Water Resources Systems Planning and Management

Water Resources Systems Planning and Management

Second Edition

Sharad K. Jain
Formerly, National Institute of Hydrology, Roorkee, Uttarakhand, India
Indian Institute of Technology, Roorkee, India

Vijay P. Singh
Department of Biological & Agricultural Engineering & Zachry
Department of Civil & Environmental Engineering
Texas A&M University College Station, TX, United States

ELSEVIER

Elsevier
Radarweg 29, PO Box 211, 1000 AE Amsterdam, Netherlands
50 Hampshire Street, 5th Floor, Cambridge, MA 02139, United States

Notices

Knowledge and best practice in this field are constantly changing. As new research and experience broaden our understanding, changes in research methods, professional practices, or medical treatment may become necessary.

Practitioners and researchers must always rely on their own experience and knowledge in evaluating and using any information, methods, compounds, or experiments described herein. In using such information or methods they should be mindful of their own safety and the safety of others, including parties for whom they have a professional responsibility.

To the fullest extent of the law, neither the Publisher nor the authors, contributors, or editors, assume any liability for any injury and/or damage to persons or property as a matter of products liability, negligence or otherwise, or from any use or operation of any methods, products, instructions, or ideas contained in the material herein.

ISBN: 978-0-12-821349-0

For information on all Elsevier Science publications visit our website at https://www.elsevier.com/books-and-journals

Publisher: Candice Janco
Acquisitions Editor: Maria Elekidou
Editorial Project Manager: Lindsay Lawrence
Production Project Manager: Kumar Anbazhagan
Cover Designer: Christian J. Bilbow

Typeset by TNQ Technologies

Contents

PART II Decision making

Preface to First Edition

Shanno devirbhistaya aapon bhavantu peetaye |
Shanyorbhi stravantu nah ||
Yajur Veda (36.12)
May beautiful waters be pleasant to us to drink
and acquire happiness,
and flow with health and strength to us.

In nature, the water of right quality in right quantity is usually not available at the right place at the right time. Sometimes, it is available at the wrong place and at the wrong time and its quality and quantity are not what they should be. Although, in nutshell, this is the gist of the problem, this seemingly innocuous, albeit naughty, interplay of words fails to convey the gravity and the magnitude of problems that are caused by the mismatch between the availability and the demand of good quality water. Nature can indeed be very harsh and ruthless at times. However, a positive outcome of this mismatch is the development of a wide range of tools and techniques for water resource system planning, development, operation, and management.

Since the publication of *Design of Water Resources Systems* by the Harvard Water Group in 1962, system techniques have been commonly applied to water resource planning, design, operation, and management. Results of numerous studies, applications, and practices have been reported in many monographs and books. Although the material presented in these books is undoubtedly relevant, praiseworthy, and authoritative, the practice of water resource management has undergone significant and far-reaching changes during recent years. In developed countries, for example, the emphasis these days is on the conservation and sustainable development of water resources, whereas the emphasis in developing countries is more on the development of untapped water resources to meet rapidly increasing demands and for the alleviation of poverty. These objectives have to be attained while simultaneously minimizing adverse impacts on the environment and the social fabric of society.

There is a growing call for involving stakeholders in decision-making at all levels. To that end, there is unanimity all over the world that water resources of a river basin have to be developed and managed by systematically integrating socio-economic, environmental, political, and engineering considerations. Furthermore, revolutions occurring in technology these days are having a profound influence on the practice of water resource engineering. Two glaring examples of this revolution are the Internet and electronic mail. These have provided a convenient, cheap, and rapid medium of communication and access to huge volumes of information. These days, people physically located far away from each other can be in close contact and can frequently exchange ideas, notes, and documents almost instantaneously. Such quick communications were unthinkable only a few years ago. These are the factors that motivated the writing of this book which provides a discussion on topics deemed relevant in the present scenario of water resource practice. Another feature, deemed important, is the employment of a real-word system as an example to illustrate the application of system concepts and techniques.

The subject matter of this book is divided into four parts. The first part, termed ***Preliminaries***, contains four chapters. Introducing the basic theme of this book, the first chapter provides an overview of the current status of water resource utilization, and the likely scenario of demands in the near future. The basic concepts, practical applications, and advantages and disadvantages of systems analysis techniques are presented. Also provided is a discussion of seven challenges for the water sector identified and debated in recent international forums. Temporal and spatial hydrologic analyses require extensive data. An understanding of how these data is measured and how the processing of these data is done is important before undertaking any water resource systems analysis. Chapter 2 presents techniques of observing and processing the data used in water resources systems. The discussion is extended to emerging techniques in Chapter 3. These techniques are Remote Sensing, GIS, Artificial Neural Networks, and Expert Systems. These techniques are versatile, highly useful, and are playing increasingly important role in water resource planning and management. Chapter 4 discusses statistical tools for deriving the desired information from these data, including commonly used probability distributions, methods of parameter estimation, regression and correlation analysis, frequency analysis, time-series analysis, and transition matrices and Markov chains.

Part 2 of this book deals with ***Decision-Making*** which is a bouquet of techniques organized in four chapters. Optimization and simulation techniques are discussed in Chapter 5. Most of the analyses dealing with planning and management of water resources systems are carried out using these techniques. Besides technical aspects, economic analysis is essential to rank the various competing projects and subsequent approval by the decision-maker. Chapter 6 dwells upon the techniques of economic analysis. During recent years, environmental and social aspects and rehabilitation and resettlement of project-affected people have come to occupy a central stage in planning as well as management of water resource projects. It is now recognized that these projects may have a lasting impact on the social fabric of the area concerned. It is, therefore, necessary that while planning as well as operating a water resources system, adequate care and precautions are exercised so that adverse impacts on the society and the environment are minimized. These issues form the subject matter of Chapter 7. Using basic analytical tools and the knowledge of environmental and social aspects, the practitioner is required to take a rational and balanced decision regarding the design and management of water resources systems. Since major inputs to a water resource system, such as rainfall, meteorological variables, etc., are not deterministic, these systems inherently have an element of risk. The concept of rational decision-making and the associated elements, such as the concept of risk, reliability, and uncertainty in water resources systems, are discussed in Chapter 8.

Part 3 of the book, comprising 2 chapters, deals with ***Water Resources Planning and Development***. Chapter 9 discusses basic concepts of planning, classification and steps of the planning process, integrated planning, institutional setup, public involvement, and planning models. Chapter 10 discusses planning and sizing of reservoirs and approaches to compute the storage capacity for conservation as well as flood control.

The fourth and last part of the book, encompassing 4 chapters, focuses on ***Systems Operation and Management***. After a reservoir has been constructed, it is essential that it is managed in the best possible way so that benefits can be maximized. The techniques of reservoir operation are discussed in Chapter 11. It includes a detailed discussion on conventional and system techniques for regulating a system consisting of reservoirs in series and parallel and the application of various systems analysis techniques, such as optimization and simulation. It also discusses real-time operation of reservoirs and the logistics required. Many dams around the world are experiencing a significant loss of their storage capacity every year due to sedimentation. This topic is discussed in Chapter 12, providing details of soil erosion and mathematical modeling of sediment transport, empirical techniques, as well as latest satellite data—based techniques to assess reservoir sedimentation, methods to prevent the sediment from entering a reservoir, and recovery of reservoir storage by flushing and dredging.

No analysis of water resources systems is complete without consideration of water quality. Chapter 13 presents fundamentals of measurement of water quality variables and modeling of water quality in rivers. A river basin is the natural unit in which water occurs and currently a lot of emphasis is being placed on holistic management of river basins. Chapter 14, the final chapter, discusses various issues related to the management of a river basin. The discussion includes integrated water quality and quantity management, water pricing, water rights, interbasin water transfer, decision support systems, privatization, and management of international river basins.

It is hoped that this book will be useful to those engaged in the practice of water resources engineering. This book is intended for senior undergraduate and beginning graduate students as well as water resource practitioners. Its purpose will be served if it provides them motivation and information in their pursuit for further search and a deeper dive in this vast reservoir of knowledge to quench their thirst.

Sharad K. Jain
National Institute of Hydrology
Roorkee, India

Vijay P. Singh
Louisiana State University
Baton Rouge, Louisiana, U.S.A.

Preface to Second Edition

This book *Water Resources Systems Planning and Management* was published in 2003. A number of advances have occurred in the water resource field during this intervening period of two decades, which have a direct bearing on the planning and management of water resource systems. It is therefore deemed appropriate to incorporate some of the advances in the second edition. All chapters have been revised, but the basic structure and layout of this book remain essentially unchanged, except for the addition of a new chapter on climate change. Hence, without repeating what is already described in the Preface to the First Edition, we discuss only additions and changes made in this edition.

As before, the subject matter of the book is divided into four parts. In the first part, termed ***Preliminaries***, Chapter 2 is beefed up with a discussion of reanalysis data which are employed in assessing the impact of climate change. Chapter 3 is significantly expanded with the inclusion of some popular GIS software, and additional AI techniques, such as a discussion of machine learning (ML), deep learning (DL), convolutional neural networks, recurrent neural networks, and comparison of Ml and DL. Chapter 4 includes multivariate probability distributions using copulas.

Part 2 of the book dealing with ***Decision-Making*** includes ecosystem services in Chapter 6, cumulative impact assessment in Chapter 7, and social impact assessment and social hydrology in Chapter 8.

Part 3 of the book on ***Water Resources Planning and Development*** includes a discussion of top-down and bottom-up planning in Chapter 9. The fourth and last part of the book, focusing on ***Systems Operation and Management***, is revised considerably. Chapter 11 discusses flood control operation of a multipurpose reservoir with an emphasis on flood limited water level (FLWL) and hedging in reservoir operation for flood control, applications of system analysis techniques for flood management, operation of a network of multipurpose reservoirs, real-time flow forecasting, uncertainties and risks in flood control operation of reservoirs, and an appendix on generalized formulations of optimization models. Chapter 12 contains a case study on sedimentation management in Sanmenxia Reservoir, China. Chapter 13 also discusses model selection. Chapter 14 is a new chapter on climate change and its impact on water resources, discussing the causes and evidence of climate change, natural internal processes, changes in external forcings, changes in composition of atmosphere due to human actions, changes in land use—land cover, study of climate change impacts on water resources, future climate scenarios and their purpose, representative concentration pathways, shared socioeconomic pathways, global climate models, downscaling climate data, statistical (or empirical) downscaling, dynamical downscaling, advantages and limitations of RCMs, adaptation and mitigation of climate change, impact of climate change on water resources, methodology to predict the impacts, impacts on quantity of water, impacts on quality of water, impacts on surface runoff and floods, impacts on droughts, impacts on groundwater, impacts on hydropower generation, and uncertainty in impact assessments. Chapter 15, the final chapter, discusses various issues related to management of a river basin.

It is hoped that the revised edition of this book will be useful to those engaged in the practice of water resource engineering, to senior undergraduate and beginning graduate students, as well as to faculty members. In this era of increasing attention to food–water–energy–health security nexus, this book may be of particular value.

Sharad K. Jain
Indian Institute of Technology
Roorkee, India

Vijay P. Singh
Texas A&M University
College Station, Texas, U.S.A.

Acknowledgments

The material for this book evolved from lecture notes prepared for various short-term and regular courses with which we were associated. A large number of colleagues and friends helped with the preparation of the book at various stages. Because of lack of space, it is not possible to mention all the names here. The authors are thankful to Dr. J. Han for his kind help with the text of Chapter 14. Some material in this book is based on studies with which we were associated and we are grateful to our coworkers in those studies. The data pertaining to the Sabarmati River system have been used in examples to illustrate various techniques throughout this book. Thanks are also due to the officers and staff of Gujarat Irrigation Department, involved with the collection of data, analysis, and management of the Sabarmati River system. The authors would also like to thank Lindsay Lawrence, Kumar Anbazhagan, and other concerned persons from Elsevier for their kind help and support during the preparation of the manuscript.

Our families (Shikha, Sanchit, Surabhi, Mukul, Chandra Mohan, Upma, Ashwini, and Parul; and late Anita, Vinay, Sonali, Vamsi, Roning, Kayden, Davin, Arti, and Alivia) created congenial atmosphere at home and allowed us to work in the evenings and on weekends on innumerable occasions. Many relatives, friends, and colleagues gave encouragement while we worked on this book. Their support and good wishes are gratefully acknowledged.

Sharad K. Jain
Vijay P. Singh

PART

I

Preliminaries

CHAPTER

Introduction to Water Resources Systems

1

Chapter outline

Water Resources Systems Planning and Management. https://doi.org/10.1016/B978-0-12-821349-0.00013-7

The objectives of this chapter are:

- to introduce the subject matter of this book;
- to explain the need and availability of water including the magnitude of the quantities involved;
- to provide an overview of systems approach; and
- to explain the challenges in the water sector in the time to come.

Earth is a blue planet whose three quarters are covered with water—a unique substance. The journey of evolution of life began in water, and it is the major constituent of the human body—an average human contains about 50 L of water. On account of its peculiar molecular structure, it is a nearly universal solvent. The movement of water on the earth has been one of the main causes of shaping the land and its topography. Water also has a high thermal inertia due to high specific heat, and therefore, the oceans have an important influence on the heat cycle of the earth. The culture, food, and life style of a place are closely knitted together and depend on the climate. It is well known that water is central to the climate of a place. Along with all the benign features, water is also a paramount vehicle to convey the fury of nature. The water-centered hazards, like floods and droughts, have caused havoc since the dawn of civilization. According to the beliefs of some religions, life on this planet is certain to be destroyed one day by a great deluge.

Since the dawn of civilization, mankind has shown a preference to settle near rivers due to the assured supply of water, facility of navigation, and fertility of river valleys. Even today, a considerable portion of the world's population lives in areas adjacent to water bodies. Every continent has a number of major cities that are located on the ocean coasts or the banks of rivers, for example, New Delhi (Yamuna), Washington DC (Potomac), London (Thames), Paris (Seine), Cairo (Nile), and Manaus (Amazon); and Tokyo, New York, Sao Paulo, and Sydney on the coasts. While the cities enjoy the benefits due to the proximity to a source of water, man has harmed the sources of water in many ways by unwise exploitation. People living close to rivers often become victims of misery and upheaval in the wake of devastating floods that have accompanied mankind throughout its history. Due to these reasons, man always had a desire to somehow tame and control nature in general and rivers in particular and use them for beneficial purposes. To fulfill this objective, man has been closely observing, measuring, and attempting to understand hydrological processes and developing techniques for the control and management of water resources. This book is an attempt to take the reader on a journey to visit some tools and developments that are the results of the quest of mankind over many centuries.

The practice of hydrologic observations dates back to historic times. The *Arthasashtra* of Cautilya (a famous administrator/economist in ancient India) discusses rainfall measurements, which were the basis of revenue collection. Varahamihira (505−587 CE), an ancient Indian philosopher, had discussed topics like the formation of clouds, signs of immediate rain, rainfall quantity, and exploration of groundwater in his treatise by the name *Brihat Samhita*. The Egyptians have been systematically measuring the flow of River Nile for a very long time. An open masonry water well was built in the first century AD at the upstream end of an island in the Nile at Cairo. The marks on the inner wall of this well constitute a water-level gauge, which was calibrated by Pliny-the-Elder (23–79 CE) in terms that illustrate the social importance of water level in the river (These were: disaster, abundance, security, happiness, suffering, and hunger. See Fig. 1.1.) Around the second century BCE, Romans had built many aqueducts on the Iberian Peninsula, whose ruins can still be seen in places like Segovia. The rainfall records of more than 100-year duration are available at many stations around the world. However, the early hydrological design practices were not very scientific, and the designs were mostly based on empiricism, thumb rules, or heuristics.

In many countries, computer-based analysis and design techniques have been in vogue since the 1950s. With the advent of computers, systems analysis techniques were introduced in the water resources area in the 1960s. One of the earliest comprehensive works on river basin simulation was published by Maass et al. (1962), who had offered the following on digital simulation: "Until digital computers of the magnitude of the (IBM 650 or Univac I) were constructed, the solution of large-scale simulation problems was not feasible. Even the storage capacities of these computers were not commodious enough .... (River basin simulation) had to await elaboration of the very large computers of the IBM 700 class, with an internal magnetic-core storage capacity of up to 32,768 words." This was the status of computer applications in the 1960s when no interactive color graphic displays were available; there were no digitizers, color laser printers, mice, disk capacities of the order of gigabytes, and the Internet. Compared to this, today, the CPU memory of a typical desktop PC is measured in

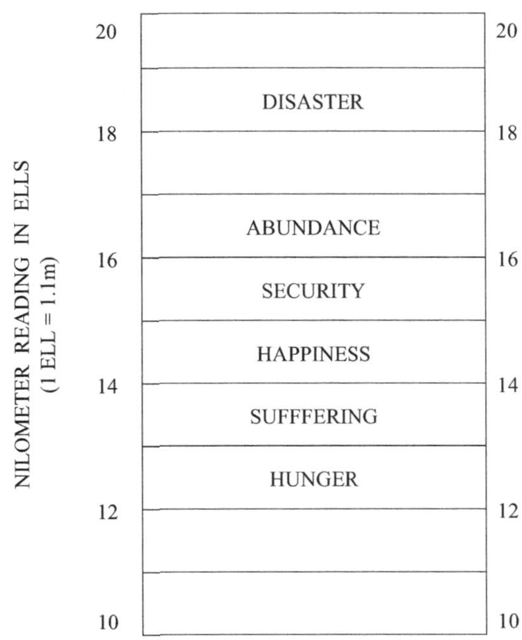

FIGURE 1.1

Nilometer built by Pliny-the-Elder.

Source: Dooge (1988). Copyright © IAHS. Used by permission.

megabytes and disk capacity in gigabytes, its processing speed is also faster thousands of times, and a variety of peripherals can be easily interfaced with it. A number of software, including compilers, editors, spreadsheets, word processing, and utilities, are loaded on each machine. The wider availability of powerful computers in the last few decades has led to a proliferation of mathematical models and studies dealing with water resource systems.

It is a bit disappointing to note that the use of modern tools for day-to-day management of water resources development projects is quite limited. There is a feeling in many quarters that despite all the developments of sophisticated models and techniques, there has not been a significant improvement in planning, design, and operation of water resources projects. Hence, the "age old problems" still remain unsolved: there are more than a billion people who do not have access to water supply, and 3 billion people still do not have adequate sanitation. The data also suggest that there is a slowdown in the growth and productivity of irrigated lands. Besides, "new problems," mostly related to environmental degradation, are cropping up. The quality of river water is degrading rapidly in many rivers despite attempts to reverse the trend. The rate of freshwater fish species extinction is five times that of saltwater species. Water diversions for irrigation are having devastating effects on water bodies, for example, the Aral Sea in Central Asia is shrinking and may soon reduce to a fraction of its original size. Deforestation/land degradation (impacts on flooding and siltation of reservoirs), salinization and waterlogging, and water contamination by chemicals, fertilizers, and human waste are all serious issues in many parts of the world. The groundwater aquifers are being mined at an unprecedented rate. About 10% of the world's agricultural food production now depends on mined groundwater. Water tables are falling as much as a meter per year in many parts of the world.

Water is a nature's gift whose availability in a region is limited by climate, geology, and topography. Traditionally, water has been considered a social commodity, being a "basic requirement" for life. In ancient times, a fundamental premise was that anybody who is thirsty should not be denied water, whatever be his income and purchasing power. This privileged social status of water was based on the doctrine of "essential service" or "public service" depending on the country. Of late, due to various reasons among which scarcity being the chief, water and related services are becoming more and more an economic asset with production and conservation costs, utilization values, opportunity costs, and demands that vary with price.

Water is required for various day-to-day activities of man. Therefore, before attempting to discuss the problems of water resources planning and management, it is necessary to examine the various uses and needs of water.

1.1 NEED FOR WATER

The need for water is derived from a variety of activities in which it is used. These activities are vital for the existence and development of human society. Because useable water is limited in its availability, it has an economic value. Furthermore, different activities require water of differing quality. Fig. 1.2 shows the sources and quality of water from the perspective of its use. For example, water of high quality is needed for domestic use, while the quality may be compromised for sanitation use. Clearly, all uses of water cannot be supported to the fullest extent, and a management policy has to be developed that can prioritize water use following established criteria. There may be conflicts and

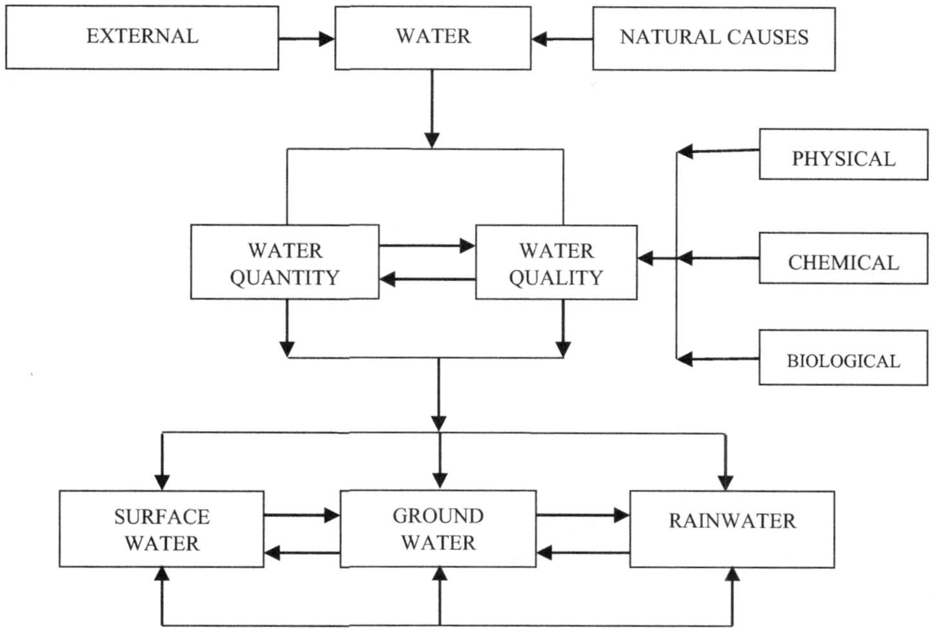

FIGURE 1.2

Sources and quality of water from the perspective of its use.

interactions among different water uses, and these, in turn, interact with water elements. The management policy has to incorporate all of these considerations.

The term "water withdrawal" refers to water removed from a source. Some of this water may be returned to the original source with changes in the quality. The term "consumptive use" refers to water that is not available after satisfying the intended purpose, for example, drinking and evaporation are consumptive uses. The term "water demand" is often used to denote the quantity of water required for a purpose. Basically, the uses of water can be divided into two categories: withdrawal (off-stream) and in-situ (in-stream). The withdrawal uses are those for which water is diverted from its natural place of occurrence. The water used for agriculture, industry, and municipal purposes are a few examples of this type. The in situ uses are those for which water need not be diverted, such as recreation, wildlife habitat, minimum flow requirement, and hydroelectric power generation (in most cases).

The utility of water lies in the various productive functions that it can perform. Traditionally, municipal and agricultural water supply have been the usual uses. These two (particularly in an agrarian economy) are viewed as social objectives and cannot be given up even if alternate uses of water give higher benefits. Other important uses are hydropower, navigation, and recreation. The objectives of water resources development are more varied now. Pollution control, repulsion of saline water, preservation of natural rivers, beautification, groundwater replenishment, and a host of other equally nonquantifiable and competitive purposes now demand equal attention along with economic purposes.

The issues of water demand and its supply are specific to time and place. There has been a continuous increase in water use over the world since the time such records are available. It has been estimated that the total water use at the beginning of the last century was about 600 km^3/year. This rate witnessed a sharp increase, and the global water use by the year 2000 is likely to be about 5300 km^3/year. Thus, there was a 10-fold rise in water use in the last century. As the population of the world is rising, the demand for water will also continue to increase in the 21st century. A number of projections of the future world population have been made, and it is estimated that the population of the world will stabilize sometime after the year 2060. The rate of urbanization will further rise, and by 2025, about 4 billion people are likely to live in urban areas.

Estimates of water use in the various continents at different times in the 20th and 21st century are given in Table 1.1.

Gleick (1997) has estimated sector-wise water requirements for the year 2025. The domestic water use has been estimated assuming that the world's entire population has access to a "basic water requirement" of at least 50 L/person/day and regions using more than that amount would implement measures that would reduce per-capita domestic water use to the present level in the developed nations which is around 300 L/person/day. In this scenario, the total domestic water needs would be approximately 340 km^3/year. In the agricultural sector, all regions are assumed to attain a minimum consumption of 2500 calories per person per day. There are likely reductions in water needed to grow these diets as a result of changes in the water-intensive components of diets, particularly meat, and changes in irrigation efficiency, cropping intensities, and irrigated areas. Taking into account the increase in population, the agricultural water consumption would be about 2930 km^3/year. The industrial water withdrawals would be at around 1000 km^3/year, and an additional 225 km^3/year will be lost due to reservoir evaporation. Thus, the total global water withdrawals are projected by Gleick (1997) to be approximately 4500 km^3 in 2025. This figure is a little low compared to the estimate of Shiklomanov (1998).

The major consumer of water in the last century was the agricultural sector, which consumed more than half of the water used. According to the data given by Biswas (1998), agriculture accounted for nearly 90% of all water use in 1900 but its share has declined to about 62% by the year 2000. This trend is likely to continue into the 21st century, even though the nutritional requirements of the rapidly increasing world population have to be satisfied. The agricultural water requirement at the beginning of the current century was about 10,000 km^3 of water per year. The demand for industrial and urban sectors was about 2500 and 240 km^3 per year; the rural sector required about 135 km^3 of water annually. In terms of per capita water demand as per the current trend, the industrial and domestic sectors require about 180 m^3 per capita per year, while the agricultural requirements are of the order of 700 m^3 per capita per year. It may be cautioned that these values are likely to change as the estimates are being continuously updated and one may come across different values from different sources. The lack of consistent and reliable data on both the supply and the use of freshwater creates serious problems in efforts to manage water resources.

Hydropower is the largest renewable source of energy and is the second largest source of electricity generation. The hydropower generation has many advantages: the plants have a high efficiency (of the order of 80%–90%), can be started and shutdown quickly, and hence are very useful as peaking plants, energy can be stored, and the reservoirs can be used for other purposes like irrigation, water supply, navigation, recreation. The operation and maintenance costs are low, and there is no environmental pollution. All these properties make hydropower a desired source of electricity generation.

Table 1.1 Historical water withdrawal, consumption, and projections.

	Historical estimates of use (km^3)								Forecasted use (km^3)		
Continent	**1900**	**1940**	**1950**	**1960**	**1970**	**1980**	**1990**	**1995**	**2000**	**2010**	**2025**
Europe	37.5	71.0	93.8	185	294	445	491	511	534	578	619
	17.6	29.8	38.4	53.9	81.8	158	183	187	191	202	217
North America	70	221	286	410	555	677	652	685	705	744	786
	29.2	83.8	104	138	181	221	221	238	243	255	269
Africa	41.0	49.0	56.0	86.0	116	168	199	215	230	270	331
	34.0	39.0	44.0	66.0	88.0	129	151	160	169	190	216
Asia	414	689	860	1222	1499	1784	2067	2157	2245	2483	3104
	322	528	654	932	1116	1324	1529	1565	1603	1721	1971
South America	15.2	27.7	59.4	68.5	85.2	111	152	166	180	213	257
	11.3	20.6	41.7	44.4	57.8	71.0	91.4	97.7	104	112	122
Australia + Oceania	1.6	6.8	10.3	17.4	23.3	29.4	28.5	30.5	32.6	35.6	39.6
	0.6	3.4	5.1	9.0	11.9	14.6	16.4	17.6	18.9	21.0	23.1
Total (rounded)	579	1065	1366	1989	2573	3214	3590	3765	3927	4324	5137
	415	704	887	1243	1536	1918	2192	2265	2329	2501	2818

Note: The row above the dotted line gives water withdrawal, and the row below the dotted line is for water consumption. Includes about 270 km^3 in water losses from reservoirs for 2025.
Source: Shiklomanov (1998).

According to the 2022 Hydropower Status Report (https://www.hydropower.org), global installed hydropower capacity was 1360 GW in 2021 and 4250 TWh of clean electricity was generated from hydropower. However, there are many "pockets" of hydropower that are yet to be developed. For example, the Zaire basin in Africa contains 20% of the hydropower potential of the world, and most of it is not harnessed. Similarly, the north-eastern region of India (chiefly the Brahmaputra River and its tributaries) has about 30% of the hydropower potential of India, and only a small fraction of it has been developed. Keeping in view the technological developments, the cost of hydroelectricity generation in the near future is expected to be in the range of 3–6 US ¢ per kWh. Furthermore, due to the development of turbines, which can efficiently operate at low heads, it may be possible to modify many existing dams which are not presently operated for hydroelectric power and use them to generate power.

Water is an important input resource for many industrial activities. The electricity generation from fossil and nuclear fuels requires water for steam, cooling, and general services. According to Herschy and Fairbridge (1998), the quantity required for condenser cooling is in the range of 0.032–0.044 m^3/s per MW. Besides, in coal-fired plants, water is also needed for ash transport. The paper industry requires 40–400 m^3 of water per ton of paper, depending on the type of raw material. If a hydraulic method of coal mining is used, about 0.08–0.14 m^3 of water is needed per ton of coal produced. Coal slurry pipelines need 0.95 m^3 of water per ton. It is estimated that about 0.163 m^3 of water per barrel (0.159 m^3) is needed to refine crude oil.

The transport of goods by water is highly fuel efficient with insignificant air pollution. The consumption of diesel per ton-km of goods by road is 0.04 L, by railway 0.011 L, and by water 0.0056 L. Navigation is not a consumptive use of water; a reservoir downstream of the navigable waterway can capture flows in the waterway and provide them for other beneficial uses.

Water is also required to sustain rivers and wetlands. River restoration has become a topic of interest in some developed countries and such projects can form part of a sustainable development plan for the river basin. The objectives of river restoration are normally to create a wider diversity of ecosystems and improve biodiversity by bringing the river into closer contact with its floodplain. A certain minimum flow is needed in a river to dilute pollution; water also washes away salts that would otherwise destroy farmlands.

1.1.1 Likely Future Trends of Water Demands

An analysis of data indicates that the rate of rise of water use has been about three times higher than the rate of increase in global population. If this figure is extrapolated, then doubling the world population would entail a sixfold increase in the total global water requirements. Coupled with increasing requirements for irrigation and industrial use, this is clearly not sustainable. Even now, water is treated as a free resource in many countries, although water deficits are being witnessed in many parts of the world. Furthermore, water is supplied either free of cost or the prices are kept artificially low in many countries due to influential lobbies and political reasons. However, as the demand versus availability worsens, it is likely that water prices will also have to be gradually increased, also as a means of controlling demands. As it becomes more and more expensive to provide one unit of water, farmers will be forced to use crop varieties that consume less water and adopt alternate means of more efficient irrigation. Water-efficient industrial processes will also have to be adopted on a large scale. The water use data (e.g., see Seckler et al., 1998) also indicate that Asia is the continent that is the highest consumer of water. This is because Asia has a very high population and agriculture, which consumes large amounts of water and is the main occupation of many Asians. In fact, there is a marked decline in the percentage of water used for agriculture from low- to high-income countries, as shown in Fig. 1.3.

Since an improved lifestyle requires more water, the improvement in the standard of living of the people is also responsible for higher water demands. Despite remarkable technical advancements, a large percentage of the world population still does not have access to clean drinking water. As governments are committed to providing clean drinking water to the entire population, urban water use is expected to rise significantly by the year 2050, which is the time by which the world population is expected to stabilize. The provision of sufficient water for all the needs will also require substantially higher investments. The reason is that the cost of developing new sources is rapidly increasing due to the higher cost of construction materials, labor, and higher provisions for rehabilitation and resettlement of the project-affected persons. Environmental laws, as well as water quality standards, are becoming more and more stringent with time, and this will also force the agencies to spend higher amounts on developing new water resources.

To meet additional land and water requirements to feed the increasing population, far more land will have to be cleared for agriculture, resulting in losses of forests, habitats, and biodiversity on a massive scale. There will also be increases in the demand for water for industry and municipal use. All this would have detrimental environmental impacts. The recent predictions about change in climate indicate that the water availability in various regions of the world is also likely to undergo changes in the future.

As the deficit of water is increasing in many parts of the world, conflicts due to water are occurring more frequently. These conflicts can be among the neighboring countries, among the adjacent states of the same country, and sometimes even among the various communities in a city.

1.1.2 Water and Ecosystems

Ecosystems are the communities of interacting organisms and the physical environment in which they live. They consist of organic and inorganic matter and natural forces that interact and change. The components of ecosystems can be divided into two groups: biotic, consisting of living beings,

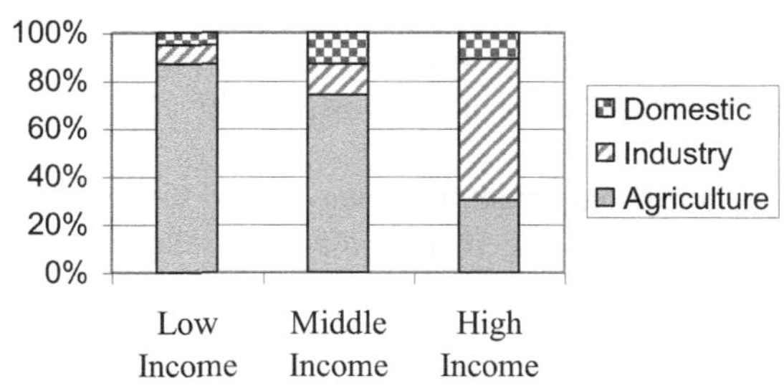

FIGURE 1.3

Annual (1999 estimates) freshwater withdrawals.

Based on data published in newspaper: The Hindu, dated June 18, 2001, New Delhi.

and abiotic, consisting of nonliving things. Humans are intimately familiar with ecosystems because they live and interact with them regularly. In fact, every part of the earth is part of an ecosystem. When studying and managing ecosystems, the scale or size is important. A small patch of forest is an ecosystem and so is a major river basin covering thousands of square kilometers. The major categories of ecosystems are coastal, forest, grassland, freshwater, and agricultural systems. Besides the divisions of the ecosystems, the linkages between them are equally important. Although human activities influence all ecosystems to some extent, such an influence on freshwater systems is very pronounced. Both excessive poverty as well as excessive industrialization jeopardize ecosystems.

At present, all nations, rich or poor, and all sections of the society, affluent or deprived, are realizing the decline in the productive and assimilative capacity of ecosystems and their impact on life. Although our understanding of ecosystems has improved dramatically in the recent past, the ability or commitment to improve their quality has not increased proportionately. If the current pattern of ecosystems' use continues, there might be a serious decline in their ability to yield the range of benefits, such as clean water and stable climate that they have been providing so far.

Historical evidence shows that water and the environment had a major role in the decline and fall of ancient civilizations. Obviously, the decline in the long-term productive capacity of ecosystems will have unhappy implications for human development and the well-being of all species. The pressure on ecosystems has tremendously increased from the latter half of the twentieth century onwards. This period has witnessed a population explosion, excessive industrialization, and a rapid rise in the consumption of resources. All this has been at a cost to ecosystems, and these pressures are likely to increase in the near future. Due to various reasons, governments worldwide are not able to control or check unsound management practices. For example, the use of water for agriculture and electrical energy is subsidized to varying degrees in most countries. Algal blooms and eutrophication of freshwater systems and waterborne diseases from fecal pollution of surface water is still a major cause of health problems in developing countries. Currently, about 40% of the world population experiences some kind of water shortage. The biodiversity of freshwater ecosystems is threatened, and a large number of species are feared to have become extinct in the last century.

To mitigate or manage the damage to ecosystems, an integrated approach that considers the entire range of services and optimizes the mix of benefits from an ecosystem and also across ecosystems is necessary. It is necessary to look beyond the traditional boundaries, since ecosystems often cross the state and national boundaries. The successful management practices that preserve or increase the capacity of an ecosystem to produce the desired benefits in the future need to be identified and implemented.

Too often in the past, water management had been limited to the sole quantitative aspect of water resources. A global approach to the management of all water-related ecosystems should be initiated in each river basin in addition to pollution control and the promotion of water recycling with a prospect of sustainable development. The root causes of water resources problems and their effects are shown in Fig. 1.4. Soil conservation, land use planning, forest management, and protection of wetlands and aquatic ecosystems are crucial for water management. The same can be said about the management of solid wastes and sludge of treatment plants whose leaching can be a source of serious, sometimes irreversible, pollution of water bodies and aquifers.

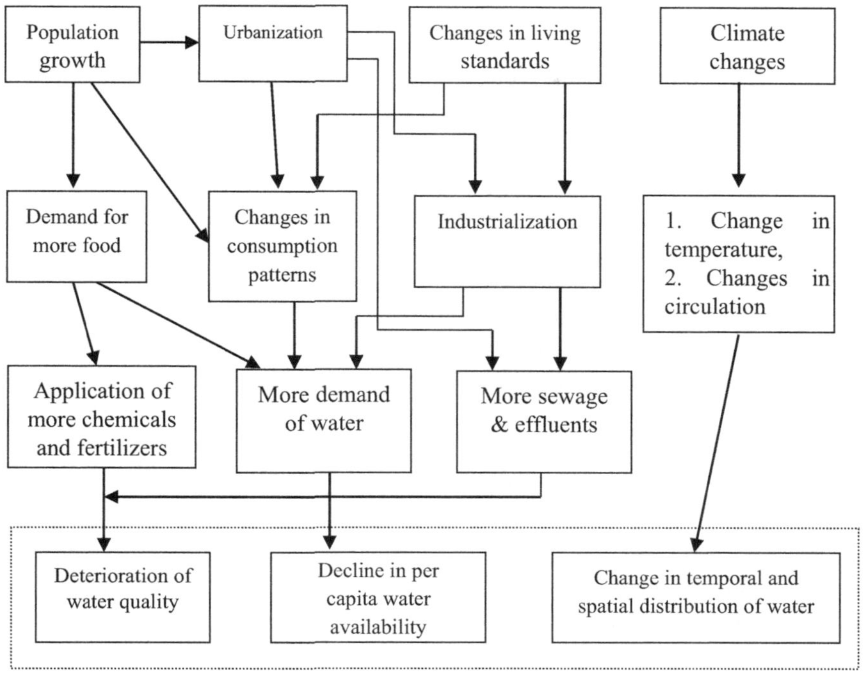

FIGURE 1.4

Major root causes of water resource problems and their effects.

1.2 AVAILABILITY OF WATER

For economic and optimum utilization, planning, design, and operation of water resources, the determination of the extent and availability of surface and groundwater is the first requisite. The distribution of water resources on the continents varies greatly in space and time. There are transitions of all kinds from one extreme to the other, that is, the water surplus in the tropical rainforest regions is in sharp contrast with the deserts, which are practically barren because of the scarcity of water. The human population, on the other hand, is distributed in quite a different manner. For example, regions with an immense abundance of water may be sparsely populated, for example, Amazon River basin, or densely settled, for example, South Asia. The deserts, which are without sufficient water, and the polar regions, where all the water is frozen, are both uninhabitable. Our ability to use water also depends on the frequency and magnitude of floods and droughts. In view of climate change, both are likely to become more severe.

The following major sources of water were identified by Cole (1998).

Natural rivers: The rivers are fed by rainfall, or snow/glacier melt, or both. Their flows vary seasonally and in upland areas streamflow varies wildly ("flashy discharge") and may be contained by impoundment in reservoirs. The flow of a river that receives groundwater inputs will be more steady. Lowland rivers are partially sustained by effluents of used water, for example, from sewage treatment works and from irrigation return flow.

Natural lakes: These usually offer a limited supply, as their range of acceptable drawdown is severely constrained.

Reservoirs: Water is stored in a reservoir behind a dam for a number of uses. In climates with distinct wet and dry seasons, water stored in a wet season is withdrawn for use in the following dry season. In many parts of the world, reservoirs may have to cater for a long series of dry years. Reservoirs may be of the following types.

- Direct supply reservoir. Most upland reservoirs supply water by pipeline or canals, often by gravity flow, and offer good quality water.
- River-regulating reservoirs, which are major water reserves and release water downstream to sustain its discharge in conditions of low natural flow.
- Pumped storage, in which river water is pumped into a (sometimes off-channel) storage, for subsequent use. This arrangement is also used in hydropower projects to provide additional power for the peak period. During the off-peak hours, water from the lower reservoir is pumped back into the higher reservoir.
- Estuary barrage, which impounds an area that otherwise would be tidal. This is the ultimate means of trapping fresh water before it enters the sea. It entails making provisions for the passage of migratory fish and providing locks for shipping.

Tube wells: These are the main sources of groundwater, usually with electric submersible pumps or with mechanical shaft-drive pumps. The yield ranges from 1 to 10,000 m^3/day, depending on the aquifer characteristics, pump capacity, and available water. Open wells (bucket and winch, or hand-pumped) are small sources of water in rural areas.

Springs: These occur where aquifers crop out above impermeable rock. They very much depend on local hydrogeological conditions. Springs may be perennial, affording a reliable minor source of supply, or only seasonal, with no year-round availability.

Saline waters: Brackish waters, which have a salt content >0.1%, are amenable to desalination by reverse osmosis systems. Although the process is expensive, it is extensively applied in arid areas (e.g., the Arabian/Persian Gulf) where fuel is relatively cheap. Even seawater can be treated; thus, it can also be desalinated by distillation processes. As explained above, recent advances in membrane technology have significantly brought down the cost.

1.2.1 Water Resources Assessment

In water resources assessment, the knowledge of hydrology, meteorology, geology, and coastal sciences is combined to provide a quantitative picture of the physical characteristics and possible variation in the availability of this natural resource. Such an assessment considers the total catchment and its meteorological inputs in various phases: (1) land phase, (2) river phase, (3) reservoir phase, (4) subsurface phase, and groundwater phase.

The relative quantities of the earth's water contained in each of the phases of the hydrologic cycle are presented in Table 1.2. Oceans contain 96.5% of the earth's water and of the remaining 3.5% on land; approximately 1% is contained in deep, saline ground waters or in saline lakes. This leaves only 2.5% of the earth's water as fresh water. Of this fresh water, 68.6% is frozen in the polar ice caps, and a further 30.1% is contained in shallow groundwater aquifers, leaving only 1.3% of the earth's freshwater mobile in the surface and atmospheric phases of the hydrologic cycle. The proportions of this

Table 1.2 World water reserves.

Item	Area 10^6 km^2	Volume 1000 km^3	Depth of run-off (m)	Percent of total water	Percent of fresh water
World ocean	361.3	1,338,000	3700	96.5	–
Ground water:	134.8	23,400[a]	174	1.7	–
Gravitational, capillary Fresh	134.8	10,530	78	0.76	30.1
Soil moisture	82.0	16.5	0.2	0.001	0.05
Antarctica ice	13.98	21,600	1546	1.56	61.7
Other ice and snow	2.25	2464	1848	0.179	7.04
Ground ice in permafrost	21.0	300.0	14	0.022	0.86
Lakes:	1.2	91	73.6	0.007	0.26
Fresh Saline	0.8	85.4	103.8	0.006	–
Marshes	2.7	11.47	4.28	0.0008	0.03
Rivers	148.8	2.12	0.014	0.0002	0.006
Biological water	510.0	1.12	0.002	0.0001	0.003
Atmospheric water	510.0	12.9	0.025	0.001	0.04
Total water	510.0	1,385,984.6	2718	100	–
Fresh water	148.8	35,029.21	235	2.53	100

[a]*Ignoring groundwater reserves in Antarctica, estimated at 2 million km^3.*
Source: UNESCO (1978), © UNESCO. Reproduced by permission of UNESCO.

water in the atmosphere, soil moisture, and lakes are similar, while that in rivers is less and that in snow and glacier ice is greater. A small amount of biological water remains fixed in the living tissues of plants and animals. The data on the earth's waters cited here are taken from a comprehensive study of world water balance conducted in the Soviet Union. These values represent only estimates and are being refined.

It is remarkable that the atmosphere contains only 12,900 cubic kilometers of water, which is less than 1 part in 100,000 of all the waters of the earth. Atmospheric water would form a layer of only 25 mm deep if precipitated uniformly onto the earth's surface (Maidment, 1993).

The biggest river basin of the world is the Amazon River basin in South America, covering an area of 6.915 million km^2; this is 38% of the total area of South America. The average discharge of this river at the mouth is 220,000 m^3/s (Herschy and Fairbridge, 1998), and the flow at Obidos can go up to 370,000 m^3/s during the wet season with flow depth of 48 m. The details of the world's rivers are available in Showers (1989). The longest river in the world is Nile whose length is 6670 km. The maximum observed discharges of some of the major rivers of the world are given in Table 1.3.

About half of the land area of the world comprises river basins (more than 200) that fall in the territory of two or more countries. The major river systems that drain a number of countries include Amazon, Congo, Danube, Ganga-Brahmaputra, Indus, Mekong, Niger, Nile, Rhine, and Zambezi. The UN (1978) has listed such river basins. The development and management of many water resources projects that involve international river basins require agreements among the concerned countries.

Table 1.3 Maximum observed discharges of selected rivers.

Country	River	Station	Catchment area (km^2)	Year-month-day	Discharge (m^3/s)
Austria	Danube	Vienna	10,700	1899/9/18	10,500
Brazil	Amazon	Obidos	46,40,300	1953	370,000
Burma	Irrawaddy	–	360,000	1877	63,700
Cambodia	Mekong	Kratie	646,000	1939	75,700
China	Changjiang	Yichang	10,10,000	1870/7/20	110,000
Egypt	Nile	Aswan	30,00,000	–	13,500
France	Rhine	Beaucaine	96,500	1856/05/31	11,640
India	Brahmputra	Pandu	404,000	1973/8/8	51,100
	Ganga	Farakka	935,340	1971/8/22	70,500
	Godavari	Dolaishwaram	307,800	1959/9/17	78,700
	Krishna	Vijaywada	251,360	1916/11/2	33,500
	Narmada	Garudeshwar	87,900	1970/9/6	69,400
Netherlands	Rhine	Lobith	160,000	1926/1/4	12,280
Pakistan	Indus	Attock	264,000	1929	23,200
USA	Mississippi	Columbus	23,87,950	1937/02/27	70,792
	Ohio	Cairo	528,300	1937/02/04	55,218
USSR	Lena	Kusur	24,30,000	1944/06/11	194,000
	Volga	Volgograd	13,50,000	1926/5/29	51,900
Zambia	Zambezi	Kariba	633,040	1958/03/05	16,990
Zaire	Zaire	Kinshasa	37,47,300	1970/12/2	67,930

Guidelines are available dealing with treaties among the basin countries. In many cases, the sharing of water is a bone of contention among the riparian states and disputes among neighboring countries due to the sharing of water resources are not unknown.

The biggest river basin of the world is the Amazon River basin in South America, covering an area of 6.915 million km^2; this is 38% of the total area of South America. The average discharge of this river at the mouth is 220,000 m^3/s (Herschy and Fairbridge, 1998), and the flow at Obidos can go up to 370,000 m^3/s during the wet season with a flow depth of 48 m. Details of the world's rivers are available in Showers (1989). The longest river in the world is the Nile whose length is 6670 km. The maximum observed discharges of some of the major rivers of the world are given in Table 1.3.

About half of the land area of the world is comprised of river basins (more than 200), which fall in the territory of two or more countries. The major river systems which drain a number of countries include the Amazon, Congo, Danube, Ganga-Brahmaputra, Indus, Mekong, Niger, Nile, Rhine, and Zambezi. The UN (1978) has listed such river basins. The development and management of many water resources projects that involve international river basins require agreements among the concerned countries. Guidelines are available dealing with treaties among the basin countries. In many cases, the sharing of water is a bone of contention among the riparian states, and disputes among neighboring countries due to the sharing of water resources are not unknown.

When one narrows down the focus from the continental stage to the national stage, the order of values also reduces. Let us consider the water balance for a country (India). The geographical area of India is 3.29 million km^2. The average annual precipitation is about 4000 billion m^3. The average annual runoff accounts for 1869 km^3, which is about 4% of the global supply. Due to topographic and other reasons, only about 690 billion m^3 of surface water is utilizable. The total replenishable groundwater potential in the country is about 432 km^3, and out of this, the utilizable quantity is about 396 km^3. Thus, the average annual utilizable water resource is 1086 km^3. The utilization in the year 1997 was: irrigation 524, domestic 30, industrial 30, power 9, and others 36, and the total being 629; all figures are in km^3. The utilization is projected to rise to 784−843 billion m^3 by the year 2025 (Table 1.4).

The data about the water balance for a country are useful for preparing national plans or when large-scale interbasin water transfer projects are to be considered. However, water resource planning is carried out basinwise, and for this purpose, the data of water balance of a basin are required. Consider the water resources of the Ganga River basin in India, whose area is 862,769 km^2. On average, this basin receives 120 cm of precipitation, and runoff is of the order of 525 km^3. The utilizable water resources of a basin depend on such factors as topography and land use, and for Ganga basin, the utilizable surface water is 250 km^3. In this basin, the total replenishable groundwater is 172 km^3.

Many programs for the assessment of water resources of the world are underway. The World Water Assessment Programme (http://www.unesco.org/water/wwap) is a UN-wide program whose goals are

Table 1.4 World water balance by continent.

Water balance elements	Europe	Asia	Africa	North America	South America	Australia	Total land area
In km^3							
Precipitation	7162	32,590	20,780	13,810	29,855	6405	110,000
Total river runoff	3110	14,190	4295	5960	10,480	1965	40,000
Groundwater runoff	1065	3410	1465	1740	3740	465	11,885
Evaporation	4055	18,500	16,455	7850	18,800	4340	70,000
Relative values							
Groundwater runoff as percent of total runoff	34	26	35	32	36	24	31
Coefficient of groundwater discharge into rivers	0.21	0.15	0.08	0.18	0.16	0.10	0.14
Coefficient of runoff	0.43	0.40	0.23	0.31	0.35	0.31	0.36

Source: Table W56 of Herschy and Fairbridge (1998). Used with kind permission of Kluwer Academic Publishers.

to: assess the state of the world's freshwater resources and ecosystems; identify critical issues and problems; develop indicators and measure progress toward achieving sustainable use of water resources; and help countries develop their own assessment capacity. The program has its secretariat at UNESCO and will publish a *World Water Development Report* at regular intervals. The Global International Waters Assessment (GIWA) is a water program led by the United Nations Environment Program. The aim of GIWA (http://www.giwa.net) is to produce a comprehensive and integrated global assessment of international waters, the ecological status of and the causes of environmental problems in 66 water areas in the world, and focus on the key issues and problems facing the aquatic environment in transboundary waters.

1.3 TECHNOLOGY FOR MEETING WATER NEEDS

Without a major technological innovation, there is little hope of meeting the ever-increasing water demands. There is no doubt that new technological changes can help improve services for millions and reduce stress on water systems around the world. At the same time, one should not and cannot ignore the wisdom of the past when dealing with water.

The traditional approach to capture rainfall has been to build dams that trap excess river flows. As a considerable volume of surface water remains to be tapped, dam construction is likely to continue to play an important role in many countries. However, the construction of new dams is becoming difficult, as most of the good sites are already developed and the remaining ones are technically difficult. In the United States, dam construction activity has been virtually nonexistent during the past quarter of a century. The social and environmental issues are becoming more crucial and will have to be successfully resolved. Therefore, careful planning is necessary before a new project is constructed. Although water is a renewable resource, the dam sites are not.

The agricultural sector witnessed a green revolution in many parts of the world during the last 4 decades. After the success of that green revolution, what is needed is a super green revolution, a revolution that is more productive as well as more "Green" in terms of conserving natural resources and the environment. To emphasize appropriate utilization of water resources, this revolution is termed as the "blue" revolution. Special emphasis is needed for drought-prone areas to attain a reasonable degree of self-sufficiency in terms of food and fodder. It can be achieved by a combination of:

(i) ecological approaches to sustainable agriculture;
(ii) optimum use of fertilizers, pesticides, and irrigation water (possibly through drip irrigation, computerized controlled sensors, and application of water and fertilizers just in time and place);
(iii) greater participation by farmers in the management of water and land resources; and
(iv) the application of improved seeds and biotechnology.

An important and promising area of innovation is biotechnology, which is undergoing a revolution. It is fueled by the groundbreaking work in modern molecular genetics and the breathtaking advances in informatics and computing. New high-yielding plants are being developed that are more environment-friendly and more drought tolerant. These plants also have increased salt tolerance. These seeds, coupled with agronomic techniques suitable for farmers with small holdings, are necessary to yield *more crop per drop* of water. The benefits of this revolution can and must be harnessed for solving the challenges of water in the interests of the poor and the environment.

Desalination of salty water is sometimes seen as the supply-side solution. Till some years ago, desalination was not considered feasible except for very moderate demands. The recent breakthroughs in membrane technology have had profound effects in many areas, ranging from waste treatment to desalination. Consequently, the cost of desalination in recent years has declined sharply to less than a dollar per cubic meter. Based on the trends of development, it is expected that by the year 2025, the cost of desalination would come down to 20 US cents/m^3 for brackish water and 40 US cents/m^3 for seawater. This technology is likely to play an important role in supplying water for municipal and industrial uses in near-coast zones. Treated wastewater can be used for agricultural purposes, and new technologies can reduce the cost of this treatment.

1.4 WATER RESOURCES PLANNING

Water resources planning and development are concerned with modifying the time and space availability of water for various purposes so as to accomplish certain basic national, regional, and local objectives. In most cases, the ability to achieve these objectives is limited by the nonuniform availability of water and other resources.

The water resources of a country are distributed unevenly in space and time, and the issue is how to best develop them. There is a need to conserve and prudently utilize the wealth of water resources. The basic motivation for a government to plan and develop water resources usually lies in the improvement of national or regional welfare, increase of the national income, national self-sufficiency, and the preservation of quality of the environment. The objective of water resources planning and management is to provide the supplies of water in accordance with the temporal and spatial distribution of demands through river regulation and distribution systems.

Depending on the system configuration, the same quantity of water can be put to various uses. The water used for hydropower generation can augment the downstream flows and can also be picked up further downstream for irrigation or other uses. Similarly, the regenerated flows from the command areas of irrigation projects in the upstream reaches can be put to further use downstream. These interlinkages and interdependence of withdrawals at various locations and use for various purposes are to be taken into account in water resources planning and management.

According to Hall and Dracup (1970), the general goals of a society can be stated in terms of the following objectives of water resources systems development:

1. To control or otherwise manage freshwater resources of the cognizant geographical or political subdivision so as to provide protection against injurious consequences of excesses or deficiencies in quantity or quality;
2. To provide or maintain water in such places and times in adequate quantity and quality for human and/or animal consumption, wildlife food production and processing, industrial production, commerce and for recreational, aesthetic, and conservation purposes as considered desirable by the body politic; and
3. To accomplish all of the above with a minimum expenditure of the available physical, economic, and human resources.

The formulation of objectives is an initial step in the planning and developmental process. The systems approach demands an explicit articulation of the objectives in the form of an objective function by which the output of the system can be determined, given the policy, initial values of the state variables, and system parameters. Systems analysis for water resources project planning is carried out with the basic objective of meeting certain fundamental requirements in an optimal way, by satisfying some economic or other criteria of optimality. The task of water resources systems planning may be:

(a) to plan new water resources projects for the least cost or optimum output, or
(b) to plan the enlargement of a system in an optimal way so as to meet its present functional requirements in a better way or to fulfill some new functions.

Detailed plans help water resources utilization in many ways. Planning helps with the assessment of the present situation in the basin, the situation desired, the gap between the two, and the means to bridge the gap. It also helps set priorities. Second, planning offers a framework and focuses on policy analysis and development and organizes public participation for each individual decision. Third, when planning processes proceed in an open and participatory mode, they result in wider public support and acceptance of the final plan. Lastly, planning has a coordinating effect because, by virtue of its nature, it forces interaction among concerned agencies.

1.5 WATER RESOURCES DEVELOPMENT

In countries with limited water resources, comprehensive and rational water resources development is a necessary condition for optimum social and economic growth. There are various ways of classifying water resources projects; in the classification based on physical nature, the broad project categories are:

1. Surface storages: reservoirs, natural lakes with artificial control of outflows.
2. Channelization: irrigation canals, navigation canals, drainage works, dykes for flood protection, and erosion control measures.
3. Diversion of water: interbasin water transfer projects.
4. Waste treatment and assimilation.
5. Groundwater extraction and artificial recharge.
6. Catchment treatment for the control of water yield and peaks.

In water resources development, there are many aspects of a question—the problem is often complex and has a multiplicity of goals and alternatives. Different persons can have diametrically opposite views on various aspects, and yet each of them may claim to be right. For example, there are widely divergent opinions about the benefits of dams. In fact, this has been the topic of passionate and often bitter debates. After a dam has been constructed, it is nearly impossible and very expensive to restore the status quo. Evidently, there is only one chance to develop the best course of action, necessitating a logical procedure that can rationally eliminate alternatives and reduce thousands of decisions to a relatively few. All this should be on the basis of a rather voluminous mass of information of divergent accuracy. The conventional methods for the development and management of water resources are often not suitable for handling complex problems of planning and management of water resources at the river basin level. Systems engineering is a powerful technique for quantitative analysis of the planning and operational problems of large river basins, as it allows consideration of complex issues in their totality.

Water is already fully allocated and exploited in many basins. Gradually, many basins are inching toward this distinction! Therefore, the elbow room for water resources development is shrinking, and in future people will have to work within tighter constraints. In this context, it is pertinent to refer to the concept of "hardcore" and "softcore" projects that were advanced by Yevjevich (1983). Referring to the development in the Nile basin, he labeled the large-scale water resources projects, such as the old Aswan dam, the High Aswan dam, and the future projects of similar size and significance as the "hardcore projects." Many other water-related activities will be needed in the basin in the form of widespread, medium or small-scale projects that will increase the available water, regulate flows by various types of storage, control sediments and salt, use water more efficiently, and solve other water-related problems. The investment for such projects will be equal to or greater than the hardcore projects if the maximum economic benefits are to be attained. These additional activities were termed as "softcore projects" by Yevjevich (1983). He noted that without the hardcore projects, many softcore projects could not be implemented. Without the effects of softcore projects, many hardcore projects may not lead to the highest efficiency in water use and to the greatest benefits. Although these observations were made with particular reference to the Nile basin, they are equally valid and relevant to every basin that is subject to large-scale development. However, in concordance with information technology jargon, more familiar terms are "hardware" to denote physical projects and "software" to denote management policies.

1.6 WATER RESOURCES MANAGEMENT

Worldwide, environmental systems are rarely managed as a cohesive whole. That is partly because air, soil, and water resources have been and continue to be managed by independent organizations having little interaction among them. For example, there are a host of organizations in the United States that have the authority to manage the nation's water resources at the local, state, regional, and national levels. The US Bureau of Reclamation and the US Army Corps of Engineers are two of the federal agencies with broad authority for water resources management. Then, there are the departments of water resources, natural resources, or environmental quality at the provincial (state) level which have the authority to manage water resources in their individual states. At the local (county or parish) level, there are water boards, commissions, and the like that manage local water resources. Similar multi-agency systems exist in most countries.

Integrated water resources management is best accomplished within a spatial unit called a river basin or watershed. Hirsch et al. (1977) coined the term *synergistic gain* to denote benefits due to the joint operation of a system of reservoirs, in excess of the benefits from optimal individual operations. These gains arise as a result of the diversity of flows in several streams used in the water supply system. These gains are realized by employing a flexible operating policy, and the ability to capture these gains is limited by physical constraints (such as canals to take water from the source to the demand center) and the ability to forecast future flows. Note that additional costs will have to be incurred to reap synergistic gains. Naturally, these should exceed the benefits.

Integrated water management can be viewed as a multi-dimensional process pivoted around the need for water, the policy to meet the needs, and the management to implement the policy. Water elements encompassing physical, chemical, and biological aspects of water quantity and quality may constitute the first dimension. Water uses, including agriculture, water supply, energy generation, industry, fish production, recreation, and transportation, may constitute the second dimension. Clearly, these water uses have to be accomplished following a well-defined management policy balancing the demand for water among different uses. The strategy to implement this policy forms the third dimension. The management must be dynamic and evolve with time, in response to changing needs and objectives. Thus, these needs and objectives may form the fourth dimension.

The water elements, their interactions, and the effects of natural as well as external constraints on them constitute the foundation upon which the edifice of integrated water management is to be built. In a river basin, a finite amount of water is available which is to be allocated for different uses, such as agriculture, water supply, industrial, energy generation, etc. (Fig. 1.5). External constraints, such as economic, demographic, transportation, and other forms of development, directly influence one or the other water elements. Likewise, climatic vagaries, climatic change, climatic extremes, and a host of natural hazards are some of the natural causes that greatly influence the water elements and have a significant impact on integrated water management.

Once a management policy is established, a strategy, including administrative infrastructure, has to be employed to undertake integrated water management as shown in Fig. 1.6. The components of

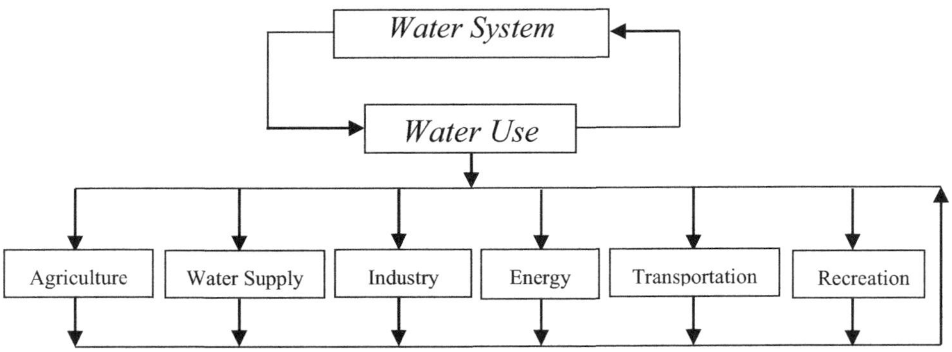

FIGURE 1.5

Elements of a water system and their interactions.

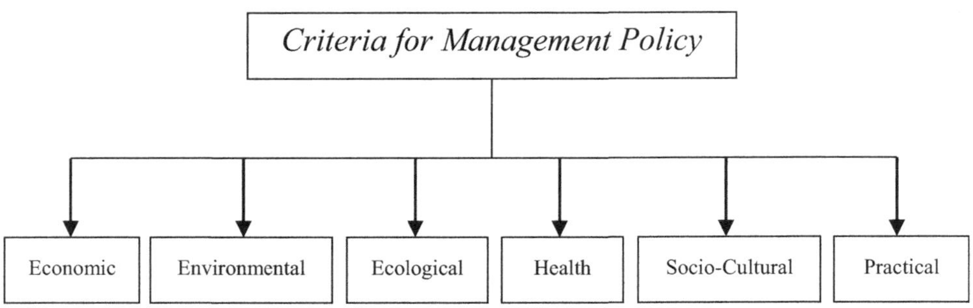

FIGURE 1.6

Criteria for foundation of a management policy.

integrated water resources management are interactive, and hence the administrative setup must be flexible and responsive to changing goals. Thus, integrated water resources management requires the integration of various components discussed earlier—physical, biological, chemical, social, economic, ecological, health, and environmental. This can be accomplished through the development and application of mathematical models. The physical, chemical, biological, environmental, and ecological components and their models must be embedded in the development of comprehensive watershed models.

The criteria, as shown in Fig. 1.6, that form the foundation of the management policy must be developed following such considerations as cost effectiveness, economic efficiency, environmental impact, ecological and health considerations, sociocultural aspects, to name but a few. The criteria must be practical, implementable, and acceptable to the society at large. Furthermore, they must be capable of responding to the changing needs of the society. The resulting policy must satisfy interactive social goals such as equity, efficiency, and environmental quality.

A model for integrated water management helps develop implementable solutions to water resources problems by combining into an optimization scheme all the essential component models. The model incorporates or accumulates all of the interactive forces or influences. Hence, it aids the decision-making process and keeps the policy results within the intersection of social goals of the management policy and the legal constraints. Such a model is shown in Fig. 1.7.

The United Nations Water Conference in Mar del Plata in 1977, the International Conference on Water and the Environment in Dublin, and the Earth Summit in Rio de Janeiro in 1992 have articulated a set of principles for good water resources management. These are referred to as the *Dublin Principles* and are:

1. The "ecological principle," which requires that water be treated as a unitary resource within river basins, with particular attention to ecosystems.
2. The "institutional principle," which recognizes that water management requires the involvement of government, civil society, and the private sector and that the principle of subsidiarity be respected. It also gives special emphasis to the role of women in water management.
3. The "instrument principle," which requires that water be recognized as a scarce economic good and that greater use be made of "user pays," "polluter pays," and other market-friendly instruments.

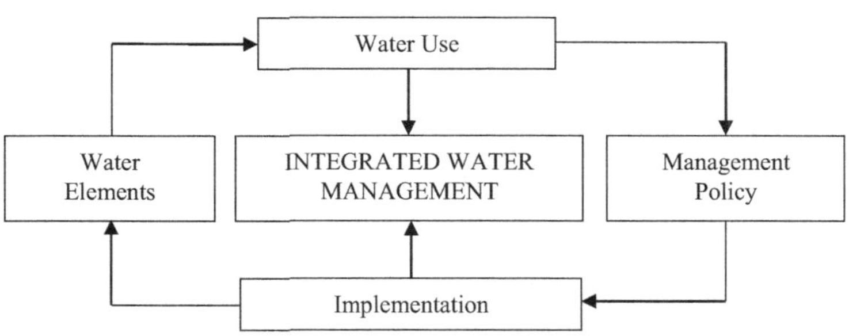

FIGURE 1.7

Integrated water management.

For any water resources, utilization to be effective and efficient, it is necessary that all the resources of a basin are managed in an integrated manner. By definition, basin-level systemic management is essential for the integrated water resources management (IWRM) principle to succeed. In other words, it is essential to take a holistic approach to IWRM. The decisions on IWRM must be participatory, technically and scientifically informed, and taken at the lowest appropriate level, but within a framework at the catchment, basin, and aquifer levels, which are the units by which nature bestows water.

Holistic river basin management is best practiced by agencies that operate at the basin level. There is a clear hierarchy in the spatial domain: basin, subbasin, etc., and many smaller units are nested within the larger one. The functions of the basin agencies must also reflect that hierarchy, with decision-making pushed down to the lowest appropriate level. It is equally imperative that decision-making must be informed and scientifically and technically sound. Effective river basin management, thus, walks on two legs: (1) institutions where users make policies and decide on the raising and spending of money; (2) competent technical agencies which provide the institutions and users with the information necessary for management.

The technology can, must, and will change to adapt to the needs of the more water-conscious world. The new technologies must be less wasteful and more sensitive to the environmental and social dimensions of decisions. It is good to draw on traditional wisdom to the extent possible while harnessing the revolutionary changes that are taking place in the electronics, information, space technology, and computer fields. But technology alone cannot bring about a more desirable future. It will require behavioral change at all levels of society everywhere. Moreover, technological change is not exogenous; it comes only if the society provides stimulation and incentives for innovation.

Despite much technological advancement, a lot remains to be achieved in the area of water resources development. Consider hydropower development, which is an environmentally friendly means to generate hydropower. In the French language, the term for hydropower is "La Houille Blanche," which means white coal. Although many areas of the world face shortages of electric power worldwide, only about 20% of the hydropower potential had been developed by 2000 (Veltrop, 1992). The cumulative hydropower capacity reached approximately 1360 gigawatts in 2020 (https://www.statista.com). The technically feasible hydropower potential of the world has been estimated at 12,900 TWhr/year, and the economically feasible potential is about 7200 TWhr/year. Table 1.5 shows the continent-wise percentage of hydropower potential developed.

The electricity supplied by hydropower far exceeds the capacity of any other renewable energy resource. Norway meets virtually the entire (99.6%) electricity demand by hydropower. Twenty-five countries worldwide depend on hydropower for more than 90% of their electricity needs. Although there is dominance of fossil fuels for electricity generation worldwide, more than 60 countries currently use hydropower for half or more of their electricity needs. Most of the installed hydroelectric capacity resides in North America, Brazil, Russia, China, and Europe.

Major improvements can and have been made in the way in which water is managed. Participatory river basin management has become a reality in some countries. Due to various measures, per capita quantity of water used has actually declined in a few cases over the past decade. While serious water problems persist in rich countries, the situation is far more challenging in developing countries. On almost all counts—service coverage, reliability of meeting demands, and water quality—poor countries have much worse conditions, and they have rapid growth in demand for municipal uses, for

Table 1.5 Continent-wise percentage of hydropower potential that has been developed.

Continent	Percentage of hydropower potential that has been developed	Percentage of electricity generated by hydropower
Africa	7	2
Asia	20	39
Australia	40	2
Europe	65	13
N. America	61	26
S. America	19	18

Source: Internet.

agriculture, for industry, and for the generation of electricity. Still worse, these countries have only a fraction of the financial resources available compared to rich industrialized countries.

Timely availability of reliable information is crucial in all management activities including those for water. The emerging information technology is helping to set up global, regional, and local databases. For example, AQUASTAT (http://www.fao.org/waicent/faoinfo/agricult/agl/aglw/aquastat/main) is a global information system of water and agriculture, established by the Food and Agriculture Organization of the UN. The objective of AQUASTAT is to provide users with comprehensive information on the state of agricultural water management across the world, with an emphasis on developing countries and countries in transition.

The technological changes alone are not enough, and it is futile to believe that just by changing institutional arrangements, the desired results can be achieved. The crux of the problem lies in creation and adoption of appropriate technologies, adequate infrastructure (including manpower), and flexible and goal-directed institutional arrangements.

Scientists from the Centre for Ecology & Hydrology (UK) and the World Water Council (http://www.worldwatercouncil.org) have developed an international *Water Poverty Index* (WPI) which grades 147 countries according to five different measures—resources, access, capacity, use, and environmental impact—to show where the best and worst water situations exist. The WPI assigns a value of 20 points as the best score for each of its five categories. *Resources* measure the per capita volume of surface and groundwater resources that can be drawn upon by communities and countries, while *access* measures a country's ability to access water for drinking, industrial, and agricultural use. Many countries have the economic capacity to provide safe water supplies and sanitation to their whole populations. *Capacity*, another WPI component, defines a country's level of ability to purchase, manage and lobby for improved water, education, and health, and *use* is a measure of how efficiently a country uses water for domestic, agricultural, and industrial purposes. In the last factor, *environment*, which provides a measure of ecological sustainability, issues included are water quality, environmental strategies and regulation, and numbers of endangered species. A country that completely meets the criteria in all five categories would have a score of 100.

According to WPI, some of the world's richest nations fare poorly in water ranking, while some developing countries score high rank. The highest-ranking country, Finland, has a WPI of 78 points.

WPI demonstrates that it is not the amount of water available that determines poverty levels in a country, but the effectiveness of how those resources are used. Some users consider the term "poverty" as being pejorative and hence attempts have been made to develop a Water Wealth Index (WWI). Sullivan et al. (2006) describe selected parameters for use in a WWI.

1.7 WATER RESOURCES SYSTEM

Before initiating a discussion on water resources systems, the basic concept of a system is briefly discussed.

1.7.1 Concept of a System

A large number of definitions of the term *system* are available. A system may be defined as *a set of objects which interact in a regular, interdependent manner.* Sinha (1991) defined a system as a collection of objects arranged in an ordered form, which is, in some sense, purpose or goal directed. According to Mays and Tung (1992), a system is characterized by: (1) A system boundary which is a rule that determines whether an element is to be considered as part of the system or of the environment, (2) statement of input and output interactions with the environment, and (3) statements of interrelationships between the system elements, inputs and outputs, called feedback.

In the context of water resources, Dooge (1973) and Singh (1988) defined a system "as any structure, device, scheme, or procedure, real or abstract, that inter-relates in a given time reference an input, cause or stimulus, of matter, energy, or information, and an output, effect, or response, of information, energy or matter."

Systems analysis, as applied to water resources, is *a rational approach for arriving at the management decisions for a particular system, based on the systematic and efficient organization and analysis of relevant information.* The use of systems techniques requires digital computers, and therefore, applications of systems analysis to water resources problems started only with the computer age in the early 1960s. Votruba et al. (1988) defined systems approach as a comprehensive method of investigation of phenomena and processes, including their internal and external relationships. When scarce resources must be used effectively, systems analysis techniques stand particularly promising.

There are many ways to classify systems. A physical system is the one that exists in the real world. A sequential system is a physical system that consists of input, output, and some working medium (matter, energy, or information) known as throughput passing through the system. Singh (1988) and Sinha (1991) classified systems as: (1) static and dynamic systems, (2) linear and nonlinear systems, (3) time-varying and time-invariant systems, (4) deterministic and stochastic systems, (5) continuous-time and discrete-time systems, and (6) lumped parameter and distributed parameter systems. The output of a static system depends only on the current inputs, while output of a dynamic system depends on the current and previous inputs. As explained by Mays (1997), the kernel of a time-invariant system does not change with time, whereas it changes with time in a time-variant system. The kernel and inputs of a deterministic system are known exactly, while those for a stochastic system are not known exactly. For a stochastic system, either the parameters in the kernel or the inputs are not known exactly and are described by statistical concepts. In a continuous time system, the time varies continuously, while the inputs, outputs, and parameter values in the kernel are known at discrete times only.

Water resources systems are generally distributed with respect to time and space. For the purpose of solution, these systems may be divided into subsystems, and each subsystem may be treated as lumped. The hydrologic system is, therefore, a physical, sequential, dynamic system. For a catchment system, the input consists of water and energy in various forms. The input-output relationship of the system may be represented mathematically (Singh, 1988) by:

$$y(t) = \Phi[x(t)] \tag{1.1}$$

where $x(t)$ and $y(t)$ are, respectively, time functions of input and output, and $\Phi[]$ is the transfer function, which represents the operation performed by the system to transform the input into output. The well-known unit hydrograph is an example of a transfer function of the catchment system.

The concept of state is basic to the systems theory. The state represents the condition of the system or is an indicator of the activity in the system at a given time. In water resources systems, the state typically may be the volume of water in the reservoir, the depth of flow of the river, or the head of groundwater at a location.

If the behavior of the system can be altered by the modification of the working of the system, the steps taken are known as exerting control. The control applied to hydrologic systems may be either natural or artificial. For example, in a catchment system, the climatic trend and cycles are the nature-applied controls that may alter the characteristics of the catchment. The man-made dams and reservoirs in river basins represent artificial controls.

Systems engineering is concerned with making decisions with respect to those aspects of a system on which some control can be applied. Water resources systems engineering can make the most significant contribution to the process of decision-making. Systems engineering was defined by Hall and Dracup (1970) as "*the art and science of selecting from a large number of feasible alternatives, involving substantial engineering content, that particular set of actions which will accomplish the overall objectives of the decision makers, within the constraint of law, morality, economic resources, political and social pressures and laws governing the physical life and other natural sciences.*" Thus, systems engineering is useful in making selections from a large number of alternatives by way of elimination. The necessity for elimination may be readily noted by a very simple example. Suppose there are 20 farming areas, and water is to be allocated to each of them. Let there be 100 alternative ways one could allocate water to any one area, each being different (say, in terms of quantity) and theoretically requiring an "evaluation" to see which might be the "best." For 20 areas, there would be $(100)^{20}$ different combinations. Suppose neither judgment nor systems analysis is used to find the "best" of these different combinations of allocations to the 20 areas. Let a high-speed computer be employed and be capable of completely evaluating one alternative in 0.001 s, compare the "worth" of that alternative to any other, and keep the best of the two. After one "earth age" (estimated by geologists as 3×10^9 years), it would be able to check out 9.45×10^{19} of the possibilities, which is only an infinitesimal fraction of 100^{20}.

Obviously, even simple problems involving just a few different values of the decisions cannot possibly be evaluated directly. In the problem cited, one could have used a bit of logic based on the knowledge of the 20 farming areas and the effects of different levels of water supply and picked a pretty good solution on the basis of judgment alone. However, even for simple problems, after applying judgment, there may still be a lot more potentially valuable alternatives that cannot be evaluated directly with the available time and money. It is for these situations that the science of systems engineering is best suited.

The theories and methods used in systems engineering are termed as systems analysis. Applied systems analysis is a general term including fields like operations research, decision theory, benefit-cost analysis, planning and scheduling, design, theory of information, application of artificial intelligence in management, and decision-making. As noted by Votruba et al. (1988), in view of the importance of systems analysis, a famous institute, namely IIASA (International Institute for Applied Systems Analysis, Laxenburg, Austria, website: www.iiasa.ac.at), has been specifically set up for this purpose. The range of fields and problems contained in applied systems analysis is apparent from the research program of this institute, which includes:

- Resources and environment: ecological problems, problems of water resources research, problems of food and agriculture, etc.
- Human settlement and services: problems of population, health, education, communication, etc.
- Management and technology: man-made artifacts, institutes, economic systems, technologies, etc.
- Systems and decision sciences: mathematical and computational problems in the analysis of large systems, etc.

1.7.2 Systems Analysis Techniques

The systems techniques can be grouped under four major categories as follows:

a. *Analytical optimization models and techniques*: This group includes optimization methods—they may be based on classical calculus and Lagrangian multipliers or mathematical programming and control theory. These modeling techniques are descriptive, that is, they usually incorporate quantitative relationships between variables of the system. These are also prescriptive in the sense that the algorithm provides the optimal solution. The mathematical programming techniques include linear, nonlinear, and dynamic programming, goal programming, and multiobjective optimization.
b. *Probabilistic models and techniques*: This group of techniques includes the techniques for analyzing stochastic system elements with appropriate statistical parameters. It encompasses all the descriptive techniques of stochastic processes to study the behavior of some aspects of the system. The important techniques in this group are the queuing and inventory theory, which are concerned with the study of queues or waiting times and inventory stocks. Such studies are associated with decisions regarding service and storage capacities. Often queuing models are combined with other optimization methods and utilize either analytical techniques or simulation and search approaches. Many reservoir problems are some type of inventory problems and have been solved using approaches that combine various techniques.
c. *Statistical techniques*: This class of techniques includes multivariate analysis and statistical inference. The techniques of multivariate analysis, including regression and correlation, factor analysis, principal component analysis, and discriminant analysis, have numerous applications in the water resources area.
d. *Simulation and Search techniques*: Simulation is a descriptive technique. A simulation model incorporates the quantifiable relationships among variables and describes the outcome of operating a system under a given set of inputs and operating conditions. Most simulation models do not contain algorithms for seeking optimal solutions. However, such models usually permit a

far less drastic simplification and approximation than is required when using an analytical optimization model. Often, a simulation model is run many times with various input and parameter data. The outputs of these runs describe the response of the system to variations in inputs and parameters. If the simulation model includes an objective function, the values of the objective for several runs generate a response surface. The model then can be combined with sampling or search techniques that explore the response surface and seek near-optimal solutions.

Systems analysis techniques do not merely deal with the engineering aspects of water resources development but also cover a multidisciplinary approach encompassing physical, social, economic, political, biological, and other characteristics of specific problems and situations. This powerful technique enables the planners, designers, and water managers to evaluate alternative development scenarios and to place before the decision-makers the effects and advantages of various alternative feasible options. Mathematical analysis of water resources problems using a systems approach is one of the most important developments in the water resources sector.

1.7.3 Characteristics of Water Resources Systems

Water resources systems can be defined as a set of water resources elements linked by interrelationships into a purposeful whole. As an example, a water supply reservoir for a small city, linked with a water distribution network, would constitute a system. In many respects, water resources systems defy rational description. To an engineer, these systems may be dams and weirs, tunnels, levees, pipelines, electrical power plants, water treatment and reclamation, spillways and similar physical works which have been constructed to provide certain benefits. An economist views them from the point of view of economic efficiency, income redistribution, and stimulation of economic growth. To a lawyer, a water resources system is a device for the implementation of water rights. To those living in an arid environment, water resources systems mean food and fiber, homes and jobs, and laws and politics. To many conservationists, water resources systems are unwanted interventions, responsible for the destruction of wild rivers, scenic beauty, and wildlife habitat. Water resources systems indeed include all these points of view, which could be physical, technological, sociological, biological, legal, geological, and agricultural. The famous water resources expert, the late Prof. V.T. Chow, coined the term "hydro-systems" to describe collectively the technical areas of hydrology, hydraulics, and water resources (Mays, 1997). This term has also been used to refer to systems for groundwater management, reservoirs, and water distribution systems.

The elements of a water resources system can be either natural (rivers, lakes, glaciers, etc.) or artificial (reservoirs, barrages, weirs, canals, hydroelectric power plants, etc.). The relationships between the elements are either real (e.g., water diversion) or conceptual (e.g., organization, information, etc.). Most of the components of these systems are artificial and are constructed to attain a certain goal. Each element of a system may also be considered as a subsystem if it has the qualifications of a system. Whether a particular object or concept is to be treated as a subsystem depends mainly on the objective of the study. The decomposition of a system is to be done carefully. According to Hall and Dracup (1970), since certain interactions internal to the entire system will become inputs and outputs between subsystems, there is considerable logic in setting subsystem boundaries so as to minimize their number. All subsystem inputs become decision variables in the subsystem optimization; hence they may control the computational feasibility of analysis. In general, the decomposition should permit a functional analysis so that subsystems can be replaced in the complete system analysis by an optimal input-output relationship. Besides, the decomposition model and the methods of analysis selected should permit a maximum of analysis of the consequences of imposing factors existing in the real system that were excluded from the analysis. This will help the decision-maker appreciate the problem in a rational manner.

1.8 ISSUES IN SYSTEMS APPROACH

Despite significant theoretical developments and the availability of infrastructure facilities, systems techniques are not routinely applied in the planning and design of water resources projects. Rogers and Fiering (1986) carried out an extensive study in which they examined more than 2500 research papers from selected reputed journals dealing with water resources. It was reported that only 38 of 723 systems-oriented papers dealt with identifiable water resources projects, only 3 of the 38 projects studied were built, and only one of these was designed according to the optimization model presented in the paper. Although considerable time has elapsed since this study, the situation may be only marginally different now. This analysis also points toward a wide gap between theory and practice in planning and design. The situation with respect to applications for management is somewhat better. Then, there are differences between developed and developing countries—there is a resistance to the use of systems analysis by many government agencies in developing countries. The resistance is not always based on blind prejudices; it may be based on some genuine and reasonable doubts. According to Rogers (1980), the following five questions need to be satisfactorily answered to convince those who are opposed to systems analysis.

Appropriateness: Most of the applications of systems techniques have been made in developed countries. Decision-makers in developing countries are quite rightly suspicious of the direct transference of such methods into situations in which the distribution of resources, the structure of society, and the governing institutions are so different. This question is best answered by successful case studies of applications in developing countries.

Reliability of database: In most developing countries, the data on economic, social, and natural resources are usually very poor in the conventional sense of time series, spatial coverage, and accuracy. Most systems analysis applications are very data intensive and very sensitive to the quality of data. In such situations, responsible administrators have every reason to be suspicious of the use of systems analysis. This question is best answered by showing how systems analysis can improve decision-making even using the shakiest data and how it can also be used to evaluate the existing data and structure additional data acquisition. It can be argued that with a less reliable database, the benefits from using systems analysis will be much greater than conventional analysis.

Model credibility: Obviously, if no models have been attempted, how can one assess their credibility? Again, the answer is to rely upon case studies of other applications which are deemed credible by their sponsors. Note that there are many levels of application and many different user groups for systems analysis models. Some may find them credible, while others may or may not.

Manpower and equipment requirements: In most developing countries, there is a shortage of skilled manpower, and this is an important hindrance to the use of systems analysis in these countries. A long-term solution is to develop indigenous expertise by introducing relevant courses in academic institutions or through foreign training. The analysis could be tailored to the available resources and reliable personnel hired whenever the need arises.

Time and money aspects: The systems analysis studies cost more than the conventional techniques and require more time. However, it should be borne in mind that the magnitude of benefits from water resources projects is very large. An improvement by a few percentages will translate into big amounts in real terms, and costs will be much smaller than the benefits accruing due to improved planning and management.

The key to convince reluctant administrators to try to use systems analysis is to provide examples drawn from applications in similar situations.

1.8.1 Potential of Systems Analysis Approach

The systems analysis approach is promising and useful for problems dealing with the management of scarce resources. Water resources are getting scarce worldwide in view of ever-increasing demands. Every year, some new regions, which earlier had surplus water, are entering the domain of water scarce or water stress areas. In such a scenario, applications of systems analysis should show a rapid increase. However, such a growth is not seen and there appears to be a resistance to the use of systems analysis by many government agencies. In developed as well as developing countries, the major responsibility for planning, development, and management of water resources projects lies with government agencies. With the developments in information technology and the availability of high-capability computers at much cheaper rates, there is undoubtedly an increasing use of computers in the analysis. However, just the use of computers will not make the development and management better. The inputs, the analysis techniques, and the implementation, all have to improve simultaneously so that extra benefits can be realized.

In many instances, the location and size of the projects are decided from political rather than technical considerations, and as a result, there is not much scope for optimizing the decision variables. Another pertinent reason is that most of the good project sites have already been developed, and now it is more of a problem of developing the best among the less suitable sites. Furthermore, a large number of agencies are usually involved in the water sector, and there is not much parallel communication and coordination among them. Due to this difficulty, the application of improved and relevant techniques is hampered.

Application of systems analysis techniques requires a huge and reliable base of hydrologic as well as economic, demographic, and social data, which are usually not available in many countries. In the absence of such a database, values are assumed based on experience, and sometimes parameters of a completely unrelated distant region are used just because they are readily available.

In many situations, there is a shortage of manpower with required skills in the concerned government agency, and this limits the possibility of the use of refined techniques. Usually, people with required skills are available only in academic and research institutes, but the necessary coordination between these institutes and field organizations may be lacking. A large number of pilot projects have been undertaken in many countries wherein advanced techniques were applied to specific projects. However, it is common to see that after the project is over, the trained manpower is gradually transferred, the consultants withdraw from the scene, and the infrastructure tends to wither away with time. Usually, within a few years, things return to the preproject stage. To overcome these problems, it would be necessary to create teams of dedicated and skilled workers, adequately motivated and compensated, within the concerned organization, and there should be a commitment at the senior level for sustained support to maintain the created infrastructure.

Increasing withdrawals from the water bodies would impose intolerable stresses on the environment, leading not only to the loss of biodiversity but to a vicious circle in which the stressed ecosystems could no longer provide services for plants and people. Unfortunately, precise figures for only limited regions are available to indicate freshwater availability and its quality. Systematically obtaining and updating such information is necessary for sound water management decisions in the future.

1.8.2 Economics in Systems Engineering

Among the major justifications of water resources projects is the role that they play as a catalyst for economic development. A geographic area may have all other necessary resources (e.g., soil, people, climate, location, etc.) for a high level of economic productivity but it is the availability of water that stimulates rapid growth. Therefore, in the water resources systems studies, economic aspects play an important role. Economic analysis provides a criterion for ranking different water development and management policies. In any scheme of water resources development, it is necessary to prepare a number of technically feasible alternatives for meeting particular objectives and assess the related costs and benefits, both tangible and intangible, for each alternative.

Each technically feasible alternative has gross benefits and costs in a given time, which are discounted to determine their net present worth. The difference between benefits and costs reflects the project's contribution to regional/national income. A multiobjective planning further expands the scope of analysis and involves broader social goals that water resources development might help attain a larger national income, a more equitable distribution of income among people and region, and environmental protection. The importance of socioeconomic factors in the execution of water resources development plans makes it essential to undertake a socioeconomic benchmark survey, which provides useful inputs for development plans. The findings can also form a benchmark against which the future improvements in the economic welfare of the population resulting from the project can be assessed.

The definition of gains and costs from water projects must include a broad spectrum of effects on social goals rather than be limited to readily quantifiable profits and expenses. There are many intangible factors that cannot be expressed quantitatively but influence the functioning of the hydroeconomic system. Economic impacts are not necessarily uniform; the gains may accrue to some who may not necessarily be the direct users, while the costs may fall on them as well on others. These inequities in costs and benefits are somewhat corrected by subsidies and taxes.

An important aspect of water management is the adoption of cost for water services. It is well known that free water leads to wastage, and consumption drops when water is metered. For example, in many irrigation systems where water is highly subsidized or free, only about 30% of the water supplied is actually used by plants. Where services are free, the result is inevitably politicization of the concerned agencies, inefficiency, and lack of accountability. Ultimately, subsidies are usually grabbed by influential groups, while the poor end up paying high prices for inferior and unreliable services. Usually, they also bear the brunt of environmental degradation.

Governments in many developing countries are not in a position to meet the investment demands for water services. The main alternative is to attract private investment for municipal water supply and sanitation, irrigation, and hydropower. However, the private sector will not enter into this venture unless they are sure of a reasonable return on their investments and there are no political risks.

Naturally, this return has to come from those who benefit from the services provided. Thus, without full-cost pricing, the present vicious cycle of wastage, inefficiency, and lack of reasonable services will continue. There will be little investment from the private sector, services will be of poor quality and rationed, and there will be little left for investing in water quality and other environmental improvements. This vicious circle needs to be converted to a "virtuous cycle" in which users pay for the services they want, and urban utilities provide these services efficiently and accountably. While doing this, the government should provide safety nets for poor people. They must also ensure that subsidies are provided to people, not to service providers.

1.9 ADVANTAGES AND LIMITATIONS OF SYSTEMS APPROACH

The systems approach for the analysis of water resources systems takes a broad and flexible view. It is a very powerful tool, capable of dealing with large-scale problems. Selection of an optimal plan is achieved through a systematic search and evaluation of various alternatives, which meet the objectives. It is flexible in suggesting modifications of the course of action to correct and compensate for any undesirable consequence of plans and operating decisions.

Essentially systems analysis consists of five steps:

(a) Define the problem.
(b) Identify the system, define its elements, and gather relevant data.
(c) Define the system objectives and constraints.
(d) Generate feasible alternatives that satisfy physical, social, political, economic, and legal constraints on the system and its management.
(e) Evaluate the alternatives for attaining system objectives and identify the most suitable among them.

As indicated in the previous section, usually, the magnitude of benefits and costs of water resources projects is quite large. Therefore, a saving of cost or increase in benefit, even by a few percentage points, translates into a big sum of money and, therefore, is worth the effort.

The systems approach has many advantages over the conventional approach in water resources planning and development. The approach focuses on the selection of definite goals and objectives and on a systematic search for alternatives. It provides the systems planner with a modern technology to analyze the system scientifically and objectively. It forces the user to identify the known and not readily known elements of the system. The approach regularly provides feedback information from each step in the analysis and thus provides flexibility for feedback correction and modification of the system. This feedback can also be used for the adoption of new system definition, system elements, and system objectives.

The high complexity of present-day problems is an important reason to apply systems techniques. In single-purpose projects, the analysis is simple, and the decision to build or not to build a project can be more easily evaluated on the basis of comparing the least-cost engineering design. But with today's increased concern for economic efficiency and environmental issues, it is now no longer a simple task to assess the allocation of water among competing demands, assess the benefits, select the best design from many alternatives, and optimally operate the system. This forces the planner, the designer, and the manager to take a wider view of the water resource system. The best in the spectrum of tools for this is the systems analysis. The following characteristics clearly bring out the demand for this approach:

(1) Water resources projects typically involve large-scale, more or less permanent, physical changes in the environment, large lakes are created, irrigation water is provided, and flood plains protected. The impact on regional economy and society are immense.
(2) The expertise of many traditional disciplines is simultaneously utilized (engineering, economics, agronomy, legal, social, and so on).
(3) The size and capital-intensive characteristics of investments in water projects, especially in the face of budgetary constraints, and the fact that they have a major effect on the economy of the region necessitates the desirability of achieving even small improvements over traditional solutions.
(4) The overall project output achieved with the approach of systems analysis is likely to be better than that achieved by an experienced staff using conventional techniques.

A word of caution here. Systems analysis is not a suitable tool in the hands of those who lack a full understanding of the water resources systems and the multiplicity of objectives. Most of the decisions related to water resources projects are quite irreversible in nature. It is equally dangerous when utilized by those who lack a reasonable appreciation of both the power and the limitations of the methods of systems analysis. A water resources systems analyst who does not recognize and integrate many quantitative and nonquantitative dimensions of the system (physical, social, economic, political, geological, etc.), to the greatest extent possible will only produce an academic exercise at best. A more likely result will be a serious mismanagement of a vital resource.

The systems approach presents some difficulties in its practical use for water resources planning and development. Most theoretical works on systems analysis are being done in universities and research agencies, which may not be concerned immediately with practical problems. The mathematics and statistics involved in the systems approach are more advanced now. Due to this, there is a big gap between theory and practice. In this scenario, the transfer of technological advances to practical use is an important task. It is most urgent and important to fill in or narrow this gap so that new developments can be soon introduced into practical applications. This requires a group of people with both theoretical background and practical experience who are willing to organize technology transfer courses.

Another important concern is mathematical computations. Most water resources systems are complex, and computations are lengthy. Fortunately, the computer hardware needed for such problems is now widely available, although software is quite expensive and may not be readily available. The availability of adequate data is another problem in the systems approach. When a complex system model is formulated, its utility is severely diminished if the requisite input data are not available.

Yet another difficult problem in analyzing the entire water resources system is dealing with intangibles. Water resource problems are not so simple that they can be expressed entirely in mathematical terms. Whereas the mathematical systems analysis is helpful in dealing with tangible factors, the overall planning and management of water resources involves intangibles, and one has to depend on judgment and experience to evaluate the intangible effects. With increased dependence on computers for decision-making, the human element should not be eliminated. Fiering (1976) has succinctly warned against overkill: "we seek optimal plans, optimal operating policies, optimal estimates of parameters, optimal anything ... It has become a new religion." The need is to have the model *serve* the decision maker rather than *the model* driving the underlying and important decisions.

1.10 CHALLENGES IN WATER RESOURCES SECTOR

The new century poses new challenges in water management. Whether a water emergency turns into a disaster depends on whether the community can take effective measures without external assistance. The long-term impacts of water disasters are usually due to the lack of prompt restoration of services and interventions. The effects of climate change and phenomena, such as the El Niño Southern Oscillation (ENSO), also require urgent attention. The recent ENSO during 1997−98 was particularly severe in causing natural disasters that, according to estimates, affected about 160 million people. The slow time scale of droughts means that its onset may not be identified until it has persisted for a considerable time.

1.10.1 Climate Change and its Impact

Water resources systems are sensitive to climate change. The mean land surface air temperature from 2006 to 2015 increased by 1.53°C as compared to the preindustrial period level (IPCC, 2019). Global warming is alleged to have contributed to increased frequency, intensity, and duration of weather extremes characterized by high temperature, extreme precipitation, wind, heat-related events, and flooding; decreased water availability; increased soil erosion; coastal degradation; permafrost degradation; and increased wildfire occurrence. These changes pose increased risk to water security, energy security, food security, as well as human health and ecosystem integrity. Fig. 1.8 shows schematically how climate change and human activities affect hydrologic changes (water quantity and quality, timing, and extreme events) and their impact and risk for human beings and ecosystems. Freshwater systems and climate change are interconnected in different ways. Wetlands in dry regions are sensitive to climate change due to decreased runoff. Land use and land cover change may amplify the effect of climate change on water resources. Climate change-induced higher water temperatures, reduced river flow, and increased precipitation intensity exacerbate water pollution, which may affect ecosystem services, biodiversity, and human health.

The impacts of climate change on hydrologic systems and water resources management can be summarized as follows: (1) Freshwater availability: renewable surface water and groundwater sources are projected to decrease in most dry subtropical regions and increase at high latitudes. (2) Freshwater storage: higher air temperatures reduce snow and ice water storage and increase evaporation from lakes, reservoirs, wetlands, glaciers, and shallow aquifers, resulting in decreased natural water storage. (3) Rainfall variability: less rainfall is projected in most dry subtropical regions and Mediterranean climates. A substantial increase in the occurrence of heavy rainfall events is expected, even in regions with decreased total precipitation amounts. In major agricultural production areas, heavy rainfall is very likely to occur (IPCC, 2021). The intensity of heavy precipitation is projected to increase as well, by about 7% per 1°C of warming at the global scale. This intensification will mainly occur over heavily cultivated regions like South and Southeast Asia and East Asia in the mid- to long term. (4) Runoff variability: annual mean runoff is projected to decrease in driest tropical regions and to increase in the

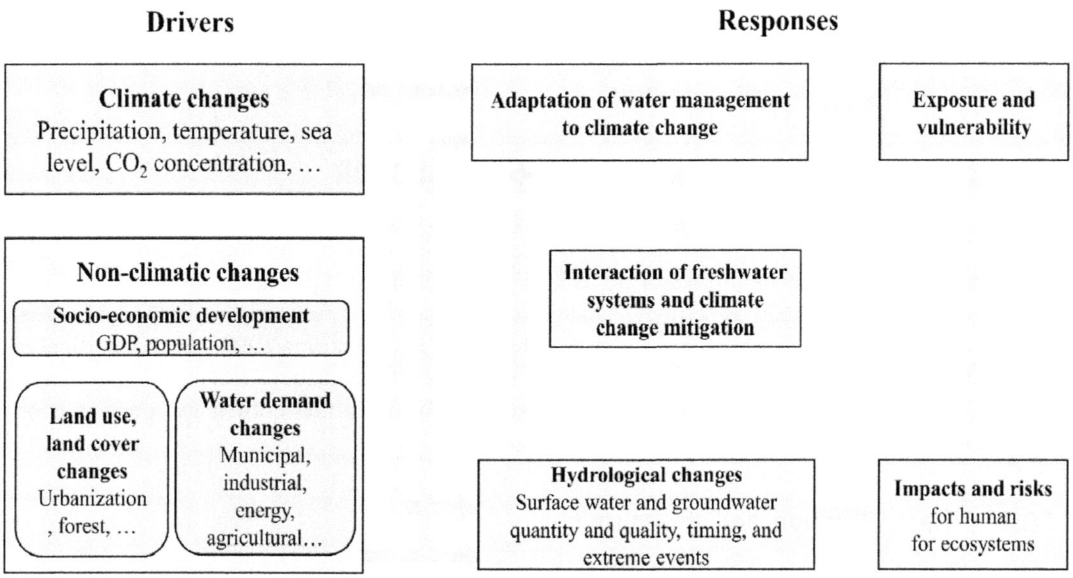

FIGURE 1.8

Framework and linkages considering impacts of climatic and socioeconomic changes on freshwater systems and consequent impacts on and risks for human beings and ecosystems.

Adapted from Jiménez Cisneros et al. (2014).

wet tropics and at high latitudes. These patterns are primarily driven by the projected changes in precipitation, temperature, and evaporation. (5) Flow variability: Climate change has reduced the maximum spring snow depth and snowmelt discharge in regions with snowfall. (6) Extreme hydrologic events: Global flood risks are projected to increase. The frequency of agricultural droughts due to decreased soil moisture, meteorological droughts due to decreased rainfall, and short hydrological droughts due to reduced surface water and groundwater are likely to increase. The increase in weather extremes likely leads to expanded warm, arid regions in many populated regions, and the impacts of climate change are more significant in arid regions than in humid regions. The most considerable impact is anticipated to occur in Africa (Lickley and Solomon, 2018). The effects of climate change impacts on hydrological changes and freshwater-related impacts of climate change on human beings and ecosystems from different studies are summarized in Table 1.6.

Further, the observed more frequent floods and droughts events, more availability in rainfall, shorter snowfall season, earlier spring snowmelt, and accelerated glacial melting tend to indicate an ongoing intensification of the hydrological cycle. This may affect water availability and increase the frequency and intensity of floods, droughts, and heavy rainfall in tropical regions.

Various organizations have prepared blueprints for water resources development, such as the World Water Vision by World Water Commission (2000). The ministerial conference of the Second World Water Forum, which took place in the Netherlands during March 2000, identified seven challenges to water security in the 21st century. The website of the *World Water Forum* (www.worldwaterforum.org) provides much useful information on this and related issues. These challenges are enumerated here.

Table 1.6 Effects of different greenhouse gas (GHG) emission scenarios on hydrological changes and freshwater-related impacts of climate change on human beings and ecosystems.

Type of hydrological change or impact	Description of indicator	References
Decrease of renewable water resources, global scale	Percent of global population affected by a decrease of more than 20% in water resources as compared to the 1990s (mean of 5 General Circulation Models (GCMs) and 11 global hydrological models).	Schewe et al. (2014)
Decrease of renewable groundwater resources, global scale	Percent of global population affected by a decrease of more than 10% in groundwater resources by the 2080s as compared to the 1980s (mean and range of 5 GCMs, population scenario SSP2)	Portmann et al. (2013)
Exposure to floods, global scale	Percent of global population annually exposed, in the 2080s, to a flood corresponding to the 100-year flood discharge for the 1980s (mean and range of 5−11 GCMs, population constant at 2005 values)	Hirabayashi et al. (2013)
Change in irrigation water demand, global scale	Change of required irrigation water withdrawals by the 2080s (on area irrigated around 2000) as compared to the 1980s (range of 3 GCMs)	Hanasaki et al. (2013)
River flow regime shifts from perennial to intermittent and vice versa, global scale	Percent of global land area (except Greenland and Antarctica) affected by regime shifts between the 1970s and the 2050s (range of 2 GCMs)	Doll and Mueller Schmied (2012)
New or aggravated water scarcity	Percent of global population living in river basins with new or aggravated water scarcity around 2100 as compared to 2000 (less than 1000 $m^3\ yr^{-1}$ of per capita blue water resources) (median of 19 GCMs, population constant at 2000 values)	Gerten et al. (2013)
Exposure to water scarcity	Population in water-stressed watersheds (less than 1000 $m^3\ yr^{-1}$ of per capita blue water resources) exposed to an increase in stress (1 GCM)	Arnell et al. (2013)
Change of groundwater recharge in the whole of Australia	Probability that groundwater recharge decreases to less than 50% of the 1990s value by 2050 (16 GCMs)	Crosbie et al. (2013)
River flow regime shift for rivers in Uganda	Shift from bimodal to unimodal (1 GCM)	Kingston and Taylor (2010)
Agricultural (soil moisture) droughts in France	Mean duration, affected area, and magnitude of short and long drought events throughout the 21st century (1 GCM)	Vidal et al. (2012)
Salinization of artificial coastal freshwater lake I Jsselmeer in the Netherlands (a drinking water source) due to seawater intrusion	(1) Daily probability of exceedance of maximum allowable concentration (MAC) of chloride (150 mg L^{-1}); and (2) maximum duration of MAC exceedance (2050, 1 GCM). As per Wikipedia, the water of this lake is now almost completely fresh.	Bonte and Zwolsman (2010)

Continued

Table 1.6 Effects of different greenhouse gas (GHG) emission scenarios on hydrological changes and freshwater-related impacts of climate change on human beings and ecosystems.—cont'd

Type of hydrological change or impact	Description of indicator	References
Decrease of hydropower production at Lake Nasser, Egypt	Reduction of mean annual hydropower production by the 2080s compared to hydropower production 1950–99 (11 GCMs)	Beyene et al. (2010)
Flood damages in Europe (EU27)	(1) Expected annual damages, in 2006; and (2) expected annual population exposed (the 2080s, 2 GCMs)	Feyen et al. (2012)

Adapted from Jiménez Cisneros et al. (2014).

1.10.2 Future Water Availability

Climate change can perturb the terrestrial hydrologic cycle and affect both water supply and demand. However, current evaluations of future climate change impacts on available water resources still remain highly uncertain, primarily due to both the large variations in rainfall projections from general circulation models (GCMs) and the uncertainties from global hydrological models (GHMs) used for the assessment of water resources. However, the influence of climate change is highly geographically dependent. Climate change may also affect irrigation potential, the differences between available renewable water and irrigation water demand. The negative value of irrigation potential indicates constrained irrigation. Projections at the end of the 21st century from 10 GHMs and six global gridded crop models (GGCM) suggest that some currently heavily irrigated regions are expected to have no irrigation potential in response to the insufficient water resources in a warming climate. However, freshwater abundance in regions like the northern/eastern United States, Europe, parts of South America, and South Asia may suffice for further expansion of irrigated cropland. In addition, climate change tends to increase potential irrigation water consumption as temperature increases in the future. As a consequence, an increase in irrigation water scarcity, the ratio of total irrigation applied to demand, is projected in most areas.

1.10.3 Meeting Basic Needs

Water is a basic human need, vital for life and health, but many of the world's poorest people cannot meet this need. Around 1.2 billion people (20%) do not have access to adequate and safe drinking water, while around half of the world's population does not have adequate sanitation. Each year between 3 and 4 million people, including more than 2 million children, die directly due to waterborne diseases. Poor sanitation also produces problems, such as worm infestations, that have a huge impact on nutrition. Bacterial infection is the main water quality problem: some areas of the world also have other water quality problems, such as fluoride, arsenic (West Bengal, India), or salinity.

Access to water is a basic human need, and the issue of satisfying basic water and sanitation needs has long been established. Despite this, the problem remains. Clearly, the traditional supply-based approaches on their own are not enough to solve it, and the appropriate and sustainable approach is to provide the poor with the means to solve the problem themselves. While "appropriate" is place dependent, a number of elements need to be part of any program to address this most basic need: secure and sustainable supplies of good quality water, education and awareness, technological innovations, integrated service provision, and community involvement. The participation and mobilization of the community are essential while not forgetting that as the custodians of health and hygiene, women hold the key to any such effort.

1.10.4 Protecting Ecosystems

Aquatic ecosystems in rivers and lakes, mangroves and other wetlands, and water-dependent ecosystems are vital sources of many other resources, are repositories of biodiversity, and are crucial for the functioning of the hydrological cycle. In many parts of the world, these are declining in area and quality, and if the current trends are not reversed, many of these will be unable to provide goods and services essential for societies to survive. Many rivers may turn into open sewers without any aquatic life forms and will transport pollutants and degrade coastal and marine ecosystems. The loss of species and habitats will dramatically reduce the world's biological diversity. The resulting declines in fish production will further jack up demands for protein from livestock production and agriculture.

The key to arrest detrimental development is to ensure that ecological processes and ecosystem integrity are deemed central to all aspects of water resources planning and management. This can only be achieved through the proper use of traditional conservation practices and sustainable management strategies. The most important measure to achieve this objective is to make conservation and rehabilitation of freshwater and related ecosystems a central tenet of water resources policies and integrated management. A balance between human needs and the intrinsic value of ecosystems needs to be established. Good governance, strong leadership, local empowerment, and universal responsibility for sustainable water resources use are some of the prerequisites. Also necessary is to create awareness, commitment, and exchange of knowledge on freshwater and related ecosystems.

1.10.5 Securing Food Supply

World food production depends on water availability. Agriculture is by far the largest user of water around the world, representing up to 90% of use in many developing countries. Foods from aquatic ecosystems, forests, etc., can be important in the diets of the poor, but farm harvests are the backbone of food production around the world. Due to population growth and changes in consumption patterns, the demand for food grains is likely to increase by 30%–40% in the next 25 years. Thus, there is going to be an ever-increasing stress on agricultural systems.

Enhancing food security is a tough challenge. Meeting irrigation demands in a sustainable manner is possible only by way of greater water productivity through *more-crop-per-drop*. Agricultural use has a huge impact on the quantity and quality of water. Improved irrigation efficiencies and advances in biotechnology will determine the future water security of the planet.

1.10.6 Sharing Water Resources

River basins are the main source of freshwater. Due to various physical interactions within a river basin, it is the logical unit for water management. There are around 300 trans-boundary river basins with shared water resources, and nearly half of the world's population lives in these river basins. There is a strong need for cooperative management of shared river basins. To achieve sustainable river basin management, riparian states should take into account the interests of other riparian states in the same basin.

The experience of cooperation in the Rhine River basin, the North American Great Lakes, the Mekong, the Danube, etc. shows that countries can and do work together to address international water issues. The basin-wide planning requires formulation of clear objectives, technical cooperation (for data collection, monitoring, etc.), and participatory decision-making on shared river basins by the public and nongovernmental stakeholders, contributing to strong river basin management. It would definitely improve decision-making through greater information and would help create a basis for implementation of policies and plans.

1.10.7 Dealing with Hazards

Too much water puts people and property at risk; too little water affects life itself. Flooding, often linked to major storms, is among the worst of natural disasters. According to estimates, more than 66 million people have suffered from storm and flood damage worldwide during the past 25 years. Population growth is forcing more people into flood-prone areas and overexploitation of aquifers. A visit to any river bank near major cities in many developing countries will readily reveal the extent of the problem. The worst sufferers are the uninsured poor, for whose benefit all the developments are planned! Coming to droughts, there are more refugees due to water shortages than due to wars, and the situation is likely to get worse.

Nonsustainable hydraulic interventions, land-use changes, and excessive water use turn the variability of natural processes into hazards that affect whole regions, often with disastrous consequences. Floods and droughts are recurrent natural phenomena that cannot always be prevented and cause large-scale suffering. Unfortunately, the damages due to floods and droughts are increasing over time. Climate change is anticipated to increase the severity of major storms, floods, droughts, and major shifts in climate patterns.

It is the responsibility of the government to provide security from floods, droughts, and other natural hazards. Sustainable flood and drought management should be based on integrated approaches that combine structural tools (embankments, dams, deepened river course) and nonstructural methods (forecasting and warning systems, contingency plans, etc.). Increasing flood damage must be prevented through town and country planning and by avoiding construction in high-risk areas, especially floodplains. Unsustainable water use should be curbed, especially in water-scarce areas. Large human settlements should not be encouraged in such areas, agriculture should economize on water use, and new technologies, such as cheaper desalination techniques, should be explored.

1.10.8 Valuing Water

Understanding and incorporating the full value of water into decision-making is an essential but as yet unrealized goal. Experience shows that wastage is less when people have to pay a price. Subsidies are widely given and can be an effective tool if they are transparent and targeted to disadvantaged groups. However, often they are "across the board." Cost recovery and pricing regimes are a problem and rarely give service providers (especially government agencies) a viable financial base. Effective water management is not possible, unless the challenges of economic valuation and financial viability are overcome. For effective water resources management, the concerned agencies must have adequate resources for investment, efficient operation, and maintenance.

It is imperative to manage water in a way that reflects its economic, social, and cultural value, and to move toward pricing water services to cover the full cost of their provision. The approach should be based on the concept of equity and the basic needs of the poor. The challenge is in two linked areas: the valuation of water resources and charges for water services. Along with water conservation and demand management, the principle of "polluter pays" needs to be practiced. These strategies will be effective only when stakeholders participate in defining charging regimes, charges are linked to the quality and reliability of services, and there is a mechanism to protect the poor. However, this requires many political and social issues to be sorted out, and in the era of coalition politics, this is a remote possibility in the near future in India.

1.10.9 Governing Water Wisely

The concept of river basin management (RBM) is widely accepted as a sound approach. But its application is not as wide because it requires a suitable institutional framework supported by appropriate laws and policies. Good governance exists where government agencies responsible for water effectively allocate and manage water resources based on legitimate policies and laws are responsive to national social and economic needs and, in the long term, to the sustainability of water resources in the country or region. To achieve good governance, it is necessary to create an enabling environment by encouraging and integrating private and public sector initiatives and to establish rules that allow clear transactions between stakeholders in a climate of trust. It is necessary to affix duties and responsibilities for the management of rivers and aquifers because these affect large populations and yet no one appears to be responsible for them.

1.10.10 Targets

The ministerial conference of the World Water Forum has also set indicative targets for the above seven challenges. In general, all the problems are proposed to be brought to the manageable limits within the next 15 years. For example, the proportion of people not having sustainable access to adequate quantities of affordable and safe water be reduced by half by 2015, or comprehensive policies and strategies for integrated water resources management be implemented in all countries by 2015. Of course, such targets were subsequently defined in the framework of Sustainable Development Goals (SDGs), discussed in Chapter 15.

It emerges from the above discussion that the main actions that need to be taken to solve the problems in the water sector are: (1) integrated management of water resources and associated resources of river basins; (2) participation of the community in the development, conservation, and management of resources; (3) improved agricultural practices; (4) an extensive and reliable database and mechanism of information dissemination; and (5) proper valuation of water. Sustainable development is now a necessity, not just a fashionable term. Water resources are a national property and should be managed accordingly. While managing them, the underlying objective should be balanced national growth rather than regional disparities and conflicts.

1.11 AN EXAMPLE OF WATER RESOURCES SYSTEM—SABARMATI SYSTEM

The real-life problems for which systems approaches are useful typically consist of many dams, weirs, and canal networks. A real-life existing system of dams and weirs is described in this section. The data of this system will be used, to the extent possible, to demonstrate the application of various techniques that will be discussed in subsequent chapters.

The Sabarmati River is one of the major west-flowing rivers in India. It rises in the Aravalli range at a north latitude of 24°40′ and east longitude of 73°20′ in the Rajasthan state at an elevation of 762 m and flows through the Gujarat state, and it outfalls into the Gulf of Cambay. The drainage basin of the river extends over an area of 21,085 sq. km and lies between longitude 71°55′E to 73°49′E and latitude 22°15′N to 24°54′N. The length of the basin is about 300 km, and it is about 105 km wide. The topography of the Sabarmati basin is hilly in the early reaches up to the Dharoi dam, after which the river flows mostly in plains. The major tributaries of Sabarmati are Wakal, Sei, Harnav, Hathmati, and Watrak. An index map of this system is shown in Fig. 1.9.

Four distinct seasons are observed in the basin. During the winter season (December to February), light rainfall occurs sometimes. The temperature during the summer season (March to mid-June) goes above 40°C and this season may have a few thunderstorms. The monsoon sets in by the end of June and continues till the end of September. About 95% of the annual average rainfall is received during

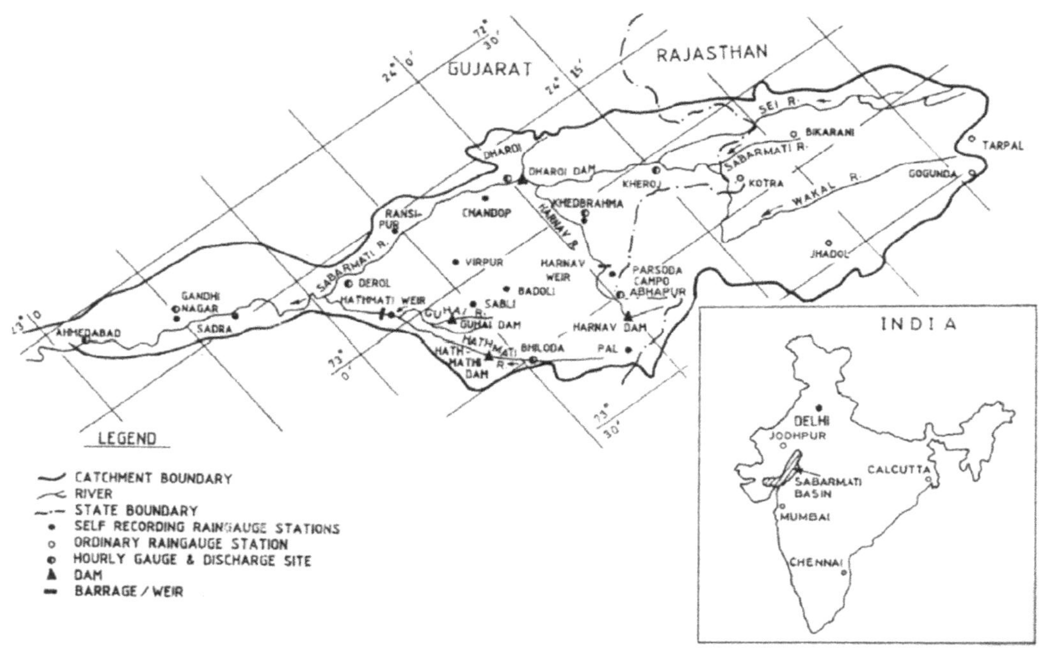

FIGURE 1.9

Index map of Sabarmati system.

monsoons. Heavy showers generally occur in association with the monsoon depression. The upper reaches of the basin receive an average annual rainfall of over 900 mm while the lower reaches receive about 650 mm; the average annual rainfall for the whole catchment is about 785 mm. A good network of hydrological and meteorological stations has been set up in the basin. At various locations, wireless stations have been established to communicate information about rainfall and discharge to a central control office, where it is used to make decisions concerning regulation.

To tap the water potential in the basin and to safeguard the downstream locations from flooding, a number of projects have been constructed on the Sabarmati River and its tributaries. One storage reservoir and three pick-up weirs have been constructed across the Harnav River to provide irrigation facilities in the command area. Harnav Dam has a dual purpose of flood control and irrigation. The catchment area at the dam site is 116 sq. km, and the live and dead storage capacities of the reservoir are 19.97 and 1.70 M m^3, respectively. The FRL and HFL of the reservoir are at levels of 332.00 and 332.25 m, respectively. The Harnav River spills excess water directly into the Dharoi reservoir.

The Hathmati dam across the Hathmati River has the dual purpose of irrigation and flood control. The catchment area at the dam site is 595 sq. km, and its live and dead storage capacities are 148.93 and 3.90 M m^3, respectively. The FRL and HFL of the reservoir are at levels of 180.74 and 183.18 m, respectively. On the Guhai River, a storage dam has been constructed for irrigation and flood control at a distance of 39 km from the source of the river. The catchment area of the Guhai River at the dam site is 422 sq. km. The dam was completed in the year 1990. The live and dead storage capacities of the reservoir are 57.04 and 5.30 M m^3, respectively. The FRL and HFL of the reservoir are at levels of 173.00 and 173.77 m, respectively. The River Guhai joins the Hathmati River between the Hathmati dam and the Hathmati weir. The purpose of Hathmati weir is to divert water for irrigation. The Hathmati River joins the Sabarmati River about 55 km downstream of the Dharoi dam.

The most important dam in the Sabarmati basin is the Dharoi dam, located at 103 km from the source of the river. The latitude and longitude of the dam are 24°00′N and 72°52′E, respectively. The dam was completed in the year 1976. The total catchment area at the dam site is 5540 sq. km, and the live and dead storage capacities of the reservoir are 775.89 and 131.99 M m^3, respectively. The purposes of the reservoir are (1) to moderate the incoming floods so that the controlled discharge at Ahmedabad city does not exceed 14,160 cumec up to the inflow rate of 21,665 cumec. Thereafter, if the inflow rate increases, the restricted outflow should be allowed up to 16,992 cumec, (2) to meet water supply requirements for the cities of Ahmedabad and Gandhinagar, and (3) to meet irrigation requirements in the command area.

The Wasna barrage is located across the Sabarmati River downstream of Ahmedabad city. The total catchment area at the barrage site is 10,619 sq. km. The barrage holds water for drinking water supply to Ahmedabad city and feeds the Fatewadi canal system. Another purpose of the barrage is to augment the underground aquifers in the Ahmedabad region.

1.11.1 Regulation of Sabarmati System

For conservation purposes, the dams and weirs are to be operated such that the water supply demands are satisfied to the maximum extent and irrigation demands are met as much as possible. As the facilities are interconnected, the integrated operation of the system will be beneficial.

At times, the Sabarmati River sends down very heavy floods, and some of these have caused devastation in Ahmedabad city, destroyed crops, changed the course of the delta channels, and filled up the harbor with silt. The highest known floods have occurred in 1875, 1941, 1950, and 1973. The peak of the design flood for the Dharoi reservoir is 27,180 cumec, while the volume of the design flood hydrograph is 3095.26 M m^3. The available storage space between FRL (189.59) and HFL (193.60) for flood moderation is 491.16 M m^3. The main cities that are located on the banks of the Sabarmati River are Gandhinagar and Ahmedabad. The safe channel capacity of the Sabarmati River at Ahmedabad is 14,160 cumec. The Dharoi reservoir should be so operated that the total flow in the river at Ahmedabad, including the flow from the catchment downstream of Dharoi, does not exceed 14,160 cumec. The inflow forecast for the Dharoi dam is issued based on observations at the Kheroj gaging site upstream of the dam.

The catchment area of the Sabarmati basin up to Dharoi dam is 5540 sq. km, and the same between Dharoi dam and Ahmedabad city (intermediate catchment) is about 5079 sq. km. Although there are some small hydraulic structures in this intermediate catchment, their capacity for flood moderation is rather limited. The Hathmati dam is ungated and cannot effectively moderate any flood. In the Guhai dam, the capacity between FRL and the MWL is only 12.5 M m^3. Therefore, no effective moderation of floods can be achieved from these structures. Hence, the entire intermediate catchment from the downstream of the Dharoi dam up to Ahmedabad city can be treated as uncontrolled. Most of this area is flat land in which agriculture is practiced, except the headwater areas of the tributaries.

1.12 CLOSURE

With the rapid growth in population and rising expectations of a better life, the limited water resources of river basins demand optimum utilization through integrated and scientific planning and management. The competing and conflicting demands of water resources call for a comprehensive and total view for planning, development, and operation of a water resources system. Systems analysis provides a rational approach for arriving at management decisions based on a systematic analysis of information. Application of systems analysis techniques to water resources problems can provide various trade-offs and suggest the best, or near best, course of action among several feasible alternatives.

For detailed modeling of a water resources system that uses large databases of varied disciplines, efficient computational devices are required. With the availability of computers in these days, the application of systems techniques to various disciplines has increased rapidly. Integration of environmental aspects into water resources planning and management is also essential in the present circumstances.

Water is everybody's concern. While the past decisions in the water sector were mainly based on technoeconomic considerations, the influence of sociopolitical factors is gradually increasing. Water management involves every person on this planet from the simple act of how he or she personally uses water to the more direct involvement of how he or she sees others use water at home or on the job. Each has a role to play as a concerned citizen and member of a community. The design and operation of a water resources project require more than technical knowledge—it involves human and societal skills and values. McCuen (1989) rightly comments that while the technical design can be standardized, it is rarely possible to standardize the value issue. He has summarized an elaborate discussion of the ethical conflicts in hydrologic practice by emphasizing that the hydrologist must consider the social impact of a design. A set of five core values has been proposed by Asmal (2002) as guidelines for future decision-making. These are: equity, sustainability, eficiency, participatory decision-making, and accountability.

Water is going to be the issue of this next century as the ultimate resource and as a potential source of conflict. As noted by *the National Geographic* magazine (Oct. 2002 supplement) "A key issue dividing Israel and its Arab neighbors, for example, is control of the tributaries of the Jordan River and West Bank aquifers ... Without oil, Middle East cannot live well; without water it cannot live." Recognizing the importance and necessity of sustainable water resources development, the UN had appropriately proclaimed the year 2003 as the *International Year of Freshwater* (details are available at www.unesco.org/water/iyfw/). To raise awareness about water and its importance, every year World Water Day is observed on 22 March. Every year, a different theme is chosen highlighting an aspect of water management. World Water Day helps focus attention to take action to tackle the global water crisis. It is a challenge for water resources experts to provide scientific and socially acceptable solutions to water-related problems. President Kennedy of the United States was quoted by Grigg (1985) to have said: *Anybody who can solve the problems of water will be worthy of two Nobel Prizes: one for peace and one for science*. Hopefully, a water specialist will qualify for this distinction in the 21st century.

References

Arnell, N.W., Lowe, J.A., Brown, S., Gosling, S.N., Gottschalk, P., Hinkel, J., Lloyd-Hughes, B., Nicholls, R.J., Osborn, T.J., Osborne, T.M., Rose, G.A., Smith, P., Warren, R.F., 2013. A global assessment of the effects of climate policy on the impacts of climate change. Nature Climate Change 3 (5), 512–519.

Asmal, K., 2002. Parting the waters. Journal of Water Resources Planning and Management, ASCE 128 (2), 87–90.

Beyene, T., Lettenmaier, D.P., Kabat, P., 2010. Hydrologic impacts of climate change on the Nile River Basin: implications of the 2007 IPCC scenarios. Climatic Change 100 (3–4), 433–461.

Biswas, A.K., 1998. Water Resources Environmental Planning, Management, and Development. Tata McGraw-Hill Publishing Company, New Delhi.

Bonte, M., Zwolsman, J.J.G., 2010. Climate change induced salinisation of artificial lakes in The Netherlands and consequences for drinking water production. Water Research 44 (15), 4411–4424.

Cisneros, J., B.E., Oki, T., Arnell, N.W., Benito, G., Cogley, J.G., Döll, P., Jiang, T., Mwakalila, S.S., 2014. Freshwater resources. In: Field, C.B., Barros, V.R., Dokken, D.J., Mach, K.J., Mastrandrea, M.D., Bilir, T.E., Chatterjee, M., Ebi, K.L., Estrada, Y.O., Genova, R.C., Girma, B., Kissel, E.S., Levy, A.N., MacCracken, S., Mastrandrea, P.R., White, L.L. (Eds.), Climate Change 2014: Impacts, Adaptation, and Vulnerability. Part A: Global and Sectoral Aspects. Contribution of Working Group II to the Fifth Assessment Report of the Intergovernmental Panel on Climate Change. Cambridge University Press, Cambridge, United Kingdom and New York, NY, USA, pp. 229–269.

Cole, J.A., 1998. Water resources: introduction. In: Herchy, R.W., Fairbridge, R.W. (Eds.), Encyclopedia of Hydrology and Water Resources. Kluwer Academic Publishers, Dordrecht.

Crosbie, R.S., Pickett, T., Mpelasoka, F.S., Hodgson, G., Charles, S.P., Barron, O.V., 2013. An assessment of the climate change impacts on groundwater recharge at a continental scale using a probabilistic approach with an ensemble of GCMs. Climatic Change 117 (1–2), 41–53.

Doll, P., Mueller Schmied, H., 2012. How is the impact of climate change on river flow regimes related to the impact on mean annual runoff? A global-scale analysis. Environmental Research Letters 7 (1).

Dooge, J.C.I., 1973. Linear Theory of Hydrologic Systems. Technical Bulletin No. 1468. USDA Agricultural Research Service, USA.

Dooge, J.C.I., 1988. Hydrology in perspective. Hydrological Sciences Journal 33 (1), 61–85.

Feyen, L., Dankers, R., Bodis, K., Salamon, P., Barredo, J.I., 2012. Fluvial flood risk in Europe in present and future climates. Climatic Change 112 (1), 47–62.
Fiering, M.B., 1976. Reservoir planning and operation. In: Shen, H.W. (Ed.), Stochastic Approaches to Water Resources, vol. II. Water Resources Publications, USA.
Gerten, D., Lucht, W., Ostberg, S., Heinke, J., Kowarsch, M., Kreft, H., Kundzewicz, Z.W., Rastgooy, J., Warren, R., Schellnhuber, H.J., 2013. Asynchronous exposure to global warming: freshwater resources and terrestrial ecosystems. Environmental Research Letters 8 (3).
Gleick, P.H., 1997. Water 2050: Moving toward a Sustainable Vision for the Earth's Fresh Water. Working Paper Prepared for the Comprehensive Freshwater Assessment for the United Nations General Assembly and the Stockholm Environment Institute, Stockholm, Sweden.
Grigg, N.S., 1985. Water Resources Planning. McGraw-Hill Book Company, New York.
Hall, W.A., Dracup, J.A., 1970. Water Resources Systems Engineering. McGraw-Hill Book Company, New York.
Hanasaki, N., Fujimori, S., Yamamoto, T., Yoshikawa, S., Masaki, Y., Hijioka, Y., Kainuma, M., Kanamori, Y., Masui, T., Takahashi, K., Kanae, S., 2013. A global water scarcity assessment under shared socio-economic pathways—part 2: water availability and scarcity. Hydrology and Earth System Sciences 17 (7), 2393–2413.
Herschy, R.W., Fairbridge, R.W. (Eds.), 1998. Encyclopedia of Hydrology and Water Resources. Kluwer Academic Publishers, Dordrecht.
Hirabayashi, Y., Mahendran, R., Koirala, S., Konoshima, L., Yamazaki, D., Watanabe, S., Kim, H., Kanae, S., 2013. Global flood risk under climate change. Nature Climate Change 3 (9), 816–821.
Hirsch, R.M., Cohon, J.L., ReVelle, C.S., 1977. Gains from joint operation of multiple reservoir systems. Water Resources Research 13 (2), 239–245.
IPCC, 2019. Climate Change and Land: An IPCC Special Report on Climate Change, Desertification, Land Degradation, Sustainable Land Management, Food Security, and Greenhouse Gas Fluxes in Terrestrial Ecosystems.
IPCC, 2021. Climate Change 2021: The Physical Science Basis. Contribution of Working Group I to the Sixth Assessment Report of the Intergovernmental Panel on Climate Change. Cambridge University Press. I.
Kingston, D.G., Taylor, R.G., 2010. Sources of uncertainty in climate change impacts on river discharge and groundwater in a headwater catchment of the Upper Nile Basin, Uganda. Hydrology and Earth System Sciences 14 (7), 1297–1308.
Lickley, M., Solomon, S., 2018. Drivers, timing and some impacts of global aridity change. Environmental Research Letters 13 (10).
Maass, A., Hufschmidt, M.M., Dorfman, R., Thomas Jr., H.A., Marglin, S.A., Fair, G.M., 1962. Design of Water Resources Systems. Harward University Press, Cambridge, Mass.
Maidment, D.R., 1993. Handbook of Hydrology. McGraw Hill Inc., New York.
Mays, L.W., 1997. Optimal Control of Hydrosystems. Marcel Dekker Inc., New York.
Mays, L.W., Tung, Y.K., 1992. Hydrosystems Engineering and Management. McGraw Hill Inc., New York.
McCuen, R.H., 1989. Hydrologic Analysis and Design. Prentice Hall, New Jersey.
Portmann, F.T., Doll, P., Eisner, S., Florke, M., 2013. Impact of climate change on renewable groundwater resources: assessing the benefits of avoided greenhouse gas emissions using selected CMIP5 climate projections. Environmental Research Letters 8 (2).
Rogers, P.P., 1980. The State of Art of Systems Analysis as a Tool in Water Policy, Planning and Management in Developing Countries. Paper No. 12. Department of City and Regional Planning, Harvard University, USA.
Rogers, P.P., Fiering, M.B., 1986. Use of systems analysis in water management. Water Resources Research 22 (9), 146S–158S.
Schewe, J., Heinke, J., Gerten, D., Haddeland, I., Arnell, N.W., Clark, D.B., Dankers, R., Eisner, S., Fekete, B.M., Colon-Gonzalez, F.J., Gosling, S.N., Kim, H., Liu, X.C., Masaki, Y., Portmann, F.T., Satoh, Y., Stacke, T., Tang, Q.H., Wada, Y., Wisser, D., Albrecht, T., Frieler, K., Piontek, F., Warszawski, L., Kabat, P., 2014.

Multimodel assessment of water scarcity under climate change. Proceedings of National Academy of Science of the United States of America 111 (9), 3245–3250.
Seckler, D., Amarasinghe, U., David, M., de Silva, R., Barker, R., 1998. World Water Demand and Supply, 1990 to 2025: Scenarios and Issues. Research Report # 19. International Water Management Institute, Sri Lanka.
Shiklomanov, I.A., 1998. Assessment of Water Resources and Water Availability in the World. Report for the Comprehensive Assessment of the Freshwater Resources of the World. Data archive on CD-ROM from the State Hydrological Institute, St. Petersburg, Russia, United Nations.
Showers, V., 1989. World Facts and Figures. John Wiley and Sons, New York.
Singh, V.P., 1988. Hydrologic Systems. In: Rainfall-Runoff Modeling, vol. 1. Prentice Hall, Englewood Cliffs, New Jersey.
Sinha, N.K., 1991. Linear Systems. John Wiley & Sons, New York.
Sullivan, C., et al., 2006. Mapping the Links between Water, Poverty and Food Security. Report on the Water Indicators workshop held at the Centre for Ecology and Hydrology, Wallingford, UK, 16 to 19 May, 2005. GWSP Issues in GWS Research, No.1. GWSP IPO, Bonn. Available at:. http:/www.gwsp.org. (Accessed 15 September 2023).
UN, 1978. Register of International Rivers. The United Nations. Pergamon Press, New York.
UNESCO, 1978. World Water Balance and Water Resources of the Earth. United Nations Educational Scientific and Cultural Organization, Paris.
Veltrop, J.A., 1992. Hydropower: needs, challenges and opportunities. In: Broch, E., Lysne, D.K. (Eds.), Hydropower'92. A.A. Balkema, Rotterdam.
Vidal, J.P., Martin, E., Kitova, N., Najac, J., Soubeyroux, J.M., 2012. Evolution of spatio-temporal drought characteristics: validation, projections and effect of adaptation scenarios. Hydrology and Earth System Sciences 16 (8), 2935–2955.
Votruba, L., Kos, Z., Nachazel, K., Patera, A., Zeman, V., 1988. Analysis of Water Resources Systems. Elsevier, Amsterdam.
Water Vision, 2000. Report of the Thematic Panel on Energy Technology and its Implications for Water Resources. Accessed at the Web-site. www.watervision.org.
World Water Commission, 2000. World Water Vision Report. Accessed from the Website. www.watervision.org.
Yevjevich, V., 1983. The Nile river basin: hardcore and softcore water projects. Water International 8, 23–34.

Further Reading

Shiklomanov, I.A., 2000. Appraisal and assessment of world water resources. Water International 2 (91), 11–32.
Singh, V.P. (Ed.), 1995. Computer Models of Watershed Hydrology. Water Resources Publications, USA.

CHAPTER

Acquisition and Processing of Water Resources Data

2

Chapter outline

Water Resources Systems Planning and Management. https://doi.org/10.1016/B978-0-12-821349-0.00003-4

Data are the foundations on which any analysis rests. The practice of hydrological measurements is very old. Kautilya initiated systematic precipitation measurements in India in the fourth century BCE. Streamflow was probably first monitored by Hero of Alexandria in the first century CE. With developments in water and hydrometeorologic sciences, there have been simultaneous developments in equipment and techniques of data collection. A number of international/national standards have been developed to ensure systematic measurements of water resources and it is necessary that observatories should conform to these standards. The Committee on Opportunities in the Hydrological Sciences (1991) has appropriately summarized the necessity of good water resources data: "Modeling and data collection are not independent processes. Ideally, each drives and directs the other. Better models illuminate the type and quantity of data that are required to test the hypotheses. Better data, in turn, permit the development of better and more complete models and new hypotheses."

The data needed for water resources are as follows: hydrometeorologic, hydrologic, hydraulic, geomorphologic, agricultural, land use and land cover, pedologic, geologic, water quality, economic, and social. Hydrometeorologic data include rainfall, snowfall, temperature, radiation, humidity, vapor pressure, sunshine hours, wind velocity, and pan evaporation. Hydrologic data include flow depth, streamflow discharge, base flow, interflow, stream–aquifer interaction, potential, water table, and drawdowns. Hydraulic data include velocity, flow depth, channel geometry, channel patterns, hydraulic works, sediment concentration, sediment load, and transport. Geomorphologic data include topographic maps showing elevation contours, river networks, drainage areas, slopes and slope lengths, and watershed areas. Agricultural data include vegetative cover, land use, treatment, and fertilizer application. Land use and land cover data include pervious and impervious areas, forest lands, urban drainage, bridges, pavements, culverts, and parks. Pedologic data include soil type, texture, and structure; soil condition; soil particle size; porosity; moisture content and capillary pressure; steady-state infiltration, saturated hydraulic conductivity, and antecedent moisture content. Geologic data include data on stratigraphy, lithology, and structural controls. More specifically, data on the type, depth, and areal extent of aquifers are needed. Depending on the nature of aquifers, these data requirements vary. For confined aquifers, hydraulic conductivity, transmissivity, storativity, compressibility, and porosity are needed. For unconfined aquifers, data on specific yield, specific storage, hydraulic conductivity, porosity, water table, and recharge are needed. Water quality data include pollutant types, pollutant concentration, pollutant load, total maximum daily load, and pollutant discharge. Economic data include the cost of water resources projects, amount and terms of investment, rate of return, and economic duration of projects. As water resources are developed for the benefit of society, social data include water-use patterns of different sections of society, use of other resources that impact water, and its quality. Each data set is examined with respect to homogeneity, completeness, and accuracy. Thus, for a water resources study, one needs a variety of data for a number of variables in the vertical as well as horizontal planes.

The activities of a hydrological service are shown in Fig. 2.1. The term *hydrological data processing* is a widely used but loosely defined term that includes a range of activities varying from simple analysis to complete modeling. The processing of hydrological data is a multistep process that begins with a preliminary checking of raw data in the field and successively higher levels of validation before it is accepted as fully validated data for further use. The passage of data from field to storage is also not a one-way process and includes several feedbacks. Sometimes, channels for feedback from data users are also maintained. Actually, processing and validation of hydrological data are not a purely statistical exercise—these require an understanding of field practices, principles of observation, and physics of the hydrological variable being measured.

Data processing also includes aggregation of data observed at a certain time interval to a different interval, for example, hourly to daily and daily to monthly. Occasionally disaggregation, that is, conversion from a long (say daily) to short (say, hourly) time step is also carried out. The computation of areal averages, for example, catchment rainfall, is also required for validation. This also provides a convenient means for summarizing large volumes of data.

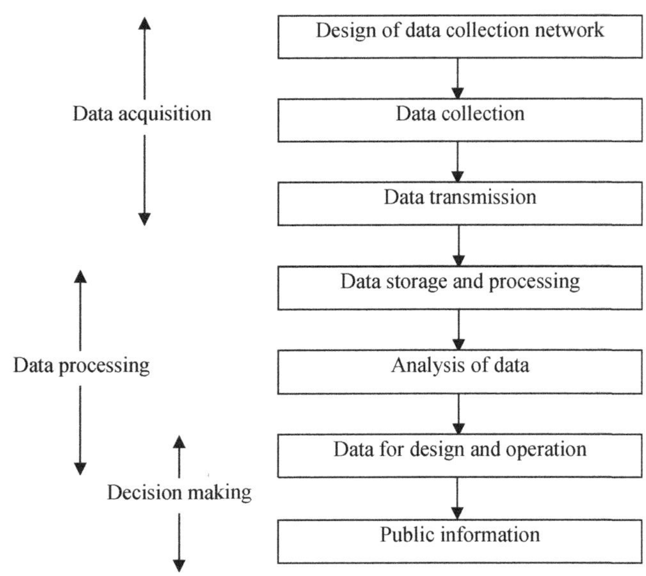

FIGURE 2.1

Activities of a hydrological service.

Adapted from WMO (2008).

Typical stages in water resources data processing are:

- scrutiny of raw data;
- data entry to computer, validation, and correction; and
- data archival, retrieval, and dissemination.

Given the central role of water resources data in planning, design, and management of water resources projects, this chapter is devoted to basic concepts of data acquisition, processing, storage, and dissemination. Some of the latest techniques of data acquisition are also discussed.

2.1 TYPES OF WATER RESOURCES DATA

There are several ways to classify water resources data. The most common way is to classify the data into three categories: time-oriented data, space-oriented data, and relation-oriented data.

The time-oriented or time-series data consist of hydrometeorological, water quantity, and water quality data that are periodically measured at a station. The time interval between observations can be constant or varying. Examples of such types of data are rainfall, river stage, and sediment concentration. Some surface water data are equally spaced in time. The space-oriented data comprise topographic maps of catchments, river networks, soil maps, etc. Traditionally, such data are stored in the form of paper maps and manually analyzed. The recent trend is to use a Geographical Information System (GIS) to input, store, and analyze such data. As described in detail in Section 3.3 in Chapter 3, different types of information, such as topographical and land use of an area, are stored in a GIS in different layers of a map, which can be overlaid and analyzed. The relation-oriented data consist of mathematical relationships between two or more variables. Typical examples are river rating curves or a spillway rating table.

Water resources data can also be classified as time-varying or time-nonvarying data. The time nonvarying or static data include most space-oriented data, which do not change with time, for example, catchment topographic map and soil map. Some features, such as river network and land use in a catchment, might gradually change with time and can be considered as semistatic. A brief description of each type of data is presented in what follows.

2.1.1 Time-Oriented Data

The values of most hydrometeorological and water quality variables change in time and such variables are classified as time-oriented data. These data can be further classified as meteorological data, hydrological data, hydraulic, and water quality data.

The time-series data include all the measurements that have an observation time associated with them and most water resources data have this property. The variable could be an instantaneous value, for example, water level in a river; an accumulated value, for example, daily rainfall; an averaged value, for example, mean daily discharge; or a statistic over a specified time period, for example, annual maximum flow. The distinction between instantaneous and accumulative values is important when the data are further processed.

Depending on the frequency of observations, the time-series data can also be classified as:

- Equidistant time series are measurements that are made at regular intervals of time (hourly, daily); the values may be instantaneous, accumulated, or averaged.

- Cyclic time series are the measurements that are made at irregular intervals of time but the irregular time sequence is repeated regularly, for example, the observation of the river stage daily at 08:30 and 17:30 h.
- Nonequidistant time series are the measurements that are made when some specified event takes place. For example, in a tipping bucket rain gauge, each tip of the bucket is recorded after a certain depth of rain has fallen.

The two most important data for hydrological analysis are precipitation and streamflow. The measurement and processing of these two will be discussed in greater detail. The time series of evaporation data forms another important input in hydrological studies. The temperature of air, soil, and water bodies is important as many processes depend on it. Other important meteorological variables include humidity, wind speed and direction, and sunshine duration.

2.1.2 Space-Oriented Data

The space-oriented data comprise catchment data (physical and morphological characteristics), river data (cross-sections, profile, bed characteristics), and lake or reservoir data (elevation—area—storage capacity). These are further discussed in Section 2.8.

2.1.3 Relation-Oriented Data

These data contain relationships between two or more variables. The variables themselves may form a time series but their relationship is of interest here. The relationship may be expressed in mathematical, tabular, or graphical form and is derived using the data. The stage—discharge rating at a site is a typical example of relation-oriented data. Note that this relation may change with time.

A mathematical relationship between two or more variables is established for a variety of purposes, such as data validation and filling-in of data gaps. The relationships between stages at two adjacent gauging stations and between the average rainfall in a catchment and the resulting outflow are some typical examples. In some instances, relationships may be established between water quality parameters and discharge to determine pollutant loads. The parameters of the relationship along with the ranges of independent variables, error statistics, and the period of applicability are required while establishing a relationship.

The stage—discharge data are the most common example of this type of data. Stage—discharge observations are the primary data to establish the relationship called the rating curve at a river-gauging station. Normally, such relationships may be expressed in parabolic or power form. More than one equation may be required to characterize the relationship that may change with time.

2.1.4 Techniques for Observation of Water Resources Data

There are many techniques in which the data that are used for water resources studies can be collected. The major techniques are described below.

Gauging Equipment

This is the most common way to observe hydrometeorological variables, such as precipitation and streamflow. A gauging site is established and is equipped with instruments that can measure the variable(s) of interest. An observer visits the site to manually measure the value of the variable and records or transmits it to the place of storage/use. These days, it is easy to transmit the data from the stations geographically spread over an area to a central office. An automated hydrologic station can measure a number of hydrometeorological variables and store/transmit the data. The equipment may be programmed to transmit the data after a selected time interval or it can be interrogated at any time to get the data. The design of networks, equipment, and methods of observation of some important variables are described in later sections.

Remote Sensing

In this technique, the data about an object are obtained without coming in physical contact with the object. The technique repetitively provides spatial data of terrain features and is discussed in Chapter 3. Weather radars are being increasingly used for the measurement of precipitation and are described in Section 2.4.1.

Chemical Tracers

In this approach, some chemicals, known as tracers, are added to the process whose data are to be obtained. The procedure to measure river discharge using tracers is described in Section 2.6.3. Tracers can also be used to determine the flow path of water or a pollutant. Nuclear or isotope techniques are employed to trace the movement of water molecules in any part of the hydrological cycle and derive information about hydrological processes. Nuclear techniques are helpful to assess the rate of sediment deposition in a water body, identify rainfall recharge, and recharge areas of aquifers, study of seawater intrusion in coastal regions, measure seepage and leakage from surface water bodies, and analyze surface water and groundwater interaction. The stable isotopes, such as oxygen-18, deuterium, carbon-13 (for C-14 dating), and N-15, are commonly used.

2.1.5 Sources of Data

The sources of water resources data can be obvious or usual as well as unusual. The usual data sources include water resources/irrigation departments, river basin/regional water authorities, experimental and research organizations, universities (for research and experimental basins, farms, etc.), and public health authorities. The unusual sources include nongovernmental organizations, airport authorities (mainly meteorological data), and municipal bodies. In any detailed study, it is advisable to search and contact unusual data sources too. This could be a painful effort but may turn out to be rewarding. A field visit is always helpful in getting supplementary information. For example, high-water marks along rivers are useful in delineating flood-prone areas and can also be used to crosscheck peak discharges.

The collection of hydrologic data involves locating the data sources, followed by inspection and evaluation of these data to establish their suitability and sufficiency for the study. Given wide differences in the practice of data storage and dissemination, the efforts required for data collection tremendously vary from country to country. In some cases, it may just mean browsing the Internet and downloading the requisite data. In others, it may mean physically visiting the concerned offices and manually copying the data from the available records.

Many times, one data-observing organization is unaware of the data collection efforts of others; even governmental agencies often know little about the data collected by other governmental and nongovernmental institutions. This poses additional problems during data collection and processing. Some of the data that are used in water resources analysis may fall under the category of confidential data and special procedures and precautions are to be followed to obtain, handle, and analyze such data, and disseminate the results.

2.2 DESIGN OF HYDROMETEOROLOGICAL NETWORKS

The information on temporal and spatial characteristics of water resources is obtained by a network of observational stations. The main purpose of planning a hydrometeorological network is to find out the hydrological characteristics of an area and gather data for planning, design, and management. Setting up a station requires an investment in equipment, in logistics, and for operation and maintenance. Scientific planning is, therefore, necessary for network design so that the desired results can be achieved with minimum cost.

The requirement for water resources data depends on their end use. Therefore, it is difficult to formulate general rules on network design. Based on spatial features, there are two types of networks: (1) areal networks, such as those for precipitation, and groundwater levels; and (2) linear networks, such as those for streamflow and river sediment. Areal networks are established to get spatial characteristics of the variables over an area, while linear networks are set up for rivers, canals, etc. On the basis of purpose, networks can be classified into three categories: basic (to get the fundamental characteristics of the variables of interest), specific (to gather data for some specific purpose, for example, a reservoir project), and temporary (which are in operation for a short period of time). While designing hydrologic networks, the following questions need to be answered:

(a) Which variables are to be measured?
(b) At which locations?
(c) How frequently?
(d) With what accuracy?
(e) For how long?

Answers to these questions determine the costs of establishment and operation of stations, costs of land acquisition, construction, equipment procurement and installation, operation, maintenance, and staffing.

The basic network is designed to provide a level of hydrological information at any location within its region of applicability that would preclude any gross mistake in water resources decision-making (WMO, 2008). In the early stages of development of a network, the first step should be to set up a minimum network. Such a network should consist of the minimum number of stations that are required to initiate planning for the exploitation of water resources in the region. The number may be based on experience or judgment.

Since hydrometeorological data networks are operated by a number of independent agencies, it is important that there is good coordination and exchange of data among them. This will reduce the overall expenditure and improve data quality. Of particular importance is the coordination between water quantity and quality data networks.

Given their central importance, precipitation and streamflow networks are discussed separately. The *Guide to Hydrological Practices* (WMO, 2008) published by the World Meteorological Organization (www.wmo.ch) contains useful guidelines to set up networks for various types of data and observe and analyze the data. Some sections in this chapter are significantly influenced by this publication. A number of other relevant WMO publications are listed in references. The handbook by ASCE (1996) and Jain & Singh (2019) also contains much relevant information.

2.2.1 Precipitation Data Networks

The optimum density of a precipitation gauge network depends on the purpose for which the observed data are to be used. For example, accurate measurements of precipitation for flood forecasting require denser networks as compared to rainfall-runoff modeling. WMO (2008) has recommended the following (Table 2.1) as minimum network densities for precipitation stations.

The optimum network should be such that it should be possible to determine required characteristics of the variable with sufficient accuracy by interpolation between values of different stations. The optimum number of rain gauge stations (N) in a network can be given by Ahuja (1960):

$$N = \left[\frac{C_V}{p}\right]^2 \tag{2.1}$$

where C_v = the coefficient of variation of the rainfall values of the existing rain gauge stations, and p = the desired maximum percentage error in the estimate of areal mean rainfall. Here, C_v is computed by

$$C_V = \sigma / P_m \tag{2.2}$$

in which σ is the standard deviation, and P_m is the mean rainfall of the existing stations. A typical value of p is 10%. Obviously, a decrease in the permissible percentage error would mean that more number of gauges would be required. Mukherjee and Kaur (1987) have proposed a small change in Eq. (2.1) by including the mean correlation (r) of precipitation over the area

$$N = \left[\frac{C_V}{p}\right]^2 (1 - r) \tag{2.3}$$

Table 2.1 Recommended minimum densities for precipitation stations.

	Precipitation station densities—minimum area (in km^2) per station	
Physiographic unit	**Nonrecording**	**Recording**
Coastal	900	9000
Mountainous	250	2500
Interior plains	575	5750
Hilly/undulating	575	5750
Small islands	25	250
Urban areas		10–20
Polar/arid	10,000	100,000

Based on WMO (2008).

Example 2.1. Six rain gauge stations have been set up in a catchment. The average annual rainfall at the stations is given in following table. Find the required number of rain gauges, if desired maximum percentage error of 5% is to be ensured in areal rainfall estimation.

Station	A	B	C	D	E	F
Rainfall (mm)	700	830	650	710	600	400

Solution. Average rainfall $= (700 + 830 + 650 + 710 + 600 + 400)/6 = 648$ mm

Standard deviation $\sigma = \sqrt{\frac{\sum (P - P_m)^2}{n - 1}} = \sqrt{\frac{103{,}484}{6 - 1}} = 143.86$ mm

$C_v = \sigma/P_m = 143.86/648 = 0.222$.

Therefore,

$N = (0.222/0.05)^2 = 19.7$, say 20.

So, 20 stations are needed if the error in rainfall estimation is to be kept below 5%.

If, suppose, an error of up to 10% is permissible, the requisite number of stations would be

$N_1 = (0.222/0.1)^2 = 4.93$, say 5.

Thus, the required number of stations significantly drops if higher error is permissible. Note that this method does not suggest the location of rain gauge stations.

2.2.2 Streamflow Networks

Every major stream should be gauged at or near its mouth. Likewise, a number of its tributaries should also be gauged. Naturally, gauging depends on the existing and likely development in the basin. According to WMO, the first gauging station is selected at the most upstream location where the drainage area is about 1300 km^2. The second station is located at a point in the downstream direction where the drainage area is approximately doubled. The WMO recommendations for a minimum density of hydrometric stations are given in Table 2.2.

Table 2.2 Recommended minimum density of streamflow stations.

Physiographic unit	Minimum densities per station (area in km^2 per station)
Coastal	2,750
Mountainous	1,000
Interior plains	1,875
Hilly/undulating	1,875
Small islands	300
Polar/arid	20,000

Based on WMO (2008).

2.3 DATA VALIDATION

The need for data validation or quality control arises because field measurements are subject to errors. Errors may also arise in data entry, during computation and (hopefully rarely) from the mistaken "correction" of "right" data (DHV, 1999). Data validation is the means by which data are checked to ensure that the corrected values are the best possible representation of the true values of the variable. Basically, data validation is carried out:

- to correct errors in the observed values where possible,
- to assess the reliability of data even though it may not be possible to correct errors, and
- to identify the source of errors to ensure that these are not repeated in future.

Measurement errors may be classified as random, systematic, or spurious in nature (Fig. 2.2). Hydrometric measurements are often subject to a combination of random and systematic errors. *Random errors*, sometimes referred to as experimental errors, are equally distributed about the mean or "true" value. The errors of individual readings may be large or small, for example, the error in a staff gauge reading where the water surface is subject to wave action. Usually, they tend to compensate with time or are minimized by taking a sufficient number of measurements. A *systematic error* or bias is a systematic difference, either positive or negative, between the measured value and the true value. Systematic errors are generally more serious and the validation process must be able to detect and

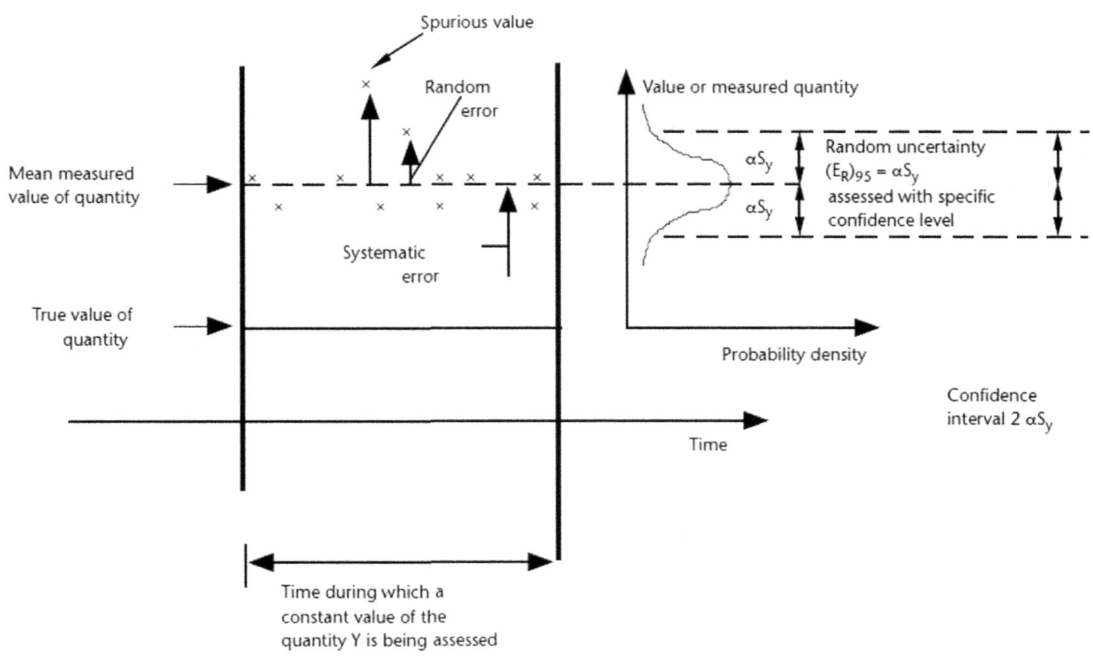

FIGURE 2.2

Description of measurement errors.

Based on WMO (2008).

correct them. Spurious errors arise due to some abnormal external cause. For example, an animal may drink water from the evaporation pan and introduce errors in the data. Sometimes, such errors may be readily detected but it may not be easy to correct them. Often, the measured data may have to be discarded.

Errors during observations typically arise due to:

- faulty equipment, for example, a current meter with worn-out parts;
- malfunction of instrument, for example, slippage of float tape in water level recorder;
- improper exposure conditions, for example, inlet of stilling well blocked so that the water level in the well differs from the river;
- personal observation errors, for example, gauge misreading;
- wrong entry of data; and
- error in field computation, for example, mistake while converting current meter rotations to velocity.

The input variables in an analysis may be directly measured (e.g., rainfall) or they may be derived using a relationship with one or more variables (e.g., discharge that has been obtained from a rating curve). In the latter case, the error in the variable (discharge) depends both on field measurements and the error in the relationship. An error may also be introduced if the relationship is no longer valid or the values are extrapolated outside the applicable range. Validation involves different types of comparisons of data and includes the following:

Single series comparison:

- between individual observations and preset physical limits;
- between sequential observations to detect unacceptable rates of change and deviations from acceptable behavior (most readily identified graphically); and
- between two measurements of a variable at a single station, for example, daily rainfall from a daily gauge and an accumulated total from a recording gauge.

Multiple stations/data:

- between two or more measurements at nearby stations, for example, flow at two sites along a river; and
- between measurements of different but related variables, for example, rainfall and river flow.

2.3.1 Levels of Validation

Ideally, data are validated soon after observation and as close to the observation station as possible. This ensures that secondary information to support validation is readily available. However, data validation at or near observation sites is not always possible due to logistics and lack of availability of trained personnel. Often a compromise must be reached—the activities that require interaction with the observers are carried out at or near the station, whereas more complex analysis is carried out in offices.

The recommended accuracy for selected hydrological variables is given in Table 2.3.

An important aspect of data validation is that none of the procedures are absolutely objective. They are basically tools to screen out suspect data that are to be further examined by other tests and

Table 2.3 Recommended accuracy (uncertainty levels) expressed at 95% confidence interval.

Variable	Uncertainty levels
Precipitation (amount and form)	3%−7%
Rainfall intensity	1 mm/h
Snow depth (point)	1 cm below 20 cm or 10% above 20 cm
Water content of snow	2.5%−10%
Evaporation (point)	2%−5%, 0.5 mm
Wind speed	0.5 m/s
Water level	10−20 mm
Wave height	10%
Water depth	0.1 m, 2%
Width of water surface	0.5%
Velocity of flow	2%−5%
Discharge	5%
Suspended sediment concentration	10%
Suspended sediment transport	10%
Bed-load transport	25%
Water temperature	0.1−0.5°C
Dissolved oxygen (water temperature is more than 10°C)	3%
Turbidity	5%−10%
Color	5%
pH	0.05−0.1 pH unit
Electrical conductivity	5%
Ice thickness	1−2 cm, 5%
Ice coverage	5% for $\geq$20 kg/m^3
Soil moisture	1 kg/m^3 $\geq$ 20 kg/m^3

Note: When a range of accuracy levels is recommended, the lower value is applicable to measurements under relatively good conditions and the higher value is applicable to measurements under difficult situations.
Based on WMO (2008).

corroborative facts. When a particular value is confirmed as incorrect, it is substituted by an alternative value that is likely to be closer to the true value of the variable; this value is flagged as corrected. Since each hydrological variable has distinct characteristics, specific validation techniques designed for each variable should be applied. It is emphasized that validation should never be treated as a purely statistical exercise; the properties and behavior of the variable under consideration should always be kept in mind.

Based on the information and techniques employed, validation can be grouped into three major categories: (1) primary validation, (2) secondary validation, and (3) hydrological validation.

2.3.2 Primary Validation

The aim of primary data validation is to highlight and, if possible, correct those data that are not within the expected range or are inconsistent. Primary validation involves comparisons within a single data series or between observations and preset limits and/or statistical range of a variable or with the expected behavior of the generating process. Sometimes, information from a few nearby stations may also be pooled. If it is not possible to definitely conclude that the suspected value is erroneous, such value is not changed but is flagged indicating that it is doubtful. All data that have been flagged as suspicious during primary validation are again screened later on the basis of additional information.

2.3.3 Secondary Validation

Secondary validation of data follows primary validation and tests the data for its expected spatial behavior as inferred from a number of neighboring observation stations. It consists of comparisons between the same variables at two or more stations. The underlying assumption is that the variable in question has adequate spatial correlation within the considered distances. This assumption must be verified on the basis of historical records and the behavior of the variable elsewhere. Some of the checks that are applied at this stage are oriented to trap specific errors known to be made by observers, while others are general and lead to the identification of spatial inconsistencies.

The spatial validation is best carried out using the data of key stations that are known to provide good-quality data. A word of caution: the key stations may not always be perfect and can also sometimes report incorrect data. If the data of all observation stations are equally reliable, then data validation becomes difficult.

When hydrological variables have a high autocorrelation, such as groundwater levels, or high correlation exists among neighboring stations, validation can be carried out with a higher level of confidence. However, some processes show a great temporal and spatial variability (e.g., convective rainfall). It is rather difficult to ascertain the behavior of such processes with the desired degree of confidence. If it is not possible to conclude whether the suspected value is erroneous, such value is not changed but is flagged as doubtful. All suspicious data are validated again on the basis of additional information.

2.3.4 Hydrological Validation

Hydrological validation consists of comparing data of (correlated) variables at nearby stations to identify inconsistencies (if any) between the time series or their derived statistics. Hydrological validation may be applied to a measured variable (water level) or to derived variables (flow, runoff). This is usually done through regression analysis or simulation modeling.

Ideally, all the data should be subjected to hydrological validation. For historical data to which no (or few) checks have been applied, hydrological validation provides an effective check on the quality and reliability of records. However, thorough hydrological validation requires a high level of professional expertize and can be time consuming. Therefore, this validation may be applied selectively both in terms of stations and tests. Finally, the validation may be able to identify a particular section of record/data item that is unreliable, but it may not always be possible to provide a correction.

2.3.5 Data Fill-in

An observed data series may contain gaps due to a fault in equipment, observer absence, etc. and these need to be filled to make the series complete for analysis. Depending on the length of the gap, the type and nature of the variable, and the information available from adjacent stations, a variety of techniques are available to fill-in gaps in a data series. For single value or short gaps in a series with high serial correlation, simple linear interpolation between known values may be adequate. Alternatively, a graph is drawn and the gap is filled by drawing a smooth line through it. Gaps in series with a high random component and little serial correlation, such as rainfall, cannot be filled in this way and must be completed with reference to neighboring stations through spatial interpolation. Longer gaps can be filled through regression analysis and statistical procedures. Large gaps in a discharge series can be filled through rainfall-runoff modeling. Note that the filled-in values will affect the statistics of the series and care is to be taken to preserve historical statistics.

2.3.6 Reanalysis Data

Reanalysis is a scientific approach to develop a comprehensive database of weather and climate data, and the primary aim is to understand how these are changing with time. The gridded database represents the state of the atmosphere at various times. It is continually updated by incorporating observations (coming from various sources, including paleo-climatic data, data obtained from tree rings, ice cores, and sediments deposits) and outputs of global climate models.

Reanalysis data typically extends over (many) centuries and/or longer and covers the entire Earth. In addition to horizontal grids, data at vertical grids would also be compiled. IPCC data repositories contain huge volumes of reanalysis data that are freely disseminated to the user; some other websites also provide this type of data. Currently, reanalysis data are mainly used for research in the field of climate sciences. But it is easy to see that such data have immense value in the field of water resources planning and management and allied fields.

2.4 ACQUISITION AND PROCESSING OF PRECIPITATION DATA

Precipitation is a primary source of fresh water. The amount, intensity, and spatial distribution of precipitation are important inputs in many hydrological studies. The total amount of precipitation that reaches the ground in a stated period is expressed as the depth to which it would cover a horizontal projection of the Earth's surface. If any part of precipitation is snow or ice, its depth when melted is included (WMO, 2008). The unit of precipitation is length and daily amounts should be read to the nearest 0.1 mm. Weekly, fortnightly, and monthly amounts should, however, be read to the nearest 1 mm at least. National standards specify the time at which daily observations of precipitation along with other meteorological variables should be made.

2.4.1 Precipitation Gauges

Precipitation is measured using a gauge that consists of a collector to delineate the area of the sample and a funnel leading to a storage device. The precipitation, thus collected, is measured by transferring the contents to a graduated measuring jar. The area of the collector and the size of the gauges vary, depending on the amount of precipitation normally received in the region where the station is located. Different types of gauges are used to measure liquid and solid precipitation.

Since the size, shape, and exposure affect the precipitation caught by a gauge, a standard gauge should be used so that observations from different gauges can be compared. Most countries have selected one particular type of precipitation gauge as the National Standard gauge that is used in the country-wide precipitation gauge network. In India, for example, Symon's rain gauge has been adopted as the standard rain gauge.

The rain gauge is usually fixed on a masonry concrete foundation sunk into the ground and the gauge is cemented into the platform such that the rim of the gauge is approximately 30 cm above the ground level. This height is necessary to prevent the splashing of water into the gauge. The rim of the gauge should be perfectly level. In a nonrecording instrument, the rainfall measured at a fixed time on any particular date is entered against that date and it is understood that the rainfall so registered has been received in 24 h preceding that time of the day of observation.

Basically, there are three types of rain gauges:

- Standard or ordinary rain gauges (ORGs) are manually read, commonly once a day at fixed hours. The readings represent the accumulated depth of rainfall.
- Self-recording rain gauges (SRRGs) record the rainfall depth in the form of a continuous plot. The data are read manually and commonly tabulated at hourly intervals.
- Automatic rain gauges with data logger. Data are available in digital form either as rainfall at fixed intervals or at timings for each event of rainfall of fixed depth (usually 1 mm).

An ordinary rain gauge (Fig. 2.3) consists of a collector and a funnel; the specifications vary with country. The precipitation caught in the gauge is poured into a graduated measuring cylinder and the depth is noted. Three types of recording gauges are in general use: weighing type, float type, and tipping bucket type. In weighing type instruments, all the precipitation falling is continuously recorded and it can measure all types of precipitation. The float and siphon recording rain gauge consists of a collector and rainfall recording mechanism mounted on a base. The recording mechanism consists of a float chamber and a siphon chamber. The recording pen is mounted on the stem of the float.

A tipping bucket rain gauge (Fig. 2.4) consists of a circular collector funnel that directs the rain into either of a pair of buckets. When a small amount (say, 0.25 mm) of rain has entered the bucket, it tilts and empties the contents. At the same time, the incoming rain is directed to another bucket that begins to fill and the process continues. An electrical pulse is generated on each tilt and is recorded. As the amount of rain to tilt a bucket is known, the instrument measures the correct amount of rainfall depth and intensity.

Precipitation Measurement by Weather Radar

The word *radar* is an acronym for radio detection and ranging. Weather radars were developed to overcome the drawbacks of the conventional measurement of rainfall using a rain gauge. Like an ordinary radar, a weather radar sends electromagnetic waves in all directions from a rotating antenna

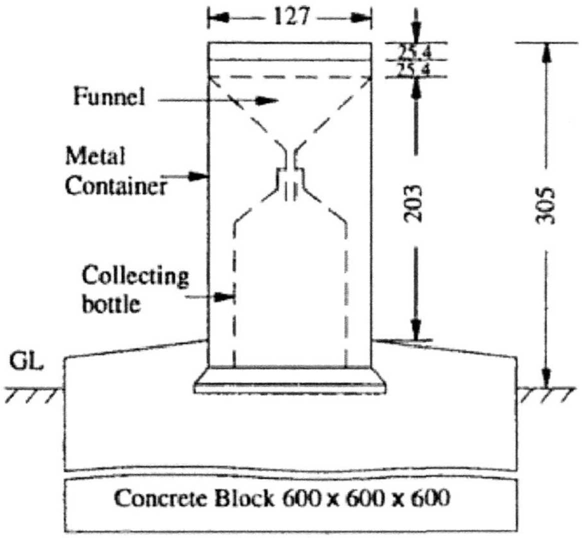

FIGURE 2.3

Standard nonrecording rain gauge.

FIGURE 2.4

Tipping bucket recording rain gauge. Right: top view with funnel removed.

(Fig. 2.5). These waves collide with raindrops and the echo of returning waves is caught by radar. The time taken by the signal depends upon the distance or range of the object and the strength of the returning signal depends on the intensity of rainfall. The direction of propagation of an electromagnetic (EM) wave is called polarization—it can be either horizontally or vertically polarized. Some weather radars are dual polarized (horizontal and vertical).

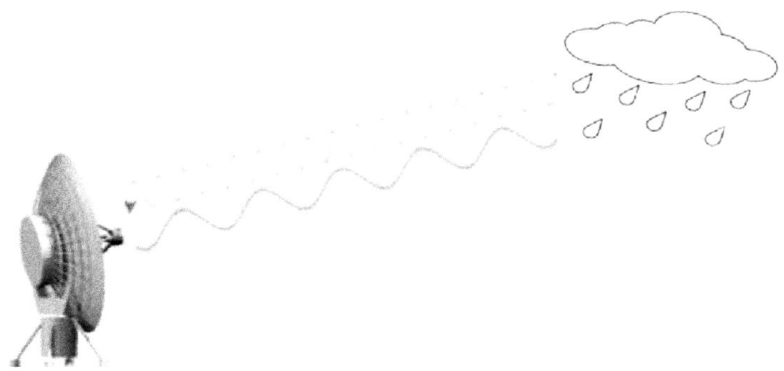

FIGURE 2.5

Principle of weather radar for measurement of precipitation.

Radar signal uses wavelengths (WLs) in the microwave region of the EM spectrum in many bands, some common are S-band WL 7.5–15 cm, frequency 2–4 GHz and C-band WL 3.8–7.5 cm, frequency 4–8 GHz. Since the attenuation of radar signal passing through a medium is inversely proportional to WL, bands with smaller WL (X-band) are not suitable in many weather applications. Thus, the bands that are used in radar depend on climatic conditions and the purpose. The long-wavelength radar does not detect light rain and snow as readily as a short-wavelength equipment.

The range of radar depends on the hardware; radars with a range of up to 200 km are common. The factors that control the measurement of precipitation are its type, size of raindrops, width of the radar beam, refraction of beam, and atmospheric attenuation. A combined use of rain gauges and radars provides much more improved rainfall estimates than what any single one of them could provide.

A radar is calibrated using the data of rain gauges and the relationship between the radar reflectivity Z (mm^6/m^3 or the sixth power of the diameter of raindrops in mm per cubic meter of the atmosphere) and rainfall intensity R (mm/h), known as Z–R relationship is established. A commonly used empirical equation is:

$$Z = AR^b \tag{2.2a}$$

where A and b are the parameters to be estimated. These two parameters can vary spatially, temporally, and on the types of precipitation (e.g., convective, stratiform, etc.). A wide range of A and b can be found in the literature. For example, for a Doppler Weather Radar in India, A = 267 and b = 1.345.

The measurement of the absolute velocity of a raindrop and its instant direction of movement require a radar that has a very precise transmitter frequency and a receiver system that is sensitive to changes in frequency induced by a moving target. Such a specialized radar is based on the Doppler effect and is known as *Doppler radar.* Although these radars are more complex and expensive, they are being installed at many places due to obvious advantages. The NEXRAD (next-generation radar) system employs Doppler radars known as WSR-88D (Vieux, 2002).

The main advantages of radar are that it can measure precipitation over areas that are difficult to access and even in bad weather, it gives areal coverage of rainfall distribution, it is possible to detect the movement of rainfall and it is capable of automatic processing. Weather radar can measure the spatial structure of precipitation fields and the data are available in near real time. Radar system allows for the measurement of localized storms, which may be missed by rain gauges, and the spatial distribution of rainfall can be simultaneously displayed on a monitor. This makes the data very useful for quantitative precipitation forecasts. Due to these reasons, the use of weather radars in precipitation measurement, forecasting, and warning has become a norm. On the negative side, the measurements from weather radar are affected by echoes from the ground or ground clutter. In some cases, these radars have been found to underestimate light rainfall and overestimate heavy rainfall. Some of these drawbacks are being removed by research and development. Of course, beyond hydrology, weather radar has many other applications.

Many books have covered this topic in greater details (see, e.g., Moore et al., 2012). Sokol et al. (2021) discuss the role of weather radar in the estimation of rainfall and its application in meteorology and hydrology.

Estimation of Precipitation by Satellites

Satellites are also being used to estimate precipitation over large areas and in near real time. Some of the concepts and techniques of analysis of satellite data have been described in Chapter 3. Images from geostationary and polar orbiting satellites along with cloud top temperature, shape, texture, and cloud history are used for the estimation of precipitation. Another possibility is to combine satellite images with radar data to obtain improved estimates.

Measurement of Snow

Many catchments receive a large amount of precipitation input in the form of snow. The accumulated snow is a natural storage of water and many major rivers of the world are snowfed. Three variables related to snow are important for water resources: snow depth, area of snow cover, and snow water equivalent.

A common method for the observation of snow depth is by stakes that are installed at an accessible location and read to assess the depth of snow. Care is necessary to measure only fresh snow. The *water equivalent of snow* is the vertical depth of water that would be obtained by the melting of snow. A snow pillow is a flat circular container that is filled with nonfreezing liquid and is used to measure the water equivalent of snow (Singh and Singh, 2001). The weight of snow on the pillow can be measured by hydrostatic techniques or pressure transducer. The snow water equivalent can also be determined by melting the snow collected in a gauge and measuring the meltwater in the same way as rainfall is measured. Isotope techniques are also used to estimate snow water equivalents.

In important catchments, snow surveys are made on permanent snow courses. A snow course is a permanently marked line where snow surveys are made. The snow courses should be carefully selected and should be representative of the area. These should be located in accessible areas where snow falls to the ground without being intercepted by vegetation, and the site should be protected from strong winds. During surveys, snow samples and depths are also measured at various places at a number of points along the course. The extent of snow cover can be assessed by aerial photography or by satellite imagery. The application of remote sensing for snow-cover mapping has been discussed in Chapter 3.

Sources and Types of Data Errors

Errors in precipitation measurement can occur due to errors in the instrument, errors while reading the instrument and transmitting or recording data, errors due to improper instrument exposure or lack of representativeness of the site, and errors that occur during the processing of data. Most of these errors could be further subclassified as systematic errors and random errors. Systematic errors are essentially due to malfunctioning of instrument, wrong exposure conditions, and/or lack of knowledge of the observer. WMO (1983a) listed the following errors for which adjustments need to be made to get a near-accurate estimate of precipitation from measured data:

i. error due to the systematic wind field deformation above the gauge orifice,
ii. error due to the wetting loss on the internal walls of the collector,
iii. error due to evaporation from the container (generally in hot climates),
iv. error due to the wetting loss in the container when it is emptied,
v. error due to blowing and drifting snow,
vi. error due to splashing in and out of water, and
vii. random observational and instrumental errors.

The first six errors listed earlier are systematic and are in order of general importance. The net error due to blowing and drifting snow and due to splashing in and out of water can be either negative or positive, while the net systematic errors due to the wind field and other factors are negative.

Random errors could arise due to the spilling of water when transferring it to the measuring jar, leakage into or out of the receiver, observational error, etc. The others could be due to the observer, such as misreading and transposing digits, and recording the data at a wrong place on the recording sheet.

2.4.2 Processing of Precipitation Data

It is common to find gaps and inconsistent values in raw precipitation data. Hence, validation and preliminary processing of precipitation data is essential before it is put to further use. The processing of data has two major objectives: to evaluate the data for its accuracy and to prepare the data in a form appropriate for subsequent analysis.

Before precipitation data are stored for further use, it is necessary to carry out validation checks. Improper registering of data includes entering data against wrong time and date, and alteration of figures. Some of the statistics that are used for checking are the values of normal rainfall, observed maximum rainfall, or value of rainfall corresponding to a 25-, 50- or 100-year return period.

Adjustment of Data

The adjustment of data has two principal objectives. The first is to make the record homogeneous with a given environment and the second is to eliminate or reduce extraneous influences by correcting for change in gauge location or exposure. An adjustment for these errors is made by *Double-Mass Analysis*.

The double-mass analysis is a graphical method to identify and adjust inconsistencies in a station's lata by comparing with the trend of reference stations' data. As the name implies, in a double-mass urve, both axes are accumulated precipitation values (see Fig. 2.6). Usually, the accumulated seanal or annual precipitation values of reference station(s) are taken as abscissa and those of the station der test as ordinate. A change in the regime of the rain gauge, such as change in exposure and the nge in location, is revealed by the change in the slope of the straight line. The older records are isted by

$$\text{Adjusted Precipitation} = \text{Raw Precipitation} \times \frac{\text{Slope of later period}}{\text{Slope of earlier period}} \tag{2.3a}$$

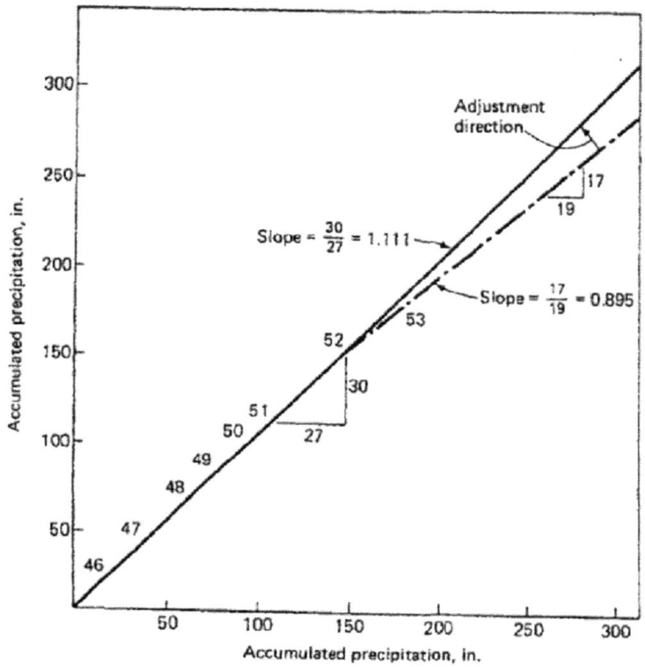

FIGURE 2.6

Double-mass analysis to check consistency of rainfall data at Blair, Nebraska. On the x-axis is the eight-station accumulated mean precipitation and on the y-axis is the accumulated precipitation at Blair.

Spatial Consistency Check

Rainfall data exhibit some degree of spatial consistency and this forms the basis of investigating the observed rainfall values. An estimate of the interpolated rainfall value at a station is obtained on the basis of the weighted average of rainfall observed at the surrounding stations. Whenever the difference between the observed and the estimated values exceed the expected limiting value, such values are considered as suspect and are flagged for further investigation and ascertaining the possible causes of departures.

Rainfall poses special problems for spatial comparisons because the correlation between stations is typically small. When rainfall is of convective type, it may rain heavily at one location, while another some distance away may have little or no rain. Over a month or season, such spatial unevenness tends to be smoothed out and aggregated totals are much more closely correlated. The spatial correlation in rainfall depends on: duration (smaller at shorter durations), distance (decreasing with distance), type of storm that has precipitation, and physiographic characteristics of the region.

The spatial consistency checks for rainfall data are carried out by relating the observations from surrounding stations for the same duration with the rainfall observed at the station. This is achieved by interpolating rainfall at the station under question with rainfall data of neighboring stations. The station being considered is called the test station. The interpolated value is estimated by computing the weighted average of the rainfall observed at neighboring stations. Ideally, the stations selected as neighbors should be physically representative of the area in which the station under scrutiny is situated. The following criteria are used to select the neighboring stations:

(a) The distance between the test and the neighboring station must be less than a specified maximum correlation distance;
(b) Too many neighboring stations should not be considered for interpolation; and
(c) To reduce the spatial bias in selection, it is advisable to consider an equal number of stations in each quadrant.

Example 2.2. Rainfall reported at a group of five stations (see Fig. 2.7) is as follows.

Station	Kundla	Idar	Virpur	Chandop	Sabli
Rainfall (mm)	132.1	12.1	103.3	125.7	149.8

During a quality control process, the data at Idar are identified as doubtful and are required to be checked for spatial consistency.

Solution. The rainfall at Idar is estimated using the distance power method and compared with the observed value. From the four quadrants around Idar (Fig. 2.7), the station nearest to each quadrant is selected for the estimation of rainfall at Idar. Using the reference coordinate system, the distance of each of the estimator stations from Idar is determined and the rainfall at Idar is estimated.

S. no	Station	Distance from Idar D_i (km)	$1/D_i^2$	R_i/D_i^2
1.	Kundla	42	5.67×10^{-4}	0.075
2.	Virpur	39	6.57×10^{-4}	0.068
3.	Sabli	75	1.78×10^{-4}	0.027
Total			14.02×10^{-4}	0.170

Rainfall at Idar $= \left[\sum(R_i/D_i^2)\right]/\left[\sum(1/D_i^2)\right] = 0.17/14.02 \times 10^{-4} = 121.25$ mm.

As the observed value is very much different from the estimated value, it is rejected and replaced by the estimated value. Note that there is a possibility that the decimal point was wrongly placed while recording the data.

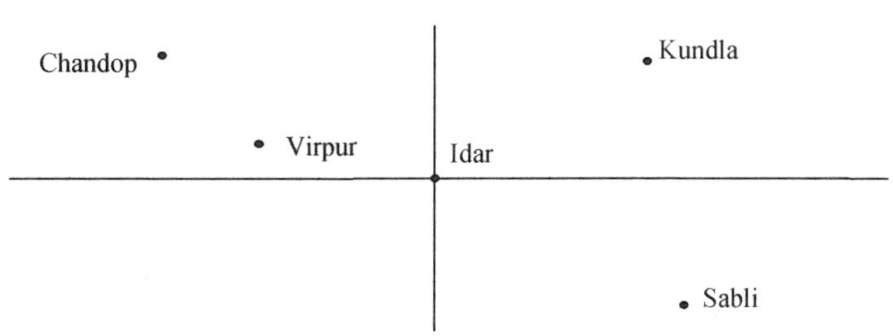

FIGURE 2.7

Location of stations for spatial consistency check.

2.4.3 Spatial Interpolation of Precipitation Data

An accurate assessment of the mean areal precipitation is needed in many hydrological analyses. Precipitation observations from gauges are point measurements. However, the precipitation process exhibits an appreciable spatial variation over relatively short distances. Numerous methods of computing areal rainfall from point measurements have been proposed. While using precipitation data, one often comes across missing data situations. Data for the period of missing rainfall could be filled using various techniques. The length of the period up to which the data could be filled is dependent on individual judgment. Due to the spatial nature of precipitation data, some type of interpolation technique is commonly adopted, making use of the data of nearby stations. The methods that are commonly used for this purpose are discussed in what follows.

Most of these methods can be used for any variable that follows a spatial behavior. The choice of any method is dependent on the quality and nature of data, required precision, availability of time and resources, and the preferences of the analyst.

Let the precipitation data be available at n stations, spread over an area, and P_i be the observed depth of precipitation at the ith station. Using a linear interpolation technique, an estimate of precipitation over the area can be expressed by

$$P^* = \sum_{i=1}^{n} P_i W_i \tag{2.4}$$

where W_i is the weight of the ith station. The spatial averaging techniques differ in the method of evaluation of these weights. Weights of an optimal interpolation technique are decided such that the variance of error in the estimation is minimum.

Arithmetic Average

The simplest technique to compute the average precipitation depth over a catchment area is to take an arithmetic average of the observed depths at gauges within the area for the time period of concern. If the gauges are relatively uniformly distributed over the catchment and the rainfall values do not differ by more than 10%, this technique yields good results. The average precipitation is

$$P = \sum_{i=1}^{n} P_i / n \tag{2.5}$$

where P is the average catchment precipitation using the data of n stations and P_i is the precipitation at station i.

Normal Ratio Method

In the normal ratio method, the rainfall P_A at station A is estimated as a function of the normal monthly or annual rainfall of the station under question and that of the neighboring stations for the period of missing data at the station under question.

$$P_A = \frac{\sum_{i=1}^{n} \frac{NR_A}{NR_i} \times P_i}{n} \tag{2.6}$$

where P_i is the rainfall at surrounding stations, NR_A is the normal monthly or seasonal rainfall at station A, NR_i is the normal monthly or seasonal rainfall at station i, and n is the number of surrounding stations whose data are used for estimation.

Example 2.3. The normal monthly rainfall at the estimator and estimated stations A, B, C, and D is known. The observed rainfall at the estimator stations B, C, and D is also known and is given in the following table. Find rainfall at station A.

Solution. The ratio NR_A/NR_i has been calculated and is given in the following table.

Station →	A	B	C	D
Normal rainfall (mm)	331.3	290.8	325.9	360.5
Event rainfall (mm)	?	98.9	120.5	110.0
NR_A/NR_i	1	1.14	1.02	0.92

The estimated rainfall at station A is

$$P_A = \frac{1.14 \times 98.9 + 1.02 \times 120.5 + 0.92 \times 110.0}{3} = 112.3 \text{ mm}$$

Distance Power Method

The rainfall at a station is estimated as a weighted average of the observed rainfall at the neighboring stations. The weights are equal to the reciprocal of the distance or some power of the reciprocal of the distance of the estimator stations from the estimated stations. Let D_i be the distance of the estimator station from the estimated station. If the weights are an inverse square of distance, the equation is

$$P_A = \frac{\sum_{i=1}^{n} \frac{P_i}{D_i^2}}{\sum_{i=1}^{n} \frac{1}{D_i^2}} \tag{2.7}$$

Note that the weights go on reducing with distance and approach zero at large distances. A major shortcoming of this method is that the orographic features and spatial distribution of the variables are not considered. The extra information, if stations are close to each other, is not properly used. The procedure for estimating rainfall data by this technique is illustrated through an example. If A, B, C, and D are the location of stations discussed in the example of the normal ratio method, the distance of each estimator station (B, C, and D) from station (A) whose data are to be estimated is computed with the help of the coordinates using the formula:

$$D_i^2 = \left[(x - x_i)^2 + (y - y_i)^2\right] \tag{2.8}$$

where x and y are the coordinates of the station whose data are estimated, and x_i and y_i are the coordinates of stations whose data are used in the estimation.

Example 2.4. Using the data of Example 2.3, estimate rainfall at station A using the distance power method.

Solution. As the coordinates of the stations are known, their distances from station A can be calculated. The weights $1/D_i^2$ are then computed for each station and the rainfall at station A is estimated as follows:

Station	Distance from station A	$1/D_i^2$	Rainfall P_i (mm)	Weighted rainfall $P_i{}^*(1/D_i^2)$ (mm)
B	28.0	1.29×10^{-3}	98.9	125.6×10^{-3}
C	17.7	3.19×10^{-3}	120.5	384.6×10^{-3}
D	42.5	0.55×10^{-3}	110.0	60.5×10^{-3}
Total		5.01×10^{-3}		570.7×10^{-3}

Therefore, rainfall at station A $= 570.7 \times 10^{-3}/5.01 \times 10^{-3} = 113.9$ mm.

Thiessen Polygon

The Thiessen polygon method is based on the concept of proximal mapping and weights are assigned to each station according to the area that is closer to that station than to any other station. This area is found by drawing perpendicular bisectors of the lines joining the nearby stations so that the polygons are formed around each station (Fig. 2.7). It is assumed that these polygons are the boundaries of the effective area that is represented by the station. The area governed by each station is planimetered and expressed as a percentage of the total area. The weighted average precipitation for the basin is computed by multiplying the precipitation received at each station by its weight and summing. The weighted average precipitation is given by

$$P = \sum_{i=1}^{n} P_i W_i \tag{2.9}$$

in which $W_i = A_i/A$, where A_i is the area represented by the station i and A is the total catchment area. Clearly, the weights will sum to unity. An advantage of this method is that the data of stations outside the catchment, if they represent some part of the catchment, may also be used. A major drawback of this method is the assumption that precipitation between two stations varies linearly and the method does not make allowance for the variation due to orography. In this method, the precipitation depth changes abruptly at the boundaries of polygons. Also, whenever a set of stations are added to or removed from the network, a new set of polygons have to be drawn.

The method fails to give any idea as to the accuracy of results. If a few observations are missing, it may be more convenient to estimate the missing data than to construct the new set of polygons. Panigrahy et al. (2009) have described algorithms for the computerized estimation of Thiessen weights.

Example 2.5. For a catchment, rainfall data at six stations for July along with their weights are given in Table 2.4. Find the weighted average rainfall for the catchment using the Thiessen polygon method.

Solution. Using the observed rainfall and station weight, the weighted rainfall at each station is computed. Summation gives the weighted average rainfall for the catchment. The computations are shown in Table 2.4.

Table 2.4 Estimation of mean areal rainfall by the Thiessen polygon method.

S. no.	Station name	Station weight	Rainfall (mm)	Weighted rainfall (mm)
1.	Sohela	0.06	262.0	15.7
2.	Bijepur	0.12	521.0	62.5
3.	Padampur	0.42	177.0	74.3
4.	Paikmal	0.28	338.0	94.6
5.	Binka	0.04	158.0	6.32
6.	Bolangir	0.08	401.6	32.1
Weighted catchment rainfall				285.6

Isohyetal Method

The isohyetal method employs the area encompassed between isohyetal lines. Rainfall values are plotted at their respective stations on a suitable base map and contours of equal rainfall, called isohyets, are drawn. In regions of little or no physiographic influence, the isohyetal contours may be drawn taking into account the spacing of stations, the quality, and variability of the data. In regions of pronounced orography where precipitation is influenced by topography, the analyst should take into consideration the orographic effects, storm orientation, etc. to adjust or interpolate between station values.

These days, computers are widely used to draw isohyetal maps. The isohyetal map for an area is shown in Fig. 2.8. The total depth of precipitation is computed by measuring the area between successive isohyets, multiplying this by the average of the two isohyets, and totaling. The average depth of

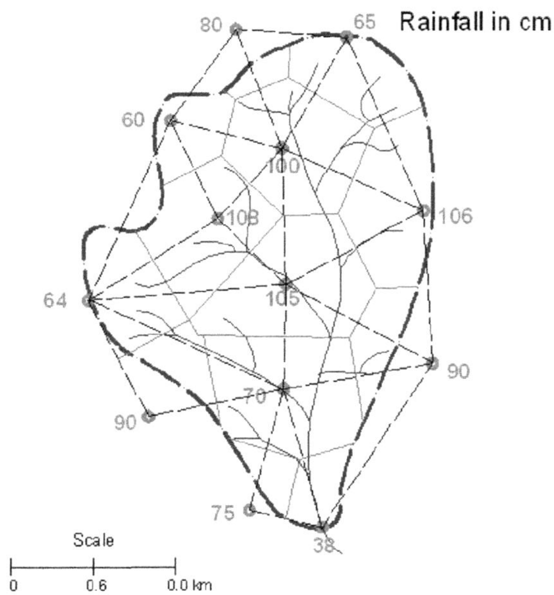

FIGURE 2.8

Thiessen polygon method for computing mean areal rainfall.

precipitation is obtained by dividing this sum by the total area. The average depth of precipitation (P_i) over this area is obtained by

$$P_A = \frac{\sum_{i=1}^{n} P_i A_i}{\sum_{i=1}^{n} A_i} \tag{2.10}$$

where A_i is the area between successive isohyets.

Example 2.6. Using the point rainfall data for a catchment, isohyetal lines were drawn, as shown in Fig. 2.9. The area enclosed by each isohyet was calculated as given in Table 2.5. Compute the average catchment rainfall.

Solution. For each isohyet, the average value is worked out (the maximum observed rainfall was 108 cm and the minimum 38 cm). This, multiplied by the area enclosed by that isohyet, gives the volume of rainfall for that isohyet. Now the volumes for different isohyetals are summed and divided by the area of the catchment to get an average catchment rainfall. The computations are shown in Table 2.5.

Kriging

For the estimation of areal averages of the variables that are considered to be realizations of stochastic processes, Matheron (1971) proposed the theory of regionalized variables. A variable, which characterizes a phenomenon varying in space and/or time and shows a certain structure, is called a *regionalized variable*. The variables, describing the depth of rainfall, water level in observation wells, and soil transmissivity, are a few examples of regionalized variables. The technique of Kriging is named after D.R. Krige who first applied this theory to gold mining in South Africa.

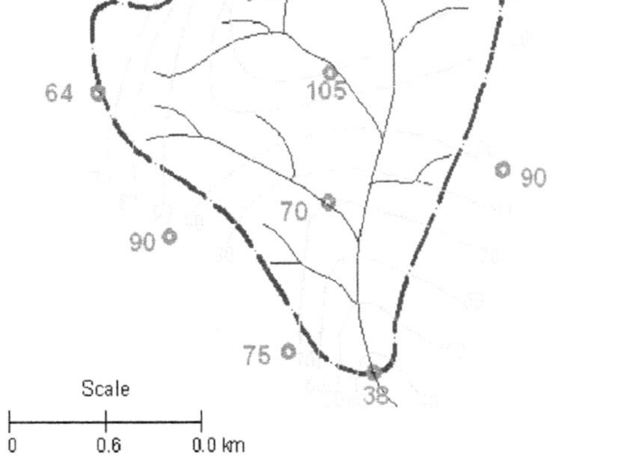

FIGURE 2.9

Isohyetal method for computing mean areal rainfall.

Table 2.5 Estimation of mean areal rainfall by the isohyetal method.

Isohyet value (cm)	Average value (cm)	Area enclosed (km^2)	Net area (km^2)	Rainfall volume (km^2-cm)
105	106.5	0.79	0.79	84.14
100	102.5	1.52	0.73	74.83
90	95	2.57	1.05	99.75
80	85	3.47	0.90	76.50
70	75	4.50	1.03	77.25
60	65	5.18	0.68	44.20
50	55	5.39	0.21	11.55
<40	39	5.41	0.02	0.78
Total			5.41	469.00
Average catchment rainfall = 469.00/5.41 = 86.69 cm				

Given the values of the variable at n observation points, p_i, i = 1, 2, …, n, the problem of Kriging is to estimate a quantity p^* which is a linear function of variables. Three types of problems may arise:

(a) To estimate the value of the variable at a point;
(b) To estimate the value of the variable over a mesh of a given area centered at a known point; and
(c) To estimate the value of the variable over a specified domain.

The first type of problem is called point Kriging, second and third types are called block Kriging. The third type is most generalized and the first two can be considered as special cases of the third type when the domain reduces to a point or a block. According to Eq. (2.4), it is required to find the set of weights that give the best possible estimation. For the estimation to be the best, the weights must be as follows: (1) unbiased, that is, there should be no systematic over or under estimation, and (2) optimal, that is, the variance between observed and computed values must be minimum.

The theory of Kriging assumes the increments of the variables to follow the weak stationarity of second order. Under this assumption, a random function is said to be stationary if the first two moments of its joint probability distribution at k arbitrary points are invariant under the simultaneous translation of all the points. The first step in the application of Kriging is the derivation of semivariogram, which is the graph of variability of the difference of the regionalized variable with respect to the distance between the data points:

$$\gamma(d_{ij}) = \text{Var}\left[p_i - p_j\right]$$
$$= \sigma^2 - \text{Cov}(d_{ij}), i, j = 1, 2 \ldots, n \tag{2.11}$$

where $\gamma(d_{ij})$ is the semivariogram, which is a function of the distance between i and j points. Fig. 2.10 shows a semivariogram.

The semivariogram is first estimated from the observed values and then one of the theoretical semivariograms is selected. For example, the exponential model of the semivariogram is

$$\gamma(d) = C_o[1 - \delta(d)] + C[1 - \exp(d / a)] \tag{2.12}$$

where $\delta(d)$ is the Kronecker delta [$= 1$ when $d = 0$, $= 0$ when $d \neq 0$]. The condition of the minimum variance and Eq. (2.11) yield a set of equations:

$$\sum_{i=1}^{n} w_i \gamma(d_{ij}) + \lambda = \gamma(d_{0j}), \quad j = 1, 2 \ldots n$$

$$\sum_{i=1}^{n} W_i = 1 \tag{2.13}$$

The solution of these (n+1) simultaneous equations yields n weights and the Lagrange multiplier λ. As mentioned earlier, the knowledge of the variogram is required for interpolation using Kriging. A number of models of the variogram have been described by Davis (2002).

Use of Station Characteristics

By the procedures described earlier, a point estimate can never be greater than the largest amount observed or less than the smallest. However, in some areas particularly in mountainous regions, precipitation patterns have known characteristics that would indicate higher or lower amounts at certain points. The use of "station characteristics" permits this to be taken into account. The station characteristic is defined as the ratio of the mean monthly precipitation at a given station to the mean monthly precipitation at a base station. A base station is one which has a long-term reliable climatological record and is representative of a large portion of the basin. The base station serves as a guide for determining the station characteristics at neighboring stations.

2.4.4 Disaggregation of Rainfall Data

Many applications, such as flood forecasting, require rainfall data of shorter duration, whereas the network of recording rain gauges (providing short-duration data) is small in comparison to that of daily rain gauges. Hence, it is often necessary to disaggregate the daily rainfall data into shorter time

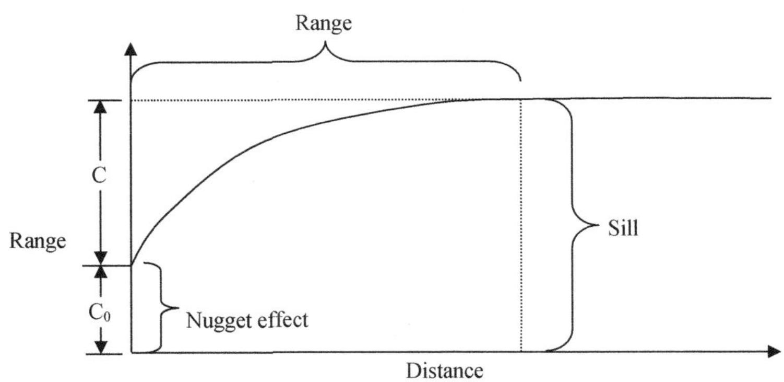

FIGURE 2.10

Plot of semivariogram.

intervals. The information on short-interval rainfall is used together with the information on daily rainfall from nearby nonrecording (daily) gauges. A common method to do this is the mass curve method.

A mass curve is a graphical display of accumulated rainfall versus time. Mass curves of accumulated rainfall at (nonrecording) daily stations and recording stations are prepared by plotting the accumulated rainfall values against time for the storm duration under analysis. Comparison of the mass curves of the recording rain gauge stations with those of the nonrecording stations helps in deciding which recording rain gauge or group of gauges is representative of which of the nonrecording rain gauge for the purpose of distributing daily rainfall into hourly rainfall.

Assume that the daily rainfall is observed at 0800 h. For converting the daily rainfall into hourly rainfall, the hourly rainfall from 0800 to 0800 h for consecutive days is accumulated and the rainfall during each hour is expressed as a ratio of the total rainfall during 24 h (0800—0800). These ratios are used to distribute the daily rainfall for the corresponding duration at those nonrecording rain gauge stations.

2.4.5 Rain Storm Analysis

While designing a dam, it is necessary that the outlet capacity is large enough to safely pass a flood of a certain magnitude. This critical flood is known as the *design flood* for the structure. The type of the hydraulic structure is the main criterion to decide the design flood hydrograph. For this purpose, the structures are classified as:

- **(i)** large or medium dams;
- **(ii)** medium structures, such as barrages, roads, and railway bridges; and
- **(iii)** small or minor structures, such as cross drainage works and minor irrigation tanks and minor road bridges.

If long-term runoff data are not available, rainfall data, which are generally available for a longer period, are used to estimate the *design storm* which is used with a suitable rainfall-runoff model to obtain a design flood. The rainstorm analysis is the first step in the design storm estimation procedure. The design storm (rainfall) is a magnitude of rainfall and its distribution which is developed for the design of specific types of structures. It has three components, namely, the rainfall amount, the areal distribution of rainfall, and the time distribution of rainfall.

The *Probable Maximum Precipitation* (PMP) is defined (WMO, 1973) as "theoretically the greatest depth of precipitation for a given duration that is physically possible over a given size storm area at a particular geographical location at a certain time of year." The Standard Project Storm is defined as a rainstorm that is reasonably capable of occurring in the region of the problem basin. It is generally the most severe rainstorm that has occurred in the region of the basin during the period of available records.

The data on volumes of precipitation during severe storms is important for examining and studying storms suitable for design purposes. Such information is generally presented in the form of tables of the maximum average depth of storm precipitation over various standard area sizes, such as 100 km^2 and 500 km^2. These data are known as the depth—area—duration data, and they could be presented in tabular as well as graphical form. WMO (2008) has described procedures to estimate the probable maximum flood and standard project flood.

2.5 ACQUISITION AND PROCESSING OF OTHER METEOROLOGICAL DATA

Besides precipitation, the other meteorological variables, such as evaporation, temperature, humidity, wind speed and direction, and sunshine hours, are also important in studies relating to water resources development and management. The acquisition and processing of these data are discussed next.

2.5.1 Pan Evaporation Data

Evaporation is an important component of the hydrologic cycle. The rate of evaporation is a function of climatic variables, such as incoming solar radiation, air and water temperature, wind speed, and saturation vapor pressure deficit. Pan evaporation provides an estimate of open water evaporation. Evaporation data from pans are frequently used to estimate evaporation from water bodies, such as lakes and reservoirs and evapotranspiration from an area. Evaporation pan readings are taken once or twice a day at fixed times.

Evapotranspiration (ET) from crop areas can be directly measured through a lysimeter (see Howell, 2005). Three types of lysimeters are used. A weighing-type lysimeter uses mechanical balance to determine the change in water content of the control volume. The hydraulics-based equipment employs hydrostatic principles of weighing, and in the volumetric-based ones, ET is measured by the amount of water added to or removed from the control volume to keep constant water content. Note that lysimeters are difficult and costly to install and maintain. Given the difficulties in the direct measurement of ET, indirect methods are generally used to estimate it. This requires measurements of meteorological variables that influence evaporation. Commonly, pan evaporation is multiplied by a coefficient to get the crop ET.

Three types of evaporation pans are in common use: the US Weather Bureau Class A pan, the GGI-3000 pan, and the 20-m^2 tank. The US pan is widely used throughout the world; the last two are widely used in Russia and some other countries. The US Weather Bureau Class A pan evaporimeter (a circular pan of 1.22 m diameter and 0.255 m deep), that rests on a white painted wooden stand, is the equipment that is used almost universally (Fig. 2.11). It is a good practice to install additional instruments along with a pan, such as an anemometer to measure wind movement over the pan, a precipitation gauge, and thermometers to measure the temperature of pan water and surrounding air. To prevent the drinking of water by birds and animals, either some chemical repellants may be added to the water (these should not pollute the water), or the pan may be covered by a wire mesh. The inner side of the pan is painted white. The water level in the pan changes due to evaporation and rainfall. A stilling well with a pointer gauge is installed in the pan. The change in the water level in the pan and the depth of rainfall at a nearby rain gauge yields evaporation losses. The amount of evaporation between two observations of water level in the pan is obtained by

$$E = P \pm \Delta d \tag{2.14}$$

where P is the depth of precipitation during the intervening period and Δd is the depth of water added (+) to or removed (−) from the pan.

Errors in the data may crop up due to observer's mistakes and instrument errors where leakage from sides or bottom is most common. Birds or animals may drink water from the pan, particularly if the covering wire mesh is damaged. The maximum value of evaporation in the region is used as the first check to screen doubtful measurements.

FIGURE 2.11

US Weather Bureau Class A land pan.

Since a pan is a small water body whose material is different from a natural body, its heat storage characteristics and air dynamics are different from a large water body. Therefore, evaporation from a pan is higher than a large open water body. An estimate of lake or reservoir evaporation is obtained by multiplying the pan evaporation by a coefficient called pan coefficient:

$$E_R = K_P\, E_{pan} \tag{2.15}$$

where K_P is the pan coefficient, E_R is the depth of evaporation from a reservoir, and E_{pan} is the pan evaporation, both in mm/day. The value of the pan coefficient depends on climate, location, season, size, and depth of the water body. This coefficient generally varies from 0.6 to 0.8. The lower values are typical of dry seasons and arid climates, while higher values are observed in humid climates. In the absence of better estimates, an average value of 0.7 is generally used.

2.5.2 Temperature and Humidity Data

Since temperature affects evaporation and snowmelt, it is needed in many water resources studies. The temperature of air, snowpack, soil, and water bodies is of interest. It is a measure of the ability of the atmosphere and water to receive and transfer heat from other bodies. The temperature of a water body is also an indicator of its quality, as it influences the amount of dissolved gases and the rate of chemical and biological activities. An accuracy of 0.1°C is enough in temperature measurements.

Temperature varies primarily with the magnitude of solar radiation and follows diurnal and seasonal cycles. It is influenced at particular times by the exchange of air masses and by cloudiness that limits incoming radiation. Temperature varies with latitude (which controls solar radiation), altitude, and proximity to oceans. Normally, temperature decreases with altitude at a rate of approximately 0.6°C per 100 m for moist air and 0.9°C for dry air. The temperature of places near a large water body, such as sea, is moderated by its influence so that the annual and diurnal range is smaller. Generally, temperatures at nearby stations are strongly correlated.

Manual observations of air temperature are made using thermometers. Four types of thermometers: dry-bulb, wet-bulb, maximum, and minimum thermometers are used. The dry-bulb thermometer gives air temperature. The wet-bulb thermometer is used to measure the temperature of the saturated air to determine the relative humidity and dew point of the surrounding air. The maximum and minimum thermometers indicate these temperatures over a span of time. Observations of temperatures are made once or twice a day at standard times. A continuous record of temperature is obtained using a thermograph in which changes in the temperature are recorded on a clock-driven revolving chart.

The atmospheric humidity has a significant influence on evapotranspiration. The commonly used term relative humidity (RH) is the ratio (in %) of the mixing ratio to the saturation mixing ratio:

$$RH = 100\, r/r_s \tag{2.16}$$

where the mixing ratio $r = m_v/m_d$. Here, m_v is the mass of water vapor and m_d is the mass of dry air. The saturation mixing ratio r_s is the ratio of the mass of water vapor in a volume of saturated air to the mass of dry air. The RH (%) can be defined as the ratio of the actual vapor pressure of the air to the saturation vapor pressure at the same pressure and temperature.

The RH does not vary drastically over a short time. Places close to the sea have higher RH and a smaller daily variation than do inland locations. The most common way of assessing the RH is the joint measurement of dry-bulb and wet-bulb temperatures. The dry bulb refers to an ordinary thermometer. The wet bulb is so called because it is covered with a clean muslin sleeve, tied around the bulb by a cotton wick that is dipped in a water container so that the wick and muslin are kept constantly moist. From these two measurements, the dew point temperature, and actual and saturated vapor pressures may also be calculated. While the actual vapor pressure may vary little during the day (except with the incursion of a new air mass), RH has a regular diurnal pattern with a minimum normally coinciding with the highest temperature (when the saturation vapor pressure is at its highest). It also shows a regular seasonal variation. RH is calculated from the wet-bulb depression (difference between dry and wet-bulb readings) using a set of tables.

RH may also be measured continuously by means of a hygrograph in which the sensor is human hair whose length varies with relative humidity. The humidity is registered on a chart on a clock-driven revolving drum and the measurement (chart) period may be either 1 day or 1 week.

Errors in the temperature data may arise due to:

- Observer error in reading the thermometer—usually it could be an error of 1°C, which is difficult to detect. Reading errors are more common in thermometers with faint graduation etchings.
- Error in registering the thermometer reading.
- Thermometer fault that results in systematic errors in temperature.

Measurement errors, if dry and wet-bulb thermometers are used, are the same as for temperature. Additionally, an error will also occur if the muslin and wick of the wet bulb are not adequately saturated or the muslin becomes dirty or is covered by slime. These defects will tend to give a high reading of the wet-bulb temperature and consequently a high value of RH. Errors in the hygrograph may result from poor calibration.

Validation of temperature data is based on location and site conditions and comparison between stations. Common errors can be detected by setting up appropriate maximum, minimum, and warning limits. For example, the summer maximum temperature can be expected not to exceed 50°C and the winter maximum temperature may not exceed 35°C. The maximum value for RH is 100%. The dry-bulb temperature should be greater than or (rarely) equal to the wet-bulb temperature.

A temperature sensor is a device to measure temperature. This can be air temperature, liquid temperature, or the temperature of solid matter. Temperature sensors have become very common these days to measure, particularly air temperature. Thermistors measure temperature by measuring the change in resistance of the electric current. Resistance temperature detectors measure ohmic resistance to determine temperature. Thermocouples use two conductors (made up of different metals) that are joined to form a junction. When this junction is heated, a voltage is produced that is directly proportional to the temperature input.

2.5.3 Wind Speed and Direction

Wind speed and direction are inputs to the calculation of evapotranspiration. Wind speeds are controlled by local pressure anomalies that, in turn, are influenced by the temperature and local topographic features. The wind speed exhibits a wide variation not only from place to place but also during the day. The wind direction may influence evaporation if the surrounding environment has different humidity values in different directions.

The wind speed is measured using an anemometer (usually a cup type). The number of rotations of the anemometer over a time interval is displayed by a counter or logged using a data logger and indicates the average speed over that interval. Normally, the wind speed over a 3-min period is considered as the instantaneous wind speed at that time. The daily wind run or the average wind speed is calculated from counter readings on successive days at the principal observation times. The wind direction is reported as 16 points of the compass either as a numerical figure or an alphabet character. Observations are made daily in the morning or twice daily in the morning and evening. Wind speed measurements may be instantaneous; these are accumulative if the wind run over a time interval is observed.

Errors in wind speed might arise as a result of the observer misreporting the counter total. Instrumental errors might arise from poor maintenance or damage to the spindle. Because of the large variability in wind speed in space and time, it is difficult to set up convincing rules to detect suspect values.

2.5.4 Sunshine Duration

The sunshine duration data is an input variable for the estimation of evapotranspiration. It is widely used for the computation of evapotranspiration in the absence of radiation measurements. The potential maximum sunshine duration depends on latitude and season; the actual sunshine hours vary due to clouds, fog, etc. The amount of bright sunshine in urban areas may be reduced by atmospheric pollution and in coastal areas it may be reduced by sea mist.

The instrument commonly employed for the observation of sunshine duration is the Campbell Stokes sunshine recorder. It is a glass sphere mounted on a section of a spherical bowl. The sphere focuses the sun's rays on a card graduated in hours, held in the grooves of the bowl that burns the card linearly through the day when the sun is shining. The card is changed daily after sunset. Hence, the sunshine recorder uses the movement of sun instead of a clock to form the time basis of the record. Different grooves in the bowl must be used in winter summer and the equinoxes, taking different card types. The lengths of burned traces on the sunshine card indicate the sunshine duration. Sunshine duration data at the required resolution is tabulated from the card.

The use of the sun as a timing device avoids timing errors. Possible errors may arise from the wrong placement of the chart may result in the burn crossing the edge of the chart and remain unregistered. The errors may also creep in while the observer notes the duration of sunshine from the chart. The values below zero or greater than the maximum possible sunshine hours for the location are rejected.

2.5.5 Automatic Weather Stations

An automatic weather station (AWS), an automated version of the traditional weather station, is a system of microprocessor-controlled hardware sensors to measure, store, and transmit data of a number of hydrometeorological variables. Its main advantage is that a number of variables are measured in one place and the whole operation of data collection, storage, and transmission is automatic. The frequency of data collection can be as small as the order of a few minutes. The variables that are measured through such a system typically include incoming and outgoing solar radiation, precipitation, sunshine duration, air and soil temperature, humidity, and wind velocity and direction.

Modern systems are capable of performing some additional actions, such as collecting additional data or initiating a set of instructions like the issuance of warnings, depending on the value of a particular variable. The systems that have facilities for telemetry can be interrogated and controlled from a distant location and commands can be issued from this control center to collect and transmit additional data as per the needs.

The energy to operate these stations is commonly provided from a battery pack that can be recharged from solar power. Thus, these stations may remain unattended for months together. A technician can visit the station after several months to download the data and do maintenance. Of course, data from a telemetry system can be downloaded as and when needed. This property of AWS makes them very useful for data collection from remote or inhospitable locations.

2.6 ACQUISITION AND PROCESSING OF STREAMFLOW DATA

The most important hydrological data for surface water analysis pertains to streamflow that has served as the lifeline for mankind and continues to do so. Its importance is also relatively more, since this source is visible in contrast to groundwater that is hidden. Streamflow records are primarily continuous records of flow passing through a particular section of the stream. Streamflow data are analyzed to determine the magnitude and variability of surface waters. These records are input for planning, design, and operation of surface water projects and are also used in the design of bridges and culverts, flood forecasting systems, and flood plain delineation.

The sections where river measurements are carried out are known as stream gauging stations. A network of these stations is established to collect data about the surface water resources of a region. The location of gauging sites is dictated by the purpose of data collection. If a particular site is needed for a specific project, the general location is in the vicinity of the project. However, if the network of gauging stations is to be established to study the general hydrology of a region, careful planning is required to identify locations so that optimum information is obtained for the resources deployed in data collection.

River-gauging stations are of three types: basic data stations, operational stations, and special stations. Basic stations collect data for a variety of uses, including planning and design of projects, and to understand the hydrological characteristics of the area. Operational stations collect data to run projects and issue forecasts. The objective of special stations is to meet specific data needs that may arise in cases, such as research, project investigation, special studies, and legal cases. Their operation is terminated when the specific need is fulfilled. Sometimes, auxiliary stations are set up to augment the records of the network. An auxiliary station may, for example, record only the peak discharge that occurs at that site during a certain period.

The number of gauging sites depends on the cost of installation and operation, the value of data, watershed size, degree of development, objective of data collection, accuracy, hydrologic characteristics, etc. Some of these factors are interrelated. For example, large watersheds involve costlier projects, and more data and higher accuracy are, therefore, needed.

The streamflow data that are of immense use in water resources are river water level or gauge stage and discharge. The terms stage and gauge height are interchangeably used to express the elevation of the river water surface with respect to an established datum. A continuous observation of the river water level or stage may be made with comparative ease and economy. At important stations, the stage is measured at short intervals and discharge is measured once or twice each day. At less important stations, only stage measurements are made regularly.

The amount of flow passing through a section per unit time is termed as discharge at that section. A continuous measurement of discharge in a natural channel is comparatively difficult, time consuming, and expensive and requires special skills. Therefore, discharge at a site is measured less often and is estimated by indirect methods. Fortunately, there exists a relation between stage and discharge at a section. This relation is termed as stage—discharge relation or rating curve that falls under relation-oriented data. This relationship is used to transform the observed stage into discharge. Note that the reliability of such discharge records is dependent on the reliability of stage data and the accuracy of the stage—discharge relation.

At many sites, discharge is not a unique function of stage; variables other than stage must also be simultaneously measured at such sites to obtain a discharge record. For example, if variable backwater occurs at a site, the information on stream slope is required. The slope can be measured by installing an auxiliary stage gauge downstream. The rate of change of stage can be an important variable where flow is unsteady and channel slopes are flat. Artificial controls are sometimes built to stabilize the stage—discharge relation. These are constructed only for low flows owing to a very high construction cost.

The International Standard Organisation (ISO) has brought out a large number of standards dealing with the measurement of liquid flow in open channels. The ISO Technical Committee TC113 deals with this theme. The publication ISO (1983) is a useful collection of standards dealing with various aspects of streamflow measurement. WMO has also brought out many publications related to streamflow measurement. Of course, the individual countries have their own standards. The topic is covered in detail in Herschey (1978, 1986, 2009).

2.6.1 Selection of Gauging Sites

After the general location of a gauging station has been determined, its precise location is selected to get the best conditions for stage and discharge measurement and to develop a stable discharge rating. For example, consider that the outflow from a reservoir is to be gauged. The general location of the gauging station will be along the stretch of the river between the dam and the confluence with the first major stream. The gauge site should be so located that it is far enough from the dam for the flow to be established fairly uniformly across the entire cross-section. On the other hand, the site should not be so far downstream that there is an appreciable flow from the intermediate catchment.

The ideal gauge site satisfies the following criteria:

(a) There should be a straight, well-defined approach channel upstream of the measuring section to ensure parallel and nonturbulent flow. The approach channel should have a uniform cross-section, free from irregularities, and the flow should have a regular velocity distribution.
(b) The river should not be braided at the gauge site and all the flow must be confined to a single stream at all stages. The site should not have a tendency to collect floating debris that may hinder measurements. Sites with high sediment deposition or scouring or high weed growth should be avoided.
(c) Banks should be permanent, high enough to contain floods.
(d) Downstream conditions should preferably be stable. Sites, which are influenced by downstream confluence with other rivers, river control structures, dams, tidal conditions, or heavy weed growth, should be avoided.
(e) Site should have unhindered access in all seasons and space to construct an office.
(f) The measurement section should be clearly visible across its width and unobstructed by trees, aquatic growth, or other obstacles.
(g) There should be a sufficient depth of flow across the whole cross-section.
(h) If the site is upstream of the confluence of two rivers, it should be located sufficiently far upstream to avoid any disturbance due to the confluence of two rivers.

An ideal site is rarely found for a gauging station and commonly judgment has to be exercised in choosing between adequate sites, each of which may have some shortcomings. Detailed guidelines for the selection of sites are given in the standard ISO-1100. A gauging station may also have hardware for real-time communication of data to a central location.

2.6.2 Measurement of Stage

Stages are measured with reference to a recognized datum, such as the mean sea level or an arbitrary datum that is selected for convenience. To eliminate the possibility of negative values of the gauge height, the datum selected should be below the elevation of zero flow. The gauge height is usually expressed in hundredths or thousandths of a meter. The water level data are measured using a variety of equipment: staff gauges, autographic water level (chart) recorders, digital water level recorders, and radars. The advantages of a nonrecording gauge are low initial cost and ease of installation. The disadvantages are the need for an observer, less accuracy, and data at larger time gaps. For long-term operation, the advantages of a recording gauge outweigh those of a nonrecording gauge. Sometimes, an automatic and a nonrecording gauge are maintained together because the electromechanical recording gauge equipment is liable to breakdowns.

Staff Gauge

Staff gauges are either vertical or inclined. Vertical staff gauges are normally porcelain enameled iron sections, graduated every 10 mm. The vertical staff gauge is used as an inside reference gauge (if installed in a well), or as an outside gauge if installed in the stream. If river stage varies over a large range during the year, observations using a single staff gauge might be difficult. In such cases, the

FIGURE 2.12

Sectional staff gauge and use of bridge piers to measure river water level.

gauge consists of stepped sections (Fig. 2.12) installed at different locations in a line normal to the flow. Each of these stepped gauges refers to the common datum and there should be sufficient overlap among them. An inclined staff gauge is usually a graduated surface attached securely to a permanent foundation. The rock outcrops on the river bank also make a good base for inclined staff gauge. Inclined gauges built flush with the stream bank are less likely to be damaged by floods, floating ice, or drift than are projecting vertical staff gauges. The gauges should be located as close to the measuring section as possible, without affecting flow conditions. Staff gauges are manually read, generally each day in the morning in lean season and at (multi) hourly intervals during high flows. In some cases, marks are placed on bridge pillars that are used to record water levels.

Water Level Recorder

A water level recorder is an instrument to sense and record water level. It consists of a time element and a gauge height element, which together produces a time series of water levels. The time element is controlled by a clock, while the gauge height element is actuated by a float or a pressure-actuated system. These recorders can be classified as either analog type or digital type, depending on the way the data are recorded. The analog-type recorders produce a graphic record of fluctuations in the water level with respect to time.

The water level recorders are generally of shaft-angular-input type, and the angular rotation of the shaft is recorded. The depth of the water surface is sensed for automatic recording by a float in a stilling well (Fig. 2.13), which follows the rise and fall of the water level. A gas-purge system that transmits the pressure head of water in a stream to a manometer is known as a bubble gauge.

A water level recorder gives a continuous record of the water level on a chart from which values are manually extracted at desired intervals. The data from a digital water level recorder can either be at equal intervals of time, usually at (fraction of) an hour interval, or at only those instants when there is a change in water level by more than a preset amount. Digital recorders store data in an electronic memory unit and these data are downloaded to a computer.

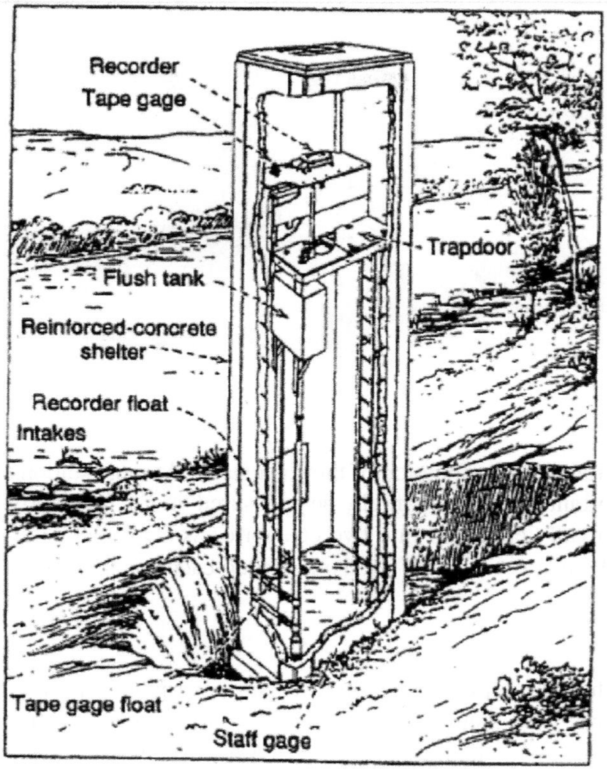

FIGURE 2.13

Stilling well for float-type recorder.

Radar Water Level Measurement

In this setup, a radar is placed above the river water surface. The radar sends an EM signal to the water surface, which is reflected by water and the signal returns to the sensor. By measuring the time taken by the signal to return to the radar and the elevation of the radar, the river water level is determined. This is a noncontact method that gives highly accurate and continuous water level data. As the instrument is not in contact with water, there is no danger of any damage to it. A drawback is the high cost of installations.

In the arrangement shown in Fig. 2.14, the radar is mounted on a ground-based platform. In satellite altimetry, radar is placed on a satellite, and the same principle is followed to measure the water-surface elevation of body, such as a river, lake, or ocean. Satellite radar altimetry measures the time taken by a radar signal to travel from the satellite to the water surface on Earth and back to the satellite. Such measurements are now being used to estimate river flows, lake water volume, etc. at present; the size of a water body is a limitation but soon technology will be able to overcome such a limitation.

2.6.3 Measurement of Discharge

Discharge is the volume of water passing through a certain section in a unit time period. It is commonly expressed in cubic meters per second (m^3/s or cumec). The discharge at a site is a function of the cross-

FIGURE 2.14

Radar river water level measurement setup.

section area and flow velocity. The cross-section area is a function of the river stage. At most stations, discharge is measured once a day; at important stations, it might be measured more frequently. Discharge measurement techniques can be broadly classified into two categories as (1) direct determination and (2) indirect determination. There are scores of methods under each category. The important ones are discussed below.

Direct Determination of Discharge

These are the methods in which either discharge is directly measured or some variable on which discharge depends is measured. The commonly used methods are: velocity–area methods, dilution techniques, electromagnetic method, and ultrasonic method. The first two are described here.

Velocity–Area Methods

The basic procedure in velocity–area methods involves measuring the flow area and velocity and these are multiplied to get discharge. Since the velocity of flow at a cross-section varies laterally and with depth, it is not enough to measure the velocity at a single point. Depending on the accuracy required, the width of the stream is divided into a number of vertical portions (Fig. 2.13). In each of these portions, the velocity is measured at one or more points along the depth to get a representative velocity in that portion. The area of the individual portion can be easily calculated if the bed profile and stage are known. The velocity may be measured by a conventional method (e.g., float or current meter) or by an advanced procedure, for example, the moving boat technique.

A *float* is a distinguishable article that floats on the water surface, such as a wooden log, a bottle partly filled with water, or branch of a tree. For a float measurement, two cross-sections sufficiently far apart on a straight reach of the channel are selected. A number of floats are introduced uniformly across the stream width a short distance before the actual upstream cross-section so that they overcome inertia and move with the velocity of water when they reach the upstream cross-section. Normally, the floats are tossed from a bridge or cableway but if there is no such opportunity, they can be launched from the stream bank. The position of each float with respect to distance from the bank is noted. A stopwatch is used to measure their travel time between the end cross-sections of the reach.

The velocity of the float is equal to the distance between the two cross-sections divided by the time taken by the float to cover this distance. The mean velocity in the vertical is equal to the float velocity multiplied by a coefficient whose value depends on the shape of the vertical-velocity profile of the stream and on the depth of immersion of the float with respect to depth. A coefficient of 0.85—0.90 is commonly used. The float method is not very reliable and its use is normally restricted to an emergency or to measure high discharges when current meter is not available or cannot be used.

Measurement of Velocity Using Current Meter

Current meter is a widely used instrument to measure the velocity of flowing water. It consists of a rotating element (rotor) whose movement is due to the reaction of the stream current. The angular velocity acquired by the rotor is proportional to the velocity of water. By placing a current meter at a point in a stream and counting the number of revolutions of the rotor during a time interval, the velocity of water at that point is determined. Current meters are of two types: those having a propeller rotating around a horizontal axis and those having a series of conical cups mounted around a vertical axis. The normal range of velocity that can be measured varies from 0.15 to 4.0 m/s.

Horizontal-axis meters consist of a propeller mounted at the end of a horizontal shaft (Fig. 2.16). These are available in a range of propeller diameters. Recently, propellers made up of plastic have been introduced. These are cheaper and respond more quickly to changes in velocity. Some current meters come with propellers of different pitch and diameter to suit various flow conditions. The horizontal-axis rotor with valves causes less disturbance to flow than vertical axis rotors. Furthermore, due to axial symmetry with the flow direction, the rotor is less likely to be entangled by debris than vertical axis rotors and the bearing friction is less compared to the vertical axis rotors. The vertical axis rotor with cups or valves can operate in lower velocities than the horizontal-axis meters, the bearings are well protected from silty water, and a single rotor serves for a range of velocities.

The current meter measurements are usually classified in terms of the means used to cross the stream during measurements, such as wading, cableway, bridge, or boat. Wading is possible in small streams of shallow depth only; the current meter is held at the requisite depth below the surface by an observer who stands in the water. In narrow well-defined channels, a cableway is stretched from bank

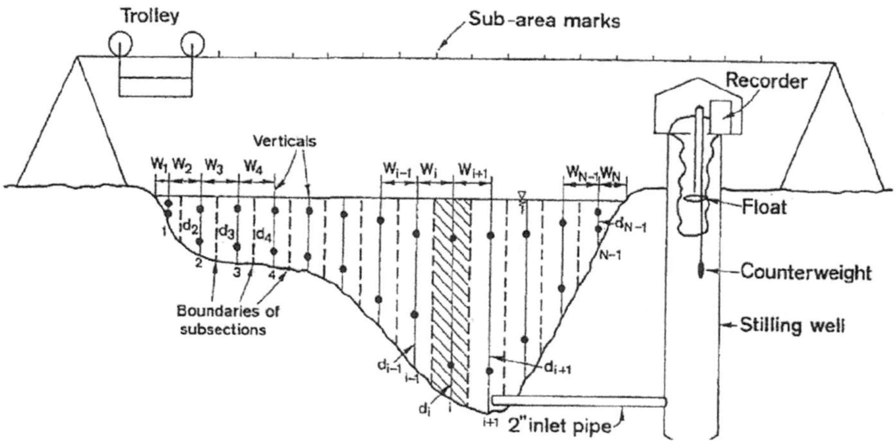

FIGURE 2.15

Schematic sketch for a velocity—area station.

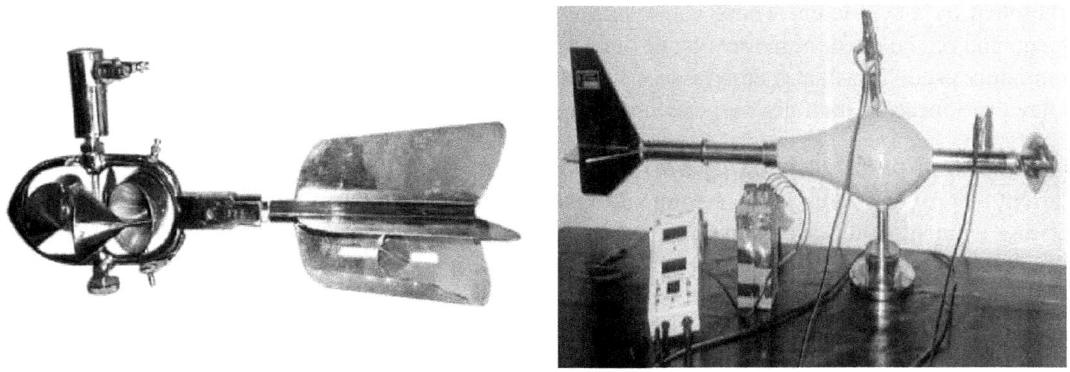

FIGURE 2.16

Cup-type current meter (left) and horizontal-axis propeller current meter (right).

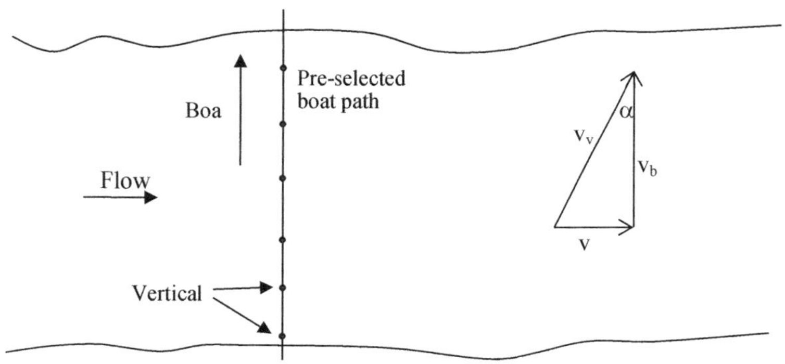

FIGURE 2.17

Moving boat method of discharge measurement.

to bank well above the flood level. A carriage moving over the cableway serves as the observation platform. Bridges are advantageous from the viewpoint of accessibility and transportation, although these are not the best locations hydraulically. The velocity measurement is performed on the downstream of the bridge to minimize the instrument damage due to drift and knock against bridge piers. Boats are most satisfactory for measurements in wide rivers.

The current meter measures the velocity at a point. However, the mean velocity in each of the selected verticals is required to estimate discharge. The mean velocity in a vertical is determined from velocity observations at one or more points in that vertical. Current meters are held down and positioned in a stable manner at the required location in flowing water by sounding weights. The weights are connected to the current meter by a hanger and pin assembly.

The section line at the gauging site is marked by permanent survey markings. The cross-section along this section line is determined by surveying with the help of sounding rods or sounding weights. When the depth of water is more or if quick and accurate depth measurements are needed, an echo sounder is used.

Moving Boat Method

On very wide streams and estuaries, the conventional methods of measuring discharge by current meter are frequently impractical or involve costly and tedious procedures. In the moving boat technique, data are collected while the observer is aboard a boat traversing the stream along a preselected path, generally normal to the direction of flow. During the traverse, an echo sounder records the geometry of the cross-section and a continuously operating current meter senses the combined stream and boat velocities. The angle between the current meter, which aligns itself in a direction parallel to the movement of the water past and the preselected path, it is also measured. Normally, data are collected at 30 or 40 observation points in the cross-section for each run. These days instruments automatically and simultaneously record the required parameters.

The velocity observed at each of the observation points in the cross-section (Fig. 2.15), v_V, is the velocity of water past the current meter resulting from both stream flow and boat movement. It is the vector sum of the velocity of water with respect to the stream bed (v) and the velocity of the boat with respect to the stream bed (v_b). The velocity of streamflow can be obtained by measuring the angle α between the selected path of the boat and a vertical vane that aligns itself in a direction parallel to the movement of water past it.

The flow velocity v, perpendicular to the boat path (true course) at each observation point 1, 2, 3, ..., can be determined from the relationship

$$v = v_V \sin \alpha \tag{2.17}$$

This equation yields that component of the stream velocity that is perpendicular to the true course even though the direction of flow may not be perpendicular.

As the current meter is usually immersed at a depth of 0.5 m from the water surface, the velocity v corresponds to the surface velocity and not the average velocity in the vertical. This surface velocity is multiplied by a coefficient ranging from 0.85 to 0.95 to obtain the average velocity of flow at the section.

Computation of Discharge

After the cross-section has been selected, the width of the stream is divided into an adequate number of partial sections to have lesser variation between two adjacent verticals. If previous measurements have shown uniformity of both the cross-section and the velocity distribution then fewer verticals may be taken. It is better if no partial section carries more than 5%–10% of the total discharge. Fig. 2.15 shows the cross-section of a river in that (n – 1) verticals are drawn. The velocity averaged over the vertical at each section is known. Considering the total area to be divided into (n – 1) segments, the total discharge is calculated by the method of mid-section as follows:

$$P = \sum_{i=1}^{n} (v_i a_i) \tag{2.18}$$

where Q is the total discharge, a_i is an individual partial cross-section area, and v_i is the mean velocity in that area. The area extends laterally from half the distance from the preceding observation vertically to half the distance to the next and vertically from the water surface to the sounded depth.

Acoustic Doppler Current Profiler

As the name suggests, the acoustic Doppler current profiler (ADCP) uses sound waves in ultrasonic range and applies the Doppler principle to measure velocity profile and compute discharge in streams. ADCP is typically towed by a boat (Fig. 2.18). Transducers are deployed beneath the water surface and point downward. Typical boat-mounted ADCPs have three or four beams pointing between 20 and 30 degrees from vertical. Three beams are needed to obtain a 3-D velocity measurement. An ADCP utilizes the Doppler principle by reflecting an acoustic signal off scatterers (small particles of sediment and other material present in water). Doppler's principle relates the observed change in frequency of a sound source to the relative velocities of the source and observer. A Global Positioning System (GPS) is used to determine the location of the boat/ADCP.

ADCP measurements are fast and accurate. ADCP can work in situations where conventional CM cannot be used or is unreliable, for example, in tidally affected flows, highly unsteady flows, and flood flows. ADCPs also provide continuous profiles of water velocity. Further, vertical-velocity distribution is not assumed but rather measured.

Dilution Technique of Streamflow Measurement

The dilution method of flow measurement, also known as the chemical method, is based on the principle of continuity applied to a tracer that is allowed to completely mix with flow. A tracer is an ion or compound that is introduced into the flow to follow its behavior. A known quantity of a tracer is introduced in the flow at an upstream section. At a downstream cross-section of the reach, the concentration of the tracer is measured at regular intervals of time.

The reach selected for measurement should be such that there is no loss or gain of water and the reach length should be sufficient to achieve complete mixing. The tracer can be introduced in two ways: (1) constant rate injection, and (2) gulp injection. In the constant rate injection method, the duration of injection should be such that a steady regime of concentration is achieved for an adequate duration (about 10–15 min) in the sampling section. In the integration method, a quantity of tracer of volume V and concentration C is added to the stream. At the sampling station, the passage of the entire tracer cloud renders a relationship between concentration and time. Common salt (NaCl) is frequently

FIGURE 2.18

An ADCP being towed along a boat for river flow measurement.

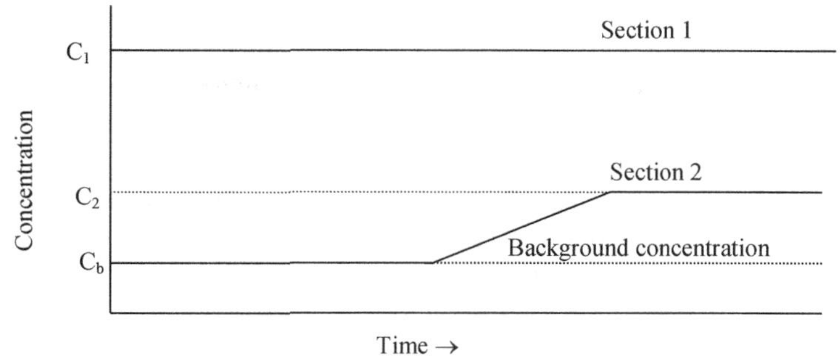

FIGURE 2.19

The concept of dilution method of discharge measurement.

used as a tracer, and it can be detected with an error of 1% up to a concentration of 10 ppm (parts per million). A cocktail of tracers may also be used.

If the tracer of concentration C_1 is injected at a constant rate Q_t at Section 1, the concentration at Section 2 gradually rises from a background value C_b to a constant value C_2 (see Fig. 2.19).

At steady state, the stream discharge Q is

$$Q = Q_t\,(C_1 - C_2)/(C_2 - C_b) \tag{2.19}$$

Example 2.7. A solution of tracer (whose background concentration in the stream was zero) with a concentration of 20 g/L was injected into a stream at a constant rate of 10 cm^3/s. At a sufficiently distant downstream section, the tracer had attained an equilibrium concentration of 4 parts per million. Estimate the discharge in the stream.

Solution. Given $C_b = 0$

Note: 1 ppm = 1 mg/L = 0.001 g/L = 1 g/m^3; 1 ppb = 10^{-6} g/L = 10^{-3} g/m^3

$C_1 = 20$ g/L $= 20 \times 10^3$ g/m^3

$C_2 = 4$ ppb $= 4 \times 10^{-3}$ g/m^3

$Q_t = 10\ cm^3/s = 10 \times 10^{-6}\ m^3/s = 10^{-5}\ m^3/s.$

$Q = 10^{-5} \times (20 \times 10^3 - 4 \times 10^{-3})/4 \times 10^{-3} = 10^{-2} \times 20 \times 10^3/4 = 50\ m^3/s.$

Indirect Determination of Discharge

These methods make use of the relationship between flow discharge and depth at specified locations. The field measurement is restricted to the measurement of depth only. Two important indirect methods are flow-measuring structures and the slope–area method.

Flow-Measuring Structures

Structures, such as notches, weirs, flumes, and sluice gates, are commonly used for flow measurement in laboratories. Sometimes, these structures are also used in field conditions, but they are limited by the ranges of head, debris, or sediment load of the stream, and the backwater effects produced by their installation. Such structures are usually built only in upper and middle reaches; in wide rivers, the size and cost of the structures are prohibitive. A typical setup consists of a reasonably straight (at least for a distance of five times the width) approach channel, a downstream channel, and the structure itself. The structure having smooth surfaces should be rigid, water-tight, normal to the flow direction, and capable of withstanding peak flows without any damage to its body. The section where the water level is measured should be at a distance of 2–4 times the maximum likely head.

The basic principle governing the use of a weir, flume, or similar flow-measuring structure is that these structures produce a unique control section in the flow. At these structures, discharge Q is a function of water-surface elevation measured at a nearby upstream location:

$$Q = f\,(H) \tag{2.20}$$

where Q is the discharge (m^3/s), and H is the head of water (m) at the structure. The equation for weirs, for example, is

$$Q = K\,H^n \tag{2.21}$$

where K and n are constants. Eq. (2.21) is applicable as long as the downstream water level is below a certain limiting water level known as the modular limit. The flows unaffected by the downstream water are known as free flows. The flow that is affected by tailwater conditions is known as drowned or submerged flow. Discharge under drowned conditions is obtained by applying a reduction factor to the free-flow discharge. For a two-dimensional weir, discharge is estimated as

$$Q = C_d \sqrt{g}\, b\, H^{1.5} \tag{2.22}$$

where C_d is the discharge coefficient, g is the acceleration due to gravity, and b is the crest width (m).

Various flow-measuring structures can be broadly considered under three categories:

(a) Thin-plate structures consist of a vertically set metal plate. The V-notch and rectangular and contracted notches are typical examples of this category.

(b) Broad-crested weirs are made of concrete or masonry and are used for large discharges.

(c) Flumes are made of concrete, masonry, or metal sheets, depending on their use and location. They depend primarily on the width constriction to produce a control section.

Slope–Area Method

In the slope–area method, discharge is estimated by observing the water-surface slope and cross-section area. It is an indirect method of discharge estimation that is used when measurement by more accurate methods, such as the velocity–area method, is not possible. Although the accuracy of slope–area method is less compared to the velocity–area method, it is sometimes necessary to use this method because the magnitude of flows is so high that the other methods of discharge estimation cannot be used.

A measurement reach is chosen for which three things are known (Fig. 2.20): (1) The cross-sectional geometry and properties at its ends, (2) the value of Manning's n, and (3) water-surface elevations at the end sections. In the selected reach, a minimum of three cross-sections are generally desirable. As far as possible, the length of the reach should be such that the difference between water levels at the upstream and downstream gauges is not less than 10 times the uncertainty in the difference. Slope is computed from gauge observations at either end of the reach, the intermediate gauge(s) are used to confirm that the slope is uniform throughout the reach.

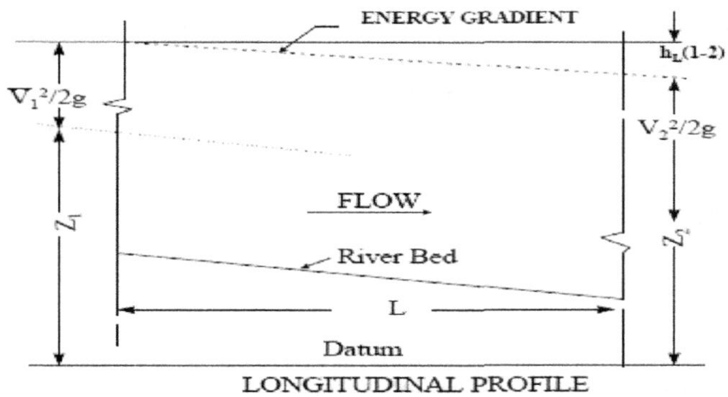

FIGURE 2.20

Channel reach for the slope–area method.

The mean velocity is established by using known empirical formulae that relate the velocity to the hydraulic mean depth, the surface slope corrected for the kinetic energy of the flowing water and roughness characteristics. Discharge is computed as the product of mean velocity and the mean cross-sectional area of flow. Contracting reaches have been found to yield consistent estimates of discharge and are preferred to expanding reaches.

The resistance equation for uniform flow in an open channel, for example, Manning's formula, can be used to relate the depths at either end of a reach to discharge. Fig. 2.17 shows the longitudinal section of a river between two sections, 1 and 2. The head at a section consists of water-surface elevation and the velocity head. The head loss is made up of two parts: (1) frictional loss and (2) energy loss due to expansion or contraction. The friction slope can be written as

$$S_f = \frac{1}{L}\left[(h_1 - h_2) + \left(\frac{V_1^2}{2g} - \frac{V_2^2}{2g}\right)(1 - k)\right] \tag{2.23}$$

where L is the reach length, k is the coefficient for energy loss; its value is 1 for contractions and 0.5 for expansions. According to Manning's formula, the mean velocity in reach 1–2 is calculated as

$$v_{1-2} = \left(\frac{1}{n}\right) R^{2/3} S^{1/2} \tag{2.24}$$

where R is the hydraulic mean depth, n is Manning's roughness coefficient, and S is the friction slope. If A is the cross-section area, then discharge Q is

$$Q = \left(\frac{1}{n}\right) A R^{2/3} S^{1/2} = K S^{1/2} \tag{2.25}$$

The term $(1/n)AR^{2/3}$ is known as conveyance (K) of the channel and it depends on channel characteristics. As the flow in the reach may not be truly uniform, the average conveyance of the reach is expressed as the geometric mean of the conveyances of the two end sections (K_1 and K_2):

$$K = \sqrt{K_1 K_2} \tag{2.26}$$

The discharge can be calculated by

$$Q = K\sqrt{S} = \sqrt{K_1 K_2 S} \tag{2.27}$$

The slope–area method can be used with some degree of accuracy in open channels with stable boundaries, or in channels with relatively coarse bed material. This method may also be used in other cases, such as alluvial channels including channels with over-bank flow or nonuniform channel cross-sections, subject to the acceptance of large uncertainties involved in the selection of the value of rugosity coefficient, such as Manning's roughness coefficient n.

Example 2.8. During a high flow, water-surface elevations of a small stream were noted at two sections A and B, 1 km apart. These elevations and other salient hydraulic properties are given below. The value of Manning's roughness coefficient can be taken as 0.020.

Section	Water-surface elevation (m)	Area of cross-section (m^2)	Wetted perimeter (m)	Remarks
A upstream	104.770	208	67.3	A is upstream of B
B downstream	103.870	203	58.5	n = 0.020

Solution. Here, an iterative approach is followed.

Slope S = (104.770−103.870)/1000 = September 0, 1000 = 0.0009.

$$\text{Conveyance of Section A } K_A = \frac{1}{0.02} \times 208 \times \left(\frac{208}{67.3}\right)^{0.667} = 22{,}075$$

$$\text{Conveyance of Section B } K_B = \frac{1}{0.02} \times 203 \times \left(\frac{203}{58.5}\right)^{0.667} = 23{,}274$$

Discharge $Q = \sqrt{22,075 \times 23,274 \times 0.0009} = 679.99\ m^3/s$

We now calculate energy loss due to constriction/expansion.

$$v_A = \frac{Q}{A} = \frac{679.99}{208} = 3.27\ m/s$$

$$v_B = \frac{679.99}{203} = 3.35\ m/s$$

$$\Delta h_v = \frac{3.27^2}{2 \times 9.81} - \frac{3.35^2}{2 \times 9.81} = -0.027\ m$$

$$\text{Sum } \Delta h + \Delta h_v \text{ to compute } S_f = \Delta h + \frac{\Delta h_v}{L} = 0.0009 + \frac{-0.027}{1000} = 0.000873$$

Updated $Q = \sqrt{22,075 \times 23,274 \times 0.000873} = 669.72\ m^3/s$

More iterations may be performed depending upon the requirement.

Special Problems in Streamflow Measurements in Arid and Semiarid Regions

Special problems associated with the measurement of streamflow in arid and semiarid regions arise from the interaction of many climatic and geological factors. The degree of difficulty varies from one region to another, depending on the combination of these factors. The main problems associated with these regions are:

1. These regions mostly have inadequate infrastructure that makes movement and communication difficult, time consuming, and expensive.

2. The harsh environment, dust, high soil erosion by strong winds, and the sediment carried by floods create problems for conventional equipment. The end result is frequent malfunctioning of equipment and loss of records.
3. The short duration and rapidity of the onset of floods, coupled with (1), imply that the team intending to measure such flows may miss them unless present at the site. The duration of peak flow may be so short that it is almost impossible to carry out measurements.
4. Soft erodible beds make sounding operations difficult when scouring may occur beneath the sounding weight as it touches the riverbed.
5. The channel may shift laterally, isolating gauge and recorder well from the flow.
6. The bed level may vary due to scour and fill, particularly during the passage of a flood. Such bed variations are difficult to measure and lead to errors in the application of cross-sectional area rendering the resultant discharge value erroneous.

Detailed planning of logistics, local knowledge, and experience are the key factors in successfully tackling the problems of such regions. New and appropriate technology in the shape of automation, telemetry, and remote sensing is necessary to overcome such difficulties. Many equipment with minor improvements could work equally well under arid and semiarid conditions. Data logging systems, with retrieval and telemetry, have proved reliable under many harsh environments and should be adapted in arid and semiarid regions. Remote sensing data can assist in planning field trips and serve as an indirect method of assessment of water resources.

2.6.4 Processing of Streamflow Data

The first check of stage data is against maximum and minimum limits. The absolute maximum and minimum limits at a particular station are carefully set such that values outside these limits are clearly incorrect. But this check does not highlight those values that are within the limits but still may be incorrect. Less extreme upper and lower warning limits are sometimes set and values outside the warning range are flagged for subsequent scrutiny. The underlying objective, while setting the upper and lower warning levels, is that the limits are violated 1–2 times every year. This would ensure that, on average, one or two extreme events are scrutinized more closely for their correctness. These limits may also be worked out using suitable statistics but care must be taken of the time interval and the length of data series under consideration.

Comparison of each data value with immediately preceding and following observations is an effective way to screen the variables that exhibit a significant serial correlation such as water level. A limit is set as the maximum acceptable positive or negative change between successive observations. What is an acceptable change in level during a rising flood hydrograph may be unacceptable during the dry season. Violations of the rise and fall limits are more readily identified from graphical plots of the hydrograph.

A visual check of the time-series data is an effective technique to detect data anomalies and must be applied to every hydrograph data. A screen display may also show the maximum and minimum limits and the upper and lower warning levels. Potential problems identified using a numerical technique can be inspected and accepted as correct or flagged as spurious or doubtful. Ideally, attempts may be made to interpret identified anomalies in terms of the performance of the observer and instruments or station, and this should be communicated to field staff for remedial actions.

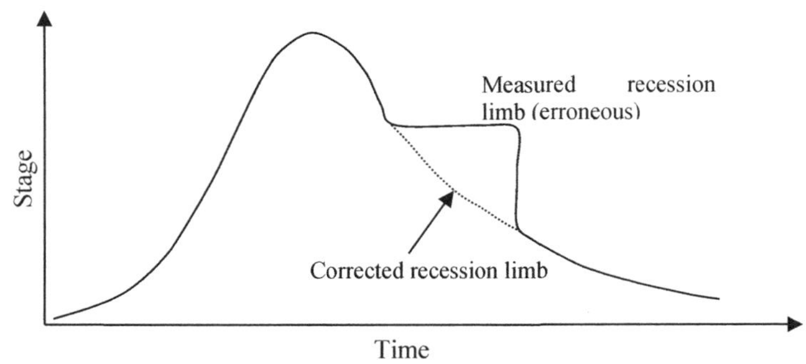

FIGURE 2.21

Erroneous recording of recession curve due to faulty equipment.

Fig. 2.18 shows a record of the recession limb of a hydrograph. The erroneous recording produced by an automatic water level recorder could be due to an obstruction in the mechanical movement that caused the float to remain hung, blockage of the intake pipe, or siltation of the stilling well. The later part of the curve showing the correct behavior was measured after removing the fault. The recession limb was corrected by drawing a smooth curve during validation (Fig. 2.21).

Some of the methods for quality control of streamflow data are:

- Test the stage or discharge of a given day in a year against the highest value and the lowest value of the same date in all the previous years.
- Apply the same test to the difference between the value on the day and the day before.
- Compare the observed data with estimates based on data from adjacent stations. The estimates may be based on regression. By transforming the data, it is possible to increase the weight on high or low values. By plotting the estimates possible errors are easily identified.
- Compare the observed discharge with estimates based on a precipitation-runoff model.
- Compare runoff at a station with runoff at an upstream station.
- Apply the double-mass curve analysis to identify shift in control.
- Apply the time-series analysis to detect changes in the homogeneity in time series.

The most common techniques to fill short data gaps are: interpolation from the adjoining values by plotting a smooth hydrograph, double-mass curve techniques, correlation with adjoining stations, and auto-correlation (for short gaps only). If a catchment has undergone significant changes, data may have to be adjusted to virgin conditions.

The processed data are used for a variety of purposes. Some of these are the computation of flow duration curves, unit hydrograph analysis, flood or low-flow frequency analysis, computation of the inflow to a reservoir, flow routing, and flood forecasting.

Stage–Discharge Relationship

The measurement of discharge at a gauging station is costly and requires trained manpower, time, and special equipment. Therefore, it is usual not to frequently measure discharge. The measurement of river stage is much easier and is carried out frequently. A relation, known as a rating curve or stage–discharge relation, exists between the river stage and discharge at a cross-section. A rating curve is developed by using the concurrent data of stage and discharge observed over a period of time. It is important that the data covers the range of stages that are likely to occur at the gauging station. As most hydrologic analyses, such as the assessment of water yield and design of projects, are based on discharge data, the rating curve has an important bearing on hydrology.

The characteristics of a stable channel do not change with time. The rating relation changes in unstable channels and, therefore, sites without a stable control should be avoided, as far as possible. The factors that influence the rating curve can be broadly classified into two groups: natural and artificial. The natural factors include the geometry of the cross-section, properties of bed and banks, alignment of the channel upstream of the gauging station, and properties of sediment being transported by the river. The artificial factors include flow regulation structures, such as a weir, channel improvement works, a bridge, land use changes, and river training works.

The combination of element(s) that control stage–discharge relation at a station is known as control. Different types of controls are: section and channel controls; natural and artificial controls; and complete, partial, and compound controls. When the geometry of a cross-section downstream of a gauge constricts the flow or there is a break in the bed slope (e.g., a fall), a section control is said to be effective. If the relation is controlled by the geometry and roughness of a reach downstream, a channel control is said to exist. A complete control governs the rating relation over the entire range of stages. This is a rare occurrence. More common is a compound control in which a section control dominates lower discharges and another channel control dominates higher discharges.

When the stage is uniquely related to discharge, this is known as simple rating curve. In a compound rating curve, more than one curve is required. To establish a rating curve, the stage and discharge data are plotted on a graph paper, as shown in Fig. 2.22, wherein stage is plotted on the Y-axis and discharge on the X-axis. Ideally, there should be a sufficient number of points, well distributed over the entire stage and discharge range at the gauging station. If the scatter of the plot is negligible or very small, a smooth curve can be drawn through the points. The scatter in the data can be due to several reasons including backwater effects, unsteady flow at the gauging site, scour of the bed and banks at the gauging site, or errors in observations.

A simple rating curve is commonly represented by a power equation that has the form

$$Q = a\,(H - c)^b \tag{2.28}$$

where Q is the discharge (m^3/s), H is the river stage (m), a and b are constants, and c is the stage (m) at which discharge is nil (known as the datum correction). This equation plots as a straight line in the log domain. The method of least squares is commonly used to estimate the coefficients of rating curves a and b in Eq. (2.28). To estimate the datum correction c, iterations are performed by varying c and the

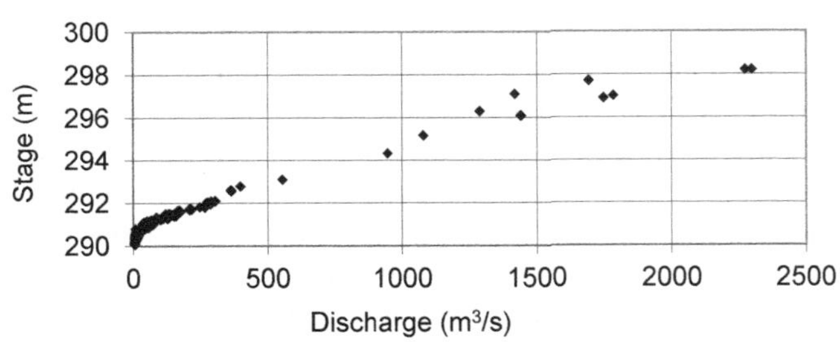

FIGURE 2.22

Plot of stage and discharge at a gauging station.

value that yields the best results is adopted. The selected value of the datum correction should be physically plausible. After establishing a stage–discharge relation, it is necessary to test it for the absence from bias and goodness of fit.

In many situations, a single curve may not be representative for the entire range of stages because the influence of roughness and boundary conditions is not the same at all stages. Up to three curves may be required—one for lower range of stages, one for middle range, and another for higher range, leading to a compound rating curve.

In large rivers with flatbed slopes and significant changes in flow rates, the effects of unsteady flow on the rating curve may be significant. The slope of the flood wavefront during the rising stage is much steeper than during the steady-state condition and the situation is reversed during the falling stage This phenomenon is known as hysteresis and it gives a loop rating curve, that is, discharge is not a unique function of the stage. Neglecting acceleration terms, Jones's formula relates discharge between unsteady and steady flow:

$$Q_{act} = Q_{steady}\sqrt{\left(1 + \frac{1}{cS_0}\frac{dh}{dt}\right)} \tag{2.29}$$

where Q_{act} = the actual (measured) discharge in channel, Q_{steady} = the steady-state discharge obtained from the rating curve, c = the wave celerity, S_0 = the energy slope (steady state), and dh/dt = the rate of change of stage.

Very few channels are stable over all stages and at all times. If the geometry and properties of the cross-section change with time due to scour or deposition, growth of vegetation, etc., this is known as a shifting control. This requires frequent updating of the rating curve. A shift in the rating curve is said to have occurred if a group of consecutive observations of discharge plots a certain percentage (say, about 5%) either to the left or the right of the established rating curve. The vertical shift of the rating curve is accounted for through a shift adjustment. In the simplest case, for a discharge measurement, the shift adjustment is obtained by subtracting the observed gauge value from the gauge value that corresponds to that discharge on the rating curve. To obtain discharge from the rating curve under shifting conditions, the shift adjustment is added to the gauge height and then the rating curve is used to read discharge.

Extrapolation of Rating Curve

Rating curves are required to be extrapolated when discharge measurements are not available over the entire range of observed stages. In simple cases, the curve may be smoothly extended. This will not be correct if, in the extended range, channel geometry changes, there is flow over the flood plain, or the roughness coefficient changes significantly. For extrapolation in low-flow ranges, it is better to plot the curve on a simple graph paper since zero discharge cannot be plotted on a log scale. A rating curve is smoothly extended. The extrapolation in high flow ranges should be attempted with utmost caution and only if indirect methods of discharge estimation such as the slope–area method cannot be used.

Rantz (1982) recommended the use of the conveyance-slope method as it is superior to the velocity–area and other methods. The conveyance (K) of a channel depends on measurable elements of a cross-section. A curve of K versus gauge height, covering the entire range, is plotted on ordinary graph paper and extrapolated as smoothly as necessary. For measured discharges, slope (S) can be obtained from Manning's equation by using the value of K. Using these data, a curve of gauge height versus slope is plotted and extended, noting that slope tends to attain a constant value at higher river stages under certain conditions. Now, discharge at the desired gauge can be computed by reading the values of K and S from respective curves. Extrapolation of discharge beyond a value exceeding twice the measured discharge should be attempted only as a last resort.

The process of establishing a rating curve is a mapping problem where stage is the input variable and discharge is the output variable. Among the emerging data analysis techniques, the artificial neural network (ANN) technique is a powerful procedure for nonlinear function mapping. Jain and Chalisgaonkar (2000) applied this technique to establish stage–discharge relations. An ANN can also successfully model a looped rating curve.

Standards have been developed by many countries for establishing the stage–discharge relation. The standard ISO 1100/2 (see ISO, 1982) deals with the determination of stage–discharge relations. Rantz (1982) has given a detailed discussion of the topic.

2.7 WATER QUALITY DATA

A large number of variables are associated with water quality information; the measurement frequency depends on the changes in the variable as well as its use. The major groups of water quality variables are organic matter, major and minor ions, toxic metals, nutrients, and sediment data. The biochemical oxygen demand, chemical oxygen demand, and dissolved oxygen are indicators of the presence of organic matter in water. The toxic metals include mercury, lead, arsenic, and nickel. The nutrients, such as phosphorus and potassium, are important from the point of view of water resources.

Water quality parameters are rarely measured continuously because of logistical reasons and are, therefore, based on sampling that may be at regular intervals or more often at irregular intervals. The recorded values represent the state of water at a particular time. Important water quality parameters or those which change rapidly with time are measured with greater frequency than those which vary slowly. For a few water quality parameters, analysis is carried out at the observation station. Most analyses require sophisticated equipment and are, therefore, carried in a laboratory.

The clean water is a clear, colorless, and odorless substance. Most natural waters are often colored by organic materials picked up from decaying plants. The physical parameters of quality include turbidity, solids, electrical conductance, color, odor, and temperature. Temperature measurements are important to understand the problem of density, viscosity, vapor pressure, oxygen saturation value, and rates of biochemical degradation. The test for residue is of great importance in the sewage treatment process to indicate the physical state of the principal constituent. The solids present in the dissolved form are related to the electrical conductivity of the water. Common measurements to assess the presence of physical impurities in water are turbidity, solids, electrical conductance, color, odor, and temperature.

A wide range of measurement techniques can be used for water quality parameters. Here, the discussion is categorized according to physical, chemical, and biological parameters.

2.7.1 Physical Parameters

The important physical properties of water from the point of view of a water quality study include temperature, density, viscosity, specific weight, and vapor pressure. The variation of these physical properties with temperature is widely available in the literature. McCutcheon et al. (1993) have tabulated range and typical concentrations of water quality parameters in streams and rivers.

Turbidity: Clear natural water allows images to be seen distinctly at considerable depths. Turbidity is measured by determining light transmission using standard light sources. The test has little meaning in relatively clear waters but is useful in defining the drinking water quality.

Solids: All contaminants of water, except dissolved gases, contribute to solids load. Solids can be classified by their size and state, by their chemical characteristics, and by their size distribution. Solids are divided into two broad groupings: dissolved (including colloidal and small suspended particles) and suspended (including settleable). The distinction is made using a membrane filter with a pore size of about 1.2 μm. Any particle passing the filter is considered dissolved, and any particle retained on the filter is considered suspended. Solids are also characterized as being nonvolatile or volatile.

The sum of dissolved and suspended (filterable and nonfilterable) solids is the total solids content. The amount of the total dissolved solids (TDSs) present in water is an important indicator of its quality for drinking, irrigation, and industrial use. As water moves on land, it picks up solids and TDS increases. TDS affects dissolved oxygen concentration and also influences the ability of a water body to assimilate wastes. Dissolved solids affect the ionic strength of water and thereby impact the mobility and transformation of metals. TDS also affects the growth and decay of aquatic life.

To determine TDS, a sample of known volume is dried in an oven and the weight of residue divided by the volume of sample gives TDS which is normally expressed in mg/L. Wide variations in the TDS concentration are observed in natural waters. The TDS concentration in rainwater is below 10 mg/L, in river water it may be of the order of hundreds of mg/L. Depending on the presence of salts, the water is categorized from saline to briny.

Color: The color of water depends on the dissolved material as well as suspended particles. Many of the colors associated with water are not true colors but the result of colloidal suspension, for example, tea. True colors result from dissolved materials, most often organisms. Most colors in natural waters result from dissolved tannin extracted from decaying plant materials. The result is a slightly brownish tint. Many industrial wastes are colored and, if not properly treated, can impart color to the receiving stream.

Odor: Odor is an indicator of the presence of pollution or toxicity in water. Pure water does not produce odor. Water usually smells due to the presence of decaying organic matter or, in the case of mineral springs, the reduction of sulfates by bacteria to hydrogen sulfide gas. The decaying organic matter may accumulate in bottom deposits large enough to provide suitable conditions for the anaerobic bacteria that produce noxious gases. Sources of the organics include plant debris washed into streams, dead animals, microorganisms, and wastewater discharge.

Temperature: Temperature affects a number of important water quality parameters. Chemical and biochemical reaction rates increase markedly with temperature. Gas solubility decreases and mineral solubility increases with temperature. The growth and respiration rates of aquatic organisms depend on temperature, and most organisms have distinct temperature ranges within which they reproduce and compete.

Lakes vary in temperature from surface to bottom, and the aquatic life varies accordingly. Cold-water species stay in deep waters, while warm-water species are found in shallow regions near the edges. The water released from the reservoir surface will be warmer and promote warm-water organisms. When the water is used for irrigation, its temperature must be high enough to induce the germination of seeds.

Hardness: Water is classified as hard or soft mainly on the basis of carbonates present in it. Normally water is classified as soft when the carbonate concentration is below 50 mg/L and is termed as very hard when it exceeds 180 mg/L. According to the guidelines of the World Health Organization, the hardness should not exceed 500 mg/L. The hardness is usually measured by titration or by measuring the magnesium and calcium ions present.

2.7.2 Chemical Parameters

The chemical characteristics of water can be classified into two categories: (a) inorganic matter, and (b) organic matter. The chemical tests of water quality include an analysis for the presence of specific ions. Gross chemical measures, such as alkalinity and hardness, are also used to define water quality. Most of the common water quality parameters reflect combinations of or interactions between ions.

Inorganic Matter

Major Ionic Species: All natural water contains a variety of dissolved salts in solution originating from rain or soil and rock with which they have been in contact. The principal chemical constituents present in most waters are cations (calcium, magnesium, sodium, and potassium) and anions (bicarbonate, sulfate, chloride, and nitrate). Typically, these ionic species are derived from the contact of the water with various mineral deposits. The most abundant species are bicarbonates, sulfates, and chlorides of calcium, magnesium, and sodium. The distribution of these species varies with geographic location and the residence time history of the water. Potassium, usually present in small amounts, is derived from soil minerals, from decaying organic matter, and from the ashes of burned plants and trees. Nitrate is usually present in small amounts. Aggregate salts are derived as TDS (mg/L). Besides, many minor ionic species, derived from the contact of water with various mineral deposits, are also present. Some of the minor constituents, such as ammonium, carbonate, and sulfide, may be present because of bacterial and algal activity.

Nonionic Species: The principal nonionic minerals found in all natural waters and groundwaters are silica, usually expressed as SiO_2. The presence of silica in water is troublesome, especially in industrial applications, where it causes severe scaling problems in boilers and heat exchangers. In addition to the major and minor ionic species, a variety of inorganic species (principally heavy metals) of anthropogenic origin may also be found. The more important of these are arsenic, barium, cadmium, chromium, lead, mercury, selenium, silver, zinc, and cyanide. These constituents are of concern primarily because of their toxicity to microorganisms, plants, and animals. Typically, these constituents come from the discharge of improperly processed industrial wastes, and high concentrations are often found in wastewater sludges.

Nitrogen and phosphorus are essential elements for the growth of plants and animals and are, therefore, often identified as nutrients. Note that both organic and inorganic forms of these constituents are of importance. Nitrogen is a complex element that can exist in seven states of oxidation. From a water quality standpoint, the nitrogen-containing compounds that are of most interest are organic nitrogen, ammonia, nitrite, and nitrate. Phosphorus is of importance in water supply systems and in the aquatic environments. Phosphorus compounds are used for corrosion control in water supply and industrial cooling water systems and in the production of synthetic detergents.

Nitrogen

Nitrogen accounts for about 80% of the gases present in the atmosphere and this huge volume maintains equilibrium of nitrogen concentration in open water bodies. The nitrogen solubility in water is very less; it is of the order of 15 mg/L. Human activities influence nitrogen in surface water in several ways. Nitrogen is present in wastes that are discharged into surface water bodies. Runoff from agriculture areas contains nutrients that contain nitrogen. It is also present in the exhaust of automobiles and industries. The key parameters for nitrogen are nitrate and ammonia. The usual range of concentration of nitrate in streams is 0.5−3 mg N/L and of ammonia in the range of 3 mg N/L. The streams that receive runoff from agricultural areas may have considerably higher concentrations of nitrates. Domestic sewage contains about 15−100 mg/L of the total nitrogen. Nitrification, which is the oxidation of ammonia and nitrite to nitrate, consumes dissolved oxygen in water. Excessive presence of nitrogen in water can lead to eutrophication.

Phosphorus

Phosphorous is an important constituent of organic matter that enters into water bodies through fertilizer, industrial waste, and rocks. Phosphorous is vital for all organisms and, in many cases, it is the nutrient that limits productivity. According to McCutcheon et al. (1993), the common venue of its concentration is 0.05 mg P/L.

pH

The pH of a solution is defined as the negative logarithm of the hydrogen ion activity, $pH = -\log(H^+)$. The range of pH is from 0 (maximum acidic) to 14 (maximum basic); pH of a neutral solution is 7. All geochemical reactions are affected by pH. Surface waters become acidic when additions of acid exceed the buffering capacity of the carbonate system. Anthropogenic sources of acidity include acid deposition and acid mine drainage. In addition, some sources of acidity arise naturally due to the oxidation of sulfide-bearing ores. Acidic waters are of concern as their low pH increases the solubility and mobility of trace metals.

Organic Matter

Organic matter is important for the health of a water body because the decomposition of organic matter draws upon the oxygen resources in water and may render it unsuitable for aquatic life. Common parameters to characterize it are: BOD (biochemical oxygen demand) and COD (chemical oxygen demand). The widely used BOD test measures the oxygen equivalence of organic matter and is the most important indicator of pollution by organic matter. Similarly, a direct measurement of dissolved oxygen is an important indication of the health of water. The absence of DO or a low level indicates pollution by organic matter. It is recorded as a percentage of saturation.

Natural Organic Compounds: Most organic compounds are composed of various combinations of carbon, hydrogen, oxygen, nitrogen, phosphorous, and sulfur. The principal organic compounds, found in wastewater, and to a much lesser degree in natural waters, include proteins, carbohydrates, and lipids.

Synthetic Organic Compounds: The presence of a large number of organic compounds in water is of concern from health, treatment, and ecological standpoints. Of greatest concern are those organic compounds that may be carcinogenic or that may cause mutation in humans and other living forms at extremely low concentrations, for example, surfactants, pesticides and agricultural chemicals, organic solvents, etc.

Toxic Metals and Organic Compounds: Toxic metals and other elements may exist naturally in waters but their concentration may be seriously increased through human activity. These metals include copper (Cu), chromium (Cr), mercury (Hg), lead (Pb), nickel (Ni), cadmium (Cd), and arsenic (As). Measurements of toxic organic substances require advanced instrumentation.

2.7.3 Biological Parameters

The surface water polluted by domestic wastewater may contain a variety of pathogenic organisms, including viruses, bacteria, protozoa and helminths. As the testing for all these organisms is costly and time consuming, the most common test is for *Escherichia coli (E. coli)* whose presence is an indicator of the potential for other pathogenic organisms. The results of testing for *E. coli* are recorded as the most probable number (MPN)/100 mL. The principal groups of microorganisms found in water may be classified as protists, plants, and animals. The most important microorganisms of concern in water and wastewater include bacteria, fungi, algae, and viruses.

Bacteria: Bacteria are single-cell protists. Although there are hundreds of bacteria, most bacteria can be grouped by form into four general categories: spheroid, rod, curved rod or spiral, and filamentous. Spherical bacteria, known as cocci (singular, coccus), are about 1–3 μm in diameter. The rod-shaped bacteria, known as bacilli (singular, bacillus), range from 0.3 to 1.5 μm in width (or diameter) and from 1.0 to 10.0 μm in length. *E. coli*, a common organism found in human feces, is about 0.5 μm in width and 2 μm in length. Curved rod-shaped bacteria, also known as vibrios, typically vary in size from 0.6 to 1.0 μm in width (or diameter) and from 2 to 6 μm in length. Spiral bacteria known as spirilla (singular, spirillum) may be found in lengths up to 50 μm. Filamentous forms can occur in the length of 100 μm and longer.

Fungi: Fungi are aerobic, multicellular, nonphotosynthetic, heterotrophic, and eukaryotic protists. Most fungi obtain food from the dead organic matter. Along with bacteria, fungi are the principal organisms responsible for the decomposition of carbon in the biosphere. They can grow in low-moisture areas and in low-pH environments. Because of these properties, fungi play an important role in the breakdown of organic materials in both terrestrial and aquatic environments. As organic materials are decomposed, fungi release carbon dioxide to the atmosphere and nitrogen to the terrestrial environment.

Algae: The name algae is applied to a diverse group of eukaryotic microorganisms that share some similar characteristics. Typically, algae are autotrophic, photosynthetic, and contain chlorophyll. The other pigments encountered in algae include carotenes (orange), phycocyanin (blue), fucoxanthin (brown), and xanthophylls (yellow). Combinations of these pigments result in the various colors of algae in nature. Algae are important microorganisms with respect to water quality. In an aquatic environment, algae form a symbiotic relationship with bacteria. If allowed to predominate, they can affect the dissolved oxygen balance by causing anaerobic conditions to exist at night.

Viruses: Viruses are parasitic particles consisting of a strand of genetic material—deoxyribonucleic acid (DNA) or ribonucleic acid (RNA)—within the protein coat. They invade living cells where the viral genetic material redirects cell activities toward the production of new viral cells at the expense of host cell growth. When the host cell dies, large numbers of viruses are released to infect other cells. They cause several types of infections, for example, the common cold. A number of viral diseases are transferred via water.

2.7.4 Sediment Data

The amount of sediment transported by a river is important for design and management of water resources projects, flood control structures, bridges, etc. The data collected includes the particle size distribution for sediment and sediment transport as bed and suspended load. The measurement of sediment is an expensive process and sediment rating curves are, therefore, widely used to indirectly assess the sediment concentration as a function of river stage or discharge.

The movement of solids transported in anyway by the flowing water is termed as sediment transport. Sediment originates from various sources, including river basin soil erosion, river bed, and bank erosion. The sediment may be temporarily stored and mobilized again, depending on its source and on the flood events. Consequently, sediment transport rates will depend on many factors, and may differ from the sediment transport capacity because of sediment availability. The total sediment transport is the sum of the suspended load and bed load. The sediment might originate as the bed material load and the wash load. The classification of sediment as per ISO (ISO 4363: 2002) is shown in Fig. 2.23.

To determine the concentration of sediment, sediment samplers are used. The type of sampler depends upon the need and the purpose. The suspended sediment samplers collect samples of water and sediment mixture in a river or lake. The sampler is lowered to the desired depth and is filled with a sample. The size of the sampler should be such that an adequate amount of sample, say at least 0.5 L, is filled. The depth integrating samplers are filled as they are lowered from surface to bed and then raised up. The rate of lowering or raising should be constant in both directions.

For sampling, the width of the cross-section is divided into subareas, depending on the variability of sediment concentration in the lateral direction. Specially designed samplers are used to measure bed-load sediments. These are lowered to the bottom and are allowed to be filled up for 5−10 min. It is necessary to do sampling at various discharges so that a rating curve relating sediment and water discharge can be prepared. Riverbed sediment sizes may change during flood events by selective erosion. The concentration of sediments is commonly expressed in g/cm^3 or kg/m^3. The international standards ISO 3716 and 4363 provide details of sediment sampling. Remote sensing data are also now being used to estimate suspended sediment in water. The change in reflectance properties of water depends on the concentration and the properties of suspended sediments and these form the basis to estimate suspended sediment. It is necessary to have adequate ground truth data to obtain reliable results.

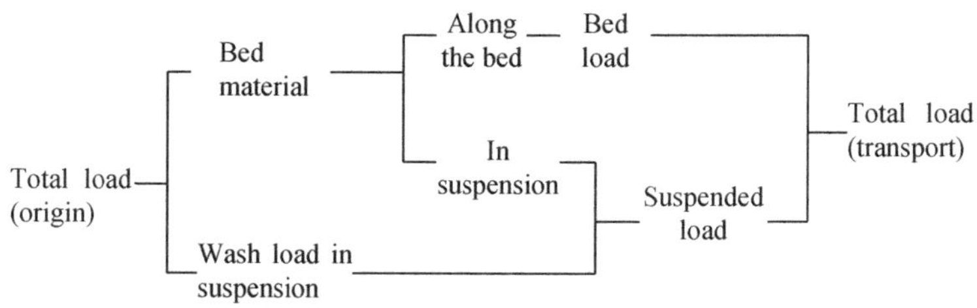

FIGURE 2.23

Definition of sediment load and transport.

Adapted from ISO 4363 (2002).

The samples are usually analyzed in laboratories. After allowing the sediments to settle down, water is carefully removed from the container and the remaining sediments are oven-dried. The particle size analysis is first carried out by sieving and then the finer sediments are analyzed using a hydrometer. The sediment transported as the suspended load is classified into three categories, depending on the particle size:

- the coarse fraction (particles above 0.2 mm diameter),
- the medium fraction (particles between 0.075 and 0.2 mm diameter), and
- the fine fraction (particles below 0.075 mm diameter).

ADCPs are being increasingly used for streamflow and suspended sediment measurements. The principle of ADCP has been discussed earlier.

2.7.5 Processing of Water Quality Data

The recording and storage of samples and data collected must follow the prescribed set of standards. For each sample, the following information should be recorded: the location of the sampling point, date of collection, purpose, and sample identification number. The location of the sampling point should be recorded in terms of district, tehsil (county or parish), and village, as well as geographical coordinates (latitude and longitude). The unique station number identifies a given sampling point in a concise form.

When a sample is collected, the information that should be recorded includes time of sampling (day, month, year, hours, and minutes), the purpose of study, and frequency of sampling. Once the water quality data have been assembled in a data storage system, the next step is to interpret the data with respect to specific questions, environmental problems, and water resource management requirements. The most frequently asked questions are:

- What is the water quality at the location of interest?
- What are the water quality trends in a region; is the quality improving or getting worse?
- How do certain parameters relate with one another?
- What is the total mass loading of materials?

Flow-Sediment Rating Relationship

A sediment rating curve is similar to a discharge rating curve, except that the relationship is established between water discharge and sediment concentration. Usually, the relation is of the form:

$$C = b\,Q^d \tag{2.30}$$

where C is the suspended sediment concentration (mg/L), Q is the discharge (m^3/s), and b and d are constants. A typical sediment rating curve is shown in Fig. 2.24. Conventionally, discharge and sediment concentration are plotted on a log—log graph paper, and a straight line is drawn. A least squares method can be used to obtain the best-fit line. Usually, the power equation is log transformed, and linear regression is applied to estimate the parameters. Typically, exponent d in Eq. (2.1) lies in the range $2 < d < 3$.

In most cases, there will be a large scatter in discharge and sediment concentration points. This scatter may span several orders of magnitude. One reason for this scatter is that soil erosion is different during different seasons of the year. If the scatter is large, it might be necessary to develop separate

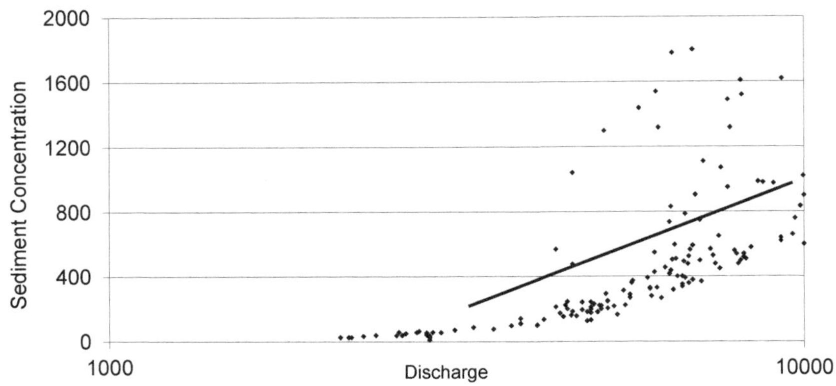

FIGURE 2.24

A typical sediment rating relationship.

rating curves for different seasons or according to streamflow generation mechanisms, such as rainfall and snowmelt, the data pertaining to rising or falling limbs of the hydrograph may also be separated.

An implicit assumption in deriving a sediment rating curve is that the sediment concentration depends on the transport capacity of flow. This holds good when most of the sediment load consists of suspended channel bed material and is not supply restricted. However, part of this material originates from the watershed and is transported as washload. Thus, sediment concentration at a given time depends on the rate of erosion within the catchment and the rate of delivery to the channel, neither of which has to be related to discharge (Pickup, 1988). The extrapolation of sediment rating curves should be carried out carefully. Sometimes, lines joining points of equal concentration, say 10%, 20% are drawn. These are used in the rating curve extrapolation.

The primary use of a sediment rating curve is to obtain the value of sediment concentration for a given discharge. The sediment rating curve along with a flow duration curve can also be used to estimate the amount of sediment being transported over a period of time, say a year. Another important use of the sediment rating curve is in the estimation of the impact of land use changes and watershed management on sediment yield. After a relationship has been developed and the scatter due to natural changes is accounted for as discussed earlier, the impact of land use changes, such as logging or watershed management, can be quantified by a shift in the relationship.

2.8 OTHER DATA

Besides the major data types discussed in the foregoing, a motley of other data are required for planning and management of water resources systems. These are discussed in the following.

2.8.1 Groundwater Data

Groundwater data can be dealt with under three headings: wells, springs, and yields/costs. Well location and hydrogeological logs are single sets of observations that do not change with time. Piezometer levels, discharge, and water quality are time-series data sets that can be validated, processed, and stored using the techniques that have been discussed previously. The well water level is most often obtained by manual dipping techniques and usually at irregular intervals, producing an irregular time-series format. The use of automatic level recorders has increased in recent years. Wide variations in the frequency of observations are noted in manual systems, ranging from three times a year to monthly/weekly observations. If the wells are pumped, the estimates of water pumped out may come from meters, from the duration of pumping, or from the quantity of power consumed. These estimates require knowledge of the pump specification and the pumping head. In addition to defining the surface drainage system of which the spring or well is a part, it is necessary to code the aquifer system(s). In the case of wells, in addition to the basic site description data, the data of well logging are also useful. The discharge of springs can be easily measured by constructing a weir or V-notch as described earlier in this chapter.

2.8.2 Reservoir and Lake Data

The time-invariant/slowly variant data of reservoirs include the elevation—area—storage capacity table, the characteristics of outlet structures, and operating policies. The time-series data include the elements of the reservoir working table, viz., inflow, outflow through various outlets, evaporation loss, and release to meet the various demands.

Lake bathymetry surveys are carried out to prepare the bed profile and estimate the volume of water in a lake. The water depth can be measured using an echo sounder. A recent development that is of immense use in field surveys is the GPS. A GPS enables the determination of coordinates of the location of an observer anywhere on the Earth, in the air, or under water using signals from a constellation of satellites. A range of equipment are available for which an accuracy of the order of cms can be obtained. This system is described in greater detail in Chapter 12. A GPS with an echo sounder is extremely helpful in lake surveys.

2.8.3 Spatial Data

Maps are the most effective means of visualizing spatial data. In water resources, both time-varying and time-invariant data are of use. The spatial data that do not vary with time include the catchment/command boundary, topography, soil map, and stream network, and geological features. The data that vary with respect to time include land use and cropping patterns. Many times, the variation of a hydrological variable is displayed through a map, for example, isohyetal map.

River channel cross-sections, longitudinal profiles, and bed characteristics are needed for many hydrological applications. These data slowly vary with time and they can, therefore, be considered as semistatic. It is important that these data are gathered for each observation station with appropriate frequency. Similarly, the longitudinal profile data at each gauging station is to be associated with a time period. River cross-sections at gauging stations are of prime importance in interpreting the stage—discharge data. The bed characteristics in the vicinity of gauging stations, including nature of the river bed and bank material, nature of mobile bed sediments and bed forms, river course stability, river type (meandering or braiding), presence of rapids upstream or downstream from the station, state of vegetation, etc., must also be available. Conventionally, such information is available in the form of paper maps, such as topographic sheets; data such as channel cross-sections are acquired through field surveys and sampling and analyzed using manual means.

The acquisition and management of spatial data using conventional means is tedious and time consuming. Due to resource constraints, it is not always possible to update these data at a desired frequency. Nowadays, this process has become considerably simpler and faster due to the availability of remotely sensed spatial data at revisit periods of the order of weeks and GIS software for processing spatial data. The current trend is to use a GIS to store and analyze spatial data. A paper map is an analog depiction of spatial data, while the GIS data are in digital form. A digital map can be prepared by digitizing a paper map or from remotely sensed data. Different types of data, such as soil and land cover, are stored in different map layers. GIS permits analysis of single or multiple layers and various layers can be overlaid, one on top of another. Once a map is available in digital form, it is easy to update or modify it.

The concept of data in the form of layers also looks attractive when storing geological data, since geologists frequently talk in terms of data layers. From a water resources point of view, spatial variation of data is important, for example, the variation of soil hydraulic properties. After the requisite data are stored in a GIS database, it is easy to answer complex queries like what the areas in the catchment are that have pine forests on shallow clay with a 3% slope? The topic of remote sensing and GIS is discussed in Chapter 3.

2.8.4 Socioeconomic and Agriculture Data

The data about population and economic activities are mostly needed for planning water resources projects. Census or gathering demographic data is a routine activity in most countries. In India, an elaborate exercise is carried out every 10 years to gather extensive demographic data, which forms an important input in national-level planning, including the water sector. The census data also form the basis to forecast the population growth for use in planning activities.

Another important input for water resources is the area under different crops and crop yield. Agricultural departments usually gather such data for individual administrative units. Since land revenues are collected, based on agricultural production, the revenue departments also gather these data. Note that the boundaries of a water resources study rarely coincide with the administrative or political boundary and it is necessary to interpolate the data before it can be used as input.

Normally, the socioeconomic and agriculture data are compiled for two conditions: "with" and "without" project. It is improper to base the analysis on a comparison of conditions before and after project construction because a decline or improvement of present conditions might occur even in the absence of the proposed project. This factor should be recognized while determining the impacts attributable to the project. Consistent assumptions should be applied to future conditions with and without the project so that comparability is assured. The net incremental benefit stream should be an accurate reflection of the project's income generating capacity, or its net contribution to real national income.

The socioeconomic data to be collected depends on the nature of the project being studied as well as the type of economy, that is, developed or developing. Sample surveys are carried out to gather requisite data. The data required for an irrigation project should cover the following aspects.

Demographic and Social Characteristics

It is essential to ascertain the available human resources, since development proceeds through and on behalf of these. The survey should determine the size of the active and potential labor force for agriculture, together with that for the tertiary sector contributing to agricultural development. The population features as well as future projections are needed to estimate the present consumption pattern and future demand of agricultural products. The level of employment and the income of the entire agricultural and nonagricultural active population must be found. The idea is to estimate the present underemployment and the labor resources available for future development. It has been established that assured irrigation is instrumental in both mechanized large farms as well as manual labor-oriented small farms. Another important factor is the migration prospects of the area, which, in turn, would lead to more reliable measurements of future population trends.

Land Use and Crop Yields

Identification of areas, earmarked for annual crops, orchards, pastures, forests, and noncultivated along with the knowledge of the system of land tenure (owner occupancy, tenancy, sharecropping), size of farms, and fragmentation, enables the preparation of preliminary blueprints on the development possibilities. These details would also show the constraints on the availability of land for cultivation. Soil properties would determine the spectrum of crops that can be grown in the area.

To assess the returns on investments in agricultural development, the rise in income resulting from the investment is needed. It is essential to know the present cropping pattern, crop rotations, varieties grown, and yields for each crop. These data are used to optimize the cropping pattern and estimate the irrigation water requirement.

Farm Budgets and Net Farm Income

Farm budget is a useful method to analyze the agricultural economy and should represent conditions anticipated for a farm typical of the area. Where wide divergence in farming patterns is anticipated, the principal patterns may be noted. The budget should cover the following factors: (1) farm investment (size of farm, value of land, building, value of farm equipment and livestock); (2) farm production (area and yield of principle crops, and production of live-stocks); (3) farm expense: details of farm expenses to produce and market the farm products; (4) gross farm income (total annual receipts from sale of crops, livestock and live stock products, and value of farm products consumed by the farm family); and (5) net farm income, which is difference between gross farm income and farm expense. By means of sample surveys, it is possible to ascertain the inputs and equipment used by farmers, such as fertilizer, seeds, labor, machinery, and breeding stock.

Agricultural Supporting Services

This covers extension services, research, storage, and processing facilities, credit, transport, etc. It will be necessary to assess the importance and the part played by public and private bodies and co-operatives in distributing the inputs and equipment commonly used. A study should be made on how the agricultural output of the area is marketed and what long-term storage capacities are. The volume and price trends of markets should be analyzed and forecasted. It is also essential to know the present agricultural credit and insurance facilities, the details of loans, and the indebtedness of farmers to deduce the amount of time needed by farmers to adapt to irrigation and intensive agriculture.

2.8.5 Water Use and Demand Data

The need for water for various sectors has been quantified in Chapter 1 and continent-wise historical water withdrawal, consumption, and projections have been given in Table 1.1. Since water is a reusable commodity, water use and water abstraction are two different things. The reuse of water is increasing globally and this is likely to grow further. Each drop of the Colorado River water is used 6–7 times before it reaches the sea (Biswas, 2000).

Water use denotes the quantity of water that is withdrawn from waterbodies such as rivers, lakes, and aquifers, for supply to fields, cities, industries, or for ecological and environmental needs. The consumptive use of water is that quantity that is evaporated, transpired, incorporated into crops, or consumed by humans or animals. This water is not immediately available for further use. For some purposes, such as hydropower generation, navigation, and recreation, water is used where it is available (river or lake) but is not withdrawn. But all purposes cannot be served this way and the off-stream uses, such as municipal water supply, irrigation, and industrial use, require that water be diverted from the source. Part of the diverted water may come back to the river, lake, or aquifer as return flow and is available for subsequent use.

The collection of water-use data is a complex task. Problems stem due to the scarcity of data, especially at the individual activity level (a farm, industrial enterprise, or household). The water-use data requirements vary according to the approach employed to represent water use in the planning effort. There are two broad approaches. The first requires data on a set of several inputs to each water-use activity (including the water itself), each activity's associated prices and costs, and a set of total outputs, including outputs of pollution, with their associated prices and costs. Such data can only come from repeated observation of the same water user over time (say, monthly totals over several years) or simultaneous observation of many users of the same sort at the same time. For self-evident reasons, the first source is known as a time series, and the second as a cross-section. Under certain conditions, the time series and cross-sectional data can be pooled, so that several inadequate data sets may be combined into one with enough size and variation to be helpful. However, under all circumstances, extreme care must be exercised in the interpretation of the available statistical information.

The second approach is determined by the process for which water is used. It requires data on what is going on within and among the many unit processes of a single water-use activity. This approach amounts to a summation of all individual water demands that can produce a large number of alternative activity designs. These designs, in turn, can be used to define water-use relations and unit water-use coefficients for specific activities, such as steel rolling, paper production, household water use, and the like.

Water resource planning needs both hydrologic and nonhydrologic data. These two broad data sets should always be mutually consistent. Spending unjustified time and resources, for instance, for refinement and improvement of the hydrologic database at the expense of the depth and scope of other nonhydrological data should always be avoided. In other words, one needs not to try too hard to improve one set of data if another set of equal importance, for whatever reason, is deficient. One of the common problems that make data acquisition difficult is that hydrologic data are always collected within the watershed boundaries, while nonhydrologic data usually refer to different spatial units that follow political and economic subdivisions of the area under consideration. Adjustments must be made to make all project data compatible in time and space.

Water demand and water use are interrelated concepts, the link being the price of water. Sometimes, these terms are erroneously used interchangeably. The traditional approach has been supply-side management but the growing water scarcity has forced consideration of demand management as a viable and desirable option. The tools and techniques of water-demand management have been classified into three categories by WMO (2008): economic, structural, and socio-political. The first category is centered on realistic water pricing. Structural techniques involve infrastructure to effect better control of water demands. The last category includes policies to encourage water conservation.

The data about municipal and industrial water supply can be obtained from the municipal water works department or from private suppliers (if the services are in private hands). The procedures to estimate water requirements for agriculture, hydropower, etc. and their future projections have been described in Chapters 9 and 10. WMO (2008) has also discussed techniques for the estimation of water requirements for many purposes.

2.8.6 Data Gathered During System Operation

It is important to note that data collection and analysis should not stop after the construction of the project. Rather data collection is a continuous activity and it should continue during the actual operation of the system, although the type of data collected may undergo change. During the operation of a water resources project, the precipitation and discharge gauging stations should continue to function to the extent feasible. Sometimes, stream gauging stations that are located in the vicinity of a dam or a barrage are relocated. The following data should be regularly measured during the operation of a project to provide useful information for improvement in the present and subsequent projects:

a. Reservoir working table: For each reservoir, a detailed working table is prepared for each time period. The time period can be daily for the low-flow season and multihourly during the high flow season. This working table gives data about the initial reservoir level and storage, inflow, release from different outlets, evaporation and other losses, spill and final reservoir level/storage. The release details can include the magnitude of demand and volume of actual release for each purpose. The table should also indicate how many outlet gates were opened and by what amount. If the project generates hydropower, the amount of power produced should also be indicated.
b. Usually a pan evaporimeter is installed at the dam site and the data are used to estimate evaporation losses from the reservoir by applying a suitable pan coefficient. It would be ideal if a meteorological station is set up at the dam site and the data of other meteorological variables, such as temperature, wind velocity, sunshine hours, and humidity, are measured. A better estimation of evaporation from a large reservoir is obtained if another pan is installed near the upstream end of the lake.
c. Crop data: This consists of the cropping pattern in the command, sowing and harvesting time, water supplied to crops, and crop yield.
d. It is advisable to carry out sedimentation surveys of major projects from time to time. The frequency of such surveys depends on the annual loss of storage due to sedimentation and normally varies from 5 to 10 years; higher sedimentation rates necessitate frequent surveys.
e. If an inflow forecasting mechanism exists, it is recommended to note the actual inflows against forecasts issued at various times. At the end of the flood season or water year, the comparison should be made to determine the efficiency of the forecast model. Suitable ways to improve this efficiency, if necessary, should be identified.
f. It has been noticed that in many command areas, the water table shows a rising tendency after surface irrigation is introduced. In all such command areas, groundwater table should be monitored at several places. If a rising tendency is noted, suitable remedial measures should be initiated without any delay.

g. For flood control projects, a detailed map should be prepared for each year showing the area inundated and the depth of flooding. An estimate of flood damage to agricultural areas and urban areas should also be made.

2.9 WATER RESOURCE INFORMATION SYSTEM

A water resource information system (WRIS) is a means to manage current and historical hydrological and related data in an organized form. The principles of a WRIS are clearly reflected in the title itself. "Water resources" indicates that the attention is limited to states, storages, and fluxes of water in space, time, and frequency phases. Hydrometry is concerned with the measurement of these states, storages, and fluxes. "Information" is data that has been manipulated and processed to give them meaning and purpose. By definition, information serves a function and is created not simply because there is something to be measured or because of curiosity. Three key features of information are: reliability, availability, and presentation. A WRIS is not simply a data collection or archive, it is a logical and structured system to collect, process, store, and disseminate water resources data.

A WRIS comprises the infrastructure of physical and human resources to collect, process, store, and disseminate data on (geo-) hydrological and hydrometeorological variables. The physical infrastructure includes observation networks, laboratories, and data processing and storage centers. Large funds and efforts are required for the operation of a WRIS. Therefore, the efficiency of the system should be such that the data are easily available to the user and the analysis leads to optimal utilization.

The primary role of a WRIS (see Fig. 2.25) is to provide reliable data sets for planning, design, and management of water resources and for research activities. The system should function in such a manner that it provides the information to users in time and in proper form. Sometimes, the scope of WRIS is extended to provide data to users on a real-time basis for short-term forecasting or operational purposes.

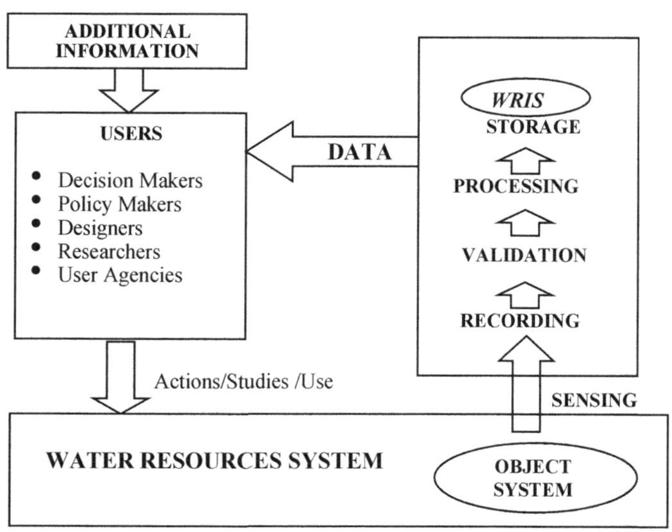

FIGURE 2.25

Schematic diagram of a water resources information system.

Adapted from DHV (1999).

The data collected for different hydrometeorological phenomena through this network is called the raw or observed data. The raw data have to be processed to ensure the reliability of the resulting information. Both raw and processed data sets have to be properly stored—processed data for dissemination and raw data to permit inspection and revalidation in response to queries from users. Note that the users have a central role in a WRIS.

As most of water-related development activities are controlled by the government sector in a majority of countries, the main agencies that observe hydrometeorological data are the data users themselves (DHV, 1999). Therefore, the network and frequency of data collection are governed by their own needs. Other data users, such as researchers, may find that data, if available, are inadequate in spatial and temporal coverage, are of varying/inadequate reliability, and are scattered at many locations. Importantly, the data that are once missed are not available later on. To obviate all these, there must be a strong linkage between data collectors and the end users and a periodic review of the WRIS.

The activities under WRIS can be broadly classified into the following categories:

i. assessing the user needs;
ii. establishing an observational network and operating it;
iii. data collection, validation, processing, and reporting;
iv. management of historical data;
v. data transmission, storage, and dissemination; and
vi. institutional and human resource development.

Item (v) is relevant to this chapter and is discussed next.

2.9.1 Data Transmission

WMO (2008) has given the classification of data transmission systems:

1. Manual with the observer at a station sending data to the central office.
2. Manual/semiautomatic system where the central office manually interrogates the automatic field stations through telephone, radio, etc. and receives the data.
3. Preprogram and time system where automatic equipment initiates the transmission of observations.
4. Automatic event indicator and the station automatically transmitting the specified change of variable to a central location.
5. Automatic system with station transmitting and central office recording data continuously.

The possible choices of transmission links are:

(a) Dedicated landlines are used when distances are short and commercial lines are not readily available.
(b) Commercial telephone lines are used wherever feasible. Two-way communication and data transmission is possible. With improvements in information technology, very good voice quality and high band rates of data transmission are possible.
(c) High-frequency radio links are used when landlines are not available or topography is difficult. The installation cost can be high.

(d) A significant development in the last few decades has been the use of satellites for data transmission. A satellite-based system consists of Data Collection Platforms (DCPs) that are installed at hydrometeorological stations. DCPs are (rechargeable) battery-operated devices that collect, encode, and communicate the data of the station to a central location through a satellite link. This system is very useful for remote and difficult-to-access locations.

The choice of a particular transmission system depends on (1) the frequency of data observation and the urgency of data; (2) the additional benefits of having forecasts based on telemetered data; (3) robustness and reliability of the system, particularly in inclement weather; and (4) availability of finances, infrastructure, and manpower to efficiently run the system.

With the growth in information technology, the trend is toward automatic observation, transmission, and storage of data, particularly in developing countries. Multiparameter data loggers can measure, store, and transmit data observed by several observation subsystems. These days, the data loggers are small, rugged, and have small power requirements. These may be battery or solar-power operated. The automatic transmission of data is usually in coded form. WMO has evolved codes related to hydrometeorological data. They have also launched an elaborate system for data observation and transmission through the World Weather Watch (WWW) program (https://public.wmo.int/en/programmes/world-weather-watch).

2.9.2 Data Storage and Retrieval

The vast amounts of data that are observed by incurring huge efforts and resources should be stored in such a way that they are easily obtainable and safe from weather and other harmful agents. Also, it is much more useful if the basic data is processed into various useful forms and kept ready to be used by the end users. This can save a lot of money and effort of the user agencies and they would be encouraged to use the data to solve different water resources problems. The archival of data is important in any field. When done in a proper manner, it enables the end users to exploit its potential of data in an efficient way and thus eliminate the tedious task of manually handling voluminous data.

Everyday, vast quantities of water resource data are collected all over the world. In this computer era, the archival of data may be accomplished in a very efficient and economic way. The basic and other processed information may be stored on computer media and the same may then be quickly made available to intended users. Hard copies in the form of data year-books may also be brought out for use by practicing engineers, planners, and managers. These water year-books can also be made available to the users on the computer. Besides the processed data, it is advisable to store the raw data as well, because it may be needed for research purposes and it might be realized at a later date that the data validation procedures had missed some aspects.

A typical setup for water resources data management is depicted in Fig. 2.26. The main components are: (1) data entry module to input data from various sources in the database; (2) user interface for data editing, display, and management; and (3) applications that can retrieve data or program the database.

Due to large volumes of water resources data, it is necessary that the data are stored such that minimum space is needed. It is estimated that the storage of graphical data on microfilms requires only about 1/300th of the storage space needed for the original data. Till a few decades ago, the digital data were archived on magnetic tapes/CD-ROMs. However, the life of magnetic tapes/CDROMs is limited and these are to be stored in a controlled environment. The current trend is to archive the data on online

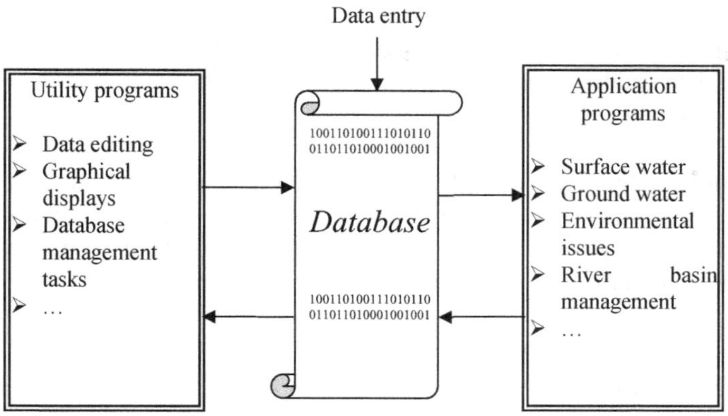

FIGURE 2.26

Management and use of hydrological database.

storage and cloud that can hold an enormous quantity of data. Just to give an idea, the data storage capacity of the popular company Google on its Cloud is more than 1 million petabytes (1 petabyte = 1.0e+15 byte) of storage (as of March 2023). The cost of hardware has dropped drastically in recent times. Of course, the volume of data that are being generated each day is also exponentially increasing.

To reduce the requirement of storage space, some type of data compression is applied. WMO (2008) has described such procedures. A number of data compression algorithms have been developed recently. For example, daily rainfall is measured to an accuracy of 0.1 mm. Rather than storing it as a real number that requires 4 bytes, the values can be multiplied by 10 and stored as an integer requiring only 2 bytes. The daily data of 1 month when rainfall was zero need not be stored as 30 zeros. An efficient way is to use the notation "30 × 0." The database files are not normal ASCII files that require larger space; these are special types of files. The suitability of a particular compression technique will depend on the characteristics of the data.

An efficient data retrieval system is also necessary so that the requisite data are quickly fetched from the database. A good retrieval system should provide the user a combination of options to select the data using criteria, such as by variable, basin, station, time period, or range of values. The user should also have a control on the format of the output, that is, tabular, plot or ASCII data files that can be directly input to another software. If graphs are displayed on the monitor, the user should have an option to print them or store on hard disk for later use. Adequate security measures should also be built in the retrieval system so that only authorized users have access to the database. Among the authorized users also, there should be various categories. Most users are given read-only access and they cannot do any modification to the database. A limited group of users are given all privileges, that is, they can read, modify, and delete data from the database. It is useful to have a log of all users who have accessed the database and the operations that they have performed so that the source can be identified in case of any mishap.

2.9.3 Data Dissemination

The basic objective of creating databases and storing data in an efficient manner is to encourage the use of data for planning, design, management, and research purposes. Therefore, there should be no hurdle in accessing the data by genuine users. Dissemination of information goes a long way to achieve this objective. The first and foremost step in the dissemination process is an up-to-date catalog of the database. WMO (2008) has outlined a data catalog format and summary reports. The catalog of data held in various databases should be updated periodically.

Many organizations routinely publish basin-wise data year-books. A typical water yearbook consists of a description of the basin, topography, soils, land use, etc., major rivers, and salient features of various water resources projects. Maps are included to illustrate all these features. The data section contains typically precipitation, streamflow, evaporation, and groundwater data. Periodically, special reports may be published giving long-term statistics of stations or highlighting special or unusual events, such as floods or droughts.

Of late, many organizations have started dispensing with paper publishing due to high costs and handling problems. With the bulk of data now available in digital form in a WRIS, the hard-copy publication is not considered as an efficient means of data dissemination. A water yearbook can be conveniently published on a computer media, such as a pen drive, which is cheaper to produce and easier to handle. A browser may also be supplied to handle data search, display, and print facility. This trend is likely to accelerate further. An important thing to remember is that the format and content of publications should depend on the needs of users. The contents should be so designed that the needs of most data users are answered and the efforts to handle data requests are reduced. The contents may also depend on the frequency of publication. The price of publication should be fixed such that it is not a burden on the organization and is affordable to the user.

Finally, in this age of computers and Internet, it is appropriate that many international organizations have established databases that can be accessed through the Internet. The Global Runoff Data Centre (GRDC) at the Federal Institute of Hydrology, Germany, has a large archive of surface water data. The center was established under the auspices of WMO and mechanisms have been evolved to supply data to a user. INFOHYDRO of WMO is metadatabase that does not actually contain any data but facilitates quick dissemination of information about institutions and agencies dealing with hydrology and catalogs of data. A similar service for climate data is INFLOCLIMA.

2.10 CLOSURE

Two basic tools for integrated management of the environment are modeling and environmental data. Both tools were available and valid in the past; however, the recent requirements for integrated environmental management have also led to a significant evolution of both modeling procedures and data management systems. Current literature provides vast amounts of studies on the modeling of different environmental processes; of late researchers are also attempting to integrate social processes with hydrological modeling. Issues related to data management systems are still not touched on in a comprehensive framework. Availability of measured data is quite uneven across the world—some regions are very well covered but others are not.

Most developed countries have well-established databases, which can be accessed easily by the users. Developing countries, on the other hand, do not have extensive data banks and the existing data may not be easily accessible to the user. There is a need for harmonization or standardization in the development of databases so that data exchange can be facilitated on regional and global levels.

References

Ahuja, P.R., 1960. Planning of precipitation network for water resources development in India. WMO flood control series 15, 106–112.

ASCE, 1996. Hydrology Handbook. ASCE Manuals and Reports on Engineering Practice No. 28. American Society of Civil Engineers, New York.

Biswas, A.K., 2000. Current perceptions and future realities in wake of water crisis. In: Goel, R.S. (Ed.), Environmental Impacts Assessment of Water Resources Projects. Oxford and IBH Publishing Co. Pvt. Ltd., New Delhi.

Committee on Opportunities in the Hydrological Sciences, 1991. Opportunities in Hydrological Sciences. National Academy Press, Washington, DC.

Davis, J.C., 2002. Statistics and Data Analysis in Geology. John Wiley and Sons, New York.

DHV, 1999. Lecture Notes of Hydrology Project Training Course on HYMOS. Hydrology Project Office, CSMRS Building. Hauz Khas, New Delhi.

Herschey, R.W. (Ed.), 1978. Hydrometry Principles and Practices. John Wiley and Sons, New York.

Herschey, R.W. (Ed.), 1986. New Techniques in Hydrometry. Adam Hilger Ltd., New York.

Herschey, R.W. (Ed.), 2009. Streamflow Measurement. Routledge, London.

Howell, T.A., 2005. Lysimetry. In: Hillel, D. (Ed.), (Editor-in-Chief). Encyclopedia of Soils in the Environment. Academic Press, New York.

ISO, 1982. Liquid Flow Measurement in Open Channels – Part 2: Determination of Stage-Discharge Relation, ISO 1100/2-1982(E). International Standards Organization, Switzerland.

ISO, 1983. Measurement of Liquid Flow in Open Channels. ISO Standards Handbook 16. International Organization for Standardization, Switzerland.

ISO, 2002. Measurement of liquid flow in open channels — Methods for measurement of characteristics of suspended sediment. ISO 4363. In: International Organization for Standardization, Geneva, Switzerland.

Jain, S.K., Chalisgaonkar, D., 2000. Setting up stage discharge relations using ANN. Journal of Hydrologic Engineering, ASCE 5 (4), 428–433.

Jain, S.K., Singh, V.P., 2019. Engineering Hydrology: An Introduction to Processes, Analysis, and Modeling. McGraw Hill Education, New York.

Matheron, G., 1971. The Theory of Regionalized Variables and its Application. Les Cahiers du centre de morphologie mathematique de Fontainbleu. Ecole Nationale Superieure des Mines, Paris, France.

McCutcheon, S.C., Martin, J.L., Barnwell, T.O., 1993. Water quality. In: Maidment, D.R. (Ed.), Handbook of Hydrology. McGraw-Hill Inc., New York.

Moore, R.J., Cole, S.J., Illingworth, A.J. (Eds.), 2012. Weather Radar and Hydrology. IAHS Proceeding and Reports, UK.

Mukherjee, K., Kaur, S., 1987. Precipitation network design. In: Jal Vigyan Sameeksha, vol. II. National Institute of Hydrology, Roorkee, pp. 35–39, 2.

Panigrahy, N., Jain, S.K., Kumar, V., Bhunya, P.K., 2009. Algorithms for computerized estimation of Thiessen weights. Journal of Computing in Civil Engineering, ASCE 23 (4), 239–247.

Pickup, G., 1988. Hydrology and sediment models. In: Anderson, M.G. (Ed.), Modelling Geomorphological Systems. John Wiley and Sons, Chicester.

Rantz, S.E., 1982. Measurement and computation of streamflow: volume 2. Computation of discharge. In: U.S. Geological Survey, Water Supply Paper # 2175. Washington, D.C.

Sokol, Z., Szturc, J., Orellana-Alvear, J., Popová, J., Jurczyk, A., Célleri, R., 2021. The role of weather radar in rainfall estimation and its application in meteorological and hydrological modelling—a review. Remote Sensing 2021, 13, 351. https://doi.org/10.3390/rs13030351.

Singh, P., Singh, V.P., 2001. Snow and Glacier Hydrology. Kluwer Academic Publishers, Dordrecht.

Vieux, B.E., 2002. Radar rainfall applications in hydrology. In: Bedient, P.B., Huber, W.C. (Eds.), Chapter 11 in Hydrology and Floodplain Analysis. Prentice-Hall Inc., New Jersey.
WMO, 1973. Manual for Estimation of Probable Maximum Precipitation. WMO - No. 332. Secretariat of the World Meteorological Organization, Geneva, Switzerland.
WMO, 1983. Guide to Meteorological Instruments and Method of Observation. WMO No. 8. World Meteorological Organization, Geneva.
WMO, 2008. Guide to Hydrological Practices. WMO No. 168. World Meteorological Organization, Geneva.

Further Reading

Gleick, P.H. (Ed.), 1993. Water Is Crisis. Oxford University Press, Oxford.
Latkovich, V.J., Leavesley, G.H., 1993. Automated data acquisition and transmission. In: Maidment, D.R. (Ed.), Handbook of Hydrology. McGraw-Hill Book Company, New York.
Reichel, G., 1998. Suspended sediment monitoring: use of acoustic Doppler current profiler. In: Herschy, R.W., Fairbridge, R.W. (Eds.), Encyclopedia of Hydrology and Water Resources. Kluwer Academic Publishers, Dordrecht.
WMO, 1972. Casebook on Hydrologic Network Design Practices. WMO No. 324. World Meteorological Organization, Geneva.
WMO, 1980. Manual on Stream Gauging. Vol. I and II, Computation of Discharge. Operational Hydrology Report No. 13, WMO No. 519. World Meteorological Organization, Geneva.
WMO, 1981. Case Studies of National Hydrological Data Banks (Planning, Development and Organisation). Operational Hydrology Report No. 17, WMO No. 576. World Meteorological Organization, Geneva.
WMO, 1982a. Concepts and Techniques in Hydrological Network Design. Operational Hydrology Report No.19, WMO No. 580. World Meteorological Organization, Geneva.
WMO, 1982b. Methods of Correction for Systematic Error in Point Precipitation Measurements for Operational Use. Operational Hydrology Report No. 21. WMO No. 589. World Meteorological Organization, Geneva.
WMO, 1988. Manual on Water Quality Monitoring – Planning and Implementation of Sampling and Field Testing. Operational Hydrology Report No. 27, WMO No. 680. World Meteorological Organization, Geneva.
WMO, 1989. Management of Ground Water Observation Programmes. Operational Hydrology Report No. 31, WMO No. 705. World Meteorological Organization, Geneva.

CHAPTER

Advanced Techniques for Data Acquisition and Systems Modeling

3

Chapter outline

Water Resources Systems Planning and Management. https://doi.org/10.1016/B978-0-12-821349-0.00015-0

The objectives of this chapter are to:

- introduce fundamental concepts of data collection through remote sensing and analysis,
- explain the concepts of a GIS and its analytical capabilities,
- introduce the ANN theory and describe its applications, and
- give an exposition of the decision support systems and expert systems.

This chapter discusses four tools which are now extensively used for the analysis of hydrosystems. These tools are labeled as advanced techniques and supplement the techniques that are conventionally applied for the analysis of water resources data. Wider applications of all these techniques began in right earnest after the 1980s or so but during this short period, they have occupied an important place in the tool kit of an analyst. The topics that have been included in this chapter are: remote sensing (RS), geographic information systems (GISs), artificial neural networks (ANNs), and decision support systems (DSSs), and expert systems (ESs). Since each of these topics is the subject matter of a complete book in itself, this chapter will only give a broad overview of the techniques and their applications to water resources problems.

3.1 REMOTE SENSING

The land phase of the hydrologic cycle is influenced and controlled by surface and near surface features of earth which have inherent spatial variability. The inputs to this hydrologic system, for example, radiation and precipitation, also have spatial attributes. In the absence of a reliable technique to measure spatial features, point values are commonly used in the analysis. The need for measurement of areal properties has thus been summarized by Klemes (1986): "It also seems that the search for new measurement methods that would yield areal distribution, or at least reliable areal totals or averages of hydrologic variables, such as precipitation, evapotranspiration, and soil moisture, would be a much better investment for hydrology." RS is a technique which has the potential to realize this necessity. In fact, it has already achieved so in many instances.

The term remote sensing (RS) implies the acquisition of information about an object without establishing any physical contact between the object and the sensing device. Although it appears to be a new technique, this way of obtaining information about an object is quite old. Photography, which has been in use for a long time, is a technique based on the same principle. Of late, the term RS technique is chiefly used to denote the acquisition and analysis of satellite data. This is a powerful technique for exploration, mapping, and management of earth resources. The main advantage of the RS technology is that it provides a broad perspective over a large area. One can "see" beyond visible electromagnetic (EM) radiation band, and data of inaccessible areas can be obtained just as easily. RS techniques have extended the scope of utilization of the EM spectrum to almost its entire range. Depending on the sensor, it is also possible to infer the characteristics of a top thin layer of the earth's surface. The interaction of EM radiation with an object can reveal a tremendous amount of information about the object: What is it? Where is it? What are some of its physical properties? What are its spatial relationships with the surroundings?

EM energy can be generated by changes in the energy levels of electrons, acceleration of electrical charges, decay of radioactive substances, and the thermal motion of atoms and molecules. Nuclear reactions within the sun produce a full spectrum of EM radiation which is transmitted through space without major changes in its character until it reaches the atmosphere. Visible light, radio waves, thermal, ultraviolet (UV), and X-rays are the familiar forms of EM radiation and propagate in accordance with the basic wave theory. The EM energy travels in a harmonic sinusoidal fashion at the velocity of light ($c = 3 \times 10^8$ m/s). The distance from one wave peak to the next is the wavelength λ, and the number of peaks passing through a fixed point in space per unit time is the wave frequency ν. The wave velocity is computed by $c = \nu\lambda$. The EM spectrum ranges from very short wavelength (10^{-7} μm) cosmic rays to very long (10^8 μm) radio waves.

In RS, the measurements of the EM spectrum are used to determine the properties of various earth features and vegetation. The basic concept is that each object, depending on its physical characteristics, reflects, emits, and absorbs varying intensities of radiation at different EM wavelength ranges. The chief source of EM radiation is the Sun. The EM spectrum can be divided into various ranges of wavelengths. Using the reflectance information from one or more wavelength ranges, it is possible to discriminate between different types of ground objects (e.g., water, dry soil, wet soil, vegetation, rocks, etc.) and map their distribution on the ground. One of the RS techniques is aerial photography which essentially makes use of only the visible part that constitutes a very small portion of the EM spectrum.

The wavelength ranges which are transmitted by the atmosphere without much loss of energy are known as *atmospheric windows*. The various ranges of the EM spectrum and the transmission of these by atmosphere are depicted in Fig. 3.1. Naturally, the wavelength ranges which are efficiently passed by the atmosphere are used in RS analysis. Useful atmospheric windows are available in the range 0.3–0.75 μm (ultraviolet, UV, and visible); 0.77–0.95 μm (near-infrared or NIR); 1.0–1.12 μm, 1.19–1.34 μm, 1.55–1.75 μm, 2.05–2.4 μm (shortwave IR); 3.5–4.16 μm, 4.5–5.0 μm, 8.0–9.2 μm, 10.2–12.4 μm, 17.0–22.0 μm (middle and thermal IR), and microwave region.

The EM radiation emitted or reflected from the objects of interest has to pass through atmosphere before the remote sensor detects it. Thus, the characteristics of atmosphere significantly determine the effective use of EM spectrum for RS. The absorption and reemission processes in the atmosphere are through changes in electronic, vibrational, and rotational quantization levels. The transmission and absorption of the energy by the gases present in the atmosphere depends on the wavelength ranges. The most important atmospheric constituents that influence the incident radiation are water vapor (H_2O), oxygen (O_2), ozone (O_3), carbon dioxide (CO_2), and aerosols. The spectral characteristics of energy sources,

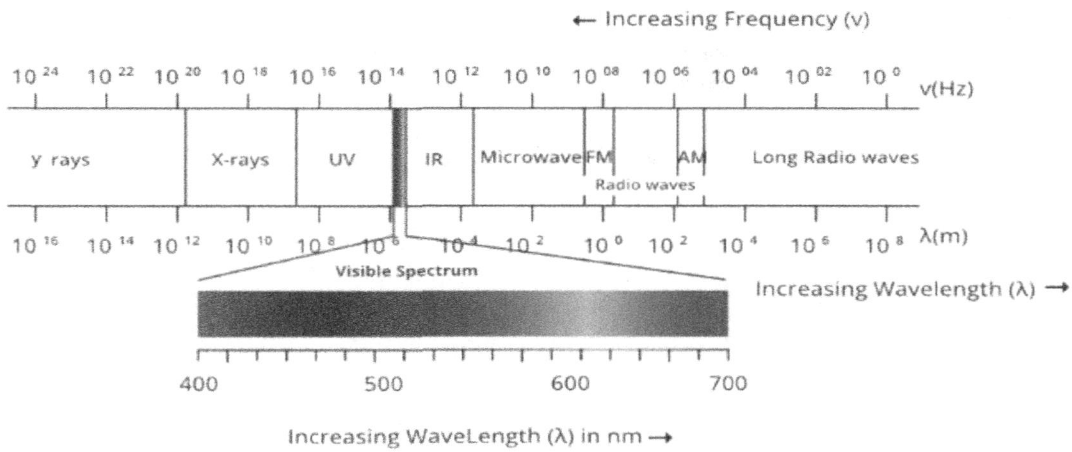

FIGURE 3.1

The electromagnetic spectrum.

Adapted from Lillesand and Kiefer (1994). Copyright © John Wiley and Sons, Inc. Used by permission of John Wiley and Sons, Inc.

atmospheric effects, and sensing systems are shown in Fig. 3.2. Note that only a small portion of EM spectrum can be detected by human eye. This *visible* range is between 0.4 and 0.7 μm. The approximate wavelengths of three primary colors are: blue 0.45–0.50 μm, green 0.5–0.6 μm, and red 0.6–0.7 μm. When an object is illuminated by visible light, its color depends upon the reflectance properties. Chlorophyll present in healthy vegetation absorbs blue and red portion and reflects green light.

Depending on the source of EM energy, the RS systems can be classified in two categories: active and passive. The sensors that generate their own energy are called *active sensors*. These sensors, such as radar, transmit energy with certain properties and record the energy that is reflected back by the features that happen to be in the signal path. The *passive sensors* use an external source of energy. Most RS systems rely on the Sun to generate the EM energy that is needed to image objects and, therefore, are classified as passive. The design of sensors is influenced by atmospheric windows, as the sensors measure energy corresponding to a particular range.

Advantages: One of the main advantages of RS techniques in natural resources management is the synoptic coverage of the earth on a periodic basis with small expenses. The advantages of RS over the conventional methods are enumerated next.

Measurement frequency: With the availability of RS data of sensors having a resolution of the order of a few meters, highly accurate maps can be prepared. For water resources modeling, the accuracy is well within the desired limits.

Speed of analysis: Great savings in time can be made with the use of the RS approach, particularly the time spent in field. The ground truth can be collected by visiting selected spots in the study area. The satellite RS data collection is as easy for a remote, inaccessible area as for a nearby area.

Sampling frequency: Currently, RS data of the same area are easily available at an interval of 15 days or less. This is a very welcome feature because the time series of changes, such as land use, can be easily prepared and used. Such data were the most difficult to gather until the advent of RS.

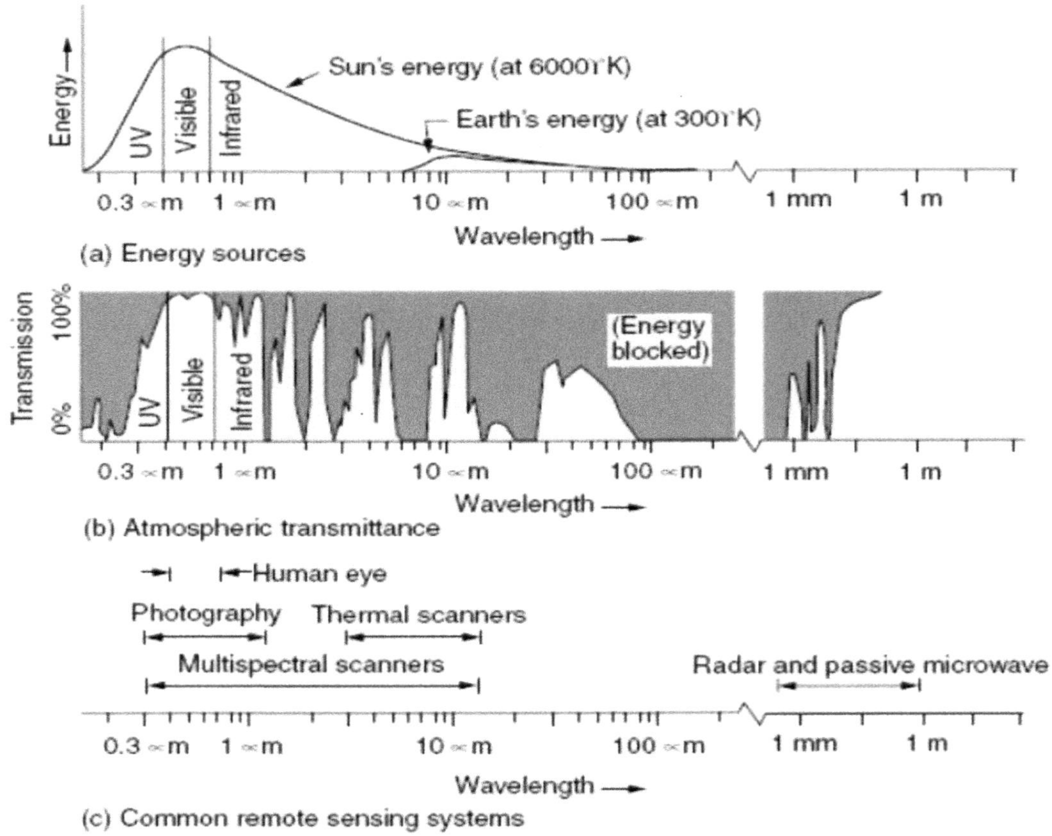

FIGURE 3.2

The spectral characteristics of (a) energy sources, (b) atmospheric effects, and (c) common sensing systems.

Source: Lillesand and Kiefer (1994) Copyright © John Wiley and Sons, Inc. Used by permission of John Wiley and Sons, Inc.

3.1.1 Basic Components of Remote Sensing Data Collection

The basic components of an RS system are shown in Fig. 3.3. The EM energy source or illumination (A) provides the energy to sense the object. The radiation energy comes in contact with the atmosphere (B) and interacts with it as it passes through. Once the energy reaches the target (C) through the atmosphere, it interacts with it in a manner that depends on the properties of both the target and the radiation. The radiation scattered or emitted from the target travels to the sensor and again interacts with the atmosphere. The sensor (D) senses the energy and either records it or transmits it (often in electronic form) to a receiver on earth (E). At the receiver, the data are stored on a computer media. The data are passed on to the users at (F) in the form of an image (digital and/or hard copy). These days most images are available in digital form; some data products may also be given in hard copy form. At the processing centers (F), the image is interpreted and analyzed by visual or digital means to extract information about the target. The final product is usually produced in the form of a map (G) and is used in planning, problem-solving, and decision-making.

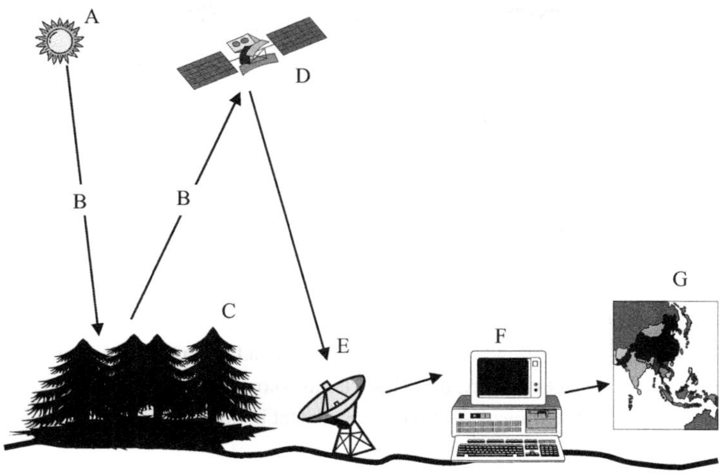

FIGURE 3.3

Basic components of a remote sensing system.

3.1.2 Remote Sensing Sensors

The commonly used sensors in RS are described in the following.

Gamma ray sensors: These sensors measure the difference between the natural gamma ray's radiation of earth and that attenuated by soil, water, or snow.

Multispectral scanners: The detectors of these sensors are sensitive to specific regions of the spectrum and measure the spectral reflectance in a narrow range of wavelengths or *bands*. Typical bands are: blue (0.45–0.50 μm), green (0.50–0.60 μm), red (0.60–0.70 μm), panchromatic (0.50–0.75 μm), and thermal infrared (TIR) (10–12.5 μm). Note that the wavelength ranges of different bands may not be exactly the same in different satellites.

Thermal sensors: The temperature of a body depends on the incident and emitted energy as well as its thermodynamic properties. The thermal sensors measure radiation in the thermal range. The thermal IR range is very useful in water resources studies.

Microwave sensors: These sensors measure the dielectric properties of various features on the earth surface. The microwaves can penetrate through clouds and for this reason, such sensors can work in cloudy weather too. They are useful for the measurement of soil moisture and mapping of areas during the flood season when the sky is usually overcast.

Applications of data of various spectral ranges in the field of water resources are indicated in Table 3.1. Chawla et al. (2020) provide lists of optical/thermal sensors and of active and passive microwave sensors.

Hyperspectral RS or imaging spectroscopy combines imaging and spectroscopy in a single system. Hyperspectral data sets may be composed of 100–150 spectral bands of narrow bandwidths (e.g., 5–10 nm). Hyperspectral sensors capture data as a set of images, each of which represents a narrow wavelength band. These images are combined to form a 3-D hyperspectral data cube which is processed.

Table 3.1 EM wavelength bands and their applications in water resources.

Band, μm	Applications
Blue (0.45–0.50)	Bathymetric mapping, land use, soil and vegetation characteristics, sediment, distinguishing soil from vegetation
Green (0.50–0.60)	Green reflectance of healthy vegetation, mapping of sediment-laden water, delineation of areas of shallow water
Red (0.60–0.70)	Vegetation discrimination because of red chlorophyll absorption
Panchromatic (0.50–0.75)	Mapping, land use
Reflective (0.75–0.90)	Biomass, crop identification, soil-crop, land-water boundaries
Near-infrared (0.7–0.8)	Vegetation boundary between land and water, and landforms
Mid-infrared (1.5–1.75)	Plant turgidity, droughts, clouds, snow-ice discrimination
Short-wave infrared 1 (1.57–1.65)	Discrimination of moisture content of soil and vegetation
Short-wave infrared 2 (2.11–2.29)	Improved discrimination of moisture content of soil and vegetation
Mid-infrared (2.0–2.35)	Geology, rock formation
Thermal infrared (10–12.5)	Relative temperature, thermal discharges, vegetation classification, moisture studies

Note: Names of bands and wavelength range can vary with satellite and sensor.
Source: Engman (1993), USGS.

3.1.3 Remote Sensing Platforms

The sensors acquiring the RS data have to be mounted on platforms from which measurements can be taken. Such platforms may be (1) ground observation platforms (towers, vehicles, etc.), (2) air-borne balloons, (3) aircrafts, and (4) satellites. The sensors mounted on ground-based vehicles are mostly used during sensor development, while verifying the design and understanding its response with respect to the characteristics of the target. If the data of a large area are to be obtained, it is necessary to take observations from a platform high up in the air. Due to this reason, satellites are the most convenient platforms. The satellites provide the coverage of all parts of earth and therefore are an ideal platform and are widely employed. The sensor may be mounted on a geostationary satellite which moves synchronously with the earth and is always positioned above the same point on the earth. Most weather or communication satellites (e.g., Meteosat: 0° longitude, 0° latitude) are of this type. Geostationary satellites provide data of limited area and may have poor resolution.

The polar orbiting satellites follow elliptical orbits (crossing near the poles) and can view a given area repetitively after a certain number of days. These satellites are sun synchronous which means that they image a particular area at the same local time in each revisit. The polar satellites provide a complete coverage of earth's surface. Some current satellites have sensors that can be steered sideways. This allows the cameras to also view and map off-nadir areas. By the use of this feature, the image of an area can be obtained more frequently than is normally possible, but the satellite has to be preprogrammed for this purpose. This feature also permits stereo coverage of an area. Some satellite systems whose data are most commonly used include the Landsat (including Thematic Mapper, TM) of NASA, the Indian Remote Sensing (IRS) satellites, the Terra and Aqua satellites of NASA, and the European Remote Sensing satellite (ERS) of the European Space Agency (ESA), GPM. The Landsat data are available since 1972 and the IRS data since 1988. The most relevant spectral bands are visible (VIS) at various colors, infrared (IR), NIR, TIR, passive microwave, and active microwave (Radar). Besides satellites, ground-based RS data collection systems are also of importance in water resources. For instance, ground-based weather radar is very useful for areal estimation of precipitation.

Currently, a number of satellite systems are operational. A few of these are described below.

Landsat Program

The Landsat program has provided perhaps the most extensively used RS data the world over. It is a joint program by NASA and the U.S. Geological Survey of USA. Its chief plank has been in delivering unrestricted global data of good geometric accuracy. Under the Landsat program, many satellites, such as Landsat 1, 2, 3, 4, and 5, were launched in the last century. Landsat 7 was launched in 1999, and Landsat 8 in 2013. Landsat 9 was launched in 2021. These satellites have been placed in near polar, near circular, sun synchronous orbit at an altitude of 700−900 km. As the satellite orbits in the north south plane, the earth below it spins around its axis, from west to east. Thus, different parts of the globe are "seen" by the satellite during different passes. The data are acquired in the descending node, that is, as the satellite moves from the north pole to the south pole.

Two important sensors on-board the Landsat satellites are Multispectral Scanning System (MSS) and TM. Both are on-line scanners and produce ground scenes of 185 × 185 km size. The MSS sensor has been a regular payload of the Landsat and has made this program a tremendous success. The TM is an advanced multispectral scanner used in Landsat 4 and 5 missions. TM operates in seven wavelength bands, out of which six are in the solar reflection region and one in thermal-IR region. The Landsat satellites offer high resolution in space (30 × 30 m pixels) and data of 7 spectral bands which are suitable for many applications. Landsat 7 has an advanced scanner, the Enhanced Thematic Mapper Plus (ETM+) operating in eight bands. Landsat 8 uses two instruments, the Operational Land Imager (OLI) for optical bands and the Thermal Infrared Sensor (TIRS) for thermal bands (total 11 bands).

The websites for Landsat-related information are: http://landsat.gsfc.nasa.gov/, www.nasa.gov, and www.usgs.gov.

Indian Space Program

Under the IRS Satellite program, the Indian Space Research Organisation (ISRO) has launched a series of land observation satellites. The operational first-generation RS satellites IRS-1A and IRS-1B were launched in 1988 and 1991, respectively. These were placed in near-polar, sun-synchronous orbit, with repetitive time of 22 days. The satellites had two linear imaging scanning sensors (LISS-I and LISS-II) for providing data in four spectral bands: VIS, IR, and NIR. Their ground resolutions were 72.5 and 36.25 m, respectively. The second generation, operational, multisensor satellites IRS-1C and IRS-1D were launched in 1995 and 1997. These were placed in near-polar, sun-synchronous orbit with a repetitive time of 22 days. These satellites have three on-board cameras. The PANchromatic (PAN) camera operates in the panchromatic region of the EM spectrum and has a spatial resolution of 5.8 m. It can be steered up to 26° across-track, thus enabling the generation of stereoscopic imagery and improved revisit capability. The Linear Imaging Self Scanner-III (LISS-III) camera operates in four spectral bands: three in visible/near infrared (VNIR) and one in short wave infrared (SWIR) region. It has a resolution of 23.5 m in VNIR bands (swath 141 km) and 70 m in SWIR band. The wide field sensor (WiFS) is a coarse resolution camera with spatial resolution of 188.3 m.

Between 1999 and 2020, about 23 satellites were launched by India for a range of applications. Recently launched satellites include Cartosat-2E, Cartosat-2F, RISAT-2B, Cartosat-3, RISAT-2BR1, and EOS-1 (RISAT-2BR2). Data from IRS satellites are available through National Remote Sensing Center https://www.nrsc.gov.in.

Copernicus Programme

Copernicus is the name of Earth observation program of the European Union. It is managed by a number of agencies/countries including the ESA. The Copernicus programme covers six main themes: atmosphere, marine, land, climate, emergency, and security. ESA is launching a fleet of Earth-observing satellites which have been named Sentinels. Seven missions will be launched under the Sentinel programme (Sentinel 1, 2, 3, 4, 5P, 5, 6). Each Sentinel mission consists of a constellation of two satellites. The Sentinel missions include radar and super-spectral imaging systems.

Besides, many other countries/consortia such as China, Japan, Brazil, etc., have thriving RS programs.

3.1.4 Resolution of Remote Sensing Data

The RS data have the following four types of resolutions.

Spatial resolution: Mather (1987) lists four criteria for the definition of spatial resolution: geometrical properties of the imaging system, the ability to distinguish between point targets, the ability to measure the periodicity of repetitive targets, and the ability to measure the spectral properties of small targets. The instantaneous field of view (IFOV) is the most common measure of geometric properties. It is the area on ground that is viewed by the instrument from a particular altitude at a given time. As shown in Fig. 3.4, it can be measured either as an angle α or as the equivalent distance on the ground. Due to various reasons, IFOV is not a useful measure of resolution and, moreover, it need not be linked to pixel size. The values can be interpolated over the cells of the image to represent any ground spacing (Mather, 1987). For example, the pixel size of IRS-1C LISS III sensor is 23.5 m. Nowadays, images with pixel sizes of the order of a meter are easily available.

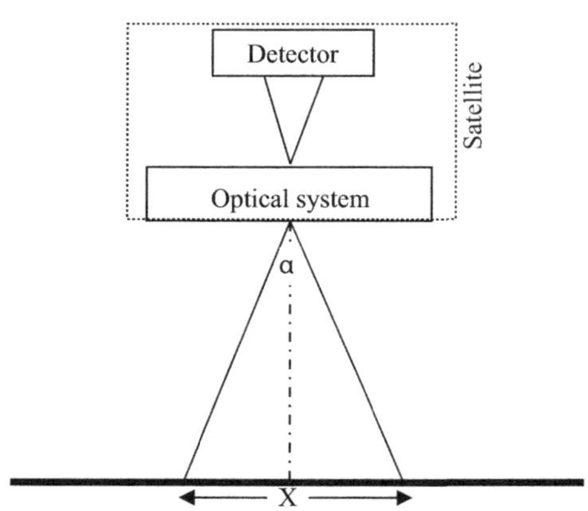

FIGURE 3.4

Angular IFOV and projection X on ground. X is the diameter of a circle.

The pixel size is chosen based on the purpose of study and the size of study area. In case of large pixel sizes, problems arise in georeferencing because ground control points with desired accuracy cannot be located.

Spectral resolution indicates the number and bandwidth of specific wavelength intervals in the EM spectrum to which a sensor is sensitive. For example, the LISS-III sensor of IRS-1C satellite operates in four spectral bands.

Radiometric resolution—In RS, the reflected radiation from an object is measured by the sensor. An electrical signal is generated whose strength depends on the radiation received by the sensor and this strength is converted and stored as digital numbers. The radiometric resolution is the number of levels which can be realized in the range of radiation levels. Typically, there are 128 or 256 levels.

Temporal resolution or revisit frequency means how often the sensor records imagery of the same area. Commonly it is of the order of two weeks. Geostationary satellites produce a very high resolution in time (say, about 30 min) but have a rather coarse spatial resolution, for example, of the order of several km. Ground-based weather radars produce a rather high resolution in time and space (e.g., 15 min in time, 1 km^2 in space). Since a large number of satellites are in orbit these days, it is possible to obtain images of an area at closer temporal gaps by compiling data from multiple satellites.

The RS data are also available as hardcopy prints of images. However, these images differ from normal photographs. In normal photographs, an object of green color is printed in green, red color in red, and so on. The resultant print is known as true color composite. The visual perception of an RS image can be altered by assigning, for example, blue, green, and red colors to observations in the green, red, and NIR bands. The product so generated is known as *false color composite (FCC).* Since vegetation is highly reflective in NIR band, it will appear as red in standard FCC. The hard copy FCC can be analyzed visually, but visual analysis is gradually becoming outdated.

The digital data contains the reflectance of each pixel in various spectral ranges. A number of hardware and software systems are available for the analysis of digital imagery or *image processing.* The current trend of software development is for PC or workstation-based systems. The following discussion is focused on analysis techniques and applications of RS to water resources.

3.1.5 Reflectance Characteristics of Earth Features

Reflectance is the ratio of energy reflected to the total energy incident on a body, expressed in percentage. An object appears green, because it reflects only wavelengths corresponding to green color in the visible spectrum. Thus, blue-colored objects absorb all light waves except those pertaining to blue color, and so on. The spectral radiant flux φ that is incident on the earth's surface is reflected, or absorbed, or transmitted. Thus,

$$\phi = \phi_r + \phi_a + \phi_t \tag{3.1}$$

where φ_r, φ_a, and φ_t are the reflected, absorbed, and transmitted parts of the flux. The relative magnitudes of these depend on the characteristics of the target on the earth's surface. This property is used to identify various features on the earth surface through measurements on their spectral properties. The remotely sensed measured signal (φ_r) as a function of the wavelength is often referred to as the *spectral signature* of the target because its analysis provides useful information about the properties of target.

Spectral reflectance is the ratio of reflected energy to incident radiation (φ_r/φ) as a function of the wavelength. The spectral reflectance characteristics of most targets depend upon the illuminating region of the EM spectrum. The MSS, TM, and LISS are some of the sensors to sense the radiances at distinct spectral bandwidths. Vegetation, for example, may reflect only 10 to 15% in the green band of the EM spectrum but as much as 40 to 60% in the NIR band. However, under certain conditions, the spectral reflectance characteristics of different objects may be the same. For example, water and wet black soil may show the same reflectance properties in a band. In such cases, it would be necessary to examine the data of other spectral bands to differentiate them. This is the basic motivation behind multispectral analysis. Fig. 3.5 shows spectral reflectance curves for water, soil, and vegetation.

The mapping and identification of vegetation is an important aspect of RS. A leaf is built-up of layers of structural fibrous organic matter. In this are pigmented water-filled cells and air spaces. The properties of pigment, leaf physiological structure, and water content affect its reflectance, absorptance, and transmittance properties. In the visible band, the green color of the leaf is due to peak reflectance at 0.54 μm. The presence of chlorophyll is responsible for very low reflectance (10%–30%) in red and blue bands. Some middle infrared (MIR) bands are absorbed by water vapor. The reflectivity rises rapidly near 0.75 μm and remains high in the NIR region in the range 0.75–1.35 μm. The peak reflectance in MIR occurs at 1.6 and 2.2 μm. Therefore, measurements in 1.55–1.75 and 2.08–2.35 μm bands give useful information about moisture content of the plant canopy.

A leaf with low pigment content gives a higher reflectance in the red region. Stress in vegetation due to diseases, insects, and nutrient deficiency affects the reflectance characteristics. A healthy plant gives a less reflectance in the red region and a high reflectance in NIR. The stress is noticeable in the NIR imagery even in the early stage of a disease. The seasonal state of maturity of a plant also influences its spectral reflectance and thus the spectral signature of a plant species usually varies during a season and in its life cycle. These spectral reflectance properties of vegetation are used to discriminate vegetated and nonvegetated areas. There can be considerable differences between plant species, and these can be used to discriminate among species.

The reflectance of soil depends on the chemical and physical properties of its components, organic matter, texture, moisture, surface roughness, and sun angle. Soils usually have a higher reflectance than

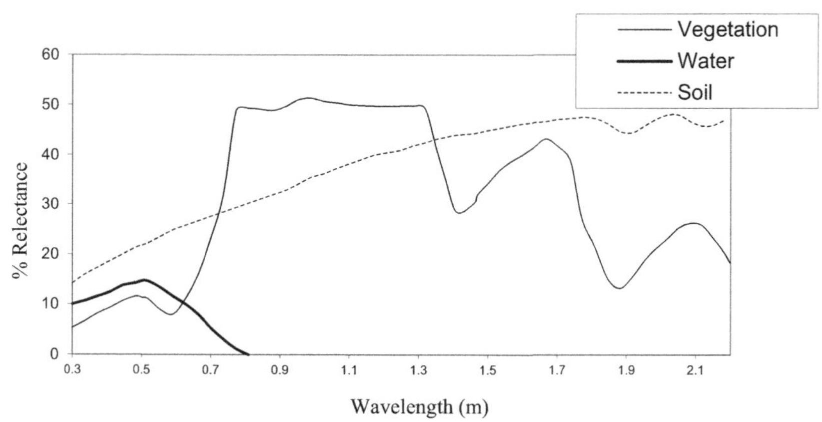

FIGURE 3.5

Typical spectral reflectance curves for water, soil, and vegetation.

plants in the visible bands. Typical curve shows rising reflectivity with increase in wavelength. In the visible band, reflectivity is affected by organic matter and moisture. In the NIR band, plants have a higher reflectance than do soils. Generally, a reduction in grain size of soil results in an increase of reflectance. In the case of rocks, the reflectance properties significantly depend on their type, chemical composition, weathering, rock outcrop, etc. The influence of weathering on spectral signatures can be either way; generally, the weathered rocks have a low reflectance.

Clear water gives low reflectance in visible and almost completely absorbs the radiation in NIR regions. The signal received by the satellite mainly depends on the conditions near the surface of the water body. Fig. 3.6 shows processes acting on solar radiation in the visible part of the spectrum over an area of shallow water. Part of solar irradiance is scattered by the atmosphere and some of it reaches the sensor. Of the irradiance reaching water surface, part is reflected. The incident solar energy that is not specularly reflected by water is refracted downward. Within the water body, radiation is either absorbed by water or dissolved substances or is backscattered by suspended particles (volume reflectance). The scatter by clear water depends on the wavelength.

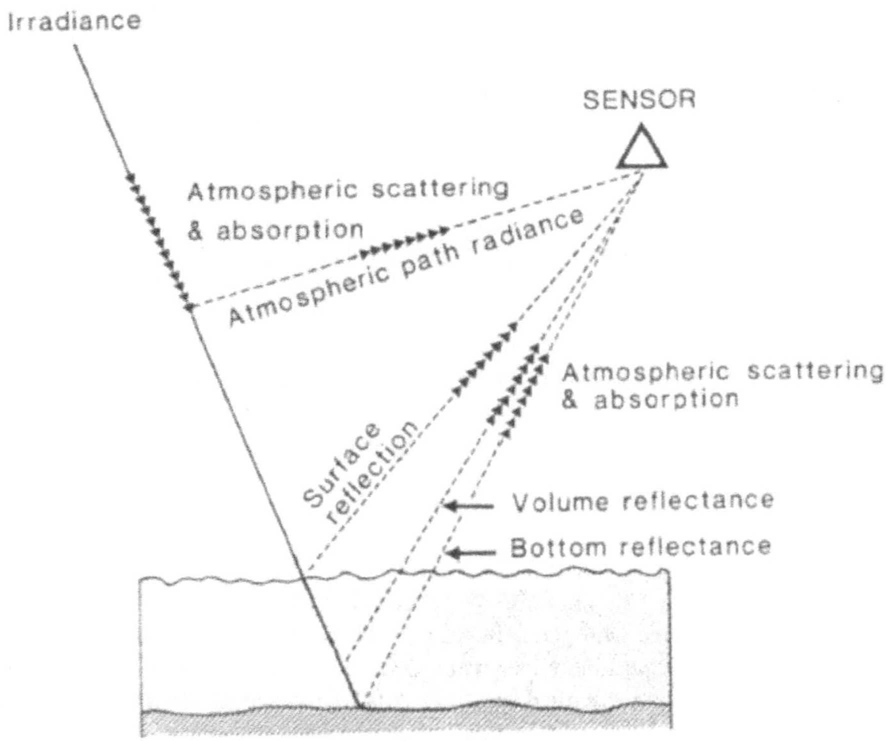

FIGURE 3.6

Processes acting on solar radiation in the visible part of the spectrum over an area of shallow water.

Source: Mather (1987). Copyright © John Wiley and Sons, Limited. Used by permission.

3.1.6 Remote Sensing Data Analysis

The satellite image can be analyzed using either the visual or the digital techniques. These are now discussed.

Visual Image Processing

The visual image processing is based on how the analyst views and interprets an image. The analyst uses his experience and knowledge about the area to interpret the hard copy of the image. Clearly, the results depend on his ability and vary from one analyst to another. The major characteristics that form the basis of image interpretation are tone/color, texture, size, shape, pattern, and association. On a standard FCC, water appears in black color, vegetation in red, etc. The water bodies have smooth texture, while cities have medium texture. Agricultural fields can be easily identified by their typical rectangular size and nestling; an airstrip will be long and thin. Shape wise, canals will be long and thin, drainage curvilinear. In terms of pattern, roads will be radial, streams in tree-like pattern, and lineaments thin, long, and crisscrossing. Examples of association are canals and agricultural fields, roads, cities, etc.

There are a few key limitations of visual interpretation. An imagery contains a huge amount of digital data, and human beings are not the best number crunchers. Computers can carry out this job much more effectively. The eye of a normal human being can only detect the difference between 8 and 16 different shades of gray and thus a lot of useful information remains unused. It is not normally possible to get consistent results from different interpreters. These days it is rare to see visual processing of images.

3.1.7 Digital Image Processing

A digital image is formed by a number of *pic*ture *el*ements or *pixel*s of the same size. Each pixel represents the spectral response of a small square-shaped area on the ground and has certain spatial and spectral properties. The spatial property defines the ground coverage of the pixel. The size of the area depends on the resolution of the on-board camera. For example, the resolution of LISS III camera of IRS-1C satellite is 23.5 m resulting in a pixel size of 23.5*23.5 m. Smaller the pixel size, more details of the target are mapped. The spectral properties define the intensity of spectral response for a pixel in various bands. Associated with each pixel is a number known as *digital number* which is the integrated radiance response of the ground covered by that pixel.

A row of pixels represents a scan line that is recorded as the sensor moves from left to right or recorded through the use of a linear array of photodetectors. An image is composed of a number of geographically ordered m scan lines, placed adjacent to one another in the direction of the y-axis. Each scan line consists of n pixels in the direction of the x-axis. If the data of only one band of the EM spectrum are recorded, an image with pixels in gray shades is obtained. Multispectral sensors detect the light reflectance in more than one band of the EM spectrum. The computer processing of an imagery allows a quantitative analysis of all spectral bands simultaneously. It is also possible to detect and interpret small differences in spectral response that a human eye cannot.

A digital image is a two-dimensional light intensity function denoted by f(i, j), where the value or amplitude of f( ) at spatial coordinate (i, j) gives the intensity (brightness) of the image at that point. Since light is a form of energy, f(i, j) must be positive and finite, and we have

$$0 < f(i,j) < u \tag{3.2}$$

where u is the upper limit of response. A continuous digital image f(i, j) is approximated by equally spaced samples arranged in the form of an n^*m array as shown below, where each element of the array is a pixel:

$$f(i,j) = \begin{bmatrix} f(0,0) & f(0,1) & \cdots & f(0,n-0) \\ f(1,0) & f(1,1) & \cdots & f(1,n-1) \\ \cdots & \cdots & \cdots & \cdots \\ f(m-1,0) & f(m-1,1) & \cdots & f(m-1,n-1) \end{bmatrix} \tag{3.3}$$

When the raw digital data are displayed on a computer screen, it is difficult to discriminate the various terrain features. Digital image processing (DIP) involves the manipulation of digital data to improve the image qualities or to enhance the features of interest with the aid of a computer. The process helps in maximizing clarity, sharpness, and details of objects of interest and leads to better information extraction. It improves the image's interpretability. Image processing operations are carried out to remove noise from the data and enhance certain features based on their spectral response. DIP is a broad subject and may involve procedures that can be simple as well as quite complex. Basically, each pixel of an image is mathematically manipulated and the operation may involve more than one image. The results of computations for each pixel are stored and form a new image. The new digital image may be subject to further manipulation, may be stored, or a hard copy may be taken.

Geometric Correction and Registration

The remotely sensed images are frequently integrated with maps. In *geometric correction*, an image is transformed so that it has the same scale and projection properties as a map. When many images of the same area are to be processed, it is often helpful if these are registered. In *registration*, the coordinate system of one image (called master image) is fit to other images of the same area.

Image Enhancement

Image enhancement algorithms are commonly applied to remotely sensed data to improve the appearance of an image and a new enhanced image is produced. The enhanced image is generally easier to interpret than the original image.

RS images are collected in multispectral bands, that is, the same scene is simultaneously scanned in several spectral bands of the EM spectrum. The radiance measured in each band is an average value over a range of wavelengths in the spectral region, termed as bandwidth. Image contrast is related to the range of gray levels (GLs) in an image, larger the range, greater the contrast and vice versa. Both linear or nonlinear contrast enhancement techniques are used for contrast enhancement. Contrast C may be computed in several ways, for example:

$$\mathrm{C} = \mathrm{GL_{Max}}/\mathrm{GL_{Min}} \tag{3.4}$$

$$\text{or } \mathrm{C} = \mathrm{GL_{Max}} - \mathrm{GL_{Min}} \tag{3.5}$$

where $\mathrm{GL_{Max}}$ and $\mathrm{GL_{Min}}$ are the maximum and minimum GLs in the image, respectively. Contrast is an indicator of the visual quality and is also a measure of the signal-to-noise ratio of an image. It is desirable to utilize the entire brightness range of the display system or hard-copy photographic film.

A contrast enhancement (often referred to as a contrast stretch) expands the original input brightness values to make use of the total dynamic range or sensitivity of the output device. The linear contrast enhancement is best applied to remotely sensed images with Gaussian or near-Gaussian histograms, wherein all the brightness values generally fall within a single, relatively narrow range of the histogram and only one mode is apparent. Unfortunately, this is a rare case, especially for scenes with extensive land and water bodies.

Nonlinear contrast enhancements can be applied to low-contrast imagery. One of the most useful techniques is *histogram equalization*. The histogram of the image is determined based on the number of output gray-scale classes into which the data are to be redistributed. Now the input data are modified to assign approximately equal number of pixels under each of the output gray-scale classes. The histogram equalization applies the greatest contrast enhancement to the most populated range of brightness values in the image. It automatically reduces the contrast in very light or dark parts of the image associated with the tail portion of a normally distributed histogram.

Principal Component Analysis

If there is a significant correlation among the data of different bands, this shows that there is redundancy in the data. Fig. 3.7 contains a plot of two highly correlated variables and AB is the best-fit line. If a plot is prepared with AB as the major axis and a line CD right angle to it as the minor axis, in most cases, the new plot will reveal more information about the structures that are present in the data. In this example, although there are two variables, the dimensionality of the data is one.

The principal component analysis (PCA), also referred to as factor analysis, is an important technique for the analysis of RS digital data. In this, the data of a number of bands are transformed into the same number of principal components (PCs) such that the first two to three PCs contain nearly 95% of the input information. The information content of higher order PCs is not much, they mostly contain noise. The ability to reduce the dimensionality from seven to two or three bands is an important consideration, especially when the transformed data contain nearly as good information as the original data. Consider a scene of Landsat TM satellite which has data of seven bands. If the dimensionality of the data is three, then an FCC can be prepared in which the three primary colors can represent three PCs. The information content of this FCC will be more than any other band combination.

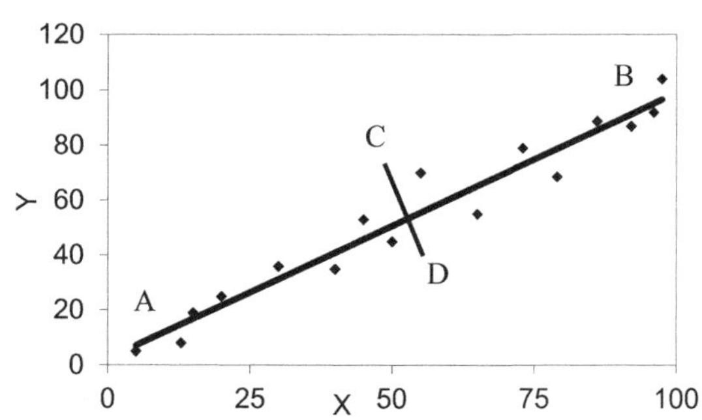

FIGURE 3.7

Variability of two high positively correlated variables.

Spectral Indices

The spectral indices are calculated using the data of two or more bands of an image and help in identifying an object or for image classification. A spectral index can be a simple ratio of two bands or a complex combination of multiple bands. Spectral indexes depend on the spectral properties of the object of interest, for example, spectral indices to identify vegetation (say NDVI) are based on the finding that a healthy vegetation will have strong reflectance in the NIR region and strong absorption in the red region. The image constructed using the computed spectral index is a new image highlighting some surface feature, for example, soil, vegetation, or water. Development and application of these indices were mostly initiated in the 1980 and 1990s. See Bastiaanssen (1998) for an early description of some indices that are routinely used in image processing. RS literature contains numerous papers on spectral reflectance indices. Ma et al. (2019) provide a comprehensive review of water-related indices.

Spectral indices to identify water-related features have numerous applications in hydrological and agricultural fields for tasks including demarcation of water bodies, crop water content, soil water assessment, snow cover, and glacier mapping. A few important indices are described below.

Normalized difference vegetation index (NDVI): This is the most commonly used index which is an indicator of the presence of chlorophyll in the vegetation:

$$\text{NDVI} = (\text{NIR}-\text{RED}) / (\text{NIR} + \text{RED}) \tag{3.6}$$

where NIR and RED are the reflectances in NIR and RED bands, respectively. The red band of EM spectrum is largely absorbed by chlorophyll and reflectance is low. On the other hand, the NIR band is largely reflected by it and has low reflectance. Therefore, healthy vegetation will have high DN value in the NIR band and low in red and consequently, large NDVI. The range of NDVI is -1 to $+1$. NDVI is a high positive number for healthy vegetation and a small positive number for a water stressed or deceased vegetation.

The modified NDWI (MNDWI) can be used to find the moisture content of plants, identify water bodies and the related features:

$$\text{MNDWI} = (\text{GREEN} - \text{SWIR}) / (\text{GREEN} + \text{SWIR})$$

where SWIR and GREEN are the reflectances in SWIR and GREEN bands, respectively.

Normalized difference wetness index (NDWI): Rao and Mohankumar (1994) proposed this index for identifying irrigated crops and suggested that this index is better than NDVI for this purpose. NDWI is defined as:

$$\text{NDWI} = (\text{SWIR}-\text{MIR}) / (\text{SWIR} + \text{MIR}) \tag{3.7}$$

Soil adjusted vegetation index (SAVI) is defined as

$$\text{SAVI} = 1.5(\text{NIR} - \text{RED}) / (\text{NIR} + \text{RED} + 0.5) \tag{3.8}$$

Many attempts have been made to relate SAVI with features of crops. These are explained in Section 3.1.8.

Bare soil index (BSI) is useful in the mapping of soils, identification of harvest, etc.

$$\text{BSI} = [(\text{RED} + \text{SWIR}) - (\text{NIR} + \text{BLUE})] / [(\text{RED} + \text{SWIR}) + (\text{NIR} + \text{BLUE})] \tag{3.9}$$

Note that the wavelength ranges for various bands vary with satellites.

Classification

Image classification is a problem of pattern recognition whose aim is to find the characteristics of objects on the earth at corresponding pixels in the image. The classification algorithms make use of the spectral reflectance values at the pixel in various bands and then try to identify the most likely class to which the pixel belongs. Thus, using the spectral reflectances, a particular pixel may be labeled as water or dense forest, etc. The aim may also be to identify all the pixels in the image that correspond to water. Two types of classification techniques are normally employed: supervised or unsupervised.

Supervised classification is carried out when the identity and location of some of the features on the image, such as urban area, wetland, and forest, are known a priori through information gathered from field visits and study of toposheets, etc. For each of these features, statistical parameters are calculated and then one can build a series of templates representing these patterns. These templates or areas are known as *training sites*, and their spectral characteristics are used to guide the classification algorithm. Next, every pixel of the image is compared with each template and is assigned the class whose properties are closest to it. In other words, it is assigned to a class of which it has the highest likelihood of being a member. The maximum likelihood classifier is the most widely used criterion.

If the identities of land features that are to be classified in a scene are not generally known a priori due to the absence of the ground truth data or other supporting data, an unsupervised classification is performed. The software then groups the pixels into different classes according to some statistical criteria. The analyst, based on his experience and familiarity with the scene, labels these clusters. If there are only a few pixels in some groups, these groups may be deleted or merged with others. Similarly, if some groups are too heterogeneous, they may be split. A combination of supervised and unsupervised classification is often helpful, particularly if adequate field data are missing.

The problem of mixed pixel occurs in classification when the ground area corresponding to the pixel has mixed features, such as part crops and part water body. The spectral properties of such a pixel show mixed behavior. Such pixels can be classified using additional ground information, supporting information from other imageries, or the judgment of the analyst. Advanced techniques, such as fuzzy classification or ANN, have also been used in image classification. Image processing techniques have been discussed in detail in texts, such as Lillesand and Kiefer (1994) and Mather (1987).

Density Slicing

Density slicing is a technique of image processing in which the histogram of the image is divided in a number of intervals or slices. All the pixels that fall in a given interval are supposed to belong to a particular category. This technique has been used in a number of studies. For example, the identification of water spread area of a surface water body. However, many times, this way of classification is considered to be inadequate and liable to errors.

Change Detection

A *change detection* of the feature involves the analysis of a series of images of different dates. In water resources studies, land cover is a most commonly studied phenomenon. The analyst might be interested in both short-term (e.g., flood inundation or snow cover) as well as long-term (e.g., deforestation or expansion of urban area) changes. It will be good if the data of the same sensor and the same resolution are used. The analysis also requires that all the images are accurately registered so that overlays can be prepared to visualize the change.

The simplest way to detect changes is by subtracting DN values of one image from the other. At places where changes are minor, the result will be close to zero. Another possibility is to take ratio of two images. If there is no change at a place, the ratio will be close to one; at places of change, the ratio will be some higher or lower value. The classification of different images can be performed singly or jointly. If the images are classified independently, the accuracy of change detection will depend on the accuracy of classification. The classification becomes complex if many images are classified together. Of course, the analyst has to decide a threshold above which the change is considered significant and below which it is insignificant or could be due to error in analysis.

Advantages of Digital Image Processing

The technique of DIP is very cost-effective for mapping large geographic areas, particularly when imageries of a number of dates are to be analyzed. It is possible to simultaneously use data of several bands for improved interpretation and modern software permit the formulation of complex interpretation algorithms which can use input data from a number of imageries. The analysis results are consistent because subjectivity is not involved. With the availability of fast computers, the interpretation and analysis can be carried out in a short time.

Although the skills and efforts required for DIP are considerably less now due to the availability of user-friendly software, it still demands sufficient experience, particularly where other than routine interpretation is to be carried out. Analysis can be expensive for small areas or for one-time interpretations as the start-up costs may be high. A good set-up requires not only a faster computer with sufficient disk storage space but also special peripheral devices, such as a laser color printer, a scanner, digitizer, data storage, etc. These days, Internet has huge data and literature repositories; some are in public domain and some are subscription based. Although the cost of satellite data has come down recently, it is still expensive, particularly the microwave data. It is tedious to evaluate the accuracy of results, more so when the target area lies in an inaccessible region.

Platforms, such as Google Earth Engine (GEE), combine multipetabyte satellite data with analytical tools/scripts. Users can analyze the images without downloading the same. Currently, GEE is free for academic and research users.

3.1.8 Applications of Remote Sensing to Water Resources Problems

Nowadays, RS data for an area are available from a number of satellites. For applications dealing with water resources, the analyst first identifies which available RS data are suitable for his problems. Then a choice is made based on considerations, such as the spatial and temporal resolution, sensors and spectral bands available, and cost of data products.

The main fields of RS application in hydrology are measurement of precipitation, land use mapping, evapotranspiration (ET) and irrigation water management, soil moisture, streamflow, snow and ice mapping, floods mapping, and water quality. The GISs aid the hydrologist to process the RS data and the derived products. The main application areas of RS technology in the field of water resources are briefly described here. This field is witnessing an exponential growth, and many innovative products will appear in future.

Precipitation estimation: Precipitation requires frequent monitoring, as it changes with space and time. Its type (solid or liquid), intensity, depth, duration, frequency, and total amounts are important in hydrology. "Measuring" precipitation from satellite data is an indirect process that estimates precipitation from active or passive signals (Levizzani and Cattani, 2019).

TRMM and GPM use active and passive radars to measure precipitation. Active radars emit signals that are reflected back to the radar by falling rain drops. The signals returning to the radar, known as radar reflectivity (Z), are an indicator of the number and size of rain/snow drops present in the atmosphere. With the known relationship between Z and R (rainfall), rainfall can be estimated from measured Z. Note that the relationship between Z and R varies with atmospheric conditions.

In RS, precipitation is estimated from the data of a single sensor, multiple sensors, or by merging data from satellites and in-situ instruments. At present, a number of missions are operating, providing data at various spatial and temporal resolutions. Sheffield et al. (2018) describe a number of missions, including Tropical Rainfall Measurement Mission Multi-Satellite Precipitation Analysis (TMPA), Global Precipitation Measurement Mission *I*ntegrated *M*ulti-satellit*E* *R*etrievals for *G*PM (GPM/IMERG), and Climate Hazards Group InfraRed Precipitation with Station data (CHIRPS).

In current times, one of the most successful operational RS systems for precipitation estimation is the Tropical Rainfall Measurement Mission (TRMM) and its successor, the Global Precipitation Measurement mission (GPM). TRMM satellite was launched in 1997 to improve the understanding of precipitation distribution and variability in tropics. The Precipitation Radar (PR) provided new insights into the structure of tropical storms. Designed for 3 years, TRMM provided precipitation data for about 17 years till 2015.

TRMM Microwave Imager (TMI) is a passive radar that measures natural thermal radiation (called brightness temperatures) from the image data. Since water surfaces, for example, oceans or lakes emit only small amounts of MW energy, they appear "cold" to a passive MW radiometer, whereas raindrops appear warm to the same radiometer. This contrast against "cold" water surface is used to infer rainfall rates. GPM core satellite was launched in 2014. GPM and its IMERG product are widely used sources of precipitation data. The GPM satellite uses the GPM Microwave Imager to measure precipitation intensities.

Products such as CHIRPS combine satellite data with gauge data. Technology to measure precipitation from space (hardware and software) is rapidly evolving, and future missions will certainly have improved capabilities.

Evapotranspiration and irrigation water management: The attempts to identify crop types using RS began in the 1980s. Over the years, a large number of RS studies related to various aspects of irrigation water management have been conducted. Irrigation is the main consumer of freshwater and in view of the water scarcity, these studies are highly relevant and useful. The attractiveness of RS stems from its ability to provide repetitive and distributed agricultural and hydrological status of a command. The major aspects that have been covered in the studies include mapping of land use, crop types, crop water requirements, soil salinity, and waterlogged areas.

ET is one of the most important processes of the hydrological cycle. Depending upon the climate, it may account for 40% or more of the moisture lost from a catchment. The conventional ground-based instruments and techniques can provide information over a very limited area, but one really needs areal ET values for atmospheric general circulation models, hydrological, and agricultural studies.

ET links water, energy, and carbon cycles. ET is hard to measure, partly because it can't be seen. It depends on a range of meteorological, soil, and vegetation properties. Many ET algorithms make uses of satellite land surface temperature (LST) measurements (Kalma et al., 2008). The attempts to identify crops by using RS began in the 1980s, and a large number of studies concerned with irrigation water management have since been conducted.

Among the satellite missions, Landsat multispectral data with a resolution of 30 m (100 m in thermal band), two sensors on-board—the OLI and the TIR—have been particularly useful for ET estimation. In fact, Landsat TIR data-based ET estimations (at a resolution of 100 m) are one of the best approaches for ET estimation. Land cover maps and leaf area estimations (LAI) derived from AVHRR (Advanced Very-High-Resolution Radiometer) at coarser but more frequent intervals are also useful in ET studies. These data are available since 1978. MODIS 2000, 1 km/6 km 1 day, 8 day products with different resolutions are available globally.

The systems to estimate ET include the algorithm developed by Mu et al. (2011) or its improvements which are widely used along with Moderate Resolution Imaging Spectroradiometer (MODIS; 1 km resolution) data. Other satellites whose data are used for ET estimation include ASTER, launched in 2000, providing data in five TIR bands; Sentinel series of satellite pairs, and geostationary satellites with coarse resolution data. Surface Energy Balance for Land Surfaces (SEBAL) is a popular algorithm to compute ET using satellite data (Bastiaanssen et al., 1998). In SEBAL, ET is calculated as a residual of energy balance. Sometimes data of multiple sensors are used to overcome the limitations of individual sensors.

There are differences among ET estimates produced by different methods and field observations. One reason is that different methods rely on meteorological data estimated from different sources which are likely to have some difference. Thus, it remains a challenge to compute ET at fine spatial and temporal resolutions that can be used for crop water management.

In addition, the identification of crops and their health is also important and many important developments in this field have taken place in the past few decades. Vegetation properties are best determined by the use of visible/infrared (VIR) sensors. Data from Sentinel-2 mission is proving to be of immense value in deriving vegetation-related information.

Soil moisture: Soil moisture is a useful variable for hydrologic studies, agriculture water and drought management, identification of flooding extent, and antecedent moisture conditions. Soil moisture is highly variable in time and space since it is driven by a number of variables, such as topography, vegetation, soil properties, and meteorological variables, that act at different scales and are measured at limited locations. RS-based soil moisture assessments are limited by the depth of penetration of the MW signal (typically top few centimeters). Note that these signals are heavily attenuated by heavy precipitation and dense vegetation.

Soil Moisture and Ocean Salinity (SMOS) satellite (launched in 2009) was the first dedicated soil moisture mission. In 2015, NASA launched Soil Moisture Active Passive (SMAP) mission that had a 10-km resolution. Data of Sentinel-1 satellites (two polar-orbiting satellites) launched by ESA are used to assess soil moisture at a resolution of 1 km. Sentinel satellites carry a synthetic aperture radar (SAR) sensor which helps in the estimation of soil moisture, surface water extent, and several other hydrological parameters.

Although, there are significant improvements in algorithms and use of the data from improved sensors, the coarse resolution and estimation limited to only top few centimeters are limiting the use of satellites to derived soil moisture.

Surface water height and extent: If surface water data can be observed from space, it will be of immense value for water resources planning and management and for the management of hydrologic extremes. Surface water can be detected from visible bands or microwave sensors (passive and radar).

Data from visible region are used to compute spectral indices, such as the normalized differences water index (NDWI, discussed earlier). Microwaves have an advantage that they can see beyond cloud and rainfall. Active MW sensors can also be used to measure lake and reservoir water levels, although currently applications are limited to larger water bodies.

To determine river discharge from satellites, river stage is measured using radar-based altimeters which can measure water levels at sufficient accuracy. However, currently use of altimeters is limited to wide rivers. The RS data are also being used to map flood-inundated areas in near real-time. With the availability of a number of satellites, the images can now be obtained at an interval of a few days. However, during flood periods, clouds may inhibit data collection and it may be necessary to use microwave data. Wang and Xie (2018) reviewed applications of RS and GIS in water resources and flood risk management.

Snow and glaciers: Fresh snow is the brightest formation on the Earth surface. The albedo or reflectance of the snow surface can be measured by RS techniques. The albedo of snow rapidly decreases with its age. The freshly fallen snow typically has an albedo of about 90%. Due to the accumulation of dust, it may reduce to 40% or so in a few days. However, as both snow and clouds have high reflectance in the visible bands, differentiation could be problematic. Delineation of snow cover areas was one of the earliest applications of satellite data in hydrology. The earlier approaches used data from visible and SWIR bands. Hence, long-term records, derived from visible band data, are available now.

Assessment of snow water equivalent (SWE) is equally important for water resources management. However, the determination of SWE is challenging, because the dialectic properties of snow crystals change as they age. The main approach to determine SWE is by the use of passive MW data. The other possibility is blending information gathered from different sensors. Problems in snow cover identification may also arise when snow is obscured by thick vegetation, bare rocks with reflectance similar to snow are present, and low sun angles generate shadows on north facing mountain slopes.

Groundwater: Since groundwater lies at a depth below the Earth surface, it cannot be directly measured from satellite data, but indirect methods can infer about the presence of ground water. An innovative use of space technology is the Gravity Recovery and Climate Experiment (GRACE) mission of NASA that consists of a pair of satellites, launched in 2002, that move at about 220 km apart in a polar orbit. As the satellite pair moves in the orbit, areas of slightly stronger gravity (higher concentration of mass) will first affect the forward satellite and will pull it away from the second satellite. As satellites continue to move in their orbit, the second satellite is pulled toward the forward satellite as it passes over the gravity anomaly. The change in the gap between the satellites is detected and measured by the precise microwave ranging system onboard the satellites. These data are used to estimate changes in total water storage, and groundwater is the major factor here. In this way, GRACE detects changes in groundwater storage over time. Using GRACE data, Rodell et al. (2009) estimated regional groundwater depletion. More details about GRACE can be seen at https://www.nasa.gov/mission_pages/Grace/spacecraft/index.html.

GRACE is a unique sensor as it does not use EM radiation. A concern with GRACE is its coarse resolution ($\sim$500 km) which hinders its use over small areas.

Water quality: The spectral response of water changes, depending upon the impurities present in it, and this property can be used to determine the quality of water by RS. By the use of reflectivity in the visible region of spectrum, parameters such as total suspended solids (TSSs), Secchi disk depth (SDD), and chlorophyll concentration can be estimated. Using the data from Landsat, MODIS,

Sentinel-2, and other satellites, statistical relationship between water quality parameters and spectral reflectance has been established.

For chlorophyll mapping, bandwidths of 675 and 700 nm have been found to be most useful and red and NIR bands turned out to be useful for turbidity and TSS mapping (Chawla et al., 2020). Apart from the use of single bands, incorporation of reflectance from multiple bands through band ratios is noted to improve water quality parameter estimates by reducing the effect of atmosphere and increasing the signal to noise ratio. The use of multispectral data along with hyperspectral data to estimate water quality parameters has been growing in recent times.

A number of new satellites with smaller pixel size and a larger number of spectral bands would be launched in the years to come. Besides many other applications, these will provide much needed useful input to water resources studies. This field is bound to see rapid development.

3.2 GEOGRAPHIC INFORMATION SYSTEMS

Conventionally, mapping, map analysis, and measurements were done manually. With the advent of computer technology, software were written to handle geographic data on computers. This has culminated in the development of GISs. A GIS, as the name suggests, is a system for input, storage, analysis, and output of geographic data or maps. While maps have been the most common conventional form of representing topography, the advent of digital maps in GIS provides an alternate method of storing and retrieving this information. The general configuration of a GIS is given in Fig. 3.8. The movement of water is linked to processes at the earth's surface and is influenced by them. In analysis and management of water projects, maps are frequently used. Besides, many other spatial data, such as the limits of community habitation and reserved areas for wild life, are important for water resources management but are not appropriately accounted for in the absence of a mechanism to do so. A GIS can link land cover data to topographic data and to other information concerning processes and properties related to geographic locations. In this way, a GIS can be effective in such studies. The effective use of GIS depends on many factors, viz., proper problem formulation, availability of data of desired accuracy, availability of software, hardware, trained manpower, etc. Some commercially

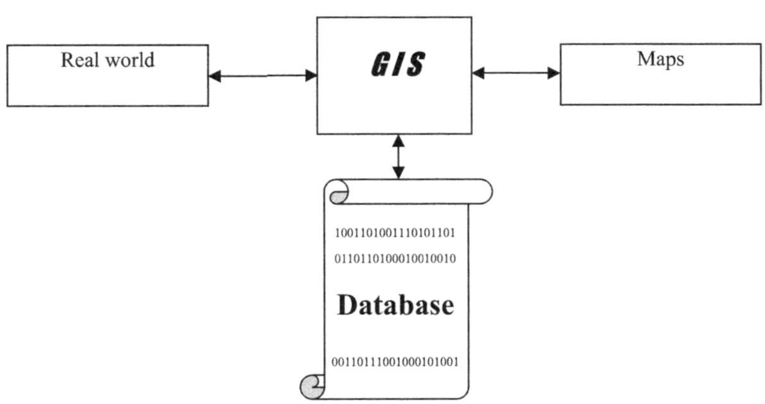

FIGURE 3.8

General configuration of a GIS.

available GISs are: Arc/Info, ERDAS Imagine, ILWIS (Integrated Land and Water Information System, etc. Johnston (1998) listed many sources of GIS data, including some Internet sites. Since Internet is a dynamic medium where new sites are regularly coming up and some old sites closing, it is helpful search for sites through search engines like Google.

Most resource management problems have a spatial dimension. In the environmental modeling domain, this is handled by spatially distributed models that describe the phenomena in one (for example, in river flow models), two (overland flow and water quality models), or three dimensions (atmosphere circulation water models). The increasing use of spatially distributed models in place of spatially lumped models is, to some extent, motivated by the wider availability of powerful and affordable computers (Loucks and Fedra, 1987). Developments in the GIS technology are important from the point of view of water resources engineering.

In GIS, the basic concept is of location, spatial distribution, and relationship; the basic elements are spatial objects. In resources modeling, the basic concept is of state, expressed in terms of numbers, mass, or energy, of interaction and dynamics; the basic elements are biological species, chemicals, and environmental media, such as air, water, or sediment. The overlap and relationship is apparent, and thus the integration of these two fields of research, technologies, or sets of methods, is an obvious and promising idea.

When applied to water resources systems, nontopographic information can include description of soils, land use, ground cover, ground water conditions, as well as man-made systems and their characteristics on or below the land surface. Description of topography is called terrain modeling, and because of the tendency of surface water to flow downhill, the hydrologic importance of terrain modeling is clear. The characteristic that differentiates a GIS from general computer mapping or drawing systems is the link to the information database. Once the database is constructed, the analysis and preparation of output maps is easy. Current GISs allow handling of large data quantities and contains a number of routines which are useful for water resources problems. The major advantage of such a GIS is that it is possible to write a routine for analysis and integrate it with GIS. This is a very welcome flexibility because an existing and tested routine can be directly used.

Spatial data can be visualized as a template composed of cells and/or points and lines, to which specific information is linked. A map can be considered to be composed of several layers (Fig. 3.9), each containing the spatial structure associated with a single category, such as channel network, soils, and land cover. If one of the layers contains the boundary of the basin, spatial operators can be used to define boundaries in the remaining layers as well as to define specific relationship among them.

To understand GIS, it is helpful to first discuss the geographic data types, data models, etc.

3.2.1 Geographic Data Types

Geographic or spatial data have certain characteristics which make them different from other data used in water resources and other areas. The spatial objects or features have some spatial *attributes* associated with them. The main features of spatial data are: geographic location, attributes, and time index. These indicate where the object is, what it is, and when it existed. The position of a terrain feature is specified with respect to a certain datum. Usually the geographic location of a point on the earth's surface is fixed by its latitude, longitude, and altitude. In addition, a cartesian coordinate system or a grid system can also be used. The position of an object can also be specified with respect to a landmark in the region.

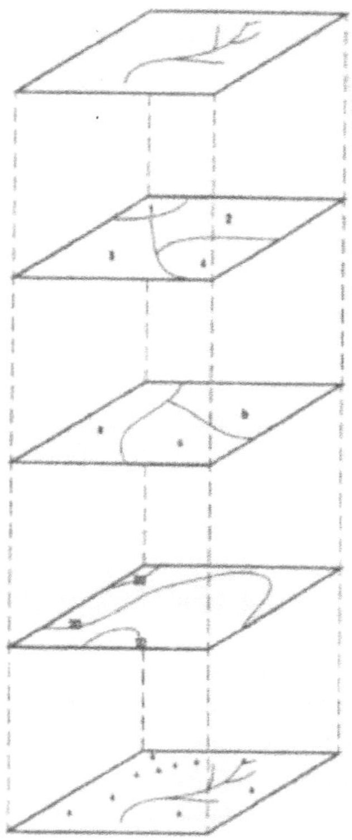

FIGURE 3.9

Categorical approach for handling spatial data.

The attributes of a spatial feature define its characteristics. The attribute can be the class to which the object belongs, for example, water body or forest; it can be an ordinal number; or it can be some measurement associated with the feature. In the last case, mathematical operations are possible on the data. The attributes of geographic data change with time and therefore it is necessary to note the time to which the attribute refers. Here, time can be specified by the calendar date, for example, April 18, 1999; it can be some time duration, for example, daily rainfall, or it can be a frequency, for example, gauge data measured twice a day.

In the GIS parlance, the real-word phenomena are *entities*, while their digital representations are *objects*. On a map, entities are represented as zero-, one-, or two-dimensional spatial objects. In an effort to standardize the data structure and storage, and facilitate and systematize data transfer among the various GISs, the Spatial Data Transfer Standard (SDTS) was developed by the National Institute of Standards and Technology (NIST, 1992). Spatial objects are graphical elements that are used to represent spatial phenomena in a map. These objects can be aggregated into more complex spatial objects. SDTS has defined several types of aggregate objects because they are effective in providing a context for the use of simple objects in a data transfer. With simple objects as the building blocks, aggregate objects denote collections of simple objects that represent real-world phenomena and, therefore, they provide a basis for defining a specific data transfer. Aggregate objects are explained in Fig. 3.10.

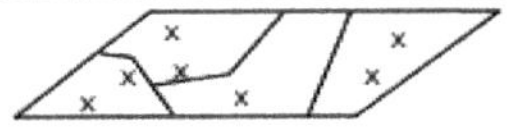	**Layer**	General term describing a collection of instances (occurrences) of spatial features in a single theme
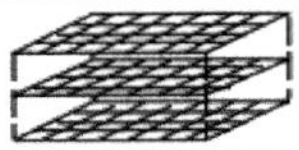	**Raster**	One or more overlapping layers for the same grid, labeled grid, voxel space, or other raster data
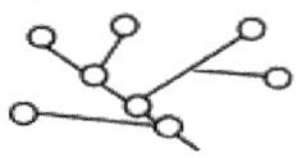	**Planar Graph**	Linear objects inter-connected on a two-dimensional surface with a node at each intersection (e.g., a topologically structured stream network or street network)
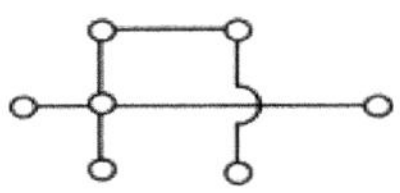	**Network**	A graph of linear objects inter-connected topologically, which, when projected onto a two-dimensional surface, allow multiple nodes at a single location or no nodes at intersection points (e.g., a gas distribution pipeline with overlapping pipes in three-dimensional space)
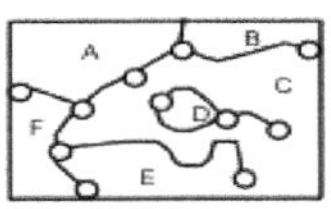	**Two-Dimensional Manifold**	A planar graph and its associated polygons which totally exhaust a surface (e.g., land cover map, parcel map)
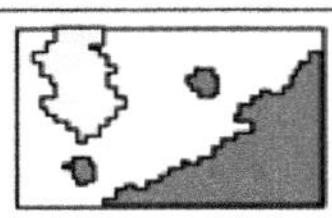	**Digital Image**	Two-dimensional array of regularly spaced pixels (e.g., unclassified satellite image, digital orthophotograph)
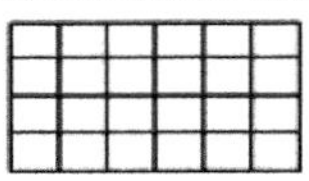	**Grid**	Matrix of cells forming a mesh with repeating pattern (e.g., grid map, digital terrain model)
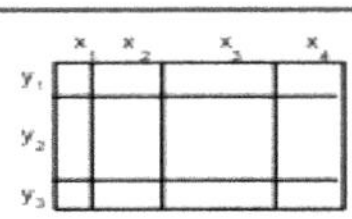	**Labeled Grid**[1]	A two-dimensional set of labeled grid cells [1] forming an irregular rectangular pattern; each labeled grid cell is identified by a spatial label
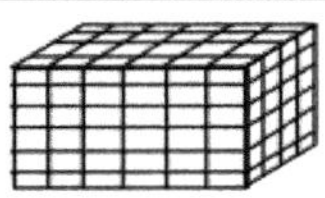	**Voxel Space, Labeled Voxel Space**[1]	Three-dimensional grids in which each voxel represents a three-dimensional volumetric unit (the three-dimensional equivalent to the grid or labeled grid)

FIGURE 3.10

Summary of aggregate spatial objects in the SDTS model.

Source: https://mapi.gov.il/gisForum/tzevet_technology/STDS-Guide%20for%20Technical%20Managers.pdf.

The geographic data can be classified into four types: points, lines, polygons, and surfaces.

Point data: A point is an object with zero-dimension which has some position but no size. Its position is specified by some coordinate system and some attributes are associated with it. A point may represent a village, a city, a stream gauging station, a rain gauge station, or a well. In a raster database, a point refers to one cell.

Line data: A line is a one-dimensional object of a certain length. A line may represent a road, a river, a canal, or an electrical power line. Depending on the object to which a line refers, it has attributes associated with it. A contour line has some altitudes, a river may have certain width, depth, and hydraulic properties. In the case of a river network, each segment may have a different order according to some system, for example, Strahler order.

Polygons: Polygons are two-dimensional objects with certain lengths, widths, and areas. These are the most common data types in a geographical database. The boundaries of a polygon may be natural features or these may be artificial. Some examples of polygons are areas of the same soil, a reservoir, a watershed, a forest, or a country. The boundaries of a catchment are natural, while the boundaries of an administrative unit, such as a district, are man-made. In case two polygons are adjacent to each other, they will have shared boundaries. A polygon may also be enclosed within another polygon, for example, a small island enclosed within a large lake. The most common polygon in water studies is a catchment. The attributes associated with such a polygon are soil types, land cover, geology, and drainage network.

Continuous surface: A continuous surface represents the altitude values of a polygon. For example, the variation of rainfall or piezometer levels can be represented as a continuous surface. A continuous surface is normally constructed using the data of point measurements. These surfaces help in better visualization of the underlying process.

3.2.2 GIS Data Structure

The approach used to represent geographic data in GIS is defined as data model. The spatial behavior of the four data types discussed above is represented using two basic models: vector and raster. The actual method of storing the spatial data on the computer is known as data structure. There are many data structures in use for vector and raster data. These are discussed below.

Vector data: A vector is a quantity which has a starting point, a magnitude, and a direction. In a vector model, objects are created by connecting two adjacent points by a straight line. A point is a one-dimensional object that specifies location and a node is a point that is a junction of two or more links. The curve joining two nodes is called an arc or a segment. A polygon is made up of arcs or segments. Areas can be created by enclosing a region by these lines to form a polygon. The conventional form of representing topography is contour line mapping. Contours can be digitally represented as a set of point-to-point vectors of a common elevation. When an entire map is stored in digital form, it is called a digital line graph (DLG).

Vector objects can be assigned several attributes, usually associated with an attribute table. Vector spatial data structures provide a compact but complex way to conceptualize geographic objects. As a result, data files are smaller in size. These are also very efficient in modeling topology. Three popular structures for handling vector data are topological, spaghetti, and TIN.

Spaghetti: This type has its origin in drawing software. In this structure, lines are drawn as seen in drawings, maps, etc. A polygon is defined as a sequence of (x, y) ordinates which represent a closed area. An example of this data structure is shown in Fig. 3.11. Strings are not interconnected, even though they may appear to be so in the map. The data structure of this model is simple. However, this structure has many drawbacks and is used for simple applications only.

Topologic data structure: Topology is defined as the transformation of geometric computations and the mathematical study of spatial relationships. In this structure, all lines are explicitly linked and the spatial relationships between entities are explicitly defined. The basic entity is a segment of straight line which is defined by the coordinates of its end points. An arc is a sequence of straight lines which begin and end at a node. The location value of starting and ending points of the areas is the same. The points where two lines join are called nodes.

FIGURE 3.11

Spaghetti data structure.

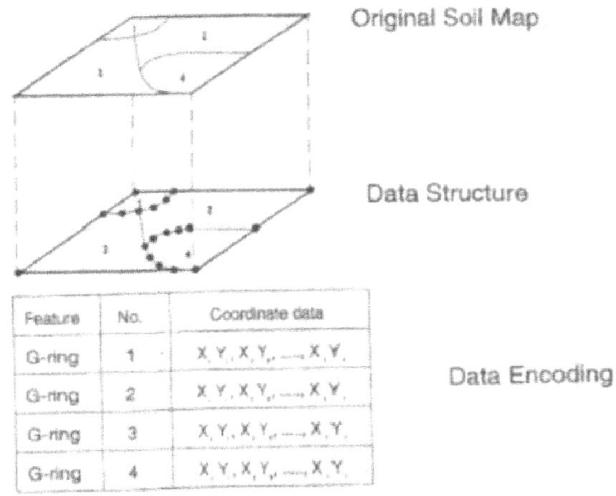

With this technique, spatial relationships among entities, mainly connectivity, are explicitly recorded. In the general case, polygons may be formed using several connected chains or arcs, as shown in the example of Fig. 3.12. The database contains information about the points in an arc, the arcs meeting at a node, and the arcs forming a polygon. The names of the polygons on both sides of the line are also stored. In the database, spatial relationships among entities, mainly connectivity, are explicitly stored. Adjacency is defined based on the polygons that surround every chain. As with other data structures, tables can be constructed to describe coordinates, connectivity of land adjacency characteristics. In Fig. 3.12, these tables are represented as summarized lists. The polygon-arc list includes the arcs associated with each polygon, even those that are common to more than one polygon. This list can be related to the arc-coordinate list to determine the coordinates of all points that constitute each polygon. The topology list defines the adjacency characteristics of all arcs.

Topologic data structures are more sophisticated than spaghetti data structures and are included in practically all vector-based GIS. Their ability to handle topology makes them well suited for contiguity and connectivity analysis (Aronoff, 1989), both of which have ample applications in hydrology. Contiguity capabilities, for example, can be applied to determine flood impact zones, correlations between soil types and land use, and others. Connectivity capabilities can be applied to drainage network and transportation analysis, emergency evacuation routes, and so on. Like spaghetti data structures, however, topologic data structures do not constitute the best solution for representing surfaces and transition zones.

Triangular irregular network (TIN): Triangular irregular networks (TINs) constitute a special case of topologic data structures in which nodes are interconnected using single links (Fig. 3.13), resulting in a set of triangular facets that cover the area of interest. The location of each node is given by its X, Y, and Z coordinates, which provide TINs with a powerful tool to simulate highly variable surfaces. The TIN topology is based on triangle adjacency and the definition of nodes that compose each triangle.

The number of possibilities for defining and locating nodes is nearly endless. As a result, many algorithms and techniques have been devised to optimize the process of TIN generation. Many GIS

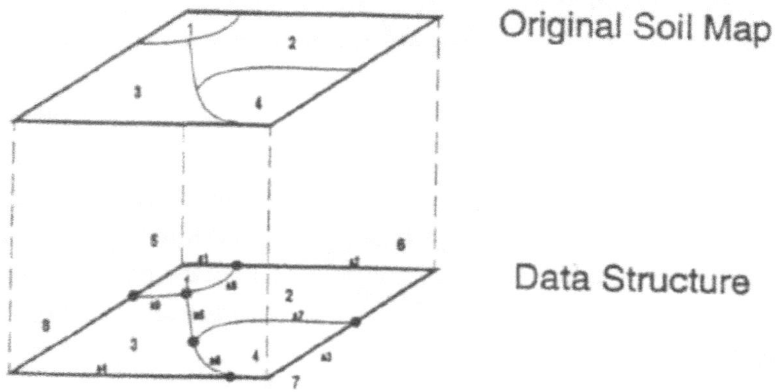

Polygon-Arc List

Polygon	Arcs
1	a1, a8, a9
2	a2, a7, a5, a8
3	a5, a6, a4, a9
4	a7, a3, a6

Arc Topology List

Arc	Left pol.	Right pol.
a1	5	1
a2	6	2
a3	7	4
a4	8	3
a5	2	3
a6	4	3
a7	2	4
a8	1	2
a9	1	3

Arc Coordinate List

Arc	X, Y pairs
a1	0,4 0,6 1,6
a2	1,6 6,6 6,3
a3	6,3 6,0 ...
a4	5,0 0,0 ...
a5	1,5 3,2
a6	3,2 4,1...
a7	3,2 3,3 ...
a8	1,5 2,5 ...
a9	0,4 1,4 1,5

FIGURE 3.12

Topologic data structure.

contain built-in routines. Customized procedures are continuously being developed for particular applications. In general, algorithms begin with a regular mesh or uniform grid of triangular elements upon which a series of point and/or linear well-defined features are drawn. Point features include peaks and wells, among others, and are maintained in the TIN as triangle nodes. Linear features include channel networks, lake shorelines, etc., and are maintained in the TIN as additional triangle edges. The algorithm then takes the location of these additional nodes and edges, and subdivides the original grid into finer triangles, causing a denser spatial resolution. The shape and spatial resolution of the original mesh may actually be the result of an interpolation process based on contour data or a digital elevation model.

TINs are well suited for representing surface spatial variability. Raster data structures can also be used for this purpose, although their applicability may not necessarily be overlapping. TINs usually provide a more compact representation for the same level of accuracy. On the other hand, raster data structures are more suited for the evaluation of parameters, such as slope and aspect, which work best when systematic sampling is used.

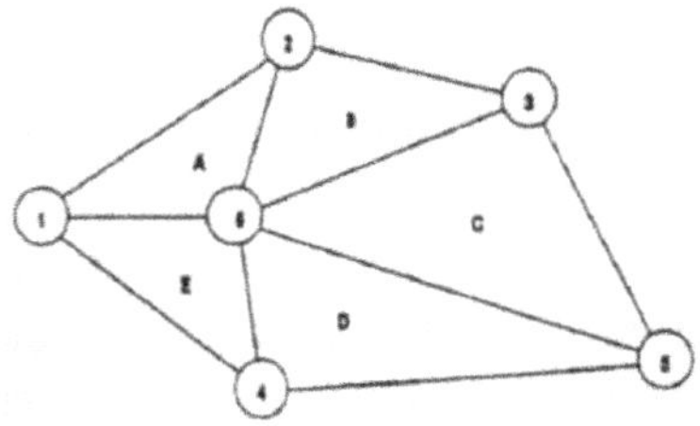

Node coordinates	
Node	Coordinates
1	X_1, Y_1
2	X_2, Y_2
3	X_3, Y_3
4	X_4, Y_4
5	X_5, Y_5

Triangle adjacency	
Triangle	Adjacent triangle
A	B, E
B	A, C
C	B, D
D	C, E
E	A, D

Triangle nodes	
Triangle	Nodes
A	1, 2, 6
B	2, 3, 6
C	3, 5, 6
D	4, 5, 6
E	1, 4, 6

Z Coordinates	
1	Z_1
2	Z_2
3	Z_3
4	Z_4
5	Z_5

FIGURE 3.13

Triangular irregular networks.

Raster Data

In the raster structure, the map area is divided into a number of cells (or pixels, in a satellite image). Each cell represents an area with uniform properties. Most raster implementations are built around square or rectangular grid cells, but other shapes, such as triangular and hexagonal, are also possible. Since grid cells are uniform, these can be referenced by their row and column numbers. Strictly speaking, raster spatial data structures are two-dimensional arrays. This means that the area that each grid cell represents can be used both to define map resolution and the number of grid cells needed to describe the spatial distribution of the attribute under study. Each pixel or grid cell may be assigned only one value, which represents the value of the attribute being mapped. Each represents a geographical region, and the value in that pixel represents some characteristic of that region.

If more than one attribute is to be handled, data layers or overlays, as shown in Fig. 3.15, need to be created. These layers can be visualized as stacked, one on top of the other.

Initial applications of GIS in water resources were mostly based on raster data. The grid is made up of regularly spaced lines, and the enclosed area of each rectangle is described in terms of its center coordinates. If the terrain is a visual image with the dots having various colors and intensities similar to a computer video screen, the use of the term raster image used for grid data as well as computer screen images is easily understood.

The size of a GIS database can be very large and several methods have been devised to reduce the storage requirement. In the regular tessellation, the size of cells is uniform which is not an efficient

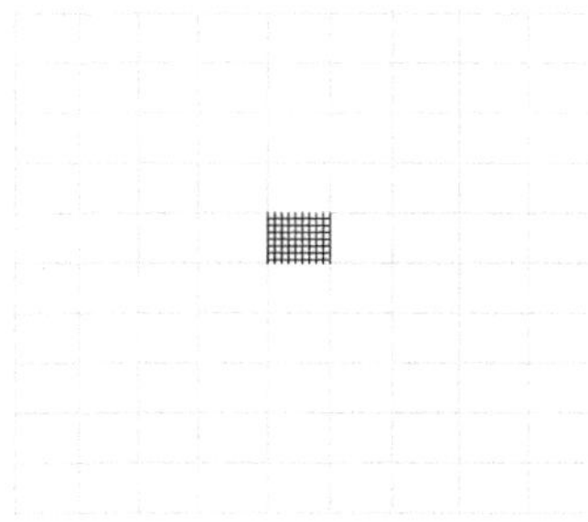

FIGURE 3.14

Raster data.

			Soil				
1	1	2	2	2	2	2	2
1	1	2	2	2	2	2	2
3	3	2	2	2	2	2	2
3	3	2	2	2	2	2	2
3	3	3	2	4	4	4	4
3	3	3	3	4	4	4	4
3	3	3	3	3	4	4	4
3	3	3	3	3	3	4	4

		Land Use					
a	a	b	b	b	b	b	b
a	a	b	b	b	b	b	b
a	a	c	c	b	b	b	b
a	a	c	c	c	b	b	b
a	a	c	c	c	c	b	b
a	a	a	c	c	c	c	c
a	a	a	c	c	c	c	c
a	a	a	a	c	c	c	c

		Topography					
19	20	20	21	20	20	20	21
20	20	21	20	20	18	20	21
20	21	20	20	18	17	20	20
21	21	20	18	15	16	20	20
21	20	20	15	13	15	18	20
20	20	1	13	11	13	15	20
18	15	10	10	10	13	15	18
15	13	10	8	10	11	13	15

FIGURE 3.15

Raster representation of the soil, land use, and topography categories shown in Fig. 3.9.

way to store data when there is less variation in the features. The important raster models of arranging data are regular tessellation, nested tessellation, and irregular tessellation. In the regular tessellation, the data are arranged in arrays. The square grid is the most commonly used arrangement, although triangular and hexagonal meshes are also used. In the square grid, raster data are arranged in arrays; each cell can be referred to by its row and column number. The referencing system of grid raster data is very convenient, and its interfacing with hardware is also simpler. In triangular tessellation, all the cells do not have the same orientation. Due to this, comparison of data is complex, but this structure is very efficient in representing topography. The size of the database can be reduced if the size of a cell depends on the variation in the surface features—the cell size can be small where the variation is more and vice versa. This concept is used in a nested tessellation model in which the cells are recursively subdivided into smaller cells of the same shape and orientation.

Raster spatial data structures are simple to conceptualize and use for overlay analysis. Further, they can be efficiently used to model high spatial variability, which makes them appealing for the manipulation of remotely sensed data and digital images. On the negative side, data files tend to be huge, although data compression techniques can reduce this burden. The size of the cell denotes the resolution of the data, bigger the size is, coarser the resolution is. Note that the size of the data files increases exponentially when higher resolution is used. The number of grid cells needed is inversely proportional to the square of the resolution and, consequently, the need to develop methods to minimize storage requirements is crucial. Most data compression methods are based on the fact that adjacent grid cells often have the same attribute value and, as a result, it is better to group them under the same category. Two such methods, quadtrees and run-length encoding, are described below.

Quadtrees: Quadtrees are perhaps the most common structure used to represent raster data. In general, the technique is based on a successive division of the map or image into quadrants until every subdivision can be assumed to be spatially homogeneous, that is, one single attribute value can be associated with it. In practical terms, grid cells can adopt variable sizes, depending on the number and spatial distribution of the original grid cells having the same attribute value. However, the analysis of this type of data is complex.

Fig. 3.16 shows a sample region and its quadtree representation. For easy visualization, a node with children is represented by a circle; a node without children and no attribute values is represented by a blank square; and a node with an attributed value is represented by a hashed square. NN, NE, SW, and

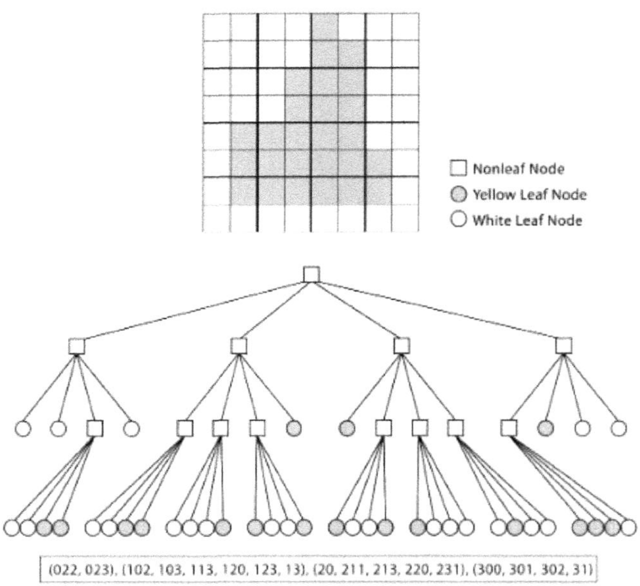

FIGURE 3.16

Sample region and its quadtree representation.

Table 3.2 Run-length encoding of the raster spatial data structure of Fig. 3.14.

Row	Value	Length	Row	Value	Length	Row	Value	Length
1	1	2	4	3	2	6	4	4
1	2	6	4	2	6	7	3	5
2	1	2	5	3	3	7	4	3
2	2	6	5	2	1	8	3	6
3	3	2	5	4	4	8	4	2
3	2	6	6	3	4			

SE are used to designate quadrants even though a numerical coding sequence, such as 1, 2, 3, is more convenient.

Run-length encoding: With this technique, grid cells having the same attribute value are organized into blocks or runs. A table can be formed to describe, for each row, the number of consecutive grid cells that are associated with the same attribute value. Table 3.2 shows, as an example, the run-length encoding that would correspond to category "soils" in Fig. 3.14.

Comparison of Raster and Vector Models

Most GISs have the ability to transform between DLGs, grid DEMs, and TIN DEMs, but contour-based methods require an order of magnitude more data storage, so that the transformation is typically from DLGs to the other forms. Both vector and raster models have some advantages as well as disadvantages. The vector data are compact, accurate, and are widely used to describe polygons. The lines and boundaries are represented in a better manner using a vector data model. The vector models are more suited for network analysis. However, the topographic analysis is difficult because the data structure is complex and each polygon has a different topological form. Further, overlaying of these maps is difficult. The structure of a raster model is simple, the analysis is also simple, because each cell has the same size and shape. There is loss of information when data are changed to a coarse resolution. Also, when raster is converted to vector, the lines and boundaries appear as a staircase. Integration of this data with satellite data is easy. Its major disadvantage is the requirement of large computer storage. However, with the development of high capacity and cheaper storage devices, this no longer appears to be a serious limitation.

It is necessary that GISs for hydrological purposes have the capability of dealing with raster as well as vector data. RS analysis involves large quantities of multitemporal raster data. The vector data are also analyzed in hydrological modeling, for example, delineation of drainage basin boundaries, automatic generation of drainage maps in a catchment, etc. Most current GIS packages have modules to convert raster data into vector data or vice versa.

Time in GIS

Most geographic databases are built assuming static geographic feature conditions. However, geographic features evolve through time. Many of these change as a result of natural processes, but they also change as a result of human intervention. Land cover, land use, land slope, and stream network geometry, including alignment, slope, and cross section, are just some examples of landscape characteristics that may change over time and that may have a direct impact on watershed hydrologic modeling.

Unfortunately, current GISs are normally atemporal in that they describe only one data state. Because most GISs in existence are based on the categorical approach, they are not capable of providing explicit linkages for objects through time. However, a reasonable goal for GIS would be to be able to respond to queries, such as what, where, when, how fast, and how often changes have taken place. Many applications in engineering, in general, and in water resources, in particular, would surely benefit from having a system capable of answering such types of queries.

3.2.3 Geographic Coordinate Systems

The position of any point on the earth's surface is commonly expressed in terms of its latitude and longitude. The latitude of a point is the angle measured at the center of the earth between the plane of the equator and the line connecting the point with the center of the earth. If the point lies in the northern hemisphere, the letter N is suffixed to the angle and the letter S is suffixed if the point lies in the southern hemisphere. The latitudes vary between 0° (equator) and 90° (poles). The meridian passing through Greenwich (UK) is the reference meridian for longitude and is termed the prime meridian. The horizontal angle between the prime meridian and the line connecting a point with the earth's center is the longitude of that point. By convention, E is suffixed for the points lying to the east of the prime meridian and W for those lying to the west. The longitudes vary between 0 and 180°. Both latitudes and longitudes are expressed in degrees, minutes, and seconds. For example, the longitude of a point may be 79° 30′ 45″ E and latitude 29° 19′ 55″ N.

Map Projection

Although the shape of the earth is considered to be spherical, actually the equatorial axis of the earth is longer than its polar axis or the earth is somewhat flattened near the poles. This shape of the earth is close to an ellipsoid or spheroid which is considered to be a better representation of the earth's shape. A spherical representation of the earth, for example, a globe, is not convenient to use and work; it cannot be included as part of a book or a report. Therefore, it is necessary to use a map projection to project the 3-dimensional surface of the earth on a 2-dimensional flat surface. Conceptually, the earth's surface is typically projected on cylinders and cones. These surfaces are "opened" and stretched to conform to the shape of a plain paper.

A number of transformations are employed in map projection which can be mathematically represented as:

$$x = f(\varphi, \theta) \tag{3.10}$$

$$y = g(\varphi, \theta) \tag{3.11}$$

where φ and θ are the latitude and longitude of the place.

There are three basic classes of map projections: the cylindrical projection, the conical projection, and the azimuthal projection. In cylindrical projection, it is visualized that a cylindrical-shaped paper is wrapped around the globe. The globe is opened up and stretched so as to conform to the cylindrical shape. If this cylinder is opened, the map of the world on a rectangular sheet will be obtained. In the case of conical projections, the globe is visualized to be wrapped by a cone and projection is taken. When the cone is opened, the fan-shaped map of the world will be obtained. In azimuthal projection, the spherical earth is projected on a plane tangent to the globe.

A map projection is named according to its class and property as well as the name of the originator and the nature of the modification of earth's surface. Some well-known projections are Mercator projection, normal cylindrical equal area (Lambert projection), polyconic projection, etc. A few commonly used map projections are described here.

Polyconic

This projection system was developed by Ferdinand Hassler in 1820. It is suitable for north-south oriented maps. It uses an infinite number of cones as developing surfaces. The parameters needed for the projection are central meridian, false easting, and northing. False northing is assigned a value of zero. The false easting is assigned a value of 500,000 m. Central meridian is generally a meridian passing through the center of the study area. The parallels are circular arcs and the meridians are radial lines. This projection is used by the Survey of India for topographic maps at the scales of 1:250,000 and 1:50,000.

Universal Transverse Mercator

The Universal Transverse Mercator (UTM) projection, developed by the US Army, is widely used in topographic maps. This projection is recommended for areas lying between 84°N and 80°S. In UTM, the earth surface is divided in 60 zones, each 6° wide in the longitudinal direction resulting in rect-angular graticule mesh (Fig. 3.17). These are numbered sequentially from west to east. The western edge of the first zone touches the 180°W meridian and the eastern edge of the 60th zone touches the 180°E meridian. In the latitude direction, each zone covers an area of 8° (except the northern most zone that covers 12°). The bottom-most zone (80°S to 72°S) is assigned letter C and the topmost letter X. The origin of each zone is located at a point at the equator where it is intersected by the central meridian of the zone. The eastings of the origin of each zone is 500,000 m. Regarding northings, for the northern hemisphere it is 0 at the equator and for southern hemisphere, it is 1,000,000 m at the equator.

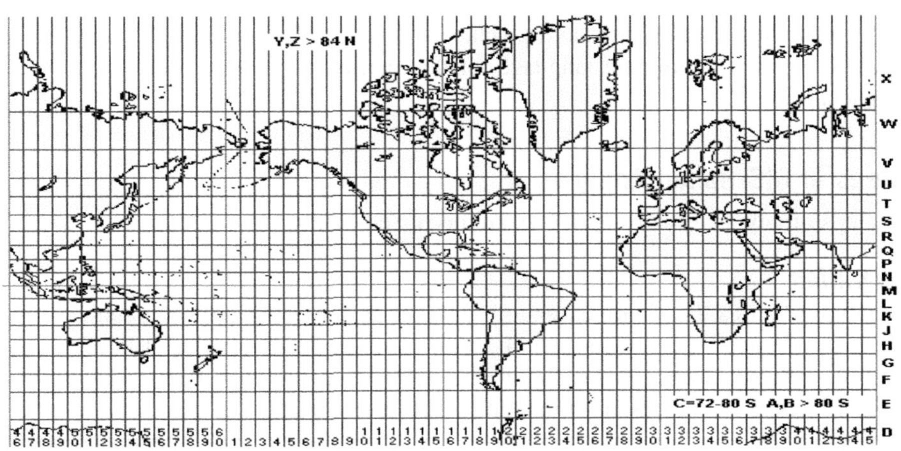

FIGURE 3.17

The UTM map projection.

Albers Equal Area Conic

This projection is used for small-scale maps. It requires two standard parallels.

Note that all maps have some kind of distortion and while choosing a map projection, the aim is to select the projection in which the distortions are within acceptable limits. The following general guidelines are followed while choosing a projection:

(a) For the countries lying in tropics, a cylindrical projection is to be used.
(b) If the country lies in temperate zone (between latitudes 23.5° to 66°), a conical projection should be used.
(c) If the region is between 66° latitude and pole, an azimuthal projection should be used.

The above is a very simplified treatment of a complex topic. For further details, a good book on geography and coordinate systems, for example, Maling (1992), may be referred to.

3.2.4 GIS-User Interface

A user interface is the way a user interacts with a GIS software to perform certain tasks. An interface is a link between computer and the user. The interface is an important part of a GIS, because a GIS is used by people with diverse backgrounds and many of them have little computer experience. Therefore, the interface should be such that it is easy and convenient for a typical user to use the system and exploit all its strengths. There are many standard interfaces, viz., menu, icon, command line, script, etc.

In a menu-based interface, the user works with cascading menu. A branch structure exists wherein from the main menu, submenus open. The actual function or operation resides at the end of the tree. In a menu, the operations are grouped based on their closeness in operations. For example, the functions dealing with a file operation (opening, creating, copying, etc.) many be grouped in one menu group. Icons are interface methods in which graphical elements are used to pictorially depict the operations that they will launch. For example,  indicates file opening under the Windows operating system. Under the command line interface, the user types in a command which is executed on pressing the return key. The optional parameters can be supplied in the command line itself. In the scripts or model mode, many commands can be written and executed as a batch. Most of the packages provide a macro language which is the programming language for that particular package and can be executed from within the package.

3.2.5 Steps of GIS-Based Analysis

The major steps of any analysis using a GIS are discussed below.

Data Preparation

Data preparation requires the identification of data and its source, scale, geographic extent of data, locating an area of interest, etc. Data identification involves deciding what data are needed and it depends on the problem. Source and scale of data will be dependent on many factors, for example, availability, storage space, processing time, computer, areal extent, etc. The geographical extent is known a priori and the boundaries are drawn on a map. It is helpful to gather requisite information for the focus area prior to the analysis in a GIS.

Data required for a GIS can be collected from ground surveys, digitizing existing maps, digitally recorded aerial photography, satellite imaging data, or combinations of thereof. A problem of the scale of accuracy arises when these data are used in combination.

In a typical GIS application, the availability of spatial database is the first problem, particularly in developing countries. In many cases, one may have to start from scratch. Note that the acquisition and compilation of the information is hardly a trivial exercise. Generally, the requisite data are available only as hardcopy maps. Even after important developments in digitizing hardware and software, the process is labor-intensive and requires skilled manpower, time, and patience.

Data Input

Data are input to GIS in many ways, namely, using digitizers, scanners, computer files, etc. Digitizers are computer devices that use a tablet and cursor to convert a paper map to a digital one. A tablet contains closely spaced mesh of wires below a flat surface over which the map is placed. The cursor is moved over the map and by pressing buttons on the cursor, coordinates of current location are sent to the computer. This is the most common method of inputting a map to a GIS but requires a lot of time and patience. While digitizing a map, many errors can occur, common among them are:

1. Dangle or dead end in segment: A polygon in a GIS should be a closed area. Error occurs when it is not so and this may be due to the presence of a dangle or dead end in a segment.
2. Intersection without node: At all intersections of lines, nodes should be formed, otherwise error will be generated.
3. Self-overlap: This error occurs by the erroneous movement of cursor causing a line to intersect itself or formation of loops.
4. Double digitization: If a line is digitized more than once by mistake, this causes multiple intersections without node.

The digitization modules have routines for minimizing errors, their checking and correction. The technique of auto snapping is used to avoid dangles. When two lines start or end closer than a specified distance, these are automatically joined and a node is created. This function can also be turned off. To delete the intermediate redundant points from a line segment, the *tunneling* function is used. In this, a tunnel of specified width is drawn between the first and the last points of consecutive sets of three points. The middle point is retained only if it lies outside the tunnel.

Scanners are similar to digitizers, except that the information is automatically input to computer. A scanner can have either a moving sensor or a fixed sensor. In the first case, a map is placed on a flat bed and sensor moves over the map. In the latter case, the map is moved over the sensor. The light reflected by the map is picked up by the sensor to form an image of the map which can be interpreted by software routines to extract the desired information. After a map is scanned, it needs to be edited to remove errors before it can be used in a GIS.

Output

The final step of GIS analysis is to present the results in the form of maps for the purpose of report or a presentation. In the output stage, various maps and information can be directed to a computer monitor or a printer. The output creation involves design of layout, contents, legend, scale, etc. The map can also include scale, graticules or grids, direction arrow, nameplate (author, date, and other information), texts, symbols, etc. Note that the design of maps, scale, contents, etc., is dependent on the objective of the problem. The required information can also be displayed on maps by various cartographic methods, such as pie charts, histograms, proportional circles, etc. Blending is an output method in which two raster maps can be superimposed to show both maps partially. This method can be used to overlay hazard maps on watershed maps.

The choice of colors in a map can be natural colors, for example, water can be shown in blue color, or an artificial color scheme can be used. Options, such as fonts, colors, and annotation procedure, may be limited by the hardware and software features. The gray shades and symbols can be selected, depending upon the variation in the output data.

3.2.6 Analysis of Geographic Data Using a GIS

In data analysis, one or more geographic data layers are manipulated to extract useful information. Many manipulations and measurements can be done on the geographic data. The GIS analytical capabilities can be broadly divided into three general groups: operations that may be performed on spatial or nonspatial data, operations performed on individual spatial data layers, and operations performed on multiple spatial data layers. Despite the differences in data structure between operations, most of these operations can be performed with either raster data or vector structure. The important operations are described here.

Data editing and management: The properties of geographic features change with time due to natural and human activities—new features might be added, the existing modified or deleted. The scope and purpose of a study may also change with time. Thus, there is a frequent need to edit the existing GIS database. Typical edit operations are: add, delete, extend, and modify. The names of objects may have to be changed. Since a user has to frequently use the edit module, it should be easy to use and powerful.

A file management function carries tasks, such as rename, delete, copy, etc. These can also be done through the operating system but are relatively safe in the GIS environment. In general, it is better to do these tasks from the GIS itself. If a certain item is to be removed, the GIS will also remove all the associated files and clean up memory. The delete operation is to be used with care, otherwise useful data can be lost forever.

Classification: It is a method of information retrieval from the data. This is somewhat similar to RS analysis discussed in Section 3.1.8. In classification, the value groups are created and the group numbers are assigned to each value in the output map. Examples of the classification operation are creating a contour map from a DEM, an isohyetal map from rainfall raster map, etc. Reclassification is a GIS operation in which another map is created by assigning new values to a map. For example, from a soil class map, soil-pH map can be prepared; from a second order land use map, first order map can be prepared, etc. The operation is also called "recoding."

Data selection and query: A GIS can be used to select data that meet certain criteria. Selection of database records may be done: (i) interactively; (ii) by specifying numerical thresholds; or (iii) by using logical operators. Interactively, data can be selected from tables or on screen using a pointing device. Quantitative data can be selected by either specifying numeral thresholds or ranges. Boolean logic uses the following four operators (Fig. 3.18) to perform on two or more data sets: AND (intersection), OR (union), NOT (negative), and XOR (exclusionary or).

Another important GIS operation is query. Using query, the location of a certain geographic object can be retrieved, or a certain geographic object can be displayed, etc. For example, in a river network, a

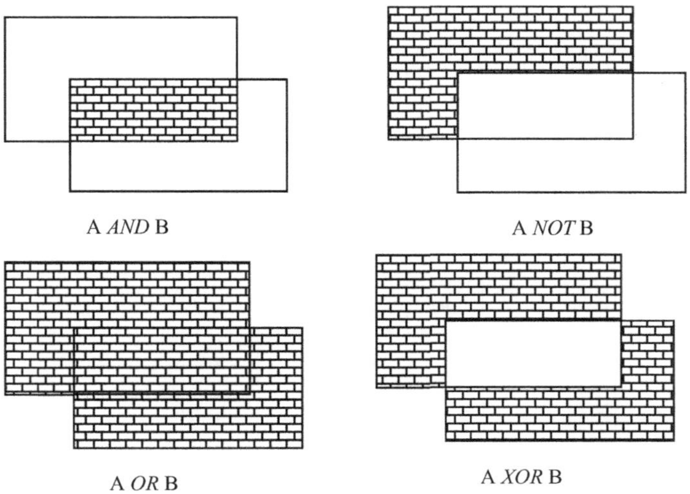

FIGURE 3.18

Venn diagram showing Boolean operations.

river with a particular name or all rivers with particular order can be highlighted. The flooded areas or particular soil or land use areas can be displayed. Two or more maps can also be overlaid, for example, flooded cropland may be displayed. The query results can also be saved as a map or different geographic data. Thus, information retrieval is also combined with query operation.

Intersection operation is used for overlay of two different types of geographic data, for example, point and area or line and area. Examples of this operation are extracting ground water wells for a basin, extracting village names for the flood affected area, extracting drainage lines in a basin, etc. The output of a union operation shows the details of both input maps.

Measurements: GISs provide a range of capabilities for measurements on spatial data depending on the type of geographic data. For quantitative attributes, the statistic, like mean pixel attribute and their standard deviations, can be computed. Area calculation is straight for both vector and raster databases. In a raster database, multiplying the number of target pixels by the area of one pixel gives the total area. Distance (d) between two points whose coordinates are (x_1, y_1) and (x_2, y_2) is:

$$d=\sqrt{(x_1-x_2)^2+(y_1-y_2)^2} \tag{3.12}$$

Other similar operations are length of string and perimeter of an area. It is also possible to compute volume between two surfaces.

Spatial aggregation and generalization: Spatial aggregation is used to reduce excess details in a raster data layer. The user defines the numbers of cells to be aggregated in each $n \times n$ window, or *kernel*. This kernel should have the same orientation as the input layer and the size should be integer multiple of cells. A GIS systematically examines all $n \times n$ nonoverlapping blocks of the input data layer, aggregates them, and assigns a summary value to it (Fig. 3.19). This summary value could be the majority value (the most frequent value), average, maximum, etc., for the kernel. In the case of tie, a mixed class may be created. Spatial aggregation reduces database size and may also be used to visualize patterns at a coarser grid than the original file.

F	F	F	A	A	A	A	A
F	F	F	F	F	F	A	A
A	F	F	F	F	F	F	A
A	A	A	F	A	F	F	F
A	A	A	A	A	A	F	F
A	A	A	A	A	A	A	A
F	F	F	A	A	F	F	A
F	A	A	A	F	F	F	A

(a)

F	F	M	A
A	F	M	F
A	A	A	M
F	A	F	M

(b)

F	F	A	A
A	F	F	F
A	A	A	A
F	A	F	A

(c)

A – Agriculture, F – Forest, M – Mixed.

FIGURE 3.19

Spatial aggregation of data of two categories in (a). To resolve tie, majority rule is followed in (b) and a mixed class is created in (c).

A somewhat similar technique is data resampling. Here, the kernel need not be an integer number of cells and may have a different orientation and size than the input cells. Resampling is performed to convert a raster data layer to another coordinate system or projection, or to match raster data layers having different cell sizes prior to overlaying them. Methods for determining which data value should be assigned to the new cell include nearest neighbor, bilinear interpolation, and cubic convolution. The nearest neighbor method assigns to new cell the value of its closest neighbor cell in the original data layer. The bilinear interpolator computes the new cell value from the values of the four cells surrounding the new cell.

A useful GIS operation is clumping which is used to aggregate patches of contiguous cells with the same attribute value and constitutes a raster database. Various groups of cells that touch at a corner or edge and have the same value are identified and merged or assigned same value.

Buffer zones: An often-used capability of a GIS is to generate a buffer zone around spatial objects, such as a flood plain zone around a river, right-of-way for road/rail, etc. The inputs are the spatial entity for which the buffer is desired and the distance or width of the buffer zone. In raster systems, buffering is done by a spreading process which computes the distance between the feature of interest and every other cell within a defined limit, resulting in halos of cells with incrementally larger distances from the central feature. In some raster GISs, additional criteria can be applied to the spread command to control the direction (uphill, downhill), resistance, and barriers to spreading (Johnston 1998). These criteria may be helpful in studies, such as routing of overland flow.

Geometric transformation: A geometric transformation manipulates a data layer to correctly overlay it on another layer of the same area. The need for this arises because maps can have different coordinate systems and/or projections, and these must be reconciled before they can be overlaid. Note that this operation involves at least two data layers. There can be mismatches even when both data layers have the same coordinate systems and projections. Such mismatches can be adjusted by techniques, such as registration by absolute position, registration by relative position, and edge matching. Most GISs can transform data among a number of projections.

Registration by absolute position: Each data layer is transformed independently using ground coordinates, such as those obtained from topographic maps. The reference points should be carefully chosen features, such as road intersections, that can be clearly identified on the digital maps. Since each layer is independently registered, errors are not propagated from one data layer to another. Even then, small errors in various data layers may cause alignment problems.

Registration by relative position: An accurately georeferenced reference layer is chosen and the other layers are transformed based upon this layer. A prerequisite here is that the reference layer should be very accurate with clear reference points which could be easily and accurately located on both data layers. A disadvantage is that the errors in the reference layer are propagated to transformed layers.

Edge matching: The idea here is that lines and polygons that cross map boundaries should match. This mismatch is quite common in the edges of adjacent maps and data layers. Most GISs have routines for (semi) automatic edge matching.

Overlay operations: This is a very useful GIS function that allows to digitally place multiple data layers on top of each other. It is also possible to overlay a vector layer over a raster data layer. Thus, one can digitize the catchment boundary from a topographic map and overlay it on the top of a satellite imagery. Similarly, on a map of catchment boundary, different layers, such as soil, land cover, slope, and channel network, can be laid. This is useful in studies, such as catchment modeling, soil erosion, etc. The maps can be manipulated using mathematical and logical functions.

Aspect: Aspect is the direction which a surface (typically hillslope) faces. It is usually expressed in degrees (0–360°) or as compass directions such as N, S, NE, NW, etc. In a TIN database, aspect is the direction of maximum slope of the triangle plane. On a hillslope, aspect controls input of solar energy which controls snow melt, ET, etc.

Line-of-sight maps: A GIS can be used to generate maps identifying the intervisibility of landscape features. A *viewshed* shows the land area that can be seen from a given point. A *visual impact* map shows all the area from which a tall object can be seen. Input data layers include elevation, height, and location of the observation point (viewshed map) or target (visual impact map). The height and location of features projecting above the land surface that could block the view (e.g., buildings, trees) are also input (Fig. 3.20). Such analysis is useful in planning communication towers.

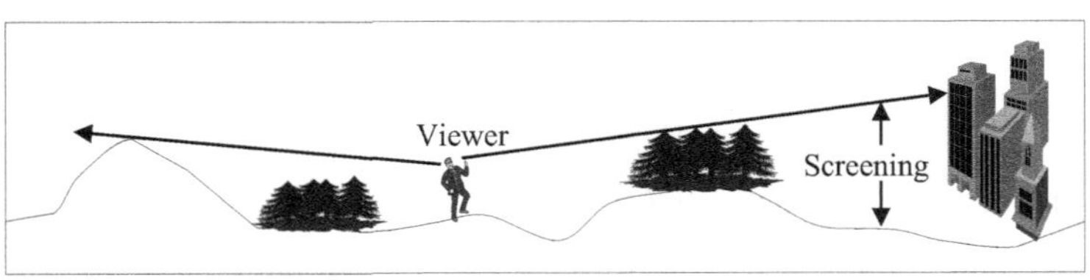

FIGURE 3.20

Pictorial presentation of viewshed concept.

3.2.7 Digital Terrain Model

The primary source of topographic information prior to the 1980s was contour maps. In these maps, elevations are represented by points and contours. Advances in digital mapping have offered essential tools to closely represent the three-dimensional nature of natural landscapes. A capability of a GIS that is most important to water resources applications is the description of the topography of a region. One such tool is the digital terrain model (DTM). DTMs are "ordered arrays of numbers that represent the spatial distribution of terrain attributes" (Vieux, 2001). These attributes include elevation, slope, slope steepness, and soil depth. A subset of DTMs which describe elevations above some arbitrary datum is the digital elevation model (DEM). Topographic variables, such as basin geometry, stream networks, slope, aspect, flow direction, can be extracted from DEMs. Three schemes for structuring elevation data for DEMs are: triangulated irregular networks (TINs), grid networks, and vector or contour-based networks (Moore and Grayson, 1991). Some spatial information is not directly described by elevation and can be described as topologic data. Topologic data define how the various pieces of the region are connected. Topology is the spatial distribution of terrain attributes.

DEMs are useful to determine physical variables, such as slope and aspects that are useful inputs in various hydrological computations, for example, overland and channel flow velocity, travel time, soil erosion, slope stability, etc. While topographic data fit within the general classification of topologic data, there are significant hydrologic attributes not related to land surface elevation. The more obvious of these are catchment areas, flow lengths, land slope, surface roughness, soil types, and land cover. These attributes help describe the ability of a region to store and transmit water.

The input data for DTM usually comes from topographic maps. Other sources could be surveys, photogrammetry, and satellite images. Special attention should be given to surface discontinuities and pits, peaks, ridges, and streams. After the data are entered into a GIS, a model is built to represent the surface behavior. The two models that are common are regular grids and TINs. In regular grids, elevation values are computed at equally spaced grid points (Fig. 3.20). In TIN, a surface is represented as a network of adjacent triangles whose vertices are the sample points.

Elevation data will rarely be available at regular grid points and interpolation will have to be carried out. DEMs are created from point and contour data using an interpolation function. The commonly used interpolation techniques include linear/bilinear interpolation, inverse weighted distance method, and kriging. The weighted average method consists of taking the weighted average of point values within a specified radius from the interpolation point. The weighted surface method fits an n-degree surface using the points within a specified radius from the point of interpolation. Kriging is also a popular interpolation technique. These have been described in Chapter 2.

The information useful in hydrologic analysis that can be derived from DTM include slope, aspect, catchment boundary, and channel network. The slope can be calculated as:

$$\tan \mathrm{S} = \left[(dz/\mathrm{dx})^2 + (dz/\mathrm{dy})^2\right]^{0.5} \tag{3.13}$$

and the aspect is defined as (Burrough, 1986):

$$\tan \mathrm{A} = -(dz/\mathrm{dy})/(dz/\mathrm{dx}) \quad -\pi < \mathrm{A} < \pi \tag{3.14}$$

where x and y are coordinates, and z is altitude.

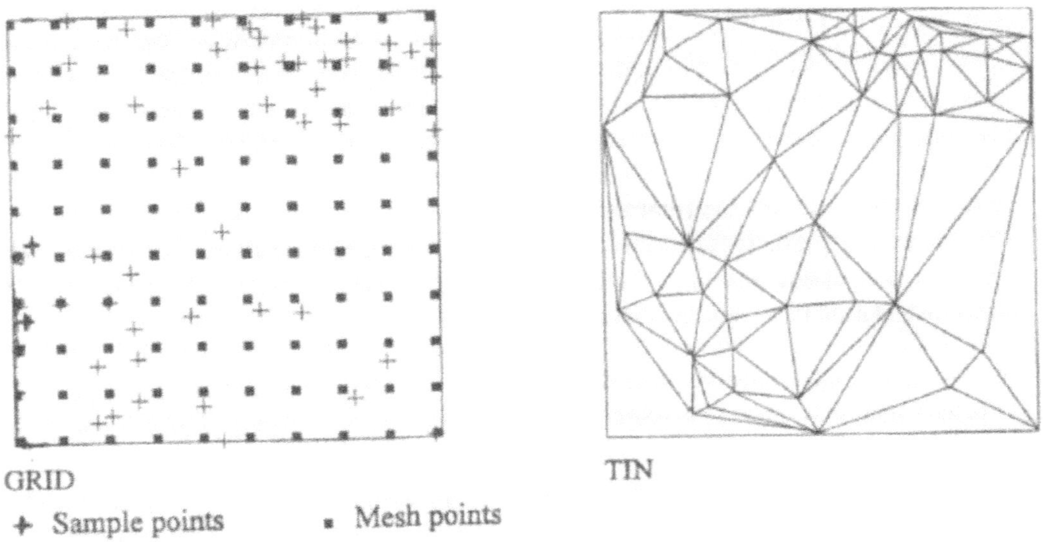

FIGURE 3.21

Regular grid and TIN models for DTM.

An inherent problem in hydrologic modeling with grid DEM data is the production of nonphysical depressions due to the noise in the elevation data, coarse contour interval, or sparse elevation data. These affect interpolation schemes resulting in an unwanted termination of drainage paths in pits. The problem is particularly acute for relatively flat areas. O'Callaghan and Mark (1984) and Jenson and Domingue (1988) have described techniques to locate and remove depressions in gridded DEM data (Fig. 3.21). Nelson et al. (2009) have given an overview of DEM data sources, their processing requirements, their advantages, disadvantages, and their applications. DEM processing may, however, be complicated by the existence of naturally pitted topography, sometimes called pothole regions. Sometimes the user has to manually alter the elevations to remove pits. The pits are filled by raising their levels iterative to nearest higher elevation points.

The algorithms to demarcate catchment boundary and channel network mainly make use of slope and aspect maps and the outlet point is given by the user. Sole and Valanzano (1996) have reviewed and described several such algorithms. BASINS (Endreny, 2002) is a toolkit for tabular and spatial data management with several modeling tools. It can be accessed at http://www.epa.gov/ost/BASINS.

DEM Visualization

Several techniques are used to display a DEM, for example, contouring, hill shading, perspective view, fly through, perspective cartography, etc. Contouring is the conventional technique of elevation representation. In hill shading, gray shades are assigned to a map based on its orientation with respect to the source of illumination. The location of the source of illumination can be specified. A perspective view is created by joining adjacent relief profiles. This is also called "wire net" technique. Profiles may be used in single directions or in two perpendicular directions. The technique is complex and uses computations of hidden points. The location of an observer can be varied to create many views.

Fly-through is a special "perspective view" technique. In this technique, many perspective views are used in an animation to create an impression as if one is flying over the terrain. The direction and the speed of flight, height, etc., can be defined interactively. In perspective cartography and texture mapping, an image or cartographic objects, for example, roads, drainage, etc., can be overlaid on the perspective view.

3.2.8 Some Popular GIS Software

Here, we briefly describe two popular GIS software that are popular these days.

QGIS (earlier known as Quantum GIS) is an open-source desktop GIS software to view, edit, and analyze geospatial data. It can process both raster and vector data; vector data are stored as either point, line, or polygon features. Multiple formats of raster images are supported, and the software can georeference images. QGIS supports many data formats, such as shapefiles, coverages, and MapInfo and integrates with many other open-source GIS software. QGIS is written in C++ language and is available for many operating systems, including Microsoft Windows and Linux/Unix. The first version of QGIS was released in 2009. QGIS has modest hardware requirements compared to commercial GIS packages.

ArcGIS is a family of GIS software developed by Environmental Systems Research Institute (ESRI). It is one of the most widely used GIS software that was first released in 1999. At that time, it was known as ARC/INFO and was a command line-based GIS system. Esri has developed the shapefile format to store and share geospatial vector. This format can represent vector data which can be points, lines, and polygons. In hydrological applications, these features can represent rain gauges, groundwater wells, canals and rivers, and catchments and reservoirs. Each of these may have associated attributes, such as the name of rain gauge or river. ArcGIS contains a large number of routines, including terrain and watershed processing, that make it easy to analyze spatial data.

3.2.9 GIS Applications in Water Resources

There are innumerable applications of GIS in water resources and the list is fast growing. This technique has been widely applied in conjunction with RS. GISs are particularly helpful in distributed modeling where they are used in managing input data and display of output. For example, long-term soil erosion from agricultural lands is estimated using the universal soil loss equation (USLE). The erosion values are given for bare standard plots under different slopes, and these are modified by certain factors. Geographic data of different factors and soils are overlaid using a GIS to obtain the mean-annual soil erosion rates. The main inputs influencing soil erosion are geology, physiography, soils, drainage density, and land use. Overlay of these maps can be used to delineate the areas that are vulnerable to soil erosion.

The SCS curve number technique is frequently used to estimate direct runoff from a watershed. Using a GIS, land use and soil maps are overlaid to get a composite map and curve numbers are assigned based on it. An average curve number can be determined or a distributed model can be applied. The SCS curve number method with GIS was used by Muzik (1996) to derive the distributed unit hydrograph. Al-Ghobari et al. (2020) attempted to estimate surface runoff using RS/GIS-based SCS curve number method.

Since surface water flows in the downslope direction, the overland and channel flow direction can be determined by using elevation data. Stream ordering is a method of assigning a numerical order to chains in a stream network based upon the number and arrangement of tributaries. Stream order has been demonstrated to be related to numerous characteristics and processes of river ecosystems. In stream ordering, headwater streams (those that receive water only from overland flow) are assigned to order number one, and chains in a downstream direction are incremented based on two different numbering methods. Terrain slope and flow direction is a local property and GIS permits a realistic representation of these in the model. Grid-based DEMs generally provide the most efficient structures for estimating terrain attributes. After a DEM map is added in ArcGIS, Spatial Analyst Tool under ArcToolbox contains routines to compute stream order parameters (such as bifurcation ratio).

Fig. 3.22 shows watershed terrain analysis using a grid model. The map of flow direction for each grid is prepared using a GIS and the equivalent network showing flow accumulation is also shown.

Catchments are defined at a point on a stream or are surrounding contributing areas for lakes. Before the advent of GIS, catchment boundaries were manually delineated. The same process can now be done automatically and quite accurately in a GIS. Water flow data layers can be used to determine the path from a known start location in the watershed to the outlet. This capability is useful in tracking

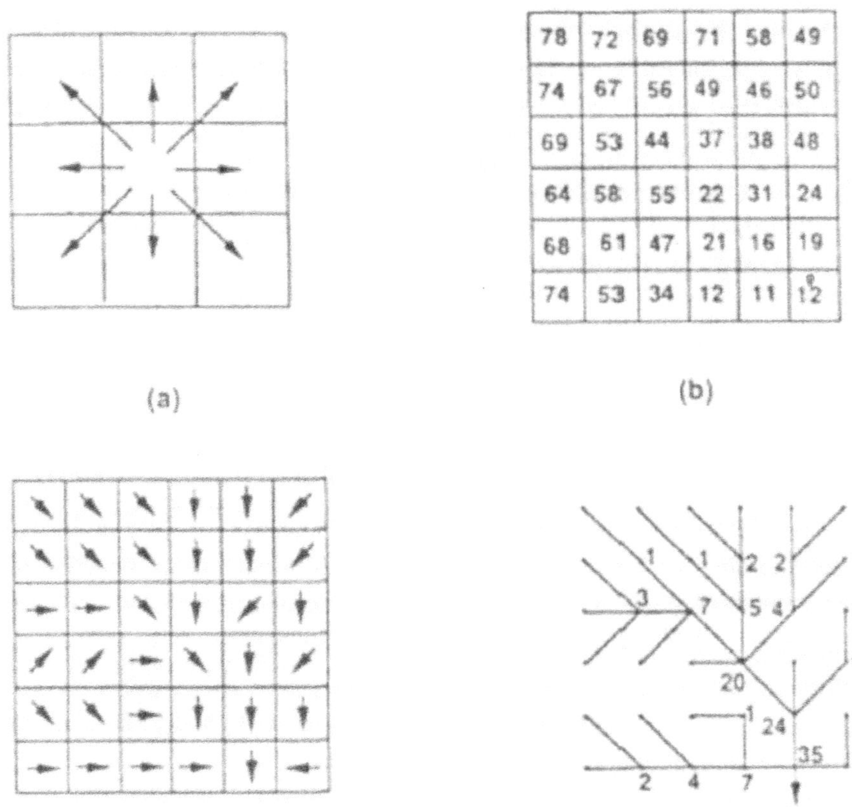

FIGURE 3.22

Watershed terrain analysis using grids; (a) the eight-direction pour-point model; (b) a grid of terrain elevations; (c) the corresponding grid of flow directions, (d) the equivalent network showing flow accumulation.

the path of a pollutant from a point or nonpoint source through the drainage network. To distinguish diffuse from channelized flow, thresholds can be applied to the flow accumulation data layer such that only cells which receive runoff from a sufficiently large drainage area are identified as channels. A network is a series of interconnected lines. Networks may be dendritic or rectilinear, as exemplified by stream and road networks. Stream networks branch hierarchically, have unidirectional flow, and are composed of network chains. Several algorithms for delineating watersheds have been covered by Vieux (2001) who has discussed distributed watershed modeling using GIS.

Hydrologic Modeling

A GIS is a powerful tool to organize and analyze spatial data. Hydrologists use a large volume of spatial and temporal data, and GIS provides a powerful framework to aid hydrological modeling.

Wider use of GIS in hydrologic modeling began around the 1980s. Many more applications of GIS and RS to water resources were described in books and papers such as DeVantier and Feldman (1993), Engman (1993), Shih (1996), Singh and Fiorentino (1996), and Islam and Sado (2002). The "Water-Ware" DSS for river basin planning developed by Jamieson and Fedra (1996) used a GIS to manage spatial data. Hydrologic literature contains a very large number of books and papers describing hydrologic models. Currently popular models/software such as Soil and Water Assessment Tool (SWAT); the suite of software developed by the Hydrologic Engineering Center, USA, including HEC-HMS, HEC-RAS, HEC-ResSIM; the Ground Water flow modeling system MODFLOW developed by the U.S. Geological Survey, the MIKE suite of software, all make use of GIS capabilities. In fact, these days, it is difficult of think of a versatile hydrologic software that does not make use of GIS.

Almost all major recent DSSs use GIS to manage spatial data.

Google Earth Engine: A recent development is GEE platform which combines a huge repository of satellite images, along with parallel computational facilities. GEE also provides an opportunity to display the results and download the results. Gorelick et al. (2017) describe the various features and facilities available on GEE. The list of technical papers describing applications of GEE in water sector as well as in many other sectors is rapidly growing.

3.2.10 Conclusions

The amount of digital data required to accurately describe the topography of even small geographic regions can be huge. This makes GIS a memory intensive and computationally intensive system. A good computer system for GIS requires large CPU memory, high-capacity disk drives, 21″ monitor, and peripherals like digitizer, scanner, color printer, etc. Even with sharp reduction in prices of hardware in recent times, the total cost of hardware and software can be quite large. Of course, all the set-up would be futile if adequate skilled manpower is not employed.

From the above discussion, it emerges that GISs are being tried in a variety of applications. The application of GIS becomes very convenient when the database exists or when it can be shared for several related purposes. Ideally, a GIS database should be created at the time of planning a project and it can be shared later on in all the studies related to the project.

In future, the use of GIS in water resources is likely to further grow rapidly. The main limitation will not be the availability of computing power but the innovative ideas and trained man-power. Further technical advancements are likely to result in improved tools for data collection, database creation, and numerical modeling. Education and awareness will also play a key role in the success of the methodology.

3.3 ARTIFICIAL NEURAL NETWORKS

Many attempts have been made to develop a technique that does not require algorithm or rule development and thus reduce the quality and complexity of the software. One such technique is known as "neurocomputing." The origin of this technique can be traced to the functioning of the human brain which contains billions of neurons and their interconnections. Due to the structure in which the neurons are arranged and operate, humans are able to quickly recognize patterns, process data, and learn from past experiences. An interesting idea that emerged in the 1940s was the possibility of emulating the processing mechanism of the brain. Although the biological unit still out-performs any man-made tool in terms of recognition, analysis, prediction, and especially learning, the alluring success from the brain-simulation models has provided enough motivation for extensive research. ANNs refer to computing systems whose central theme is borrowed from the analogy of biological neural networks. ANNs represent highly simplified mathematical models of our understanding of the biological neural networks. They include the ability to learn and generalize from examples, to produce meaningful solutions to problems even when input data contains error or are incomplete, and to adapt solutions over time to compensate for changing circumstances and to process information rapidly. ANNs are also referred to as "neural nets," "artificial neural systems," and "parallel distributed processing systems."

The ANN approach is faster compared to its conventional compatriots, robust in noisy environments, flexible in the range of problems it can solve, and highly adaptive to the newer environments. An ANN has the ability to learn from examples, to recognize a pattern in the data, to adapt solutions over time, and process information rapidly. Due to these established advantages, by now ANNs have numerous real-world applications, such as image processing, speech processing, performing general mapping from input pattern (space) to output pattern (space), grouping similar patterns, solving constrained optimization problems, robotics, and stock market predictions. Mathematically, an ANN is often viewed as a universal approximator (ASCE, 2000a). Applications of ANNs to water resources problems are rapidly gaining popularity due to their immense power and potential in mapping of nonlinear system data.

Some of the reasons why ANNs have become an attractive modeling tool are:

1. They are able to "learn" the relation between input and output variables even when the underlying physical laws are unknown or not precisely known.
2. They are a useful tool in modeling complex processes.
3. The mathematics is simple and one need not solve complex partial differential equations with attendant problems like instability of algorithm.
4. They work well even when the training sets are incomplete or contain noise.
5. They are able to adapt to solutions over time.
6. Once a network is trained, it is easy to use.

A water resources system may be nonlinear and multivariate, and the variables involved may have complex interrelationships. Often, the problems are ill-defined and solutions are difficult to come by using physically based methods. Such problems can be efficiently solved using ANNs. In many cases, the existing knowledge is far from perfect and, therefore, empirical models are used. ANNs are handy in such cases too. Because of their built-in mechanism of growing "wiser" with "experience," ANNs are capable of adapting their complexity and their accuracy increases as more and more input data are made available to them. They are capable of extracting the relation between inputs and outputs of a process without any knowledge of the underlying principles. The processes involving several parameters are easily amenable to neuro-computing. Because of the generalizing capabilities of the activation function, one need not make any assumption about the relationship (linear, nonlinear, etc.) between input and output as in the case of regression analysis. All these properties make ANNs an attractive tool for water resources practitioners.

3.3.1 Definition

The origin of ANNs lies in the quest to mimic the functioning of human brain, the definition and jargon can hardly defy this connection. There are many definitions of an ANN. The definition proposed by Govindraju and Rao (2000) is one of them: "A neural network is a massively parallel distributed processor that has a natural propensity for storing experimental knowledge and making it available for use. It resembles the brain in two respects:

1. Knowledge is acquired by the network through a learning process.
2. Interneuron connection strengths, known as synaptic weights, are used to store the knowledge."

3.3.2 Structure and Classification of ANNs

An ANN is a network of parallel, distributed information processing system that relates an input vector to an output vector. It consists of a number of information processing elements called neurons or nodes, which are interconnected via unidirectional, weighted signal channels called connections. The networks with a large number of neurons are frequently used for practical problems. The way these neurons are interconnected determines how computations proceed.

The most common way of classifying ANNs is based on the number of layers: single layer, bilayer, and multilayer. Another classification is based on the direction of data flow through the network. The networks where information passes one way (forward) are known as feed-forward networks. The information is received by the input layer nodes which process and pass it on to the next (hidden) layer. The hidden layer(s) nodes also process it and pass to the next layer till the final output layer. In a recurrent ANN, the information flows through the nodes in both directions. To achieve this, the previous network outputs are recycled as current inputs. In fully connected networks, each node is connected to every other node.

The most widely used network structures in water resources area are the multilayer, feed-forward networks. The remaining discussion is focused only on such networks.

3.3.3 Feed-Forward ANNs

A feed-forward ANN has an input layer, an output layer, and one or more hidden layers between the input and output layers. Each of the neurons in one layer is connected to all the neurons of the next layer and the neurons in one layer are only connected to the neurons of the immediate next layer. The strength of the signal passing from one neuron to the other depends on weights of the interconnections. The intermediate layers enhance the network's ability to model complex functions. The optimal architecture of an ANN is the one that yields the best performance in terms of error minimization while retaining a simple and compact structure (ASCE, 2000a).

A three-layer feed-forward ANN is shown in Fig. 3.23. The input to the network is received by the neurons in the input layer. The data passing through the connections from one neuron to another are manipulated by weights which control the strength of a passing signal. When these weights are modified, the data transferred through the network change and the network output alters. The neurons in a layer share the same input and output connections, but do not interconnect among themselves. Each layer performs specific functions. All the nodes within a layer act synchronously, meaning at any point of time, they will be at the same stage of processing. The activation levels of the hidden nodes are transmitted across connections with the nodes in the output layer. The level of activity generated at the output node(s) is the network's solution to the problem presented at the input nodes.

A typical neuron is shown in Fig. 3.24. Depending on the layer in which the neuron is located, its input may be an input causal variable or outputs of neurons of the previous layer. Every neuron receives signals from each neuron in the previous layer. At each neuron in hidden and output layer, every input is multiplied by its weight, the product is summed and passed through a transfer function to produce its result. The weights leading to the *j*th neuron in a layer form a weight vector $W_j = (w_{1j}, \ldots w_{ij}, \ldots, w_{nj})$, where w_{ij} represents the weight of the connection from the *i*th neuron in the previous layer to the current neuron.

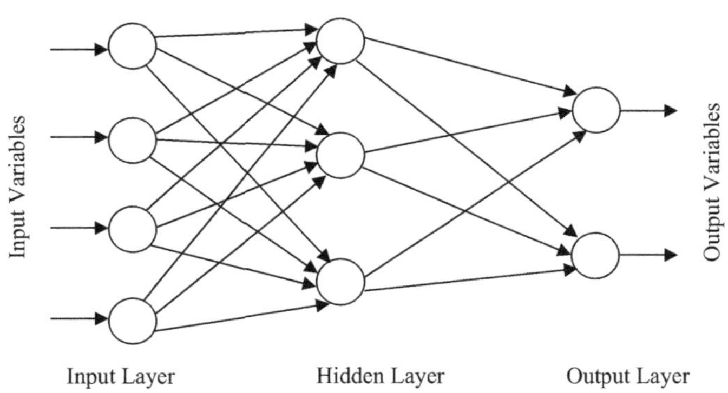

FIGURE 3.23

Schematic representation of a three-layer feed-forward ANN.

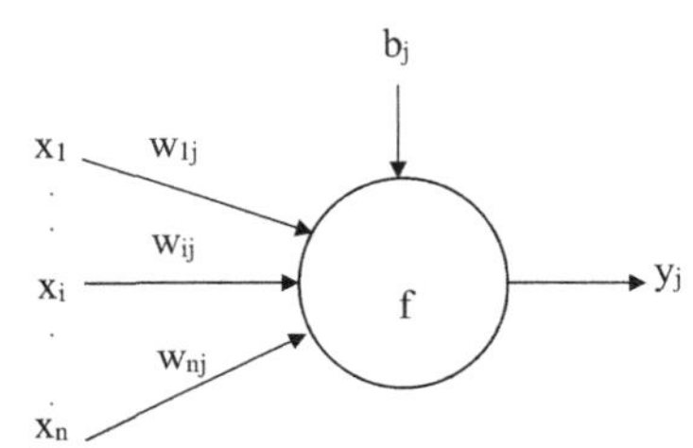

FIGURE 3.24

A typical processing node of an ANN.

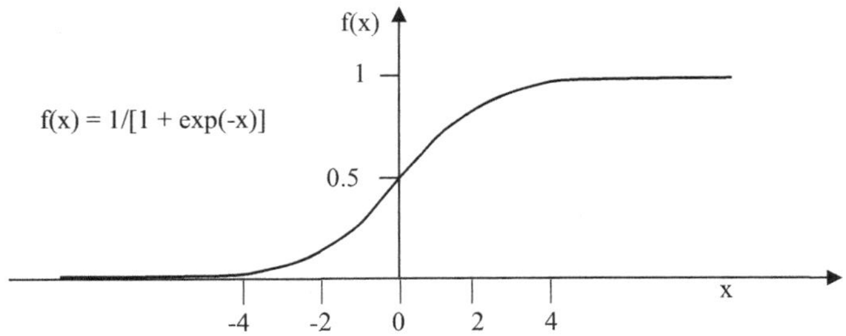

FIGURE 3.25

Sigmoid function.

The most commonly used transfer function is a steadily increasing S-shaped curve, called a *sigmoid or logistic function*. The basic characteristics of the sigmoid function are that it is continuous, differentiable everywhere, and is monotonically increasing and these make it suitable for use with an ANN.

The sigmoid function is shown in Fig. 3.25. The input to the function can vary between $\pm\infty$ and output y_j is always bounded between 0 and 1. The attenuation at the upper and lower limbs of the "S" constrains the raw sums smoothly within fixed limits. The transfer function also introduces a nonlinearity that further enhances the network's ability to model complex functions.

To obtain the output of a node j (y_j), the weighted sum of input signals is taken. This should exceed the threshold or bias of the neuron (b_j) before the neuron can fire. The resulting value is passed through the activation function to get the output. The sigmoid function is:

$$y_j = 1 / [1 + \exp(-z)] \tag{3.15}$$

and the *tan h* function is

$$y_i = (e^z - e^{-z}) / (e^z + e^{-z}) \tag{3.16}$$

where $z = \Sigma\, w_{ij}\, x_i$.

The potential of feed-forward neural networks can be attributed to three main factors (Kothari and Agyepong 1996): (1) multilayered feedforward neural networks do not need an explicit mathematical equation relating inputs and outputs; (2) a feed-forward network with a single hidden layer with an arbitrary number of sigmoidal hidden nodes can approximate any continuous function; and (3) a feedforward network with a single hidden layer of m sigmoidal nodes achieves an integrated squared error of $O(1/m)$, while a linear combination of a set of m fixed basis functions achieves an integrated squared error of $O(1/m^{2/d})$, where d is the dimension of the input (Barron, 1993).

3.3.4 Designing an ANN

The ANN design consists of finding a simple architecture which yields the desired performance. There is no analytical solution to determine an optimal ANN architecture and a unique solution cannot be guaranteed. Since the numbers of input and output nodes are problem-dependent, the designer has to determine the number of hidden layers and the number of nodes in each hidden layer. According to Hsu et al. (1995), three-layer feed-forward ANNs can be used to model real-world functional relationships that may be of unknown or poorly defined form and complexity. Therefore, in such networks, the problem reduces to finding the optimal number of nodes in the hidden layer. Generally, a trial-and-error approach is used. This number should be chosen carefully since the performance of a network critically depends on it—a network with too few nodes will give poor results, while it will overfit the training data if too many nodes are present. Maren et al. (1990) recommend the geometric mean of the numbers of neurons in input and output layers as a good starting guess.

Among the automatic algorithms, there are two major variants. The pruning algorithms, as the name suggests, begin with a large network and systematically remove the nodes whose contribution is minimal. The other variant, the growing algorithms, begin with a small network and add nodes till the improvement in performance is insignificant.

3.3.5 Training of ANN

The knowledge of an ANN is contained in its weights. The objective of training/learning is to determine the set of weights and thresholds so that the ANN gives the desired output. The process is similar to calibration of a watershed model. In general, it is assumed that an ANN does not have any prior knowledge about the problem before it is trained. At the beginning of training, weights are initialized either with a set of random values or based on some previous experience. When the network weights are altered, the data transfer through the ANN changes and the network performance alters. The learning algorithm adjusts the weights such that for an input signal, the ANN output is close to the desired output. Several learning examples are presented to the network, each contributing to the optimization of weights. The results of an ANN keep on improving as more and more data are made available to it, because it has a built-in mechanism of growing "wiser" with "experience." This adjustment can be continued recursively until a weight space is found, which results in the smallest overall prediction error. At this stage, when an ANN has learned enough examples, it is considered trained. The final weight matrix of a successfully trained neural network represents its knowledge about the problem. Note that the aim of learning is to get a network that generalizes the relationship between input and output rather than the one that memorizes it.

There are primarily two basic learning strategies for ANNs—supervised and unsupervised. The supervised training algorithm uses a large number of inputs and outputs patterns. The inputs are cause variables of a system, and the outputs are the effect variables. This training procedure involves the iterative adjustment and optimization of connection weights and threshold values for each of nodes. The primary goal of training is to minimize the error function by searching for a set of connection strengths and threshold values that cause the ANN to produce outputs that are equal or close to targets. In contrast, an unsupervised training algorithm uses only an input data set. The ANN adapts its connection weights to cluster input patterns into classes with similar properties. Supervised training is most common in water resources applications.

In supervised training, the available data set is generally partitioned into two parts: training set and validation set. The training data set should contain sufficient input and output pairs and the entire range of inputs should be included so that the network can adequately learn the underlying relationship between input and output variables.

Caution is to be exercised in supervised learning process so that it does not end up in *overtraining a network* or *overfitting*. This happens when the network "learns" a data set too well, that is, while trying to capture the underlying principles, it also tries to fit the noise that is present in the data set (Fig. 3.26). Note that the network is supposed to learn trends in the data set and not remember individual patterns. In this eventuality, the ANN results will be very good for the training data set but poor for others.

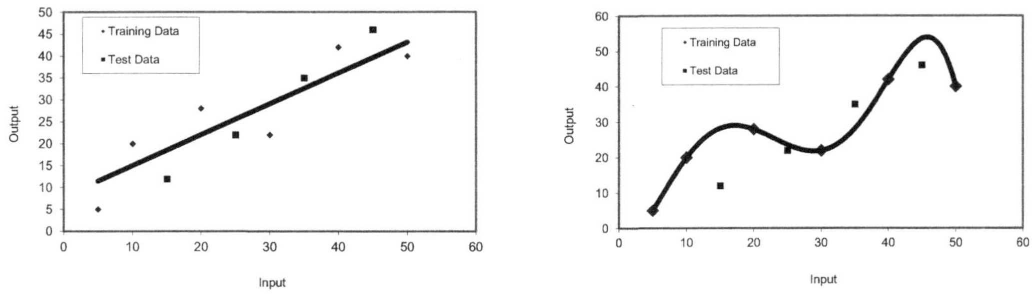

FIGURE 3.26

A well-trained ANN (left) and an overfit model (right).

ASCE (2000a) has recommended a cross-training procedure to overcome the overfitting problem. The goal of this procedure is to stop training when the network begins to overtrain. A portion of the data set is reserved for cross-training purpose. After the adjustment of network parameters with each epoch, the network is used to find the error for this data set. Initially, errors for both the training and cross-training data sets go down. After an optimal amount of training has been achieved, the errors for the training set continue to decrease, but those associated with the cross-training data set begin to rise. This is an indication that further training will likely result in the network overfitting the training data. Training is stopped at this stage, and the current set of weights and thresholds are assumed to be optimal.

The validation of a trained ANN is performed by subjecting it to patterns that it has not seen during training. Some error criterion, such as MSE or a plot of ANN output versus desired response, can form the basis to assess its performance.

There are many algorithms to train a network. Hsu et al. (1995), proposed a Linear Least Squares SIMplex (LLSSIM) algorithm for training a three-layer feed-forward network and demonstrated its application. Error back propagation (BP) algorithm is perhaps the most popular algorithm to train multilayer feed-forward network and is discussed next.

Data Standardization

Before applying ANN, the input data need to be standardized so as to fall in the range [0,1]. A typical variable, say discharge (Q), which can vary between zero and some maximum value Q_{max} can be standardized by the following formula:

$$Q_s = Q/Q_{max} \tag{3.17}$$

where Q_s is the standardized discharge. A different formula will be more suitable for a variable that varies within a certain range. Minns and Hall (1996) have rightly emphasized the importance of the correct standardization. There is, however, some danger of losing information in standardization.

3.3.6 Error Back Propagation Algorithm

The error BP algorithm based on the generalized delta rule was proposed by Rumelhart et al. (1986) to adjust the interconnection weights during training. In this algorithm, a set of inputs and outputs is selected from the training set and the network calculates the output based on inputs. The actual output is subtracted from the target output to find the output-layer errors. The weights of all neurons are adjusted by an amount that is proportional to the strength of the signal in the connection and the total measure of the error. The total error at the output layer is then reduced by redistributing this error backwards through the hidden layers until the input layer is reached. This backward propagation of errors gives the algorithm its name. This process continues for a number of prescribed sweeps or until a prescribed error tolerance is reached. The mean square error (MSE) over the training samples is a typical objective function. If all possible sets of weights are plotted against the corresponding sum-of-squares of errors, the result is an error surface shaped like a bowl. Its bottom marks the set of weights with the smallest sum-of-squared error. The goal during the training is to find the bottom of the bowl or the best set of weights. A typical error function can be given as

$$E = \sum_{p=1}^{N} \sum_{n=1}^{m} \left(T_{pn} - O_{pn}\right)^2 \tag{3.18}$$

where T_{pn} is the target value of the nth neuron for the pth pattern, O_{pn} is the output value of the nth neuron for the pth pattern, N is the total number of patterns, and m is the total number of output neurons. The increment of weights connecting node i to j at the nth pass, $\Delta w_{ij}(n)$, is given by

$$\Delta w_{ij}(n) = -\varepsilon^* \frac{\partial E}{\partial w_{ij}} + \alpha^* \Delta w_{ij}(n-1) \tag{3.19}$$

where α and ε are known as the momentum factor and the learning rate, respectively. The momentum factor controls the speed of training and helps prevent the oscillations in weights. The algorithm can be trapped in local minima and learning rate can be adjusted to increase the chance of avoiding the same.

The BP training algorithm involves two steps. The first step is a forward pass, in which the effect of the input is passed forward through the network to reach the output layer. This is compared with the desired output, and the error is computed. In the second step, the error is propagated back toward the input layer and the weights are modified according to Eq. (3.4). The BP algorithm is based on the steepest descent method. The problem of local minima is faced in most nonlinear optimization problems. It can be addressed to some extent by adjusting the step size and choosing different starting points. Besides, the closer the initial guesses are to the optimum point, the faster the training is, but there is no definite way of making a good initial guess of the weights.

The computations begin with the initialization of weights. The steps of algorithm, following ASCE (2000a), are as follows:

- Step 1. Do Steps 2–9 till the stopping condition is met.
- Step 2. For each training pair of data set, perform Steps 3–8.

Feed-forward:

- Step 3. Each input neuron (X_i, $i = 1, 2, \ldots, n$) receives input signal x_i and sends it to all units in the next (hidden) layer.
- Step 4. Each neuron in the hidden layer (Z_j, $j = 1, 2, \ldots, p$) sums its weighted input signals.

$$\mathrm{Zin_j} = \mathrm{v_{oj}} + \sum \mathrm{x_i v_{ij}} \ \text{for } \mathrm{i} = 1, 2, \ldots, \mathrm{n} \tag{3.20}$$

where v_{ij} is the connection weight and v_{oj} is the bias value. The activation function is applied to compute the output signal:

$$Z_j = f(Zin_j) \tag{3.21}$$

This signal is sent to all units in the output layer. Typically, "*f*" is the sigmoidal nonlinear function, defined in Eq. (3.15) or tanh function defined in Eq. (3.16).

- Step 5. Each output neuron (Y_k, k = 1, 2, …, m) sums its weighted input signals

$$Yin_k = wo_k + \sum z_i w_{ij} \text{ for } j = 1, 2, \ldots, p \tag{3.22}$$

Again, the activation function is applied to compute the output signal:

$$Y_k = f(Yin_k) \tag{3.23}$$

Back-propagation of error:

- Step 6. Each output neuron, (Y_k, k = 1, 2, …, m), computes its error using the target pattern corresponding to the input training pattern

$$\delta_k = (t_k - y_k)f'(Yin_k) \tag{3.24}$$

calculates its weight correction term (to update w_{jk})

$$\Delta w_{jk} = \partial \delta_k Z_j \tag{3.25}$$

calculates its bias correction term (to update w_{ok} later)

$$\Delta w_{ok} = \alpha \delta_k \tag{3.26}$$

and sends δ_k to nodes in the previous layer.

- Step 7. Each hidden unit (Z_j, j = 1, 2, …, p) sums its delta inputs (from units in the next layer)

$$\delta in_j = \sum \delta_k w_{jk} \text{ for } k = 1, 2, \ldots\ldots, m \tag{3.27}$$

multiplied by the derivative of its activation function to calculate its error information term

$$\delta_j = \delta in_j f'(Zin_j) \tag{3.28}$$

calculates its weight correction term (used to update v_{ij} later):

$$\Delta v_{ij} = \partial \delta_j x_i \tag{3.29}$$

and calculates its bias correction term (used to update v_{oj} later):

$$\Delta v_{oj} = \alpha \delta_j \tag{3.30}$$

Update weights and biases:

- Step 8. Each output neuron, Y_k, k = 1, 2, …, m, updates its bias and weights (j = 0, 1, …, p):

$$w_{jk}(new) = w_{jk}(old) + \Delta w_{jk} \tag{3.31}$$

Each hidden node (Z_j, j = 1, 2, …, p) updates its bias and weights (i = 0, 1, …, n):

$$V_{ij}(new) = v_{ij}(old) + \Delta v_{ij} \tag{3.32}$$

- Step 9. Test if the stopping condition is satisfied.

Although BP is a powerful algorithm, it has several drawbacks that can lead to problems during training. The most common problems are long and slow training process, network paralysis, and local minima.

The training process usually starts with a long step size which is gradually reduced till the desired convergence is achieved. There are no definite guidelines to determine the step size. Too small a step size will increase the training time abnormally, while a large step size may skip the optima. Jain et al. (1999) found that learning rates between 0.4 and 0.8 give good results. The problem of moving target arises as the weights have to continuously adjust their values from one output to another for successive patterns. The change in a weight during one training pass may be nullified in the next pass because of a different pattern and this may also slow down training.

When there are large adjustments of weights in the initial epochs, this may lead to *network paralysis*. When all the nodes produce large outputs, the derivative of the activation function can become small and this will slow down the learning process (change in weights), because the change in weights is proportional to the derivative of activation function. This problem is usually avoided by reducing the step size or learning rate.

Example 3.1 Train the following 3-layer network (Fig. 3.27) by the BP method: Assume the learning rate to be 0.25. The input pattern is (0, 1) and the output 0.

Solution To begin with, the weight values are set to random values: 0.6, 0.4, 0.5, −0.2 for weight matrix 1, and 0.3 and 0.8 for weight matrix 2. The input signal set to the neurons I1 and I2 of the input layer which just passes the signal to the hidden layer. Now consider the working of the hidden layer.

Input of hidden neuron H1: $0 * 0.6 + 1 * 0.5 = 0.5$
Output of hidden neuron H1: $1/[1 + \exp(-0.5)] = 0.6225$
Input of hidden neuron H2: $0 * 0.4 + 1 * (-0.2) = -0.2$
Output of hidden neuron H2: $1/[1 + \exp(+0.2)] = 0.4502$

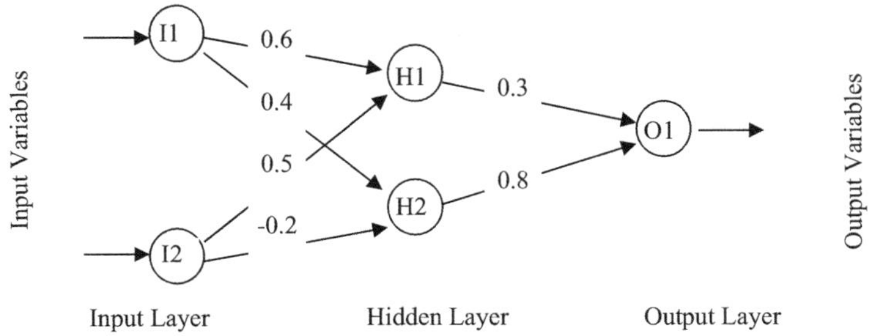

FIGURE 3.27

Example 3-layer ANN.

The signal now reaches the output layer.

Input of output neuron O1: 0.6225* 0.3 + 0.4502* 0.8 = 0.5469
Output of output neuron O1: 1/[1 + exp(−0.5469)] = 0.6334

Since the target output is 0.

Error at the output neuron = 0−0.6334 = −0.6334

To modify the weights, we first calculate

$$\delta = \text{Error}^*(\partial \text{Out}/\partial x) = (\text{Target}-\text{Out})^*\text{Out}^*(1.0-\text{Out})$$

Change in weight

$$\Delta w_{pq,k} = \eta \delta_{q,k}\ \text{Out}_{p,j}$$

and $w_{pq,k}(n+1) = w_{pq,k}(n) + \Delta w_{pq,k}$

First, change the weights in weight matrix 2:

$\delta = (-0.6334) * 0.6334 * (1-0.6334) = -0.1470.$
$\Delta w_{11,3} = 0.25^*(-0.1470)^*0.6225 = -0.0229$
$\Delta w_{21,3} = 0.25^*(-0.1470)^*0.4502 = -0.0165$
New value of weight 1: 0.3 + (−0.0229) = 0.2771
New value of weight 2: 0.8 + (−0.0165) = 0.7835

Now consider the weights in weight matrix 1:

Change in weight 1: 0.25 * (−0.6334) * 0* 0.6225 * (1−0.6225) = 0
Change in weight 2: 0.25 * (−0.6334) * 0* 0.4502 * (1−0.4502) = 0
Change in weight 3: 0.25 * (−0.6334) * 1* 0.6225 * (1−0.6225) = −0.0372
Change in weight 4: 0.25 * (−0.6334) * 1* 0.4502 * (1−0.4502) = −0.0392

Hence,

New value of weight 1: 0.6 + 0 = 0.6 (not changed)
New value of weight 2: 0.4 + 0 = 0.4 (not changed)
New value of weight 3: 0.5 + (−0.0372) = 0.4628
New value of weight 4: 0.2 + (−0.0392) = −0.2392.

3.3.7 Cascade Correlation Algorithm

This algorithm starts training with a minimal network, that is, without any node in the hidden layer. During training, the network grows by adding new hidden units one by one, maximizing the impact of the new node on the network error, creating a multilayer structure. If the output error is greater than the desired value, a node is added to the hidden layer. Once a new hidden node has been added to the network, its input-side weights are frozen. The hidden nodes are trained so as to maximize the correlation between output of the nodes and output error. A training cycle is divided into two phases. First, the output nodes are trained to minimize the total output error. Then, a new node is inserted and connected to every output node and all previous hidden nodes. The new node is trained to correlate with the output error. The addition of new hidden nodes is continued until the maximum correlation between the hidden nodes and error is attained.

The training or weight updating is done for any two layers at a time, and the weights are optimized with the help of a gradient ascent method. In this method, the correlation between the hidden unit output and residual network error is maximized. A noteworthy property of this algorithm is that the determination of architecture is part of training. The steps for this algorithm are summarized as follows (Thirumalaiah and Deo, 1998):

1. Consider only the input and output nodes.
2. Train directly the input-output weights over the entire training set by the delta rule.
3. Add one new hidden node. Connect it to all input nodes as well as to all other existing hidden nodes. Take all training sets one by one and adjust the input weights of this new hidden node after each training set so as to maximize the overall correlation S between the new hidden node's value and the residual error defined as:

$$S = \sum_{o} \left| \sum_{p} \left(V_p - \overline{V}\right)\left(E_{p,o} - \overline{E}\right) \right| \tag{3.33}$$

where V_p is the output of the new hidden node for training pattern p; E_{po} is the network output error for output node o on pattern p; and bars denote the averages of the respective quantities. Pass the training data set one by one and adjust input weights of the new node after each training set unit S does not change appreciably.

4. Once the training of the new node is done, freeze its input weights and the output side weights are trained once again using the delta rule.
5. Go to step 3, and repeat the procedure until the specified minimum error is reached or a specified maximum number of iterations is over.

Thirumalaiah and Deo (2000) have reported that this algorithm significantly reduced the training time. However, in some instances, it is possible to get better results (in terms of a higher coefficient of correlation and efficiency) with the help of a different training algorithm.

3.3.8 Modular Neural Network

In applications in which the training data are fragmented or a discontinuous representation of the process having a significant variation over the range of the inputs, the normal three-layer ANNs may not give good performance. For example, the relationships between rainfall and runoff are likely to be quite different for low and high-flow events because different catchment properties govern the flow in these ranges. To a conventional ANN, these data may appear to be inconsistent and noisy, and if this is the case, the ANN will not be able to properly learn the data behavior. A modular neural network (MNN) can adequately handle such problems. An MNN is based on the assumption that a complex nonlinear problem can be divided into several subproblems.

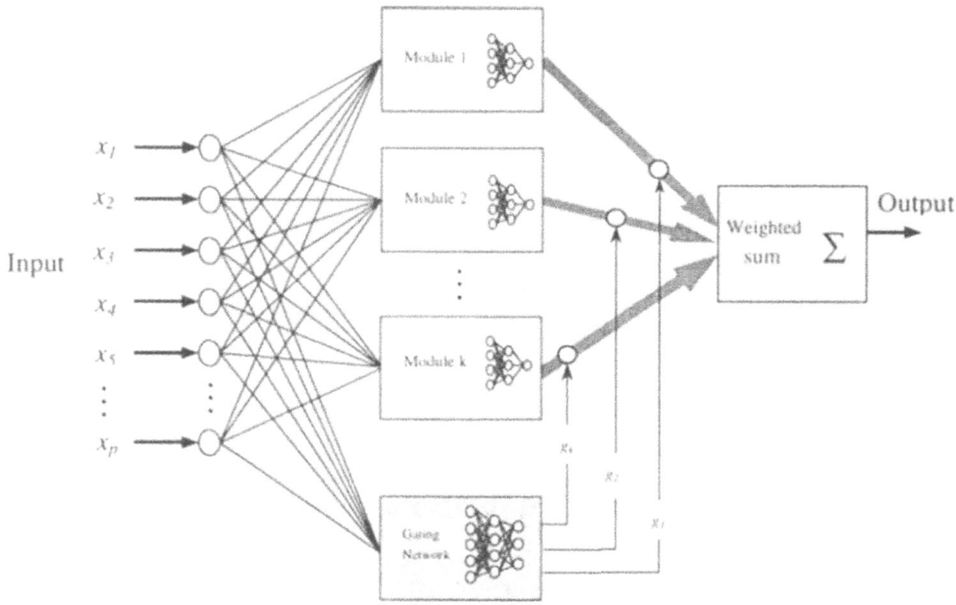

FIGURE 3.28

Schematic diagram of a modular neural network (MNN).

Source: Govindaraju and Rao (2000).

A schematic diagram of an MNN is shown in Fig. 3.28. It consists of several expert feed-forward networks, each of which receives the input vector. The input is also received by a gating network which produces scalar outputs that are partitions of unity at each point in the input space. Corresponding to the input vector, each expert network produces certain response and the weighted sum of the responses is the network output. Here, the weights are assigned by the gating network. According to Zhang and Govindaraju (2000), the gating network output can be regarded as the probability that an input vector is attributed to a particular expert. The number of expert modules is equal to the number of partitions in the input and is dependent on the problem. The data of three medium sized watersheds were used for monthly runoff predictions based on rainfall and temperature data. In this study, the input was partitioned in three subsets representing low, medium, and high flows and three expert networks were used. The authors showed that the performance of the MNN was better or comparable to the performance of a three-layer ANN.

3.3.9 Radial-Basis Function Networks

The back-propagation algorithm of a multilayer feed-forward ANN is a gradient descent algorithm that may terminate at a local optimum. This problem is overcome in radial-basis function (RBF) networks by incorporating the nonlinearity in the transfer functions of the nodes of the hidden layer and thus parameter optimization becomes a linear search. The theory and structure of the RBF networks is based on the theory of interpolation in multidimensional spaces. However, in case of exact interpolation of every data point, a very large number of nodes in the hidden layer will be required. As most water resources data contain a noise element too, the network would also try to mimic the noise. To overcome these problems, the number of RBFs (M) employed in the RBF networks is much smaller than the number of training patterns (N). The number of nodes in the hidden layer M is determined during training.

The RBF networks consist of three layers, viz., a transparent input layer, a hidden layer, and an output layer. The input layer passes the variables to each node in the hidden layer. A transformation function at each node in the hidden layer transforms the incoming values. The transfer function in RBF networks is a radially symmetric basis function and hence this name of the networks. An RBF has a center μ where the function value is the highest and a spread σ that indicates the radial distance from the center within which the function value is significantly different from zero. There is a wide choice of such functions; the Gaussian RBF is the most commonly used function. The responses from the hidden layer are multiplied by the weights of links connecting hidden and output layers and then summed up to yield the network output. The diagram of a typical network is shown in Fig. 3.29.

The response y_j of the jth node of the hidden layer for the pth input pattern is given by

$$y_j = f\left(\frac{\|X^p - \mu_i\|}{2\sigma^2}\right) \tag{3.34}$$

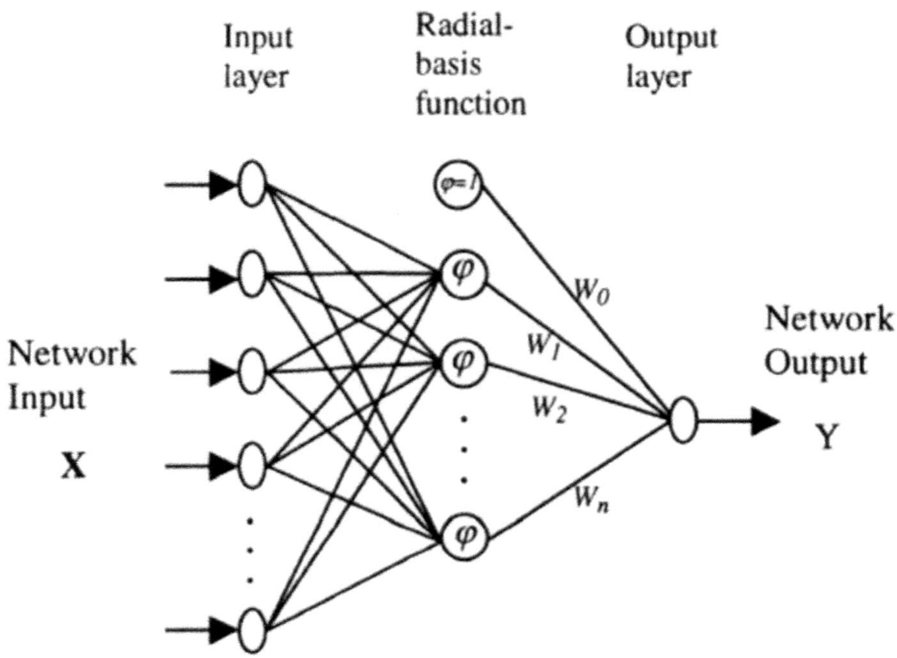

FIGURE 3.29

A schematic diagram of RBF network.

After Govindaraju and Zhang (2000).

where ||. || denotes Euclidean distance. The output of the kth node of the output layer becomes

$$Z_{p,k} = \sum_{j=1}^{M} y_j w_{k,j} \tag{3.35}$$

where $w_{k,j}$ is the weight of the link from the jth node in the hidden layer to the kth node in the output layer.

The training of RBF networks is a two-step process. In the first step, the input patterns are clustered in groups and the center and spread of each group are determined. In the next step, the interconnection weights are obtained. Govindaraju and Zhang (2000) describe several training strategies of these networks.

The RBF network and BP mainly differ in terms of handling of nonlinearities. In BP networks, nonlinearity is handled through the activation (e.g., sigmoid) function; in the RBF network, the Euclidean distance between the center and input is computed by a nonlinear function. The results are combined at the output layer in a linear summation fashion.

3.4 ARTIFICIAL INTELLIGENCE AND MACHINE LEARNING

Artificial intelligence (AI) is a branch of knowledge whose aim is to make machines that think and act like human beings. Computers are excellent machines at applying rules and carrying out computations. Humans can "easily" complete many tasks that appear to be very tough, but there are tasks that are relatively straightforward or easy for a human being but may be very difficult or complex for a computer. For example, picking up a fragile object such as an egg and place it on a hard surface is a trivial task for a human being, but it is not so easy for a machine to perform. Likewise, driving a car through a crowded and bumpy road is not very difficult for a human being, but it is tedious for a machine and involves transmission and rapid processing of a large volume of data.

AI techniques that involve analysis of large volumes of data to learn patterns help attain this objective. A schematic diagram explaining the domains of AI, machine learning (ML), and deep learning (DL) is given in Fig. 3.30.

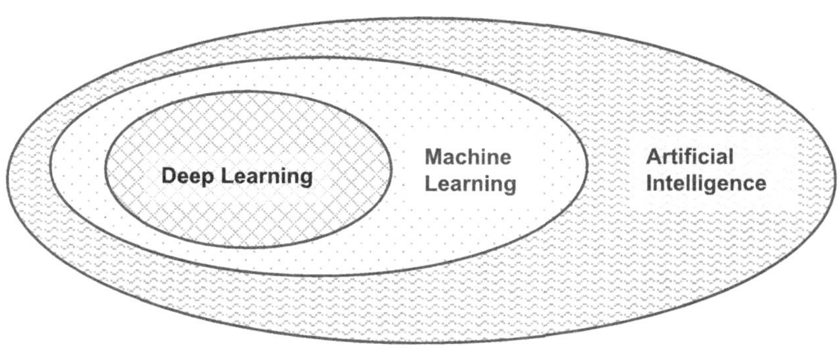

FIGURE 3.30

A schematic diagram explaining domains of artificial intelligence, machine learning, and deep learning.

3.4.1 Machine Learning

ML falls within the domain of AI whose main goal is to train computers so that they are able to perform various tasks without the need of programming. For this purpose, computers are provided with data, (mostly structured) and they learn the patterns in the data. With time, computers are expected to become "familiar" with data and are in a position to take decisions on their own without the need for human intervention.

Similar to DIP, a machine can learn in two ways: Supervised learning and unsupervised learning.

In supervised learning, a machine/model learns with the help of human inputs. Training data are input to the machine/model with the aim to teach it how it should respond to the input data. Once the machine is adequately trained, its performance is evaluated to see if it can correctly respond to the data. Based on outcome, the machine may be declared adequately trained or may require more training.

In unsupervised learning, the machine/model is provided with input data and it attempts to find patterns and associations in the data. A commonly used method is clustering in which the computer tries to classify data based on attributes.

Besides the applications in hydrology, ML tools are used in many day-to-day tasks. For example, e-mailing platforms such as Google employ these to screen spam. Voice-activated devices such as Alexa use ML for voice recognition, and so on.

3.4.2 Deep Learning

DL models introduce sophisticated approach to ML so that the computers are able to perform complex tasks such extracting data from an image and initiate actions based on it. This is achieved by use of "deep neural networks" that have multiple layers. Data are passed between nodes that are highly interconnected. Passing through the network, the inputs are transformed in a highly nonlinear manner to produce the desired output. Thus, ML is about using the computers to perform complex tasks without an explicit need for programming. Note that DL, a type of ML, is a subset of AI. The needs of data to build a DL system are typically large.

Some of the algorithms that are employed in DL are briefly described here.

3.4.3 Convolutional Neural Networks

Convolutional neural networks (CNNs) are special algorithms, designed to efficiently process images. Here, a weight-based filter is applied to each element of an image. This allows the computer to understand the elements present in the picture. Applications of CNNs include identifying a particular feature, for example, water in the image to determine the extent of flooding. The technology using computer for image analysis and comprehension is known as computer vision.

3.4.4 Recurrent Neural Networks

The conventional neural networks do not have memory which is an integral part of neural networks of living beings. Recurrent neural networks (RNNs) bring in memory in the network architecture. RNNs have the ability to consider past data and past decisions when taking a decision at the current time step. On account of this unique ability, RNNs are being applied in many tasks where the context of problem-solving becomes important, for example, natural language processing.

3.4.5 Comparison of ML and DL

(1) ML is concerned with the ability of computers to think and perform tasks with minimal human intervention. DL is concerned about providing computers the ability to learn and think based on simplified modeled structures of human brain.
(2) ML requires considerable human intervention. In contrast, DL systems are more complex but once satisfactorily set up, they require very less intervention.
(3) Requirement of computing power is less in ML, but DL requires less human intervention.
(4) After training, DL systems can analyze videos, images, and unstructured data and extract useful information from them, but ML can't do this task easily.
(5) But hardware and resource requirements of DL systems are typically high. ML programs are much less complex than DL algorithms; these can run on conventional computers also.
(6) ML systems can be set up easily and begin operation rapidly, but the results may have limited power. DL systems take time to set up but once set, they begin producing results after a short time. Also, the quality of results is likely to improve with time as more and more data become available and are used.
(7) ML typically requires structured data and uses conventional algorithms. DL employs ANNs and is designed to consider large volumes of unstructured data.

3.4.6 Applications of ANNs in Water Resources

An ANN is a network that relates the inputs and outputs of a system, and these networks have been successfully used to map nonlinear input and output relationships in a wide range of areas. The immense success with which ANNs have been used to model the nonlinear system behavior indicates that this approach can be useful in the field of water resources. ANNs have been used for flow predictions, flow/pollution simulation, parameter identification and to model complex nonlinear input-output time series.

A brief review of selected applications follows. A set of two papers published by the ASCE task committee on the application of ANNs in hydrology (ASCE, 2000a,b) contains an exhaustive review of theory and applications of ANN in water resources.

French et al. (1992) developed a three-layer feed-forward ANN to forecast rainfall intensity in space and time and compared the results with two other methods of short-term forecasting. Since an ANN relates the pattern of inputs to the pattern of outputs, volume continuity is not a constraint. However, care must be taken to avoid the presentation of contradictory information to the ANN. Chang and Tsang (1992) compared the multiple regression and ANN approaches to model SWE from multichannel brightness temperatures and reported that the results of the ANN approach were better.

Hsu et al. (1995) have shown that the ANN model approach provides a better representation of the rainfall-runoff relationship of a medium-sized basin than the ARMAX approach or the Sacramento soil moisture accounting model. Raman and Sunilkumar (1995) investigated the use of ANNs for synthetic inflow generation and compared the model performance with that of a multivariate time-series (ARMA) model. Smith and Eli (1995) used a neural network model for simulating runoff using data of a synthetic watershed. They trained a BP network to predict the peak discharge and the time to peak. Minns and Hall (1996) reported a series of numerical experiments in connection with the application of ANN to rainfall-runoff modeling and concluded that the ANNs are capable of identifying usable relationships between discharges and antecedent rainfalls.

Carriere et al. (1996) designed a virtual runoff hydrograph system based on ANN by training a recurrent BP neural network. They obtained good correlation between observed and predicted data. An advantage of using ANN for rainfall-runoff modeling is that parameters relating to the catchment can be avoided in the input and virtually no model parameter needs to be manually calibrated. Raman and Chandramauli (1996) derived reservoir operating policies for a dam in India by two approaches: (i) a DP and neural network procedure (DPN model) and (ii) DP and a multiple linear regression procedure (DPR model). Based on comparison of performances, it was concluded that the DPN model provided better performance than other models. Dawson and Wilby (1998), while using an ANN for river flow forecasting, have given an overview of ANNs, their training, and data standardization. Based on the results of an application study, they have highlighted the ability of ANN to cope with missing data and to learn from the event currently being forecast in real-time. They have also emphasized the need for a thorough investigation into the relationship between the training period length and the hydrological realism of the ANN forecast.

Fernando and Jayawardena (2000) applied RBF networks for runoff forecasting. In general, the performance of RBF networks was found to be as good or better than multilayer feed-forward networks. They also mention that the efforts needed in the case of an RBF network with OLS algorithm are considerably less. For reservoir inflow prediction, time-series analysis and ANNs were used by Jain et al. (1999). It was found that high flows were modeled better through the ANN. For reservoir operation, the ANN model performance was found to be the best among other models. Birikundavyi et al. (2002) also found that a simple ANN can achieve accuracy superior with that of ARMAX and deterministic models for 7-day ahead forecasting.

A very useful application of ANNs is to stage-discharge relationship at a gauging site. Jain and Chalisgaonkar (2000) applied ANN for this purpose and demonstrated that an ANN can also successfully model a loop rating curve (hysteresis effect). This is not possible using the conventional technique which can only fit an average or steady-state curve. Besides streamflow, the phenomenon of hysteresis is noticed in other branches of hydrology too, for example, soil moisture retention curve and ANNs can be conveniently used in these areas also. Jain (2001) further extended this work by applying the ANN concept to establish the relation between river stage, discharge, and sediment discharge. The input to the ANN consisted of the river stage at the current and previous time periods, and the water discharge and sediment concentration at previous time periods. Such an ANN has two output nodes, one corresponding to the water discharge and the other corresponding to the sediment discharge. It was found that for ANNs, the SSE is about an order of magnitude smaller and the correlation coefficient is much higher than is for the conventional method.

Minasny and McBratney (2002) found ANN to be a significantly improved tool to model parametric pedotransfer functions of soils. Kumar et al. (2002) concluded that the ANN can predict reference crop ET for an area better than the Penman–Monteith method. ANNs have also been applied to groundwater remediation problems, identification of pollution sources, and infilling streamflow data. Some researchers have a feeling that the ANN could perhaps be regarded as the ultimate black-box model.

Recent times have seen many excellent reviews of ANN/AI in water resources. Cigizoglu (2008) wrote a review on ANNs in water resources. Kratzert et al. (2018) wrote a review on rainfall-runoff modeling by using long short-term memory (LSTM), and Shen (2018) published a review on DL research with special focus on water resources. Sit et al. (2020) reviewed applications of deep learning in hydrology and water resources.

3.4.7 Issues in ANN Applications

By now, ANNs are firmly established as a viable black-box modeling tool. However, along with numerous advantages, ANNs have some disadvantages too. The first and foremost is the requirement of adequate data of desirable quality and quantity. To be fair to ANNs, this is a crucial requirement with all modeling techniques and ANN cannot be an exception. Presently clear guidelines are not available except that the entire range of likely inputs should be covered. Guidelines to select network architecture for a given type of problems are also badly missing. In this context, ASCE (2000b) has raised the following very pertinent questions which need to be resolved.

(a) Can ANNs be made to reveal any physics? The application of ANNs can get a boost if some physical explanation of their functioning is available. This will also help in selecting the appropriate type of network and learning algorithm for a given problem. Some neural networks can provide statistical interpretations in terms of conditional probabilities. For instance, a feed-forward network can learn the posterior probability of a classification. This problem is receiving attention of many researchers.

(b) Can an optimal training set be identified? ANNs cannot learn without data—they are data-intensive and poor training data will result in poor learning. An optimal data set should fully represent the modeling domain, has minimum required data points, and there should not be repetition of data. So far, there are no guidelines about these.

(c) Can ANNs improve time-series analysis? The time-series models are based on extracting the correlation and dependence structure of the data. While ANNs have been shown to work better than a time-series model for river stage and discharge prediction, they have not given any insight into the process. It would be welcome if ANNs can bring out the relationships among the variables and highlight those features of input data that are not revealed by other techniques.

(d) Can training of ANNs be made adaptive? The most time-consuming part of ANN application is training. As new data become available, the previously trained ANN has to be retrained. Since catchment properties change with time, it is important to incorporate the new information in the model. The ANN applications will be immensely benefited if the training can be made adaptive, that is, the new information is incorporated into the models without the necessity of complete retraining.

(e) Are ANNs good extrapolators? Many studies have shown that ANNs work well in the range of input data that were used for training and their performance deteriorates if during application, the input data are out of this range. This aspect is receiving attention of hydrologists. Imrie et al. (2000) have presented a methodology for training ANNs to produce models that generalize well on new data and can extrapolate beyond the range of values included in the calibration range. They claimed good results with the data from the catchment of the River Trent and a modified cascade-correlation algorithm.

A thinking and solution to the above issues will certainly help in a better understanding of the ANNs. Of course, ANNs cannot be considered as a panacea for all types of problems of water resources or an alternative to other modeling approaches. Gupta et al. (2000) have aptly commented: "*We do not advocate that the ANN approach be generally used in place of the conceptual modeling approach, because the conceptual methodology provides the strengths that the ANN approach does not—in particular, the conceptual approach has the potential to be applied to ungaged watersheds or to simulate the potential behavior of a watershed under land use changes ... Further, implementation of the ANN approach does not require the considerable amount of expertise and data required to calibrate a conceptual watershed model.*" Logically then, ANN should be viewed as an alternative to conventional computing techniques.

3.5 EXPERT SYSTEMS AND DECISION SUPPORT SYSTEMS

Ever since the computers have become an integral part of the emerging technology, efforts are on to use their capabilities in decision-making. One of the outcomes of these efforts is the emergence of "Artificial Intelligence" (AI). AI is concerned with efforts in making computers think and do things intelligently. According to Barr and Feigenbaum (1981), "*Artificial intelligence is that part of computer science which is concerned with designing intelligent computer systems, that is, systems that exhibit the characteristics we associate with intelligence in human behavior.*" Based on the AI concepts, a wide spectrum of areas are being developed, viz., robotics, natural language processing, speech recognition, ANN, computer-aided intelligent instructions, and knowledge-based ESs. In earlier periods, AI was considered only as a research topic, but later on it found more applications in specialized programs.

Intelligence requires software to be knowledgeable not only about their own possibilities and constraints but also about the application domain and about the user, that is, the context of its use. Defaults and predefined options in a menu, sensitivity to context and history of use, built-in estimation methods, or alternative ways of problem-solving that depend on the user can all be achieved by the integration of ESs technology in the user interface and in the system itself.

The development of ES involves separation of problem-solving techniques and domain-dependent knowledge so that existing problem-solving frameworks may be used in new domains with appropriate knowledge base. This led to knowledge engineering which is the process of constructing an ES. Basically, the expert knowledge is stored in the computer in an organized manner. This knowledge base is used to provide advice and, if necessary, logic behind a particular decision.

ES follows the reasoning process that a human decision-maker would go through to arrive at a decision. The most prominent problem-solving assets of an expert, particularly in the domain of water resources, are theoretical and practical experience, archived data, and operational expertise. The expert must have sound theoretical knowledge that has been acquired through studies, thinking, and discussions. An expert gathers experience by solving a range of practical problems and in this process, he also builds a large database. His involvement in operational tasks further helps in refining his tools and techniques.

At the core of an ES is the knowledge about a specific problem domain. An ES is a combined human-computer system designed to solve problems that normally require logical consideration of both facts and heuristics, or rules-of-thumb, to arrive at a decision. An ES can be defined as an "*interactive computer program that incorporates experience, judgment, rules of thumb, intuition and other expertise to provide knowledgeable advice about a variety of tasks.*" Simonovic (1991) defined a water resources ES as "a computer application that assists in solving complicated water resources problems by incorporating engineering knowledge, principles of systems analysis, and experience, to provide aid in making engineering judgments and including intuition in the solution procedure."

Professional engineers using computer-aided analysis and design have incorporated expertise in the program in the form of constraints and limitations, assumptions and approximations, interactive inputs and outputs, including graphics. Some people consider the results more as advice rather than answers.

The important characteristics and advantages of an ES are:

(a) When experts retire, the valuable knowledge that they have acquired over the years goes with them; ES helps in preserving such a perishable knowledge.
(b) ES can also help an expert either as an assistant, or a partner.
(c) ES provides justification and explanation to the conclusions that it has arrived at.
(d) ES can be designed to interact with humans in a suitable way.
(e) ES incorporates knowledge associated with humans which is otherwise scarce and mobile.
(f) ES can help in the reduction of expenses due to mediocrity.
(g) ES can provide an efficient solution for complex decision problems requiring large amounts of information and many possible outcomes.

A rule-based approach can also be a substitute for a numerical model, even in the socioeconomic domain. An example is environmental impact assessment (EIA) based on a checklist of problems, which can be understood as a diagnostic or classification task. A qualitative label is assigned to potential problems based on the available data on environment and planned action, and a set of generic rules assessing and grading the likely consequences. One of the famous early ES was a software named as MYCIN which was developed by the Stanford Research Institute in 1976 to help a physician in his diagnosis and prescription. It successfully used domain-dependent information. The first attempt to develop an ES for water resources problems was by Gasching et al. (1981) who developed a consultation system called HYDRO. This system was intended to provide advice comparable to that of an expert hydrologist in selecting parameter values characteristic of the watershed under consideration.

ESs are attractive because they offer an excellent way of organizing rules. However, note that ESs provide no new intrinsic validity to the use of rules or uncertainty in the process of forming a decision. While some problem domains are very complex but deterministic in nature, most experientially derived rules do not perfectly capture the relationships within a problem domain. The ESs are helpful because they deal with two types of uncertainties: those of imperfect rules and imperfect information. Imperfect rules may result from a poor understanding or intentional simplification of the system. Rather than saying that "If A then B," one might express the degree to which he believes that B follows from A. The answer might be: "If A then probability of B = 0.8." Here probability represents the chance of premise B following premise A. Such a rule is termed an uncertain rule because of its imperfect association. Imperfect information may either be incomplete or it may have noise.

An effective ES must provide the following which are of direct concern to a system manager (Palmer and Holmes 1988): (a) Consider only operating rules that are feasible and that follow existing management procedures. (b) Consider operating rules that are appropriate for a wide range of conditions. (c) Assume that managers consider the same criteria for each decision. (d) Provide quantitative information (such as the probability of yield and failure for a wide variety of initial conditions and overall historic flow regimes) that is not readily available otherwise. (e) Operate on a computer available in a typical office. (f) Serve as a training tool for future managers.

3.5.1 Expert Systems Architecture

A typical ES consists of (Fig. 3.31): (i) a knowledge base; (ii) a working memory; (iii) an inference engine; (iv) system analysis, graphics, and other software; and (v) a communication module including interfaces to the developer and the user. The knowledge base consists of declarative knowledge which are facts about the domain, and procedural knowledge which are scientific, analytical, or heuristic rules from the domain. The rules in turn a have trigger part which, based on facts, triggers the body part of the rule which are actions for processing instructions on the knowledge base, input-output instructions, or control instructions. They may include metarules which are rules about rules. The working memory is the current active set of the knowledge base and may include a knowledge management module. The problem-solving module is the inference engine which may also provide justification for the advice from ES. Graphic, numeric analysis, statistical analysis, system analysis, and simulation software may also be available. Data communication module provides an exchange of information among the various modules and may also provide interfaces to the developer and user.

ESs differ from traditional computer programs in the following respects:

(a) ESs are knowledge-intensive programs;
(b) ESs mimic the decision-making and reasoning process of human experts. They can provide advice, answer questions, and justify their conclusions;
(c) In ES, expert knowledge is usually divided into a number of separate rules;
(d) The rules forming expert knowledge base are separated from the methods (inference mechanism, reasoning mechanism or rule interpreter) for applying the knowledge to the current problem;
(e) ESs are highly interactive; and
(f) ESs have a friendly/intelligent user interface.

Knowledge is driven by a mechanism called inference engine, which needs data in the context space. An explanation module is sometimes provided and the user interface, as the name suggests, assists in interaction between the user and context space. A knowledge acquisition facility is provided to manage (add or edit) the knowledge base.

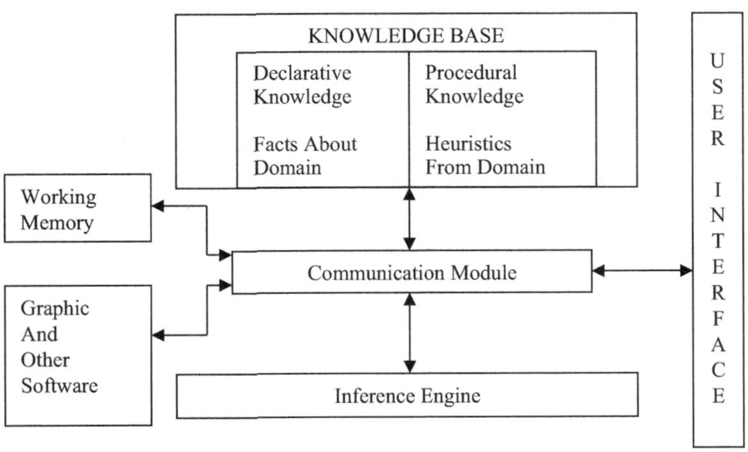

FIGURE 3.31

Architecture of a typical expert system.

3.5.2 ES Development

Five sequential stages in the development of ES were enumerated by Hayes-Roth et al. (1983):

(i) Identification—determining characteristics of the problem,
(ii) Conceptualization—finding concepts to represent knowledge,
(iii) Formalization—designing structures to organize the knowledge,
(iv) Implementation—formulating rules embodying the knowledge, and
(v) Testing—validating the rules.

While building an ES, the gathering and structuring of rules involve significant efforts and resources. Therefore, it is necessary to approach in an organized manner. The first step is to precisely define the problem domain. The next step is to document the expert decision-making process, most commonly by watching or asking the expert. Rouhani and Kangari (1987) have recommended the following steps:

(a) Observe the expert without interruption.
(b) Informally discuss representative problems with him.
(c) Ask the expert to describe the basis for his judgment and empirical rules. Organize a knowledge base.
(d) Ask the expert to solve a series of problems. Focus on the decision process.
(e) Solve a range of problems with the rule base.
(f) Verify the accuracy of judgments. Iterate.
(g) Ask other experts.

Another good approach is the "thinking-aloud" experiment wherein the expert is asked to think through the problem, while speaking out loudly and a verbatim transcript is taken. This is then followed with a "cross-examination" period where the expert is asked in-depth questions about specific topics. The transcripts of those two sessions are then broken into short meaningful phrases used in the expert's analysis. From those phrases, premises and their relations are identified and formed into rules. Another source of rules is simulation models. While these models can be integrated into the system model, it may pose a computational/burden to do so. In such cases, heuristics that generalize the results of a number of simulation runs are developed. Cuena (1983) describes how rules for a flood-control problem are generated in this manner.

To take full advantage of the ES features, a good coordination between the knowledge engineer and the expert is necessary. At times, this requires patience and understanding because an expert will be a busy person and may not always be willing to share his expertise. One individual expert should not be given too much importance because this will give a strong personal bias. A review and critical examination by other experts will help overcome this bias.

3.5.3 ES Techniques

The order of execution of the rules and/or procedures in an ES is governed by the inference engine in terms of the problem-solving strategy used. Maher (1986) considers two approaches:

The derivation approach: A solution is derived that is most appropriate for the problem from a list of predefined solutions stored in the knowledge base of ES. It includes forward chaining (which works from an initial state of known facts to the goal state), backward chaining (which works from a hypothetical goal state to the facts in terms of subgoals which are preconditions for the goal stated. If the hypothesis is not supported by facts, it tests for another goal state and so on in a predefined order of goals), and the mixed initiative which combines forward and backward chaining strategies.

The formation approach: A solution is formed from eligible solution components stored in the knowledge base. It includes problem reduction (factoring the problem into subproblems), plan-generate-test (which generates all possible solutions, prunes inconsistent solutions, and tests the remaining solutions), and agenda control (assigning a priority rating to each task in the agenda and perform tasks with higher priority before those of lower priority). These may be combined with other techniques for hierarchical planning and least commitment, backtracking, and constraint handling.

3.5.4 ES Tools

A wide variety of development tools and environments are available for ES. They include programming languages like PASCAL, C++, Java, and Python. ES shells provide a framework and tools to build a system. They contain all the modules required for an ES except that the knowledge base is hollow. If this base is filled with knowledge in specified syntax, then it becomes an ES. Although the criteria for the selection of a tool solely depend on the type of problems, a few general principles can guide a knowledge engineer in selecting an approximate shell. For an engineering problem, the prototype of ES must be able to perform hybrid chaining, that is, forward and backward. They must be able to handle uncertainty, and provide interface to other software. The hardware requirements must be such that the shell is highly portable and can be used on a PC. ES shells make the development of these systems easier than previously possible. Many such shells are commercially available. The use of a good tool helps in lessening the effort to develop an ES and the developer can pay more attention on acquiring knowledge and refining the rule base.

An ES also has a module for editing the knowledge base, that is, adding, deleting, or modifying knowledge. The user interface acts as an intermediary between the end user and inference engine and conceals the complexity of the system from end user by helping him in a friendly manner. The ES must be friendly with a national language interface, should be able to support user-defined functions, and must not be too complicated.

Some ESs also have an explanation facility. This module has the knowledge to explain how the system has arrived at particular answers. The explanation includes displaying the inference chains and explaining the rationale behind the use of each rule in chains. The ability to examine the reasoning process and explain their operation are truly innovative and important qualities of ESs. This facility becomes very handy in debugging the system when its performance is not as expected.

Every ES shell may not have all these facilities. Depending upon the type of the problem, the knowledge engineer has to choose the shell that best suits the requirement. For example, a backward chaining rule is suitable with an ES in medical field, while for an engineering design, a system with forward chaining is better.

3.5.5 Knowledge Base

A knowledge base is that module which contains domain-specific knowledge, which must be of high quality as the ES behavior critically depends on this. There are mainly three ways of representing knowledge, viz., rule, taxonomy (semantic nets), and frames. Representation of knowledge in the form of rules is the most popular formalism with "IF-condition-THEN-action" statements. A set of rules specifies the program to react and is useful when the knowledge is in the form of condition action. It has its inherent simplicity, understandability, and ease of modification. The rule base contains facts and heuristics that are specific to the problem being addressed. An example of a rule for the management of a storage reservoir in water-scarce region in high flow season is:

IF	Demand	>	Average
AND	Storage	≥	Capacity/2
AND	Date	≥	August 15
THEN	Curtail		Irrigation releases to half

Typically, such a rule will be one of many in a rule base relating system conditions and management action.

Representation of semantic (implying meaning) nets is used when knowledge is a subset of some other bigger set. A semantic network consists of points (nodes) connected by links (arcs) that describe the relation between nodes. The biggest advantage of this is that it is possible to represent hierarchical information. Frames refer to a special way of representing common concepts and situations. A frame is a network of nodes and relations organized in a hierarchy, where topmost nodes represent general concepts and lower ones more specific instances. A node is a collection of attributes called slots, and their values are stored in the respective slots. Frames are useful in giving specification details.

Knowledge Acquisition

Knowledge in any specialty is of two types: public and private. The body of information that is widely shared and agreed upon is public knowledge. Pump efficiencies, pipe roughness coefficients, and channel roughness coefficients are some examples of public information in a water resource setting. Private knowledge consists of rules of thumb or heuristics developed by human experts through experience. These heuristics are subjective rules of good judgment that allow human experts to make educated guesses and deal with incomplete data. Perceived public resistance to water restrictions, estimated political liabilities, and the risk avoidance characteristics of an operator are examples of private information. This characterization between public and private knowledge is of particular value in water resources planning. Often modelers devote a significant effort in capturing public information but devote little attention to private information.

Knowledge acquisition is obtaining of problem-solving skills from an individual and incorporating these in a computer program. These techniques establish information as specialized facts, procedures, and judgmental rules about a narrow domain area (Hayes-Roth et al., 1983). Knowledge acquisition is extremely important in ESs development and is often a significant obstacle. The success of an ES depends upon the quality of the knowledge gathered and its effective assimilation into a rule base. Unfortunately, knowledge acquisition is an iterative process and many revisions of the system rule base may be necessary. Knowledge acquisition is difficult for several reasons. Domain experts often are unaccustomed, or unable to formulate and articulate their problem-solving methods. They may forget portions of information, introduce inconsistencies, or not express their problem-solving algorithms explicitly. They may fear losing their jobs to computers or exposing their techniques to the scrutiny of coworkers and the public. Finally, they may simply mistrust or feel uncomfortable with the process.

3.5.6 Inference Engine

An inference engine is computer software that examines the knowledge base and answers the questions posed by the user. Simple logic schemes based upon the status of system variables and on the rule base accomplish this task. The inference engine is the most crucial component of an ES since it makes and manipulates the database for problem solution. It is a mechanism that derives the knowledge, that is, a sophisticated system guiding the selection of a proper response to a specific situation. This is known as pruning. Three formal approaches used in this case are production rules, structured objects, and predicate logic. Production rules consist of a rule set, a rule interpreter which specifies when and how to apply the rules, and a working memory that holds data, goals, or intermediate results. Structured objects use vector representation of essential and accidental properties. Predicate logic uses propositional and predicate calculi. Context is the work space for the problem constituted by the inference mechanism from information provided by the user and meta rules in the knowledge base. The inference engine can work in two ways, viz., forward chaining and backward chaining.

Forward chaining starts with known initial state and proceeds in the forward direction until the goal state is arrived at. From the given information, the inference engine searches the knowledge base for rules whose precedence matches the given current state and fires those rules adding more information to the working memory. The basic steps are:

(1) The system is presented with one or more conditions.
(2) For each condition, the system searches the rules in the knowledge base for those rules that correspond to the condition in IF part.
(3) Each rule can in turn generate new conditions from the conclusions of the invoked THEN part, which are added to the existing ones.
(4) If any condition is added to the system in step (3), it will be again processed from step (2). If there are no new conditions, the session ends.

Backward chaining engines (goal-driven engines) involve reasoning in backward direction. The system selects a goal state and reasons in the backward direction and establishes the initial state conditions necessary for that goal state to be true. If the given initial state conditions match with the arrived initial state conditions, then that goal is the solution. Otherwise, the system selects another goal and repeats the process. This process is very similar to the backward chaining associated with dynamic programming. The steps are:

(1) Select a goal state and rules whose THEN portion has that goal state as conclusion.
(2) Using IF portion of the selected rules, establish subgoals to be satisfied for the goal state to be true.
(3) Using steps (1) and (2), establish initial conditions necessary to satisfy all subgoals.
(4) If the given initial state corresponds with required initial state, then the selected goal is one solution. If not, select another goal state.

3.5.7 Applications in Water Resources

A large amount of domain-dependent expertise is available in planning, design, construction, and integrated operation of water resources systems and they are often heuristic in nature. Hence, water resources engineering has good potential for ES applications. The role and context of application of an

ES in water systems management is shown in Fig. 3.31. The following examples demonstrate the variety of issues in water resources management that need and can be tackled effectively through the judicial use of ES.

Some of the application areas of ES in water resources are detailed below (Fig. 3.32).

System modeling and analysis: Mathematical modeling and analysis of water resources systems is an area that is well suited for ES application. ES can be useful in data analysis (hazard evaluation; Wilson, 1986); hydrologic modeling and parameter estimation (Engman et al., 1986; Delleur, 1988) and in formulation of the management problem, choice of model to be used, preparation of input data, etc., say, for a reservoir system (Savic and Simonovic, 1989); water-supply system operations (Shepherd and Ortolano, 1996), and for an urban storm sewer system (Lindberg and Nielsen, 1986). An ES for the calibration of SWMM model was developed by Baffaut and Delleur (1990).

Reservoir operation: Integrated management of a system of multipurpose reservoirs and conjunctive use of surface and groundwater are complex problems requiring multidisciplinary knowledge and experience and, therefore, are suitable for ES applications. An ES for dam operations would use available data on current reservoir levels and inflows, downstream channel conditions, storage capacities, precipitation forecasts, data on demands, etc., and suggest the best operational strategies to maximize overall benefits. In a multireservoir system, such data on all the structures can be jointly analyzed to formulate the best policy for integrated operation. Some applications of ES for this purpose include ES for real-time reservoir operations by Armijos et al. (1990), ES for real-time operation of a multipurpose reservoir system by Fischer and Schultz (1991), and ES for short-term reservoir operation by Simonovic et al. (1992). Since a number of factors and data are involved in real-time operation, a DSS that logically organizes data and models can be of great value. Dam managers are motivated to use DSS by their successful application in many engineering/management tasks and wider availability of computers with user-friendly graphic interfaces, but some are awed by their complexity. Many studies describe the development and application of DSS to real-life problems with the aid of RS, GIS, graphical tools integrated to multiple criteria decision-making (Ahmad and Simonovic, 2006; Qi and Altinakar, 2011; Werner et al., 2013; Şensoy et al., 2018).

Flood management system: ES can be very effective in overall flood management planning and flood forecasting. A flood management planning system would contain knowledge of the basin like topography, channel network, soil properties, flood damage centers, meteorological characteristics of the region, etc. These data could be used in preparing a blueprint for flood management using structural and nonstructural measures. Some relevant applications are: ES for dam site selection by Engel and Beasley (1991); ES to select methods to calculate design flood flows by Varas and Von Chrismar (1995), and for flood frequency analysis by Chow and Watt (1990). Ghobadi and Kaboli

FIGURE 3.32

Role and context of an ES in water management.

Some of the application areas of ES in water resources are detailed below.

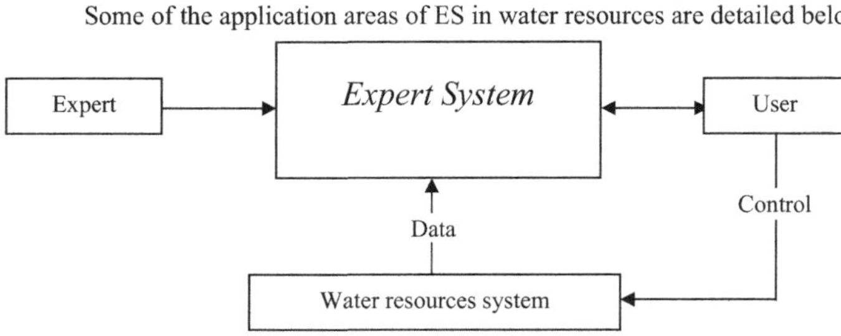

(2020) presented a web-based DSS for flood management by reservoirs with gated spillways. Zhang et al. (2021) have systematically described the steps of flood control operation including predepletion for various forecast horizons and types of floods.

A river stage or discharge forecasting system could use the data, such as stages at various locations in the reach, lateral flow data, current and forecasted precipitation data, cross-section data, longitudinal-section data, characteristics of flood plains, and roughness properties, and use these data in a mathematical model. The results along with the knowledge base can be used to formulate improved flood forecasts. In a similar way, a system for drought forecasting can also be developed. Raman et al. (1992) report an ES for crop planning during droughts.

Water quality monitoring: A water quality monitoring system would take data readings and determine possible causes of poor water quality, aid in the location of pollution sources, and suggest remedial actions to improve water quality. Wishart et al. (1990) developed ESs for the interpretation of river quality data. Sponemann and Fahs (1986) describe the control of an urban storm sewer system using ES.

The MEXSES system developed by Fedra et al. (1991) for the Lower Mekong basin is a rule-based ES, using hierarchical checklists to perform screening level EIA. The system is geared for the assessment of water resources development projects, such as dams and reservoirs, hydropower and irrigation schemes, flood control, navigation, aquaculture, etc. The indicators used to assess a given project are based on checklists of items specific to a project type, covering environmental as well as selected socioeconomic topics. Each indicator is rated on a qualitative scale, from "not significant" to "major." A system of hierarchical checklists was used with a rule-based deduction process including a recursive explanation function and a knowledge base browser. MEXSES uses a knowledge representation, combining an object-oriented design for the descriptors, the basic elements in the inference procedure, with near natural language rules.

Jamieson and Fedra (1996a) described a DSS called "WaterWare" that has, among other modules, an ES. The capabilities of this product were demonstrated through two applications to Thames River basin in England and Rio Lerma in Mexico by Jamieson and Fedra (1996b). In recent times, much work has been done to develop DSS; this topic is discussed in a later chapter.

3.6 CLOSURE

Knowledge-based ESs aim at bringing together the expertise of individuals gathered over long periods of professional practice and the power of the digital computers, especially the personal computers. This can be placed at the disposal of scientists and field engineers for more effective management of the scarce water resources particularly in extreme events like floods and droughts.

The current ESs simulate the thinking of an expert only in a gross manner. Still much work is needed to include a number of important aspects of human thinking process like perceiving significance, reaching intuitive conclusions, examining a single issue from different perspectives, understanding basic principles, and generalization and breadth of knowledge. Hence, compared to human experts, they appear narrow and shallow. However, in the current form, they have the potential to at least partly relieve an expert of a difficult task and give timely advice to beginners and nonexperts. As ES applications increase, it is expected that the vast body of heuristic knowledge and expertise available with water resources professionals will be successfully compiled, organized, formalized, and codified into ES packages for efficient management of scarce resources. There are reasons to believe that ES will have commonplace applications in water resources in near future and these will be an integral part of most DSSs.

An advantage of ESs over traditional modeling approaches is its ability to incorporate human intuition and experience into the modeling process. This argument rests on the ability of ESs to capture human decision-making expertise and represent this expertise as a series of rules and facts. Another major advantage is in bridging gaps between scientific developments and their practical application or reducing the time gap between "lab-to-land" transition.

References

Ahmad, S., Simonovic, S.P., 2006. An intelligent decision support system for management of floods. Water Resources Management 20 (3), 391−410.

Al-Ghobari, H., Ahmed, D., Abed, A., 2020. Estimation of surface water runoff for a semi-arid area using RS and GIS-based SCS-CN method. Water (MDPI) 12 (7), 1924. https://doi.org/10.3390/w12071924.

Armijos, A., Wright, J.R., Houck, M.H., 1990. Bayesian inferencing applied to real-time reservoir operations. Journal of Water Resources Planning and Management, ASCE 116 (1), 38−51.

Arnoff, S., 1989. Geographical Information Systems: A Management Perspective. WDL Publications, Ottawa, Canada.

ASCE Task Committee on Application of Artificial Neural Networks in Hydrology, 2000a. Artificial neural networks in hydrology, I: preliminary concepts. Journal of Hydrologic Engineering, ASCE 5 (2), 115−123.

ASCE Task Committee on Application of Artificial Neural Networks in Hydrology, 2000b. Artificial neural networks in hydrology, II: hydrological applications. Journal of Hydrologic Engineering, ASCE 5 (2), 124−137.

Baffaut, C., Delleur, J.W., 1990. Calibration of SWMM runoff quality model with expert system. ASCE Journal of Water Resources Planning and Management 116 (2), 247−261.

Barr, A., Feigenbaum, E.A. (Eds.), 1981. The Handbook of Artificial Intelligence, vol. 1. Morgan Kaufman.

Barron, A.R., 1993. Universal approximation bounds for superpositions of a sigmoidal function. IEEE Transactions on Information Theory 39 (3), 930−945. https://doi.org/10.1109/18.256500.

Bastiaanssen, W.G.M., 1998. Remote Sensing in Water Resources Management: The State of the Art. International Water Management Institute, Colombo, Sri Lanka.

Bastiaanssen, W., Pelgrum, H., Wang, J., Ma, Y., Moreno, J., Roerink, G., Van der Wal, T., 1998. A remote sensing surface energy balance algorithm for land (SEBAL): Part 2: validation. Journal of Hydrology 212, 213−229.

Birikundavyi, S., Labib, R., Trung, H.T., Roisselle, J., 2002. Performance of neural networks in daily streamflow forecasting. Journal of Hydrologic Engineering, ASCE 7 (5), 392−398.

Burrough, P.A., 1986. Principles of GIS for Land Resource Assessment. Oxford Clarendon Press, London.

Carriere, P., Mohaghegh, S., Gaskari, R., 1996. Performance of a virtual runoff hydrograph system. Journal of Water Resources Planning and Management, ASCE 121 (6), 421−427.

Chang, A.T.C., Tsang, I., 1992. A neural network approach to inversion of snow water equivalent from passive microwave measurements. Nordic Hydrology 23, 173−182.

Chawla, I., Karthikeyan, L., Mishra, A.K., 2020. A review of remote sensing applications for water security: quantity, quality, and extremes. Journal of Hydrology 585 (2020), 124826.

Chow, K.C.A., Watt, W.E., 1990. A knowledge-based expert system for flood frequency analysis. Canadian Journal of Civil Engineering 17 (4), 597−609.

Cigizoglu, H.K., 2008. Artificial neural networks in water resources. In: Gonca Coskun, H., et al. (Eds.), Integration of Information for Environmental Security. Springer, pp. 115−148. Chapter in.

Cuena, I., 1983. The use of simulation models and human advice to build an expert system for the defense and control of river floods. Proceedings of Eighth IJCAI 246−249.

Dawson, C.W., Wilby, R., 1998. An artificial neural network approach to rainfall-runoff modelling. Hydrological Sciences Journal 43 (1), 47–66.

Delleur, J.W., 1988. A front end expert system for the calibration of SWMM runoff block. In: Annual Water Res. Conf., WRP&M Div. ASCE, Norfolk, VA.

DeVantier, B.A., Feldman, A.D., 1993. Review of GIS applications in hydrologic modelling. Journal of Water Resources Planning and Management, ASCE 119 (2), 246–261.

Endreny, T.A., 2002. BASINS toolkit for hydrological monitoring, modelling, and assessment. Hydrological Processes 16, 1331–1335.

Engel, B.A., Beasley, D.B., 1991. DSS: dam site selector expert system for education. ASCE Journal of Irrigation and Drainage Engineering 117 (5), 774–782.

Engman, E.T., 1993. Remote sensing. In: Maidment, D.R. (Ed.), Chapter 24 in Handbook of Hydrology. McGraw-Hill, New York.

Engman, E.T., et al., 1986. An expert system for snowmelt runoff modeling and forecasting. Proceedings of Water Forum 86, 174–180. ASCE.

Fedra, K., Winkelbauer, L., Pantulu, V.R., 1991. Expert Systems for Environmental Screening. An Application in the Lower Mekong Basin. RR-91-19. International Institute for Applied Systems Analysis. A-236l Laxenburg, Austria.

Fernando, D.A.K., Jayawardena, A.W., 2000. Runoff forecasting using RBF networks with OLS algorithm. Journal of Hydrologic Engineering, ASCE 393, 203–209.

Fischer, H., Schultz, G.A., 1991. An expert system for real-time operation of a multi-purpose multi-unit reservoir system. In: Proceedings of the Symposium on Hydrology of Natural and Manmade Lakes. IAHS, pp. 151–157. Pub. No. 206.

French, M.N., Krajewski, W.F., Cuykendall, R.R., 1992. Rainfall forecasting in space and time using a neural network. Journal of Hydrology 137 (1–4), 1–31.

Gasching, J., Reboh, R., Reiter, J., 1981. Development of Knowledge-Based Expert System for Water Resources Problems. SRI Project 1619. SRI International, USA.

Ghobadi, M., Hesam, S.K., 2020. Developing a Web-based decision support system for reservoir flood management. Journal of Hydroinformatics 22 (3), 641–662. https://doi.org/10.2166/hydro.2020.185.

Gorelick, N., Hancher, M., Dixon, M., Simon, I., Thau, D., Moore, R., 2017. Google earth engine: planetary-scale geospatial analysis for everyone. Remote Sensing of Environment 202 (2017), 18–27.

Govindaraju, R.S., Rao, A.R., 2000. Artificial Neural Networks in Hydrology. Kluwer Academic Publishers, Dordrecht.

Govindaraju, R.S., Zhang, B., 2000. Radial-basis function networks. In: Govindaraju, R.S., Rao, A.R. (Eds.), Artificial Neural Networks in Hydrology. Kluwer Academic Publishers, Dordrecht.

Gupta, H.V., Hsu, K.-L., Sorooshian, S., 2000. In: Govindaraju, R.S., Rao, A.R. (Eds.), 'Effective and Efficient Modeling for Streamflow Forecasting' in Artificial Neural Networks in Hydrology. Kluwer Academic Publishers, Dordrecht.

Hayes-Rolth, F.D., Waterman, D.A., Lenat, D.B., 1983. Building Expert Systems. Addison-Wesley, Reading, Mass.

Hsu, K.-L., Gupta, H.V., Sorooshian, S., 1995. Artificial neural network modeling of the rainfall-runoff process. Water Resources Research 31 (10), 2517–2530.

Imrie, C.E., Durucan, S., Korre, A., 2000. Riverflow prediction using artificial neural networks: generalisation beyond the calibration range. Journal of Hydrology 233, 138–153.

Islam, M.M., Sado, K., 2002. Development priority map for flood countermeasures by remote sensing data with geographical information system. Journal of Hydrologic Engineering, ASCE 7 (5), 346–355.

Jain, S.K., 2001. Development of integrated sediment rating curves using artificial neural networks. Journal of Hydraulic Engineering, ASCE 127 (1), 30–37.

Jain, S.K., Chalisgaonkar, D., 2000. Setting stage-discharge relations using ANN. Journal of Hydrologic Engineering, ASCE 594, 428–433.

Jain, S.K., Das, A., Srivastava, D.K., 1999. Application of ANN for reservoir inflow prediction and operation. Journal of Water Resources Planning and management, ASCE 125 (5), 263–271.

Jamieson, D.G., Fedra, K., 1996. The 'WaterWare' decision-support system for river basin planning. 1. Conceptual design. Journal of Hydrology 177, 163–175.

Jamieson, D.G., Fedra, K., 1996a. The 'WaterWare' decision-support system for river-basin planning. 1. Conceptual design. Journal of Hydrology 177 (3–4), 163–175.

Jamieson, D.G., Fedra, K., 1996b. The 'Water Ware' decision-support system for river-basin planning 3. Example applications. Journal of Hydrology 177 (3–4), 199–211.

Jenson, S.K., Domingue, J.O., 1988. Extracting. topographic structure from digital elevation data for geographic information system analysis. Photogrammetric Engineering and Remote Sensing 54.

Johnston, C.A., 1998. Geographic Information Systems in Ecology. Blackwell Science Ltd., London.

Kalma, J.D., McVicar, T.R., McCabe, M.F., 2008. Estimating land surface evaporation: a review of methods using remotely sensed surface temperature data. Surveys in Geophysics 29 (4), 421–469.

Klemes, V., 1986. Dilettantism in hydrology: transition or destiny? Water Resources Research 22 (9), 177S–188S.

Kothari, R., Agyepong, K., 1996. On lateral connections in feed-forward neural networks. In: Proceedings of IEEE International Conference on Neural Networks, 13-18. Institute of Electrical and Electronics Engineers, New York.

Kratzert, F., Klotz, D., Brenner, C., Schulz, K., Herrnegger, M., 2018. Rainfall–runoff modelling using long short-term memory (LSTM) networks. Hydrology and Earth System Sciences 22, 6005–6022, 2018.

Kumar, M., Raghuwanshi, N.S., Singh, R., Wallender, W.W., Pruitt, W.O., 2002. Estimating evapotranspiration using artificial neural network. Journal of Irrigation and Drainage Engineering, ASCE 128 (4), 224–233.

Levizzani, V., Cattani, E., 2019. Satellite remote sensing of precipitation and the terrestrial water cycle in a changing climate. Remote Sensing 11, 2301. https://doi.org/10.3390/rs11192301, 2019.

Lillesand, T.M., Kiefer, R.W., 1994. Remote Sensing and Image Interpretation. Wiley, New York.

Lindberg, S., Nielsen, J.B., 1986. Modeling of Urban Storm Sewer Systems in 31, pp. 687–696.

Loucks, D.P., Fedra, K., 1987. Impact of changing computer technology on hydrologic and water resource modeling. Review of Geophysics 25 (2). March.

Ma, S., Zhou, Y., Gowda, P.H., Dong, J., Zhang, G., Kakani, V.G., Pradeep, W., Chen, L., Colton Flynn, K., Jiang, W., 2019. Application of the water-related spectral reflectance indices: a review. Ecological Indicators 98, 68–79. March 2019.

Maher, M.L., 1986. Problem solving using expert system techniques. In: Kostem, C.N., Maher, M.L. (Eds.), Proceedings of Symposium Expert Systems in Civil Engineering. Seattle, USA.

Maling, D.H., 1992. Coordinate Systems and Map Projections. Pergamon Press, Oxford.

Maren, A.J., Hartson, C.T., Pap, R.M., 1990. Handbook of Neural Computing Applications. Academic Press, San Diego.

Mather, P.M., 1987. Computer Processing of Remotely Sensed Images, An Introduction. John Wiley and Sons, Chicester.

Minasny, B., McBratney, A.B., 2002. The neuro-m method for fitting neural networks parametric pedotransfer functions. Journal of Soil Society of America 66, 353–361.

Minns, A.W., Hall, M.J., 1996. Artificial Neural Networks as rainfall runoff models. Hydrological Sciences Journal 41 (3), 399–418.

Moore, I.D., Grayson, R.B., 1991. Terrain-based catchment partitioning and runoff prediction using vector elevation data. Water Resources Research 27 (6), 1177–1191.

Mu, Q., Zhao, M., Running, S.W., 2011. Improvements to a MODIS global terrestrial evapotranspiration algorithm. Remote Sensing of Environment 115, 1781–1800. https://doi.org/10.1016/j.rse.2011.02.019.

Muzik, I., 1996. Flood modeling with GIS-derived distributed unit hydrographs. Hydrological Processes 10, 1401–1409.

Nelson, A., Reuter, H.I., Gessler, P., 2009. Chapter 3 DEM production methods and sources. In: Hengl, T., Reuter, H.I. (Eds.), Developments in Soil Science, vol 33. Elsevier.

NIST, 1992. Spatial Data Transfer Standards (SDTS), FIPS Pub. 173. US Department of Commerce, Computer Systems Laboratory, Maryland. SuDocs C13.52:173.

O'Callaghan, J.F., Mark, D.M., 1984. The extraction of drainage networks from digital elevation data. Computer Vision, Graphics, and Image Processing 28, 232–344.

Palmer, R.N., Holmes, K.J., 1988. Operational guidance during droughts: expert system approach. ASCE Journal of Water Resources Planning and Management 114 (6), 647–666.

Qi, H., Altinakar, M.S., 2011. A GIS-based decision support system for integrated flood management under uncertainty with two dimensional numerical simulations. Environmental Modelling and Software 26 (6), 817–821.

Raman, H., Chandramauli, V., 1996. Deriving a general operating policy for reservoir using Neural Network. Journal of Water Resources Planning and Management 122 (5), 342–347.

Raman, H., Sunilkumar, N., 1995. Multivariate modeling of water resources time-series using artificial neural networks. Hydrological Sciences Journal 40 (2), 145–163.

Raman, H., Mohan, S., Rangacharya, N.C.V., 1992. Decision support for crop planning during droughts. Journal of Irrigation and Drainage Engineering, ASCE 118 (2), 229–241.

Rao, P.P.N., Mohankumar, A., 1994. Cropland inventory in the command area of Krishnarajsagar project using satellite data. International Journal of Remote Sensing 15 (6), 1295–1305.

Rodell, M., Velicogna, I., Famiglietti, J.S., 2009. Satellite-based estimates of groundwater depletion in India. Nature 460, 999–1002. https://doi.org/10.1038/nature08238.

Rouhani, S., Kangari, R., 1987. Expert systems in water resources. In: Water for the Future: Hydrology in Perspective. Proceedings of Rome Symposium. IAHS. Publication No. 164.

Rumelhart, D.E., McLelland, J.L., the PDP Research Group, 1986. Parallel Distributed Processing, Explorations in the Micro Structure of Cognition, vol. I. Foundations, The MIT Press, Cambridge.

Savic, D.A., Simonovic, S.P., 1989. Introduction of fuzzy reasoning in intelligent decision support for reservoir analysis. In: Maksimovic, C., Radojkovic, M. (Eds.), Computational Modelling and Experimental Methods in Hydraulics, Proceedings of HYDROCOMP '89. Elsevier, pp. 404–414, 1989.

Şensoy, A., Uysal, G., Şorman, A.A., 2018. Developing a decision support framework for real-time flood management using integrated models. Journal of Flood Risk Management 11, S866–S883.

Sheffield, J., Wood, E.F., Pan, M., Beck, H., Coccia, G., Serrat-Capdevila, A., Verbist, K., 2018. Satellite remote sensing for water resources management: potential for supporting sustainable development in data-poor regions. Water Resources Research 54, 9724–9758. https://doi.org/10.1029/2017WR022437.

Shen, C., 2018. A transdisciplinary review of deep learning research and its relevance for water resources scientists. Water Resources Research 54, 8558–8593. https://doi.org/10.1029/2018WR022643.

Shepherd, A., Ortolano, L., 1996. Water-supply system operations: critiquing expert-system approach. ASCE Journal of Water Resources Planning and Management 122 (5), 348–355.

Shih, S.F., 1996. Integration of remote sensing and GIS for hydrological studies. In: Singh, V.P., Fiorentino, M. (Eds.), Geographical Information Systems in Hydrology. Kluwer Academic Publishers, Dordrecht.

Simonovic, S.P., 1991. Knowledge-based systems and operational hydrology. Canadian Journal of Civil Engineering 18 (1), 1–11.

Simonovic, S.P., Venema, H.D., Burn, D.H., 1992. Risk-based parameter selection for short-term reservoir operation. Journal of Hydrology 131 (1–4), 269–291.

Singh, V.P., Fiorentino, M., 1996. Geographical Information Systems in Hydrology. Kluwer Academic Publishers, Dordrecht.

Sit, M., Demiray, B.Z., Xiang, Z., Ewing, G.J., Sermet, Y., Demir, I., 2020. A comprehensive review of deep learning applications in hydrology and water resources. Water Science and Technology 82 (1), 2635–2670. https://doi.org/10.2166/wst.2020.369.

Smith, J., Eli, R.N., 1995. Neural Network models of rainfall-runoff process. Journal of Water Resources Planning and Management 121 (6), 499–508.

Sole, A., Valanzano, A., 1996. Digital terrain modeling. In: Singh, V.P., Fiorentino, M. (Eds.), Geographical Information Systems in Hydrology. Kluwer Academic Publishers, Dordrecht.

Sponemann, P., Fuchs, L., 1986. Use of artificial Intelligence for real time control in. In: Sriram, D., Adey, R. (Eds.), Applications of AI in Engineering Problems, vol. II. Springer Verlag, pp. 415–422.

Thirumalaiah, K., Deo, M.C., 1998. River stage forecasting using artificial neural networks. Journal of Hydrologic Engineering, ASCE 3 (1), 63–67.

Thirumalaiah, K., Deo, M.C., 2000. Hydrological forecasting using neural networks. Journal of Hydrologic Engineering, ASCE 5 (2), 180–189.

Varas, E.A., Von Chrismar, M., 1995. Expert system for the selection of methods to calculate design flood flows. Hydrological Sciences Journal 40 (6), 739–749.

Vieux, B.E., 2001. Distributed Hydrologic Modeling Using GIS. Kluwer Academic Publishers, Dordrecht.

Wang, X., Xie, H., 2018. A review on applications of remote sensing and geographic information systems (GIS) in water resources and flood risk management. Water (MDPI) 10 (5), 608. https://doi.org/10.3390/w10050608.

Werner, M., Schellekens, J., Gijsbers, P., van Dijk, M., van den Akker, O., Heynert, K., 2013. The Delft-FEWS flow forecasting system. Environmental Modelling and Software 40, 65–77.

Wilson, J.L., et al., 1986. GEOTOX: a knowledge based system for hazard site evaluation. In: Sriram, D., Adey, R. (Eds.), Applications of AI in Engineering Problems, vol. II. Springer Verlag, pp. 661–671.

Wishart, S.J., Lumbers, J.P., Griffiths, I.M., 1990. Expert systems for the interpretation of river quality data. Journal of the Institution of Water and Environmental Management 4 (2), 194–202.

Zhang, B., Govindaraju, R.S., 2000. Modular neural networks for watershed runoff. In: Govindaraju, R.S., Rao, A.R. (Eds.), Artificial Neural Networks in Hydrology. Kluwer Academic Publishers, Dordrecht.

Zhang, J., Min, Y., Feng, B., Duan, W., 2021. Research and application of key technologies for dynamic control of reservoir water level in flood season. Water 13, 3576. https://doi.org/10.3390/w13243576.

Further Reading

Ambast, S.K., Keshari, A.K., Gosain, A.K., 2000. Estimation of regional evapotranspiration for a part of WYC command using remote sensing technique. In: Proceedings of International Conference on Integrated Water Resources Management for Sustainable Development, New Delhi.

Bausch, W.C., 1995. Remote sensing of crop coefficients for improving irrigation scheduling of corn. Agricultural Water Management 27 (1), 55–68.

Bausch, W.C., Neale, C.M.U., 1987. Crop coefficients derived from reflected canopy radiation: a concept. Transactions of the American Society of Agricultural Engineers 30 (3), 703–709.

Choudhury, B.J., Ahmed, N.U., Idso, S.B., Reginato, R.J., Daughtry, C.S.T., 1994. Relations between evaporation coefficients and vegetation indices studied by model simulations. Remote Sensing of Environment 50, 1–17.

Demers, M.N., 1997. Fundamentals of Geographic Information Systems. John Wiley and Sons, New York.

DHI, 1998. MIKE-SHE Users Guide and Technical Reference Manual. Danish Hydraulic Institute, Denmark.

Dozier, J., 1989. Spectral signature of Alpine snow cover from the Landsat thematic mapper. Journal of Remote Sensing of Environment 28, 9–22.

Dugdale, G., Milford, J.R., 1986. Rainfall estimation over the Sahel using Meteosat thermal infrared data. In: Rolfe, E., Battrick, B. (Eds.), ISLCP. Parametrisation of Land-Surface Characteristics: Use of Satellite Data in Climate Studies. European Space Agency, Paris.

Engman, E.T., Gurney, R.J., 1991. Remote Sensing in Hydrology. Chapman and Hall, London.

Goel, M.K., Jain, S.K., 1997. Automated Mapping of Snow Cover Using IRS-1C Data. Report No. TR(BR)-6/96-97. National Institute of Hydrology, Roorkee.

Hall, D.K., Martinec, J., 1985. Remote Sensing of Ice and Snow. Chapman and Hall, London.

Hsu, K., Gao, X., Sorooshian, S., Gupta, H.V., 1997. Precipitation estimation from remotely sensed information using artificial Neural Networks. Journal of Applied Meteorology 36 (9), 1176–1190.

Jayaseelan, A.T., Rao, D.P., 1997. Satellite contribution to natural disaster management in India. In: Conference on Geomatics in Disaster Management. Dehradun, India.

Karunanithi, N., Grenney, W.J., Whitley, D., Bovee, K., 1994. Neural networks for river flow prediction. Journal of Computing in Civil Engineering, ASCE 8 (2), 210–220.

Kite, G.W., Ellehoj, E., Dalton, A., 1996. GIS for large-scale watershed modelling. In: Singh, V.P., Fiorentino, M. (Eds.), Geographical Information Systems in Hydrology. Kluwer Academic Publishers, Dordrecht.

Maksimovic, C., Radojkovic, M., 1989. Computational Modelling and Experimental Methods in Hydraulics, Proc. HYDROCOMP '89. Elsevier.

Mendicino, G., 1996. Design of GIS for hydrological applications. In: Singh, V.P., Fiorentino, M. (Eds.), Geographical Information Systems in Hydrology. Kluwer Academic publishers, Dordrecht.

Nagarajan, R., 2000. Temporal orbital satellite images in the drought assessment and early warning decisions. Proceedings of International Conference on Integrated Water Resources Management for Sustainable Development, New Delhi.

Pauwels, V.R.N., Hoeben, R., Verhoest, N.E.C., Troch, F.P.De, Troch, P.A., 2002. Improvement of TOPLATS-based discharge prediction through assimilation of ERS-based remotely sensed soil moisture values. Hydrological Processes 16 (5), 995–1013.

Quinn, P.F., Beven, K.J., Lamb, R., 1995. The ln(a/tanβ) index: how to calculate it and how to use it in the TOPMODEL framework. Hydrological Processes 9, 161–182.

Rango, A., Martinec, J., 1999. Modeling snow cover and runoff response to global warming for varying water years. World Resource Review 11 (1), 76–91.

Richards, F., Arkin, P., 1981. On the relationship between satellite-observed cloud cover and precipitation. Monthly Weather Review 109, 1081–1093.

Sabins, F.F., 1987. Remote Sensing Principle and Interpretation. W.H. Freeman and Co., San Francisco.

Schultz, G.A., 1993. Application of GIS and remote sensing in hydrology. HydroGIS 93: Application of Geographic Information Systems in Hydrology and Water Resources. IAHS Pub. No. 211, pp. 127–140.

Seidel, K., Schaper, J., Martinec, J., 2000. Mapping of snow cover and glaciers with high resolution remote sensing data for improved runoff modeling. In: Proceedings of International Conference on Integrated Water Resources Management for Sustainable Development, New Delhi.

Sell, P.S., 1985. Expert System—A Practical Introduction. Macmillan, London.

Simonovic, S.P., 1989. Expert system design for intelligent decision support for surface water quality data management. In: Maksimovic, C., Radojkovic, M. (Eds.), Computational Modeling and Experimental Methods in Hydraulics, Proceedings of HYDROCOMP '89. Elsevier, Dordrecht, pp. 383–393, 1989.

Simonovic, S.P., 1990. An expert system for the selection of a suitable method for flow measurement in open channels. Journal of Hydrology 112 (3–4), 237–256.

Smith, J.A., 1993. Precipitation. In: Maidment, D.R. (Ed.), Chapter 3 in Handbook of Hydrology. McGraw-Hill, New York.

Tokar, A.S., Markus, M., 2000. Precipitation-runoff modeling using artificial neural networks and conceptual models. Journal of Hydrologic Engineering, ASCE 5 (2), 156–161.

Turban, E., 1995. Decision Support System and Expert System: Management Support System. Prentice-Hall Inc., Englewood Cliffs, New Jersey.

CHAPTER

Statistical Techniques for Data Analysis

4

Chapter outline

Water Resources Systems Planning and Management. https://doi.org/10.1016/B978-0-12-821349-0.00009-5

The objectives of this chapter are to:

- explain the basic concepts of statistical analysis of data,
- describe the frequency distributions that are commonly used in water resources, and describe methods of estimation of their parameters,
- discuss regression and correlation analysis, and
- briefly discuss time series analysis.

The observed behavior of many water resources variables is chance-dependent and cannot be adequately explained in terms of known physical laws. This could be because of: (a) a poor or incorrect understanding of the underlying complex processes, (b) nonavailability of sufficient data, and (c) inherent randomness of the variable. In such cases, techniques for statistical analysis are summoned to make inferences about the behavior of variables.

Statistics deals with methods to draw inferences about the properties of a *population* based on sample data. Population refers to a collection of objects. It can be finite or infinite, for example, the collection of all flow data of a river at a given site. Often, measurements of the entire population are not available and what is generally available is a limited number of observations or a finite *sample*. Based on this sample, properties of the population are determined by assuming them to be unbiased estimates of the properties of population.

A variable whose value at any time is not influenced by the value at earlier time(s) is known as a *random variable*. Such a variable can be discrete which can take on only a finite set of values, such as number of rainy days at a place in a year. It can also be continuous and can take on any value, for example, the water level of river at a gauging site or the magnitude of rainfall at a place.

For many problems, the sample data consist of measurements on a single random variable; the techniques of analysis are called univariate analysis and estimation. Univariate analysis is carried out by using measurements of the random variable, which is called sample information, to discern the statistical properties of population from which sample measurements are likely to have come. After the underlying population has been identified, one can make probabilistic statements about the future occurrences of the random variable values, this represents univariate estimation. It is important to remember that univariate estimation is based on the assumed population, not the sample, and the sample is used only to characterize the population.

The following are the main steps of statistical analysis of data:

(i) It is always useful to first plot the sample data.

(ii) Select a set of probability distribution functions.
(iii) Fit the selected distributions to the sample data. Common methods of parameter estimation are least squares (LS) method; methods of moments (regular moments, partial moments, geometric moments, cumulative moments), probability weighted moments (PWMs), and linear moments; entropy-based methods; and method of maximum likelihood (ML).
(iv) Select the best fit distribution using the goodness-of-fit tests.
(v) Use the best-fit probability distribution to make inferences about the likelihood of occurrence of the magnitudes of the random variable.

If all the values of a random variable and the corresponding probabilities are known or found, the relation between these values and probabilities is described by a probability distribution. Knowing this distribution, the probability of any value of the random variable can be determined.

In statistical analysis of multivariable data, the functional forms of the relationships are studied. Linear regression analysis is one of the ways to develop a suitable form of the multiple-variable model, wherein a dependent variable takes on values caused by variations in one or more independent or predictor variables.

4.1 BASIC CONCEPTS

Let X denote a random variable and x be a particular value of X. The cumulative distribution function (CDF), $F_X(x)$, is the probability that the random variable X is less than or equal to x:

$$F_X(x) = \mathrm{P}(X \leq x) \tag{4.1}$$

The probability density function (PDF) describes the relative likelihood that a continuous random variable X takes on different values and is the derivative of the CDF:

$$f_X(x) = d\,\{F_X(x)\}/dx \tag{4.2}$$

The PDF and CDF of a random variable are shown in Fig. 4.1.

At this stage, it is useful to state some of the properties of $f(x)$ and $F(x)$ for continuous random variables.

1. The probability of a random variable cannot be negative

$$F(x) \geq 0, -\infty < x < \infty \tag{4.3}$$

2. The sum of probabilities of all possible outcomes is equal to 1, that is, the area under the PDF is unity.

$$\int_{-\infty}^{\infty} f(x)dx = 1 \tag{4.4}$$

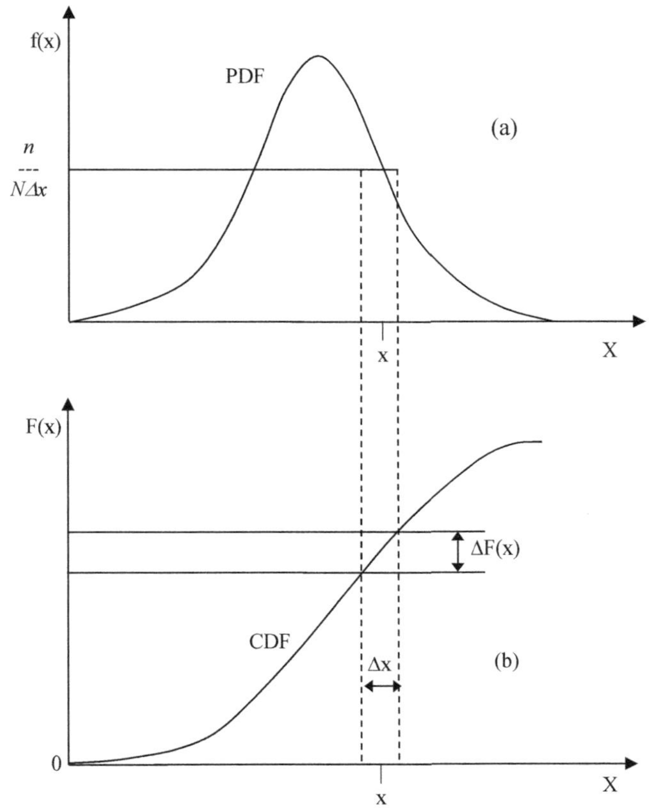

FIGURE 4.1

PDF and CDF of a random variable.

$$3.\ P(X \leq x) = F(X \leq x) = F(x) \int_{-\infty}^{x} f(x)dx \tag{4.5}$$

If a and b are any real numbers such that $a < b$, the events $X \leq a$ and $a < X \leq b$ will be mutually exclusive. Then $P(X \leq b) \geq P(X \leq a)$ or $F(X \leq b) \geq F(X \leq a)$, and

$$P(X \leq b) = P(X \leq a) + P(a < X \leq b)$$

$$= \int_{-\infty}^{a} f(x)dx + \int_{-a}^{b} f(x)dx = \int_{-\infty}^{b} f(x)dx = F(X \leq b)$$

This yields

$$P(a < X \leq b) = P(X \leq b) - P(X \leq a)$$

$$= \int_{-\infty}^{b} f(x)dx - \int_{-\infty}^{b} f(x)dx$$

$$= \int_{a}^{b} f(x)dx, for\ a < b \tag{4.6}$$

4. The probability that X (continuous variable) assumes a particular value is zero, that is, $P(X = a) = F(x = a) = 0$,

$$\int_{a}^{a} f(x)dx = F(a) - F(a) = 0 \tag{4.7}$$

$$5.\ F(+\infty) = \lim_{x \to -\infty} F(x) = 1 \tag{4.8}$$

Also

$$F(-\infty) = \lim_{x \to -\infty} F(x) = 0 \tag{4.9}$$

This can be verified from the area under the PDF.
For discrete random variables, analogous statements can be made:

$$1.\ \sum_{i} f(x_i) = 1 \tag{4.10}$$

where $f(x_i)$ represents the probability of $X = x_i$ in the sample space if the observations are finite in the sample. Thus, this can be replaced by $p(x_i)$.

$$2.\ P(a \le x \le b) = \sum_{x_i \ge a}^{x_i \le b} p(x_i) \tag{4.11}$$

$$3.\ P(X \le x_k) = \sum_{i=1}^{k} p(x_i) \tag{4.12}$$

4.1.1 Distribution Characteristics

There are four principal moments for characterizing probability distributions:

(i) the central tendency or the value around which all other values are clustered,
(ii) the spread of the sample values around mean,
(iii) the asymmetry or skewness of the frequency distribution, and
(iv) the flatness of the frequency distribution.

These characteristics are expressed in terms of the parameters of distributions, and the parameters can themselves be expressed in terms of moments. These parameters are estimated from the observed sample data and are then used as estimates of the parameters of the population distribution.

Measures of Central Tendency

In statistics, various measures of location are described. The important measures are the following.

(i) Arithmetic mean: If x_1, x_2 ..., x_n represent a sequence of observations, the mean of this sequence is the ratio the of sum of values and the number of values:

$$\bar{x} = \frac{1}{n}\sum_{i=1}^{n} x_i \tag{4.13}$$

where $\bar{x}$ represents the sample mean; population mean is generally represented by μ.

(ii) Mode: It is the value in the sample (or population) which occurs most frequently. It is the peak value of the PDF. Note that a sample or population may have more than one peak.

(iii) Median: It is the middle value of the ranked values for a sample (or population). The median divides the distribution in two equal parts.

Measure of Dispersion or Variation

Some of the important measures of dispersion or variation include:

(i) Variance: It represents the dispersion of data about the mean and is expressed as:

$$s^2 = \frac{1}{n}\sum_{i=1}^{n} (x_i - \bar{x})^2 \tag{4.14}$$

(ii) Standard deviation: The unbiased estimate of population standard deviation (s) from the sample is given as the square root of the variance, that is,

$$s = \left[\frac{1}{n}\sum_{i=1}^{n} (x_i - \bar{x})^2\right]^{1/2} \tag{4.15}$$

For $n < 30$, the unbiased estimate of s is found by replacing n by $n-1$ in this equation.

(iii) The coefficient of variation C_V is a dimensionless dispersion parameter and is equal to the ratio of the standard deviation and the mean:

$$C_v = s/\bar{x} \tag{4.16}$$

The variance has the square of the units of the original data. The standard deviation has the same dimensions as of the data. The coefficient of variation is a dimensionless quantity.

Measures of Symmetry

If the data are exactly symmetrically displaced about the mean, then the measure of symmetry should be zero. If the data to the right of the mean (larger) are more spread out from the mean than those on the left then, by convention, the asymmetry is positive and vice versa for negative asymmetry.

The third moment of the sample data about the mean is given by:

$$M_3 = \frac{1}{n}\sum_{i=1}^{n} (x_i - \bar{x})^3 \tag{4.17}$$

This moment is zero if the data are symmetrical. Otherwise it is positive or negative.

Coefficient of skewness: It is a nondimensional measure of the asymmetry of the distribution of the data. An unbiased estimate of the coefficient is given by:

$$C_S = \frac{n \sum_{i=1}^{n} (x_i - \overline{x})^3}{(n-1)(n-2)S^3} \tag{4.18}$$

Symmetrical frequency distributions have very small or negligible sample skewness coefficient C_s, while asymmetrical frequency distributions have either positive or negative coefficients. Often a small value of C_s indicates that the frequency distribution of the sample may be approximated by the normal distribution since $C_s = 0$ for this function. The symmetrical and skewed distributions are shown in Fig. 4.2.

Note that because the third central moment has dimensions equal to the cube of the data, it is not of direct use while comparing different data sets. The coefficient of skewness does not have this disadvantage and is, therefore, preferred.

Measures of Peakedness or Flatness

The kurtosis coefficient measures the peakedness or the flatness of the frequency distribution near its center. An unbiased estimate of this coefficient is given by:

$$C_k = \frac{n^2 \sum_{i=1}^{n} (x_i - \overline{x})^4}{(n-1)(n-2)(n-3)S^4} \tag{4.19}$$

The kurtosis of a normal distribution is 3.

Standard Errors of Sample Statistics

Because of the short length of most records, the statistics calculated from the sample are only estimates of the true or population values which would be available if very large samples were available. The reliability of statistics calculated from the sample can be judged from the standard error of estimate

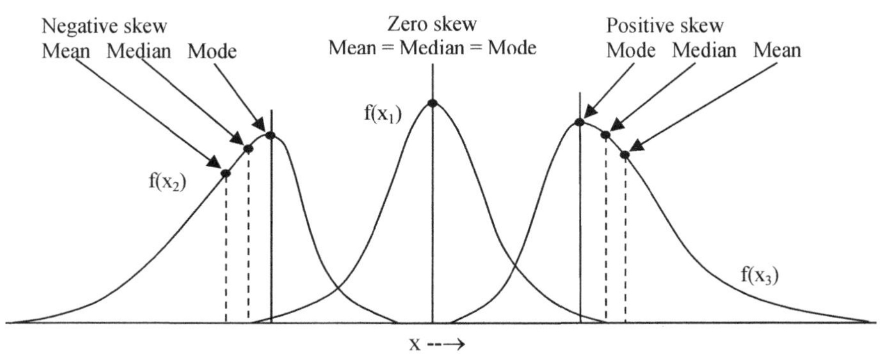

FIGURE 4.2

Symmetrical and skewed distributions.

(SEE). According to the statistical theory, the probability that the true or population value of each statistic is within one SEE of the value calculated from the available data is about 68%.

The standard errors of mean, standard deviation, and coefficient of skewness are, respectively, given below:

$$S_e(\bar{x}) = s/\sqrt{n} \tag{4.20}$$

$$S_e(S) = s/\sqrt{2n} \tag{4.21}$$

$$S_e(C_s) = \sqrt{6n(n-1)/[(n+1)(n+2)(n+3)]} \tag{4.22}$$

Clearly, the SEE of each parameter becomes smaller, as the length of record used in the analysis becomes longer.

Graphical Presentation of Data

For graphical presentation of data in the form of histograms and cumulative histograms of frequency (or relative frequency or probability), a frequency table is prepared. The range of data is divided into a number of intervals of convenient size, and the number of frequencies of values occurring in each interval is entered alongside. This table provides a valuable summary. The selection of class interval can affect the appearance of a frequency histogram. If the class intervals are very large, the table is compact but loses detail. If the intervals are too small, the table may be too bulky and not succinct enough. For the choice of class interval, the following guidelines may be considered:

(a) Brooks and Carruthers' rough guide:

$$\text{No of classes} \le 5 \log(\text{sample size}) \tag{4.23}$$

(b) Charlier's rule of thumb:

$$w = (\text{maximum value} - \text{minimum value})/20 \tag{4.24}$$

where w is the size of class interval. Number of classes is generally 15 to 25.

(c) According to Sturges (1926), the class interval can be estimated by:

$$m = 1 + 3.3 \text{ Log } (n)$$

where m is the number of classes, and n is the number of observations.

The frequency table can be prepared using the following steps:

(i) Arrange the variable (X_i) in increasing or decreasing order of magnitude.
(ii) Decide the number of class intervals (NC) and thereby the size of the class interval ΔX following the above guidelines.
(iii) Divide the ordered observations X_i into NC intervals (or groups).
(iv) Determine the absolute frequency n_j by counting the observations that fall within the jth class interval for $j = 1, \ldots,$ NC.
(v) Determine the corresponding relative frequencies as n_j/n, $j = 1, \ldots,$ NC.

(vi) Compute the cumulative relative frequencies F_j, $j = 1, \ldots$ NC. These cumulative frequencies approximate the probabilities as:

$$F_j = F(X \leq x) \text{ if order is increasing, or}$$

$$F_j = F(X > x) \text{ if order is decreasing.}$$

(vii) Prepare the plots for the relative frequencies as well as cumulative relative frequencies on simple graph papers, taking the group interval as abscissa and the relative frequencies or cumulative relative frequencies as ordinate.

Example 4.1. The annual flow of Sabarmati River at Dharoi is plotted in Fig. 4.3 for the period 1868–1965. Find the statistical parameters of these data and plot the histogram.

Solution. The histogram of the annual flow of Sabarmati River at Dharoi for the period 1868–1965 is plotted in Fig. 4.4.

Mean of the data $\bar{x} = 65206/98 = 665.37$ million cubic m.

Variance $\sigma^2 = 11841713/98 = 120833.8$ (million cubic m)^2.

Standard deviation $\sigma = (120833.8)^{0.5} = 346.9$ million cubic m.

Coefficient of variation $C_V = 346.9/665.37 = 0.521$.

FIGURE 4.3

Annual flow of Sabarmati River at Dharoi.

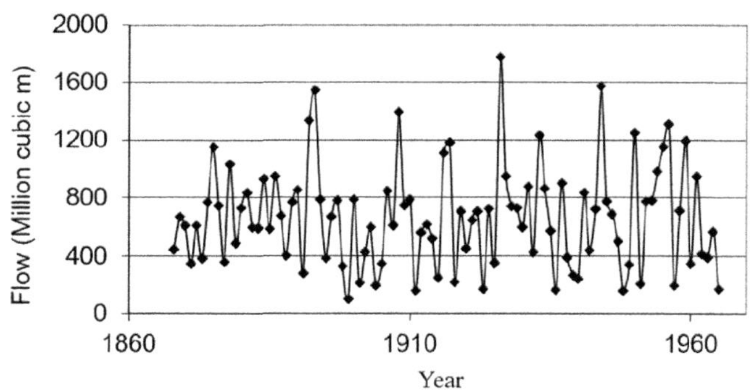

FIGURE 4.4

Cumulative histogram of annual flows of River Sabarmati.

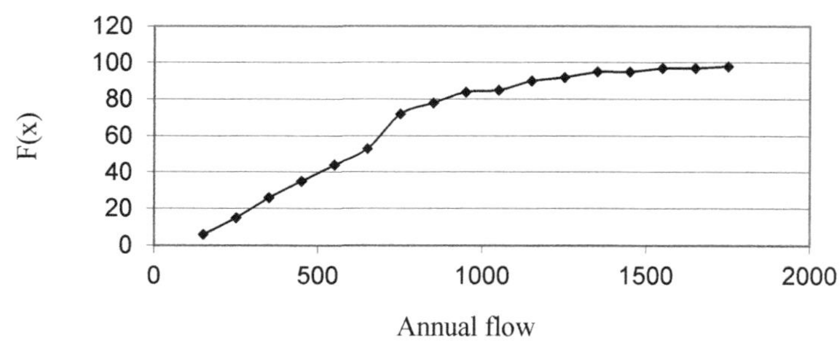

Coefficient of skewness $C_s = 0.76$ (positively skewed).
Kurtosis $C_k = 3.65$.
Standard error of mean $= 346.9/(98)^{0.5} = 35.042$.
Standard error of SD $= 346.9/(2*98)^{0.5} = 24.85$.
Standard error of $C_s = [6*98*97/(99*100*101)]^{0.5} = 0.239$.

4.2 PROBABILITY DISTRIBUTIONS

A distribution is an attribute of a statistical population. It describes the relation between the random variable and the probabilities. The distribution gives important information about the data, whether they are bunched together or spread out, and whether they are symmetrically disposed on the X-axis or not. Distribution also tells the relative frequency or proportion of various X values in the population in the same way that a histogram gives that information about a sample.

The distributions that are commonly used in water resources problems are described in the following. A summary of the distributions is provided in Table 4.1.

4.2.1 Continuous Probability Distributions

The commonly used continuous distributions are the normal, log normal, extreme value type-1 (Gumbel or EV1), gamma, Pearson type-III, log-Pearson type-III, Weibull, and logistic distributions. The probability density functions (PDFs), CDFs, and other properties of these distributions are given below.

Normal Distribution

Also known as Gaussian or error distribution, the normal distribution is a symmetrical bell-shaped probability density function. When a hydrologic variable, integrated over a large time period, is used in analysis, the variable is expected to follow a normal distribution. The normal distribution has two parameters, mean μ and standard deviation σ, and its PDF can be expressed as

$$f(x)\frac{1}{\sigma\sqrt{2\pi}}\exp\left[-\frac{(x-\mu)^2}{2\sigma^2}\right] \quad -\infty < x < \infty \tag{4.25}$$

Integrating Eq. (4.25), the CDF of the normal distribution is:

$$F(x) = \frac{1}{\sigma\sqrt{2\pi}}\int_{-\infty}^{x}\exp\left[-\frac{(u-u)^2}{2\sigma^2}\right]du \tag{4.26}$$

For the normal distribution, the reduced variate is $Z = (x-\mu)/\sigma$. The mean of the reduced variate is 0, standard deviation $\sigma_z = 1$, and its coefficient of skewness is 0. Fig. 4.5 shows the normal distribution and the area under the distribution curve for three values of the standard variate.

The normal distribution is the most widely used distribution and is employed in the analysis of variance, estimation of random errors of hydrologic measurements, hypothesis testing, generation of random numbers, etc. A random variable that is made up of the sum of many small independent effects is expected to follow a normal distribution. Many hydrologic variables are not normally distributed,

Table 4.1 Summary of distributions commonly used in hydrology.

Distribution	Probability density function	Range	Mean	Variance
Binomial		$0 \le x \le n$	np	$np(1-p)$
Geometric	$P(x) = pq^{x-1}$, $q = 1-p$	$1 \le x \le \ldots.$	$1/p$	q/p^2
Poisson	$P(x) = \frac{\lambda^x \exp(-\lambda)}{x!}$	$0 \le x$	Λ	Λ
Exponential	$f(x) = \lambda \exp(-\lambda x)$	$0 \le x \le \infty$	$1/\lambda$	$1/\lambda^2$
Gamma	$f(x) = \frac{\lambda^n x^{n-1}}{(n-1)!} \exp(-\lambda x)$	$0 \le x \le \infty$	n/λ	n/λ^2
Normal	$f(x) = \frac{1}{\sigma\sqrt{2n}} \exp\left[-\frac{(x-\mu)^2}{2\sigma^2}\right]$	$-\infty < x < \infty$	μ	σ^2
Log-normal ($y = \ln x$)	$f(x) = \frac{1}{\sigma x\sqrt{2n}} \exp\left[-\frac{(\ln x-\mu_y)^2}{2\sigma_y^2}\right]$	$0 < x < \infty$	μ_y or $\exp\left(\mu_y + \sigma_y^2/2\right)$	σ_y^2 or $\mu_x^2\left[\exp\left(\sigma_y^2\right) - 1\right]$
Gumbel	$f(x) = \alpha \exp\{-\alpha(x-\beta) - \exp[-\alpha(x-\beta)]\}$	$-\infty < x < \infty$	$\mu + \gamma/\alpha$ $\gamma = 0.5772$	$n^2/6\alpha^2$
Pearson type III	$f(x) = \frac{1}{a\Gamma(b)}\left(\frac{x-c}{a}\right)^{b-1} \exp\left(-\frac{x-c}{a}\right)$	$-\infty < x < \infty$	$ab + c$	a^2b
Log Pearson type III ($y = \ln x$)	$f(x) = \frac{1}{ax\Gamma(b)}\left(\frac{\ln x-c}{a}\right)^{b-1} \exp\left(-\frac{\ln x-c}{a}\right)$	$0 < x < \infty$	$\mu_y = c + ab$	$\sigma_y^2 = a^2b$

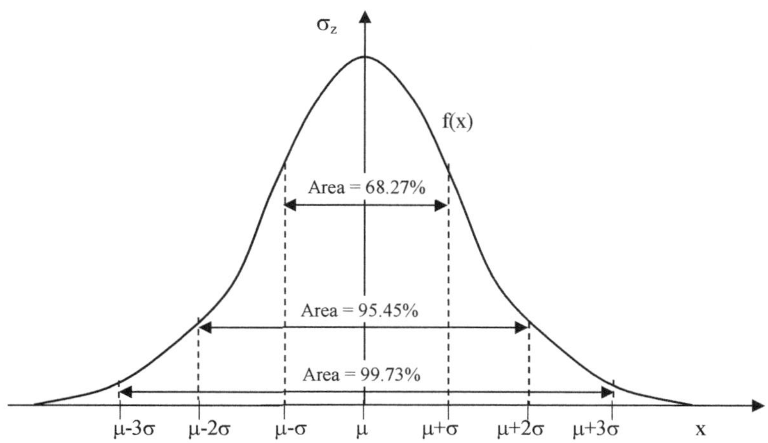

FIGURE 4.5

The normal distribution and the area for three values of the standard variate.

but transformations can, in many cases, make them approximately normally distributed. When the time interval over which a hydrologic variable is measured increases, the variable approximately follows a normal distribution because the number of causative effects increases.

Log-Normal Distribution

When a random variable is the resultant of the product of many small effects, its logarithm is made up of the sum of logarithms of these small effects. The logarithm of such a random variable can be expected to follow a normal distribution. Hence, if the variable is transformed to the log domain, it is likely to follow the normal distribution. Let $Y = \ln X$. If Y is normally distributed, then X is log-normally distributed. The PDF of the log-normal distribution is

$$f(x) = \frac{1}{x\sigma_y\sqrt{2\pi}}\exp\left[-\frac{\left(\ln x - \mu_y\right)^2}{2\sigma_y^2}\right] x > 0 \tag{4.27}$$

The parameters of log-normal distribution are μ_y and σ_y which can be estimated by transforming all x_i's to y_i's by

$$y_i = \ln x_i \tag{4.28}$$

Extreme Value Type 1 Distribution (EV1)

Let a series of large number of (N) observations of random variable be subdivided into n subsamples of size m each, such that $N = nm$. Each subseries shall have two extreme values: one maximum and one minimum corresponding to, for example, floods and droughts. Gumbel (1958) showed that the n largest values of subsamples asymptotically follow an extreme value type 1 (EV1) distribution. This distribution, also known as the Gumbel distribution or double negative exponential distribution, is widely used for frequency analysis of floods, maximum rainfall, etc. This distribution is essentially a log-normal distribution with constant skewness (approximately 1.14). Its PDF and CDF are as follows:

$$f(x) = \alpha \exp\{-\alpha(x-\beta) - \exp[-\alpha(x-\beta)]\} - \infty < x < \infty; -\infty < \beta < \infty; \alpha > 0$$

$$F(x) = \exp\{-\exp[-\alpha(x-\beta)]\} \tag{4.29}$$

where α and β are the scale and location parameters. The estimates of parameters using the method of moments are:

$$\widehat{\alpha} = \frac{1.283}{s}; \quad \widehat{\beta} = \bar{x} - 0.45s \tag{4.30}$$

Example 4.2. For the Sabarmati River data of Example 4.1, find the value of parameters of EV1 distribution.

Solution. The mean and standard deviation of the data are 665.37 million cubic m and 346.9 million cubic m, respectively. Therefore, the method of moment estimates is:

$$\alpha = 1.283/346.9 = 0.00367.$$

$$\beta = 665.37 - 0.45 * 346.9 = 508.$$

Gamma Distribution

The probability density function of the gamma distribution, with λ and n as parameters, is given by

$$f(x) = \frac{\lambda^n x^{n-1}}{(n-1)!}\exp(-\lambda x), x > 0, \lambda > 0, n = 1, 2, 3\ldots \tag{4.31}$$

This distribution is extensively used in hydrology though it is not widely used in frequency analysis. It can be used to determine the time to nth event which is the time to the first event T_1 plus the time interval between the first and second events T_2, plus the time interval between the second and third events T_3, and so on, or $T_1 + T_2 + \ldots + T_n$, because the time interval between events is described by the exponential distribution, the gamma distribution

The mean and the variance of the gamma distribution are

$$E(X) = n/\lambda \tag{4.32}$$

$$\mathrm{Var}[X] = n/\lambda^2 \tag{4.33}$$

Example 4.3. The average time interval between floods in some parts of Gujarat is 2 years. Compute the probability that there will be a period less than or equal to 10 years for the occurrence of five floods.

Solution. Here, $\lambda = ½ = 0.5$.

$$F(x \le 10) = \int_0^{10} \frac{0.5^5 x^4}{4!} \exp(-0.5x)dx$$

Hence,

$$= \frac{1}{768} \int_0^{10} \exp(-0.5x)dx$$

$$= \left[\frac{\exp(-0.5x)}{768} (-2x^4 - 16x^3 - 96x^2 - 384x - 768) \right]_0^{10}$$

$$= \frac{1}{768} \{\exp(-5)[-2(10)^4 - 16(10)^3 - 96(10)^2 - 384(10) - 768] - \exp(0)(0 - 0 - 768)\}$$

$$= -0.433 + 1 = 0.567.$$

Pearson Type-III Distribution (PT3)

The PT3 is a three-parameter gamma distribution and is widely used in hydrology. Its parameters are related to mean, standard deviation, and skewness.

$$f(x) = \frac{1}{a\Gamma(b)} \left(\frac{x-c}{a} \right)^{b-1} \exp\left(-\frac{x-c}{a} \right) \tag{4.34}$$

where *a, b,* and *c* are the scale, shape, and location parameters, respectively, and $\Gamma(b)$ is a gamma function. If $c = 0$, this distribution becomes a two-parameter gamma distribution. Parameters *a, b,* and *c* are related to mean, standard deviation, and coefficient of skewness as (method of moment estimates)

$$a = \sigma / b \tag{4.35a}$$

$$b = (2/C_s)^2 \tag{4.35b}$$

$$c = \mu - \sigma\sqrt{b} \tag{4.35c}$$

Example 4.4. For the Sabarmati River data of Example 4.1, find the parameters of the PT3 distribution.

Solution. The estimates of parameters using the method of moments are

$$b = (2/0.76)^2 = 6.93 \text{ million cubic m.}$$

$$a = 346.9/\sqrt{6.93} = 131.78 \text{ million cubic m.}$$

$$c = 665.37 - 346.9^* \sqrt{6.93} = -247.84.$$

Log-Pearson Type-III Distribution (LPT3)

If the random variable $Y = \ln X$ follows a PT3 distribution, then X follows the LPT3 distribution. This distribution was recommended by the U.S. Water Resources Council for adoption as the standard distribution to be used in flood frequency analysis. Its PDF is given by

$$f(x) = \frac{1}{ax\Gamma(b)}\left(\frac{\ln x - c}{a}\right)^{b-1} \exp\left(-\frac{(\ln x - c)}{a}\right) \tag{4.36}$$

It is a very versatile distribution and can accommodate a variety of shapes. The mean, standard deviation, and coefficient of skewness of LPT3 distribution are given by

$$\mu_y = c + ab \tag{4.37a}$$

$$\sigma_y = a\sqrt{b} \tag{4.37b}$$

$$\gamma_y = 2/\sqrt{b} \tag{4.37c}$$

Transformation Techniques

In many instances, it is better to transform the data to a particular distribution of known characteristics instead of assuming that a known distribution fits the data. Since the properties of normal distribution are completely defined, the given data are transformed to a normal distribution. Of the several transformations that are available, the power transformation is most commonly used (Jain and Singh, 1986a,b):

$$y = (x^{\lambda} - 1)/\lambda, \text{for } \lambda \neq 0 \tag{4.38}$$

$$= \ln x \quad \text{for } \lambda = 0$$

The reciprocal and square root transformations can be obtained as special cases of Eq. (4.38).

4.2.2 Discrete Probability Distributions

The use of discrete probability distributions is restricted generally to those random events in which the outcome can be described as success or failure, that is, there are only two mutually exclusive events in an experiment. Moreover, the successive trials are independent and the probability of success remains constant from trial to trial. The binomial or Poisson distributions can be used to find the probability of occurrence of an event r times in n successive years.

Binomial Distribution

This distribution arises in Bernoulli processes where in any trial, the event may or may not take place. The probability of occurrence of the event is the same from one trial to another. This distribution usually occurs while dealing with complementary events. A common example is tossing of coins in which the probability of head appearing is the same in each trial. The occurrence of wet and dry days over a given time interval is also a complementary event. The probability of occurrence of the event r times in n successive years is given by:

$$P_{r,n} = n\, C_r P^r q^{n-r} = \frac{n!}{r!(n-r)!} p^r q^{n-r} \tag{4.39}$$

where $P_{r,n}$ is the probability of a random event of a given magnitude and exceedance probability P occurring r times in n successive years. The probability of the event not occurring at all in n successive years is:

$$P_{o,n} = q^n = (1-p)^n \tag{4.40}$$

The probability of an event occurring at least once in n successive years:

$$P_1 = 1 - q^n = 1 - (1-p)^n \tag{4.41}$$

Example 4.5. An analysis of data on the maximum one-day rainfall depth at a station indicated that a depth of 280 mm had a return period of 50 years. Determine the probability of a one-day rainfall depth equal to or greater than 280 mm occurring (a) once in 20 successive years, and (b) two times in 15 successive years.

Solution. Here, $P = 1/50 = 0.02$.

(a) In the first case, $n = 20$, $r = 1$. Therefore, from Eq. (4.39)

$$P_{1,20} = \frac{20!}{19!1!} * (0.02) * (0.98)^{19} = 0.272.$$

(b) In this case, $n = 15$, $r = 2$. Therefore,

$$P_{2,15} = \frac{15!}{13!2!} * \left(0.02^2\right) * (0.980)^{13} = 0.0292.$$

Example 4.6. What is the probability that a 5-year flood will not occur at all in a 10-year period?

Solution. Here, $p = 1/5 = 0.2$, $n = 10$, and $r = 0$. Hence, the probability is

$$P_{0,10} = \frac{10!}{0!10!} * 0.02^0 * (0.8)^{10} = 0.1074.$$

Poisson Distribution

The Poisson distribution is a limiting form of the binomial distribution when p is very small and n is very large, and np tends to a constant value λ. This may happen when the interval over which the Bernoulli process is defined gets smaller and smaller and the number of trials becomes greater and greater, keeping np constant. The Poisson distribution has only one parameter λ that denotes the expected mean frequency of occurrence of some event in a given time t. The probability distribution of the number of events in a given time is

$$P(X = x) = \frac{\lambda^x \exp(-\lambda)}{x!} \quad \lambda > 0, \mathrm{x} = 0, 1, 2, \ldots \tag{4.42}$$

The CDF of the Poisson distribution is

$$\mathrm{P}(X \le x) = \sum_{i-0}^{x} \frac{\lambda^i \exp(-\lambda)}{i!} \tag{4.43}$$

The conditions for the application of Poisson distribution are: (a) the number of events is discrete, (b) two events cannot coincide, (c) the mean number of events per unit time is constant, and (d) events are independent. Thus, it can be applied to the following situations with p relatively small and n relatively large to determine the probability of:

(i) droughts in a given time period,
(ii) number of rainy days at a given location,
(iii) probability of rare flood events, and
(iv) probability of reservoir being empty in any one year out of a long period of record.

4.3 METHODS OF PARAMETER ESTIMATION

A number of methods are available to estimate parameters of hydrologic models. Some of the popular methods used in hydrology include (1) method of moments (Nash, 1959; Dooge, 1973; Harley, 1967; Singh, 1988); (2) method of PWMs (Greenwood, et al., 1979); (3) method of mixed moments (Rao, 1980, 1983; Shrader et al., 1981); (4) L-moments (Hosking, 1986, 1990, 1992); (5) maximum likelihood estimation (MLE) (Douglas, et al., 1976; Sorooshian et al., 1983; Phien and Jivajirajah, 1984); and (6) LS method (Jones, 1971; Snyder, 1972; Bree, 1978a, 1978b). A brief review of these methods is given here.

4.3.1 Method of Moments for Continuous Systems

The method of moments is frequently utilized to estimate parameters of linear hydrologic models (Nash, 1959; Dooge, 1973; Singh, 1988). Nash (1959) developed the theorem of moments which relates the moments of input, output, and impulse response functions of linear hydrologic models. Moments of functions are amenable to the use of standard methods of transform, such as the Laplace and Fourier transforms. The method of moments has been used to estimate the parameters of frequency distributions. Wang and Adams (1984) reported on parameter estimation in flood frequency analysis. Ashkar et al. (1988) developed a generalized method of moments and applied it to the generalized gamma distribution. Kroll and Stedinger (1996) estimated moments of a lognormal distribution using censored data.

Let X be a continuous variable (it may or may not be a random variable) and $f(x)$ its function satisfying some necessary conditions. The rth moment of $f(x)$ about an arbitrary point is denoted as M_r^a (f). Here M denotes the moment, the subscript ($r \geq 0$) denotes the order of moment, the superscript denotes the point about which to take the moment, and the quantity within the parentheses denotes the function, in normalized form, whose moment is to be taken. Thus, the rth moment of the function $f(x)$ can be defined as

$$M_r^a(f) = \int_{-\infty}^{\infty} (x-a)^r f(x)dx \tag{4.44}$$

This is the definition used normally in statistics. If the area enclosed by the function $f(x)$ does not add to unity, the definition of Eq. (4.44) becomes

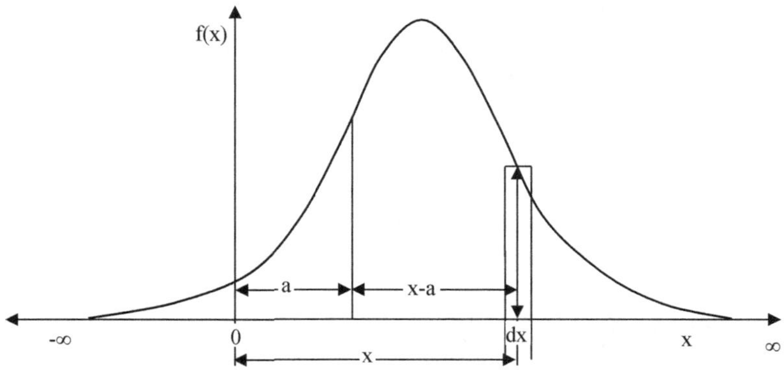

FIGURE 4.6

Concept of moment of a function $f(x)$ about an arbitrary point.

$$M_r^a(f) = \frac{\int_{-\infty}^{\infty} (x-a)^r f(x)dx}{\int_{-\infty}^{\infty} f(x)dx} \tag{4.45}$$

As the denominator in Eq. (4.45) defines the area under the curve which is usually unity or made to unity by normalization, the two definitions are numerically the same. In this text, the definition of Eq. (4.44) is used with $f(x)$ normalized beforehand. It is assumed here that the integral in Eq. (4.44) converges. There are some functions which possess moments of lower order; some do not possess any moment except of zero order. However, if a moment of higher order exists, moments of all lower orders must exist. Fig. 4.6 shows the concept of moment of a function about an arbitrary point.

Moments are statistical descriptors of a distribution and reflect on its qualitative properties. For example, if $r = 0$, then Eq. (4.44) yields

$$M_0^a = \int_{-\infty}^{\infty} (x-a)^0 f(x)dx = \int_{-\infty}^{\infty} f(x)dx = 1 \tag{4.46}$$

Thus, the zero-order moment is the area under the curve defined by $f(x)$ subject to $-\infty < x < \infty$. If $r = 1$, then Eq. (4.44) yields

$$M_1^a = \int_{-\infty}^{\infty} (x-a)^1 f(x)dx = \mu - a \tag{4.47}$$

where μ is the centroid of the area or mean. Thus, the first moment is the weighted mean about the point a. If $a = 0$, the first moment gives the mean. When $a = \mu$, the rth moment about the mean is

$$M_r^\mu = \int_{-\infty}^{\infty} (x-\mu)^r f(x)dx \tag{4.48}$$

Henceforth, for simplicity of notation, we will drop the superscript if the moment is taken about 0. The descriptive properties of the moments with respect to a specific function can be summarized as follows:

$M_0 =$ Area
$M_1 =$ Mean
$M_2^\mu =$ Variance, a measure of dispersion of the function about the mean
$M_3^\mu =$ Measurement of skewness of the function
$M_4^\mu =$ Kurtosis, a measure of the peakedness of the function

4.3.2 Method of Moments for Discrete Systems

For a discrete function, represented as f_j, $j = -\infty, \ldots, -1, 0, 1, \ldots, \infty$, the rth moment about any other arbitrary point can be defined in a manner analogous to that for continuous functions. When the arbitrary point is the origin, the rth moment is defined as

$$M_r = \sum_{m=-\infty}^{\infty} m^r f_m \tag{4.49}$$

When f_m is normalized:

$$\sum_{m=-\infty}^{\infty} f_m = 1 \tag{4.50}$$

Otherwise,

$$M_r = \sum_{m=-\infty}^{\infty} m^r f_m \Big/ \sum_{m=-\infty}^{\infty} f_m \tag{4.51}$$

It can be noticed that Eqs. (4.49) and (4.51) are analogous to Eqs. (4.44) and (4.45). Fig. 4.7 explains the concept of moment of a discrete function.

Example 4.7. The histogram of annual flows of Sabarmati River is given Table 4.2 (A plot is available in Fig. 4.4). Find the mean and variance of the data using the method of moments.

Solution. The first moment of data = (150*6 + 250*9 + 350*11 + ... + 1750*1)/98 = 664.2857 cumec.

This is the mean of data.

The second moment about the mean will give the variance.

Second moment = $[(150-664)^2 * 6 + (250-664)^2 * 9 + \ldots + (1750-664)^2 * 1]/98 = 120000$ cumec^2.

4.3.3 Method of Probability Weighted Moments

Greenwood et al. (1979) introduced the method of PWMs and showed its usefulness in deriving explicit expressions for parameters of distributions whose inverse forms $X = X(F)$ can be explicitly defined. They derived relations between parameters and PWMs for Generalized Lambda, Wakeby,

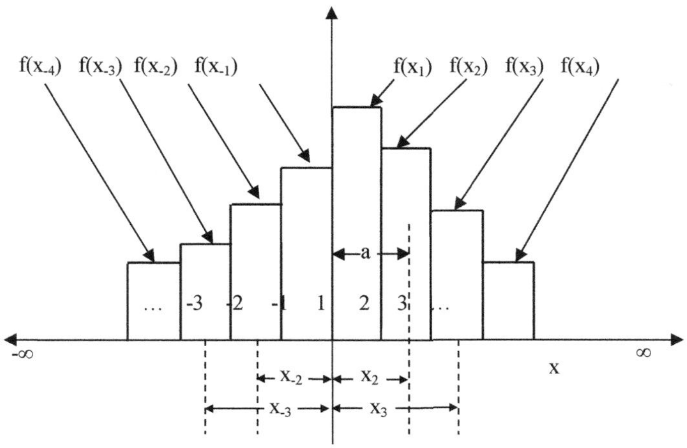

FIGURE 4.7

Concept of moment of a discrete function about an arbitrary point.

Table 4.2 Histogram of annual flows of Sabarmati River.

Discharge range	Frequency	Discharge range	Frequency
100–200	6	200–300	9
300–400	11	400–500	9
500–600	9	600–700	9
700–800	19	800–900	6
900–1000	6	1000–1100	1
1100–1200	5	1200–1300	2
1300–1400	3	1400–1500	0
1500–1600	2	1600–1700	0
1700–1800	1		

Weibull, Gumbel, Logistic, and Kappa distributions. Hosking (1986) developed the theory of linear moments and applied it to estimate parameters of several distributions. For flood frequency analysis, Haktanir (1996) modified the conventional method of probability-weighted moments for the estimation of parameters of any distribution without the need to use a plotting position formula. Wang (1997) defined partial PWMs and derived them for extreme value type I and III distributions. He applied these moments to lower bound censored samples.

Let a PDF be denoted as $F = F(X) = P[X \le x]$. The PWMs of this function can be defined as

$$M_{i,j,k} = E\left[x^i F^j (1-F)^k\right] = \int_0^1 [x(F)]^i F^j (1-F)^k dF \tag{4.52}$$

where $M_{i,j,k}$ is the PWM of order (i, j, k), E is the expectation operator, and i, j, and k are real numbers. If $j = k = 0$ and i is a nonnegative integer, then $M_{i,0,0}$ represents the conventional moment of order i

about origin. If $M_{i,0,0}$ exists and X is a continuous function of F, then $M_{i,j,k}$ exists for all nonnegative real numbers j and k.

For nonnegative integers j, k, we can express

$$M_{i,0,k} = \sum_{j=0}^{k} \binom{k}{j} (-1)^j M_{i,j,0} \tag{4.53a}$$

$$M_{i,j,0} = \sum_{k=0}^{j} \binom{j}{k} (-1)^k M_{i,0,k} \tag{4.53b}$$

If $M_{i,0,k}$ exists and X is a continuous function of F, then $M_{i,j,0}$ exists. When the inverse $X = X(F)$ of the distribution $F = F(X)$ cannot be analytically defined, it may, in general, be difficult to derive $M_{i,j,k}$ analytically.

We normally work with the moments $M_{i,j,k}$ into which x enters linearly. In particular, the PWMs for hydrologic applications are defined as

$$a_r = M_{1,0,r} = E\left[x\{1 - \mathrm{F}(x)\}^r\right], r = 0, 1, 2, \ldots \tag{4.54a}$$

$$b_r = M_{1,r,0} = E\left[x\{F(x)\}^r\right], r = 0, 1, 2, \ldots \tag{4.54b}$$

Here $a_{k-1} = E[x_{1:k}]$ and $b_k = E[x_{k:k}]$ are the expected values of extreme order statistics.

In general, a_r and b_r are functions of each other as

$$a_r = \sum_{k=0}^{r} (-1)^k \binom{r}{k} b_k \tag{4.55a}$$

$$b_r = \sum_{k=0}^{r} (-1)^k \binom{r}{k} a_k \tag{4.55b}$$

Therefore,

$$\begin{aligned} a_0 &= b_0 & b_0 &= a_0 \\ a_1 &= b_0 - b_1 & b_1 &= a_0 - a_1 \\ a_2 &= b_0 - 2b_1 + b_2 & b_2 &= a_0 - 2a_1 + a_2 \\ a_3 &= b_0 - 3b_1 + 3b_2 - b_3 & b_3 &= a_0 - 3a_1 + 3a_2 - a_3 \end{aligned} \tag{4.56}$$

A complete set of these a or b probability-weighted moments characterizes a distribution.

4.3.4 Methods of Mixed Moments

Rao (1980, 1983) proposed the method of mixed moments (MIXM) which is applicable to any log-probability distribution. As the name suggests, the MIXM method is based on mixing the moments of real and logarithmically transformed data. Thus, only the first two moments (mean and variance) of the data are used. For example, if it is desired to fit the log-Pearson type (LP) III distribution to a given set of data, then its parameters can be estimated in two ways: (1) the first method uses the mean ($\bar{x}$) and variance S_x^2 of real data and the mean of logarithmically transformed values ($Y = \log X$). (2) The second method uses the mean of real data ($\bar{x}$) and the mean and variance S_y^2 of logarithmically transformed data ($Y = \log X$). Using Monte Carlo experiments, Rao (1980) showed that the first method possessed superior statistical properties as compared to the second method.

4.3.5 Method of L-Moments

The probability-weighted moments characterize a distribution but are not meaningful by themselves. L-moments were developed by Hosking (1986) as functions of PWMs which provide a descriptive summary of the location, scale, and shape of the probability distribution. These moments are analogous to ordinary moments and are expressed as *linear* combinations of order statistics, hence the name. They can also be expressed by linear combinations of probability-weighted moments. Thus, the ordinary moments, the PWMs, and L-moments are related to each other. L-moments are known to have several important advantages over ordinary moments and have less bias than ordinary moments because they are linear combinations of ranked observations. As an example, variance (second moment) and skewness (third moment) involve squaring and cubing of observations, respectively, which compel them to give greater weight to the observations far from the mean. As a result, they result in substantial bias and variance.

If X is a real value ordered random variate of a sample of size n, such that $x_{1:n} \leq x_{2:n} \leq \ldots \leq x_{n:n}$ with the cumulative distribution $F(x)$ and quantile function $x(F)$, then the rth L-moment of X (Hosking 1990) can be defined as a linear function of the expected order statistics as:

$$L_r = \frac{1}{r}\sum_{k=0}^{r-1}(-1)^k\binom{r-1}{k}E\{X_{r-kr}\}, r = 1, 2, \ldots \tag{4.57}$$

where E{. } is the expectation of an order statistic and is equal to

$$E\{X_{jr}\} = \frac{r!}{(r-j)!j!}\int x\{F(x)\}^{j-1}\{1-F(x)\}^{r-j}dF \tag{4.58}$$

As noted by Hosking (1990), the natural estimator of L_r, based on an observed sample of data, is a linear combination of the ordered data values, that is, an L-statistic. Substituting Eq. (4.58) in Eq. (4.57), expanding the binomials of $F(x)$ and summing the coefficients of each power of $F(x)$, one can write

$$L_r = E\left[xP_{r-1}^*\{F(x)\}\right] = \int_0^1 x(F)P_{r-1}^*(F)dF, \quad r = 1, 2, \ldots \tag{4.59}$$

where $P_r^*(F)$ is the r-th shifted Legendre polynomial expressed as

$$P_r^*(F) = \sum_{r}^{k}(-1)^{r-k}\binom{r}{k}\binom{r+k}{k}F^k \tag{4.60}$$

Eq. (4.60) can simply be written as

$$P_r^*(F) = \sum_{k=0}^{r} P_{r,k} F^k \tag{4.61}$$

and

$$P_{r,k} = (-1)^{r-k} \binom{r}{k} \binom{r+k}{k} \tag{4.62}$$

The shifted Legendre polynomials are related to the ordinary Legendre polynomials $P_r(u)$ as $P^*{}_r(u) = P_r(2u-1)$ and are orthogonal on the interval (0,1) with a constant weight function.

The first four L moments are

$$L_1 = E(x) = \int x dF \tag{4.63}$$

$$L_2 = \frac{1}{2} E(x_{22} - x_{12}) = \int x(2F-1)dF \tag{4.64}$$

$$L_3 = \frac{1}{3} E(x_{33} - 2x_{23} + x_{13}) = \int x(6F^2 - 6F + 1)dF \tag{4.65}$$

$$L_4 = \frac{1}{4} E(x_{44} - 3x_{34} + 3x_{24} - x_{14}) = \int x(20F^3 - 30F^2 + 12F - 1)dF \tag{4.66}$$

4.3.6 Method of Maximum Likelihood Estimation

The MLE method is widely accepted as one of the most powerful parameter estimation methods. Asymptotically, the ML parameter estimates are unbiased, have minimum variance, and are normally distributed, while in some cases, these properties hold for small samples. The MLE method has been extensively used for estimating parameters of frequency distributions as well as fitting conceptual models.

Let $f(x;\, a_1, a_2, \ldots a_m)$ be a PDF of the random variable X with parameters a_i, $i = 1, 2, \ldots, m$, to be estimated. For a random sample of data $x_1, x_2, \ldots x_n$, drawn from this probability density function, the joint PDF is defined as

$$f(x_i, x_2, \ldots. x_n; a_1, a_2, \ldots. a_m) = \pi_{i=1}^{n} f(x_i; a_1, a_2, \ldots. a_m) \tag{4.67}$$

Interpreted conceptually, the probability of obtaining a given value of X, say x_1, is proportional to $f(x;\, a_1, a_2, \ldots a_m)$. Likewise, the probability of obtaining the random sample $x_1, x_2, \ldots x_n$ from the population of X is proportional to the product of individual probability densities or the joint PDF. This joint PDF is called the likelihood function, denoted by L:

$$L = \pi_{i=1}^{n} f(x_1; a_1, a_2, \ldots a_m) \tag{4.68}$$

where parameters a_i, $i = 1,2, \ldots m$, are unknown.

By maximizing the likelihood that the sample under consideration is the one that would be obtained if n random observations were selected from $f(x;\, a_1, a_2, \ldots a_m)$, the unknown parameters are determined, and hence the name of the method. The values of parameters so obtained are known as MLE estimators. Since the logarithm of L attains its maximum for the same values of a_i, $i = 1,2, \ldots m$, as does L, the MLE function can also be expressed as

$$\ln L = L^* = \ln \pi_{i=1}^{n} f(x_1; a_1, a_2, ...a_m) = \sum_{i=1}^{n} \ln f(x_1; a_1, a_2, ...a_m) \tag{4.69}$$

Frequently ln[L] is maximized, for it is many times easier to find the maximum of the logarithm of the ML function than that of the normal L.

The procedure for estimating parameters or determining the point where the MLE function achieves its maximum involves differentiating L or $\ln L$ partially with respect to each parameter and equating each differential to zero. This results in as many equations as the number of unknown parameters. For m unknown parameters, we get

$$\begin{aligned} \frac{\partial L(a_1, a_2, \ldots a_m)}{\partial a_1} &= 0 \\ \frac{\partial L(a_1, a_2, \ldots a_m)}{\partial a_2} &= 0 \\ &\vdots \\ \frac{\partial L(a_1, a_2, \ldots a_m)}{\partial a_m} &= 0 \end{aligned} \tag{4.70}$$

These m equations in m unknowns are then solved for the m unknown parameters.

Applying the method of ML, the parameters of EV1 distribution for the Sabarmati data are $\alpha = 0.00354$ and $\beta = 503.6$. Recall that the estimates using the method of moments were $\alpha = 0.00367$ and $\beta = 508$.

4.3.7 Method of Least Squares

The method of least squares (MOLS) is one of the most frequently used parameter estimation methods in hydrology. Natale and Todini (1974) presented a constrained MOLS for linear models in hydrology. Williams and Yeh (1983) described MOLS and its variants for use in rainfall-runoff models. Jones (1971) linearized weight factors for LS fitting. Shrader et al. (1981) developed a mixed-mode version of MOLS and applied it to estimate parameters of the log-normal distribution. Snyder (1972) reported on fitting of distribution functions by nonlinear LS. Stedinger and Tasker (1985) performed regional hydrologic analysis using ordinary, weighted, and generalized LS.

Let there be a function $Y = f(X; a_1, a_2, \ldots a_m)$, where a_i, $i = 1, 2, \ldots m$, are parameters to be estimated. The MOLS involves estimating parameters by minimizing the sum of squares of all deviations between observed and computed values of Y. Mathematically, this sum D can be expressed as

$$D = \sum_{i=1}^{n} d_1^2 = \sum_{i=1}^{n} [y_0(i) - y_c(i)]^2 = \sum_{i=1}^{n} [y_0(i) - f(x; a_1, a_2, \ldots a_m)]^2 \tag{4.71}$$

where $y_0(i)$ is the ith observed value of Y, $y_c(i)$ is the ith computed value of Y, and $n > m$ is the number of observations. The minimum of D in Eq. (4.71) can be obtained by differentiating D partially with respect to each parameter and equating each differential to zero, for example,

$$\frac{\partial \sum_{i=1}^{n} [y_0(i) - f(x_i; a_1, a_2, ...a_m)]^2}{\partial a_1} = 0 \tag{4.72}$$

The resulting m equations, usually called the normal equations, are then solved for the estimation of m parameters. This method is frequently used to estimate parameters of a linear regression model (see Section 4.7).

4.4 CONCEPT OF ENTROPY

Entropy can be considered as a measure of the degree of uncertainty or disorder associated with a system. Indirectly it also reflects the information content of space-time measurements. Entropy is viewed in three different but related contexts and is hence typified by three forms: thermodynamical entropy, statistical entropy, and information-theoretical entropy. In water resources, the most frequently used form is the information-theoretical entropy.

The concept of entropy provides a quantitative measure of uncertainty. To that end, consider a probability density function (PDF) $f(x)$ associated with a dimensionless random variable X. The dimensionless random variable may be constructed by dividing the observed quantities by its mean value, for example, annual flood maxima divided by the mean annual flood (MAF). As usual, $f(x)$ is a possible function for every x in some interval (a, b) and is normalized to unity such that

$$\int_a^b f(x)dx = 1 \tag{4.73}$$

The most popular measure of entropy was first mathematically given by Shannon (1948) and has since been called the Shannon entropy functional (SEF), denoted as $I[f]$ or $I[x]$. It is a numerical measure of uncertainty associated with $f(x)$ in describing the random variable X and is defined as

$$I[f] = I[X] = -k \int_a^b f(x) \ln\left[\frac{f(x)}{m(x)}\right] dx \tag{4.74}$$

where $k > 0$ is an arbitrary constant or scale factor depending on the choice of measurement units, and $m(x)$ is an invariant measure function guaranteeing the invariance of $I[f]$ under any allowable change of variable, and provides an origin of measurements of $I[f]$. Scale factor k can be absorbed into the base of logarithm and $m(x)$ may be taken as unity so that Eq. (4.74) is often written as

$$I[f] = I[x] = -\int_a^b f(x) \ln[f(x)]dx; \quad \int_a^b f(x)dx = 1 \tag{4.75}$$

We may think of $I[f]$ as the mean value of $-\ln[f(x)]$. Actually, $-I$ measures the strength, $+I$ measures the weakness. SEF allows choosing that $f(x)$ which minimizes the uncertainty. Note that $f(x)$ is conditioned on the constraints used for its derivation. Singh (1988, 1998, 2013) has described the theory of entropy and has given expressions of SEF for a number of probability distributions.

4.4.1 Principle of Maximum Entropy

According to the principle of maximum entropy (POME), "the minimally prejudiced assignment of probabilities is that which maximizes entropy subject to the given information." Mathematically, it can be stated as follows: given m linearly independent constraints C in the form

$$C_i = \int_a^b y_i(x) f(x) dx, i = 1, 2, ..., m \tag{4.76}$$

where $y_i(x)$ are some functions whose averages over $f(x)$ are specified. The maximum of I, subject to the conditions in Eq. (4.76), is given by the distribution

$$f(x) = \exp\left[-\lambda_0 - \sum_{i=1}^{m} \lambda_i y_i(x) \right] \tag{4.77}$$

where λ_i, $i = 0,1, \ldots, m$, are the Lagrange multipliers and can be determined from Eqs. (4.76) and (4.77) along with the normalization condition in Eq. (4.73).

4.4.2 Entropy-Based Parameter Estimation

The general procedure for deriving an entropy-based parameter estimation method for a frequency distribution involves the following steps: (1) Define the given information in terms of constraints. (2) Maximize the entropy subject to the given information. (3) Relate the parameters to the given information. More specifically, let the available information be given by Eq. (4.76). Since POME specifies $f(x)$ by Eq. (4.77), inserting Eq. (4.77) in Eq. (4.75) yields

$$I[f] = \lambda_0 + \sum_{i=1}^{m} \lambda_i C_i \tag{4.78}$$

In addition, the potential function or the zeroth Lagrange multiplier λ_0 is obtained by inserting Eq. (4.77) in Eq. (4.78) as

$$\int_a^b \exp\left[-\lambda_0 - \sum_{i=1}^{m} \lambda_i y_i \right] dx = 1 \tag{4.79}$$

resulting in

$$\lambda_0 = \ln \int_a^b \exp\left[-\sum_{i=1}^{m} \lambda_i y_i \right] dx \tag{4.80}$$

The Lagrange multipliers are related to the given information (or constraints) by

$$-\frac{\partial \lambda_0}{\partial \lambda_i} = C_i \tag{4.81}$$

It can also be shown that

$$\frac{\partial^2 \lambda_0}{\partial \lambda_i^2} = \mathrm{var}[y_i(x)]; \quad \frac{\partial^2 \lambda_0}{\partial \lambda_i \partial \lambda_j} = \mathrm{cov}\left[y_i(x), y_j(x)\right], i \neq j \tag{4.82}$$

With the Lagrange multipliers estimated from Eqs. (4.81) and (4.82), the frequency distribution given by Eq. (4.77) is uniquely defined. It is implied that the distribution parameters are uniquely related to the Lagrange multipliers. Clearly, this procedure states that a frequency distribution is uniquely defined by the specification of constraints and application of POME.

Example 4.8. Consider an exponential distribution of random variable X as $f(x) = a\exp(-ax);\ a > 0, x > 0$. Determine the parameter a using the entropy theory.

Solution. The method of parameter estimation involves (1) specification of constraints, (2) method of Lagrange multipliers, (3) construction of the zeroth Lagrange multiplier, (4) relation between Lagrange multipliers and constraints, (5) relation between Lagrange multipliers and constraints, and relation between parameters and constraints. These steps are now followed here.

Specification of constraints: For the definition of entropy, we first take the logarithm of the PDF:

$$\ln f(x) = \ln a - ax \tag{4.83}$$

The entropy of the exponential PDF can be expressed as

$$-\int_0^\infty f(x)\ln f(x)dx = -\ln a\int_0^\infty f(x)dx = a\int_0^\infty xf(x)dx \tag{4.84}$$

The constraints for the exponential distribution function can be expressed as

$$C_1 = \int_0^\infty f(x)dx = 1 \tag{4.85}$$

$$C_2 = \int_0^\infty xf(x)dx = \bar{x} \tag{4.86}$$

where $\bar{x}$ is the first moment of the distribution about the origin or mean, and C_1 and C_2 are constraints.

Method of Lagrange multipliers: The Lagrangian function L can be expressed as

$$L = -\int_0^\infty f(x)\ln f(x)dx - (\lambda_0 - 1)\int_0^\infty f(x)dx - \lambda_1\int_0^\infty xf(x)dx \tag{4.87}$$

In order to maximize entropy for determining unbiased $f(x)$, the Lagrangian function can be differentiated with respect to $f(x)$ and equated to zero:

$$\frac{\partial L}{\partial f(x)} = -\ln f(x) - 1 - (\lambda_0 - 1) - \lambda_1 x = 0 \tag{4.88}$$

where λ_0 and λ_1 are the Lagrange multipliers. Eq. (4.88) leads to the unbiased $f(x)$:

$$f(x) = \exp(-\lambda_0 - \lambda_1 x) \tag{4.89}$$

Construction of zeroth Lagrange multiplier: Inserting Eq. (4.6) in Eq. (4.2), one obtains

$$\int_0^\infty f(x)dx = \int_0^\infty \exp(-\lambda_0 - \lambda_1 x)dx = 1 \tag{4.90}$$

Eq. (4.90) yields

$$\exp(-\lambda_0) = \int_0^\infty \exp(-\lambda_1 x)dx \tag{4.91}$$

Eq. (4.91) defines the zeroth Lagrange multiplier as

$$\lambda_0 = -\ln \lambda_1 \tag{4.92}$$

Eq. (4.91) can also be expressed as

$$\lambda_0 = \ln \int_0^\infty \exp(-\lambda_1 x)dx \tag{4.93}$$

Relation between Lagrange multipliers and constraints: Eq. (4.92) can be differentiated with respect to λ_1 as

$$\frac{\partial \lambda_0}{\partial \lambda_1} = -\frac{1}{\lambda_1} \tag{4.94}$$

Differentiating Eq. (4.91) with respect to λ_1 as

$$\frac{\partial \lambda_0}{\partial \lambda_1} = -\frac{\int_0^\infty x\exp(-\lambda_1 x)dx}{\int_0^\infty \exp(-\lambda_1 x)dx} \tag{4.95}$$

Eq. (4.95) can be written as

$$\frac{\partial \lambda_0}{\partial \lambda_1} = -\frac{\int_0^\infty x\exp(-\lambda_0 - \lambda_1 x)dx}{\int_0^\infty \exp(-\lambda_0 - \lambda_1 x)dx} = -\int_0^\infty x\exp(-\lambda_0 - \lambda_1 x)dx = -\bar{x} \tag{4.96}$$

Equating Eq. (4.94) to Eq. (4.96), one gets

$$\frac{1}{\lambda_1} = \bar{x} \tag{4.97}$$

Relation between Lagrange multipliers and parameter: using Eq. (4.92) in Eq. (4.89), the entropy-based $f(x)$ can be written as

$$f(x) = \exp(\ln \lambda_1 - \lambda_1 x) = \lambda_1 \exp(-\lambda_1 x) \tag{4.98}$$

Comparing the exponential PDF and Eq. (4.98),

$$\lambda_1 = a \tag{4.99}$$

Relation between parameter and constraint: Equating Eqs. (4.97) and (4.99), one obtains

$$a = \frac{1}{\bar{x}} \tag{4.100}$$

4.5 MULTIVARIATE PROBABILITY DISTRIBUTIONS USING COPULAS

Water resources systems encounter complex environmental and hydrological processes, such as floods, droughts, rainstorms, windstorms, weather extremes, and tides which are characterized by two or more correlated random variables. For example, drought is characterized by severity, areal extent, duration, and interarrival time. Likewise, a flood is characterized by peak flow, duration, volume, and interarrival time. These processes are multivariate, and their treatment needs multivariate frequency analysis. For multivariate frequency analysis, there are three methods that are commonly applied: (1) statistical method, (2) entropy theory, and (3) copula method. Traditional statistical methods normally require the marginal frequency distributions of individual random variables to be of the same type and the same type should be relatively simple for the derivation of multivariate distribution, but in real life, individual random variables follow different frequency distributions which can be simple or complicated. Thus, these methods are too restrictive and are difficult to apply to those multivariate random variables that have different frequency distributions. The entropy theory is applicable to random variables having different marginal frequency distributions but requires an explicit expression of correlation structure and the *mathematics* becomes complicated even for two random variables to obtain an analytical form of the multivariate distribution. The copula method is free of these restrictions and is the most viable method for deriving multivariate distributions. In recent years, this method has been extensively applied in water engineering (Zhang and Singh, 2019).

The copula method is founded on the theorem proposed by Sklar (1959) according to which the multivariate distribution is a function of marginal distributions. Let $X_1, X_2, \ldots, X_n$ be random variables with marginal cumulative distributions $F(x_1)$, $F(x_2)$, ..., $F(x_n)$, respectively, where x_i, $i = 1, 2, \ldots, n$, are values of X_i, $i = 1, 2, \ldots, n$. Then, the relation between cumulative joint distribution function $F(x_1, x_2, \ldots, x_n)$ and copula $C(u_1, u_2, \ldots, u_n)$, where $u_i = F(x_i)$, i-1, 2, ..., n, can be expressed as

$$\begin{aligned} F(x_1, x_2, \ldots, x_n) &= P(X_1 \le x_1, X_2 \le x_2, \ldots, X_n \le x_n) \\ &= C(F_1(x_1), F_2(x_2), \ldots, F_n(x_n)) = C(u_1, u_2, \ldots, u_n) \end{aligned} \tag{4.101}$$

$$F(x_1, x_2, \ldots, x_n) = P(X_1 \le x_1, X_2 \le x_2, \ldots, X_n \le x_n) = C(F_1(x_1), F_2(x_2), \ldots, F_n(x_n)) = C(u_1, u_2, \ldots, u_n)$$

Copula can also be thought of as

$$C(u_1, u_2, \ldots, u_n) = F\left(F_1^{-1}, F_2^{-1}, \ldots, F_n^{-1}\right) \tag{4.102}$$

in which $x = F_i^{-1}(u_i)$. Thus, copula is essentially a function that captures the essential features of the dependence of multivariate random variables by connecting the multivariate distribution to its marginals. Thus, the determination of the CDF of correlated random variables, denoted as H, amounts to the determination of C and the dependence function. If $c(u_i, u_2, \ldots, u_n)$ denotes the density function of copula $C(u_i, u_2, \ldots, u_n)$, then

$$c(u_1, u_2, \ldots, u_n) = \frac{\partial C^n(u_1, u_2, \ldots, u_n)}{\partial u_1 \partial u_2 \ldots \partial u_n} \tag{4.103}$$

The mathematical relation between copula density function and multivariate probability density function, $f(x_1, x_2, \ldots, x_n)$ can be expressed from Eq. (4.103) as

$$c(u_1, u_2, \ldots, u_n) = \frac{f(x_1, x_2, \ldots, x_n)}{f(x_1)f(x_2)\ldots f(x_n)} \tag{4.104}$$

This two-step method enables the derivation of multivariate probability distributions with different marginal distributions regardless of their dependence structure. Central to this method is the determination of the dependence structure that is represented by a copula. Different families of copulas have been proposed and have been described by Zhang and Singh (2019). The Archimedean copula family is commonly employed for hydrologic analyses because of its ease of determination, a large variety of copula families belonging to it, and its application to both positively and negatively correlated hydrologic variables. For this reason, the one-parameter Archimedean copulas are applied for determining the joint probability distribution of correlated flood variables. The method is illustrated using an example of two random variables using one-parameter Archimedian copula.

Let X and Y be two random variables with their CDFs, respectively, as $F(x)$ and $F(y)$ and let $U = F(x)$ and $V = F(y)$. Then, U and $V$are uniformly distributed random variables; and u will denote a specific value of U, and v will denote a specific value of V. The one parameter Archimedean copula, denoted as C_θ, where θ is the copula parameter, can be expressed as

$$C_\theta(u, v) = \varphi^{-1}(\varphi(u) + \varphi(v)) \tag{4.105}$$

where φ is the copula generator which is a convex decreasing function satisfying $\varphi(1) = 0$; and $\varphi^{-1}() = 0$ when $v \geq \varphi(0)$.

We consider the Gumber–Hougaard copula family for which the generating function is expressed as

$$\varphi(t) = (-\ln\ t)^\theta, t = u\ or\ v \tag{4.106}$$

Now Eq. (4.105) can be expressed as

$$C_\theta(u, v) = C_\theta(F(x), F(y)) = H(x, y) = \exp\left\{ - \left[(-\ln u)^\theta - (-\ln v)^\theta\right]^{\frac{1}{\theta}} \right\} \tag{4.107}$$

It may be noted that

$$\varphi(u) = (-\ln u)^\theta; \quad \varphi(v) = (-\ln v)^\theta \tag{4.108}$$

Also, parameter θ is related to Kendall's coefficient of correlation, τ between X and Yas

$$\tau = 1 - \theta^{-1} \tag{4.109}$$

The relationship between Kendall's τ and the generating function parameter given by Eq. (4.109) indicates that only the positive dependence structure of the bivariate data can be analyzed for the Gumbel–Hougaard copula. In flood hydrology, flood peak and volume, and flood duration and volume are positively correlated and therefore their bivariate distributions can be derived using this copula.

The other popular copula family is Ali–Mikhail–Haq family. The generating function is

$$\varphi(t) = \ln\left[\frac{1 - \theta(1 - t)}{t}\right] \tag{4.110}$$

$$\tau = \left(\frac{3\theta - 2}{\theta}\right) - \frac{2}{3}\left(1 - \frac{1}{\theta}\right)^2 \ln(1 - \theta) \tag{4.111}$$

Therefore, Eq. (4.107) can be written as

$$C_\theta(u, v) = C_\theta(F(x), F(y)] = H(x, y) = \frac{uv}{1 - \theta(1 - u)(1 - v)} \tag{4.112}$$

The Ali–Mikhail–Haq copula applies to both negatively and positively dependent random variables. This means that joint distributions of flood peak and volume, flood peak and duration, and flood volume and duration can be derived using the Ali–Mikhail copula. However, this copula is not suitable if the dependence structure represented by Kendall's τ approaches a very high value of positive or very low value of negative correlation.

Example 4.9. Consider two random flood variables X and Y that have exponential distribution functions as

$$F(x) = 1 - \exp(-x); \quad F(y) = 1 - \exp(-y) \tag{4.113}$$

Determine the joint distribution of X and Y.

Solution. It is assumed that the joint distribution of X and Y, $H(x, y)$, can be expressed by the Ali–Mikhail–Haq copula function. Here $u = F(x) = 1 - \exp(-x)$; $v = F(y) = 1 - \exp(-y)$. Then, using Eq. (4.94), the joint distribution can be expressed as

$$H(x, y) = C_\theta(u, v) = \frac{[1 - \exp(-x)][1 - \exp(-y)]}{1 - \theta \exp(-x)\exp(-y)} \tag{4.114}$$

where θ is the copula parameter which can be determined from its relationship with Kendall's tau τ as given by Eq. (4.111).

Example 4.10. Consider two random variables with different marginal distributions. Let a random flood variable X follow EVI or Gumbel distribution and another flood variable Y follow the exponentially distribution which can be expressed as

$$F(x) = \exp[-\exp(-x)]; F(y) = 1 - \exp(-y) \tag{4.115}$$

Solution. The copula function is assumed to be given by the Ali–Mikhail–Haq family. Here $u = F(x) = \exp[-\exp(-x)]; v = F(y) = 1 - \exp(-y)$. Then, the joint distribution of these two random flood variables, H(x, y), can be given by Eq. (4.114) as

$$H(x, y) = C_\theta(u, v) = \frac{\exp[-exp(-x)]\,[1 - \exp(-y)]}{1 - \theta\{1 - \exp[-\exp(-x)]\exp(-y)\}}$$

in which θ is the copula parameter which can be estimated from the relationship with Kendall's τ as given by Eq. (4.111).

4.6 PROBLEMS OF PARAMETER ESTIMATION

The parameters of a distribution function are estimated from sample values. There are, of course, myriad ways by which to obtain parameter estimates. The sample data may contain errors, the hypotheses underlying the method of parameter estimation may not yield accurate estimates, and there may be truncation and round-off errors. These sources of errors may result in errors in parameter estimates. Each estimate of a parameter is a function of sample values which are observations of a random variable. Thus, the parameter estimate itself is a random variable having its own sampling distribution. An estimate obtained from a given set of values can be regarded as an observed value of the random variable. Thus, the goodness of an estimate can be judged from its distribution.

Some important questions arise here. How should we best use the data to form estimates? What do we mean by the best estimates? Also, are these estimates unique? How do we select the best parameter estimator if there is one? A number of statistical properties are available by which to address the above questions. These are discussed below.

4.6.1 Bias

Let the parameter be a and its estimate a_c. The estimate a_c is called an unbiased estimate of a if $E(a_c) = a$. In general, an estimate will have a certain bias $b(a)$ depending on a so that

$$E(a_c) = a + b\,(a) \tag{4.116}$$

Obviously, $b(a) = 0$ for an unbiased estimate. It should, however, be noted that an individual a_c is not equal to or even close to a even if $b(a) = 0$. It simply implies that the average of many independent estimates of a will be equal to a.

The bias in a given quantity is usually measured in dimensionless terms and is often referred to as standardized bias (or BIAS). Thus, BIAS is defined as

$$\text{BIAS} = \frac{E(\widehat{a}) - a}{a} \tag{4.117}$$

where $\widehat{a}$ is an estimate of parameter or quantile of a. In Monte Carlo experimentation, large numbers of samples of different sizes are generated from a given population. For each sample, then, an estimate of a is obtained. If there are, say, 1000 samples of a given size generated, then there are 1000 values of parameter a. Thus, $E(a)$ is the average of the 1000 estimates of a for a given sample size and is estimated as

$$E(\widehat{a}) = \sum_{i=1}^{n} \widehat{a}_i \,/\, n \tag{4.118}$$

where n is the number of samples generated or the number of values of the a estimate. The value of a in Eq. (4.118) is the true value of a or the value of parameter a of the population.

4.6.2 Consistency

Let there be a sample of size n. The estimate a_c is called a consistent estimate of a if it converges to a with probability one as n tends to infinity. Because many unbiased estimates have variances of the type

$$\text{Var}\,(a_c) \cong C/(n)^{0.5} \tag{4.119}$$

where C is constant. The condition of consistency is satisfied in most cases. In practice, it is desirable to have Var (a_c) as small as possible. This would imply that the probability density function of a_c would be more concentrated about a.

4.6.3 Efficiency

An estimate a_c of a is said to be efficient if it is unbiased and its variance is at least as small as that of any other unbiased estimate of a. If there are two estimates of a, say a_1 and a_2, then the relative efficiency of a_1 with respect to a_2 is defined as

$$e = \frac{E(a_1 - a)^2}{E(a_2 - a)^2} \leq 1 \tag{4.120}$$

if $E(a_2 - a)^2 > E(a_1 - a)^2$, then $e \leq 1$. An efficient estimate has $e = 1$. If an efficient estimate exists, it may be approximately obtained by the use of the MLE or entropy method.

4.6.4 Sufficiency

An estimate a_c of a is said to be sufficient if it uses all of the information that is contained in the sample. More precisely, let a_1 and a_2 be two independent estimates of a. Now, a_1 is considered a sufficient estimate if the joint probability distribution of a_1 and a_2 has the property.

$$f(a_1, a_2) = f\ (a_1)f(a_2|a_1) = f(a_1)\mathrm{K}(x_1, x_2, \ldots x_n) \tag{4.121}$$

in which $f(a_1)$ is the distribution of a_1, $f(a_2|a_1)$ is the conditional distribution of a_2 given a_1, and $\mathrm{K}(x_1, x_2, \ldots x_n)$ is not a function of a but only of x_i's. If Eq. (4.121) holds, then a_2 does not produce any new information about a which is not already contained in a_1. In this case, a_1 is a sufficient estimate.

4.6.5 Standard Error

Another dimensionless performance measure frequently used in hydrology is the standard error (SE), defined as

$$\mathrm{SE} = \sigma(\widehat{a}) / a \tag{4.122}$$

where σ (.) denotes the standard deviation of a and is computed as

$$\sigma(\widehat{a}) = \left[\frac{1}{n-1}\sum_{i=1}^{n}\{\widehat{a}_i - E(\widehat{a}_i)\}^2\right]^{1/2} \tag{4.123}$$

where the summation is over n estimates $\widehat{a}$ of a. In Monte Carlo experiments, referred to as above, for each sample size, a value of SE is obtained. Thus, this measure is similar to the coefficient of variation.

4.6.6 Root Mean Square Error

The root mean square error (RMSE) is one of the most frequently employed performance measures and is defined for parameter a estimate as

$$\mathrm{RMSE} = E\left[(\widehat{a} - a)^2\right]^{1/2} / a \tag{4.124}$$

where $E[.]$ is the expectation of [.]. It can be shown that RMSE is related to BIAS and SE as

$$\text{RMSE} = \left[\frac{n-1}{n}\text{SE}^2 + \text{BIAS}^2\right]^{1/2} \tag{4.125}$$

4.6.7 Robustness

Kuczera (1982a,b,c) defined a robust estimator as the one that is resistant and efficient over a wide range of population fluctuations. Two criteria for resistant estimator are mini-max and minimum average RMSE. According to the mini-max criterion, the maximum RMSE for all population cases should be minimum. Thus, for a resistant estimator, the average RMSE as well as the maximum RMSE should be minimum.

4.6.8 Relative Mean Error

Another measure of error in assessing the goodness of fit of hydrologic models is the relative mean error (RME) defined as

$$\text{RME} = \frac{1}{N}\left(\sum_{i=1}^{N}\left[\frac{Q_0 - Q_c}{Q_0}\right]^2\right)^{0.5} \tag{4.126}$$

in which N is the sample size, Q is the observed quantity of a given probability, and Q_c is the computed quantity of the same probability. Also, used sometimes is the relative absolute error defined as

$$\text{RAE} = \frac{1}{N}\sum_{i=1}^{N}\left|\frac{Q_0 - Q_c}{Q_c}\right| \tag{4.127}$$

4.7 HYPOTHESIS TESTING

Many times, while analyzing water resources data, questions arise such as: Does the flow at a given site follow a normal distribution? Is the quality of water in the river violating relevant standards? Is there significant correlation between two given variables? If the results of the model are very close to the observed one, a conclusion may be reached without using a statistical test, but sometimes the difference could be such that well-articulated tests are needed to arrive at a conclusion. Statistical procedures known as hypothesis testing are followed in these situations. The hypothesis tests can be broadly divided into two categories: parametric tests and nonparametric tests. The tests that require the specification of the type of distribution of data are termed as parametric test. Examples are the t-test and the tests related to linear regression. In most cases, it is assumed that the data follow a normal distribution. The nonparametric or distribution free tests do not require the specification of the distribution of data. These tests do not require that the data should follow a specified distribution. Examples of these tests are Kendall's Tau test and Kruskal—Wallis test.

The main steps in conducting statistical tests are: (i) formulation of the hypothesis that is to be tested, (ii) formulation of an alternate hypothesis, (iii) formulation of the test statistic and significance level, (iv) determination of the distribution of test statistic, and (v) to conduct the test.

The first step in parametric tests is to formulate a hypothesis which is termed as null hypothesis and is denoted by H_0. Usually, this is the hypothesis of no change or no difference. For example, the distributions of flow at two stations are identical or there is no correlation between two given variables. The characteristics of the data that are to be examined are quantified by the test statistic. The test checks whether the behavior of the test statistic is similar to what is expected (leaving aside the possibility of this to happen due to chance alone) if H_0 were true. This null hypothesis may be mathematically written as:

$$H_0: \mu_1 = \mu_0 \tag{4.128}$$

This statement says that the statistical measure (in this case, mean) of the parent population from which the sample was drawn is not different from the mean μ_0 of a population.

After choosing the null hypothesis, an alternative hypothesis is formulated such that the null and the alternate hypothesis are mutually exclusive and all inclusive. The alternate hypothesis is a statement of some departure from the null hypothesis that might be expected. The alternate hypothesis can be stated as

$$H_a: \mu_1 \neq \mu_0 \tag{4.129}$$

This states that the mean of population from which the sample was drawn is not equal to the specified population mean. Some typical examples of null hypothesis are: the distributions of flow at two stations are different or the two variables under examination are related to each other.

Since the rejection of null hypothesis implies that the specific assumption is not true, the alternate hypothesis must be sufficiently general. If the null hypothesis is rejected, the true behavior lies somewhere in the vast set of possibilities stated in the alternate hypothesis. Sometimes the outcome of hypothesis testing is stated as rejection of the null hypothesis versus the failure to reject the null hypothesis.

After the hypothesis has been formulated, statistical analysis is carried out to either accept it or reject it. The hypothesis may be true or false and it might be accepted or rejected. This produces four possible combinations which are indicated in Table 4.3.

If a correct hypothesis is accepted or a wrong hypothesis is rejected, this is a right decision. If, however, a null hypothesis which is true is rejected, this leads to an erroneous conclusion and type I error is said to have been committed. In statistical jargon, the probability of making type I error is termed as level of significance. This probability is to be specified before carrying out the test.

Table 4.3 Hypothesis testing: possible outcomes and their probabilities.

	Hypothesis is correct	Hypothesis is incorrect
Hypothesis is accepted	Correct decision. This outcome has a probability $1-\alpha$	Type II error with a probability β
Hypothesis is rejected	Type I error with a probability of α	Correct decision with probability $1-\beta$

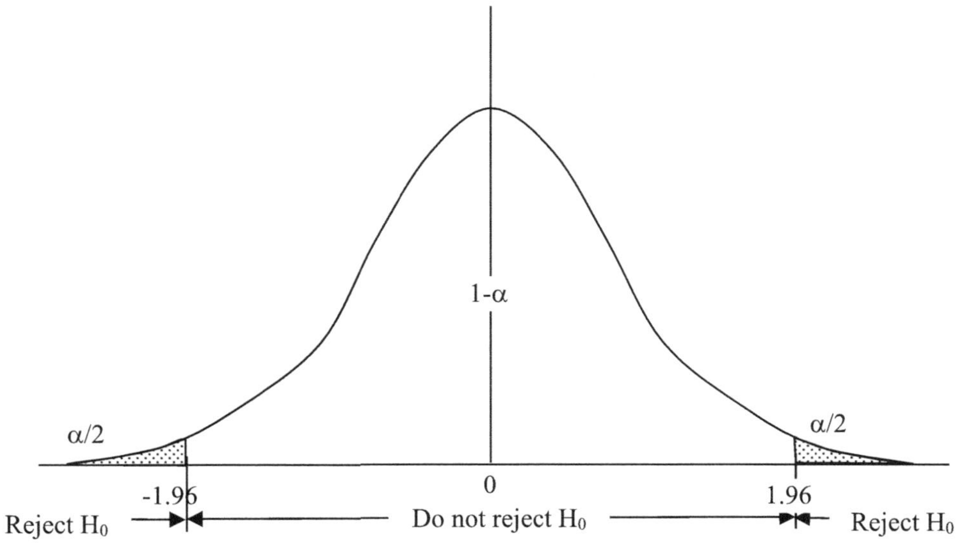

FIGURE 4.8

Two-sided test of hypothesis.

In hydrology, most commonly the significance level of 0.05 (1 in 20) or 0.01 (1 in 100) is adopted. If a level of 0.05 is chosen, it implies that the decision of statistical test may be in error about one time out of 20. In terms of distribution properties, this corresponds to 5% of the area under the curve. This concept is illustrated for a two-sided test in Fig. 4.8 in which the test statistic under null hypothesis is normal and each shaded area near the two tails contains 2.5% of the total area. This shaded area is termed as the area of rejection or the critical region. Since the alternative hypothesis in Eq. (4.96) is of inequality type, the null hypothesis is rejected if the test statistic falls in the critical region either because it is too high or too low. The significance level that is chosen in a particular circumstance depends upon the risks associated with a wrong decision.

If an incorrect hypothesis is accepted, this leads to a type II error, denoted by β. The null hypothesis is formulated with the intention that it will be rejected because this eliminates the possibility of a type II error. The probability of making type II error is not known; this error increases as significance level α increases.

4.7.1 The *t*-Test

The uncertainty in the estimates of parameters of a probability distribution, such as mean and standard deviation, can be studied using Student's *t*-distribution. This distribution is similar to the normal distribution, but its shape is dependent upon the size of sample; the exact shape depends upon the number of observations in the sample. The *t*-distribution approaches the normal distribution when there are infinite observations in the sample.

In statistical tests, the same sample is used to estimate the parameters of the distribution and perform the test and thus there is multiple use of observations. The concept of degrees of freedom is used to overcome this limitation. The term *degree of freedom* is defined as the number of observations in a sample less the number of parameters being estimated. The tables of *t*-distribution list the value of *t* statistics corresponding to various levels of significance and the degrees of freedom (ν).

Table 4.4 contains the values of *t*-distribution for selected degrees of freedom and significance level. The value of the *t*-distribution can be read from the row corresponding to ν and the column corresponding to the significance level α. For example, for $\nu = 10$, the t value is 1.81 for 5% significance level. This implies that 95% of the area of the curve lies to the left of value 1.81. Since the *t*-distribution is symmetric, 5% of the area in the left tail is to the left of the t value of −1.81 for $\nu = 10$. For the case where one is interested in 95% of the area but with 2.5% in each tail, the critical t value for $\alpha = 2.5$ is 2.23 for $\nu = 10$.

Example 4.11. Consider a test of hypothesis about sample mean. Assume that the annual runoff data at a site follows normal distribution and 21 observations for this data are available. For these observations, the mean x_m is 15.0 mm and the standard deviation s_d is 5.0 mm. The question now is whether, at 5% significance level, the mean annual runoff can be considered to be drawn from a population whose mean is 17.0 mm.

Solution. To test this, the null hypothesis is H_0: $\mu = 17.0$ mm against the alternate hypothesis H_a: $\mu \neq 17.0$ mm.

The test statistic is

$$t = [n\,(x_m - \mu)/s_d]^{0.5} = [21 * (15.0 - 17.0)/5.0]^{0.5} = -1.83.$$

Since this is a two-tailed test, the null hypothesis will be rejected if the test statistic is either too high or too low and hence the value of t is needed for $\alpha/2$ and $\nu = n-1$. From Table 4.4, $t_{\alpha/2,n-1} = t_{2.5,20} = 2.09$. Since $|t| = 1.83$ which is less than 2.09, the statistic does not fall in the region of rejection and the null hypothesis is accepted.

Table 4.4 Values of *t*-distribution for selected degrees of freedom and significance level.

	Significance level α (%)			
Degrees of freedom ν	**10**	**5**	**2.5**	**1**
1	3.08	6.31	12.71	31.82
2	1.89	2.92	4.30	6.96
3	1.64	2.35	3.18	4.54
4	1.53	2.13	2.78	3.75
5	1.48	2.02	2.57	3.36
8	1.40	1.86	2.31	2.90
10	1.37	1.81	2.23	2.76
15	1.34	1.75	2.13	2.60
20	1.32	1.72	2.09	2.53
30	1.31	1.70	2.04	2.46
60	1.30	1.67	2.00	2.39
120	1.29	1.66	1.98	2.36
∞	1.282	1.645	1.96	2.32

4.7.2 Chi-Square Distribution

Another distribution that is frequently used in hypothesis testing is the chi-square distribution. Let there be a sample of size n and values are taken from a normal population having a mean μ and standard deviation σ. The observations can be standardized using the relation

$$Z = (X - \mu)/\sigma \tag{4.130}$$

If the standardized values are squared and added, they follow a new statistic:

$$Y = \sum_{i=1}^{n} Z_i^2 \tag{4.131}$$

The variable Y follows a chi-square (χ^2) distribution with n degrees of freedom. The chi-square distribution is a special case of the gamma distribution. Similar to the t-distribution, this distribution also has a single parameter. However, this distribution is not symmetric and is always positive. The chi-square tests are single-tailed and the region of rejection is near the right tail. Table 4.5 lists the chi-square values for the selected degrees of freedom. For example, for 10 degrees of freedom, 5% of the area in the right tail (region of rejection) from χ^2 values is from 18.31 to ∞.

The goodness of fit test determines whether it is appropriate to use a particular distribution for the given sample data. Visual judgment is one way in which the data are plotted on an appropriate probability paper to check whether the match is acceptable or not. The chi-square test is also widely used for this purpose. The test procedure consists of dividing the sample into a number of segments or classes depending upon the data range. For each segment, the actual number of observations and the expected according to the distribution under test are computed. The test statistic is

Table 4.5 Values of Chi-square distribution for selected degrees of freedom and significance level.

	Significance level α (%)			
Degrees of freedom ν	**10**	**5**	**2.5**	**1**
1	2.71	3.84	5.02	6.63
2	4.60	5.99	7.38	9.21
3	6.35	7.82	9.35	11.34
4	7.78	9.49	11.14	13.28
5	9.24	11.07	12.83	15.09
8	12.02	14.07	16.01	18.48
10	15.99	18.31	20.48	23.21
15	22.31	25.00	27.49	30.58
20	28.41	31.41	34.17	37.57
30	40.26	43.77	46.98	50.89
40	51.81	55.76	59.34	63.69
50	63.17	67.50	71.42	76.15
100	118.50	124.34	129.56	135.81

$$\chi_c^2 = \sum_{i=1}^{k} (O_i - E_i)^2 / E_i \tag{4.132}$$

where O_i and E_i are the observed and expected number of observations in the ith segment and k is the total number of segments. If p parameters are estimated from data, χ^2_c follows a chi-square distribution with ($k-p-1$) degrees of freedom. If the difference between the actual and expected observations in the segments is large, it implies that samples were not drawn from the assumed distribution. Therefore, the null hypothesis that the observations follow the assumed distribution is rejected if $\chi^2_c > \chi^2_{1-\alpha,k-p-1}$.

Example 4.12. In a goodness of fit test, the data were divided in 10 classes and the value of χ^2_c came out to be 10.44. If two parameters of the distribution were computed, test whether the chosen distribution is appropriate for the data at a significance level of 0.1?

Solution. The degree of freedom is $10-2-1=7$. From the table of chi-square values, $\chi^2_{10,7} = 12.02$. Since this value is greater than 10.44, the null hypothesis cannot be rejected. It is, therefore, concluded that the chosen distribution properly describes the behavior of data.

4.8 LINEAR REGRESSION

It is an approach which is widely used to describe linear cause and effect relations between two variables. The objective is to predict a dependent variable based on an independent variable. The linear regression equation is:

$$y_i = a + bx_i + \epsilon_i i = 1, 2, \ldots n \tag{4.133}$$

where y_i is the ith value of the dependent or regressed variable and x_i is the ith value of the independent or regressor variable. The regression line crosses the y-axis at a point a (the intercept) and has a slope b, and ϵ_i is the random error term for the ith data point. The variables involved in regression should be chosen carefully and there should be a logical reason behind this choice. A scatter plot of y versus x should be made to ascertain the dependence structure. Sometimes, a transformation of x, such as a power or log transformation, improves the regression relation.

4.8.1 Parameter Estimation

The regression coefficients (a and b) are estimated by minimizing the sum of squares of deviations of y_i from the regression line. For a point x_i, the corresponding $\widehat{y}_i$ given by the regression equation will be:

$$\widehat{y}_i = a + bx_i \tag{4.134}$$

The residual error at this point is $e_i = y_i - \widehat{y}_i$. It provides a measure of how well the least-squares line conforms to the raw data. If the line passes exactly through each sample point, the error e_i would be zero. The sum of square of errors is:

$$S_{se} = \sum_{i=1}^{n} e_i^2 = \sum_{i=1}^{n} (y_i - \widehat{y}_i)^2 \tag{4.135}$$

Minimizing S_{se} leads to the following values of parameters:

$$b = S_{xy}/S_{xx}$$

and

$$a = \overline{y} - b\overline{x} \tag{4.136}$$

where

$$S_{xx} = \sum_{i=1}^{n} (x_i - \overline{x})^2$$

$$S_{yy} = \sum_{i=1}^{n} (y_i - \overline{y})^2 \tag{4.137}$$

$$S_{xy} = \sum_{i=1}^{n} (x_i - \overline{x})(y_i - \overline{y})$$

4.8.2 Goodness of Regression

The goodness of regression is measured by the variability of the dependent variable that is explained by the regression relation. Referring to Fig. 4.9, one can write,

$$y_i = \overline{y} + (\widehat{y}_i - \overline{y}) + (y_i - \widehat{y}_i)y_i = y_i - \overline{y} - \widehat{y}_i + \overline{y} + \widehat{y}_i or\ y_i - \widehat{y}_i = (y_i - \overline{y}) - (\widehat{y}_i - \overline{y}) \tag{4.138}$$

Squaring both sides and summing up

$$\sum (y_i - \widehat{y}_i)^2 = \sum (y_i - \overline{y})^2 - 2\sum (y_i - \overline{y})(\widehat{y}_i - \overline{y}) + \sum (\widehat{y}_i - \overline{y})^2 \tag{4.139}$$

Since

$$\widehat{y}_i = a + bx_i \quad (\widehat{y}_i - \overline{y}) = b(x_i - \overline{x})$$

Further, from Eq. (4.139),

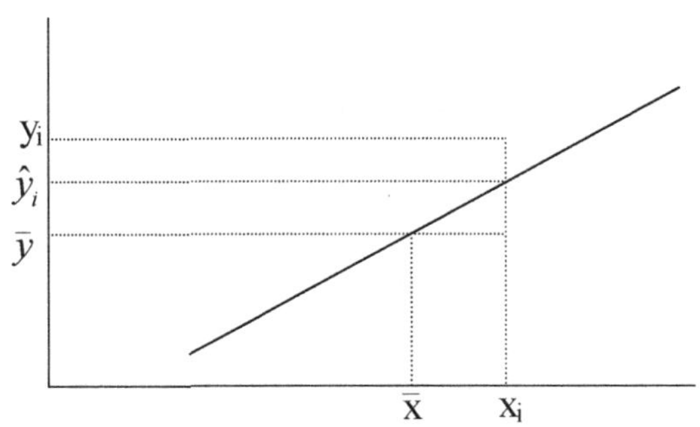

FIGURE 4.9

Graphical representation of regression related variables.

$$b\sum(x_i-\bar{x})^2=\sum(x_i-\bar{x})(y_i-\bar{y})$$

Therefore, substituting and simplifying,

$$\sum(y_i-\hat{y}_i)^2=S_{se}=\sum(y_i-\bar{y})^2-\sum(\hat{y}_i-\bar{y})^2 \tag{4.140}$$

However,

$$\sum(y_i-\bar{y})^2=\sum y_i^2-n\bar{y}^2$$

and

$$S_{sr}=\sum(\hat{y}_i-\bar{y})^2$$

Therefore,

$$S_{se}=\sum y_i^2-n\bar{y}^2-S_{sr}$$

or

$$\sum y_i^2=n\bar{y}^2+S_{se}+S_{sr} \tag{4.141}$$

Thus, the total sum of squares $\Sigma y^2{}_i$ consists of three components. The first is the sum of squares due to mean, the second is the sum of squares of errors, and the third is the sum of squares due to regression S_{sr}. The quantity S_{yy} is also termed as sum of squares about the mean or sum of squares corrected for mean.

Some important indicators of the goodness of regression are:

$$\text{Mean square error (mse): } s^2=S_{se}/(n-2) \tag{4.142a}$$

$$\text{Standard error of regression } s=(\text{mse})^{0.5} \tag{4.142b}$$

$$\text{Correlation coefficient } r=S_{xy}/\left(S_{xx}\,S_{yy}\right)^{0.5} \tag{4.142c}$$

$$\text{Coefficient of determination } R^2=1-S_{se}/S_{yy} \tag{4.142d}$$

The coefficient of determination represents the fraction of variance that is explained by regression. The closer this ratio is to unity, the "better" is the regression relation.

4.8.3 Inferences on Regression Coefficients

The variances of coefficients a and b are needed to determine their confidence bands. From Eq. (4.136),

$$b=\frac{S_{xy}}{S_{xx}}$$

$$=\sum(x_i-\bar{x})(y_i-\bar{y})\,/\,S_{xx}=\sum y_i(x_i-\bar{x})\,/\,S_{xx}$$

So,

$$\text{var}(b)=\sum(x_i-\bar{x})^2\text{var}(y_i)\,/\,S_{xx}^2=S_{xx}s^2\,/\,S_{xx}^2$$

Hence,

$$\sigma_b^2 = s^2/S_{XX}$$

or $\sigma_b = s/\sqrt{S_{XX}}$

Thus, the standard error of b, $S_b = s/\sqrt{S_{XX}}$.

The variance of coefficient a

$$\text{var}(a) = \text{var}(\bar{y} - b\bar{x}) = \text{var}(\bar{y}) - \bar{x}^2\text{var}(b)$$

$= s^2/n + s^2\bar{x}/S_{XX}$

So, the standard error of a

$$S_a = s\sqrt{\frac{1}{n} + \frac{\bar{x}^2}{S_{xx}}} \tag{4.143}$$

***Test of Hypothesis concerning* a *and* b**

Hypothesis H_0: $a = a_0$ versus H_a: $a \neq a_0$ is tested by computing $t = (a - a_0)/S_a$ which has a t distribution with $n-2$ degrees of freedom. H_0 is rejected if $|t| > t_{(1-\alpha/2),\,(n-2)}$.

Hypothesis H_0: $b = b_0$ versus H_b: $b \neq b_0$ is tested by computing $t = (b - b_0)/S_b$. H_0 is rejected if $|t| > t_{(1-\alpha/2),(n-2)}$.

Hypothesis H_0: $b = 0$ is tested by computing $t = (b-0)/S_b$. H_0 is rejected if $|t| > t_{(1-\alpha/2),\,(n-2)}$ and in this case the regression equation is able to explain a significant amount of variation in y.

4.8.4 Confidence Intervals

The confidence interval at the $\alpha\%$ significance level indicates that in repeated applications of the technique, the frequency with which the confidence interval would contain the true parameter value is $(100-\alpha)\%$. A typical value of α is 0.05 which corresponds to $(1-0.05)^*100\% = 95\%$ confidence limits. These intervals are defined if the true relationship between the variables is linear and the residuals e_i are independent, normally distributed random variables with constant variance. If the model is correct, then a/S_a and b/S_b should follow the t distribution with $(n-2)$ degrees of freedom. Hence, for coefficient a, the lower and upper limits are:

$$(l_a, u_a) = \left\{\text{a} - \text{t}_{(1-\alpha/2),(n-2)}\text{S}_\text{a},\ \text{a} + \text{t}_{(1-\alpha/2),(n-2)}\ \text{S}_a\right\} \tag{4.144}$$

For coefficient b, the lower and upper limits are:

$$(l_b, u_b) = \left\{\text{b} - \text{t}_{(1-\alpha/2),(n-2)}\text{S}_\text{b},\ \text{b} + \text{t}_{(1-\alpha/2),(n-2)}\ \text{S}_\text{b}\right\} \tag{4.145}$$

where $t_{(1-\alpha/2),(n-2)}$ represents Student's t values corresponding to the probability of exceedance $\alpha/2$ and $(n-2)$ degrees of freedom.

Confidence Intervals on Regression Line

These depend on the variance of $\widehat{\overline{y}}_k$ which is the predicted mean value of $\widehat{y}_k$ for a given x_k:

$$\widehat{\overline{y}}_k = a + bx_k \tag{4.146}$$

Then,

$$\text{var}\left(\widehat{\overline{y}}_k\right) = \text{var}(a) + x_k^2\text{var}(b) + 2x_k\text{cov}(a, b)$$

$$= s^2\left[\frac{1}{n} + \frac{(x_k - \overline{x})^2}{S_{xx}}\right] \tag{4.147}$$

Hence, the standard error of $\widehat{\overline{y}}_k$ would be

$$S_{\hat{\overline{y}}_k} = s\left[\frac{1}{n} + \frac{(x_k - \overline{x})^2}{S_{xx}}\right]^{1/2} \tag{4.148}$$

So, the lower and upper confidence limits on the regression line are:

$$(L, U) = \left[\widehat{\overline{y}}_k - S_{\hat{\overline{y}}_k}t_{(1-\alpha/2),(n-2)}, \widehat{\overline{y}}_k + S_{\hat{\overline{y}}_k}t_{(1-\alpha/2),(n-2)}\right] \tag{4.149}$$

Confidence intervals on an individual predicted value of y are:

$$S_{\hat{y}_k{}'} = S\left[1 + \frac{1}{n} + \frac{(x_k - \overline{x})_2}{\sum(x_i - \overline{x})^2}\right]^{1/2} \tag{4.150}$$

Example 4.13. Precipitation and runoff for a catchment for the month of July are given below in Table 4.6. (a) Develop the rainfall-runoff relationship in the form: $y = a + bx$; where y represents the runoff and x represents the precipitation. (b) What percent of the variation in runoff is accounted for by the developed regression equation?

Solution. (a) The various variables required to calculate a and b are computed in Table 4.6. Here, $\overline{x} = 687.05/16 = 42.94$, $\overline{y} = 234.04/16 = 14.63$. The regression coefficients are:

$$\text{b} = \text{S}_{\text{xy}}/\text{S}_{\text{xx}} = 369.423/570.0559 = 0.648$$

and $a = \overline{y} - b\overline{x} = 14.63 - 0.648{*}42.94 = -13.195$

Hence, the regression equation is: $y = -13.195 + 0.648\ x$.

(b) The percent of variation in y that is accounted for by the regression is computed as the coefficient of determination (r^2) multiplied by 100. The value of S_{se} has been computed in Table 4.7.

Coefficient of determination $R^2 = 1 - S_{se}/S_{yy} = 1 - 123.668/363.0714 = 0.659$.

Thus, 66% of variation in y is explained by the regression equation. The remaining 34% variation is due to unexplained causes. Spreadsheet softwares such as Microsoft Excel have functions to compute regression parameters. The function "Slope(known_ys, known_xs)" gives the slope of regression line

Table 4.6 Precipitation runoff data and calculations.

SN	Year	Precipitation (x)	Runoff (y)	$x-\bar{x}$	$y-\bar{y}$	$(x-\bar{x})^*(y-\bar{y})$	$(x-\bar{x})^2$	$(y-\bar{y})^2$
1	1953	42.39	13.26	−0.55	−1.37	0.7535	0.3025	1.8769
2	1954	33.48	3.31	−9.46	−11.32	107.0872	89.4916	128.1424
3	1955	47.67	15.17	4.73	0.54	2.5542	22.3729	0.2916
4	1956	50.24	15.50	7.3	0.87	6.3510	53.2900	0.7569
5	1957	43.28	14.22	0.34	−0.41	−0.1394	0.1156	0.1681
6	1958	52.60	21.20	9.66	6.57	63.4662	93.3156	43.1649
7	1959	31.06	7.70	−11.88	−6.93	82.3284	141.1344	48.0249
8	1960	50.02	17.64	7.08	3.01	21.3108	50.1264	9.0601
9	1961	47.08	22.91	4.14	8.28	34.2792	17.1396	68.5584
10	1962	47.08	18.89	4.14	4.26	17.6364	17.1396	18.1476
11	1963	40.89	12.82	−2.05	−1.81	3.7105	4.2025	3.2761
12	1964	37.31	11.58	−5.63	−3.05	17.1715	31.6969	9.3025
13	1965	37.15	15.17	−5.79	0.54	−3.1266	33.5241	0.2916
14	1966	40.38	10.40	−2.56	−4.23	10.8288	6.5536	17.8929
15	1967	45.39	18.02	2.45	3.39	8.3055	6.0025	11.4921
16	1968	41.03	16.25	−1.91	1.62	−3.0942	3.6481	2.6244
	Total	687.05	234.04	0.01	−0.04	369.4230	570.0559	363.0714

Table 4.7 Regression computations.

SN	Year	Precipitation (x)	Runoff (y)	$\hat{y}$	$S_{se} = (y-\hat{y})^2$
1	1953	42.39	13.26	14.2737	1.0276
2	1954	33.48	3.31	8.5000	26.9365
3	1955	47.67	15.17	17.6952	6.3764
4	1956	50.24	15.50	19.3605	14.9036
5	1957	43.28	14.22	14.8504	0.3975
6	1958	52.60	21.20	20.8898	0.0962
7	1959	31.06	7.70	6.9319	0.5900
8	1960	50.02	17.64	19.2180	2.4900
9	1961	47.08	22.91	17.3128	31.3282
10	1962	47.08	18.89	17.3128	2.4874
11	1963	40.89	12.82	13.3017	0.2321
12	1964	37.31	11.58	10.9819	0.3577
13	1965	37.15	15.17	10.8782	18.4195
14	1966	40.38	10.40	12.9712	6.6113
15	1967	45.39	18.02	16.2177	3.2482
16	1968	41.03	16.25	13.3924	8.1656
	Total	687.05	234.04	234.0884	123.6680

and the function "Intercept(known_ys, known_xs)" gives its intercept. The **LINEST** function fits multiple linear regerssion line and computes related statistics.

The coefficient of correlation $(r) =$ square root of coefficient of determination

$$= \sqrt{0.66} = 0.81.$$

Example 4.14. Using the data of Example 4.9, (a) compute the 95% confidence interval on a and b and test the hypothesis that $a = 0$ and the hypothesis that $b = 0.500$ for the above regression; (b) calculate the 95% confidence limits for the regression line. Calculate the 95% confidence interval for an individual predicted value of y.

Solution. (a) Computation of 95% confidence intervals on a and b.

Mean square error (Table 4.7): mse $= S_{se}/(n-2) = 123.668/14 = 8.83$.

Standard error of regression: $s_r = \text{mse}^{0.5} = 8.83^{0.5} = 2.97$. This is a very useful indicator of the quality of regression relationship.

$$\text{Standard error of } b\ (S_b) = s_r/\sqrt{(S_{xx})} = 2.97 / \sqrt{570.0559} = 0.125.$$

$$\text{Standard error of } a\ (S_a) = s_r\sqrt{\frac{1}{n} + \frac{\bar{x}^2}{S_{xx}}} = 2.97\sqrt{\frac{1}{16} + \frac{42.94^2}{570.0559}} = 5.39$$

From the t-table, for $\alpha = 0.05$, $n-2 = 14$, $t_{(1-0.025),(16-2)} = t_{0.975,\ 14} = 2.14$.

So, 95% confidence intervals on a:

$$(l_a, u_a) = \left\{ a - t_{(1-\alpha/2),(n-2)} \cdot S_a,\ a + t_{(1-\alpha/2),(n-2)} \cdot S_a \right\}$$
$$= (-13.1951 - 2.14 \text{ x } 5.39,\ -13.1951 + 2.14 \text{ x } 5.39)$$
$$= (-24.73,\ -1.66).$$

Similarly, 95% confidence intervals on b:

$$(l_b, u_b) = \left\{ b - t_{(1-\alpha/2),(n-2)} \cdot S_b, b + t_{(1-\alpha/2),(n-2)} \cdot S_b \right\}$$
$$= (0.648 - 2.14 \text{ x } 0.125, 0.648 + 2.14 \text{ x } 0.125)$$
$$= (0.38, 0.92).$$

(ii) Testing the hypothesis H_o: $a = 0$ versus $a \neq 0$.

Here, $t = (a-0.00)/S_a = -13.1951/5.39 = -2.44$. Since $t_{(1-\alpha/2),(n-2)} = t_{0.975,14} = 2.14$, $|t| > t_{0.995,14}$. Hence, the hypothesis H_0: $a = 0$ is rejected.

Hypothesis H_0: $b = 0.5$ versus H_a: $b \neq 0.5$.

In this case, $t = (b-0.5)/S_b, = (0.648-0.50)/0.125 = 1.184$. Since $|t| < 1.184$, the hypothesis H_o cannot be rejected.

From the above tests, it is concluded that the intercept is significantly different from zero. However, the slope is not significantly different from 0.5. The significance of the overall regression can be evaluated by testing H_o: $b = 0$. Under this hypothesis,

$$t = \frac{b - 0.00}{S_b} = \frac{0.648}{0.125} = 5.184$$

Since $|t| > t_{0.975,14}$, we reject H_o. The regression equation is able to explain a significant amount of the variation in Y.

4.8.5 Extrapolation

An extrapolation of a regression equation beyond the range of X used in estimating a and b is discouraged for two reasons. First, the confidence intervals on the regression line become very wide as the distance from $\bar{x}$ is increased. Second, the relation between Y and X may be nonlinear over the entire range of X and only approximately linear for the range of X investigated. An example of this is shown in Fig. 4.10.

4.9 MULTIPLE LINEAR REGRESSION

The association of three or more variables can be investigated by multiple linear regression and correlation analysis. If all the variables (dependent and independent) are in linear form, the regression is referred to as the multiple linear regression. Often a nonlinear association between the variables is handled by transforming the variables to linear form and applying multiple regression as it is easier to treat linear equations. The general form of the multiple linear regression equation is:

$$y_i = b_1 x_{i,1} + b_2 x_{i,2} + \ldots + b_p x_{i,p} \tag{4.151}$$

When set of n observations is available on the dependent and each of the p independent variables, there shall be n equations for p unknowns.

$$\begin{aligned} y_1 &= b_1 x_{1,1} + b_2 x_{1,2} + \ldots + b_p x_{1,p} \\ y_2 &= b_1 x_{2,1} + b_2 x_{2,2} + \ldots + b_p x_{2,p} \\ &\ldots\ldots \\ y_n &= b_1 x_{n,1} + b_2 x_{n,2} + \ldots + b_p x_{n,p} \end{aligned} \tag{4.152}$$

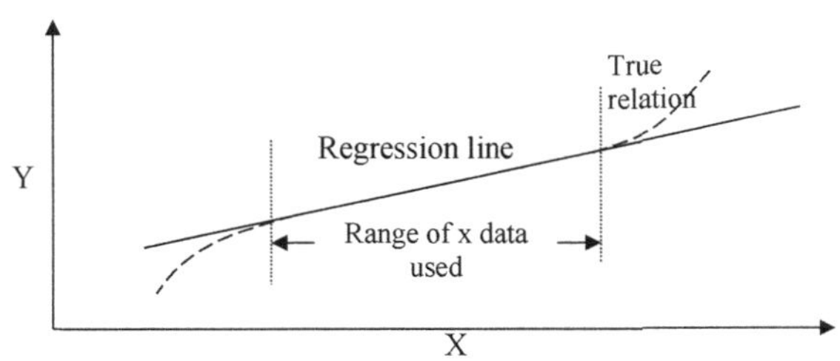

FIGURE 4.10

Regression line and extrapolation.

where y_i is the ith observation of the dependent variable; and $x_{i,1}$, $x_{i,2}, \ldots x_{i,p}$ are independent variables.

If a regression equation with p parameters is fitted to a set of n data points of variables, the number of degrees of freedom will be $n-p$. If the number of parameters is equal to the sample size, the regression equation will pass through all the points as there is no degree of freedom. It cannot be used for prediction as the errors of parameters are inversely proportional to the number of degrees of freedom.

In designing the multiple linear and nonlinear regression relations, the selection of dependent and independent variables is very important. The dependent variable is defined by the problem itself. The independent variables are selected due to the following two reasons:

(i) They have been observed in the past concurrently with the dependent variable so that the regression equation may be established. In future, they may be used to predict the dependent variable.

(ii) An analysis of physical phenomenon indicates a cause-and-effect relation between dependent and independent variables.

These criteria are necessarily subjective. The variables that are known to have no effect on the dependent variable are neglected. In matrix notation, Eq. (4.152) can be written as:

$$\frac{Y}{n \times 1} = \frac{X}{n \times p} \frac{B}{p \times 1} \tag{4.153}$$

4.9.1 Estimation of Regression Coefficients

The difference between observed and predicted (using regression) values of y or the error $= y_i - \hat{y}_i$. The regression coefficients are obtained by minimizing the sum of squares of errors using the equation:

$$\frac{b}{(px1)} = \frac{(X'X)^{-1}}{(pxn)(nxp)} \frac{X'}{(pxn)} \frac{Y}{(nx1)} \tag{4.154}$$

where X' is the transpose of matrix X.

Coefficient of determination (R^2)

$$\text{Let } Z_{i,j} = \left(X_{i,j} - \overline{X}_j\right)/S_j \tag{4.155}$$

where $\overset{j}{\bar{x}}$ and S_j are the mean and standard deviation of the jth independent variable. The correlation matrix is:

$$R = Z'Z/(n-1) = \left[R_{i,j}\right] \tag{4.156}$$

where $R_{i,j}$ is the correlation between the ith and jth independent variables. R is a symmetric matrix since $R_{i,j} = R_{j,i}$. The coefficient of determination is defined as:

R^2 = Sum of squares due to regression/Sum of squares about mean

$$\text{or } R^2 = \left(b'X'Y - n\overline{Y}^2\right)/\left(Y'Y - n\overline{Y}^2\right) \tag{4.157}$$

Here b' is the transpose of vector b of size $(1xp)$, and Y' is the transpose of vector Y of size $(1xn)$. Let the residual error be $\varepsilon = Y - X\,b$. The variance of the error, var (ε), is

$$\mathrm{Var}(\varepsilon) = S^2 = \sum (y_i - \hat{y}_i)^2 \,/\, (n-p) \tag{4.158}$$

4.9.2 Inferences on Regression Coefficients

(i) Standard errors of b_i $\left(S_{\hat{b^i}}\right)$

Let $C = X'\,X$ and C_{ii}^{-1} is the ith diagonal element of $(X'X)^{-1}$. Then

$$\mathrm{Var}\left(\hat{b}_i\right) = S_{\hat{b}_i}^{\,2} = C_{ii}^{-1} S^2 \tag{4.159}$$

$$S_{\hat{b}_i} = \left|\sqrt{C_{ii}^{-1} S^2}\right| \tag{4.160}$$

(ii) Confidence intervals on b_i

If the model is correct, the quantity $\hat{b}_i \big/ S_{\hat{b^i}}$ follows a t-distribution with $(n-p)$ degrees of freedom.

The confidence intervals on b_i are given as

$$\left(L_{\hat{b}i}, U_{\hat{b}i}\right) = \left(\hat{b}_i - t_{(1-\alpha/2)/(n-p)} S_{\hat{b}i}, \hat{b}_i + t_{\left(1-\frac{\alpha}{2}\right)/(n-p)} S_{\hat{b}i}\right) \tag{4.161}$$

(iii) Test of hypothesis concerning b_i

The hypothesis that the ith variable is not contributing significantly to explaining the variation in the dependent variable is equivalent to testing the hypothesis H_o: $b_i = b_o$ versus H_a: $b_i \neq b_o$. The test is conducted by computing:

$$t = \left(\hat{b}_i - b_o\right) / S_{\hat{b}_i} \tag{4.162}$$

H_o is rejected if $|t| > t_{(1-\alpha/2),(n-p)}$. If this hypothesis is accepted, it is advisable to delete the concerned variable from the model.

Significance of the Overall Regression

The hypothesis H_o: $b_1 = b_2 = \ldots\ b_p = 0$ versus H_a: at least one of these b's is not zero is used to test whether the regression equation is able to explain a significant amount of variation of Y or not. The ratio of the mean square error due to regression to the residual mean square has an F distribution with $p-1$ and $n-p$ degrees of freedom. Hence, the hypothesis is tested by computing the test statistic:

$$F = \frac{\left(b'X'Y - n\overline{Y}^2\right)/(p-1)}{\left(Y'Y - \widehat{b}'X'Y\right)/(n-p)} \tag{4.163}$$

H_o is rejected if F exceeds the critical value $F_{(1-\alpha),\ (p-1),\ (n-p)}$.

Confidence Intervals on Regression Line

To put the confidence limits on $Y_k = X_k b$, it is necessary to estimate the variance of $\widehat{y}_k$. This is given by

$$S_{\hat{Y}_k}{}^2 = S^2 X_k (X\prime X)^{-1} X\prime_k \tag{4.164}$$

where

$$(L, U) = \left\{\widehat{Y}_k - t_{(1-\alpha/2),(n-p)}\ S_{\hat{Y}_k},\ \widehat{Y}_k + t_{(1-\alpha/2),(n-p)} S_{\hat{Y}_k}\right\}$$

***Confidence Intervals on Individual Predicted Value of* Y**

$$\widehat{Y}_K = X_K\ \widehat{b} \tag{4.165}$$

$$(L', U\prime) = \left\{\widehat{Y}_k - t_{(1-\alpha/2),(n-p)} S_{\hat{Y}_k{}'}, \widehat{Y}_k + t_{(1-\alpha/2),(n-p)} S_{\hat{Y}_k{}'}\right\} \tag{4.166}$$

$$S'_{\hat{Y}_k}{}^2 = S^2\left[1 + X_k (X'X)^{-1} X'_k\right]$$

4.9.3 Stepwise Regression

The most common procedure for selecting the best regression equation is stepwise regression. The regression equation is built with one variable at a time, adding at each step the variable that explains the largest amount of the remaining unexplained variation. After each step, all the variables in the equation are examined for significance and any variable that no longer explains a significant amount of variation is discarded. The steps of stepwise regression are:

(i) The variable which has the highest simple correlation with the dependent variable is picked up as the first independent variable.

(ii) The variable which explains the largest of the residual variation in the dependent variable after the first step is added as the next variable.

(iii) Test the significance of the new variable and retain or discard depending on the results of this test.

(iv) Repeat steps (ii) andand (iii) until all of the variables not in the equation are found to be insignificant and all the variables in the equation are significant.

Care must be exercised to see that the resulting equation is rational. Of course, the real test of the regression model is its ability to predict those observations of the dependent variable that were not used in estimating the regression coefficients. Transformation of independent variables may significantly improve the regression relationship.

4.10 CORRELATION ANALYSIS

Correlation is a mathematical measure of the strength of relationship between two variables or within the same series. A measure of the relationship is a dimensionless coefficient, called correlation coefficient. However, note that correlation is not an evidence of a causal relationship between two variables. If one variable drives the other, they may be correlated, as rainfall and runoff. The variables may also be correlated if they share the same cause. Examples include dependent variables, such as river discharge, concentration or transport rates of sediment, and concentration or transport rates of substances that are transported in association with suspended sediment (Hirsch et al., 1993).

4.10.1 Cross-Correlation

The correlation between two time series or cross-correlation $r_{x,y}$ is given by

$$r_{x,y} = s_{x,y} / \left(s_x s_y\right) \tag{4.167}$$

where $s_{x,y}$ is the sample covariance between X and Y, and s_x and s_y are the sample standard deviations of X and Y, respectively. As with autocorrelation, $r_{x,y}$ can also range from -1 to 1; $r_{xy} = +1$ or -1 implies a perfect linear relationship between X and Y and $r_{x,y} = 0$ implies linear independence, although there may be other types (say, nonlinear) of dependence. If observations in a time series are correlated, this must be kept in mind while drawing any inferences about the data or when modeling the process that has produced the time series. For example, the correlation for the precipitation-runoff data of Table 4.6 is 0.812 which indicates a high correlation between these two variables. Note that r_{XY} is also called Pearson correlation coefficient.

A high correlation between two variables need not necessarily be due to a cause-and-effect relation between them. If there is a correlation between some variables that do not have a cause-and-effect relation, this is termed as spurious correlation. The monthly flows of two adjacent streams may be highly correlated, but this could be because the influencing external causes are the same. A high correlation in this case does not mean that a change in flow of one stream will force the other stream's flow to change. It is to be noted that independent variables are uncorrelated, but uncorrelated variables are not necessarily independent. The dependence in correlated variables is a stochastic dependence and is not always physical or cause-and-effect dependence. Any apparent correlation between variables that are in fact uncorrelated is termed as spurious correlation.

4.10.2 Serial or Autocorrelation

The autocorrelation or serial correlation of a series is defined as linear correlation between a time series and the same series at a later interval of time. Assume that in a time series, observations are equally spaced in time and that the statistical properties of the process do not change with time. The autocorrelation of a time series (having n observations) at lag k (r_k) is given by

$$r_k = \frac{\left[\sum_{i=1}^{n-k} x_i x_{i+k} - \sum_{i=1}^{n-k} x_i \sum_{i=1}^{n-k} x_{i+k}\right] / (n-k)}{\left[\sum_{i=1}^{n-k} x_i^2 - \left(\sum_{i=1}^{n-k} x_i\right)^2 / (n-k)\right]^{0.5} \left[\sum_{i=1}^{n-k} x_{i+k}^2 - \left(\sum_{i=1}^{n-k} x_{i+k}\right)^2 / (n-k)\right]^{0.5}} \tag{4.168}$$

Here the lag is the amount of offset when comparing the values of the series. The autocorrelation of lag 1 is determined by computing the correlation between elements 1, 2, ..., (n-1) of a series and the elements 2, 3, 4, ... n of the same series. From Eq. (4.168), it is clear that r_0 is unity. Note also that as k increases, the number of pairs of observations used in estimating r_k decreases, since all of the summations contain $n-k$ terms. Therefore, serial correlation should only be estimated for k sufficiently smaller than n; usually correlation at lags exceeding 20 is not much useful.

A purely random process will have $r_k = 0$ for all k, indicating that all of the observations in the sample are independent of each other. The elements of a sample of data possessing serial correlation are not random elements. The plot of autocorrelations at various lags is known as a correlogram. A typical correlogram begins at a value of +1.0 at 0 lag, and then decays at higher lags. At lags of near coincidence of the elements, the correlogram shows a rise; it falls otherwise. Correlograms help reveal the characteristics of a time series and disclose intervals of time or distance at which the time series has a repetitive nature.

The annual flows of Sabarmati River at Dharoi were shown in Fig. 4.3. The correlogram of this series up to a lag of 20 is plotted in Fig. 4.11. It can be seen from the correlogram that there is very poor autocorrelation in the series. The autocorrelation at lag 1 is 0.0064 and at lag 2, it is −0.0295.

4.10.3 Inferences on Correlation Coefficients

For uncorrelated variables, $r_{x,y} = 0$ and for correlated ones, $r_{x,y} \neq 0$. Even in uncorrelated populations, the sample correlation coefficient will rarely be zero. This deviation from zero is likely to be due to chance. Thus, statistical tests are needed to determine if the deviation of the sample correlation coefficient from zero may be ascribed to chance or not. This test can be conducted by making a hypothesis H_o: $r_{x,y} = 0$ or H_o: $r_{x,y} = r^*$ where r^* is known.

Assume that X and Y are random variables from a bivariate normal distribution. The population correlation coefficient is denoted by ρ. If $\rho = 0$, then the quantity

$$t = r\left[(n-2)/\left(1-r^2\right)\right]^{0.5} \tag{4.169}$$

has a t-distribution with (n−2) degrees of freedom. The hypothesis H_o: $\rho = 0$ is rejected if $|t| > t_{1-\alpha/2,n-2}$.

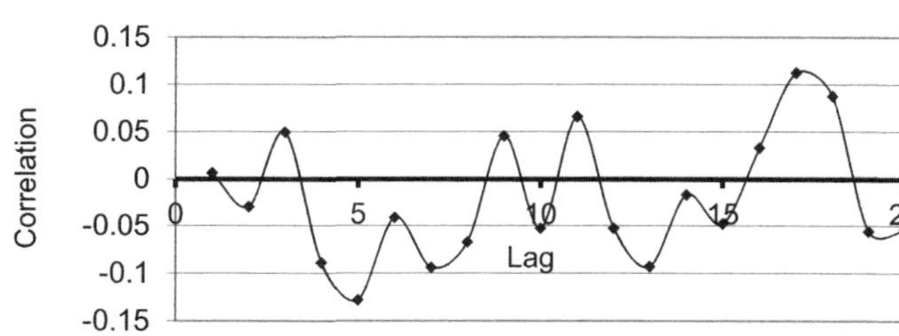

FIGURE 4.11

Correlogram of annual flows of Sabarmati River.

Kendall's Rank Correlation Test

Also known as Kendall's τ test, this is an effective and general measure of correlation between two variables. As it is a rank-based nonparametric procedure, it overcomes the problems due to the effect of extreme values and to deviations from a linear relationship. Thus, it is well suited to use for dependent variables for which the variation around the general relationship exhibits a high degree of skewness or kurtosis. Examples include dependent variables, such as river discharge, and concentration or transport rates of sediment.

The steps to conduct Kendall's test for correlation (the null hypothesis H_o is that the distribution of dependent variable Y does not change as a function of independent variable X) are as follows.

1. The n data pairs $(x_1, y_1), (x_2, y_2), \ldots (x_n, y_n)$ are indexed according to the magnitude of the x value, such that $x_1 \le x_2 \le \ldots \le x_n$. Any pair of observations (x_i, y_i) and (x_j, y_j), where $i < j$, are regarded as concordant if the sort order of (x_i, x_j) and (y_i, y_j) agrees: that is, if either both $x_i > x_j$ and $y_i > y_j$ holds or both $x_i < x_j$ and $y_i < y_j$; otherwise they are regarded as *discordant*.
2. Kendall's correlation coefficient τ is defined as

$$\tau = \frac{\text{Number of concordant pairs} - \text{number of discordant pairs}}{\text{Total number of pairs}} \tag{4.170a}$$

3. One can also compute τ as

$$\tau = \frac{2}{n(n-1)} \sum_{i<j} \text{sgn}(x_i - x_j)\,\text{sgn}\left(y_i - y_j\right) \tag{4.170b}$$

where $<0\text{sign}(\theta) = 1 \text{ if } \theta > 0 = 0 \text{ if } \theta = 0 = -1 \text{ if } \theta$

4. Compute Z statistic for hypothesis testing (Abdi, 2007):

$$Z = \frac{3\tau\sqrt{n(n-1)}}{\sqrt{2(2n+5)}} \tag{4.170c}$$

For $n > 10$, the test is conducted using a normal approximation (Hirsch et al., 1993). The standardized test statistics Z is computed as follows:

5. Replace S by τ below.

$$z = \begin{cases} \dfrac{S-1}{\sqrt{Var(S)}} & S > 0 \\ 0 & S = 0 \\ \dfrac{S+1}{\sqrt{\text{Var}(S)}} & S < 0 \end{cases} \tag{4.170d}$$

where $\text{Var}(\tau) = \frac{2(2n+5)}{9n(n)}$

6. The null hypothesis is rejected at a significance level α if $|Z| > Z_{(1-\alpha/2)}$, where $Z_{(1-\alpha/2)}$ is the value of the standard normal distribution with a probability of exceedance of $\alpha/2$. If some of the x and/or y values are tied, the formula for Var(S) is modified. If the sample size is less than 10, then it is necessary to use tables for the S statistic.

Spearman's Rank Correlation

Spearman's rank correlation measures the strength and direction of association between two ranked variables. It gives the measure of monotonicity of the relation between two variables, that is, how well the relationship between two variables could be represented using a monotonic function.

Spearman's rank correlation coefficient is computed by

$$\rho = 1 - \frac{6\sum d_i^2}{n(n^2-1)} \tag{4.171}$$

where ρ = Spearman's rank correlation coefficient, d_i = the difference between the two ranks of each observation, and n = the number of observations.

Spearman Rank Correlation varies between +1 and −1:

- A value of +1 means a perfect association of ranks
- A value of 0 means that there is no association between ranks
- A value of −1 means a perfect negative association of ranks

Example 4.15. Consider the marks of six students in two subjects given in the table. Compute Spearman's rank correlation coefficient.

Student	Subject A	Subject B
A	71	61
B	92	77
C	83	95
D	88	73
E	79	90
F	87	93

Solution. Step 1: Rank both the data in descending order. The highest marks will get a rank of 1 and the lowest marks will get a rank of 6.

Step 2: Calculate the difference between the ranks (d) and the square value of d.

Step 3: Sum all d^2 values.

Student	Subject A	Rank	Subject B	Rank	Difference in ranks (d)	d^2
A	71	6	61	6	0	0
B	92	1	77	4	−3	9
C	83	4	95	1	3	9
D	88	2	73	5	−3	9
E	79	5	90	3	2	4
F	87	3	93	2	1	1
					Total	32

Step 4: Insert the values of d and n in the formula:

$$\rho = 1 - \frac{6\sum d_i^2}{n(n^2-1)} = 1 - \frac{6 \times 32}{6(6^2-1)} = 1 - 0.914 = 0.086$$

The Spearman's Rank Correlation for the given data is 0.086. The value is near 0, which means that there is a very weak correlation between the two ranks.

4.11 FREQUENCY ANALYSIS

Frequency analysis is performed to determine the frequency of the likely occurrence of hydrologic events. This information is required to solve a variety of water-resources problems, for example, design of reservoirs, floodways, bridges, culverts, levees, urban drainage systems, irrigation systems, stream-control works, water-supply systems, and hydroelectric power plants, floodplain zoning, setting of flood-insurance premiums, etc. Although the frequency analysis of virtually every component of the hydrologic cycle is required, the emphasis here will be on frequency analyses of streamflow extremes and rainfall only.

The hydrologic data to be analyzed for frequency analysis must be treated in light of the objectives of the analysis, length and completeness of record, randomness of data, and homogeneity. The length of record should be more than 25 years for the derived distribution to be acceptable. The hydrologic data must have been controlled by a uniform set of hydrologic and operational factors. For example, the factors causing a winter rain flood are quite different from those during a spring snowmelt flood or a local cloudburst flood. These two types of floods should not be combined into a single record. Sometimes a hydrologic record may have gaps. Missing data may sometimes be estimated using regional analysis or by correlation with other hydrologic data in the region.

Hydrologic data are generally presented in chronological order constituting the complete duration series (CDS). For frequency analysis, CDS is seldom used, because the hydrologic design of a project is normally dictated by only a few critical events. Therefore, hydrologic data can be selected in two ways: (1) partial duration series (PDS) and (2) annual duration series (ADS). PDS is composed of the data exceeding a specified base level. In ADS, one value (usually the highest) is selected from each year. The two series are comparable if the record is longer than 10 years and either can be used.

4.11.1 Point Frequency Analysis

The frequency distributions presented earlier can be fitted to the data. Two commonly used methods of fitting are: (1) the graphical method and (2) frequency factors.

Graphical Method

This method involves fitting of an assumed probability distribution to observed data. The sample data are arranged in either ascending or descending order of magnitude and each data point is assigned a rank. If these are arranged in descending order of magnitude, then the highest value will

Table 4.8 Some commonly used plotting-position formulas.

		Values for m = 1 and N = 10	
Method	**Formula $P_m(X > X_m)$**	P_m	T_m
Hazen (1914)	$(m-0.5)/N$	0.05	20.0
California (1923)	(m/N)	0.10	10.0
Weibull (1939)	$(m)/(N+1)$	0.091	11.0
Beard (1943)	$(m-0.31)/(N+0.38)$	0.066	15.0
Chegodayev (1955)	$(m-0.3)/(N+0.4)$	0.067	14.9
Blom (1958)	$(m-0.375)/(N+0.25)$	0.0609	16.4
Gringorten (1963)	$(m-0.44)/(N+0.12)$	0.055	18.1
Cunnane (1978)	$(m-0.4)/(N+0.2)$	0.58	17.2
Adamowski (1981)	$(m-0.25)/(N+0.5)$	0.071	14.1

be assigned the rank of 1, the second highest the rank of 2, and so on; the lowest value will have the rank of N, where N is the number of data points in the sample. This arrangement gives an estimate of the exceedance probability, that is, the probability of a value being equal to or greater than the ranked value. If the values are arranged in ascending order, then an estimate of the nonexceedance probability, that is, the probability of a value being less than or equal to the ranked value, is obtained. These data points are plotted on probability paper, with their positions determined from a plotting-position formula.

Many plotting-position formulas are available; some commonly used ones are given in Table 4.8. Adamowski (1981) has shown that all of these formulas are special cases of

$$P_m = \frac{m-a}{N+b} \tag{4.172}$$

where a and b are constants, P_m is the exceedance probability of the mth observation, and m is the mth value of N ordered observations, such as $P_1 < P_2 < \ldots < P_N$. A commonly used plotting-position formula in hydrology is

$$Pm = \frac{m}{N+1} \tag{4.173}$$

Clearly, the return period of the mth data point, T_m, is

$$T_m = (N+1)\,/\,m \tag{4.174}$$

The observed values and their exceedance probabilities are plotted on the probability paper corresponding to the assumed probability distribution. On the ordinate of the graph paper are observed values and on the abscissa the probabilities or return periods. The objective of using the probability paper is to linearize the distribution so that plotted data can be represented by a straight line. A best-fit straight line is then drawn through the plotted points. The line is assumed to give the probabilities of all values beyond the observed range.

4.11.2 Frequency-Factor Method

Chow (1951) proposed the use of a frequency factor for hydrologic frequency analysis. If a hydrologic variable X is plotted chronologically in time, then a particular value x is found to be composed of two parts: namely, the mean, $\bar{x}$, and the departure from the mean Δx:

$$x = \bar{x} + \Delta x \tag{4.175}$$

The variable Δx can be expressed as the product of the standard deviation S and the frequency factor K. Therefore,

$$\Delta x = S\,K \tag{4.176}$$

$$x = \bar{x} + S\,K \tag{4.177}$$

where K depends on the return period T and the PDF of X; K literally means the number of standard deviations above and below the mean to achieve the desired quantile. For a distribution, a relation between K and T can be derived. For two-parameter distributions, K varies with T. For skewed distributions, it varies with the coefficient of skewness (C_s) and is very sensitive to the length of record. The frequency factor for some commonly used distributions is given here.

Normal distribution: Recall the definition of the standard normal variate, $Z = (X-\mu)/\sigma$, where μ is the population mean, and σ is population standard deviation of the variable X. Its observed values are expressed as

$$z = (x - \bar{x})\,/\,S \tag{4.178}$$

$$\text{or } x = \bar{x} + S\,z \tag{4.179}$$

Thus, for the normal distribution, K is the standard normal variate, which can be obtained from the tables of standard normal distribution.

Log-normal distribution: If a variable follows log-normal distribution, its logarithms will follow normal distribution and then the formula for normal distribution can be applied.

Gumbel distribution: If the reduced variate is Y, the frequency factor for this distribution is

$$K = (y - 0.577)\,/\,1.283 \tag{4.180}$$

$$\text{where } y = 1.283(x - \bar{x})\,/\,S + 0.577 \tag{4.181}$$

Log-Pearson Type 3 Distribution: For the Log-Pearson Type 3 Distribution, K is a function of both the return period and C_s. Values of K for log-transformed data are given by the Water Resources Council (1967). To fit this distribution, transform the data, xi (annual floods), to their logarithmic values, y_i. Compute the mean, standard deviation S_y, and C_s for the log values. Get the value of K for the desired T from the tabulated values. If C_s falls between -1 and $+1$, an approximate value of K is obtained from

$$K = \frac{2}{C_s}\left[\left\{\left(z - \frac{C_s}{6}\right)\frac{C_s}{6} + 1\right\}^3 - 1\right] \tag{4.182}$$

Compute y from $y = \bar{y} + KS_y$. Then, compute $x = \exp(y)$ for the desired T value.

Example 4.16. The mean annual flood of a river is 32,100 cumec and the standard deviation of flood peaks is 6000 cumec. Assume that the peaks follow the Gumbel distribution. What is the probability of a flood of magnitude of 40,000 cumec occurring in the river within the next 5 years?

Solution. Given $\bar{x} = 32{,}100$ cumec, $S = 6000$ cumec, and $x = 40{,}000$ cumec.

$$\text{Now } y = 1.283(40000-32100)/6000 + 0.577 = 2.266.$$

$$F(y) = \exp\left[-\exp(-2.266)\right] = 0.901$$

$$P = 1-F(y) = 1-0.901 = 0.0985$$

The return period of the flood event is $1/P = 1/0.0985 = 10.15$. The probability of a 40,000 cumec flood occurring in the next five years $= 1-0.901^5 = 1-0.594 = 0.406$.

4.11.3 Confidence Limits

A value of the variate estimated from a probability distribution for a given return period is usually in error due to the limited sample size. Therefore, a statement indicating the limits about the estimated value within which the true value is contained with a specific probability is needed. This statement is made by constructing confidence limits, which are also called the confidence intervals, confidence bands, error limits, or control curves. The confidence interval indicates the limits about the estimated value and the probability with which the true value will lie between those limits. This statement accounts for the sampling errors only.

Let the confidence probability be α. The confidence interval of the variate x corresponding to a return period T is bounded by values x_1 and x_2 (Nemec, 1973) as

$$x_{1,2} = x \pm G(\alpha)\, S_e \tag{4.183}$$

where $G(\alpha)$ is a function of the confidence probability α and can be determined by using the table of normal variates. As an example,

α (%)	50	68	80	90	95	99
$G(\alpha)$	0.674	1.00	1.282	1.645	1.96	2.58

S_e is the probable error expressed as

$$S_e = \left(1 + 1.3K + 1.1K^2\right)^{0.5} S_{N-1} / \sqrt{N} \tag{4.184}$$

in which K is the frequency factor of the distribution under consideration, S_N is the standard deviation of the sample, and N is the sample size. By using this method, confidence limits can be placed above and below the fitted distribution curve. If the Gumbel distribution is considered, then for a given sample and T, 80% confidence limits are about twice as big as 50% ones, and 95% confidence limits are about thrice as big as 50% limits.

4.11.4 Regional Frequency Analysis

For many watersheds, streamflow data are either insufficient or nonexistent at the sites of interest. The methods of frequency analysis using data from a single site will have then limited predictive value because of large sampling errors. To overcome data deficiency, a regional frequency analysis is performed. By defining a region that is hydrologically similar in terms of the variable to be studied, data from several gaging sites within this homogeneous region are pooled together into a single regional frequency analysis. Examples of regional frequency analysis are estimation of design flood from rainfall-runoff relationship, prediction of flood peaks from the relation between observed values and drainage-basin characteristics, and estimation of rainfall depths and frequencies in ungaged areas from characteristics at well-gaged sites in the same area.

The first step in a regional analysis is to define the region itself. The definition of a region depends on the quantities to be estimated. Many methods are available to define a region that is homogeneous. For mean annual precipitation, large physiographic regions can be used, whereas for peak flow, the regions may be confined to drainage basins of certain sizes. Regional boundaries can be defined in terms of similarity of flood-frequency curves or flow curves. Homogeneity tests are used to check if flood-frequency curves in a region can be considered homogeneous.

4.11.5 Index Flood Method

The index flood (IF) method, developed by the U.S. Geological Survey (Dalrymple, 1960; Benson, 1962), is widely used to perform regional flood-frequency analysis. The basic premise of this method is that a combination of streamflow records maintained at a number of gaging stations will produce a more reliable, not a longer, record, and thus will increase the reliability of frequency analysis within a region. There are two major parts of the IF method. The first is the development of basic dimensionless frequency curves representing the ratio of the flood of any frequency to an index flood (the MAF). The second is the development of relations between geomorphologic characteristics of drainage areas and the MAF by which to predict the MAF at any point within the region. By combining the MAF with the basic frequency curve, a regional frequency curve is produced.

Basic Frequency Curve

In large regions that are homogeneous with respect to flood-producing characteristics, individual streams whose drainage areas vastly differ in size have frequency curves of approximately equal slope if discharge is expressed as a ratio of the mean. The flood peaks at each gaging station are divided by an index flood (which is often taken as the MAF at the station) and are thus reduced to dimensionless ratios. The individual curves plotted using the flood ratios can be superimposed and will nearly coincide.

These curves will pass through the recurrence interval of 2.33 years at the ratio of 1.0, but may have different slopes. The variation of these slopes may be used to test the homogeneity of the region. Within the homogeneous region, the ratios are compiled for all stations and then the median ratio for each is obtained. By plotting the median ratios against recurrence interval, a regional frequency curve is obtained. This is supposedly the best representation of a flood-frequency relation obtained by combining all dimensionless curves.

Homogeneity Test

The homogeneity test was developed by Langbein (Dalrymple, 1960) and can be used to test if a region is homogeneous to permit combining individual frequency curves with confidence to form a regional curve. The question is: Do the records differ from one another by amounts attributable to chance? The answer to this question lies in computing these differences and setting limits that will be acceptable statistically.

The SEE S_e of the reduced variate y for the Gumbel distribution (EV1) can be written as

$$S_e = \exp(y)\{1/\{(T-1)N]\}^{0.5} \tag{4.185}$$

where T is the recurrence interval, and N is the number of years of record. If a normal distribution of the estimates is assumed, then 95% of the estimates of the T-year flood will lie within $2S_e$ of their most probable value of T. The test employs the 10-year flood because this is the longest recurrence interval for which most records will give dependable estimates. For $T = 10$ years,

$$2S_e = 0.666 \exp(y) \,/\, N^{0.5} \tag{4.186}$$

and y for the EVl distribution is 2.25. Therefore, the confidence limits are specified by

$$2.25 \pm 6.33/N^{0.5}$$

Dalrymple (1960) listed values of y corresponding to T, and lower and upper confidence limits with the corresponding T for various values of N.

Mean Annual Flood

The MAF is defined as the value of the graphical frequency curve at the recurrence interval of 2.33 years. Benson (1962) confirmed experimentally that MAF has a magnitude equivalent to the flood of a 2.33-year recurrence interval. It is dependent on many factors that can be classified as either physiographic or meteorologic. The physiographic factors influencing MAF at a given point are: (1) drainage area and shape; (2) stream slopes; (3) natural storage in lakes, swamps, or channels; (4) land slope; (5) land use; (6) geology; (7) stream density; (8) stream pattern; (9) elevation; (10) aspect; (11) orographic position; (12) basin relief; and (13) soil cover.

The meteorologic factors are connected with the magnitude and distribution of precipitation received by a drainage area, and include: (1) storm types; (2) type of region, whether humid or arid; (3) storm pattern; (4) storm direction; (5) precipitation intensities; (6) snowmelt; (7) storm volume; and (8) extent of ice jams. Evaluation, treatment, and use of some of these factors are difficult. The most commonly used factor is the mean annual precipitation.

Sometimes factors, representing the composite effect of the above factors, have been used in flood frequency studies. The mean annual runoff reflects the effect of precipitation and basin characteristics. Another factor is the lag time, which is the time difference between the centers of rainfall and runoff. It represents the composite effect of all topographic factors.

Median Flood Ratios

The ratios of several floods of different recurrence intervals to the MAF are tabulated for each station. Enough recurrence intervals are selected to define the curve appropriately. By tabulating the flood ratios, the median ratio for each recurrence interval is computed. This median of a recurrence interval is the mid-value of its flood ratios if their number is odd or the mean of the two central ratios if their number is even.

Regional Frequency Curve

Each median flood ratio is plotted against its recurrence interval on the probability paper. An average frequency curve is then drawn. This is the regional frequency curve, showing flood discharge in a ratio to MAF, is based on all significant discharge records available, and represents the most likely flood-frequency values for all areas in the region.

Some weaknesses of the IF method are:

1. The flood ratios for comparable streams may differ due to large differences in IF. If IF is not typical and is obtained from a short period of record, the remainder of the frequency curve may be faulty.
2. The homogeneity test cannot be applied at a level much higher than that of the 10-year flood because many individual records are too short to adequately define the frequency curve at higher levels. In many cases, individual curves show wide and sometimes systematic differences at higher levels.
3. Within a flood frequency region, frequency curves are combined for all sizes of drainage areas, excluding the largest. Recent studies using ratios of less frequency floods have shown in all cases that the ratio of any specified flood to MAF varies inversely with the drainage area. In general, the larger the drainage area, the flatter the frequency curve. The effect of drainage area is relatively greater for floods of higher recurrence intervals.

4.11.6 Multiple-Regression Method

The relation of flood peaks of selected recurrence intervals to basin and climatic parameters is determined by multiple-regression methods. The resulting relation is of the form

$$Q_T = b\prod_{i=1}^{N} x_i^{a_i} \tag{4.187}$$

where Q_T denotes the T-year flood; b is the regression constant; x_i, i = 1, 2, …, N, are independent basin and climatic parameters; a_i, i = 1, 2, …, N, are regression coefficients. Many studies have used models similar to Eq. (4.187) to estimate the flood magnitude of recurrence interval T. The basin and climatic characteristics are normally evaluated from topographic, geologic, and climatic maps. Although the basin and climatic factors to be used in regression analysis vary from one region to another, the most important factors are drainage area, main-channel slope, and mean annual precipitation. Many of these factors are interrelated. For example, in general, as drainage area increases, slope and rainfall intensity decreases.

Some of the advantages of using a multiple-regression analysis are: (1) It provides a mathematical relation between Q of a specified value and T the independent variables. (2) It provides an evaluation of the independent variables that best define the dependent variable. (3) It provides a measure of the accuracy of the equation in terms of the SEE, and tests the significance of the coefficients of each independent variable. (4) It evaluates the relative significance of each independent variable by indicating those variables that have a coefficient that is significantly different from zero at a particular percent confidence level. (5) It provides an easy evaluation of the coefficients when the dependent and independent variables are transformed to their logarithms and used in a linear regression.

4.12 TIME SERIES ANALYSIS

A *time series* is a set of observations generated sequentially in time. If the set is continuous, the series is said to be *continuous*; if it is discrete, the time series is said to be *discrete*. Here only discrete time series where observations are made at some fixed interval h will be considered. The observations in a discrete series made at equidistant time intervals $\tau_0 + h,\ \tau_0 + 2h,\ \ldots\ \tau_0 + th,\ \ldots\tau_0 + Nh$ may be denoted by $z(\tau_1),\ z(\tau_2),\ \ldots z(\tau_t),\ \ldots z(\tau_N)$. For many purposes, the value of τ_0 and h are unimportant; these are needed if the observation times are to be defined exactly. Of course, the information content of a time series is affected by the choice of h, particularly if the series has rapid changes. If τ_0 is adopted as the origin and h as the unit of time, z_t can be regard as the observation at time t.

A discrete time series may arise in two ways:

(a) By sampling a continuous time series; for example, the continuous river flow at a station may be sampled at hourly intervals.

(b) By accumulating a variable over a period of time; for example, rainfall may be accumulated over a period of a day.

A hydrologic time series can be divided in two basic groups: (1) univariate (single) time series, for example, monthly streamflow at a point, and (2) multivariate (multiple) series of different kinds at one point. The examples of the second type are series of flow and water quality variables at a station. If a time series, for example, daily precipitation, is composed of nonzero and zero values, it is known as intermittent series. A time series whose values have been observed at regular intervals, such as each day or each hour, is termed as regularly spaced time series.

Time series analysis is useful for many applications, such as forecasting, detecting trends in records, filling-in missing data, and generation of synthetic data. Time series analysis is a subject in itself and only a brief introduction of it is given in the following.

4.12.1 Components of a Time Series

A time series can be divided into a number of components. The main components of a hydrologic time series are: trends and other deterministic changes, cycles or periodic changes and autocorrelation, almost periodic changes, such as tidal effects, and components representing stochastic or random variations. Trend and jumps are introduced in a time series due to gradual or sudden changes in the major factors of the process that is responsible for the time series. For example, a runoff time series will have a trend as a result of major land use changes in the upstream catchment; a water quality time series may exhibit trends if a new factory upstream begins to discharge its effluent in the river. The closure of a diversion dam will lead to a jump in the series because the flow will be reduced due to diversion. The turning point test and Kendall's rank correlation tests are commonly employed to test randomness and trend in a series.

The time series of water resource variables which are measured or accumulated at subannual time intervals normally have periodic patterns. Such patterns can be seen in time series, for example, monthly rainfall, daily runoff, daily volumes of urban water demands, and these series are said to have seasonal or periodic patterns. Due to seasonality, the statistical properties of time series vary with time (week or month, etc.). In harmonic analysis, the periodic component of a time series is represented using a series of sine functions. For a series without trend, the harmonic equation is:

$$z_t = \mu + \sum_{i=1}^{L} \lambda_i \sin\left(\frac{2\pi t i}{T} + \varphi_i\right) + \varepsilon_i \tag{4.188}$$

where μ is the population mean, λ_i and φ_i are the amplitudes and phases of the wave, and i/T is the frequency of the wave.

4.12.2 Stationary Stochastic Processes

A special class of stochastic processes, called *stationary processes*, is based on the assumption that the process is in a particular state of statistical equilibrium. A stochastic process is said to be strictly stationary if its properties are unaffected by a change of time origin; that is, if the joint probability distribution associated with m observations z_{t1}, z_{t2} ... z_{tm}, made at any set of times t_1, t_2, ... t_m, is the same as that associated with m observations z_{t1+k}, z_{t2+k}, ..., z_{tm+k}, made at time $t_1 + k$, $t_2 + k$, ... $t_m + k$. Thus, for a discrete process to be strictly stationary, the joint distribution of any set of observations must be unaffected by shifting all the times of observation forward or backward by any integer amount k. The statistical properties of a nonstationary time series are time-dependent.

Assuming that the stationarity assumption holds true, the joint probability distribution $p(z_{t1}, z_{t2})$ is the same for all times t_1, t_2, which are a constant interval apart. Therefore, the nature of the joint distribution can be inferred by plotting a scatter diagram using pairs of values (z_t, z_{t+k}), of the time series, separated by constant interval or lag k. The covariance between z_t and z_{t+k} is called the autocovariance at lag k and is calculated by

$$\gamma_k = \text{cov}[z_t, z_{t+k}] = E[(z_t - \mu)(z_{t+k} - \mu)] \tag{4.189}$$

For a stationary process, the variance at time ($t + k$) is the same as at time t. The estimate of the kth lag autocovariance γ_k is

$$c_k = \frac{1}{N} \sum_{t=1}^{N-k} (z_t - \bar{z})(z_{t+k} - \bar{z}), k = 0, 1, 2..., K \tag{4.190}$$

The estimate of lag k autocorrelation is obtained by

$$r_k = c_k / c_0 \tag{4.191}$$

which implies that $r_0 = 1$.

A common cause of autocorrelation or dependence in many hydrologic time series is the storage effect. In the case of river flow series, for example, this storage might be at the catchment surface, in unsaturated zone, or in ground water zone.

4.12.3 Time Series Models

A mathematical model representing a time series or stochastic process is called a *time series model.* The model has a certain structure and a set of parameters. The important categories of time series models are as follows.

Autoregressive (AR) models: Autoregressive (AR) models are extremely useful to represent certain practical series. Let the values of a process at equally spaced times t, $t-1$, $t-2$, ... be Y_t, Y_{t-1}, Y_{t-2}, ... and let z_t, z_{t-1}, z_{t-2} ... be the deviations from the mean μ; for example, $z_t = Y_t - \mu$. In an AR model, the current value of the process is expressed as a finite, linear aggregate of previous values of the process and a shock a_t. Thus

$$z_t = \varphi_1 z_{t-1} + \varphi_2 z_{t-2} + \ldots + \varphi_p z_{t-p} + a_t \tag{4.192}$$

is called an AR process of order p and is denoted by AR(p). Introducing a *backward shift operator* B, defined by $Bz_t = z_{t-1}$, Eq. (4.173) can be written as:

$$\left(1 - \varphi_1 \mathrm{B} - \varphi_2 \mathrm{B}^2 - \ldots - \varphi_p \mathrm{B}^p\right) z_t = a_t \tag{4.193}$$

$$\text{or } (\mathrm{B})\ z_t = a_t \tag{4.194}$$

Here $\varphi(B) = 1 - \varphi_1 \mathrm{B} - \varphi_2\,\mathrm{B}^2 - \ldots - \varphi_p\,\mathrm{B}^p$ is termed as an AR operator of order p. The AR models have been extensively used in water resources, because this form has an intuitive type of time dependence and the AR models are simple to use.

Moving average (MA) models: Another kind of model of great practical importance in the representation of observed time series is the finite moving average (MA) process. Here z_t is linearly dependent on a finite number q of previous a's. Thus,

$$z_t = a_t - \theta_1 a_{t-1} - \theta_2 a_{t-2} - \ldots - \theta_q a_{t-q} \tag{4.195}$$

is called an MA process of order q and is denoted by MA(q). Similar to the AR operator, an MA operator of order q can be written as

$$\theta(B) = 1 - \theta_1 \mathrm{B} - \theta_2 \mathrm{B}^2 - \ldots - \theta_q \mathrm{B}^q \tag{4.196}$$

and the MA(q) process can be written as

$$z_t = (\mathrm{B})\ a_t \tag{4.197}$$

Autoregressive-moving average models: Greater flexibility in fitting time series models is achieved by including both autoregressive and MA terms in the model. This leads to the mixed autoregressive-moving average ARMA(p, q) model:

$$z_t = \varphi_1 z_{t-1} + \ldots + \varphi_p z_{t-p} + a_t - \theta_1 a_{t-1} - \ldots - \varphi_q a_{t-q} \tag{4.198}$$

$$\text{or } z_t - \varphi_1 z_{t-1} - \ldots - \varphi_p z_{t-p} = a_t - \theta_1 a_{t-1} - \ldots - \theta_q a_{t-q}$$

$$\text{or } (\mathrm{B}) z_t = (\mathrm{B})\ a_t \tag{4.199}$$

which employs $p + q + 2$ unknown parameters μ, $\varphi_1, \ldots, \varphi_p, \theta_1, \ldots, \theta_q, \sigma_a^2$, that are estimated from the data. The simplest member of ARMA (*p*, *q*) family is the ARMA (1, 1) model which can be written as

$$z_t - \varphi_1 z_{t-1} = a_t - \theta_1 a_{t-1} \tag{4.200}$$

The combination of AR and MA models makes it possible to simulate many hydrologic processes by using a small number of parameters. For example, the flow in a stream results due to a number of causes such as precipitation and groundwater effluence. This mixed behavior can be conveniently modeled by ARMA models. In practice, an adequate representation of actually occurring stationary time series can be frequently obtained with AR, MA, or mixed model, in which *p* and *q* are not greater than 2 and often less than 2. Note that ARMA (*p*,0) model is the same as AR (*p*) and ARMA (0,*q*) model is the same as MA (*q*). In Eq. (4.165), z_t and a_t may represent time-dependent discharge (output) and rainfall (input).

The ARMA models are suitable for stationary hydrologic series. In the case of nonstationary series, the periodic or seasonal fluctuation can be removed by taking the differences and the ARMA model can be applied to the resultant series. The resultant model is termed as autoregressive integrated moving average (ARIMA) model. Consider a time series that is homogeneous except in level, that is, the various segments of the series look identical, except the difference in level about which it changes. Such a series can be adequately represented by a model of the form:

$$\varphi(\mathrm{B})\ \nabla z_t = (\mathrm{B})\ a_t \tag{4.201}$$

where ∇ is the *backward difference operator* defined as

$$\nabla z_t = z_t - z_{t-1} = (1 - \mathrm{B}) z_{t-1} \tag{4.202}$$

Thus, ARIMA(*p*, *d*, *q*) is an ARMA model that is fitted to the data after taking the *d*th difference of the series:

$$\varphi(\mathrm{B})\nabla^d\ z_t = (\mathrm{B})a_t \tag{4.203}$$

where ∇^d indicates that the series is differenced *d* times. The notation $\nabla_n = 1{-}\mathrm{B}^n$ indicates differencing with lag of *n*. The first-order differencing (eq. 4.167) is helpful in removing the trend of a series or nonstationarity in the mean. Two consecutive differencing operations are necessary to remove nonstationarity in the mean and slope. However, it may not always be possible to remove nonstationarity by differencing alone, other transformations may also be needed.

4.12.4 Partial Autocorrelation Function

The partial autocorrelation function is another way of representing the time dependence structure of a series. It is useful for the identification of the type and order of the model of a given time series. Let φ_{kj} denote the *j*th coefficient in an AR process of order *k*; φ_{kk} being the last coefficient. Now, φ_{kk} satisfies the set of equations

$$\rho_j = \varphi_{k1}\ \rho_{\mathrm{j}-1} + \varphi_{k(k-1)}\ \rho_{j-k+1} + \varphi_{kk}\ \rho_{\mathrm{j}-\mathrm{k}} j = 1, 2, \ldots k \tag{4.204}$$

leading to the set of equations known as the Yule—Walker equations. These are written as

$$\begin{bmatrix} 1 & \rho_1 & \rho_2 & \cdots & \rho_{k-1} \\ \rho_1 & 1 & \rho_1 & & \rho_{k-2} \\ \vdots & & & & \\ \rho_{k-1} & \rho_{k-2} & \rho_{k-3} & \cdots & 1 \end{bmatrix} \begin{bmatrix} \varphi_{k1} \\ \varphi_{k2} \\ \vdots \\ \varphi_{kk} \end{bmatrix} = \begin{bmatrix} \rho_1 \\ \rho_2 \\ \vdots \\ \rho_k \end{bmatrix} \tag{4.205}$$

or

$$P_k \, \varphi_k = \rho_k \tag{4.206}$$

Solving these equations for $k = 1, 2, 3, \ldots$, successively, the values of φ_{11}, $\varphi_{22} \ldots$ are obtained as a function of ρ. The quantity φ_{kk}, regarded as a function of the lag k, is called the *partial autocorrelation* function. For an AR(p) process, the partial autocorrelation function φ_{kk} will be nonzero for k less than or equal to p and zero for k greater than p. Stated in another way, the partial autocorrelation function of an AR(p) process has a cutoff after lag p. Conversely, the autocorrelation function of an MA(q) process has a cutoff after lag q, while its partial autocorrelation tails off. If both autocorrelations and partial autocorrelations tail off, a mixed process is suggested.

Partial autocorrelations may be estimated by successively fitting AR processes of orders 1, 2, 3, ... by LS and picking out the estimates φ_{11}, $\varphi_{22} \ldots$ of the last coefficient fitted at each stage. Alternatively, approximate Yule—Walker estimates of the successive AR processes may be employed. The estimated partial autocorrelations can then be obtained by substituting estimates r_j for the theoretical autocorrelations in Eq. (4.204) to yield

$$r_j = \widehat{\varphi}_{k1} r_{j-1} + \widehat{\varphi}_{k2} r_{j-2} + \ldots + \widehat{\varphi}_{k(k-1)} r_{j-k+1} + \widehat{\varphi}_{kk} r_{j-k}, j = 1, 2, \ldots k \tag{4.207}$$

4.12.5 Fitting of ARMA Models

Initially, one does not know the order of the ARMA process for fitting to an observed time series as well as the order of differencing that is required, if any. Therefore, the model is built iteratively, that is, a set of models is identified using the characteristics of the data and its adequacy is tested. Depending on the results, the model may be adopted or another candidate model is identified. While fitting a time series model, at least 50 observations should be used. If sufficient observations are not available, one proceeds by using experience to build a preliminary model. This model may be updated as more data becomes available. While applying ARMA models, the main stages are:

- *Model identification* which involves the use of data and any information on how the series was generated to identify a subclass of parsimonious models worthy to be considered.
- *Parameter estimation* which involves an efficient use of the data to make inferences about parameters conditional on the adequacy of the considered model.
- *Diagnostic checking* involves checking the fitted model in its relation to the data with the intent to reveal model inadequacies and to achieve model improvement.

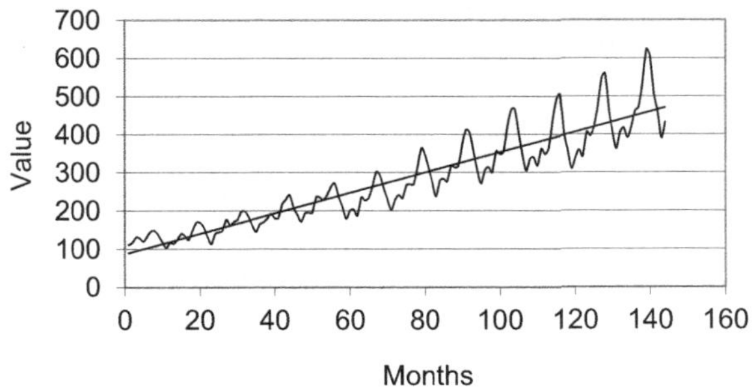

FIGURE 4.12

Plot of a periodic time series.

In any time series modeling, it is worthwhile to first plot the data. A visual inspection of the data always gives useful information about the behavior of the process. The autocorrelation function of the series also gives useful information. If the autocorrelation function fails to die out rapidly, it suggests that the series may be nonstationary and may require differencing to obtain a stationary series. Fig. 4.12 shows a plot of a periodic time series (Box and Jenkins, 1976, airlines data series). The rising trend of the series points to the need of differencing of the series. The correlogram of this time series is plotted in Fig. 4.13.

Having tentatively decided the order of differencing d, the general appearance of the estimated autocorrelation and partial autocorrelation functions of the appropriately differenced series are studied. These provide clues about the choice of the orders p and q for the AR and MA operators. In doing so, the characteristic behavior of the theoretical autocorrelation function and of the theoretical partial autocorrelation function for AR, MA, and mixed processes is used. For example, if φ_1 of an AR(1) model is +ve, ρ_k decays exponentially to zero, while for φ_1 −ve, it oscillates in sign. The autocorrelation function of AR(2) model has different forms, depending on the values parameters take on. The properties of autocorrelation and partial autocorrelation functions of a few model types are described in Table 4.9.

To illustrate, the series in Fig. 4.12 was differenced at lag 1 and the differenced series is plotted in Fig. 4.14. The correlogram of the differenced series is plotted in Fig. 4.15. Although the differenced series has considerably small correlations, these are still quite high at lags of the multiple of 12 and this shows that further differencing is needed.

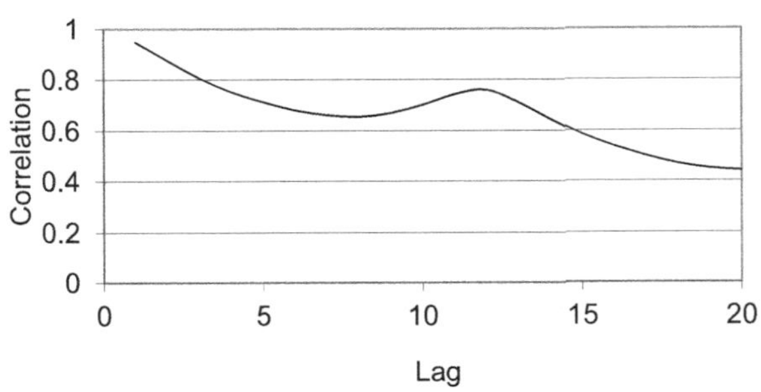

FIGURE 4.13

Correlogram of periodic time series of Fig. 4.12.

Table 4.9 Properties of autocorrelation and PAC functions of some time series models.

Model type	Autocorrelation function	Partial autocorrelation function
AR(1), first-order autoregressive	Decreases exponentially	$\varphi_{1,1} \neq 0$ $\varphi_{i,i} = 0$ for $i = 2,3,4, \ldots$
AR(p), pth-order autoregressive	Mixed type of damping from lag 1	$\varphi_{i,i} \neq 0$ for $i \leq p$ $\varphi_{i,i} = 0$ for $i > p$
MA(1), first-order moving-average	$\rho_1 \neq 0$ $\rho_i = 0$ for $i = 2,3,4, \ldots$	Decreases exponentially
MA(q), qth-order moving-average	$\rho_i = 0$ for $i > q$ $\rho_i \neq 0$ for $i \leq q$	Mixed type of damping from lag 1
ARMA (1, 1), autoregressive moving-average	Decreases exponentially after lag 1	Decreases exponentially after lag 1
ARMA (p, q), autoregressive moving-average	Mixed type of damping after lag $q+1$	Mixed type of damping after lag $p-q$

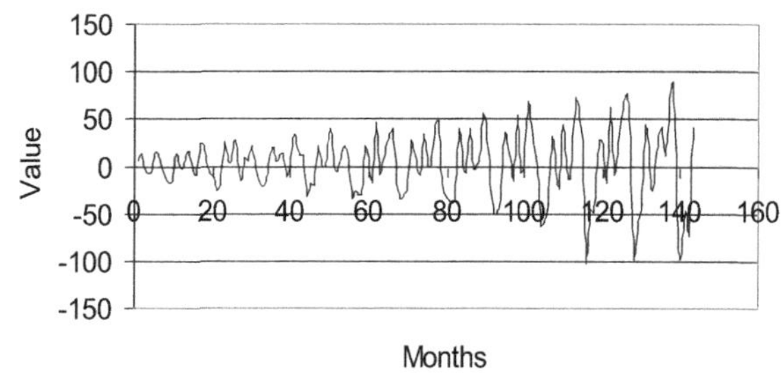

FIGURE 4.14

Time series of Fig. 4.12 after differencing at lag 1.

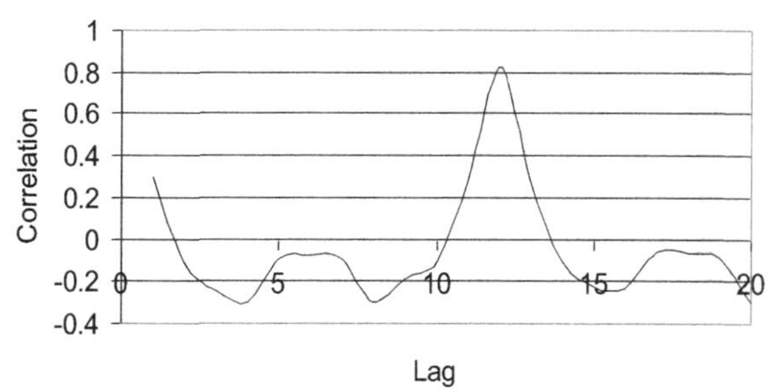

FIGURE 4.15

Correlogram of the time series of Fig. 4.14.

For most monthly hydrological series, it is often helpful to first standardize the series by subtracting the mean and dividing by the standard deviation of the corresponding month. A first-order differencing of the resultant series is often adequate to yield a stationary series that can be modeled by the ARMA class of models.

The common techniques to estimate the parameters of a time series model are the method of moments, the MOLS, and the method of ML. Mostly, the parameters obtained by the method of moments are used as the first approximation and are refined by other methods.

After an ARMA model has been fitted, it is necessary to apply statistical tests to check its adequacy and suitability. The tests which are used for this purpose include *the Porte Manteau Lack of Fit Test*, *the Akaike Information Criterion (AIC)*, and the test of correlogram. The statistic AIC is calculated by

$$\mathrm{AIC}(p,q) = N \ln\left(\underset{\in}{\overset{\wedge}{\sigma}} 2\right) + 2(p+q) \tag{4.208}$$

where N is the sample size, and $\underset{\in}{\overset{\wedge}{\sigma}} 2$ is the MLE of the residual variance. The model which gives the minimum AIC is selected.

Examination of the residuals (difference between observed and computed values of a dependent variable) of a model is always helpful. The residuals of an adequate model should resemble white noise, the lag-one serial correlation should be close to zero, and they should have small variance. A test to check whether the residuals of a model are independent or not is the *Porte Manteau lack-of-fit* test. In this, the Q statistic is determined by

$$Q = N \sum_{k=1}^{L} r_k^2 \tag{4.209}$$

where r_k denotes the autocorrelation of residuals at lag k and L is the number of lags considered. Q approximately follows a chi-square distribution with (L-p-q) degrees of freedom.

The technique of overfitting, in which a more elaborate model is fitted to the data and then the results are compared, has also been recommended. Box and Jenkins (1976), Salas et al. (1980), and Salas (1993) are good references for time series models.

The water resources literature contains innumerable applications of time series modeling to a wide range of problems. The ARMA models are frequently used in rainfall-runoff modeling. A number of well-known hydrologic models are special cases of the ARMA model. For example, the Muskingum model of flood routing is obtained by setting certain parameters of this equation to zero. The ARMA model parameters do have a physical interpretation. As the precipitation event occurs, the MA parameters influence the rising limb of the hydrograph and these parameters will, therefore, depend on the basin physiographic characteristics and its state. Similarly, the AR parameters will more heavily influence the recession limb of the hydrograph. These parameters can vary from storm to storm.

4.13 MARKOV MODELS

Many data sequences consist of a succession of mutually exclusive states, for example, annual streamflow data at a site. The future states of any stochastic process, for example, annual flow at a particular location, cannot be predicted with certainty. However, based on past data, it is possible to assess the probability of any particular outcome. Consider a stream in which the annual flow varies from 50 to 90 million m^3 and the interest lies in the nature of transitions from one state to another. To construct transition matrix, the data are classified into several mutually exclusive states. Let the data be

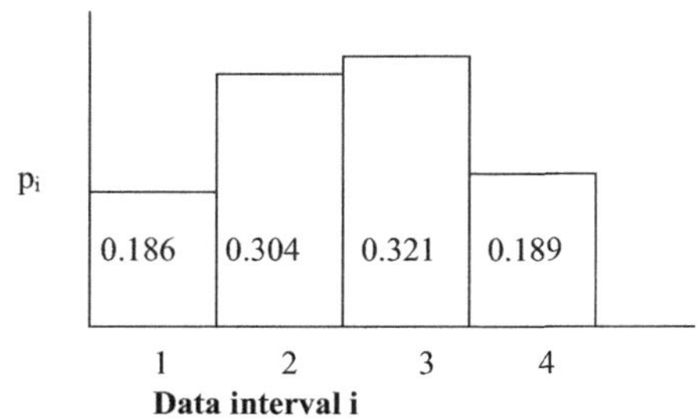

FIGURE 4.16

Histogram of probabilities.

classified into four classes of intervals of 10 million m^3 each: 50–60, 60–70, 70–80, and 80–90 million m^3, designated as A, B, C, and D. Now it is possible to determine the percent of time the past annual flows were within each of the four intervals. Here it is assumed that the probability distribution of annual streamflows does not significantly change with time and past data provide a reasonable basis to define probability distributions. Denote p_i as the fraction of time the annual flow was in the interval i where $i = 1, 2, 3, 4$. Then,

$$p_i = [\text{Number of years when flow was in interval i}]/[\text{Total number of years of record}] \quad (4.210)$$

The following is an example of vector p of probabilities that may be obtained:

$$p = \{p_1, p_2, p_3, p_4\} = \{0.186, 0.304, 0.321, 0.189\} \quad (4.211)$$

The probabilities of Eq. (4.176) are shown in a histogram in Fig. 4.16. Thus, the probability of an annual streamflow in a year falling in the range from 50 to 60 mcm is 0.186, from 60 to 70 mcm is 0.304, and so on. These streamflow intervals can be thought of as discrete states of the streamflow system. If all possible states are defined by these four intervals, then

$$\sum_{i=1}^{4} p_i = 1 \quad (4.212)$$

The next question is the tendency for one state to succeed another. Since there are four states, a 4*4 matrix can be constructed, each box showing the number of times a given state is succeeded by another. If each element in the *i*th row is divided by the total of the *i*th row, each row in this matrix will sum to one. This matrix is called a transition probability matrix. Let P_{ij} be the conditional probability of the current annual streamflow being in interval j given that last year's streamflow was in interval i.

$$Pij = \frac{\text{Number of events in interval j following those in interval } i}{\text{Total number of events in interval } i}$$

$$\text{Then, } \sum_{j=1}^{4} p_{ij} = 1 \text{ for all rows } i \quad (4.213)$$

A typical transition probability matrix is shown in Table 4.10.

Table 4.10 Matrix of streamflow transition probabilities.

		Streamflow state j in year $y+1$			
		A (50–60)	B (60–70)	C (70–80)	D (80–90)
Streamflow state i in year y	A	0.40	0.25	0.25	0.10
	B	0.20	0.30	0.40	0.10
	C	0.10	0.40	0.30	0.20
	D	0.10	0.20	0.30	0.40

The matrix shows, for example, that when state A occurs in a year, it is followed by state A 40% of the time, by B 25% of time, and so on. Note that this probability is independent of the total number of times that A occurs. Each row of the matrix is the conditional probability vector corresponding to an initial stream flow state. Each element P_{ij} in the matrix is the probability of a transition from streamflow i in one year to streamflow j in the next year. These conditional probabilities are called transition probabilities. If the transition from one discrete value to the next can be described by a matrix of transition probabilities as that shown in Table 4.10, this stochastic streamflow process is called a discrete Markov process and the matrix is called a first-order Markov chain. These probability estimates improve as more observations are available.

Stochastic processes in which the probability of a future state is dependent only on the present state and not on any of the past states are said to follow a first-order (lag one) Markov chain. A useful property of Markov chains in describing ergodic processes, such as streamflows, is that there exists a stationary or steady-state probability distribution that is independent of the initial state. This implies that although next year's flow probabilities are dependent on this year's flow, the distribution of streamflows far in future, say 20 years from now, will be independent of this year's flow.

Coming back to the transition probability matrix of Table 4.10, assume that in year y streamflow was 64 million m^3. Since this is in the interval between 60 and 70, the state of streamflow in year y was 2 or B. The initial probability vector $p^{(y)}$ for year y is {0, 1, 0, 0}. Knowing $p^{(y)}$, it is easy to determine the probabilities $p_j^{(y+1)}$ of each of the four possible streamflow states j that can occur in year $(y+1)$. In Table 4.10, the probability vector or row for state corresponding to B is {0.2, 0.3, 0.4, 0.1}. This vector can be calculated by using the fact that in year $(y+1)$, the probability of being in state j is equal to the sum of the probabilities of being in each state i in year y times the probability of a transition from state i in year y to state j in year $(y+1)$. If $p_j^{(y+1)}$ represents the unconditional probability of being in state j in year $y+1$, this can be computed by:

$$p_j^{(y+1)} = p_1^{(y)} p_{1j} + p_2^{(y)} p_{2j} + p_3^{(y)} p_{3j} + p_4^{(y)} p_{4j} = \sum_i p_i^{(y)} p_{ij} \forall j \tag{4.214}$$

where p_{ij} is the probability of a transition from state i to state j; here it is assumed to be independent of year y. Denoting the matrix of p_{ij}'s by P and the probability vector for year y by $p^{(y)}$, Eq. (4.192) can be written in a compact way:

$$p^{(y+1)} = p^{(y)} P \tag{4.215}$$

Table 4.11 Successive streamflow probability vectors.

	Streamflow state probabilities			
Year y	p_1^y	p_2^y	p_3^y	p_4^y
Y	0.000	1.000	0.000	0.000
$y+1$	0.200	0.300	0.400	0.100
$y+2$	0.190	0.320	0.320	0.170
$y+3$	0.189	0.306	0.323	0.183
$y+4$	0.187	0.305	0.321	0.187
$y+5$	0.187	0.304	0.321	0.188
$y+6$	0.186	0.304	0.321	0.189
$y+7$	0.186	0.304	0.321	0.189

Likewise, the probability of each streamflow state in year $(y+2)$ can be obtained knowing $p^{(y+1)}$:

$$p^{(y+2)} = p^{(y+1)}\ P \tag{4.216}$$

In a similar manner, it is easy to compute the probabilities of each possible streamflow state for year $(y+1)$, $(y+2)$, $(y+3)$, and so on. The probability vectors for the first seven years are listed in Table 4.11.

Note that as y increases, the probabilities tend toward a limiting value, that of $(y+7)$ in this example. These are the unconditional steady-state probabilities of having any one of the possible streamflow states. These are the same probabilities that are shown in the histogram of Fig. 4.16.

The annual streamflow correlation will not be as strong as assumed here. However, the shorter the time duration of streamflow data, for example, monthly, weekly, and daily, the autocorrelation will increase.

4.14 CLOSURE

Statistical analysis of hydrological variables provides useful information about the nature of distribution that the data tend to follow. This can be used to predict the magnitude and associated frequency. Regression techniques are widely used to develop prediction equations for different hydrological variables. These prediction equations are useful to estimate the dependent hydrological variables, which are difficult to monitor, based on the data of independent variables which can be easily monitored.

References

Abdi, H., 2007. The Kendall Rank Correlation Coefficient. In: Neil Salkind (Ed.), Encyclopedia of Measurement and Statistics. Sage, Thousand Oaks (CA).

Adamowski, K., 1981. Plotting formula for flood frequency. Water Resources Bulletin 17 (2), 197–202.

Ashkar, F., Bobee, B., Leroux, D., Morisette, D., 1988. The generalized method of moments as applied to the generalized gamma distribution. Stochastic Hydrology and Hydraulics 2, 161–174.

Beard, L.R., 1943. Statistical Methods in Hydrology. U.S. Army, Corps Eng. Civ. Works Invest. Proj. Rep. CW-151.

Benson, M.A., 1962. Evolution of Methods for Evaluating the Occurrence of Floods. U.S. Geological Survey, Water Supply Paper 1580-A, https://doi.org/10.3133/wsp1580A.

Blom, G., 1958. Statistical Estimates and Transformed Beta-variable. John Wiley & Sons, New York.

Box, G.E.P., Jenkins, G.M., 1976. Time Series Analysis—Forecasting and Control. Holden Day Inc., San Francisco.

Bree, T., 1978a. The stability of parameter estimation in the linear model. Journal of Hydrology 37, 47–66.

Bree, T., 1978b. The general linear model with prior information. Journal of Hydrology 39, 113–127.

California, 1923. Flow in California Streams. Calif., Dep. Public Works, Div. Eng. Irrig., Bull. 5.

Chegodayev, N.N., 1955. Formulas for the calculation of the confidence of hydrological quantities. In: Chow (Ed.), Handbook of Applied Hydrology. McGraw-Hill, New York.

Chow, V.T., 1951. A general formula for hydrologic frequency analysis. Transactions—American Geophysical Union 32, 231–237.

Cunnane, C., 1978. Unbiased plotting position — a review. Journal of Hydrology 37 (3–4), 205–222.

Dalrymple, T. 1960. Flood-Frequency Analysis. U.S. Geological Survey, Water Supply Paper 1543-A.

Dooge, J.C.I., 1973. Linear Theory of Hydrologic Systems. Technical Bulletin No. 1468. Agricultural Research Service, U.S. Department of Agriculture, Washington, D.C, p. 327.

Douglas, J.R., Clarke, Newton, S.G., 1976. The use of likelihood functions to fit conceptual models with more than one dependent variable. Journal of Hydrology 29, 181–198.

Greenwood, J.A., Landwehr, J.M., Matalas, N.C., Wallis, J.R., 1979. Probability-weighted moments: definition and relation to parameters to several distributions expressible in inverse form. Water Resources Research 15, 1049–1054.

Gringorten, I.I., 1963. A plotting rule for extreme probability paper. Journal of Geophysical Research 68, 813–814.

Gumbel, E.J., 1958. Statistics of Extremes. Columbia University Press, New York.

Haktanir, T., 1996. Probability-weighted moments without plotting position formula. Journal of Hydrologic Engineering 1 (2), 89–91.

Harley, B.M., 1967. Linear Routing in Uniform Open Channels. M. Engg. Sc. Thesis, University College, Cork, Ireland.

Hazen, A., 1914. Storage to be provided in impounding reservoirs for municipal water supply. Transactions of the American Society of Civil Engineers Paper.

Hirsch, R.M., Helsel, D.R., Cohn, T.A., Gilroy, E.J., 1993. Statistical analysis of hydrologic data. In: Maidment, D.R. (Ed.), Handbook of Hydrology. McGraw-Hill Inc., New York.

Hosking, J.R.M., 1986. The Theory of Probability-Weighted Moments. Technical Report RC 12210. Mathematics. IBM Thomas J. Watson Research Center, Yorktown Heights, New York.

Hosking, J.R.M., 1992. Moments or L-moments. An example comparing two measures of distributional shape. The American Statistician 46 (3), 186–189.

Hosking, J.R.M., 1990. L-Moments: analysis and estimation of distribution using linear combination of order statistics. Journal of the Royal Statistical Society: Series B 52 (1), 105–124.

Jain, D., Singh, V.P., 1986a. A comparison of transformation methods for flood frequency analysis. Water Resources Bulletin 22 (6), 903–912.

Jain, D., Singh, V.P., 1986b. Comparing methods of transformation for flood frequency analysis. In: Shen, H.W., Obeysekara, J.T.B., Yevjevich, V., DeCoursey, D.G. (Eds.), Multivariate Analysis of Hydrologic Processes. Colorado State University, Fort Collins, Colorado, pp. 755–767.

Jones, L.E., 1971. Linearizing weight factors for least squares fitting. Journal of the Hydraulics Division, ASCE 97 (HY5), 665–675.

Kroll, C.N., Stedinger, J.R., 1996. Estimation of moments and quantiles using censored data. Water Resources Research 32 (4), 1005–1012.

Kuczera, G., 1982a. Combining site-specific and regional information: an empirical Bayes approach. Water Resources Research 18 (2), 306–314.

Kuczera, G., 1982b. Robust flood frequency models. Water Resources Research 18 (2), 315–324.

Kuczera, G., 1982c. On the relationship between the reliability of parameter estimates and hydrologic time series data used in calibration. Water Resources Research 18 (1), 146–154.

Nash, J.E., 1959. A systematic determination of unit hydrograph parameters. Journal of Geophysical Research 64 (1), 111–115.

Natale, L., Todini, E., 1974. A Constrained Parameter Estimation Technique for Linear Models in Hydrology. Publication No. 13. Institute of Hydraulics, University of Pavia, Pavia, Italy.

Nemec, J., 1973. Summary: Interaction between reservoirs and the atmosphere and its hydrometeorological elements. In: Man-Made Lakes: Their Problems and Environmental Effects, pp. 398-405.

Phien, H.N., Jivajirajah, T., 1984. Fitting the S_b curve by the method of maximum likelihood. Journal of Hydrology 67, 67–75.

Rao, D.V., 1980. Log-Pearson type 3 distribution: method of mixed moments. Journal of Hydraulics Division. ASCE 106 (HY6), 999–1019.

Rao, D.V., 1983. Estimating log Pearson parameters by mixed moments. Journal of Hydraulic Engineering 109 (8), 1118–1132.

Salas, J.D., Delleur, J.R., Yevjevich, Y., Lane, W.I., 1980. Applied Modeling of Hydrologic Time Series. Water Resources Publications, Colorado.

Salas, J.D., 1993. Analysis and modeling of hydrologic time series. In: Maidment, D.R. (Ed.), Handbook of Hydrology. McGraw-Hill Inc., New York.

Shannon, C.E., 1948. A mathematical theory of communications, I and II. Bell System Technical Journal 27, 379–443.

Shrader, M.L., Rawls, W.J., Snyder, W.M., McCuen, R.H., 1981. Flood peak regionalization using mixed-mode estimation of the parameters of the log-normal distribution. Journal of Hydrology 52, 229–237.

Singh, V.P., 1988. Hydrologic Systems, Vol. I: Rainfall-Runoff Modeling. Prentice-Hall, Englewood Cliffs, New Jersey.

Singh, V.P., 1998. Entropy-based Parameter Estimation in Hydrology, Water Science and Technology Library, vol 30. Kluwer Academic Publishers, Dordrecht.

Singh, V.P., 2013. Entropy Theory and its Application in Environmental and Water Engineering. John Wiley, New York.

Sklar, A., 1959. Functions de Repartition an Dimension Set Leursmarges. Publications de L'In-stitut de Statistique de L'Universite de Paris.

Snyder, W.M., 1972. Fitting of distribution functions by nonlinear least squares. Water Resources Research 8 (6), 1423–1432.

Sorooshian, S., Gupta, V.K., Fulton, J.L., 1983. Evaluation of maximum likelihood parameter estimation techniques for conceptual rainfall-runoff models: influence of calibration data variability and length on model credibility. Water Resources Research 19 (1), 251–259.

Stedinger, J.R., Tasker, G.D., 1985. Regional hydrologic analysis: 1. Ordinary, weighted and generalized least squares compared. Water Resources Research 21 (9), 1421–1432.

Sturges, H.A., 1926. The choice of a class interval. Journal of the American Statistical Association 21, 65–66.

Wang, S.X., Adams, B.J., 1984. Parameter Estimation in Flood Frequency Analysis. Publication 84-02. Department of Civil Engineering, University of Toronto, Toronto, Canada.

Wang, Q.J., 1997. LH moments for statistical analysis of extreme events. Water Resources Research 33 (12), 2841–2848.

Water Resources Council, 1967. A uniform technique for determining flood flow frequencies. Washington, D.C.
Weibull, W., 1939. A statistical theory of the strength of material. Ing. Vetenskapa Acad. Handlingar 151, 1–45.
Williams, B.J., Yeh, W.W.G., 1983. Parameter estimation in rainfall-runoff models. Journal of Hydrology 63, 373–393.
Zhang, L., Singh, V.P., 2019. Copulas and Their Applications in Water Resources Engineering. Cambridge University Press, Cambridge, England.

Further Reading

Davis, J.C., 1986. Statistics and Data Analysis in Geology. John Wiley and Sons, New York.
Haan, C.T., 2002. Statistical Methods in Hydrology. Iowa State Press, Ames, U.S.A.
Kite, G.W., 1977. Frequency and Risk Analysis in Hydrology. Water Resources Publications, Colorado.
Landwehr, J.M., Matalas, N.C., Wallis, J.R., 1979a. Probability-weighted moments compared with some traditional techniques in estimating Gumbel parameters and quantiles. Water Resources Research 15, 1055–1064.
McCuen, R.H., 1993. Statistical Hydrology. Prentice-Hall, New Jersey.
McGuess, R.H., Snyder, W.M., 1985. Hydrologic Modelling, Statistical Methods and Applications. Prentice-Hall, Englewood Cliffs, New Jersey.
Rao, A.R., Hamed, K.H., 1994. Frequency analysis of upper Cauvery flood data by L-moments. Water Resources Management 8, 183–201.
Singh, V.P., 1992. Elementary Hydrology. Prentice-Hall, Englewood Cliffs, NJ.
Yevjevich, V., 1972. Probability and Statistics in Hydrology. Water Resources Publications, Fort Collins, Colorado.

PART

II

Decision making

CHAPTER

Systems Analysis Techniques

5

Chapter outline

Water Resources Systems Planning and Management. https://doi.org/10.1016/B978-0-12-821349-0.00004-6

The objectives of this chapter are to:

- introduce systems analysis techniques;
- explain optimization techniques that are used in water resources;
- describe stochastic optimization techniques with special reference to water resources; and
- discuss the simulation tool.

The basic philosophy of an objective in planning or operation is the concept of change for the better; here "better" is defined by person(s) or society having both the desire and the authority to take action. Two important characteristics of change are the direction and the magnitude of desired change. Since the direction of change cannot be easily specified, except saying that it should be toward improvement, the word objective is more or less exclusively used to measure the magnitude of change. The magnitude can be indicated by either some cardinal or an ordinal measure, although a cardinal measure is desirable from an analytical point of view. Further, the magnitude of change may be described by a single measure or it may require a set of measures. If there is more than one measure, the level of achievement of any one objective is usually dependent on the level of achievement of the remaining objectives. If there are a number of objectives, there may be some common indices. One can attempt to aggregate all the common indices under different heads till no further aggregation is possible. The objectives left at this stage are termed as noncommensurate objectives. This implies that these objectives cannot be further aggregated without ignoring features that are of concern to the decision-makers.

5.1 SYSTEMS ANALYSIS TECHNIQUES

The development of systems analysis techniques can be traced to the 17th century when Newton developed the differential calculus method of optimization. The operations research techniques were developed during the second world war when the allied powers had to fight war on many fronts and they were required to deploy limited resources in the best possible manner. Since the techniques were aimed at getting the best results from military operations, these were known as operations research techniques. Dantzig developed the simplex method of linear programming (LP) in 1947. The work by Kuhn and Tucker in 1951 provided the impetus for the development of nonlinear programming (NLP) techniques. In 1957, Bellman developed the principle of optimality which laid the foundation for important developments in dynamic programming (DP) techniques. The goal programming and multiobjective optimization saw growth and applications during the decade of the 1960s and onwards. A firsthand account of the history of LP has been provided by Dantzig and Thapa (1997).

The popular operations research techniques include optimization methods, simulation, queuing theory, network flow theory, and game theory. Among these, optimization and simulation are extensively used in water resources problems. Therefore, only these two techniques will be covered in this chapter.

5.2 OPTIMIZATION

In many engineering problems, there are a number of possible solutions. It is, therefore, required to evaluate each alternative solution and then choose the best from the point of view of interest, say economic or convenience. Optimization is the science of choosing the best among a number of possible alternatives. The driving force in the optimization models is the objective function (or functions in multiobjective optimization). The term optimal solution essentially refers to the best from the solution of the mathematical model under all assumptions and constraints, whether explicitly stated or implicitly included in the formulation. Clearly, the optimal solution indicated by the model may be far from the actual system's optimal solution. Dantzig and Thapa (1997) defined mathematical programming (or optimization theory) as "that branch of mathematics dealing with techniques for maximizing or minimizing an objective function subject to linear, non-linear, and integer constraints on the variables." The word programming should not be related to computers; here it means "scheduling," the setting of an agenda, or creating a plan of activities (ReVelle et al., 1997).

Any "optimal" solution derived is clearly dependent on the assumptions and criteria and their associated uncertainties. Some of these uncertainties might be derived from the selection of model structure, parameters, scope, or focus. Others might be related to data, the optimization techniques used to solve the mathematical models, and the inability to account for many nonquantitative and nontangible considerations in the model.

An optimization problem can be stated as:

$$\text{Maximize (or Minimize)} \quad f(X) \tag{5.1}$$

subject to

$$g_j(X) \geq 0, j = 1, 2, \ldots, m \tag{5.2}$$

$$h_j(X) = 0, j = m + 1, m + 2, \ldots, p \tag{5.3}$$

where X is a vector of n-variables which are known as decision variables, g(X) are the inequality constraints, and h(X) are the equality constraints. To solve an optimization problem, the value of decision variables is systematically changed. The range over which a decision variable can be changed is known as its feasible range. The decision-maker evaluates the available alternatives on the basis of some prescribed criterion function. This criterion function, denoted by f(X), is known as the objective function. Its choice depends on the problem. While formulating the optimization problem, one should carefully decide the objective function which should properly reflect the preference of the decision-maker. In the beginning of analysis, objectives are often unclear or loosely stated. Considerable efforts may be needed to clarify them. Typically, the objective function may represent benefits which are maximized or it may represent costs which are to be minimized.

The availability of resources is usually limited and is expressed with the help of constraints. Here, g and h are the inequality and equality constraints. These constraints restrict the range over which the decision variables can change and thus affect the optimum solution. The number of decision variables and the number of constraints depend on the problem. If the number of constraints is zero, then the problem is known as the unconstrained optimization problem.

In most practical problems, the surfaces of objective functions have more than one peak or trough. The graph in Fig. 5.1 shows the variation of two objective functions with the decision variable for a problem which has only one decision variable. The objective function shown with solid lines (Z_1) has only one extreme point. If the second objective function shown by dotted lines (Z_2) is to be minimized, points A and B are known as local optimum, because the value of the objective function is the lowest only in the vicinity of these points. At point C, the value of the objective function is the lowest among

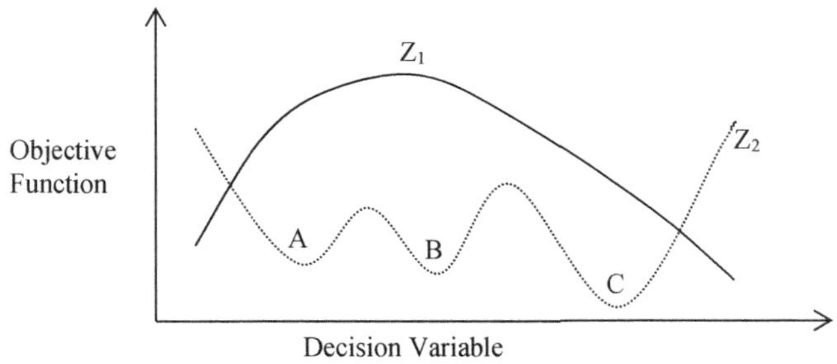

FIGURE 5.1

Variation of two objective functions with decision variables.

all the points and hence this point is termed as the global optimum. The value of the objective function at a local optimum is more than the global optimum in case of a minimization problem and vice versa for the maximization problem. In problems where the objective function has this type of behavior, the solution algorithm may end up at a local optimum. In an optimization problem, the constraints force the solution to lie within a limited region which is known as the *feasible region*. It is to be noted that usually real-life problems have a large number of decision variables and constraints. Therefore, in such problems, it is helpful to understand the nature of the objective function and constraints.

5.2.1 Classification of Optimization Techniques

Optimization techniques are also known as *mathematical programming* techniques. They can be classified in several ways, such as on the basis of the existence of constraints, the nature of the problem, the nature of the equations involved, the permissible values of the design variables, the deterministic nature of the variables involved, the separability of the functions, the number of objective functions involved, etc. The usual way of classifying optimization techniques is based on the nature of the problem or equations involved. These techniques can be classified as LP, NLP, goal programming (GP), DP, integer programming (IP), etc. This classification is useful from a computational point of view, since many methods have been developed solely for the efficient solution of a particular class of problems.

In the classification which depends on the existence of constraints, the problem can be classified as constrained optimization or unconstrained optimization. The input variables to a problem of water resources could be either deterministic or stochastic and depending upon that, the technique can be classified as deterministic optimization or stochastic optimization. According to Yeh and Becker (1982), stochastic optimization is useful for planning purposes, while deterministic optimization is a viable approach for real-time reservoir operation (see Section 11.7) with frequent updating of streamflow forecasts. Recently, many new optimization techniques have been used in studies dealing with water resources. Genetic algorithm (GA) is one such approach whose use began in the 1970s (see Goldberg, 1989; Dandy et al., 1996; Wardlaw and Sharif, 1999). Another technique that has become popular in systems control is fuzzy programming (see Pedrycz, 1993; Russell and Campbell, 1996).

Although optimization encompasses a very wide range of subjects, keeping in view the current status of the application of optimization techniques in water resources, the discussion in this chapter is limited to LP, NLP, and DP only.

5.3 LINEAR PROGRAMMING

Optimization problems in which the objective function and constraints are linear functions of decision variables and the decision variables are nonnegative are termed as LP problems. Dantzig and Thapa (1997) defined LP as a technique that "is concerned with maximization or minimization of a linear objective function in many variables subject to linear equality or inequality constraints." An optimization problem can be classified as an LP problem if it meets the following conditions:

- The decision variables of the problem are nonnegative, that is, positive or zero.
- The criterion function or objective function is described by a linear function of the decision variables, that is, a mathematical function involving only the first powers of the variables with no cross products.
- The operating rules governing the processes, commonly known as constraints, are expressed as a set of linear equations or linear inequalities.

The LP type of optimization problem was first recognized in the 1930s by economists while developing methods for optimal allocation of resources. During the World War II, the United States Air Force sought more effective procedures to allocate resources and this led to the development of LP. G.B. Dantzig, who was a member of the Air Force Group, formulated the general LP problem and devised the simplex method of solution in 1947. This was a significant step in bringing LP into wider usage. Since then, LP models have been widely used to solve a variety of military, economic, industrial, social, engineering, and hydrological problems. The number of applications of LP has grown immensely in the past few decades.

The LP models have been extensively used to solve water resources problems. Although the objective function and the constraints are not linearly related with the decision variables in many real-life water resources problems, these can be approximated by linear functions and the LP technique can be used to obtain the solution.

5.3.1 Assumptions in LP

Four basic assumptions are implicitly built into LP models:

(a) Proportionality assumption

The contribution of a decision variable to the objective function and its usage in various resource consuming activities is directly proportional to its values.

(b) Additivity assumption

This assumption indicates that the total usage of resources and contribution to the overall measure of effectiveness are equal to the sum of the corresponding quantities generated by each activity conducted by itself at a given level.

(c) Divisibility assumption

According to this assumption, the fractional values of the decision variables are permissible.

(d) Deterministic assumption

The parameters of the LP model are assumed to be known with certainty.

5.3.2 Mathematical Representation of an LP Problem

There are many ways to represent an LP problem. A general LP problem can be expressed in the conventional way. A compact way of representation of a linear program is a matrix form. Solutions are obtained by converting the general LP problem to a standard form. Each form is described in what follows.

Conventional Form

An LP problem consists of a linear objective function and a set of linear constraints. The constraints may be expressed in terms of inequalities or equalities. In most cases, especially in real-life water resources problems, constraints appear as inequalities. In many cases, it is observed that a linear program may consist of both types of constraints, that is, some constraints may be of equality type and others may be of inequality type.

A general LP problem can be written in conventional form as:

$$\text{Minimize (or maximize)}: Z = f(x) \tag{5.4}$$

subject to:

$$g_j(x) \geq 0; \quad j = 1, 2, \ldots, m_1 \tag{5.5}$$

$$h_j(x) \leq 0; \quad j = 1, 2, \ldots, m_2 \tag{5.6}$$

$$l_j(x) = 0; \quad j = 1, 2, \ldots, m_3 \tag{5.7}$$

where Z is the objective function; X is an n-dimensional decision vector; $g_j(x)$ and $h_j(x)$ are inequality constraints; $l_j(x)$ are equality constraints; and m_1, m_2, and m_3 denote the number of constraints for these types, respectively. The objective function is a linear function of the decision variables.

Standard Form

An LP problem with m constraints and n variables can be represented in standard form as follows:

$$\text{Minimize (or maximize)}: Z = c_1 x_1 + c_2 x_2 + \ldots + c_n x_n \tag{5.8}$$

subject to:

$$\begin{aligned} a_{11} x_1 + a_{12} x_2 + \ldots + a_{1n} x_n &= b_1 \\ a_{21} x_1 + a_{22} x_2 + \ldots + a_{2n} x_n &= b_2 \\ \cdot \quad \cdot \quad \ldots \quad \cdot \quad &\cdot \\ \cdot \quad \cdot \quad \ldots \quad \cdot \quad &\cdot \\ \cdot \quad \cdot \quad \ldots \quad \cdot \quad &\cdot \\ a_{m1} x_1 + a_{m2} x_2 + \ldots + a_{mn} x_n &= b_m \end{aligned} \tag{5.9}$$

$$x_i \geq 0, i = 1, 2, \ldots\ldots n. \tag{5.10}$$

$$b_i \geq 0, i = 1, 2, \ldots\ldots m. \tag{5.11}$$

where Z represents the objective function; x_i's are the decision variables; and c_i's are the cost (or benefit) coefficients representing the cost (or benefit) incurred by increasing the x_i decision variable by one unit. The right-hand side of constraint equations represents the resource availability. These arise due to limited availability of a particular resource, say, water. The a_{ij} coefficients are called technological coefficients and quantify the amount of a particular resource i required per unit of the activity j.

The standard form of an LP problem is solved algebraically. The main features of the standard form are:

- The objective function is either of the maximization or minimization type.
- All the constraints are expressed as equations, that is, equality-type constraints, except the nonnegativity constraints associated with the decision variables.
- All the decision variables are restricted to be nonnegative.
- The RHS constant of each constraint is nonnegative.

Matrix Form

In matrix notation, the standard LP problem can be expressed in a compact form as:

$$\text{Minimize (or maximize) } Z = C^T X \tag{5.12}$$

subject to:

$$A X = b \tag{5.13}$$

$$X \geq 0 \tag{5.14}$$

$$b \geq 0 \tag{5.15}$$

where A is an $m \times n$ matrix, X is an $n \times 1$ column vector, b is an $m \times 1$ column vector, C is an $n \times 1$ column vector, and Z represents the objective function. Superscript T in Eq. (5.12) refers to the transpose operation. Thus, one can write:

$$X = [x_1, x_2, \ldots\ldots\ldots, x_n]^T$$

$$C = [c_1, c_2, \ldots\ldots\ldots, c_n]^T$$

$$b = [b_1, b_2, \ldots\ldots\ldots, b_m]^T$$

$$A = \begin{bmatrix} a_{11} & a_{12} & \cdots & a_{1n} \\ a_{21} & a_{22} & \cdots & a_{2n} \\ \cdots & \cdots & \cdots & \cdots \\ a_{m1} & a_{m2} & \cdots & a_{mn} \end{bmatrix}$$

5.3.3 Formulation of an LP Model

The three basic steps in constructing an LP model are:

- Identify the decision variables and represent them in terms of algebraic symbols.
- Identify all the restrictions or constraints in the problem and express them as equations or inequalities which are linear functions of the decision variables.

- Identify the objective or criterion which is to be either maximized or minimized, and represent it as a linear function of the decision variables.

These three basic steps will be clear when one formulates a number of linear programs. Here, one may note that the model building is not a science but is primarily an art and comes mainly by practice. Depending on the experience, skill, and scientific knowledge about the system under consideration, the devèloped model will meet the realism and fulfill the intended objectives. Any discrepancy in the model formulation will yield an erroneous result, and sometimes may even give physically meaningless solution. Hence, it is necessary to work out many exercises on problem formulation before handling a real-life problem.

5.3.4 Reduction of a General LP Problem to a Standard Form

The simplex method for solving an LP problem requires the problem to be expressed in the standard form. But not all LP problems appear in the standard form. In many cases, some of the constraints are expressed as inequalities rather than equations; at least it is most often true in the case of water resources problems. In some problems, all the decision variables may not be even nonnegative. Hence, the first step in solving an LP problem is to convert it to the standard form. The procedure to convert a general program to the standard form is outlined below:

- Convert all inequalities to equalities.
- Convert all decision variables unrestricted in sign to strictly nonnegative.
- Make all the right-hand side constants of the constraints nonnegative.

Handling Inequality Constraints

An inequality constraint of the type $\leq$ can be converted to the equality type by introducing a new nonnegative variable called a slack variable. This new variable is added to the left-hand side of the constraint. Hence, the constraint

$$a_{11}\, x_1 + a_{12}\, x_2 + \ldots. + a_{1n}\, x_n \leq b_1 \tag{5.16}$$

can be written as:

$$a_{11}\, x_1 + a_{12}\, x_2 + \ldots. + a_{1n}\, x_n + s_1 = b_1, s_1 \leq 0 \tag{5.17}$$

Here, s_1 is a slack variable.

Similarly, an inequality constraint of $\geq$ type can be converted to the equality type by introducing a new nonnegative variable called a surplus variable. This new variable is subtracted from the left-hand side of the constraint. Thus, the constraint

$$a_{11}\, x_1 + a_{12}\, x_2 + \ldots. + a_{1n}\, x_n \geq b_1 \tag{5.18}$$

can be written as:

$$a_{11}\, x_1 + a_{12}\, x_2 + \ldots. + a_{1n}\, x_n - s_2 = b_1, s_2 \geq 0 \tag{5.19}$$

Here s_2 is a surplus variable. Furthermore, a constraint of $\geq$ type, if desired, can be easily converted to one of $\leq$ type by multiplying by -1 throughout the equation.

Handling Variables Unrestricted in Sign

In some situations, it may become necessary to introduce a variable in the LP model that can have both positive and negative values. Since the standard form requires all the variables to be nonnegative, a variable unrestricted in sign must be transformed. The unrestricted variable is replaced by the difference of the two nonnegative variables. If x_1 is unrestricted in sign, it can be replaced by $x_1 = x_2 - x_3$, where x_2 and x_3 can have only positive values.

Handling Constraints Having Negative Right-Hand Side Constant

Since the right-hand side constant of each constraint must be nonnegative, the constraints having negative right-hand side constants are multiplied by -1 throughout to get the constraint in the standard form. Thus, the constraint

$$3x_1 - x_2 - 2x_3 = -5 \tag{5.20}$$

will take the form

$$-3x_1 + x_2 + 2x_3 = 5 \tag{5.21}$$

in the standard form. It is important to note here that if the inequality type constraints are being multiplied by -1, their nature will reverse.

Interchanging the Nature of the Objective Function

The nature of the objective function, if desired, can be changed by putting a negative sign with the prescribed expression for the objective function. That means that a maximization problem is equivalent to a minimization problem with the negative of the objective function, that is,

$$\text{Max } [Z] = \text{Min } [-Z] \tag{5.22}$$

5.3.5 Canonical Form of an LP Problem

A system of equations that possesses at least one basic variable in all equations is called a canonical system. A variable is said to be a basic variable in a given equation if it appears with a unit coefficient in that equation and is absent in all other equations. This system is useful to obtain the optimal solution and forms a basis for the simplex method. To get a canonical system, a sequence of pivot operations is performed on the equations such that there is at least one basic variable in each equation.

Pivot Operations to Convert a System of Equation into Canonical Form

Consider the following system of equations.

I_0	$2x_1 + x_2 - x_3 = 3$	(5.23)
II_0	$2x_1 - x_2 + 4x_3 = 17$	
III_0	$3x_1 + x_2 + x_3 = 13$	

To convert this system of equations into canonical form, we first convert the coefficient of x_1 as unity in the first equation and zero in the remaining two equations as follows.

$I_1 = I_0/2$	$x_1 + x_2/2 - x_3/2 = 3/2$	Step 1.1: divide I_0 by 2	(5.24a)
$II_1 = II_0 - 2I_1$	$0x_1 - 2x_2 + 5x_3 = 14$	Step 1.2: $II_0 \rightarrow 2x_1 - x_2 + 4x_3 = 17$ $2I_1 \rightarrow 2x_1 + x_2 - x_3 = 3$	
$III_1 = III_0 - 3I_1$	$0x_1 - 1/2x_2 + 5/2x_3 = 17/2$	Step 1.3: $III_0 \rightarrow 3x_1 + x_2 + x_3 = 13$ $3I_1 \rightarrow 3x_1 + 3/2x_2 - 3/2x_3 = 9/2$	

Next, we convert the coefficient of x_2 as unity in the second equation and zero in the remaining two equations as follows.

$I_2 = I_1 - II_2/2$	$x_1 + 0x_2 + 3/4x_3 = 5$	Step 2.2: $I_1 \rightarrow x_1 + 1/2x_2 - 1/2x_3 = 3/2$ $II_2/2 \rightarrow 0x_1 + 1/2x_2 - 5/4x_3 = -7/2$	(5.24b)
$II_2 = II_1/(-2)$	$0x_1 + x_2 - 5/2x_3 = -7$	Step 2.1: Divide II_1 by (-2)	
$III_2 = III_1 + II_2/2$	$0x_1 + 0x_2 + 5/4x_3 = 5$	Step 2.3: $III_1 \rightarrow 0x_1 - 1/2x_2 + 5/2x_3 = 17/2$ $II_2/2 \rightarrow 0x_1 + 1/2\ x_2 - 5/4x_3 = -7/2$	

Next, we convert the coefficient of x_3 as unity in the third equation and zero in the remaining two equations as follows.

$I_3 = I_1 - II_2/2$	$x_1 + 0x_2 + 0x_3 = 2$	Step 3.2: $I_2 \rightarrow x_1 + 0x_2 + 3/4x_3 = 5$ $III_3{}^*(3/4) \rightarrow 0x_1 + 0x_2 + 3/4x_3 = 3$	(5.24c)
$II_3 = II_1/(-2)$	$0x_1 + x_2 + 0x_3 = 3$	Step 3.3: $II_2 \rightarrow 0x_1 + x_2 - 5/2x_3 = -7$ $III_3{}^*(5/2) \rightarrow 0x_1 + 0x_2 + 5/2x_3 = 10$	
$III_3 = III_2/(5/4)$	$0x_1 + 0x_2 + x_3 = 4$	Step 3.1: Divide III_2 by (5/4)	

Eq. (5.24c) is now a canonical system and the solution is $x_1 = 2$, $x_2 = 3$, $x_3 = 4$.

The number of basic variables is decided by the number of equations in the system. The variables that are not basic are called nonbasic variables. By applying the pivot operations, a given variable can be made a basic variable. The solution obtained from a canonical system by setting the nonbasic variables equal to zero and solving for the basic variables is called a basic solution. A basic feasible solution is a solution in which the values of the basic variables are nonnegative. The basic feasible solution satisfies all constraints. A basic feasible solution that provides the minimum (or maximum) value of the objective function is called an optimum solution. It may be noted that the feasible region of a properly formed LP problem is a convex set. A set is convex if it is not possible to find two points such that not all points on the line joining them belong to the set.

5.3.6 Graphical Solution of an LP Problem

The graphical method is a simple way to solve LP problems. This method is also very useful in conceptual understanding of the solution technique. However, it can be used to solve the LP problems involving at most two decision variables. In the following, an LP problem having two decision variables will be discussed. The feasible region and constraints are graphically shown in Fig. 5.2.

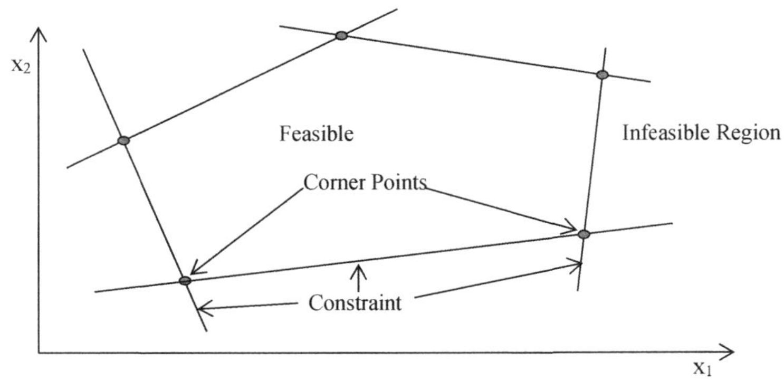

FIGURE 5.2

Definition sketch of feasible region and constraints.

Example 5.1. Solve the following problem using the graphical method of LP.

$$\text{Max } Z = 2\,x_1 + x_2 \tag{5.25}$$

subject to:

$$2x_1 - x_2 \leq 8 \tag{5.26}$$

$$x_1 + x_2 \leq 10 \tag{5.27}$$

$$x_2 \leq 7 \tag{5.28}$$

$$x_1, x_2 \leq 0 \tag{5.29}$$

Solution. In Fig. 5.3a, the constraints are plotted against the coordinate axes x_1 and x_2. To plot the first constraint, $2x_1 - x_2 \leq 8$, plot a straight line $2x_1 - x_2 = 8$. Similarly, plot lines $x_1 + x_2 = 10$ and $x_2 = 7$ to mark the second and third constraints. The nonnegativity constraints are plotted as the axes themselves. The feasible region can be easily delineated and is shown in Fig. 5.3a by the bounded pentagonal region formed by the lines of each constraint, including nonnegativity. The solution begins with an arbitrary value of the objective function, say 8 and the line $2x_1 + x_2 = 8$ is plotted (Fig. 5.3b). Both x_1 and x_2 should be in the feasible region. There are infinite points on the objective function line inside the feasible region, and each of these points is a solution to the problem. Since it is a maximization problem, the objective function line is shifted to the right as far as possible while ensuring that at least one point lies in the feasible region. It can be seen that the farthest point up to which we can go is the point (6, 4). Beyond this point, although the value of the objective function increases, there is no feasible solution. Hence, this is the optimum solution of the problem with $x_1 = 6$ and $x_2 = 4$, and the optimal value of the objective function is 16.

Importance of Corner Points

A closer inspection of Fig. 5.3 shows that the feasible region is compact and continuous and the optimum point is always a corner point. Further, if a constraint passing through this corner point is

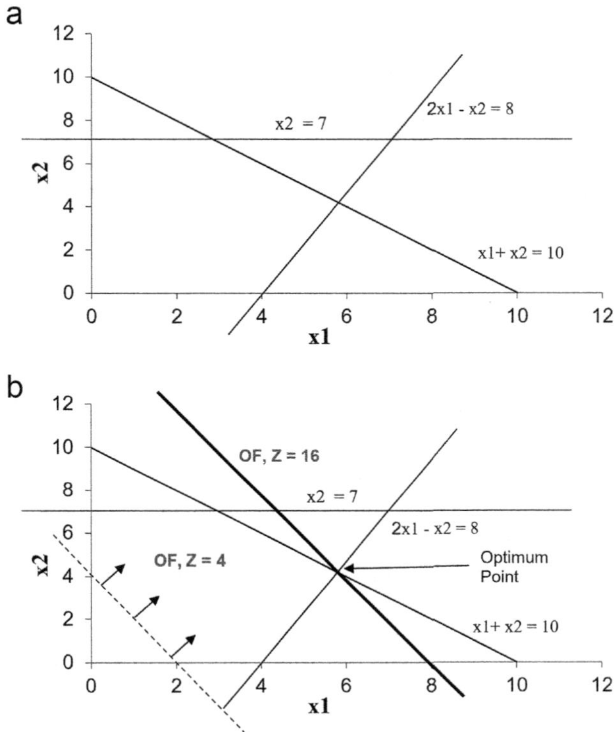

FIGURE 5.3

(a) Feasible region for Example 5.1, (b) Graphical solution of illustrative Example 5.1.

parallel to the objective function line, all the points falling on this constraint (or objective function) will have the same (optimum) value of the objective function. In this case, the problem will have infinite solutions. Such LP problems are said to have alternative or multiple optimal solutions. Some important properties of the corner points are:

- If there is only one optimal solution to an LP problem, it must be a feasible extreme point. If there are multiple optimal solutions, at least two must be adjacent feasible extreme points.
- In every problem, there are only a finite number of feasible extreme points.
- If a feasible extreme point is better than all its adjacent feasible points, then it is better than all other feasible extreme points. This property holds if the feasible region is convex. Based on this property, one need not enumerate all the extreme points, and the status of one extreme point can be ascertained to determine whether the optimal solution has been reached or not.

In some problems, the feasible region may not be a closed convex polygon, and hence it may be possible to increase the objective function value continuously and still be inside this region. Such types of problems are termed as having an unbounded solution. One may also notice in Fig. 5.3 that the constraint $x_2 \leq 7$ has no influence on the optimal solution. Such constraints are known as nonbinding constraints. The constraints which are met with equality at the optimum point (e.g., $x_1 + x_2 \leq 10$) are called binding constraints.

5.3.7 Simplex Method of LP

Motivation: A given system of simultaneous equations can be converted to canonical form and a basic solution can be obtained. One can move to another basic solution by choosing a different set of basic and non-basic variables. So, the optimal solution of a LP problem can be found by generating all solutions, retaining feasible solutions, and comparing them to find the optimal value of the objective function.

If a problem has m constraints and n decision variables with $n \geq m$, a basic solution can be obtained by setting any $n - m$ variables equal to zero. The number of basic solutions to be inspected is thus equal to the number of ways in which m variables can be selected from a set of n variables. There will be ${}^{n}C_{m}$ such solutions. Suppose that in a problem, number of decision variables $n = 30$, and number of constraints, $m = 20$, then number of possible basic solutions will be ${}^{30}C_{20} = 30!/[(30-20)!20!] = 30{,}045{,}015$. Hence, to solve this problem, one has to obtain 30,045,015 solutions and compare them – a formidable task even with the help of a fast computer !! Therefore, a computational scheme is needed that generates a series of basic feasible solutions and systematically moves from one feasible solution to the next until the optimal solution is reached. Computational efficiency is important.

The simplex method, developed by Dantzig, is an efficient method to solve LP problems. It is an iterative procedure to solve problems which are in the standard form. The simplex method requires that the constraint equations be expressed as a canonical system from which a basic feasible solution can be readily obtained. Once a basic feasible solution is available, attempts are made to improve it until the optimal solution is obtained.

The equations containing only the slack variables can be automatically considered as canonical system with the slack variables as basic variables. However, in many cases, finding a canonical system with a basic feasible solution is not an easy task. One way to obtain a basic feasible solution is to arbitrarily choose the basic variables and use a technique, such as Gaussian elimination, to get the solution. A systematic approach to get a canonical system starting from a basic feasible solution is the use of artificial variables. The artificial variables are added in those equations in which no basic variables appear by inspection. An auxiliary objective function is formed which is equal to the sum of artificial variables. This method is called two-phase simplex, as there are two objective functions. The first phase aims at the minimization of the auxiliary objective function. If, as a result of this phase, this function cannot be made zero, then the problem is infeasible and the algorithm is terminated. When the auxiliary objective function becomes zero, the optimization of the main function is taken up.

Computational Steps of the Simplex Method

The computational steps of the simplex method in tableau form are as follows:

1. Express the problem in standard form.
2. Start with an initial basic feasible solution in canonical form and set up the initial tableau.
3. Use the inner product rule to find the relative profit (or cost) coefficients ($\overline{C}_j$). This rule states that the relative profit (or cost) coefficient of a variable x_j ($\overline{C}_j$) is obtained by subtracting the product of the row matrix consisting of profit (or cost) coefficients (C_j) of basic variables and the column matrix consisting of transformation coefficients (a_{ij}) corresponding to x_j in the canonical system from the actual profit (or cost) coefficient corresponding to variable x_j. If all the relative profit coefficients are negative or zero, the current basic feasible solution is optimal for a maximization problem. For a minimization problem, all $\overline{C}_j$ should be positive or zero at the optimal point.

4. If the solution is not optimum, select the nonbasic variable with the most positive $\overline{C}_j$ value (highest value) to enter the basis in a maximization problem. In a minimization problem, the nonbasic variable with the most negative $\overline{C}_j$ value (lowest value) is selected to enter the basis. The decision is arbitrary in the case of a tie. Let this variable be x_r. The value of the objective function can be also computed by multiplying the row matrix consisting of profit (or cost) coefficients of basic variables and the column matrix consisting of right-hand side constants.
5. Apply the minimum ratio rule to determine the basic variable to leave the basis. The minimum ratio rule is that for this variable (x_r), take b_i/a_{ir} ratio for each constraint row i (for those constraints only which have +ve a_{ir} values), and the minimum ratio determines the row in which the basic variable will have unit coefficient. The corresponding variable from this row (which was a basic variable) will leave the basis. The constraint row corresponding to the entering basic variable is known as pivot equation, and the element located at the intersection of the entering column and pivoting row is known as the pivot element.
6. Perform the pivot operation to get the new tableau in canonical form, and get a new basic feasible solution.
7. Go to step 3 and repeat the steps until an optimal solution is found.

If, during the simplex iterations, the value of one or more basic variables becomes zero, it is termed as a degenerate solution. In such an event, there is no assurance that the solution will improve further. Sometimes, the degeneracy is temporary, and the solution improves after a few iterations.

The simplex method, when used for a large problem requires considerable computer time and storage. Some techniques have been developed which require less time and storage. Among these techniques, the revised simplex method is the most popular in which the time and storage are saved by manipulating only selected entries of the simplex tableau. A new algorithm has been recently developed by Karmarkar (1984) which moves through the interior of the feasible region to attain the optimum.

Example 5.2. Solve the problem of Example 5.1 using the simplex method.

Solution. The problem is written in canonical form by introducing slack variables, x_3, x_4, and x_5:

$$\text{Max } Z = 2\,x_1 + x_2 \tag{5.30}$$

subject to:

$$2x_1 - x_2 + x_3 = 8 \tag{5.31}$$

$$x_1 + x_2 + x_4 = 10 \tag{5.32}$$

$$x_2 + x_5 = 7 \tag{5.33}$$

$$x_1, x_2, x_3, x_4, x_5 \geq 0 \tag{5.34}$$

Here, the values of n and m are 5 and 3, respectively. Hence, there will be three basic variables which can be chosen arbitrarily. If the variables x_3, x_4, and x_5 are considered basic variables, the problem is in the canonical form. This finishes Step 1 and paves the way to Step 2. The initial basic feasible solution will be $x_1 = 0$, $x_2 = 0$, $x_3 = 8$, $x_4 = 10$, and $x_5 = 7$. The initial simplex tableau is formed as follows:

Here, one should note that the contents below the dotted line are not part of the initial simplex tableau. The bottom row separated by the dotted line shows the value of the relative profit or cost coefficients corresponding to the variable x_j ($\overline{C_j}$). It is computed as stated in Step 3. Since two values of relative profit coefficients are not either negative or zero, this is not the optimal solution and hence this solution is to be improved. For this, we see that the value of the relative profit coefficient is maximum for x_1 variable, thus x_1 will enter the basis. This completes Step 4. Moving to Step 5, since row 1 gives the minimum (b_i/a_{ir}) ratio, the variable x_3 will leave the basis. Thus, the row 1 is the pivot row and the number 2 in this row is the pivot element. This concludes Step 5 (Tableau 1)

Coming to Step 6, all the coefficients in the pivot row are divided by the pivot element, and x_r is eliminated from all rows except row 1. To make the coefficient of x_1 unity, row 1 is divided by 2. To eliminate x_1 from the second row, row 1 is multiplied by $-1/2$ and is added to the second row. In row 3, already the x_1 variable does not appear. After performing these operations, a new simplex tableau is formed as follows:

This new table shows the constraints in canonical form, and thus the improved basic feasible solution is $x_1 = 4$, $x_2 = 0$, $x_3 = 0$, $x_4 = 6$, and $x_5 = 7$. This completes Step 6, and now we move to Step 7. Again, from Tableau 2, it is clear that this solution is not optimal because all $\overline{C_j}$ are not either negative or zero. Thus, we again repeat the process. A close inspection of Tableau 2 shows that variable x_2 will enter the basis and variable x_4 will leave the basis. Therefore, a new tableau is again formed with row 2 as the pivot row and the number 3/2 in this row as the pivot element. The new tableau is shown below:

Tableau 3 shows that all relative profit coefficients are either negative or zero. Thus, the optimal point has been reached and the computations are terminated. The optimal value of the objective function is 16, and the optimal solution is $x_1 = 6$, $x_2 = 4$, $x_3 = 0$, $x_4 = 0$, $x_5 = 3$, and $Z = 16$.

Tableau 1

C_j	2	1	0	0	0	RHS
	x_1	x_2	x_3	x_4	x_5	
	2	−1	1	0	0	8
	1	1	0	1	0	10
	0	1	0	0	1	7
$\overline{C_j}$	2	1	0	0	0	Z = 0

Tableau 2

C_j	2	1	0	0	0	RHS
	x_1	x_2	x_3	x_4	x_5	
	1	−1/2	1/2	0	0	4
	0	3/2	−1/2	1	0	6
	0	1	0	0	1	7
$\overline{C_j}$	0	2	−1	0	0	Z = 8

Tableau 3						
C_j	2	1	0	0	0	RHS
	x_1	x_2	x_3	x_4	x_5	
	1	0	1/3	1/3	0	6
	0	1	−1/3	2/3	0	4
	0	0	1/3	−2/3	1	3
$\overline{C_j}$	0	0	−1/3	−4/3	0	Z = 16

Matlab Script

Software packages, such as Matlab, provide commands to solve LP problems. The script to solve the above problem (minimize of −ve of the OF) by using Matlab is as follows:

```
clear all
f = [-2 -1];
A = [2 -1; 1 1; 0 1];
b = [8 10 7];
Lb = zeros(3,1);
Aeq=[];
beq=[];
[x,fval] = linprog(f,A,b,Aeq,beq,lb)
```

After running this script, Matlab responds:

```
Optimal solution found.

x =
    6.0000
    4.0000

fval =
   -16
```

Interpreting Simplex Tableau

The final simplex tableau, besides giving information about the optimal solution, also contains other useful information. From the simplex Tableau 3 above, one can readily determine that the values of basic variables are $x_1 = 6$, $x_2 = 4$ and the value of the objective function is 16. The value of nonbasic decision variables at the optimum point is zero, except x_5 which is basic. The values of optimum slack variables are ignored, because they do not affect the decision. However, if a slack variable is a basic variable at an optimal stage, the corresponding constraint is nonbinding and the corresponding resource is abundant. Otherwise, the constraint is binding and the resource is scarce. Note that the RHS coefficients can be viewed as resource constraints.

The simplex tableau also contains information about the per unit worth of a resource, which is also known as its shadow price. This information is useful while fixing priorities about the allocation of funds for various resources. The shadow price of a nonbinding resource is zero, while it is nonzero for a binding constraint. The per-unit worth of a resource is given by $\partial Z/\partial b_i$, $i = 1, 2, \ldots, m$. Any change in the availability of the resource corresponding to the binding constraints will change the optimum solution. The value of per-unit worth of a resource can be obtained from the final tableau in the objective function row under the starting basic feasible variables. In the example above, the per-unit worth of resources for the constraint number 1, 2, and 3 are 1/3, 4/3, and 0, respectively. This implies that an increase of availability of resource 1 by one unit will lead to an increase in the objective function value by 1/3 units.

5.3.8 Duality in LP

Associated with every LP problem (termed as the primal problem), there exists another problem known as the dual problem. The dual problem is formulated by transposing the rows and columns of the primal problem including the right-hand side and the objective function, reversing the inequalities, and maximizing the objective function instead of minimizing.

Consider the following problem (primal problem):

$$\text{Minimize } Z = C^T x \tag{5.35}$$

subject to:

$$A\,x \geq b \tag{5.36}$$

$$x \geq 0 \tag{5.37}$$

Then, its dual problem can be stated as

$$\text{Maximize } Z_1 = b^T y \tag{5.38}$$

subject to:

$$A^T y \leq C \tag{5.39}$$

$$y \geq 0 \tag{5.40}$$

Here y is a column vector ($m \times 1$). To write a dual problem, it is necessary to write the primal problem in a particular way. In a minimization problem, all the constraints must be written in $\geq$ form and all the constraints of a maximization problem must be written in $\leq$ form. For example, let the primal be:

$$\text{Min} \quad z = x_1 + x_2$$

$$\text{subject to: } x_1 + 2x_2 \geq 5$$

$$2x_1 + x_2 \geq 4$$

$$x_1, x_2 \geq 0$$

then the dual is

$$\text{Max} \quad z_1 = 5y_1 + 4y_2$$

$$\text{subject to} \quad y_1 + 2y_2 \leq 1$$

$$2y_1 + y_2 \leq 1$$
$$y_1, y_2 \geq 0$$

Relationship Between Primal and Dual Problems

Some interesting relations exist between a primal problem and its dual. These are:

1. The dual of the dual is the primal.
2. If the primal is a minimization problem, then the dual is a maximization problem.
3. If dual has a finite solution, then the primal also has a finite solution.
4. For each variable in the primal, there exists a constraint in the dual and vice versa.
5. If any variable in the primal is unrestricted in sign, then the corresponding constraint in the dual is an equality constraint and vice versa.
6. If the primal has an unbounded solution, the dual has either an unbounded solution or is infeasible.
7. In the final solution (if it exists), if any constraint in the primal problem is satisfied as equality, the corresponding dual variable will have a value greater than zero, and vice versa.

The dual variables, y, are also termed as simplex multipliers, Lagrange multipliers, shadow prices, marginal costs, or opportunity costs. If a constraint is viewed as a resource constraint, the dual variable gives the marginal value of relaxing the constraint. It shows the change in the objective function per unit change in the RHS at the optimum, all other things remaining the same. Furthermore, if the problem contains a large number of constraints, it is more efficient to solve the dual since, in general, an additional constraint requires more computational effort than an additional variable. The solution procedure for a dual problem is on similar lines as for a primal. However, in the case of primal simplex, the feasibility is maintained throughout, while the dual simplex starts with an infeasible solution while maintaining optimality.

The dual problem of cost minimization may be viewed as a product maximization problem, where the product is maximized by varying the y variables, the imputed cost of each constraint. These implicit values are the shadow prices of constraints. They define the marginal value of the contribution of each constraint to the objective function, say how much more output can be obtained by relaxing a constraint by one unit. If the price of a resource is less than its shadow price, it is desirable to buy that resource and expand the production. The value of the slack variable at the optimum solution indicates cost (in terms of lowering the output) of using any activity which is not included in the optimum solution.

5.3.9 Postoptimality Analysis

Many times it is necessary to study the variation in the optimal solution resulting from the change in the various parameters, such as the cost coefficients, technological constants, or due to the addition or deletion of variables or constraints. The study of change in the optimal solution due to these changes is known as the postoptimality analysis. The following changes affect the optimal solution:

- Changes in the RHS constants of constraint equations,
- Changes in the objective function coefficients,
- Changes in the coefficients in the constraints,

- Addition of new variables, and
- Addition of new constraints.

Due to these changes, the optimal solution may change in the following way:

- The optimal solution may remain unchanged,
- The basic variables remain the same, but their values change, or
- The basic variables as well as their values change.

In most cases, it is not necessary to solve the problem from the beginning, and the final simplex tableau can be used to get the required answer.

5.3.10 Important Classes of LP Problems

Many day-to-day LP problems have some unique features which allow the use of special techniques for solution. Some of these problems are briefly discussed below:

Transportation Problem

In many real-life situations, a product is manufactured at a number of locations and it is required to transport it to a number of destinations. For example, a big company may have a number of production and demand centers, spread out geographically. The objective of a transportation problem is to devise a schedule of movement which minimizes the cost of transportation. This problem can also be formulated as a regular LP problem and solved using the simplex method. However, its special structure allows a more efficient and convenient procedure for its solution.

Network Flow Problems

A network is a configuration which consists of nodes joined by directed arcs. Each arc can support a flow. Associated with each arc are three parameters: a lower bound on the flow, an upper bound on the flow, and a cost value which represents the expenses of moving one unit of flow along the arc. A circulation is a set of flows in the network which preserves conservation of flow at each node. This means that the total flow into a node must equal the total flow out of the node.

The objective of a network flow optimization algorithm is to generate a circulation in the network which minimizes the total system cost (defined as the sum of flows in each arc times the arc expense) subject to the capacity restrictions on the flow in each arc. Thus, the problem can be written as:

$$\text{Minimize} \sum C_i x_i \tag{5.41}$$

$$\text{subject to} \quad l_i \le x_i \le u_i \tag{5.42}$$

$$x_{in} = x_{out} \tag{5.43}$$

where x_i = flow in arc i, C_i = cost value for arc i, l_i = lower bound on arc i, u_i = upper bound on arc i, x_{in} = sum of flows into node j, and x_{out} = sum of flows out of node j.

The out-of-kilter algorithm (OKA) is a general method to generate the optimum circulation in a capacitated cost network. OKA begins with a circulation in the network. Now, maintaining a circulation, the flows are changed to achieve the objective without violating any of the continuity and limit constraints. OKA establishes a pricing system which assigns a price to each node. This technique has been described in detail in many books, such as Jensen and Barnes (1980). The surface water allocation model (AL-V) by Martin (1981) makes use of network flow optimization approach.

Other topics of interest in LP include the transportation problem and the assignment problem. Very efficient computer packages are available to solve LP problems which make this technique very attractive to use. For more on the LP and other optimization techniques, the reader may refer Rao (2009) and Taha (2017).

5.4 NONLINEAR PROGRAMMING

An optimization problem in which either the objective function and/or one or more constraints are nonlinear functions of decision variables is termed an NLP problem. An NLP problem can be stated in the general form as:

$$\text{Minimize (or maximize) } f(x) \tag{5.44}$$

subject to

$$h_j(x) = 0, j = 1, 2, \ldots, m \tag{5.45}$$

$$g_j(x) \leq 0, j = (m+1), \ldots, p \tag{5.46}$$

where f(x) denotes the objective function, $h_j(x)$ are the equality constraints, $g_j(x)$ represent the inequality constraints, and $x = [x_1, x_2 \ldots, x_n]$ is a vector of decision variables. Since $g_j(x) \geq 0$ can be written as $-g_j(x) \leq 0$, the inequality constraints can be denoted by $g_j(x) \leq 0$. Here m and p are nonnegative integers. If $m = p = 0$, then the problem is said to be unconstrained. The problem reduces to an LP problem if $h_j(x)$, $g_j(x)$, and f(x) are all linear functions of decision variables.

Before proceeding further, let us define some properties of f(x). Let f(x) be a continuous and continuously differentiable function. The vector of first partial derivatives of the function with respect to various variables is called the gradient of the function:

$$\nabla f(x) = \left(\frac{\partial f}{\partial x_1}, \frac{\partial f}{\partial x_2}, \ldots, \frac{\partial f}{\partial x_n}\right)^T \tag{5.47}$$

The gradient vector at a given point represents the direction along which the change in the function values at the maximum rate. The matrix of second partial derivatives of the function (if it exists) is known as the Hessian matrix. This is a square and symmetric matrix.

$$H(x) = \nabla^2 f(x) = \begin{vmatrix} \frac{\partial^2 f}{\partial x_1^2} & \frac{\partial^2 f}{\partial x_1 \partial x_2} & \cdots & \frac{\partial^2 f}{\partial x_1 \partial x_n} \\ \frac{\partial^2 f}{\partial x_2 \partial x_1} & \frac{\partial^2 f}{\partial x_2^2} & \cdots & \frac{\partial^2 f}{\partial x_2 \partial x_n} \\ \vdots & \vdots & \vdots & \vdots \\ \frac{\partial^2 f}{\partial x_n \partial x_1} & \frac{\partial^2 f}{\partial x_n \partial x_2} & \cdots & \frac{\partial^2 f}{\partial x_n^2} \end{vmatrix} \tag{5.48}$$

The solution of an NLP problem will be global if the objective function and feasible region are convex. In a convex region, a line joining any two points within the region will always be in the domain of the function. Mathematically, a function is said to be convex if for any two points x_1 and x_2, and for all α, $0 \le \alpha \le 1$:

$$F[\alpha x_1 + (1-\alpha)\, x_2] < [\alpha F(x_1) + (1-\alpha)F] \tag{5.49}$$

The concept of a convex function is geometrically represented in Fig. 5.4. The convexity or concavity of a function f(x) can also be ascertained by the following:

Function f(x) is concave	if H(x) is negative semidefinite
Function f(x) is strictly concave	if H(x) is negative definite
Function f(x) is convex	if H(x) is positive semidefinite
Function f(x) is strictly convex	if H(x) is positive definite

The convexity and concavity of function f(x) depends on the properties of the Hessian matrix. The Hessian matrix can be classified into following types:

Type of Hessian matrix	Condition to be satisfied
Positive definite	$x^THx > 0$ for all $x \neq 0$
Negative definite	$x^THx < 0$ for all $x \neq 0$
Indefinite	$x^THx > 0$ for some x and >0 for some other x
Positive semidefinite	$x^THx \geq 0$ for all x
Negative semidefinite	$x^THx \leq 0$ for all x

A function of two variables may have partial derivatives of both variables being zero at a point, and in this case, the Hessian matrix will be neither positive nor negative definite. Such a point is called a saddle point.

5.4.1 Lagrange Multipliers and Kuhn–Tucker Conditions

The concepts of Lagrange multipliers and Kuhn–Tucker conditions are important and useful for constrained NLP problems. The Lagrange multiplier method converts an NLP problem with equality constraints to an unconstrained problem by developing an augmented objective function. For the

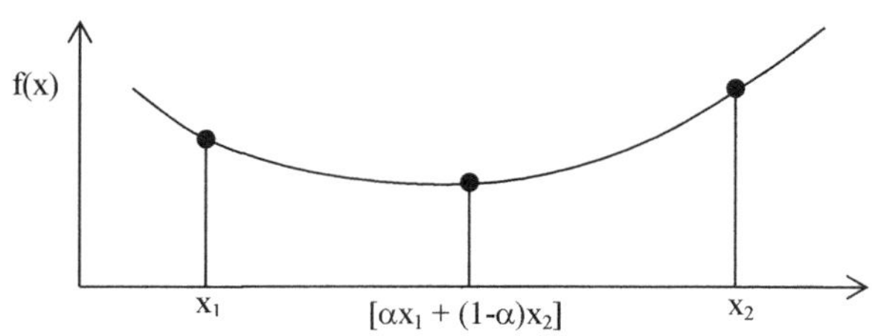

FIGURE 5.4

Geometric representation of a convex function.

constrained problem given by Eq. (5.44) with equality constraints given by Eq. (5.45), the Lagrange function (minimization problem) is:

$$L(x, \lambda) = f(x) + \lambda\, g(x) \tag{5.50}$$

where λ is the vector of Lagrange multipliers. If the original problem has n variables and m constraints, the augmented objective function will have (n + m) variables. These variables can be obtained by setting the partial derivative of L(x, λ) to zero:

$$\frac{\partial f}{\partial x_i} + \sum_{j=1}^{m} \lambda_j \frac{\partial g_j}{\partial x_i} = 0, i = 1, \ldots n \tag{5.51}$$

$$\lambda_j\, g_j = 0, j = 1, 2, \ldots\, m \tag{5.52}$$

The Lagrange multipliers are also known as dual variables (see Section 5.3.8), shadow prices, opportunity costs, etc. The jth Lagrange multiplier represents a marginal change in the value of the objective function in the vicinity of the optimal solution with respect to the right-hand side of the jth constraint. The Lagrange multipliers indicate how much the value of the objective function at the optimal point will change for a small change in the right-hand side of the constraint.

A necessary condition for a local optimum is that the first derivative of the function is zero; this is also a sufficient condition for a convex or concave function. These conditions need extension when there are inequality constraints. Consider the minimization problem given by Eq. (5.44) with m inequality constraints given by $g_i(x)$ which is differentiable and let $x \geq 0$. According to the Kuhn–Tucker conditions, an optimal solution x^* to this problem will exist only if there exist $\lambda_1, \lambda_2, \ldots, \lambda_m$ such that the following conditions are satisfied:

$$1.\ \text{If } x_j^* > 0 \text{ then } \frac{\partial F}{\partial x_j} + \sum_{i=1}^{m} \lambda_i \frac{\partial g_i}{\partial x_j} | x_j^* = 0; \quad j = 1, 2, \ldots, n \tag{5.53a}$$

$$2.\ \text{If } x_j^* = 0 \text{ then } \frac{\partial F}{\partial x_j} + \sum_{i=1}^{m} \lambda_i \frac{\partial g_i}{\partial x_j} | x_j^* \geq 0; \quad j = 1, 2, \ldots, n \tag{5.53b}$$

$$3.\ \text{If } \lambda_i > 0 \text{ then } g_i(x_1^*, x_2^*, \ldots, x_n^*) = b_1; \quad j = 1, 2, \ldots, m \tag{5.53c}$$

$$4.\ \text{If } \lambda_i = 0 \text{ then } g_i(x_1^*, x_2^*, \ldots, x_n^*) \leq b_1; \quad j = 1, 2, \ldots, m \tag{5.53d}$$

$$5.\ x_j^* \geq 0; \quad j = 1, 2, \ldots, n \tag{5.53e}$$

$$6.\ \lambda_i \geq 0;\ i = 1, 2, \ldots m \tag{5.53f}$$

The condition given by Eq. (5.53a) is the necessary condition for an optimum provided the stationary point is not at the boundary. The condition specified by Eq. (5.53b) is the supplementary condition when the optimum may be at the boundary. The condition given by Eq. (5.53c) suggests that the inequality constraints introduced into the Lagrangian function is binding (i.e., it suggests equality). According to Eq. (5.53d), the Lagrangian multipliers of the constraints that are not binding vanish. The conditions given by Eqs. (5.53e) and (5.53f) ensure nonnegativity of the decision variables and Lagrangian multipliers.

5.4.2 Classification of Nonlinear Programming Methods

Depending on the nature of the problem, the subclasses into which the solution techniques can be divided are unidirectional search methods, unconstrained optimization techniques, and constrained optimization techniques. A brief discussion of these methods follows.

5.4.3 Unconstrained Nonlinear Programming Methods

An optimization problem without any constraint is called an unconstrained optimization problem. Although unconstrained optimization problems are rare in real life, in several nonlinear optimization techniques, the constrained problem is converted into an unconstrained problem which is subsequently solved. All the unconstrained optimization algorithms are iterative in nature. The computation starts at an initial point, and advancement is made toward the optimum point in a systematic manner using some property of the function. A suitable criterion is chosen to terminate the computation when no further improvement is possible. These methods can be further classified into two categories, depending on whether the derivatives of the objective function are used or not.

Unidirectional search methods

In these methods, the objective function is optimized with respect to one variable only. The objective function is assumed to be unimodal, that is, it has only one maxima or minima in a given interval. If the region where the optimal solution lies is known, a search is made in the direction in which the objective function is expected to improve from an initial point, using a fixed step size (which is decided based on the required accuracy). This method is simple but inefficient. An improvement is to increase the step size if the objective function improves in the search direction. A variant is to evaluate the objective function at predetermined equally spaced points in an exhaustive search over the feasible space.

Dichotomous (dichotomy—the division of two things that are completely different) search is a sequential search method in which a new search is made based on the outcome of the previous searches. The function value is determined at two points close to the center of the search region. Depending upon the relative values of the objective function at the two points, nearly half of the search region can be eliminated from further scrutiny. Successive searches are carried out by following this scheme till a solution with desired accuracy is obtained.

The Fibonacci method is used to find the minimum value of an unimodal function over the search region. It uses the Fibonacci numbers (F), defined as

$$F_0 = 0,\ F_1 = 1$$

$$F_n = F_{n-1} + F_{n-2},\ n = 2, 3, 4 \ldots$$

This yields the sequence of Fibonacci numbers, F_n.

n	0	1	2	3	4	5	6	7	8	9	...
F_n	0	1	1	2	3	5	8	13	21	34	...

Fibonacci was an Italian mathematician who is credited for developing Fibonacci numbers. However, some sources mention that Indian mathematicians had described these numbers in the sixth century. Consider that the search region for variable x is $a \le x \le b$, and n function evaluations are to be performed to find the optimum. If $L_0 = (b - a)$, a quantity $S_1 = (F_{n-2}/F_n)^* S_0$ is computed. Two function evaluations are performed at points x_1 and x_2 that are S_1 distance away from either end. So, the position of these points will be

$$x_1 = a + S_1 = a + (F_{n-2}/F_n)^* S_0$$

$$x_2 = b - S_1 = b - (F_{n-2}/F_n)^* S_0$$

Now, the function is evaluated at x_1 and x_2, and depending upon the relative values at these points and remembering the unimodal assumption, the region of length S_1 near a or b is discarded from further consideration. The revised search region now is

$$L_1 = L_0 - S_1 = L_0 - (F_{n-2}/F_n)^* S_0 = (F_{n-1}/F_n)^* S_0$$

For the next round, we already have an evaluation point in the search region and the function will be evaluated at one more point so that the two points are at a distance $S_2 = (F_{n-3}/F_{n-1})^* L_1$ from each end of the reduced search region. The process of discarding a part of the search region and evaluating the objective function continues and after n rounds, the search region will be $L_n = (F_1/F_n)^* S_0$ and we will have the optimum point with this accuracy. Thus, in the Fibonacci method, the number of iterations is decided based on the desired accuracy in locating the optimum.

The ratio of two Fibonacci numbers F_n/F_{n-1} as $n \to \infty$ is $\varphi = 1.618$. This number is also called as the Golden ratio. It has the property: $\varphi = 1 + 1/\varphi$. Ancient Greek architects and scientists such as Pythagoras and Euclid studied this number extensively. Greek architects believed that if the building has sides a and b that satisfy the relation $(a+b)/a = a/b = 1.618$, the building will have the most pleasing appearance. In the golden Section method, the function values are evaluated at a distance $(1 - 1/\varphi)^* L_0 = 0.382^* L_0$ from either end. Now, depending upon the relative values of the objective function, the region from the point of higher function value to the end of the region is discarded and the next round of computations begins. The computations are terminated when the search region is sufficiently small or the solution with the desired accuracy is obtained.

These techniques may not be useful for real-life problems, because mostly such problems have more than one variable. However, in some multidimensional problems, optimization is performed by systematically conducting unidirectional searches until the optimum is found. Constrained problems involve the optimization of an objective function subject to one or more constraints and are much harder to solve than unconstrained problems with a comparable number of independent variables and the degree of nonlinearity, because of the additional requirement that the solution must satisfy the constraints.

Direct Search Methods

The methods that use only function values to guide the search for the optimum value are either search methods constructed from geometric intuition or theoretically based techniques which have a mathematical foundation. These methods are also known as pattern search methods and do not require the gradient of the objective function. Let f(x) be the function to be optimized. It is assumed that f(x) is continuous, and $\nabla f(x)$ may or may not exist but is not available, and f(x) is unimodal in the domain of interest. In the case of multimodal functions, the solution may terminate at a local minimum.

Among the methods which do not need the derivative of the objective function, the most frequently used techniques are the pattern search methods (such as Powell's or Hooke and Jeeves' method), Rosenbrock's method of rotating coordinates, and the simplex method. In pattern search methods, starting from an initial point, several favorable unidirectional moves are made and the local behavior of the function is established. Using these, the pattern direction is determined as the most favorable direction of movement and a base point is established by moving in this direction. These two steps are repeated until the optimum value is found. The main motivation of moving in the pattern direction is that convergence can be considerably slow if movements are made only in coordinate directions. The cyclic use of searches in coordinate or any fixed set of directions is inefficient and may fail to converge to a local optimum. The Hooke and Jeeves method is a combination of one-at-a-time exploratory move (to understand the local behavior of the objective function), and the pattern moves (to take advantage of the pattern direction) are based on some heuristic rules. Powell's method is a theoretically based method that uses the history of iterations to build up directions for acceleration. The algorithm was devised assuming a quadratic objective function, and it will converge in a finite number of iterations for such functions. A quadratic function in an n-dimensional real space is defined by:

$$F(x) = a + b^T x + x^T Q x \tag{5.54}$$

Another commonly used method is the Rosenbrock's method of rotating coordinates. Here, the coordinate system is rotated at each stage of optimization in such a manner that the first axis is oriented toward the locally estimated direction of the valley and all other axes are made mutually orthogonal and normal to the first axis. Because the coordinate system can be rotated, depending on the need, this method can follow curved and steep valleys.

Computationally, direct search methods are relatively uncomplicated; hence, their algorithms are easy to implement. On the other hand, they can be, and often are, slower than the derivative-based methods. Often, the objective function is the only reliable information that is available in a practical engineering problem and therefore, the direct methods are important.

Gradient-Based Methods

A major drawback of direct methods is that they require an excessive number of function evaluations to locate the solution. Then there may be a need to seek stationary points (the points where the first derivative of the objective function is zero) and thus a motivation to use the methods that employ gradient information. The gradient of a function can be computed using Eq. (5.47). The gradient-based methods are iterative, since the elements of the gradient are, in general, nonlinear functions of the decision variables.

The function value changes fast by moving in the direction of the gradient. Hence, this direction is also called the steepest ascent direction and the maximization methods which use the first derivative are also known as the steepest ascent methods. The unconstrained minimization methods which use the derivatives of the objective function are also called the steepest descent methods. However, the steepest ascent direction is a local property and it changes as one moves along the objective function surface. It is, therefore, necessary to evaluate the gradient at many points and thus the method is iterative.

It is assumed here that the function f(x) and its first and second derivatives $\nabla f(x)$ and $\nabla^2 f(x)$ exist and are continuous. Some methods use the first derivative and some require both first and second derivatives. It is assumed that the elements of the gradient are available in closed form or can be reliably approximated numerically. The gradient-based methods employ an iterative procedure:

$$x^{(k+1)} = x^{(k)} + \alpha^{(k)} s\{x^{(k)}\} \tag{5.55}$$

where $x^{(k)}$ is the solution at the kth step, $\alpha^{(k)}$ is the step-length parameter, and $s\{x^{(k)}\}$ is the search direction in the N-dimensional space of decision variables. The various methods differ in the manner in which s(x) and α are determined at each iteration. Usually $\alpha^{(k)}$ is selected so as to minimize f(x) in the s $\{x^{(k)}\}$ direction. Therefore, efficient single-variable minimization algorithms are required to implement these methods. Since the gradient is the direction of the steepest descent, a simple gradient search algorithm is:

$$x^{(k+1)} = x^{(k)} - \alpha \nabla f\{x^{(k)}\} \tag{5.56}$$

It remains to find a suitable value of α. The algorithm becomes slow near the minimum point as $\nabla f(x)$ tends to zero. The Cauchy method determines the step length α such that the function value f $\{x^{(k+1)}\}$ is minimum along the gradient direction. The Newton method that makes use of a second-order gradient is an improvement over this method. The iterative scheme of Newton's method is

$$x^{(k+1)} = x^{(k)} - \alpha \nabla^2 f\ \{x^{(k)}\}^{-1} \nabla f\{x^{(k)}\} \tag{5.57}$$

Marquardt Algorithm

This procedure was proposed by Marquardt (1963) by using the strengths of both Cauchy's and Newton's methods. Also known as Levenberg—Marquardt method, it requires second-order information and allows for convergence with relatively poor starting guesses for the unknown variables. In general, a steepest descent procedure would be expected to converge for poor starting values but requires a lengthy solution time. The Gauss Newton method, on the other hand, will converge rapidly for good starting estimates. In this method, a least-square objective function is utilized. The search direction is given by

$$s\{x^{(k)}\} = -\left[H^{(k)} + \lambda^{(k)} I\right]^{-1} \nabla f\{x^{(k)}\} \tag{5.58}$$

where I is the identity matrix and λ is used to control the search direction as well as the step-length. When λ is very large, the search is in the gradient direction. When λ equals zero, the algorithm reduces to the Newton method. In the Marquardt algorithm, the initial values of λ are large and decrease toward zero as the optimum is approached.

Conjugate Gradient Methods

The conjugate gradient methods exploit the conjugacy concept by using gradient information. To understand the conjugate property of ellipses, consider an ellipse shown in Fig. 5.5. According to the conjugate property, the line AB joining the points of contact of two parallel tangents to an ellipse must pass through the center of the ellipse.

FIGURE 5.5

The concept of conjugate directions.

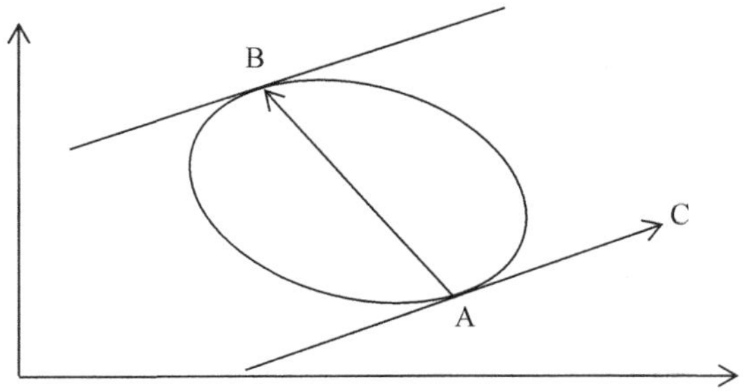

In Fig. 5.5, directions AB and AC are conjugate directions. Mathematically, for an n-dimensional case, if A is an n × n symmetric positive definite matrix, then a set of directions $\{S_i\}$ is said to be conjugate if

$$S_i \, A \, S_j = 0 \text{ for all } i \neq j \quad \text{and } i = 1, 2, \ldots, n; \quad j = 1, 2, \ldots n \tag{5.59}$$

The methods, that locate the minimum of a quadratic function of n variables in steps whose number is related to n, are known as quadratically convergent. The convergence of a steepest descent method can be significantly improved by changing it to a conjugate gradient method. To complete the function search along n conjugate directions, n minimization cycles are necessary. After these n cycles, all the search directions are mutually conjugate (search has also been done along coordinate or nonconjugate directions) and the minimum of the quadratic function would have been found. The Fletcher–Reeves method is a conjugate gradient method in which the search direction is established as a linear combination of all the previous search directions and newly determined gradient.

A class of methods uses the following scheme to generate search directions at the kth iteration

$$s\{x^{(k)}\} = -H^{(k)} \, \nabla f\{x^{(k)}\} \tag{5.60}$$

where $H^{(k)}$ is an n*n matrix, termed as metric. When H changes at each iteration, it is called as variable metric. This is a positive-definite matrix and at the first iteration, it can be an identity matrix. The best variable metric method is the Davidson–Fletcher–Powell (DFP) method in which the first-order derivatives are used to get the approximation of the Hessian matrix of the function. If the optimal step length is $\lambda^{(k)}$, then the iterations in the DFP method proceed as follows:

$$x^{(k+1)} = x^{(k)} - \lambda^{(k)} \, H^{(k)} \nabla f\{x^{(k)}\} \tag{5.61}$$

This method is stable and quadratically convergent and is used extensively. Rao (2009) found it to be the best general-purpose unconstrained optimization technique that uses derivatives.

5.4.4 Constrained Nonlinear Programming Methods

The methods of constrained nonlinear optimization can be classified into two groups: direct methods and indirect methods. In direct methods, the constraints are handled explicitly. Most popular in this category are the methods of feasible directions. In these methods, the current solution is improved by moving in usable feasible directions.

$$x^{(k+1)} = x^{(k)} + \lambda S \tag{5.62}$$

Here S is the direction along which a small step of size λ can be taken without leaving the feasible domain and at the same time improving the objective function value. There are two important steps at each stage of iteration: finding a usable feasible direction at the given point and determining the length of step along this direction. Except for convex problems, the algorithm may terminate at a local optimum. Zoutendijk's method and Rosen's gradient projection method come under this category. Basically, the methods differ in the ways in which the usable feasible directions are found.

In indirect methods, the constrained problem is solved by solving a sequence of unconstrained problems. Penalty function methods, which come under this category, follow this strategy. Following Rao (2009), the problem given by Eqs. (5.44) and (5.45) is converted to the following unconstrained minimization problem:

$$F = f(x) + r_k \sum_{j=1}^{p} G_j\left[g_j(x)\right] \tag{5.63}$$

where G_j is some function of the constraint g_j and r_k is the penalty parameter. The second term on the right hand side is termed as penalty term. Two variations of penalty function methods are used. In the interior penalty function method, a feasible starting point is obtained and the sequence of iterations coverages to the constrained minimum. In the exterior penalty function method, a slightly different form of Eq. (5.63) is used and the function F increases as some power of the amount by which the constraints are violated. The sequence of computations converges to the desired solution from the exterior of the feasible region.

The generalized reduced gradient (GRG) method is similar to the simplex method of LP in the sense that the n decision variables are partitioned into m basic (x_B) and (n − m) nonbasic (x_N) variables. The problem is then expressed as:

$$\text{Min (or max) } f(x_B, x_N) \tag{5.64}$$

subject to

$$g(x_B, x_N) = 0 \tag{5.65}$$

and bounds on the variables. The basic variables can be expressed in terms of nonbasic variables as $x_B(x_N)$. The objective function, expressed in terms of nonbasic variables, is known as the reduced objective function and its gradient, the reduced gradient. In the GRG method, a sequence of reduced problems is solved by a gradient search method. These methods are especially very successful in problems where the constraints are nearly linear.

Another powerful technique is the successive linear programming (SLP) technique. The SLP algorithm also solves NLP problems by solving a sequence of linear programs. These algorithms are attractive for large sparse nonlinear optimization problems where usually only some variables appear nonlinearly in the objective function and/or constraints.

Based on the concepts drawn from principles of natural biological evolution, (Duan et al., 1992) developed a global optimization strategy called the "shuffled complex evolution (SCE-UA)" method. SCE-UA begins with a population of points sampled randomly from the feasible space. This population is partitioned into several communities, each of which contains (2n + 1) points, n is the dimension of the problem. Each community is made to evolve based on a statistical "reproduction" process that uses the "simplex" geometric shape to direct the search in an improvement direction. At periodic stages in the evolution, the entire population is shuffled and points are reassigned to communities to ensure information sharing. As the search progresses, the entire population tends to converge toward the neighborhood of global optimum, if the size of initial population is sufficiently large. SCE-UA algorithm has been widely used in various fields, including water resources.

Vrugt et al. (2003) presented a modified version of the SCE-UA algorithm by merging the strengths of the Metropolis algorithm, controlled random search, competitive evolution, and complex shuffling. They claimed that their algorithm, called SCEM-UA, significantly reduces the number of model simulations needed to infer the posterior distribution of the parameters. Markov Chain Monte Carlo (MCMC) methods are popular for estimating the posterior probability distribution of parameters of hydrologic models. The Metropolis-Hastings (MH) algorithm is a general MCMC algorithm.

5.4.5 Some Common Problems in NLP Applications

While solving an NLP problem, several things require attention. First, an algorithm may fail because the problem was wrongly formulated. Unfortunately, this error may not be revealed directly, but rather through the failure of some portion of the algorithm.

Overflow in the user-defined function is a common problem. Overflow may occur when the optimization problem has an unbounded solution since in this case the function value gradually becomes large. An unbounded problem is also indicated in a minimization problem, when there is a consistent decrease in the function with no sign of convergence. The appropriate remedy depends on the reason for the unboundedness.

If the program is self-developed, there might be errors in programming, for example, in computation of step-length or the gradient or Hessian of the objective function (inaccurate finite-difference approximation). Computation of the gradient of the function is necessary in most NLP techniques. Numerically, the gradient can be approximated up to sufficient accuracy by the central difference or the forward difference formula. A poor scaling of variables can also cause instability in numerical procedures. There may be an imbalance between the values of the function and changes in decision variable x; the function values may change little even though x changes significantly. Conversely, the function may change extremely rapidly, even though x changes hardly at all.

Overly stringent accuracy requirements may also cause a failure of the algorithm. In some instances, an algorithm may indicate that it has been unable to terminate successfully, whereas, in fact, the solution has already been found. A good programming practice is to input the maximum number of times the problem functions are evaluated, and/or an upper bound on the number of iterations. Such bounds are useful in several situations and serve as a protection against errors in formulation that would otherwise not be revealed. In particular, the upper bound on the number of function evaluations may be reached when an optimization problem has an unbounded solution. This failure will also occur if a large number of iterations are performed without any significant improvement in the objective function.

A detailed discussion of these techniques is available in many textbooks, such as Rao (2009), Wagner and Himmelblau (1972), and Reklaitis et al. (1983). Many books contain steps of the various solution algorithms. One can develop a working, if not efficient, program by carefully following these steps. R is a programming language and environment for scientific computing and graphics. It provides a wide variety of statistical and graphical tools including optimization. A big advantage is that R is available as Free Software that runs on a variety of UNIX, Windows and Mac platforms. Software such as Matlab ® contain routines for non-linear optimization.

5.5 DYNAMIC PROGRAMMING

DP is an enumerative technique developed by Richard Bellman in 1953. This technique is used to get the optimum solution to a problem which can be represented as a multistage decision process. The entire DP formulation is based on the Bellman principle of optimality. According to this principle, an optimal policy has the property that whatever the initial state and decisions are, the remaining decisions must constitute an optimal policy with respect to the state resulting from the first decision. The proof of this theorem can be obtained by contradiction. In Fig. 5.6, let the optimal path for going from A to D be ABCD. According to Bellman's theorem, the optimal path from B to D will be BCD and not BED. If the optimal path from B to D is BED, then the optimal path from A to D will be ABED and not ABCD.

DP is not a class of optimization techniques, but as an algorithm, it is a powerful procedure to solve sequential decision problems. Many problems in water resources involve a sequence of decisions from one period to the next period and are known as sequential decision problems. Such problems can be decomposed into a series of smaller and easily solvable problems that can be conveniently handled by DP. For example, the operation of a reservoir proceeds in a sequential manner from one time period to another. An important feature of DP is that nonlinearities and constraints can be readily accommodated. In fact, constraints serve to reduce the region to be covered in computations and are helpful in that sense. In a DP problem formulation, the dynamic behavior of the system is expressed by using three types of variables:

State variables—define the condition of the system. For example, the amount of water stored in the reservoir may represent its state. If a problem has one state variable per stage, it is called a one-dimensional problem; a multidimensional problem has more than one state variable per stage. Thus, the optimization of operation of a system of two reservoirs will have two state variables, one for each reservoir.

Stage variables—define the order in which events occur in the system. Most commonly, time is the stage variable. There must be a finite number of possible states at each stage.

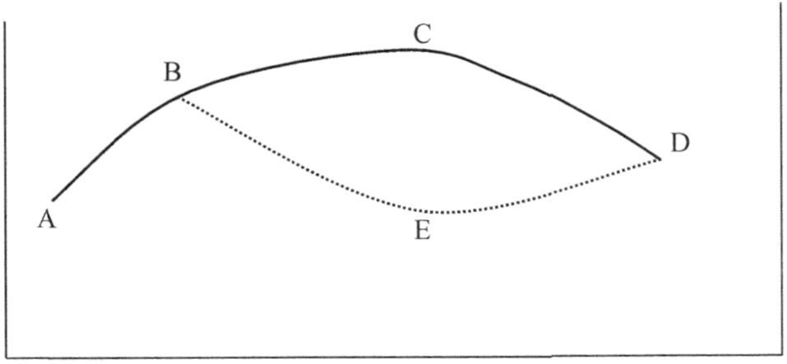

FIGURE 5.6

Illustration of the principle of optimality.

Control variables—represent the controls applied at a particular stage and transform the state of the system. For a reservoir operation problem, the release of water from the reservoir is a typical control variable.

The dynamic behavior of the system is expressed by an equation known as the system equation. It can be written in discrete form as:

$$s(t+1) = f[(s(t), u(t), t] \quad t = 1, 2, \ldots, N \tag{5.66}$$

where s(t) is the state variable at time t, u(t) is the control applied at time instant t, which lasts for a finite duration, and f[.] is the given functional form. The state of the system at any stage should lie in the domain of admissible states at that stage; the controls should also lie in the admissible domain at that stage:

$$s(t) \, \varepsilon \, S(t), u(t) \, \varepsilon \, U(t) \tag{5.67}$$

where S(t) and U(t) are the domains of admissible states and controls at stage t. The function f[.] should be invertible, that is, it must be possible to express the decision variable as an explicit function of state variables:

$$u(t) = f^{-1}[s(t+1), s(t), t] \tag{5.68}$$

For an invertible system, the order of the state vector is equal to the order of the control vector. Thus, the knowledge of stage variables enables one to compute the decision variables. For instance, in reservoir regulation problems, the mass balance equation (which is also the state equation) is invertible.

With each state transformation, a return is associated, which may either represent benefits or costs. Typically, the benefits are maximized and the costs are minimized. The optimal decision made at a particular stage is independent of decisions made at the previous stage, given the current state of the system. It is necessary that the objective function of a DP problem should be separable. It should be possible to write individual objective functions at each stage as functions of state and/or decision variables at that stage. Likewise, the constraints should also be separable or each constraint should be associated with an individual stage only. For a multidimensional problem, it would be necessary to evaluate the objective function for all discrete combinations of state variables.

A set of decisions for each time period is called a policy, and the policy which optimizes the objective function is called the optimal policy. The set of states resulting from an application of the policy is called the state trajectory. For example, the volume of water stored in a reservoir can be considered to be its state. The state of a reservoir is transformed due to inflows and can be controlled by releasing water from the storage. This water can be used for some useful purpose (e.g., irrigation) to yield monetary returns or it may also cause flood damage downstream and a cost is associated with this damage. A problem of optimizing the operation of a reservoir could be to find the releases (controls) which yield the best returns.

5.5.1 Recursive Equation of DP

Let R[s(t), u(t), t] be the return from operating a system which is at state s(t) and the control u(t) is applied at stage t. Further, let F[s(N),N] be the sum of returns from the application of controls from the initial stage at $t = 0$ to the final stage at $t = N$. The objective of maximizing the sum of returns from the system can be expressed as

$$\text{Max } F[s(N), N] \tag{5.69}$$

Let the state of the system at t = 0, s(0) ε S(0) be known and the returns F[s(0),0] be also known. Let F*[s(0),0] be the optimum value of these returns. Now, consider the first stage (of duration Δt). The optimal return for this period is given by

$$F^*[s(1), 1] = \underset{u(0)\ \varepsilon\ U(0)}{\text{Max}} \quad R[s(0), u(0), 0] + F^*[s(0), 0] \tag{5.70}$$

Here, R[s(0),u(0),0] indicates the returns that are obtained when at stage 0, the system is in state s(0) and control u(0) is applied [U(0) is the domain of admissible states]. This equation is solved for each discrete level of state at t = 0 as a function of control variables u(0). To do this, the state is discretized into a number of discrete levels. Now a particular lattice point is chosen and all the admissible levels of decision variables which lead to this state are chosen. For each of these decision variables, the return F[s(1),1] is calculated. The maximum among these returns gives the value of F* [s(1),1]. This computation is repeated for each discrete value of s(1) and the results are stored. The progress of computations in a forward algorithm is shown in Fig. 5.7.

Computations are performed in a similar fashion for stage 2, 3 …., N. Note that there can be more than one decision variables at a given state. The recursive equation for any stage t can be written as:

$$F^*[s(t), t] = \underset{u(t-1)\ \varepsilon\ U(t-1)}{\text{Max}} \quad R[s(t-1), u(t-1), t-1] + F^*[s(t-1), t-1] \tag{5.71}$$

Finally, at the end of stage N, the values of F*[s(t),t], t = 1,2 … N are available. The optimal value of control variables or the optimal policy is obtained by tracing back the values of returns, corresponding to those states which satisfy the initial and final values and the constraints. The optimal state trajectory can be determined by using the system equation, once the optimal policy is known. This is the method of enumeration or brute-force optimization. The tree of computations generated in enumeration is shown in Fig. 5.8.

The above computational scheme of DP is known as the forward algorithm, since computations start at the initial value of the state variable at stage 1 and move forward stage-by-stage. In contrast with this, computations can also commence at the final value of the state variable at the last stage and move backwards. The optimal policy is retrieved by tracing forward from the returns. This algorithm is termed as the backward algorithm.

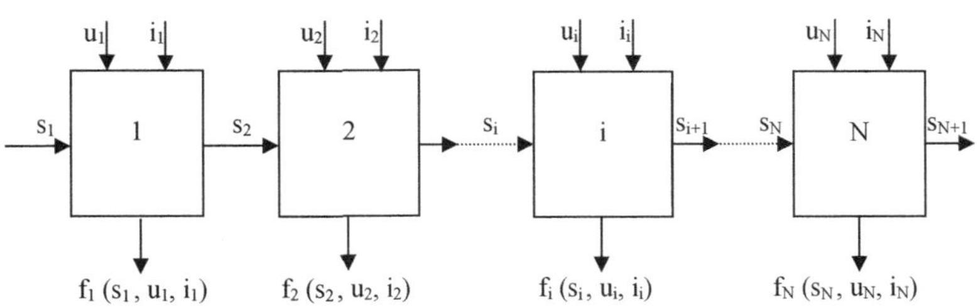

FIGURE 5.7

Progress of DP computations in a forward algorithm.

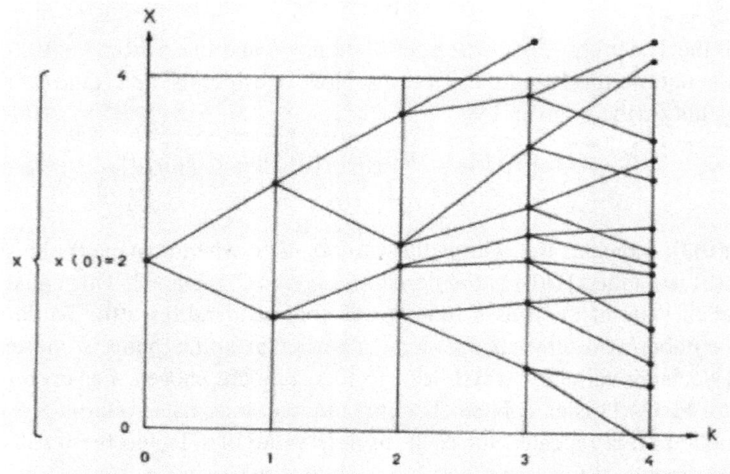

FIGURE 5.8

Tree generated by enumeration in DP.

Example 5.3. A system of three reservoirs is to be constructed. The yield versus cost at the reservoir sites is given in the following table.

	Cost		
Yield	**Reservoir 1**	**Reservoir 2**	**Reservoir 3**
0	0	0	0
20	15	10	20
40	30	35	40

Find the minimum cost combination to get a total system yield of 60 and 80.

Solution. This problem can be solved using enumeration method of DP by carrying out computations in various stages. Here, stage refers to the number of reservoirs in the system at any time. At stage 1, only reservoir 1 is considered. Therefore, the cost of providing various yields (in the range 0–40) at stage 1 will be same as the cost of providing yields by reservoir 1. At stage 2, the reservoir 2 is added and the combination of reservoirs 1 and 2 is considered. The computation of cost to get various yields in the range of 0–80 from reservoir 1 and reservoir 2 is given in the following table:

	Yield from		Cost of		
Total yield	**Reservoir 1**	**Reservoir 2**	**Reservoir 1**	**Reservoir 2**	**Total cost at stage 2**
0	0	0	0	0	0
20	20	0	15	0	15
	0	20	0	10	10[a]

40	40	0	30	0	30
	20	20	15	10	25[a]
	0	40	0	35	35
60	20	40	15	35	50
	40	20	30	10	40[a]
80	40	40	30	35	65[a]

[a]*indicates optimal solution for that yield.*

Proceeding to stage 3, all the three reservoirs are considered. Now, for a total system yield of 60 and 80, the possible combinations and corresponding costs are given in the following table:

	Yield from		Cost of		
Total yield	**Reservoir 3**	**Stage 2**	**Reservoir 3**	**Stage 2**	**Total cost**
60	40	20	40	10	50
	20	40	20	25	45
	0	60	0	40	40[a]
80	20	60	20	40	60[a]
	40	40	25	40	65

[a]*indicates optimal solution for that yield.*

The last column of the above table gives the total cost of providing various yields. It can be seen from the table that for a total system yield of 60, the minimum cost is 40 units. Now, the optimum solution can be traced backward. To get a yield of 60, reservoir 3 should not be constructed and a yield of 60 units should be obtained from stage 2. From computations of stage 2, one can note that reservoir 1 should give a yield of 40 and a yield of 20 units must be obtained from reservoir 2.

Similarly, for a total system yield of 80, a yield of 20 must be planned from reservoir 3, and a yield of 60 from stage 2. The table for stage 2 shows that a yield of 40 should be obtained from reservoir 1 and reservoir 2 must provide a yield of 20.

It is clear from this example that computations are quite simple in the enumeration method. However, as the number of variables and their discretization increase, one encounters a major computational problem because the computer memory and time requirements increase exponentially. For example, consider a two-reservoir problem. If reservoir 1 takes on 40 feasible states and reservoir 2 takes on 20 feasible states, the DP recursive equation would have to be evaluated at 800 points in each period. In general, if there are n state variables at each stage and each state variable has m discrete values, then one needs to evaluate the objective function at m^n points. Each of these values and equal number of values of state variables need to be stored in computer memory at each stage. One can easily imagine the consequences as the number of state variables increases or as a finer discretization is used. This problem arising due to storage and comparison of abnormally large number of variables was termed by Bellman as the *curse of dimensionality* (see Bellman and Dreyfus, 1962; Buras, 1966).

Several procedures have been developed to overcome the curse of dimensionality. Intuitively, the number of variables to be stored can be reduced by adopting a coarser grid for initial computations.

After the optimal solution is located, a finer grid can be constructed in the vicinity of this solution. However, in this scheme, one may miss the global optimum and the solution may converge to a local optimum. Another attractive procedure to alleviate the curse of dimensionality is the discrete differential DP (DDDP).

5.5.2 Discrete Differential DP

The technique which uses the concept of increments for state variables was introduced by Larson (1968) and termed as state increment DP (SIDP). Heidari et al. (1971) used this concept for reservoir operation studies and referred it as discrete differential dynamic programming (DDDP). The major difference between Larson's SIDP and DDDP is the time interval used in computations, which is variable in the former and fixed in the latter. In fact, DDDP is a generalization of SIDP.

The DDDP procedure starts with an assumed trial state trajectory, which is a sequence of feasible state vectors resulting in a corresponding initial policy, and an initial value of the objective function. The DP recursive equation is then used to examine a restricted set of values of the state variables or the neighboring states that are one small increment above and below the trial state trajectory. This subdomain is called a corridor (see Fig. 5.9), and the trial trajectory lies at the center of the corridor, although this is not a necessary condition. More than one discrete state on either side of the trajectory may be chosen, but the choice of three quantized states at each stage is most suitable for computational efficiency. Now, DP computations are performed within this restricted corridor and a neighboring trajectory that gives a better value of the objective function is found. This new trajectory replaces the trial state trajectory and the procedure continues. The procedure is assumed to have converged to a local optimum when the trajectories in two successive iterations are the same and a better value of the objective function cannot be found. This can be interpreted as a sort of successive approximation

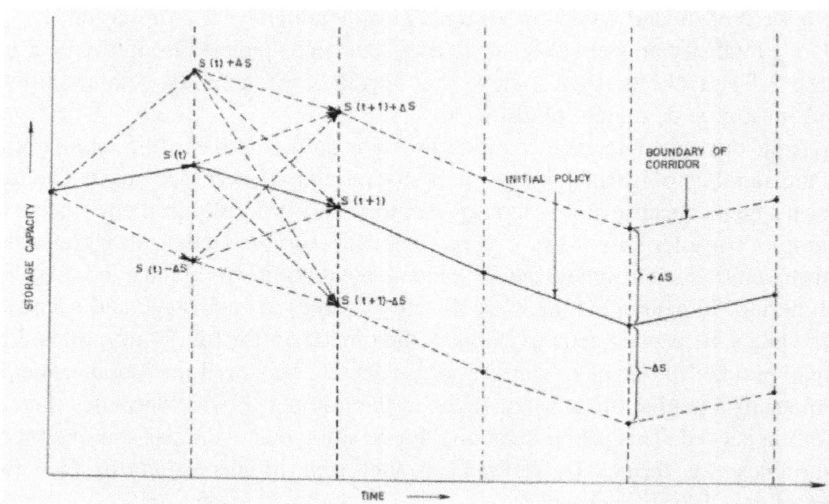

FIGURE 5.9

Subdomain of DDDP computations.

scheme. An initial estimate of the policy is made, and this is used to construct an improved estimate. The scheme cannot assure the global optimum and may converge to a local optimum. However, by starting from different initial solutions, the possibility of finding the global optimum is increased. This technique is particularly suitable for invertible systems. The water resources systems are mostly invertible. For example, assuming that the inflows to a reservoir are known, the releases from it can be determined if the states of the reservoir at different times are known.

When a real-life problem is solved by DDDP, the number of state transitions is considerably reduced, and computer time and storage requirements are drastically reduced. To obtain quick convergence, two procedures have been suggested to compute the increments of state variables. The first is to keep the increments small and constant throughout an iteration. The second is to reduce the size of increments as the iterations proceed. Goulter and Tai (1985) recommend that at least 5 to 10 discrete states should be used at each stage. In general, strong correlation is found between the number of iterations required for convergence and the size of increments used at each iteration. Furthermore, several iterations with a small increment of state variables should be allowed at the end of computation to improve the value of the objective function.

5.5.3 Advantages and Disadvantages of DP

DP is essentially an enumerative technique which is specially suited to multistage decision problems. Some of the advantages of using this technique for problems of water resources systems are:

(i) The DP formulation is the same for linear as well as nonlinear problems. Thus, no extra effort is required for nonlinear problems. This property is very useful since many problems cannot be realistically linearized.

(ii) The incorporation of constraints in LP and NLP problems is more difficult than in DP problems. In DP, the constraints serve a useful purpose. They limit the feasible region and thus may lead to reduction in the computational time requirement.

(iii) The stochastic nature of a problem can be easily considered in the DP formulation. The algorithm developed for a deterministic problem does not have to be significantly changed to incorporate stochasticity. This is in contrast with other techniques where incorporation of stochasticity requires too much change in the algorithm and significant increase in computational time.

Along with the above advantages, there are also some disadvantages. The major disadvantage is that DP is not basically tailored in such a fashion that generalized programs can be written using it. Thus, a new computer program has to be developed or an existing program has to be significantly modified and tested for each new application. On the other hand, standard computer programs are widely available for LP.

5.6 STOCHASTIC OPTIMIZATION

The system behavior and input of water resources projects display stochasticity which needs to be appropriately accounted for. Depending on the way the stochastic nature of the system and inputs are treated in an optimization formulation, the solution techniques are classified as implicit stochastic optimization (ISO) or explicit stochastic optimization (ESO). In ISO, the system and the stochastic nature of the input are represented by statistical models and these models are used to generate realizations of the input time series over the operation horizon. A suitable deterministic optimization technique is applied to find the optimum decision variables for each input realization. Since data generation techniques and simulation are used, the problem need not be solved analytically. This approach is known as Monte Carlo technique.

In this approach, inputs are represented by a time series model or probability distribution and the system behavior is modeled. Inputs are transformed into system outputs, and statistical characteristics of the outputs are gathered. This enables estimates of various output probabilities to be made which can be related to risk. A regression analysis is carried out to establish a relationship among system inputs, state variables, outputs, and optimum decisions for all of the generated sequences. This relationship can be used to take operation decisions when the future is unknown. This strategy was used by Young (1967) to derive reservoir operation rules. Since deterministic DP was used, the procedure was termed as Monte Carlo dynamic programming (MCDP).

A drawback of ISO is that it requires considerable computational time and effort to generate a large number of synthetic input sequences, solve a large number of optimization problems, and do multivariate regression analysis. Furthermore, one may not always get a good relationship between the variables involved.

In contrast to ISO, ESO directly uses the probability distributions of inputs in optimization. The objective function is the sum of benefits over all states, stages, inputs, and decisions multiplied by the probability that these conditions occur. Thus, the objective function represents the expected total benefit for the system. Computations are performed to find that set of probabilities which maximizes the expected total benefit. These probabilities are then used to calculate the conditional probabilities of making a decision given that the system is in a certain state at a given stage and it receives certain inputs. Ideally, the solution should yield a pure strategy, that is, one decision should have a probability of unity and all other decisions should have zero probability. Unfortunately, one does not always obtain pure strategies; "mixed" strategies are occasionally obtained and are suitably used to arrive at a decision. In the ESO, if the optimization technique used is LP, this problem is known as stochastic LP. The linear decision rules are used in LP to disallow the possibility of mixed strategy.

The advantage of ESO procedures over ISO is that the results obtained from ESO are based on a conditional probability distribution at each stage. Therefore, more information is utilized for the choice of a decision at each stage. Instead of just a single estimate, the probability distribution of inputs is used. The method involves a great deal of computation time and storage so that its application to complex systems is severely limited. The common assumption is that the inputs at each stage have a steady-stage probability distribution and the system can be represented by a cyclic (repetitive) operation. Thus, only one cycle needs to be analyzed, and the system properties can be represented using a small number of discrete values for states, inputs, and decisions.

5.6.1 Chance-Constrained Linear Programming

The future inputs to a water resources project, for example, a reservoir, are random and hence the resulting states and decisions, such as storage and releases, will also be random. In some optimization problems, the constraints that limit the range of values of these specify the percentage of time that these ranges can be violated. The constraints that explicitly define these limits are termed as chance constraints. Chance constraints are used in mathematical programming to constrain the optimization to those decisions that represent a failure probability smaller than the constraint value. Basically, a constraint on the probability of failure is transformed into its deterministic equivalent. A chance constraint that ensures that some variable $\widehat{x}$ is no greater than the value of a random variable X at least some fraction α of time is written as (Loucks and Dorfman, 1975):

$$\text{prob}[\widehat{x} \leq X] \geq \alpha \tag{5.72}$$

where *prob* denotes probability.

In reality, many problems, such as reservoir sizing and operation, require a nonlinear optimization formulation, but the use of linear decision rules permits their framing as an LP problem. ReVelle et al. (1969) were the first to present a chance-constrained formulation to solve reservoir design and operation problems. They used a linear decision rule which allows simple formulation of chance constraints for problems dealing with reservoirs and the probability distribution of inflows can be easily considered. The linear decision rule relates the release from a reservoir to storage as:

$$R_t = S_t - b_t \tag{5.73}$$

where R_t is the release from the reservoir, S_t is the initial storage, and b_t is a decision variable, all for period t. The objective of their study was to minimize the capacity of the reservoir and determine the optimum coefficients in the decision rule, subject to chance constraints on freeboard, storage, and releases. The freeboard consideration requires that at least a volume V be available at the end of period t for temporary storage of flood peaks. In other words, storage S_t at the end of the period should not be greater than the reservoir capacity C minus the freeboard requirement V, at least 100α % of the time:

$$P(S_t \leq C - V) \geq \alpha \tag{5.74}$$

By substitution of the linear decision rule from Eq. (5.73) and by using the continuity equation:

$$S_t = S_{t-1} + I_t - R_t = I_t + b_t \tag{5.75}$$

The chance constraint can be formulated as

$$P[I_t + b_t \leq C - V] \geq \alpha \tag{5.76}$$

$$F_{It}\ (C - V - b) \geq \alpha \tag{5.77}$$

or

$$C - V - b \geq i_\alpha \tag{5.78}$$

where i_α is the α quantile point from the distribution of I_t, $F_{It}\ (i_\alpha) = \alpha$. The above use of the linear decision rule, chance constraints, and optimization in an LP formulation yields the chance-constrained LP. Though the use of the rule results in conservative designs, the model is useful in multireservoir studies and preliminary screening studies. It enables simpler application of chance constraints in an LP problem.

5.6.2 Stochastic Dynamic Programming

The DP formulation where the stochastic nature of variables is not considered is known as deterministic DP. Since many water resources variables are stochastic in nature, the DP approach is frequently modified to account for this stochasticity. The DP formulation which takes into account the stochastic nature of variables is known as stochastic dynamic programming (SDP). DP can be adapted in two ways to handle stochastic input data. The first of these is called MCDP. The basic idea of this procedure (Monte Carlo techniques are discussed later in Section 5.9.2) is to generate a number of synthetic streamflow sequences which match the properties of observed inflow series. For each of these series, a DP formulation is used to get the optimum policy and so there will be as many policies as the number of synthetic sequences. These optimum policies are then used in a regression analysis to determine the causal factors influencing the optimal policy.

An alternative to this procedure is to formulate the problem as a true SDP problem and use the policy iteration and the policy improvement routines. The SDP approach for the development of an optimal operation policy of water resources systems was first used in the 1950s. In this method, it is necessary to analyze streamflows on a time period (usually monthly) basis and express the relation between these as transition or conditional probabilities of period-to-period flows. The computation of transition probability matrix has been described in Chapter 4.

Where the probability of various values of a variable is dependent on the value of that variable in a previous time period, the sequence of events so described is called a Markov chain. When the probability of being in a given state after another given state is a fixed quantity, it is termed as constant or stationary conditional probability. Many hydrologic variables exhibit this property.

An examination of monthly river flow data at a number of sites shows that the conditional probability connecting monthly flows is not a stationary quantity. Therefore, the sequence of monthly flows can be regarded as connected by 12 sets of different conditional (or transitional) probabilities to form a nonstationary (cyclic) Markov chain. The derivation of the optimal policy is based on the assumption that the system described is ergodic. For an ergodic system, the final system state is independent of the starting state. For example, in a reservoir operation problem, this is equivalent to stating that no matter what the state of reservoir at the start of computations is the steady state of the system will be independent of that starting state.

Consider that a system is to be managed for one-time period only and at the end of this time period, its state is of no value. Let this be the end point of the study period. The calculations step backward in time. The optimum decision r for this last time period is obtained by the following equation:

$$f_1(s_1, I_2) = \max_{r} R(r) \tag{5.79}$$

where $f_1(s_1, I_2)$ is the expected return from the optimal operation of a system which has 1 time period to the end of the operation horizon; s_1 is the state of the system at the start of the first calculation time period; I_2 is the input to the system in the first time period; and R(r) is the return obtained consequent to applying controls in this period. The decision should satisfy the applicable constraints. Thus, for each discrete value of state variable and for each discrete value of input, there will be a value of r which will give the maximum R(r).

Now, consider the period before the last time period (which is the calculation time period number 2, as the counting is backward). According to Bellman's principle of optimality, an optimal policy for these last two time periods must include one of the policies already determined for time period number 1. relating to the state that the system attains when the optimal decision is made for time period number 2. Often, the next state that will be occupied is not a deterministic function of the current state and decision and may depend on uncertain events, such as rainfall, streamflow, or political decisions. Transition probabilities are used to account for such situations. One can then write

$$f_2(s_2, I_3) = \max_{r} \left[R(r) + \sum_{I_3=0}^{I_3=I_3\ \max} P(I_2|I_3)\ f_1(s_1, I_2) \right] \tag{5.80}$$

where $P(I_i|I_{i+1})$ is the transition probabilities connecting the input in the ith time period I_i with input in the $(i+1)$th time period I_{i+1}. This equation states that for all values of s_2, the state at the start of the second computational time period (or the second last in the study period) and all possible values of input during the preceding period (I_3), there exists a value of control (r) which maximizes the objective function. The value of r is chosen from its feasible domain. The right-hand side of this equation expresses that r will be chosen such that the return from decision r in the current (second) time period, together with the worth of the system state at the start of the succeeding time period, will be a maximum. Note that the value of the system state at the start of the next time period for all possible states of that system will be known from the preceding calculations. Eq. (5.80) uses the fact that one period's input is related to the preceding period's input by the conditional probabilities $P(I_2|I_3)$. Thus, for each value of I_3, it is possible to assign a probability to I_2 so that the probability of a situation which derives from I_2 (such as s_1) can similarly be assigned that probability. Thus, it is possible with given values of s_2 and I_3 to range over all possible values of r and determine both the current period returns and the expected value of the resulting state of the system. The value of r is chosen to maximize the sum of these returns.

This procedure is repeated for all possible values of s_2 and I_3 and function f_2 is completely evaluated. Similarly, f_3 is evaluated from f_2 and so on, leading to the general formulation:

$$f_i(s_i, I_{i+1}) = \max_{r} \left[R(r) + \sum_{I_{i+1}=0}^{Ii+1=Ii+1\ \max} P(I_i|I_{i+1})\ f_{i-1}(s_{i-1}, I_i) \right] \tag{5.81}$$

In table below, the temporal progress of the various computations is presented.

Time period index for computations		$i+1$	I	$i-1$	...	2	1	End of study period\start of computations
Input I		I_{i+1}	I_i	I_{i-1}	...	I_2	I_1	
System state at the beginning of the period	s_{i+1}	s_i	s_{i-1}	...	s_2	S_1		
Return function		$f_{i+1}(s_{i+1}, I_{i+2})$	$f_i(s_i, I_{i+1})$	$f_{i-1}(s_{i-1}, I_i)$	...	$f_2(s_2, I_3)$	$f_1(s_1, I_2)$	
Progress of time ⟶								

Starting at some time in the future and using the transition probabilities between the input in one-time period and that in the adjacent time period, it is possible to calculate the values of r for each time period as a function of state variables s_i and I_{i+1}. Put together, these r's form an optimal policy for the control of the system. Under certain circumstances, this policy converges when the values of r that are used to evaluate the function $f_i(s_i, I_{i+1})$ repeat for all values of i, as i becomes large. This method of determination of an optimal policy is termed as policy-iteration routine, and it is an effective way to develop an optimal policy.

By starting this procedure at an arbitrary future time period, and stepping backward one period at a time, it is easy to find the optimum control at any state of the system. The decision to be made in early time periods will be affected by the conditions specified at the end of the study period, which is the start of the calculation procedure. However, by carrying the calculations sufficiently far from the end, the results will be free from the influence of the starting conditions, and an optimum policy can be established. The result of this calculation is a set of matrices of decisions to be made under all states of the system, that is, for all values of s_i and I_{i+1} in each period, say a month, of the year.

A major advantage of this formulation is that the developed policy can be used for design as well as for actual operation. The state of the system is described by two variables, the input in the preceding period I_{i+1} and the system state s_i. These quantities are available at the time when the decision is to be taken. This formulation can also be used to assess a given design. A model of this kind also allows to explore the sensitivity of the return and the uncertainties associated with it to variation in input parameters (correlation, uncertainty, etc.), constraints on the system operation, and return functions.

5.7 MULTIOBJECTIVE OPTIMIZATION

Water resources systems are usually characterized by multiple objectives, multiple decision-makers, and multiple constituencies. Before discussing this topic, it is useful to differentiate between multipurpose and multiobjective. According to Major (1977), the term multiobjective refers to multiple economic, social, environmental, and other objectives of water development; and multipurpose refers to multiple functions like navigation, flood control, water supply, recreation, etc. Clearly, these two terms are not synonymous—purposes can vary and still be aimed at the same objective, and one purpose can fulfill more than one objective.

A fundamental characteristic of multiobjective water resources problems is that the various objectives are often noncommensurate and may be in conflict. In view of this, a multiobjective analysis has a central role in water resources planning and management. In fact, the area of water resource planning and management is responsible for many developments in the field of multiobjective optimization.

Trade-offs are an inherent part of negotiation, of reaching consensus, and of compromise solutions. Thus, if two solutions are compared, may be that the first solution achieves higher levels of certain objectives and lower levels of other objectives when compared to the second solution. For example, the use of a reservoir for flood control purposes may be achieved at the expense of reducing benefits from conservation operation. Moreover, as environmental and other socioeconomic aspects now dominate and influence policy decisions, the importance and the need for a multiobjective analysis have become more critical and evident. The multiobjective analysis should be viewed not only as a methodological approach but also as a philosophy. However, analysts and decision-makers should be aware of the properties, efficacy, and limitations of the multiobjective analysis.

While formulating and screening alternative plans, the objectives of the project should be given explicit and quantitative consideration. This is especially important in the planning of river basins, where there are likely to be several conflicting and noncommensurate objectives. The objectives that can't be expressed in a common unit are noncommensurate. For example, one may want to maximize irrigation benefits (measured in monetary units) and environmental quality (measured in units of pollutant concentration). Traditionally, only one objective (economic efficiency) is considered and the other objectives are included either as constraints or as commensurate with the primary objective in some way. This limitation can be overcome in multiobjective analysis.

A multiobjective analysis is "vector optimization." The essential difference between a scalar optimization problem and a vector optimization problem is that the objective function is a vector instead of a scalar. To some persons, the term vector optimization is contradictory in itself, since one cannot optimize a vector. Fundamental to multiobjective analysis is the concept of Pareto optimum, which is also known as the noninferior solution. Qualitatively, a noninferior solution of a multiobjective problem is one where any improvement of one objective function can be achieved only at the expense of degrading another. The Pareto optimal solutions and the associated trade-off values help decision-makers select an acceptable level of assurance and the corresponding cost. The decision-makers can make known their preferences with respect to the level of assurance against uncertainties in the model's prediction at the expense of a degradation (reduction) in the model's optimal solution. Some authors prefer the term "best-compromise solution" to suggest that a noninferior solution so identified is optimal only in terms of a particular set of value judgments.

In a single-objective model, the decision-makers' preferences are assumed to be known and expressed as a single-objective function F(X) whose value for a particular solution vector X gives the utility of X for the decision-maker. A general multiobjective analysis problem involves the selection of a k-dimensional vector of decision variables, $X = (x_1, x_2, \ldots x_k)$, which may represent project outputs, the allocations of scarce resources as inputs, and operation policies. There might be some relationships among variables which are expressed through constraints, $g_i(X) \le 0, i = 1, 2, \ldots, m$. Each quantifiable objective can be described by a function $F_j(X)$ that should be either maximized or minimized. Using these notations, the vector optimization problem can be stated as:

$$\text{Maximize } Z(x) = \left\{f_1(x), f_2(x), \ldots, f_p(x)\right\} \quad (5.83)$$

$$x \in X$$

subject to

$$g_i(x) \le 0 \quad i = 1, 2, \ldots m \quad (5.84)$$

where Z(x) is the p-dimensional objective function, that is, there are p objectives; and $g_i(x)$ are the constraints. The set of all feasible solutions X is defined as:

$$X = \{x |\ g_i(x) \le 0, i = 1, 2, \ldots, m\} \quad (5.85)$$

Every feasible solution to the problem implies a value for each objective.

It is to be noted that without information about preferences which provide a rule for combining, the objectives may be incomparable. Such an incomplete ordering, which is characteristic of multiobjective planning problems, implies that in the absence of preference information, an optimal solution cannot be found to the problem since all feasible solutions are not comparable. A complete ordering, which is characteristic of scalar (single-objective) optimization problems, can be obtained for a vector optimization only by introducing value judgments into the solution process.

5.7.1 Noninferior Solutions

The trade-offs between separate and noncomparable objectives can be explored by following anyone of several approaches. If there are only a few objective functions and decision variables in the vector X, it may be easy to enumerate all feasible combinations of decision variables. If there are two objective functions and two decision variables, the values of decision variables and objective functions can be visualized by plotting them in the decision and objective space. An enveloping curve of these values can be easily drawn and this would indicate the efficient vectors X and the trade-offs that are possible among these efficient combinations. This concept is shown in Fig. 5.10.

A solution x^* of a multiobjective problem is termed as noninferior if and only if there does not exist another x' such that $f_i(x') \geq f_i(x^*)$, $i = 1,2,\ldots n$, with strict inequality holding for at least one i. The collection of all noninferior solutions is referred to as the set of noninferior solutions.

Each point within the feasible region in Fig. 5.10 represents a particular set of values for the decision variables in the vector X that satisfy the constraints. Each point on the transformation curve gives the maximum value of objective $F_1(X)$ given a particular value of objective function $F_2(X)$. Note that a point M located in the feasible region represents a solution that is not preferred because one can either get a better value of the objective function $F_2(X)$ by moving to point B or a better value of the objective function $F_1(X)$ by moving to point C. The trade-offs defined by certain combinations of feasible and efficient decision vectors are of interest in decision-making. Each decision vector on the DBCE portion of the feasibility frontier is efficient because in that segment of the curve, there can be no increase in the value of one objective without a decrease in the value of the other objective. The solutions on the AD segment of the curve are inferior solutions because one can obtain a better combination of the values of both objective functions (assuming that the values are to be maximized) elsewhere on the curve. Inferior solution vectors are of interest only if some of the objectives are to be minimized. When all the objectives are to be maximized, only the efficient solutions need be considered.

FIGURE 5.10

Transformation curve and indifference curves for two-objective case.

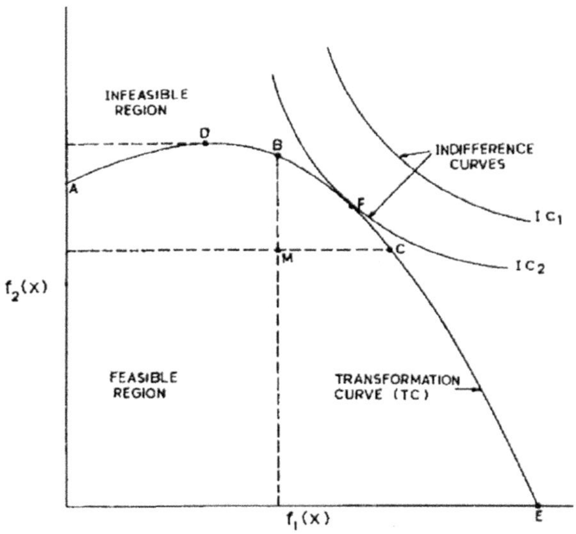

The goal in the planning stage is the identification of plans which lie on the transformation curve, since this is the set of efficient trade-offs. Any point on the transformation curve corresponds to a specific trade-off or marginal rate of substitution between the objectives. This rate equals the slope of the curve at that point. For example, the solutions D, B, and C in Fig. 5.10 correspond to three different marginal rates of substitution between objective functions $F_1(X)$ and $F_2(X)$.

The indifference curves IC_1 and IC_2 are the curves of equal preference of a decision-maker. It is possible that different individuals or groups may have different sets of indifference curves. The optimal plan for any particular policy maker is the plan on the objective transformation curve which achieves the highest level of preference of the decision-maker, for instance, point F in Fig. 5.10. The optimal trade-off or marginal rate of substitution is given by the slope of the transformation curve at that point F.

Example 5.4. The following two-objective two-decision variable problem by Cohon and Marks (1975) is used to illustrate the concepts of multiobjective optimization.

$$\text{Max } Z(x) = [Z_1(x), Z_2(x)] \tag{5.86}$$

$$Z_1(x) = 5x_1 - 2x_2 \tag{5.87}$$

$$Z_2(x) = -x_1 + 4x_2 \tag{5.88}$$

subject to

$$g_1(x):\ -x_1 + x_2 - 3 \leq 0 \tag{5.89a}$$

$$g_2(x):\ x_1 + x_2 - 8 \leq 0 \tag{5.89b}$$

$$g_3(x):\ x_1 - 6 \leq 0 \tag{5.89c}$$

$$g_4(x):\ x_2 - 4 \leq 0 \tag{5.89d}$$

$$g_5(x):\ x_1 \geq 0 \tag{5.89e}$$

$$g_6(x):\ x_2 \geq 0 \tag{5.89f}$$

Solution. In this example with two decision variables and two objectives, the feasible region in the objective space can be found by enumeration of all extreme points and computation of the values of each objective at each of these corner solutions. These points and the values of the objective functions are listed in Table 5.1.

Table 5.1 Extreme points and values of objective functions.

$\overline{X}$	x_1	x_2	Z_1	Z_2
1	1	4	−3	15
2	4	4	12	12
3	6	2	26	2
4	6	0	30	−6
5	0	0	0	0
6	0	3	−6	12

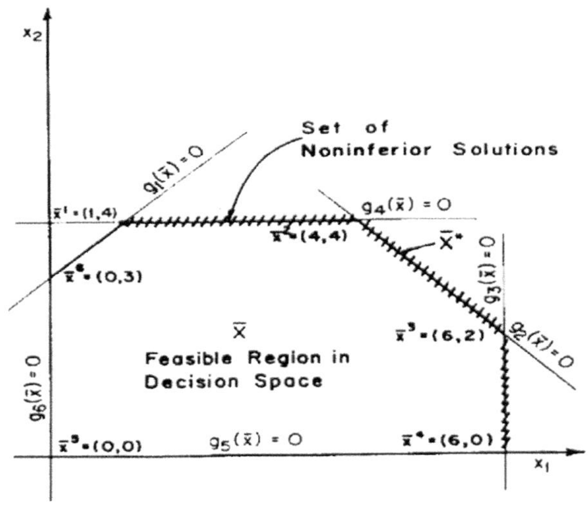

FIGURE 5.11

The feasible region in decision space X and the set of noninferior solutions X*.

After Cohon and Marks (1975).

The noninferior set Z(X*) can be found by applying the definition of noninferiority. The set of noninferior solutions contains four extreme points: $Z(x^1)$, $Z(x^2)$, $Z(x^3)$, and $Z(x^4)$. The point $\overline{X}$ 2 (4,4) is the best-compromise solution. Note that this enumeration procedure is computationally feasible only for very small problems. The feasible region in the decision space and the set of noninferior solutions are shown in Fig. 5.11. Fig. 5.12 shows the feasible region and the noninferior set Z(X*) in the objective space Z(X).

In the absence of preference information, no particular noninferior solution can be identified as preferable to any other noninferior solution. Of course, if preferences are known as represented by an indifference surface, then one of the noninferior solutions can be identified as the best-compromise

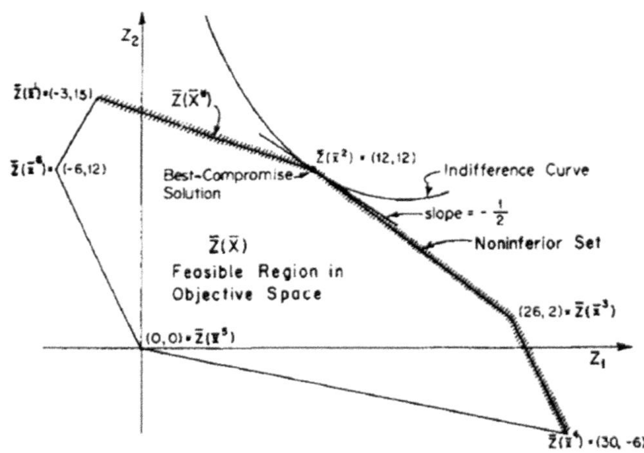

FIGURE 5.12

Feasible region in objective space Z(X), the noninferior set Z(X*), and the best-compromise solutions.

After Cohon and Marks (1975).

solution. The term "best-compromise solution" indicates that a noninferior solution so identified is optimal only in terms of a particular set of value judgments.

The use of enumeration method as a means of defining the feasibility frontier and the corresponding trade-offs between each efficient decision vector cannot be adopted as the number of variables and objectives increases. For these reasons, optimization techniques are usually suggested as a means of estimating feasible and efficient decision vectors X.

5.7.2 Classification of Multiobjective Optimization Techniques

The multiobjective analysis techniques can be classified in three groups (Cohon and Marks, 1975):

(a) Generating techniques,
(b) Techniques based on prior articulation of preferences, and
(c) Techniques based on progressive articulation of preferences.

5.7.3 Generating Techniques

These were among the first techniques that were developed to solve multiobjective optimization problems. The purpose of these techniques is to identify the set of noninferior solutions in the decision space as well as objective space within which the best-compromise solution will lie. The solution contains the maximum information from the model without preference information from the decision-maker. Two popular methods in this category are the weighting method and the constraint method.

Weighting Method

The weighting method was the first technique developed to solve multiobjective optimization problems. It is based on the premise that noninferior solutions can be obtained by solving a scalar optimization problem in which the objective function is a weighted sum of the components of the original vector-valued objective function. Cohon and Marks (1975) found that the solution to the following problem is, in general, noninferior:

$$\text{Max} \sum_{k=1}^{p} w_k Z_k(x) \tag{5.90}$$

subject to

$$x \in X \tag{5.91}$$

where $w_k \geq 0$ for all k and strictly positive for at least one objective. Thus, the noninferior solutions can be obtained through the use of an LP package (if all functions and constraints are linear or can be linearized) by parametrically varying the weight w_k in the objective function.

Constraint Approach

This method replaces (p − 1) objective functions with (p − 1) lower bound constraints:

$$\text{Maximize } Z_r(x) \tag{5.92}$$

subject to

$$Z_k(x) \geq L_k \quad \text{for all } k \neq r \tag{5.93}$$

where L_k is a lower bound on objective k; its value can be varied parametrically to evaluate the impact on the single objective function $Z_r(x)$. The noninferior solution can be identified by solving p problems and taking a different objective function each time.

In the generating techniques, the trade-offs among the objectives are explicitly considered. These methods are intuitively appealing since results can be graphically presented. However, the essence cannot be graphically captured for larger problems. The computational burden also grows tremendously with the increase in the number of objective functions. Therefore, these techniques remain attractive only for small problems with two or three objective functions.

Techniques Based on Progressive Articulation of Preferences

As the name suggests, these techniques consist of finding a noninferior solution, getting decision-maker's reaction to this solution, modify the problem, and repeat the steps until a satisfactory solution is attained. The STEp Method (STEM) is one such method. The method requires construction of a pay off table. This method is not very suitable for water resources problems.

Techniques Based on Prior Articulation of Preferences

In these techniques, the information about preferences of the objectives is used to do partial ordering of the objectives and eliminate some noninferior solutions. The aim is also to reduce the intensive computational load of generating techniques. The goal programming is one such technique. It is discussed in detail in Section 5.8.

The *Electre* method attempts to structure partial ordering of alternatives which is stronger than the incomplete ordering implied by noninferiority. In this method, a specific outranking relationship is developed for the set of noninferior solutions. This method is not applicable to water resources problems, since it is not computationally attractive and tradeoffs are obscured by the analysis.

5.7.4 Surrogate Worth Trade-off Method

All the multiobjective optimization methods are directed toward evaluating, in some order of magnitude, the relative worth or utility of each of the objectives so that they can be treated as if there was only one composite objective. It might be called the commensurate approach, since it is directed to the measurement and summation of the worth of all objectives in a common set of units. In the case of water resources systems, the decision-maker may not be interested in evaluating the relative true worth of all the combinations of objectives, but rather in the evaluation of the relative true worth of changes that might be incurred in these objectives due to changes in the set of decisions. The interest is also in the values of those incremental changes for the decision sets that are already at a Pareto optimum. The order of magnitudes of values upon which the trade-off of the decision-maker depends can be used as a substitute for the unknown true worth ratios of the marginal gain and losses between objectives. For this reason, these are called as surrogate worth. The term surrogate worth is defined to be a positive number whenever the true worth of Δf of the numerator in the trade-off ratio between any two objectives is considered to be higher than the true worth of the denominator of the ratio. It will be a negative number when the opposite is true and will be zero when the decision-maker cannot distinguish between their relative worths.

The surrogate worth trade-off (SWT) method was developed by Haimes and Hall (1974) who proposed that the choice of optimal weights should be made with the knowledge that trade-offs depend on the levels of objectives. The SWT method recognizes that the optimization theory is usually more concerned with the relative value of additional increments of the various noncommensurate objectives, at a given value of each objective function, than it is with their absolute values. Further, when the values of objective levels attained are known, it is easier for decision-makers to assess the relative value of the trade-off of marginal increases and decreases between any two objectives than it is to assess their absolute average values. The distinguishing feature of the SWT method is the generation of "trade-off functions" which show the relationship between a weight on one objective (when another objective is the numeraire) and the values of that objective. A set of trade-off functions may be interpreted as a disaggregated noninferior set, in which the objectives are considered in pairs.

If one objective function is considered primary and all others at minimum satisfying levels are considered constraints, the Lagrange multipliers related with the (p − 1) objectives as constraints will be zero or nonzero. The Lagrange multiplier for a constraint that limits the optimum is nonzero. The nonzero Lagrange multipliers correspond to the noninferior set of solutions, whereas the zero Lagrange multipliers correspond to the inferior set of solutions. Furthermore, the set of nonzero Lagrange multipliers represents the set of trade-off ratios between the principal objective and each of the constraining objectives. These Lagrange multipliers are functions of the optimal level attained by the principal objective function as well as the level of all other objectives satisfied as equality (binding) constraints. Consequently, these Lagrange multipliers form a matrix of trade-off functions. It is assumed that the objective functions are differentiable functions of the right-hand side levels of the constraints ($\in_j$).

Next, the worth ratios are selected. Since the worth ratios only represent relative worth (not the absolute level of worth of objectives), any surrogate ratio that varies monotonically with correct ratios will suffice. The distinguishing feature of the SWT method is the generation of "trade-off functions" which show the relationship between a weight on one objective (when another objective is the numeraire) and the values of that objective. A set of trade-off functions may be interpreted as a disaggregated noninferior set, in which the objectives are considered in pairs. The computational procedure is first to transfer the multiobjective problem into

$$\text{Maximize } Z_r(x) \tag{5.94}$$

subject to

$$x \in X \tag{5.95}$$

$$Z_k(x) \geq L_k \tag{5.96}$$

where L_k is the lower bound on the kth objective for all $k \neq r$. One of the objectives expressed as a constraint, say, objective s, is then varied over the k values of L_s, keeping the other objectives, all $k \neq r$, s, fixed at L_k. The problem in Eqs. (5.87)–(5.89) is solved for each value of L_s, producing at most k noninferior solutions. The dual variable associated with the constraint for the sth objective when the rth objective is in the objective function is T_{rs} which is the trade-off between objectives r and s:

$$T_{rs}(x) = df_r(x)/df_s(x) \tag{5.97}$$

where

$$df_r(x) = \sum_{k=1}^{p} \frac{\partial f_r(x)}{\partial x_k} dx_k \tag{5.98}$$

There are p values of T_{rs} generated by solving the modified problem with the k values of L_s. The trade-off T_{rs}, taken as a function of $Z_s(x)$, is the previously referred trade-off function.

With generated values of T_{rs} and $Z_s(x)$, regression analysis is used to get the function $T_{rs}[Z_s(x)]$. Next, another of the $k \neq r$ objectives is selected as objective s. This process is repeated until $T_{rk}[Z_k(x)]$ is generated for all $k \neq r$. The next step is to replace the rth objective and repeat the procedure until all $T_{jk}[Z_k(x)]$ are generated for all $j = 1, 2, \ldots, p$, and all $k = 1, 2, \ldots., j-1, j+1, \ldots, p$. The result is a set of functions which relate the weights to the levels of the objectives (these can be displayed graphically). The number of trade-off functions is in general equal to p^2. However, since $T_{kk} = 1$ and $T_{jk} = 1/T_{kj}$, the number of trade-off functions is little less.

The trade-off functions give the analyst the required information to extract "surrogate worth functions" W_{jk} from the decision-maker. There is one surrogate worth function for every trade-off function; thus the intent of constructing W_{jk} is to attach values to the previously computed trade-offs. The W_{jk} functions are ordinal, varying between -10 and $+10$, with some arbitrary but predetermined value which indicates an acceptable ("optimal") trade-off. The set of optimal trade-offs or weights found by this method are then used to identify the best-compromise solution.

The SWT method provides more information than other methods, although less than the maximum information associated with the generating methods is available. The information supplied is not complete in the sense that the trade-off functions are generated between two objectives, assuming fixed values for all of the remaining objectives. Thus, the variation of trade-offs with the level of objectives is captured in only a limited sense.

The SWT method can be a powerful tool when there are difficulties in evaluating trade-offs. Its greatest utility is for problems with several objectives ($p > 3$), since it leads decision-makers through a systematic comparison of objectives, two at a time. This approach may decrease the confusion associated with high dimensionality in the objective space when it is administered properly. Unfortunately, this method is vulnerable to its computational sensitivity to the number of objectives which is a generic characteristic of multiobjective solution techniques.

Cohon and Marks (1975) evaluated various multiobjective optimization techniques and suggested that when there are less than four objectives, a generating technique, such as the weighting method or constraint method, should be used. When there are four or more objectives, a technique which restricts the size of the feasible region, such as the SWT method, may be more suitable to use.

The purpose of multiobjective techniques is to assist the decision-maker in estimating efficient alternative solutions and trade-offs that may be required to obtain an acceptable solution. The iterative process of proposing a solution and having it accepted or rejected by the decision-maker is one means of focusing on trade-offs that are considered acceptable (need not be optimal) by the decision-maker. But sometimes, the decision-maker may not be adequately aware of the issues and implications and may not be able to clearly and consistently state his preferences.

5.8 GOAL PROGRAMMING

Most real-world decision problems involve multiple and often conflicting goals which cannot be maximized or minimized simultaneously within the constraints. In such a context, one has to think in terms of deviations from the optimum of individual objectives and thus try to achieve a balance between various objectives. When LP is used to solve decision problems, all constraints have equal importance and the optimum solution must satisfy all constraints. However, this assumption is not realistic and all constraints may not have equal importance. Such problems can be efficiently solved using the GP technique which can solve problems with a single or multiple goals. These goals may be noncommensurate, meaning that they cannot be measured on the same-unit basis. Thus, there is a need to establish a hierarchy of importance among these conflicting goals so that low-order goals are considered only after the higher-order goals are satisfied or have reached the point beyond which no further improvements are desirable.

In many cases, the management does not try to "optimize," instead it tries to "satisfice." An optimizer usually seeks the best possible outcome for a given objective, such as profit maximization in LP. A satisficer, on the other hand, attempts to achieve a satisfactory level of multiple objectives. GP is an appropriate technique for decision analysis. If management can provide an ordinal ranking of goals in terms of their contributions or importance to the organization and all relationships of model are linear, GP can be used to solve the problem. GP was devised for situations wherein the decision-maker proposes to seek maximization or minimization of the weighted absolute deviations or departures from the individual optimum. The main aim of GP is to establish a specific numerical goal for each objective, formulate an objective function for each goal, and then seek a solution that minimizes the (weighted) sum of deviations of these objective functions from their respective goals.

There are three possible types of goals:

- A lower, one-sided goal sets a lower limit that should not be underachieved (but exceeding the limit is acceptable).
- An upper, one-sided goal sets an upper limit that should not be exceeded (but falling under the limit is acceptable).
- A two-sided goal sets a specific target that should not be missed on either side.

In GP, instead of trying to maximize the objective criterion directly, the deviations among goals and what can be achieved within the given set of constraints are to be minimized. Such type of variable is represented in two dimensions, positive and negative deviations from each goal. The objective function becomes the minimization of these deviations based on the relative importance or priority assigned to them.

To understand the problem formulation of a goal program, it is necessary to define some notations. Any variable x_j, positive, zero or negative, can be expressed as the difference of two positive variables, that is,

$$x_j = x_j^+ - x_j^- \tag{5.99}$$

where

$$x_j^+ = x_j \text{ if } x_j \geq 0 \tag{5.100}$$

$$x_j^+ = 0 \text{ if } x_j \leq 0 \tag{5.101}$$

Similarly,

$$x_j^- = 0 \text{ if } x_j \geq 0 \tag{5.102}$$

$$x_j^- = -x_j \text{ if } x_j \leq 0 \tag{5.103}$$

Thus, x_j^+ and x_j^- are both positive and represent the positive and negative components (only components without sign). Clearly, the product $x_j^+ \times x_j^- = 0$ is always satisfied with one of the components or both components equal to zero. Furthermore,

$$|x_j| = x_j^+ + x_j^- \tag{5.104}$$

The matrix used in GP is composed of two types of constraints: goal and nongoal. Each goal constraint may be assigned a positive or negative deviational variable or both. These variables are shown in Fig. 5.13. In this figure, the line labeled, "Goal," indicates the complete goal attainment. If more than the desired goal level is achieved, there is positive deviation from the goal (d^+). Underachievement (d^-) means there will be a negative deviation (d^-) from the goal. An optimal solution is obtained when the sum of nonattainment of goals is minimized according to the priority structure.

5.8.1 Goal Programming Model

Goal programming is a linear mathematical model in which the optimum attainment of multiple goals is sought within the given decision environment. The objective function is composed of either a pair or a single deviational variable for each goal constraint. If overachievement is acceptable, the positive deviation (d_i^+) can be eliminated from the objective function. On the other hand, if underachievement is satisfactory, the negative deviation (d_i^-) should not be included. The exact achievement of a goal requires both negative and positive deviations to be represented in the objective function to achieve the ordinal solution.

An optimal solution is obtained when the sum of nonattainment of goals is minimized according to the priority structure established by the decision-maker. To achieve the goals according to their importance, GP provides a means by which the negative or positive deviations about the goal may be ranked according to an ordinal priority ranking scale in order of preference of each goal level. Weights are assigned to priority factors for minimizing the deviational variables. They are only assigned to deviational variables which have been assigned the same priority levels. The deviational variables and the ordinal priority factors are always present in each objective function.

A GP problem in standard format is

$$\text{Minimize } Z = d^- + d^+ \tag{5.105}$$

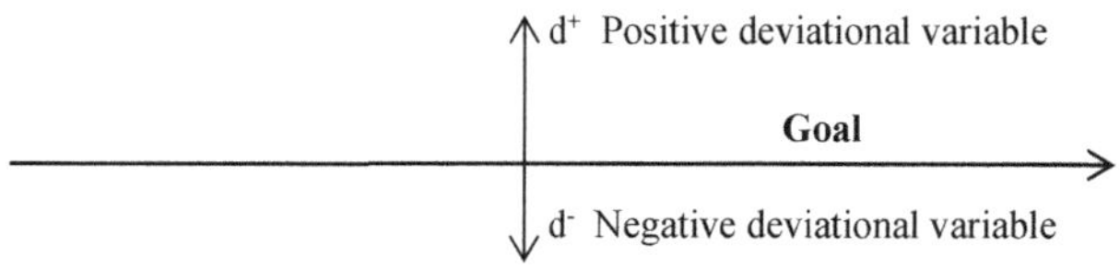

FIGURE 5.13

A two-dimensional goal space.

subject to

$$BX + d^- - d^+ = h \tag{5.106}$$

$$AX \leq b \tag{5.107}$$

$$X, d^-, d^+ \geq 0 \tag{5.108}$$

where B is a (1 × n) row vector of objective function coefficients, X is an (n × 1) column vector of real variables, b is an (m × 1) column vector of right-hand side constant, A is an (m × n) matrix of technological coefficients, d^+, d^- are the deviational variables in positive and negative directions, and h is the goal level set by the decision-maker. A goal constraint can be less than, greater than, or equal to type. Table 5.2 shows these three possibilities and how they are handled in GP formulations.

While solving the GP problems, the following points must be considered:

(i) In GP, the objective is to minimize the total nonattainment of goals. This is achieved by minimizing deviational variables through the use of preemptive priority factors and differential weights. There is no profit maximization or cost minimization per se in the objective function. Therefore, preemptive factors and differential weights take the place of C_j used in LP.

(ii) The objective function is expressed by assigning priority factors to certain variables. These preemptive priority factors are multidimensional, since they are ordinal rather than cardinal values.

(iii) The GP solution will allow some lower priority goals to go unsatisfied in order that higher priority goals, which may conflict with lower priority ones, achieve the targets. The ranking of deviational variables is the most important step in formulating a GP problem. The highest priority factor is assigned to deviational variables of the most important goal. The lowest priority factor is assigned to deviational variable of the least important goal. Thus, the low-order goals are considered only after higher order goals are achieved. The priority factors have the relationship of $P_j >>> P_{j+1}$ which means that P_j always takes priority over P_{j+1}.

The basic assumption in formulating the initial tableau of GP is identical to that of LP. One can assume the initial solution to be at the origin where values of all decision variables are zero. Considering linear problems, one can visualize situations of cost functions having different cost slopes for positive and negative deviations.

Table 5.2 GP model formulations and inclusion of the deviational variable in the objective function.

S.no.	Goal or constraint type	Processed goal or constraint	Deviational variable to be minimized in the objective function
1.	$f_i(X) \leq b_i$	$f_i(X) + d_i^- - d_i^+ = b_i$	d_i^+
2.	$f_i(X) \geq b_i$	$f_i(X) + d_i^- - d_i^+ = b_i$	d_i^-
3.	$f_i(X) = b_i$	$f_i(X) + d_i^- - d_i^+ = b_i$	$d_i^- + d_i^+$

Priority Ranking

After assigning the deviational variables, the next step is to assign the ordinal priority factors. The negative and/or positive deviations about the goal are ranked according to an ordinal priority ranking scale in order of preference of each goal level.

If deviations from the specified target have different cost slopes, then the objective function can be written as

$$\text{Minimize } z = \sum_{k=1}^{K} \left(w_k^+ d_k^+ + w_k^- d_k^-\right) \tag{5.109}$$

subject to a set of constraints. Here w^+ and w^- represent the cost slopes for positive and negative deviations.

Based on the priority levels, GP can be classified as:

(i) preemptive goal programming, and
(ii) nonpreemptive goal programming.

When there is a hierarchy of priority levels for goals in which one or more of the goals are far more important than others, it falls under preemptive GP. When dealing with goals on the same priority level, the approach is nonpreemptive GP. It is possible to specify the order in which objectives are to be satisfied in a lexicographic context in which case the weights are referred to as preemptive weights. In this case, the objective function is generally written as

$$Z = \left[p_1 h_1 \left(d^+, d^-\right), p_2 h_2 \left(d^+, d^-\right), \ldots, p_k\, h_k\left(d^+, d^-\right)\right] \tag{5.110}$$

One-sided goals: In this case, g_k for the kth goal represents the bound on that goal rather than a specific amount that should be attained if possible. If g_k is a lower bound goal, then

$$\sum_{j=1}^{n} c_{jk} x_j \geq g_k \tag{5.111}$$

In this case, any attainment over g_k is fine but any deviation below is to be avoided, if feasible. The change that this causes in the formulation of the objective function is that only negative deviations are incorporated into the objective function. Both types of deviations are still there in the constraints as before as both can still occur. Similarly, only positive deviations are incorporated in the objective functions in the case of upper bounds.

Noncommensurable Objectives

Suppose there are k objectives of the following type which cannot be combined into a single objective function:

$$z_1 = \sum_{j=1}^{n} c_{j1} x_j \tag{5.112}$$

The overall objective function for the model becomes

$$\text{Maximize } z = \text{Minimize } \left[z_1, z_2, \ldots, z_k\right] \tag{5.113}$$

An optimal solution to this problem is the one that makes the smallest z_k as large as possible. This model is not in the LP format. However, it is equivalent to the following LP model:

$$\text{Maximize } Z = z$$

subject to

$$\sum_{j=1}^{n} c_{j1}x_j - z \geq 0 \quad \text{for } k = 1, 2, \ldots., k \tag{5.114}$$

$$x_j \geq 0 \quad \text{for } j = 1, 2, \ldots, n$$

and any other constraints in the original model. The maximum feasible value of the new variable z in this model must equal the smallest $z_k = \Sigma\, c_{jk}\, x_j$, so an optimal solution for (x_1, x_2, …, x_n) will make this smallest z_k as large as possible.

When the objectives are to be minimized rather than maximized, the overall objective function for the original model would change to

$$\text{Min } Z = \max\{z_1, z_2, \ldots, z_k\} \tag{5.115}$$

and the corresponding LP model is

$$\text{Min } Z = z \tag{5.116}$$

$$\sum_{j=1}^{n} c_{jk}x_j - z \geq 0 \quad \text{for } k = 1, 2, \ldots, k \tag{5.117}$$

$$x_j \geq 0 \quad \text{for } j = 1, 2, \ldots, n \tag{5.118}$$

and any other constraints in the original model.

Example 5.5. Consider two objectives of maximizing food production in two different regions (x_1, x_2 are the numbers of projects to be undertaken):

$$\text{Objective 1: Max } z_1 = 2000\, x_1 \tag{5.119a}$$

$$\text{Objective 2: Max } z_2 = 3000\, x_2 \tag{5.119b}$$

subject to

$$x_2 \leq 4 \quad \text{Equipment constraint} \tag{5.120a}$$

$$x_1 + 2x_2 \leq 10 \quad \text{Experts constraint} \tag{5.120b}$$

$$60x_1 + 20x_2 \leq 300 \quad \text{Money constraint} \tag{5.120c}$$

$$x_1, x_2 \geq 0$$

Solution. The equivalent LP problem is

$$\text{Max } Z = z \tag{5.121}$$

subject to

$$2000x_1 - z \geq 0 \quad (5.122a)$$

$$3000x_2 - z \geq 0 \quad (5.122b)$$

$$x_2 \leq 4 \quad (5.122c)$$

$$x_1 + 2x_2 \leq 10 \quad (5.122d)$$

$$60x_1 + 20x_2 \leq 300 \quad (5.122e)$$

$$x_1 \geq 0, x_2 \geq 0, z \geq 0$$

The solution to this problem is

$$x_1 = 45/11, \text{so } z_1 = 8182.$$

$$x_2 = 30/11, \text{so } z_2 = 8182.$$

Hence, $z = 8182$.

Example 5.6. A farmer produces two crops: rice and wheat. The farmer has a production capacity of 40 ton of crops. Because of limited sale opportunity, he can sell a maximum of 24 tons of rice and 30 tons of wheat. The gross margin from the sale of 1 ton rice is 80 units and for wheat, it is 40 units. Find the optimal production.

Solution. Let the optimal production of rice and wheat be x_1 and x_2 tons. If the farmer had only the single goal of profit maximization, the decision problem could be easily formulated as an LP problem, as illustrated below:

$$\text{Maximize} \quad Z = 80x_1 + 40x_2 \quad (5.123)$$

subject to

$$x_1 + x_2 \leq 40 \quad (5.124a)$$

$$x_1 \leq 24 \quad (5.124b)$$

$$x_2 \leq 30 \quad (5.124c)$$

$$x_1, x_2 \geq 0 \quad (5.124d)$$

The optimum solution is $x_1 = 24$, $x_2 = 16$, and the total profit $Z = 2560$ units.
This profit maximization problem can also be solved by the GP approach:

$$\text{Minimize } Z = p_1\left(d_1^+ + d_2^+ + d_3^+\right) + p_2 d_4^- \quad (5.125)$$

subject to

$$x_1 + x_2 + d_1^- - d_1^+ = 40 \quad (5.126a)$$

$$x_1 + d_2^- - d_2^+ = 24 \quad (5.126b)$$

$$x_2 + d_3^- - d_3^+ = 30 \quad (5.126c)$$

$$80x_1 + 40x_2 + d_4^- - d_4^+ = 10000 \quad (5.126d)$$

The objective function of the GP formulation indicates that the highest priority (p_1) is assigned to the minimization of d_i^+. Since in the LP model the first three constraints are "less than or equal to" inequalities, the solution must be within the region that satisfies these three constraints. By assigning the highest preemptive priority factor to the minimization of positive deviations (d_i^+) in the first three constraints, the same restrictions are met. The second priority factor of GP is assigned to the minimization of d_4^-, that is, minimization of the underachievement of some high profit goal. Arbitrarily, a limit of 10,000 units is set, knowing that such a high profit will never be achieved. The minimization of underachievement of the profit goal will drive the values of x_1 and x_2, within the area of feasible solution, as close to 10,000 as possible. The solution of the GP problem is the same as the LP solution: $x_1 = 24$, $x_2 = 16$, and $Z = 2560$ units.

Example 5.7. Consider that the farmer in Example 5.6 has set the following goals as arranged in order of their importance:

1. He wants to avoid any underutilization of the production capacity.
2. He wants to sell his crop as much as possible. Since the gross margin from the sale of rice is twice the amount from wheat, he has twice as much desire to sell more rice as wheat.
3. The farmer wants to minimize the extra production from the farm.

How each of these goals can be incorporated in the model?

Solution. The consideration of each of these goals is discussed below.

Production capacity: Since the goal regarding extra production was given the lowest priority, it is quite possible that the production of both rice and wheat may be more than 40 tons. The production capacity restriction can be expressed as:

$$x_1 + x_2 + d_1^- - d_1^+ = 40 \tag{5.127}$$

where d_1^- and d_1^+ represent the under- and overutilization of the production capacity. Depending on the actual utilization, at least one of these variables will be zero.

Sales capacity: Due to the limited sale opportunity, the farmer can sell a maximum of 24 tons of rice and 30 tons of wheat. One can omit positive deviations from the constraints which can be written as:

$$x_1 + d_2^- = 24 \tag{5.128}$$

$$x_2 + d_3^- = 30 \tag{5.129}$$

where d_2^- and d_3^- represent the underachievement of sale goals for rice and wheat, respectively.

In addition to the variables and constraints stated above, the following preemptive priority factors are defined in order to pursue the stated goals:

p_1: The highest priority is assigned to minimize the underutilization of production capacity (i.e., d_1^-).
p_2: The second priority factor is assigned to minimizing the underachievement of sale goals (i.e., d_2^- and d_3^-).
However, the farmer also wants to assign differential weights (not preemptive but ordinary numerical weights) to the achievement of the sale goal according to the gross margin ratio between rice and wheat. Hence, the farmer assigns twice the weight to d_2^- as assigned to d_3^-.

p_3: The lowest priority factor is assigned to minimizing the extra production (i.e., d_1^+).

The objective function can now be stated as:

$$\text{Minimize } Z = p_1 d_1^- + 2p_2 d_2^- + p_2 d_3^- + p_3 d_1^+ \quad (5.130)$$

The objective is to minimize deviations from the goals. The value of the deviational variable associated with the highest preemptive priority must be minimized first to the fullest possible extent. When no further improvement is possible or desired at the highest goal, the next attempt is to minimize the value of the deviational variables associated with the next highest priority factor, and so forth. Therefore, the complete model is:

$$\text{Min } Z = p_1 d_1^- + 2p_2 d_2^- + p_2 d_3^- + p_3 d_1^+ \quad (5.131)$$

$$\begin{aligned} \text{subject to} \quad & x_1 + x_2 + d_1^- - d_1^+ = 40 \\ & x_1 + d_2^- = 24 \\ & x_2 + d_3^- = 30 \\ & x_1, x_2, d_1^-, d_2^-, d_3^-, d_1^+, d_2^+, d_3^+ \geq 0 \end{aligned}$$

Applications of GP

An important property of GP is its ability to handle multiple incompatible goals according to their importance. A GP model performs three types of analysis:

(1) It determines the degree of attainment of defined goals with given resources;
(2) It determines the input requirements to achieve a set of goals; and
(3) It provides the optimum solution under the varying input and goal structures.

The biggest advantage of GP is its flexibility, which allows model simulation with numerous variations of constraints and goal priorities. GP can be applied to almost unlimited managerial and administrative decision areas. Allocation planning and scheduling, and policy analysis are the most readily applicable areas of GP. Some references for further study are Ignizio (1976), Lee (1972), and Loganathan and Bhattacharya (1990).

5.9 SIMULATION

Simulation is the process of duplicating the behavior of an existing or proposed system. It consists of designing a model of the system and conducting experiments with this model either for better understanding of the functioning of the system or for evaluating various strategies for its management. The essence of simulation is to reproduce the behavior of the system in every important aspect to learn how the system will respond to conditions that may be imposed on it or that may occur in the future. The main advantage of simulation models lies in their ability to accurately describe the reality. If a simulation model can be developed and is shown to represent a prototype system, it can provide insights into how the real system might perform over time under varying conditions. Thus, proposed configurations of projects can be evaluated to judge whether their performance would be adequate or not before investments are made. In a like manner, operating policies can be tested before they are implemented in actual control situations. Hufmschmidt and Fiering (1966) describe the simulation technique for the design of water resources systems. James and Lee (1971) have noted that simulation is the most powerful tool to study complex systems.

Usually, the structure or behavior of the system being simulated is so complex that its analytical expression is not possible. A simulation model of a water resource system duplicates its operation with a defined operational policy, using the parameters of physical and control structures, time series of flows, demands, the variables describing water quality, etc. The evaluation of design parameters or operation policy is through the objective function (flow or demand related measures or economic indices) or some measure of reliability. Since simulation models do not use an explicit analytical procedure for the determination of the best combination of controlling variables, it is necessary to proceed by trial and error or follow a strategy of parameter sampling.

Since models are abstractions of reality, they usually do not describe all the features that are encompassed by a real-world situation. Only those aspects of the system that are relevant to the objective of the study are modeled so that solution is obtained at a reasonable cost and within a prescribed time frame. If the simulation model has to reproduce all the complexities of the prototype, it will be as complex as the prototype. Therefore, the model builder should attempt to model the detailed functioning of individual components to the extent necessary so as to meet the overall accuracy requirements, while not making it unnecessarily complicated. To illustrate, if the objective is the design of a large storage reservoir for irrigation and municipal water supply, it is quite unnecessary to model the complete runoff process. On the other hand, a monthly flow-generation model is entirely unsuited for modeling peak discharges. An important aspect of model building in the context of simulation is to find the best permissible simplifications. When, for example, should the engineer responsible to issue flood forecasts use a simple routing model and when he should employ a dynamic wave model, using the complete St. Venant equations? The difference in efforts and computer time for the two methods is very large. The main reasons for searching for a simple model may be a lack or low quality of data. For example, consider that there are only a few rainfall and discharge data stations in a large catchment. In this scenario, there is no justification to set-up a detailed model which requires huge data, long time to calibrate and run, and skilled manpower.

The main advantage of simulation models lies in their ability to closely describe the reality. If a simulation model can be developed and is shown to represent the prototype system realistically, it can provide insight about how the real system might perform over time under varying conditions. Thus, proposed configurations of projects can be evaluated to judge whether their performance would be adequate or not before investments are made. In a like manner, operating policies can be tested before they are implemented in actual control situations. Simulation is widely believed to be the most powerful tool to study complex systems.

5.9.1 Classification of Simulation Models

Simulation models may be physical (a scale model of a spillway operated in a hydraulics laboratory), analog (a system of electrical components, resistors and capacitors, arranged to act as analog of pipe resistances and storage elements), or mathematical (a compilation of equations and logical statements that represent the actions of a system's elements). Mathematical simulation models are very useful and popular in the field of water resources. The platform that is used to operate models of this type is the digital computer. Only simulation models of this type are discussed here.

Simulation models can also be classified as static or dynamic. Dynamic models take into account the changing parameters of the system (structures and facilities) and the variations in their operation. The simulation model of a water resource system is considered a dynamic model if the operational policy can be dynamically changed with time and if such changes correspond to the system demands and related changes in system parameters. These are assumed as fixed in static models. The development and application of dynamic models is a more involved exercise and often static models give acceptable results.

Many hydrological variables are stochastic in character. Deterministic and stochastic simulation models are distinguished by the way this stochasticity is accounted for. A time-series of gauged flows represents a sample of the stochastic process. Under certain conditions, deterministic simulation models can be used with confidence. For example, if measured monthly flows for a period of 40 years are used as input in a study, a deterministic model may be adequate. If the process is stationary, the sample can be considered a reasonably good characterization of the stochastic process. Note that as the inflow series will not repeat exactly, the future performance of the system will differ from that obtained by the model.

Two principal methods are used to account for stochastic properties in the simulation model:

- The synthetic flows generated by methods of stochastic hydrology are used as inputs, or
- The simulation model is combined with other models that permit a stochastic solution (e.g., the chance-constrained model).

Besides the above, the degree of aggregation can be used to classify simulation models. A model with a high level of details is suitable to investigate the operation of an existing water resource system. The typical objective of applying such a model is the improvement of the system operation. On the contrary, simulation models with much more aggregated data are appropriate for the design of water resource systems. Jacoby and Loucks (1972) were among the first to use this strategy to investigate a system in combination with analytical optimization models. Aggregation simplifies modeling but should not introduce significant deviations from reality. Simplification can also be achieved by neglecting variables that do not impart a decisive effect on the system behavior. If the output is not sensitive to the variation of certain variables, these can be considered as constant.

5.9.2 Monte Carlo Simulation

Design of real-world systems is generally based on observed historical data. For example, the observed streamflow data are used in sizing a reservoir, historical traffic data are used in the design of highways, observed data are used in the design of customer services, etc. However, frequently the historical records are not long enough and the observed pattern of data is not likely to repeat exactly. The performance of a system critically depends on the extreme values of input variables and the historical data may not contain the entire range of input variables. There are many instances when a flood with peak value exceeding the historical records entered a reservoir.

An important conclusion from the above is that one does not get a complete picture of the system performance and risks involved when historical data are used in evaluation. Thus, for instance, the planner cannot determine the risk of a water supply system failing to meet the demands during its economic life, because this requires a very large sample of data which are not commonly available.

For many systems, some or all inputs are random, system parameters are random, initial conditions may be random, and boundary condition(s) may also be random in nature. The probabilistic properties of these are known. For such systems, simulation experiments are conducted using a set of inputs which are synthetically (artificially) generated. While generating inputs, it is ensured that the statistical properties of the random variables are preserved. Each simulation experiment with a particular set of inputs gives an answer. When many such experiments are conducted with different sets of inputs, a set of answers is obtained. These are statistically analyzed to understand or forecast the behavior of the system. This approach is known as Monte Carlo simulation, and using it, planners get better insights into the working of the system and can determine the risk of failures, for example, chances of a reservoir running dry or a customer service center failing to provide services within promised time.

The main advantages of Monte Carlo simulation are that the system, its inputs, outputs, and parameters can be easily described. All the critical parameters of the system can be included in its description. The other advantages include saving in time and expenses. It is important to remember that the synthetically generated data are no substitute of the actual observed data, but this is a useful pragmatic tool which allows the analyst to extract detailed information from the available data.

Random number generation is an important part of Monte Carlo analysis. In the early days of mathematical simulation, mechanical means were employed to generate random numbers. The techniques that were used to generate random numbers were drawing cards from a pack, drawing numbered balls from a vessel, reading numbers from a telephone directory, etc. Printed tables of random numbers were also in use for quite some time. The Monte Carlo techniques have got this name because roulette wheels similar to those in use at Monte Carlo were used to generate random numbers. The current approach is to use a computer to generate random numbers and this is discussed in Appendix 5A.

5.9.3 Time Management in Simulation

Modeling of a continuous process by a discrete model requires the assumption that the continuous changes during a defined period take place instantaneously at the end or at the beginning of the period. The decision-making process in water resources systems is discrete; simulation models are also discrete models. The real-life process, however, is continuous. Therefore, the time step size is an important aspect of the model and should be chosen carefully. This choice depends on the degree of aggregation and the time variability of inputs.

Event scanning and periodic scanning are two common ways of time management in simulation models. In the event scanning approach, the clock is advanced by the amount of time which is required for the occurrence of the next event. In many natural phenomena, the periods of high activity are separated by long periods in which the system lies inactive. This approach is suitable for these types of situations and can save computational time. Note that it requires some scheme to determine the time when the events take place.

In periodic or fixed time scanning, the whole simulation horizon is divided into smaller time periods. The system time is incremented by the predetermined step and simulation is performed. This procedure is repeated till the end of the period of analysis. A judicious choice of time increment is necessary in the periodic scanning approach. This increment should be small enough so that no significant information is lost. The fixed-time scanning approach is commonly used in water resources problems.

A simulation model of a water resources system commonly mimics the behavior of the system in discrete time steps using arithmetical and logical procedures (algorithms) by a series of "snapshots." These algorithms try to reproduce the way inputs to the system are acted upon and transformed into outputs. The size of this time step depends on the purpose of the study. In planning studies dealing with irrigation, hydroelectric power generation, water supply purposes, minimum flow maintenance, etc., monthly or 10-day periods are generally used. These periods make it possible to reflect the seasonal variability of demand and hydrological inputs. During floods, however, the condition of a surface water system changes rapidly and, therefore, weekly or 10-day time steps are too long to give any meaningful result. Hence, for the purpose of flood-related simulation studies, multihourly or shorter time steps are used.

5.9.4 Design of Sampling Strategy

Sampling strategy is a combination of methods whose basic object is to find the nature of the response surface in a problem and the location of extreme points. A good strategy should minimize the computational efforts and save time. The first few samples are exploratory with the aim to know the nature of the response surface and to find out the kind of variation it has—whether it is smooth or there are jagged edges. These data help in the identification of the region of response surface which requires further detailed examination. The range of decision variables over which further experiments need to be conducted is also identified. Next, extensive sampling is carried out in the identified region to get the desired information. The sampling methods are diagrammatically shown in Fig. 5.14.

The basic sampling strategies when using simulation models are: systematic sampling, random sampling, and the gradient methods (e.g., the steepest ascent method). In systematic sampling, a survey of the response surface is carried out at points systematically distributed over the range of input variables. The method of uniform grid is suitable when a large number of alternatives are to be examined. In this method, a grid of uniform spacing is drawn in the feasible region and sampling is carried out at various nodes of the grid. The size of the grid is decided keeping in view the number and range of variables. The objective function will have to be evaluated at many points if the grid size is small. In case a large grid size is chosen, the chances of missing the extreme points is higher, particularly if the response surface is jagged.

An alternative sampling scheme which is known as a single factor method involves experiments in which one variable is changed at a time. The process is repeated and each variable is evaluated over its range. This strategy is suitable when the variables involved are independent, but it requires the objective function evaluation at a large number of points to gather desired information about the response surface. An improvement of this method consists of changing two or three variables at a time to find out the behavior of the response surface and further experiments are directed in the direction of

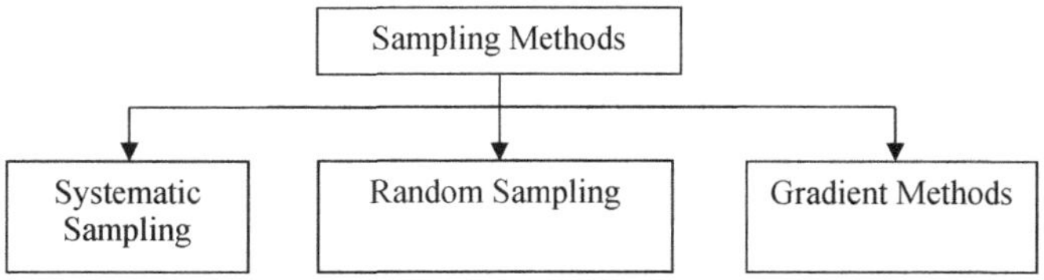

FIGURE 5.14

Sampling methods used in simulation.

the desired change. If the number of variables is small and the variables that have a major influence on the objective function can be easily identified, this method is useful. As the number of combinations in systematic sampling could be large, a combination of this sampling with heuristic sampling is often used. Obviously inefficient combinations of parameters are excluded in advance. This method is effective in situations where the system behavior can be guessed beforehand.

The random sampling method is suitable at the beginning of computation when the system behavior is completely unknown. This strategy is helpful in preliminary identification of the zone of response surface which requires further refinement. Some points are chosen at random from a systematic grid and the objective function is computed at these points. Note that as the number of sampling points increases, better information of response surface is obtained and the chances of finding the extreme points also increase.

In a detailed investigation of the behavior of a system and its response to input parameters, the method of steepest ascent is often used. By small variations of parameters, the direction of steepest ascent on the response surface (i.e., the surface formed by the values of the objective function in n-dimensional space) is identified. The parameters are changed in the direction of the steepest ascent and a new value of the objective function is computed. The algorithm is repeated until the region of the optimum is reached.

5.9.5 Steps in Simulation Modeling

The following are the steps in development and application of a simulation model:

(a) Define the problem;
(b) Describe the water resources system and its hydrological relationships;
(c) Decide the model structure, input, and output;
(d) Design the model;
(e) Test the model, if it is not suitable, go to step (c); and
(f) Apply the model to the problem.

After the model of a system is developed, experiments are conducted with it to verify the analytical results or to answer the question "*WHAT IF?*" The simulation models are much helpful in understanding the consequences or implications of changing one or more of the decision variables.

In the application of simulation models, it is often advantageous to subdivide a large problem into subsystems by any one of the following methods:

1. The first approach is based on the flow of water particles. Subsystems are defined on the basis of the origin of flow or its modification. If some groups of elements show little relationships to other elements but the interactions inside these groups are relatively strong and numerous, a subsystem consisting of this group of elements can be formulated. For example, the delineation of a basin in various subbasins is based on the flow paths of water.
2. The second method of identifying subsystems is the functional approach. This method is adopted if the flow approach is not suitable due to typical needs and requirements of the system functions. For example, consider the simulation of operation of a multipurpose reservoir for conservation and flood control purposes. The length of the time period for conservation simulation is usually a month or a week. No flood routing calculations are performed when such a long time step is used.

However, to simulate flood control regulation, a multihourly time step is used. A short-term operation has some unique characteristics. Flood routing is very important and the model may also have a component for forecasting.

5.9.6 Inputs to Simulation Models

The requirement of input data depends on the objective of using a simulation model. Broadly, there are two main types of input data for the simulation model: (1) the variables describing the system, its configuration, and parameters of structures, for example, reservoir capacities, characteristics of outlets, and aquifer properties; and (2) time series of flows, demands of the system (either existing or projected), evaporation depths, operation policy, etc. The observed hydrological long time-series data will require preprocessing before it can be used in the model. For instance, the model may require the average precipitation in the catchment and this will require the determination of station weights if the Thiessen polygon method is used. The periods of observation often do not coincide for all the stations relevant to the system, and the stations may not be located at points where the data are needed by the model. If the records are interrupted and the observation periods are shorter in some stations, the records should be completed by filling-in the missing values. The procedures for processing of hydrometeorological data are described in Chapter 2 as well as in many books such as Singh (1992).

The parameters of the system include the storage capacities of reservoirs, carrying capacities of canals, river, the acceptable minimum releases from reservoirs, diversion of flows within the basin or interbasin diversion, characteristics of aquifer, requirements for water quality, etc. The input data also include the requirements at the demand centers and diversion points.

The economic inputs include costs of storage, irrigation diversion, power plants, recreation facilities, etc. The economic output values include costs of operation, maintenance and replacement of facilities, benefits associated with water supply for municipal, industrial, agricultural use, hydropower, recreation, reduction of flood damage, and low-flow augmentation. The broad data requirements for analysis of water resources systems have been discussed in Chapters 2 and 9.

5.9.7 Outputs of Simulation Models

The output of a simulation model can be in terms of hydrological and/or economic variables. The hydrologic results comprise values of various design variables, operation policies, working tables, hydrographs at important locations, etc. The design variables are, for example, the optimum storage capacity of the reservoir. Typical hydrological variables for a surface water project comprise, for various time periods, the reservoir storage, inflows, demands, release, spill, power generated, and deficit in release for various purposes. Reliability indices are evaluated in terms of volume or time deficits.

Display or print of input data in the initial model runs helps in detecting the input errors. A good simulation model should allow the user to control how much detailed output he wants. The user should carefully decide what variables and how much detailed results are needed; unnecessary and long tables are difficult to study and the useful information may remain unnoticed. The interpretation of results is simpler if these are also available graphically.

Water resources systems at certain periods fail to meet target outputs. Therefore, the simulation model outputs should include the deficits for various targets, energy deficits; reservoir level fluctuations; and the duration, magnitude, and total volume of the deficits. Reliability indices are helpful in assessing the system performance. The economic consequences of these deficits are evaluated by an economic loss function. The economic output data typically include the time-series of benefits and costs, cash flows, and benefit/cost ratio.

The interpretation of output is an important step in the analysis. A careful browse is necessary for a proper choice of variables to be changed for the next step of computation. The sensitivity of system performance to various decision variables can be ascertained through a few model runs. To verify the results of simulation models, "common sense" based on prior experience is often used. If the results of simulation model fall outside the expected limits, detailed print out containing, for example, the reservoir contents at the end of each month, releases from reservoirs, flow at important points of the system, can be taken. This will help locate possible errors, if any.

5.9.8 Pros and Cons of Simulation Models

Simulation and other types of models can be valuable aids in decision-making, but their advantages and disadvantages must be weighed for the circumstances of concern. Some advantages of the use of simulation models include:

(a) They impose a logic and structure to analyses;
(b) They provide insights into system behavior;
(c) Their structure is ideally suited to experimental work;
(d) They may be designed to accommodate many options;
(e) Projections into the future are facilitated with their use; and
(f) They can aid in communications between analysts and policy makers.

Some common problems associated with simulation models are:

(a) Some simulation models are data hungry and their data requirements cannot be easily met;
(b) The handling of intangibles is difficult in simulation models;
(c) There are high costs of development and/or use of simulation models, particularly the complex ones;
(d) Trained and experienced man-power is needed to interpret their output; and
(e) The users may not always be ready to accept the models and their results.

5.10 CLOSURE

Since optimization and simulation models are used for a common end, many times the analyst is in a dilemma as to which type of model to use. Simulation models are effective tools for problems, such as evaluating the performance of various configurations of reservoir and hydroelectric power plant capacities, water use allocation targets, operating policies, and the like. But they are not a very effective means for choosing or defining the best configurations or combinations of capacities, targets, and policies. For these problems, optimization models have proven to be effective, at least, for eliminating the worst solutions from further consideration.

Due to several reasons, optimization models cannot determine the exact optimal solution to water resources management problems. The first reason is that their solution algorithms often require some simplifying assumptions which may not be realistic and may be limiting. Nonlinear cost, benefit and loss functions, and nonlinear expressions required for defining evaporation losses, hydroelectric energy production, and some combinations of flood control alternatives are often approximated by piecewise linear functions. The analyst should be aware of the implications of these limitations.

The modeling of even relatively small surface water systems may require a large number of mathematical statements, constraints, and variables. Even though very high-capability desktop computers are easily available nowadays, software, trained man power, and finances may sometimes be a limiting factor. It is often necessary to make simplifying assumptions to reduce the size of the problem to manageable limits.

A third and perhaps the most limiting aspect of optimization models is the conceptual difficulty associated with the quantification and specification of a criterion for evaluating each possible management alternative. Public policy objectives are a mixture of monetary and nonmonetary goals which make quantification difficult. Furthermore, during the preliminary planning phase of water resources projects, the goals and objectives are not always clear and there may not be an agreement among decision-makers on what these goals should be and the extent to which they are to be satisfied.

There are other important limitations related to the quantification of hydrologic, technologic and economic uncertainties, and inaccurate/incomplete data. All of these mathematical, computational, conceptual, and data limitations restrict the use of optimization models to preliminary screening. Those alternative investments and operating policies that survive the preliminary screening process should be further analyzed, evaluated, and improved using simulation or other techniques where appropriate. While simulation models share many of the same conceptual and data limitations, they are far less restrictive mathematically and computationally. Hence, they are usually better suited for evaluating more precisely the alternatives defined by the optimization or preliminary screening models.

To take full advantage of the strengths of systems techniques, it is necessary that due care and attention are paid to problem formulation. The objective(s) and constraints should be carefully designed. Usually, great effort of the analyst goes in to reduce the system to a manageable representation without compromising its essential features and relationships. After listing all the alternatives in the beginning, those that are clearly inferior may be discarded so that the potential and competing solutions can be examined in detail. Sensitivity analysis gives further insight into the solution; it also highlights the variables which should be carefully monitored.

Systems analysis may be applied for structuring a water resources project. To that end, a block diagram of the system is drawn and the elements are connected by logical statements. In this form of representation, it is easier to see how different components interact within the system and with its environment. By isolating subsystems, their performance can be tested and analyzed separately.

APPENDIX 5A

5A.1 Generation of Random Numbers

Most modern compilers have built-in routines to generate uniformly distributed random numbers between 0 and 1. A number of arithmetic techniques are used for this purpose, such as midsquare method, the congruence method, the composite generators, etc. The most popular of these is the congruence method. The random number generators have a mathematical expression that is used as a recursive equation to generate numbers. To start the process, a number known as "seed" is input to the equation which gives a random number. This number is input to the equation to generate another number and so on. When this process is repeated n times, n random numbers are obtained. The recursive equation used in the *congruence method* is:

$$R_i = (aR_{i-1} + b)\ (\text{modulo d}) \tag{5A.1}$$

where R_i are the random variables; and a, b, and d are positive integer constants which depend upon the properties of the computer. The word "modulo" denotes that the variable to the left of this word is divided by the variable to the right (in this case d) and the remainder is assigned the value R_i. The initial value of the random variable (R_0) in Eq. (5A.1) is called the seed. The properties of the generated numbers depend on the values of constants a, b, and d, their relationships, and the computer used. The value of constant a needs to be sufficiently high; low values may not give good results. Constants b and d should not have any common factors. In computer generation, the sequence of random numbers is repeated after a certain lag and it is desirable that the length of this cycle should be more than the numbers that are needed for the study. This lag increases as d increases and therefore a large value of d should be chosen. Normally, d is set equal to the word length (the number of bits retained as a unit) of the computer; a typical value being $2^{31} - 1$. This technique of random number generation is "deterministic," that is, the generated numbers can be duplicated again. Therefore, these numbers are called *pseudorandom numbers*. The generated random numbers should be tested to ensure that these are not serially correlated and are uniformly distributed.

Example 5A.1. Generate uniformly distributed random numbers using Eq. (5A.1) with a = 5, b = 3, and d = 7. The seed R_0 can be assigned a value of unity.

Solution. Eq. (5A.1) is rewritten as $R_i = (5 * R_{i-1} + 3)$ (modulo 7)

i	R_{i-1}	$(5 * R_{i-1} + 3)$ (modulo 7)	R_{i-1}
1	1.00	(5 * 1.00 + 3) mod 7 = 8.00 mod 7	0.14
2	0.14	(5 * 0.14 + 3) mod 7 = 3.70 mod 7	0.53
3	0.53	(5 * 0.53 + 3) mod 7 = 5.65 mod 7	0.81
4	0.81	(5 * 0.81 + 3) mod 7 = 7.05 mod 7	0.01
5	0.01	(5 * 0.01 + 3) mod 7 = 3.05 mod 7	0.43
6	0.43	(5 * 0.43 + 3) mod 7 = 5.15 mod 7	0.73
7	0.73	(5 * 0.73 + 3) mod 7 = 6.65 mod 7	0.95
8	0.95	(5 * 0.95 + 3) mod 7 = 7.75 mod 7	0.11
9	0.11	(5 * 0.11 + 3) mod 7 = 3.55 mod 7	0.51

It may be noted that the numbers are (nearly) repeating after a cycle of 7.

5A.2 Transformation of Random Numbers

The input to the prototype system will have a certain probability distribution. The input random variables in Monte Carlo simulation should follow the same probability distribution. Therefore, the uniformly distributed random numbers are converted to follow the desired probability distribution. The variables involved may either be continuous or discrete random variables.

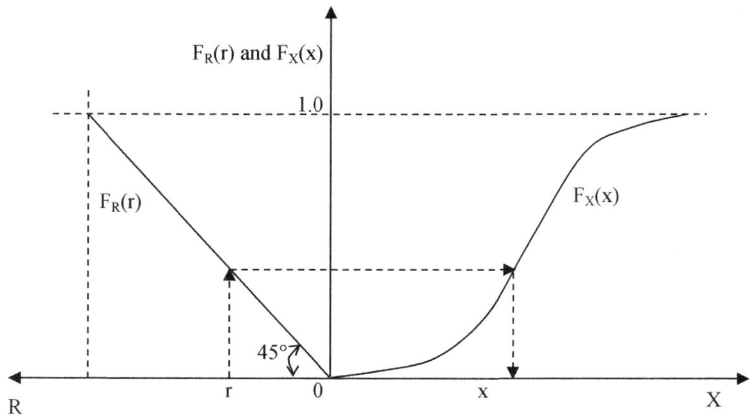

FIGURE 5A.1

Determination of random number x with desired distribution from uniformly distributed random number r.

If the inverse form of a distribution can be expressed analytically, the inverse transformation is the simplest method to generate random variables that follow a given (Vrugt, et al., 2003) distribution. In this method, first a uniformly distributed random number r_i in the range [0,1] is generated. If $F_Q(q)$ is the desired cumulative distribution function of random variable Q, then Q can be generated as

$$Q = F_{Q^{-1}}[r] \tag{5A.2}$$

where F_Q^{-1} is the inverse of the cumulative distribution function of random variable Q. This is a simple and computationally efficient method, graphically illustrated in Fig. 5.15.

Example 5A.2. Generate exponentially distributed random numbers with parameter $\lambda = 2.3$.

Solution. The cumulative distribution function of an exponential distribution is

$$F_X(x) = 1 - e^{-\lambda_x}$$

Its inverse can be written as

$$x = F_{X^{-1}}[r] = \ln(1 - r)/\lambda \tag{5A.3}$$

Since $(1 - r)$ is uniformly distributed, this can be replaced by r which is also uniformly distributed. Hence, exponentially distributed random numbers with the desired property can be generated by

$$x = -\ln(r)/2.3$$

If the first uniformly distributed random number $r_1 = 0.89$, the corresponding number x_1 will be

$$x_1 = -\ln(0.89)/2.3 = 0.05067.$$

The other common methods to generate random variables are the composition method and the function-based method.

References

Bellman, R.E., Dreyfus, S., 1962. Applied Dynamic Programming. Princeton University Press, Princeton, New Jersey.

Buras, N., 1966. Dynamic programming in water resources development. In: Chow, V.T. (Ed.), Advances in Hydroscience, vol 3. Academy Press, New York.

Cohon, J.L., Marks, D.H., 1975. A review and evaluation of multi-objective programming techniques. Water Resources Research 11 (2), 208–220.

Dandy, G.C., Simpson, A.R., Murphy, L.J., 1996. An improved genetic algorithm for pipe network optimization. Water Resources Research 32 (2), 449–458.

Dantzig, G.B., Thapa, M.N., 1997. Linear Programming 1: Introduction. Springer-Verlag, New York.

Duan, Q., Soroosh, S., Gupta, V., et al., 1992. Effective and efficient global optimization for conceptual rainfall-runoff models. Water Resources Research 28 (4), 1015–1031.

Goldberg, D.E., 1989. Genetic Algorithms in Search, Optimization and Machine Learning. Addison-Wesley, Reading, Mass.

Goulter, I.C., Tai, F.-K., 1985. Practical implications in the use of stochastic dynamic programming for reservoir operation. Water Resources Bulletin 21 (1), 65–74.

Haimes, Y.Y., Hall, W.A., 1974. Multiobjectives in water resources system analysis: the surrogate worth trade-off method. Water Resources Research 10 (4), 615–623.

Heidari, M., Chow, V.T., Koktovic, P.V., Meredith, D.D., 1971. Discrete differential dynamic programming approach to water resources systems optimization. Water Resources Research 7 (2), 273–282.

Himmelblau, D.M., 1972. Applied Nonlinear Programming. McGraw Hill Book Company, New York.

Hufmschmidt, M.M., Fiering, M.B., 1966. Simulation Techniques for Design of Water Resources Systems. Harvard University Press, Cambridge, Massachusetts.

Ignizio, J.P., 1976. Goal Programming and Extensions. D C Death and Company, Lexington Books.

Jacoby, H.D., Loucks, D.P., 1972. Combined use of optimization and simulation models in river basin planning. Water Resources Research 8 (6), 1401–1414.

James, L.D., Lee, R.R., 1971. Economics of Water Resources Planning. McGraw-Hill Book Co., New York.

Jensen, P.A., Barnes, J.W., 1980. Network Flow Programming. McGraw-Hill Inc., New York.

Karmarkar, N., 1984. A new polynomial-time algorithm for linear programming. Combinatorica 4, 373–395.

Larson, R.E., 1968. State Increment Dynamic Programming. Elsevier, New York.

Lee, S.M., 1972. Goal Programming for Decision Analysis. Auerbach Publishers Inc., Philadelphia.

Loganathan, G.V., Bhattacharya, D., 1990. Goal programming techniques for optimal reservoir operations. Journal of Water Resources Planning and Management, ASCE 116 (6), 820–838.

Loucks, D.P., Dorfman, P., 1975. An evaluation of some linear decision rules in chance constraint models for reservoir planning and operation. Water Resources Research 11 (6), 777–782.

Major, D.C., 1977. Multiobjective Water Resources Planning, Water Resources Monograph No. 4. American Geophysical Union, Washington, DC.

Marquardt, D.M., 1963. An algorithm for least squares estimation of non-linear parameters. Journal of the Society for Industrial and Applied Mathematics 11, 431–441.

Martin, Q.W., 1981. Surface Water Resources Allocation Model, AL-V, Program Documentation and Users Manual. Texas Department of Water Resources, USA.

Pedrycz, W., 1993. Fuzzy Logic and Fuzzy Systems. John Wiley, New York.

Rao, S.S., 2009. Optimization, Theory and Practice. Wiley Eastern, New Delhi.

Reklaitis, G.V., Ravindran, A., Ragsdell, K.M., 1983. Engineering Optimization – Methods and Applications. John Wiley & Sons, New York.

ReVelle, C., Jores, E., Kirkby, W., 1969. Linear decision rule in reservoir management and design 1: development of stochastic model. Water Resources Research 5 (4), 767–777.

ReVelle, C., Whitlatch, E.E., Wright, J.R., 1997. Civil and Environmental Systems Engineering. Prentice Hall, New Jersey.

Russell, S.O., Campbell, P.F., 1996. Reservoir operating rules with fuzzy programming. Water Resources Research 122 (3), 165–170.

Singh, V.P., 1992. Elementary Hydrology. Prentice-Hall, New Jersey.

Taha, H.A., 2017. Operations Research – an Introduction. Pearson Education Limited, Harlow, England.

Vrugt, J.A., Gupta, H.V., Bouten, W., Sorooshian, S., et al., 2003. A Shuffled Complex Evolution Metropolis algorithm for optimization and uncertainty assessment of hydrologic model parameters. Water Resources Research 39 (8), 1201. https://doi.org/10.1029/2002WR001642.

Wardlaw, R., Sharif, M., 1999. Evaluation of genetic algorithms for optimal reservoir system operation. Water Resources Research 125 (1), 26–33.

Yeh, W.W.-G., Becker, L., 1982. Multi-objective analysis of multireservoir operations. Water Resources Research 18 (5), 1326–1336.

Young, G.K., 1967. Finding reservoir operation rules. Journal of Hydraulics Division, ASCE 93 (HY6), 297–321.

Further Reading

Fletcher, R., 1980. Practical Methods of Optimization, vol 1. John Wiley & Sons, New York.

Neufville, R. de, Stafford, J.H., 1971. Systems Analysis for Engineers and Managers. McGraw-Hill Book Company, New York.

ReVelle, C., 1999. Optimizing Reservoir Resources. John Wiley & Sons Inc., New York.

Votruba, L., Kos, Z., Nachazel, K., Patera, A., Zeman, V., 1988. Analysis of Water Resources Systems. Elsevier, Amsterdam.

CHAPTER

6 Economic Considerations in Water Resources Management

Chapter outline

Water Resources Systems Planning and Management. https://doi.org/10.1016/B978-0-12-821349-0.00020-4

The objectives of this chapter are:

- to explain the basic economic concepts,
- to present the discounting techniques that are commonly used,
- to discuss the optimal allocation of water from the point of view of economics, and
- to describe the procedure of allocation of cost in a multipurpose project.

Water is a renewable resource whose supply is uncertain in nature and depends on space and time. Water is a mobile resource whose availability at one place is affected by its use at other places. For this reason, the problems related to water supply are site-specific. When its availability is less, droughts occur, which adversely affect agriculture, municipal life, and cattle. An excess availability of water leads to flooding, which also has high damage potential. Water is also a very good solvent, and this property makes it vulnerable to pollution by a number of pollutants that adversely affect its economic value. Highly polluted water is a liability rather than an asset.

Rarely, water is completely consumed while being used to meet the requisite demands. A fraction of the water withdrawn (mostly of degraded quality) comes back to the source as a return flow. Affected by this phenomenon are the users in the downstream reaches of the sources because the quantity, quality, and timing of water available to them are impacted by the upstream usage. These interdependencies are known as externalities.

Due to the site-specific nature of water supply and demand, the problems of water resources management are also site-specific. The projects which are constructed to utilize raw water resources are huge and have important impacts on the regional economy and environment. These projects provide significant employment opportunities and also support ancillary activities. Another feature of water management is that the projects display *economy of scale*, that is, the cost per unit of water decreases as the project size increases.

Water is a unique economic commodity. It has a high economic value, yet it is available freely. In an extensive study in the United States, Frederick et al. (1996) found that the value of water is the highest in the drier, water-scarce regions (USA $191 per acre foot) and is the lowest in humid regions ($7 for Great Lakes region). Regarding use, the water value is the highest for industrial processing ($282) and domestic uses ($194), and the lowest for recreation and waste disposal ($3).

Water was also the central subject of an interesting enigma in economics that came to be known as the *water–diamond paradox*. This paradox can be stated as follows. Although water is necessary for human survival and consequently has higher value in terms of use, its price is very low. In contrast, diamonds are not at all essential for human survival but their value is very high. This paradox could be explained when the marginal and total values of these two commodities are considered. It is well known that the total utility of water exceeds the utility of diamonds. However, the marginal utility of diamonds exceeds the marginal utility of water because diamonds are scarce and water is much more easily available. This scarcity-based higher marginal utility of diamonds is the root cause of the much greater price of diamonds than that of water.

After a water resources development project is made operational, it produces a time pattern of economic consequences. As large sums of money and other resources are involved, it is necessary that the consequences must be carefully predicted, evaluated, and compared before the project is taken up for execution. For example, consider that a dam is to be constructed in a basin to provide water for irrigation, municipal and industrial use, generate hydropower, and control floods. The reservoir may also submerge villages and agricultural and forest areas, and people may have to be rehabilitated and resettled. Different locations and different heights of the dam will produce different patterns of desirable and undesirable effects. All the benefits and costs must be suitably considered to determine the best project configuration.

An improvement in the quality of life and enhancement of public welfare through accelerated economic development is the most common goal of national governments. The impact of a project on economic welfare can be evaluated in terms of the relative values of associated benefits and costs. If benefits are more than costs, it implies that the nation would be better off by taking up the project. On the other hand, an excess of costs over benefits indicates wastage of money. In this case, compensating the affected people by providing relief measures or rehabilitating them might be a better policy. Of course, it is also necessary to consider factors like environmental sustainability, employment generation, and national security. Logic would guide that usually a project should be taken up only if its benefits exceed costs.

6.1 BASIC PRINCIPLES OF PROJECT ECONOMICS

Engineering economics is concerned with applying economic criteria to select the best alternative from a group of feasible engineering designs or to evolve the best economic policy for planning, operation, or management of an engineering project. The principles of engineering economics guide the ranking of alternatives so that they may be compared to determine which alternative should be selected. The project evaluation process requires the prediction of consequences expected to result from picking an alternative, estimation of the magnitude of each consequence, and converting the consequences into commensurable units to facilitate comparison.

An important economic concept is that of opportunity costs. It refers to the benefits foregone when a resource is used for one purpose instead of its next best alternative use. Another concept is marginalism. This emphasizes the importance of considering incremental gains related to incremental costs. The law of diminishing returns implies that from the producers' point of view, an increase in the use of a given input, when all other inputs are held constant, leads to decreasing increments of products. The responsiveness of demand to changes in prices is measured in terms of the elasticity of demand. This term is defined as the percentage by which the quantity taken changes in response to a 1% change in the variable. The "price elasticity of demand" for water is a measure of the willingness of the consumers to reduce water use due to a rise in price or the tendency to use more if the price reduces.

Some important frequently used concepts of economic analysis are discussed below.

6.1.1 Cash Flow Diagram

After the physical consequences of each alternative are identified, only the relevant and important consequences are considered for further analysis. The consequences that do not influence the decision-making process are eliminated. Each consequence is assigned a monetary value. After assigning the monetary values to all the relevant consequences, the cash flow diagram is drawn to convert the time stream of monetary value into an equivalent single number. The intangible values are properly treated and quantified.

The graphic representation of each monetary value with time is called a *cash flow diagram*. In a cash flow diagram, the horizontal axis denotes time. The benefits are represented by arrows pointing upward, while costs are represented by arrows pointing downward. The length of the arrow need not be scaled although attempts are made to make them proportional to the cost or benefit. For convenience of analysis and with little loss in accuracy for long-lived projects, all cash flows during a year are by convention combined into lump sums occurring at the end of the year. Fig. 6.1 shows a cash flow diagram for a hypothetical irrigation project. A large sum of expenditure is made in the beginning and, thereafter, benefits are received every year. The diagram also shows expenditures that are made for maintenance every few years. In a strict sense, annual benefits and costs will not be constant every year

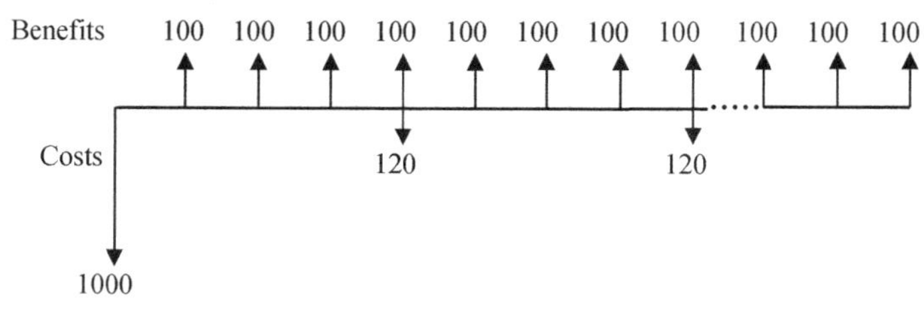

FIGURE 6.1

Cashflow diagram for a hypothetical project.

but will vary depending on crop production and market price of inputs and produce. However, only expected average values are normally predicted in advance, even though the random component could also be introduced.

6.1.2 Discount Rate

A discount rate is the expression of the time value of the capital that is used in equivalence calculations to compare alternatives. The rate is essentially a value judgment based on a compromise between the present consumption and capital formation from the viewpoint of the decision maker. For public works planning, this means the viewpoint of the society as a whole. The ideal discount rate would achieve a rate of capital formation maximizing the total social welfare. Many viewpoints are encountered in the selection of the best discount rate. Depending on the viewpoint, the choice of a discount rate also changes. The economic feasibility of a project greatly depends on the discount rate. The discount rate differs from the interest rate if they are defined precisely. The interest rate expresses the time value of the capital; interest is the fee paid by an entity to use the capital of another, for example, the rate at which banks make payments to depositors. In some way, it is rent paid to use someone else's money just like the rent paid to use a car or an apartment. The interest rate is determined by the capital market. An enterprise seeking to improve its own economic health will try to earn money by investing it in projects that give returns exceeding the borrowing rate of capital.

A low discount rate is favorable to the construction of public works projects and the groups that benefit from project construction. Sometimes, allocation for current needs may have to be sacrificed for future benefit. However, excessive diversion of resources to the public sector is detrimental in the long run. Generally, a lower value may be chosen for justifying projects of greater national importance, and a higher value may be chosen for projects in private sectors. However, too low a rate has serious adverse consequences for national economic growth. Since in many countries, all water resources projects are owned by the government or public undertakings and there is hardly any involvement of the private sector, a smaller rate of interest may be chosen by the decision maker. Finally, the discount rate changes with time and from one public body to another.

When choosing a discount rate, the lower bound should be at least as big as the available risk-free interest rate. Thus, no investment is worth undertaking unless it is just as productive as any of the comparable investments that are commonly available, such as long-term government bonds (de Neufville and Stafford, 1971). It will be better for any rational government to pay back its debts rather than invest in any project whose returns are less than the interest rate of its bonds. Most large-scale water projects are taken up by governments and for such projects, social discount rates should apply.

6.1.3 Discount Factors

Before a comparison can be made, all components or consequences of a project under consideration are to be expressed in a common unit. The most convenient unit in the present era is a money unit. However, the value of money changes with time and therefore, even though the consequences are expressed in monetary units, amounts at different times cannot be compared or combined. To make them comparable, all monetary values are converted to equivalent amounts at some definite time. This conversion is made by using *discount factors*.

The major obstacles in comparison to alternatives are the differences in kind and in time. The two basic principles: equivalence of kind and equivalence of time are employed to facilitate the comparison. The following example will help in clarifying these concepts.

Example 6.1. Consider two irrigation projects having the same construction costs. The first project provides water for rice and will yield 40 tons of rice and the second for sugarcane to produce 50 tons of sugarcane. If one has to invest money for only one project, which one should be selected?

Solution. This question cannot be answered with the given information because the outputs from the two irrigation projects are not in commensurable units. To answer this question, some additional information is needed—whether 40 tons of rice is more valuable than 50 tons of sugarcane or vice versa. If both crops are converted into monetary values, the comparison will be easy. The project whose output has more monetary value will evidently be selected.

Example 6.2. Assume that there are two options A and B for the production of rice in Example 6.1. Option A will provide water immediately and will cost X. Under option B, the cost will be Y and the water supply will start after 5 years from now. Which option should be selected?

Solution. It is easy to see that 40 tons of rice this year will not have the same value compared to the value 5 years from now. If $X = Y$, people would be more inclined to invest money to produce 40 tons of rice now (option A) than 5 years from now (option B). However, the two options cannot be compared with the available data because of differences in time and costs. To compare, information about the price of rice and discount rate is needed.

We begin with the simplest case. Assume that some principal amount P units (say Indian rupees) are invested at an annual interest rate of i percent. After 1 year, the amount will grow to become $P(1 + i)$. After one more year, it will be $P(1 + i)^2$. In this way, after n years, the final amount will be $P(1 + i)^n$. If this final amount is denoted by F then

$$F = P(1 + i)^n \tag{6.1}$$

In this case, the principal amount is paid just once and it gets compounded to become F after n years. The final amount is obtained by multiplying P by the factor $(1 + i)^n$. This factor is termed as single-payment compound amount factor. The *rule of 72* has emerged from the tables of this factor. It states: *the value of a sum doubles when the product of i (in %) and n is about 72*. Thus, a certain amount invested at 12% interest rate will double in 6 years ($12 \times 6 = 72$). Also, at $i = 8\%$, the present worth of an amount that will be received 9 years later is half.

Depending upon the payment of principal and compounding interest, many discount factors are used in economic analysis. A summary of these is given in Table 6.1.

Example 6.3. A bank launches a new scheme in which an investor has to deposit Rs. 1000 each year for 10 years and he will get Rs. 14,500 at the end of 10 years. What is the rate of interest that an investor will be getting upon joining this scheme?

Solution. Applying the equation for the series compound amount factor (Table 6.1)

$$F = A[(1 + i)^n - 1] / i$$

$$14,500 = 1000^* [(1 + i)^n - 1] / \mathrm{i}$$

Solving by trial and error, i is about 0.08. So, the investor will get return at 8% per annum.

Table 6.1 Summary of discounting factors.

Type of discount factor	Description	Formula
Single-payment factors		
Compound—amount factor	An amount P invested at the beginning of the first year grows to F after the end of n years.	$F = P(1+i)^n$
Present worth factor	It gives the present value of a future amount F that might be available after n years (inverse of above).	$P = F/(1+i)^n$
Uniform annual series factors		
Sinking-fund factor	The amount A that should be invested at the end of each year, for n years, to yield F at the end of n years.	$A = Fi/[(1+i)^n - 1]$
Capital-recovery factor	The amount A that will be received at the end of each year, for n years, if amount P is invested at the beginning of the first year.	$A = Pi(1+i)^n/[(1+i)^n - 1]$
Series-compound amount factor	The amount F that will be obtained at the end of n years if an amount A is invested at the end of each year, for n years.	$F = A[(1+i)^n - 1]/i$
Series present worth factor	The present value of P if an amount A is invested at the end of each year, for n years.	$P = A[(1+i)^n - 1]/i(1+i)^n$

Example 6.4. A friend of Mr. B offers him a bond for Rs. 6000. This bond will yield him Rs. 150 per month for the next 50 months. Mr. B has the money and it can otherwise give him a return @ 1% per month. Should he accept the offer?

Solution. To make a decision, the present worth of a payment of Rs. 150 for 50 months @ 1% is to be determined. Using the formula for the series present worth factor,

$$P = A[(1+i)^n - 1] / [i(1+i)^n]$$

$$= \frac{150[(1+0.01)^{50} - 1]}{[0.01(1+0.01)^{50}]}$$

$$= 5879$$

As the present worth of the bond is less than the asking price, the offer should be rejected.

6.1.4 Sunk Cost

While comparing projects, or analyzing the justification for alterations or expansions in a project, it is necessary to see how much money has already been spent. Obviously, the money already spent cannot be made available for investment now nor can it be recovered. Therefore, such an investment is not taken into consideration for an economic study. Past expenditure, which has no economic relevance in deciding future alternatives, is termed as *sunk cost*. Sunk costs may be disregarded while comparing alternatives except for their influence on future cash flows. However, sunk costs may influence decisions for two main reasons. The decision maker may have some prior commitment to continue a policy so that the past efforts are not wasted. Also, an undepreciated book value for assets that have no economic worth may restrict the freedom to make a new investment. In any case, past mistakes are not legitimate and plausible reasons to continue with an unjustifiable policy.

6.1.5 Intangible Values

An economic study seeks to evaluate all consequences in commensurable monetary units. However, there are many items that cannot be quantified in terms of monetary units. An old temple, for instance, which will be submerged in the reservoir or a valley with natural beauty cannot be assigned a money value. Each such matter that cannot be expressed in monetary terms is called an *intangible.* Such matters do not have direct effects on human beings physically through the loss of health or life, emotionally through the loss of national prestige or personal integrity, or psychologically through environmental changes. Moreover, monetary values cannot measure the achievement of many extra economic goals, such as income redistribution, increased economic stability, or improved environmental quality.

The inability to express a value in economic units does not necessarily preclude its evaluation in other units. All intangible values should be quantified as precisely as possible because these do influence decisions. While weighing whether a given sacrifice in economic value is worthwhile to achieve a goal, the decision maker should have access to the best possible data on the nature of the intangible consequences in addition to the economic consequences.

6.1.6 Salvage Value

Some project elements may have economic or physical lives shorter than the period of analysis. Such elements need periodic replacement. However, all alternative projects must be evaluated over the same period of analysis. If the period of analysis is not an even multiple of life of elements, the service life of some replaced elements will extend beyond this period. In this case, an adjustment must be made through a negative cash flow equal to the value of the element at the end of the period of analysis. The value of the unused life of an element at the end of the period of analysis is its salvage value. The straight-line depreciation may be used to estimate the salvage value (S) of an element by the following formula:

$$S = I(1 - U / L) \tag{6.2}$$

where U is the years of unused life, L is the years of total life, and I is the initial value.

Example 6.5. A pump requires replacement every 20 years and is to be used in a project where the economic study is based on a period of analysis of 70 years. What salvage value should be used if the initial cost is Rs. 25,000? Also, determine the total number of pumps that will be required during the economic life of the project which is also 70 years.

Solution. As a new pump will be installed at the interval of 20 years, the fourth pump will be required to be installed in the year 60. This pump will have 20 – (70 – 60) = 10 years of useful life remaining at the end of the period of analysis. Therefore, $U = 10$ years, $L = 20$ years, and $I =$ Rs. 25,000. From Eq. (6.2)

$$\text{Salvage value, } S = 25{,}000\ (1 - 10/20) = \text{Rs. } 12500$$

$$\text{Total number of pumps required} = 70/20 = 3.5 = 4 \text{ (rounded to next higher integer)}$$

6.1.7 Marginal Returns

The rate of change of the output of a process with respect to changes in individual resources is known as marginal product with respect to that resource:

$$MP_i = \Delta P\ /\Delta x_i \tag{6.3}$$

where ΔP is the change in the product, and Δx_i is the change in input x_i. As more and more quantity of a resource is used, its marginal product eventually decreases, leading to the law of diminishing marginal return. At the optimum solution, the marginal product per unit cost of the resource must be the same for all resources.

6.1.8 Planning Horizons

The planning horizon is the most distant future time considered in an engineering economic study. The uncertainty inherent in predicting the more distant future favors short planning periods. However, the water resources projects have long life necessitating an analysis of the long-term effects of projects even though they might be constructed to meet immediate requirements. Actually, three different time horizons must be considered in any economic analysis: (1) the economic life, (2) the physical life, and (3) the period of analysis. These are explained below.

Economic life: The time span after which the benefits from the operation of a project are less than the cost of operation is called the economic life. After the economic life is over, the project no longer gives benefits that exceed the operation and maintenance cost. The economic life of different project components may be different, for example, the spillway gates and turbines in a hydropower plant will have different life spans.

Physical life: This is the length of time beyond which the project can no longer physically perform its intended functions. The economic life can be greater or equal to the physical life, because of obsolescence and changing demands for services. Many factors cause economic life to be shorter than physical life. As an example, due to rapid developments in electronics, it may be economical to get a defective computer replaced rather than repaired.

Period of analysis: This is the length of time over which the project consequences are considered in the analysis. The project's economic life is the upper limit of this period. For water resources projects, the period of analysis may be in the range of 50–100 years.

6.2 DEMAND AND UTILITY OF WATER

Consider that a consumer has to purchase some commodity to meet his needs and a number of alternatives are available to him. These alternatives mainly differ in terms of price and the consumer is aware of this. The behavior of a consumer and his decision-making process can be explained using some concepts that are explained in this section.

6.2.1 Water Demand and Cost

A user gets many types of benefits from water that are (Young, 1986): (1) commodity benefits, (2) waste assimilation benefits, (3) public and private aesthetic and recreational benefits, (4) species and ecosystem preservation, and (5) social and cultural values. The first three can be treated as economic considerations, because they are characterized by increasing scarcity and the associated problems of allocation among competing uses to maximize economic benefits.

The demand for water can be for in situ or withdrawal uses. For in situ or in-stream uses, such as hydropower generation, recreation, and waste assimilation, water need not be removed from its source. The withdrawal uses, such as municipal and industrial uses, and irrigation, water needs to be removed from its source. The willingness of consumers to pay for water varies inversely with the amount being procured. The following functional relationship relates the demand to the cause variables

$$D_w = f(P_w, P_a, I, P, Z) \tag{6.4}$$

where D_w is the demand for water over a given time; P_w is the price of water; P_a is the price of an alternative source of water; I is the income of the consumer; P refers to the overall price index such as the consumer price index; and Z represents the effect of other relevant factors.

The demand function of a consumer is used to determine the quantity of a commodity that he is willing to buy as a function of price and income. A typical demand function is plotted in Fig. 6.2. As the quantity of a commodity that is demanded increases as the price is reduced, the graph of the demand function slopes toward the right. The consumption decreases as the price increases and this is known as the substitution effect.

Three components that add up to make the cost of water were mentioned by Rogers et al. (1998). These are as follows: the full supply cost, the full economic cost, and the full cost. Each of these is composed of separate elements. Fig. 6.3 schematically shows the components. The full supply cost includes the costs associated with the supply of water to a consumer without consideration of either the externalities imposed upon others or of the alternate uses of water. For a water supply system, full supply costs are composed of operation and maintenance (O&M) cost and capital charges. O&M costs include the price of raw water, electricity for pumping, expenditure for treatment plants, wages, and expenditure for maintenance of the components. Capital charges should include depreciation charges and interest costs. The sum of the full supply cost, the opportunity cost associated with the alternate

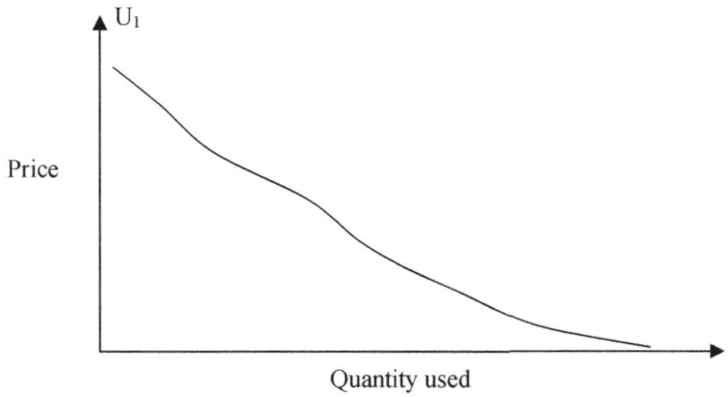

FIGURE 6.2

A typical demand curve for water.

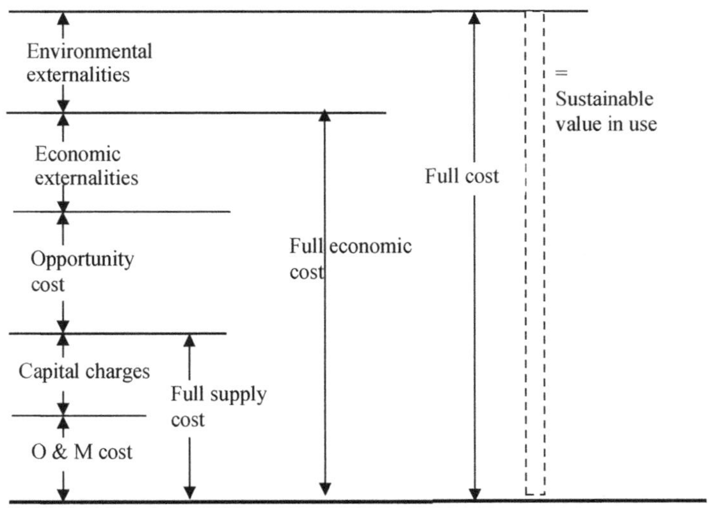

FIGURE 6.3

General principles for cost of water.

From Rogers et al. (1998). Copyright © Global Water Partnership. Used by permission.

use of the same water resource, and the economic externalities imposed upon others due to the consumption of water by a specific sector gives the full economic cost of water. The opportunity cost arises because by consuming water, another user of water is being deprived of water. If there is no shortage of water, the opportunity cost is zero.

6.2.2 Elasticity of Water Demand

The elasticity of demand is a measure of the responsiveness of consumers' purchases to variations in price. It is defined as the percentage by which the quantity used changes in response to a 1% change in price. The price elasticity of demand for water is a measure of the willingness of a consumer to reduce the use of water in response to the rise in price or use more water if the price falls. If the quantity demanded does not change much with price, then the demand is said to be inelastic and it is elastic when there are large changes in demand with price. A change in the price of water does not produce immediate reactions but the effects are felt in the long run. The uses of water for outdoor activities, such as lawns and gardens, show an elastic response, while the use in homes is inelastic.

As the elasticity of a commodity is determined by the demand curve, it is influenced by all those factors that influence demand. Typically, the more elastic a commodity is and the more substitutes are available, the wider is the range of uses and the larger is the proportion of consumer's income that is spent on the commodity.

6.2.3 Utility Theory

An intelligent consumer always chooses that particular alternative that gives him the maximum level of satisfaction. Utility functions provide information about the satisfaction that a consumer gets by consuming a certain quantity of given commodities during a specified span of time. Let the quantity of *n* commodities that a consumer uses be c_1, c_2, ... c_n. His utility function can be expressed as

$$U = f(c_1, c_2, \ldots, c_n) \tag{6.5}$$

A utility function is a continuous function. Suppose that the consumer chooses a number of consumption combinations of these commodities such that the level of satisfaction remains the same. A locus of these points is known as an *indifference curve*. Note that the indifference curves do not intersect because a given combination of commodities will yield a unique level of satisfaction. Such a curve when there are only two commodities is shown in Fig. 6.4. In this figure, different curves represent different values of utility, viz., U_1, U_2, U_3 ...

It can also be seen from Fig. 6.4 that the end segments of the indifference curves are asymptotic to the axes because as less and less of one commodity is consumed, more and more of the other commodity must be used to derive the same level of satisfaction. The change in utility function with respect to a change in consumption of a commodity at a point on the indifference curve is its marginal utility. The differential of utility function Eq. (6.5) can be written as

$$\mathrm{dU} = (\partial \mathrm{U} / \partial \mathrm{C}_1)\mathrm{dC}_1 + (\partial \mathrm{U} / \partial \mathrm{C}_2)\mathrm{dC}_2 + \ldots + (\partial \mathrm{U} / \partial \mathrm{C}_1)\mathrm{dC}_1 \tag{6.6}$$

Here, quantity $(\partial U/\partial C_1)$ is the marginal utility of commodity C_1, and so on.

The utility theory has not found wide applications in water resources systems engineering. The main reason behind this (Loucks et al., 1981) is the requirement that the decision-maker quantifies his attitudes toward risk. This is usually problematic because it is difficult for someone to articulate how he values an investment whose return is a random variable.

FIGURE 6.4

Indifference curves for two commodity cases.

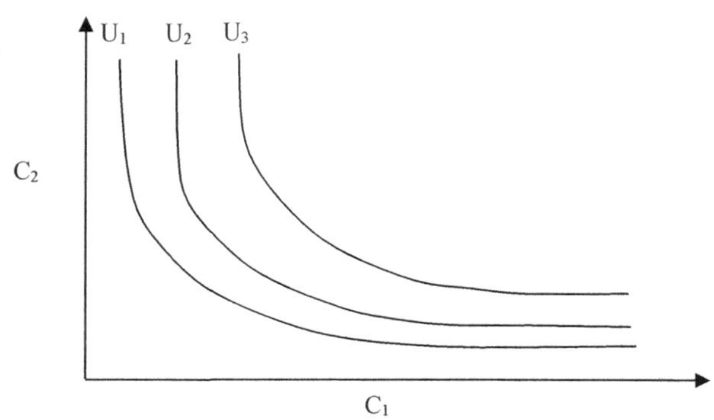

6.3 PROJECT ECONOMICS AND EVALUATION

Project economics involves the identification and quantification of all kinds of costs and benefits associated with a project and the economic analysis of the proposed plans. Project evaluation involves testing the project for all types of feasibilities, its evaluation, and the assessment of implications of changes in inputs. These studies are important before adopting a project plan or policy because water resources projects directly affect the development of a state, or region and the living standard of the people therein.

6.3.1 Project Cost

The initial cost of a project includes construction cost, engineering and administration cost, right-of-way cost, the cost of relocating facilities, and other minor costs. Construction cost is the amount spent on physical works outlined in the plans and specifications. The engineering cost is the amount spent on preparing the plans; structural and hydrologic design, and specifications; inspecting construction work; conducting special investigations, such as structural and hydraulic model studies, geologic explorations; technical review of engineering details, etc. The administration costs include salary and perks of staff, office expenses (paperwork, audit), and legal costs. The right-of-way cost is the payment to use the land required for project purposes. The land cost, which may still be used by the original owner but is secured against possible threats due to the project is termed as easement cost. Relocation cost is the expense to move or modify the existing facilities, such as bridges, roads, railroads, and powerlines located on the project right-of-way. Other costs include payments for water-rights acquisition (if any), publicity through media, etc.

After installation, the project requires continuous expenditure for operation, maintenance, and replacement. Operation costs include those costs that are associated with the control of outlets/gates, supervision of hydroelectric plants, purchasing power for pumping, and other activities required to produce project output on a continuing basis. Maintenance costs are those associated with the preventive measures to reduce breakdowns, repair of works, etc. Major repairs of civil works are usually needed after large floods. Replacement includes periodic changing of parts, such as pumps, well casings, or machinery whose useful life is less than that of the project as a whole, and normal wear and tear. The capital cost of a project is compiled by adding costs for items, such as investigation and planning, land, building, works, tools and plants, work-charged staff, maintenance during construction, and contingency.

6.3.2 Project Installation Cost

The construction cost can be estimated by determining the quantity of each input required to complete the project and multiplying the quantity by a unit cost. The unit cost can be based on contract bid prices for that item in similar recent construction projects. A contingency estimate of about 15% of the sum of the item costs is normally added to the total to allow for unforeseen developments during construction. The contingency allowance should be larger for preliminary planning estimates where the analysis has been less detailed. Note that the norms will vary with country and region.

Engineering and administration costs are estimated as a fixed percentage of the construction cost. Extra costs may be added when unusual design or construction problems are likely to be encountered or when the project is located in difficult terrain.

The price for a right-of-way is estimated based on the judgment of professional appraisers from their experience with the real-estate market. Adjustments to this estimate are needed because the market value of the property is based on an implicit capitalization of the expected future income at a discount rate well in excess of that used in project planning. An unwilling seller places a value on a property in excess of the market price since he wants to derive a value from his property, based on sentimental attachment, effect on his way of life, and established relationships with others in his community. In some cases, the resistance to sell depends on the degree of threatened change in his life and the seller would like to be compensated for this.

The adjustment for the right-of-way cost due to the first kind is made by expressing the market value, M_v, in terms of the economic value, E_v, as:

$$E_v = i\, M_v / d \tag{6.7}$$

where d is the planning discount rate, and i is the implicit market discount rate. If this adjustment is not made, the use of two discount rates will severely distort the project evaluation. The above expression assumes that the property has an indefinitely long life. A typical value for i is about 7%–8%. The economic value for the second type of adjustment is difficult to quantify.

Easement costs are taken as a percentage of the purchase cost determined by the degree the easement restricts private land use. However, easement costs are not too popular because their cost is usually close to the full land value. Severance damages must be paid when the right-of-way separates parcels under common ownership or interferes with land use or access. The acquisition cost, legal fees, and administrative expenses can be estimated as a fixed sum for each parcel that must be purchased.

6.3.3 Rehabilitation and Resettlement Costs

Rehabilitation and resettlement (R&R) of the population displaced or affected due to a project has been a hotly debated issue in recent years. In countries with high population density, a large number of people may be affected by a project. Therefore, this is a very important issue and in the absence of a satisfactory resolution, the progress of a project may be completely halted. The packages for rehabilitation at the new site consist of infrastructure facilities, compensation for the land submerged, and plots for building houses at the new site at nominal costs. Free facilities for transport of personal effects and loans, etc. are provided to the population displaced. In many projects, the R&R packages offered by the project authorities have been rejected by the beneficiaries as being inadequate. This topic is dealt with in detail in Chapter 7.

Recent experience shows that the R&R expenses are a significant part of the total project budget. However, there is a tendency on the part of planners to earmark a small amount for this purpose so that the benefit–cost (BC) ratio, and thereby the justification of the project is not adversely affected. However, such projects usually face funding problems when the work is at an advanced stage. It is, therefore, important that a realistic estimate of R&R costs is made by considering the current market (or even slightly higher) prices. A reasonable amount may be set aside to provide soft loans to the affected people to start a new business or vocation.

6.3.4 Operation, Maintenance, and Replacement Costs

The operation costs are estimated from expected salaries, fringe benefits, and overhead costs of hiring the required operating personnel, providing them with the required equipment and supplies, and the market cost of electricity or fuel consumed in operation. The number and type of operational personnel, supplies, and equipment required can be estimated from experience in operating similar projects. The annual cost of project administration should also be included.

The annual maintenance cost may be estimated as a percentage of the construction cost for each group of items. Typical percentages are 3.0 for earth channels, 0.5 for earth dams, and 0.1 for concrete structures. An alternative method is to estimate these costs from a maintenance program in which the required quantity of personnel, equipment, and supplies is estimated, and its cost is determined from prevailing prices.

Replacement costs are developed from the estimated dates and costs of replacing assets whose lives are shorter than those of the project. Annual replacement cost, A_c, over the project life is determined as follows:

$$A_c = (P - L)\left[(A/P, \%i, L_a) - (A/P, i\%, L_p)\right] \quad (6.8)$$

where P, L, L_p, and L_a are the costs of the new asset, the salvage value of the old, the project life, and the asset life, respectively. The notation (A/P, $i\%$, L) denotes the capital recovery factor (see Table 6.1).

6.3.5 Total Annual Cost

The total annual cost of a project is the sum of the annual recovery cost and annual expenditure on operation, maintenance, and replacement. The annual recovery cost is obtained by multiplying the capital investment with the capital recovery factor. The capital recovery factor (*CRF*) indicates the number of rupees one can withdraw in equal amounts at the end of each of N years if Rs. One is initially deposited at $i\%$ interest per annum. It is a factor by which the capital investment or the present value at the beginning of a project's life should be multiplied to get an equivalent annual recovery cost or fixed annual figure sufficient to repay exactly the present sum in N years with interest rate i. The *CRF* is explained in Table 6.1. If a fixed initial cost (capital investment) of a project is C_o, and a constant annual operation, maintenance, and repair cost is OMR, then the total annual cost (*TAC*) is given by

$$TAC = (CRF)C_o + OMR \quad (6.9)$$

Example 6.6. The construction of a hydroelectric project would cost Rs. 70 million. The project has an annual operation and maintenance cost of Rs. 4 million and a 50-year life. What is the annual cost of the project if a planning discount rate of 3% is to be used?

Solution. The CRF is computed by

$$CRF = \left[0.03(1 + 0.03)^{50}\right] / \left[(1 + 0.03)^{50} - 1\right] = 0.0388654$$

$$TAC = (0.0388654)^{*}\ 70 + 0.40 = 3.1205846 = \text{Rs. } 3.12 \text{ million/year.}$$

6.3.6 Project Benefits

Benefits from a water resources project can be classified into two categories: direct and indirect. Direct benefits refer to those that are produced on account of direct physical effects of projects, such as increased agricultural production, hydro-power generation, reduction of flood damage, and benefits from municipal and industrial water supply. Benefits arising out of technological and economic linkages of the effects of a project are regarded as indirect, for example, increased benefits due to better quality of life (because of the availability of sufficient water) or increase in agricultural productivity on account of flood control measures. These benefits may stem from or be induced by direct benefits.

Normally, the benefits that are taken into account for project formulation belong to the category of direct benefits only. Indirect benefits are not included because there is no well-accepted way of quantifying them. Project promoters are often eager and include these benefits to improve the benefit–cost ratio. As indirect costs also enter into the estimates along with benefits, they tend to reduce the overall benefits. However, whatever that may be, there is no justification for neglecting the indirect effects altogether. These benefits and costs constitute an important aspect of project impact and must be appropriately included in the benefit–cost analysis.

The procedures for estimating annual benefits are different for flood control, drainage, and anti-erosion works. In the case of flood control works, say, embankments, an estimate of annual benefits is made by finding out the average monetary value of annual flood damages of a few years before the construction of the project. In some cases, the benefit is estimated by multiplying the area affected by some assumed value of the damage per hectare for unirrigated and irrigated lands. From this, an estimate of the average annual damage after the construction of the project is deducted. There is also provision for appropriate adjustments for the beneficial value of silt deposition, if any. However, there are no standard guidelines to estimate these and the practice varies from country to country and may be even within the same country.

6.4 DISCOUNTING TECHNIQUES

Water resources planning and management involves choices among physically feasible alternatives which are governed by many factors, such as technical, economic, social, financial, environmental, and political. An engineering alternative is a course of action that is physically capable of achieving the design objectives. The alternatives are called mutually exclusive if only one from a set can be selected, which may be due to conflicting site requirements, limited financial resources, or inputs. Sometimes, it may be practical to implement two or more alternatives to arrive at the solution. A properly defined alternative must be specified with sufficient clarity so that its economic and intangible consequences can be evaluated.

An economic analysis is performed in a series of steps as given below:

1. Each promising alternative is identified and explicitly defined in physical terms.
2. The physical consequences of each alternative are expressed in terms of monetary estimates, including the dates as well as the magnitudes of the receipts and disbursements.
3. A period of analysis is decided.
4. The estimates of the lives and salvage values, if any, of the structures and other assets are prepared.
5. A discount rate is selected and applied to convert the predicted time stream of monetary values into an equivalent single number.
6. The alternatives are compared on the basis of equivalent monetary values.

The procedures in which discounting factors are systematically applied to compare alternatives are known as discounting techniques. The alternatives may be either different projects or different sizes of the same project. Discounting factors are applied to convert cash flows to an equivalent single number at some definite time. The present value of either benefit or cost is obtained by applying suitable discounting factors to cash flow. It can also be estimated as:

$$PV = \sum_{t=1}^{T} (1+i)^{-t} A_t \tag{6.10}$$

where *PV*, *T*, *t*, *i*, and A_t denote the present value (of benefit or cost), the planning horizon, the index for time, the discount rate, and the amount (benefit or cost) at time *t*, respectively. The discounted amounts for costs and benefits, or net benefits are analyzed by one of the discounting techniques to arrive at the best alternative.

The commonly used discounting techniques are as follows: (1) the benefit–cost ratio method, (2) the present worth method, (3) the rate-of-return method, and (4) the annual-cost method. If properly applied, each method should lead to more or less the same conclusion. These methods are discussed in what follows.

6.5 BENEFIT–COST RATIO METHOD

The principles of welfare economics are based on the assumption that each individual is the best judge of his or her own welfare, and the welfare of the society is based on that of its citizens. The gross welfare of the society increases if the welfare of one individual increases without a corresponding reduction for individuals. This is called a *Pareto improvement.*

A fundamental concept of the benefit–cost (BC) analysis is the principle of Pareto superiority, which was defined by Vilfredo Pareto (an Italian social scientist) as follows: *Economic state 1 is to be judged socially superior to economic state 2 if at least one person individually judges 1 superior to 2, and no one judges 2 superior to 1.* In other words, a Pareto improvement is a change in economic situation that makes one or more members of society better off without making any one worse off. Although this concept sounds good, is difficult to realize because in practice, a policy aimed at making some one better might make some others worse. Therefore, a modification in terms of a majority rule is desirable. This leads to what has been termed as *Potential Pareto Superiority* rule which states: "Economic state 1 is socially superior to economic state 2 if those who gain by the choice of 1 over 2 could compensate those who lose so that if compensation were paid, the final result would be that no one would be worse off than he would be in state 2." A Pareto optimum is a solution in which it is not possible to increase welfare of any individual without decreasing the welfare of some other person.

The BC ratio is defined as the ratio of the present worth of benefits and the present worth of costs. The benefits are positive effects of a project for which the beneficiaries are willing to pay. The costs are the value of opportunities forgone because of the commitment of resources to the project or the willingness to pay to avoid detrimental effects. The present worths of costs and benefits are computed separately, and then the ratio is determined. The annual values can be used as an alternative without affecting the ratio. The term *opportunity costs* refers to the benefits that are foregone when a scarce resource is used for a purpose instead of its next best alternative use.

Note that the BC criterion requires that those who gain are able to compensate losers and still be better off. This is the rationale of the BC analysis. However, the criterion does not demand that the compensation actually be paid, only that it is possible that suitable compensation exists to leave no one worse off.

Although it does not reflect the real productivity, the BC ratio has a lot of relevance under certain conditions. This is when several independent projects are to be chosen, given some capital constraints. Then, it is appropriate to rank the projects by their respective BC ratios, implementing successively lower projects until the BC ratio of the marginal project reaches unity.

6.5.1 Steps of BC Analysis

As stated, the BC ratio (R) is the ratio of the present value of all benefits to the present value of all costs. It can be calculated as

$$R = \frac{B}{C} = \sum_{t=0}^{n} \frac{B_t}{(1+i)^t} \Big/ \sum_{t=0}^{n} \frac{C_t}{(1+i)^t} \tag{6.11}$$

where B_t and C_t are the monetary values of benefits and costs incurred at time t respectively, i is the discount rate, and n is the life of the project. In some cases, the sum of the discounted net benefits in those years when gross benefits exceed the total cost (discounted positive net benefits) divided by the sum of discounted net benefits in those years when total costs exceed gross benefits (discounted negative benefits) is computed. This discounting method is also termed as the *net benefit-investment ratio*.

While computing BC ratios, it is important to use uniform computational methods to achieve comparable results. A different and lower value of the BC ratio is obtained using gross benefits and gross costs. In other words, moving a cost from the denominator and subtracting it from the numerator (as by omitting the production cost from the gross cost stream and instead deducting it from the gross benefit stream) will change the ratio value. Clearly, any manipulation that reduces the size of the denominator will increase the ratio. It is common not to compute the BC ratio using gross costs and gross benefits but to compare the present worth of net benefits with the present worth of investment cost plus operation and maintenance costs.

The following steps are followed to choose the best alternative by this method.

1. Calculate the BC ratio for each alternative.
2. Choose all alternatives having a BC ratio exceeding unity. Reject the rest. If the sets of mutually exclusive alternatives are involved, proceed to steps 3, 4, and 5.
3. Rank the alternatives in the set of mutually exclusive alternatives in order of increasing cost. Calculate the BC ratio by using the incremental cost and the incremental benefit of the next alternative above the least costly alternatives.
4. Choose the more costly alternative if the incremental BC ratio exceeds unity. Otherwise, choose the less costly alternative.
5. Continue the analysis by considering the alternatives in order of rank.

While performing the above steps, all BC ratios should be computed by using the same discount rate and the same period of analysis. The BC ratio does not distinguish between a big project and a small project. Being a ratio, it does not reflect the actual productivity. This ratio is particularly favorable to projects where most of the cost is the initial investment and the year-to-year expenses are minimal. It is important to recognize that the best project has the greatest net benefits, not the largest benefit–cost ratio. Sometimes, the BC ratio method leads to different decisions than do other techniques. However, this conflict occurs when the incremental-cost principle of steps 3–5 is neglected.

The BC ratio can also be obtained by dividing the net annual benefits estimated as the difference between the "with" and "without" scenarios by the annual cost. In India, for example, the latter includes annualized interest on capital cost at 10%, depreciation at 2%, and operation and maintenance expenses. The discount rate used is usually the opportunity cost of capital.

The decision on whether particular cash flows should be considered costs or negative benefits is sometimes arbitrary and affects the BC ratio. There is considerable variation among agencies in the definition of cost in the BC ratio. Most agencies include the entire government cost on the cost side, but there are differences in the treatment of associated costs by individuals, which in some instances are considered among costs, in others as negative benefits. Since these costs are rarely a high percentage of all project costs, the error is relatively small. The practice understates BC ratios, since the subtraction of a negative constant from the numerator plus its addition to the denominator of a ratio greater than one always acts to reduce the result. Thus, if benefits are 12, government costs 7, and associated costs 2, the BC ratio according to economists is $(12-2)/7$ or 1.43, while engineers may calculate it as $12/(7+2)$ or 1.33.

The BC ratio method is the conventional technique that is widely used to analyze public works proposals. After eliminating technically unsound alternatives, some other criteria are applied for further screening of projects. This has led to the use of the BC ratio as the major deciding criterion. Generally, only those schemes that show a favorable BC ratio are taken up for further examination or execution. In special cases, strong justification, other than economic criterion based on the basic needs of the specific area, is taken into account. For example, a project may be taken up in a drought-prone area, even though it might not have a favorable BC ratio.

Some economists have advanced the concept of a social BC ratio. In this approach, the benefits and costs are social benefits and social costs, and a social discount rate is used. These are different than the actual benefits and costs. The reason is that the value of a unit output generated by a project in a poor or backward region is more than the same output in a prosperous region. How much more this value will be, will depend on the sociopolitical conditions of the country.

6.5.2 Incremental Benefit and Cost

Any change in the proposed project plan results in changes in associated benefits and costs. The terms *incremental benefit* and *incremental cost* indicate the changes that occur in the benefit and the cost, respectively, due to alterations in the project plan. According to the incremental cost principle, the change in benefits and costs resulting from a given decision determines the merit of that decision. Each project segment should be judged on its own merit. The decision to increase the size of a project should be justified by extra benefits and the incremental benefit should be more than the incremental cost. Analysis based on the total cost and total benefit may give different solutions to those obtained from the incremental cost principle.

Example 6.7. Two alternative projects are under consideration. The estimated cost of the first proposal is Rs. 40 million, whereas that for the second is Rs. 45 million. If the benefits from these proposals are Rs. 45 and 48 million, respectively, which proposal should be adopted?

Solution. The BC ratio for proposal 1 = 45/40 = 1.125. The BC ratio for proposal 2 = 48/45 = 1.067. Since both proposals have BC ratios greater than 1, both will be analyzed using the incremental principle as per step 3 described earlier. The first proposal will have rank 1, and the second one will have rank 2. Thus, incremental cost (second proposal over first proposal) = 45−40 = Rs. 5 million. Incremental benefit (second proposal over first proposal) = 48−45 = Rs. 3 million. Incremental BC ratio = 3/5 = 0.6. As the incremental benefit−cost ratio is less than 1, the less costly alternative will be selected. Therefore, proposal 1 will be selected.

Example 6.8. A project costing Rs. 2.0 million is expected to produce the benefits of Rs. 3.0 million. Before starting the project, it is observed that the adoption of some advanced technology can increase the benefits to Rs. 3.2 million. If the consultation charge for the advanced technology is Rs. 0.5 million, state whether the adoption of advanced technology is justified?

Solution. BC ratio without advanced technology = 3.0/2.0 = 1.5.

Incremental cost of advanced technology = 0.5 million rupees.

Incremental benefit from advanced technology = 3.2−3.0 = 0.2 million rupees.

B/C ratio with advanced technology = 3.2/(2 + 0.5) = 3.2/2.5 = 1.28.

As the incremental cost of advanced technology is more than the incremental benefit, the adoption of advanced technology is not justified. Further, the BC ratio without the advanced technology is higher than with advanced technology. Hence, it is advisable to continue with the original plan.

Example 6.9. An amount of Rs. 1000 is invested each in projects A and B. Project A returns Rs. 200 at the end of the year for 10 years while Project B returns Rs. 130 at the end of the year for 20 years. Rank the projects using the BC ratio method if the discount rate is 4%. Also, rank them if the discount rate is 11%. This example illustrates the role of the discount rate in project ranking.

Solution. The computations are shown in the table below. The present value of benefits can be obtained using the series present worth factor (see Table 6.1).

			For r = 4%		For r = 11%	
Project	**Initial cost**	**Benefits**	**Present value of benefits**	**BC ratio**	**Present value of benefits**	**BC ratio**
A	Rs. 1000	Rs. 200 per year for 10 years	Rs. 1622	1622/1000 = 1.622	Rs. 1177	1177/1000 = 1.177
B	Rs. 1000	Rs. 130 per year for 20 years	Rs. 1766	1766/1000 = 1.766	Rs. 1035	1035/1000 = 1.035
Evidently, at r = 4%, project B scores over A due to higher BC ratio. But when r = 11%, project A has higher ranking than B.						

6.5.3 Economic Rationale of Benefit−Cost Analysis

The rationale for BC analysis is based on two fundamental economic concepts: scarcity and substitution. The first of these implies that the resources, for example, water, housing, food, education, or good environment, are limited and should be used efficiently. The concept of substitution indicates that individuals, social groups, and institutions are generally willing to trade off a certain amount of one objective for more of another. It implies that the limited resources could be put to alternate uses to obtain the maximum benefit. The general economic problem is how to use available scarce resources to maximize the resulting human welfare. The scarcity is registered in the marketplace by price. In regard to water resources development, the maximization of the net benefit objective requires the efficient allocation of water to various uses (for example, hydropower, irrigation, water supply, flood control, and navigation) over time and space.

The water resources projects take considerable time in realizing the maximum possible benefits and this must be reflected in their ranking. To clearly bring out the advantages of quicker implementation, the discounted cash flow method may be used. This method better reflects the time value of money.

Different types of benefits and costs are generated by a given pattern of resource use. For a BC analysis, the benefits and costs are to be expressed in monetary units. For that, it is required to estimate the monetary value of irrigation water, hydropower production, navigation, flood control, etc. An important source of prices used is the current market trend.

Some benefits and costs are correctly registered in market prices, and some are registered in no markets, although simulated market values (i.e., what users would pay if a market existed) can be computed, and for others, it may be nearly impossible to have an adequate market valuation. The market prices are assumed to be an accurate reflection of marginal social values. But if the markets are not competitive, the prices are not appropriate indicators of costs and benefits and should be adjusted appropriately. In some cases, the market prices are not registered, for example, for the use of water for environmental preservation or recreation and the analyst has to estimate these prices.

Even when benefits and costs are registered in market prices which could be measured as the aggregate net willingness-to-pay of those affected, it may still not be possible that the beneficiaries could compensate the loser in all cases and that everyone would benefit from the project. While the owners of many resources required for the project, such as land, are compensated, those people who lose their favorite scenic sites or suffer a decrease in the value of their property because of the project are seldom compensated. Likewise, the host population (those who receive the resettlers) do not generally receive any compensation. The compensation criterion also ignores the resultant income redistribution which should be considered during the plan selection process. The compensation criterion implies that the marginal social value of income to all affected parties is the same. Although the analysis will determine the amount of income stream generated over and above the costs of labor and other inputs, it does not specify who actually receives it.

A real-world economic system, even when operating under fairly competitive conditions, seldom generates a socially acceptable distribution. As things stand, the common feeling is that the poor become poorer while the rich, richer possibly because the poor do not have resources to exploit the new opportunities. Maybe the distribution of income is not socially considered as adequate but is the outcome of the lack of knowledge of how to correct it because of the ability of the economically powerful groups to prevent corrective measures. In such a case, market prices cannot be taken as an acceptable measure of the value for the evaluation of projects or policies that affect the economically disadvantaged segments of the population. This is particularly true for developing countries.

The market prices reflect economic values when these are fair and competitive or they will be shadow prices. The term *shadow price* denotes the value assigned by the public to goods and services that are not marketed. Market prices do not always reflect the relative scarcity of different resources because these may be modified to differing degrees by taxes, subsidies, and exchange rates (if imported products are involved). For some final goods and services, however, the concept of the opportunity cost is not applicable because it is the consumption value that sets the economic value, not the value in some alternative use. In such instances, the "willingness to pay" criterion will need to be applied. This is because the ultimate objective of all economic activity is to satisfy the consumption wants, all opportunity costs are derived from consumption values, and thus from the willingness to pay. The topic is covered in detail by Howe (1983a,b) and Griffin (1998).

Economic efficiency in resource allocation is obtained when no individual can be made better off without making someone else worse off. This condition is termed as Pareto optimality. The beneficial effects of a project increase positive utility or remove existing negative utility. A drawback of the benefit–cost analysis approach is that a potential Pareto improvement treats all users equally. In other words, efficient allocations are not necessarily fair or just. Another criticism is that the conventional measures of benefits and costs are dependent on the existing distribution of wealth, and may serve to enshrine the status quo (Young, 1986).

6.6 OTHER DISCOUNTING METHODS

The other important discounting methods are discussed in this section.

6.6.1 Present Worth Method

While selecting a project, an intuitive approach can be to choose the project that has the largest present worth of the discounted algebraic sum of benefits minus costs over its life. It is the present worth method. This technique is also called the net present worth or the net present value because this method reduces a stream of costs and benefits to a single number. The net worth (benefit—cost) for each year is computed and discounted to the present. A sum of these gives the net present value (*NPV*):

$$NPV = \frac{B_0 - C_0}{(1+i)^0} + \frac{B_1 - C_1}{(1+i)^1} + \ldots + \frac{B_t - C_t}{(1+i)^t} + \ldots + \frac{B_n - C_n}{(1+i)^n} \quad (6.12)$$

where C_t is the monetary value of costs and B_t is the monetary value of benefits, both at time t, i is the discount rate, and n is the life of the project. The determination of the appropriate discount rate is important in this method. Sensitivity analysis of this value (i) will, to some extent, reduce the arbitrariness in selecting this value.

The steps involved in this method are enumerated below.

1. Calculate the present worth of each alternative using appropriate discounting factors.
2. Choose all the alternatives having a positive present worth. Reject the rest. If no sets of mutually exclusive alternatives remain, stop. Otherwise, step 3 or 4, as appropriate, is adopted to choose the best alternative.
3. In a set of mutually exclusive alternatives, choose the one that has the greatest present worth.
4. If in a set of mutually exclusive alternatives, some have benefits that cannot be quantified but are approximately equal, choose the alternative having the least cost.

The following rules are followed while calculating the present worth of a project.

Rule 1. Compute all present worths to the same time base, irrespective of the initial time of different alternatives.

Rule 2. Compute all present worths by using the same discount rate, even if the alternatives are being financed from different sources.
Rule 3. Base all present worths on the same period of analysis, irrespective of differences in economic life.

The discounting rate used for the analysis is either the social rate of discount or the opportunity cost of capital. The social rate of discount is the rate at which society's weight on increments to consumption declines over time. This rate of discount depends on many factors, including (1) society's present level of consumption, (2) the expected growth of consumption, (3) the expected growth of population, (4) the rate at which marginal utility of consumption diminishes, and (5) society's time preference. The opportunity cost of capital refers to the *marginal productivity of capital* (MPC), which, in turn, is the same as the internal rate of return of the marginal project. The internal rate of return will be discussed in a later section.

Example 6.10. There are three mutually exclusive alternatives for a water supply project. The present worth of costs and benefits associated with these alternatives is given below:

	Present worth (million Rs.)		
Item	**Alternative A**	**Alternative B**	**Alternative C**
Construction cost	42	38	40
Operation and maintenance cost	10	12	11
Benefit	60	59	56

Compare the alternatives and suggest the best. Assume that all present worth figures have been obtained using the same discount rate and period of analysis.

Solution. Net present worth for alternative A = 60 − (42 + 10) = Rs. 8 million.
Net present worth for alternative B = 59 − (38 + 12) = Rs. 9 million.
Net present worth for alternative C = 56 − (40 + 11) = Rs. 5 million.
Thus, alternative B should be chosen, since its net present worth is the maximum.

6.6.2 Rate-of-Return Method

The rate of return is the discount rate at which the present worth of the discounted algebraic sum of benefits minus costs over the project life equals zero. This rate, which is found by trial and error, is the rate of return that equates to the initial cost and the sum of discounted future net benefits. At this rate, the benefit−cost ratio is close to one. This rate represents the average rate of interest at which a project pays back the investment over its life time. It, therefore, is a criterion for comparing alternative investment opportunities. Some writers call this technique as the *internal rate-of-return* (IRR) method.

Mathematically, IRR is some discount rate r such that the initial cost C_0 is equal to the present worth of benefits:

$$C_0 = \frac{B_1 - C_1}{(1+r)^1} + \frac{B_2 - C_2}{(1+r)^2} + \ldots + \frac{B_n - C_n}{(1+r)^n} \tag{6.13}$$

Alternatively, it is the discount rate r that would make the NPV of the project equal to zero. Newnan (1983) has defined the rate of return as the interest rate earned on the unrecovered investment such that the payment schedule makes the unrecovered investment equal to zero at the end of the life of the investment. The higher the rate, the more attractive is the project. A project with an IRR exceeding some predetermined level (e.g., the social discount rate) is deemed acceptable. For a given stream of benefit and cost, the determination of IRR is by trial and error. For assumed values of the discount rate, one can calculate net benefit, which is equal to the present value of benefits minus the present value of costs. Fig. 6.5 illustrates the concept of IRR.

Example 6.11. The initial cost of a project is Rs. 1000 and it gives benefits @ Rs. 250/- per year for 5 years. Find the IRR. What will be the IRR if benefits are obtained at the same rate for 6 years?

Solution. Application of Eq. (6.13) yields

$$1000 = \frac{250}{(1+r)^1} + \frac{250}{(1+r)^2} + \frac{250}{(1+r)^3} + \frac{250}{(1+r)^4} + \frac{250}{(1+r)^5}$$

This equation can be solved by trial and error to yield $r = 7.9\%$. Note that IRR should not be construed as one will get an annual return @ 7.9% of Rs. 1000. Each payment of Rs. 250 implies a return @ 7.9% on the unrecovered investment plus the partial return of the investment. For the first year, out of Rs. 250, 79 is the return on Rs. 1000 (at 7.9% per annum) and the balance $250 - 79 = 171$ is repayment of the investment. This leaves unrecovered investment of Rs. 829 at the beginning of 2nd year. The uncovered investment reduces each year and at the end of 5th year, it will be nil.

If the benefits are obtained for 6 years, there will be one more term in the equation:

$$1000 = \frac{250}{(1+r)^1} + \frac{250}{(1+r)^2} + \frac{250}{(1+r)^3} + \frac{250}{(1+r)^4} + \frac{250}{(1+r)^5} + \frac{250}{(1+r)^6}$$

The solution gives $r = 13\%$. This shows that the IRR of the venture has increased considerably. The following steps are involved in comparing alternatives by this method.

1. Calculate the rate of return for each alternative.
2. Choose all alternatives having a rate of return exceeding the minimum acceptable value, known as the minimum attractive rate of return (MARR). Reject the rest. If sets of mutually exclusive alternatives are involved, proceed to steps 3, 4, and 5.

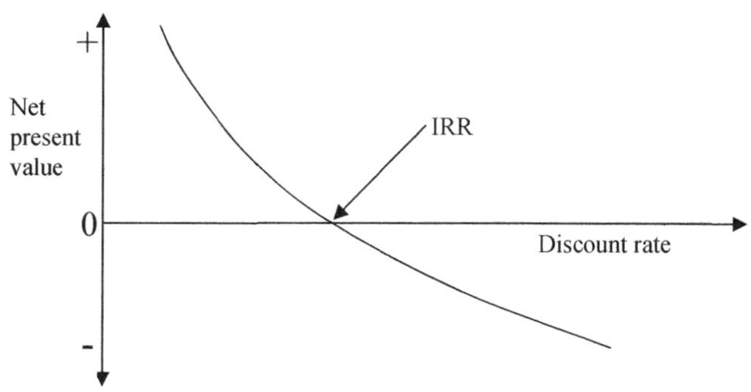

FIGURE 6.5

Illustration of internal rate of return.

3. Rank the alternatives in the set of mutually exclusive alternatives in order of increasing cost. Calculate the rate of return on the incremental cost and incremental benefits of the next alternative above the least costly alternative.
4. Choose the costlier alternative if IRR exceeds the minimum acceptable discount rate. Otherwise, choose the less costly alternative.
5. Continue the analysis by considering the alternatives in order of rank.

Other decision rules are also used in this method. All alternatives are compared over the same period of analysis. The rate-of-return method will not lead to the same decision as the present worth method unless the incremental analysis of steps 3–5 is used in place of selecting the mutually exclusive alternative with the highest rate-of-return. This method must be applied with caution because more than one rate of return exists when annual costs exceed annual benefits in years after annual benefits first exceed annual costs. However, this method, using steps 3–5, still gives consistent answers even when dual solutions exist. The water resources planner needs to be alert to this problem when comparing alternatives by the rate-of-return method.

While using this method, it is necessary to select an MARR be desirable to work out a minimum cut-off rate. This rate varies with the country and depends on the economic conditions, inflation, etc. For a large country with wide disparities, it will not be advisable to follow a uniform rate throughout. Somewhat lower values have been recommended for drought-prone regions, chronically flood-prone areas, and hilly regions. A lower rate of return would also be appropriate in a water-scarce basin most of whose flow has already been utilized. According to Newnan (1983), MARR should be equal to the highest of the three: the cost of borrowing money, the cost of capital, and the opportunity cost. The cost of borrowed money is the interest rate at which money can be borrowed. The cost of capital depends on many factors. Basically, the money a firm uses for investment is drawn from the various components of the overall capitalization and their rate of return determines it. Further, in the event of limited money, a person or a firm may not be able to use all the investment opportunities. After choosing better opportunities, the rate of return of the best-rejected alternative is the opportunity cost.

The main advantage of this method is that it gives a measure of the strength of a project in terms of a single, well-understood number. Also, the discount rate need not be assumed as in other criteria. Two problems may be encountered with this method.

(i) The rate r that is the solution of Eq. (6.13) is not necessarily unique. Newnan (1983) explains how to resolve this situation.
(ii) The criterion implicitly assumes a single discount rate over the life of the project.

Example 6.12. A person has Rs. 10 million to invest. The available alternatives with required investment and returns are as follows. (1) Rs. 3 million; 15%; (2) Rs. 4 million, 20%; (3) Rs. 1 million; 18.5%; (4) Rs. 2 million, 19%; (5) Rs. 3 million, 18%; (6) Rs. 4.5 million, 16%; (7) Rs. 2 million, 17%. Find the opportunity cost of money for him?

Solution. If the alternatives are ranked according to the rate of return, the following ordering is obtained: (1) Rs. 4 million, 20%; (2) Rs. 2 million, 19%; (3) Rs. 1 million, 18.5%; (4) Rs. 3 million, 18%; (5) Rs. 2 million, 17%; (6) Rs. 4.5 million, 16%; (7) Rs. 3 million, 15%. Clearly, the person would invest in the first four alternatives. Among the rejected ones, the best alternative can give a return @17% and this is the opportunity cost for him.

6.6.3 Comparison of BC Ratio and IRR Method

Let O be the annual operation, maintenance, and replacement (OMR) costs that are assumed uniform for the sake of simplicity. If K denotes the fixed investment at the beginning, i the discount rate, and T the life of the project in years, the present value of the total cost will be

$$PV_C = \sum_{t=1}^{T} \frac{o}{(1+i)^t} + K \tag{6.14}$$

Let B be benefits received annually. The present value of total benefits is

$$PV_B = \sum_{t=1}^{T} \frac{B}{(1+i)^t} \tag{6.15}$$

and the BC ratio (R) is

$$R = \frac{PV_B}{PV_C} = \sum_{t=1}^{T} \frac{B}{(1+i)^t} \left[\sum_{t=1}^{T} \frac{o}{(1+i)^t} + K \right]^{-1} \tag{6.16}$$

Dividing the numerator and denominator $\sum_{t=1}^{T} \frac{1}{(1+i)^t}$ by, the ratio becomes

$$R = \frac{B}{O + K\left[\sum_{t=1}^{T} \frac{1}{(1+i)^t} \right]^{-1}} \tag{6.17}$$

Define

$$a_{iT} = \left[\sum_{t=1}^{T} \frac{1}{(1+i)^t} \right]^{-1} \tag{6.18}$$

Actually, a_{iT} is the annual capital charge per unit investment, representing both interest and amortization. It can be calculated, given i and T. Now, Eq. (6.17) becomes

$$R = B \,/\, [O + a_{iT}K] \tag{6.19}$$

Eq. (6.19) expresses R, that is., the BC ratio, in terms of annual costs and annual benefits. Referring to Eq. (6.13), the rate of return r is given by

$$K = \sum_{t=1}^{T} \frac{B - O}{(1+r)^t} \tag{6.20}$$

On the pattern of a_{iT}, we define a_{rT} as

$$a_{rT} = \left[\sum_{t=1}^{T} \frac{1}{(1+r)^t} \right]^{-1} \tag{6.21}$$

Hence, Eq. (6.20) becomes

$$K = (B - O)/a_{rT} \tag{6.22}$$

or

$$B = a_{rT}K + O \tag{6.23}$$

and the *BC* ratio R is

$$R = (a_{rT}K + O)\,/\,(a_{iT}K + O) \tag{6.24}$$

Solving for a_{rT},

$$a_{rT} = [R\,(a_{iT}\,K + O) - O]\,/\,K = (a_{iT} + O\,/\,K)R - O\,/\,K \tag{6.25}$$

or

$$a_{rT} = a_{iT}\,R + (O\,/\,K)(R - 1) \tag{6.26}$$

This enables computation of a_{rT}, knowing *R, O, K*, and *i*. Given a_{rT}, *r* can be readily found. If there are no current costs (*O* is zero), the two criteria coincide and Eq. (6.26) becomes

$$a_{rT} = a_{iT}\,R$$

Here, whatever the interest rate is, a higher *R* will result in a higher value for a_{rT} and hence higher *r*. If *R* is 1, the two criteria become identical as the second term in Eq. (6.26) becomes zero.

The degree to which the use of *R* affects the ranking of projects as compared to the rate-of-return method depends on the range of values of *O*/*K*. Projects of similar type, such as hydroelectric or irrigation projects, will obviously have similar values of *O*/*K*. But there will be much larger differences in projects of different purposes.

6.6.4 Annual Cost Method

In this method, all benefits and costs are converted into equivalent uniform annual figures. Decision rules resemble those for the present worth method because each annual cost is present worth times a constant capital recovery factor. The steps involved are as follows:

1. Calculate the net annual benefit of each alternative.
2. Choose all alternatives having a positive net annual benefit. Reject the rest. If sets of mutually exclusive alternatives are involved, proceed to step 3 or 4, as appropriate.
3. Choose the alternative in a set of mutually exclusive alternatives having the greatest net annual benefit.
4. If the alternatives in the set of mutually exclusive alternatives have benefits that cannot be quantified but are approximately equal, choose the alternative having the least annual cost.

Example 6.13. The annual benefits from projects A and B are Rs. 4.2 and 4.1 million, respectively, and the annual costs of these projects are Rs. 3.95 and 3.6 million, respectively. State which project is more beneficial? Also, find the net annual worth in percentage by which the chosen project is more favorable over the other.

Solution. Net annual worth of project A = 4.2−3.95 = Rs. 0.55 million.

Net annual worth of project B = 4.1−3.60 = Rs. 0.50 million.

As project A has more net annual worth, it will be more beneficial.

Difference in the net annual worths (A over B) = 0.55 − 0.50 = 0.05.

Percentage difference by which project A is more favorable over B = 0.05 × 100/0.50 = 10%.

6.6.5 Comparison of Discounting Techniques

Each of the four discounting techniques will yield the same solution if used correctly. However, each technique has advantages and disadvantages associated with ease of calculation, presentation, and understanding of the results. These are important considerations while selecting a method. The present worth method is simpler, safer, easier, and more direct. It does not require an additional set of computations to apply the incremental-cost principle. The direct expression of the net present worth is conceptually straightforward and easily presented. The weak point is that the method involves working with large numbers which may be hard to visualize. Besides, the present worth method cannot be used to rank projects in order of economic desirability unless all of them require equal investment. Being an absolute measure, the present worth is not effective to compare the profitability of the alternative investment. It is an important complement to IRR particularly to compare projects that are mutually exclusive or that have a high rate of return.

The rate-of-return technique is a valuable analytical tool because it does not require a preselected discount rate. The rates of return are intuitively meaningful to many investors and the resulting rates can be compared with those for many other types of investment. This method is criticized because:

(1) ambiguous answer may occur due to dual solutions,

(2) it necessitates the calculation of incremental rate-of-return for interdependent projects,

(3) there is a danger of people's accepting the overall return in contrast with incremental rates of return as indicators of rank, and

(4) the solutions via trial-and-error are a bit difficult to obtain.

The rate-of-return technique has distinct advantages over other discounted measures of projects worth in a sense that the calculation does not depend on assumptions about the opportunity cost of capital. Unlike the net present worth, a relative measure can be used to compare the profitability of projects.

The benefit−cost ratio method is widely used by water resources planners. However, this method, without applying the required incremental BC analysis, can lead to serious errors. Interdependent projects cannot be ranked according to their BC ratios, because each enlargement must pass the incremental BC ratio test.

As the annual cost technique uses constant multiplies of the present worth method, it has the same advantages and disadvantages, except for the use of smaller numbers. The annual cost method is sometimes preferred because more people are accustomed to thinking in terms of annual costs than of present worths. As far as the selection of a method is concerned, it depends primarily on the purpose of the analysis. It is not possible to use the benefit−cost and rate-of-return techniques when benefits cannot be evaluated. Costs alone must be compared by using the present worth or annual cost method. More calculations are involved in the rate-of-return and BC ratio methods. A comparison of three measures of present value is given in Table 6.2.

Table 6.2 Comparison of three measures of present value.

	Net present value	Internal rate of return	Benefit–cost ratio
Selecting or ranking rule for independent projects			
No constraint on costs	Select all projects with NPV > 0; project ranking not required	Select all projects with IRR greater than cut-off rate of return; project ranking not required	Select all projects with B/C > 1; project ranking not required
Constraint on costs	Not suitable for ranking projects	Ranking all projects by IRR may give incorrect results	Ranking all projects by B/C where C is defined as constrained cost will always give correct ranking
Mutually exclusive projects (no constraint on costs)	Select alternative with largest NPV	Selection of alternative with highest IRR may give incorrect results	Selection of alternative with highest IRR may give incorrect results
Discount rate	Appropriate discount rate must be adopted	No discount rate required, but cut-off rate of return must be adopted	Appropriate discount rate must be adopted

Adapted from Gittinger (1982).

6.7 PROJECT FEASIBILITY AND OPTIMALITY

The development, utilization, preservation, and management of most water resources projects involve political and social objectives in addition to the specific objectives of the project. All decisions are concerned with the quality of life and its distribution to society now and in the future. Before adopting a project or going for economic analysis, it must be ensured that all projects pass feasibility criteria. Each project must pass six feasibility tests: Engineering, Economic, Financial, Political, Social, and Environmental (or EEFPSE).

6.7.1 Engineering Feasibility

If the proposed project is physically capable of performing its intended objective(s), it is called technically feasible. The engineering design must be confined within the technologically feasible region. Engineering analysis will show the combination of outputs that can be physically produced and those that cannot be produced by a project. All other feasibility tests are carried out only for the alternatives that pass engineering feasibility tests.

6.7.2 Economic Feasibility

The project is said to be economically feasible when the benefits from the project exceed the costs. It is important to note that the comparison should be between *with* and *without* rather than *before* and *after* because many of the after-effects may occur even without the project. Thus, the use of before and after concepts is not proper in project justification.

6.7.3 Financial Feasibility

Even though a project may clear all other feasibility tests, it may remain unimplemented if funds are not available. The test of financial feasibility is determined by examining potential sources of available funds and is passed if sufficient funds can be raised to pay for project construction and operation. A project may be economically feasible but financially infeasible because the benefits are insufficiently concrete for the beneficiaries to appreciate their true value or are thinly distributed among too many beneficiaries or a large area. A project may be economically infeasible but financially feasible (although rare) because someone is willing to pay for the fulfillment of noneconomic goals. Financial feasibility also depends on local interests who want the likely economic benefits to the degree that they are willing to arrange their portion of the required funds.

6.7.4 Political Feasibility

The political feasibility of a project is achieved when the required political approval can be secured. Political feasibility is determined by an analysis of how key decision makers assess the favorable and adverse effects of the project, the direction of popular feelings, and the project potential for obtaining widespread public support. Ordinarily, political support follows economic and engineering feasibility. Sometimes, despite economic infeasibility, political pressure for a project is sufficiently strong to eventually pave the way for its construction. Conversely, strong groups who feel that they may be adversely affected, are at times able to stop a project in spite of all other favorable indicators.

6.7.5 Social Feasibility

If the target beneficiaries respond favorably to project construction, it is said that the project is socially feasible. This feasibility is determined by assessing the change that the project is expected to impart to the lives of the beneficiaries and evaluating the willingness of those affected to adopt. The opinion of the people, who will be beneficiaries or affected by the project, must be positive regarding the project launching. Otherwise, successful completion of a project within the scheduled time may not be possible and this will change the time stream of costs and benefits. Land acquisition, rehabilitation and resettlement, employment, income redistribution, project output distribution, and environmental deterioration are some of the issues on which different groups in the project region may have diametrically opposite viewpoints. This is not a good omen and is likely to put hurdles in project completion.

The success of a project depends on the willingness of the users to realize project benefits. The more drastic are the changes that a project requires in the lives of the beneficiaries, the greater is the likely resistance to change. The infusion of productive capital does not automatically transform a society.

6.7.6 Environmental Feasibility

All project proposals considered for evaluation must pass environmental feasibility. The project should not generate adverse environmental consequences. The issues related to ecological disturbances may raise a number of constraints during execution. The growing awareness of ecological considerations over the last few decades has led to the involvement of many diverse decision makers and a number of additional institutional and social constraints. It will ultimately affect the cash flow pattern and delay the completion.

The water resources development projects are viewed by some as having a destructive influence on the environmental quality as they often lead to the submergence of forests and displacement of people. The conflict can be mitigated by introducing a design that is in consonance with the environment, is aesthetically pleasing, and by choosing a plan that leads to minimum deforestation and resettlements of people.

The environmental aspects have been discussed at length in Chapter 7.

6.7.7 Cost and Benefit Curves

Economic evaluation of production alternatives is based on the variation in total production cost with the level of production output, and the variation in the resulting benefit with the level of production output. The total cost curve is developed by summing the required input costs for a series of outputs. Similarly, the total benefit curve is developed by summing values received by output users. The total cost includes fixed costs and variable costs. The fixed costs remain constant regardless of the output, while the variable costs vary with the level of the output. The average cost or average benefit curves are developed from total cost or total benefit curves by dividing the total value by the level of output. Average cost curves are usually U-shaped. The marginal cost or marginal benefit curves represent the slope of the total cost or total benefit curves. The slope represents the change in the total cost or the total benefit associated with a one-unit change in the output. Like average curves, marginal curves are generally U-shaped but justified more to the left.

The common approach to estimate benefits from flood management is the "property damage avoided" approach. This approach measures the present value of the expected property damages which are avoided by a project or policy. The replacement and maintenance costs to the affected structures are estimated for a number of water levels with and without the project. The estimated annual benefit for a given flow level is the difference between these costs.

6.7.8 Optimal Allocation of Water to Individual Users

In the problem of optimal water allocation to individual users, the usual assumption is that water is supplied to each individual user by an existing supplier. Naturally, the private supplier would like to allocate the water among the users so that his benefits are the maximum. However, from the point of view of society, it is important to achieve economic efficiency and equity (recall the concept of Pareto superiority). The requirement to provide services much beyond the design limits is the root cause of the degradation of services in many cases.

The problem of optimal allocation can be easily understood with the help of demand and cost curves. In Fig. 6.6, the horizontal axis represents the quantity of water used per unit time while the vertical axis shows prices. The curve labeled MW depicts the users' marginal willingness to pay for an additional unit of water per unit time. This curve slopes downwards to the right as the marginal willingness goes on decreasing as additional units are made available. The curve MC shows the incremental or marginal cost of providing one unit of water per unit time. Clearly, the marginal cost increases as additional quantities are made available. The two curves intersect at point O^* where the marginal cost and marginal benefit are equal. Therefore, this is the optimal point where the marginal

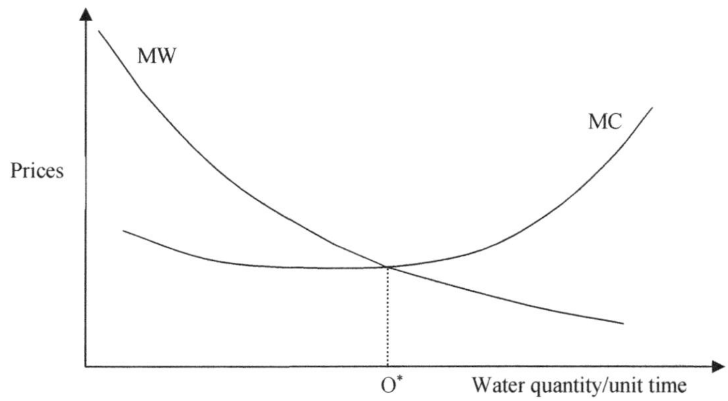

FIGURE 6.6

Optimal allocation of water to individual users.

willingness to pay for each additional unit of water is equal to the marginal cost of supplying that unit. Note that if supply is more than this amount, then the marginal benefit will be less than the marginal cost of supply. Conversely, when a lesser quantity is supplied then the marginal benefits will exceed the marginal costs.

6.7.9 Optimal Allocation of Water Among Different Uses

In the case where limited water is to be allocated among different sectors in a region, the optimal solution from an economic point of view would be the one that maximizes the benefit. According to the *equimarginal principle* (Young, 1986), the net benefit function is maximized when the net marginal benefits per unit of water used are equal in all use sectors. Consider Fig. 6.7 for the case of two uses,

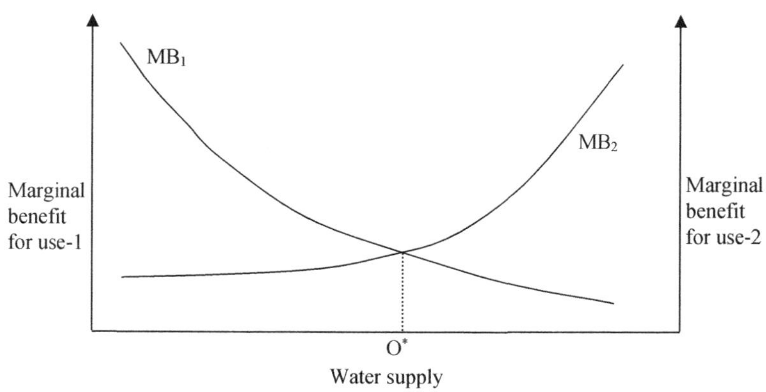

FIGURE 6.7

Optimal allocation of water to two different uses.

agriculture and urban. The curve MB_1 represents the marginal net benefit for supplies to use-1 and slopes downwards to the right indicating diminishing returns as more quantity is supplied. The curve MB_2 shows marginal benefits for use-2 and here it is drawn in reverse form from the right-hand axis just to illustrate the underlying concept. The intersection of these two curves at O^* represents the optimal balance between the two uses.

6.7.10 Allocation of Ground Water

The problem of groundwater allocation is quite different from that of surface water allocation. The major difference is that the velocity of groundwater flow is much slower than that of surface water. Also, the aquifers are not very extensively measured and subsurface investigations are fairly expensive. Usually, the optimal allocation is ascertained by balancing the diminishing returns to the present period use to the increased pumping cost and the discounted future benefits. The enforcement of groundwater allocation is difficult, although very much necessary. In the absence of such enforcement, many regions in the world are witnessing rapid declines in the groundwater table, which is also responsible for a number of environmental problems and water quality problems.

6.7.11 Project Optimality

Water resources planning, development, or management can be thought of as a production process. In planning a production process for the public sector, many valuable insights can be gained from analyzing how economic forces would act to order production under ideal conditions. The basic purpose of production is to convert resources (input) into more useful forms (output). A water resources project is constructed to produce such desired outputs as irrigation water, reduced flood damage, a navigable channel, or electric power from a set of inputs such as earth, concrete, steel, pumping energy, and natural streamflow.

After obtaining an optimal solution, it is necessary to examine how sensitive the project's net present worth, rate of return, BC ratio, etc. are to likely changes, such as increased construction cost, extension of the construction period, and changes in output prices. An analysis to find what happens under these changed circumstances is called sensitivity analysis. As the outputs of the projects are subject to uncertainty, sensitivity analysis should form an integral part of economic analyses of all projects.

In general, project outputs are sensitive to changes in four principal areas, namely prices, delay in implementation, cost overrun, and yield. As regards prices, every project should be examined to see what happens if the assumptions about the sale prices of the project's product prove wrong. The delay in implementation affects many projects. The cause could be the administrative delay in ordering or receiving equipment and stores, delay due to agitation against the project, nonavailability of funds, etc. Determination of the effects of delay on the net present worth, the financial and economic rates of return, and the net benefit—investment ratio of a proposed agricultural investment is an important part of the sensitivity analysis. Every project should be tested for sensitivity to cost overrun, especially for construction because so often the costs are incurred early in the project when they weigh heavily in the discounting process, and also for facilities that must be completed before any benefit can be realized. It may also be necessary to test a proposed project for sensitivity to errors in the estimated yield. There is a tendency in many agricultural projects to be optimistic about potential yields based mainly on data from experimental fields, especially when new cropping patterns are proposed. It has also been found in many projects that the cropping patterns being followed by the farmers were quite different than what was assumed, resulting in serious deviations in water demands as well as returns.

6.8 VALUATION OF ECOSYSTEM SERVICES

Millions of species live on earth. An ecosystem is a group or community composed of living and nonliving things and their interactions with each other. Every organism depends on other organism(s) for growth and survival. This dependence of organisms on one another and their surroundings forges an interacting system called an ecosystem. It is a dynamic complex of biotic and abiotic components that are interactive and maintain an equilibrium in the ecosystem, and we as humans are an integral part thereof. The interactions among different components of ecosystems are fundamental to a healthy environment. Natural ecosystems include agricultural systems, forests, grasslands, and aquatic habitats. Estuarine and coastal ecosystems fall under the broad category of marine ecosystems.

Healthy ecosystems render a range of services and provide various goods that are useful to society, such as clean water and air, regulating climate and moderating extremes of weather, natural pollination of crops, and human mental and physical well-being. Put together, these benefits are termed as "ecosystem services." Ecosystem functions refer to the habitat, biological or system properties, or processes of ecosystems.

6.8.1 Classification of Ecosystem Services

The Millennium Ecosystem Assessment (MEA) grouped the ecosystem services into four broad categories (Fig. 6.8): provisioning (e.g., food to eat, timber to be used for domestic/commercial purposes, and water), regulatory (e.g., regulating climate and disease), supporting functioning of the ecosystems (such as nutrient cycles and oxygen production), and cultural (e.g., local rituals, religious practices, or supplies for traditional clothing, recreation). Ecosystem goods (such as food) and services (such as waste assimilation) represent the benefits human populations derive, directly or indirectly, from ecosystem functions. Damages due to flood and landslide represent harms from ecosystems.

A brief description of the four categories of services follows.

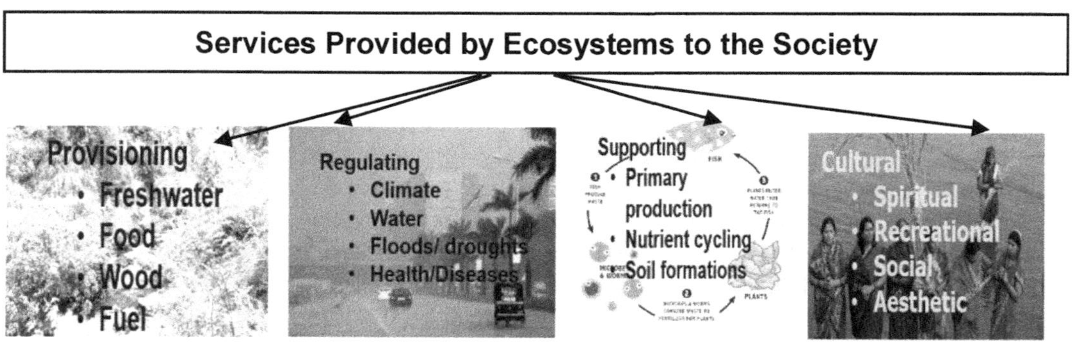

FIGURE 6.8

Four categories of services provided by ecosystems.

Provisioning Services

Ecosystems are sources of freshwater, food, wood, fuel, natural gas, oil, raw material for cloth-making, medicines, and other resources. They provide suitable conditions (raw materials or energy) for these products to grow and be available to the user in sufficient quantity and good quality.

Forests produce a variety of timber products, firewood, charcoal, and wood chips. In addition, forests also produce nonwood products, such as fodder, aromatic and medicinal plants, and wild foods. Of course, the importance of forests in controlling the concentration of greenhouse gases is well known. Studies have also found that forests can mitigate small- and medium-magnitude floods. Likewise, a river in which good quality water flows provides water for a range of uses by the society, is a place for religious and cultural rituals/events, impacts the microclimate of the place, and provides many spiritual benefits.

Regulating Services

Regulating services refer to the benefits provided by ecosystems that moderate natural phenomena. Through these services, ecosystems help maintain balance in ecology. They help in regulating climate, and water, including water-related extremes, such as floods and droughts. Forests, which are components of ecosystem, help maintain the concentration of greenhouse gases in the atmosphere, check soil erosion, and increase infiltration of water. Bees pollinate flowers and bacteria decompose waste material. Biodiversity (or "biological diversity" is the variety of life on Earth) of ecosystems work as natural controllers of health and diseases.

Supporting Services

Supporting services help other services perform their jobs. Three such services are shown in Fig. 6.8, viz., primary production, nutrient cycling, and soil formation. Here, primary production refers to the production of organic matter by way of processes, such as photosynthesis and chemosynthesis (chemosynthesis is the biological conversion of one or more carbon-containing molecules, usually carbon dioxide or methane, and nutrients into organic matter using the oxidation of inorganic compounds, for example, hydrogen gas, hydrogen sulfide, or ferrous ions as a source of energy, rather than sunlight, as in photosynthesis. Source: Wikipedia). On average, a human consumes about 378 L of pure oxygen per day.

Nutrient cycling is the movement of nutrients by biotic and abiotic processes through an ecosystem. Since all living beings require nutrients to survive and function, the service provided by nutrient cycling impacts many ecosystem services.

Supporting services also help ecosystems sustain themselves. For example, the water cycle, carbon and nitrogen cycles, nutrient cycle, photosynthesis, soil formation, etc., all also help ecosystems. Without supporting services, the ecosystems will not be able to provide the other three services sustainably.

Cultural Services

Humans have two-way interaction with nature. They impact nature and get, in turn, impacted by nature which influences cultural, social, and intellectual development. Many old customs and traditions function around nature. These include religious ceremonies and social events/festivals. Many people derive spiritual rejuvenation when they come in contact with elements of nature, for example, flowing water, natural forests, etc. In India, people in large numbers come to rivers to take bath and many ceremonies are performed by the side of flowing water. Numerous tourist places have been developed around water bodies. Note that cultural services are nonmaterial benefits—one can only "feel" them.

Because ecosystem services are not fully accounted for in market-based economics or adequately quantified in terms of economic services, they are often ignored or given inadequate importance in policy decisions. This "overlook" may ultimately compromise the sustainability of humans in the biosphere. The global economy would rapidly halt without the services of ecological life-support systems.

6.8.2 Valuing Ecosystem Services

Healthy ecosystems provide both tangible and intangible benefits; the benefits derived can be direct or indirect. Although the benefits provided by nature have been recognized for a long time, only in the recent years attempts have been made to estimate the benefits in monetary terms.

Ecosystem valuation attempts to estimate the costs of making healthy ecosystems and the benefits derived therefrom. In a simplified framework, the benefits from the four types of services described earlier can be categorized into two groups: direct use and indirect use. The direct-use value is the direct yield of the ecological system if sold in a commodity market, for example, the selling price of timber, fish, water, or other commodities. The indirect use value is the benefit derived from the indirect utilization of ecosystem services. These values are the benefits derived from ecosystem services outside the system. For example, a catchment may naturally filter water for the benefit of people living downstream; and a forest may provide protection to the downstream communities against floods. A range of methods are used to calculate the market values and estimate nonmarket values of services. Where a market does not exist, valuations can be estimated by asking people for their willingness to pay for the particular ecosystem service.

Ecosystem valuation typically assigns a monetary value to the services provided by an ecosystem. Such valuation attempts to capture the range of benefits and costs in the complex natural sphere with a range of economic methodologies. It needs to be acknowledged that such valuations are estimates and involve large uncertainties as the process involves a range of nonmarket costs and benefits. Many past studies have attempted to estimate the value of a wide variety of ecosystem services. The benefits provided by natural ecosystems are widely recognized yet poorly understood. Nevertheless, this is a concept whose time has come.

Economists utilize a variety of methods to calculate the market values and measure nonmarket values. Where market prices do not exist, prevalent environmental economic methods are used to assign a monetary value to ecosystem services. These methods include the stated preference methods and the revealed preference methods. Stated preference methods perform a survey of stakeholders and ask them how much they are willing to pay for a certain ecosystem (service). The contingent valuation method is an example of such methods. Revealed preference methods typically use a relation with a market good or service to estimate the willingness of the users to pay for the service.

Costanza et al. (1997) grouped ecosystem services into 17 major categories. The major ones are as follows: climate regulation, water regulation, water supply, soil formation, nutrient cycling, waste treatment, and food production. Costanza et al. (1997) completed a worldwide estimate of the worth of ecosystem services, amounting to US $33 trillion annually. The present value of these services would be about $44 trillion.

de Groot et al. (2012) synthesized more than 300 works evaluating the 10 main biomes of the world: Open oceans, Coral reefs, Coastal systems, Coastal wetlands, Inland wetlands, Lakes, Tropical forests, Temperate forests, Woodlands, and Grasslands. Their results show that the total value of ecosystem services is considerable and ranges between 490 International $/ha/year (int$/year) for the total bundle of ecosystem services that can potentially be provided by an "average" hectare of open oceans to almost 350,000 int$/year for the potential services of an "average" hectare of coral reefs. This potential benefit can take many forms depending on the degree of exploitation and such exploitation's sustainability, but can result, for example, in large ecotourist revenues for local communities, protection from storm destruction, or profit for an international lumber company. To illustrate at the country level, the total value of ecosystem services for India in 2011 was reported to be US $1.8 trillion/year (Chopra et al., 2022).

Valuations of ecosystem services are being increasingly used to formulate sustainable and environment-centric policies Although, as of now, the valuation of ecosystem services has numerous uncertainties due to knowledge and data gaps, it is helpful to prepare such estimates to make the engineers and decision makers aware of this concept and to let them know the magnitude of the value of such services.

6.8.3 Concerns Regarding Ecosystem Valuations

Although the values of ecosystem services are being estimated and new understandings are being developed, many knowledge gaps remain. Many natural resources, such as clean air, quiet surroundings, and biodiversity, are not purchased/sold in markets, and it is very difficult to assign a price to them.

After estimating the ecosystem costs and benefits, in some instances, payments for ecosystem services have been made, for example, to landlords to preserve forests, to improve conditions in watersheds, etc. A question is who should pay for ecosystem benefits and where these benefits should go?

6.9 ALLOCATION OF PROJECT COST

In multipurpose water resources projects, it is necessary to divide the project cost among several uses and/or different groups of beneficiaries to examine the viability of uses and fix charges for services. For instance, a dam might be serving for irrigation and hydropower and the cost of providing each service will be needed to levy fair charges for each. Similarly, the share of flood control in the cost of a multipurpose project must be distinguished from that for irrigation to properly assess net benefits. The problem assumes significance when the costs and benefits of a project are spread over more than one administrative unit which is quite often the case. For example, two nearby cities may decide to jointly build a facility for water supply. An objective basis of cost allocation is needed to avoid interstate or intergroup disputes. The procedure for dividing the total cost among responsible heads is called cost allocation. Once a formula to allocate costs is established, it may have to be incorporated into a legally binding cost-sharing agreement. Cost allocation falls under financial analysis.

If there are many user groups, cost allocation requires the division of costs among these groups. Each entity that is assigned a cost is called a cost center. The methods that are commonly used in practice are as follows: (1) allocation of costs in proportion to some numerical criterion, such as population and water use, or (2) allocation of costs like marginal costs directly and dividing the remaining cost based on some criterion.

A project element is a distinct physical part of the project. Some project elements serve only one cost center and their costs are termed as direct costs. If an element serves more than one cost center, the difference in its cost with and without serving a cost center is the *separable cost* of that element with respect to that cost center. This cost can be determined from project designs with the two scenarios. *Nonseparable cost* is the difference between the total project cost and the sum of separable costs. Thus, in a project serving irrigation and hydropower demands, the separable cost of hydropower is the cost of the project less the cost of an equivalent pure irrigation project. Normally, the sum of separable costs for all the purposes will be less than the total project cost due to economies of scale. Nonseparable costs include joint costs and common costs. If a project element contributes to the production of more than one output, its cost is joint cost. Common costs are those indirect or other fixed costs that must be incurred but cannot be associated with any specific purpose such as the wages of common-pool administrative staff.

In a fair system, the allocation of cost to any cost center should neither be less than the additional cost of including that center in the plan nor more than the total benefits of the center. The allocation should not burden any objective with a greater investment than the fair capitalized value of the annual benefit of that objective. It should not result in charging any objective with a greater investment than would be necessary for its development at an alternate single-purpose site. Mathematically, this can be represented as (Heaney and Dickinson, 1982):

$$x(i) \leq \min[b(i), c(i)] \, \forall \, i \in N \tag{6.27}$$

where $x(i)$ is cost allocated to group i, $b(i)$ is benefit of group i, $c(i)$ is alternative cost if group i acts independently, and N is set of all groups, $\{1, 2, \ldots \text{i}, \ldots n\}$. Further

$$\sum_{i \in S} x(i) \leq c(S) \, \forall \, S \subset N \tag{6.28}$$

where $c(S)$ is the alternative cost if the subset S acts independently and S is any subset of the master set N.

The allocation of costs by following the methods in the first category is simple. The commonly used methods in the second category are the separable costs, remaining benefits (SCRB) method, alternative justifiable-expenditure method, and use-of-facilities method. In the SCRB method (James and Lee, 1971), the joint costs of a project are apportioned among the purposes using the ratio of remaining benefits to the total benefits. The remaining benefits are the lesser of benefits of costs of the least-cost alternative minus separable cost. The total remaining benefits are the benefits of independent action with less separable cost. Following Heaney and Dickinson (1982), separable costs are

$$\text{sc}(i) = c(N) - c[(N) - \{i\}] \, \forall \, i \in N \tag{6.29}$$

where $\text{sc}(i)$ is the separable cost to group i, $c(N)$ is the total cost for the grand coalition of n groups, and $c[(N) - \{i\}]$ is the total cost for the grand coalition with group i excluded. After separable costs are assigned to each group, the remaining costs are nonseparable costs (nsc)

$$nsc = c(N) - \sum_{i \in N} \text{sc}(i) \tag{6.30}$$

The prorated share of the nonseparable costs to each cost center is based on the ratio of their remaining benefits to the total remaining benefits. The prorated share is given by

$$\beta(i) = [\min\{b(i), c(i)\}] - sc(i)^* \left\{ \sum_{i \in N} \{[\min(b(i), c(i))] - \mathrm{sc}(i)\} \right\}^{-1} \tag{6.31}$$

The sum of the prorating factors will be unity. Thus, in the SCRB method, the total charge to the ith cost center is

$$x(i) = \mathrm{sc}(i) + \beta(i)^* \ \mathrm{nsc} \tag{6.32}$$

The following points are important while allocating costs:

1. The allocation process should be straightforward and simple enough to be easily understood. Conflicting interests are resolved on the basis of their best understanding.
2. The sum of the allocations to all the cost centers normally equals the total project cost. Most of the methods require this condition to be fulfilled but this has not been followed in many instances. Heaney and Dickinson (1982) quote cost allocation for TVA projects where it was concluded that it is not possible to meaningfully assign joint costs; these were viewed as sunk costs that need not be recovered.
3. Joint facilities should be operated in accordance with cost allocation. It is not equitable to allocate most of the cost to a center having a low service priority in facility operation.
4. Costs nonseparable to any single or group of cost centers must be allocated among all centers; and costs separable to a group but not to a single cost center must be allocated among the centers composing that group.
5. As cost allocation directly affects economic and social efficiency, that allocation method should be used which does the most to promote the desired social goals.

The essence of cost allocation is a successful resolution of conflicting interests among parties in a fair manner. Strictly speaking, there is no unique method of cost allocation in conflicting situations, because each party tends to influence the allocation by shifting the largest share of joint costs to others. The SCRB method has a flaw in that it is not monotonic in total costs, that is., an increase in total costs may result in some participants having to pay less than before due to the way the marginal costs are introduced. Allocation of joint costs is complex and some authors believe that there is no economically justifiable way for this. Besides, most allocation schemes use information about demands and the optimal scale of development that in practice may be unreliable or nonexistent. Young et al. (1982) cite a case from Sweden in which the cost of a water supply system was to be allocated among six groups of municipalities. Of all techniques, the scheme based on the size of the population was actually adopted.

According to Votruba et al. (1988), the practice in Chechoslovakia was to allocate costs in proportion to benefits and side effects expressed in monetary units. If there are significant side effects among inputs and outputs of the multipurpose water engineering project, which cannot be evaluated by shadow prices, these effects may be evaluated by the initial and OMR costs of the substitute or they may be determined by agreement among the users.

6.9.1 Cost Allocation Practices in India

According to the Damodar Valley Corporation Act of 1948, the joint costs are allocated to different purposes in proportion to the expenditure that would have been incurred in constructing a separate structure solely for that purpose, less any amount that is solely attributable to the object. In the case of the Hirakud Project, the same principle of the alternative justifiable-expenditure method was originally followed and accordingly the allocation of the costs of storage capacities between flood control, irrigation, and power was in the ratio of 38:20:42. In 1952, this approach was replaced by a new method based on the ultimate utility of water for various purposes. Flood control as a separate purpose was eliminated. Subsequently, at a seminar organized by the Government of India in 1961, it was decided that joint costs be allocated to various purposes in proportion to the reservoir capacity or quantity of water utilized for each purpose. In light of this, flood control was to share 25% of the reservoir cost on account of the consideration that the reservoir operates for flood control for 3 months in a year.

For the Kosi project in Bihar, the cost of a barrage is deemed to be a common facility for flood control and irrigation, and its cost is divided equally between them. In the case of the Rengali Project in Orissa, the cost of Stage I of works was allocated equally between flood control and irrigation, since the same storage capacity could be used equally for both functions. In April 1967, the Government of India recommended the adoption of the facilities used method for allocation of joint costs of multipurpose river valley projects. The alternative justifiable-expenditure approach was recommended by the Rashtriya Barh Ayog in 1976 to allocate costs in multipurpose projects. A discussion on the economic aspects of irrigation projects in India was given by Navalawala (1993).

6.9.2 Funding Needs in Water Sector

The water sector projects are highly capitalistic. The construction of large infrastructure to manage river basins, interbasin transfers, canal networks, water supply, and wastewater collection and treatment plants, all require significant funding. The amortization of this large amount is only possible over a very long period and may be extending to several decades. It is a fact that many governments are not in a position to bear all costs and that public funding has reached its limits. Except in a few, the funding requirements greatly exceed the abilities of national or regional public budgets to sustain services.

Every year various users pay large sums for water consumption and treatment. According to estimates, the amount is about 1% of the World Gross Product or about US$ 300 billion per year (1990s estimate). In terms of the volume of water consumed, the urban consumers use about 10%–15% of the total but pay most of the above amount while agricultural users consume about 70% of the total quantity but are highly subsidized. Paradoxically, many of the poor living in arid regions pay a huge price for water, typically 5–10 times that which would be necessary to pay the full costs of a well-designed water supply system. It is instructive to note that:

- In many developing countries, poor people often have to purchase water in small quantities (often of dubious quality) to meet their basic needs by paying very high prices for it. Alternately, some of these people have resorted to costly individual means to pump water.
- The time and efforts of the women who have to walk over long distances every day to fetch water are rarely taken into account.
- If the individual expenses to get water for domestic use (cisterns or roof-top tanks, boreholes, and pumping) are summed up for a sizable society, the totals may reach very high amounts that could be better used by the community.

In some countries, water for agriculture is provided free of cost or the prices are so low that these do not even cover the amount spent on operation and maintenance of infrastructure. This is an important reason behind the degradation of infrastructure in many places. The unwillingness of private investment in this sector also largely stems from this. A private supplier will charge not only for water as such, but also the expenses that he has to incur to make it available where the user needs it, with the required reliability in terms of quantity and quality. However, reliable services can be provided at affordable cost. For instance, the price of treatment and supply of a cubic meter of drinking water (including taxes) is nearly the same as that of a "soft drink" in a bar!

The role of governments in water resources development is crucial. The governments' primary responsibilities are: (1) create and implement a legislative and regulatory framework that governs water, (2) create national and basin-level authorities, (3) plan, construct, or operate water resources development projects alone or in association with private parties or create conditions that are conducive for the participation of private parties in these activities, and (4) provide the enabling framework for community action that empowers the poor, women and the minorities. Unfortunately, traditional public subsidies, mainly sourced from general budgets, have reached their limit and it is becoming increasingly difficult almost everywhere to meet the needs of the water sector with traditional public budgetary means.

The money from the private sector will flow to this sector only if investors are sure of getting competitive returns and there is an efficient, strong, and transparent regulatory framework that protects the interests of both investors and consumers. Undoubtedly, the private sector can considerably improve the dismal technical and financial performance that characterizes most public utilities in developing countries. It is, however, pertinent to note that participation of the private sector is not a panacea for all problems and the experiences so far have been mixed.

6.9.3 Funding Needs for Climate Action

Climate change has emerged as the biggest environmental problem facing the world and all the countries are going to be affected. However, the people who have small role in causing climate change may be the first and the worst sufferers. Most of the greenhouse gas emissions are coming from the wealthier sections of the people. Poorer nations have fewer industries and burn lesser fossil fuels. Huge finances are required to adapt and take actions to combat climate change. In the global talks, there is agreement to mobile US $100 billion for developing countries to take climate action – they will require to adapt to climate change and take steps to cut emissions of greenhouse gases. But so far, this target has not been achieved. Funding required to climate adaptation is going to rise further as impacts become more severe and are felt over larger areas. According to some recent reports, countries may have to spend up to $300 billion a year by 2030 and $500 billion by 2050 to address adverse climate change impacts. Of course, these estimates will be revised in future but it is being felt that the current allocation and spending is much less than what is required. Water sector will account for a substantial part of total allocation for climate action. Many climate actions in water sector are considered "no regret" actions meaning thereby that in any case, these will not cause any harm. For example, if wastage of water or leakage from water utility infrastructure is reduced, it will improve water use efficiency and help reduce risk due to water scarcity.

6.9.4 Case Study of Dharoi Project

In this section, the main heads of expenditure that were used to estimate the project cost and the apportionment of cost among purposes will be explained. As per the project report (Govt. of Gujarat, 1976), the revised estimates of the cost of works related to the Dharoi project pegged the total expenditure at Rs. 500.184 million. While preparing the estimates, the expenditures were grouped under three heads: dam and appurtenant works, Fatewadi barrage (near Ahmedabad city) and canal system, and main canal and branches.

The total cost of the dam and the appurtenant works was Rs. 325.951 million. The main heads under this expenditure were subdivided into preliminary surveys and investigations, acquisition of lands (submergence, for offices and residential colonies), structures likely to be submerged, and rehabilitation. The key construction activities for which funds were needed were masonry dam and spillway, earthen dam, dykes, head regulator for canal, and instrumentation for earth dam and dykes. The expenditure on residential and nonresidential buildings that were required for the housing of staff and offices of the project, their maintenance, and construction of approach roads, etc. were also grouped under this head.

As water supply for irrigation was one of the main purposes of the Dharoi project, the construction of canals and water distribution system was an important activity. The heads of expenses under this broad heading included preliminary investigations and tests, land acquisition for canals and branches, and construction cost of regulators, earthwork, lining of the canals, falls, cross-drainage works, bridges, and escapes. It was also necessary to construct some buildings for office purposes for the canal division, service roads, etc. The total cost under this head was estimated at Rs. 94.049 million. As part of this project, a barrage known as Fatewadi barrage was also constructed downstream of the city of Ahmedabad to divert water for agriculture. The expenditure on barrage and canal system was estimated at Rs. 80.184 million. It may be noted that these estimates were prepared in the mid-1970s.

While allocating the cost of the project for various purposes, the following benefits were identified.

- **a.** Drinking water supply to the city of Ahmedabad and Gandhinagar.
- **b.** Irrigation in the command area.
- **c.** Partial flood control, and
- **d.** Improvement of road communication due to the dam and additional roads constructed for the project purposes.

It was argued that the flood control benefit is a result of the build-up of storage in the reservoir above FRL and hence it is an incidental benefit for this project. Similarly, the improvement in road communication was also considered to be incidental in nature. Therefore, the apportionment of the project cost was made for irrigation and water supply purposes. The committee constituted to identify the method for this apportionment recommended the following two methods.

1. Reservoir storage space used for each purpose, and
2. Quantity of water utilized for each purpose.

In the reservoir space method, the idea is that each purpose should be charged on the basis of the reservoir storage space that is used to hold water for that purpose. The space required for municipal water supply needs in post-monsoon season as well as storage reserved for water supply (to ensure 90% reliability) were entirely charged for water supply use. Likewise, the storage space to meet irrigation requirements in the command in post-monsoon period was charged to irrigation. The carryover space was allocated between water supply and irrigation as per expected use. The evaporation loss was divided between these two purposes in the ratio of storage space occupied for each purpose. The space allocated for sediment accumulation was equally divided between two purposes.

When all these requirements for two purposes were summed up and percentages were determined, these turned out to be 59.5% for water supply and 40.5% for irrigation. Therefore, of the total project cost, 59.5% was charged to the water supply purpose and 40.5% to irrigation. The expenditure for water supply was charged to Ahmedabad and Gandhinagar city in proportion to the projected requirements. Accordingly, 92.6% of the cost was allocated to the city of Ahmedabad and 7.4% to the city of Gandhinagar.

6.10 CLOSURE

The economic analysis of water resources projects is an evaluation of various alternatives, plans, or policies. It can be used to identify the best and most feasible alternative as per economic criterion to logically resolve the conflicts and to assess the sensitivity of the project outputs to changes in project configuration and economic conditions. Economic analysis is intended to guide decisions on the use of scarce water resources and economic conditions and to provide criteria for ranking different water development and management policies.

In any scheme of water resources development, it is necessary to prepare a number of technically feasible alternatives to meet particular objectives and estimate costs and benefits, both tangible and intangible, for each alternative. Each of the technically feasible alternatives will produce a unique time pattern of consequences. To compare alternative plans or policies, all consequences are expressed in monetary values, and cash flows representing the variation of costs and benefits with time stream are discounted to find out their net present worth.

Sensitivity analysis and multiobjective planning further expand the scope of analysis and involve broader social goals that water resources development might help attain, like a larger national income, a more equitable distribution of income between people and region, and environmental protection. Fundamental to all development planning is the availability of reliable and adequate data about water and related land resources of a region meant to serve. Socioeconomic factors play a vital role in planning, evaluating, and implementing the water resources development plans. The economic analysis involves the decision makers while testing the projects for their feasibility and allocating the cost in multipurpose projects.

References

Chopra, B., Khuman, Y.S.C., Dhyani, S., 2022. Advances in ecosystem services valuation studies in India: learnings from a systematic review. Anthropocene Science 1, 342–357. https://doi.org/10.1007/s44177-022-00034-0.

Costanza, R., d'Arge, R., de Groot, R., Farber, S., Grasso, M., Hannon, B., Limburg, K., Naeem, S., Robert, V., O'Neill, Paruelo, J., Raskin, R.G., Sutton, P., van den Belt, M., 1997. The value of the world's ecosystem services and natural capital. Nature 387, 253–260. https://doi.org/10.1038/387253a0.

de Groot, R., Brander, L., van der Ploeg, S., Costanza, R., Bernard, F., Leon, B., Christie, M., Crossman, N., Ghermandi, A., Hein, L., Hussain, S., Kumar, P., McVittie, A., Portela, R., Rodriguez, L.C., ten Brink, P., van Beukering, P., 2012. Global estimates of the value of ecosystems and their services in monetary units. Ecosystem Services 1 (2012), 50–61.

de Neufville, R., Stafford, J.H., 1971. Systems Analysis for Engineers and Managers. McGraw-Hill Book Company, New York.

Frederick, K., VandenBerg, T., Hanson, J., 1996. Economic Values of Freshwater in the United States. Resources for the Future, Washington, DC, p. 1996.
Gittinger, J.P., 1982. Economic Analysis of Agricultural Projects. Johns Hopkins University Press, Baltimore.
Govt. of Gujarat, 1976. Sabaramati Reservoir Project at Dharoi (Revised), I. Public Works Department, Govt. of Gujarat, Gandhinagar.
Griffin, C.G., 1998. The fundamental principles of cost-benefit analysis. Water Resources Research 34 (8), 2063–2071.
Heaney, J.P., Dickinson, R.E., 1982. Methods for apportioning the cost of a water resources project. Water Resources Research 18 (3), 476–482.
Howe, C.W., 1983a. Benefit-cost analysis for water systems planning. In: AGU Water Resources Monograph 2. Washington, DC.
Howe, C.W., 1983b. Impact of price on residential water demand: new insights. Water Resources Research 18 (5), 713–716.
James, L.D., Lee, R.R., 1971. Economics of Water Resources Planning. McGraw-Hill Book Co., New York.
Loucks, D.P., Stedinger, J.R., Haith, D.A., 1981. Water Resources Systems Planning and Analysis. Prentice Hall Inc., New Jersey.
Navalawala, B.N., 1993. Economic Aspects of Irrigation and Pricing of Water. Irrigation Water Management, Water Management Forum, Institution of Engineers, Ahmedabad, India.
Newnan, D.G., 1983. Engineering Economic Analysis. Engineering Press Inc., San Jose, California.
Rogers, P., Bhatia, R., Huber, A., 1998. Water as a Social and Economic Good: How to Put the Principle into Practice. Global Water Partnership/Swedish International Development Cooperation Agency, Stockholm, Sweden.
Votruba, L., Kos, Z., Nachazel, K., Patera, A., Zeman, V., 1988. Analysis of Water Resources Systems. Elsevier, Amsterdam.
Young, H.P., Okada, N., Hashimoto, T., 1982. Cost allocation in water resources development. Water Resources Research 18 (3), 463–475.
Young, R.A., 1986. Chapter 3: Water economics. In: Mays, L.W. (Ed.), Water Resources Handbook. McGraw-Hill, New York.

Further Reading

Lund, J.R., 1992. Benefit-cost ratios: failures and alternatives. Journal of Water Resources Planning and Management, ASCE 118 (1), 94–100.
ReVelle, C., Whitlatch, E.E., Wright, J.R., 1997. Civil and Environmental Systems Engineering. Prentice Hall, New Jersey.

CHAPTER

7

Environmental and Social Considerations

Chapter outline

Water Resources Systems Planning and Management. https://doi.org/10.1016/B978-0-12-821349-0.00012-5

The objectives of this chapter are to:

- explain the role of water in the environment,
- provide an overview of environmental impacts of water resources projects,
- explain the procedures for the assessment of environmental impact of water resources projects and mitigation,
- discuss rehabilitation and resettlement aspects, and
- describe environmental and social issues associated with Sardar Sarovar Dam, a major multipurpose project in India.

The word environment means "surrounding conditions influencing development or growth." The surroundings encompass the whole complex of factors—including all the living beings (the flora and the fauna); the various life-supporting systems, such as land, air, and water; as well as society. The study of interactions among the various elements of the environment is the subject of ecology. Ecology is "the branch of science concerned with the relationships between organisms and their environment." Ecohydrology is a closely related interdisciplinary study area that deals with the interactions between hydrologic and ecological systems. It can be viewed as a subfield of hydrology with an emphasis on ecological aspects. Hydrologic and ecological systems interact in water bodies (such as lakes, rivers, and ponds) and in forests and terrestrial ecosystems. Hydrologic systems have an important role in the evolution and sustenance of ecosystems.

Environment can be considered a natural resource. When this resource is overutilized, problems usually arise. Overexploitation of resources is common in both developing and developed countries, albeit due to different reasons.

The environmental "hot spots" are the locations that have critical deficiencies in assimilative or supportive capacity. Water, both quantitatively and qualitatively, is the limiting resource in most regions and the future planning should be based on this reality. Among natural resources development projects, water-related projects have the most profound impact on the environment. The interest and involvement of the general public in environmental issues began to grow in the 1960s. This, of course, does not mean that these problems were not recognized or the issues were not addressed in past projects. Nevertheless, the attention and the resources earmarked to address these issues have grown with public concern. The Environmental Impact Assessment (EIA) became an integral part of each important project in the 1990s. Currently, for many projects, social and environmental problems are proving to be more difficult to handle than technical problems. Now, along with technical and financial feasibility, it is necessary to examine the social and environmental feasibility of a water resources project.

A major man-made intervention, like a storage reservoir or extensive groundwater pumping, will change the existing balance and force the environment to seek a new stable state. The new state does not get established immediately, it is developed over time; compared to the current, the new state might be better and more acceptable or worse and distressing. Thus, the developers need to ensure that any development does not augment the processes that are undesirable or harmful to the society. The magnitude of intervention is crucial. Any intervention that will adversely affect the environment with consequential irreversible harm to life is to be checked. Further, no intervention is not always the right strategy.

As is true for many other natural processes, the environment is not a static concept. It is an ever-evolving and dynamic entity. This concept is elaborated in the next section.

7.1 DYNAMISM OF ENVIRONMENT

Environment is in dynamic balance with its elements. Any major change in one element upsets this balance and depending on the quality and intensity of the destabilizing action, the environment regains its balance or attains a new dynamic balance over a period. Consider this: in a dense forest, chopping off a few branches of a few trees disturbs the balance locally but this will not cause noticeable adverse impacts on the forest as the trees will grow again in a short period of time. Controlled exploitation of forests also disturbs the environmental balance but a new balance with small degradation is attained in a reasonable period of time. However, after the uncontrolled plunder of forests, the balance and biodiversity may suffer a big loss forever, leading to progressive degradation of the environment.

The environment in a place keeps on evolving due to natural processes, changing climatic conditions, as well as human activities. There is no unique steady state for the environment at a place on a time scale longer than a century. For example, the Indo-Gangetic plains in north India are the result of natural processes that have taken place over millions of years. The place where the mighty Himalayas stand today was occupied by the ocean in the geologic past. Likewise, many climatic and hydrologic variables follow a long-term cycle. Therefore, when nature itself is not in a permanent state, the life of human-made systems cannot be unlimited and it is futile to talk of their permanency. What needs to be ensured is the longevity of these structures, harmony with the environment, and their ability to meet expected future requirements.

While managing the environment, depending on the malaise, the remedy may be simple, rapid, and painless or it may be complex, slow, and painful. The shortage of water in an area may be overcome by a few wells or it may require long-distance transfer of water. Flood mitigation may necessitate the construction of a large reservoir. The analogy is similar to human illness—a sick person may need a small tablet to get well or may even require a major surgery to regain health. In cases of extensive treatment, expert supervision coupled with the necessary follow-up and monitoring is needed to avoid adverse side effects. Equally important is the reversibility of the system. For some environmental problems, if a remedial action is not initiated in time, consequences may be irreversible and it may be impossible to restore the system to its previous state. People have to understand this in addressing the challenges due to climate change.

While assessing the degree of intervention that is likely to be bearable, the natural variability in the environment is an important factor. Environmental systems are stretchable to some extent; the range of stretch is different under different natural settings. It is not possible to quantify this inherent property of nature and provide some index in view of the large number of variables and uncertainties. The idea can only be expressed in a somewhat qualitative manner. Nevertheless, the systems should not be stretched so much that they collapse or breakdown.

7.2 WATER IN ENVIRONMENT

Water has a central and key role in the environment, but the contribution and importance of water in the environment are not the same in all geographical areas or all year round in the same area. Depending on the location, topography, geology, and precipitation characteristics, some areas have higher water availability than others. The role of water in the environmental system also depends on other factors. Many times, forests, pastures, agricultural cultivation, and urban areas provide different sets of environmental parameters. A change in land use brings in a new "environmental state" for water. The developmental potential also varies considerably from area to area. Many ills of the development programs for water have their roots in the inadequate appreciation of these differences and attempts to apply similar solutions for dissimilar environmental situations. Vegetation is a good indicator of the availability of water in an area. Plants are an indicator of the quality of water in the ecosystem; all tree species do not survive well in saline or waterlogged areas. Water management in arid regions must be different than in humid areas, and so on.

Most environment-centered discussions and writings tend to be influenced by emotions and ignore the hard ground realities. Apparently, people form their opinion about the environment mostly based on what they see around the picnic spots or in movies. However, the natural environment is not beautiful and pleasant everywhere or at all times. At times, it can be extremely harsh, dangerous, ruthless, and furious. Imagine the conditions in the deserts, droughts, avalanches, cyclones, hurricanes, or a river in spate. Those who have been through these can appreciate what these situations mean. Even seemingly innocuous weather with temperatures close to 40°C causes several hundred deaths each year. Besides, the environment of a place, which is pleasant in one season, may become quite hostile in other seasons. After all, the migratory birds from near-polar regions move to warmer places in winter because their habitats are no longer fit for living or there is not enough food. Natural water is also not always pristine and healthy as is widely believed. Ocean water is too salty to consume. Pure rainwater, after falling on the earth and mixing with pollutants, may no longer be fit for human consumption. This is the reason why in case of flooding of the countryside, paradoxically, water may spread epidemic, and providing clean drinking water to the population is a serious challenge.

In adverse situations, only human effort and ingenuity can mellow the harshness of the environment, control threats, and make life comfortable. Places, such as Israel and Kuwait, are testimony to this. However, new challenges are emerging because of the increasing population pressure, climate change, etc., and people are settling in unsafe areas. Providing life-support systems to them, including water and other amenities and protection against disasters, is necessary. Water is the dominant element in the ecology that can bring about a desired change. Obviously, land and air cannot be as easily manipulated as water. Hence, water has an additional role to play, namely improving the environment, and is to be managed and developed in harmony with its surroundings. The principal aim of water management must be to create this harmony where it does not exist by changing the availability pattern in a dry area or protecting an area from water excesses. While dealing with water, the objective should not be just environmental protection but also improvement.

In many regions, because of poverty, lack of resources, and increasing pressure of population, the environment has been degraded and forest wealth has been plundered. Water management in such areas has, therefore, an additional important role—regenerating and restoring the lost plant life and biodiversity, which, in turn, can support other life and lead to improvement in the environment.

Every effort to develop water resources results in some modification of the environment. Sometimes, the impact is confined mainly to the river course, aquifer, or lake itself. In other instances, effects are much more widespread and may result in considerable alterations in land resources, forests, or fisheries. Beyond this, water resources development (WRD) may have major impacts on human settlements and economic activities. The extent of these impacts depends on the ability of various physical, natural, and human systems to absorb them.

Sustainable development and environmental protection are two related aspects of an optimal resource utilization strategy. Due to the sheer size and consequent developmental activities, the WRD projects do result in significant impacts on the regional environment—these impacts can be positive as well as negative. To have proper development of the resources without destroying the ecological balance, environmentally sound planning is needed.

Environmental issues were not understood or viewed with seriousness till a few decades ago but the awareness and knowledge have considerably increased in recent times. There is a growing concern about the adverse social and environmental impacts of water development projects in many countries. The reason is the result of studies carried out all over the world, which have demonstrated the significance of environmental impacts. Therefore, it has become imperative to evaluate any development plan, especially WRD projects, from the social and environmental standpoint and take adequate mitigation measures right in the beginning.

7.3 ENVIRONMENTAL IMPACTS OF WATER RESOURCES PROJECTS

WRD activities date back to thousands of years. However, only during the last 4–5 decades, both the size and number of projects have increased significantly, with concomitant impact on the environment. The water resources-related activities that lead to major environmental impacts are as follows: construction of dams and canals, submergence due to reservoirs, irrigation, and waterlogging, excessive withdrawal of groundwater over a large area, and interbasin transfer of water. Although the environmental impacts could be beneficial or adverse, the impacts that are adverse need detailed analysis and investigation.

The interaction between the various environmental components is a polymorphous task. Humans are the cause of many environmental disturbances and also the hub around which the impact assessment should be addressed. Assuming that the environmental system is in "equilibrium" stage (i), the development of a water resources project will induce an action into the system. This action will produce a reaction and dynamic interaction between the environmental components involved. After a quasi-steady state has been reached, a new "equilibrium" stage ($i + 1$) of the environmental system will be evolved. A schematic representation of the man-nature interaction is given in Fig. 7.1. The purpose of EIA is to predict stage ($i + 1$), compare with stage (i), and improve decisions, planning, implementation, and monitoring of the project. A trade-off between the conditions of the human component and the ecological system at stage ($i + 1$) is possible. During the EIA, emphasis should be placed on the analysis of the impacts of the proposed water project; however, when applicable, the effects of natural evolution processes must be taken into account.

There are many ways to classify environmental impacts as discussed in the following.

7.3.1 Primary and Secondary Impacts

Environmental impacts can be generally classified as primary and secondary. Primary impacts are caused directly by the project works, such as loss of forests due to submergence, change of a river regime due to the construction of a dam, etc. These impacts are relatively easy to measure. Secondary impacts are caused by the project outputs, such as flow regulation, channelization, and water supply. These are indirect consequences of project outputs. For example, a project may begin supply of water for irrigation and its effect on the fertility of the irrigated soil can be termed as a secondary impact. Secondary impacts could be equal or more pronounced than the primary impacts and, unfortunately, often more difficult to predict and measure. The distinction between primary and secondary impacts is often arbitrary.

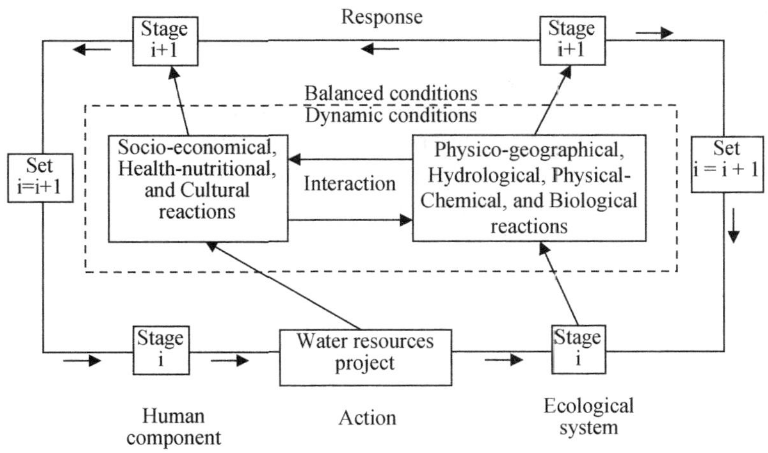

FIGURE 7.1

Human–nature interaction resulting from a water resources project.

In an ecosystem, impacts are usually complex and one impact often leads to another, resulting in a chain reaction. For instance, deforestation could contribute to increased soil erosion that would accelerate reservoir siltation, leading to a loss in fishery, which, in turn, may cause malnutrition. A major impact may often arise due to a combination of factors. The causal linkages between impacts may not be direct or clear-cut. Various types of water-related activities, such as land clearance, construction, water impoundment, water channelization, and changes in land-use patterns, can cause beneficial or adverse impacts on the environment.

7.3.2 Long- and Short-Term Impacts

Environmental impacts can also be classified as short term and long term. Short-term impacts occur during investigation, construction, and immediate postconstruction phases. After these activities are over, the impacts also subside. Long-term impacts stem from permanent large-scale changes, such as creation of a large lake, development of perennial irrigation instead of seasonal irrigation, construction of a large canal, and large-scale deforestation.

7.3.3 Acceptable and Unacceptable Impacts

Environmental impacts can also be broadly categorized as follows:

1. Totally unacceptable impacts, for example, submergence of important places. Such a place could be a large city, an important historical monument, a national park, or a forest inhabiting rare species.
2. Conditionally acceptable impacts, such as dislocation of population, submergence of forest land, submergence of mineral deposits, and waterlogging due to canal irrigation.
3. Neutral impacts that are neither desired nor adverse, such as the submergence of unproductive wasteland.

Environmental impacts can be further classified as quantifiable or nonquantifiable. The impacts that can be measured and mathematically expressed as functions of decision variables are quantifiable. For example, the area of submergence of forest land can be measured and, in turn, can be expressed as a function of the storage capacity of the reservoir, a decision variable. An increase in the incidence of malaria because of the construction of the reservoir is an example of a nonquantifiable impact. Even though a nonquantifiable impact may be important, it cannot be satisfactorily incorporated into mathematical models of the system. Its impacts can be gauged only through qualitative studies and judgment.

7.3.4 Adverse Impacts

Adverse impacts of WRD projects need to be identified during their planning and investigation phase as well as the implementation and operation phases. Although these phases appear sequentially in time, in practice, there could be a considerable gap between them. It is necessary to anticipate the adverse impact of all the project phases in advance so that measures to eliminate or mitigate them, where practicable, are initiated in time. Every effort needs to be made to enhance the beneficial impacts to the extent possible. As adverse impacts need to be examined very carefully, they are listed first.

Planning and Investigation Phase

The major activities that are liable to affect the environment are as follows: construction of access roads; foundation investigations, drilling and blasting; locating quarries for construction materials; layout of construction plants, etc. Sometimes it is necessary to construct access roads or tracks in the mountains during the project investigation phase. In fact, most such tracks are small clearings where land is leveled so that vehicles can pass. Creating access roads or even motorable tracks in otherwise inaccessible areas does not have significant environmental consequences. Sometimes these paths are used for undesirable and illegal plunder of forest wealth. Closing the tracks or barricading the entries after the investigations usually checks the problem.

Forest cutting in the investigation stage is generally not substantial and the adverse impact is essentially temporary. If the period between the investigation and implementation phases is large, the forest is expected to rejuvenate itself. An adverse impact of drilling and blasting in the investigation phase is expected to be minor, as these operations are carried over small areas. Similarly, areas of quarries and construction plants, etc. are generally quite small compared to the catchment area and the influence during the investigation stage is not important. Of course, the news that a project is likely to come up in the area may trigger new economic activities.

Construction Phase

During the construction phase, the environment near the project site faces extensive disturbance but this phase is of very short duration compared to the life of a project. In the case of dams or weirs/barrages, the project site covers a small area; the areal coverage could be large for an interbasin water transfer and canal project. While some disturbance is unavoidable, timely actions could largely mitigate the long-term adverse impacts. In the case of canals, the construction activity is spread over a large geographical area. But there is not much impact on the environment, except for the chopping of trees and minor changes in the topography due to cutting or filling. Similarly, in the case of hydropower projects, trees may have to be felled to construct dam/diversion structure and powerhouse and lay transmission lines.

Construction of dams/hydropower projects involves the transportation of heavy machinery, gates, turbines, penstocks, and so on. If the existing roads and bridges are inadequate to carry heavy loads, either these will have to be suitably strengthened and/or new roads may have to be constructed to reach the project site, which is generally away from the highways/cities. Road construction causes disturbance to the environment. Measures would be required to mitigate the impacts due to preparatory activities.

To minimize the adverse impacts on the environment, the construction equipment should meet applicable norms for sound and emissions. The sound levels should be kept within specified limits and blasting operations should not coincide with nesting periods of birds. If new evidences of archeological importance are encountered during digging, the work at that place should be suspended and expert organizations should be immediately approached for advice. In mountain terrains, one may come across geological surprises—subsurface conditions are quite different than what was assumed. In some projects, this has resulted in revising tunnel alignment, revised construction plans, and cost and time overruns. Construction of tunnels has also resulted in an unwanted impact—the drying of springs due to obstructions created in the movement of water in the subsurface zone.

During the construction of tunnels, large volumes of muck are produced and it should be safely disposed. At times, the muck is thrown into rivers, causing problems in downstream areas. Detailed guidelines about muck disposal have been prepared and should be followed to minimize adverse impacts on the environment. Likewise, construction materials and equipment should be transported such that this activity does not harm the environment.

In large projects, a large number of workers are engaged in construction work, and temporary housing near the project site is provided to them. The influx of a large number of temporary visitors in an area usually results in environmental degradation but the impacts are usually small and temporary.

Immediately after construction is over, all efforts should be made to restore the environment in the vicinity of project areas to its original or even better state.

Operation Phase

During this phase, the impacts are either due to the diversion of water from an area to another, which may also involve trans-basin diversion, or applying water to an area that is facing problems due to inadequate water. In all diversion projects, only the excess water after considering the need of that region is diverted. Hence, this should not lead to any undesirable consequences.

The environment will suffer if the incoming floods for a hydro project are not moderated properly. Also, projects should be operated such that the environmental flow requirements in the downstream river are always met. Groundwater should be utilized such that the water tables do not fall rapidly and excessively.

The negative environmental impacts of a WRD project are summarized in Table 7.1.

7.3.5 Beneficial Impacts

To take a holistic view of WRD, it is necessary to examine the beneficial impacts of WRD projects. It is interesting that of all the registered large dams in the world, only about 5% are in Africa where most of the countries facing severe water scarcity are located. About 55% of large dams are in North America and Europe and largely because of this, there are not likely to be severe water shortages (Keller et al., 2000). The world's biggest dam, The Three Gorges Dam, has been constructed on the Yangtze River in

Table 7.1 Summary of negative environmental impacts of WRD projects.

Phase	Impacts on physical environment	Impacts on socioeconomic environment	Impacts on biotic environment
Planning and investigation	Minor increase in pollution.	Small increase in economic activities.	Some forest cutting.
Construction	Higher erosion and pollution at construction site. Temporary change of flow patterns and sediment load of rivers. Blasting operations. Impacts due to unsafe muck disposal.	Law and order problems due to influx of "outsiders." Loss of cultural and historical monuments. Dislocation of people from construction area. Influx of people with different cultural backgrounds.	Workers' health. Destruction of flora and fauna. Deterioration in downstream water quality. Waterborne diseases.
Operation	Possibility of waterlogging in command. Changes in river flow regime. Less water in downstream areas due to diversion. Less flooding of wetlands with silt. Changed flow regime in coastal zone. Saline water intrusion. Sedimentation of reservoirs.	Displacement of people from submerged area. Loss of income from the area impounded. Possibility of spread of waterborne diseases. Population increase. Ill effects of tourism and recreation.	Loss of flora and fauna, wildlife habitat. Loss of natural reserves. Changes in water quality and sediment load (eutrophication). Damage to ecology due to tourism. Spread of aquatic weeds in water bodies. Spread of diseases.

China. Reservoir behind this dam is 600 km long and has gross capacity of 39.3 km^3. Its power station has an installed capacity of 22,500 MW.

We now describe the beneficial impacts of water projects.

Irrigation Benefits

In many places, irrigation is the only source of providing water to crops at the right time. Increase in agricultural production is the most important beneficial impact of irrigation projects. In the global context of food production, there is no alternative to self-sufficiency in food production on a sustained basis. Consider the case of India with about one-third area being drought prone and about one-eighth flood prone. This coupled with erratic rainfall and the need to feed over one billion people, the importance of irrigation is self-evident.

For many countries, irrigation is the key to rural and national development. With assured irrigation, agricultural production increases, diversified crops are grown, per capita food availability is better, and food and nutrition intakes are higher. Higher availability of water also promotes increased livestock holding; this can be a source of assistance in farming and increased availability of animal protein. In addition to the improved food and nutrition, the improvements in education, health facilities, the status of women, and general advances in the overall quality of life further advance the health status of the rural people. With increased agricultural production, a number of agroindustries are setup yielding multiple benefits—higher income to farmers, reduced wastage of crops, better nutrition to people, etc.

Based on a study of an irrigation project, Shah (1993) concluded that gross receipts per man-day in irrigation areas increased by more than 100% in all economic activities. Irrigated farming also increased permanent employment of the order of 30–50 man-days per ha. This is a very significant improvement in view of the large-scale unemployment in many countries. Not only this, the growth of agricultural output also triggers growth in other sectors of the economy. For example, it leads to increased demand for fertilizers, pesticides, farm machinery, water pumps, motors, transportation, cold storage, etc. Of course, there is a need for repair and maintenance of farm equipment and this generates more employment.

International Water Management Institute (IWMI) carried out a multicountry study to examine in detail the benefits and potential harms of irrigation (Hussain, 2007). They found that irrigation benefits vary widely across systems, and depend on many factors that include local conditions, governance, irrigation policy, and broader economic and political factors. It was found that through the multiplier effect, indirect irrigation benefits could be larger than direct benefits. The distribution of irrigation benefits also varies widely by the type of benefit and the socioeconomic status of beneficiaries. The direct benefits generally accrue to landholders, while a significant part of the indirect benefits accrue to the landless and small farmers. Further, integrated management of systems and services increases overall benefits manifold. Irrigation can also lead to some adverse social, health, and environmental impacts, such as displacement of people due to new projects, health risks from water-related diseases, land and water degradation, loss of biodiversity, and river degradation due to water withdrawals. Often, negative social and environmental consequences adversely affect the poor more than others. Most adverse impacts are due to inadequacies and ineffectiveness of institutions and governance. Many potential adverse impacts can be avoided/minimized by effective planning, design, and management. Regarding secondary and tertiary benefits, in India, the average receipts for households with irrigation are found to be more than double than of those without. As it happens, families with higher income spend more on nonfood items, say cloths, TV, entertainment, mobile phones, houses, vehicles, and so on. The sale of cement, consumer durable (washing machine, TV), and automobiles has been found to be higher during the years of good crops when the farmers have higher disposable income.

Hydropower Benefits

Many large water projects also generate hydropower. The generation of clear energy from a renewable resource where no fossil fuel is burned is a boon for air quality. The operation and maintenance costs of hydropower plants are very low, of the order of 1% of capital cost, and these can be started and stopped in a few minutes. The water is not consumed in the process and is immediately available for other uses. According to estimates, about 16% of the world's electricity comes from hydro sources and nearly 60% of all renewable electricity is generated by hydropower. Since no harmful gases are emitted in hydropower, this sector is becoming a preferred choice to mitigate climate change.

Flood Control

The storage of excess flood waters behind a dam or confining the river flow in between embankments are two major strategies of flood control. The amelioration of the distress from flooding and the consequent increase in food production from the areas protected from the ravages of floods is, by no means, a small benefit. The benefits of the Tennessee Valley development in the United States are well known. A series of dams built in the Damodar basin in India has substantially controlled flooding in a basin that was earlier termed "the sorrow of Bengal." The Three Gorges Project is the backbone of the flood control system to protect the areas in the middle and lower reaches of the Yangtze River in China. The average annual savings from flood damage alone are likely to be equal to a significant part of the cost of this project. The influence of the High Aswan Dam (HAD) on the economy of Egypt is well known. The stored water saved Egypt from famine in 1972–73 and again in nine successive drought years 1979–87. Lake Nasser (reservoir behind HAD with a gross capacity of 168.9 billion cubic meters) has protected the Nile valley from major floods in 1964, 1975, and 1988. It has brought 0.324 million ha of additional land under irrigation and has converted 0.283 million ha from flood to perennial irrigation.

Municipal and Industrial Water Supply

A number of metropolitan cities depend on dams and canals for domestic water supplies. Sometimes water is supplied over large distances, for example, to the city of Mumbai (from dams in the Western Ghats) and Delhi (Upper Ganga Canal and Ramganga dam). Many industries, for example, steel making, thermal power generation, petroleum refining, etc., critically depend on water. The Rihand dam provides water for several mega-thermal power plants in India. Reservoirs have been specifically constructed to meet industrial water needs, for example, the Tenughat dam in India for the water needs of the Bokaro Steel plant.

Water resources projects provide a dependable source of drinking water. The very availability of drinking water leads to the improvement of health and sanitation. Improved economic status due to these projects also raises the general standard of living.

Fisheries, Flora, and Fauna

Large reservoirs also provide a good site for the development of fisheries. Reservoirs in many countries have helped fisheries development by stocking them with fingerlings of sweet water fish varieties. The reservoirs are also known to have attracted aquatic birds from long distances in the migration season and may be developed as aquatic bird sanctuaries. Many wildlife and bird sanctuaries around the world are by-products of river valley projects.

Reservoirs also have positive impacts on the climate of the nearby area. In contrast to the creation of heat islands by mega cities, large industries, or thermal power plants, reservoirs create cool islands. Cool winds coming from the lake give a refreshing and relaxing feeling.

Recreation

Water has a central role in outdoor recreation. Large water bodies are tourist attractions and places for recreation because water-based sports, such as swimming, boating, and skiing, are immensely popular. Many reservoir sites are used or are being developed as recreational centers around the world.

Table 7.2 Summary of positive environmental impacts of WRD projects.

Phase	Impacts on physical environment	Impacts on socioeconomic environment	Impacts on biotic environment
Planning and investigation	Minor	Some reduction in unemployment Escalation of real estate prices	No impact
Construction	Diversion of river flows More dust and air pollution Development of infrastructure, means of communication such as roads, etc.	Higher employment and incomes More commercial activities More labor and visitors to the area	Trees may be cut Land submergence may begin in later stages Compensatory afforestation may be carried
Operation	Conservation of water Regulated stream flow. Generation of electric power Improved quality of water Reduced downstream flows and silting Raising groundwater levels Navigation facilities Catchment area improvement Command area development More pleasant climate	Higher and diversified grain and fiber production Lesser damages due to floods Reliable domestic and industrial water supply Diversification of economic activities Better healthcare More avenues of recreation Higher earnings from tourism Poverty alleviation	More species and quantity of aquatic life More visits by migratory and seasonal birds Diversity of flora

Vrindavan Garden at Krishnaraj Sagar in Karnataka (India) is a world-famous example. The High Aswan Dam and its surroundings are tourist attractions and the number of tourists per annum has increased from 80,000 in 1960 to 300,000 in the 1990s (Abu-Zeid et al., 1997). Sailing along the Nile from Cairo to Aswan in winter has become a popular sport.

Besides the above, there are many other benefits like navigation, wildlife preservation, and microclimatic improvements due to water bodies. The positive environmental impacts of WRD projects are summarized in Table 7.2.

Asmal (2002) presented a balanced opinion about dams by stating: "... dams are neither the problem nor the solution. They are merely one tool that society may collectively select or reject to improve their lives. ... it is safe to say that dams have delivered many benefits. It is also safe to say that in too many cases, the price paid to secure these benefits has been unacceptable and often unnecessary."

7.4 ENVIRONMENTAL IMPACTS OF RESERVOIRS

The creation of reservoirs results in far-reaching changes in the ecosystem. The major effect is on the land, which is inundated and on the aquatic environment. Soil erosion, sedimentation, stratification, adverse effect on fish, and proliferation of aquatic weeds are some of the major disruptions in the ecosystem that may involve economic loss. The effects of reservoirs on the terrestrial environment are felt on forests, wildlife, groundwater, climate, and agriculture. Human environment is affected in respect of alterations in human settlements, occupational patterns, and waterborne diseases. But, it would be unfair to blame dams for all the ills—they are just one of the options or tools that society may choose to manage river waters.

An increasing constraining factor in dam construction is the conflict between the concern for the environmental effects of reservoirs and the growing need to manage water resources for irrigation, electric energy demands, flood control, etc. Compared to big dams, small dams are said to be less destructive to aquatic life and cause less damage to the general environment and other aesthetic factors. However, they can impart limited control on streamflow. A series of small dams cost more than one big dam of equivalent capacity and are more difficult to manage.

An understanding of the interrelationship between reservoirs and natural ecosystems of the region is essential to not only preserve the existing environment but also to further improve its quality. Such goals as the improvement of natural conditions and environments are adequately achieved by the construction of multipurpose reservoirs. The long-term use of reservoirs has clearly proved that in a variety of climatic zones, the reservoirs very well harmonize with the natural environment. The environmental impact of reservoirs and remedial measures are given in the following.

7.4.1 Physical Impacts

Most of these impacts have been discussed in Section 7.3. A few important impacts that are specifically due to reservoirs are discussed here.

Submergence of Land

As water is stored behind a dam, land is submerged. Usually, the dam site is selected to keep the submergence as small as possible, subject to the project objectives, and technical requirements of the dam foundation, abutment, and the reservoir. Usually, submergence is less than 10% of the benefited area. Although no hard and fast rule can be laid down, such submergence should not be more than 20% of the extent of the benefited area, because with larger submergence, problems of resettlement and rehabilitation become intractable. It may be noted that, generally, the displaced population is about 2%–4% of the population benefited from the project. This issue is discussed in detail later in Section 7.10.

While large dams lead to the submergence of the land area, they are not a major cause of deforestation. For example, even if all the envisaged dams in the Ganga basin are constructed, the submerged area of the forest will be less than 2%. Poor people in many countries depend on wood for domestic fuel requirements. The availability of cheap electrical energy by hydroelectric projects will help in the reduction of deforestation because some fuel requirements can be met by electricity. In fact, a large number of trees are cut every year to meet the needs of packaging industries, timber for building construction, furniture and transport sector, etc. Indiscriminate grazing is also harmful to forests.

As most dams are constructed in the upper reaches of a river, the submergence of forests and wasteland cannot be avoided. Clearly, there is a trade-off between submergence of cultivable and forest lands, or submergence of lands at one site against another site. However, the forest cover lost on account of WRD projects is a very small part of the total forest loss in a country. In any case, to reduce this loss, permission to cut forest is usually given with the stipulation of compensatory afforestation. Various countries have evolved different norms. In India, for example, an area equivalent to five times the area of forest likely to be affected (submerged or cut for any of the project features, such as canal) is to be afforested and maintained for 5 years at project costs. The project estimates should provide about 1% of the project costs for this purpose. These provisions, if properly followed, should more than compensate for the ensuing deforestation.

A reservoir may also submerge private agricultural or nonagricultural lands; such lands are required to be acquired under the prevailing land acquisition act. Experience shows that land acquisition is commonly resisted by owners and hence, it usually takes time. It is not always feasible to strictly apply the legal provisions to such acquisition. Many matters are better sorted through dialogue and persuasion rather than confrontation.

At times, reservoirs may submerge structures of public importance, such as temples and mosques, which are owned by the community. As a rehabilitation policy, such structures are usually recreated at a somewhat improved scale of facilities at the new rehabilitation sites. The project may also submerge structures of architectural importance or of cultural heritage. Such structures may have to be shifted block-by-block to a new safe site. At times, such submergence involves sites occupied by past civilizations (e.g., Nagarjunkonda behind the Nagarjunsagar reservoir in India). Of course, in some cases, it may be impractical to shift the ruins and reconstruct them at safe places except at prohibitive costs. It might then be appropriate to create a museum at project costs with extensive photographic records of the past civilization. The important monuments recovered from such excavations could be exhibited so that future generations are made familiar with the culture of the past civilization. While constructing the High Aswan Dam, 17 temples that were considered as important monuments were shifted to higher elevations. These include Abu-Simbel, Philae, and Kalabsha, and these are now attractions for tourists (Abu-Zeid et al., 1997).

It is important to ensure that the proposed project does not submerge or affect adversely any endangered species of flora or fauna. Therefore, it is desirable that botanical and zoological survey of the submergence area is done by experts from related fields. The remedial steps, if necessary, should be initiated in the planning stage itself.

Reservoir-Induced Seismicity

One important objection to large WRD projects is reservoir-induced seismicity (RIS). RIS is the incidence of an earthquake triggered due to the impoundment of water behind a dam. Many people believe that reservoirs trigger earth tremors due to the load of water. However, a reservoir, at worst, can only advance an earthquake that would have occurred otherwise too. The magnitude of forces associated with an earthquake is several orders bigger compared to the additional load of water in the reservoir. The change in stresses due to water load is too small to cause a fracture in the earth's crust (Srivastava, 1993). Therefore, the presence of a reservoir does not increase the severity of an earthquake. While RIS may be associated with some of the large dams and reservoirs, there is adequate evidence to demonstrate that some of the large dams have not shown RIS (e.g., Bhakra, Beas, Pong, Ramganga, and Tarbela projects, which are located in the Himalayas). In some areas, earthquakes have occurred after the reservoirs were created but there is no conclusive evidence that these were caused by the reservoirs. Yet, another view is that the water seeped from the dam provides a lubricant effect and triggers small quakes. In fact, this is a positive feature as it helps in the release of energy in small shots that are less damaging than a big earthquake. While there is no complete agreement on this issue, a general belief is that earthquakes are caused by the filling of reservoirs at sites where natural stresses in the underlying rock mass have developed to a state close to rupture. Perhaps, reservoirs might advance the earthquake, which would otherwise have occurred sometime later.

Earth or rockfill dams have a better capacity to resist earthquakes than a rigid dam due to their inertia, high damping, and the ability to undergo large strains without cracking. The dam design can take care of earthquakes by choosing an appropriate type of dam and a reasonable safety factor. The designs of project elements, such as intakes, powerhouse should provide for earthquake resistance. Modern technology has enabled designers to build safe dams in highly seismic regions in central Asia, the United States, Japan, etc. Many tall dams have successfully withstood large earthquakes.

Siltation

The river water deposits silt into the reservoir that, in turn, triggers a number of environmental problems. Silting may be reduced by placing outlets in the dam at such points that allow some of the silt to migrate to the downstream channel. Siltation of reservoirs due to sedimentation from the catchment area is incorrectly attributed as an adverse impact of WRD projects. Increased siltation should be attributed to indiscriminate activities in the watershed areas, such as shifting cultivation, overgrazing, and faulty cultivation practices. Reservoir sedimentation has been discussed at length in Chapter 12.

7.4.2 Biological Impacts

Some of the major impacts in this category are as follows.

Flora and Fauna

Deforestation of the reservoir or canal submerged area and consequent displacement of wildlife and population are inevitable. Adverse impacts include the removal of feeding areas, some loss of habitat, and limitations of movement. The area submerged in a reservoir constitutes an insignificant percentage of its catchment area and canals occupy a small part of the command area. Moreover, mature ecosystems, such as primary forests, are not much affected by the reservoir and its surroundings. However, to enhance flora and fauna, it is essential to develop national parks, game reserves, and forest reserves. Many times the data on forest loss are not correct. In India, of the reported 3% loss of the forest area due to river valley projects, much loss has occurred in areas that are often described as forests but do not have the required forest cover (Hasan et al., 2000).

Fishery

Some species of fish migrate upstream for breeding. The creation of reservoirs without fish ladders obstructs this movement. Therefore, whenever possible fish ladders should be provided. However, if the dam height is large, say more than 25–30 m, the fish ladder becomes long and expensive. Free passage of migratory fish to and from their spawning grounds is also disturbed due to the obstruction created by the construction of the dam. The change in river regime downstream of the dam may be harmful to the fish growth. LaBounty (1984) noted that the construction of The Three Gorges Dam on the Yangtze River would result in a loss of habitat for many kinds of aquatic life, including the unique Chinese Sturgeon and there is a possibility of reducing nutrient input to the downstream and estuarine fisheries. On the benefit side, the creation of a reservoir provides a conducive environment favoring the reproduction of several fish species. If the river flows below the reservoir are higher during the dry season than under natural prestorage conditions, this creates favorable conditions for the development of fisheries in the downstream river reach.

Waterborne Diseases

Waterborne diseases are illnesses that are caused by microscopic organisms, such as viruses and bacteria. These organisms are ingested by consuming contaminated water or by coming in contact with feces. Large aquatic bodies are prone to give rise to waterborne diseases, such as malaria, filaria, and schistosomiasis. Among the waterborne diseases, diarrhea is the major killer; it is the second leading cause of death for children under the age of 5. In developing countries, a significant number of children die every year due to waterborne diseases. Therefore, it is extremely important that precautionary steps are taken so that these adverse impacts are minimized. Over 95% of waterborne diseases are preventable, and the ways to practice sanitation and treat drinking water are well known, a lack of finances and other resources, the right policies, etc. prevent millions of people from access to the tools. Climate change is expected to increase the incidence of waterborne diseases.

Aquatic Nuisance Plants

Proliferation of aquatic nuisance vegetation is associated with reservoirs. Concern, however, is limited to those aquatic plants that are larger than microscopic algae. Early planning and action can avoid some of the hazards posed by aquatic plants to public health, fisheries intake structures, and navigation. Of particular importance is the vegetation menace of water hyacinth or bull rush. Water hyacinth is a floating weed that has a prolific growth and in a relatively short period, it can cover large areas of water bodies like a carpet. This carpet harbors mosquitoes, depletes dissolved oxygen, and affects fish life. If not checked in time, this problem can attain menacing proportions. In one instance, a hydropower plant had to be shut down temporarily because the weeds had choked the intakes. Existence of water hyacinth could be an indicator of the organic pollution of water. Removal of weeds is difficult and mechanical means have to be employed. Weeds have been used for biogas generation with some success.

Water Quality

The quality of water is an important consideration for all WRD projects, as it affects all aspects of water use—for humans, animals, crops, and even industries. Construction of a project does not degrade the quality of water—the reckless and indiscriminate use of chemical fertilizers, pesticides, and disposal without proper treatment does that. Of course, building a dam over a river deprives the downstream land of nutrients normally brought by the river in the form of silt. After the construction of a dam, river flow is stored in the reservoir and this alters the supply of freshwater to the downstream area. There are instances where to get the "maximum" benefit from the reservoir; the flow in the channel downstream of the dam has been reduced to almost nil. In the areas near the river mouth, this can lead to saltwater intrusion into low-land or estuarine areas. In any case, the annual release of water from a reservoir is lower than that in the preconstruction period, because of water losses through evaporation from the reservoir and diversions.

To overcome the adverse impacts of hydro projects on river health, certain flows should always be left in rivers to sustain aquatic species and ecosystems. Such flows are termed as environmental flows (EF). The principle governing EFs recognizes that these flows are necessary to maintain downstream ecosystems and the communities that depend on them. According to the widely quoted Brisbane Declaration (2007), "Environmental flows describe the quantity, timing and quality of water flows required to sustain freshwater and estuarine ecosystems and the human livelihoods and well-being that depend upon these ecosystems" (Arthington, 2012). A detailed discussion on this topic is given in Jain et al. (2021).

7.4.3 Small Dams versus Big Dams

Although this is not an environment-related issue, it is discussed here because an oft-repeated suggestion in many discussions related to the environment is that small dams are more suitable than big dams. Basically, there are three places to store water: soil profile, surface storages, and aquifers. The storage in soil profile is very important for agriculture but only small quantities of water can be stored for a short period. A comparative analysis of advantages, limitations, and key issues associated with groundwater, a small reservoir, and a large surface reservoir is given in Table 7.3.

The main argument against big reservoirs is that they submerge large areas compared to many small reservoirs. Any dam, big or small, needs a suitable site. One just cannot build a dam wherever one wants. If a major dam is to be replaced by a number of small dams, there must be a number of suitable sites on the same river. Keller et al. (2000) argued that it is very difficult to construct safe small dams. Of course, one will also have to build dams in lower reaches, which will mean more

Table 7.3 Comparative advantages, limitations, and key issues associated with groundwater, small surface reservoir, and a large surface reservoir.

	Groundwater storage	Small surface reservoirs	Large surface reservoirs
Advantages	Negligible evaporation loss Ubiquitous distribution Operational efficiency Available on demand Good water quality	Ease of operation Multiple use Groundwater recharge	Multipurpose Large, reliable yield Carryover storage Low cost per m^3 water stored Groundwater recharge
Limitations	Slow recharge rate Groundwater contamination Cost of extraction Recoverable fraction	High evaporation loss fraction Relatively high unit cost Absence of over-year storage	Complexity of operations Sedimentation High initial investment Large gestation period Requires good sites
Key issues	Declining water levels Rising water levels Regulation of use Groundwater pollution	Sedimentation Population displacement Submergence Environmental impacts	Social and environmental impacts Rehabilitation and resettlement

Adapted from Keller et al. (2000).

submergence (often of good agricultural land) due to flat slopes and more population displacement due to higher population density.

It is useful now to examine some real cases. Consider Britain's largest reservoir Quoich on Quoich River near Fort Augustus in the Scottish Highlands. The lake behind the 38 m high earth-rockfill dam has a storage capacity of 3828 million m^3. According to Robbroeck (1996), if all British reservoirs are arranged in an ascending order of size and their volumes and surface areas aggregated, the total volume of the 327 smallest reservoirs would be needed to replace the volume of the largest, and the total submerged area would be 6705 ha, 3.5 times the area of Quoich. A similar analysis for South Africa shows that 433 small reservoirs would be needed to replace the volume (5246 million m^3) of the Gariep reservoir, with an aggregate area 222 times larger. A comparison was also made by Shah (1993) in India between the proposed Girna dam in the Mahanadi basin in Orissa and a smaller Girna dam plus eight satellite storages making up the same volume. In the case of smaller dams, the cost would be 150% higher, 60% more land would be submerged, considerably less energy would be generated, and evaporation would be 50%–60% higher.

Although social and environmental problems are probably not in direct proportion to the area submerged, it can be safely deduced that a large number of small reservoirs will be far less acceptable from that point of view. Economics would also be much worse: loss of advantage of scale, more site establishment, more spillways, and diversion and outlet works. Silt accumulation is also substantially less, as the United States Department of Agriculture figures show: reservoirs smaller than 10 acre-feet (ac-ft) silted up at an average rate of 3.5%/year, smaller than 100 ac-ft at 2.7%/year and smaller than 1,000,000 ac-ft at 0.16%/year. This alone is a powerful argument against a large number of small reservoirs. A comparative study of some key characteristics of three sizes of structures is made in Table 7.4.

Table 7.4 Contrast of characteristics of the High Aswan Dam, Dharoi reservoir, and a minor tank in Sri Lanka.

Characteristic	High Aswan Dam	Dharoi reservoir	Typical minor tank in Sri Lanka
Storage capacity	168.9 km^3	1.321 km^3	4.1 ha-m
Surface area	6500 km^2	138 km^2	5.0 ha
Net irrigated area	2,648,000 ha	36,827 ha	5.0 ha
Storage fraction of area times depth	0.29	NA	0.4
Annual evaporation loss	14 km^3	0.15 km^3	2.0 ha-m
Annual evaporation depth	2.7 m	2.458	1.0 m
Dam height	111 m	45.87 m	2 m
Crest length	3.830 m	1207 m	170 m
Embankment volume	44,300,000 m^3	NA	2600 m^3
Command area	3.4 million irrigated hectares	773,778 ha	<10 ha

Adapted from Keller et al. (2000).

While dealing with hydropower projects, comparisons must look at impacts per unit of output. The impacts of a single large hydro project must be compared with the cumulative impacts of several small projects yielding the same power and level of service. The most fundamental determinant of the nature and magnitude of impacts of hydropower projects are the specific site conditions, not the scale of the project (www.hydropower.org).

Clearly, the degree of regulation and reliability that is provided by a large dam is not possible with a small tank. Small dams tend to dry up fast during droughts, as the surface area is large. In such periods, major and medium projects are the mainstay of water supply. One has to consider this aspect also when planning projects in drought-prone and arid regions. To conclude, one must go for the optimal size of the project rather than getting bogged down in the small versus big controversy.

7.5 ENVIRONMENTAL PROBLEMS IN COMMAND AREAS

The adverse environmental impacts at the level of commands are essentially a human-induced problem; irrigation per se is not responsible for environmental degradation. The problems arise due to mismanagement of irrigation water and flawed policies that do little to check its injudicious use and wastage. The affected areas are, on average, about 3%−5% of the benefited areas. Moreover, the problems are not present in all commands and the affected command is not completely spoiled.

The canal water charges are kept at a very low level in many countries. Usually, neither the cost of irrigation system management nor the water productivity is a basis for deciding irrigation charges. In many countries, a strong political lobby is behind fixing irrigation charges and payment of dues, and concerns, other than economic, may be the dominating factor. Mismanagement of irrigation water and defective policies are largely responsible for several environmental problems in command areas. The main problems are discussed next.

Two problems of considerable importance in surface irrigated areas are waterlogging and soil salinity. Waterlogging occurs when roots cannot respire due to excess water in the soil profile. The problems arise because water and chemicals are applied in excess. Also, the drainage aspects of agricultural areas have not been given due attention. In many irrigation projects, adequate drainage is not planned initially because it will jack-up the project cost. The problem surfaces only after many years of operation. These problems have spread to many fertile and irrigated lands, particularly in arid and semiarid regions. As per estimates, waterlogging affects approx. 10% of the global land area and significantly reduces crop yield.

In some instances, seepage from canals is responsible for raising the groundwater table (GWT) to rather undesirable levels or for salinity and alkalinity problems in command areas. It is not possible to completely prevent seepage from canals unless the entire section is lined. However, such linings are quite expensive and their maintenance is a problem. Excess seepage from canals is due to deficiencies in design, construction, or maintenance. But seepage from canals is not all that bad because it also helps recharge groundwater. In many parts of the world, properly designed canals are in use for more than a century without any adverse effects.

GWT can be controlled through horizontal and vertical drainage. Horizontal drains are deep drains that take away the excess water from fields to natural drainage channels. In vertical drainage, groundwater is pumped out by vertical wells (shallow or deep tubewells depending on the situation) and either using this water for irrigation (with or without dilution depending on its quality) or draining it through natural surface drainage.

Degradation of soil fertility by waterlogging is a common and serious impact of irrigation. Many irrigation commands suffer from waterlogging to some degree. In arid regions, more irrigation water than needed for evapotranspiration must be applied to soil to avoid accumulation of salts in the root zone. As a result, groundwater is easily contaminated by fertilizers and pesticides percolating with irrigation water. This problem is more acute if groundwater is extensively used for irrigation and has a high dissolved solid content.

7.5.1 Lowering of Water Table

Overexploitation of groundwater is a serious problem in many parts of the world. Electric or fossil fuel-powered pumps and tubewells find extensive use to exploit this vital resource. Groundwater mining is the withdrawal of water from the aquifers over a period of time at rates that exceed the recharge rate. It is also called overdraft or mining of aquifers. As there is no state regulation for the use of groundwater in many parts of the world, GWTs are drastically falling in intensively cropped areas as withdrawals exceed recharge. The main reasons for the excessive use of groundwater are as follows: (1) inadequate availability or absence of surface water, (2) no disincentive to extract larger quantities of water, (3) subsidized or flat rates of electricity to extract groundwater, and (4) low rainfall. The decline in the water table results in an increase in pumping costs over time. Other problems could be land subsidence, induction of rock chemicals in water, and seawater intrusion in coastal areas.

7.5.2 Building-Up of Nitrate and Pesticide Residues in Groundwater

Irrigation projects influence groundwater in various ways. Irrigation has induced the use of more fertilizers and pesticides. Fertilizer consumption has increased because it has, in conjunction with irrigation, augmented production, while more pesticides are being used to control more diseases and pests. Application of more irrigation water without proper drainage leads to an increase in soil salinity and alkalinity. As water evaporates, it leaves behind salt particles on top soil layers. This damages vegetation and disturbs the ecosystem balance.

Important problems related to fertilizer vis-a-vis environmental quality are nitrate pollution of groundwater, eutrophication of lake and river water increased emission of gaseous nitrogen, and metal toxicities. The fertilizer-related pollution is rapidly increasing in many countries. It is affecting agricultural production and deteriorating the quality of land and water. There are reports of nitrate pollution of groundwater and surface flows through saline seeps originating from feed lots or high fertilizer fields.

Agricultural sector is the major user of pesticides. The injudicious use of chemicals deteriorates the land quality and contaminates water, food, and the environment. The spurt in pesticide use has resulted in secondary pest outbreaks. The pesticide residues in soil may create a variety of hazards. Soil microorganisms that cause breakdown of cellulose, nitrification, turnover of organic matter, and other biological materials may be adversely affected by pesticides. Pesticides and chemicals inhibit the microbial population in the soil, thereby resulting in reduced nitrogen fixation by symbiotic bacteria. There may be a serious decline in the population of earthworms due to pesticide residues and this affects crop yields.

7.5.3 Problem of Weeds

The problem of weed infestation is common in canal irrigation projects. An adequate and reliable supply of water, sunlight, and nutrients provides optimal conditions for weed growth. Bullrush, a type of grass, grows in canals, ditches, drains, and in waterlogged lands. It disturbs the flow of water causing obstruction to canal flow or drainage. Manual recourse of weed removal is expensive. It is better to drain the area where these weeds grow, which limits their growth.

7.5.4 Health Hazards

The spread of some diseases that are injurious to human health is another side effect of canal irrigation. The major waterborne diseases observed in rural areas are dysentery, cholera, malaria, and filariasis. Although huge amounts of money are spent on disease eradication and health improvement programs, the incidence of waterborne diseases shows no sign of abatement in rural areas in some developing countries, but only a small part of it can be attributed to WRD projects.

As the causes of human-induced environmental degradation vary from command to command, the remedial measures are also different. Basically what is needed is the judicious use of irrigation water for sustained agricultural development. Some of the ways to do this are as follows: better on-farm management, drainage of agricultural land, conjunctive use of surface and groundwater, environmental friendly input—output pricing policy, and formation of water users association.

7.6 ENVIRONMENTAL IMPACT ASSESSMENT

The purpose of an EIA is to determine, before implementation, the environmental impacts of a proposed action so that unintended or undesired consequences can be reduced or eliminated. EIA is a process to identify, understand, evaluate, and predict the influence of some action of man on the environment. These influences have spatial and temporal dimensions and the comparison is made with the scenario if the action had not taken place. It has been explained in Fig. 7.2. EIA should be part of policy analysis. It defines and assesses a proposed project's physical, biological, and socioeconomic effects, bringing together all aspects in a form that permits a rational decision. Negative environmental impacts are exposed, thus allowing their alleviation through the identification of possible alternative sites and the construction process. In many countries, EIA has been incorporated into legislation.

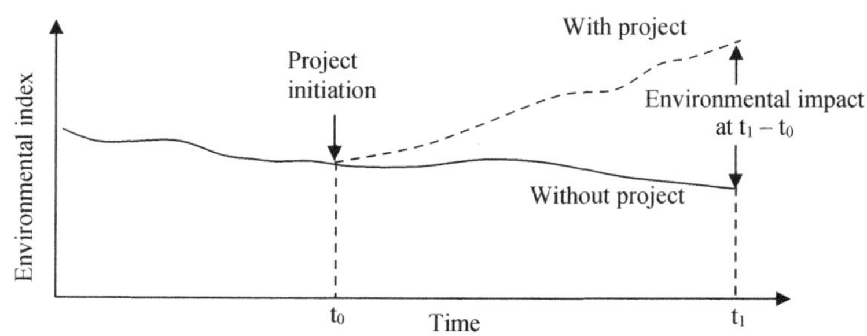

FIGURE 7.2

Conceptual representation of environmental impacts.

A large investment in any sector, including water, precludes funds and resources from going to other sectors in any given economy. International funding agencies evaluate such aspects before they fund large-scale water resources projects. The impact analysis component is particularly important in the planning process. Besides the environment, a water project can have an impact on many aspects of society, for example, commerce, disposable income, consumption amount and patterns, construction activities, population distribution, and health. Therefore, every project document should be accompanied by an analysis of these so that all future project consequences (both favorable and unfavorable) can be identified, assessed, quantified (to the extent possible), and integrated into the decision-making process.

The scope of an EIA study can be very large. Rees (1981) has suggested that the following details may be included:

- a description of the nature and characteristics of the proposed development;
- a description of the existing bio-physical and socioeconomic environment;
- an assessment of significant types of environmental impacts during site preparation, construction, and operation;
- integration of the expected bio-physical impact with indirect socioeconomic consequences and community response;
- review of the compatibility of the proposed development with approved land and water management objectives, and environmental standards and quality criteria for the area(s) likely to be impacted;
- reasons for choosing the particular location and project specification and operation from among possible alternative, adverse impacts that cannot be avoided; and
- a summary suitable for decision-makers and other interested parties.

A multidisciplinary approach is a must to analyze the previous details properly.

7.6.1 Environmental Impact Assessment Procedure

A comprehensive EIA procedure consists of many stages that depend upon the requirements of the regulators or funding agencies. Typically, it consists of the following stages.

(a) Screening: Here, based on the broad parameters of the project, its potential social and environmental impacts, and the spatial extent of these impacts, the project is assigned a particular category. The projects whose impacts are likely to be larger are processed more vigorously compared to those whose impacts are likely to be localized.

(b) Scoping: This stage identifies the key issues and scope of the EIA study. Terms of reference for the EIA are also decided at this stage.

(c) EIA study: At this stage, a detailed study to determine the environmental impact is carried. It often involves the collection of relevant data. The study identifies the likely environmental and social impact of the proposed project and wherever the impacts are likely to be adverse and significant, a plan called Environmental Management Plan (EMP) is prepared to mitigate the impacts.

(d) Public consultation: This is an important part of the EIA process in which the project documents are shared with the people who are likely to be impacted by the project and presentation is made by the project proponents before the public and their views are solicited. If concerns are raised by the public, these are carefully addressed.

(e) Appraisal: In this stage, the EIA study report is presented to the decision-making body. Generally, the report is examined by a body of experts from different domains who carefully scrutinize different parts. Depending upon the outcome, the report may be accepted or modifications are ordered. When the decision-maker is convinced that the project satisfies the laid down norms, it may be taken up for implementation.

(f) Monitoring: The regulators monitor the construction activities to ensure that the applicable guidelines are not violated. Periodic monitoring is also carried out postconstruction to ensure that the project is not causing any adverse impact. In case of violations, appropriate remedial measures are initiated.

While assessing the impacts of the proposed development, sufficient details should be available to give a clear picture in respect of the following aspects:

- general location, specific siting on a detailed map, and project layout;
- size/magnitude of operation;
- site preparation and construction;
- transportation/communications requirement;
- possible damages due to natural hazards and preventive strategy;
- dangers due to hazards, such as spill of poisonous chemicals, leakage of dangerous gases, fire, etc., and safety setup;
- waste treatment and disposal; and
- monitoring and surveillance systems.

Before analyzing the likely effects on the environmental system by the proposed development, it is worthwhile to assess the nature and characteristics of the existing environment. For this purpose, the existing environment will need to be described in terms of its present characteristics, especially the ones that are likely to prevail for the entire duration of the proposed development. Such an evaluation may require the initiation of large-scale surveys and/or long-term monitoring programs, and therefore, needs sufficient time for completion. The effects on the environmental system may be evaluated by knowing the effects on various elements, for example, physical and biological resources, socioeconomic development, etc.

7.6.2 Techniques of Environmental Impact Assessment

EIA is helpful to identify the possible negative impacts of the project and the design can be appropriately modified to ensure environmental protection. Fig. 7.3 shows a flowchart of preliminary and detailed assessment procedures for WRD projects. In a technique known as the Delphi process, selected experts first individually rank the projects, say, on a 1−100 scale. Next, statistics of these rankings are presented to the experts who are asked to come up with a new individual ranking. A rapid convergence is obtained and finally, the mean of impacts is determined. This process is repeated for all alternatives and these are ranked in order of increasing impacts. The advantages and disadvantages of this method have been described by Westman (1985). Some of the methodologies that are used in EIA studies are described in what follows.

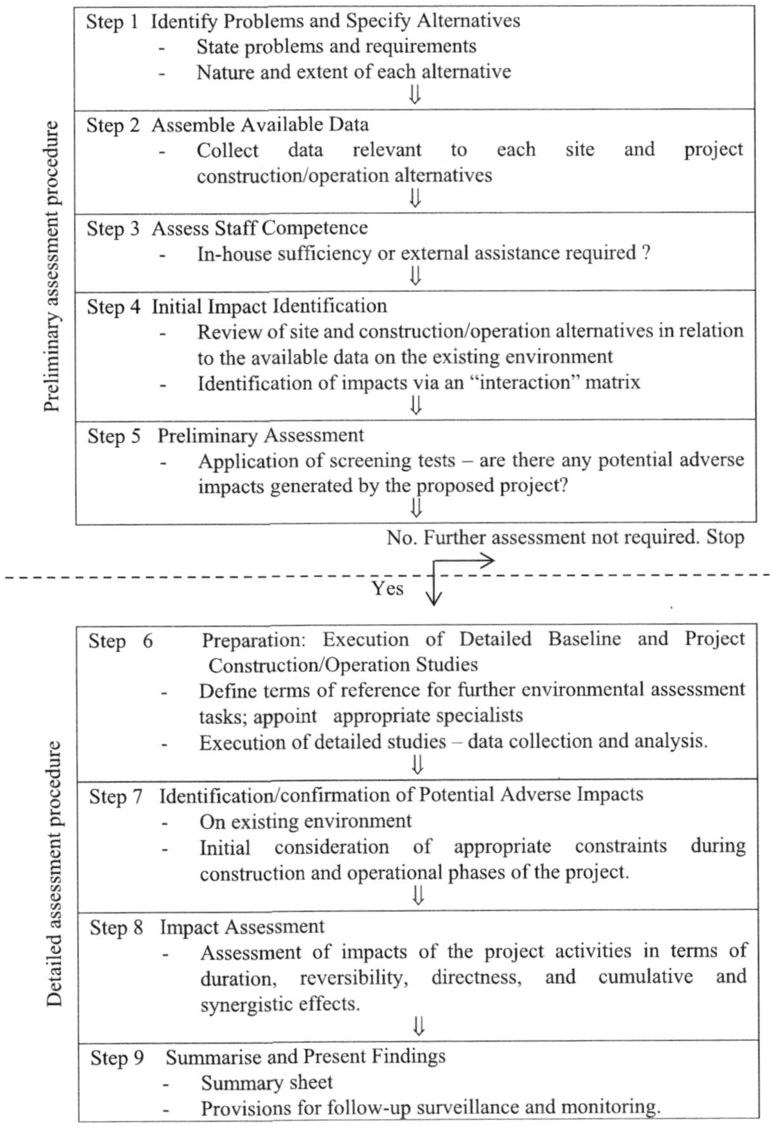

FIGURE 7.3

Flowchart of preliminary and detailed assessment of a WRD project.

Ad Hoc Method

This technique of EIA is quantitative by nature and gives information in comparative statements for different development alternatives of a WRD project. The method is simple and can be easily understood by decision-makers as well as laymen. It is not based on expert opinion and can be completed

Table 7.5 An example of ad hoc method.

		Alternative sites			
S. no	**Item**	**A**	**B**	**C**	**D**
1.	Submerged area (Ha)	10,000	3,000	5,000	8,000
2.	Irrigation potential (Ha)	50,000	10,000	20,000	5,000
3.	Power generation (MW)	11,000	10,000	7,000	3,000
4.	Soil erosion	Severe	Mild	Mild	Insignificant
5.	Displacement (number of people)	25,000	15,000	10,000	5000
6.	Weed growth	Yes	No	Yes	No
7.	Fish culture	No	Yes	Yes	Yes
8.	Water quality impacts	Yes	Yes	No	No

quite fast. In this method, the area and nature of the expected impacts are identified. A team of technical experts representing related disciplines reach a consensus in qualitative terms. The ad hoc method is not much preferred, as the results are subjective and lack consistency. Its use is discouraged these days.

Example 7.1. Four alternate sites are available for a project. Determine the environmental impacts of these sites using the ad hoc method.

Solution. The quantitative and qualitative assessment of the consequences of the alternate project sites is given in Table 7.5.

Environmental Indices

An index is a quantified representation of a specific element, in this case, an environmental consequence. The quantification can be made by assigning a numerical value or a simple "yes" or "no." Depending on the project, a certain number of indices must be incorporated into the impact study. The indices can be subdivided as follows:

a. Resources indices: These types of indices show the change in the potential of the system or its subsystem.
b. Ecological indices: These types of indices show the change of abiotic and biotic environmental components.
c. Socioeconomic indices and cultural indices: These types of indices show the changes toward the improvement of living conditions.

Example 7.2. This example of application of the environmental indices method is based on Sahu et al. (1992). The task is to evaluate the impact of some components of the Narmada Sagar project in India.

Solution. Table 7.6 gives evaluation of environmental impacts in the catchment, reservoir, and at the dam site by following the environmental indices method.

Table 7.6 EIA of some components of Narmada Sagar project.

S.no	Item	Impact	
		−ve	+ve
A. Catchment area			
1.	Change in the erosion and deposition pattern	−1	
2.	Some loss of forest cover and grasses due to fuel needs and grazing cattle	−2	
	Total	−3	+0
B. Reservoir area			
1.	Scenic spots, archeological sites, temples ruins	−1	
2.	Mines of semiprecious materials, rocks, boulders, and sand would be submerged	−2	
3.	Flora will be affected	−1	
4.	Wildlife will be affected	−2	
5.	Sedimentation in the reservoir would occur, causing subsidence of late bottom	−2	
6.	Lake eutrophication	−1	
7.	Water quality degradation	−1	
8.	Weeds development in lakes	−1	
9.	Rehabilitation and resettlement of displaced persons	−8	
10.	Hazards due to water infestation to human and cattle health	−1	
11.	Railway and highway dislocation and recounting	−3	
12.	Fishery development in lakes		+7
13.	Recreation and tourist resorts		+4
14.	Fore-shore bed will be available for cultivation		+2
15.	Microclimatic improvement due to low temperature		+3
	Total	−23	+16
C. Dam site and related area			
1.	Communication facility over bridge on spillway		+4
2.	Ornamental garden on top of the dam, the toe of the dam on the downstream side		+2
3.	Flood control benefits to areas situated on river banks of towns and villages		+3
4.	Seepage from dams may produce beneficial effects		+1
5.	Fish mortality downstream of the spillway	−1	
6.	Seismic activity may trigger due to impoundment of water in the lake	−1	
7.	Township and community development in the area		+7
8.	Rise of water table in wells thereby reducing pumping lifts		+4
	Total	−2	+21

The environment indices are described in detail in Section 7.6.3. The use of such indices can be seen in the ad hoc method explained earlier. Proper identification, selection, and classification of environmental indices are essential for a successful impact assessment study. Most EIA studies contain a list of indices ranging between 50 and 200 with the majority having 50–100 indices. The number of indices used in a particular study depends on the quantity and quality of data available. While developing or identifying indices for a new WRD project, the baseline data and the data of an existing project in the vicinity are very useful. The ad hoc method and indices methods have many common features.

Matrix Assessment Method

ICOLD (1982) developed a matrix method to analyze the environmental impact of dams. This method is based on the action–response relations that are expressed using a matrix. The matrix of environmental variables and project activities identify a cause-effect relationship between specific activities of the project and environmental impacts. During analysis, a matrix is constructed in terms of actions causing impacts (rows) and effects (columns). Expected impacts are incorporated and quantified in this matrix. Note that sometimes, the construction effects may prove to be more damaging than the operational effects. Such a procedure helps with the identification of activities that need more attention and scrutiny. Note that there are different methods to compute the total in the table. Table 7.7 shows an example of the matrix method. Whenever a detailed appraisal of certain aspects included in the initial matrix is needed, an expanded matrix is developed (Pendse, 1987).

Example 7.3. The Tallahala development is a water impoundment project on the Tallahala River, southern Mississippi, constructed to provide water supply, flood control, and water quality control. An earthfill dam will create an artificial lake with 80.76 million m^3 of storage for flood control, 80.76

Table 7.7 Illustrative example of matrix method of EIA.

Priority values	Proposed action → Impact on ↓	Dam construction	Reservoir filling	Relocation of population	Toxic discharge	Total
10	Health	3 5	4 6		5 7	12 18
9	Forestry	4 4	6 7			10 11
5	Archeological sites	2 5	4 7			6 12
3	Tourism		7 6			7 6
4	Fishery	2 6			4 7	6 13
7	Social life	2 4		9 8		11 32
6	Navigation		5 5			5 5
8	Downstream water quality		6 7		7 8	13 15
Total		13 24	32 38	9 8	16 22	

First number in a cell—magnitude of likely impact, Second number—importance of likely impact.

million m^3 of storage for water supply, 16 million m^3 of storage for water quality regulation, and 7.2 million m^3 for sedimentation. The lake will have a 64 km shoreline with a maximum length and width of 9.6 and 3.2 km, respectively, at normal water levels. The total land required for satisfactory implementation of the project is 62.82 sq. km.

The area to be affected by the project is a low-population, low-income area. The major urban zones are the cities of Laurel (24,000 inhabitants), Bay Springs, and Heidelburg, which are showing declining population. Recreation activities are limited in the area. The water quality of the Tallahala River is acceptable north of Laurel, but dramatically polluted downstream of Laurel, where one major industrial installation produces an average loading of about 2600 lb/day of biochemical oxygen demand (BOD). Two to three times a year, the land experiences extensive flooding in both urban and agricultural areas with a negative impact on the economy. Due to environmental conditions, mosquitoes are a problem, and malaria is still a concern. About 70% of the land in the Tallahala drainage basin is devoted to agriculture and 30% is woodland. About 80% of the agricultural land is utilized for cotton and 20% for pasture. Woodland areas consist of more than 40 wood plant species. Principal wildlife present in the project area comprises deer, wild turkey, squirrel, rabbit, and quail. Habitat is available within the basin for either rare or endangered species. Historical or archeological sites have not been documented.

Solution. Based on the interaction matrix system developed by the ICOLD, the impact assessment is presented in Table 7.8. Positive impacts are denoted by (+), while negative by (−). Explanation of the ICOLD matrix system symbols is provided in Table 7.9. From Tables 7.8, it is evident that most of the actions will have beneficial effects on the economy of the area. Health and ecology will be subjected to gains and losses so that a detailed quantitative analysis should be conducted. Deforestation and possible industrialization have the most detrimental impacts.

Checklist Method

This is a quantitative method that facilitates rapid assessment of the impacts on the environment. Checklist methodologies range from a simple listing of environmental elements to sophisticated techniques where a weighing factor is assigned to each element according to its importance, and then scaling techniques evaluate the impact of each alternate solution. A large number of checklists have been formulated Canter (1981) developed a checklist of factors related to environmental quality. The checklists can be of different types: simple checklist, descriptive checklist, scaling checklist, or weighing checklist.

An example of the simple checklist method is given in Table 7.10.

A simple example of a weighting checklist is the water quality index (WQI) developed by the National Sanitation Foundation of the United States. WQI is the aggregated result of nine environmental factors and is computed by

$$\mathrm{WQI} = \prod_{i=1}^{9} I_i^{W_i} \tag{7.1}$$

where I_i is the subindex value of environmental factors and w_i is the assigned weight importance. The nine factors and their weights are given in Table 7.11. The subindex value depends on the measured value of the factor through functional relationship. WQI ranges between 1 and 100 and classifies the

Table 7.8 ICOLD interaction matrix for Tallahala River dam (For symbols, see Table 7.9).

	Actions											
Effects	**101**	**103**	**104**	**105**	**106**	**108**	**201**	**202**	**207**	**302**	**303**	**305**
101				+								
102	+			+	+							
103		+	+		+	+		+				
104	+									+		
105		+	+		+			+				
106	+											
107			±	±								
108	+	+	+	+	+		±		−	+	−	
109					+	+		+		+		
112		+										
113			+							+		
115			+									
116		+		−		+					−	
202	−		+						−			
203	−		+						−			
205	−								−			
206									−			
302	−			−	−							
501									−	+		
502	+									+		
504	+		+							+		
505				−								
507				−			+					−
508				−					−			
601				−					−	+		
602				−					−	+		
603				−					−	+		
604			−	−		+						−
605				−		+						−
606				−								−
607				−			+					−
608				−		+						
609				−								

system accordingly as follows: 0–25, very bad; 26–50, bad; 51–70, medium; 71–90, good; 91–100, excellent. Canter (1981) provides a thorough discussion of checklist methods.

Example 7.4. River Iskar is 368 km long, it springs from Rila mountain in southwestern Bulgaria and it flows north where it joins the Danube River. The river is a recipient of industrial and domestic wastes

Table 7.9 ICOLD matrix system: explanation of symbols.

Actions	
A 101: Irrigation	A 201: Presence of the dam
A 103: Potable water	A 202: Reservoir
A 104: River control	A 207: Deforestation
A 105: Industrial use	A 302: Reservoir shores
A 106: Navigation	A 303: Fluctuation zone
A 108: Fishing	A 305: Down river of reservoir
Effects	
E 101: Industrialization	E 302: Physics and chemistry
E 102: Employment	E 501: Forest
E 103: Tourism	E 502: Pasture
E 104: Agriculture and livestock	E 504: Cultivation land
E 105: Communication	E 505: Superior plants
E 106: Commerce or trade	E 507: Phytoplankton
E 107: Relocation and land value	E 508: Rare and threatened species
E 108: Social acceptance	E 601: Mammiferous
E 109: Recreation	E 602: Birds
E 112: Domestic water supply	E 603: Insects
E 113: Purchase of land	E 604: Economic fish species
E 115: Protection against natural hazards	E 605: Other species
E 116: Health	E 606: Macroinvertebrate
E 202: Erosion	E 607: Microorganism
E 203: Material in suspension	E 608: Zooplankton
E 205: Sedimentation	E 609: Rare and threatened species
E 206: Stability	

Table 7.10 Checklist method of EIA.

Item	Likely impacts					
	Beneficial			Harmful		
	Short term/long term	Reversible/ irreversible	Local/ widespread	Short term/long term	Significant/ normal	Local/ wide
1. *Atmospheric*						
a. Air pollution						
b. Climate						
c. Temperature						
2. *Land use*						
a. Forest						
b. Grazing						
c. Residential						
d. Industrial						
e. Agriculture						
3. *Water*						
a. Lake						
b. Rivers						

Table 7.11 Water quality index and environmental impact of Iskar River, Bulgaria.

Water quality factors	Weight w_i	Site I			Site II			Site III		
		Measured quantity	$I_i^{(+)}$	$I^w_{i\,i}$	Measured quantity	$I_i^{(+)}$	$I^w_{i\,i}$	Measured quantity	$I_i^{(+)}$	$I^w_{i\,i}$
DO (%)	0.17	1	2	1.125	104	98	2.180	101	99	2.184
BOD (mg/L)	0.10	48	2	1.072	8	35	1.427	4	60	1.506
Suspended solids (mg/L)	0.08	580	20	1.271	420	45	1.356	500	35	1.329
Nitrates (mg/L)	0.10	2.40	88	1.565	0.70	98	1.582	0.05	99	1.583
Fecal coliforms (N/100 mL)	0.15	2*	90	1.964	2*	90	1.964	2*	90	1.964
pH	0.12	7*	90	1.716	7*	90	1.716	7*	90	1.716
Phosphate (mg/L)	0.10	0.3*	80	1.550	0.3*	80	1.550	0.3*	80	1.550
Turbidity (Jtu)	0.08	5*	90	1.433	5*	90	1.433	5*	90	1.433
Temperature variation °C	0.10	2*	83	1.556	2*	83	1.556	2*	83	1.556
Remarks		Bad		27.942	Good		77.131	Good		80.6
Loss in % per year										
Fish economy		12.9			0			0		
Recreation		32.2			0			0		
Irrigation		6.5			0			0		
Inert material yield		41.9			0			0		
Flora in flooded areas		6.5			0			0		

*Note: * denotes assumed values; + the subindex value was estimated from Ott (1978).*

(purified and unpurified) from many factories and urban areas. As a result of this pollution, detrimental environmental effects have been documented. To assess the impact of river water quality on the environment, a number of data were collected from three river sites. These data were dissolved oxygen, biological oxygen demand, oxydizability, dissolved and undissolved substances, NH_4, Fe, Mn, and phenols (Ivanov, 1984). Assessment of environmental detriment was done according to losses in fish economy, recreation, irrigation, inert material yield, and flora in flooded areas. Compute the water quality index using the method developed by the National Sanitation Foundation (NSF) of the United States.

Solution. For the application of the NSF water quality index method, five additional water quality factors should be provided. These factors are fecal coliforms, pH, phosphates, turbidity, and temperature. As no information was available, it is assumed that these factors are of minor importance and therefore moderate values can be assigned for the study. In Table 7.11, the water quality index is estimated for the three river sites. At site one, the water quality is bad and polluted, while at sites two and three water quality is good and acceptable. From the same table, it is evident that bad water quality causes an alarming rate of environmental deterioration.

Network Method

The environmental components interact with human elements and hence are very complicated and complex. These represent the impact causes and consequences through an integrated network system. The network is generally shown as a tree, sometimes known as "impact tree." To arrive at the tree structure, one has to answer a series of questions related to each of the project activities, such as primary impact area, secondary impact area, etc. The main limitation of the network approach is that this provides inadequate information on the technological treatment of the problem. An example of this method is given in Table 7.12. It is nothing but a way to present factual information. However, the network can become very complex for a real-life problem.

Overlay Method

In this procedure, environmental impacts are assessed by using cartographic techniques. The project area is depicted by physical, social, and ecological characteristics of the environment. Maps are superimposed on each other to assess the environmental characteristics within the project boundary. The method draws its strength from the graphical display of types of impacts, the extent of these, and the geographical area affected. This method is useful for selecting a site among various alternatives. The base map of the project is prepared on a transparency and the existing environmental features and their boundaries are marked. The severity of impacts can be depicted using color codes. Various transparencies for different impacts give an overall picture of the degree, severity, and extent of the impacts. Fig. 7.4 illustrates the method with a simple example.

The availability of powerful GIS packages has considerably improved the ease and effectiveness of this method. CBIP (1995) contains a number of Environmental Impact Assessment case studies. Many international organizations, such as FAO and UNEP (Hussein et al., 2004), have developed manual/guidelines for EIA. Many countries have issued detailed procedures to be followed for appraisal of infrastructure projects from the environmental angle. India, for example, has a comprehensive set of laws and regulations for this purpose.

Table 7.12 Network method of EIA.

Specific activity	Create a reservoir			
Basic affected resources	**Land**		**Water**	
Change in land use	Decrease in forest area	Increase in built-up area	Decrease in stream length	Increase in lake area
Physical effects	Faster flow velocities Higher erosion	Reduced infiltration	Change in flow regime.	Change in evaporation Change in groundwater regime Sedimentation
Biological effects	Higher crop production Loss of biodiversity	Less biomass production Loss of biodiversity		More fish production More aquatic plants
Social impacts	Decrease in tranquility	Higher population density More conflicts		More recreational opportunities
Importance of terminal effects	High	Medium	Low	High

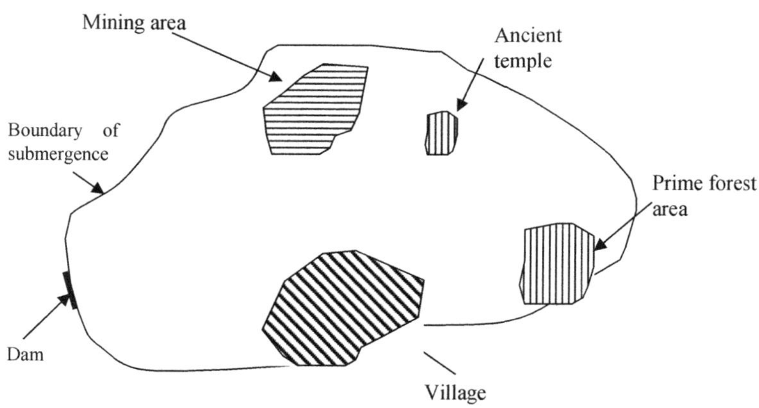

FIGURE 7.4

Illustration of overlay method of EIA.

7.6.3 Indices for Environmental Impact Assessment

An index is a means devised to reduce a large quantity of data down to its simplest form, retaining essential features to answer the questions that are being asked. In short, an index is designed to simplify. In the process of simplification, of course, some information is lost. However, if the index is designed properly, the lost information will not seriously distort the interpretation. Indices are also defined as the quantified limits for indicators. Quantification in this context means, but is not restricted to, a certain amount or concentration that can be the uppermost or the lowest value permitted. Hence, a "yes" or "no" answer would also fall under this quantification. Indices are intended to be protective measures for man or for man's direct sources for food and water. The indices have been developed for the short-term protection of nature. In terms of near future, limits have to be set as to the degree to which the existing nature should be protected. However, as nature is a composite whole, standards have to be found for selected indicators, thus defining and transforming the selected indicators into indices.

Indices of environmental impacts are a means to quantify and evaluate the quality of the environment with regard to some particular element. Environmental monitoring data consist of routine measurements of physical, chemical, and biological variables that are intended to give insights into environmental conditions. These data often provide an important yardstick to judge the effectiveness of regulatory programs in improving environmental quality. From a purely conceptual point of view, environmental monitoring data serve as a feedback loop to evaluate the effectiveness of regulatory activities. Once the requisite data are collected, there is a further need to translate these into a form that is easily understood. After the indices are developed, they should serve as "indicators" to examine trends, to highlight specific environmental conditions, and to help governmental decision-makers in evaluating the effectiveness of regulatory programs. Of course, indices are not the only source of information that forms the basis of decisions. Decision-making is based on many other considerations besides indices and monitoring data.

Six basic uses of environmental indices are.

- **(a)** Resource allocation—indices may be applied to environmental decisions to assist managers in allocating funds and determining priorities.
- **(b)** Ranking of allocations—indices may be applied to assist in comparing environmental conditions at different locations or geographical areas.
- **(c)** Enforcement of standards—indices may be applied to specific locations to determine the extent to which legislative standards and existing criteria are being met.
- **(d)** Trend analysis—to determine the changes in environmental quality.
- **(e)** Public information—to inform the public about environmental conditions.
- **(f)** Scientific research—indices are a means to reduce a large quantity of data to a form that gives insights into the environmental phenomenon and helps in scientific research.

In each of these applications, an index helps in conveying information about the nature and state of the environmental process. Because the questions being asked are different in each application, the index may differ in terms of the variables included, the basic structure, and the manner in which it is applied. Since different users have different data-reporting needs, identification of the users should be a critical part of the development and application of environmental indices. Some of the indices are discussed next (see Unesco, 1984).

Resource Indices

The potential use of a system or a component within a system may be referred to as a resource and the index showing the status of resource's potential, as it is developed, is known as the resource index. This index regulates human health requirements by setting limits to the useable potential of the resource or its components. These resources may include potable and industrial water supply, food production resources, energy creation resources, mineral resources, and resources for recreation.

Indices for resources for water supply may cover hydrological and quality properties. The indices would depend on the source of water such as surface and underground, the amount available from each source with its variations, and the amount and quality of water returned after use. The indices for food supply sources may depend on the type of seeds, fertilizers, pesticides, soil, climate, quantity and quality of irrigation water, and water management practices. Water is also used to generate energy, the quantum depending on the volume and head of water. The relevant index will depend on these factors as well as their natural variation.

Ecological Indices

A basic understanding of the ecological system is needed to identify indicators that are significant for protecting the ecological system. Based on the experience gained in protecting subsystems, such as rivers, lakes, and wetlands, limits have been set for the import of organic and inorganic pollutants as well as for plant nutrients. These indices are being evolved. Unesco (1984) has suggested the following ways to develop indices.

Changes of Abiotic Components

The hydrological components to be used could be a percentage of surface runoff, a percentage of groundwater supply, evaporation over the course of the year, etc. The climate component to be used could be a change of velocity and direction and a change of mean temperature over the years. The physical components to be used could be the change of flow, change of temperature, stratification, etc. The chemical components to be used could be the quality of water.

Changes of Biotic Components

The pattern of natural and cultivated land or water bodies, diversity and variety of species, food web chain, metabolism of the system, vulnerability to ecosystem, toxic effects of components in the system, probity, and tropic status are some of the biotic components. As all man-made changes and human actions induce stress on an ecosystem, the total amount of energy import could be used as a general index.

Valued Ecosystem/Environmental Components (VECs)

The concept of valued ecosystem/environmental components (VECs) has been widely employed in impact assessments (IAs). A VEC is a fundamental element of the physical, biological or socio-economic environment that has significance from scientific, economic, social or cultural perspective. VECs were introduced to avoid the inefficient approach of 'count everything' that was common in the past and to focus the attention to a holistic view of ecological health and integrity. In EIA, those VECs are included that may be affected by a project's activities. For example, in case of a water resources project, riverflows, soils, forests, wildlife, fish, fish-habitats, biodiversity, population, infrastructure, economy, family and community life may be affected by a proposed project and may be considered as VECs. It can be seen that VECs are mostly selected from cultural, ecological, economic, and social areas.

VEC selection is challenging in IA studies. In selection of VECs for infrastructure projects, project proponents along with related government engineers initially identify overall environmental concerns for IA. Political leaders and location of the project also influence selection of VECs. The extent of geographical area considered depends on the size and nature of the project. The initial list is screened by the subject experts, NGOs and inputs may be formally or informally obtained from the general public and the local/affected community. Legislation and regulations pertaining to environment significantly influence selection of VECs. Ayodele and Gunn (2015) noted that 'the ecosystem approach is being circumvented by just focusing on certain parts of the environment or certain species or issues' and that VEC selection is based on a great deal of 'speculation or judgment calls', with little attention given to holistic ecosystem considerations.

While selecting VECs, availability of data should also be given a thought. If a VEC is selected for which not much data is present or the data is biased or unverified, the purpose of IA will not be served and in fact, the results may be misleading. Many governments or their agencies, for example Government of Canada, have issued guidelines for the selection of VECs and assessment of potential impacts of projects.

Other Indices

All man-made and man-induced changes have to be measured and quantified with respect to social, economical, cultural, or a combination of one or more of these goals. However, the time frame is of great importance in such goal setting. The benefits should be passed on to the future generations rather than be limited to the present generations.

7.6.4 EIA Procedures

The assessment of environmental impact became a necessary requirement in many countries from the 1970s onwards. However, the practices of EIA have not undergone major improvement during the past 30 years or so. The only change has been that more parameters and data are now being analyzed with the wider availability of computers. According to Biswas (1998), there are at least three fundamental problems with the current EIA techniques.

1. The linkages between environmental assessment and the socioeconomic aspects of water development at the policy level are not clear. Even when attempts to link have been made, the linkages were, at best, descriptive.
2. Although some progress has been made in the application of EIA at the project level, commensurate progress at the policy and program level has been missing. The useable EIA techniques for water resources are still in their early stages and much progress is needed before water policy and programs can be analyzed and made environmentally sound before implementation.
3. EIA studies have followed more or less the same philosophy irrespective of the location of the application area. The studies have concentrated on what is not environmentally more sound WRD than what it is. In this way, attention has been focused only on one part of problem. A holistic approach to the issue should be to identify the factors that contribute to the sustainable development of water resources.

Biswas (1998) cites the example of the High Aswan Dam of Egypt to highlight the political and ideological factors behind the negative publicity of this dam in respect of environmental impacts. This dam was completed in 1968 and this was the period when the environmental movement was initiated.

After the United States withdrew from this project, it was completed with technical and financial support from the Soviet Union. This project was technically and economically sound and, therefore, it was possible to criticize it only on environmental grounds in the era of the cold war. Although no scientific evidences were available, a series of articles in the popular media at that time caught the imagination of the public in the west. Many myths surrounding the High Aswan Dam have been repeated so many times that these have come to be accepted as facts, according to him, this is a remarkably successful dam without which, Egypt would have been in dire economic straits. Since not building the dam would have been a catastrophic decision for Egypt, the issue should have been what steps should have been taken to maximize the positive environmental impacts and reduce the negative ones.

7.6.5 Cumulative Impacts Assessment

A conventional project and site-specific approach to environmental assessment has its limitations when it comes to assessing potential cumulative impacts on environmental resources. This is because the impact of a particular project on an environmental resource may be considered insignificant when assessed in isolation, but may be significant when evaluated in the context of the combined effect of all past, present, and reasonably foreseeable future activities that may have or have had an impact on the resources in question. Cumulative impact assessment (CIA) also provides valuable and important inputs, particularly in monitoring of environmental sustainability impacts. For these reasons, the explicit assessment of cumulative impacts is desirable in EIA practice.

Cumulative impacts generally refer to impacts that are additive or interactive (synergistic) in nature and result from multiple activities over time, including the project being assessed. Cumulative impacts can be defined as "the impacts on the environment that result from the incremental impact of the action when added to other past, present, and reasonably foreseeable future actions regardless of what agency (federal or nonfederal) or person undertakes such other actions." Cumulative impacts:

(a) are caused by the aggregate of past, present, and future actions;
(b) are the total effect, including both direct and indirect effects, on a given resource, ecosystem, and human community of all actions taken, no matter who has taken the actions;
(c) need to be analyzed in terms of the specific resource, ecosystem, and human community being affected;
(d) cannot be practically analyzed beyond a reasonable boundary; the list of environmental effects must focus on those that are meaningful;
(e) rarely correspond to political or administrative boundaries;
(f) may result from the accumulation of similar effects or the synergistic interaction of different effects;
(g) may last for many years beyond the life of the project that caused the effects; and
(h) should be assessed in terms of the capacity of the affected resource, ecosystem, and/or human community to accommodate additional effects.

The process of analyzing cumulative impacts is an enhancement of the traditional environmental assessment components: (1) scoping, (2) describing the affected environment, and (3) determining the environmental consequences. Generally, it is also critical to incorporate the analyses of cumulative effects into the development of alternatives for an environmental assessment, since it is only by reevaluating and modifying alternatives in the light of the projected cumulative effects that adverse consequences can be effectively avoided or minimized.

In many ways, scoping is the key to analyzing cumulative impacts, as it provides the best opportunity for identifying important cumulative impact issues, setting appropriate boundaries for analysis, and identifying relevant past, present, and future actions. By evaluating resource impact zones and the life cycle of effects, rather than projects, the boundaries of CIA can be defined. Scoping can also facilitate the interagency cooperation needed to identify agency plans and other actions whose effects may overlap those of the proposed action.

Determining the cumulative environmental consequences of an action requires delineating the cause-effect relationships between multiple actions and resources, ecosystems, and human communities of concern. The successful analysis of cumulative effects ultimately depends on the careful application of individual methods, techniques, and tools to the environmental assessment. The unique requirements of CIA (i.e., focus on resource sustainability, expanded geographic and time boundaries) must be addressed by developing an appropriate conceptual model using a suite of primary methods: questionnaire, interviews, and panels; checklists; matrices; networks and systems diagrams; modeling; trends analysis; and overlay mapping and geographic information systems.

In most cases, it will be beyond the scope of an EIA to include a full-fledged CIA. However, it is important that environmental assessment, where appropriate, includes a preliminary assessment of the potential for cumulative effects and specific recommendations on the need for, and the conduct of, a CIA. The key issues in considering a CIA relate to defining appropriate analytical boundaries in time and space, identifying and predicting future resource use and impacts, and evaluating the significance of cumulative effects that are predicted to take place. The spatial boundaries are best defined flexibly to allow adjustments during the assessment process, as the nature of linkages becomes better defined. The setting of temporal boundaries will be largely a function of data availability as well as the levels of uncertainty and confidence in prediction. A concern is to determine the minimum data requirements that will allow defensible and robust impact predictions. The criteria for judging the significance of cumulative effects are not different from those for other types of environmental assessment, but threshold effects and irreversible changes in the use of critical resources will generally be key concerns.

7.7 INTEGRATION OF ENVIRONMENTAL ASPECTS IN WATER RESOURCES PLANNING

For a long time, environmental issues were given a minor importance in WRD and operation. This perspective has completely changed in recent times, mostly based on the results of the studies carried out all over the world, which have demonstrated, beyond doubt, that environmental impacts of major WRD projects are significant. Without rallying to extreme views advanced by some environmentalists that any change in natural environment is detrimental, a balanced approach is that such impacts have to be adequately assessed for each project to enable their incorporation as an integral part in the planning process. The ultimate goal of WRD is to ensure that water of acceptable quality is available in sufficient quantity at the right location, at the right time, at the right price, on a reliable basis. Another developmental goal could be the protection against excess water, that is, flood management and control. Because of the uncertainties inherently associated with all phases of water resources management, all this can be achieved only within some limits of assurance, reliability, and cost.

It is necessary that multipurpose WRD projects should take into account, right at the planning stage, appropriate steps for protection, preservation, and development of the environment. Environmentally sound water management implies that:

(a) development be controlled to ensure that the resource itself is maintained and that adverse effects on other resources are considered and where possible, ameliorated;
(b) options for future development are not foreclosed; and
(c) efficiency in water use and in the use of capital are key criteria in strategy selection.

The beneficial and harmful aspects of WRD projects have been discussed earlier. Learning from the results of case studies of existing projects, the planners are now better aware of the lacuna in the previous planning approaches. Such studies have brought out the consequences of unbalanced development that were earlier difficult to foresee. Many of the environmental studies that have been carried out assess environmental impacts through indicators. However, most models that have been developed for EIA concentrate on limited aspects, such as water pollution, and not on overall river basin development.

There is a significant difference in the socioeconomic conditions, availability and usage of resources, and technology in the developing and developed countries. The priorities that developing countries set for their own environment will not necessarily be those that people in richer countries might want to adopt. Thus, although some cultures in poor countries may value their natural heritage strongly, most developing country governments are likely to give lower priority to amenity damage as long as basic human needs remain unmet (Herschy et al., 1998). Consequently, the approach and the methodology of environmental studies differ from country to country. An automatic transposition of methodologies from one place to the other is likely to produce unrealistic results. Generally, studies try to assess the environmental impacts through selected commensurate units, termed environmental impact units (EIU). Differences in the EIU scores describe the condition of environmental factors with and without the project.

There are many constraints that limit the potential application of available knowledge by water professionals and decision-makers for reducing environmental disruptions to a minimum. The methodologies are still being developed and the baseline data are usually not available. The framework currently used to consider various environmental impacts is overwhelmingly biased toward assessing only the negative impacts. Realistically, any reasonable water development project will have discernible positive impacts, among others, on environment in varying degrees. Thus, while considering impacts, one should not ignore positive benefits, which are the real motivation for development projects. Technology developments aim to achieve balanced environmental development. On one hand, more water, food, and other commodities have to be supplied to each person in an ever-expanding population. On the other hand, the area of cultivable land is reducing and water is becoming scarcer each day. This necessitates a need for striking a balance between the environment and development.

Water resources planning calls for consideration of possible impacts on the environment, both beneficial and adverse. Water resources planners assess, quantify, and present them to the decision-maker who chooses an appropriate plan from the various competing alternatives. Any policy in isolation does not help in preventing the deterioration of environmental quality. A blend of technical, economic, social, and political aspects should form part of the developmental strategy to achieve the twin objectives of increasing production and improving the environment.

7.7.1 Optimization Methods

It is required that the adverse environmental impacts from the development are minimized, while the optimum level of development is achieved. This can be accomplished through the use of optimization techniques, such as multiobjective or goal programming. Suppose there are two objectives: minimization of the total cost (TC) and minimization of population displaced (P). If C is a vector of reservoir capacities, the problem can be stated as follows:

$$\text{Min } z = f[TC(C), P(C)] \tag{7.2}$$

subject to usual constraints. The techniques to solve an optimization problem have been described in Chapter 5.

The optimized matrix approach can be useful in cases where the number of objectives is large. This approach is based on search techniques and can provide a set of feasible noninferior alternatives in matrix form. The set of solutions, while satisfying the upper bounds fixed on cost, and impact, provides optimal plans for both impact and cost. The problem formulation is as follows:

$$\text{Minimize (cost, environmental impacts, population displacement)} \tag{7.3}$$

subject to

$$C(Y_k) \le C \quad k = 1, 2 \ldots n \text{ (cost constraint)} \tag{7.4}$$

$$P(Y_k) \le P \quad k = 1, 2 \ldots n \text{ (population displacement constraint)} \tag{7.5}$$

$$A(Y_k) \le A \quad k = 1, 2 \ldots n \text{ (area submergence constraint)} \tag{7.6}$$

$$F(Y_k) \le F \quad k = 1, 2 \ldots n \text{ (forest submergence constraint)} \tag{7.7}$$

where Y_k is the yield of the kth reservoir; and P, C, A, and F are the upper limits fixed on population, cost, area of submergence, and forest submergence, respectively.

The matrix approach works out the total cost and the total magnitude of each type of impact for each combination. If any of the upper bounds is violated, the combination is rejected, thus leaving only the feasible alternatives for consideration in the next step. The feasible alternatives are compared with each other to locate the inferior alternatives. The model provides a set of feasible noninterior plans. These alternatives include the minimum-cost and minimum-impact plans and help in detailed trade-off studies between cost and impact and between various impacts.

7.8 ENVIRONMENTAL CONSIDERATIONS IN RESERVOIR PLANNING AND OPERATION

Water resources projects influence and transform the environment to a degree and over a range that varies from project to project. In addition to the classical criteria of technical, economic, and financial feasibility, dam projects have to satisfy a fourth and particularly stringent criterion, namely social and political acceptance. An important factor for such an acceptance is compatibility with the environment. The solution is to be found by striking a balance between divergent, and sometimes contradictory, goals.

Dams are essential because of the benefits that their reservoirs offer all over the world, by storing water in times of surplus and dispensing it in times of scarcity. Dams prevent or mitigate devastating floods and catastrophic droughts. Thus, dams are an integral part of engineered infrastructure. Still, more dams will be needed in the future for the adequate management of the world's limited, unevenly distributed, and in many places, acutely scarce water resources. Simultaneously, there is a need to protect and improve the natural environment as the supporter of all life. Then, there is the social side to the comprehensive concept of the environment: the people, their land and settlements, their economy, and traditions. The impact of dams and reservoirs on this environment is inevitable and undeniable; land is flooded, people are resettled, the continuity of aquatic life along a river is interrupted, and its runoff modified and often reduced by diversions.

Thus, dam engineers find themselves confronted with the basic problems inherent in the transformation of the natural world into a human environment. In their never-ending quest to provide a growing number of people with a better life, the need to develop natural resources, including water, means that the natural environment cannot be kept completely unchanged. However, great care must be taken to protect the environment from all avoidable harm or interference. We must cooperate with nature's inherent fragility, its dynamism without overtaxing its powers of regeneration, and its ability to adapt to a new equilibrium that is beneficial. Importantly, all this is to be attained while ensuring that the people directly affected by a project are better off than before.

The contribution of dam engineers to the development of water resources is based on proven technology which is testified by over 39,000 large dams. This technology continues to benefit from ongoing refinement and a steady growth of knowledge and experience, in particular with regard to its social and environmental consequences.

7.8.1 Guidelines of ICOLD

The International Commission on Large Dams (ICOLD), a premier international organization, has attached great importance to the environmental and social aspects of dams and reservoirs. It has emphasized that these aspects should be addressed with the same concern which has made the question of dam safety a predominant concept. With the aim of balancing the need for the development of water resources with the conservation of the environment in a way that will not compromise future generations and to enhance the awareness of the environmental issues of dam engineering, ICOLD (1998) has brought out a *Position Paper on Dams and Environment* (available at its website). This paper emphasizes the following aspects of environmental policy:

(a) Concern for the environment, including both natural conditions and social aspects, must be manifest from the first planning steps, throughout all phases of design and implementation, and during the entire operating life of a project. Dam promoters must be aware of the fact that although dams are the most important means of making surface water available at the place and time of demand, there are also other, nonstructural means of increasing water utilization that can be applied in addition to dams or as an alternative, such as the tapping and recharging of groundwater or desalination of seawater.

(b) In the past it has been the hallmark of our very best engineers to see the natural environment as one of their responsibilities too, which is why many dams and reservoirs harmonize so well with their environment. Today, however, the enormous increase in human knowledge, including that in the field of environmental science, means that a whole team of specialists is needed to access and utilize that knowledge for a WRD project.

(c) The larger the project, the greater the effects on the natural and social environment to be expected, and the wider the scope of the multidisciplinary, holistic studies that they require. Large-scale development demands integrated planning for an entire river basin before the implementation of the first individual project(s). Where a river basin is part of more than one country, such planning presupposes international cooperation.

(d) Projects must be judged everywhere and without exception by the state-of-the-art of technologies involved and by current standards of environmental care. The scope for reducing any detrimental impacts on the environment through alternative solutions, project modifications in response to particular needs, or mitigating measures should be thoroughly investigated, evaluated, and implemented.

(e) The decision on what is usually a very considerable investment for a dam project must be based on unequivocally realistic economic analysis, especially in the case of a large project in a developing country that would tie down a major share of its financial resources for many years. Any tendency to overstate the benefits and understate the costs must be strictly avoided. This also requires taking the impacts on the natural and social environment into account. Despite proposals put forward by international financing institutions and a growing literature on the subject, some such impacts are difficult to quantify or express in monetary terms. In such cases, they must be incorporated into the decision-making process at a higher level of judgment.

An important item on the benefit side is the useful life of the reservoir. The available live storage volume must be estimated according to reliable data on the transportation of solids according to realistic assumptions on reservoir sedimentation.

(f) Involuntary resettlement must be handled with special care, managerial skill, and political concern based on comprehensive social research, and sound planning for implementation. The associated costs must be included in the comparative economic analyses of alternative projects but should be managed independently to make sure that the affected population will be properly compensated for. For the population involved, resettlement must result in a clear improvement of their living standard, because the people directly affected by a project should always be the first to benefit instead of suffering for the benefit of others. Special care must be given to vulnerable ethnic groups.

(g) Even if there is no resettlement problem, the impact of WRD projects on local people can be considerable during both construction and operation. All such projects have to be planned, implemented, and operated with the clear consent of the public concerned. Hence, the organization of the overall decision-making process, incorporating the technical design as a subprocess, should involve all relevant interest groups from the initial stages of project conceptualization, even if existing legislation does not (yet) demand it.

(h) A complete postconstruction audit of an entire project or at least a performance analysis of major impacts should be carried out to determine the extent to which the environmental objectives of the project or of certain mitigating measures are being achieved.

(i) As soon as a project becomes operational, its impact on the environment should be assessed at regular intervals, based on data and sources resulting from adequate preconstruction monitoring. Depending on the individual situation, certain critical parameters should be monitored as a basis for a subsequent performance analysis of the project.

(j) There is also a need for more ecological research on dams and reservoirs that have already seen many years of service. Mistakes and shortcomings could be avoided, many of the recurring controversies relating to the ecological impacts of new dam projects could be prevented and the problems involved could be clarified and solved more easily if our latent store of long-term experience with the operation of so many dams and reservoirs were to be collected, processed, evaluated and published in the framework of research projects based on carefully directed investigations.

7.8.2 Guidelines for Planning

The following is a list of do's and don'ts for those who are involved in the planning and development of water resources projects. These steps can go a long way in overcoming and controlling the resistance and hatred toward the projects.

1. Before taking up a project, the promoters should look at other means of meeting the objectives, such as demand management, minimization of wastage, more effective use of existing developed resources, combating of pollution, recycling, and improved irrigation practices. This is also necessary to preempt the accusations by the opponents.
2. The final plan of any project should contain a realistic and credible balance sheet setting out the direct and indirect costs and the direct and indirect benefits. Environmental and social costs and benefits must be established by experts in their field and be given their due weight.
3. The balance sheet must clearly reflect the cost and benefits of the various uses that the water will be put to. An engineering department alone cannot prepare such a compilation. A multidisciplinary team of engineers, geologists, economists, agronomists, biologists, botanists, sociologists, etc. is needed.
4. It is evident that a WRD project cannot be evaluated in isolation. The alternative uses of the resources required for the project should always be considered.
5. In the past, the developers have been on the defensive. Armed with the true facts and a well-conducted public relations and participation campaign, it will be easier to put across their viewpoint. Based on the experiences of some recent projects, the expectations and aspirations of the public at large are well known now.
6. The views of the people and agencies in one country, though expressed with good intentions, may be out of place for another country. Many countries possess limited resources, which are often urgently needed for social upliftment: better education, health, housing, clean water, and proper sanitation. These countries need to prioritize the application of these resources and decide what percentage should be earmarked for development works and what should be set aside for the protection of the environment.
7. Water resources specialists are increasingly being called upon to answer questions on dam safety, the stability of slopes of reservoirs, earthquake-resistant design, erosion below the dam, reservoir sedimentation, etc. These arguments are viewed with suspicion and people feel that they give conservative estimates of likely costs, damage to the environment, land submerged and people affected, water quality as affected by storage and the like, while exaggerating the benefits due to dams, flood control, hydropower, etc. Serious and patient efforts are needed to improve the image and establish credibility.

8. If people do not believe in a project then either the people are right or the promoters have not been successful in putting across their arguments. A free and frank discussion helps in bringing forward the truth and a better solution to the problem.
9. The technologists can predict the extent to which a reservoir can protect a valley from likely floods, but they cannot put in place a legislation to prevent people from settling in a flood plain. The success of many projects depends on forming appropriate rules and implementing them.

7.9 SUSTAINABLE DEVELOPMENT

Earth is essentially a single system with various biological, chemical, and physical interactions occurring within its environs. Based on the analyses of environmental systems, their trends, and experience, two things are clear: (1) environmental problems arising today are largely caused by various human activities, and (2) environmental problems transcend all boundaries and thus affect all nations and communities. In view of these, it is considered necessary all over the world that all developmental activities should be planned such that current demands are appropriately met without unduly limiting the various options of future generations. In fact, this is a refined form of the concepts, such as "eco-development," which were prevalent in the early 1970s.

The term *sustainable development* (SD) became fashionable in the 1980s. It rapidly captured the imagination of development practitioners and analysts. This term was popularized mainly by the Brundtland Commission report, *Our Common Future* (WCED, 1987). Although it may appear to be so, the concept itself is not new. As noted by Biswas (1994), the general philosophy behind the sustainability concept was expounded centuries, if not millennia, earlier. Similar thoughts on living in harmony with nature can be found in ancient Indian religious texts, such as *Rig-Veda*.

Today, the foremost issue in the minds of political leaders, planners, engineers, and social scientists is the issue of SD. Basically, SD aims at maintaining equilibrium between human needs and economic development, while preserving environmental conservation through the efficient use of natural resources. It emphasizes the need to review environmental protection and economic growth with parallel compatibility. Now, it is considered to be the most reasonable way of combining the current growth with the planning of future projects.

Sustainability issues are not new issues, nor is sustainability a new concept. The current interest in sustainable water resources management has come from a realization that some of the past or current activities have or could cause irreversible damage to the ecosystem. This damage may adversely affect not only our own lives but also the lives of our successors. To address these questions, it is helpful to differentiate growth from development. According to Loucks et al. (2000), growth involves making the pie bigger, building new capacity in new places, improving the standard of living, changing land use, etc. Development involves capacity expansion in situ, redistribution of existing resources, more efficient use of scarce resources, water quality management, and the like.

Water resources management is the vector sum of a progression of legislation, policies, regulations, engineering practices, and institutional traditions. In many instances, natural calamities like floods and droughts are the motivation for changes in the way water resources are managed. These changes could be directed toward the attainment of the objective of sustainable development.

7.9.1 Defining Sustainable Development

There are many ways in which the term "sustainable development" has been defined. It was defined by WCED (1987) as the "development that meets the needs of the present without compromising the ability of the future generations to meet their own needs." However, this definition has been criticized as vague, simplistic, and internally inconsistent (Biswas, 1994). claimed that one could easily identify more than 100 definitions of sustainable development without much difficulty. Even an organization like the United Nations does not have a uniform and acceptable definition for use by its various component organs and the definitions used by UNEP, FAO or ILO differ in significant ways. With specific reference to water resources, ASCE Task Committee for Sustainability Criteria (1998) has defined the concept as follows: "Sustainable water resource systems are those designed and managed to fully contribute to the objectives of society, now and in the future, while maintaining their ecological, environmental, and hydrological integrity." In the opinion of Loucks et al. (2000), sustainable management implies managing for the long term. Water resources systems that are able to satisfy, to the extent possible, the changing demands placed on them over time without system degradation, can be called "sustainable."

It is important to note that sustainable development is not a top-down but a bottom-up approach. It requires that development efforts are decentralized and local people are involved at all levels of planning, design, and implementation. These days, the notion of sustainability is applied at all levels and in ecological, sociological, and economic terms. The popularity of this term emanates from the fact that the current development trends appear to be unsustainable in a variety of ways. Given concerns with economic decline, population growth, and heavy resource depletion during the past 2 decades, many analysts have made pessimistic predictions about the future possibilities for the continued growth of the global economy and the ability of developing countries to attain the economic levels reached by developed industrial societies.

Some societies seem to have exceeded the carrying capacity of the land and resources to provide food and basic needs. For example, the results of recent investigations indicate that the rise of sea level has averaged 1–2 mm per annum during the last century. Acceleration in this rate arising from global warming could trigger ocean thermal expansion or the recession of glaciers at a rate of 8–29 cm by 2030. This could drastically increase the frequency of flooding on islands and low-lying coastal areas and a reduction in the reserves of freshwater due to an increase in saltwater encroachment. Here, it is pertinent to note that there is a lot of uncertainty in the interpretation of recent climate data. Kite (2000) has shown that if the climate data from all climate stations in California are averaged, there is an upward trend in temperature. However, if the stations are divided into urban and rural stations, two different trends are seen: the curve of urban stations shows an upward trend while the curve for rural stations shows a downward trend. Clearly, more research is needed before making a prophetic warning.

While discussing sustainability, the ideas of renewability, resilience, and recoverability are also important. Extensive literature is available on these concepts. See Clark et al. (1994) for definition and discussion. Renewability signifies the rate at which a resource can be replaced so that sustainability is achieved by restricting the level of use to something at or below the rate of replacement. Resilience signifies the ability to withstand stress without long-term or irreversible damage. Recoverability is a concept that accepts that detrimental impact may take place but concentrates on the rate or frequency of impact in relation to the inherent rate of recovery.

Xu et al. (2002) have defined an index, named sustainability index (SI), as the ratio of aggregated possible water deficit to the corresponding supply in the same region:

$$SI = \begin{cases} (S - D)/S, & S > D \\ 0, & S \leq D \end{cases} \tag{7.8}$$

where D is the water demand, and S is the available supply. SI values greater than 0.2 correspond to low or no stress of water supply (demand ≤80% of the potential water supply), whereas those smaller than 0.2 reflect vulnerable conditions (demand >80% of the potential water supply). SI = 0 indicates an unsustainable water supply, that is, water demand equals or exceeds all available local water resources. Note that this index is based on arbitrary thresholds and considers only a small part of the much wider scope encompassed in the definitions given earlier.

Many people view SD from ethical and moral angles. According to Hussain (1992): "… the question of SD involves the question of morality. … I feel that economic growth without concern for morality becomes unsustainable. But at the same time morality without economic growth becomes an empty word." This idea can be further developed by arguing that SD and the enforcement of human rights are interdependent and mutually supporting principles. Thus, any development that violates human rights cannot be sustainable. The preservation of natural wealth, its flora, fauna, and pristine environment ensures that the options and rights of future generations are safeguarded.

The origins of the SD concept lie in a criticism of the other paradigms of development. On account of the support this term has got from many individuals and institutions, it has tremendous drive and force. Thus, SD is argued to be the best and any other kind of development is considered bad. Given the widespread acceptance of the term, it is almost obligatory for all sincere developers to ensure that their practices are sustainable.

The definition of sustainability completely directs attention to those aspects that cannot be sustained. By trying to define sustainable water development in terms of only those factors that could contribute to unsustainability, the entire attention is focused on one part of the equation. The other part that has been completely ignored could possibly be as important as the negative aspects, if not more. Sustainable water development, as it is analyzed at present, focuses only on what it is not, and then attempts to ameliorate the potential negative effects. To take a holistic approach, consideration should first be given to what is sustainable water development, and then move on to what is unsustainable.

Fisher (1997) pointed out that the use of this term evokes more questions than answers about what a new paradigm of development might be. Despite detailed discussions, it has not yet been possible to identify a development process that can be planned and implemented, and which would be inherently sustainable; however, this may be defined. Biswas (1994) concluded that there is more success in identifying certain aspects of development that are unsustainable—then taking appropriate remedial steps to reduce or even eliminate those undesirable effects—than in devising a holistic process that is intrinsically sustainable right from the very beginning. In his opinion, it is easier to agree upon what is unsustainable than what is sustainable.

Coming to hydrological processes, by nature these have fluctuations that could be so great that statistically significant data would be very expensive to obtain to categorically conclude whether such variations are natural or not. If additional factors, such as potential climatic changes, are superimposed on already complex issues, the degree of uncertainty in terms of detecting or predicting the transition process greatly increases (Abu-Zeid et al., 1992). One is then confronted with the difficult issue of even identifying the direction of any change, let alone the degree of change. The critical question then is: what early warnings could indicate the beginning of a transition process from sustainable to unsustainable? The present knowledge is inadequate even to identify the parameters that could indicate the passage from one stage to the other. Thus, currently, we really cannot accurately detect, much less predict, the transition of any such sustainable system to an unsustainable one.

7.9.2 Issues in Sustainable Development

The following issues about the concept of SD have been compiled from recent technical literature.

1. What exactly is to be sustained and how long does it have to be sustained? At what level does sustainability operate: at the level of the individual, specific cultural groups, a region, a country, or the world? And over what spatial scales should sustainability considerations apply?
2. What are the appropriate temporal scales when considering the sustainability of a specific project? One of the challenges of measuring sustainability is to identify the appropriate temporal scales in which those measurements should be made.
3. How would a WRD project or a watershed development strategy that is planned under SD principles compare to one planned under the current (unsustainable) practices?
4. How can sustainability be measured? Is it an ecological concept, a social concept, or something else?
5. Is sustainability an ability to adapt to changing circumstances or is it the maintenance of interrelationships through the suppression of change? Is the ideology based upon conservative foundations or is it progressive?
6. We do not know what future generations will want from us. They may not appreciate and interpret our actions the way we did for our forefathers. How can we identify what our descendants would like us to do?
7. We do not even know with certainty what all the short-term, let alone the long-term, impacts of our current management decisions will be. How to put sustainability in operation?
8. Is it appropriate to try to satisfy the present needs even if they overstress the system designed to meet them? If not, what are, if any, other sustainable ways to meet these needs? Is it really feasible to increase the benefits derived from water resources and at the same time increase or maintain the sustainability of those systems?
9. According to Biswas (1994), no attempt has been made to define or even discuss what is meant by the long term. Does sustainability cover 50 years, or 100, 500, 1000, or even more? Some people vaguely speak of "several" generations. Even if one considers the lowest figure of 50 years, there is a fundamental dichotomy as to its use in the real world. For example, considering irrigated agriculture, generally, the economic planning horizon of farmers extends to one cropping season or at most two. The overriding philosophy of nearly all farmers has been to maximize economic returns from their agricultural activities within this time frame. Thus, the mind-set is inherently based on maximizing profits over a continual series of short-term periods. Although short-term benefits could have long-term costs, generally short-term considerations have won over the long-term implications.

10. How to allocate, over time and space, renewable as well as nonrenewable resources, for example, the waters that exist in many deep groundwater aquifers that are not being replenished by nature? Loucks et al. (2000) note that to preserve nonrenewable resources now for the use of our descendants in the future, the interests of sustainability would imply that those resources should never be consumed as long as there is a future. If permanent preservation seems unreasonable, then how much of a nonrenewable resource might be consumed, and when?
11. How to determine the correct strategy or the optimal future? How to account for the inevitable and profound effects of future technological developments that may mitigate many of the adverse effects of current unsustainable practices? With the exception of the loss of species, what other resources are vulnerable to irreversible decisions?
12. The available resources and demands balance is completely different in developing countries compared to developed ones. For instance, due to the increase in population, the nutrition demands are rising in developing countries. Due to financial reasons, small farmers in developing countries (most of whom are poor), are forced to consider only the short-term economic implications for their survival. Up to what level agricultural activities can be intensified without sacrificing sustainability? Clearly, the sustainable development strategies in these two situations will be different. But in what ways?

This list is not exhaustive by any means. There are many unanswered questions related to the sustainable development and management of any renewable or nonrenewable natural system. These issues are unresolved, variously defined, or openly contested to hinder the sustainable development of a practical guide for developers. A better understanding of these questions will clarify the ramifications of sustainable development and only then the true implementation of sustainable development principles and goals may begin. A number of studies related to different sectors have been carried out to analyze and identify the relevant factors. For example, Bassam (1999) identified five major elements for sustainability in agricultural production systems: policy and management, energy and input, genetic resources, climate, soil, and water.

On account of the impending crisis, crucial and important planning and management decisions in various regions of the world have to be taken. No decision-maker can wait until all the questions are answered. At the same time, they need to work toward increasingly sustainable levels of development and management. This includes learning how to get more from limited resources and how to minimize wastage. New ideas and technology will have to be developed to achieve increased economically efficient recycling. Management approaches that are more nonstructural and compatible with the environmental and ecological life-support systems must be identified. Better ways of planning, developing, upgrading, maintaining, and paying for the infrastructure that permits effective and efficient resource management and provides needed services must also be defined. No single discipline, profession, or stakeholder group has the wisdom to know what will be sustainable and what is right for all of us, living now and in the future. Such decisions can only be made through a process involving all the relevant disciplines and all the interested and affected stakeholders.

Change over time is certain; the direction is uncertain. Sustainability requires that public and private institutions also change over time in ways that are responsive to the demands of societies. The aspirations of societies change with time. Earlier, people wanted to control and manage natural water systems to meet social requirements. Now many societies are moving toward "conservation." A number of national, state, and local restoration projects are underway in Australia, Europe, and North

America that are an evidence of changing expectations and values. To achieve higher levels of sustainability of renewable water resource systems, their renewing capacity, that is, the ability to produce the desired amounts and qualities of water and to support the environment and ecosystems must be preserved and enhanced. This is a necessary prerequisite to enable such systems to satisfy to the maximum extent possible, the needs of future generations.

Professionals must work within the social infrastructure of a community or region and in collaboration with an informed and involved public. This can certainly lead to more socially compatible, creative, appropriate, and hence sustainable, uses of technology and resources for addressing a community's or region's water resource problems or needs. From the perspective of water resources management in developing countries, two things are important: (1) the development process must not be impeded, and (2) the environment must be protected simultaneously. It is not an easy task because of the complex technical, economic, social, and political factors involved. But then there is no other choice. Equally, in terms of environmental management, it is essential to move beyond the current negative-reactive approach to a proactive-creative one. For the sake of a better quality of life for the millions of people in developing countries, a more optimistic approach is necessary.

The United Nations Conference on Environment and Development (UNCED) was held in 1992, In Rio de Janeiro, Brazil. It was convened to address urgent problems of environmental protection and socioeconomic development. The conference adopted Agenda 21, a plan to achieve sustainable development in the 21st century. To ensure effective follow-up, the Commission on Sustainable Development (CSD) was created in December 1992. The CSD is a functional commission of the UN Economic and Social Council. Details about this commission are available on the Internet at http://www.un.org/esa/sustdev/csdgen.htm.

7.10 SOCIAL IMPACTS

The aim of social impact analysis is to study the influence of water resources projects on the social and cultural life of the region. The people who are affected by a project were classified into three categories by Scudder and Biswas (1998a): those who must be relocated because of project work and inundation due to reservoirs (the relocatees); the communities, or the host population who receive the relocatees; and the other project-affected people. The last category includes the people who live in the vicinity of the project, those who live near the reservoir periphery but do not require relocation, and those who live downstream of the projects and whose lives are affected by, say changes in river regimes, etc. Usually, the number of people in the third category is far more than the relocatees and the hosts. It is, however, not easy to determine who are the people in the last two categories and to find their exact numbers except in simple cases where, say, relocatees are settled near a town.

Since the relocatees are the people who are most affected by a project and their problems are "visible," they get the most attention, while the other two categories do not get a fair share. Importantly, most discussions on the social impacts of water resources projects are biased toward those who are adversely affected; the beneficiaries are usually not highlighted. A possible cause is that while the adversaries are usually vocal and organize protests, strikes, etc., the beneficiaries choose to remain invisible.

It is true that resettlement leads to a number of stresses, at least in the initial years. The availability of land, other resources, and employment opportunities to the displaced people is gradually reducing.

The stresses that project-affected people have to undergo are physiological, psychological, and sociocultural. The main cause of these stresses is that the people have to leave behind their ancestral place and settle in a new climate (to which they may not be habitual) and adjust to a strange society whose values, cultures, and traditions they may not like. They may feel intimidated by jeering the host population and may have to adopt a profession for which they have no liking and experience. According to Cernea (1990), forced population displacement caused by dam construction is the single most counter developmental with social consequences of water resources development.

The degree of stress faced by the relocatees also depends on the ability of the national economy to provide gainful employment to the resettlers. In the absence of employment, resettlers may be worse off because their earlier sources of daily necessities, etc. are lost. Besides, the population densities as well as government authority and control tend to increase after resettlement. Strangely, political leaders may be caught in a "no win" situation (Scudder and Biswas, 1998b). If they support migration, they lose their constituency, as people generally resist change and do not wish to move. If they oppose removal, their clout is adversely affected when displacement occurs.

Another important social problem that needs careful attention is the economic and cultural difference between displaced people and host society. Naturally, the attempts to shift the people to completely different climatic conditions, for example, hilly people to plains, have also been resisted and rejected. Ideally, both should have similar ethnic and linguistic composition to avoid conflicts. Sometimes, social problems crop up because of a lack of knowledge or concern by the planners and implementing agency. For example, in the Mahaweli Project in Sri Lanka, ethnically opposite groups were made to settle in a common area leading to communal disturbances (Scudder and Biswas, 1998b). It was reported that due to ethnic problems caused by Ghana's Kpong dam, fighting broke out between resettlers and the host population, killing many people.

Accurate data on the number of people that are affected by WRD projects are not readily available except for the projects whose impacts are confined to a well-defined area or to a small population whose number is already known. Among the three groups discussed earlier, one may get data about relocates only. Very little is known about the projects that were completed more than a decade ago. The number of affected people varies from thousands for a smaller project to more than a million for a project like The Three Gorges Dam on Yangtze River in China. According to the World Bank (1994) estimates, the displacement toll of 300 large dams that, on average, enter into construction every year is estimated to be above 4 million people. Unfortunately, governments worldwide have been accused of deliberate underestimation of the number of relocatees as well as the range and extent of impacts. Yet, another cause of difficulty in understanding the social impacts is the nonavailability of data on preproject status against which the impacts due to the project can be compared. What is required is a carefully chosen sample of at least 1% of families who are tracked and periodically interviewed beginning with the initiation of detailed planning. The focus should not be limited to just the major projects; medium and minor projects should also be covered.

It is generally assumed that the impact of a river valley project on the people living in the downstream areas is positive, that is, they get benefited by flood control aspects but lose nothing. However, there are cases where the construction of levees and other flood control mechanisms coupled with urbanization led to a reduction in wetlands. This reduction was responsible to lessen outputs from them and reduce biodiversity depriving benefits for many local people who depend upon the productive use of these wetlands.

The moment a dam project is approved and the area for submergence is earmarked, the government as well as private parties stop making investments for improvement and, in many cases, for routine maintenance. During the time between project approval and initiation of settlement, the life of relocatees is already adversely affected and their living standards drop. Most large WRD projects have seen significant cost overruns and in this eventuality, it is common to see that the funds allocated to overcome social problems are diverted to meet construction and establishment expenses.

7.10.1 Social Impact Assessment

Social impact assessment (SIA) provides a framework to collect, compile, analyze, and incorporate social information into the design and delivery of developmental interventions. SIA ensures that development interventions are well informed and take into account the key relevant social issues. To be relevant, SIA should incorporate a participatory strategy for involving a wide range of stakeholders.

For a water management project, stakeholders are those who have some interest in the project from social, economic, cultural, or other perspectives. Stakeholders may also have direct or indirect involvement in project preparation, construction, and operation.

Stakeholders for a hydro project may be divided into several groups.

(a) *Beneficiary groups*: they directly or indirectly benefit from the construction or operation of the project. These people may be the consumers of the project output such as electric power or those who earn by providing project services, such as planning, investigations, designing, supervising, and consulting. Some beneficiaries may be supplying labor, materials, equipment, and logistics to the project. Beneficiaries may also be getting indirect benefits. For example, transport connectivity or the environment of a city may have improved due to a project.

(b) *Sufferer groups*: whose interests may be compromised because of a project. For example, people adversely impacted by the acquisition of their land or house for the project, people whose houses are damaged due to construction-related activities, etc. At times, people suffer due to a project if construction and operation are not carried out by following standard operating procedures, for example, carelessly disposing of muck and waste.

(c) *Developers and investors*: Promoters, managers, and operators of water development projects.

(d) *Other stakeholders*: Political leaders and officers at different levels—National, State, and Local governments; office-bearers from related sectors and regulators.

(e) *NGOs, civil society*: they monitor the activities of the developers, help vulnerable groups, and liaise between different groups.

All the members in the beneficiary or sufferer group do not reap benefits or harms to the same degree; these differ from stakeholder to stakeholder. Beneficiary and sufferer groups may not always be directly associated with the project. Further, the beneficiary at a time may become a sufferer at a different time and vice versa. Not all the residents in an area will benefit/suffer due to the project and to the same extent.

Stages in SIA

The following are the main stages in SIA:

- Identify the area of influence of the project and the baseline conditions for the society and human environment.
- Develop an effective mechanism to involve all potential stakeholders from the public.

- Describe the proposed project (including studied alternatives) to the public in clear terms, using the language that the public understands.
- Identify and describe all the probable social impacts of the proposed project.
- Ascertaining the responses of the public to the impacts, particularly the harmful impacts, especially to the vulnerable sections of the society.
- Developing monitoring plans and mitigation measures.

Many countries have established procedures to carry out SIA and SIA is an integral part of providing approvals to proceed with the project. In many countries, SIA pays particular attention to directly impacted weaker sections and the vulnerable sections of the society in the project area.

7.10.2 Sociohydrology

As the water crises are becoming more complex and challenging and knowing that water resources link different sectors, the need for a more integrated paradigm in managing water resources by involving all stakeholders is rising. However, even now water management is dominated by approaches that largely emphasize technical solutions. These approaches have worked well in the short term but at times, they may result in unintended outcomes in the long term.

An emerging science, sociohydrology, the science of people and water, is aimed at understanding the dynamics and coevolution of coupled human-water systems (Sivapalan et al., 2012). Sociohydrology is focused on developing a generalizable understanding of the interactions and feedbacks between natural, technical, and social processes, with the aim to improve water management. Sociohydrology and the field of ecohydrology have considerable similarities. Ecohydrology explores the coevolution and self-organization of vegetation in the landscape in relation to water availability (Eagleson, 2002).

In traditional hydrology, human-induced water resources management activities are prescribed as external forcings in the water cycle dynamics, under the assumption of stationarity. In sociohydrology, humans and their actions are considered part and parcel of water cycle dynamics, and the aim is to predict the dynamics of both (Di Baldassarre et al., 2019). The inclusion of the discipline of sociohydrology in mainstream hydrology is expected to broaden the latter, as the world is changing due to human influence.

As per Di Baldassarre et al. (2019), work in sociohydrology has built on a history of works in three related fields: (1) water resources systems analysis where the focus is mainly on decision support by pursuing an optimization route; (2) integrated water resources management (IWRM). IWRM was introduced in the 1990s. It is concerned with problem-solving but the concept is not widely implemented so far; and (3) development of interdisciplinary frameworks that explore the influence of society and natural systems (including water) on each other. Examples include social-ecological systems and coupled human—nature systems. Ross et al. (2020) found important overlap as well as a considerable difference between two similar-sounding subfields: sociohydrology and hydrosocial research. As per them, while sociohydrology focuses on power and scale in studying water demand, hydrosocial research concentrates on practical responses to climate extremes. It is hoped that further developments in sociohydrology will lead to environmentally and socially sound projects and improved management of water resources.

7.10.3 Rehabilitation and Resettlement

Worldwide, a large number of people migrate to new areas of habitat every year. The displacement of people for nation-building or development is often considered unavoidable. There are various types of resettlement: spontaneous resettlement, facilitated spontaneous resettlement, sponsored voluntary resettlement, and involuntary resettlement. Here our attention is confined to involuntary resettlement consequent to a WRD project. Rehabilitation and Resettlement (R&R) of the population displaced or affected by a WRD project is a hotly debated topic of recent times.

A major reason for opposition to WRD projects in recent years is the displacement of the population due to these projects. It is now a concern because of the realization that displacement, by its very nature, results in the breakdown of family and community networks, and causes social and economic distress. Nevertheless, the governments have carried out national development plans and the displaced people have been compensated for in some way. It is important to note that the number of people that are subject to involuntary resettlement due to reasons, such as natural calamities (floods and droughts), political suppression, law and order problems, war, terrorist acts, and search for jobs, etc., is much larger than the water resources projects. Among the development-induced activities, the proportion of WRD projects is coming down because fewer and fewer of these projects are being constructed, while the number of people affected by infrastructural development works is increasing.

With the increasing depletion of natural resources and the competition for land, the issues associated with the displacement of people through infrastructure development have become more complex, requiring more attention, more resources, in short, a new paradigm. When the majority of the project-affected people are poor and unskilled, besides relocation, rehabilitation of these people becomes equally important. Can a project be termed as a developmental activity if it leads to the dislocation of vast populations? To avoid social tensions, involuntary resettlement should be planned such that the resettlers and the host population both benefit. This could be in terms of higher income, improved living standards, and health of the affected people.

The problem of resettlement attains gigantic proportions in densely populated or resource-starved nations. In recent times, a number of dams have been constructed or are being constructed in developing countries. Table 7.13 gives the number of people that were or will be subject to displacement due to large water resources projects. The Gezhouba project in China displaced more than 20,000 people and The Three Gorges Dam in the same region is likely to result in displacement of over a million people. The Saguling dam in Indonesia necessitated the resettlement of about 65,000 people. The Srisailam dam in South India has displaced about 100,000 people and the Sardar Sarovar project on Narmada will affect even more number of people. According to Herschy et al. (1998), the water resources projects constructed in China in the past 30 years have displaced more than 10 million people, the corresponding number for the last 4 decades in India is 20 million. It is also reported that in many cases, the actual displacement is much more than estimated. The feasibility report of the Kiambere dam in Kenya estimated that some 1000 people would be displaced but during construction, this number turned out to be 6000. The same authors also report that against the estimate that 200 people will be affected by the Ruzizi II hydropower project, the actual number turned out to be 16,000.

A detailed and up-to-date baseline socioeconomic database is essential to prepare R&R plans. One of the major criticisms in many projects has been that some planners come out with different figures at different times and the numbers given by different agencies do not match. The confusion can go up to the extent that the number of affected villages/towns may not be consistent. A well-organized database

Table 7.13 The number of resettlers from some major projects.

Project	Country	Number of resettlers
Three Gorges	China	1,250,000
Danjiangkou	China	383,000
Sanmenxia	China	319,000
Xinjiang	China	306,000
Dongpinghu	India	278,000
Upper Krishna II	India	220,000
Xiaolangdi	China	180,000
MCIP III Irrigation	India	168,000
Andhra Pradesh Irrigation II	India	150,000
Gujarat Med. Irrigation II	India	140,000
Sardar sarovar	India	127,000
Tehri	India	105,000
Aswan High dam	Egypt	100,000
Subarnarekha Group	India	100,000
Srisalam	India	100,000
Kossou	Ivory Coast	85,000
Akosombo	Ghana	84,000
Longtan	China	73,000
Shuikou I and II	China	67,000
Mahaweli I–IV	Sri Lanka	60,000
Saguling	Indonesia	65,000
Kariba	Zambia and Zimbabwe	67,000
Cirata	Indonesia	56,000
Sobradinho	Brazil	55,000
Paulo Af. IV	Brazil	52,000
Yacyreta	Argentina and Paraguay	50,000
Itaparica	Brazil	50,000
Kainji	Nigeria	44,000
Yantan	China	40,000

Based on Scudder and Biswas (1998a) and others.

will be necessary to clearly identify the project-affected persons. There are numerous cases where rogue elements try to force or bribe for the inclusion of their names in such lists in order to enable them to claim the declared benefits. A database will also help in arriving at a realistic outlay for R&R and thereby a realistic estimate of the project cost. There is a need for a class-benefit analysis of the project to understand who stands to gain and who are the losers. Although detailed plans exist for the physical structure of a project, a detailed blueprint for R&R does not exist even for all major projects.

At present, not many countries have well-defined policies for resettlement. Even where such policies exist, the political and administrative will to implement them in a human-centric approach is generally missing. Due to immense variations in policies and socioeconomic factors, the R&R programs vary considerably from one country to the other. In many countries, such programs are still being evolved and modified in light of experiences. While doing that, the basic principles of fairness, transparency, and equity should be adhered to in all cases. In India which is a large country, each state has its own R&R policy. The packages for rehabilitation at the new site generally consist of infrastructure facilities, compensation for the land submerged, and plots for building houses at the new site at nominal costs. Free facilities for the transport of perennial effects and loans, etc. are provided to the population displaced. A successful implementation also requires that R&R plans are properly phased in time so that there is no 11th-hour shifting. A rush through these activities generates tension and heart burning.

While considering and evaluating various project alternatives, appropriate weight should be given to minimize the displacement of the population, because the higher the magnitude of displacement, the more will be human misery, expenditure, and delay. It is important to perceive and propagate unavoidable resettlement as a development opportunity. It is noted that many R&R programs are rejected by the project-affected people on the ground that these are inadequate and unacceptable. While a certain element of greed and desire to get the maximum out of it cannot be ruled out, many of the packages prepared by the government agencies were clearly inadequate and were improved after agitation. To avoid complications and agitation, it would be better (although difficult) if the resettlement programs were prepared by involving the affected population. Such a close working relationship will also help overcome the problems due to the arrogance of officials. The extra time and efforts spent should be considered as an investment as it will eliminate many future problems.

The requirement of funds for resettlement also increases day by day. Scudder and Biswas (1998b) reports that the resettlement component of 20 projects financed by the World Bank between 1986 and 1993 was 9% of the total cost. In countries where population densities are higher, this amount will be considerably more. For example, the Three Gorges Project of China will require one-third of project funds for resettlement if the Chinese policy guidelines were followed. The finances required for the Sardar Sarovar project in India are also quite high.

Often resettlement is considered inadequate, because the government agencies may lack awareness of the complexity and range of issues, the goals may be improperly perceived, and in some cases, they may pretend to ignore the reality because this will weaken the justification for the project. Of course, one can imagine that finding land and employment for over one million people who will be affected by the Three Gorges Project is a real herculean task.

R&R of tribals or hill-area people whose lifestyles and culture are radically different from the people of plains needs careful attention. The tribals are used to living in hills and forests form an integral part of their life. They practice their own way of agriculture. Their relocation even in command areas of nearby canals can inflict a cultural shock on them and could be a cause of avoidable social conflict. It would be desirable to relocate them within the forests on the fringes of the reservoir. Side-by-side, measures for their health and economic uplift should also be initiated so that they are brought into the national mainstream while retaining their own identity.

The displaced people usually find that while they lose much, they gain very little from the projects that displaced them. As pointed by Gill and Fisher (199) the result is that those who are made to sacrifice for the "national good" are not those who ultimately benefit from the development projects that displace them. It does not seem to be merely coincidental that a major proportion of the displaced are the poor and the powerless. However, the justification is that some people will have to sacrifice in the "national interest." But the equally important questions are: who sacrifices and who benefits? What happens to the displaced? Do they benefit or gain from the process that displaces them? And what is their role in the whole process? These questions should be addressed in the right earnest.

All said and done, resettlement can also be viewed as an opportunity in the long run because it can remove cultural constraints to the developmental and entrepreneurial initiative, and political and other restrictions. It can also bring the tribal people into the national mainstream, although there are likely to be different opinions on this matter. Every human being has a right to development, but for it to be just and sustainable, it must be development for all—not development for some at the cost of others. The process of development must not violate the principles of democracy and the human rights of the people involved. Instead, it must ensure, as far as possible, people's participation in planning and implementation, and it must bring a share of the benefits to them. Recognizing this, Article 1 Clause 1 of the United Nations General Assembly Declaration on the Right to Development states: "The human person is the central subject of development and should be the active participant and the beneficiary of the right to development." The same article in Clause 3 goes on to add, "States have the right and the duty to formulate appropriate natural development policies that aim at the constant improvement of the well-being of the entire population of all individuals."

Importantly, R&R is not an administrative problem; it is a human problem. Since R&R deals with people, pragmatic policy decisions with the involvement of the affected people at appropriate levels and stages is necessary. Many people, particularly the old, who may have been born and brought up on that land, consider it as their ancestral property and have an emotional and nostalgic attachment. Such cases should be dealt with care, respect, and patience that they deserve. The use of force should always be avoided and people should be motivated to move by giving them higher standard of living and sustainable means to lead a dignified life. It is also necessary to ensure that the displaced people are made partners in prosperity due to the project and do not turn into adversaries.

Earlier, a common feeling was that the compensation paid for acquiring land for projects was inadequate and nontransparent, and due to this, most people were unwilling to give their lands for development projects. Addressing these concerns, the Government of India has enacted "The Right to Fair Compensation and Transparency in Land Acquisition, Rehabilitation and Resettlement Act, 2013." Provisions of this Act are applicable to land acquisition, compensation, rehabilitation, and resettlement, for the cases when the Government acquires land for its own use, hold, and control, including for Public Sector Undertakings and for public purposes. A key point is that whenever government intends to acquire land for a public purpose, it shall consult the concerned Panchayat (village administrative body), Municipality or Municipal Corporation, and shall carry out a SIA study in consultation with them. The value of the land to be acquired shall be based on the market value of the land or the average sale price for similar types of nearest lands. After this act has come into force, there are much less difficulties in land acquisition, although the cost of projects has risen.

7.11 CASE STUDY—SARDAR SAROVAR PROJECT, INDIA

The Sardar Sarovar Project in the Narmada basin, India, is one the most ambitious and equally controversial projects of recent times. The salient features of the project, its benefits, environmental impacts and opposition are described in the following. The material is based on (Chitale, 1997; Fisher, 1997; SSNNL, 2000) Internet, newspaper reports, etc.

Narmada, the largest westward flowing river in India, traverses over 1310 km. It rises in the state of Madhya Pradesh in central India and passes through the states of Maharashtra and Gujarat on its way to the Gulf of Khambhat. The Narmada drainage basin covers 98,796 square km. About 85% of the Narmada's average annual flow of 50 km^3 occurs during 4 months of monsoon rains from June through September (Chitale, 1997). Floods are intense and sudden, reaching their peaks in a short time. Harvesting this flow would require a large system of reservoirs. Hence, the Narmada River is the subject of one of the largest basin development schemes in the world that would include, when completed, 30 major, 135 medium, and about 3000 minor dam projects in the Narmada basin.

Tapping the resources of the Narmada has been the dream of political leaders and development planners for decades. Large parts of Gujarat and Rajasthan face recurrent droughts and there have been instances when the water had to be transported by trains to save the people from famine. The idea of constructing dams on the Narmada River was first suggested in 1946. For quite some time, this idea could not materialize because the states did not agree on the distribution of the river water. The then Prime Minister of India, Mr. Jawaharlal Nehru, laid the foundation stone for the Sardar Sarovar dam, a multipurpose project that is the terminal dam of the basin-wide scheme in 1961. This project was delayed and in 1965, a committee was appointed by the Government of India to prepare a detailed plan for the development of the Narmada basin. The committee recommended the construction of a dam and a canal in Gujarat and 12 major projects in Madhya Pradesh. The two principal dams proposed were the Indira Sagar Dam and the Sardar Sarovar. The recommendations of the committee were endorsed by the Government of Gujarat but rejected by the Governments of Madhya Pradesh and Maharashtra. Subsequently, in 1969, the issue was referred to the Narmada Water Disputes Tribunal, which was established under India's Interstate Water Disputes Act of 1956. The Tribunal considered the issues for a decade and made its final award in 1979.

This award, which provides for the diversion of 11,718.25 million m^3 (9.5 million acre-feet, MAF) of water from the reservoir into a canal and irrigation system, has formed the basis for the construction of the current Sardar Sarovar Project. Finances for this ambitious project were secured in 1985 when the World Bank entered into credit and loan agreements with the Governments of Gujarat, Madhya Pradesh, and Maharashtra. It provided US$ 450 million for the construction of the dam and the canal. The construction of the dam began in earnest in 1987. Another major project under construction upstream of Sardar Sarovar is the Indira Sagar project.

7.11.1 The Project

Sardar Sarovar is an ambitious and technologically complex irrigation scheme, which is to draw upon the flow of the Narmada River to alleviate the water needs of large areas of the state of Gujarat. The project, which is one of the largest water resources projects ever undertaken in India, includes a dam, a riverbed powerhouse, a main canal, a canal powerhouse, and an irrigation network. Its projected impact extends over a large area, and it will potentially affect 25–40 million people. The components of the project are designed to irrigate vast areas of Gujarat and Rajasthan (although not a basin state, was also later allocated a share of its waters), and to provide drinking water to areas of central and northern Gujarat. The water is to be delivered by creating a storage reservoir on the Narmada River with a full reservoir level of 138.684 m, along with an extensive canal and irrigation system. Fig. 7.5 shows the Sardar Sarovar Project Area, including the river basin, the submergence area, and the canal system.

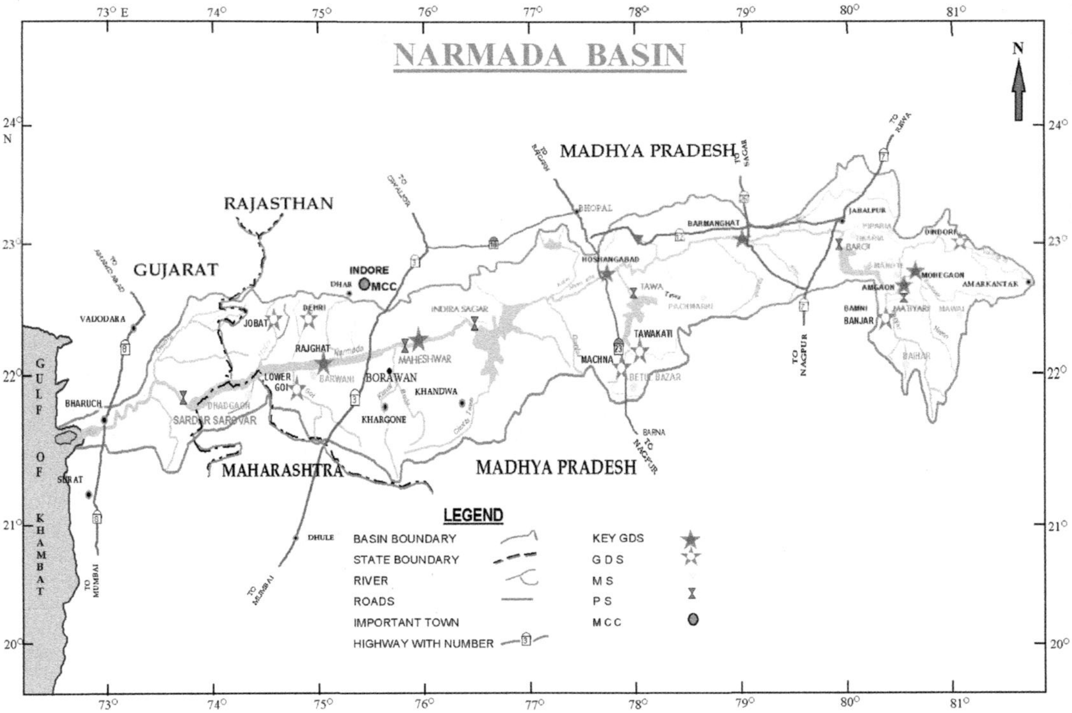

FIGURE 7.5

Narmada basin, India.

Narmada Control Authority.

The main dam is 1210 m long and 163 m high from the deepest foundation level (Fig. 7.6). The live storage capacity of the reservoir is 5860 MCM. Project will irrigate about 1.905 million ha land, benefiting 1 million farmers, providing drinking water to 9490 villages and 173 towns (30 million people), and will generate about 1 billion kWh of electricity every year from a powerhouse having 1450 MW installed capacity.

The Narmada Main Canal will be the largest of its kind in the world, extending 458 km to the Rajasthan border and crossing 19 major rivers and 244 railway lines or roads. With 31 branch canals, the aggregate length of the distribution system will be 75,000 km (within Gujarat). The main canal will be 250 m wide at the head and 100 m wide at the border: the capacity of this canal system is such that it will be able to empty the proposed reservoir storage in less than 2 months. The canal will also transport Narmada water to Saurashtra and Kutch region of Gujarat, which are drought-prone areas. Many wildlife sanctuaries and parks will get water from the project.

The dam has been constructed in a hilly region, and the reservoir created behind the dam extends to over 200 km upstream, submerging approximately 37,690 ha of land in three states: Gujarat, Maharashtra, and Madhya Pradesh. Out of this, 11,279 ha is agricultural, 13,542 ha is forest land, and 12,869 ha is river bed and wasteland. While the full impact of the project remains in dispute, a total of

FIGURE 7.6

View of Sardar Sarovar Dam from downstream.

Rupeshsarkar, https://commons.wikimedia.org/w/index.php?curid=93105527.

245 villages have been submerged, mostly partially, affecting about 110,000 people. Many of these people, especially in Gujarat and Maharashtra, are "tribals" and have no formal title to their land. A large number of farmers, about 140,000 according to an estimate, will lose land to the canal and irrigation systems. In addition, thousands of people living downstream find their lives affected by the project. Weigh this up against the benefits described earlier. The area that will be submerged is about 1.65% of the area that will get benefits. The ratio of the population displaced to the population benefited is 1:37. Generation of wealth in an area also contributes to the general economic development of the area. For different levels of irrigation efficiency, the internal rate of return of the project could be between 16.77 and 21.88 on economic prices of the inputs and outputs and the corresponding benefit–cost ratios between 1.59 and 3.29 (Chitale, 1997). The acceptable values are 9 and 1, respectively.

7.11.2 Controversy

The complexity of the project allows for considerable dispute about how to best calculate and compare the projected costs and anticipated benefits. The project's proponents emphasize the enormous benefits it is expected to bring to millions of people at the cost of displacing comparatively few. They ask that these projected benefits—the provision of drinking water to millions of people and irrigation of millions of hectares of land—be weighed against the relatively small number of people who will be displaced from their land. From their perspective, this balance tilts heavily in favor of the project.

The opponents of the project question both the projected benefits and the cost of the project. They argue that the irrigation benefits have been vastly overestimated and that adequate drinking water may never reach the neediest drought-prone areas of Gujarat, such as Kutch. According to them, the economic costs are based on unrealistic figures and have been grossly underestimated. Also, the human and environmental costs have been vastly underestimated or ignored by misreporting the number and the extent to which people will be affected by the project and disregarding the costs of cultural disruption that will occur when tribal people are moved from their traditional lands. In their opinion, the number of people to be affected must include those living in the submergence area, those displaced by the construction of infrastructure, those affected by the canal, and those living downstream whose lives and livelihoods will be affected. The reservoir submergence has ignored the adverse effect on the forest cover and Narmada water will never reach far-flung drought-prone Kutch area, an oft-repeated dream of the dam builders.

In the Sardar Sarovar Project, the major concern about compensation has focused on the category of people identified as "oustee." An oustee is an individual "whether landed or landless, who since at least one year prior to the date of publication of the notification under the relevant Indian Land Acquisition Act, has been ordinarily residing, or cultivating land, or carrying on any trade, occupation, or calling or working for gain in Gujarat, Madhya Pradesh, and Maharashtra, who would be displaced from his usual habitat due to the carrying out of the Project." Two factors affect the compensation for which the Sardar Sarovar Project oustee is entitled. One is the assessment of an oustee's right to compensation, which is complicated by the disparity between the way the government administers, registers, and taxes land and the way people conceive of and use resources. Secondly, the R&R policies of the three states of Gujarat, Madhya Pradesh, and Maharashtra vary in the compensation they give to oustees. With the new R&R policy and other efforts, the matter has been largely resolved although there are some unresolved topics.

The three states have different norms for the treatment of "major" sons (sons over the age of 18), encroachers (those residing on and cultivating land to which they do not have legal title), and the landless. The Gujarat policy makes no distinction between landed and landless oustees and offers full benefits to major sons. According to the award of the Narmada Water Disputes Tribunal, all people displaced by the Sardar Sarovar reservoir have the right to settle in Gujarat, if they so desire. Naturally, so long as Gujarat's resettlement and rehabilitation benefits are significantly better than those offered by the other two states, most displaced people are apt to take this option. Hence, the Madhya Pradesh resettlement plans anticipate that only 10% of the displaced people from that state will remain in the state. An unequal compensation is seen to violate the spirit of the Tribunal Award, which provides that all oustees may remain in their home states. To do so under current policies would entail a financial sacrifice for some, while relocation to Gujarat would mean for many "a long cultural journey." Therefore, while the right of choice still exists in principle, the disparity in benefits means a choice between migration to another state or a lower standard of living.

The concern of compensation dominates the discussions of R&R; there are disagreements about what constitutes full, fair, and appropriate compensation, and further disagreements about whether all the oustees will or can be fully, fairly, and appropriately compensated for under prevailing circumstances. The proponents argue that displacement should be treated as a development opportunity and that project-affected people should not only regain their standard of living but also be treated as the first beneficiaries of the project. Clearly, high costs attached to R&R reduce the cost-benefit ratio, making it more difficult to raise political and financial support. Even in cases where project benefits make it possible to offer attractive R&R packages, high compensation is opposed by the decision-makers fearing that this will set a precedent of high awards that could not be met by all future projects. Note that the oustees mostly belong to marginal and disempowered communities and R&R requires that all people affected by the project must "improve or at least regain the standard of living they were enjoying prior to their displacement."

7.11.3 Protests

Meanwhile, a number of Indian Non-Governmental Organizations (NGOs) began opposing the project, mainly on the grounds of environmental, human (tribal) rights, skewed economics of the project, etc. The *Narmada Bachao Andolan* (NBA) is the main opponent. They were later joined by several foreign NGOs. Some NGOs, such as *Action Research in Community Health Association* (ARCH), who were originally opposed to the project because of insufficient compensation for the affected population, later accepted the improved measures and supported the continuation of construction.

Originally, the campaign against the Sardar Sarovar Project revolved around resettlement issues. Earlier disagreements between the states, mainly about water allocation and the height of the proposed dams, had been resolved amicably by the middle of 1974. After certain clarifications, this was made an award by a Tribunal under the Inter-State Water Disputes Act at the end of 1979. This Tribunal also set out a resettlement and rehabilitation scheme, which, at that time, was considered very liberal though the World Bank insisted on even higher standards. Landless farm laborers and so-called "encroachers" were to receive 2 ha compensation, often more than the area possessed by the communities among whom they were to be resettled. It had been planned to resettle some of the displaced people on degraded forest land. In the meantime, a new "Forest Conservation Act" was passed in 1980, forbidding any forest land to be diverted to other purposes, unless specifically sanctioned by the Central Government. This led to a shortage of land needed for resettlement.

Gradually, NGOs started to employ radical protest techniques, such as marches, hunger strikes, traffic disruptions, and intimidation of those wanting to be resettled. They even instituted a "mass drowning rather than relocate" campaign at the first village threatened by the rising waters. Much was made of the fact that many of the so-called "oustees" were still tribal. A lot of incitation of these relatively primitive people took place. The proponents, however, claim that the tribal families have shown a desire to avail themselves of this development opportunity. The NGOs also started canvassing for foreign support through newspaper articles, talks, and petitions to heads of governments, the UN, and donor agencies. All this led to Japan withdrawing its financial support to the project. The World Bank appointed an Independent Review Mission, the first of its kind in the history of the Bank. The report of this mission was interpreted variously by various people; it was criticized by many, for example, Alagh et al. (1997), and apparently further complicated the matter. After the World Bank withdrew from the project, the Government of India decided to proceed without external help.

The opponents managed to mobilize a substantial number of people in India and elsewhere against the project, but a large-scale popular support for it was shown by about a million people turning up for a proproject demonstration. The public issue of bonds was floated by the Government of Gujarat to mobilize funds for the project. All the issues were oversubscribed, showing public support for the project.

7.11.4 Project Status

The construction of the dam has been completed. Water level at the Sardar Sarovar dam reached the level of 138.7 m on 15 Sept. 2019. Most project components have also been completed and are providing benefits. Water is already flowing through the Narmada Main Canal and has reached the faraway places in Gujarat and Rajasthan. Many websites contain information about this project, for example, Sardar Sarovar Narmada Nigam Ltd. (www.sardarsarovardam.org) and Narmada Control Authority (http://nca.gov.in/). Water resources planners and decision-makers who have worked on this project or followed its progress have received invaluable experience from this project.

Just downstream of the dam on the bank of the Narmada River, near the Kevadiya colony, a 182 m tall statue of Sardar Vallabhbhai Patel, the first deputy prime minister and home minister of India, has been constructed (completed in 2018). This statue adds to the charm of the Sardar Sarovar Dam and millions of tourists visit this statue every year.

7.12 CLOSURE

Sometimes it is fashionable to denounce all technologies and raise a slogan for going "back to nature." But one should not forget the risks faced by our ancestors. Even today, fishermen who face the periodic onslaught of cyclones need to be provided with advance warnings and damage-proof shelters for protection during cyclones. The traditional mud houses in the villages cannot provide the required safety. Similarly, during an extended drought, the people and cattle have to be provided water at least for survival. While facing the furies of nature, the role of high technology can hardly be ignored. The technology is a result of human ingenuity, efforts, and skills. It is, therefore, not appropriate to denounce them mindlessly.

Most people do not prefer the "natural" and raw living environment, but developed and managed conditions; not the natural water from the ground may be polluted but treated water, free from impurities. It is wrong to believe that nature has provided all waters as healthy. Natural lakes can get heavily polluted by excessive decay of the vegetative matter flowing in them. It is for human endeavors to arrest such excessive decay and/or to transform these waters into consumable ones by utilizing proper technology.

In the present situation, it is necessary to filter "environmental noise" from a few serious efforts for true environmental awakening. It is true that efforts of the so-called environmentalists, as distinct from environmental scientists, have succeeded in retarding the pace of harnessing additional waters for irrigation, drinking, or hydropower or for putting up new flood control works. It is not in the interest of the developing societies to allow this unsettled situation to continue for a long time. A determined effort is necessary to resolve the issues and move ahead quickly with refined and improved policies in light of new environmental data and proper analysis. Discussion on an issue, beyond a limit, is counter productive. It will be very unfortunate if WRD gets bogged down in "paralysis by analysis!"

7.13 APPENDIX 7A

7.13.1 Report of World Commission on Dams

The World Commission on Dams (WCD) launched its Final Report *Dams And Development: A New Framework for Decision-Making*, in 2000 (World Commission on Dams, 2000). The Commission tried to gather the views of all parties in the increasingly confrontational debate about the role that 45,000 large dams worldwide have played in development. Estimates suggest that some 40–80 million people have been displaced by dams worldwide, while the livelihoods of many more living downstream were affected but not recognized. This report will have an impact on the future role of the $42 billion dam industry. Expectedly, the report was appreciated by a section of people, while it was rejected by another group.

7.13.2 Main Findings of the Review of Large Dams

The WCD report provides a mammoth review of the technical, financial, and economic performance of dams as well as their environmental and social performance. Together with its assessment of potential alternatives to dams and the study of decision-making processes, it offers an insight into one of the most controversial development debates of our time.

WCD conducted detailed reviews of eight large dams in Turkey, Norway, the United States, Zambia and Zimbabwe, Thailand, Pakistan, Brazil, and South Africa. It also prepared country reviews for India and China, as well as a briefing paper on Russia and the newly formed Commonwealth of Independent States. A survey of 125 large dams was also undertaken, along with 17 thematic reviews on social, environmental, and economic issues; on alternatives to dams; and on governance and institutional processes.

The commission concluded that dams have made an important and significant contribution to human development, but in too many cases, the social and environmental costs have been unacceptable and unnecessary. A new framework for decision-making that moves beyond simple cost-benefit trade-offs to introduce an inclusive "rights and risks approach," which recognizes all legitimate stakeholders in negotiating development choices was suggested.

The commission found that

- Dams deliver significant development services in more than 140 countries.
- On a global scale, hydropower dams account for 19% of electricity generated and for an estimated 12%–16% of global food production.
- About 12% of large dams supply domestic and industrial water, and large dams provide flood control services in more than 70 countries.
- Large dams display a high degree of variability in delivering predicted water and electricity services—and related social benefits—with a considerable portion falling short of physical and economic targets, while others continue to generate benefits after 30–40 years.
- Large dams have demonstrated a marked tendency toward scheduled delays and cost overruns.
- Large dams have led to the loss of forests and wildlife habitat and the loss of aquatic biodiversity of upstream and downstream fisheries. The efforts to counter the ecosystem impact of large dams have met with limited success.
- The negative social impacts reflect a pervasive and systematic failure to assess and account for the range of potential negative impacts on displaced and resettled people as well as downstream communities.

7.13.3 Alternatives to Dams for Water and Energy Resources Development

The commission examined the alternatives for meeting energy, water, and food needs and found that while there is far greater scope for utilizing these, no universal formula applies as local and national conditions are central to determining viable options. Obstacles, such as market, institutional, intellectual, and financial barriers, limit the adoption rate of alternatives. Improved system management, particularly in the irrigation sector but also through a reduction in water losses, more efficient system management, and an upgrading of distribution technology, can alleviate demand for new supply sources. Demand-side management has significant potential and provides a major opportunity to reduce water stress and power requirements.

7.13.4 Key Recommendations

The commission argued that it is not necessary to trade-off one person's gain against another's loss. Rather, by negotiating outcomes through multicriteria analysis—technical, environmental, economic, social, and financial—the development effectiveness of water and energy projects will be improved, unfavorable projects will be eliminated at an early stage, and the options chosen will be what key stakeholders agree best meets the needs in question.

- A set of five core values should be adopted for future decision-making: equity; sustainability; efficiency; participatory decision-making; and accountability.
- Seven strategy priorities for water and energy resources development: gaining public acceptance; comprehensive options assessment; addressing existing dams; sustaining rivers and livelihoods; recognizing entitlements and sharing benefits; ensuring compliance; and sharing rivers for peace, development, and security.
- National governments should review existing procedures and regulations concerning large dam projects and time-bound licenses for all dams.
- Bilateral aid agencies, export credit agencies, and multilateral development banks should establish procedures that ensure that approved financing for dam projects emerges from an agreed process of ranking alternatives.

References

Abu-Zeid, M.A., Biswas, A.K., 1992. Some major implications of climatic fluctuations on water management. In: Abu-Zeid, M.A., Biswas, A.K. (Eds.), Climatic Fluctuations and Water Management. Butter-Heinemann, Oxford.

Abu-Zeid, M.A., El-Shibini, F.Z., 1997. Egypt's high Aswan dam. Water Resources Development 13 (2), 209–217.

Alagh, Y.K., Buch, D.T., 1997. The Sardar Sarovar project and sustainable development. In: Fisher, W.F. (Ed.), Toward Sustainable Development: Struggling over India's Narmada River. Rawat Publications, Jaipur, India.

Arthington, A.H., 2012. Environmental Flows: Saving Rivers in the Third Millennium. University of California Press, Berkeley, p. 424.

ASCE Task Committee for Sustainability Criteria, 1998. Sustainability Criteria for Water Resource Systems. Division of Water Resources Planning and Management, ASCE, Reston, USA.

Asmal, K., 2002. Parting the waters. Journal of Water Resources Planning and Management, ASCE 128 (2), 87–90.

Bassam, N.El, 1999. Developments in Agriculture – Global Key Issues, Natural Resources and Development, vol. 51. Institute for Scientific Cooperation, Tubingen, Germany, pp. 39–57.
Biswas, A.K., 1994. Sustainable water resources development: some personal thoughts. Water Resources Development 10 (2), 109–116.
Biswas, A.K. (Ed.), 1998. Water Resources Environmental Planning, Management, and Development. Tata McGraw Hill Publishing Company Ltd., New Delhi.
Canter, L.W., 1981. Methods of Environmental Impact Assessment: Theory and Applications (Emphasis of Weighting-Scaling Checklists and Networks). Environmental Impact Assessment, NATO ASI Series, Toulouse, France.
CBIP, 1995. Environment Impact Assessment Studies (Case Studies). Central Board of Irrigation and Power, New Delhi.
Cernea, M.M., 1990. Poverty Risks from Population Displacement in Water Resources Development. Harvard Institute for International Development, Cambridge, USA. Development Discussion Paper 355.
Chitale, M.V., 1997. The Narmada project. Water Resources Development 13 (2), 169–179.
Clark, M.J., Gardiner, J., 1994. Strategies for handling uncertainty in integrated river basin planning. In: Kirby, C., White, W.R. (Eds.), Integrated River Basin Development. John Wiley and Sons, Chichester.
Di Baldassarre, G., Sivapalan, M., Rusca, M., Cudennec, C., Garcia, M., Kreibich, H., et al., 2019. Socio-hydrology: scientific challenges in addressing the sustainable development goals. Water Resources Research 55, 6327–6355. https://doi.org/10.1029/2018WR023901.
Eagleson, P.S., 2002. Ecohydrology: Darwinian Expression of Vegentation Form and Function. Cambridge University Press, Cambridge, England. https://doi.org/10.1017/CBO9780511535680.
Fisher, W.F. (Ed.), 1997. Toward Sustainable Development: Struggling over India's Narmada River. Rawat Publications, Jaipur, India.
Gill, M.S., 1997. Resettlement and rehabilitation in Maharashtra for the Sardar Sarovar Narmada project. In: Fisher, W.F. (Ed.), Toward Sustainable Development: Struggling over India's Narmada River. Rawat Publications, Jaipur, India.
Hasan, Z., Goel, R.S., 2000. Water resources development in Himalayan region and environmental impacts. In: Oxford, R.S.G. (Ed.), Environmental Impacts Assessment of Water Resources Projects. IBH Publishing Co. Pvt. Ltd., New Delhi.
Herschy, R.W., 1998. Environment priorities for development: water. In: Herschy, R.W., Fairbridge, R.W. (Eds.), Encyclopedia of Hydrology and Water Resources. Kluwer Academic Publishers, Dordrecht.
Hussain, A., 1992. Working towards Sustainable Development. Conference Report. Southern Asian Institute, New York.
Hussein, A., Bisset, R., Sadler, B., 2004. Environmental Impact Assessment and Strategic Environmental Assessment: Towards an Integrated Approach. United Nations Environment Programme (UNEP), Geneva.
Hussain, I., 2007. Direct and indirect benefits and potential disbenefits of irrigation: evidence and lessons. Irrigation and Drainage 56, 179–194. https://doi.org/10.1002/ird.301.
ICOLD, 1982. Dams and Environment. Bulletin 35. The International Commission on Large Dams, Paris.
ICOLD, 1998. Position Paper on Dams and Environment. The International Commissionon Large Dams. Available at: http://genepi.louis-jean.com/cigb/chartean.htm.
Ivanov, K., 1984. A method for estimation of the chemical indices and its application to project 'Iskar' in Bulgaria. In: Hydro-environmental Indices: A Review and Evaluation of Their Use in the Assessment of Environmental Impacts of Water Resources Projects. Working Group on IHP-II, Project A3.2. Unesco, Paris.
Jain, S.K., Mishra, P.K., Sharma, A., 2021. In: Mukherjee, A. (Ed.), Environmental Flows and Their Assessment, Chapter 1 in the Book Riverine System: Hydrology, Hydrosocial and Hydro-Heritage Connectivity. Springer/ Capital Publication, ISBN 978-3-030-87066-9.
Keller, A., Sakthivadivel, R., Seckler, D., 2000. Water Scarcity and the Role of Storage in Development. Research Report 39. International Water Management Institute, Colombo, Sri Lanka.

Kite, G.W., 2000. Climate change and water resources planning: a somewhat skeptical view. In: Proceedings of International Conference on Integrated Water Resources Management for Sustainable Development. ICIWRM, New Delhi, 2000.
LaBounty, J.F., 1984. Assessment of the environmental aspects of construction of the Three Gorges project on the Yangtze river. Water International 9, 10−17.
Loucks, D.P., Stakhiv, E.Z., Martin, L.R., 2000. Editorial: sustainable water resources management. Journal of Water Resources Planning and Management, ASCE 43−47 (Mar).
Olagunju, A.O., Gunn, J.A.E., 2015. Selection of valued ecosystem components in cumulative effects assessment: lessons from Canadian road construction projects. Impact Assessment and Project Appraisal 33 (3), 207−219. https://doi.org/10.1080/14615517.2015.1039382.
Ott, W.R., 1978. Environmental Indices − Theory and Practice. Ann Arbor Science Publishers, Michigan.
Pendse, Y.D., 1987. 'Environmental Aspects of Water Resources Development in India, Commemorative Volume. CBIP Diamond Jubilee, New Delhi.
Rees, C.P., 1981. Guidelines for environmental impact assessment of dam and reservoir projects. Journal of Weed Science and Technology 13, 57−71.
Robbroeck, T.van, 1996. Reservoirs: Bane or boon. Geoffrey Binnie Lecture. British Dam Society, London.
Ross, A., Chang, H., 2020. Socio-hydrology with hydrosocial theory: two sides of the same coin? Hydrological Sciences Journal 65 (9), 1443−1457. https://doi.org/10.1080/02626667.2020.1761023.
Sahu, H., 1992. Evaluation impacts on environment of Narmada Sagar project, India. In: Parr, N.M., Charles, J.A., Walker, S. (Eds.), Water Resources and Reservoir Engineering. The British Dam Society, London.
Scudder, T., 1998a. Social impacts. In: Biswas, A.K. (Ed.), Water Resources Environmental Planning, Management, and Development. Tata McGraw Hill Publishing Company Ltd., New Delhi.
Scudder, T., 1998b. Resettlement. In: Biswas, A.K. (Ed.), Water Resources Environmental Planning, Management, and Development. Tata McGraw Hill Publishing Company Ltd., New Delhi.
Shah, R.B., 1993. Role of major dams in the Indian economy. Water Resources Development 9 (3), 319−336.
Sivapalan, M., Savenije, H.H.G., Blöschl, G., 2012. Socio-hydrology: a new science of people and water. Hydrological Processes 26, 1270−1276.
Srivastava, L.S., 1993. Hydrodynamic Behaviour of Water Reservoirs. National Seminar of Hydrological Hazards − Prevention and Mitigation, Indian Association of Hydrologists, Roorkee, India.
SSNNL, 2000. Meeting the Challenges of Development - Sardar Sarovar Project on River Narmada. Sardar Sarovar Narmada Nigam Ltd., Govt. of Gujarat, Gandhinagar.
Unesco, 1984. Hydro-environmental Indices: A Review and Evaluation of Their Use in the Assessment of the Environmental Impacts of Water Resources. Technical Document in Hydrology, Working Group Report on IHP-II.
WCED, 1987. Our Common Future (The Brundtland Report). World Commission on Environment and Development, Oxford University Press.
Westman, W.E., 1985. Ecology, Impact Assessment, and Environmental Planning. John Wiley and Sons, New York.
World Bank, 1994. Resettlement and Development. The Bankwide Review of Projects Involving Resettlement 1986-1993. The World Bank, Washington D.C.
World Commission on Dams, 2000. Dams and Development: A New Framework. Earthscan, London.
Xu, Z.X., Takeuchi, K., Ishidaira, H., Zhang, X.W., 2002. Sustainability analysis for Yellow river water resources using the system dynamics approach. Water Resources Management 16, 239−261.

Further Reading

Biswas, A.K., 1997. Water development and the environment. Water Resources Development 13 (2), 141−167.
CWRC, 1997. Three Gorges Project. Changjiang Water Resources Commission, Wuhan, China.

CHAPTER

Rational Decision Making

8

Chapter outline

Water Resources Systems Planning and Management. https://doi.org/10.1016/B978-0-12-821349-0.00006-X

The art of decision-making started with the beginning of life when the very survival depended on the decisions made and actions taken. With the progress of civilization came reason and experience which led to the development of scientific methods for problem-solving. In day-to-day life, one makes many decisions which may involve certain amounts of risk. People's intuitive judgment, cognitive ability and availability of data are some of the factors that influence these decisions. While the first two aspects are the subjects of psychology, the last pertains to water resources. The uncertainty of future events greatly affects all stages of design and operation of water resource systems. Despite the immense variability in nature which has to be coped with and the availability of limited information, decisions have to be taken and implemented. The decision theory attempts to provide a systematic approach to making rational decisions. It can be shown (Klir, 1991; Shackle, 1961) that the necessity of decision-making results totally from uncertainty. In other words, if the uncertainty did not exist, there will be no need for decision-making. The overall philosophy and the various dimensions of decision-making have been very aptly illustrated by Haimes and Stakhiv (1985), as shown in Fig. 8.1.

When a number of decisions are possible, the general approach is to first eliminate the inferior alternatives till only a few of them remain. Then, considering the available information, the alternative that is expected to have the highest value or the least risk is chosen. Of course, there is no guarantee that this alternative will turn out to be the best; there might be situations wherein the outcome will not be up to the expectation. The optimization methods that help eliminate alternatives have been discussed in Chapter 5. Optimization, that is, searching for the extremum of a function, does not involve the necessity of real decision-making, although the variables which are changed to arrive at the

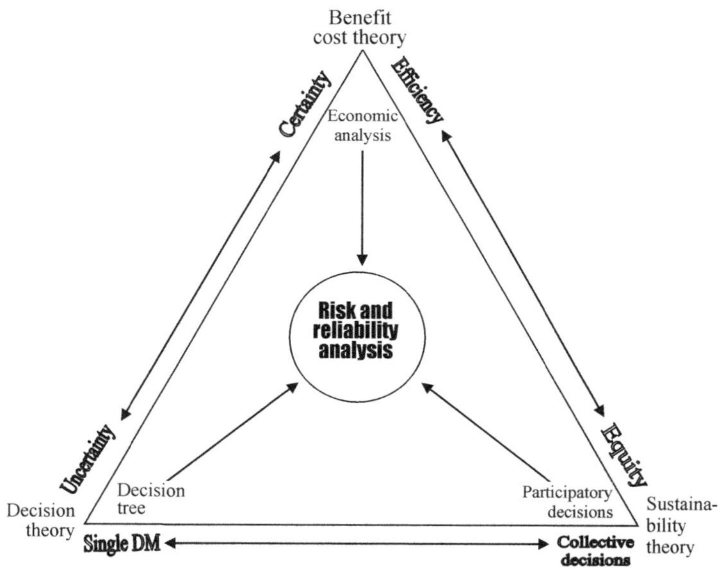

FIGURE 8.1

The decision triangle.

Adapted from Haimes and Stakhiv (1985).

extremum are called decision variables. In operations research and control theory, the availability of full knowledge needed to achieve the maximum gain is assumed. In fact, decisions are made by defining an objective function and a set of constrains. These methods are primarily concerned with the operational ordering of policies. The present chapter is focused on rational decision-making in real-life situations.

Risk and uncertainty are central to rational decision-making. A risky decision exposes the decision-maker to the possibility of some type of loss, but there are many situations when he has to make such a decision. It is not possible to completely avoid risk, one has to choose between risks. Naturally, the primary source of risk is the vagaries of nature. A decision to build a project may be risky because a large flood or earthquake may occur and endanger the structure with the resulting loss of life and property. Second, a decision may be risky because the natural phenomena are not clearly understood. Sometimes, a risky choice has to be made if the cost of an alternative that can control the risk is more than the ability or willingness to pay for it. The ultimate goal is to reduce uncertainty and thereby risk. At this stage, it is useful to understand what is implied by rationality in decision-making.

8.1 CONCEPT OF RATIONALITY

The presence of uncertainty notwithstanding, decisions about environmental systems, such as water storage reservoirs, irrigation systems, water purification systems, and so on should be rational. To judge whether a decision is rational or not, the decision-making process itself needs to be considered along with the criterion of rationality. There are three main criteria of rationality. First, the decision must be aimed at an objective that is to be achieved. Consider, for example, that the life of an overhead water tank is over and now it is to be replaced. The typical objectives might be to design a tank that meets some projected demands is economical, is durable, and has a pleasing appearance. Failure to meet any of these objectives in a new design may make it unacceptable, and the underlying decision will be irrational.

In a real-life situation, there might be many designs that meet the stated objectives. This leads to the second criterion of rationality which states that the decision-maker must identify and study enough alternatives to ensure that the best solution is among them. This does not imply an exhaustive search, but he must have an open mind for alternative solutions. For example, the tank designer may be a specialist in steel structures, but that should not prevent him from considering the advantages of reinforced concrete. Each alternative has its pros and cons and has a certain amount of risk.

The third criterion of rationality is that the choice between risky alternatives is made in accordance with an objective evaluation process. In the tank example, each alternative has associated benefits and costs and all the consequences must be considered when ranking the alternatives. A difficulty, however, is encountered with the evaluation of those aspects for which there is no market value, for example, service interruption or aesthetic value. The answers in such cases are not very precise and have an element of subjectivity. Nevertheless, one can make a guess and obtain some upper and lower limits which will suffice to rank alternatives and find the best among them.

Assume that the initial cost of a design is expressed in terms of a monetary unit. Maintenance requirements and useful life can also be evaluated in the same unit. One can calculate what it will cost to paint the tank, etc. The only problem there is that these expenditures are not immediate like initial cost but occur at specified intervals and money has a time value. The money markets have established interest rates that express this difference in value. One can calculate with little difficulty the present value of, say, a payment of 200,000 units every 25 years. It is the amount of money, which invested at the current rate of interest, will yield in 25 years 200,000 units plus the original investment (see Chapter 6 for a detailed treatment of this subject).

A common argument is that the factor of safety as used in structural design is adequate to take care of adverse impacts of hydrologic uncertainty as well. Note that this factor of safety can, at best, be considered sufficient only with respect to the longevity of the structure against influences due to hydrologic uncertainty. Hydrologic design and operation of projects requires explicit risk analysis and management. However, it is not uncommon to find resistance to apply risk and uncertainty analysis as an additional decision-aiding tool. The tendency is magnified when dealing with low-probability high-consequence events, such as dam failure.

To summarize, a rational decision-making involves (1) defining an objective, (2) identifying alternative means of achieving the objective and attendant risks, and (3) applying a ranking procedure to determine the best alternative. In the real world, often little planning is done and little rationality is invoked even for important decisions. One of the common errors is the failure to recognize alternatives. All too often, the course of action is determined by precedence, tradition, accident, prejudice, or shortsightedness. Another common error is that important decisions are too often postponed till there is hardly anytime for anything but to continue the present practice. A third common error is the presumption in the value judgment as to what is important and what is not. Logically, the judgment of the client or the community to be served must be paramount.

8.2 RISK ANALYSIS AND MANAGEMENT

The concept of risk combines a probabilistic measure of the occurrence of an event with a measure of the consequences of that event. The occurrence may include the intensity, time of start, or duration. Consider a detention pond for flood control in an urban area. The dam may be overtopped and breached thereby causing damage in the urban area. The risk would be the probability of overtopping and the specified damage or harm in a given period (Kaplan and Garrick, 1981). Thus,

$$\text{risk} = \text{uncertainty} + \text{consequence} \tag{8.1}$$

This definition overcomes the limitations of the approach when risk is defined as probability times consequences. A drawback of this definition is that it equates a low-probability high-damage scenario with a high-probability low-damage scenario. Following Kaplan and Garrick (1981), risk R is defined by a set of triplets:

$$R = \{< s_i, p_i, x_i >\}, \qquad i = 1, 2, \ldots, N \tag{8.2}$$

Here s_i is scenario description, p_i is the probability of that scenario, and x_i is the consequence of that scenario. This definition answers the three critical questions related to risk, viz., what can go wrong? What are the chances of it? What are the consequences thereof?

Sometimes, risk is simply computed as the probability that a particular adverse event occurs during a stated period of time. In this case, risk is conceived as being equivalent to the probability of exceedance or nonexceedance of critical large or critical small values of a random variable, respectively. Risk exists objectively in nature, in society, or in technology-based systems, regardless of whether an investigator has properly conceived the controlling random variables or has collected sufficient data on them.

In planning and operation of water resources, risk can be measured in many ways. Corley (1979) listed the following ways to measure the parameters of risk in water resources systems operations: (i) probability of occurrence of a specified undesirable outcome, (ii) the number of occurrences of a specified length of time, (iii) the expected number of occurrences during a specified period, and (iv) investment required to prevent the occurrence of risk. While evaluating the performance of a hydrosystem, it is common to consider the number of failures over a specified time horizon. Such a characterization of performance has at least two unsuitable assumptions (Moy et al., 1986):

(1) All failures are given equal importance; there is no difference between failures by magnitude and consequence.
(2) Failures are treated as though their effects are independent of one another.

Although these assumptions have been widely used, these are not realistic, since all failures will not be of the same magnitude and importance. A failure with a deficit of 20 units from a 100-unit target is not as damaging as a 50-unit deficit from the same target. Moreover, the effect of failures is not independent. In fact, the shortage in a consecutive deficit period is certainly of greater damage than that in which the shortage periods are separated by periods of adequate supply. From the viewpoint of the users of a water supply system, four periods of failure within one year with each shortfall followed by a period of no shortage are likely to be more acceptable compared to four consecutive deficit periods during the same time span. Three commonly used measures of system performance are: (i) how often the system is in a satisfactory state (reliability), (ii) how quickly the system returns to a satisfactory state once a failure occurs (resiliency), and (iii) how significant the likely consequences of a failure may be (vulnerability). The commonly used performance indices for water resources projects are defined below.

Reliability: Reliability is defined as the probability of a system being in a satisfactory state. Denote the state of system by random variable X_t at time t, where t takes on discrete values 1, 2, ..., n. Then the possible X_t values can be partitioned into two sets: S, the set of all satisfactory outputs, and F, the set of all unsatisfactory outputs. The reliability of the system can be expressed as (Hashimoto et al., 1982):

$$\alpha = P\{X_t \in S\} \tag{8.3}$$

For a water supply system, a failure is said to have occurred when supply is less than the demand. Therefore, the reliability is the ratio of nonfailure periods to total periods in the operating horizon.

Resiliency: Resiliency describes the capacity of a system to return to a satisfactory state from a state of failure. Therefore, it may be defined as the conditional probability of the system being in a satisfactory state in period t when it was not in a satisfactory state in period t−1:

$$\beta = P\{X_t \in S / X_{t-1} \in F\} \tag{8.4}$$

By the fundamental theorem of probability, Eq. (8.4) may be expressed as

$$\beta = P\{X_{t-1} \in F, X_t \in S\} \big| P\{X_{t-1} \in F\} \tag{8.5}$$

Let Y_t be a zero-one integer variable such that $Y_t = 1$ when $X_t \in F$ and $Y_t = 0$ when $X_t \in S$. The total time in the state F may be expressed as

$$T_F = \sum_{t=1}^{n} Y_t \tag{8.6}$$

Similarly, another zero-one integer variable Z_t is employed to indicate a transition from an unsatisfactory state to a satisfactory state:

$$\begin{aligned} Z_t &= 1, X_{t-1} \in F, X_t \in S \\ &= 0 \quad \text{otherwise} \end{aligned} \tag{8.7}$$

The probability that a system is in state F in period t−1 and enters state S in the following period may then be expressed as:

$$p_s = \frac{1}{n}\sum_{t=1}^{n} Z_t \tag{8.8}$$

Now, resiliency (β) can be computed as

$$\beta = p_s/T_F = \sum_{t=1}^{n} Z_t \Big/ \sum_{t=1}^{n} Y_t \tag{8.9}$$

Clearly, a resilient system is capable of quick recovery from a deficit state to normal operation.

Vulnerability: Even though the probability of failure of a system may be very low, it is necessary to examine the damage due to a possible failure. In real life, few systems can be made so large or so safe that failures are impossible and even when it is possible to provide such a level of security, the cost is likely to be prohibitive. Logically then, efforts should be made to ensure that the damages by a failure are not severe. The vulnerability is an important criterion to describe the severity of failure for a system. In order to create a quantitative indicator of the system vulnerability, let the numerical indicator of the severity for the ith failure be s_i and the corresponding occurrence probability be p_i. Then, the system vulnerability can be expressed as (Jinno et al., 1995)

$$\gamma = \frac{1}{NF}\sum_{i=1}^{NF} p_i s_i \tag{8.10}$$

in which γ is the vulnerability of a system and NF is the total number of failures.

In risk analysis of hydrologic systems, the deficit of water may be taken as the severity of failure. Assuming that each failure (with different water deficits) occurs with the same probability, Eq. (8.10) can be simplified as:

$$\gamma = \frac{1}{NF}\sum_{i=1}^{NF} def_i \tag{8.11}$$

in which def_i is the ith water deficit. Eq. (8.11) states that the average water deficit may be regarded as the vulnerability of a hydrologic system.

The measurement of the average water deficit by itself is insufficient or inadequate for comparison among different systems or different periods for the same system. Therefore, Eq. (8.10) is normalized by the demand target during the corresponding period. The vulnerability used here is the average deficit during the whole water supply period divided by the average water demand during this period.

$$\gamma = \sum_{i=1}^{NF} def_i \Big/ \sum_{i=1}^{NF} VD_i \tag{8.12}$$

in which VD_i is the water demand during the ith deficit period; NF is the total number of failures. In general situations, γ is less than one and greater than zero. A value of $\gamma = 1$ means that the system is in its most vulnerable state. The situation $\gamma = 0$ shows that the system is in its satisfactory state and no deficit occurs. In general, the larger the deficit, the larger is the vulnerability.

Resilience is a measure of how quickly the system recovers from failure and vulnerability is a measure of the magnitude or consequences of failure, should it occur. Both these terms can be expressed in a variety of ways (Hashimoto et al., 1982). Since random variables are involved, it is possible to determine the values of those measures that exceed with a specified probability. A measure of resilience is the partial and total derivatives of the system response. The partial derivatives of the system response with respect to a decision variable measure the sensitivity of response to that variable alone. If the partial derivative is small, the system is robust with respect to such changes. If a system failure tends to persist after it has occurred, it may have serious implications even though such failures do not occur frequently and the reliability is high. The associated operating policy may be less desirable than a policy which results in a lower reliability but a higher resilience. In this context, Moy et al. (1986) noted that the desirability to have a reliable reservoir with the least number of operational failures is commonly accepted by public and decision-makers. But this commonly accepted notion of reliability is a short-sighted view of reservoir operation. On account of this undue emphasis on meeting demands whenever possible, little or no consideration is given to extreme and infrequent events. It is the extreme event failures to which a highly reliable reservoir is most subject.

8.2.1 Risk Index

When an environmental system includes many subsystems and there is a need to compare them, the use of reliability, resiliency, and vulnerability alone is inconvenient. To overcome this, an integrated index can be used. Such an integrated index as a weighted function of reliability, resiliency, and vulnerability is helpful to compare the performance of different systems. For instance, a drought risk index can be defined as

$$\mu = w_1 \times (1 - \alpha) + w_2 \times (1 - \beta) + w_3 \times \gamma \tag{8.13}$$

where w_1, w_2, and w_3 are weights such that $\sum w_i = 1.0$. In the simplest situation, all weights can be assumed to be equal, that is, $w_1 = w_2 = w_3 = 1/3$.

8.2.2 Classification of Risks

There are three types of risks:

i. Risks for which statistics of identified casualties are available,
ii. Risks for which there may be some evidence, but where the connection between suspected cause and injury to one individual cannot be traced, and
iii. Best estimates of probabilities of events that have not yet occurred.

All risks are conditional and the conditions are often implied by the context; these are not explicitly stated. For example, the risk of death due to a flood will significantly differ from one place to the other and from one country to the other. There are many risks that we wish to minimize, such as those relating to human health, environmental pollution, flooding, etc. That indeed is the basis of the insurance policy. Insurance is the baseline against which people are prepared to take risks and risk is traded off in exchange for payment.

Risks are also classified as external risk and manufactured risk. An external risk is experienced as coming from outside. A manufactured risk refers to risk situations of which we have very little experience of confronting, for example, environmental risks of global warming, climate change, etc. From the earlier days of human civilization up to the threshold of modern times, risks were primarily due to external sources—floods, famines, plagues, etc. Very recently, the focus has shifted from what nature does to us to what we have done to nature. This marks the transition from the predominance of external risk to that of manufactured risk. Much of what used to be natural is not completely natural anymore. Consequently, natural phenomena, such as floods, droughts, diseases, extreme weather, etc., are no longer entirely natural; they are being influenced by human activities too.

In these circumstances, there are two extremes to characterizing risk. On one hand, if the risk is real, there must be an explicit statement to that effect and the risk must be emphasized even at the cost of scare mongering. On the other hand, if the risk turns out to be minimal, there will indeed be accusations of scare mongering. Furthermore, if the risk is not emphasized and it turns out to be significant, there will then be accusations of cover up. If the risk is not real (imaginary), it may be emphasized. No action is needed on a risk that is not real and is not emphasized, see Fig. 8.2.

One way to cope with the rise of manufactured risk is to employ the "precautionary principle." It presupposes action about issues even though there is insufficient scientific evidence about them. Sometimes this principle is not helpful in coping with risk and responsibility. Sometimes, one needs to be bold rather than cautious in supporting innovation or other forms of change.

The economic risk is the risk of financial loss associated with a product or system due to potential hazards causing loss of production, damage, or other financial consequences. Economic risks are related to financial losses that represent the commercial consequences of a hazard. Risks to human safety can also have economic consequences. The loss can be partial or total, temporary or permanent. The loss can be financial or related to safety. Costs can be capital, operating, maintenance, and/or life cycle costs. An environmental risk is a measure of potential threats to the environment which combines the probability that events will cause or lead to degradation of the environment and severity of that degradation.

Hazard is viewed as a source of danger. Safeguards are introduced to contain hazard within acceptable risk because

$$\text{Risk} = \text{hazard}/\text{safeguards}$$

In other words, one can reduce risk by increasing hazard, though it is not possible to make it zero, howsoever big the safeguard may be.

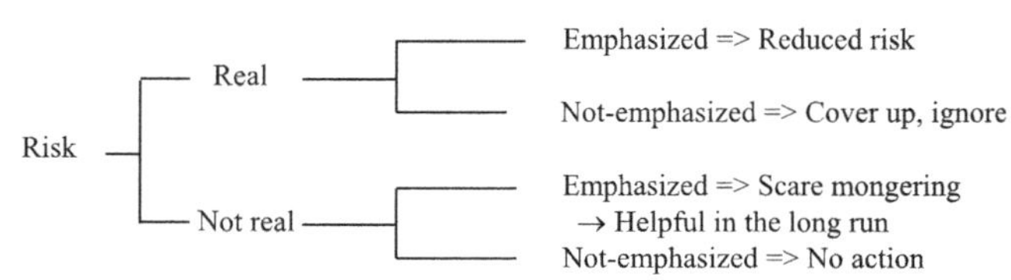

FIGURE 8.2

Classification of risk.

Example 8.1. Consider a watershed where flooding is not uncommon and when flooding does occur, it is hazardous to the ecosystem. It is assumed that the probability of hazard is the same as the probability of flooding. Some safeguards have been emplaced in the watershed to mitigate the severity of flooding and the consequent damage. Based on streamflow data available for the watershed, it is determined that the potential for flooding in a year of concern is 0.01. The probability that the safeguards will be effective is found to be 0.8. Compute the risk due to flooding in that year.

Solution. In this case, the risk is simply the ratio of 0.01–0.8, which is 0.0125. Thus, the risk is 1.25%, which is pretty low. This suggests that the safeguards are effective.

8.2.3 Sources of Risk

There are a number of complex sources of risk to water resource systems; often there are interactions among these. Yevjevich (1985) identified four sources: nature, technological, socioeconomic, and human. The sources under each of these are shown in Fig. 8.3. Among these sources, technological causes are largely controllable to a reasonable degree and so are human factors. The risks affecting the environment should not be confused with risks caused by environmental effects, either natural or man-made. The natural and socioeconomic sources are yet to be fully understood.

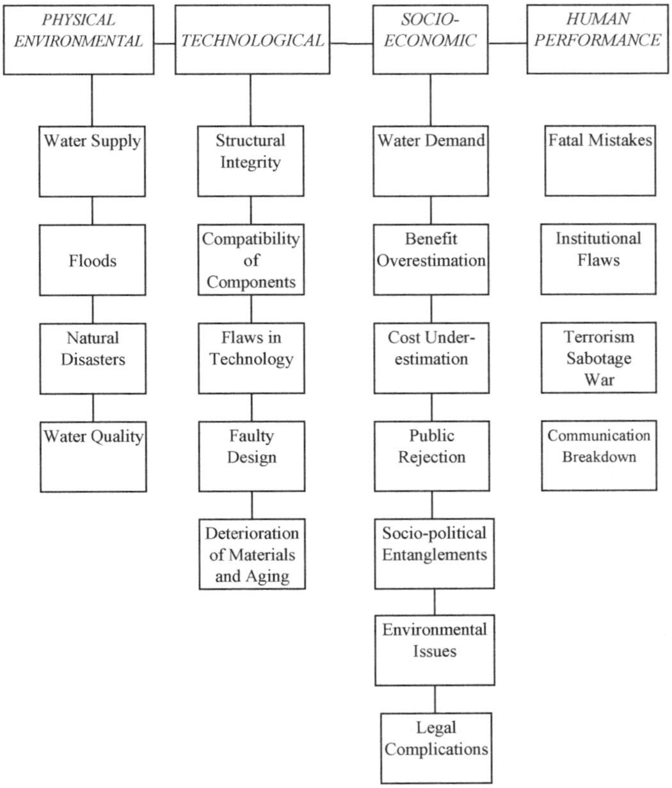

FIGURE 8.3

Types and sources of risk and uncertainty in water resources.

Adapted from Yevjevich (1985).

For a detailed risk analysis, one needs to collect and analyze huge amounts of data for a number of variables. Therefore, the majority of applications so far have been to simple problems, such as flood risk, drought risk, etc. Two data related aspects are important in decision-making:

(a) Imperfect data, either due to bad equipment, measurement errors, inadequate coverage or due to variability and complexity of the underlying system lead to nonoptimal decisions.
(b) Meteorological, hydrologic, social, and economic processes are highly influenced by random factors. The performance of a project will be affected by interactions of all these factors. The likely effect of these factors cannot be analyzed, unless there is adequate database.

8.2.4 Estimation of Risk

All systems have a probability of failure and complete avoidance of risk of calamitous failure is not possible. The engineering approach to quantify risk is to investigate possible failure mechanisms or modes and their analysis. Engineering limits the risk to an acceptable level for individuals and society. This involves quantification of the reliability of various components and examination of the systematic failure. This helps establish the overall reliability of the system. Estimation of risk for a management action is incomplete till the risk of nonaction is also estimated.

Although an examination of past events helps understand failure modes, in view of the disastrous consequences of engineering failures, it is not acceptable to wait for disasters to occur so as to build up a body of case histories as a basis for policy decisions. An anticipatory approach based on judgment and experience is required. Risk estimation provides this by methods based on a systematic analysis of complex plans into its component subsystems. Further analysis of failure mechanisms follows, and then the risk is synthesized by drawing together models of the individual subsystems. This procedure requires a wide range of data on past failures and knowledge about the various processes. The results are subject to substantial uncertainties due to inadequacies in the data and insufficient accuracy of scientific knowledge.

The deterministic approach of risk estimation can be illustrated by the use of the factor of safety, which is the ratio of design strength to design stress. Various factors affect these parameters. In practice, there will be a distribution of stresses and strengths. Assuming that the mean stress will be smaller than the mean strength, one can show that failure will occur where the upper end of the stress distribution encounters the lower end of the strength distribution. This leads to definitions of safety factors and safety margins in probabilistic terms.

In terms of decision-making, the deterministic approach incorporates an implicit value judgment as an acceptable standard of practice and is derived from an extension of past practice and experience which may be inadequate to deal with rapidly changing technology. In contrast, the probabilistic approach describes hazards in terms of risk of failure and their associated consequences.

One of the difficulties in risk assessment is in defining an "acceptable risk." One cause of the difficulty is that risks are not linearly comparable. The acceptability cannot be stated in isolation; it depends on the context in which to assess risk and the attendant costs and benefits (not only financial). A person may be willing to take a large risk if he stands to gain a large benefit or avoid large cost. When the cost of avoidance of a risk is small, most people will be willing to incur that cost.

The risk of failure and calamitous consequences that may result are greatly influenced by management. The absence or lack of adequate management and auditing of safety are important contributing factors in major disasters. There are considerable difficulties in determining and quantifying public and political risks, and especially in defining an acceptable social risk. The "rule of thumb" for acceptable level of risk is usually a round number with low chances of occurrence (to provide a heightened sense of safety), say one-in-ten thousand. This is a common but arbitrary social standard by which the outcomes of risky events are judged.

The return period, for which a hydraulic structure, such as a dam, should be designed, is calculated based on the acceptable risk. If for a time invariant hydrologic system, the probability of occurrence of an event, x, greater than the design event, x_o, during a period of n years is P, then the probability of nonoccurrence, q is 1-P. If this design event has a return period of T years and a corresponding annual probability of exceedance of p, then:

$$p = 1/T \tag{8.14}$$

The probability of nonoccurrence in any one year is:

$$q = 1 - 1/T \tag{8.15}$$

The probability of nonoccurrence in n years (design life) is:

$$q_n = (1 - 1/T)^n \tag{8.16}$$

Hence, the probability that x will occur at least once in n years or the risk R is:

$$R = 1 - (1 - 1/T)^n \tag{8.17}$$

Identification of hazard is the first step in risk estimation and involves human elements—cultural, organizational, group, and individual—which are the most frequently contributing causes of disasters. It depends on the knowledge, experience, engineering judgment, and imagination. The hazard classifications are based on some combination of measures of (a) the population at risk, (b) the likely loss of life, (c) economic losses, and (d) the impact on infrastructure and other structures of importance. The problem of analyzing both high-frequency failures (local flood control) and low-frequency failures (dam safety) by using a common evaluation paradigm seems to reflect the engineering analog of the social scientists' dilemma of the intransitivity in choice, while aggregating individual preferences to collective social choices. It is doubtful whether decision theory can provide much assistance to resolve this dichotomy. Hence, the engineering profession's resort to conservative design standards may well be justified.

Some of the issues related to risk analysis that need consideration are:

- **(i)** Risk analysis requires estimates of the exceedance probability of extreme hydrologic events. These probabilities are highly variable and are likely to affect the choice of alternatives.
- **(ii)** Many intangible factors cannot be measured in economic terms (loss of life, social dislocation, and environmental effects).
- **(iii)** The expected-value approach of annualizing damages due to a one-time low-probability catastrophic event is questionable.
- **(iv)** Reliability of various hydrologic models has not been firmly established, thus putting a question mark on many critical inputs.

Example 8.2. A hydraulic structure has a design life of 100 years. What is the risk involved if it is designed for a 50-year return period flood?

Solution. Here, $n = 100$ years, $T = 50$ years and substituting these values of n and T in Eq. (8.17):

$$R = 1 - (1 - 1/50)^{100} = 0.867 \text{ or } 86.7\%$$

Based on the acceptable risk, the return period for which the structure should be designed can be ascertained. The next example illustrates this.

Example 8.3. For what return period must a highway engineer design a critical underpass drain if he is willing to accept only 10% risk that flooding will occur in the next five years?

Solution. Here, the risk involved (R) is $= 0.10$, $n = 5$ years. The return period to be adopted may be computed by substituting these values in Eq. (8.17):

$$R = 1 - (1 - 1/T)^{n}$$

$$\text{or} \quad 0.1 = 1 - (1 - 1/T)^{5}$$

T turns out to be 47.6 years. This means that there is a 10% chance that a 48-year return period flood will occur once in the next five years.

8.2.5 Risk Management

Risk involves a number of unknowns and it is necessary to manage them; merely taking a negative attitude toward risk is not healthy. Risk management is the process of making decisions to accept a known or an assumed risk and/or the implementation of actions to reduce the probability of occurrence of an event. Risk management is also concerned with the mitigation of those risks that are derived from unavoidable hazards through the optimum specification of warning, safety devices, and risk control procedures, such as contingency plans and emergency actions. Risk management requires the following tasks (Haimes, 1981):

(1) Risk identification: It involves identification of the nature, type, and source of risk and uncertainties. The major types of risks are environmental, financial, and technological.
(2) Risk quantification: It entails formulating appropriate measures of risk and estimating the likelihood of occurrence of all consequences associated with risky events as well as their magnitude. This may involve developing probabilities of future events.
(3) Risk evaluation: It includes evaluation and optimizing trade-off of impacts of risk and its aversion. It requires decision-making regarding both an acceptable level of risk and its equitable distribution. This phase also involves the development of measures to reduce or prevent risk.
(4) Risk management: It involves decision-making to reduce or prevent risk, the formulation of policies, the development of risk-control options, and execution of such policy options.

Risk cannot be perceived simply as a one-dimensional objective concept, such as the product of the probabilities and consequences of any event. The perception of risk is inherently multidimensional and personalistic. A particular risk means different things to different people and different things in different contexts. Given the conditional nature of all risk assessment, these are derived from social and institutional assumptions and processes, that is, risk is socially constructed. Public attitudes favoring less risk are compatible with the desire to develop better technology and greater regulation of technology.

8.3 UNCERTAINTY ANALYSIS

Environmental systems are subject to uncertainty, but their planning, design, operation, and management are often done without accounting for it. According to Chow (1979), uncertainty can be defined in simple language as the occurrence of events that are beyond man's control. In respect of water resources projects, the uncertainties can be natural, model, parameter, data, and operational uncertainties. Natural uncertainties are associated with the intrinsic variability of the system. This implies that the performance indicators of the system will vary for different sets of equally likely input sequences. In this case, the system performance must be treated as a random variable. The model uncertainties arise when the model is not able to closely represent the true behavior of the system. The uncertainties that are associated with construction, maintenance, and management of the system are of operational type.

However much one may like, uncertainties cannot be completely eliminated. At best, one can reduce them by better equipment, standard data collection procedures, dense network of stations, better models, and maintenance. Uncertainty analysis is carried out to determine the statistical properties of output as a function of stochastic input parameters. This helps find the contribution of each input parameter to the overall uncertainty of the model output and can be used to reduce the output uncertainty.

8.3.1 Classification of Uncertainties

Uncertainty can be inherent or *intrinsic*, caused by randomness in nature; or can be *epistemic*, caused by the lack of knowledge of the system or paucity of data. Environmental phenomena exhibit random variability, and this variability is reflected when observations are made and samples analyzed. For example, there is inherent randomness in the climatic system and it is impossible to precisely predict what the maximum rainfall would be in a given city in a given year, even if there is a long history of data. Similarly, there is no way to precisely predict the amount of sediment load that a given river will carry during a given week at a given location. Likewise, the maximum discharge of a river for a given year cannot be predicted in advance. Because randomness is an inherent part of nature, it is not possible to reduce the inherent uncertainty. One of the common ways to deal with the first source of uncertainty in water resources projects is the use of a design flood whose return period is larger than the design life of the project.

Following Pat-Cornell (1996), the intrinsic uncertainty can be divided into inherent uncertainty in time and inherent uncertainty in space. A stochastic process expressed as a time series of a random variable exhibits uncertainty in time. Thus, the time series of three-hour maximum annual rainfall at a given station and annual instantaneous maximum discharge of a river at a station are examples of inherent uncertainty in time. Environmental systems also exhibit uncertainty in space. For examples, hydraulic conductivity of soils varies with location and direction, and a space series along a given transect can be formed.

Epistemic uncertainty is extrinsic and is knowledge-related. Knowledge relates to the environmental system as well as to data. Thus, this type of uncertainty is caused by a lack of understanding of the causes and effects occurring in the system. If the system is fully known, the uncertainty may be caused by a lack of sufficient data. For example, using laboratory experimentation or computer simulations, it may be possible to construct a mathematical model for an environmental system, but it will be impossible to determine the parameters for the vast range of conditions encountered in nature. Likewise, in case of open-channel flow, the flow dynamics is reasonably known, but it is not possible to estimate the shear stress for the range of flow and morphologic conditions that are found in practice. Epistemic uncertainty changes with knowledge and can be reduced with increasing knowledge and longer records of good quality data. In general, knowledge can be increased by gathering data, research, experience, and expert advice.

The epistemic uncertainty can be divided into statistical uncertainty and model uncertainty. Statistical uncertainty combines parameter uncertainty and distributional uncertainty (Vrijling and van Gelder, 2000). Parameter uncertainty is caused by either lack or poor quality of data, or inadequate method of estimation. It is also not always clear which type of a distribution a particular environmental random variable follows. For example, the annual maximum instantaneous discharge of a river can be described by many distributions, such as the log-Pearson type 3 distribution and the 3-parameter lognormal distribution. These two types of uncertainties are not always independent and distinguishable. For example, the identification of a correct distribution model depends very much on the accuracy with which its parameters can be estimated.

Usually, the variance of parameter estimation is proportional to $1/n$ where n is the data length and the precision is proportional to $1/n^{0.5}$ (Burges and Lettenmaier, 1982). This means that to improve the precision of parameters by a factor of 2, the required data length will be four times. But the data themselves may have associated uncertainties which could arise due to measurement errors, inconsistency, errors during recording of data, and inadequate representation of the variable due to limited samples in spatial and temporal domains.

Vrijling and van Gelder (2000) suggested division of the statistical uncertainty in time and space. Consider a time series of the annual maximum instantaneous discharge of a river. In most cases, the time series is short to determine the discharge of a long recurrence interval, say, 500 years. This is the case of scarcity of information. The same applies to droughts, minimum flows, and a host of environmental variables. Although an estimate of a 500-year flood can be made using any of the standard techniques, this estimate is subject to uncertainty. This is an example of statistical uncertainty in time or statistical uncertainty of variations in time. This uncertainty can be reduced by strengthening the database.

The spatial mapping of an environmental variable is subject to uncertainty as enough data are usually not available. This is termed as statistical uncertainty in space or statistical uncertainty of variations in space. Generally, very little information is available to map spatial variability of variables like erosion rates and hydraulic conductivity of an aquifer.

Environmental models are imperfect because of uncertainties. Consider, for example, an air pollution model that describes the pollutant concentration in space and time. The model is imperfect because there are many gaps in our knowledge about pollutant dispersion in the atmosphere or the model is simplified for practical reasons. In any case, the model is subject to uncertainty.

8.3.2 Sources of Uncertainties

Uncertainties in environmental analysis arise due to (1) randomness of physical phenomena or (2) errors in data and modeling. Modeling and data errors are of two types: systematic and random. A stochastic or random phenomenon is characterized by the property that its repeated occurrences do not produce the same outcome. For example, when the same rainfall occurs at different times over a watershed, it produces different hydrographs.

Systematic errors are characterized by their deterministic nature and are frequently constant. For example, a rain gage located near a building tends to produce biased rainfall measurements. A gage operator tends to operate the gage in a biased manner. Changes in experimental conditions also tend to produce systematic errors. For example, if the location of a rain gage is changed, it will produce biased measurements.

8.3.3 Analysis of Errors

Errors are determinate if a logical procedure can be employed to evaluate them. Random errors can be determined using statistical tools. Systematic errors are often determinate because they can be evaluated using subsidiary experiments or other means. Sometimes, some determinate errors can be removed using correction factors. For example, the U.S. National Weather Service pan is used to determine evaporation in an area by applying a correction factor to the pan measurement.

If experimental data are subject to small errors, then the terms "precision" and "accuracy" are often employed. An experiment is said to have high precision, if it has small random error. It is said to have high accuracy, if it has small systematic error. This means that precision relates to the repeatability of measurements and accuracy to the deviation between the true value and the estimator. Thus, there are four possibilities for characterizing experiments: (1) precise and accurate, (2) precise and inaccurate, (3) imprecize and accurate, and (4) imprecize and inaccurate. Our objective is to reduce both systematic and random errors as much as possible. However, for economy of effort, one must try to strike a balance between the sources of error, giving greater weight to the larger of the two.

An important but difficult aspect of data errors is mistakes and rejection of data. When data of natural phenomena are collected, anomalies are seemingly found. These anomalies and their causes should be carefully analyzed. The causes of these anomalies may be random or systematic. Under no circumstances should observed data be discarded, unless there is strong compelling reason to do so. For example, in frequency analysis of the annual instantaneous maximum discharge of a river, the so-called outliers or inliers are usually encountered. These must be dealt with, not discarded away. Most often, these anomalies are expected. If the normal probability law (or chi-square criterion) indicates that these anomalous values are expected, they must be retained. Consider another case. If the deviation of an anomalous value is too large and has a small chance of occurring, the Chauvenet criterion may be used as a guide for accepting or rejecting the value or more appropriately flagging the suspicious situation. According to this criterion, if the probability of the value deviating from the mean by the observed amount is 1/2N or less (N = number of observations), then there is reason for suspicion or rejection.

Example 8.4. A watershed has a network of rain gages for measurement of rainfall. It is ascertained that if a rain gage measures rainfall within 2%–5% of the true value, then the rain gage is said to be accurate and precise. Based on an analysis of rainfall observations, it has been determined that one of the rain gages, called A, always measures rainfall about 10%–15% away from the true value. There is another rain gage, called B, which is found to measure rainfall within 2% away from the true value. Another rain gage, called C, is found to measure rainfall somewhat unpredictably, that is, sometimes 15% away from the true value, sometimes 20% away from the true value and sometimes very close to the true value. Another rain gage, called D, measures rainfall sometimes 5% higher, sometimes 5% lower, sometimes 2% higher and sometimes 2% lower, but always within 5% away from the true value. What can be said about the measurements of these rain gages?

Solution. The measurements of rain gage A are precise but inaccurate, because A has the repeatability quality. Rain gage B is accurate and precise because it has repeatability quality and its observations are close to the true values. Rain gage C is inaccurate and imprecize because its measurements are neither repetitive nor accurate. Rain gage D is imprecize but accurate because its observations are close to the true values, but its characteristic is not repeatable. These characteristics are shown in Fig. 8.4.

Measures of Errors

Experimental data errors can be investigated using the criterion of repeatability of measurements. In observing natural phenomena, however, there is no way to repeat the measurement. In laboratory experimentation, if a measurement is repeated, say N times, and the measured values do not exactly match the "true" value, the differences are analyzed. It should however be noted that the true value is never known and only an estimator can be obtained. If N becomes large, the arithmetic average of the measured values approaches a constant value and if this is the case, the estimator approaches a constant value and is qualified as a "consistent" estimator. Thus, consistency is one measure of experimental data error and is tied to the sample size. Ideally, the estimator should be consistent and without bias. However, the estimator, although consistent, may be biased if N is too small, that is, the estimator may be either too large or too small. Thus, bias is another measure of experimental error which is not tied to the sample size.

Not all statistics are unbiased estimators. Some are consistent estimators because they converge to the parent population as sample size increases, but for finite sample size, they need correction to become unbiased best estimate. For example, for N identical independent measurements, the sample

FIGURE 8.4

Measurement of rainfall by four rain gages. Gage A is precise, inaccurate; gage B is precise, accurate; gage C is imprecise, inaccurate; gage D is imprecise, accurate.

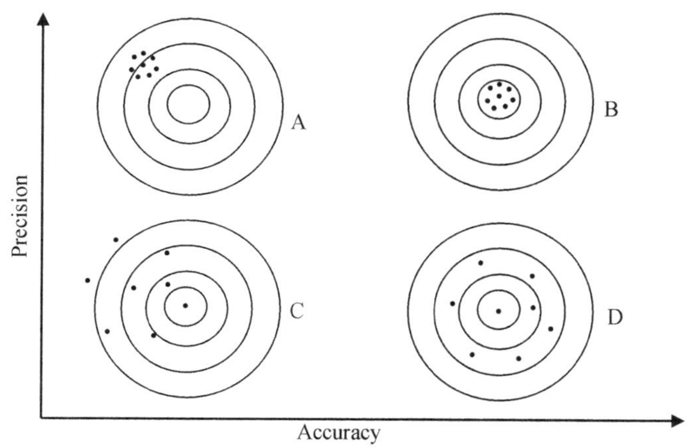

mean and sample variance are consistent statistics, but only the sample mean is an unbiased estimator. To obtain an unbiased estimate of the variance, the sample variance is multiplied by a factor N/(N−1).

Extraction of Information

Data are a source of information. It is the only medium of communication with nature. The purpose of data analysis is, therefore, to extract the maximum information contained in it. Statistical concepts used to analyze data are threefold: (1) aggregate characteristics, (2) variation of individual values from aggregate properties, and (3) frequency distribution of individual values. In the first case are mean (arithmetic, median, mode, harmonic, and geometric), deviation (mean and standard), variance, coefficient of variation, and higher order (such as skewness, kurtosis, etc.) or other types of moments (such as probability weighted, linear, and geometric). The moments and frequency distribution are interconnected. While computing these statistics, issues relating to rounding off and truncation have to be dealt with.

Propagation of Errors

Consider a function $z = f(x, y, \ldots)$ of two or more quantities x, y, ... If there are errors in x, y, ..., there will be error in z, whose amount is denoted as dz. Let these errors in x, y, ... be denoted as dx, dy, ... The errors dz, dx, dy, ... must, of course, be of the same kind, that is, all average errors, all standard deviations, etc. For propagation of errors, two cases can be distinguished: (1) independent errors and (2) nonindependent errors. In the case of independent errors, there is the possibility of compensation or counterbalancing. For example, if the error in x causes z to be large, the error in y may cause it to be small. Thus, the net result would be that the total error in z would be less than the sum of individual errors. This result does not hold in the case of nonindependent errors where the errors actually add algebraically.

Let the error be specified by variance whose square root yields the standard deviation. It is assumed that the function varies sufficiently slowly so that it can be represented by the first few terms in the Taylor series expansion. Let $\bar{z}, \bar{x}, \bar{y}, \ldots$ be the mean values of z, x, y, ..., respectively. Their variances are defined, respectively, as $\sigma_z^2 = (z-\bar{z})^2, \sigma_x^2 = (x-\bar{x})^2, \sigma_y^2 = (y-\bar{y})^2, \ldots$ Let a point P be defined by $(\bar{x}, \bar{y}, \ldots)$. Expanding z in Taylor series at P

$$z = f(\bar{x}, \bar{y}, \ldots.) + \left.\frac{\partial f}{\partial x}\right|_P (x-\bar{x}) + \left.\frac{\partial f}{\partial y}\right|_P (y-\bar{y}) + \ldots. \tag{8.18}$$

where the derivatives of f are evaluated at pint P. Eq. (8.18) yields $\bar{z} = f(\bar{x}, \bar{y}, \ldots.)$, which is expected.

Consider the variance of z, which can be expressed with use of Eq. (8.18) as

$$\sigma_z^2 = \left(\left.\frac{\partial f}{\partial x}\right|_p\right)^2 (x-\bar{x})^2 + \left(\left.\frac{\partial f}{\partial y}\right|_p\right)^2 (y-\bar{y})^2 + \ldots. + 2\left(\left.\frac{\partial f}{\partial x}\right|_p\right)\left(\left.\frac{\partial f}{\partial y}\right|_p\right)(x-\bar{x})(x-\bar{y})^2 + \ldots.. \tag{8.19}$$

or

$$\sigma_z^2 = \left(\left.\frac{\partial f}{\partial x}\right|_p\right)^2 \sigma_x^2 + \left(\left.\frac{\partial f}{\partial y}\right|_p\right)^2 \sigma_y^2 + \ldots. + 2\left(\left.\frac{\partial f}{\partial x}\right|_p\right)\left(\left.\frac{\partial f}{\partial y}\right|_p\right)\sigma_{xy} + \ldots. \tag{8.20}$$

where σ_{xy} is the covariance of x and y.

Eq. (8.20) can be generalized for independent and nonindependent or correlated errors. Let $z = f(x_1, x_2, \ldots, x_n)$. Then, the variance of z is:

$$\sigma_z^2 = \sum_{i,j=1}^{n} \left(\frac{\partial f}{\partial x_i} \bigg|_P \right) \left(\frac{\partial f}{\partial x_j} \bigg|_P \right) \sigma_{x_i x_j} \tag{8.21}$$

It should be noted that covariances are always symmetric, that is, $\sigma_{x_i x_j} = \sigma_{x_j x_i}$, and vanish for independent errors.

Example 8.5. Consider the case of two independent variables x and y, and $z = f(x, y)$. Assume that the errors are independent.

Solution. The standard deviation of z is

$$\sigma_z = \sqrt{\left(\frac{\partial f}{\partial x}\right)^2 \sigma_x^2 + \left(\frac{\partial f}{\partial y}\right)^2 \sigma_y^2} \tag{8.22}$$

If $z = xy$, then

$$\sigma_z = \sqrt{(\bar{y})^2 \sigma_x^2 + (\bar{x})^2 \sigma_y^2} \tag{8.23}$$

Eq. (8.23) is more meaningfully expressed in terms of the coefficient of variation (standard deviation divided by the mean), ε, as

$$\varepsilon_z = \sqrt{\varepsilon_x^2 + \varepsilon_y^2} \tag{8.24}$$

If $z = x/y$, then Eq. (8.24) also holds. If $z = x + y$ or $z = x - y$, then

$$\sigma_z = \sqrt{\sigma_x^2 + \sigma_y^2} \tag{8.25}$$

Eq. (8.25) shows why in environmental analysis, computations based only on the water balance equation are not popular. An example is the significant error obtained when evaporation from a lake or a watershed is computed based on water balance.

Another case is $z = x^m y^n$. In this case,

$$\varepsilon_z = \sqrt{m^2 \varepsilon_x^2 + n^2 \varepsilon_y^2} \tag{8.26}$$

Eq. (8.26) shows the error multiples in a nonlinear case. If $z = x^m$, then $\varepsilon_z = m\,\varepsilon_x$. If $z = cx$, where c is constant, then $\varepsilon_z = \varepsilon_x$.

It is now possible to determine the best value of a quantity x from two or more independent measurements whose errors may be different. Intuitively, a measurement with less error should carry more weight. However, how exactly the weighting should be done is not quite clear. To that end, the principle of least squared error may be invoked. Consider two independent measurements of x as x_1 and x_2, with their respective (plus or minus) errors as σ_1 and σ_2. It may be reasonable to assume an estimate of x as

$$\overline{x}_{12} = ax_1 + (1 - a)x_2 \tag{8.27}$$

Eq. (8.27) is similar to the Muskingum hypothesis used in flow routing. It can be shown that

$$x_{12} = \frac{\frac{x_1}{\sigma_1^2} + \frac{x_2}{\sigma_2^2}}{\frac{1}{\sigma_1^2} + \frac{1}{\sigma_2^2}} \tag{8.28}$$

$$\sigma_{12}^2 = \left[\frac{1}{\sigma_1^2 + \sigma_2^2}\right]^{-1} \tag{8.29}$$

Eqs. (8.28) and (8.29) can be generalized as

$$\overline{x} = \frac{\sum_{i=1}^{n}\left(\frac{x_i}{\sigma_i^2}\right)}{\sum_{i=1}^{n}\left(\frac{1}{\sigma_i^2}\right)} \tag{8.30}$$

and

$$\sigma_z = \left\{\sum_{i=1}^{n}\left(\frac{1}{\sigma_i^2}\right)\right\}^{-1/2} \tag{8.31}$$

Now consider the case when errors are correlated. Let $E = A/(A + B)$, where A and B are independent measurements, with their mean and variances, respectively, denoted as $\overline{A}$, σ_A^2 and $\overline{B}$, σ_B^2. It can be shown that

$$\sigma_E^2 = \frac{(\overline{B})^2}{(\overline{A} + \overline{B})^2}\sigma_A^2 + \frac{(\overline{A})^2}{(\overline{A} + \overline{B})^2}\sigma_B^2 \tag{8.32}$$

Eq. (8.32) can also be derived by expressing $E = A/U$, where $U = A + B$, and then applying the Taylor series. If $z = x + y$ or $z = x - y$, then it can be shown that

$$\sigma_z = \sqrt{\sigma_x^2 + \sigma_y^2 \pm 2\sigma_{xy}} \tag{8.33}$$

Eq. (8.33) contains the covariance term. Similarly, if $z = xy$, then

$$\varepsilon_z = \sqrt{\varepsilon_x^2 + \varepsilon_y^2 + 2\frac{\sigma_{xy}}{(\overline{x}\,\overline{y})}} \tag{8.34}$$

If $z = x/y$, then

$$\varepsilon_z = \sqrt{\varepsilon_x^2 + \varepsilon_y^2 - 2\frac{\sigma_{xy}}{(\overline{x}\,\overline{y})}} \tag{8.35}$$

Eqs. (8.34) and (8.35) are similar, except for the sign of the covariance term. If $z = x^m y^n$, then it can be shown that

$$\varepsilon_z = \sqrt{m^2\varepsilon_x^2 + n^2\varepsilon_y^2 + 2\frac{mn\,\sigma_{xy}}{(\bar{x}\,\bar{y})}} \tag{8.36}$$

Eqs. (8.33) to (8.36) contain covariance terms and should be calculated.

8.3.4 Analysis of Uncertainty

The subsequent analysis assumes that the uncertainty can be measured quantitatively, at least in principle. For example, if an unbiased coin is tossed, it is not known in advance whether head will turn up. But it is known that the event "head will turn up" has a probability of ½ or 50% each time the coin is tossed. Similarly, it is possible to assign a probability to the event that the peak flow in the Red River in any given year will exceed 3000 m^3/s. One can also determine the probability that the compressive strength of concrete, manufactured in accordance with given specifications, will exceed 25,000 kN/m^2 or that the maximum number of cars that have to wait at a railway crossing will exceed 20 on any given working day. The kind of uncertainty involved here is the uncertainty associated with the randomness of the event.

To investigate the uncertainty in the conclusions reached about uncertain events is beyond the scope here. For example, one may calculate that the probability p that the peak flow in the Red River exceeds 3000 m^3/s in any given year is 3.3%. But there is an element of uncertainty in estimation of p, which depends very much on the length and quality of data. The question thus arises: What is the probability that p lies within a given range $p \pm \Delta p$? In each case, there are events that can be analyzed to a degree that reasonable people, using reasonable procedures, come up with reasonably close probability assessments. Recall that the goal is only to deal rationally with uncertainty, not completely eliminate it.

To deal with uncertain events in the decision-making process, events that are certain to occur, or conclusions that are certainly true, must, be fully taken into account. These can be given a weight of 1 (certain events). Impossible events, on the other hand, are disregarded in decisions and these are given the weight of 0. Any in-between event is given a weight equal to the probability of its occurrence. Thus, the more likely an event, the more the weight it gets and the greater is its relative effect on the outcome or the decision.

An evaluation of safety and reliability requires information on uncertainty, which is determined by the standard deviation or coefficient of variation. Questions of safety or reliability arise principally due to the presence of uncertainty. Thus, an evaluation of uncertainty is an essential part of the evaluation of engineering reliability. The uncertainty due to random variability in physical phenomena is described by a probability distribution function. For practical purposes, its description may be limited to (a) central tendency and (b) dispersion (e.g., standard deviation) or coefficient of variation.

To deal with uncertainty due to prediction error (estimation error or statistical sampling error and imperfection of the prediction model), the coefficient of variation or the standard deviation is normally employed, which represents a measure of the random error. In effect, the random error is involved whenever there is a range of possible error. One source of random error is the error due to sampling, which is a function of the sample size. The random sampling error can be expressed in terms of the coefficient of variation (CV) as $\Delta_1 = \mathrm{CV}/\sqrt{\mathrm{N}}$, where N = sample size.

Consider, for example, the mean annual rainfall for Baton Rouge to be 60.00 inches. Conceivably, this estimate of the true mean value would contain error. If the rainfall measurement experiment is repeated and other sets of data obtained, the sample mean estimated from other sets of data would most likely be different. The collection of all the sample means will also have a mean value, which may well be different from the individual sample mean values, and a corresponding standard deviation. Conceptually, the mean value of the collection of sample means may be assumed to be close to the true mean value (assuming that the estimator is unbiased). Then, the difference (or ratio) of the estimated sample mean (i.e., mean value of 60 in.) to the true mean is the systematic error, whereas the coefficient of variation, or standard deviation of the collection of sample means represents a measure of the random error. In effect, random error is involved whenever there is a range of possible error. One source of random error is the error due to sampling, which is a function of the sample size.

Example 8.6. Consider the mean annual rainfall for Baton Rouge, which is given above as 60 in. based on a sample of data. The mean rainfall estimated by the arithmetic mean method is about 5%—10% higher than the true mean. Taking the sample standard deviation of 15 in. and the number of observations in the sample as 25, compute the total random error in the estimated mean value.

Solution. The corresponding CV is 15 in. / 60 in. = 0.25. Assuming the random sampling error (expressed in terms of CV) would be

$$\Delta_1 = 0.25/\sqrt{25} = 0.05$$

The systematic error or bias may arise due to factors not accounted for in the prediction model that tends to consistently bias the estimate in one direction or the other. With this information, a realistic prediction of the mean rainfall requires a correction from 90% to 95% of the corresponding mean (arithmetic) rainfall. If a uniform probability density function (pdf) between this range of correction factors is assumed, then the systematic error in the estimated arithmetic mean rainfall of 60 in. will need to be corrected by a mean bias factor of

$$\mathrm{e} = (0.9 + 0.95)/2 = 0.925$$

whereas the corresponding random error in the estimated mean value, expressed in terms of CV, is

$$\Delta_2 = \frac{1}{\sqrt{3}}\left(\frac{0.95 - 0.90}{0.90 + 0.95}\right) = \frac{1}{\sqrt{3}}\left(\frac{0.05}{1.85}\right) = \frac{0.027}{1.73} = 0.016$$

The total random error in the estimated mean value is

$$\Delta = \left(\Delta_1^2 + \Delta_2^2\right)^{0.5} = \left[(0.05)^2 + (0.016)^2\right]^{0.5} = (0.0028)^{0.5} = 0.053$$

The systematic error is a bias in the prediction or estimation and can be corrected through a constant bias factor. The random error, called standard error, requires statistical treatment. It represents the degree of dispersiveness of the range of possible errors. It may be represented by the standard deviation or coefficient of variation of the estimated mean value. An objective determination of the bias as well as the random error requires repeated data on the sample mean (or median), which are hard to get.

For quantification of uncertainty measures, we confine ourselves to the error of prediction or the error in the estimation of the respective mean values, that is, the systematic and random errors will refer to the bias and standard error, respectively, in the estimated mean value of a variable (or function of variables). It is important that the uncertainty measures are credible, for the validity of a calculated probability depends on credible assessments of individual uncertain measures. Methods for evaluating uncertainty measures depend on the form of the available data and information.

For a set of observations, the mean value is

$$\overline{x} = \frac{1}{n}\sum_{i=1}^{n} x_i \tag{8.37}$$

and the variance is

$$\sigma_x^2 = \frac{1}{n-1}\sum_{i=1}^{n}(x_i - \overline{x})^2 \tag{8.38}$$

The uncertainty associated with the inherent randomness is given by

$$CV = \frac{\sigma_x}{\overline{x}} \tag{8.39}$$

The above estimated mean value may not be totally accurate relative to the true mean (especially for small sample size n). The estimated mean value given above is unbiased as far as sampling is concerned; however, the random error of the above $\overline{x}$ is the standard error of $\overline{x}$, which is

$$\sigma_{\overline{x}} = \sigma_x/\sqrt{n} \tag{8.40}$$

Hence, the uncertainty associated with random sampling error is

$$\Delta_x = \sigma_{\overline{x}}/\overline{x} \tag{8.41}$$

This random error in $\overline{x}$ is limited to the sampling error only. There may, however, be other biases and random errors in $\overline{x}$, such as the effects of factors not included in the observational program.

Often, the information is expressed in terms of the lower and upper limits of a variable. Given the range of possible values of a random variable, the mean value of the variable and the underlying uncertainty may be evaluated by prescribing a suitable distribution within the range. For example, for a variable x, if the lower and upper limits of its values are x_l and x_u, the mean and CV of x may be determined as follows:

$$\overline{x} = \frac{1}{2}(x_l + x_u) \tag{8.42}$$

$$CV = \frac{1}{\sqrt{3}}\left(\frac{x_u - x_l}{x_u + x_l}\right) \tag{8.43}$$

where the pdf of the variable is uniform between x_l and x_u.

Alternatively, if the pdf is given by a symmetric triangular distribution and is prescribed within the limits x_l and x_u, the corresponding CV would be

$$CV = \frac{1}{\sqrt{6}}\left(\frac{x_u - x_l}{x_u + x_l}\right) \tag{8.44}$$

With either the uniform or the symmetric triangular distribution, it is implicitly assumed that there is no bias within the prescribed range of values for x. On the other hand, if there is bias, the skewed distributions above may be more appropriate. If the bias is judged to be toward the higher values within the specified range, then the upper triangular distribution would be appropriate. In such a case, the mean value is:

$$\overline{x} = \frac{1}{3}(x_l + 2x_u) \tag{8.45}$$

and CV is

$$CV = \frac{1}{\sqrt{2}}\left(\frac{x_u - x_l}{2x_u + x_l}\right) \tag{8.46}$$

Conversely, if the bias is toward the lower range of values, the appropriate distribution may be a lower triangular distribution, with the mean value as

$$\overline{x} = \frac{1}{3}(2x_l + x_u) \tag{8.47}$$

and CV as

$$CV = \frac{1}{\sqrt{2}}\left(\frac{x_u - x_l}{x_u + 2x_l}\right) \tag{8.48}$$

Another distribution may be a normal distribution where the given limits may be assumed to cover $\pm 2\sigma$ from the mean value. In such cases, the mean value is

$$\overline{x} = \frac{1}{2}(x_u + x_l) \tag{8.49}$$

and CV is

$$CV = \frac{1}{2}\left(\frac{x_u - x_l}{x_u + x_l}\right) \tag{8.50}$$

The seemingly different types and sources of uncertainty can be analyzed in a unified manner. Consider a variable x (true value) whose prediction is given as $\widehat{x}$. Let there be a correction factor N to account for error in $\widehat{x}$. Therefore, the true x may be expressed as

$$x = N\widehat{x} \tag{8.51}$$

For random variable x, the model $\widehat{x}$ should be a random variable. The estimated mean value $\widehat{x}$ and variance $\sigma_{\widehat{x}}^2$ (e.g., from a set of observations) are those of $\widehat{x}$. Then, $CV = \sigma_x/\overline{x}$ represents the inherent variability. The necessary correction N may also be considered a random variable, whose mean value "e" represents the mean correction for systematic error in the predicted mean value $\overline{x}$, whereas CV of N, Δ, represents the random error in the predicted mean value $\overline{x}$. Assuming N and $\widehat{x}$ to be statistically independent, the mean value of x, is

$$\mu_x = e\overline{x} \tag{8.52}$$

The total uncertainty in the prediction of x then becomes

$$\Omega_x \cong \sqrt{CV_x^2 + \Delta_x^2}, \qquad CV_x = \sigma_x/\overline{x} \tag{8.53}$$

The above analysis pertains to a single variable. If Y is a function of several random variables $x_1, x_2, ..., x_n$, that is,

$$Y = g(x_1, x_2, ..., x_n) \tag{8.54}$$

the mean value and associated uncertainty of Y are of concern. A model (or function) $\hat{g}$ and a correction N_g may be used, so

$$Y = N_g \hat{g}(x_1, x_2, ..., x_n) \tag{8.55}$$

Thus, N_g has a mean value of "e_g" and CV of Δ_g. Using the first-order approximation, the mean value of Y is

$$\mu_y \cong e_g \hat{g}\left(\mu_{x1}, \mu_{x2}, ..., \mu_{xn}\right) \tag{8.56}$$

where e_g is the bias in $\hat{g}\left(.,.,---,.\right)$ and $\mu_{xi} = e_i \bar{x}_i$. Also, the total CV of Y is

$$\Omega_y^2 = \Delta_g^2 + \frac{1}{\mu_g^2} \sum_i \sum_j \rho_{ij} c_i c_j \sigma_{xi} \sigma_{xj} \tag{8.57}$$

in which $c_i = \partial g / \partial x_i$ evaluated at (μ_{x1}, μ_{x2},... μ_{xn}), ρ_{ij} is correlation coefficient between x_i and x_j.

8.3.5 Value of Information

Many water resources decisions are made without adequate data. The loss in net benefits resulting from imperfect project design due to limited historic record length also marginally decreases as the sample size increases. The "worth of data" measures explain the loss in benefits due to uncertainty or shortage of data. The value of hydrologic data can be related to benefits foregone because of insufficient data. The marginal value of streamflow data shows a decreasing trend, and it approaches the marginal cost at record lengths beyond 100 years.

The incremental value of data is the greatest in the initial stages when information is still at its lowest level. At this level, even a limited addition to information is of great value. As the volume of data on a specific variable increases, returns diminish. Klemes (1977) assessed the value of information in reservoir optimization with numerical experiments and demonstrated an insensitivity of results to the input distribution as measured by the information content. As either hydrologic or economic uncertainties grow, the standard linear operation policy often is the optimum policy. Bayesian decision theory approaches also utilize uncertainty related to the sample length and either update the information with new available data or take expectations over the risk functions to make decisions.

8.4 UTILITY THEORY

With the advancement of economic theory came the acceptance of insurance against various forms of risk. The rapid technological advancements necessitated continuous decisions. At times, natural and man-induced events seem to occur in unaccounted ways, and hence there was an urgent need to quantify uncertainty.

The expected benefit criterion that is normally used for ranking projects has a drawback. It gives no indication of undesirable outcomes and fails to consider the decision-maker's aversion to risk. The gambling theory illustrates the need for a methodology to incorporate risk aversion in project appraisal. When gambling, most people seem to prefer a small prize with a good chance of success compared to a large reward with a high risk. Studies of such behavior led to the hypothesis of expected utility. Basically, utilities are numbers which a person assigns to the consequences of an action; the preference for a particular action is shown by its higher expected utility.

It is important to note that the face value of wealth is not the same as what it is really worth to a person. For instance, 100 rupees may be of great importance to a poor man but is comparatively worthless to a millionaire. This means that a graph of utility against monetary worth is not linear. For instance, a risk avoider's curve, which is more common, is concave downwards, and this is in direct contrast with the curve of a risk seeker, as shown in Fig. 8.5.

The utility theory, which embodies indifference curves, values, and preferences, is an important concept to find solutions to uncertainty and risk. Utility is a measure of the worth of various outcomes and forms part of an individual's nonlinear psychological variables. It is not easy to determine the form of a person's utility curves, but once these are found, it is easy to use them. This topic has been covered in Chapter 6.

The utility theory tries to express the preference of a rational decision-maker through a utility function. Consider a water resources project which is capable of giving maximum benefits of X. The worst-case scenario corresponds to zero net benefits. Now, the utility function U(B) of net benefits B equals the probability p at which the decision-maker is indifferent to a project that yields net benefits B for certain and to a project that yields the net benefits of X with probability p and net benefits of 0 with probability 1−p (Loucks et al., 1981).

The need for a methodology to incorporate a risk preference into project evaluation is illustrated by the problem of *gambler's ruin*. Let a gambler with A rupees to wager goes for betting. As per the rules, if the gambler wins any game, he receives twice what is bet, and if he loses, the entire amount placed on the bet is lost. The gambler somehow knows that the probability of winning any game is 70%. If the

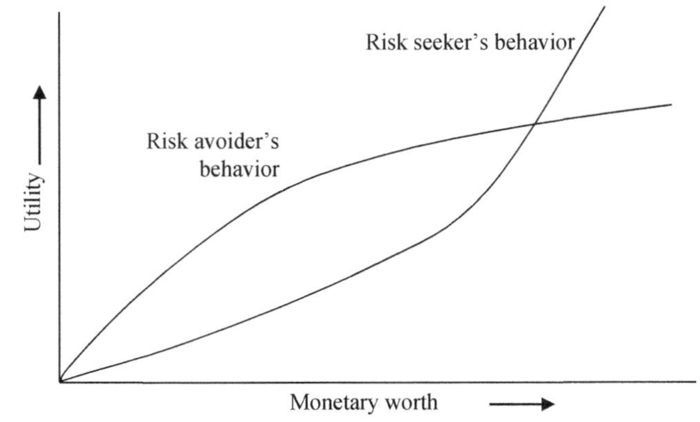

FIGURE 8.5

Risk avoider's and seeker's utility functions.

gambler wishes to maximize expected earnings, he should bet all that he has on each game because the expected winnings when x rupees are bet (maximum bet amount is A) is

$$0.30 \times 0 + 0.70 \times 2x = 1.4x \tag{8.58}$$

which clearly increases with x. Assume that the gambler starts with x rupees and bets all the earnings on each of n games. Evidently, the expected earning after n games is $(1.40)^n$ x. But hidden herein lies the gambler's ruin because the probability that the gambler has any earning after n games is the probability of winning all n games, or $(0.70)^n$. For illustration, if $n = 20$, $(0.70)^{20} = 0.08\%$ and this implies that his chances of winning all 20 games are very small and eventually, he is almost certain to loose everything. Of course, most people will not follow this strategy. For instance, after winning 19 games in a row, a few individuals will risk loosing the whole earning in the next game, since the chances of winning so many games uninterrupted are too meager.

Another decision strategy is the max-min strategy. As the name suggests, the max-min decision strategy aims to maximize the smallest benefits that can be obtained under given circumstances. The decision vector is obtained by solving

$$\max_{x_i} \left[\min_{w_j} NB\left(x_i \middle| w_j\right) \right] \tag{8.59}$$

where x_i is a decision vector and w_i are the set of parameters. The max-min decision criterion is a simple but pessimistic criterion for making decisions. In the gambling example, this approach would suggest not to indulge in betting, since the minimum gain is nil. However, it may not appear to be a reasonable policy in this instance, particularly in view of favorable chances of winning.

While venturing for a risky alternative, most people weigh two things: the severity of the alternative outcomes and their likelihood. This requires assigning probabilities to each of the parameter sets. In general, there is no single best way of estimating probabilities of uncertain events. For an event that has taken place in the past, and for which past outcomes can be used to predict the likelihood of future outcomes (e.g., rainfall or streamflows), a probability distribution based on past observations may be derived. If these objective probabilities cannot be determined, one may proceed by first projecting the limits of the range of possible outcomes, and subdividing this range into subranges. Then, weights (probabilities) reflecting one's assessment of the relative likelihood of each subrange can be assigned. Alternatively, a reasonable probability distribution can describe the distribution of uncertain parameters based on subjective estimates of the distribution's mean, variance, or various quantiles. Estimates of probabilities or probability distributions should not be biased by considerations of risk preference or risk aversion. The estimation of probabilities should reflect what unbiased experts believe to be the actual likelihood of possible outcomes. The Bayesian approach (discussed in Section 8.6) is helpful in these circumstances.

A frequent trade-off in systems design is between projects which maximize the expected benefits and those which maximize the benefits of the worst outcome but in so doing achieves the smallest expected benefits. Typically, should one choose the project A which gives the net benefits of 100 with certainty or the project B with expected benefits of 150 and also the risk of receiving net benefits of only 50? There are no clear-cut answers to such questions. The utility theory helps in eliciting the decision-maker's preferences and make a choice.

8.4.1 Expected Monetary Value of a Decision

The expected monetary value (EMV) of a decision is the weighted sum of the values of the possible outcomes of that decision. In other words, the money value of each possible outcome is multiplied by the probability of the outcome and all weighted money values are added up. This procedure is simple when the number of possible outcomes is finite and a definite probability can be associated with each. But sometimes the number of possible outcomes is infinite. Assume that one has to determine EMV of the annual flood damage along a river. Each possible river stage above a critical level corresponds to a certain amount of damage. But the number of possible river stages is infinite.

In this case, one can divide the range of possible river stages into a finite number of intervals, each, say, 0.5 m higher than the previous one. One then assigns an average flood damage to each interval. Since a definite probability can now be associated with each interval, it is easy to assign a definite probability for each of the finite number of average damages. The weighted sum of the flood damages or the EMV can be determined. The concept is further illustrated with an example.

Example 8.7. A contractor bids for an excavation job for a project. The only time available for the excavation is the month of August and it is clear that on rainy days the contractor cannot work. To ensure compliance with the time clause in the contract, there is a condition that the contractor forfeits his payment if he does not complete the work in time. The contractor enters a bid for Rs. 1,000,000.

The contractor has three options: (a) He calculates that he can complete the work in 25 days at a cost of Rs. 500,000 with his own equipment. But a study of precipitation records reveals that there is a 30% chance of having less than 25 rain free days in August. (b) The contractor can buy additional equipment and can then finish the work in 20 days at a cost of Rs. 600,000. But there is a 10% probability of having fewer than 20 rain free days in August. (c) The contractor can also make a deal with another contractor and finish the work in 15 days at a cost of Rs. 800,000. A study shows that there is virtually no chance of having fewer than 15 rain free days in August. What is the reasonable course of action?

Solution. It is helpful to draw the decision tree with various possible outcomes of the alternative courses of action, because the consequences are dependent on the decisions and the random factor (the precipitation in the month of August). The possible outcomes, profit or loss, can be entered with the appropriate probability weights. Together, they determine the EMV of each decision. The preferred decision is the one with the greatest EMV.

It appears from Fig. 8.6 that alternative B involving the purchase of additional equipment yields the highest EMV and would be the preferred decision.

Alternatives		Status of the job	Benefits /loss	Net return
	A	Complete the job	0.7 × 500,000	200,000
		Can't complete the job	0.3 × (– 500,000)	
	B	Complete the job	0.9× 400,000	300,000
		Can't complete the job	0.1× (– 600,000)	
	C	Complete the job	1.0× 200,000	200,000
		Can't complete the job	0.0× (– 800,000)	

FIGURE 8.6

Decision tree for example problem.

A few comments are in order. In the first place, EMV has little to do with the actual expectations in any concrete situation. It stems from the time when the theory of probability was used primarily to evaluate strategies in games of chance. A gambler might consistently place a bet in which the chances of winning, say Rs. 100, were 20%. He would then expect to win "in the long run" an amount of Rs. 20 per bet and would call Rs. 20 the expected value of the bet.

In the second place, one might note that the decision policy leading to the result shown in Fig. 8.6 may be rational, but that it is not necessarily the one that every contractor would want to adopt. In particular, note that alternatives A and C have the same EMV. This would be interpreted as indifference as to the choice between the two. Yet many people would express strong preference for either one or the other in a concrete situation. Of course, this observation does not invalidate the procedure.

8.4.2 Improving the Decision Policy

A common criticism of this decision procedure is that the calculated EMVs do not necessarily identify the "best" decision. While most people would agree that alternatives A and C in the previous example should not necessarily be rated equally desirable, opinions would vary as to which of the two would be preferred and why. It appears, therefore, that EMV does not adequately sum up the consequences of each possible outcome, which are different for different people. Indeed, a 30% risk of losing Rs. 500,000 may well be wholly unacceptable for a small contractor since it could put him out of business. Even the 10% chance of losing Rs. 600,000 may make alternative C more attractive to him than alternative B. A large firm, on the other hand, would not consider the prospect of such losses completely unacceptable.

This nonlinear property of the value of losses is the basis for taking out insurance. A loss of say Rs. 100,000 due to the destruction of property by fire may have an annual probability of occurrence of only 0.0001. EMV of this loss is Rs. 10 per year. But it is not irrational to protect oneself against such a crippling loss by means of insurance which may cost Rs. 50 per year. The value of gains is not linear either. A person would undoubtedly get more satisfaction out of a gain of Rs. 800,000 than out of a gain of Rs. 400,000. But the satisfaction, or more precisely, the value to him is not necessarily twice as much.

To express the nonlinearity of gains and losses, one might use a value measure which is called the "utility value" to distinguish it from the monetary value. It will, by definition, allow the linear operations of weighting and addition necessary in evaluating the consequences of any outcome. It can be related to real money; however, the relationship could be different for different persons depending on their circumstances. Fig. 8.7 shows such a relationship.

The curve shown in Fig. 8.7 has a 45° slope near the origin, indicating that for relatively small gains and losses, the two units are identical. Large gains are discounted somewhat and large losses are given greater negative utility value because of their crippling effect. Performing the decision calculations in utility value has a profound effect on the relative desirability of alternatives A and B as can easily be checked.

Needless to say, the determination of a utility relationship, such as shown in Fig. 8.7, is a difficult task. Furthermore, decision-makers that have very large assets compared to the gains and losses involved are likely to identify EUV with the EMV.

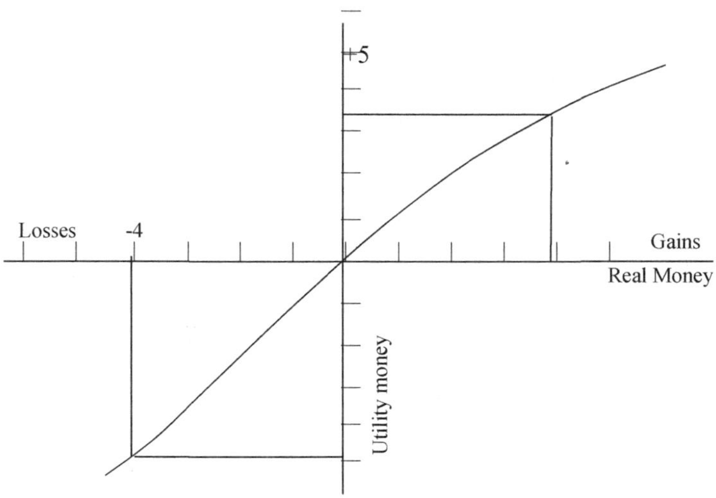

FIGURE 8.7

Effective monetary value and effective utility value.

8.4.3 Sensitivity Analysis

The effect of uncertainty of parameters on system performance can be estimated by varying the value of the uncertain parameter(s) and comparing the results in the changed circumstances. Such an analysis is very frequently performed, for example, in catchment modeling to identify the parameters to which the model response is sensitive. Depending on the cost and effort needed to gather additional data that are needed to reduce the uncertainty and improve system performance, a logical follow-up action can be planned.

A simple capacity expansion model of Loucks et al. (1981) can be used to illustrate the techniques of sensitivity analysis and to present three alternative measures of the cost of errors in model parameters. A capacity expansion model for water and sewage treatment plants, pipelines, canals, and other structures, which have long lives and show significant economies of scale, can be easily formulated. The cost of a structure, including initial capital costs and the present value of operation and maintenance costs, with capacity x is often well approximated by the power function:

$$K(x) = ax^b \tag{8.60}$$

where x is a measure of size of structure and b is the elasticity of costs defined as

$$b = \frac{x}{K}\frac{dK}{dx} \tag{8.61}$$

Its range is [0,1] and it reflects the economies of scale. Eq. (8.61) can also be written as

$$\frac{dK}{K} = b\frac{dx}{x} \tag{8.62}$$

According to Eq. (8.62), if $b = 0.50$, then a 20% change in capacity will result in only 10% change in costs. Typical values of b for water treatment and wastewater treatment facilities range from 0.60 to 0.80. Parameter b plays an important role in scheduling projects. For $b < 1$, it is economically advantageous to build ahead of the anticipated demand to capture the economies of scale that result from construction of large facilities. However, these savings must be weighted against the opportunity cost of committing resources to build capacity which may not be needed for many years.

The parameters for water resources projects are selected after analyzing demographic, economic, and engineering data. In many countries, the quality and quantity of such data are less than desired. Sensitivity analysis helps assess the cost and the effect of errors in design parameters. Assume that the set of "true" parameters is known and is denoted by vector w. The true parameter values are not known in practice. Let the best available estimates be denoted as w_e. The knowledge of true parameter values allows the determination of optimal design capacity x. The use of "non-true" w_e values yields nonoptimal design capacity of the facility, denoted by x_e.

The minimum cost of facility that is achievable with parameter value w_e is $C(x_e|w_e)$; the real cost of a nonoptimal design x_e evaluated with true parameter values w is $C(x_e|w)$. Mis-specification of a parameter may result in a significant loss of economic efficiency (LEE), which is the amount by which the actual costs of implementing the nonoptimal decision exceeds the costs that would result if the optimal design x^* was selected. It can be expressed as a ratio of the cost of optimal design:

$$LEE = [C(x_e|w) - C(x^*|w)] / C(x^*|w) \tag{8.63}$$

Because the design x^* achieves the minimal value of $C(x|w)$, LEE is never less than zero. A large value of LEE indicates that large resources would be wasted by using nonoptimal design parameters. Consider a situation in which parameters of a model can take on a range of values represented by n discrete alternative parameter vectors $w_1, \ldots, w_n$. Associated with each parameter vector w_i is a decision vector x_i which maximizes the system's net benefits $NB(x_i|w_i)$ so that

$$NB(x_i|w_i) = \text{Max } NB(x|w_i), x \in X$$

where X is the set of feasible decisions. The cost of choosing other than the optimal decision could be measured by the loss of economic efficiency, $LEE(x_i|w_j)$, also called regret. The term regret is defined as the nonnegative amount by which the net benefits could have been increased by the selection of the optimal decision vector x_j for parameter vector w_j rather than decision x_i:

$$LEE(x_i|w_j) = \max_k [NB(x_k|w_j)] - NB(x_i|w_j) = NB(x_j|w_j) - NB(x_i|w_j) \geq 0 \tag{8.64}$$

8.5 SYSTEMS ANALYSIS TECHNIQUES FOR RATIONAL DECISION MAKING

When a decision is to be taken under risk, the degree of ignorance about the data is expressed in terms of a probability density function. Therefore, it will not be possible to give the exact value of benefits in such a case. Commonly, the expected value of benefits (say, agricultural benefits) is maximized or the expected value of costs (say, flood damages) is minimized.

Generally, any management policy is not absolutely "safe", since there is some nonzero probability of an undesirable outcome. The policy may, however, be modified to reduce this risk, but the resulting benefits will be smaller. Therefore, there is a trade-off between the risk level and the expected benefits. Usually, the consequences of an undesirable outcome govern the selection of an appropriate risk level.

8.5.1 Stochastic Optimization

Deterministic approaches fail to account for the uncertainties inherent in the prediction of economic, hydrologic, technologic, and other factors that affect the performance of water resources systems. Hence, optimization techniques must be extended to incorporate mathematical descriptions of various random processes. The general form of an optimization problem is

$$\min f(X) \tag{8.65}$$

$$\text{subject to} \quad g_i(X) = b_i, i = 1, 2 \ldots n \tag{8.66}$$

where vector X contains the decision variables.

The risk is considered either through an implicit approach or through an explicit approach. If the latter approach is used in an optimization framework, it involves optimizing the expected value of the performance criterion and the use of inflow probability distribution in the constraints. The risk level can be a variable in the constraints or it can be a decision variable. The stochastic optimization techniques are an extension of deterministic counterparts that have been discussed in Chapter 5.

The uncertainty may arise in a problem through the objective function or one or more constraints. The uncertainty in the objective function arises from imprecise knowledge or inability to provide the right value of the future benefits and costs resulting from alternative decisions. One way to handle this uncertainty is to substitute the expected value for the uncertain net benefit function. The use of the expected value of the objective function is satisfactory if the alternatives are not too extreme so that the expected value can be substituted for the expected utility of a decision (Loucks et al., 1981).

If the uncertainty in constraints is small, it may be satisfactory to use the expected values by following the same logic as above. However, when there are large variations in some quantities, the use of the expected values for random quantities is not proper as it will result in significant violations of the constraints. When only the right-hand side b_i of one or more inequality constraints is random, chance constraints can be written that define the probability P_i that the constraint can fail:

$$\Pr[g_i(X) \le b_i] \ge 1 - P_i \tag{8.67}$$

This equation indicates that the constraints can be violated no more than 100 P% of the time. Suppose $g_i(X)$ is the demand of water and b_i represents the total quantity of the available water which is uncertain. If the demand should be met at least 95% of the time, Eq. (8.67) becomes

$$\Pr[g_i(X) \le b_i] \ge 0.95 \tag{8.68}$$

which indicates that 95% of the time, demand $g_i(X)$ should be less than the availability or a 5% probability of a water shortage is acceptable. The major advantage of these types of chance constraints is that they can be converted into deterministic equivalents, given the knowledge of the distribution function $F_i(b_i)$. By definition,

$$\Pr[B_i \le b_i] = F_i(b_i) \tag{8.69}$$

which implies that

$$\Pr\left[B_i \le F_i^{-1}(b_i)\right] = P_i \tag{8.70}$$

Thus, there is a 100 P_i % chance that the value of B_i will be less than or equal to the quantity $F^{-1}{}_i(P_i)$ which is denoted $b_i^{(Pi)}$. Thus, for a fixed decision X, if

$$g_i(X) \le b_i^{(Pi)} \tag{8.71}$$

then the constraint $g_i(X) \le b_i$ will be violated at most the 100 P_i% of the time.

When the risk is incorporated into the constraints, the approach is known as reliability constrained programming. This approach has been a subject of several investigations. The stochastic DP technique is a very efficient means of deriving optimum policies for water resources systems. The technique is normally used to maximize the expected net benefits. But in doing so, it may well derive a policy that, despite its economic advantages, allows the system to fail more frequently than is acceptable. A formulation that includes a penalty term overcomes this problem. The recursive equation of probabilistic dynamic programming is (Askew, 1974):

$$F_n\{S(n)\} = \underset{R(n)}{\max} \sum_{I(n)} P\{I(n)\}[B\{R(n) + F_{n-1}(S(n-1))\,/\,(1+r) - W] \tag{8.72}$$

where S(n), B{R(n)}, and P{I(n)} are the initial storage of the reservoir, the net benefits associated with a release R(n), and the probability of inflow I(n), all for the period n, respectively. The term r denotes the discount rate. Further, W is a penalty function whose value is greater than zero for failure and zero otherwise. The effect of this penalty function is to reduce the expected net benefit in the event of a failure. To get a graph of trade-off between the expected net benefits and the risk level, the operation of the system is simulated using a range of penalty function values and the ratio of the expected net benefit and the maximum possible benefits is plotted against the average number of failures. However, the use of a penalty function to represent the reluctance to failure does not always guarantee an optimum solution. It is better that the decision space is constrained to those decisions which satisfy the constraint rather than changing the objective function. Some investigators suggest that the future failures should not be considered as important as the present ones. The discount rate can also be looked upon as a risk premium-higher premium is to be paid for riskier ventures and vice-versa.

8.5.2 Stochastic Simulation

The simulation technique was discussed in detail in Chapter 5. Simulation models may be either deterministic or stochastic. Stochastic simulation provides planners with a powerful tool for evaluation of the probability distribution of the performance indices of complex stochastic water resources systems.

When simulating any system, the modeler designs and conducts mathematical experiments. He has to collect inputs and specify initial conditions. The initial conditions can be based on the most likely situation or can be random conditions. In case of stochastic simulation, a number of runs of the simulation model are made by varying the initial conditions or inputs to the system. The variation of inputs may also require synthetic generation of data. The synthetic data should have the same statistical properties as the observed data and a number of techniques and software packages are available for this purpose. It is important to plan model runs, what outputs are to be collected for the system performance, and how these are to be presented. The outputs are studied to analyze system performance indices. Typically, one has to see if there are statistically significant differences in the indices and are these changes in performance indices due to modifications of the system's design or operating policy or these are just sampling fluctuations.

Consider that a reservoir is operated to meet water supply demands. A number of equally likely synthetic streamflow sequences are generated. The operation of the reservoir is simulated using two operation policies, and a number of runs are taken for each policy using the generated inflow sequences. Let the performance indicator be the failure of reservoir to meet the demand. Denote by F_{1i} and F_{2i} the failure rates using the first and second policy and the ith synthetic streamflow sequence. If the difference Z_i ($=F_{1i} - F_{2i}$) is positive more frequently, it indicates that there are more failures using the first policy. One can plot Z_i and see its variation about the zero line.

One can also attempt to find if the difference between failures using two policies is statistically significant—can the difference between the failure rates be attributed to the fluctuations that occur in the average of any finite set of random variables? The significance of the difference between the mean failure rates can be tested by first finding the mean and standard deviation of Z_i. If the sample size n is sufficiently large, the variable t is defined as

$$t = [Z_m - \mu_Z]/\left[s_Z/n^{0.5}\right] \tag{8.73}$$

where Z_m is the mean of all Z_i and s_Z is the standard deviation. If Z_i are normally distributed, then t has a Student's t-distribution. The value of t obtained from Eq. (8.73) can be compared with standard tables. The expected number of failures is the same in both cases, $\mu_Z = 0$.

8.6 BAYESIAN DECISION MAKING

A practical way in which uncertainty can be quantified is through probabilities. The decision theory deals with optimal actions, based on the past and current information, in addition to inferences and conclusions. In its application, one has to deal with possible states of nature and choose between different possible actions. The main argument for justifying the decision-making of this type is that in most practical situations, knowledge of probabilities is incomplete. For instance, in contrast with the games of chance and lotteries, the probability that a horse wins a race is subjective, that is, it depends on personal knowledge. These subjective probabilities that quantify personal knowledge could be incorporated in Bayesian decision theory. Because of the inclusion of the subjective element, the approach is often referred to as decision-making under uncertainty. On the other hand, by the same definition, decisions made under risk are based on objective probabilities.

8.6.1 Bayes' Theorem

In many decision-making problems, statistical parameters of the underlying processes are unknown and estimates of these parameters are crucial to the decision process. For such inferential purposes, Bayesian analysis is a useful practical tool to the decision-maker. In application of the decision theory, Bayesian probabilities are used to weigh the utilities of feasible actions. This leads to the Bayes solution which is based on the optimum expected utility.

The Bayes theorem follows from the definition of conditional probability and is regarded as a fundamental theorem to revise probability through evidence. Bayes' theorem involves a prior (or a priori) distribution of a variable which may be based on observed data, some theoretical reason, or on the investigator's judgment about the likely behavior of the variable. This distribution contains all the relevant information about the variable before additional data becomes available. Given the a priori distribution, the posterior distribution can be evaluated when a new set of data becomes available.

Let α_i, i = 1, 2, …, n, denote all possible values of a variable which is subject to uncertainty. Furthermore, let x represent a forecast or estimate of the variable. The prior probabilities estimated before the receipt of new data of the variable are denoted by $P_o(\alpha_i)$ and the conditional probabilities of x subject to the properties of variables α_i are denoted by $P(x|\alpha_i)$. The posterior probabilities $P_1(\alpha_i|x)$ represent the probabilities of different values α_i, given the sample x. The joint probability $P(\alpha_i, x)$ of α_i and x, based on conditional probabilities, is estimated by

$$\begin{aligned} P(\alpha_i, x) &= P_o(\alpha_i)\, P(x|\alpha_i) \\ &= P(x)\, P_1(\alpha_i|\, x) \end{aligned} \tag{8.74}$$

Now, the marginal probability of x is the sum of joint probabilities for all i:

$$P(x) = \sum_{i=1}^{n} P_0(\alpha_i) P(x|\alpha_i) \tag{8.75}$$

Substituting the value of P(x) from Eqs. (8.75) to (8.74), one obtains the Bayes theorem:

$$P_1(\alpha_i|x) = P_0(\alpha_i)P(x|\alpha_i) / \sum_{i=1}^{n} P_0(\alpha_i)P(x|\alpha_i) \tag{8.76}$$

According to Eq. (8.76), the posterior probability of a value α_i, conditional to a sample x, is the joint probability of α_i and x (that is, the product of the prior probability of α_i and the probability of x conditional to α_i divided by the marginal probability of x). Note that the sum of the posterior probabilities (for all i) equals unity.

Using the posterior distribution, one can make better inferences on the variable. Of course, the inferences depend on the assumptions made regarding the probability model. While making a decision, a rational individual uses some prior information that is available with him. The Bayesian method can help revise probabilities.

Although the mathematical soundness of the Bayesian approach has been widely accepted, there is some controversy with regard to the choice of prior distribution. An objection to the use of Bayes' theorem is that prior distributions are either unknown or not easily obtained. It is often suggested that the Bayesian theory should be used only when the prior probabilities can be expressed clearly. Nevertheless, many people believe that the Bayesian decision theory is a useful tool in situations where uncertainties could only be quantified through experience and professional expertise. Several water resources planning and management problems fall in this class.

Prior probabilities can be determined by following two approaches: (1) data-based sources—the information is obtained from past data samples, and (2) theoretical and other considerations—the information is derived from nondata-based sources. Clearly, as more data become available, the relative influence of prior probabilities on posterior probabilities reduces.

The Bayesian decision rule requires the estimation of optimum expected utility which is commonly referred to as the Bayes risk. The procedure is to assign numbers or utilities (which could be in the form of gains or losses) to the possible actions and to weight these by the posterior probabilities. The optimum action is then found by comparing the expected utilities. The corresponding expected utility is the Bayes risk.

8.6.2 Application of Bayes Theorem

The Bayes theorem has numerous applications in water resources, for example, to update the weather forecasts, reservoir operation, etc. Using this theorem, the previously enumerated probabilities can be updated in the light of current information. The application is illustrated through two examples.

Example 8.8. The weather at a place can be described by two states: wet (α_1) and dry (α_2). Based on observed data, the prior probabilities of these states are $P_o(\alpha_1) = 0.2$ and $P_o(\alpha_2) = 1 - P_o(\alpha_1) = 0.8$. Also, let x_1 and x_2 be the forecasts of wet and dry weather based on current data. Estimation from previous forecasts suggests that the conditional probabilities of success of these forecasts are $P(x_1|\alpha_1) = 0.80$ and $P(x_2|\alpha_2) = 0.70$. Estimate the posterior probability of rain given dry weather.

Solution. Clearly the probabilities of failure are $P(x_2|\alpha_1) = 1 - P(x_1|\alpha_1) = 0.20$ and $P(x_1|\alpha_2) = 1 - Pr(x_1|\alpha_1) = 0.30$. According to Bayes theorem, the posterior probability of rain when it is forecasted is given by

$$\begin{aligned} P_1(\alpha_1/x_1) &= P_o(\alpha_1)P(x_1|\alpha_1)/\{P_o(\alpha_1)P(x_1|\alpha_1) + P_o(\alpha_2)P(x_1|\alpha_2)\} \\ &= 0.2 \text{ x } 0.8 / [0.2 \text{ x } 0.8 + 0.8 \text{ x } 0.3] = 0.4. \end{aligned} \tag{8.77}$$

The probability of dry weather is

$$P_1(\alpha_2|x_1) = 1 - 0.4 = 0.6.$$

The posterior probability of dry weather is

$$\begin{aligned} P_1(\alpha_2/x_2) &= P_o(\alpha_2)P(x_2|\alpha_2)/\{P_o(\alpha_2)P(x_2|\alpha_2) + P_o(\alpha_1)P(x_2|\alpha_1)\} \\ &= 0.8 * 0.7/(0.8 * 0.7 + 0.2 * 0.2) = 0.933 \end{aligned} \tag{8.78}$$

Hence, $P_1(\alpha_1|x_2) = 1 - 0.933 = 0.067$.

The procedure is shown diagrammatically in Fig. 8.8. Predictions of rain, however, have only an approximately 40% chance of success. One could expect that, out of 100 forecasts of dry weather, 93 are correct.

Prior probabilities	Conditional probabilities	Joint probabilities	Posterior probabilities
Wet α_1, Dry α_2	Wet x_1, Dry x_2		
$P_o(\alpha_1) = 0.2$	$P_1(x_1|\alpha_1) = 0.8$	$P_o(\alpha_1)\,P(x_1|\alpha_1) = 0.16$	$P_1(\alpha_1|x_1) = 0.4$
	$P_1(x_2|\alpha_1) = 0.2$	$P_o(\alpha_1)\,P(x_2|\alpha_1) = 0.04$	$P_1(\alpha_1|x_2) = 0.067$
$P_o(\alpha_2) = 0.8$	$P_1(x_1|\alpha_2) = 0.3$	$P_o(\alpha_2)\,P(x_1|\alpha_2) = 0.24$	$P_1(\alpha_2|x_1) = 0.6$
	$P_1(x_2|\alpha_2) = 0.7$	$P_o(\alpha_2)\,P(x_2|\alpha_2) = 0.56$	$P_1(\alpha_2|x_2) = 0.933$

FIGURE 8.8

Revision of probabilities of weather forecasting.

Adapted from Kottegoda (1980).

Now consider a case when prior information is vague and the prior distribution is assumed to be uniform, that is, $P(\alpha_1) = P(\alpha_2) = 0.5$. If the conditional probabilities are the same, it follows that $P_1(\alpha_1 |x_1) = 0.73$, $P_1(\alpha_2 |x_1) = 0.27$, and also that $P_1(\alpha_2|x_2) = 0.778$ from which $P_1(\alpha_1|x_2) = 0.222$. One could see that, on account of the diffused prior information, the posterior probabilities are close to the original conditional probabilities. One may question the necessity of this procedure when more definite prior information is available. The justification for combining prior and current information is relevant when the estimated probabilities are based on a limited number of observations and therefore are not reliable.

Finally, consider an extreme situation when $P_0(\alpha_1) = 0.1$ and $P_0(\alpha_2) = 0.9$; the conditional probabilities remaining the same. This example is merely given to illustrate concepts because, in this case, weather forecasts are not really needed. It is found that $P_1(\alpha_1|/x_1) = 0.203$, $P_1(\alpha_2|x_2) = 0.969$. Clearly, almost definite prior information has a strong influence on the posterior probabilities.

Example 8.9. The following simplified application of Bayesian decision-making is based on discrete probabilities for reservoir operation (see Kottegoda (1980) for a similar example). This reservoir is designed for irrigation purposes. Forecasts of inflows are made, and these form the basis for optimal actions before the commencement of irrigation season. Let three discrete water yields, α_1, α_2, and α_3, be available from this reservoir. The prior probabilities of these yields, assessed from rainfall, runoff, and catchment properties, are: $P_0\ (\alpha_1) = 0.2$, $P_0\ (\alpha_2) = 0.6$, and $P_0\ (\alpha_3) = 0.2$.

Based on current hydrologic information on the catchment and the reservoir storage at the start of the year, a forecast x_j, $j = 1, 2, 3$, of water available during the coming year is issued prior to the release for irrigation. The conditional probabilities $P(x_j|\alpha_i)$ of each of the three forecasts relative to a given reservoir yield are as follows. Using these forecasts x_j, the *a posteriori* probabilities $P(\alpha_i|x_j)$ of the true reservoir yield are to be evaluated using Bayes' theorem.

	Conditional probabilities for forecast $P(x_j\|\alpha_i)$			
Reservoir yield	x_1	x_2	x_3	**Sum**
α_1	0.6	0.2	0.2	1.0
α_2	0.1	0.7	0.2	1.0
α_3	0.2	0.1	0.7	1.0

Based on the forecasts, the decision-maker has to decide the area A_i, $i = 1,2,3$, that can be irrigated. The annual net benefits (Rupees) from farming different extents of areas A_i and for each of the yields α_i, $i = 1, 2, 3$, are given in the table below.

	Estimated annual net benefit by farming area		
Reservoir yield	A_1	A_2	A_3
α_1	500	400	300
α_2	600	700	600
α_3	700	800	900

(a) If the decision-maker acts using the Bayes solutions based on forecasts, find the expected annual benefit from irrigation?

(b) Find the expected annual income in the two limiting cases: (i) when the forecasts are 100% accurate, and (ii) when the forecasts are totally unreliable and actions are taken solely on the prior probabilities.

Solution. To solve the problem using the Bayes theorem, it is necessary to first compute the joint probabilities of reservoir yield and forecast, that is, $P_0(\alpha_i)\ P(x_j|\alpha_i)$

	Joint probability for forecast		
Reservoir yield	x_1	x_2	x_3
α_1	$0.2 \times 0.6 = 0.12$	$0.2 \times 0.2 = 0.04$	$0.2 \times 0.2 = 0.04$
α_2	$0.6 \times 0.1 = 0.06$	$0.6 \times 0.7 = 0.42$	$0.6 \times 0.2 = 0.12$
α_3	$0.2 \times 0.2 = 0.04$	$0.2 \times 0.1 = 0.02$	$0.2 \times 0.7 = 0.14$
Marginal probabilities	0.22	0.48	0.30

The posterior probabilities $P(\alpha_i|x_j)$ are obtained by dividing the joint probabilities by the marginal probabilities. Their sum is 1.0.

	Posterior probability for forecast		
Reservoir yield	x_1	x_2	x_3
α_1	0.545	0.083	0.133
α_2	0.273	0.875	0.400
α_3	0.182	0.042	0.467
Total	1.0	1.0	1.0

For each forecast x_i, $i = 1, 2, 3$, either of three actions A_j, $j = 1, 2, 3$, is possible. To find the action that will be given the highest expected return, the annual assessed returns of the net income are weighted by the corresponding posterior probabilities and are then added for each pair of x_i and A_j. The expected returns from actions taken are given in the following table.

	Estimated annual net benefit		
Forecast	A_1	A_2	A_3
x_1	0.545×500 | 0.273×600 | $= 563.7$ 0.182×700 |	0.545×400 | 0.273×700 | $= 554.7$ 0.182×800 |	0.545×300 | 0.273×600 | $= 491.1$ 0.182×900 |
x_2	0.083×500 | 0.875×600 | $= 595.9$ 0.042×700 |	0.083×400 | 0.875×700 | $= 679.3$ 0.042×800 |	0.083×300 | 0.875×600 | $= 587.7$ 0.042×900 |
x_3	0.133×500 | 0.400×600 | $= 633.4$ 0.467×700 |	0.133×400 | 0.400×700 | $= 706.8$ 0.467×800 |	0.133×300 | 0.400×600 | $= 700.2$ 0.467×900 |

With forecasts x_1, x_2, and x_3, the Bayesian solution is that optimal actions A_1, A_2, and A_3 should be taken with maximum expected incomes of 563.7, 679.3, and 706.8, respectively. To determine the long-term average annual income from these optimum actions, the maximum expected incomes are multiplied by marginal probabilities of respective forecasts. Therefore, the expected annual income is $563.7 \times 0.22 + 679.3 \times 0.48 + 706.8 \times 0.3 = 662.12$.

(b) If the forecasts are 100% accurate, the conditional probabilities $P(x_j|\alpha_i)$ form an identity matrix:

	Conditional probability for forecast		
Reservoir yield	x_1	x_2	x_3
α_1	1.0	0.0	0.0
α_2	0.0	1.0	0.0
α_3	0.0	0.0	1.0

These conditional probabilities are equal to the posterior probabilities $P(\alpha_i|x_j)$. The marginal probabilities of forecasts are equal to the corresponding prior probabilities. The expected returns from various decisions are the same as the second table of the example:

	Expected return for action		
Forecast	A_1	A_2	A_3
x_1	500	400	300
x_2	600	700	600
x_3	700	800	900

The expected annual income is $500 \times 0.2 + 700 \times 0.6 + 900 \times 0.2 = 700$. When only the prior probabilities are known, the expected returns from actions are as given below:

	Expected return for decision		
Reservoir yield	A_1	A_2	A_3
α_1	$500 \times 0.2 = 100$	$400 \times 0.2 = 80$	$300 \times 0.2 = 60$
α_2	$600 \times 0.6 = 360$	$700 \times 0.6 = 420$	$600 \times 0.6 = 360$
α_3	$700 \times 0.2 = 140$	$800 \times 0.2 = 160$	$900 \times 0.2 = 180$
Sum	600	660	600

As per this table, action A_2 should be taken invariably and the expected annual income will be 660.

To summarize, the expected annual income values are: if the forecasts are available, the expected annual income from optimum action is 662.12; if perfect forecasts are available, the income is 700; and if only the prior information is available, the income is 660. Thus, the expected value of forecast is $662.12 - 660 = 2.12$, and the expected value of perfect information is $700 - 660 = 40$. Apparently, the expected value of forecasts is almost negligible and the perfect forecasts give only about 7% additional benefits. Of course, these values are highly dependent on the estimated benefits and there can be wide variations from case to case. The limitations in such procedures are due to likely errors in the assessment of benefits or costs of action and in the estimation of prior and conditional probabilities.

8.7 CLOSURE

To conclude, one can state: (1) Environmental data are subject to both intrinsic and epistemic errors which can be random or systematic. (2) The sources of errors are many and are the root cause of uncertainty. (3) Two main measures of errors are bias and consistency. (4) Errors propagate in analyses and the rate of propagation depends primarily on the nature of the function describing the system and the dependence among errors. (5) Errors can be minimized but not eliminated completely.

A rational decision-making involves defining an objective, identifying alternative means of achieving the objective and attendant risks, and applying a ranking procedure to determine the best alternative. In the real world, often little planning is done and little rationality is invoked even for important decisions. Often, the common cause is the failure to recognize alternatives. All too often, precedent, tradition, accident, prejudice, or shortsightedness determines the course of action. Frequently, important decisions are postponed till there is hardly any choice but to continue the present practice. A third common error is the presumption in the value judgment as to what is important and what is not.

Design and management of a water resources project involves interaction of water resources engineers with many other disciplines. In particular, the interaction with structural and geotechnical engineers is important. Within the respective engineering professions, an inherent resistance to risk analyses is noticed and this is magnified when dealing with low-probability high-consequence events, such as dam failure or extensive, large-scale water supply shortages. There is potential for substantial savings as well as improved planning and operation if people from the various disciplines work in close interaction. Since many factors responsible for risk and uncertainty have a common cause and origin, it would be better to evolve an integrated safety assessment and management procedure rather than apply independent factors of safety.

References

Askew, A.J., 1974. Optimum reservoir operating policies and the imposition of a reliability constraint. Water Resources Research 10 (1), 51–56.

Burges, S.J., Lettenmaier, D.P., 1982. Reliability measures for water supply reservoirs and the significance of long-term persistence. In: Unny, T.E., McBean, E.A. (Eds.), Decision Making for Hydrosystems: Forecasting and Operation. Water Resources Publications, Colorado, USA (1982).

Chow, V.T., 1979. Risk and reliability analysis applied to water resources in practice. In: McBean, E.A., Hipel, K.W., Unny, T.E. (Eds.), Reliability in Water Resources Management. Water Resources Publications, Colorado, USA.

Croley, T.E., 1979. Risk in reservoir design and operation: a state-of-the-art review. In: McBean, E.A., Hipel, K.W., Unny, T.E. (Eds.), Reliability in Water Resources Management. Water Resources Publications, USA.

Haimes, Y.Y., 1981. Risk/benefit Analysis in Water Resources Planning and Management. Plenum Press, New York, USA.
Haimes, Y.Y., Stakhiv, E.Z. (Eds.), 1985. Risk-based Decision Making in Water Resources. ASCE, New York, USA.
Hashimoto, T., Stedinger, J.R., Loucks, D.P., 1982. Reliability, resiliency, and vulnerability criteria for water resource system performance. Water Resources Research 18 (1), 14–20.
Jinno, K., Zongxue, X., Kawamura, A., Tajiri, K., et al., 1995. Risk assessment of a water supply system during drought. Water Resources Development 11 (2), 185–204.
Kaplan, S., Garrick, B.J., 1981. On the quantitative definition of risk. Risk Analysis 1 (1), 11–27.
Klemes, V., 1977. Value of information in reservoir optimization. Water Resources Research 13 (5), 837–850.
Klir, G.J., 1991. Measures and principles of uncertainty and information. In: Proceedings of NATO Advanced Study Institute on Information Dynamics. Plenum Press, New York.
Kottegoda, N.T., 1980. Stochastic Water Resources Technology. The Macmillan Press Ltd., London.
Loucks, D.P., Stedinger, J.R., Haith, D.A., 1981. Water Resources Systems Planning and Analysis. Prentice Hall Inc., New Jersey.
Moy, W.-S., Cohon, J.L., ReVelle, C.S., 1986. A programming model for analysis of the reliability, resilience, and vulnerability of a water supply reservoir. Water Resources Research 22 (4), 489–498.
Pat-Cornell, M.E., 1996. Uncertainties in Risk Analysis: Six Levels of Treatment. Reliability Engineering and System Safety, No. 54. Elsevier Science Limited, Northern Ireland.
Shackle, G.L.S., 1961. Decision, Order and Time in Human Affairs. Cambridge University Press, Cambridge.
Vrijling, J.K., van Gelder, P.H.A.J.M., 2000. Policy implications of uncertainty integration in design. In: Wang, Hu (Eds.), Stochastic Hydraulics 2000. Balkema, Rotterdam, pp. 633–646.
Yevjevich, V., 1985. The risk factor in water resources. In: Haimes, Y.Y., Stakhiv, E.Z. (Eds.), Risk-based Decision Making in Water Resources. ASCE, New York, USA.

Further Reading

Unny, T.E., McBean, E.A. (Eds.), 1982. Decision Making for Hydrosystems: Forecasting and Operation. Water Resources Publications, Colorado, USA.

PART

III

Water resources planning and development

CHAPTER

Water Resources Planning

9

Chapter outline

Water Resources Systems Planning and Management. https://doi.org/10.1016/B978-0-12-821349-0.00005-8

River basin planning for water resources development and management has been practiced in many parts of Asia and Africa for at least 9000 years. The oldest recorded practice of irrigated agriculture has been traced in Jericho in 7000 BCE (Saha and Barrow, 1981). There are also recorded histories of sophisticated engineering works for water regulation in China, Egypt, and Iraq, which date back to several thousand years. An intricate system of basin irrigation involving longitudinal dikes parallel to the main channel of the Nile to regulate flood waters and a network of cross dikes and canals to conduct flood waters into predesignated basins was evolved in Egypt as early as 3400 BCE (Hamdan, 1961). About 250 BCE, the Chinese had built a sophisticated system, called Duijiangyang Irrigation Project, which is still in use. This system uses dikes and spurs to divert river water but has no dams. In India, the Grand Anicut (Kallanai Dam) was constructed by the Chola King Karikalan during the first century A.D. on the River Cauvery. The dam was built to protect the downstream populations against floods and provide water for irrigation in the Cauvery delta. The Grand Anicut is the world's oldest still-in-use dam; it is also credited with being the fourth oldest dam in the world and the first in India. By the seventh century AD, the Chinese had developed a highly sophisticated network of structures for irrigation, making a balanced use of ground and surface water resources. They had organized a system of administrative authority to ensure a high state of maintenance of these structures and perfected a land-use pattern that maximized the use of available irrigation facilities. Detailed accounts of canal irrigation systems in Iraq dating as far back as 4000 BCE are also available. An important milestone of the modern form of river basin planning was the initiation of developments in the Tennessee Valley through the Tennessee Valley Authority (TVA) in the 1930s in the United States.

A water resource can be defined as any aspect of water that has value or that is exploited by the user to get a certain benefit. A benefit is any tangible or intangible output that is valued or desired by producers or individuals. The different aspects of a water resource of interest are its quantity and quality, potential energy, flow depth, surface area, aesthetic value, waste assimilating capacity, health benefits, its biological productivity, etc. Typical bulk users of water are as follows: municipalities and industries, irrigated agriculture, forestry, hydroelectric power plants, thermal power plants, commercial navigators, recreational water users, and fish and wild life. Integrated water resources management is a way to manage an area's water resources taking into account the fact that the aspects such as quantity and quality are facets of the same physical water system. It recognizes the physical interconnections among components of ecosystems and the relationship among the users.

Society needs to prepare plans to develop water resources because at most places/times, either we have too much water, or too less water, or water quality is poor. In addition, water may also be needed for the rejuvenation of the environment. Planned development of a resource leads to more value for money spent and minimizes unwanted outcomes.

Different water users need different amounts and types of water. The principal categories of water use are shown in Table 9.1. On average, the total consumption of water by all users should not exceed

Table 9.1 Principal categories of water use.

Infrastructure F Agriculture, Forestry and Aquaculture A Industry I	F	Drinking	W	W Withdrawal N In-stream use O On-site **(a)** Highly consumptive use **(b)** Heavy impact on water quality
	F	Domestic uses	W	
	F	Public uses in settlements	W	
	A	Rain-fed agriculture	O	
	A	Livestock	W	
	A	Fish and wildlife	N	
	A	Forestry	O	
	A	Irrigation (a)	W	
	F	Navigation	N	
	I	Hydropower	N	
	I	Thermal power	W	
	I	Mining (b)	W	
	A	Swamp and wetland habitat	O	
	I	Cooling	W	
	I	Manufacturing (b)	W	
	I,	Waste disposal (b)	N	
	F	Recreation	N	
	F	Aesthetic enjoyment	N	
	A	Utilization of estuaries	N, O	

Adapted from UN (1976).

the rate of its replenishment by the hydrologic system in a region. Hence, the availability of water in a region and the amount and nature of requirements determine how many water users can be supported. As the rate of resource replenishment is finite and limited, the water users may be in competition with each other.

A water resources (WR) development project is a set of structural or nonstructural activities to develop or improve water resources for the benefit of society. A physical WR system is a collection of various elements (e.g., reservoirs, canals, pipelines, etc.), which interact in a logical manner and are designed in response to social needs. The ultimate goal of WR planning and management is to serve the public by ensuring that water of required quantity and quality is available at the right location and at the right time. The aim is also to protect society from the harmful effects of water. All this must be achieved within accepted levels of assurance.

Planning was defined by Weiss and Beard (1971) as *the process by which the society directs its activities to achieve goals it regards as important*. According to UN (1972), "Planning aims at optimal use of available resources. WR planning involves estimation of short term and long term needs and ways to meet these needs. It involves a comparative evaluation of alternative solutions with respect to their technical, economic and social merits. Planning needs looking into the future and looking from a broad spectrum of disciplines." The US Water Resources Council (WRC) issued a set of "Principles and standards for planning water and related land resources" in 1973. These were revised in 1979 and 1980 and specified that the overall purpose of planning should be to improve the quality of life through contributions to:

(a) National economic development,
(b) Environmental quality,
(c) Regional economic development, and
(d) Other social effects.

WR planning is a logical course of actions leading to the selection of the best acceptable project in response to an identified need. Because of wide variations in the distribution of surface water and groundwater resources over a region, WR planning is always broad in scope. Such planning is needed at different levels and for different purposes of water management. It, therefore, requires that many different uses of water are considered and evaluated, leading to the articulation of trade-offs among conflicting and competing objectives. It requires that decisions are made at many different levels, ranging from national or even international water plans to regional or local projects and involving experts and decision-makers who have varied backgrounds and who are often not water-cognizant: politicians, lawyers, and social scientists. The spectrum of objectives that are considered important for a particular water resources project by such a motley group may differ very widely.

WR planning requires a well-coordinated team of qualified professionals with clear objectives and scope of the project, who can draw a plan that is acceptable to those who are impacted by the project and to the decision-maker. This is an involved task because water resources are subject to natural variations, and future changes in demography and economy are difficult to predict. Consequently, elements of uncertainty enter the process. The other noteworthy aspect is that many WR decisions are more or less irreversible. For instance, a dam that has been built in a river valley exists practically forever, regardless of whether there is a need for it or not. It will never be possible to restore the site to its original condition, even if a dam that is no longer needed is carefully decommissioned.

To appreciate the scope of the water resources planning process, the following characteristics may be noted:

(1) WR systems most often have multiple objectives and functions.
(2) While technical aspects of the problem provide the foundations, institutional, social, and other considerations are also important and may often be the deciding factors.
(3) Multiple decision-makers, representing various constituencies, needs, and aspirations, are commonly involved in the planning process.
(4) The projects have a significant influence on society and regional economy.
(5) Elements of risk and uncertainty characterize almost all WR systems.
(6) The planning and decision-making processes are hierarchical in nature.
(7) Activities, such as problem definition and formulation, data collection, processing, and analysis, constitute the dominant effort in the planning process, at least in the initial stages.
(8) The wide scope of WR planning requires experts from many different disciplines, such as hydrology, engineering, agriculture, economics, and social sciences.

9.1 INTEGRATED PLANNING

With the best project sites already utilized in many countries, multipurpose goals must be adopted to make the best use of remaining limited sites. Due to ever-increasing demand of water for agricultural, domestic, industrial, and other purposes, a great deal of emphasis needs to be laid on optimum utilization of water resources. A wise exploitation of WR calls for integrated planning which is the planning of water, land, and other associated resources with coordination among geographical, functional, and procedural aspects.

It is important to note that uncoordinated planning activities are likely to lead to an imbalance in resource use because the availability of one resource in natural ecology is closely related to the use of another. In the absence of coordinated planning and sustainable development, the natural balance existing among different resources will be disturbed leading to harmful results. The two basic requirements that must be met in basin-wide integrated planning are as follows: improved coordination of a diverse variety of human activities, and integration and utilization of large amounts of information. An integrated plan must explicitly identify the factors and interrelationships that form the basis to plan and implement activities for the use of resources to achieve desired goals. Planning should be organized in such a way that all decisions are optimum and the relationship between resource use and outcome is foreseen.

The concept of integrated planning has its roots in regional and comprehensive planning. In regional planning, the activities are project-oriented and concerned with economic development in a region. Of late, the environmental and social issues have also become important. As the awareness of the significance of interrelationships among diverse projects has increased, it is recognized that planning should more explicitly take into account a large number of variables and functions. This led to the concept of comprehensive planning. In the late 1960s, it was felt that planning activities must become more closely coordinated and interrelated. Additional efforts for interfacing different planning activities and their close coordination led to the concept of integrated planning. These days, it is essential to examine environmental protection schemes, afforestation, catchment area treatment, soil conservation, conjunctive use, command area development, etc. while preparing project feasibility reports. These measures orient the project toward overall development.

9.2 STAGES IN WATER RESOURCES PLANNING

The scope of WR planning process can vary from broad-based preliminary planning of a new project to a detailed evaluation of a selected physical project (a feasibility study). The broad flow of activities of WR planning activities is shown in Fig. 9.1. The project may be a large (and sometimes) internationally financed activity or a small project that is financed by a small city or a private party.

Stephenson and Peterson (1991) identified three levels of planning, associated with three areas of different geographic extent: national, regional or river basin, and local areas. Based on the time sequence of the planning process, the entire life cycle of a project can be divided into four phases (Haimes et al., 1987): (1) planning, (2) construction/implementation, (3) operation, and (4) management.

Phase 1, which is relevant to this chapter, can be categorized into five stages:

Stage 1. The project initiation stage: It begins with the statement of needs and includes preliminary planning, feasibility, and field investigations.
Stage 2. The data collection stage: Detailed data are gathered for analysis and decision-making.
Stage 3. Project configuration stage: A large number of alternatives are investigated and a small number of promising alternatives are selected for detailed analysis.
Stage 4. Detailed planning stage: The design parameters, operation rules, costs, benefits, etc., of the alternatives selected in stage 3 are determined, and the final project configuration is selected. Actually, this phase represents a detailed form of stages 2 and 3.
Stage 5. The design stage: The final configuration is translated into a detailed structural design.

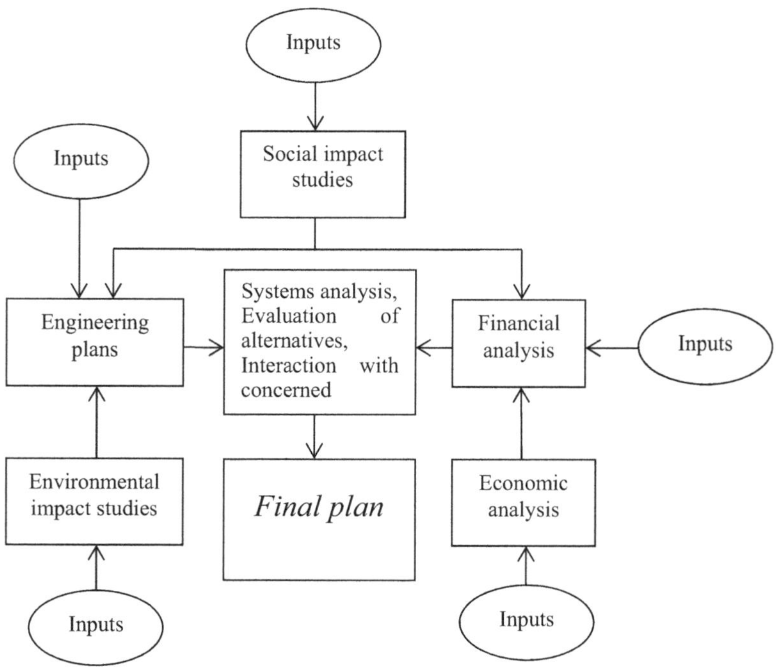

FIGURE 9.1

Flow of activities in the planning process.

Adapted from Grigg (1985).

The fifth stage is not a direct part of WR planning, as it mostly involves structural design and financial aspects of a project. This stage is beyond the scope of this book.

The earlier classification into five stages is only one of many similar classifications. The UN (1970) has outlined a four-phase program for integrated river basin development consisting of: (1) preliminary investigation and organization; (2) general reconnaissance of existing conditions; (3) initial phase of implementation, including actual start of small-scale projects; and (4) construction and operation of major structures. Simonovic (1989) outlined a four-step planning process being followed in Yugoslavia: (1) inventory, forecast, and analysis of available water resources; (2) inventory, forecast, and analysis of water demand; (3) formulation of alternative solutions for satisfying water demands from available water resources; and (4) comparison and ranking of alternative plans. According to the US WRC guidelines, the planning process consists of six major steps:

(a) Specification of the water and related land resources problems and opportunities associated with the federal objective and specific state and local concerns.
(b) Inventory, forecast, and analysis of water and related land resource conditions within the planning area relevant to the identified problems and opportunities.
(c) Formulation of alternative plans.
(d) Evaluation of the effects of alternative plans.
(e) Comparison of alternative plans.
(f) Selection of recommended plan based on the comparison of alternative plans.

It can be seen that the steps/stages given by the various agencies are more or less identical. In the following discussion, the earlier five-stage classification will be elaborated. The subdivision into five stages is convenient, since each stage is a logical precursor of the previous stage. However, the situation may be significantly different in some countries. Also, not all the proposed projects clear all stages. Unfortunately, one may find water resources projects that are stalled at various stages, including the construction stage. There are also instances where the project configuration was finalized in a court based on legal rather than technical considerations.

The definition of the five stages of planning yields a conceptual model of the planning process which is shown in Fig. 9.2. Here, the stages are a part of a sequential decision process, in which the tasks to be executed in each stage are represented by boxes and the connecting lines denote decisions to be taken. Arrows indicate the direction of the information flow from one stage to the next. Of course, there will always be many formal and informal linkages that cannot be shown in such a diagram. The planner must, therefore, allow enough flexibility for later adjustments, because most operation procedures are developed on the basis of some assumptions, and it is very likely that the real world will not function as predicted during planning.

9.2.1 Relationship Among Stages

The process of planning a WR project can also be visualized as a hierarchical structure of subsystems and decisions. A considerable overlap is bound to be present when dividing a complex process, such as WR planning in stages. In fact, this is nothing but a logical continuum between them. For example, stages 2 and 3 may be combined to form a preliminary feasibility stage. Some aspects of stages 2 and 4 may be combined into a feasibility study that provides the basis for the final financial decisions that are made before the project is designed and executed. Stages 2 and 4 may also be coupled in terms of data development and improvement. Furthermore, planners at stage 4 itself may need the design of components that is normally carried out at stage 5 to have more accurate cost estimates. In certain cases,

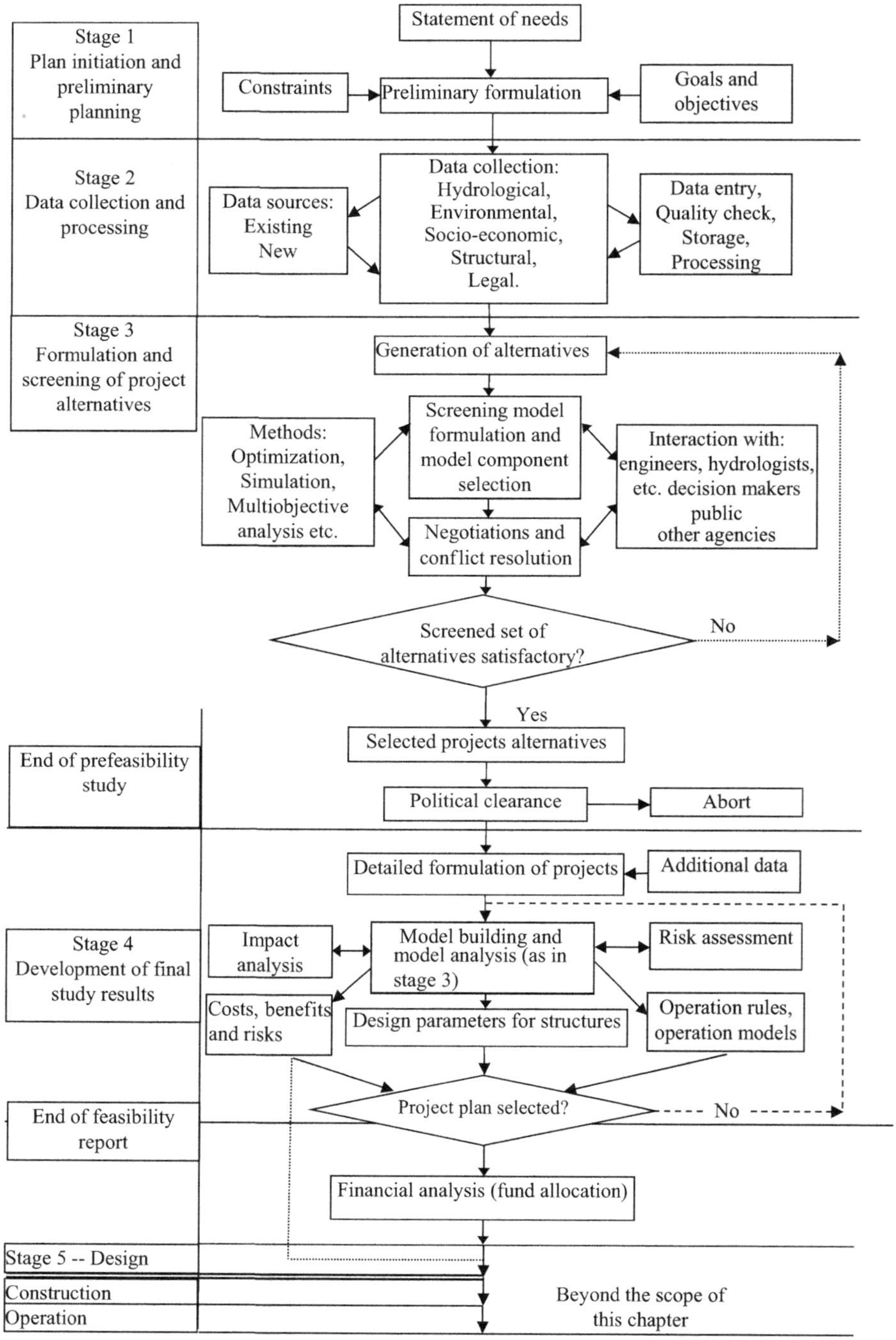

FIGURE 9.2

Stages in the water resources planning process.

Reproduced from Haimes et al. (1987). Copyright © UNESCO. Reproduced by permission of UNESCO.

only one plan may be selected, and a detailed analysis is made for that plan alone. Sometimes, construction activities, such as excavation for the dam or building a coffer dam, may begin before the design of certain components, for example, spillway gates is being finalized.

Stage 3 is basically a preliminary screening, while stage 4 brings the project very close to its final configuration. Thus, stage 4 requires more resources and the use of sophisticated techniques. Depending on the circumstances, there may be a need for detailed analysis of some portions of stage 3, with perhaps the generation of a few additional alternatives.

During stage 4, the project is analyzed in detail, including the generation of one or more suitable integrated models. This will require the following steps:

(a) quantitative definition of all variables and terms;
(b) quantification (to the extent possible) of final objectives, constraints, input—output relationships and measures (structural and nonstructural);
(c) identification and evaluation of available candidate models;
(d) evaluation of the database needed for step c;
(e) construction of new models;
(f) integration of newly developed models with existing models, as appropriate; and
(g) model calibration, validation, and application.

The planning and policy options developed through the use of models and their associated trade-offs and impacts are discussed in detail at this stage. Ideally, this requires the participation of all concerned decision-makers, stakeholders, constituencies, and agencies. The final result of this stage should be an optimal plan.

A brief discussion of data collection and processing is given next. As this topic has been discussed in Chapter 2, here the attention is focused on those aspects that are related to planning only.

9.3 DATA COLLECTION AND PROCESSING

Data collection and processing constitutes stage 2 of the planning process although this activity may continue till the final design is prepared. In stage 2, the data needed for the project should be collected and their quality and quantity evaluated. Depending on the availability of data, decisions need to be made about the collection of additional hydrometeorological data, water demand and quality data as well as demographic, economic, and ecological information.

The data collection process requires identifying the sources of data, exploration of these sources, inquiries about other possible data sources, evaluation of data quality, and computerization of data for processing. This process involves a lot of fieldwork and planning because the data collected must be purpose-oriented. The purpose must govern the type, the accuracy, and the time horizon of the data. The acquisition and processing of water resources data have been discussed in Chapter 2. The purpose of this section is to explain main aspects related to planning.

9.3.1 Specifications and Sources of Data

When the results of the preliminary planning exercise indicate that the project will come into being, a detailed analysis is initiated. By now, some data are already available from preliminary analyses but in most cases these are insufficient for a detailed investigation. At this stage, a broader database is needed. The first effort should be to compile whatever data are already available and pertinent to the project. This may require visits to many offices scattered all over the project area. In many instances, the data may have to be manually copied and then entered into the computer. If the available data are limited and further observations and measurements are required, additional field sampling and investigation programs are initiated. It is common to set up additional rain gages and stream gaging stations specifically to collect data for project planning.

Due to the high level of automation and modern means of information dissemination, data collection is a simpler exercise in developed countries. But in many countries, much of the data are still in manuscript form and are scattered in various branches of a data collection agency. Usually, no data inventory is available in a central office and it may be necessary to visit each branch office and copy and computerize the data. Also, there may not be a common format for storing the data and the quality and reliability may widely vary from agency to agency and sometimes across the same agency. Therefore, sufficient time and funds should be allocated for data collection. The problem may be slightly simpler because many times the planning agency may also be involved in the observation of a few variables. The accessibility of data banks through the Internet is becoming a common practice now in many countries. One should also inquire about the other data sources as sometimes nontraditional departments may also be observing pertinent data.

In many organizations, data collection is just one of the responsibilities, sometimes the least important. Therefore, the personnel assigned to data collection works are frequently moved to more urgent tasks. The maintenance of equipment may also not be as prompt as it should be. The observer might not have sufficient training and may not be adequately motivated and aware of the importance of the data that he is producing.

9.3.2 Data Adequacy

The adequacy of data is defined with respect to the purpose for which the data are to be used and the consequences of using inadequate data. Virtually any hydrologic database can be considered to be inadequate in some respects. Therefore, the assessment of data sufficiency should be based on sampling and parameter uncertainties and on an evaluation of how sensitive the key project parameters are to changes in the data accuracy. From an economic standpoint, the data can be considered adequate when the marginal cost of obtaining the additional data is equal to the marginal benefits from this data. However, this concept is difficult to apply because of uncertainties in the evaluation of future benefits.

The analytical methods that can be used in a specific situation should be commensurate with the quality and coverage of the database. There is no point in subjecting unreliable data to intensive processing. The reliability and representativeness of the data must be evaluated before choosing analytical tools. A common feeling is that the experts always press for more and more detailed data while even the available data are not fully used. However, the mathematical models cannot be effective without adequate data for model calibration, validation, and application. This is why the data-gathering process should be directly related to the needs of analysis and tools. The analyst must remember that his methods can hardly be used to give appropriate results if they are not based on an adequate and realistic description of the system.

In many instances, particularly in developing countries, WR projects are planned based on hydrologic and nonhydrologic databases, which are far smaller than those desired for effective analysis. However, the planning activities cannot be suspended until detailed data are available. Of course, in some instances special stations are installed during early stages of planning to collect more data and analysis is refined after desired data are available. Sometimes, plans are hurriedly prepared, because a project is to be pushed through due to political pressure or the existence of problems requiring immediate action. In such situations, it is worth considering the possibilities of implementing the project in stages, if this is technically feasible, although this always entails additional costs. The other possibility is to design the project in such a way that eventual losses due to the use of imperfect data are minimized. The strategy of "wait and watch," hoping that uncertainty about some of the crucial factors influencing the project will be reduced, is not generally helpful. Waiting does not always improve things and in fact new uncertainties may crop up with time.

9.3.3 Data Quality Control

The term data quality control denotes preliminary checking that is undertaken to weed out obvious errors and inconsistencies from the raw data. The methods include preliminary checking, plotting, and removal of errors by spatial and temporal consistency checks. The preliminary checking must ensure the overall correctness of the indicative information; simple spatial and statistical checks should be applied to see if the data provide a reasonable long-term picture of water availability and use in the region under consideration. Preliminary processing may also include the identification of data gaps as well as filling such gaps by suitable techniques. These days, all analyses are performed on computers.

It is also important that the data are representative of the current hydrological conditions in the basin. This is especially important for streamflow series in a basin subject to large-scale man-made changes, for example, deforestation or upstream withdrawals. Although relatively long records may be available, these may no longer be representative unless man-made changes in the basin are appropriately accounted for.

9.3.4 Data Systems

Database systems include data stored on computer media and the collection of software for operations such as data input, search, retrieval, updating, and deletion. Of late, many organizations have started calling their data files as databases without giving attention to properties, such as exclusion of data redundancy, provision for data independence and protection, and precise definition of mutual relations among different data. The importance of these aspects has grown considerably with the expansion of databases and the development of better software for data processing and management.

WR planning requires handling a large amount of data, and the task of data processing, compiling results, and report writing cannot be coped with conventional methods. Adequate attention to develop a database in the initial stages will save considerable time, effort, and headaches in the long run. The type of information system that is advisable for a given plan depends very much on the level of effort, availability of computer facilities, data management experts, etc.

Depending on the purpose, data requirements vary considerably. For long-range planning, monthly or annual data are needed. For operational purposes, short-time data are needed. Therefore, a data series that serves a function may not be able to serve another. With the expansion of the Internet, the common system configuration consists of a database server and online users who access the database from remote hosts. With the development of software for the management and analysis of geographical data, the task of planners has become relatively easy. The GIS tool has been described in Chapter 3.

9.4 ESTIMATION OF FUTURE WATER DEMANDS

The use of water can be divided into two categories: consumptive use, in which water is an end to itself, and nonconsumptive use, in which water is a means to an end. The first category includes the use for municipal, agricultural, industrial, and mining purposes. The nonconsumptive uses are in-stream uses, such as hydropower, transportation, and recreation. Consumptive uses are modeled using consumptive functions and nonconsumptive uses, using production functions. The water use refers to the amount of water applied to achieve various ends so that it is a descriptive concept. Water demand is the scheduling of quantities that consumers use per unit of time for a particular price of water, which is an analytical concept. Modern water resources projects are mostly multipurpose, catering for flood-damage reduction, irrigation, hydroelectric power generation, domestic and industrial water supply, navigation, recharging of groundwater, conservation and improvement of soil and sediment abatement, low flow augmentation and water quality control, etc. The sequence of these uses is shown in Fig. 9.3.

The ability to plan and design new water resource facilities is directly linked to the ability to predict future water requirements. In the context of planning, the future could refer to years or decades depending on the particular problem. Demand forecasting is an estimation of water use in the future based on the previous water use, socioeconomic and climate parameters of the past and present water use, and projections of population and economic growth. Because of the size and capital intensiveness of most water resources projects, the time scale in water demand forecasting generally is 15–25 years for medium-range forecasting and about 50 years for long-range forecasting. According to UN (1976), the time horizon must be fixed in relation to periods that are characterized by the adoption of new technical means and methods by far-reaching social and economic changes or by a radical substitution of raw material resources. This will lead to the elaboration of a number of water resources development scenarios. It will be possible to select from these the most probable scenario, or that which is preferable by elimination.

The projections of future population are important inputs in planning because the quantities, such as the demand for water and energy, crucially depend on these. Most countries have specialized agencies to provide these projections. International organizations, such as the UN, also make these projections from time to time. Sometimes, estimates are made for different scenarios of population growth. For example, the population of India by the year 2050 is projected to be 1581 million in the high growth scenario and 1346 million in the low growth scenario (NCIWRD, 1999).

A forecast is an estimate of the future state of a variable that has four dimensions: quantity, quality, time, and space. In planning, the major factors determining the project cost are the quantity of water that must be stored, treated, and supplied and the quality of wastewater to be collected, treated, and

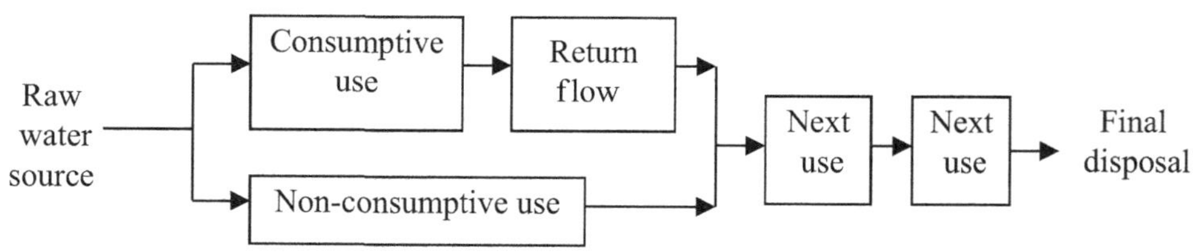

FIGURE 9.3

Steps in use of water resources.

disposed. Water demand and use exhibit hourly, daily, monthly, seasonal, and annual variations. The type, size, and timing of engineering facilities largely depend on these quantities. Forecasts of water demand should also reflect technological changes in production processes, product outputs, raw materials, and waste treatment methods, and public policies with respect to water use and development. An explicit inclusion of these factors is important in medium and long-range forecasts. Otherwise, forecasts would be of limited value to decision-makers. Therefore, simplistic methods, such as a linear extrapolation of past water demand (called projection), are generally not appropriate for long-term forecasting. Besides the magnitude, the variation of demands is also equally important to examine how far the various uses are compatible to each other.

A broad database is needed for forecasting and this should include gross and per capita demands, water use, effluents discharged with and without treatment, and the quality of sources of water. Additional data include water charges and effluent treatment cost. This information would be most suitable if it is available for each planning unit, say basin or subbasin. Jain et al. (2018) provide a brief review of flood forecasting techniques and their applications.

In what follows, water requirements for municipal and irrigation are described in detail. The water use for hydroelectric power is discussed in Chapter 10. Some industries, such as petroleum refining, chemicals and steel manufacturing, textiles, food processing, and pulp and paper mills, are such intensive water users that they must be identified and studied separately, the basic procedure remaining the same.

9.4.1 Municipal Water Use

Municipal water use can be divided into categories of residential (houses and apartments), commercial (businesses and stores), institutional (schools and hospitals), industrial, and other water use (park watering, swimming pools, fire-fighting). Unlike the water use in agriculture, where water is an input into a production system, municipal water use is mostly for meeting human needs. To the water delivered for these uses (or consumption) must be added the loss due to leakage from the distribution system to determine the amount of treated water. Now, adding the amount consumed by the treatment processes, the water required for the city is obtained.

Forecasting the municipal water demand is an important task for water utility agencies, involving three interrelated activities. The first activity is supply management which refers to forecasting the water demand so that new supply facilities can be designed, sequenced, and timed. A second interrelated activity is the demand management to determine the impact of water price changes, conservation measures, and rationing. The third activity is the collection of wastewater, its treatment, and disposal.

The total annual water use is a result of the combined effect of many factors. The per capita approach is commonly used to examine the relation between the total annual water use and population. Naturally, the annual water use would increase as the population grows. If such a relationship can be established, planners can predict the amount of water use for the anticipated population growth. Regression analysis is one of the most frequently used statistical techniques to estimate future water requirements. Here, the first task is to identify the factors that might affect the water use and to include the population size, price of water, average income, and annual precipitation.

The first step to establish such a relationship is to plot the population size (on the horizontal axis) versus the corresponding water use (on the vertical axis) in the form of a scatter diagram. The next step is to develop a mathematical function to describe this upward trend of water use with respect to the population size. It may be assumed that the water use, Q, is linearly related to population, POP, which can be approximated by the following equation:

$$Q = b_0 + b_1 \, POP + e \qquad (9.1)$$

in which b_0 is the intercept, b_1 is the slope of the line, and e is the error term denoting the discrepancy between the observed water use and that estimated by the straight line equation. As data points will not exactly fall on a straight line, the error (e) accounts for the failure of the proposed model to exactly fit the observed data. If *e* is zero, Eq. (9.3) is a deterministic model in which the water use (Q) is uniquely determined by population (POP). This is a simple linear regression model containing only one independent variable. A general extension of Eq. (9.1) involving more than one independent variable yields a multiple linear regression model which can be expressed as

$$Q = b_0 + b_1 \, x_1 + \ldots + b_k \, x_k + e \qquad (9.2)$$

Once the form of the model is finalized, the next phase of analysis is to estimate the regression coefficients. Regression analysis is an iterative process; the success in developing a reasonable model depends largely on the analyst's ability to interpret the resulting model and to correlate the model behavior with the process under investigation. The regression analysis has been discussed in Chapter 4.

Due to the ever-changing nature of social, economic, and political environments in a region, there are numerous uncertainties in any forecast. Errors in water use forecasts may arise from inappropriate assumptions made in determining the model parameters. If the population projections are too high, naturally the demand estimates will be high. An unreasonably high estimate of economic growth or too high an emphasis on past trends without a basic understanding of the reasons behind this trend and their sustainability will also produce a high estimate. Planners, due to their overenthusiasm in promoting the projects, may also adopt unreasonable values. Whatever the cause, errors in forecasting produce excess economic costs which may be avoided through the use of improved approaches. Additionally, improved methodologies for forecasting water demands are needed to account for: (1) growing number of conflicts among water uses and water users; (2) increasing realization of interrelationships among different outputs from WR systems; and (3) increasing scope and scale of WR development.

9.4.2 Water Requirements for Irrigation

Almost every plant process is directly or indirectly affected by the supply of water. Water constitutes about 80%–90% of most plant cells and tissues in which there is active metabolism. The crop water need (ET crop) is defined as the depth (or amount) of water needed to meet the water loss through evapotranspiration. In other words, it is the amount of water needed by the various crops to grow optimally (https://www.fao.org/3/s2022e/s2022e07.htm).

The three important factors for estimating irrigation requirements are as follows: evaporation, transpiration, and consumptive use. Transpiration is the process by which water vapor leaves plant life and enters the atmosphere. The term consumptive use includes transpiration, evaporation, or the water evaporating from the adjacent soil surface, the surface of plant leaves, and the water used by the plant for its metabolic activities. As the water used in the metabolic process is very small (less than 1% of the ET requirements), the terms consumptive use and actual ET are interchangeably used.

The three major factors that determine the amount and timings of the irrigation water are as follows: (1) crop water requirements, (2) availability of water, and (3) capacity of the soil root zone to store water. The water requirement of a crop is *the quantity of water required by a crop in a given period of time for its normal growth under field conditions.* It can be calculated by

$$WR = ET + AL + SR \tag{9.3}$$

where WR is the crop water requirements, ET is the evapotranspiration of the crop, and AL is the application losses, which include losses during conveyance and application in the field, and SR is the special requirements. As rainfall and moisture present in the root zone of soil may meet a part of the water requirement, the irrigation requirement (IR) can be calculated as follows:

$$IR = WR - (ER + S) \tag{9.4}$$

where ER is the effective rainfall, and S is the contribution of soil profile.

As described in the Irrigation and Drainage Paper 56 (Allen et al., 1998), the FAO has adopted the Penman–Monteith (PM) equation (named here *FAO56-PM*) as the standard technique to compute reference ET or ET_0. The PM equation is physically based because it attempts to incorporate the physiological and aerodynamic characteristics of the reference surface. The FAO has recommended the use of the PM method to compute reference ET from a grass surface and has specified a grass reference ET equation (Allen et al., 1998). The reference surface is a hypothetical grass reference crop with specific characteristics. ET_0 is affected by climatic parameters only and ET_0 is a climatic parameter that can be computed from weather data. ET_0 expresses the evaporating power of the atmosphere at a specific location and time of the year and does not consider crop characteristics and soil factors. A number of methods are available for its determination. These include the pan evaporation method, energy budget method, temperature-based methods, aerodynamic profile methods, and the combination method.

The FAO56-PM equation for the daily time step is stated as

$$ET_0 = \frac{0.408\Delta(R_n - G) + \gamma \dfrac{900}{T + 273} u_2(e_s - e_a)}{\Delta + \gamma(1 + 0.34u_2)} \tag{9.5}$$

where ET_0 = the grass reference ET in mm/day, Δ = the slope of vapor pressure versus temp curve in kPa/°C, R_n = the net radiation at the crop surface in MJ/m^2/day, G = the soil heat flux density in MJ/m^2/day, T = the mean daily air temp at 2-m height in °C, u_2 = the mean daily wind speed at 2-m height in m/s, e_s = the saturation vapor pressure in kPa, e_a = the actual vapor pressure in kPa, and γ = the psychrometric constant in kPa/°C. FAO has also provided an equation to compute ET_0 for hourly time steps (in mm/hour).

Actual ET can be obtained by multiplying PET with the appropriate crop coefficients that account for crop characteristics. The crop coefficient relates PET to ET of a disease-free crop grown in large fields under optimum soil water and fertility conditions and achieving full production potential under the given growing environment. The crop coefficients vary seasonally but an average value can be used for rough calculations. Allen et al. (1998) have given procedures for the selection of appropriate values that take into account crop characteristics, time of sowing, stage of crop development, and general climatic conditions. They have also discussed various aspects of crop water requirements in detail and have provided typical values of ET crop and crop coefficients for various climatic conditions.

FAO has programmed the methodology to estimate crop water requirements in a software named CROPWAT. It can be freely downloaded from https://www.fao.org/land-water/databases-and-software/cropwat/en/.

9.5 PLAN INITIATION AND PRELIMINARY PLANNING

This is the first stage of the water resources planning process. Usually, a plan for a water resources project is initiated to meet existing needs or those that are likely to arise in the future and this is followed by formulation of the plan. Sometimes these needs are clearly identifiable and sometimes not. For example, a project may be built to provide irrigation water in an area that is drought prone or economically backward. Utmost care and efforts are required to analyze, in the broadest possible terms, their real nature, since this is a very important phase of the planning process. There are no golden rules for the successful identification of needs. A beginning should be made with broad thinking. Extensive groundwork and contact with local people is necessary to clearly understand the aspirations and expectations of society. It is of utmost importance that the voice of the user for whom the project is being contemplated is properly heard and taken into account in planning. After all, the mission of the project is to serve the people, not the people involved in its planning. The planning team should strive hard to clearly understand the problem in the beginning rather than redefining it at later stages of the project planning. According to Grigg (1985), problem identification is one of the most important parts of any management process, especially problem-solving.

One of the biggest difficulties in WR planning is that the needs are sometimes not clearly identified in the initial stages. In some cases, one may have to proceed even though the needs have not been defined as well as they should be. This issue also poses a problem while formulating the objectives because the aims may not be clear and the project boundaries and the data needed to even initiate the work may not be available or nonexistent. There may be various groups with divergent interests lobbying to orient the development for their benefit. In a democratic set-up, elections to various bodies are the time when people commonly raise demands for various projects. At that time, the politicians are forced to look into them; they usually agree and make commitments. Obviously, many of these projects are not well thought of and only a few pass the initial screening. The screened projects are included in the list of appropriate bodies, depending on the scope and extent.

The levels of planning differ in character and scope from one country to another, but all of them require that water problems be formulated in the context of the overall economic and social aspirations of a given region or nation. The level at which decisions are taken at the first stage also varies. In some cases, there may not be any regional representation in the identification of plans and the local people may not be enthusiastic about the project. Hence, it is always good and necessary to have some representation of regional and subregional levels with its extent, depending on the disaggregation of problems. In a federal set-up, the decision-making level depends on the project size. Smaller projects are approved at a local level while bigger projects or those that involve interstate basins need clearance from progressively higher and usually some federal authority.

The main tasks under the problem formulation are determining the goals or objectives, deciding the investigations to be carried out, and chalking out the approach for analysis. The clarity of the problem formulation decreases as the scope increases. For example, a nationwide water resources plan will be less specific than a regional irrigation project.

In the first stage, the problem should not be viewed only from the perspective of water, but a broad view should be taken. For example, an area may have a shortage of electric power but this should not immediately mean that a hydroelectric power project needs to be constructed. A better solution might be a thermal power plant. The solution also depends on the attitudes and stage of development of the society. A society that is affluent, nature-loving, and conservation-oriented may not be willing to construct a dam to meet water requirements. However, such concerns will not be prominent in another situation, for example, when a dam contributes to the water supply to a drought-prone area or the production of badly needed food and fiber. However, local concerns are not always the deciding factors. At times, vocal and influential outsiders are able to stall or even thwart a project, may be out of conviction or due to other interests.

Translation of needs into problem formulation is an iterative process. It undergoes changes in time and becomes more and more focused and pinpointed with time. The problem formulation depends on the nature and scope of the problem, the planning level, various constraints (technical, economic, political, etc.) that must be taken into account, and above all, the project objectives. The problem formulation is also subject to several constraints. Administrative and hydrologic boundaries rarely intersect, time and funds allocation for the problem are often limited, and various regulations narrow the range of planning options. Moreover, skilled professional personnel may not be available. These are some of the constraints that influence the problem formulation. The appropriate consideration and resolution of these constraints is necessary to search for and implement a viable solution.

Once the preliminary formulation of a plan nears completion, a clear statement of the project objectives should emerge. Such a statement cannot be prepared without detailed data about the cost of achieving each of these objectives. Various alternatives to meet each objective should be considered and the adverse or beneficial impacts of each decision on other objectives should also be considered. It is necessary to have realistic information about the cost of meeting each objective and the benefits likely to arise. For example, in many projects, faced with opposition, the promoters began quoting those objectives also which were not initially part of the project or highlighted those objectives that were a minor part of the project but had higher emotional value. This was done to buttress the justification for the project but caused confusion and problems later on.

9.5.1 Dependency of Water Sector Plan on Other Sectors

For this discussion, "other" implies nonwater sectors, for example, energy, transportation, rural development, poverty alleviation, etc. Every development scheme that has a component concerned with water should not be considered to be a WR project. Unless the solution has substantial control and management of water, the project need not be in the water domain. Therefore, all competing proposals of other nonwater sectors of the economy should be considered before the formulation of a water plan. The issues that may arise at this stage typically are hydropower versus other sources of power development, navigation versus rail/road transportation. This also requires studying substitution and trade-offs. The interface and interdependence of water and related land resources with other sectors of the economy should be recognized in a preliminary project formulation. Water resources planners may have inadequate information concerning other sectors of the economy. For example, the assumptions about the future cropping patterns are frequently not realistic because farmers follow that particular cropping schedule which is the best from their point of view rather than from the project viewpoint.

The extent to which water resources development is treated as an individual sector of the national economy depends on the major income-generating sectors of the economy. For example, in India, where agriculture was the main occupation till 3–4 decades ago, the resource allocation for the water sector was made under the broad heading of irrigation. In fact, the nodal ministry in the water sector was known as the Ministry of Irrigation. Recently, this Ministry has been named as Ministry of Water Resources but the allocations under the national plan continue to be under the heading of irrigation. The responsibility for the management of water resources is shared among a number of ministries, for example, Ministry of Agriculture, Ministry of Energy, and Ministry of Rural Development. Many times, there is considerable overlap in the responsibilities of these ministries resulting in confusion and suboptimal WR management.

9.5.2 Articulation of Project Objectives

The goals and objectives are stated differently at various planning levels. The ones at the national level tend to be global (e.g., to enhance national economic development, social well-being, and regional economic development). Moreover, they do not detail the conflicting issues. They are intentionally as encompassing and as comprehensive as possible to ensure broad support by various constituencies and stakeholders. After an agreement about the general project objectives is reached, more focus should be centered on specific objectives and their translation into design criteria. These criteria require the definition of measures that will be used to assess the degree to which individual objectives have been met.

One of the most important parts of preliminary plan formulation is a clear statement of project objectives. In most practical situations, objectives cannot be taken as given. A difficulty may arise because the objectives may be conflicting and the alternative means of satisfying any one objective may produce substantial adverse effects on another. It is usually impossible to appropriately define objectives without having detailed information about the feasibility and cost of achieving them. Only a rigorous quantitative analysis can indicate whether a particular objective is feasible or not and how much it will cost to achieve it. Such analysis and ultimate choice of socially relevant project objectives requires judgment on the part of the water resources planner as well as other participants in the planning process, for example, the politician. The perception of project objectives by the public at large and other constituencies is also equally important.

The specific project objectives usually coincide with some water management aspects, such as water supply, protection against floods, hydropower production, and development of navigation. An analysis of project objectives shows that depending on the character and the scope, objectives can be stated in very different ways. While formulating objectives, many assumptions are also to be made. This is a difficult task and many times the planners make such assumptions that help justify the project. An important and crucial assumption in irrigation projects is about the cropping pattern in the command area. An analysis of the existing projects shows that in many cases, the actual cropping pattern is completely different from the one assumed in the planning stage. Before the project, the farmers were growing crops that were best suited as per the water availability at that time. However, when adequate water became available, they switched to the crops that gave them the best returns.

9.5.3 Project Constraints

The enumeration of project objectives and constraints go together. The constraints are helpful in the sense that they restrict the feasible alternatives and reduce their number. However, from an evaluation point of view, constraints often have a function similar to objectives. Reasons like topography and dense population make some projects so narrowly constrained that only a few feasible options are left.

All constraints on the projects should be clearly identified and listed at the initial stage. At this stage, the constraints should not be viewed as an absolute restriction and these need not be treated as inviolable. Of course, some constraints depend on physical factors, for example, water available at a particular site and these cannot be violated. A problem may arise when there are differing views of the experts on the subject particularly if adequate data are not available. For example, different experts may arrive at different yields at the given site and the planner will be at a loss while choosing the right number. Such types of problems, however, can be minimized if detailed data are available. Although the constraints considered in water resources planning vary widely, most attention is paid to technical and economic constraints. Also, important and often overlooked or underestimated are the institutional and cultural constraints that rule out certain project alternatives. In general, all constraints should be explicitly specified and open to debate in the plan initiation phase to avoid controversies that may surface at later stages. Often planners may also be constrained to use only the existing data, irrespective of how inadequate and unreliable the database is.

The availability of funds clearly imposes a constraint on the magnitude of development. However, such constraints can be overcome to various degrees and should not be viewed as sacrosanct. There are instances when the political decision-makers decided to take up a project even though the full funding was not arranged. The perception of funding agencies also changes with time due to developments in the global economy, the progress of the project, and the views of influential groups.

The earlier discussion suggests that the constraints must be scrutinized from many points of view as analysis proceeds and technical possibilities emerge, and their roles may be subject to change. Some constraints are permanent and can never be violated, while others are binding in the short run and may be changed with the passage of time or removed by invention or technological improvement. However, irrespective of the nature of particular constraints, the systems analyst should examine their influence on the marginal cost and project outcomes. Analyses should also bring out the consequences of violating or relaxing a constraint.

9.5.4 Planning for Operation

Planning for operation, which generally leads to the generation of operational rules for the project, is an important step in the planning process. It is also the most intensive systems analysis step. Although the operation policies are developed for the entire planning horizon at this stage, they are not followed in reality. These policies considerably change when the real-life operation commences. However, they serve the following important objectives in the planning process:

(a) Provide an analytical mechanism with which to develop design criteria and thus optimize the project design.
(b) Enable the planner to better understand the couplings among the various subsystems (reservoirs, rivers, groundwater systems, etc.) and consequently account for the system constraints and attributes.
(c) Enable the agencies responsible for operating the project to initiate contractual agreements with, for example, electric power or water supply utilities.
(d) They lead to the identification of major gaps in data, if any. If needed, a new data collection program can be started.

(e) Help uncover early signs of conflicts with other agencies and/or water resources operating entities. In this case, a process of negotiation may be initiated and/or some of the project design may be altered to accommodate these newly discovered institutional or organizational constraints.
(f) Provide a useful training medium for those who will be responsible to operate and manage the project when it is completed.
(g) Assist in the development of an appropriate cost-sharing formula for the project.

These and other objectives associated with operational rules dictate that the planning team should develop a reasonable planning-for-operation policy as part of the planning process.

9.5.5 Conjunctive Use Planning

Integrated WR planning must deal with the availability of water from all sources, the use to which water is to be put in the best possible manner (including the amount and timing), the various impacts of uses on its quality and on the environment, constraints in water development and management, and socioeconomic needs. One option to solve water resources problems is the conjunctive use of water, which is the coordinated management of surface water and groundwater. It should be attempted whenever possible.

Various advantages of conjunctive use are worth mentioning here. Operation of both surface and groundwater reservoirs provides for larger water storage and hence greater water conservation. Greater utilization of groundwater leads to smaller surface distribution systems. As the pumping well would act as a vertical drainage and aid in controlling the water table, a basin where conjunctive use is practiced would require a small drainage system. In conjunctive use planning, canal lining can be reduced, as seepage from canals provides recharge to groundwater. Surface reservoirs can store water during the period of excess flows to control floods. This water can be released later for artificial recharge. The conjunctive use leads to lesser evapotranspiration loss. Surface water is a source of electric energy and this energy can be transmitted to a faraway place and used to withdraw groundwater.

The problem of selecting the best strategy for conjunctive use of surface and groundwater in a complex system where conflicting interests compete for limited natural and financial resources can be solved by the systems approach. Therefore, the systems approach is being increasingly used to solve various problems associated with conjunctive use planning more so with the advent of digital computers. Basically, the problems have been solved in two frameworks: optimization and simulation. The topic of conjunctive use planning has been the theme of innumerable technical papers starting with the classic work of Buras (1963, 1972) and Rogers and Smith (1970). Willis and Yeh (1987) also provide a detailed discussion on this topic. Singha (2014) has presented an overview on issues and methods of the conjunctive use of surface water and groundwater resources for sustainable irrigated agriculture.

9.6 INSTITUTIONAL SET-UP

An administrative structure that guarantees careful operation and maintenance of completed systems and that has sufficient flexibility to adjust to changing needs is necessary for efficient water resources planning and operation. The establishment of a well-functioning water administration with strong powers of regulation and a well-trained maintenance staff hardly requires emphasis. Countries, that have a well-developed WR administration that has evolved with time and experience, have been able to reap large benefits. The organizations that are involved in water management commonly belong to the federal government, state government, and local/municipal authorities. In addition, there are river basin authorities, private utilities (for water, sanitation, and energy), nongovernmental organizations (NGOs), and water user associations/societies.

The organization that is commonly assigned the responsibility to initiate planning for a new WR project is often an existing institution or group. Most commonly, an existing structure fits the objectives of the project; in rare cases, a new planning entity may be created. If a national agency charged with water resources planning is already in existence, it may, in all probabilities, take the lead and prepare the plan. There will be a local set-up for a small-scale project.

To ensure a sufficiently comprehensive plan and to evaluate several project options, a mix of agencies is preferable. This is because government agencies in many countries have developed a specific mission over the years. One of them, usually the biggest, is entrusted with the leadership and coordination responsibilities. Typically, this will be a water resources or an irrigation department. There might be a federal organization with a clear mandate to technically examine and approve all the projects before they are included in the national plans. Sometimes, more than one such organization examines the project from specific angles, say water resources, environment, pollution, etc. The involvement of external experts from academic or research institutes in the planning process is quite common; specific research projects to study important aspects may also be occasionally funded.

9.6.1 Top-Down and Bottom-Up Planning

Two general approaches are followed in planning and management activities: the top-down approach and the bottom-up approach.

In the top-down approach, the various activities related to plan preparation are carried out by expert government institutes, staff therein, and consultants. These experts/institutes carry out all the steps necessary for the preparation of water resources development plans and prepare the required reports/drawings. It is assumed that these people have the required background and understanding of the issues and tools to prepare the plans. In bottom-up planning, the concerned citizens work with the government institutes and/or NGOs for the preparation of plans. Experience shows that the plans developed primarily by professionals without significant citizen involvement may be technically flawless, but they have little chance of success if they do not take into consideration the concerns of affected local stakeholders and do not have their support (Loucks and van Beek, 2017).

These days, most WRD plans are prepared by the significant involvement of various stakeholders at different stages. When the stakeholders participate in making plans, they have a sense of ownership of the projects which means that the resistance is less and they also make efforts to see that the plans work. Many times, the public comes up with useful ideas that result in more efficient plans. However, to obtain useful input from the public, they need to be properly informed. This topic is discussed in detail in the next section.

9.6.2 Involvement of Experts

The multidisciplinary nature of water resources planning necessitates that interdisciplinary interaction should take place at each stage. The contributions of experts from different fields are necessary to ensure the best results. Usually, the results are more multidisciplinary than interdisciplinary, meaning that although there is interaction, it tends to take on the shape of a presentation of results by individual experts as seen in the light of their own expertise.

The major group of experts involved in the plan initiation and preliminary planning phase consists of systems analysts. They must interact with other groups who have expertize in municipal and industrial water supply, irrigation, hydropower production, forestry, etc. It is crucial that the systems analysts and disciplinary experts fully understand project purposes and objectives. This is especially important when the disciplinary experts identify constraints that reduce the modeling freedom of the systems analyst. It is also important to consult experts from other fields, such as lawyers, biologists who investigate rare species, archaeologists, etc. to ensure that these issues are also suitably addressed.

The locally available expertize and the scope of the project influence the selection of experts. Since water projects are closely linked to social and economic life, the consultants should be familiar with these facets of society and not impose alien practices.

9.6.3 Decision-Making Levels

Many water resource projects are large, and huge sums of public money are invested. The allocation of money for these projects depends upon the needs of the other sectors of the economy. Therefore, decision-making chains are long, and final decisions are made at political levels. Natural disasters, like floods or droughts, often provide a push for initiation, formulation, approval, and implementation of water resources projects.

The basis for a decision on a WR project is a detailed report in which the project objectives are outlined as well as the means by which they are to be accomplished and associated costs. The consequences of the project in terms of benefits and adverse impacts are also detailed. WR planning is the sum of all activities that lead to such a detailed project report. The larger the project and the more intensive the use of WR, the broader becomes the scope of the planning process. It is, therefore, necessary to evolve a hierarchy of levels for water resources planning, beginning at a level where all possible projects are considered in the context of a general national master plan. A national water plan must spawn subplans, which cover more details for a smaller area.

Typical of such a planning cascade is a division into various levels. Different countries have different planning levels, but in general one can identify three levels, and these are often associated with different planning authorities.

(i) Level A is a reconnaissance study or a general framework study. The temporal horizon is about 30–50 years. The purpose is to identify major current or prospective problems. The geographical coverage is generally very large. This level may involve international agreements on the allocation of water from a river that flows through two or more countries. These agreements are reached on the basis of water resources development as well as many other national interests, including strategic aspects. The decisions on a national or international level are of great consequence as they set the strategy for development.

(ii) Level B is a comprehensive planning effort for a smaller region. This level should follow Level A, where problems have already been identified. The time horizon is about 15 years. The purpose of water resources planning on this level is to set priorities for the long-term development of a country. Its decision level is largely political and involves technical inputs only on a limited scale, usually as financial data or constraints. Often, this is the level at which the decision whether to proceed with the planning for a project or not is taken and funds earmarked.

(iii) Level C is implementation planning, where specific project designs are developed. Generally, Level C should follow Level B, because specific plans or recommendations from the Level B effort are implemented here. This level is regional; its results are incorporated into a regional water plan that identifies WR projects within the context of different requirements imposed by alternative development plans of a region. The objective of such a study is to set priorities and make recommendations for the allocation of WR to different users.

9.6.4 Compatibility Among Agencies

Even though cooperation with agencies during the planning process is essential, this will not necessarily lead to a coordinated output. All the agencies may not feel the need for cooperation. There may be interagency or personal disagreements or dislikes. Usually, there is no clearly identified person for coordination and it is common to see that different persons or a person without much interest, background, responsibility, or authority is assigned the duty of coordination. It is also important that a water resources plan does not get "drowned" by the activities of another agency's project, for example, a water quality improvement plan poisoned by the diversion of polluted flows by another agency. Although these things look very simple and logical, there may be complications. Frequent interaction and coordination are the best ways to avoid waste of time and resources and save effort.

The possible ways to enhance cooperation include jointly working on problems, mutual visits, and exchange of personnel. With the developments in information technology, it has become much easier to share and exchange data and documents. Although it is important to put in place a proper mechanism, the actual coordination depends on persons—unless the concerned persons are enthusiastic about it, the things just do not work.

9.6.5 Capacity Building

This is not part of the planning process but has its relevance and importance for the overall development. The shortage of trained manpower is a major hindrance in applying improved technology, particularly in developing countries. This shortage can be overcome in two ways. As an immediate solution, some funds can be earmarked to train a few personnel on the techniques that will be used in the analysis. These people may also impart training to their coworkers. The second and long-term solution is to initiate advanced courses in local universities that can be taken by the prospective candidates. At the same time, job positions at an appropriate level with commensurate benefits should be created in the planning department. Another option is to hire experts for the training of needed personnel.

A hierarchy of objectives of education programs for water resources engineers, planners, and managers was presented by Dyck (1990). A number of sample course structures of different durations based on the background of the target audience have also been presented in this publication. In a training program for young water professionals in India, situation understanding and improvement (SUIP) concept was applied, combined with mentoring and experiential learning (Maheshwari et al., 2023). This novel experiment was found to benefit the participants' professional outcomes significantly.

9.7 PUBLIC INVOLVEMENT

Evaluations of many past development projects have shown that poor identification of the needs of local communities and inadequate assessment of the social impacts is a key reason for project failures. This experience has forced planners to search for ways to improve projects. An important remedy is a more rigorous preproject analysis of social and cultural conditions and more interaction with people and consulting local communities during project design and implementation in "open" planning. Furthermore, local residents with deep knowledge about the area can sometimes provide better inputs than "an external planner." At the same time, local communities may not be able to appreciate the new ideas unless they understand the thought process that has generated these ideas. Thus, there is a necessity for a framework within which all stakeholders can access and analyze information, establish priorities, and develop plans. The United Nations (UN, 1997) has laid special emphasis on public involvement in WR projects. Since WR projects are capital intensive, it is common to approach international financing and donor agencies for funding. The major financing agencies use the term *Public Involvement* to denote the activities through which the concerned stakeholders are integrated into the decision-making process.

Public involvement (PI), also termed as public participation, is the process through which the views of all interested parties (stakeholders) are integrated into project decision-making. Here *public* refers to persons or groups having an interest in the project. Public participation refers to the involvement of such individuals or groups in decision-making or trying to influence decisions. The basic premise behind such participation is that the officers of a governmental agency or the engineers of a water authority may not know fully well what the public wants and what is best for them. An important implication of this finding is that public participation should not be limited to just ascertaining different views. It should also make the project-affected people conversant about the decisions being made and their implications on their life and environment. At the same time, it is important to ensure that expression and consideration of public viewpoints do not improperly impede the decision-making process. While coordinating public participation, planners should be aware of (1) the limitations inherent in public participation, (2) the requirements that must be met to ensure adequate anticipation, and (3) how that participation should be structured.

The term *stakeholders* includes all individuals and groups with an interest in a project. They can be divided into four categories: government agencies, directly affected parties, indirectly affected parties, and other parties (including NGOs). Some stakeholders are easily identified; to find others, field visits, studies, and discussions are necessary.

In several countries, WR planning and management is the responsibility of specialized agencies that represent the interests of all water users. The public is indirectly represented in the planning process through their political representatives who are members of government administrative agencies. Except through politicians and nongovernmental organizations, generally there is no worthwhile public participation in WR planning activities in many countries. In some cases, this has led to serious problems during the execution of projects. It is necessary to acquire land for reservoir and canal projects; there may also be displacement of population due to land inundation. A large-scale displacement disturbs the social setting and usually old people have a nostalgic attachment with the land. There might be resentment against the project among the project-affected people if they feel that they have not been adequately compensated for. Public participation in project planning is helpful largely in overcoming such problems, although it does not always work that way. In many countries, public involvement in a project is ensured through public hearings where planners present and discuss their plans to the general public.

The main aim of PI is to create openness and dialogue from the outset of a project to improve decision-making. It is important to note that development should be based on a process of sharing knowledge and values, rather than attempting to impose new values. Also, there are no prescriptive methods for involving the public in decision-making. While the underlying principles of PI are applicable to all countries (and all natural resources projects), the mechanism and degree to which the public is permitted to participate will vary considerably from one society to another (according to sociopolitical and cultural context) and from one project to another.

Although PI may appear to be time consuming and costly at first, the long-term benefits far exceed initial costs. Nevertheless, it is an iterative (repetitive) and flexible process that should, ideally, take place throughout the lifetime of a project. As far as possible, the PI process should be integrated into project planning, and in particular into the environmental assessment process. This process should start at the earliest stage possible in decision-making; better even when the project is being conceived. As things are, it would be wonderful if all stakeholders accept PI as the normal way in which to plan and implement projects.

9.7.1 Advantages of Public Involvement

The concept of PI has evolved over time through learning and experience. The key advantages of public involvement in decision-making are:

- PI reduces the risk of project failure by improving the quality of planning and decision-making.
- By bringing a diverse range of values and opinions to the table, PI can improve problem-solving.
- PI helps in the development of a feeling of partnership with local communities. It thereby overcomes the local resistance and provides a conducive work environment.
- PI significantly reduces conflicts between individuals, groups, and organizations.
- PI helps in improving the project performance by using the technical expertize of the public.
- The poorer the people, or the scarcer the resource, the more important it is that local communities take part in project planning and decision-making, because their survival and well-being may critically depend on it. In some cases, the project in question may be the most crucial chance of development for a generation or more.

Specifically, the benefits that PI can bring to all stakeholder groups, including the government, developers, and the affected parties, are:

Advantages to the Government

- Increased credibility, legitimacy, and positive image through transparent decision-making, particularly when decisions are controversial.
- Improved coordination between governmental departments as per the needs of the PI process.
- Higher level of commitment of all stakeholders to decisions made.
- Reduced risk of serious confrontation, thereby minimizing project costs and delays.
- Development of a sense of belonging and responsibility among local communities toward projects.

Advantages to Financing Agencies

- Realistic information about the needs and preferences of local communities.
- Better project database right from the early planning.
- Improved technical design of projects, thereby reducing costs.
- Increased market share by virtue of positive image.

Advantages to Affected Parties

- Improved understanding of the project and its impact on their lives.
- A project that meets their actual needs.
- Higher chances of success of the project and thereby improvement in the living standards.
- A platform for local communities to voice their concerns at all levels of government.
- Better targeting of benefits.
- Increased levels of accountability of government and developer to local communities.

9.7.2 Activities in Public Involvement Process

The PI includes a number of activities where each step is a prerequisite and leads to the next. Considerable preplanning is needed and PI has to be carefully handled to gain the most out of it. For example, one cannot have participation without consultation and consultation cannot occur without information dissemination. Goodman (1984) classified the various public participation and education techniques into five categories. These are large group meetings, small group meetings, organizational approach, media, and community interaction. He analyzed 23 PI techniques and discussed the advantages and disadvantages of each of these.

The four main activities of PI are enumerated:

1. *Information gathering:* It is a systematic analysis of existing social, cultural, and economic conditions about directly affected groups of stakeholders (such as farmers or indigenous minorities). Data are obtained through surveys, questionnaire, site visits, and polls. These are analyzed to identify key issues, key people and organizations, and their level of interest.
2. *Information dissemination:* It is a process of providing information about a project to stakeholders. A variety of tools can be used depending on-site conditions, level of literacy, infrastructure, and attitudes of the communities: electronic and print media, exhibitions, conferences, and seminars. Trained field staff is useful for this purpose. The public should be kept informed regularly—during active periods as well as during "quiet times." To the extent possible, the information should be disseminated in local language without too much technical jargon. Clearly, the stakeholders can participate and be useful only if they are fully informed.
3. *Consultation:* This is the step in which decision-makers listen to the views of other stakeholders to improve project design before implementation or to make necessary changes during implementation. Consultation should involve the government, affected parties, lending agencies, and NGOs through workshops, round table discussions, and seminars, preferably at a project site. It is important to classify stakeholders because not all of them will interact at the same level. Irrespective of their knowledge and skills, all the members of the public should be treated with dignity and respect.

4. *Participation:* This is basically an extension of consultation where directly affected groups become joint partners in the design and implementation of projects. They participate in "making" decisions. It is a good idea to involve district/county-level authorities in this effort. Finally, the public should be informed about the decision and reasons thereof.

It is important that all stakeholders, including those who may be affected unknowingly, are involved and given a fair opportunity to influence the design and implementation of a project. Indirectly affected parties (often poor or marginalized communities) are the most difficult to identify and involve in the PI process. Planners should take a proactive role in facilitating and monitoring the PI process from early consultation to the post-decision period. Needless to say, even the best techniques fail, unless the agency responsible for overseeing the PI processes has the required sincerity, integrity, and commitment.

In some countries, public hearings and public participation are required by law before the final approval of any major WR project that involves public funds. These public hearings have a great advantage because the public concerns, objections, and views (other than those of directly interested agencies) are heard, and subsequent modifications are often incorporated into these plans. There is a need to systematize this participation to the extent possible and integrate it with the entire planning and screening process. The development (and design) of questionnaires that can articulate public preferences in a cogent way is also important. Also, a preliminary education of the public on the issues at stake and the preparatory steps for public hearings and evaluation of these alternative plans should be planned well in advance.

Detailed guidelines for public participation have been prepared by many countries, for example, the United States, and the United Kingdom (https://www.gov.uk/government/publications/water-resources-planning-guideline/water-resources-planning-guideline, accessed on July 22, 2022), India, and European Union; funding agencies, for example, The World Bank and the Asian Development Bank (ADB).

Examples

(i) It was reported by UN (1997) that during the early 1980s, the National Irrigation Administration (NIA) in the Philippines changed its approach from the construction of large-scale irrigation systems to assisting in the development of small-scale community systems. A program of new projects was initiated, building on indigenous methods and improving these methods via participatory projects with local farmers. Additionally, NIA embarked upon an extensive training program for farmers and its own staff and assisted farmers in organizing themselves into formal irrigation associations. By the mid-1980s, in participatory irrigation schemes, crop yields were generally 20% higher than in nonparticipatory schemes, and farmers' satisfaction with the canals and structures was considerably higher. The levels of cost recovery were between 5 and 7 times higher, and water distribution between farmers was more equitable.

(ii) A Brazilian water and sanitation project for low-income communities involved communities and was able to design systems at a cost of US$ 50 per capita, well within the government-stated ceiling of US$ 120 per capita. Both the funding agencies and the government benefited from PI.

(iii) In India, some states are encouraging the transfer of responsibilities of management of irrigation systems to Water User's Associations. The state of Maharashtra is one of the leaders in this aspect. The results of case studies have shown that the transfer of rights and responsibilities has led to improved performance in terms of maintenance of structures, equity in water distribution, and higher productivity. Simultaneously, the adoption of volumetric pricing has resulted in higher water use efficiency. The recovery of water use charges is also better.

9.8 FORMULATION AND SCREENING OF ALTERNATIVES

This is the third stage in the water resources planning process. The main activities at this stage are as follows: classification of project alternatives, the actual generation of project alternatives, and screening of the project alternatives. The outcome after this stage is the formulation of selected alternative projects and the evaluation of their relative advantages and disadvantages.

9.8.1 Classification of Alternatives

In the planning process, all plausible project alternatives should be considered: feasible and infeasible, structural and nonstructural, and water and nonwater. Although some may view the study of infeasible alternatives as wasteful, important and valuable information might be gained from such an effort. For example, a sensible measure or plan that happens to be at the time politically or institutionally infeasible can shed light on the cost associated with existing institutional impediments and might indicate specific ways for removing or alleviating such obstacles. Nonwater alternatives often constitute an integral part of a water alternative package; for example, land transportation might be considered as an alternative to navigation. Technical constraints also govern the selection of alternatives. For example, a dam site might have an excellent rock foundation but would require major work in relocating people or rerouting transportation routes, while another location might require extensive foundation work.

Furthermore, it must be realized that there are often different alternatives to accomplish the same objective. For example, flood protection can be achieved by retention structures, flood levees, or zoning to prevent settlements in flood-prone areas. Similarly, alternatives for water supply include the use of ground or surface water or both. Hydropower generation should be considered within a broader economic scale, with nuclear or fossil-fire generating units considered as part of the system. Such considerations commonly lead to the use of pumped storage plants, where excess energy during low-consumption periods is used to pump water into a temporary storage at a high elevation, from which it is released through turbines to generate energy during peak hours.

9.8.2 Generation of Alternatives

Stage 3 of the planning process involves the formulation and screening of project alternatives. Depending on the number of alternatives to be examined, there are several possibilities for their screening. These are as follows:

(i) If the number of alternatives is small, the screening step can be eliminated.
(ii) If there are many but not too many alternatives, it is helpful to systematically screen them using mathematical models.
(iii) If the number of alternatives is large, some type of hierarchical screening in stages is helpful. An increasing rigidity of selection and/or exclusion criteria can be adopted as screening proceeds till a small number of alternatives remain.

Stage 3 has a considerable overlap with stage 4 (development of final study results). Therefore, ideally, the same personnel should be involved. The planner who is engaged in the generation and analysis of alternative projects should carefully select the decision variables and their feasible range. Alternatives are generated by selecting the various possible sites for projects and then various sizes and configurations of the components. The planner and the decision-maker also decide which of the systems' objectives should be kept as such and which should be considered as constraints.

9.8.3 Techniques for Screening Alternatives

There are two principal options to deal with the screening problem. In the first option, alternatives are screened using some mathematical technique or judgment based on prior experience; a simplified representation of the system is chosen. Application of LP as a screening tool at this stage is a fairly common practice. In the second option, a set of potential alternatives is developed (often based on experience or experts' recommendations), and these are then evaluated and ranked according to some criteria. In both options, a single criterion or a multitude of criteria may be employed. Screening of alternative plans is an iterative process. The techniques that are commonly used for screening have been described in Chapter 5.

Screening techniques may range anywhere between the rule of thumb and formal analysis, depending on the type of the problem and the level of screening at which ranking procedures are used. By means of optimization methods, a large number of alternatives can be evaluated, but while using these models, a detailed description of all the alternatives is not possible. By contrast, a simulation model allows for a very detailed description of the system, but only a few alternatives can be investigated due to limitations of time and other logistics. The choice of the right systems analysis tool depends on the individual problem and the preference of the analyst.

9.8.4 Evaluation of Alternatives and Finalization

As planning deals with the future, the planners have to make predictions about the conditions when the project will be operational. The future cannot be predicted with certainty and there is always an element of uncertainty. In addition, decision-makers must analyze the various alternatives that often involve conflicting economic, societal, environmental, and political forces. With ever-increasing public awareness and influence in the decision-making process, the planning task has become even harder. The main steps of the evaluation process are as follows:

To elaborate and quantify:

- the scope and extent of the system,
- the structural and nonstructural components of the system,
- constraints—topographical, hydrological, structural, financial, institutional, etc.
- target demands for various purposes, for example, irrigation, municipal and industrial, hydropower, water quality maintenance, recreation, etc., and
- flood management aspects, if any.

To analyze:

- evaluation of alternative plans that have been identified at earlier stages,

- economic analysis, and
- assessment of environmental impact of alternative plans.

The prefeasibility study is considered to be over when the potential alternative plans that are suitable for implementation have been selected. In many countries, a public hearing is held at this stage to explain the plans and elicit public views and comments. The decision-makers at various levels also participate in the discussions. Based on the outcome of these, the plans are suitably modified. The planning agency may also be asked to carry out additional or more elaborate analyses during the scrutiny of the project proposal. This process culminates in the decision to continue or discontinue the planning activity. This decision is usually taken at the political level.

9.9 MODELS FOR WATER RESOURCES PLANNING

Because of the complexity of the issues involved in WR planning and because of the significant impacts of such projects on the environment and regional/national economy, appropriate planning methods must be employed that can handle the problems satisfactorily. The systems analysis approach is a set of tools for the analysis of water resources projects. In systems analysis, the system and its components are described by means of mathematical models. The objective of the analysis during planning is to find the system design with the best possible combination of elements to meet the desired objective.

A model is a simplified representation of the real system and the utility of modeling results depends on how well the modeler is able to perceive actual relationships and capture them in the model. While physical models are mostly used to design various physical components of a project, for example, spillway, mathematical models are useful in analysis, such as hydrologic, economic, and impact assessment. Mathematical models are also very handy in evaluating the consequences of alternative plans. Clearly, one cannot conduct experiments on a prototype project. The use of models is often less expensive and convenient than conducting comprehensive surveys or other conventional approaches.

Mathematical models describe the physical processes, such as the movement of a flood wave in a river channel in a simplified manner through arithmetic and logical statements. These models have assumed a unique role in the planning and management of water resources. Models are currently used to investigate virtually every type of WR problems for small- and large-scale studies and projects and at all levels of decision-making. The use of models has made it feasible to quantitatively compare the likely effects of alternative decisions. At the cost of redundancy, it needs to be stressed that systems engineering is not a substitute for the decision-making process. Rather, it is a set of tools that help with these tasks.

9.9.1 System Decomposition

A necessary condition for the successful use of systems methodologies for WR planning is the ability to develop a model that takes care of various objectives, constraints, and input–output relationships of the system being modeled. Only if this condition is met, the results of the model will be meaningful and implementable.

A real-life WR system possesses most of the following characteristics:

(i) multiple noncommensurable objectives as well as multiple decision-makers;
(ii) a large number of variables and parameters;
(iii) a large spatial and temporal database;
(iv) a large number of subsystems; and
(v) relations among variables that are nonlinear, space and time-dependent, and stochastic.

Working under such a complex situation, the analyst may be bogged down in the plethora of models, analyses, and results. With this motivation, the concept of the hierarchical approach was presented by Haimes (1977) and is based on the decomposition of large-scale and complex systems and subsequent modeling of the system into "independent" subsystems. This decentralized approach utilizes the concepts of levels and layers and enables the modeler to analyze the behavior of subsystems at a lower level. The results are then transmitted to higher levels. When applying the hierarchical approach to water resources systems, several combinations of hierarchical structures are possible. The four major decompositions are as follows.

Temporal Decomposition

The planning horizon for water resources projects is of the order of 100 years. However, the conditions of the system change with time; the storage capacity of a reservoir changes with time; the development of a command area proceeds with time; the area to be protected from floods undergoes changes; and so on. Hence, an intermediate term of 10–15 years, often referred to as planning-for-operation, is usually embedded in this long-term planning. There may be another short term of 2–5 years further nested into it. All these plans have to be compatible with each other and well coordinated as they relate to the same system.

Physical-Hydrological Description

A river basin is the natural unit for planning and operation of water resources projects although administrative units are adopted in some cases. A regional water resources development program may cover several river basins. Each basin can be further divided into subbasins and so on. The topic of river basin modeling has been discussed in Chapter 14.

Political-Geographical Description

More than 200 river basins in the world are transboundary basins, that is, they fall under the territory of two or more nations. Within the same nation also, the project area may fall under different political or administrative units, for example, state, district, or county. The analyst may consider either political or natural boundary as a criterion for decomposition.

Goal-Oriented or Functional Description

Water resources systems can be analyzed with respect to their economic and functional goals. The models following this pattern typically are demand and supply models and models for hydroelectric power generation, irrigation, municipal and industrial use, etc.

9.9.2 Selection of Systems Analysis Tools

Although a spectrum of systems analysis tools, varying from very sophisticated to simple, is now available, the choice should be to go for "appropriate" technology. In other words, the technique chosen should be commensurate with available data, equipment, expertise, and site conditions. According to Miser (1982), analytic tools should be chosen considering six principles:

1. They should be appropriate to the problem and to the prospective solutions that may emerge.
2. They should match appropriately the available data. A method may be very attractive but if it requires data that are not available for the focus system, it cannot yield trustworthy results.
3. They should be internally consistent (the sophisticated analysis of one part should not be bludgeoned by hazy speculation in another).
4. They should be balanced in detail and accuracy (if one enters with order-of-magnitude estimates, one is seldom entitled to five—figure accuracy in the results, or if accurate estimates are combined with very questionable estimates, this fact should be reflected in how the results are presented).
5. They should be appropriately interdisciplinary in the light of an appreciation of the problem with which the work began and is being continued.
6. They should be appropriate, if at all possible, to the process of presenting the findings that will emerge at the end of the planning study (the client will surely not want to poke into details, but some understanding of the analytic tools has a persuasive value for many users of systems analysis results).

Generally, the techniques of analysis for a particular project are not completely chosen in the beginning nor they are finalized in one go. The selection also depends on the preferences of the analyst and the requirements of the funding agency, if any. In the early stages, design may be based on thumb rules, and sophisticated techniques come into the picture with the involvement of experts. There are many instances where tools of widely varying complexities have been applied in the same study. It is not uncommon to see that the output of a crude analysis forms the inputs to a sophisticated analysis! Frequently, a tool is employed just because it is available or a person trained for that tool is a member of the planning team, although more relevant tools might be available.

It is advisable to use simple models in the earlier stages of planning. The reason being that not much data may be available at these stages, a large number of alternatives may have to be tried and the available time may be limited. There is a trade-off between accuracy and expediency; the balance often tips toward the latter to provide quick and reasonable results. But in later stages, the balance should gradually be moved in favor of greater accuracy and more detail. Actually, the requirements at later stages necessitate the use of the best possible tools to refine the analysis.

An important requisite for the viable use of models in the planning process is the perception (by the planners and the public) of their credibility. The entire study can lose the participatory support of the concerned agencies if the models and procedures used are perceived as lacking in credibility and scientific footing.

The systems analyst has to find the tools that will best meet the planning needs. An approach, which allows the aggregation of several models, can be very helpful. In complex planning efforts, simulation is usually the preferred approach. Jacoby and Loucks (1972) were among the first to demonstrate how the optimization and simulation models can be effectively used together in river basin planning. Often, simulation is coupled with some kind of single-objective or multiobjective optimization, for example, for optimization of water allocation at each simulation step.

Many times, one comes across the debate whether optimization is a better tool or simulation is. This debate is unfortunate at best because the prime objective should always be to solve the problem in the best possible way rather than promote a particular technique. Simulation, when used as a search technique, is often preferred in complex problems. Many analysts prefer to take a number of simulation models by changing the input to get a feel for the behavior of the system or to do "manual optimization." In this approach, the analyst can effectively use his experience and judgment.

9.9.3 Use of Multiobjective Analysis

Multiobjective analysis is especially pertinent in river basin planning due to the presence of several conflicting and noncommensurable objectives. For example, economic efficiency is measured in monetary units, while environmental quality may be measured in units of pollutant concentration. However, society is placing an increasing importance on nonpecuniary objectives that are difficult to quantify monetarily. As pointed earlier, Major (1977) noted that the term "multiobjective" refers to multiple economic, social, environmental, and other objectives of water development, and "multipurpose" refers to multiple functions of water projects like navigation and flood control. These are not synonymous: purposes can vary and still be aimed at the same objective, and one purpose can fulfill more than one objective.

Traditionally, benefit–cost analysis has dominated the planning process. In this, only one objective (economic efficiency) is considered, with the other objectives being included either as constraints or as being somehow commensurate with the primary objective. In fact, this analysis is a simplified case of multiobjective analysis in which all objectives are expressed in terms of benefits and costs.

Fundamental to multiobjective analysis is the Pareto optimum, also known as the noninferior solution. The generation of Pareto optimal plans can be invaluable in identifying specific characteristics and attributes of a basin's subarea as well as in quantifying the complex interrelationships among the many components in the planning process. The noninferior solutions and the associated trade-off values help the decision-maker select an acceptable level of assurance and the corresponding cost. In other words, decision-makers can make known their preferences with respect to the level of assurance against uncertainties in the model prediction at the expense of degradation in the model's optimal solution. While formulating and screening alternative plans, multiple purposes, which are often in conflict and competition, must also be given explicit and quantitative consideration to the extent possible. For example, the use of reservoirs for flood control purposes may be achievable at the expense of reducing hydropower generation.

Multiobjective analysis should be viewed not only as a new method but also as a philosophy. Trade-offs are an inherent part of negotiation, of reaching consensus, and of compromise solutions. Thus, the use of multiobjective and trade-off analysis in the development of final plan results can be a natural step in this phase. The role of multiobjective analysis is particularly critical in addressing nonstructural plans, in which the cost, benefits, and risks cannot be easily quantified in monetary terms as is the case in more structured plans. Furthermore, as environmental and other socioeconomic aspects dominate and influence policy decisions, the importance and need of multiobjective analysis become more and more critical and evident.

For most water resources systems, decisions are not made by a single individual but rather by a group of individuals. These may be legislative bodies, the river valley authorities, etc. In every case, each member of the group has a personal view of the significance, importance, and relative value of the various objectives being considered. Each decision-maker may have a different constituency to whom he or she is responsible. This means that the decision-maker must integrate the relative influence and views of the segments of this constituency into the evaluation of the merits of alternatives. The critical influence of these decision-makers and stakeholders must be recognized throughout the planning process.

Multiobjective planning involves an interaction between planners and decision-makers and requires efficient communication for its success. According to Major (1977), there are four aspects of communication in multiobjective analysis: (1) between planning leaders and those who develop models and techniques for the production of alternate plans, (2) between planners and participants in the political process, (3) communication with planners in other sectors, and (4) communication after delivery of a report or the completion of the planning process. In view of this, communication activities must be carefully planned and periodically reviewed to ensure proper flow of information. These days, decision-support systems (DSSs) are being employed to help in multiobjective analysis. The North Atlantic Regional Water Resources Study (Major and Schwarz, 1990) was a large-scale application of multiobjective planning.

9.9.4 Object-Oriented Modeling Approach

The object-oriented (o-o) approach is frequently used in software development. An object is a "black box" which receives and sends messages. A black box actually contains code (sequences of computer instructions) and data. Traditionally, code and data have been kept apart. For example, in the C programming language, units of code are called functions, while units of data are called structures. In the C language, functions and structures are not formally connected—a function can operate on more than one type of structure, and more than one function can operate on the same structure. However, in the o-o programming, code and data are merged into a single indivisible entity, which is an object. This has many advantages. The main characteristics of the o-o approach are the following (Simonovic et al., 1997):

Identity: It implies that data are organized into discrete, recognizable entities called objects. These objects can be real-world things, such as a reservoir, a canal, or a conceptual thing, such as an operation policy.
Classification: An object is defined via its class, which determines everything about an object. Objects are individual instances of a class. For example, one may create a class "river" and then have objects like rivers. The choice of class depends on the application.
Polymorphism: It implies that the same operation may behave differently on different classes. It is an action that the object performs or is subject to.
Inheritance: Let there be a class that can respond to a group of different messages and a new similar class is required with a few more messages. All that is needed is to create a subclass of the original class, which inherits all the behavior of the original class. The original class is called the parent class of the new class. Inheritance also promotes reuse.

Many o-o programming languages are available, such as Java and C++, which permit easy use of this modeling approach. These days, the Python language is gaining popularity for this purpose.

9.9.5 Model Credibility

The credibility of a model refers to the faith in the model and acceptance of its results by the concerned parties. The lack of model credibility and acceptance constitutes one of the most common reasons for misgivings about models. This lack of faith in the minds of decision-makers and the public is an important, although often neglected, issue and needs to be dealt with at all stages of the planning process. One way out is to have a presentation about the model and its applications in a similar set-up to the decision-makers in the early stages of analysis. One should, however, not rush to conclude that all decision-makers are averse to model use. In fact, the reaction of a decision-maker very much depends on his background and exposure; nowadays, it is not uncommon to find a decision-maker who is well versed with the modeling terminology and understands and appreciates the analysis.

Models are only as good as the perception of their creators. Because of this, they should be only one part of the decision process. Before the results of a model are accepted, it has to be proved or verified using the data of the same or a nearby system. In many instances, the results of an application of the model to a faraway place, maybe in a different country, are cited to support the model. Often, such justifications fail to convince the target audience. Therefore, it should be ensured that the model is applicable to the present case and the assumptions are not violated.

9.10 SENSITIVITY ANALYSIS

Sensitivity analysis is an important part of any analysis; the same is true for the planning process too. The aim of this analysis is to understand how sensitive the various plans are to the characteristics of decision variables and input data. For example, if it is found that an irrigation project can cater for 10% more command area and the marginal benefit–cost ratio is more than unity, the decision-maker may like to bring the additional area under the ambit of the project. Furthermore, if it is realized that the project design is not sensitive to a particular input variable, there is no need to put too much effort in collecting data about that variable.

9.10.1 Risk and Uncertainty Analysis

Planning is concerned with the future, which is always uncertain. Risk analysis provides an assessment of trade-offs among the beneficial and adverse consequences that result from adopting a particular risk level. The existence of these trade-offs requires that risk management should be an integral part of the decision-making process. The efficacy of risk analysis in WR planning and management can be gauged by the assistance it provides to planners and decision-makers in the following ways:

(a) It identifies the sources of risk and uncertainty associated with exogenous variables and events derived from hydrological, environmental, and social factors.
(b) It quantifies the input–output relationships with respect to the randomness of these exogenous variables and events.
(c) It quantifies, as far as possible, the probable impacts that risk and uncertainty and their associated trade-offs will have on alternative policy decisions.
(d) It enables decision-makers to make scientific use of information about risk and uncertainty.

A relevant example of risk analysis is the dam-break modeling. This study is carried out to understand and quantify the damages that will take place if a dam holding water fails. There have been many instances of dam failures in the past. Different types of dams, for example, concrete dams, and earth and rockfill dams, have different failure behavior and fail due to different reasons. In the case of concrete dams, 29% of the cases are due to overtopping, 53% of the cases are due to foundation, and 18% of the cases are due to other reasons, including a shortfall in construction/design, material degradation, gate failures, and seismic events. The main causes of the failure of earth and rockfill dams are overtopping, piping, or earthquakes. The list of dams that have failed in the past include the Teton Dam (USA) due to hydraulic fracture, the Machhu Dam-II (India) due to overtopping, and the Sheffield Dam (USA) due to liquefaction. The Koyna dam (India) had suffered damages due to an earthquake. Since the failure of dams is associated with huge losses of life and property, it is important that the risks associated with a dam failure are properly examined before constructing a dam. About 2000 people were killed due to the failure of the Machhu dam in 1979 (Herschy and Fairbridge, 1998). In one of the most devastating dam failures of recent time, two dams in Derna, Libya, collapsed in September 2023 due to heavy rainfall from the Storm Daniel. The flood water caused havoc in the city of Derna and killed upwards of 5000 people.

A number of mathematical models are available that can be used to delineate the downstream area likely to be submerged and the maximum water level reached in case a dam fails. Many funding agencies and governments have made dam failure analysis a precondition before giving clearance for major and medium dam projects. Also, if an important project, such as a nuclear powerhouse, is to be constructed downstream of a dam, a dam-break analysis is necessary to evaluate risks. Wurbs (1987) has carried out a comparative evaluation of several dam breach flood wave models. Singh (1996) has discussed the theory and mathematical modeling of dam failure and its applications in detail. The topic of risk and reliability analysis has been dealt with in greater detail by Singh et al. (2007).

9.10.2 Uncertainties Associated with Objectives and Constraints

After the goals and objectives are adopted by the planning team, they become the dominant force that drives the planning process. Goals are positive attributes or characteristics that individuals or society try to attain. Goals of individuals and society are an unbounded set, that is, any stated goal is included within at least one more encompassing goal, and there is a set of more narrowly defined goals within it. Two major sources of uncertainty related to planning goals and objectives should be identified and addressed at the appropriate stage of the planning process. These are (1) perceptions of long-term societal goals and objectives and (2) perceptions of the long-term availability of technological and nontechnological means with which the planning goals and objectives can be achieved.

Societal goals and objectives are intrinsically hierarchical. Hence, uncertainty associated with each subgoal and subobjective contributes to the uncertainty of the overall societal goals and objectives. The input data may have errors in it. The model may have errors and the assumptions may not be realistic. Moreover, these errors and uncertainties are associated with all levels of planning. Then there are uncertainties associated with the perception of the availability of long-term technological and non-technological means of achieving the planning goals and objectives. This is particularly true for the assessment of future technology and its cost, reliability, and acceptability. Thus, the planning team should assess and evaluate the uncertainties associated with the goals on which the selected plan(s) are based and with the ways and means of realizing these goals. The approaches to rational decision-making discussed in Chapter 8 will be helpful to this end.

9.10.3 Postevaluation of Projects

Although this is not strictly part of planning, a study of some recently constructed projects could be a very good learning exercise. The aim should not be to find mistakes in those projects but to learn what methodology was used and why, and how the same can be useful for the project under question. At the same time, the weak points of analysis should also be noted down so that these could be improved upon. It will also be useful to have interaction and personal rapport with the planners associated with the study of those projects. There are many minor points and experiences, which could be very important for people working on similar problems. Many such things may not be available in a formal report; these can only be communicated and appreciated through personal interaction.

9.11 INTERACTION BETWEEN ANALYST AND DECISION-MAKER

In cases where there are different groups of planners and analysts, in the initial stages of interaction, the planners develop blueprints of several alternative plans. The analyst then uses mathematical models that incorporate a simplified model of the plans, the various internal relationships, objectives, and constraints to put these alternative plans in quantitative form. The planners reevaluate the alternatives and modify them as appropriate, using the quantitative information generated by analysts. Following several iterations among planners and analyst(s), alternative plans are ready for evaluation by the public and/or policy analysts and decision-makers. In many cases, there may not be distinct groups of planners and analysts or only one team may be working on the problem. Naturally, in these cases, the process of zeroing on the alternatives will not be so clear-cut although the final outcome will be the same.

The analyst, while consulting decision-makers, such as politicians and senior bureaucrats, must be aware that they may try to exclude alternatives that compete with those that they favor. Decision-makers can also identify political and institutional constraints that would exclude some alternatives. Detailed outlines of the plans must be presented to politicians, bureaucrats, and affected agencies early enough so that they will not be taken by surprise and react by completely rejecting the plan. Thorough discussions will help make them amenable to accept the plan and to identify themselves with it; this will also increase the probability that they may ultimately become its advocates.

The planners' objectives should be the development and/or formulation of a final plan that can enhance the social well-being of the people in the region, can ultimately be accepted by the public and other policy or decision-makers, and can also be implemented. In the screening process, this objective should guide the planners toward a compromise plan or solution that is viable and that has a good chance of acceptance. It should never appear at any stage that the planners want to impose their decision on the public and it would be very unfortunate if, in the end, the target beneficiaries become the worst critics or opponents of the project.

9.11.1 Presentation of Results

A good study loses its value if the decision-maker is not convinced or satisfied with the utility of the results. Therefore, decision-maker(s) should be involved as much as possible in the study through interim progress reports, discussions, presentations, etc. Sometimes this is difficult because the decision-maker may change during the course of the study and the interaction with a new person will have to be commenced from scratch.

It is quite possible that people representing a wide spectrum of backgrounds and interests will review the final study reports. A common practice is to present the reports in three segments.

(i) An executive summary that is suitable for politicians and senior bureaucrats.
(ii) The main text that is suitable for experts and professionals, who may have to scrutinize the work.
(iii) Appendices containing input data, results, figures, drawings, computer listings, etc. The material must be presented to enable the technical personnel to check and verify the analysis and results.

There are, of course, different angles from which decisions must be made—technical, political, etc. The decision-maker will be interested in the extent to which the planner has considered various options. Along with the absolute values of levels of objectives that can be attained by the project (such as the value of hydropower generated, average annual flood damage saved, etc.), it is also useful to include the percentage improvement in various indicators. This will be helpful to the decision-maker to judge and appreciate the contribution of the project. Finally, the use of computer graphics and other tools has made it easier to quickly prepare and present the results in an attractive format.

If funding for a project is being sought from an international financing or donor agency, such as the World Bank, a detailed project report will have to be prepared according to the norms of the concerned agency. In the World Bank, a project is examined from a number of different aspects like technical, economic, environmental, financial, institutional, organizational, and managerial aspects. In the past, the development assistance and landing operations focused mainly on construction and infrastructure facilities without giving much attention to water quality, environmental issues, and operation and maintenance. The approach by the World Bank (1993) recognizes that a comprehensive policy framework is needed, management should be decentralized, and water be treated as an economic good. Great emphasis is placed on the pricing and participation of stakeholders and the importance of the project for the regional or national economy, its impact on the improvement of general welfare, and the standard of living. Besides, organizational arrangements for the construction, operation, and management skills are also given weightage and improved methods have been devised to determine social impacts on lower-income groups of society. After completion, a report is prepared in which the success of the implementation of the project and difficulties that were encountered are included.

Likewise, each major agency has its own guidelines for the preparation of detailed project reports. Most such agencies get the project appraisal done by their own staff or consultants. Many countries/agencies have issued detailed guidelines for the preparation of project reports. For example, the Government of India (1980) has issued such guidelines for the preparation of detailed reports for river valley projects.

9.12 WATER RESOURCES PLANNING—CASE STUDIES

Major (1977) has described four applications of multiobjective planning: the North Atlantic Regional Water Resources Study (discussed in detail by Major and Schwarz, 1990); the Rio Colorado study (described later); the Big Walnut study; and the Managua study. Several case studies related to specific aspects, such as flood control planning, groundwater planning, and wastewater management planning, have been illustrated by Goodman (1984) and Grigg (1985). French et al. (1979) described a software for interactive water resources planning using computer graphics.

Major and Lenton (1979) applied WR planning techniques to the Rio Colorado basin, Argentina. This river rises from the snowmelt runoff in the Andes and flows in a generally easterly direction for about 1100 km through arid countries to its mouth in the Atlantic. Rio Colorado is one of the largest rivers in Argentina and is used for irrigation and hydroelectric power. The river flows through several provinces and the interests of each of these riverine provinces are different from others and the national government. This basin has about 13 sites for hydropower generation, 20 sites for irrigation, and the irrigable area is of the order of 800,000 ha. The goals of this study were to: (1) illustrate the modern river basin planning methods by an application and (2) enable the decision-maker to understand the physical, economic, and social trade-offs involved in the choice of a development scheme for the river basin and thus assist them in choosing among the alternative development plans. A large number of experts from the Massachusetts Institute of Technology (MIT), USA and professionals from Argentina took part in this study.

Three types of models were used in the Rio Colorado study. For screening purposes, a mixed integer programming model was used. The most promising sites were analyzed using the simulation approach. The system configuration that was analyzed through the simulation model was further studied by a sequencing model. The objective was to schedule the development optimally in four future time periods taking into account benefits over time, budget constraints, constraints on the number of farmers available to work in new areas, and the project interrelationships such as the necessity that the irrigated area is not developed before the construction of dam to supply water to it. The process of model use of this study is shown in Fig. 9.4.

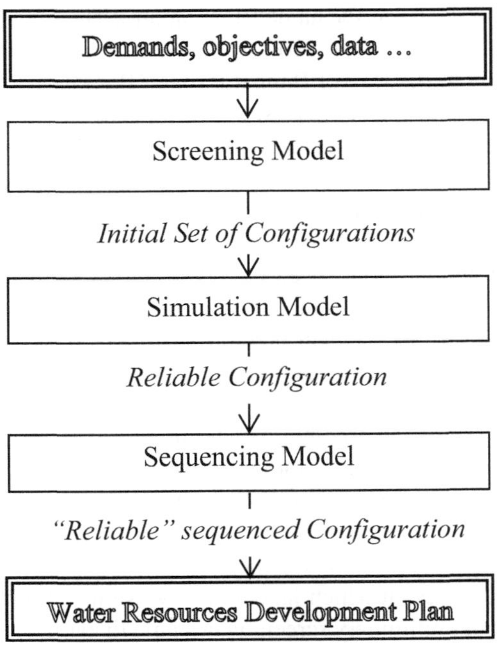

FIGURE 9.4

Process of model use in water resources planning.

Adapted from Major and Lenton (1979).

9.12.1 Ganga–Brahmaputra–Barak Basin Study

Chaturvedi and Rogers (1985) have presented the results of extensive studies on the Ganga–Brahmaputra–Barak basin in the Indian subcontinent. The following discussion is based on their book detailing the study. This basin lies in four countries: India, Nepal, China, and Bangladesh, the major portion being in India. They named this system as *Greater Ganga system*, and this is the second-largest international river basin in terms of runoff, second only to the Amazon basin in South America. It drains an area of 1.38 million square kilometers and the peak outflow from the system at its estuary is 141,000 m^3/s (cumec). It carries about 127.61 m-ha-m of water in the Bay of Bengal each year of which about 80% is in the monsoon season. The north-eastern boundary of this system is formed by the Himalayas, which are geologically young mountains and it is bounded by Vindhyachal mountains in the south. A low watershed separates the basin from the Indus basin in the west. The extensive alluvial plains of the Ganga basin are part of this system. The delta of the Greater Ganga system covering an area of 56,700 sq. km is one of the biggest in the world.

The Ganga River rises in the Himalayas at an elevation of 7010 m above the mean sea level. It flows for a distance of 2500 km which makes it the 15th longest river in Asia and the 39th in the world. It drains about one-third of the geographical area of India and supports more than 40% of its population. The monthly mean flow of the Ganga River at its tail end reaches up to 57,000 cumec.

The Brahmaputra (the son of Brahma, the Creator of universe in the Hindu mythology) River rises in the great glacier in the northern-most chain of Himalayas in the Kailash range just south of a lake called Konggyu Tsho. After flowing for a distance of 1700 km through southern Tibet parallel to the main Himalayan range where it is known as Tsangpo, it enters India as the Dihang River. After its confluence with some streams, it attains the name Brahmaputra. The Majuli island, which is the largest river island covering an area of more than 1200 sq. km is part of the Brahmaputra. This river joins Ganga and together they flow to the Bay of Bengal through Bangladesh.

The Greater Ganga system has wide diversities in physiographic-geographical characteristics, topography, soil, and land use as well as socioeconomic aspects. The basin is like an elongated bowl with very high steep mountains in the north, comparatively low mountains in the south and east, and a very flat fertile alluvial plain in between. The region has also witnessed rapid growth in population over the last few decades; there has been tremendous urbanization, and demands for water have risen rapidly. The development in the basin has been largely on an ad hoc basis. Although the region has huge water potential, due to various reasons including its international character, most of this potential remains unutilized. More than 80% of annual precipitation takes place in 4 months of monsoon; the area receives solid as well as liquid precipitation. The physiographic and meteorological characteristics of the system coupled with monsoon concentrated precipitation lead to heavy floods. As the Himalayas are geologically young and erodible mountains and have very steep slopes, the high flows also carry high sediment loads. Most of the storage sites and hydroelectric potential lie in Nepal and the North-East part of India. The emphasis on WR development in the past was on flow diversions limited to low flows.

The schematic diagram of the system used in the coordinating model is given in Fig. 9.5. There are 49 river-schematic nodes of interest at which junctions, diversions, groundwater pumping, and return flows occur. In the first stage of preliminary screening, a single linear model was constructed to explore and coordinate all the demands placed on the system. This model was used as a tool to explore the various goals to be obtained and the constraints on development. The idea was that once the system has

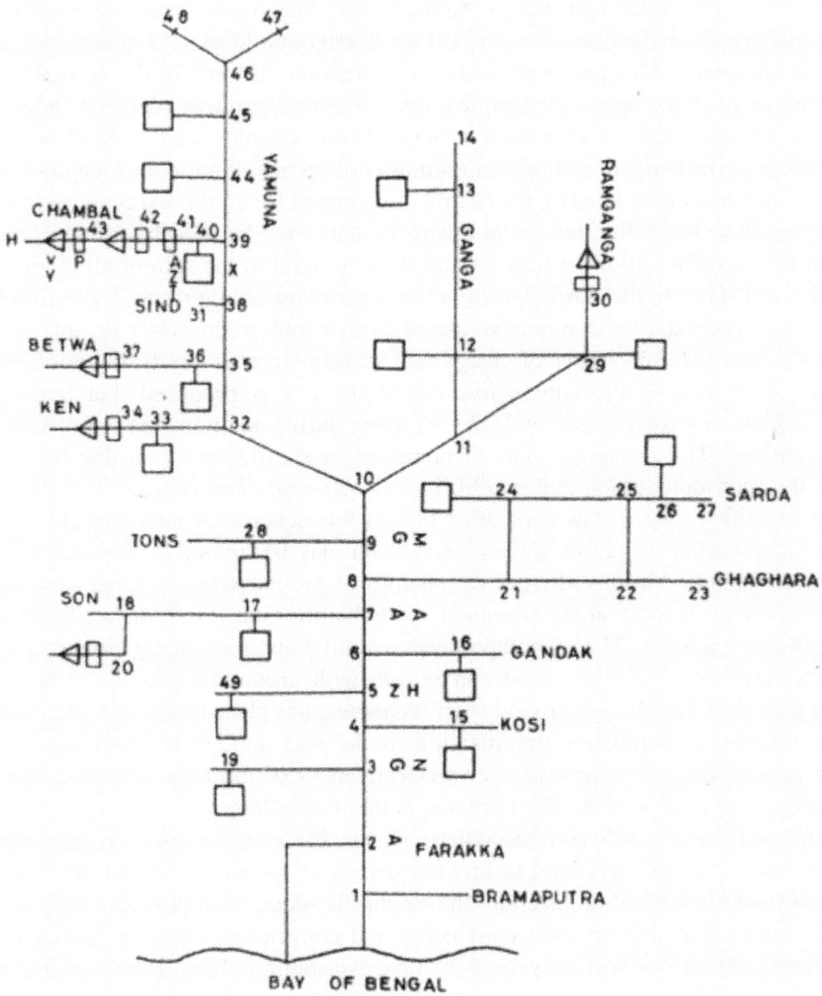

FIGURE 9.5

Schematic diagram of Greater Ganga system.

Reproduced from Chaturvedi and Rogers (1985). Copyright © Indian Academy of Sciences. Used by permission.

been fully explored by this inexpensive model, it can be broken up into smaller, more manageable pieces for a more complete analysis. The model considered five reservoirs and 75 irrigation works as already developed. In the second stage, the entire basin was decomposed into smaller systems. Two types of decomposition were planned. In the first case of hydrologic decomposition, the system was divided into nine subbasins. The operation of each subbasin was optimized by varying the irrigation level under the given water releases and energy target levels. Each subbasin then reports to the central organization, its optimal irrigation level and search for energy production, shadow prices, and their effective ranges on the water and energy target levels. The master problem was then solved by maximizing incremental irrigation areas and the total irrigation level, and energy production for the whole basin is computed. This process is done in an iterative fashion. In the second type of decomposition, the basin was decomposed by political units. The problem was analyzed under two schemes: flow quota scheme and resource allocation scheme. These schemes were worked out by three algorithms. The first algorithm requires that the minimum flow at any point in the Ganga should be greater than the sum of the flow quota fixed by the central government for all the states above that point. Under the second algorithm, water that leaves an upstream state may be used by a downstream neighbor, provided that this neighbor allows a flow greater than or equal to the sum of all the upstream states' quota into the next state. In the third algorithm, the restriction of algorithm 2 was lifted.

The authors emphasized the conjunctive use of surface and groundwater for this system pointing out that groundwater is a major user of energy and surface water is a rich source of energy through hydroelectric generation. In the conjunctive use study, it was proposed that infiltration may be increased during the monsoon season by heavy pumping during nonmonsoon season and thus creating groundwater storage. The extent and date of pumping were estimated such that it is replenished in 120 days of the monsoon season and equilibrium is achieved. The areas suitable for groundwater storage were identified, and it was concluded that in this scheme of underground storage of flood waters, the total potential irrigation in the Ganga basin may be limited by the area of irrigable land rather than water supply. Various alternative schemes of groundwater recharge were proposed. The first involves pumping heavily along perennial rivers before monsoon to lower the water table and induce groundwater recharge. The second proposes a similar approach along nonperennial rivers. The third involves irrigation during the monsoon season with groundwater lowered adequately in the nonmonsoon period so that enough groundwater recharge takes place to provide adequate supplies for nonmonsoon months. A simulation-optimization model was applied to study the surface water–groundwater interaction and comparative cost-effectiveness of the three alternate approaches. The sensitivity analysis showed that the third scheme is the most attractive.

Chaturvedi and Rogers (1985) concluded that a reasonable approach for such large systems is that a programming model may be first used to find out the range for which simulation studies should be carried out, particularly taking into account the stochastic nature of inputs and outputs. For detailed modeling, the simulation will be most convincing and convenient. However, simulating the entire system in diverse conditions will be extremely time consuming. They also emphasize that trained manpower is the most important prerequisite for WR development.

9.12.2 Water Resources Planning in Egypt

Egypt is said to be the gift of the Nile River. Though the country has three sources of water, the Nile River, rainfall, and groundwater, the Nile River is the most important because of its large flow in comparison to the other two sources. Probably, no other country is so completely dependent on a single source of water as Egypt is on the Nile. The river sustains more than 98% of the population on a narrow green band bisecting a land that is otherwise nearly barren (see Fig. 9.6). Recognizing the importance of the Nile River flow for irrigation, hydropower, municipal needs, and navigation, Egyptians have devoted large efforts to regulate its flow. An index map of the Nile River is given in Fig. 9.6. Planning for WR in Egypt has aimed at the following:

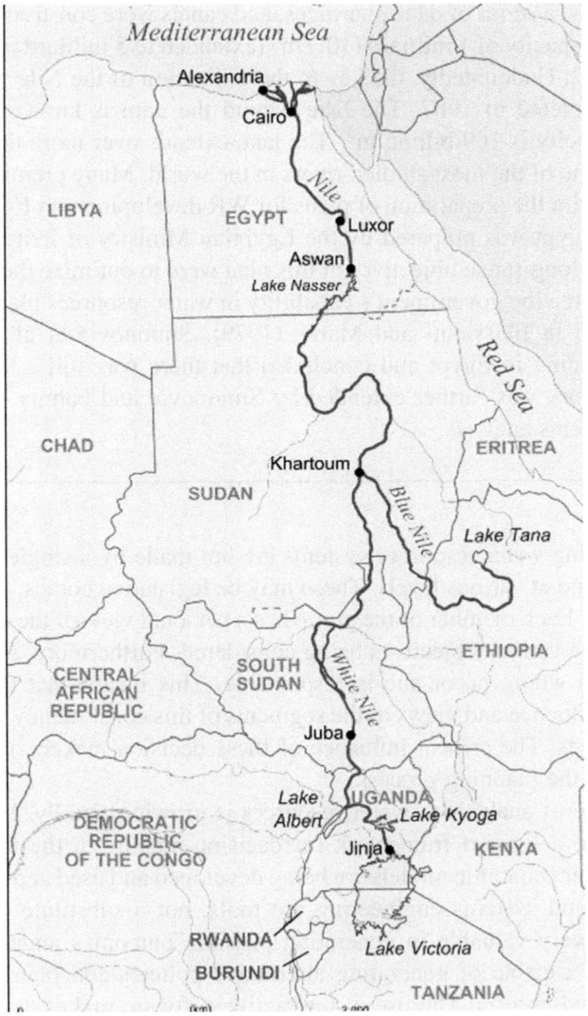

FIGURE 9.6

Index map of the Nile River basin.

Source: https://en.wikipedia.org/wiki/Nile.

(i) Control of the river's waters, its preservation, and development,
(ii) Securing navigation all the year through,
(iii) Production of hydroelectric power,
(iv) Laying down appropriate programs necessitated by the development policy for the reasonable use of all water resources,
(v) Generalization of perennial irrigation for the provision of increasing water requirements for accelerating food requirements.

To realize these aims, a series of dams, barrages, and canals were constructed. The Aswan dam was initiated in 1902 with capacity of 1 milliard (10^9) m^3 (extended to 5 milliard m^3 in 1923) and Zefta and Assiut barrages in 1902. Undoubtedly, the key to the utilization of the Nile water is the High Aswan Dam, which was completed in 1967. The lake behind the dam is known as Lake Nasser and its maximum storage capacity is 169 billion m^3. The lake extends over more than 500 km.

The Nile River is one of the most studied rivers in the world. Many premier institutes of the world have been associated with the preparation of plans for WR development in Egypt. The Master Plan for WR development in Egypt was prepared by the Egyptian Ministry of Irrigation in association with experts from MIT. The long-range objectives of this plan were to optimize the development and use of Egypt's WR and reinforce the government's capability in water resources planning. The results of this project were described in Elassiouti and Marks (1979). Simonovic et al. (1997) applied the o-o approach for WR planning in Egypt and concluded that there was still a lack of integral national-level planning. This work was further extended by Simonovic and Fahmy (1999) by the integrated use of the o-o and systems analysis.

9.13 CLOSURE

The decisions concerning water resources systems are not made by a single individual but rather by groups of individuals and at various levels. These may be legislative bodies, river valley authorities, a government set-up, etc. Each member of the group has a personal view of the significance, importance, and relative value of the various objectives being considered. Furthermore, each decision-maker may have a constituency to whom he or she is responsible. This means that the decision-maker must integrate the relative influence and views of the segments of this constituency into the evaluation of the merits of the alternatives. The critical influence of these decision-makers and stakeholders must be recognized throughout the planning process.

Application of systems analysis to water resources is growing rapidly. Mathematical models are gradually evolving into a support framework for decision-making in the form of DSSs. Improved system models and better economic models are being developed and used at different levels. Of course, mathematical models and systems engineering are tools, not a substitute for the decision-making process. They can be very valuable in generating possible outcomes under certain conditions and assumptions. They are capable of generating alternative policies and plans that are optimal under specific assumptions and criteria. The use of interactive software makes decisions more transparent and helps in accomplishing two-way communication with the systems analyst.

There are continuous changes in the planning process, dictated by the need for ever-increasing complexities of economic and social institutions as WR projects are important components of the infrastructure. For example, at the local level, a hydropower project is part of a regional electric supply grid; at the national level, it becomes part of the national grid. There is also a continuing need to review the planning decisions of yesterday. This is necessary in light of developments and evolution of the social and economic fabric of the country and of the needs and demands that are placed on the water resources of the region.

It may be emphasized that the procedures discussed here do not guarantee the quality of the results of planning; these will ultimately depend on the correctness of the data describing the system, the appropriateness of the tools, and the skills of the planning team. It is necessary to obtain all the needed information on each of the system elements before any detailed analysis is performed. For this, checklists are sometimes used. However, even the best checklists and planning schedules can only be a guide and must be used with care and discretion. They can supplement, but not replace, the skill and intuition of an experienced and creative planner. Finally, no hard-and-fast rules exist on what planning procedures are to be used to find the best approach to address the planning process.

References

Allen, R.G., Pereira, L.S., Raes, D., Smith, M., 1998. Crop Evapotranspiration: Guidelines for Computing Crop Requirements. Irrigation and Drainage Paper No. 56. Food and Agriculture Organization (FAO), Rome.

Buras, N., 1963. Conjunctive operation of dams and aquifers. Journal of the Hydraulics Division, ASCE HY6, 111−131.

Buras, N., 1972. The Scientific Allocation of Water Resources. Elsevier Book Company, New York.

Chaturvedi, M.C., Rogers, P., 1985. Water Resources Planning – Some Case Studies for India. Indian Academy of Sciences, Bangalore.

Dyck, S. (Ed.), 1990. Integrated Planning and Management of Water Resources (Guidance Material for Course for Engineers, Planners, and Decision-Makers). International Hydrological Programme, UNESCO, Paris.

Elassiouti, I.M., Marks, D.H., 1979. Water resources planning in Egypt. In: Proceedings of the International Conference, (held in Cairo, Egypt).

French, P.N., Johnson, L.E., Loucks, D.P., Greenberg, D.P., 1979. Interactive Water Resources Planning Using Computer Graphics. Department of Environmental Engineering, Cornell University, Ithaca, New Jersey.

Government of India, 1980. Guidelines for Preparation of Detailed Project Reports of Irrigation and Multipurpose Projects. Working Group Report. Government of India, New Delhi.

Goodman, A.S., 1984. Principles of Water Resources Planning. Prentice-Hall Inc., Englewood Cliffs, New Jersey, USA.

Grigg, N.S., 1985. Water Resources Planning. McGraw-Hill Book Company, New York.

Haimes, Y.Y., 1977. Hierarchical Analyses of Water Resources Systems: Modeling and Optimization of Large-Scale Systems. McGraw-Hill Inc., New York.

Haimes, Y.Y., Kindler, J., Plate, E.J., 1987. The Process of Water Resources Project Planning: A Systems Approach. Studies and Reports in Hydrology # 44. UNESCO, Paris.

Hamdan, 1961. Evaluation of irrigation agriculture in Egypt. In: Stamp, L.D. (Ed.), A History of Land Use in Arid Regions. UNESCO, Paris.

Herchy, R.W., Fairbridge, R.W. (Eds.), 1998. Encyclopedia of Hydrology and Water Resources. Kluwer Academic Publishers, Dordrecht.

Jacoby, H.D., Loucks, D.P., 1972. Combined use of optimization and simulation models in river basin planning. Water Resources Research 8 (6), 1401−1414.

Jain, S.K., Pankaj, M., Sanjay K., J., Pavithra, P., Singh, V.P., Tullos, D., Kumar, S., Agarwal, S.P., Dimri, A.P., 2018. A Brief review of flood forecasting techniques and their applications. International Journal of River Basin Management 1−16. https://doi.org/10.1080/15715124.2017.1411920.

Loucks, D.P., van Beek, E., 2017. Water Resource Systems Planning and Management: An Introduction to Methods, Models, and Applications. Deltares, UNESCO-IHE, and Springer Nature, Switzerland.

Maheshwari, B., Atkins, D., Hagare, D., Spencer, R., Dillon, P., Jain, S., Rollason, R., Reynolds, J., Dollin, J., Batelaan, O., Packham, R., Patel, J. B., Purkait, M. K., Bhaduri, A., Mailapalli, D., Ashok, A., Prasad, J., 2023. Mentoring in the young water professionals training program: lessons for effective capacity development. World Water Policy 9 (3), 334−348. https://doi.org/10.1002/wwp2.12129.

Major, D.C., 1977. Multiobjective Water Resources Planning. Water Resources Monograph No. 4. American Geophysical Union, Washington, DC.

Major, D.C., Lenton, R.L., 1979. Applied Water Resources Planning. Prentice-Hall Inc., Engelwood Cliffs, New Jersey.
Major, D.C., Schwarz, H.E., 1990. Large-scale Regional Water Resource Planning: The North Atlantic Regional Study. Water Science and Technical Library. Kluwer Academic Publishers, Dordrecht.
Miser, H.J., 1982. Lessons from Experience in the Practice of Operations Research. Professional Papers, International Institute for Applied Systems Analysis, Laxenburg, Austria.
NCIWRD, 1999. Integrated Water Resources Development – A Plan for Action. Report of the National Commission for Integrated Water Resources Development. Government of India, Ministry of Water Resources, New Delhi.
Rogers, P., Smith, D.V., 1970. An algorithm for planning irrigation projects. ICID Bulletin 10, 15–30.
Saha, S.K., Barrow, C.J., 1981. River Basin Planning: Theory and Practice. John Wiley & Sons, New York.
Simonovic, S., 1989. Application of water resources systems concept to the formulation of a water master plan. Water International 14, 37–50.
Simonovic, S., Fahmy, H., El-Shorbagy, A., 1997. The use of object-oriented modeling for water resources planning in Egypt. Water Resources Management 11 (4), 243–261.
Simonovic, S., Fahmy, H., 1999. A new modeling approach for water resources policy analysis. Water Resources Research 35 (1), 295–304.
Singh, V.P., 1996. Dambreak Modeling Technology. Kluwer Academic Publications, Dordrecht.
Singh, A., 2014. Conjunctive use of water resources for sustainable irrigated agriculture. Journal of Hydrology 519 (Part B), 1688–1697. https://doi.org/10.1016/j.jhydrol.2014.09.049.
Singh, V.P., Jain, S.K., Tyagi, A., 2007. Risk and Reliability Analysis: A Handbook for Civil and Environmental Engineers. American Society of Civil Engineers, USA, ISBN 978-0-7844-0891-9, p. 783.
Stephenson, D., Peterson, M.S., 1991. Water Resources Development in Developing Countries. Development in Water Science # 41. Elsevier, Amsterdam.
UN, 1970. Integrated River Basin Development, Report of a Panel of Experts. Department of Economic Affairs, The United Nations, New York.
UN, 1972. Water Resources Project Planning, Water Resources Series No. 41. The United Nations, New York.
UN, 1976. The Demand for Water. Natural Resources – Water Series No. 3. The United Nations, New York.
UN, 1997. Public Involvement Guidelines for Natural Resources Development Projects. The United Nations, New York.
Weiss, A.O., Beard, L.R., 1971. A multi-basin planning strategy. Water Resources Bulletin 7 (4), 750–764.
Willis, R., Yeh, W.W.-G., 1987. Groundwater Systems Planning and Management. Prentice-Hall, Engelwood Cliffs, New Jersey.
World Bank, 1993. Water Resources Management; A World Bank Policy Paper. The World Bank, Washington, DC.
Wurbs, R.A., 1987. Dam breach flood wave models. Journal of Hydraulic Engineering, ASCE 113 (1), 29–46.

Further Reading

Chaturvedi, M.C., 1987. Water Resources Planning and Management. Tata McGraw-Hill Publishing Company Ltd., New Delhi.
Meta Systems Inc., 1975. Systems Analysis in Water Resources Planning. Water Information Center Inc., New York.
US Water Resources Council, 1973. Water and Related Land Resources, Establishment of Principles and Standards for Planning. Washington, DC.

CHAPTER

Reservoir Sizing

10

Chapter outline

Water Resources Systems Planning and Management. https://doi.org/10.1016/B978-0-12-821349-0.00014-9

Dams are constructed for two main functions. The first is to store water in the lake behind the dam to even out the fluctuations in river flow and match the availability with demand. If surplus water is available in a stream at a time earlier than its demand, it has to be carried over through storage from the period of surplus to the period of deficit. The purpose of regulation is to match the releases from the dam with the demand or to temporarily store high flows in the river to reduce flood damage. The second function of the dam is to create a hydraulic head of water (difference in height between the surface of a reservoir and the river downstream) in the reservoir upstream of the dam so that water can be diverted into a canal and flow due to gravity. The creation of storage and head allows dams to generate electricity; to supply water for agriculture, industries, and households; to control flooding; and to assist river navigation by providing regular flows and drowning rapids. Other reasons for building large dams include fisheries and recreation activities, such as boating.

It is believed that the early Mesopotamians were the first dam builders. The Sumerians had built networks of irrigation canals in the plains along the lower Tigris and Euphrates rivers about 6500 years ago. Remains of dams that were built around 3000 BCE as part of an elaborate water supply system in Jordan have been found. Earth and rock-fill dams had been built around the Mediterranean, in the Middle East, China, and Central America by the late first millennium BC. The remains of impressive dams and aqueducts in Spain are fine examples of the ingenuity of Roman engineers. South Asia has a long history of dam building. Long earthen embankments were built to store water for irrigation. The remains of the Indus Valley Civilization which flourished 4000–5000 years ago show that these people were familiar with many hydrologic principles and had built well-planned networks of water supply and drainage works. The landscape of South India is dotted with a large number of very old small artificial ponds known as "tanks," which are still in use as an important source of irrigation water. According to Morris and Fan (1998), the oldest reservoir in operation today is the Aftentang reservoir (storage 100 million m^3), constructed west of Shanghai during 589–581 BCE. The oldest continuously operating dam still in use is the Kofini flood control diversion dam and channel constructed in 1260 BCE on the Lakissa River upstream of the town of Tiryns, Greece, which it continues to protect. Another noteworthy very old dam is the Kallanai dam (also known as the Grand Anicut), built across the Cauveri River in Tamil Nadu, India. It was built during the reign of Chola king Karikalan in about 150 AD. The structure is still in use.

The 19th-century dams were mainly earthen embankments designed largely on the basis of thumb rules. Dam builders in the 19th century had little streamflow or rainfall data, and few statistical tools to analyze whatever hydrologic data had been gathered. As a consequence, some of these dams failed. After the turn of the 19th century, there were important developments in civil engineering. Consequently, the size of dams and power stations being built began to increase rapidly. The improvements in dam engineering allowed the high dams to be built and progress in turbine design increased the head at which the turbines could operate.

The International Commission on Large Dams, has created a database of the world's largest dams (http://www.icold-cigb.org). There were more than 58,000 dams by 2022. The Asian continent accounts for more than 64% of all dams, and China has built most of these. Engineers of the former USSR have built most of the large reservoirs; they are followed by engineers from Canada. Earth dams predominate for some 65% of all reported dams.

A dam contains a number of structural features other than the main wall itself. Spillways are used to discharge water when the reservoir level threatens to become dangerously high. Dams built across broad plains may include long lengths of ancillary dams and dikes. Weirs and barrages are constructed to divert river flow; they do not have significant storage and cannot effectively regulate flows. A weir is normally a low masonry or concrete wall. A barrage is a bigger (usually metallic) structure, often extending for hundreds of meters across a wide river.

10.1 NEED FOR RESERVOIRS

An important aspect of water resources development projects is planning and operation of reservoirs, which are the most important components of a water resources project. Using a reservoir, the natural streamflow can be regulated so that the outflow follows the desired pattern. In day-to-day life, the concept of reservoirs is frequently used. For example, in cities where the municipal water supply is erratic, residents use vessels to store water whenever there is running water in the taps, and the stored water is withdrawn and used for various domestic needs when the taps go dry.

Due to the large size, the reservoir projects are capital intensive, that is, they require huge amounts of money, manpower, land, and other resources. Furthermore, these projects significantly affect the environment, population, and economy of the region in which they are constructed. As the financial resources available are usually limited, these should be carefully used to impart the maximum possible benefit to the national economy. Moreover, once the dams are in place, it is not easy to undo or partially offset their harmful impacts. Recently, some of the dams have been involved in various controversies that have resulted in frequent reviews, changes in design, and delays in construction resulting in huge cost overruns. Due to these reasons, it is necessary that these projects are planned with utmost care and detailed examination of the issues involved. Now, most countries have put in place procedures to examine dam project proposals by experts from a range of disciplines so that all the concerns and objections by different stakeholders are resolved before the construction is taken up.

The commonly used terms that are relevant to a reservoir are defined in Appendix A. A schematic diagram of a reservoir is given in Fig. 10.1.

The time unit that is adopted for analysis of data for long-term planning is 1 year. To define year, one can follow the concept of calendar year, financial year, or water year. In water resources analysis, usually the concept of water year is used. The starting month of a water year varies from country to country, depending on the geographic location and climatic pattern. In general, a water year begins during the time of low flows. From a statistical point of view, the starting month should be chosen such that the coefficient of variation of annual streamflows is maximized while the autocorrelation is minimized (McMahon and Mein, 1986).

The principal function of a reservoir is the regulation of natural streamflow by storing surplus water in the high-flow season to control floods and releasing the stored water in the dry season to meet various demands. Generally, the major part of the annual streamflow is available during a few months

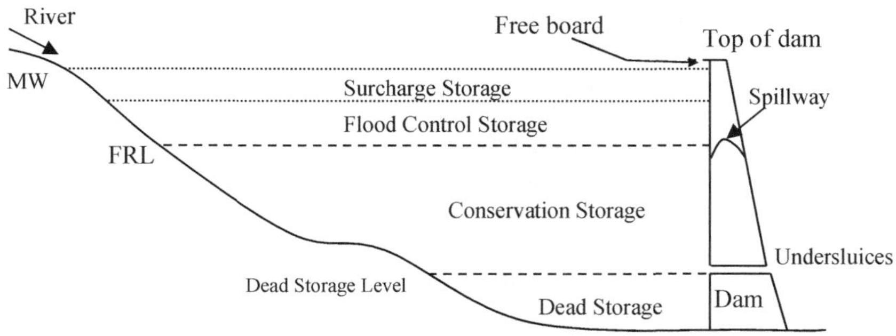

FIGURE 10.1

Schematic diagram of a reservoir.

of the rainy season. However, the demands for water arise all year round, and therefore it is necessary to store the excess water in the rainy months so that it can be used when the natural streamflow is not sufficient to meet the demand. The water stored in a reservoir may be diverted by means of pipes or canals to faraway places where it is needed; this diversion results in spatial changes of water availability. The water may also be kept in the reservoir and released later for beneficial uses resulting in temporal changes. In short, the aim of a reservoir is to match the temporal and spatial availability of water with demands. Depending on the magnitude of natural inflows and demands at a particular time, the reservoir storage is either built up or water is supplied from the storage.

As a result of storing water, a reservoir provides a water head that can be used for the generation of electric power. The reservoir also provides an empty storage space for moderating inflow peaks. A reservoir also provides a pool for navigation to negotiate rapids, habitat for aquatic life, and facilities for recreation and sports. It enhances scenic beauty, promotes afforestation, and supports wild life.

10.1.1 Classification of Reservoirs

Various classifications of reservoirs are possible depending on the purpose, size, and storage space available in it. These are briefly discussed next.

Classification Based on Purposes Served

Depending on the number of purposes that a reservoir serves, a reservoir may be classified either as single purpose or multipurpose. A single-purpose reservoir serves only one purpose. This purpose may be either a conservation purpose like water supply for domestic and industrial purposes, irrigation, navigation, generation of hydroelectric power, and recreation. The flood control purpose is a nonconservation in nature. A multipurpose reservoir is designed and operated to serve a combination of these purposes.

Classification Based on Size

Depending on the size, reservoirs are classified as major, medium, or minor. These norms, however, vary from country to country. In India, if the gross capacity and the hydraulic head of the reservoir exceed 60×10^6 m^3 and 30 m, respectively, the reservoir is classified as a major reservoir. If the gross capacity lies between 10 and 60 × 10^6 m^3 and the hydraulic head lies between 12 and 30 m, a reservoir is classified as a medium reservoir. Minor reservoirs have a gross capacity of less than 10 million m^3 and a hydraulic head of less than 12 m.

Classifications Based on Storage Capacity

Based on the storage space provided, a reservoir may be classified as a seasonal storage or over-year storage. A seasonal storage reservoir is designed to serve conservation purposes for periods of low flows. These reservoirs fill and spill frequently and are constructed on small tributaries to serve relatively small areas. An over-year storage reservoir is designed to serve for periods exceeding more than a water year. The storage in an over-year storage reservoir at the end of a water year is carried over to the next year. These reservoirs may neither be completely full nor dry every year.

10.2 CHARACTERISTICS AND REQUIREMENTS OF WATER USES

As a thumb rule, the bigger a reservoir is, the more are the purposes that it can serve. The major purposes for which a reservoir is used and the functional requirements for these are discussed next. The irrigation requirements have been described in greater detail in Chapter 9. The hydropower generation is discussed in greater detail in a later section.

10.2.1 Irrigation

Irrigation demands are consumptive and only a small fraction of the water supplied is available to the system as return flow. These requirements have direct correlation with rainfall in the command area. Irrigation requirements are seasonal in nature and the variation largely depends on cropping patterns in the command area. In general, demands will be small during the wet season and large during winter and summer months. The average annual demands remain more or less steady unless there is an increase in the command area or a large variation in the cropping pattern. The safety against droughts depends on the available water in the reservoir, and hence it is desirable to keep as much water in storage as possible consistent with current demands.

Food and Agriculture Organization (FAO) has adopted the Penman–Monteith (PM) equation as the standard technique to compute reference ET. The procedure to compute crop water requirement has been described in the report FAO-56 (Allen et al., 1998). FAO has also developed a decision support tool called CROPWAT, which is software to calculate crop water requirements and irrigation requirements based on soil, climate, and crop data. CROPWAT also allows the development of irrigation schedules for different management conditions and the calculation of scheme water supply for varying crop patterns. The software can be freely downloaded from https://www.fao.org/land-water/databases-and-software/cropwat/en/.

10.2.2 Municipal and Industrial Water Supply

Water stored in many reservoirs is used to meet domestic and industrial needs. Generally, water requirements for municipal and industrial purposes show small changes throughout the year, more so when compared with irrigation and hydroelectric power demands. In most developing countries, water requirements for these purposes are seen to increase with time due to growth and expansion. The seasonal peak of the demand for domestic uses is observed in summer. For the purpose of design, a target value is arrived at by projecting population and industrial growth. The supply system for such purposes is designed for a high level of reliability. Stringent quality norms are placed on water supplied for drinking and cooking use, not many people directly consume tap water supplied to homes due to concerns about its quality.

10.2.3 Hydroelectric Power

Water is a renewable source of energy. The hydroelectric power generation is a nonconsumptive use of water because after passage through the power plant where its mechanical energy is converted to electric energy, the same water can again be utilized for other uses downstream. Due to this feature, hydroelectric projects are frequently multipurpose. As a result of research and development in turbine technology, efficient turbines with capacities varying from several hundreds of MW to a few MW have been developed. Therefore, one now comes across mega hydropower projects, providing power to a big region, to microprojects, catering to the needs of a small village. It is estimated that one-quarter of the electrical energy generated in the world is from hydropower. Some advantages of hydropower generation are:

- This is a renewable source of energy, the sun being the prime mover of the water cycle. As no payment is made for the input, the production is free from inflation.
- The hydropower plants do not require much outlay on account of operation and maintenance and have a long life.
- The hydropower generation does not pollute the environment; no heat is produced and no harmful gases are released. These days, there is renewed interest in energy generation by hydropower since the world community has set targets to limit global warming below 2 degrees, mainly by controlling the emission of greenhouse gases.
- The hydropower power plants work at a very high efficiency (say up to 90%), whereas the thermal power plants work at a comparatively low efficiency.
- The plant can be started or shut down in a short time, with no wastage of water.

The electricity demands usually vary seasonally and to a lesser extent daily (domestic and industrial demands during weekends are less compared to weekdays) and even hourly. The degree of fluctuation depends on the type of loads being served, viz. industrial, municipal, and agricultural. For example, in the case of municipal areas, the hydroelectric demands are at maximum during the peak summer months. Furthermore, during the course of a day, two demand peaks are observed, one in the morning and another in the evening. The hydroelectric power plants are usually part of a regional or national grid and their operation is governed by their role in the grid. The hydropower generation aspect is discussed in greater detail in Section 10.5.

10.2.4 Flood Control

Flood control reservoirs are designed and operated to moderate flood flows that enter into them. Flood moderation is achieved by storing a part of flood flows in the reservoir and releasing the balance. The degree of flood attenuation or moderation depends on the empty storage space available in the reservoir when the flood impinges on it. The achievement of this purpose requires the availability of empty storage space in the reservoir. As far as possible, the releases from the storage are kept smaller than the safe capacity of the downstream channel.

10.2.5 Navigation

Storage reservoirs may also be operated to maintain a stretch of downstream river navigable. The requisite depth of flow in the navigation section is maintained by releasing water from the dam. The demand for water for this purpose depends on the type and volume of traffic in the navigable waterways. The water requirements for navigation show a marked seasonal variation. There is seldom any demand during the wet period when sufficient depth of flow is available in the channel. The demands are at a maximum in the dry season when large releases are required to maintain the required depth.

10.2.6 Thermal/Nuclear Power Generation

Water is also an important input in thermal/nuclear power generation where it is used for cooling purposes. The simplest arrangement, known as once-through cooling, consists of diverting the water from the source to the power plant where it is used to cool the condensers and the heated water is returned to the source. Although this method has many advantages, the main being the low cost of construction, it is being gradually discarded due to thermal pollution of the receiving waters. Many new power plants use evaporative cooling tower systems. In such systems, a tall cooling tower is constructed and the water, after cooling the condensers, is passed through an air stream, cooled, and recycled to the condensers. In fact, cooling towers with recirculating water are typical visual identity of power plants.

10.2.7 Recreation

The benefits from recreation are derived when the reservoir is used for swimming, boating, skiing, and other water sports and picnics. Usually, the recreation benefits are incidental to other uses of the reservoir, and rarely a reservoir is constructed solely for recreation. The recreation activities are best supported by a reservoir, which remains nearly full during the recreation season. Large and rapid fluctuations in the water level of a reservoir are harmful from a recreational point of view because they can create marshy land near the rim of the reservoir. The famous Brindavan Gardens is located in the Indian State of Karnataka, adjoining the Krishnarajasagara Dam built across the river Cauveri. Water drawn from the Krishnarajasagara Dam meets the requirements of the Garden.

10.2.8 Minimum Flow Maintenance

Many times, it is necessary to release a certain minimum amount of water in the river below the reservoir from water quality (dilution of pollution) or environmental considerations. The release under this head varies seasonally and may get the highest priority.

10.3 RESERVOIR PLANNING

As reservoirs are capital-intensive projects, it is necessary that these projects are carefully planned and executed. All the available data should be analyzed and if necessary, further information should be gathered so that the best decision is taken with respect to location, size, and type of structure and auxiliary facilities. First, the best site for the dam and reservoir is selected and then a number of investigations are carried. These are described in what follows.

10.3.1 Site Selection Criteria for a Reservoir

The following factors should be kept in mind while selecting the site for a reservoir:

1. The reservoir site should be such that the leakage of water through the ground is minimum. Sites having permeable rocks reduce the water tightness of the reservoir. The rocks that allow less passage of water include shales, slates, schists, gneiss, and crystalline igneous rocks such as granite.
2. A suitable site for the dam must exist. The dam should be founded on a sound watertight rock base, and percolation below the dam should be minimum. The cost of the dam depends on the suitability of a site and is often a controlling factor in the site selection.
3. The reservoir basin should have a narrow opening in the valley so that the length of the dam is the least possible.
4. The cost of the real estate for the reservoir, including road, railway, rehabilitation, and resettlement, must be as small as possible.
5. The topography of the reservoir site should be such that it has adequate storage capacity without submerging excessive land and other properties.
6. The site should be such that a deep reservoir is formed. A deep reservoir is preferable to a shallow one because of the lower cost of the land submerged per unit of capacity, less evaporation losses due to a smaller water spread area, and less likelihood of weed growth.
7. The reservoir site should be such that it avoids or excludes water from those tributaries that have a high concentration of sediments in water.
8. The submergence area should not contain, to the extent possible, sites of archeological importance, major towns, forested area, and habitats of rare species.
9. The soil and rock mass at the reservoir site must not contain harmful minerals and salts.

It is often not possible to find a site that satisfies all of the earlier conditions. In such cases, the planner uses discretion to choose the site that best meets the project objectives.

10.3.2 Investigations for Planning a Reservoir

The various investigations required for reservoir planning are described next:

Engineering Surveys

The area around the potential dam site is surveyed in detail and a map with a small contour interval (of the order of a few meters) is prepared. From the map, the following are prepared:

(a) area-elevation curve,
(b) storage-elevation curve,
(c) details of the land and property likely to be submerged, and
(d) suitable site to locate the dam.

Conventionally, the topographic maps are prepared by carrying out a plain-table survey of the area. However, these days the maps can be quickly prepared using remote sensing techniques. This technique saves time and cost is very handy for areas which are difficult to access. This technique has been described in Chapter 3. Currently, LIDARs are finding increasing deployment in applications that require precise elevation data. LIDAR stands for "light detection and ranging". It is a method to determine distances by laser scanning. LIDAR sensor is typically mounted on an aircraft.

Area-Elevation and Storage-Elevation Curves

Once the site of the dam is finalized, a map on the reservoir area with a small contour interval, say 1 m or so, is prepared. Starting with the lowest contour, the areas enclosed by the successive contours can be determined with a planimeter. Clearly, as elevation of a contour increases, the enclosed area also increases. Thus, a curve may be drawn with elevation on the *X*-axis and area on the *Y*-axis. Such a curve for a reservoir is shown in Fig. 10.2. The contour plan also shows the water spread corresponding to the maximum water level in the reservoir. This information is used to determine the area likely to come under submergence.

The reservoir capacity or the volume of storage corresponding to a given water level may be calculated by the trapezoidal formula. Thus, if A1 and A2 are the areas between two successive contours, and h is the contour interval, the intermediate storage volume V can be calculated using the formula:

$$V = (A1 + A2)h/2 \tag{10.1}$$

The total reservoir capacity at a given elevation is computed by adding the incremental volumes up to that elevation. The storage volumes corresponding to various water surface elevations may be calculated and a curve, called capacity curve, may be plotted between elevation and storage as shown in Fig. 10.2.

Geological Investigations

In all major civil engineering projects, geological surveys are critically important. The geological investigations are required to give detailed information about the following:

1. hydro-geology of the area;
2. suitability of foundations for the dam;
3. geological and structural features, such as folds, faults, and fissures, in the reservoir basin;
4. water tightness of reservoir basin;

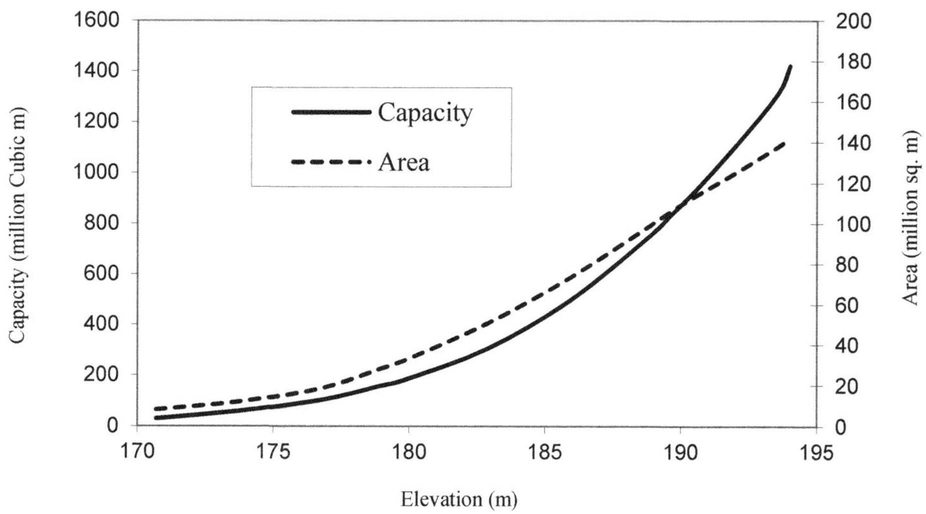

FIGURE 10.2

Elevation—area—capacity curves of Dharoi reservoir.

5. location and extent of permeable and soluble rocks, if any;
6. groundwater conditions in the region; and
7. location of quarries for materials required for dam construction and quantities available from them.

The geological investigations cost little in comparison to the total cost of the project; typically, it may amount to between 0.25 and 1% of the project cost. This relatively small amount represents a valuable insurance against difficulties that might arise during construction. There have been instances where the discovery of adverse geological features during construction led to disruption of work and time plus cost overruns.

An important requirement for the reservoir site is that there should be no danger of large leakage when the ground is under pressure from the full head of water in the reservoir. The geology of the dam site is important from the point of view of a suitable foundation for the dam. The nature of subsurface geology should be explored by trial bores or various means of geophysical exploration. The geology of the catchment area should also be studied since it affects runoff and percolation.

Hydrological Investigations

These investigations are important in reservoir planning as the reservoir size, height of dam, capacity of the irrigation canals, the installed capacity of the power house, etc. depend on the available water. The hydrological investigations can be divided into the following heads:

a. study of runoff pattern at the proposed dam site to determine the water availability and storage capacity required to meet the target demands,
b. determination of the hydrograph of the worst flood for spillway design,
c. estimation of evaporation and other losses from the reservoir particularly in an arid area,
d. sedimentation studies to determine the sediment inflow into the reservoir and its impact on the reservoir performance, and
e. simulation studies to study the performance of the reservoir under a given inflow series and demand pattern.

For reservoir planning, the first step is the correct assessment of water availability at the site. This requires a sufficiently long sequence of data; the data length depends on the type of storage, type of project, and variability of flows. An analysis carried using a longer period of data will give more reliable results. A longer data series would be required for over-the-year-storages. A comparatively shorter length will suffice for within-the-year storage where the spill occurs almost every year and the critical period is of the duration of a few months. Different agencies have issued guidelines regarding the minimum length of data required, and one such guideline is given in Table 10.1.

The flow sequences required for the planning of projects need to be prepared for an appropriate time unit so that the simulation studies can have the desired resolution. The number and type of additional hydro-meteorological stations to be set up in the catchment and the command areas, if any, are decided during hydrologic investigations. While fixing the location of additional stations, future requirements for the operational stage of the project should also be considered.

After assessing the data required and availability, various techniques are used to extend/generate long-term flow sequences (if necessary) for proper evaluation of water availability and project planning. Since rainfall data are normally available for a longer period than runoff data, it is common to extend the runoff data using rainfall data. The water availability analysis is described in a later section.

Table 10.1 Minimum length of data required for design of various projects.

Type of project	Minimum data length for use in analysis
Diversion projects	10 years
Within-the-year storage projects	25 years
Over the year storage projects	40 years
Systems involving a combination of the above	Depending on the predominant element

Reconnaissance (Preliminary) Investigations

The main purpose of such investigations is to screen out the inferior alternatives and to decide on further data that need to be collected for detailed feasibility investigations of the remaining selectable alternatives. A reconnaissance survey will identify the scope of a project plan with respect to its geographical location, project functions, approximate size of its various components, likely problem areas, and time and cost of conducting feasibility investigations.

Actually, a reconnaissance investigation is a preliminary version of a feasibility investigation carried out in a short time and with less accuracy. It considers all the physical, engineering, economic, environmental, and social aspects related to the project. It is usually conducted with the available data. Collection of some new data, if considered necessary for reconnaissance, is made by surveys. These may include the simple cross-section (instead of detailed topography) of the stream at the dam site, surface investigations of geological conditions at the dam site, subsurface explorations for dam foundation, quality and quantity of available construction materials, and so forth. Preliminary designs are made by using short-cut methods (e.g., empirical curves, tables, and previous experience). Cost and benefits of the project are also estimated. Based on the results of the reconnaissance, a set of project plans is selected for subsequent detailed investigation.

Feasibility Investigation

The aim of feasibility investigations is to ascertain the soundness and justification, or otherwise, of different alternative plans chosen after carrying out preliminary investigations. The analyses need to be of high accuracy and dependability so that the reliability of results, on the basis of which the final selection of the project plan is made, is not questioned. However, the completion of the feasibility investigation does not mean the end of the project planning. Minor changes may be required for various reasons before construction and even during construction.

The first step in the feasibility investigation is to collect or update the basic data of different types. The accuracy and reliability levels of these data must be consistent with the degree of accuracy required for feasibility justifications. The basic data for dams and reservoirs include topographic surveys of sites, information on streamflow and design flood, land costs, reservoir clearing costs, communication facilities, climatic conditions affecting construction, fishery, and wildlife to be preserved, construction material, foundation conditions of dam site and reservoir area, availability of trained manpower and environmental and other considerations. The facilities and appurtenances necessary for the functioning of the project must be specified and considered while making cost estimates.

Preconstruction Investigations

These investigations are carried out if the time elapsed between the feasibility investigation and the commencement of construction is large. It is essential that any new information that might have become available during the intervening time is incorporated into final designs. For example, a flood of large magnitude might have occurred during this intervening period and this may necessitate changes in the design flood for the project and consequently the design of the spillway and related structures. Best (1998) has provided a discussion on investigations for dam construction.

10.4 ESTIMATION OF WATER YIELD USING FLOW DURATION CURVES

A popular method to study streamflow variability is the flow duration curves. A flow duration curve of a stream is a plot of discharge against the percent of time the flow was equaled or exceeded. This curve is also known as the discharge–frequency curve. To prepare it, streamflow data are arranged in the descending order of discharges using class intervals. The data used can pertain to any time step, daily, weekly, 10-daily, or monthly. If N number of data points are used, the plotting position of any discharge (or class value) Q is

$$P = m/(N + 1)100\% \tag{10.2}$$

where m is the order number of the discharge (or class value), and P is the percentage probability of the flow magnitude being equaled or exceeded. The plot of the discharge Q against P is the flow duration curve (Fig. 10.3). An arithmetic scale, a semilog scale, or log–log graph may be used, depending on the range of data and use of the plot. The flow duration curve represents the cumulative frequency distribution and shows the streamflow variation of an average year. The ordinate Q_P at any percentage probability P represents the flow magnitude in an average year that can be expected to be equaled or exceeded P% of the time and is termed as P% dependable flow. In a perennial river, Q_{100} (100% dependable flow) is a finite value, for example, it is about 60 units in Fig. 10.3. In an intermittent or ephemeral river, streamflow is nil for a finite part of a year, and as such Q_{100} is zero.

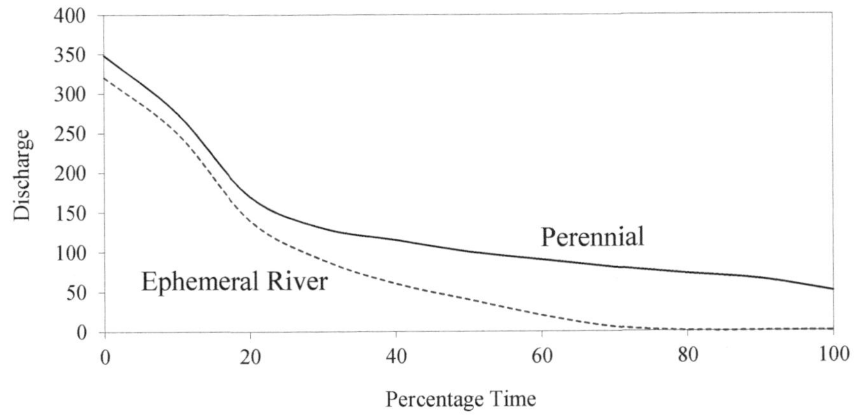

FIGURE 10.3

Typical flow duration curves.

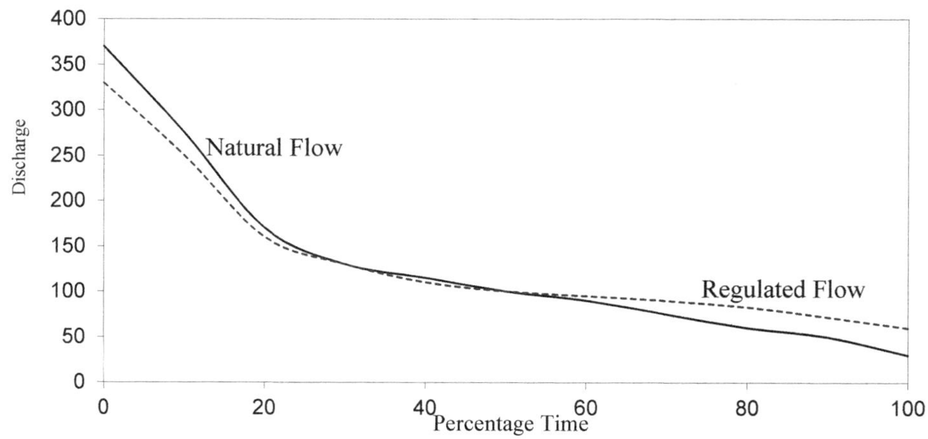

FIGURE 10.4

Reservoir regulation effect on flow duration curve.

Some important characteristics of a flow duration curve are:

1. The slope of a flow duration curve depends on the time interval of the data selected. For example, a daily streamflow data gives a steeper curve than a curve based on the monthly data for the same stream. This is due to the smoothening of small peaks in monthly data.
2. The presence of a reservoir on the stream considerably modifies the virgin-flow duration curve depending on the nature of flow regulation. Fig. 10.4 shows the typical reservoir regulation effect.
3. The virgin-flow duration curve plots as a straight line on a log probability paper at least over the central region. From this property, various coefficients expressing the variability of flow in a stream can be developed to describe and compare different streams. A steep slope of the flow duration curve indicates a stream with a highly variable discharge. On the other hand, a flat slope indicates a slow response of the catchment and a small variability. At the lower end of the curve, a large flat portion indicates considerable base flow. A flat curve on the upper portion is typical of river basins having large flood plains and also of rivers having large snowfall.
4. The chronological sequence of occurrence of the flow is masked in the flow duration curve. A discharge of say, 1000 cumec, in a stream will have the same percentage P whether it occurred in January or June. This aspect, a serious handicap, must be kept in mind while interpreting a flow duration curve.

10.4.1 Procedure to Prepare a Flow Duration Curve

The Institute of Hydrology (1980) has outlined the procedure for estimating the flow duration curve depending on the availability of data at or near the site of interest. The guidelines suggested for a given length of record are given in the following:

More than 10 years of records: Such records need no adjustment or standardization as this period of data will probably provide a sufficiently accurate flow duration curve.

Two to 10 years of records: For this length of records, divide the daily flow data by the average flow over the period of record before analysis. This overcomes to a great extent the departures due to wet or dry years. The conversion to the long-term flow duration curve is made using an estimate of the long-term average flow.

Less than 2 years: This length of record may be treated as short and some indirect approaches are used for flow duration curve computations. These approaches are based on the use of catchment characteristics.

Preparing Flow Duration Curves from Daily Flow Data

The flow duration curves from daily flow data may be prepared using the following steps:

i. Choose a class interval (CI) such that about 25–30 classes are formed.
ii. Assign each day's discharge to its appropriate CI.
iii. Count the total number of days in each CI.
iv. Cumulate the number of days in each CI and get the number of days above the lower limit of each CI.
v. Compute the probabilities of exceedance by dividing the quantities obtained from step (iv) by the total number of days in the record (for example, 365 if the 1-year record is being used to construct the flow duration curve).
vi. Multiply the probabilities of exceedance obtained from step (v) by 100 to get the percentage exceedance.
vii. Plot the probabilities of exceedance in percent against the corresponding lower bound of CI on linear graph paper. Sometimes the flow duration curve better approximates as a straight line if the log normal probability paper is used in place of linear graph paper.

Sometimes a flow duration curve is prepared for duration other than the base duration for which data are available. For example, the daily data at a site might be available but the flow duration curve of 10-day flows might be needed. The flow duration curves for durations other than the base duration can be prepared as follows:

i. Derive a hydrograph whose values are not simply daily discharges but are average discharges over the previous D days (D-duration of flow duration curve). It is equivalent to the outcome of passing a moving average of D-day duration through the daily data. Generally, 1, 5, 7, 10, 30, 60, 90, 180, and 365 days are adopted as the standard values for D.
ii. Plot the flow duration curve from the data of discharge hydrograph derived in step (i) using the procedure described previously. Fig. 10.5 illustrates typical flow duration curves for different durations.

10.4.2 Use of Flow Duration Curves

The flow duration curves are used to estimate the dependable flows of a river for various reliabilities. In various countries, guidelines have been formulated for the planning of river valley projects for different purposes. For example, according to the practice in India, irrigation projects are planned using 75% dependable flow. Hydropower and drinking water projects are planned with 90% and 100% dependable flows, respectively. The 90% dependable flow is also used as a measure of ground water

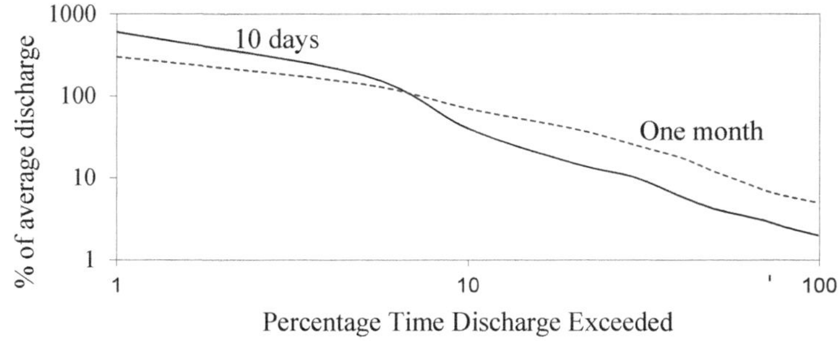

FIGURE 10.5

Flow duration curves for different durations.

contribution to stream flow. This same value can also be used as a measure of run-of-the-river hydropower potential.

Other important uses of flow duration curves are:

1. to evaluate the characteristics of the hydropower potential of a river,
2. to design drainage systems,
3. for flood control studies,
4. to compute the sediment load and dissolved solids load of a stream, and
5. to compare adjacent catchments with a view to extend streamflow data.

10.5 HYDROPOWER GENERATION

Since time immemorial mankind has been using the energy of water falling from a height to perform useful works, for example, water wheels to grind grains. Hydroelectric energy is produced by converting the mechanical energy of falling water into electrical energy by a turbine and generator. Hydropower is a renewable and clean power source that does not produce any air or water pollution and is one of the most efficient ways to generate electricity. In multipurpose projects, the generation of hydroelectric energy is combined with other uses, such as irrigation and water supply, resulting in tremendous savings of money and other resources. Another advantage of hydropower over other forms of electricity generation is that reservoirs can store water during times of low demand and can quickly start generating during the peak hours of electricity use. Thermal power plants take much longer to start up from cold than hydropower plants.

With respect to types of site development, there are four major classifications of hydroelectric projects: storage, barrages, run-of-river, and pumped storage. Storage projects usually have heads in the medium to high range (greater than 25 m) and have provisions to store relatively large volumes of water during periods of high streamflow to provide water for power generation during periods of deficient streamflow. The power house is commonly located at the toe of the dam, although in some cases it might be away from the dam.

Peaking operation is frequently associated with storage projects and this requires large and sometimes rapid fluctuations in releases of water through the generating units. It is often necessary to provide facilities to even out the fluctuations in the discharge if rapid changes of discharge below the project are not desired. For example, such an arrangement exists in the Bhakra Nangal project in India where a small barrage (Nangal barrage) has been constructed at some distance downstream of the Bhakra dam (a major power project).

A barrage, also known as pondage, has a very small storage capacity. It can regulate the flow only up to minor extent and generate power according to the weekly or daily variation of the load. Hence, the tail water fluctuations are usually quite large, particularly in peaking operations. If the cycle of peaking operation is a single day, the pondage requirements are based on the flow volume needed to sustain generation for 12 h. If more storage capacity is available and large fluctuations in the reservoir level are permissible, a weekly cycle of peaking operation may be considered. Since industrial and commercial consumption of power is significantly lower on weekends than on week days, an "off-peak" period is created from Friday evening until Monday morning. During this period, water can be accumulated in the pondage for later use.

Run-of-river plants have little or no storage and, therefore, must generate power from streamflow as it occurs with little or no benefit from at-site regulation. These projects generally have productive heads in the low to medium range (5–30 m) and are quite frequently associated with navigation or other multipurpose developments. For a base-load run-of-river project to be feasible, the stream must have a relatively high baseflow. Sometimes, the falls in irrigation canals are also used to generate energy. Because of (near) absence of storage, there is usually very little operational flexibility in these projects. The existence of upstream storage project(s) may make a run-of-river project in the lower part of the basin feasible where it would not otherwise be. However, the storage projects must provide a regulated outflow that is useable.

10.5.1 Pumped Storage Schemes

Pumped storage projects consist of a high-level forebay where inflow or pumped water is stored until it is needed for power generation and a low-level afterbay where the power releases are stored. These projects depend on pumped water as a partial or total source for generating electric energy. The pumping and generation are done by units composed of reversible pump turbines and generator motors connecting the forebay and afterbay. The water is pumped from the afterbay to the forebay when the normal power demand is low and released from the forebay to the afterbay to generate power when the demand is high. Such projects derive their usefulness from the fact that the demand for power is generally low at night and on weekends and therefore, pumping energy at a very low cost will be available from idle generating facilities. The feasibility of pumped storage developments arises from the need for relatively large amounts of peaking capacity, the availability of pumping energy at a cheap rate, and a load with an off-peak period long enough to permit the required amount of pumping.

There are three types of pumped storage development: diversion, off-channel, and in-channel. The diversion type of development usually consists of pumping in one basin to a forebay on or near the divide between that basin and an adjacent basin and it does not recirculate the water between the forebay and afterbay. The water is released through generating units into an afterbay located in the adjacent basin. The advantages are that it may be possible to pump against a head that is very small in relation to the head for generating, provided a source of water for pumping can be located at an elevation that is not too far below the forebay. This scheme has the disadvantage that separate pump-motor and turbine-generator units are required, whereas a single reversible unit is used in the other two types of pumped storage development.

The off-channel type of pumped storage development is most suitable when a forebay site exists on a hill above a stream where an afterbay can be constructed. The head differential should be large and the forebay site should be close to the afterbay to avoid head loss and reduce construction costs. The water requirement to support this type of development is not large after the initial supply has been provided. Since the system primarily recirculates water, it is necessary to provide water only to replace losses due to evaporation and leakage.

In the in-channel type of pumped storage development, the reservoir of a conventional power project is used as a forebay. The afterbay could be a reservoir from a downstream project or a reservoir provided solely to serve as an afterbay. This type of development is more attractive if the cost of the afterbay is shared with other purposes. In an in-channel pumped storage project, the maximum possible amount of water is pumped back from the afterbay to the forebay during the low-flow period. During the less severe dry periods, only a part of the water that is used to generate power is pumped back into the forebay. During periods of high streamflow, none of the water is pumped back. This type of project is most valuable when there is a large difference between the streamflow quantities in low-flow periods and the average conditions because the reversible mechanism can add substantial firm energy with attendant pumping costs that are far less than what would be incurred with other types of development.

These days, pumped storage schemes have become very popular. The system is considered as a giant "green" battery because it stores power (in the form of water) and releases it when needed.

10.5.2 Components of Hydropower Projects

In a storage project, the reservoir behind the dam stores water that is used to generate electric power. The portion of the reservoir that is immediately upstream of the intake structure is known as forebay. The water is withdrawn from the forebay through an intake structure and is carried to the power house through penstocks. A penstock is a conduit to carry water from the forebay to turbines and gates and valves are installed to control the water flow. Depending on the site conditions, open channels or tunnels may instead be used. A surge tank is constructed to handle the problems of water hammer. The turbines and generators are installed in the power house. The water comes out of turbines through the draft tube and joins the tail water.

The term "static head" (see Fig. 10.6) denotes the difference between the water surface elevation in the forebay and the elevation of tail water (TWL). The "net head" is a static head with less losses in the penstock:

$$\text{Net head} = \text{Static head} - \text{Losses} = \text{Water surface elevation in forebay} - \text{TWL} - \text{Penstock losses} \tag{10.3}$$

Although the head is usually related to the dam height, a low dam can yield a high head if the turbines and generators of the powerhouse are located some distance downstream of the dam. The amount of power generated is a function of discharge and the hydraulic head. It can be computed as

$$P = 9.817QH\eta \tag{10.4}$$

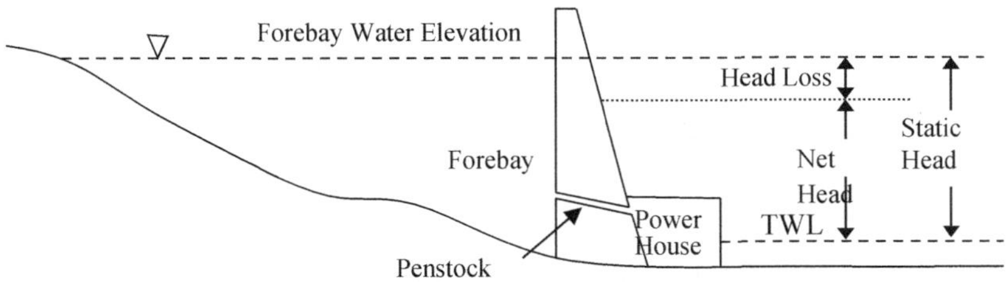

FIGURE 10.6

Terms related to head in a hydropower plant.

where P is the electric power in kW, Q is the discharge through the power plant in m^3/s, H is the net head in m, and η is the overall efficiency of the power plant expressed as a ratio (usually about 0.85). The overall efficiency of the power plant is obtained by multiplying the turbine efficiency with the generator efficiency. Hydroelectric power generation depends on the volume of water passing through turbines and the effective head. Thus, the same amount of power can be produced by releasing more water at a low head or less water at a high head. Furthermore, it is better to construct these plants in hilly areas where steep slopes provide high heads.

Load (Demand) of Hydropower Projects

With respect to the type of load served, hydropower projects can be classified into two categories: base-load plants and peaking plants. Base-load projects generate power to meet the base-load demand (the demand that exists 100% of the time). Usually, the base load is served by thermal generating facilities. But if there is a relatively abundant supply of water with a high degree of reliability and fuel is relatively scarce, hydroelectric projects may also be constructed to serve the base load. In this respect, the functionality of a project may rapidly change depending upon the supply and demand situation of electricity.

Peaking plants supplement base-load generation during periods of peak power demands. These plants must have sufficient capacity to satisfy peak demands of a system and water should be sufficient to provide the peaking support for as long and as often as is needed. In general, a peaking hydroelectric plant is desirable in a system that has thermal generation facilities to meet base-load demands. The hydroelectric generating facilities are particularly adaptable to the peaking operation because their output can be changed rapidly. Further, the seasonal variations in streamflow can be easily offset by providing storage.

The demand for electrical energy is known as load. The ratio of the average power demand to peak power demand for the time period under consideration is known as a load factor and this is computed on a daily, weekly, monthly, or annual basis. Thus,

$$\text{Load factor} = [\text{Average power demand}]/[\text{Peak power demand}] \tag{10.5}$$

an appropriate time unit is chosen in this equation.

The generation of energy is limited by the installed capacity of the power plant. The term firm energy is used to denote the energy that is available with 100% reliability. The concept is similar to firm power. The energy that is available over and above the firm energy is known as secondary energy.

The generation of energy depends on the installed capacity of the power plant. Depending on the configuration of the system of which the hydropower plant is a part, it may be operated to meet the base load or the peak load. It is more efficient to operate the hydropower plant to meet the peak load because these plants can be put on and off at short notice and there is no wastage of resources. However, there might be other considerations in operation.

10.5.3 Estimating Hydropower Potential and Demand

Analogous to the concept of firm water, firm power is the maximum quantity of power that can be guaranteed to be delivered 100% of the time according to some prescribed distribution. The hydroelectric power potential is determined on the basis of the critical period as indicated by the historical streamflow record. The critical period is a function of the power demand, streamflow, and available storage. If a project serves more than one purpose and if, in serving another purpose, some of the storage or streamflow is not available for power production, the streamflow data should be adjusted to reflect the "loss" Losses, such as evaporation, leakage, and station use, must also be deducted from the available flow before calculating the potential energy. The amount of power generated over a time, or energy, is expressed in kilowatt-hour (kW-hr). It can be computed as follows:

$$KWHR = 9.817QHT\eta \quad (10.6)$$

in which KWHR is the hydropower generated during the period in kW-hr and T is the number of hours in the period.

While computing the average head, the tailwater elevation should represent average conditions during the time when power generation actually occurs. For example, in a peaking project that usually generates power at or near the installed capacity for a short duration, the tailwater elevation should correspond to the discharge at the installed capacity rather than to the average discharge. Likewise, if there are releases that do not pass through the generating units but significantly affect the tailwater, the tailwater elevation should reflect the combination of power releases and other releases.

Two methods are used to estimate the hydropower potential at a given site: The flow duration curve method and Sequential Streamflow Routing (SSR) method. In the first method, the flow duration curve at the site is the basic input. The net head for various discharges is estimated. Using the data of the useable range of flow duration curve and head versus discharge data, a head-duration curve is developed. The hydropower equation is used to estimate the power generated at many points on the flow duration curve and a power duration curve is developed. The average annual energy (AE in kW-hr) and the dependable capacity (DC in kW) can now be calculated:

$$AE = 87.6 \int_0^{100} Pdp \quad (10.7a)$$

$$DC = 0.01 \int_0^{100} Pdp \quad (10.7b)$$

where P is power in kW and p is the percent of time. The major advantage of this method is that it is simple and fast. However, it cannot take into account the installed capacity and key project features, such as intake and power plant characteristics.

In the SSR method, the time step size and period of analysis are chosen and the operation of the reservoir is simulated. For each time period, the reservoir outflow is computed using the continuity equation and other constraints on operation. The amount of energy generated corresponding to this outflow and head is calculated using Eq. (10.6). The process is repeated for all the time steps. Now the average annual, monthly, etc. generation as well as firm energy can be computed. An advantage of this method is that it can take into account the reservoir characteristics as well as power plant features. The results of this method are more realistic as compared to the flow duration curve method.

Knowing the average head, the volume of water needed to generate the desired amount of energy for a given time period can be computed using Eq. (10.6). The summation of volumes for various periods will give the total volume required.

Example 10.1. Compute the volume of water required to generate 4.75 MW-hr of electric energy for 1 day if the average head is 95.0 m and efficiency of the power plant is 0.85.

Solution. The energy to be generated is 4750 kW-hr and $T = 24$. From Eq. (10.6)

$$9.817 \times {}^{*}Q \times 95.0 \times 24 \times 0.85 = 4750$$

$$\text{or} \quad Q = 0.2497 \text{ cumec.}$$

$$\text{Volume of water required for one day} = 0.2497 \times 24 \times 3600 = 21574.08 \text{ m}^3.$$

10.6 RESERVOIR LOSSES

A portion of the water stored in a reservoir is lost and is not available for beneficial use due to various processes. The major causes of the loss of water are evaporation, leakage through the body of the dam, and groundwater flow.

10.6.1 Evaporation Losses

The loss of water due to evaporation depends on the nature of the evaporating surface and meteorological factors (see Section 2.5.1 for detail). The factors affecting the evaporation process are radiation, wind speed, and vapor pressure of the air overlying the surface. The amount of evaporation also varies with latitude, season, time of day, humidity, and condition of the sky. It is difficult to categorically express the relative effect of the controlling meteorological factors. If radiation exchange and all other meteorological elements were constant over a shallow lake for a considerable time, the temperature of water and evaporation would become constant. If the wind speed were suddenly doubled, the rate of evaporation would also be double for some time.

The quality of water in a reservoir also affects evaporation to a small extent. This reduction takes place because dissolved solids reduce the vapor pressure for evaporation, the temperature of water rises and this partially offsets the effect of reduction in vapor pressure. Moreover, any foreign material which affects the reflective property of the water surface affects evaporation.

A pan evaporimeter is commonly used to estimate evaporation from a lake. The pans can be installed in three ways: on the land surface, sunken in ground, and floating on water surface. The pans installed on or above the ground surface show a little higher evaporation as extra heat is absorbed by the side walls. The main advantages of surface pan are economy and ease of installation and maintenance. An estimate of the depth of evaporation can be obtained by multiplying the pan evaporation by the pan coefficient.

The evaporation from a reservoir can be best approximated by a pan floating on the lake's surface. However, the installation and maintenance expenses are quite large. Observation of evaporation data is difficult and many times, splashing takes place which renders the records unreliable. Due to these reasons, these pans are not very common in use.

Evaporation losses from a reservoir depend on the water spread area and the rate of evaporation. These are expressed in terms of water depth. The depth of water lost by evaporation may typically be 100–200 cm annually. Evaporation in hot months is 2–5 times that in winter months. In humid areas, low values, say about 50 cm per annum, may be common while evaporation can go up to about 350 cm per annum in arid climates.

Due to the significant loss of water in arid climates, extensive efforts have been made to control the same. In India, efforts have been made to control evaporation losses by laying layers of monomolecular chemical films on the surface of the reservoir. These films try to retard evaporation by curtailing the exchange of energy between water and atmosphere. However, these films have not been very successful because they are expensive to lay and easily break due to wind and waves.

Example 10.2. The water spread area of a reservoir is 15 sq. km and the evaporation from a pan (pan coefficient = 0.7) at the dam for a day is 1.0 cm. Compute evaporation loss from the reservoir.

Solution. Evaporation loss = $15 \times 1000 \times 1000 \times 0.01 \times 0.7 = 105000\ m^3$.

Empirical Equations to Estimate Evaporation from Reservoirs

A number of empirical equations are in use for the estimation of evaporation from reservoirs. Most of these are some form of the generalized Dalton's Law of Evaporation

$$E_L = C(e_w - e_a) \tag{10.8}$$

where E_L = evaporation from lake (mm/day), C = a coefficient, e_w = saturated vapor pressure at the water surface temp in mm of Hg, e_a = actual vapor pressure of overlying air at a specified height, in mm of Hg.

The general form of the equations to estimate evaporation based on meteorological data is as follows:

$$E_L = K\, f(u)(e_w - e_a) \tag{10.9}$$

where f(u) = wind speed correction function, K is a coefficient.

Meyer's formula is a popular empirical method to compute lake evaporation

$$E_L = K_M \left(1 + \frac{u_9}{16}\right)(e_w - e_a) \tag{10.10}$$

where K_M = coefficient to account for influencing factors, = 0.36 for large, deep lakes; 0.5 for small, shallow ponds; u_9 = monthly mean wind velocity in km/h at 9 m above ground = $u_1(9)^{1/7}$.

10.6.2 Seepage Losses

A reservoir also exchanges flow with aquifers though the magnitude is small compared to the surface water inflow. The amount of this flow depends on the physiographical features and soil characteristics in the vicinity and the position of the water table. Assuming homogeneous conditions, the flow can be computed by the Darcy law.

Seepage losses are difficult to estimate and are not important in most cases. Seepage is generally more when the reservoir is underlain by porous strata having ample outlets beneath the surrounding hills or under the dam. Absorption losses may be significant in the initial stages but gradually reduce as the soil pores become saturated. Generally, the reservoir banks are not so permeable as to cause significant leakage. Where the banks have continuous seams of porous strata or are made of fractured rock formations, pressure grouting is done to seal the fractured rock. The water lost from the reservoir to the ground water through seepage and passages in the reservoir bed is not amenable to measurement. However, an assessment of the losses can be made by considering the inflow, the outflow, and precipitation over the reservoir and evaporation loss and then labeling the unaccounted loss of water due to seepage and absorption. Evidently, the accuracy of the determination depends on the precision with which other variables of the equation are estimated.

10.6.3 Leakage through Dam

This component of outflow consists of the loss of water from the reservoir on account of leakage through the body of the dam as well as through gates and spillways. It is not easily possible to relate these losses with a measurable quantity. For example, the losses through the gates or valves of undersluices depend on their design, installation, and maintenance. A simplifying assumption is that these losses linearly vary with the reservoir level. In general, the amount of water lost due to these reasons varies between 0.5% and 4% of the total discharge through the structure.

10.6.4 Water Balance of a Reservoir

The water balance equation for a reservoir is nothing but the mass balance or continuity equation. This equation states that the sum of inflow and outflow components and change in storage (with appropriate signs) must be zero over a given time interval. This equation can be expressed as

$$I_S + I_G + P - E - Q - L - \Delta S \pm \delta = 0 \tag{10.11}$$

where I_S is the surface water inflow into the reservoir; I_G is the ground water inflow into the reservoir; P is the precipitation on the surface of a reservoir; R is the release from the reservoir; E is the evaporation from the reservoir; L is the storage loss including seepage; ΔS is the change in reservoir storage during the period of computation; and δ is the error term. All the terms are expressed either in volume units or depth units.

The water mainly enters a reservoir through surface inflow and direct precipitation; the water that leaves the reservoir comprises releases through outlets and spillways, evaporation, and losses due to seepage. To the extent possible, all the components of the water balance equation should be independently measured or estimated. The error term δ represents the net effect of errors in the estimation of different components. In practice, it is likely that errors will be present while measuring or computing various terms of the water balance equation. A large value of δ represents a significant error in estimating different variables involved in Eq. (10.9). However, a small value of δ does not necessarily indicate that errors are small; may be the errors of opposite sign balance themselves.

Components of reservoir water balance are associated with many hydrological processes, such as precipitation directly falling over the reservoir, evaporation, inflows from the upstream catchment, and outflow through outlets including spillways. However, often the only observations available for water balance computations are weather data and reservoir water levels. Regarding surface inflows, data are commonly limited to the mainstream (s); flows in numerous small tributaries are very rarely measured and their contributions are typically estimated on their catchment area and rainfalls.

The water balance equation may be applied for any time interval. The mean water balance is a term normally used for computations that are spread over an annual cycle, for example, a water year. Sometimes this term is also used for seasonal water balances. The computations of the mean water balance are simplest in nature; with the shortening of the computational period, a more detailed accounting procedure is required. The additional factors that are to be included in the computations include bank storage during reservoir filling, water loss due to water and ice left on the banks when the reservoir is drawn down and the return of this water to the reservoir later on. Sokolov and Chapman (1974) provide a detailed discussion on water balance computations.

10.7 RANGE ANALYSIS

This is an important component of storage analysis. Let x_i, $i = 1, 2, \ldots n$, represent a time series of flows at a particular site on a stream. This time series can be a daily, weekly, or monthly series. Assume that these flows are being fed into a big reservoir and an amount equal to the mean of the series (x_m) is being taken out. Let

$$S_1 = \Delta x_1 = x_1 - x_m \tag{10.12}$$

be the increase or decrease (depending on the relative magnitude of x_1 and x_m) in the reservoir content at the end of the first time period. In this manner,

$$S_i = \Delta x_1 + \Delta x_2 \ldots + \Delta x_i = \sum_{j=1}^{i} \Delta x_i \tag{10.13}$$

represents such a change at the end of the ith time period. The maximum of n values of S_i, denoted by S_n^+ is called the maximum surplus, or surplus or maximum partial sum of deviates. The minimum of n values of S_i represented by S_n^- is called minimum deficit or deficit or minimum partial sum of deviations. The sum of magnitudes of surplus and deficit

$$R_n = S_n^+ + |S_n^-| \tag{10.14}$$

is called the *range*. The terms surplus, deficit, and range are graphically represented in Fig. 10.7.

The statistic range is always greater than zero and represents the storage capacity required in a reservoir to maintain an outflow equal to x_m if the x_i were inflows. However, this implies that there is no loss of water due to spill or other reasons.

The statistic range depends on the properties of the series as well as its length. As the length of the series increases, the range will either increase or remain the same. So far, the outflow from the

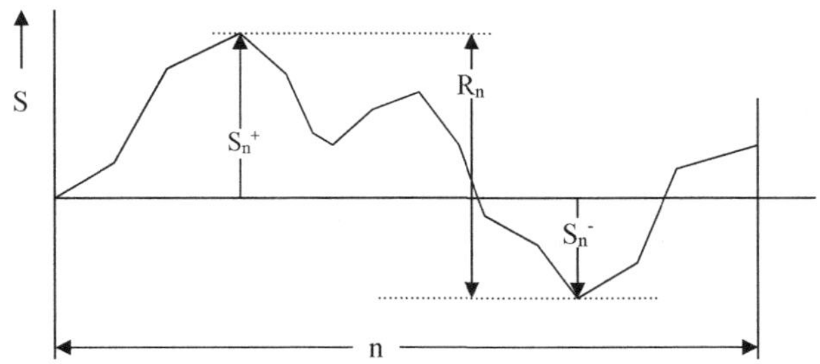

FIGURE 10.7

Definition diagram of surplus, deficit, and range.

reservoir is assumed to be equal to the mean of all inflows. Different values of the parameters under discussion are obtained if the outflow is not x_m but is equal to the mean of the subseries of size n. In such cases, the adjective adjusted is used with the parameters and they are called adjusted surplus, adjusted deficit, and adjusted range.

10.7.1 Hurst Phenomenon

Extensive investigations on the properties of the range were carried out by Hurst et al. (1965). It was concluded that the rescaled range, R/σ, where σ is the standard deviation of the data, increases with the length of the series. It can be proved if a series follows a normal distribution and its members are independent of each other then, for a large value of n

$$R/\sigma = 0.5\sqrt{n\pi} = 1.25\sqrt{n} \tag{10.15}$$

However, Hurst found that, for the natural phenomenon, the relationship between R/σ and the length of the series is given by

$$R/\sigma = (n/2)^H \tag{10.16}$$

where H is a variable. This equation was derived based on an analysis of 75 phenomena and 690 portions of these. It was found that the variable H is normally distributed with a mean of 0.73 and a standard deviation of 0.09. Of the above two equations for rescaled range, Eq. (10.15) expresses that it increases with a 0.5 power of n while Eq. (10.16) shows that it increases with a 0.73 power of n. This discrepancy in the value of exponent H is termed as "*Hurst phenomenon*."

After the discovery of this unusual behavior of natural variables by Hurst, a tremendous amount of research has been conducted to study the causes of this behavior and a number of models have been developed to reproduce this phenomenon in the time series models. Before going for a discussion of the causes of the Hurst phenomenon, the dependence structure of hydrologic time series is discussed.

10.7.2 Dependence in Hydrologic Time Series

In a dependent time series, an element is influenced by its predecessors or the past of the series shapes the present. The dependence of a series can be analyzed either by correlogram analysis or range analysis. Assume that the hydrologic time series being studied is stationary. Such a series is generated by a stationary process whose probability laws do not change with time. Generally, the geophysical, biological, and other natural processes are nonstationary but these can be assumed stationary for a relatively short time span.

A hydrologic time series displays two types of dependence: long-term and short-term. It is observed that the autocorrelation coefficient of a hydrologic time series dies out as the lag increases. This implies that a particular value of the variable is influenced only by the recent past values of the series, the distant values do not affect the present. This type of dependence is termed as short-term dependence and the process is said to be a short-memory process. In these processes, the incidents tend to fade from the system memory as time passes. Although this also looks intuitively correct, this observation fails to explain the Hurst phenomenon.

The Hurst phenomenon can be satisfactorily explained by the long-term dependence which implies that the process has infinite memory. The long-term dependence is associated with the failure of the correlogram to die at high lags. However, it is difficult to explain this physically. For example, it is hard to figure out as to how the discharge of a particular day is affected by the discharge of say 100 past days and by what mechanisms the impact of a hydrological event is carried over for years together. These counter arguments have given rise to a controversy about the appropriate explanation of the Hurst phenomenon. The short-term dependence, although physically believable, cannot explain the Hurst phenomenon. On the other hand, long-term dependence can explain this feature but it cannot be physically explained. Two types of models have been developed in hydrology corresponding to these two types of dependence: the long-memory models and the short-memory models.

The concept of stationarity is also a prerequisite for proper interpretation of the Hurst phenomenon since the rescaled range is also a function of deviations from the mean. Klemes (1974) conducted a number of simulations using white noise and the mean level was changed in different ways. It was shown that the value of H increased with this type of nonstationarity.

Another explanation, put forward to explain the Hurst phenomenon, was that the length of the available record is not long enough for H to attain a value of 0.5. It has been argued that if a sufficiently long series of observations is available, H would tend to attain an approach of 0.5. A plausible explanation of the value of H higher than 0.5 is the persistence, the higher values being the effect of dependence on observed natural series. The dependence can be taken into account in Markovian models but these models cannot reproduce $H > 0.5$. Hence, if the dependence is considered to be the likely cause of the Hurst phenomenon, then long-memory models are required. The short-term persistence is caused by the storage effect.

Kumar (1982) pointed out that in geophysical processes, the memory manifests itself mostly through the conservation of mass and momentum and it has the Markovian property that the past influences the future only through its influence on the present. Thus, once the present state has been arrived at, it is no longer significant, from the point of view of future development, as to how it was arrived at.

Short-memory models have been used in hydrology to generate synthetic data. For many hydrologic studies, particularly those concerned with the design and management of water resource systems, sufficiently long data series are required to determine the operating rules and test the system under various conditions to evaluate their performance. It may be mentioned that a longer generated series does not contain any further information than its parent series. However, its use helps with greater extraction of the information already contained in the parent series. The autoregressive models (see Chapter 4) are a class of short-memory models that have been extensively used in hydrology.

Salas-La Cruz (1972) has derived relationships for the determination of the range of periodic-stochastic processes may be employed to obtain the required storage capacity.

10.7.3 Sensitivity of Reservoir Storage to Inflow Statistics

The minimum reservoir capacity that is required to meet a given demand depends on the generating mechanism of inflows, apart from the nature of demands themselves. Wallis and Matalas (1972) conducted simulation experiments to determine the sensitivity of the reservoir capacity to various parameters of an inflow sequence, including the Hurst exponent H. They used two approaches to generate inflows: the Markov process and the fractional Gaussian noise process. The most important parameter of the Markovian process is ρ_u which is lag u serial correlation coefficient. The fractional Gaussian noise model was proposed by Mandelbrot and Wallis (1969).

Wallis and Matalas (1972) generated a number of sequences using these two models and the mean storage for various levels of development was determined using the sequent peak algorithm. It was found that over certain ranges of values of the level of development (α), ρ_u, and H, the mean storage was insensitive to the inflow-generating process. For $\alpha < 0.80$, the mean storage depends on ρ_u. For $\alpha \geq 0.80$, the mean storage mainly depends on H. Klemes et al. (1981) conducted simulation experiments to determine differences in reservoir performance reliability when inflows are generated using long-memory models and short-memory models. The reliability of the reservoir was characterized in three different ways:

- Occurrence-based reliability R_a, which is the number of nonfailure years expressed as a percentage of the total number of years in a given period.
- Time-based reliability R_t, which is the total duration time of all nonfailure intervals expressed as a percentage of the total length of the given period.
- Quantity-based reliability R_v, which is the actual amount of water supplied expressed as a percentage of the total demand during the given period.

Usually, at least some years contain shorter or longer periods of nonfailure and during most failure periods, the outflow is not reduced to zero. This leads to the condition $R_a \leq R_t \leq R_v$. Based on simulation experiments, the following observations were made by Klemes et al. (1981):

- The short-memory model leads to overestimation of reservoir performance reliability as compared to the long-memory model.
- Overestimation is, in general, highest for annual reliability, lower for time-based reliability, and lowest for quantity-based reliability.
- Overestimation of all three reliability characteristics is very small for reservoirs with storage coefficients up to about one for any draft ratio. For reservoirs with a storage coefficient greater than 1, it is small for draft ratios up to about $D = 0.6-0.8$ and over $D = 1.1$, depending on the inflow parameters, such as the coefficient of variation and lag-one serial correlation coefficient.
- Overestimation is maximal for draft ratios close to one and increases with reservoir coefficients up to about 2 or 3. For higher values of this coefficient, no further increase was detected.
- Overestimation slightly increases with the variability of inflows and decreases with the increase of the lag 1 correlation coefficient.

Thus, it was concluded that the use of short or long-memory model makes no difference in the reservoir performance reliability if the draft ratio is either very low or very high. The length of the memory in the annual inflow series is irrelevant if the regulation itself has no over-year memory. For example, if the draft is low, the reservoir fills up every year and if it is high, the reservoir empties every year; in both cases, only seasonal regulation is involved.

It was further pointed out that the decision-makers with high-risk aversion would prefer long-memory models; those with low-risk aversion, short-memory models. The replacement of a short-memory streamflow model with a long-memory model amounts to the incorporation of a small safety factor into the reservoir performance reliability. However, in most practical cases, this factor is much smaller than the accuracy with which the performance reliability can be assessed.

10.8 REGULATION REGIME FUNCTION

The storage reservoirs are the most effective way of regulating the natural flow of a stream. The reliability of getting a specified amount of release mainly depends on, inter alia, the storage capacity of the reservoir. The more is the storage capacity, the higher is the reliability of supplying a given amount of water or the higher will be the yield for a specified reliability. This is so because the reservoir basically provides the storage space to carry inflows from a period of excess to that of deficit. A study of the storage–reliability–yield relationship is essential to provide preliminary estimates of capacity or yield of a reservoir. The relationship relates inflow characteristics, reservoir capacity, release, and reliability. This analysis is the main aim of the stochastic theory of storage. The methods of storage–yield analysis can be broadly classified into sequential and nonsequential methods. Among the two, sequential methods that use historical inflow series are more popular; the most common example is the mass curve method. The simulation technique is widely used these days.

The inflow process, the reservoir storage, and the outflow process constitute the streamflow regulation system. The operation regime of this system is specified by the storage capacity of the reservoir (S), yield (q), and a measure of reservoir reliability (r). The relationship among these three is designated by a function (Klemes, 1981):

$$\varphi = \varphi(S, q, r) \tag{10.17}$$

where $S > 0$, $q > 0$, $0 < r \leq 100\%$. This function is called the regulation regime function, or regime function, or storage–yield function, or storage–draft function. Any two of the three variables involved in Eq. (10.15) can be regarded as independent and the third one following an appropriate functional form:

$$S = S(q, r) \tag{10.18a}$$

$$q = q(S, r) \tag{10.18b}$$

$$r = r(S, q) \tag{10.18c}$$

Klemes (1981) pointed out that the available methods based on the stochastic storage theory can directly handle only Eq. (10.18c). This equation can be regarded as the basic equation by means of which the regime function can be evaluated. Eqs. (10.18a) and (10.18b) can be solved only for the finite deterministic inflow series and $r = 100\%$ with the aid of the mass curve method. The nature and shape of the regime function depend on the statistical properties of inflows. The shape also depends on the length of the time period used in the analysis. The required storage size reduces with an increase in the time period due to the averaging out of fluctuations. The storage capacity arrived at using the mean annual flows is termed as long-term storage and the one using mean monthly (or of durations of the same order) gives a very close estimate of the storage capacity required to meet the given demands. Seasonal or within-year storage is the difference of these two storage values.

10.8.1 Development of Components of Regime Function

As pointed out earlier, an iterative search using simulation is best suited for developing different components of the regime function or determining the value of the third variable knowing the other two. The different components of the regulation regime function for Dharoi reservoir are presented in graphical form in Figs. 10.8a–10.8c, which correspond to Eqs. (10.18a–10.18c), respectively. The yield is expressed as a ratio of the mean inflow and it is termed as the degree of regulation, level of development, or draft ratio. Similarly, the storage capacity is also expressed as a ratio of the mean annual inflow total and can be termed as the storage ratio or storage coefficient. For the Dharoi reservoir, the mean annual inflow was 86.808×10^7 cubic meter. A distribution of yield among different months was adopted, based on the irrigation demand, and an annual reliability measure was adopted. It is readily apparent from Fig. 10.8a that the value of the reservoir yield decreases rapidly with an increase in the reliability r. Similarly, Fig. 10.8b shows that after a certain limit, the marginal requirement of the storage space for small increases in the storage required is quite large, more so for smaller values of the yield. In other words, it means that smaller degrees of regulation for the reservoir can be obtained with quite small storage ratios.

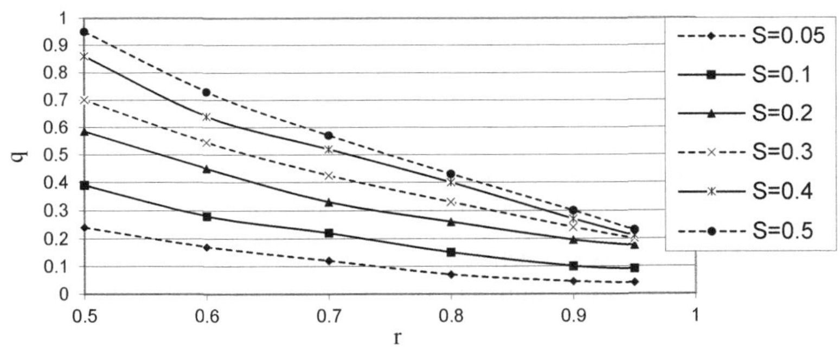

FIGURE 10.8A

Regulation regime function S = S(q,r).

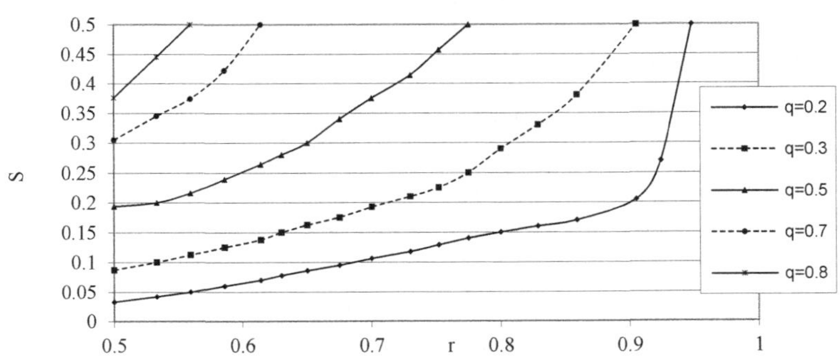

FIGURE 10.8B

Regulation regime function q = q(S,r).

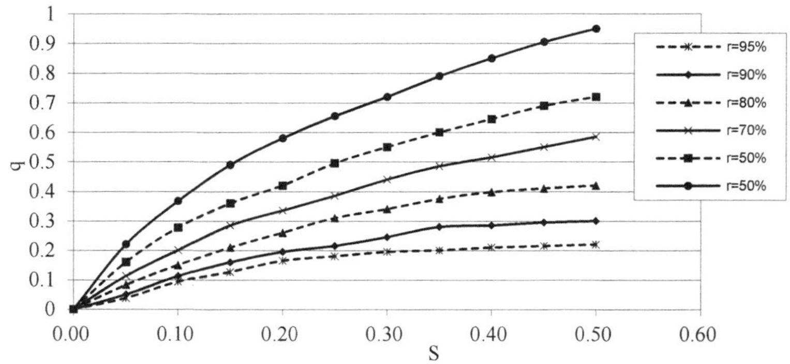

FIGURE 10.8C

Regulation regime function r = r(S,q).

10.9 RESERVOIR CAPACITY COMPUTATION

Having estimated the water requirements for an intended project and having assessed the available water at a prospective site, a planning engineer is faced with one of the three situations:

a. The rate at which water is available is always in excess of the requirements.
b. The total available water over a period of time is greater than or equal to the overall requirements, but at times, the demand exceeds the availability.
c. The total available water is much less than the requirements.

In the first case, water can be used directly from the stream as and when needed. A storage reservoir is the solution to the second case. In the third case, a supplemental source or an alternative site has to be explored.

Once it is decided that a storage reservoir is required at a particular site, the next important decision is to finalize the capacity of the reservoir. The required storage to meet given demands depends on three factors: the variability of streamflows, the size of demands, and the reliability of meeting the demands. The procedures for estimating the storage capacity needed to meet given demands or the possible yield from a given project design and data constitute the *storage–yield (SY) analysis.* The selection of a suitable method mainly depends on the following factors:

- The level of study: More accurate methods should be selected in the design stage while an approximate method can be adopted in the planning stage.
- Reliability of input data: An accurate method is useful only if the input data are reliable.
- Time and facilities available for study: A sophisticated method can be adopted if sufficient time and facilities are available.

The storage capacity of a reservoir is divided into a number of zones based on the useful purposes the reservoir is required to serve. These zones are discussed next.

10.9.1 Storage Zones in a Reservoir

For ease in analysis and operation (see Section 10.4), the entire reservoir storage space is conceptually divided into a number of zones by drawing imaginary horizontal planes at various elevations (see Fig. 10.1). The lowest zone is the dead (or inactive) zone and its storage capacity is denoted by S_{min}. The bulk of the storage capacity for conservation purposes is provided in the conservation (or active) storage zone and is denoted by S_{active}. The top level of the conservation zone is termed as full reservoir level (FRL) or normal pool level. If the storage space above FRL is exclusively reserved for flood control, the maximum storage capacity is $S_{max} = S_{min} + S_{active}$. In some cases, the flood control space may be temporarily used to store water for use in the dry season. The highest level up to which the water is allowed to rise in a reservoir is known as the maximum water level (MWL). When there is flow over the spillway, a temporary surcharge storage is created above the flood control pool.

Dead Storage Zone

Dead storage is provided in a reservoir to serve two purposes:

(a) Most rivers carry a sizable amount of sediment either as suspended or bed load. Upon entering a reservoir, the velocity of flow reduces and hence its carrying capacity is lost. So the sediment settles down and keeps on accumulating. On account of this accumulation, the effective storage capacity of the reservoir goes on reducing with time. This phenomenon is termed as reservoir sedimentation and is discussed in Chapter 12.

(b) For efficient working of turbines in a hydropower project, it is necessary that the head variation must be within a specified range and a minimum head must always be available.

The storage provided in the dead zone is the greater of the above two factors.

The bulk of the storage space is provided by the conservation zone. The methods for its estimation are discussed next.

10.10 STORAGE REQUIREMENT FOR CONSERVATION PURPOSES

A number of techniques are available to compute the storage capacity for conservation purposes, such as irrigation, municipal and industrial water supply, the hydropower generation. Depending on the type of data and the computational technique used, the reservoir capacity computation procedures are classified into the following categories:

10.10.1 Critical Period Techniques

The techniques based on the critical period concepts are the earliest techniques of storage—yield analysis. The critical period is defined as the period in which an initially full reservoir, passing through various states (without spilling), empties. One such method, known as the *Mass Curve Method* was the first rational method proposed to compute the required storage capacity of a reservoir. The other popular method in this group is the *Sequent Peak Method* proposed by Thomas and Bourdon. Some analytical methods, such as the Alexander Method and Dincer Method, are also based on the critical period concept, but these are not commonly used and are of academic interest only.

10.10.2 Simulation/Optimization Techniques

Among optimization techniques, those based on linear programming (LP) and dynamic programming have been found to be particularly suitable. The simulation approach can be used as stand-alone or it can be used to further modify and test the results of critical period or optimization methods.

10.10.3 Probability Matrix Methods

These methods use statistical laws to analytically solve the storage—yield problems. Some of the well-known methods in this class are Moran's and Gould's methods. Moran formulated a system of simultaneous equations involving time and water volume. The methods could be applied to any streamflow distribution. However, the solution of the problem was possible under simplified assumptions only and hence the utility of this method for real-life problems is rather limited. These methods are not commonly used for real-life problems. A comparative analysis of selected methods of storage—yield analysis is given in Table 10.2.

Kottegoda (1980), McMahon and Mein (1986), Klemes (1981), and Nagy et al. (2002) have discussed methods such as Moran's method, Alexander's method, Gould's method, and Phatarford's method. It is clear from Table 10.2 that simulation is the best method for SY analysis.

10.10.4 Mass Curve Method

Also known as the Rippl mass curve method, it is a simplified method commonly used in the planning stage. The method considers the most critical period of recorded flow. In the methods based on the critical period concept, a sequence of streamflows containing a critical period is routed through an

Table 10.2 Comparison of selected methods of storage—yield analysis.

	Mass curve	Sequent peak	Simulation	Dincer	Alexander	Moran	Modified gould
Assumes initially full reservoir	Yes	Yes	Yes	Yes	Yes	No	No
Gives reservoir contents	Yes	Yes	Yes	No	No	No	Yes
Accounts for evaporation	No	No	Yes	No	No	Yes	Yes
Variables demands	No	Yes[a]	Yes	No	No	No	Yes[a]
Reliability measures							
R_a	No	No	Yes	No	No	Yes	Yes
R_t	Yes	No	Yes	No	No	Yes	Yes
R_v	Yes	Yes	Yes	Yes	Yes	Yes	Yes
Computational load	Low	High	High	Low	Low	High	High

[a]*Provided annual demand remains constant.*

initially full reservoir in the presence of specified demands. The reservoir capacity is obtained by finding the maximum difference between cumulative inflows and cumulative releases. Define a function X(t) as:

$$X(t) = \sum_t x(t)\, dt \tag{10.19}$$

where x(t) may represent monthly flows. The graph of X(t) versus time is known as the mass curve. In the mass curve method, the storage capacity can be determined either graphically or analytically. The method proposed by Rippl in 1883 to determine the storage capacity of a reservoir, is a graphical technique. According to Klemes (1974, 1979), Rippl plotted a residual mass curve Z of reservoir inflows relative to draft q:

$$Z_t = \int_0^t (x - q)d\tau = \int_0^t x d\tau - \int_0^t q d\tau = X_t - Q_t \tag{10.20}$$

and this was used to determine the smallest capacity of the reservoir that is necessary to ensure the release at the desired rate without failure throughout the whole period under consideration. This reservoir will be empty only in the most critical period. The concept is illustrated in Fig. 10.9.

The mass inflow curve and mass demand (here demand includes the total water demand and evaporation) curve are accumulated separately. For a constant draft, the yield mass curve is a straight line having a slope equal to the draft rate. At each high point on the mass inflow curve, a line is drawn parallel to the yield curve and extended until it meets the inflow curve. The maximum vertical distance between the parallel yield line and the mass inflow curve represents the required storage. The mass curve of inflows for Dharoi reservoir is plotted in Fig. 10.10. The line AB is the mass curve of demands. Two lines parallel to line AB, namely line CD and EF, are drawn so that they are tangent to the mass curve of inflows at points C and E. The maximum vertical distance between the mass curve of inflows and line CD and EF is noted. The maximum of these is the required storage.

The mass curve technique, although simple and straightforward, is not without shortcomings. This method is suitable when the draft is constant. It is not possible to consider evaporation losses in a meaningful way in this method which could be significant in arid climates. As with all deterministic methods of analysis, the particular set of stream flow figures used is just one sample from a large population and hence conclusions based on that sample will include a sampling error of unknown magnitude. The method has the implicit assumption that the storage which would have been adequate in the past will also be adequate in the future. Although this is not clearly true, the error caused just on this count is not really serious, particularly if sufficiently long flow series has been considered. However, this problem will arise in any other method since the true future is not known and Rippl's method is not tied to the use of the historical record Klemes (1974, 1979). Some methods address this problem by explicitly considering the stochasticity of inflows.

One more drawback of the mass curve that is quoted in the literature is that an explicit economic analysis cannot be done with this technique. The storage size cannot be related to the economic life of the project and usually an estimate of the storage increase with the increase in the length of the record used. The objective of this analysis is not economic; there is no comparison of costs and benefits when

FIGURE 10.9

Definition sketch for Rippl's mass curve method: (a) Hydrographs of inflow x and draft q, (b) mass curves of inflow X and draft Q, and (c) residual inflow mass curve Z defined in Rippl's sense Eq. (10.18) and his procedure of determining the storage capacity K necessary for nonfailure reservoir operation during the period T.

Courtesy: Klemes (1979).

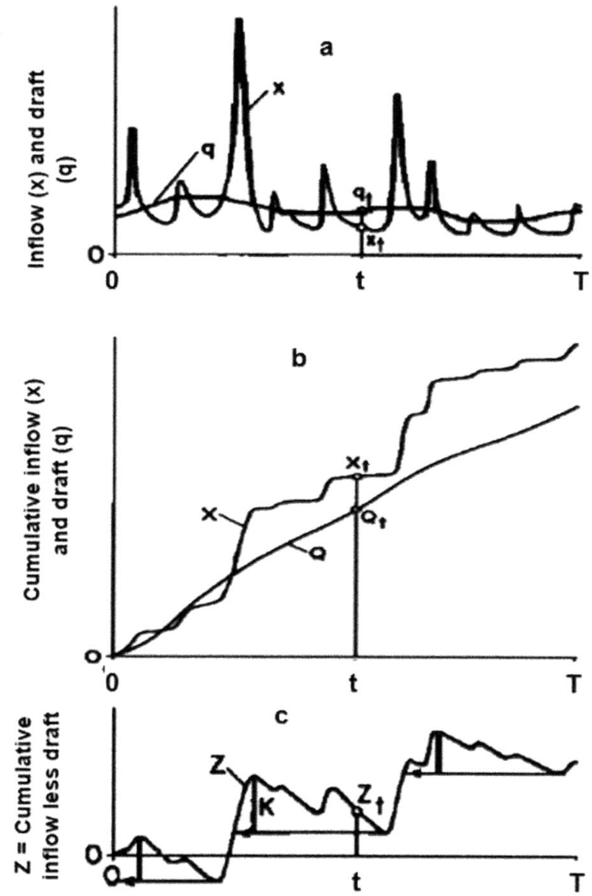

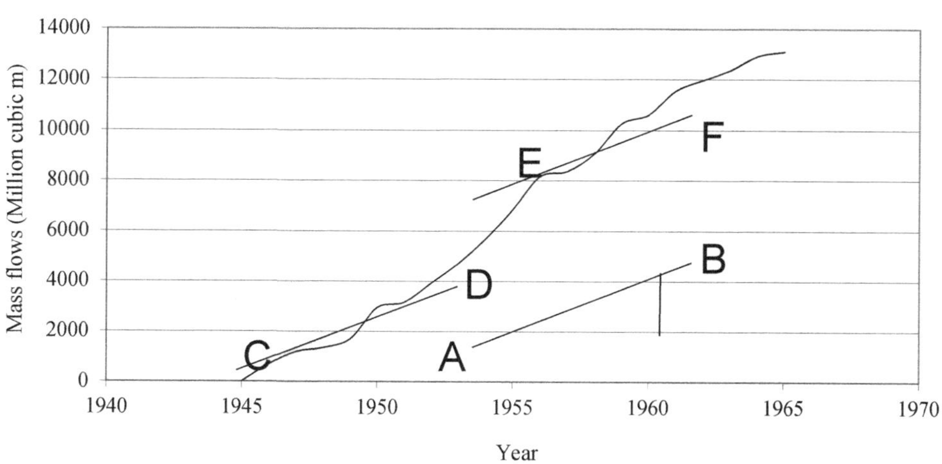

FIGURE 10.10

Mass curve for storage analysis.

using the mass curve method. Furthermore, the size of storage cannot be computed for a particular level of reliability. Klemes (1979a,b) provided a systems-analytic interpretation of the mass curve method and noted that it represents a backward-moving, forward-looking recursive maximization when the method is written as

$$K_i = \max(k_{i-1}, C_i), i = 1, 2, \ldots n \tag{10.21}$$

where subscript i runs backward in time, C_i is the minimal fill-up necessary for the ith dry season, and K_i is the minimal storage capacity necessary for the period from the beginning of the ith dry season to the end of the whole period T. Klemes (1974, 1979) also provided an economic interpretation of the mass curve method. Assuming a convex loss function, the regulation based on a firm value of the target release is optimal under conditions of extreme hydrologic and economic uncertainty about the future. Further, the regulation that is aimed at the greatest possible equalization of outflow is optimal under conditions of perfect knowledge of future streamflow combined with minimum economic uncertainty.

10.10.5 Sequent Peak Algorithm

An analytical solution of the mass curve method is given in the sequent peak algorithm. This method was proposed to circumvent the need to choose the correct starting storage which is required in the mass curve procedure. The computations are quite simple and can be carried out as follows. Let I_t be the inflow to the reservoir in the period t, R_t be the release from the reservoir, and S_t the storage at the beginning of t. The reservoir is assumed to be empty in the beginning. The mass curve of the cumulative net flow volume (inflow − outflow) against time is used. This curve will have many peaks (local maxima) and troughs (local minima). For any peak P_i the next following the peak of magnitude greater than P_i is called a sequent peak. To take care of the case when the critical period falls at the end of the record, computations are performed for twice the length of the inflow record assuming that the inflows repeat after the end of the first cycle. The variable S_t is calculated as follows:

$$S_t = \begin{cases} S_{t-1} + R_t - I_t & \text{if positive} \\ 0 & \text{if negative or zero} \end{cases} \tag{10.22}$$

The required storage capacity is equal to the maximum of S_t values.

Example 10.3. The monthly inflow data for a reservoir are available for 56 months. The required constant release is 34.0 million m^3. Find out the required storage using the sequent peak method.

Solution. Following Eq. (10.20), computations are shown in Table 10.3. As seen from Table 10.3, the required reservoir capacity would be 675.8 million m^3 which is the maximum of the last column in Table 10.3. Here the calculations need not be repeated for the second cycle because the storage at the end of the first cycle is close to zero and therefore the second cycle would be identical to the first cycle.

The sequent peak algorithm can consider variable release from the reservoir. The reliability of the reservoir can be obtained indirectly. Since the reservoir would be able to meet the worst drought from the record, the implied probability of failure would be $1/(N+1)$. The algorithm is very fast and easy to program. A single historical record is used to compute the storage and hence the method is limited in that sense. It is also not possible to exactly consider the losses, but these can be approximately included in the releases.

Table 10.3 Illustration of sequent peak algorithm (all data in million m^3).

Period (t)	Storage S_{t-1}	Inflow I_t	Release R_t	Storage S_t
1	0.00	17.10	34.0	16.90
2	16.90	47.20	34.0	3.70
3	3.70	76.70	34.0	0.0
4	0.00	2.60	34.0	31.40
5	31.40	0.70	34.0	64.70
6	64.70	0.0	34.0	98.70
7	98.70	0.0	34.0	132.70
8	132.70	0.0	34.0	166.70
9	166.70	0.60	34.0	20.10
10	20.10	0.10	34.0	234.0
11	234.00	0.70	34.0	267.30
12	267.30	0.0	34.0	301.30
13	301.30	6.20	34.0	329.10
14	329.10	10.10	34.0	353.0
15	353.00	40.7	34.0	346.30
16	346.30	0.6	34.0	379.70
17	379.70	0	34.0	413.70
18	413.70	0	34.0	447.70
19	447.70	0.3	34.0	481.40
20	481.40	0.2	34.0	515.20
21	515.20	0.4	34.0	548.80
22	548.80	0	34.0	582.80
23	582.80	0.3	34.0	616.50
24	616.50	0	34.0	650.50
25	650.50	8.7	34.0	675.80
26	675.80	184.9	34.0	524.90
27	524.90	527.2	34.0	31.70
28	31.70	48.1	34.0	17.60
29	17.60	17.10	34.0	34.50
30	34.50	47.20	34.0	21.30
31	21.30	76.70	34.0	0.0
32	0.00	2.60	34.0	31.40
33	31.40	0.70	34.0	64.70
34	64.70	0.0	34.0	98.70
35	98.70	0.0	34.0	132.70
36	132.70	0.0	34.0	166.70
37	166.70	0.60	34.0	20.10
38	20.10	0.10	34.0	234.0
39	234.00	0.70	34.0	267.30
40	267.30	0.0	34.0	301.30

Table 10.3 Illustration of sequent peak algorithm (all data in million m^3).—cont'd

Period (t)	Storage S_{t-1}	Inflow I_t	Release R_t	Storage S_t
41	301.30	6.20	34.0	329.10
42	329.10	10.10	34.0	353.0
43	353.00	40.7	34.0	346.30
44	346.30	0.6	34.0	379.70
45	379.70	0	34.0	413.70
46	413.70	0	34.0	447.70
47	447.70	0.3	34.0	481.40
48	481.40	0.2	34.0	515.20
49	515.20	0.4	34.0	548.80
50	548.80	0	34.0	582.80
51	582.80	0.3	34.0	616.50
52	616.50	0	34.0	650.50
53	650.50	8.7	34.0	675.80
54	675.80	184.9	34.0	524.90
55	524.90	527.2	34.0	31.70
56	31.70	48.1	34.0	17.60

10.10.6 Stretched-Thread Rule

One of the objectives of streamflow regulation problems with the aid of the mass curve analysis is to determine regulation aimed at the greatest possible equalization of the reservoir outflow (Klemes, 1981). The need for this regulation arises from a desire to reduce the losses that could arise due to flows being either too high or too low. Such regulation can be considered to have the maximum economic effect and has also been termed as ideal. A simple method to determine such a regulation policy was introduced in Europe in 1923 by Varlet and is known as *The Stretched-Thread Rule* (Klemes, 1979).

In this method, the mass curve of inflows is drawn as shown in Fig. 10.11. Now this mass curve is shifted upwards by a distance equal to the reservoir storage capacity. Let a thread be stretched between these two mass curves whose lower end is either midway between the two curves or zero; the upper end is also fixed the same way. The shape of this thread represents the shortest path between two opposite ends of the corridor formed by two mass curves. This thread is the mass curve of the reservoir outflow which gives the greatest possible equalization of outflow. The lower mass curve represents the line of full reservoir and the upper, the line of empty reservoir. The volume of water in storage at any time is given by the vertical distance between the stretched-thread and the upper mass curve.

Klemes (1979) detailed a procedure for numerical computation of the shortest path when the corridor is enclosed within two broken lines consisting of straight-line segments and the end points of the shortest path are specified. The lower broken line is given as a series $\{X_t^*\}$, $t = 0, 1, \ldots, N$, the upper broken line as $\{X_t^{**} = X_t^* + K_t\}$, where K_t is the width of the corridor so that $K_t > 0$ for all t. In the storage reservoir problem, a variable width of the corridor arises if constraints on storage additional to those of empty and full reservoir are specified, for instance, variable requirements on freeboard and minimum storage level throughout the year. The steps suggested are as follows:

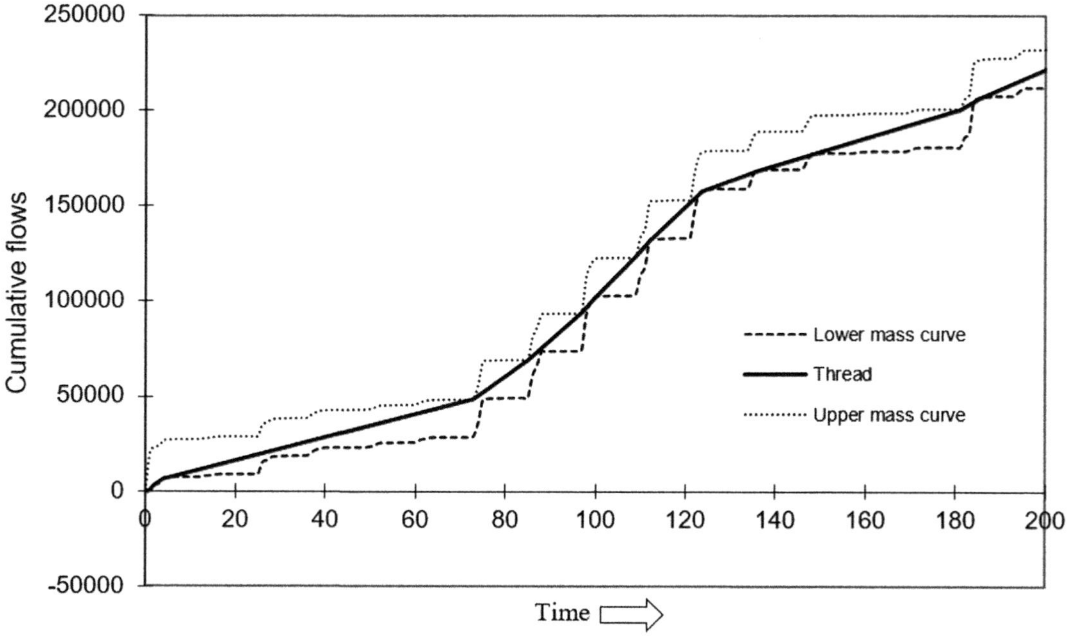

FIGURE 10.11

Stretched-thread analysis for Dharoi reservoir.

1. A straight line connecting the end points A_o and A_N is computed.
2. The corridor boundaries are checked to see whether any of them are crossed by the line A_oA_N. If no crossing is recorded, the line A_oA_N is the desired shortest path. If X^* crosses A_oA_N, the point of maximum distance of X^* above A_oA_N is identified as a corner point A_t of the shortest path; if X^{**} crosses A_oA_N, the point of maximum distance of X^{**} below A_oA_N is identified as another corner point A_j.
3. The corner point closest to the starting point A_o, in this case A_j, is regarded as an end point of the shortest path in the period (0,j).
4. Steps one to three are repeated with A_j replacing A_N.
5. If no additional corner points are identified in the interval (0, j), the straight line A_oA_j is the first segment of the shortest path and the search moves to the next interval (j, i) with A_j and A_i representing the starting and the end points, respectively. If, however, an additional corner point, say A_k, is identified in the period (0, j), the search moves to the interval (0, k).
6. In general, the search always moves forward in time only after the shortest path in the whole past period has been found.

Fig. 10.11 shows the stretched-thread diagram for the Dharoi reservoir.

10.10.7 Storage–Yield Analysis

The storage–yield (SY) analysis is carried out to determine the smallest volume of storage required to meet given demands with a stated reliability. It is also used to reassess the water demand which can be satisfied by an existing reservoir. The required storage depends on the volume of demand, reliability of meeting them, and reservoir inflows. The SY analysis problems can be of two types: given the storage, calculate the yield; or given the yield, calculate storage. Both optimization and simulation techniques are used in SY analysis.

Optimization Techniques

Among the various available optimization techniques, LP and dynamic programming (DP) have been extensively used for storage–yield analysis. Here, an LP-based simplified formulation is discussed. The problem formulation is essentially the same for DP.

Consider that a reservoir is to be constructed at a particular site. Monthly inflow data for the past *n* months are available. The projected demand for water during a critical year is known along with its distribution each month. The losses from the reservoir are neglected for the time being. The problem is to find out the minimum capacity of the reservoir which will supply the required quantity of water without failure. Let D be the annual water demand from the reservoir and α_i, i = 1,2, ... 12 be its fractions for different months. Hence, the demand in a particular month will be α_i D. Let I_t be the inflow to the reservoir during the tth month and R_t be the water actually released from the reservoir. The objective is to find the minimum capacity of the reservoir, which can meet the demand.

$$\text{Min C} \tag{10.23}$$

Representing the storage content of the reservoir at the beginning of month t by S_t, the continuity equation (neglecting losses) is

$$S_t + I_t - R_t = S_{t+1} \quad t = 1, 2 \ldots n \tag{10.24}$$

This equation has to be satisfied for each of the n months and hence there will be n such constraints in the formulation. The value of S_1 is given as input.

It is also required that the amount of water actually released from the reservoir must be more than or equal to the amount demanded. This can be mathematically expressed as follows:

$$R_t \geq \alpha_i D, t = 1, 2 \ldots n \tag{10.25}$$

The index i = 1, 2, ... 12 represents the calendar month. Since this constraint also must hold for each month, there will be *n* such constraints. If the capacity of the required reservoir is C, then in any month, from a physical point of view, the storage content of the reservoir must be equal to or less than this value. Hence,

$$S_t \leq C, t = 1, 2 \ldots n \tag{10.26}$$

Besides the storage S_t, capacity C, annual demand D, and release R_t can take on only positive values. This completes the problem formulation. The problem is quite easy to solve particularly due to the availability of standard package programs.

The computations for firm power can also be performed in a similar manner. The maximum firm power output can be obtained by

$$\text{Max}\left[\min HE_t\right], t = 1, 2, \ldots n \tag{10.27}$$

where HE_t is the amount of hydropower produced during period t. The maximum possible firm power that can be generated depends on the site conditions, hydrology of the area, and the capacity of generating equipment. The lower bound of firm power is zero.

10.10.8 Simulation Method (Behavior Analysis)

Simulation is essentially a search procedure. It is one of the most widely used techniques to solve a large variety of problems associated with the design and operation of water resources systems. The reason is that this approach can be realistically and conveniently used to examine and evaluate the performance of a set of alternative options available. Furthermore, serial correlation of inflows, seasonality, etc., are easy to account for. Also, it is easy to present the technique and its results to nontechnical persons.

Assume that a site has been identified for the construction of a dam. The reservoir has to cater for irrigation for a nearby area and the target demand of water for different months is given. The problem is to find the required capacity of the reservoir. The elevation–area–capacity table for the site is available. A sufficiently long series of streamflows at the site is available and the reliability to meet the demands has been specified.

The computational steps using a binary search method are as follows:

(a) At the beginning of computations, the upper bound of the storage capacity is decided based on the water availability, site conditions, etc. The lower bound of storage is taken as dead storage S_{min}.

(b) The reservoir is initially assumed to be full. The effect of the initial storage value will not be significant if the operation of the reservoir was simulated using a long inflow series.

(c) A trial value of storage as the average of upper and lower bounds is assumed.

(d) The operation of the reservoir is simulated for the whole period of record. At each time interval t, an attempt is made to satisfy the demand to the extent possible. If there is not enough water in the reservoir to meet the demand during any period, the demand is met to the extent possible and the period is treated as failure. If the available water in the reservoir in a period is less than S_{min}, no release is made. The storage is depleted by evaporation only and the reservoir is assumed to have failed during that period. If there is so much water during the period that there is no place to keep it even after meeting all the demands, the extra water over the storage capacity is spilled. The storage at the end of the period is computed using the continuity equation:

$$S_{t+1} = S_t + I_t - E_t - R_t \tag{10.28}$$

Step (d) is repeated for all the n periods.

(e) Now, the reliability of the reservoir (REL) is computed by

$$REL = 1.0 - FAIL/n \tag{10.29}$$

where FAIL is the number of failures (number of periods when release $R_t < \text{demand } D_t$).

(f) If this reliability is less than the desired value, it means that the trial value of the reservoir capacity is small. Hence, the present value is adopted as the lower bound for the next iteration.

The feasible region below this lower bound is discarded and the trial value for the next iteration is chosen midway between the upper bound and the new lower bound. Go to step (d).

(g) If, however, the reliability comes out to be higher than the requirement, the trial size of the reservoir is bigger than what is necessary and hence the region between the current value and the upper bound is discarded for further examination. The present capacity value becomes the new upper bound. Again, the trial value for the next iteration is chosen as the mean of new upper bound and old lower bound. Go to step (d).

(h) The iterations are performed till the desired value of the reliability is achieved.

This method converges quite rapidly as the feasible region is halved every time. It may be seen that in this method, the generation of hydroelectric power can also be easily considered. The evaporation losses can be easily considered if information about the depth of evaporation is available. The evaporation loss E_t is a function of both S_t and S_{t+1}.

Undoubtedly, a sufficiently long and reliable inflow data series is the prime requirement for SY analysis. Although the method to be adopted for a particular problem will depend on the available data, simulation has been rated as the best method for SY analysis. McMahon and Mein (1986) recommend that the results of the preliminary analyses should be further refined using behavior analysis.

Many times the available inflow data length is less than desirable or the data may have gaps. In these situations, one has to resort to the synthetic generation of stream flow sequences. The purpose of synthetic generation is to have a number of streamflow sequences that are equally likely to occur in the future. A large number of statistical/techniques are available for synthetic data generation; many software packages are also available. After the synthetic sequences are available, the operation of the system is simulated using these different sequences. Naturally, a range of storage values will be obtained and this variation indicates the sampling error that is associated with the use of the short-period data. As the length of the sequences increases, it will be noticed that the variation of the storage value will reduce. The estimate of the required storage increases with the length of the sequence.

The computations to arrive at the preliminary assessment of the live storage for the Dharoi reservoir are given in Appendix B.

10.10.9 Reservoir Screening

Reservoir screening is meant to select the reservoir(s) that should be incorporated into development schemes. Each reservoir with a certain storage capacity is capable of giving a certain yield and therefore, a relation between the storage capacity and yield can be developed. As there is a relation between storage capacity and cost, one can develop a relationship between cost and yield:

$$C = f(Y) \tag{10.30}$$

where C is cost and Y is yield. If the total yield of the system is to be Y_T, the purpose of reservoir screening is to identify the combination of reservoir(s) that provides this yield Y_T at the minimum cost.

DP is a powerful tool to solve this problem. The screening process using a DP model involves two stages. In the first stage, the yield from a reservoir is estimated for various values of storage capacities by simulating each reservoir individually. These results are used in the second stage where the selection of reservoirs and their sizes is finalized by DP. This method has two advantages. First, it can use any number of years of input data at the simulation stage, and second, it can handle nonlinear functions easily. An example of the application of DP for reservoir screening is presented in Chapter 5.

10.11 FLOOD CONTROL STORAGE CAPACITY

The requirement of storage space for flood control is in conflict with the requirements for conservation needs. The conservation requirements, such as water supply and hydropower generation, require the storage space to be full while the flood control aspect requires the availability of empty storage space.

From the point of view of analysis, the demands for water supply and hydroelectric power are relatively deterministic in nature, while the demand for flood control storage is completely stochastic. Furthermore, the time period for analysis is usually of the order of 1 month for conservation purposes while for flood control purposes, it is of the order of a few hours. The requirement of storage space for flood control is estimated by using the design flood hydrograph. An initial reservoir level is assumed at which this flood hydrograph impinges the reservoir. The maximum level attained by the reservoir is computed by routing the hydrograph through the reservoir. The maximum height of the dam is obtained after adding the free board to this level. To begin with, the top of the conservation pool (FRL) is a good choice for initial storage for computations. Before discussing the steps to determine flood control space, it is useful to briefly describe the design flood for a reservoir.

10.11.1 Reservoir Design Flood

A design flood is a hypothetical flood (peak discharge or hydrograph) adopted as the basis in the engineering design of project components. Some of the common purposes are:

(i) Design floods adopted for the safety of structures against failure by overtopping, etc. during floods. For example, the design flood adopted for dams to decide the spillway capacity.
(ii) Design floods adopted for flood control and drainage works to provide safety to downstream areas against flooding.

Since the design flood adopted often marks the difference between safety and disaster, utmost attention has been given the world over to select and estimate the design flood that is most appropriate for a given case. Economic, social, and other nonhydrologic considerations influence the philosophy of protection, and hence the selection. Policies have been laid down by most organizations for various applications and are followed unless there are compelling local factors for deviation in the particular case.

Many approaches are used to estimate design floods. The rational formula was widely used in early times and is still used in some countries. The other popular approach is based on the unit hydrograph. The design storm for the project is determined using meteorological data and then the unit hydrograph is applied to determine the corresponding flood hydrograph. When a sufficiently long flow series is available, frequency analysis is carried out to determine the floods of various frequencies. For important projects, the results of various approaches are compared to obtain the design flood.

Design Flood to Determine Spillway Capacity

The design criteria for flood control schemes have evolved over the years, pooling the experiences and practices followed by various organizations and individuals. Law (1992) provided an overview of spillway design flood standards and freeboard requirements in Europe. The prevalent hydrologic design criteria for determining the spillway capacity for India have been detailed in Indian Standard IS 11223-1985, "Guidelines for fixing Spillway Capacity" According to these guidelines the inflow design floods that need to be considered for various functions of spillways are given.

Inflow Design Flood for Dam Safety

It is the flood for which the dam should be safe against overtopping and structural failure. The criteria for classification of dams are based on size and hydraulic head (see Section 10.1.4). There is no single universally accepted criterion, but a general consensus is to adopt PMF as the design flood for large dams that have high hazard potential. According to the guidelines being followed in India, the following types of spillway design flood are recommended for various sizes of dams (Table 10.4).

For minor structures, a flood of 50- or 10-year frequency is adopted depending on the importance of the structure. Floods of larger or smaller magnitudes may be used if the hazard involved is high or low, respectively. The relevant parameters to be considered in judging the hazard in addition to the size would be:

(i) distance and location of the downstream areas. Due consideration is given to the likely future developments, and

(ii) the maximum carrying capacity of the downstream channel at a level at which catastrophic damage is not expected.

Inflow Design Flood for Efficient Operation of Energy Dissipation Works

The energy dissipation arrangements for the spillway may be designed for the best efficiency for a smaller inflow flood than the inflow design flood for the safety of the dam.

Inflow Design Flood to Check the Extent of Upstream Submergence

The inflow design flood to check the extent of the upstream submergence depends on local conditions, the type of property and the effects of its submergence. Except for very important structures in the upstream like power houses and mines for which the levels corresponding to SPF or PMF may be used, smaller design floods and levels attained under these may suffice. In general, a 25-year return period flood for land acquisition and a 50-year return period flood for the built-up property acquisition may be adopted.

Table 10.4 Inflow design flood for various types of dams.

Type of dam	Storage capacity (million m^3)	Hydraulic head (m)	Inflow design flood
Small	0.5–10	7.5–12	100 year return period
Intermediate	10–60	12–30	SPF[b]
Large	>60	>30	PMF[a]

[a]*Probable maximum flood (PMF): it is the theoretically largest flood resulting from a combination of the most severe meteorological and hydrologic conditions that could possibly occur in a given area. PMF is used as a design flood for large hydraulic structures whose failure may lead to catastrophic floods.*

[b]*Standard project flood (SPF): the flood that would result from a sever combination of meteorological and hydrological factors, reasonably applicable to the region. Extremely rare combinations of these factors are excluded. Magnitude of SPF is smaller than the PMF.*

Inflow Design Flood for the Extent of Downstream Damage in the Valley

The inflow design flood to check the extent of downstream damage depends on local conditions, the type of property, and the effects of its submergence. For important facilities like power houses, the outflows under the inflow design flood for the safety of dams and all gates operating conditions are relevant. Normally, the discharge relevant to check the acceptability of the downstream submergence may be smaller than that for power houses at or near the toe of the dam.

For important projects, dam-break studies may be undertaken as an aid to terminate the design flood. Where professional judgment or studies indicate an imminent danger to present or future human settlements, the PMF should be used as the design flood. Besides the available guidelines, location-specific factors should also be considered when choosing a particular type of flood.

Design Flood for Fixing Freeboard

The design of spillways and the size of the flood control pool are determined using reservoir routing.

Return Level and Return Period

The terms return level and return period are relevant for the determination of the design flood of a project. These terms are relatively easily understood given stationarity but become ambiguous in the nonstationary setting. A stochastic process is defined as stationary if its mean, variance, and auto-covariance are time independent.

The term return period can be interpreted in different ways. Some authors equate the return period with the "average recurrence interval" defining it as the average time elapsed between two successive realizations of the event. The difference between the two definitions arises because of the probability of having more than one exceedance in a given year. In fact, in some places, the use of the terms "recurrence interval" and "return period" has been criticized as these may lead to confusion in the minds of some decision-makers. It is better, therefore, to express the rarity of an event in terms of annual exceedance probability (AEP). AEP is defined as: "the probability that a given flow value will be exceeded in any 1 year."

The term m-year return level is defined as the high quantile for which the probability that the annual maximum exceeds this quantile is 1/m. Under the stationarity assumption, the return level is the same for all years, and this gives rise to the notion of the return period. Under stationarity, the return period of a particular event is the inverse of the probability that the event will be exceeded in any given year. Thus, the m-year return level is associated with a return period of m years. In the stationary case, there is a one-to-one relationship between a return level (the quantile) and a return period (the associated time interval).

Under the nonstationarity conditions, let the return period be interpreted as the expected waiting time. Let the m-year return level r_m be defined as the level for which the expected waiting time until an exceedance occurs is m years. So, r_m is the solution to Eq. (10.31) (Cooley, 2013):

$$m = 1 + \sum_{i=1}^{\infty} \prod_{y=1}^{i} F_y(r_m) \tag{10.31}$$

If the return period is interpreted as the expected number of events, the m-year return level r_m can be obtained from the solution to Eq. (10.32) (Cooley, 2013):

$$1 = \sum_{y=1}^{m} \left(1 - F_y(r_m)\right) \tag{10.32}$$

10.12 RESERVOIR ROUTING

The passage of a flood hydrograph through a reservoir is an unsteady flow phenomenon. The routing of a flood wave through a reservoir is known as reservoir routing. It is an important part of the reservoir analysis whose major applications are: fixing MWL during reservoir design, design of spillway and outlet works, and dam-break flood wave analysis. A reservoir can be either controlled or uncontrolled. The controlled reservoirs have a spillway with gates to control the outflow. The spillway of an uncontrolled reservoir does not have gates.

The continuity equation is the governing equation in all hydrologic routing methods. It essentially states that the difference between the inflow and outflow is equal to the rate of change of storage:

$$I - Q = dS/dt \qquad (10.33)$$

where I is inflow, Q is outflow, S is storage, and t is time.

Over a small time interval Δt, the difference between the total inflow and outflow volumes is equal to the change in storage. Hence, Eq. (10.33) can be written as follows:

$$I_m \Delta t - Q_m \Delta t = \Delta S \qquad (10.34)$$

where I_m, Q_m, and ΔS denote the average inflow, the average outflow, and the change in storage during time period Δt, respectively.

For the sake of clarity, it is necessary to introduce the frequently used terms, viz., translation and attenuation characteristics of the flood wave propagation. The translation is the time difference between the occurrence of inflow and outflow peak discharges, and attenuation is the difference between the inflow and outflow peak discharges. Having known the peak discharge of the outflow hydrograph, the attenuation is

$$\text{Attenuation} = [\text{Inflow peak discharge}] - [\text{Outflow peak discharge}] \qquad (10.35)$$

$$\text{and Translation} = [\text{Time-to-peak of inflow}] - [\text{Time-to-peak of outflow}] \qquad (10.36)$$

10.12.1 Reservoir Routing Techniques

Reservoir routing requires the relationship between the reservoir elevation, storage, and discharge to be known. This relationship is a function of the topography of the reservoir site and the characteristics of the outlet facility. Using the basic Eq. (10.33), several methods for routing a flood wave through a reservoir have been developed, namely:

The mass curve method The modified puls method The Goodrich method The coefficient method	The Puls method The Wisler–Brater method The Steinberg method

A schematic depiction of reservoir routing is given in Fig. 10.12. Some of these methods are described in the following. Singh (1988) has discussed these methods in detail.

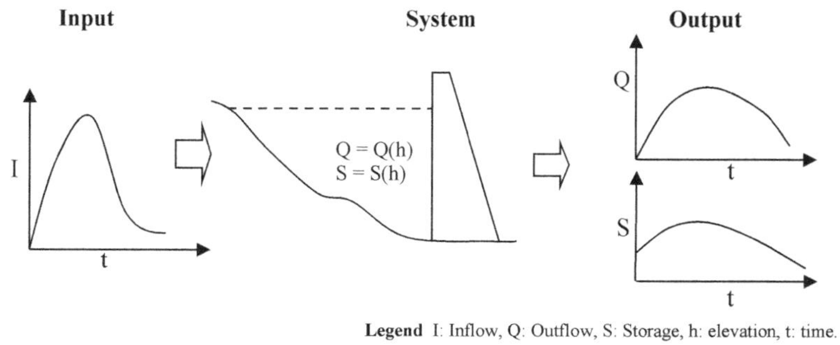

FIGURE 10.12

Schematic representation of reservoir routing.
h, elevation; *I*, inflow; *Q*, outflow; *S*, storage; *t*, time.

10.12.2 Mass Curve Method

This is one of the most versatile methods of reservoir routing, various versions of which include: (i) direct, (ii) trial and error, and (iii) graphical. Here the trial and error version is described.

For solution by trial and error method, Eq. (10.33) can be rewritten as follows:

$$M_{t+\Delta t} - (V_t + Q_m \Delta t) = S_{t+\Delta t} \quad (10.37)$$

where $M_{t+\Delta t}$ is the accumulated mass inflow at time $t + \Delta t$, and V_t is the accumulated mass outflow at t. A storage–discharge relationship and the mass curve of inflow should be prepared before obtaining the trial and error solution. The time step size Δt is chosen and the mass inflow hydrograph is computed. The steps of the trial and error solution are as follows:

(a) For the current time period, the mass outflow is assumed. It may be a function of the accumulated mass inflow.
(b) The reservoir storage is computed by deducting mass outflow from mass inflow.
(c) For this storage, compute reservoir elevation and then corresponding outflow.
(d) Using this outflow and the outflow for the previous period, compute average outflow, and then new mass outflow.
(e) Compare the mass outflow of step (d) with that of step (a). If they are not within a desired proximity, another mass outflow is assumed and steps (a) to (d) are repeated.
(f) If the two values agree, increment the time step and go to step (a).

10.12.3 Modified Puls Method

This method is also referred to as the Storage–Indication method. The basic law used in the modified Puls method states: The inflow minus outflow is equal to the rate of change in storage. Assuming $I_m = (I_1 + I_2)/2$, $Q_m = (Q_1 + Q_2)/2$ and $\Delta S = S_2 - S_1$, Eq. (10.33) is written as follows:

$$(I_1 + I_2)\Delta t/2 - (Q_1 + Q_2)\Delta t/2 = S_2 - S_1 \quad (10.38)$$

where suffixes 1 and 2 denote the beginning and the end of the time interval Δt, and Q may incorporate the controlled discharge as well as uncontrolled discharge. Separating the known quantities from the unknown ones and rearranging:

$$(I_1 + I_2) + (2S_1/\Delta t - Q_1) = (2S_2/\Delta t + Q_2) \tag{10.39}$$

Here, the known quantities are I_1 (inflow at time 1), I_2 (inflow at time 2), Q_1 (outflow at time 1), and S_1 (storage in the reservoir at time 1). The unknowns are S_2 and Q_2. Since one equation with two unknowns cannot be solved, a relation between storage, S, and outflow, Q is needed. As the outflow from the reservoir during floods takes place through the spillway, the discharge passing through the spillway can be related to the reservoir elevation which, in turn, can be related to the reservoir storage. Such curves are invariably available for any reservoir. The outflow through the spillway can be computed from the following equation:

$$Q = C_d\, L\, H^{1.5} \tag{10.40}$$

where Q is the outflow discharge (cumec); C_d is the coefficient of discharge; L is the length of spillway (m); and H is the depth of flow above the spillway crest (m). Thus, a curve/table of elevation versus discharge can be prepared. The reservoir storage also depends on the elevation. Therefore, we have

$$S = S\,(Y) \quad \text{and} \quad Q = Q\,(Y) \tag{10.41}$$

where Y represents the water surface level. The right side of Eq. (10.39) can be written as follows:

$$2S/\Delta t + Q = f(Y) \tag{10.42}$$

Before developing the $[(2S/\Delta t) + Q]$ versus outflow relation, it is necessary to select a time interval Δt such that the actual nonlinear shape of the inflow hydrograph, particularly the crest segment, can be closely linearized within this interval. For smoothly rising hydrographs, a minimum value of $t_p/\Delta t = 5$ is recommended, where t_p is the time to peak of the inflow hydrograph. A computer-based calculation would normally use a much greater ratio, say 10–20. Once Δt is fixed, the relations of Eq. (10.42) can be used to prepare curves or tables of $(2S/\Delta t \pm Q)$ versus Q since for a given elevation, Q and S are known.

The computations are performed as follows. At the beginning, the initial storage and outflow discharge are known. In Eq. (10.39), all the terms on the left-hand side are known at the beginning of the time step Δt. Hence, the value of $(S_2 + Q_2\Delta t/2)$ at the end of the time step is calculated by Eq. (10.39). The outflow can be calculated using the relation between $(S_2 + Q_2\Delta t/2)$ and Q. This procedure is repeated to cover the full inflow hydrograph.

10.12.4 Coefficient Method

In the coefficient method, the reservoir is represented by a single conceptual storage element assuming storage S to be directly proportional to outflow Q:

$$S = K\,Q \tag{10.43}$$

where K is a proportionality factor equal to the reciprocal of the slope of the storage curve that can be a constant or a variable function of outflow. If K is constant, then the reservoir is linear; otherwise, the reservoir is nonlinear.

For flood routing, a finite difference approximation is normally employed. Eqs. (10.39) and (10.43) can be combined and written as follows:

$$\Delta t\ (I_1 + I_2)/2 - (Q_1 + Q_2)\Delta t/2 = K(Q_2 - Q_1)$$

or

$$Q_2 = Q_1 + C(I_1 - Q_1) + C(I_2 - I_1)/2 \tag{10.44}$$

where

$$C = \Delta t/(K + \Delta t/2) \tag{10.45}$$

If K is variable, then C can be derived and plotted as a function of Q. For each routing period, the appropriate value of C must be obtained corresponding to the outflow under consideration. Then, routing can be performed by using Eq. (10.44).

10.12.5 Reservoir Routing with Controlled Outflow

Most of the big dams have gated spillways and the gates are raised or lowered to control the reservoir outflow. The dam may also have undersluices to release water control for irrigation, water supply, etc. The operation of the spillway gates and undersluices depends on the state of the reservoir, the level of demands, and the operation policy. In gated dams, reservoir outflow can be either (1) controlled, (2) uncontrolled, and (3) partly controlled. Incorporating controlled outflow, the continuity equation can be written as follows:

$$(I_1 + I_2)/2 - (Q_1 + Q_2)/2 - Q_c = (S_2 - S_1)/\Delta t \tag{10.46}$$

where Q_c is the mean controlled outflow from the reservoir during the time interval Δt. Rearranging the terms, Eq. (10.46) can be written as follows:

$$2S_2/\Delta t + Q_2 = I_1 + I_2 + 2S_1/\Delta t - Q_1 - 2Q_c \tag{10.47}$$

When the controlled outflow Q_c is known, the solution can be obtained on the same lines as the modified Puls method. The solution of Eq. (10.47) is simple if the entire outflow is controlled.

10.12.6 Major Applications of Storage Routing

The storage routing has numerous applications. The major ones are discussed next.

Determination of Capacity of Flood Control Pool

The storage routing of floods entering a reservoir is employed to determine the maximum levels attained for given reservoir characteristics, operation policy, and initial conditions. The data are used to demarcate the area likely to be submerged and to fix the heights of the dam. The routing of flood waves of various return periods provides inputs for economic and risk analyses of flood control storage capacity; it also helps in developing a policy for the operation of flood control pools.

Sizing Capacities of Outlet Structures

The larger the outlet capacities, the smaller will be the maximum reservoir level for a given shape of the incoming flood and the reservoir level at the beginning of this flood. The storage routing yields the maximum reservoir level attained for the given design of outlet works. The decision variables of spillway design include the type, width, height, and number of spillway openings (either controlled by gates or uncontrolled); the shape of spillway crest; and properties of gates. Optimal design of outlet structures can be obtained by systematically changing the relevant parameters.

Rates of Change of Reservoir Levels

The stability of banks along the reservoir shores, wave erosion, and landslides into the reservoir are important in reservoir management. An important variable that affects these is the permissible rate of change of reservoir water level. This rate depends on soil and rock properties and is expressed in meters/day or cm/hour. The maximum permissible rate of change is used in storage routing for the design of outlets and fixing the MWL.

Effects of Reservoirs on Downstream Floods

The extent of flooding in river reaches downstream of a reservoir can be assessed by knowing the outflow from the reservoir. Storage routing provides the requisite information along with the probability estimates.

Computation of Dam Breach Outflows

To study the dam breach problems, such as the dynamics of breach openings and outflow hydrographs through breaches, storage routing is employed. The outflow hydrograph resulting from a dam break, and the extent of flooding are input in preparing development plans for areas downstream of a reservoir.

10.12.7 General Comments

The selection of a proper routing time interval Δt is important in all reservoir routing problems. Its value should be neither too long nor too short. If it is too long, the variability of the inflow hydrograph, particularly the crest segment, may not be properly accounted for. If it is too short, it takes more effort to perform flood routing. Further, Δt is assumed so that the inflow and outflow are approximately linear during this period. Usually, Δt should be one-third to one-half of the travel time through the reservoir. Furthermore, the routing interval Δt need not be constant. It can be kept large when there are small changes in inflow and small when there are large changes therein.

The routing operation performed by the trial and error solution of the mass curve method is simple and easily done. This can be efficiently adapted to complex routing problems. The Puls method and the modified Puls method both have two shortcomings. First, the assumption that the outflow begins at the same time as the inflow implies that the inflow passes through the reservoir instantaneously regardless of its length. Second, it is difficult to choose an appropriate Δt because the negative outflow occurs during recession whenever $\Delta t > 2S_2/Q_2$ or $Q_2/2 > S_2/\Delta t$. The former drawback is not serious if the ratio of $T_t/T_m \leq 0.5$, where T_m denotes the time to peak of the inflow hydrograph and T_t denotes the travel time. T_t is defined as L/u, with L being the length of the reach and u being the average steady-state velocity. The latter weakness can be circumvented by plotting discharge versus $[(2S/\Delta t) + Q]$ curve on a log–log paper and comparing the plot with the line of equal values. If the plotted values lie above the line of equal values, the drawn figure must be abandoned and a new value of Δt must be selected. Further negative outflow can be avoided usually by taking Δt less than T_t.

The Wisler–Brater method requires an observed basis for routing computations so this can best be simulated in controlled conditions. Hence, its use is less for practical purposes. The Steinberg method requires K-curves and their superimposition over storage curves and thereby involves graphical work before actual routing is carried out.

To solve a reservoir routing problem, the following data are needed:

(a) water surface elevation versus storage volume and elevation versus outflow discharge curve/table,
(b) inflow hydrograph,
(c) initial values of storage, inflow, and outflow, and
(d) for the coefficient method, the value of proportionality constant K, which is the reciprocal of the slope of the storage curve, is also needed.

Example 10.4. The spillway of a dam is ogee crested whose crest is at an elevation of 101 m. Route the hydrograph given in Table 10.5 through the reservoir. Reservoir water elevation versus outflow, storage data are given in Table 10.6. Initially, the reservoir level is at an elevation of 102 m and the initial outflow is 27 cumec. Take time interval $\Delta t = 1$ h.

Route the flood through this reservoir.

Solution. The calculations for the Modified Puls method are shown in Table 8:

- Column 1: water surface elevation, from 101 to 107 m.
- Column 2: the head above the spillway crest.

Table 10.5 Inflow hydrograph for Example 10.4.

Time (h)	1	2	3	4	5	6	7	8
Inflow (m^3/s)	27	40	60	100	160	240	180	130
Time (h)	9	10	11	12	13	14	15	16
Inflow (m^3/s)	90	60	40	35	30	27	27	17

Table 10.6 Reservoir water elevation versus outflow, storage table.

(1)	(2)	(3)	(4)	(5)	(6)
Elevation (m)	Head (m)	Outflow (m^3/s)	Storage (10^6 m^3)	Storage S (m^3/s)-h	$[(2S/\Delta t) + Q]$ (m^3/s)
101	0	0	0.00	0.00	0.00
102	1	27.00	1.50	416.67	860.33
103	2	58.08	2.50	694.44	1446.97
104	3	98.33	3.50	972.22	2042.77
105	4	166.00	4.50	1250.00	2666.00
106	5	220.07	5.50	1527.78	3275.63
107	6	299.85	6.50	1805.56	3910.96

- Column 3: the outflow (given or calculated).
- Column 4: the storage volume in cubic meters above the spillway crest elevation.
- Column 5: storage volume in (cumec-h), computed by dividing the values in Column 4 by $\Delta t \times 3600$.
- Column 6: the storage indication quantities $[(2S/\Delta t + Q]$, in m^3/s.

Illustration:

$$\text{For storage} = 1.5 \times 10^6\ m^3,\ \text{storage}\ (m^3/s)\text{-h},\ 1.5 \times 1{,}000{,}000/\ (1 \times 3600) = 416.67$$

and

$$[(2S/\Delta t) + Q] = 2 \times 416.67/1 + 27.0 = 860.33\ m^3/s.$$

The routing is summarized in Table 10.7 in which (Col. 1) shows time; (Col. 2) shows the inflow hydrograph; (Col. 3) shows elevation, and so on.

Here, the initial outflow is 27 m^3/s; the initial storage indication value is 860.33 m^3/s (corresponding to the initial elevation of 102 m in Table 10.6. The recursive procedure should continue until the outflow has substantially declined. Sample calculations are shown in Table 10.7.

Details of calculations.

First time step

$$[2S/\Delta t + Q]\ (\text{Column 7}) = 806.33 + (40 + 27) = 873.33$$

$$\text{Corresponding outflow (Column 8)} = 27.69\ m^3/s$$

Second time step

$$[2S/\Delta t - Q]\ (\text{Column 6}) = 873.33 - 2 \times 27.69 = 873.33 - 55.4 = 817.96$$

$$[2S/\Delta t + Q]\ (\text{Column 7}) = 817.96 + (40 + 60) = 817.96 + 100 = 917.96$$

$$\text{Corresponding } Q\ (\text{Column 8}) = 30.05\ m^3/s$$

And so on.

After completing the computations, it will be seen that the peak of the outflow hydrograph will be less than the peak of the inflow hydrograph and the peak of the outflow hydrograph will occur after the peak of inflow. One can also observe (Fig. 10.13) in Eq. (10.29) that when $dS/dt = 0$, $I = Q$. Thus the peak of the outflow hydrograph will fall on the falling limb of the inflow hydrograph.

10.13 FIXING TOP OF DAM

The top of a dam is fixed a little above the high flood level by providing extra space known as freeboard. Freeboard is the vertical distance between the MWL and the top of the dam. Freeboard is

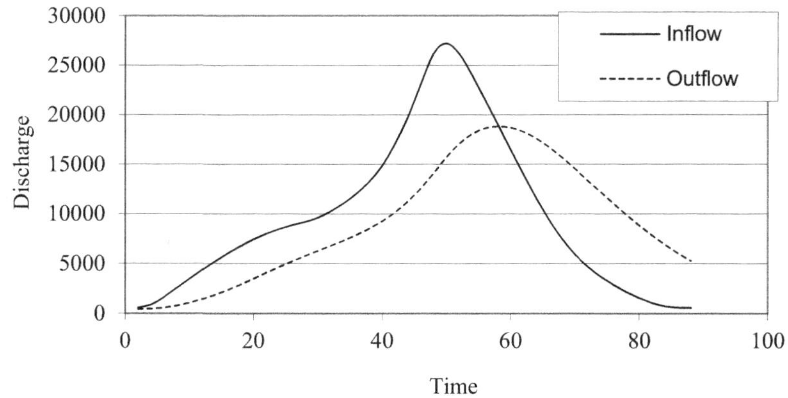

FIGURE 10.13

Plot of Dharoi reservoir inflow and outflow for a flood event.

provided to ensure the safety of the dam against overtopping in the event of an adverse combination of hydro-meteorological variables, such as strong winds during a major storm which may push up the reservoir water level. The size of the freeboard depends on the design flood for freeboard, wind set-up, and the wave run-up that is expected depending on the length of the reservoir, prevailing winds, etc. When the wind blows over the surface of a reservoir, water is piled up at the leeward end and is lowered at the windward end. The set-up is more pronounced in shallow reservoirs and can be calculated by (WMO, 1994):

$$H_s = ku^2\, l\, n\cos/gd \tag{10.48}$$

where H_s is the height of set-up above still pool level, u is the wind speed measured at an elevation of 10 m above the still pool level, l is the fetch length (straight length of unobstructed water surface exposed to wind action), n is a dimensionless coefficient dependent upon the configuration and hydrology of lake, θ is the angle between wind direction and line along which fetch is measured, g is acceleration due to gravity, d is average water depth along the direction of wind, and k is a dimensionless shear-stress coefficient. For rectangular lakes of uniform depth, $n = 1$ and $k = 1.45 \times 10^{-6}$ provided $u \le 880\ d\ (g/l \cos \theta)$.

When wind blows across reservoirs, waves are generated which run up on the upstream face of dams. This run-up is critical only when the reservoir water level is near the top. Adequate freeboard is required to prevent the overtopping of a dam by waves.

10.14 APPENDICES

10.14.1 Definitions

(a) Dead storage zone: This is the bottommost zone in a reservoir and the corresponding storage is also termed as inactive storage. Generally, this zone is provided to cater to the sediment entering the reservoir, to provide a minimum head for hydropower plants, or to provide a minimum pool for recreation facilities. Usually, all the outlets are located above this zone. The withdrawals from this zone, if any at all, are made only in extremely dry conditions. The entire reservoir storage which lies above the inactive storage is called live or active storage.

(b) Buffer zone: This is the storage space on the top of the dead storage zone and the reservoir level is brought down to this zone under extreme drought situations. When the reservoir is in this zone, the release from the reservoirs caters only to the essential needs.

(c) Conservation zone: Water is stored in this zone to cater to various conservation requirements like irrigation and water supply hydropower generation. This zone normally accounts for most of the storage space available in a conservation reservoir.

(d) Flood control zone or surcharge zone: This zone is located on the top of the conservation zone. The storage space of this zone is exclusively earmarked for absorbing or moderating floods impinging the reservoir. Depending on the flood volume and downstream release constraints, water is stored in this zone to attenuate a flood peak. After the flood peak has passed, this zone is emptied as soon as possible to prepare for subsequent flood events.

(e) Spill zone: This storage space above the flood control zone corresponds to the flood rise during extreme floods and spilling. This space is occupied mostly during high flows and the releases are at or near maximum.

(f) Full reservoir level (FRL): This is the highest level of the reservoir at which water is intended to be held for various conservation uses, including part or total of the flood storage without allowing any passage of water through the spillway.

(g) Maximum water level (MWL): It is the highest level to which the reservoir water will rise while passing the design flood with the spillway facilities in full operation. This level refers to the top of the spill zone.

(h) Active storage: The storage space in a reservoir that is above the dead storage is termed as active storage. It is also called live storage.

(i) Within-year storage: Some reservoirs are operated to provide water over a short period of low flows only and therefore, these may fill up and empty several times a year. The storage required in such a reservoir is known as within-year or seasonal storage (see Fig. 10.A1).

(j) Carryover storage: When the water stored in a reservoir at the end of a year is carried over to the next year, this is called as carryover storage. This storage is estimated using annual data (ignoring seasonal fluctuations).

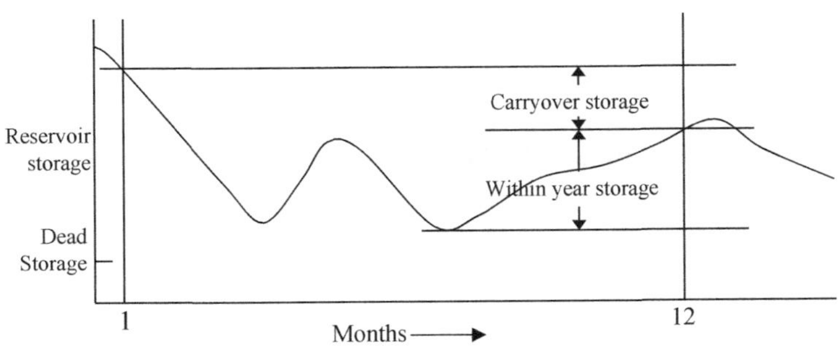

FIGURE 10.A1

Carryover and within-year storages.

FIGURE 10.A2

Variation of yield and firm yield of a reservoir.

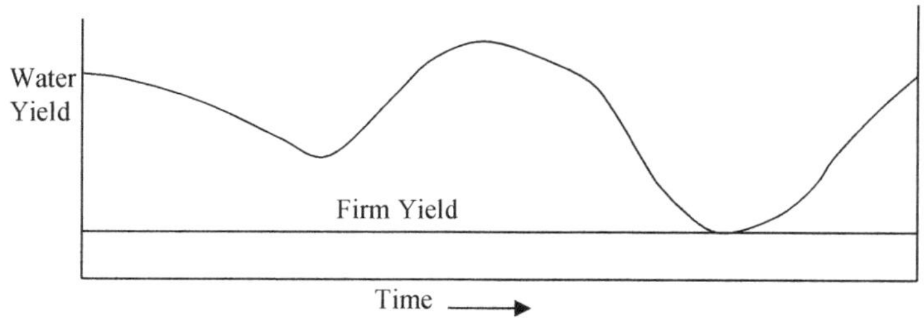

(h) Release: Release or draft is the amount of controlled outflow from a reservoir during a given time interval to satisfy various demands. Release from the reservoir is also expressed as a ratio or percentage of mean inflow. This ratio is usually below 0.9; the low values are for regions where evaporation losses are higher and high values are for regions where these losses are small.

(i) Yield: For the reservoirs serving for conservation purposes, the amount of water released for these purposes is called the reservoir yield. For the reservoirs where the stored water is used to generate hydroelectric power, the yield is defined as the amount of power delivered during a time interval.

(j) Firm yield: Firm water yield from a reservoir is defined as the maximum quantity of water that can be guaranteed to be delivered with 100% reliability (See Fig. 10.A2). The firm power yield of a reservoir can also be described in a similar manner. For example, if the yield of a reservoir for the months of January to December is 35, 42, 55, 67, 90, 75, 52, 33, 37, 40, 36, and 41 units. Then, the firm yield will be the minimum of these values, that is, 33 units.

(k) Reliability: Reliability of a system (reservoir) is described by the probability α that the system is in a satisfactory state. A reservoir is in a satisfactory state if it meets all the demands. The reliability is given by

$$= \text{Prob}[X_t S] \qquad (10.A1)$$

where X_t is the state of the system at time t and S is the domain of admissible states. Further, risk, which is the probability of failure, is $(1 - \alpha)$. Thus, if a reservoir meets all the demands in 9 months of a year, its reliability will be 9/12 = 0.75 and the risk = 1.0−0.75 = 0.25.

Volume reliability (R_v) is the ratio of the volume of water supplied to that demanded:

$$R_v = \text{Volume of water supplied/Volume of water demanded} \qquad (10.A2)$$

10.14.2 Fixing Live Storage Capacity of Dharoi Reservoir

The computations for preliminary fixing the live storage capacity of the Dharoi reservoir are given in this appendix.

1. Irrigation (direct demand)	
a. October requirement	030.838 Mm^3
b. Fair weather (Nov. to June) requirement	145.557 Mm^3
Total	176.395 Mm^3
2. Water supply demand for 9 months (October to June)	197.365 Mm^3
3. Fair weather (October to June) lake losses	088.197 Mm^3
Total	461.957 Mm^3
4. Less post-monsoon yield in the design year (1877)	013.939 Mm^3
Balance	448.018 Mm^3
5. Add carry over	196.995 Mm^3
Total	645.013 Mm^3
6. Add water supply reservation to ensure 99% reliability	137.662 Mm^3
7. Live storage	782.675 Mm^3
8. Add silt pocket of 131.988 Mm^3 at RL 175.87 m	131.988 Mm^3
9. Gross storage required	914.663 Mm^3
9a. Gross storage provided	907.878 Mm^3
10. Corresponding FRL	RL 189.60 m

The gross storage of the reservoir is computed to be 907.87 Mm^3 with FRL 622.0 m as detailed:

Details of project components		Requirement (Mm^3)
A. Gross utilization		
a. Irrigation (396.606 Mm^2)		
i. Direct command requirement		218.335 Mm^3
b. Water supply demand		
i. Ahmedabad	7.8674 cumec	197.365 Mm^3
Gandhinagar	0.5660 cumec	
Downstream riparian rights	0.8490 cumec	
Total	9.2824 cumec	
ii. Water supply requirement of 9 months (October to June) less the availability from the run of the river below Dharoi		
c. Lake losses		144.323 Mm^3
d. Gross utilization		560.023 Mm^3
e. Net utilization (560.023−144.323)		415.70 Mm^3
f. Design carry over		196.995 Mm^3
g. Reservation for water supply		137.662 Mm^3
h. Design year 1877 having maximum carryover of 196.995 Mm^3		196.995 Mm^3
B. Storage provision		
1. Irrigation (direct demand)		
a. October requirement		030.838 Mm^3
b. Fair weather (Nov. to June) requirement		145.557 Mm^3

Continued

Total	176.395 Mm^3
Say	197.365 Mm^3
2. Water supply demand for 9 months (October to June)	197.365 Mm^3
3. Fair weather (October to June) lake losses	088.197 Mm^3
Total	482.927 Mm^3
4. Less post-monsoon yield in the design year (1877)	013.939 Mm^3
Balance	468.988 Mm^3
5. Add carry over	196.995 Mm^3
Total	665.983 Mm^3
6. Add water supply reservation to ensure 99% reliability	137.662 Mm^3
7. Live storage	803.645 Mm^3
8. Add silt pocket of 131.988 Mm^3 RL. 577.0	131.988 Mm^3
9. Gross storage required	935.633 Mm^3
9a. Gross storage provided	907.878 Mm^3
10. FRL required.	RL 622.20

The average utilization of water for water supply and irrigation is worked out as under on the basis of the monthly reservoir working table of design year 1877.

Details of project	Requirement in (Mm)
i. Annual water supply utilization for October to June (reliability 99%)	197.365^3
ii. Water supply reservoir (to ensure 99% reliability)	137.662^3
iii. Annual irrigation utilization (reliability 75%)	218.335^3
iv. Lake losses	144.323^3
v. Carry over	196.995^3

References

Allen, R.G., Pereira, L.S., Raes, D., Smith, M., 1998. Crop Evapotranspiration - Guidelines for Computing Crop Water Requirements. FAO Irrigation and Drainage Paper No. 56. Food and Agriculture Organization (FAO), Rome.

Best, E., 1998. Dams. In: Herschy, R.W., Fairbridge, R.W. (Eds.), Encyclopedia of Hydrology and Water Resources. Kluwer Academic Publishers, Dordrecht.

Cooley, D., 2013. Return periods and return levels under climate change. In: Book: Extremes in a Changing Climate, pp. 97–114. https://doi.org/10.1007/978-94-007-4479-0_4.

Hurst, H.E., Black, R.P., Simaika, Y.M., 1965. Long Term Storage. Constable, London.

Institute of Hydrology, 1980. Low Flow Studies, Wallingford, UK.

ICOLD, 1988. World Register of Large Dams. International Commission on Large Dams, Paris.

Klemes, V., 1974. The Hurst phenomenon: a puzzle? Water Resources Research 10 (4), 675–688.

Klemes, V., 1979. Storage mass-curve analysis in a system-analytic perspective. Water Resources Research 15 (2), 359–370.

Klemes, V., 1981. Applied stochastic theory of storage in evolution. In: Chow, V.T. (Ed.), Advances in Hydroscience, vol. 12. Academic Press, New York.
Klemes, V., Srikanthan, R., McMahon, T.A., 1981. Long-memory flow models in reservoir analysis: what is their practical value? Water Resources Research 17 (3), 737–751.
Kottegoda, N.T., 1980. Stochastic Water Resources Technology. The Macmillan Press Ltd., London.
Kumar, A., 1982. Hydrologic Time Series Modelling – an Overview. Report No. SA-1. National Institute of Hydrology, Roorkee.
Law, F.M., 1992. A review of spillway flood design standards in European countries, including freeboard margins and prior reservoir levels. In: Parr, N.M., Charles, J.A., Walker, S. (Eds.), Water Resources and Reservoir Engineering. The British Dam Society, London.
Mandlebrot, B.B., Wallis, J.R., 1969. Computer experiments with fractional Gaussian noises, 2. Rescaled ranges and spectra. Water Resources Research 5 (1), 242–259.
McMahon, T.A., Mein, R.G., 1986. River and Reservoir Yield. Water Resources Publications, Colorado, USA.
Morris, G.L., Fan, J., 1998. Reservoir Sedimentation Handbook. McGraw Hill Book Company, New York.
Nagy, I.V., Asante-Duah, K., Zsuffa, I., 2002. Hydrological Dimensioning and Operation of Reservoirs, vol. 39. Water Science and Technology Library, Kluwer Academic Publishers, Dordrecht.
Salas-La Cruz, J.D., 1972. Range Analysis for Storage Problems of Periodic-Stochastic Processes. Hydrology Paper, 3. Colorado State University, Fort Collins, Colorado.
Singh, V.P., 1988. Hydrologic Systems: Rainfall-Runoff Modeling, vol. I. Prentice Hall, New York.
Sokolov, A.A., Chapman, T.G., 1974. Methods for Water Balance Computations. The Unesco Press, Paris.
Wallis, J.R., Matalas, N.C., 1972. Sensitivity of reservoir design to the generating mechanism of inflows. Water Resources Research 8 (3), 634–641.
WMO, 1994. Guide to Hydrological Practices. WMO No. 168. World Meteorological Organization, Geneva.

Further Reading

Box, G.E.P., Jenkins, G.M., 1976. Time Series Analysis: Forecasting and Control. Holden-Day Inc., San Francisco.
FAO, 2023. CROPWAT accessed from: https://www.fao.org/land-water/databases-and-software/cropwat/en/. on 10 Jan 2023.
Hall, W.A., Dracup, J.A., 1979. Water Resources Systems Engineering. Tata McGraw-Hill Publishing Company, New Delhi.
Linsley, R.K., Franzini, J.B., Freyberg, D.L., Tchobanoglous, G., 1992. Water-Resources Engineering. McGraw Hill Inc., New York.
McCuen, R.H., 1989. Hydrologic Analysis and Design. Prentice Hall, New York.
Salas, J.D., Delleur, J.W., Yevjevich, V., Lane, W.L., 1980. Applied Modelling of Hydrologic Time Series. Water Resources Publications, USA.
Starosolszky, O., 1987. Applied Surface Hydrology. Water Resources Publications, Colorado, USA.
Subramanya, K., 1987. Engineering Hydrology. Tata McGraw Hill, New Delhi.
WMO, 1986. Manual for Estimation of PMP, Operational Hydrology Report No 1. World Meteorological Organization, Geneva.
Wurbs, R.A., 1996. Modeling and Analysis of Reservoir System Operations. Prentice Hall PTR.

PART

Systems operation and management

IV

CHAPTER

Reservoir Operation

11

Chapter outline

Water Resources Systems Planning and Management. https://doi.org/10.1016/B978-0-12-821349-0.00010-1

The objectives of this chapter are:

- to explain the problem and issues in reservoir operation,
- to explain conventional and rule curve-based approaches of reservoir operation,
- to explain the procedures for operation of a system of reservoirs for multiple purposes, and
- to illustrate the systems analysis technique with a real-life example.

After structural facilities, such as dams, barrages, hydropower plants, etc., come into being, the benefits that could be reaped depend to a large extent on how these facilities are operated and managed. The efficient use of water resources requires not only judicious design but also proper management after construction. By far, most good dam sites have already been developed in many countries. Due to various reasons, there is slowdown in the construction of new projects and the construction cost of new projects is increasing with time. Biswas (1991) estimated that the unit cost of water from the next generation of municipal water supply projects would usually be two to three times higher than from the present generation. It is, therefore, imperative that all projects are managed in the best possible manner. A conceptual depiction of the need of regulation to meet the requirements of the society is given in Fig. 11.1.

Guidelines for operation of a reservoir have to be developed in the planning stage of a project. Later on, these are refined on the basis of actual operational experience. Such schedules range from rigid rules which are to be exactly followed to flexible guidelines that permit considerable leverage to the operator. Such schedules can be in graphical, tabular, or narrative form or a combination of these.

Reservoir operation is an important component of water resources planning and management. After construction, detailed guidelines are given to the operator to enable him to take appropriate decisions. A reservoir operation policy specifies the amount of water to be released from storage at any time depending on the state of the reservoir, level of demands and any information about the likely inflow to the reservoir. The operation problem for a single-purpose reservoir is to decide the releases to be made from the reservoir so that the benefits for that purpose are maximized. For a multipurpose reservoir, additionally, it is also required to optimally allocate the release among purposes. The complexity of the problem of reservoir operation depends on the extent to which the various intended purposes are compatible. If the purposes are compatible, less effort is needed for coordination. At this stage, it is helpful to discuss the conflicts among various purposes.

11.1 CONFLICTS IN RESERVOIR OPERATION

While operating a reservoir that serves more than one purpose, a number of conflicts arise among demands for various purposes. The conflicts that arise while operating a multipurpose reservoir may be classified as follows:

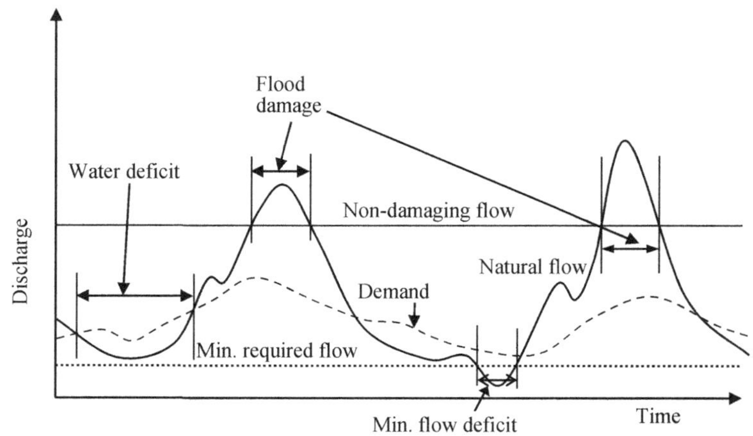

FIGURE 11.1

The need of regulation to meet the requirements of the society.

11.1.1 Conflicts in Reservoir Space

These conflicts occur when a reservoir (of limited storage) is required to satisfy divergent purposes, for example, water conservation and flood control. If the geological and topographic features of the dam site and the funds available for the project permit, the dam of sufficient height can be built and storage space can be clearly allocated for each purpose. However, this seldom being the case, multipurpose reservoirs with shared space are developed. The satisfaction of conservation purposes requires the reservoir to be filled to the maximum extent possible, whereas the objective of flood control is best met when sufficient vacant space is available in the reservoir. Therefore, while regulating such a reservoir, the crucial decision is whether to fill the reservoir or keep it vacant. A full reservoir allows reaping higher benefits by satisfying conservation purposes, but at the same time, there is a higher risk of flood damages. On the other hand, an empty reservoir can moderate floods more effectively, but if the flows are not up to the expected volume, the reservoir may remain vacant and consequently the conservation purposes will suffer.

11.1.2 Conflicts Among Purposes

Within the conservation purposes also, conflicts can arise when the pattern of water use is different and the requirement of storage space for one purpose is not in conformity with the other purpose. For example, irrigation demands may show one pattern of variation, depending on the crops, season, and rainfall, while the hydroelectric power demands may have a different variation. The water required for consumptive uses like irrigation, municipal water supply can't be shared with any other use. The conflicts in daily discharge are also experienced in a reservoir which serves more than one purpose. If a reservoir is for a consumptive use and hydroelectric power generation, the releases for the two purposes may vary considerably in the span of a day. Similarly, the benefits from the use of a reservoir for recreation are high during summer, but the irrigation demands may also be the highest during this period requiring drawdown of the reservoir level.

11.1.3 Conflicts within the Same Purpose

A deficit of water can be distributed over time in different ways. A typical decision is whether the supply should be cut now so that there is a small deficit for a longer period or postpone the cut for the future and risk a bigger shortage albeit for a shorter time. The impact of these two decisions will be different in different situations and will also depend on the use of water, viz., irrigation, municipal water supply, and so on.

In a multireservoir system, the decision-maker also has to allocate releases as well as deficits among reservoirs. Evidently, none of the common uses of water are 100% compatible with each other. The operation policy should guide the operator in satisfactorily resolving these conflicts. It may be emphasized here that the key word in regulation of a multipurpose reservoir is "compromise."

11.2 CRITICAL ISSUES IN RESERVOIR OPERATION

According to James and Lee (1971), the following six issues need to be optimally resolved while developing an operation policy of a reservoir:

1. *Use of flood storage*: Whether flood inflows should be stored to reduce current damages or released to provide additional storage space in case new rains produce even greater flows.
2. *Use of total storage*: Whether storage space should be filled to save water for beneficial use or emptied to contain potential floods.
3. *Release of stored water*: Whether water stored within the reservoir should be released for present use or retained for use during possible future droughts.
4. *Release by reservoir*: How much of the water to be released for beneficial use should come from each reservoir in which water is stored.
5. *Use of available water*: How the water released from the reservoir should be divided among various potential uses.
6. *Release elevation*: Whether the released water should be taken from near the surface or from some elevation deeper within the reservoir.

The above issues are briefly discussed in what follows.

11.2.1 Use of Flood Storage

The two main objectives of flood management are: (i) to minimize the downstream damages and (ii) to ensure dam safety. An important decision while regulating a flood control reservoir is whether flood flows should be stored in the reservoir to control the current flood or be let out to provide additional storage space in case a bigger flood occurs. This question arises when the storage level in the reservoir is in the flood control zone. The controlling parameters are the available flood storage, the current and forecasted inflows, the safe carrying capacity of the downstream channel, and the status of other reservoirs in the system. While making releases, the current flow at the damage center and the likely contribution from the catchment downstream of the reservoir up to the damage center should also be considered. It might be prudent to release at a rate equal to the safe capacity of the downstream channel less local flows, if the reservoir is in the flood storage zone. If the forecast indicates the possibility of larger floods, releases slightly exceeding safe carrying capacity of the downstream channel can be made to avoid severe damages subsequently. There is an economic trade-off between the increase in the downstream damages caused by larger releases and the increase in the expected value of future damages caused by less storage space available to contain subsequent flows. To properly manage floods, the operator should know the damages which high reservoir releases can cause in the downstream area.

Reliable precipitation and inflow forecasts are not always available to the dam operators. Therefore, the flood control regulation schedule for a reservoir is normally developed based on the information that is likely to be available with the operator at the dam site, viz., the current inflow rate, reservoir elevation and rate of rise/fall, and volume of inflow that can be expected in a flood.

11.2.2 Use of Total Storage

The second important issue is whether the storage space should be filled by storing water for some future beneficial use or be emptied to absorb likely floods. This question primarily deals with the operation of a reservoir in wet season when the objective may be to moderate the potentially dangerous floods whenever there is a significant probability of their occurrence and to store water to the capacity of the reservoir for beneficial use in the dry season. The controlling parameters are the amount of water currently in storage, the vacant space for flood control, the value of stored water, and the risk of flood.

The usual practice is to keep the reservoir empty in the first few weeks of the flood season and gradually fill it up as the season progresses. The rule curves for a multipurpose reservoir are designed so that the release from the dam should not be very and high reservoir is full at the end of the filling season. However, if the flows in the wet season are not as expected, the reservoir may not be completely filled. Clearly, a better strategy would be to fill the reservoir sufficiently at the first available opportunity. If a larger flood is expected subsequently, the storage could be depleted to create the requisite vacant space and the storage is refilled again when the flood begins to recede. This procedure can be effectively implemented if reliable forecasts of rainfall and inflows are available.

The economic trade-off is between the value of the additional water stored within the reservoir and the additional flood damages if vacant space is not available when the flood occurs. Water can be stored in the joint-use space of a reservoir for other purposes as long as the flood control operation is not hindered. The water stored for conservation purposes is released based on demands. Thus, the main operating decisions that require the operator's attention are the amount of flood control space that ought to be available at any time and size of the releases required to create additional space, if necessary.

11.2.3 Release of Stored Water

The third question is important for operation during a dry season to ensure that the water would be available to meet the demands during the most critical year. There is a trade-off between the benefit received from additional water when put to present use and its expected value in future. The trade-off is in terms of the probability of wastage of water if the reservoir spills versus the probability of a severe drought which would make the stored water very valuable.

11.2.4 Release by Reservoir

The fourth question pertains to the apportionment of release among different reservoirs in a system. How the water to be released for beneficial purposes should be divided among reservoirs? For flood control purposes, this will depend on the vacant storage space available in the reservoirs, conditions in the downstream reaches of each of the reservoirs as well as the likely inflow to each of the reservoirs. If the reservoirs serve conservation purposes, intuitively more water should be drawn from the reservoir that is likely to receive more inflow. For the best results from the operation of a reservoir system, it is necessary that the operation policies are jointly developed.

11.2.5 Use of Available Water

This issue pertains to the allocation of water among different uses. James and Lee (1971) suggest a division of water among the various uses until its marginal value for each is equal.

11.2.6 Release Elevation

Primarily due to temperature variations, reservoirs tend to stratify into zones of different density. The warmer upper layer of a lake is known as epilimnion (it might be colder during winters). The temperature changes rapidly with depth in the intermediate metalimnion zone. The lowest zone where water is usually densest and coldest has the lowest dissolved oxygen concentration and the largest concentration of sediments is known as the hypolimnion zone. These three zones are well defined only during warmer summer months; the reservoir is nearly isothermal during other seasons.

The important water quality parameters from a reservoir operation point of view are temperature, biochemical oxygen demand, dissolved oxygen, and suspended sediments. If the quality of released water is an important consideration, it is necessary to have outlets that can draw water from different elevations. The operator can then let out water from the appropriate outlet(s) to ensure that the water of the desired quality is supplied.

With the above background, the various approaches to develop policies for operation of reservoirs will be discussed in the following.

11.3 BASIC CONCEPTS OF RESERVOIR OPERATION

The drawdown refill cycle of a reservoir is usually 12 months long except when the reservoir capacity is large in relation to streamflows. The cycle may extend over many years in arid regions. In many regions of the world, the refill periods (when inflows are more than the demands and therefore extra water is stored in the reservoir for later use) and drawdown periods (when inflows are smaller than the demands and therefore water is withdrawn from storage to meet various demands) are distinctly separated. For example, in monsoon climate, high flows occur during certain calendar months only. Many reservoirs receive a significant portion of their annual flows through snowmelt. In addition, benefits from operation of reservoirs considerably improve when reliable weather and inflow forecasts are available.

A reservoir is operated according to a set of rules or guidelines to store and release water depending on the purposes it is required to serve. The decisions regarding releases in different time periods are made in accordance with the available water, inflows, demands, time of the year, etc. Many operation rules are based on intuition and common sense. For example, in a multireservoir operation, the consumptive demand may be met from the reservoir that is nearest to the demand point so as to minimize transit losses and wastage. Likewise, in irrigation operation, the manager may release water to save the standing crop from serious damage and take the risk of shortage of water for a future crop.

For reservoirs which are designed for multiannual storage, the operation policy is based on long-term targets. The estimates of water availability are made using long-term data. The requirements for conservation uses are worked out by projecting the demand data. The magnitudes of releases for the uses which are to be served from storage on a long-term basis are determined and the reservoir is operated accordingly. In periods of droughts, based on prespecified priorities, the supply for some uses is curtailed keeping in view the minimum demands of each purpose. Consideration is given to the maintenance of essential services even if it is at the cost of agriculture and industrial production.

Many basins in cold countries experience floods when the snow melts. It is possible to fairly accurately predict the runoff volume during this period by using snow surveys data and storage may be allocated to ensure desired flood protection. One can be quite sure that this space will be filled by the time flood season is over. However, such a long-range forecasting with desired reliability is not possible for rain-fed rivers in monsoon climate, and a calculated risk is taken while allocating storage space for flood control.

11.3.1 Long-Range Planning Schedules

The long-range schedules are developed during the planning stages typically to estimate the project reliability, likely benefits or the type and extent of demands that can be met. Such schedules are also developed during the system expansion, that is, when a new reservoir is to be added to an existing configuration. For conservation purposes, long-term (monthly or annual) data are commonly used for this type of analysis. Suitable assumptions are made to keep the problem tractable. For example, normal values of the amount of evaporation may be used to estimate evaporation losses from the reservoir. Despite their crudeness, the results of such studies can provide useful insights into the problem and form the basis of refined and detailed studies in the latter stages of planning and eventual operation.

11.3.2 Rigid Operation Schedules

Rigid operation schedules are needed for the eventuality when decisions are to be taken urgently, but the required detailed data are not available or there is not enough time to analyze them. Such a situation is most likely to arise in flood management. An example of extreme rigidity in operating schedule is a single purpose flood control reservoir with ungated spillway(s). In a way, the operation policy is built into the structure through the elevation of the spillway crest and its shape. Rigid schedules for flood control operation of a gated reservoir are needed for use by nontechnical staff at the dam site. Such schedules are formed on the basis of the study of the design flood or the probable maximum flood. Therefore, these schedules are based on data, such as the reservoir elevation, river stage at a downstream point, reservoir inflow, and the rate of change of reservoir elevation which are available to the operator even when all means of communication breakdown.

As too much rigidity is a hindrance in realizing the maximum benefits, some flexibility in operation is preferred. While there are no two opinions that the operator should be provided with a detailed set of instructions covering all the situations that are likely to arise, he should have enough flexibility to fine tune the decisions based on specific at-site conditions. Usually, the day-to-day reservoir operation is based on current and forecast of streamflow, status of demands, and precipitation outlook. A crucial factor in effective operation, particularly in case of floods, is the availability of reliable forecasts. Many day-to-day decisions are taken based on judgment and supported by knowledge gained by the study of past events.

11.3.3 Standard Linear Operating Policy

The simplest of the reservoir operation policies is the standard linear operating policy (SLOP), graphically represented in Fig. 11.2. According to this policy, if in a particular period, the amount of water available in storage is less than the target demand, all the available water is released. If the available water is more than the target demand but less than target demand plus available storage capacity, the release equal to the target demand is made and the excess water is stored in the reservoir. In case, even after making releases equal to the target demands, there is no space to store the excess water, all the water in excess of the maximum storage capacity is released.

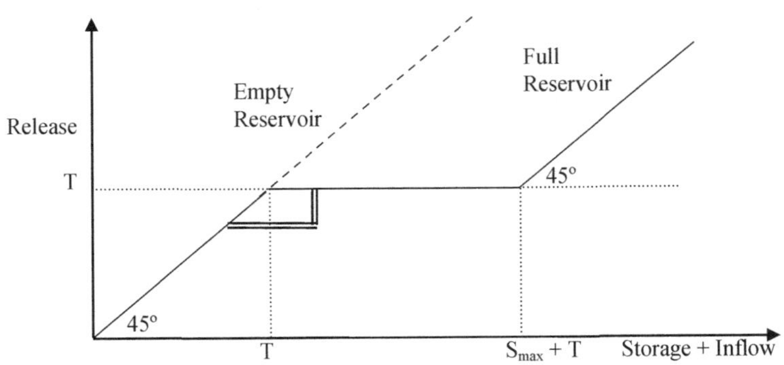

FIGURE 11.2

Graphical representation of the standard linear operation policy.

Let A_w represent the available water and T the target demand. Mathematically, the SLOP can be expressed as:

$$\begin{array}{lll} \text{If} & A_w \le T, & \text{Release} = A_w \\ \text{If} & T < A_w \le S_{max} + T, & \text{Release} = T \\ \text{If} & A_w > S_{max} + T, & \text{Release} = A_w - S_{max} \end{array} \tag{11.1}$$

The reservoir will be empty in the first case and full in third. The zone of feasible releases lies between the lines of full and empty reservoir. The SLOP is a one-time operation policy without relation to the release of water at any other time. This type of time isolated releases of water is neither beneficial nor desirable. The water beyond the target demand in any period has no economic value. Although this policy is frequently used in planning studies, it is not used in day-to-day operation due to its rigidity. Some of its drawbacks can be minimized by introducing rationing in the event of a deficit. For instance, if the available water is less than the demands for the current and next three months, the release may be reduced by some amount, say 25%. Depending upon the circumstances, rationing may be introduced in several stages; the number of stages and the extent of rationing can be decided by simulation. The double line in Fig. 11.2 shows one such possibility. The resulting rule was termed *rationing rule* by ReVelle (1999).

11.4 RULE CURVES

A rule curve or rule level specifies the storage to be maintained in a reservoir as closely as possible during different times of the year, while trying to meet various demands. The rule curves are generally derived by operation studies using historic or generated flows. Here the implicit assumption is that a reservoir can best satisfy its purposes if the storage levels specified by the rule curve are maintained in the reservoir at different times. The rule curve as such does not give the amount of water to be released from the reservoir. This amount will depend on the inflows to the reservoir and the demands for various purposes. Different rule curves may be developed for different purposes such as municipal water supply, irrigation, hydropower generation and for flood control.

11.4.1 Derivation of Rule Curves

The derivation of rule curves depends on the type of the reservoir and the purposes to be served. A reservoir may be classified either as a seasonal reservoir or a multiannual reservoir. The storage of a seasonal reservoir is utilized to carry water from the wet season to the dry season, whereas multiannual reservoir storage is used to carry water from a wet period to a subsequent dry period which could occur several years later. Consider the case of a reservoir with seasonal storage serving conservation needs. If this reservoir is able to meet the demands during the critical year, it will be able to do so in all other years.

The streamflow of a river during the driest year on record and the water requirements have been plotted in Fig. 11.3a. Assume that the reservoir is full at time A. From A to B, the demands exceed the natural inflow and hence the reservoir will deplete and will be empty at B. From time at A onwards, the inflow and demand curves diverge and the cumulative difference is maximum at B. This difference represents the required storage capacity. The mass curves of inflows and demands have been plotted in

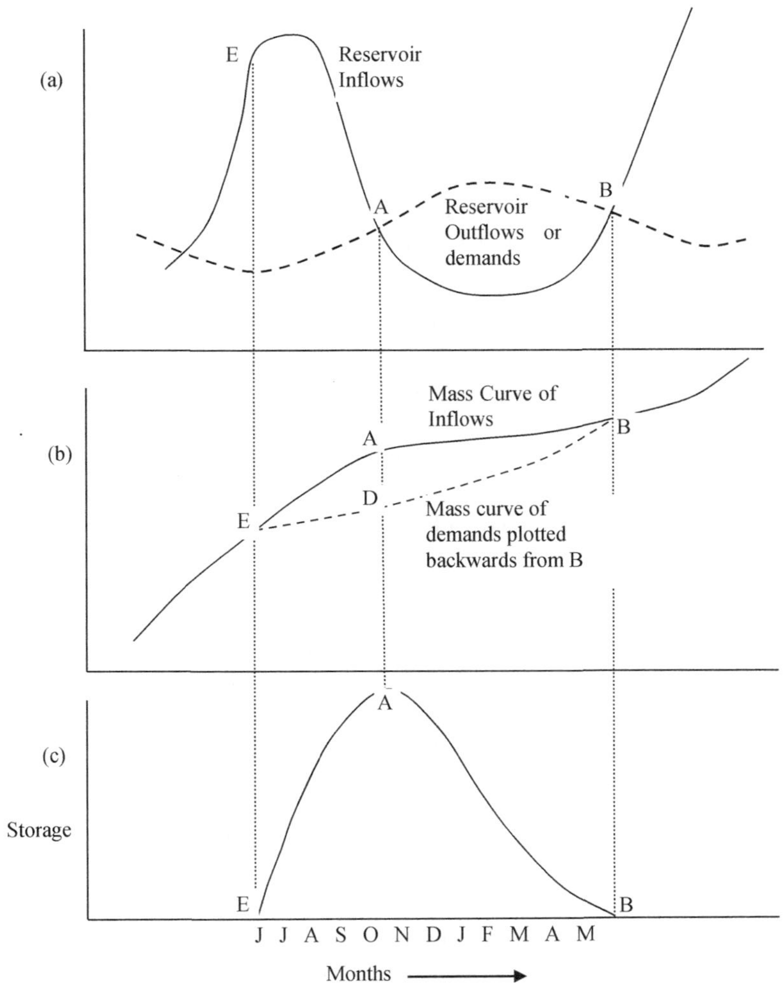

FIGURE 11.3

Development of rule curve for conservation operation.

adapted from Kuiper (1965).

Fig. 11.3b. In this figure, at point B, the reservoir is empty. From this point, the demand mass curve is plotted backwards in time and curve BE is obtained which is nothing but the mass curve AB of demands extended to the left. The vertical ordinates between the inflow mass curve EAB and demand mass curve EDB represent the volume of water which is in storage during the period from E to B. These vertical ordinates have been plotted against time in Fig. 11.3c and the resulting curve is the rule curve. This rule curve represents, in a reverse order of time, the accumulation of the deficiency between demands and available streamflow during the critical period. Since this analysis has been performed for the driest year on record, it can be safely concluded that whenever there is more water in the reservoir than specified by the rule curve, there is no danger of failure of the reservoir.

Since a rule curve depends on the flow pattern in a critical year, it would be desirable that rule curves for other near-critical years are also prepared. When these curves are plotted on the same graph, these will cross one another at several places. Finally, a smooth enveloping curve is drawn and is the requisite rule curve. The rule curves for carry-over storage can also be prepared in a similar way. The above is the graphical technique for preparing rule curves. A computer-based approach for the preparation of rule curves is explained in a later section.

11.4.2 Operation of a Reservoir Using Rule Curves

While operating a reservoir with the help of rule curves, there are several possibilities. If the water level at any time is above the elevation stipulated by the rule curve at that time (i.e., enough water), releases are made to meet all conservation demands. If the available storage is in the vicinity of that indicated by the rule curve, the release of water should be restricted such that the storage does not fall appreciably below the rule curve level. If for some reason, the level in the reservoir is much below the rule curve, the release should be curtailed with attempts to return to the rule curve level at the earliest.

The rule curves implicitly reflect the trade-off among various reservoir objectives in the long run and serve as a guide for short-term operation. Thus, the operator has flexibility to decide the releases so that the long-term objectives are fulfilled to the maximum possible extent. A reservoir operation schedule which gives some leverage to the operator to use his judgment, and experience is termed as flexible schedule. Note that the rule curves only specify the ideal levels to be maintained and the operators can use their experience and judgment to distribute excess or deficit over space and time to maximize benefits. The release decision may also incorporate the relative priority among various uses and in case of deficits, the higher priority demands are met first.

To provide further flexibility in operation, different rule curves may be specified for different circumstances. For example, there can be three different rule curves—one for a normal year in which reservoir inflows are close to the average flow (within $\pm 20\%$ of the average), another for a dry year (inflow below 80% of the average), and the third for a wet year (inflow greater than 120% of the average). In situations where it is possible to forecast floods or snow melts, conditional rule curves can be defined. These rule curves may be presented either in the form of tables or graphs and show the desired reservoir levels as a function of the expected inflows. Conditional rule curves may be defined for the entire water year or a part thereof.

Many times due to various reasons, such as low inflows, minimum requirements for demands, etc., it is not possible to adhere to the rule curve. In case of deviations, there are several ways to return to the rule levels. One way is to return to the rule curve by curtailing the release beyond the minimum required if the deviation is downward or making releases at higher rates if the deviation is upwards.

The rule curves are developed using past streamflow and demand data which will not be repeated in future. Therefore, there is a possibility of improvement over rule curve-based operation. Of course, this requires detailed input data, better models, infrastructure, and trained personnel. But the significant amount of larger benefits which can be reaped by improved regulation makes their use attractive for major reservoirs, particularly when the resources are limited and have to be allocated among a number of competing users. Several such approaches are discussed later in the chapter.

11.4.3 Concept of Storage Zoning

As discussed in Section 10.9.2, the entire reservoir storage space can be conceptually divided in a number of zones by drawing imaginary horizontal planes at various elevations. The sizes of these zones need not be constant and can vary with time as shown in Fig. 11.4.

During the actual operation, the reservoir managers are expected to maintain the reservoir level in the specified zones. This conceptual division of a reservoir into a number of zones and the rules governing the maintenance of storage levels in a specified range are based on the assumption that at a specified time, an ideal storage zone exists for the reservoir and benefits can be maximized by keeping the storage in this zone. This concept is in some way akin to a rule curve with the added advantage that this approach gives more flexibility to the decision-maker who can carefully steer the storage level within the specified zone to maximize gains. In addition, the rule governing the maintenance of the reservoir level in a particular zone may be conditioned upon the hydrologic state of the system. Thus, the reservoir operator may be advised to keep the level in one zone if streamflow is X and in another zone if the flow is Y.

The common operation policy is to release as much water as possible irrespective of the damages in the downstream area when the reservoir is in the spill zone, to release at the maximum nondamaging rates when the reservoir is in flood control zone, and to bring the reservoir to the top of the conservation zone at the earliest possible time. The release from the conservation zone depends on the requirements of water for various purposes intended to be met by the stored water and the day-to-day releases may be adjusted based on the anticipated inflow and the future requirements up to the end of the operating horizons. When the available water is expected to be less compared to the demand, releases may be curtailed. Broadly, a rule curve also aims at this type of operation.

In the context of multiple reservoirs, zoning offers some flexibility in operation of individual reservoirs. Sometimes, defining subzones within the conservation zone provides further flexibility. While managing multiple reservoirs, attempt is made to balance the storage level in different reservoirs, that is, at any time all the reservoirs are maintained in the same zone to the extent possible. This type of operation is necessary to restore balance among reservoirs after an unexpected or extreme hydrologic event.

There are three approaches for such balancing of reservoir contents. The first, known as the "equal function" policy, is to maintain keeping all reservoirs at their same zonal position, that is, at a level

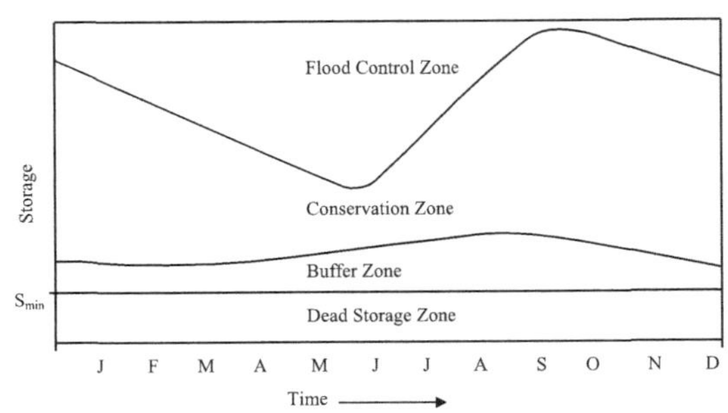

FIGURE 11.4

Variation of reservoir zones with time.

where the percentage filling of the zone is equal for all the reservoirs. The second one is based on a reservoir ranking or priority concept. Each reservoir is assigned a priority. The entire water in a zone of the lowest priority reservoir is utilized before drawing water from the next lowest priority reservoir, and so on. The third concept is based on a "storage lag" policy. Withdrawals from some reservoirs are begun before drawing water from the same zones of other reservoirs. After a certain volume has been released from the initial group of reservoirs, releases are made from all reservoirs, maintaining the percentage difference of available zone volume. This policy is followed to provide a readily available reserve of water if corrections in interreservoir balancing are needed after an unexpected or extreme hydrologic event.

Conditional Rule Curves

Conditional rules have also been used to regulate multiple-reservoir systems. These policies define reservoir releases not only as a function of the existing storage volumes and the time of the year but also as a function of the expected natural inflows into the reservoirs for some prespecified time period in the future. Such policies can be described as functions, in tabular form, or as a diagram. For the reservoirs that receive substantial snowmelt, the winter snow depth is a typical input.

The multiple zones and subzones and operating rules are prescriptive in character as compared to the simple curve. Defining flow ranges further guides in operation. These ranges for the individual channels downstream of the reservoirs can be defined as a function of the upstream storage volume. Loucks and Sigvaldason (1980) defined three ranges as shown in Fig. 11.5. The *normal flow range* is considered ideal and the flow should be in this range as long as all the upstream reservoirs are within their respective ideal zones. The *extended range* is the enlarged range of flows that could be utilized if one or more upstream storage volumes are either in flood control or buffer zone. The *extreme range* is

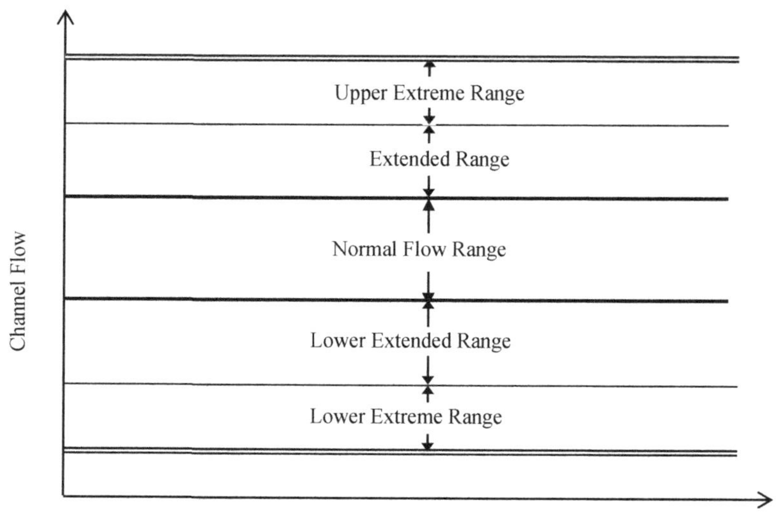

FIGURE 11.5

Channel flow ranges.

Adapted from Loucks and Sigvaldason (1980).

the further enlarged range of flows that could occur if one or more of the upstream storages are in either the spill or inactive zone. The size and extent of these ranges can be a function of time. With multiple zoning for storage volumes and flow ranging for channel flows, there is less need for operator judgment when balancing reservoir levels with channel flows and keeping the system within the restrictions imposed by these zones or levels and flow ranges.

Sometimes, the water level of a reservoir is fluctuated over a small range to reduce the incidence of diseases. Such fluctuations destroy breeding habitat of mosquitoes.

11.5 OPERATION OF A MULTIRESERVOIR SYSTEM

The discussion so far was limited to operation procedures for a single reservoir. It is well known that the benefits from the joint operation of a system of reservoirs can be substantially larger than the sum of benefits obtained from the operation of individual reservoirs. A system may consist of reservoirs in series, in parallel, or a combination. Approaches to develop operation policies for a system of reservoirs are discussed in the following. Some of these operational policies are developed by intuition and are anticipatory. But this does not diminish their utility and effectiveness.

11.5.1 Reservoirs in Series

Consider a system of two reservoirs in series as shown in Fig. 11.6. A complex system can be decomposed into this simple configuration. The diversion demand D_1 can be met only by reservoir 1 while demands D_2 and D_3 can be satisfied by both reservoirs. The rules for refill and drawdown of reservoirs in series for various purposes are given in Table 11.1.

The reservoirs shown in Fig. 11.6 can serve conservation demands best by minimizing the uncontrolled outflow of water from the system. The spill from any reservoir, except the lowest, can be captured by a downstream reservoir. Thus, the most upstream reservoir should be filled up first (subject to the availability of inflows), followed by the reservoir just downstream to it, and so on. This strategy permits capture of spills from the upstream reservoirs in the system itself.

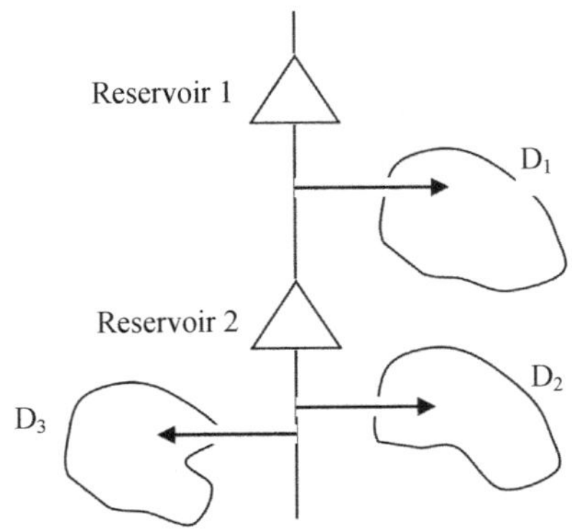

FIGURE 11.6

A system of two reservoirs in series.

Table 11.1 General rules for operation of reservoirs in series.

Purpose	Refill period	Drawdown period
Water supply	Fill upstream reservoirs first	Withdraw from downstream reservoirs first
Flood control	Fill upstream reservoirs first	Withdraw from downstream reservoirs first
Energy storage	Fill upstream reservoirs first	Withdraw from downstream reservoirs first
Hydropower production	Maximize storage in reservoirs with greatest energy production per unit of water	Maximize storage in reservoirs with greatest energy production per unit of water
Recreation		Equalize marginal recreation improvement of additional storage among reservoirs

adapted from Lund and Guzman (1999).

During the drawdown season, where the natural streamflows are small in comparison with the demands, the most downstream reservoir should be drawn down first and so on. The demands at a location are met by the immediately upstream reservoir before using any other upstream reservoir. This rule can be bypassed if due to various reasons, such as topography, system configuration, it is not possible to meet all the demands by all reservoirs. The relative magnitudes of water loss due to evaporation and seepage from various reservoirs should be considered while applying these rules.

If these rules are applied to the system shown in Fig. 11.6, demand D_1 is met from reservoir 1 (meeting this demand from reservoir 2 may involve pumping) and demands D_2 and D_3 are met from reservoir 2. When there is not enough water in reservoir 2, water from reservoir 1 is released (assuming that it has enough water) to meet demands D_2 and D_3. This rule ensures minimization of spills from reservoir 2. The spills of reservoir 1 can be captured by reservoir 2.

11.5.2 Hydropower Reservoirs

Water stored in a hydropower reservoir provides assured supply as well as hydraulic head. During the filling season, the aim usually is to have as much energy (in form of water) as possible stored in the system at the end of the season. Since the water stored in an upstream reservoir (higher elevation) has higher potential energy, the upstream reservoirs in a series should be filled first. After generating at an upstream reservoir, the water can be captured in a downstream reservoir where it again generates energy. The same logic also holds good for any spill from an upstream reservoir. Of course while storing water, one has to also examine the compatibility of uses of water. The storage of water for energy generation and other conservation uses may be compatible but will have conflict with the flood control purpose.

During the drawdown period, the objective of operation is to maximize hydropower production for a given total storage amount vis-à-vis the demands. Recall that a reservoir system can generate the maximum amount of power when all the reservoirs are full because the hydraulic heads will be the highest in this case. If the available water in the system is limited, it should be allocated among the reservoirs in such a way that the hydropower production is maximized. The governing variables here are storages, inflows, installed capacities, and efficiencies of power plants. Note that in a small reservoir, the rate of increases of head per unit volume of additional water is higher as compared to a large reservoir (with more

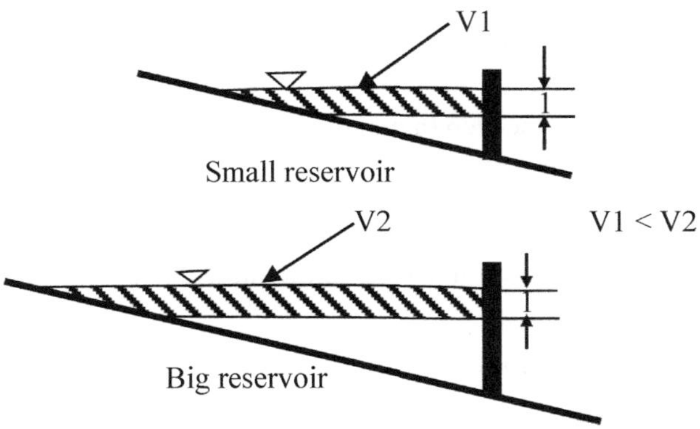

FIGURE 11.7

Volumes of water needed for a unit change in head in a big and small reservoir.

adapted from Lund and Guzman (1999).

surface area), all other things remaining the same. As shown in Fig. 11.7, the volume of water needed to increase the head by 1 unit in a smaller reservoir (V1) is less than the volume needed (V2) to increase the head in a larger reservoir by the same increment.

The important variables in hydropower production, namely, reservoir capacity, volume of inflows, and efficiency of power plant, determine the overall suitability and ranking of the reservoirs. Thus, if the storage can be increased, it should be in reservoirs with the greatest ability to produce power and vice versa. Following this reasoning, Lund and Guzman (1999) suggested a measure that can be used to rank the reservoirs:

$$V_i = a_i \eta_i \left(\sum_{j=1}^{i} I_j \right) \tag{11.2}$$

where V_i is the increased power production per unit increase in the storage, a_i is unit change in hydropower head per unit change in storage, and η_i is the power generation efficiency, all for reservoir *i;* and I_j is direct inflows and releases into reservoir *j*; and the summation is for all reservoirs upstream of reservoir *i*. The most upstream reservoir in the series is numbered 1 and so on. The reservoirs are ranked according to the V_i values. The filling begins at the highest value of V_i and proceeds in the descending order.

A variant of this rule is the *storage effectiveness index* method developed by the US Army Corps of Engineers (USACE, 1985).

Storage Effectiveness Index Method

This method was developed by USACE to maximize firm hydropower production during the drawdown season. For each reservoir, a storage effectiveness index is calculated for each time-step, using forecast inflows and power demands for the current time-step and remaining time-steps in the drawdown season. Lund and Guzman (1999) gave the following computational steps of this method:

Step 1. Find the firm energy requirement for the current time-step E_f.

Step 2. Estimate the shortfall of firm hydropower production due to insufficient inflows to the system.

$$S_f = E_f - 720\sum_{i=1}^{n} I_{Ui}H_i(S_i)\eta_i \tag{11.3}$$

where S_f is the energy shortage (kW-hr) for the current time-step (a month consisting of 720 h), I_{Ui} is the inflow upstream of reservoir i during the current time-step (m^3/s), and H_i is the hydropower head (m) for reservoir i as a function of current reservoir storage S_i. Eq. (11.3) assumes that all flows can be passed through turbines to generate power.

Step 3. For each reservoir, the volume of water required for that reservoir to individually eliminate the shortfall (ΔS_i) is estimated as

$$\Delta S_i = S_f/(720 * \ H_i\ \eta_i) \tag{11.4}$$

where H_i is the average head at which the volume ΔS_i is released.

Step 4. For each reservoir, the energy loss in the remainder of the drawdown season due to the release of ΔS_i during this time-step is estimated.

Step 5. The storage effectiveness ratio (SER) for reservoir i is

$$\text{SER}_i = E_{Li}\ /\ S_f \tag{11.5}$$

where E_{Li} is the drawdown season power loss due to drawdown of reservoir i by ΔS_i units. This ratio is calculated for each reservoir and the reservoirs with the lowest ratios are to be drawn down first.

11.5.3 Reservoirs in Parallel

The simplest configuration of parallel reservoirs is shown in Fig. 11.8. It consists of two reservoirs located on two different streams which join downstream of the reservoirs. The direct demands of reservoirs 1 and 2 are D_1 and D_2, respectively. Either or both of the reservoirs can meet demand D3. An important difference in operation of series and parallel reservoirs is that the release from an upstream reservoir cannot be captured by a downstream reservoir. Therefore, balancing of the operation is important in such cases. Rules to operate parallel reservoirs are summarized in Table 11.2.

In this configuration, the commonly followed procedure is to discharge water first from the reservoir with larger drainage area or potential inflows per unit storage capacity. To that end, the

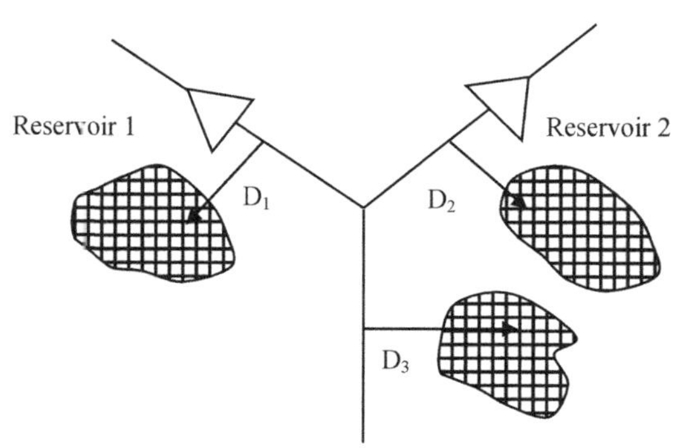

FIGURE 11.8

A system of two parallel reservoirs.

Table 11.2 General rules operation of reservoirs in parallel.

Purpose	Refill period	Drawdown period
Water supply	Equalize probability of spill among reservoirs	Equalize probability of emptying among reservoirs
Flood control	Keep more vacant space in reservoirs likely to receive bigger floods	Not applicable
Energy storage	Equalize EV of energy spill among reservoirs	For the last time-step, equalize EV of seasonal energy spill among reservoirs
Hydropower production	Maximize storage in reservoirs with greatest energy production per unit of water	Maximize storage in reservoirs with greatest energy production per unit of water
Recreation	Equalize marginal recreation improvement of additional storage among reservoirs	Equalize marginal recreation improvement of additional storage among reservoirs

Note: EV = expected value.
adapted from Lund and Guzman (1999).

drainage areas to storage volume capacity ratios for two reservoirs are compared (assuming the runoff per unit of drainage area is the same). The reservoir with the larger ratio will supply water for demand D_3 before the other reservoir is drawn down. Discharging water first from the reservoir having the largest drainage to the storage volume capacity ratio will usually result in a reasonable conservation of water.

Several types of rules have been developed for conservation operation of parallel reservoirs during refill periods. A typical objective of such system is to minimize expected shortages. These rules prescribe ideal releases or storage levels for reservoirs in parallel to avoid the inefficient condition of having some reservoirs full and spilling, whereas other reservoirs have unused storage capacity. The severity of shortages is reduced by minimizing uncontrolled spills from the system.

New York City Rules

The New York City (NYC) rules are helpful for a parallel system of reservoirs that have a single demand downstream of all reservoirs. Clark first stated these rules in 1950 for the NYC water supply system. In operating this system, an attempt was made to have the storage in each of the watersheds, at all times, fall on the same percentage year. The reservoirs are operated to minimize expected shortages by using the probability of spills. Clark noted that the physical spill is minimized when the probabilities of spill at the end of the refill season are the same for each reservoir. The deficit in meeting the target is minimized when the physical spill is minimum.

Application of the NYC rule requires prediction of inflows. Evidently, better accuracy of predictions would result in lesser spillage of water. A high degree of accuracy in predicted inflows is not critical in the early periods of the refill season. The releases are recalculated at each period, and the reliability of flow forecasts becomes progressively more important as one reaches the end of the refill season. The NYC rule has been found to be optimal or near-optimal in a variety of operating conditions and system configurations. The optimality of rules also depends on the coefficient of variation of mean monthly flows and the correlation between flows on adjacent streams.

Originally, NYC rules assumed that the unit value of water is the same in each reservoir. But this may not be always the case. In a water supply system, the quality of raw water affects its unit value because poor quality water would entail higher treatment costs. For hydropower generation, a unit volume of water in a reservoir having greater head can generate more energy and will, therefore, have higher value. The NYC rules were later modified to handle situations where the unit value of water varies between reservoirs but is constant in any individual reservoir.

The general form of the NYC rule equates the probabilities of spill at the end of the refill season adjusted by the unit value of water for each reservoir

$$h_i \, \Pr\left[CQ_i \geq K_i - S_{fi}\right] = \lambda, \text{ for all } i \tag{11.6}$$

where h_i is the unit value of water in reservoir i, CQ_i is the cumulative inflow to reservoir i from the end of the current period to the end of the refill season, K_i is the storage capacity of reservoir i (assumed to be the same in every period), S_{fi} is the storage at the end of the current period for reservoir i, and λ is constant across all reservoirs in parallel. Note that if the unit value of water is the same among reservoirs that are supplying water, h_i can be incorporated into constant λ in Eq. (11.6).

Generally, historical data are used to estimate cumulative inflows CQ_i. The release from each reservoir for the current period is found by knowing the initial storage, expected inflow, and the end-of-period storage that satisfies Eq. (11.6). The sum of these releases should equal the total downstream target release. A trial-and-error procedure may have to be adopted to find the releases. If the purpose of reservoirs is energy generation, the probabilities of spill of the potential energy are equated.

Hydropower Rules

For steady-state hydropower production at reservoirs in parallel, the storage effectiveness of parallel reservoir j can be defined as

$$V_j = \eta_j \, a_j \, I_j \tag{11.7}$$

where subscript j refers to an individual parallel reservoir. While emptying parallel reservoirs sequentially, those with the smallest V_j are emptied first. The filling should take place in the reverse order. Sheer (1986) and Lund and Guzman (1999) have derived rules for complex cases.

11.5.4 Other Rules

Some other useful rules for reservoir operation are discussed in this section.

Space Rule

The space rule seeks to leave more space in reservoirs where greater inflows are expected, or where inflows with greater potential energy are expected in the case of energy storage. The space rule, proposed by Maass et al. (1962), seeks to minimize the volume of spills. An inefficient condition arises in operation of multiple reservoirs when some of them are full and spilling, and others are unfilled. The spill-minimizing objective implies that this rule is especially suitable in the system's refill season. This requires monitoring storage volumes and estimating future inflows. When parallel reservoirs are operated by this rule, the objective is to equalize the probability that the reservoirs will have filled at the end of the drawdown refill cycle. In that event, all the reservoirs will be full and spilling, full and not spilling, or partly full, the unoccupied storage space being proportioned to inflows during the drawdown refill cycle. Mathematically, the rule is:

$$\frac{S_{\max j} - S_{Ijk} - Q_{jk} + R_{jk}}{\sum_{j}^{m}\left(S_{\max j} - S_{Ijk} - Q_{jk}\right) + R_T} = \frac{Q_{j,n-k}}{\sum_{j}^{m} Q_{j,n-k}} \tag{11.8}$$

where S_{maxj} is the full capacity of the *j*th in a series of *m* parallel reservoirs; S_{Ijk} is the initial contents of the *j*th reservoir in the *k*th month of a series of *n* months; Q_{jk} is the flow into the *j*th reservoir in the *k*th month; R_{jk} is the release from the *j*th reservoir in the *k*th month; R_T is the sum total of target releases required; and $Q_{j,n\text{-}k}$ is the predicted flow into the *j*th reservoir for the remaining *n-k* months of the drawdown refill cycle. Solving the above equation for R_{jk}, the release from the *j*th reservoir in the *k*th month is:

$$R_{jk} = \left[\sum_{j}^{m}\left(S_{\max j} - S_{Ijk} - Q_{jk}\right) + R_T\right] \times \left(Q_{j,n-k} / \sum_{j}^{m} Q_{j,n-k}\right) + S_{Ijk} + Q_{jk} - S_{\max j} \tag{11.9}$$

subject to the constraint

$$0 < R_{jk} < \left(S_{Ijk} + Q_{jk}\right) \tag{11.10}$$

The space rule is useful in situations where inflow forecasting is reliable as in the case of runoff from snowmelt. For other types of streamflows, the effectiveness of the space rule would be a function of the coefficient of variation of the mean monthly flows, the correlation between flows on the adjacent stream and the reliability of flow forecasts. Although this rule aims at minimizing the water spilled in the system during the remainder of the drawdown-refill cycle, it does not guarantee minimum system spill. The rule also does not provide the total system yield R_T which has to be computed separately. ReVelle (1999) has proposed a few variants of the space rule.

The following numerical example illustrates the use of space rule to determine releases from two parallel reservoirs.

Example 11.1. While operating two parallel reservoirs for irrigation for a given month, the target output is 400×10^3 m^3. The unregulated flow from the catchment below the reservoirs and above the point of irrigation diversion is nil. For the data given in Table 11.3, find the release from each reservoir.

Solution. The computations following Eq. (11.9) are shown in Table 11.3. To meet the target output of 400×10^3 m^3, the release from the reservoir number 1 would be 36.4×10^3 m^3 and from the reservoir number 2, it will be $363.8 * 10^3$ m^3. The releases from the individual reservoirs are determined based on the storage in them and the expected inflows.

The space rule can be modified and used to apportion releases among reservoirs for flood control, based on short intervals of time. It is also valid when each unit of water is of equal value in a given reservoir but not in different reservoirs. However, the space rule must be modified in its form to deal with this situation. Unequal values of water may be present in a system, for instance, when there is a power-plant downstream of reservoir B to generate electric energy from irrigation releases, but no such plant exists below reservoir A. Space rule can also be applied to maximize the economic value of the system output by minimizing spills of the higher valued water from reservoir B at the expense of spills of downstream valued water from reservoir A. This rule provides flexibility in operation by increasing

Table 11.3 Determination of reservoir releases using space rule (Example 11.1).

Row	Details	Reservoir No. 1 (10^3 m^3)	Reservoir No. 2 (10^3 m^3)	Total for reservoir 1 and 2
A	Maximum storage capacity	200.0	2000.0	2200.0
B	Storage at the beginning of month	100.0	1000.0	1100.0
C	Empty storage space at the beginning of month, *Row A* − *Row B*	100.0	1000.0	1100.0
D	Inflow during the month	50.0	500.0	550.0
E	Target irrigation release for the month	To be computed		400.0
F	Total space that would be available at the end of current month $C - D + E$	To be computed		950.0
F	Predicted inflow between end of current month and end of refill cycle	100.0	1000.0	1100.0
H	Fraction of required space at end of month	100/1100	1000/1100	1
I	Allocation of space at the end of month = $F \times H$	86.4	863.6	950.0
J	Storage contents at end of month $A - I$	113.6	1136.4	1250.0
K	Release for the current month $B + D - J$	36.4	363.6	400.0

release toward the end of the drawdown-refill cycle to free reservoir space for predicted inflows that might otherwise spill.

Pack Rule

The pack rule uses streamflow forecasts during the last few months of drawdown-refill cycle and tries to avoid spills by additional releases of water in advance (say, for secondary energy generation). The rule was so named by Maass et al. (1962) because the expected future spill is as tightly packed as possible into future spare turbine capacity. Mathematically,

$$R_d = Q_{n-k} - (S_{\max} - S_{Tk}) - P_{n-k} \tag{11.11}$$

where R_d denotes the additional releases for the current month k for the generation of dump energy; $Q_{n\text{-}k}$ is the predicted flow into the reservoir for the remaining n-k months of the drawdown refill cycle; $S_{\max}$ is the full reservoir capacity; S_{Tk} is the reservoir contents in the current month after current flows have been added and releases made to meet the target output for energy, and $P_{n\text{-}k}$ is the useful water capacity of turbines for the remaining n-k months of the drawdown refill cycle. If the right hand side of the equation is not positive, $R_d = 0$ and the equation is further subject to constraint: $P_c \geq R_d \leq S_{Tk}$ where P_c is the useful water capacity of turbines in the current month after releases have been made through turbines to meet the target output for energy.

The pack rule can be applied whenever releases beyond the specified output requirements are of value. In Fig. 11.9, the operation of a reservoir which also generates hydropower is depicted. Assume

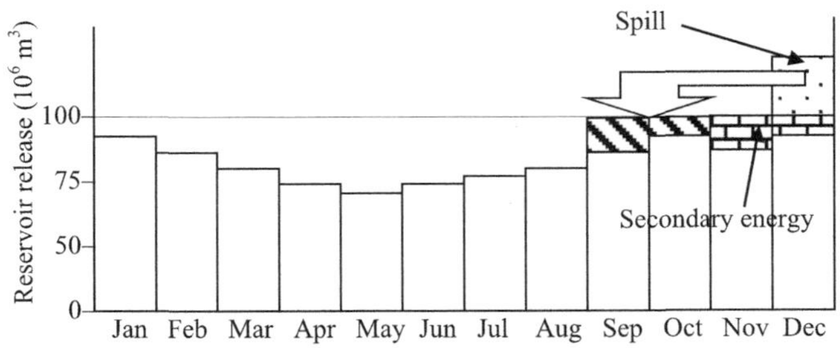

FIGURE 11.9

Operation of a reservoir. Shown here is spill that would result from a rigid rule and generation of extra secondary energy using this spill by applying pack rule.

that the drawdown-refill cycle begins in the month of Jan and ends in Dec. By applying the pack rule, secondary energy was generated during the last two months of the cycle, shown by rectangles with brick-shaped hatching. In the last month of this drawdown refill cycle, there is spill of water of the order of 25×10^6 m^3, shown by dotted rectangle. If the pack rule is applied, additional secondary electric energy, shown by rectangles hatched with slanting lines, can be generated during the months of Sep and Oct. Clearly, the application of the pack rule has two advantages. It minimizes spill of water which has no value and may as well cause some unwanted consequences in the downstream areas. At the same time, additional secondary energy is generated by the system to increase the overall benefit.

Hedging Rule

The aim of hedging is to distribute the anticipated shortage uniformly so that its severity is reduced. It is sometimes economical to accept a small current deficit in releases so as to decrease the probability of more severe water or energy shortage at a later date. The effect of hedging is brought out in Fig. 11.10 for a reservoir operating for irrigation/water supply. In Fig. 11.10a, the total volume of release is 240 million m^3 and the volume of deficit is 55 million m^3. The largest deficit of 30 million m^3 occurred in the month of Jan followed by a deficit of 20 million m^3 in Feb. This deficit was distributed by following the hedging rule and the revised release schedule is shown in Fig. 11.10b. After the released were revised, the largest deficit is 15 million m^3 which occurred in the months of Jan and Feb. This severity is just half of the severity of the first case and is expected to be less damaging although the total volume of deficit stays the same.

The economic justification of the hedging rule arises from nonlinear benefit or loss functions of the proposed uses of water. If the marginal values of water for specific uses are constant, the economic losses from shortages will be linear. It is well known that the streamflows are stochastic and to some extent the demands, particularly those arising from agriculture areas, too are stochastic. These two facts will suggest that it is better to postpone shortages as long as possible. However, the marginal values of water for a specific use are not constant. Since large deficits are more damaging than smaller ones, is preferable to avoid the possibility of suffering heavy deficits.

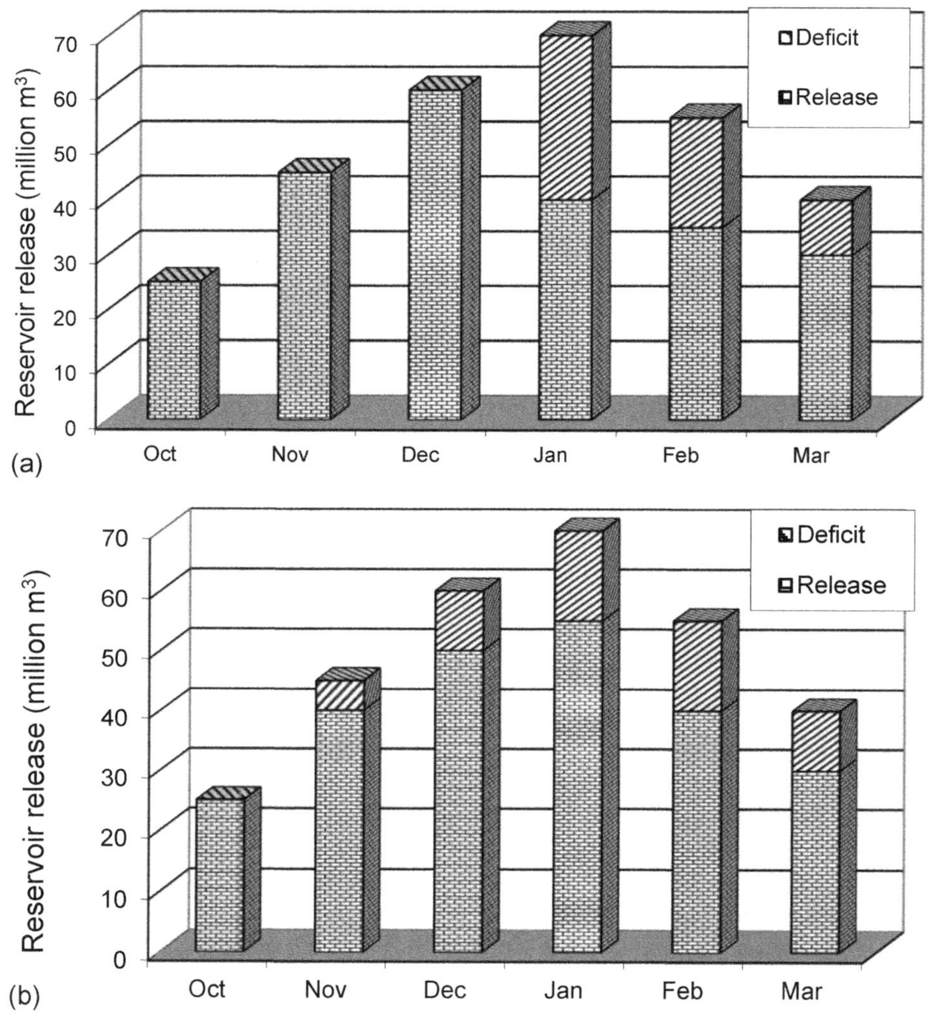

FIGURE 11.10

(a) Hedging rule—monthly outputs and shortages without hedging. (b) Hedging rule—monthly outputs and shortages with hedging.

11.5.5 Selective Withdrawal

The quality of supplied water is very important for many uses. For example, cold water is not good for temperature-sensitive crops, such as rice. The temperature and dissolved oxygen content of water released from a dam affect biological life in the stream. Due to taste and odor problems, it is not desirable to supply poor quality water for domestic use. Since the quality of water stored in a reservoir varies with depth, in some instances, it is required to supply water from a particular layer. This mode of operation is termed as selective withdrawal. For this purpose, an intake that can withdraw water from the desired layer(s) is a prerequisite. Many reservoirs have conduits to allow rapid release of turbid water during flood events. Fontane et al. (1981) presented a methodology combining optimization and simulation for determining operational guidelines for selective withdrawal structures to meet downstream water temperature objectives.

11.6 RESERVOIR OPERATION FOR FLOOD CONTROL

Among the measures of flood control, a storage reservoir with gates to control the outflow is perhaps the most effective means. A reservoir helps moderate floods by reducing the peak, delaying its arrival, and extending the hydrograph base. The moderation of a flood through storage is achieved by storing a part of flood volume in the rising phase of the hydrograph and releasing the same gradually in the receding phase of the flood. The degree of moderation or flood attenuation depends on the empty storage space available in the reservoir when the flood impinges on it. The flood control pool must be emptied as quickly as the downstream flooding conditions allow; this will reduce the risk of highly damaging future releases, should a major flood occur in quick succession. Operation of a multipurpose reservoir during high flows can be divided in three periods (Fig. 11.11). Here, ① is the predepletion stage where, if necessary, vacant space is created by releasing stored water; ② is period of very high flows when reservoir is regulated to maintain outflows below safe channel capacity, ensuring safety of dam. After the high flow period is over, reservoir is filled in stage ③. Correct identification of beginning and termination of various stages is important.

The reservoir regulation consists of storing peak flows over and above the safe (nondamaging) carrying capacity of the channel at the damage point in the reservoir. The reservoir is emptied after the passage of the flood to make space for control of subsequent floods. In Fig. 11.12, ABCDE represents the inflow hydrograph. The line ZZ represents the nondamaging carrying capacity of the river channel

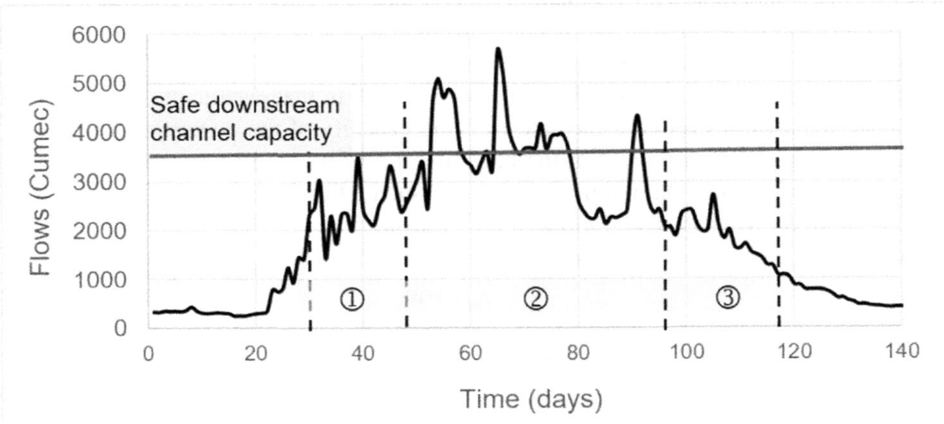

FIGURE 11.11

Stages in operation of a multipurpose reservoir during the high flow period.

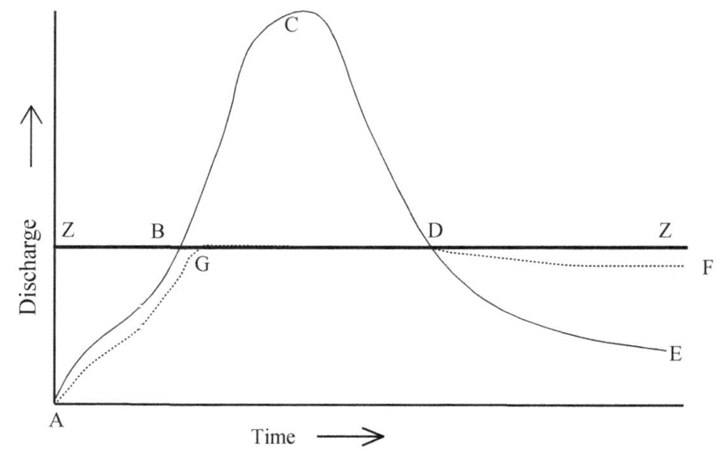

FIGURE 11.12

Ideal operation of a reservoir for flood control.

downstream of the reservoir. From point B to point D, the natural flow in the river exceeds its safe carrying capacity. If there were no reservoir, from the time corresponding to point B up to point D, the flood water will have spilled over the channel banks and cause damage. The moderated release from the reservoir under the ideal operation is given by the dotted curve AGDF. As soon as the inflow begins to increase, the release is gradually increased till some point G where the release equals the safe carrying capacity. While the inflows from point B to D exceed the safe carrying capacity of the downstream channel, the release is maintained within safe range by storing the volume in the segment BCD in the flood control zone of the reservoir. After point D, the inflows continue to fall rapidly, but the release, while still in safe zone, exceeds inflow so that the reservoir is quickly emptied.

The above operation is a near-ideal situation which is possible only if the perfect fore-knowledge of the hydrograph is available. In absence of such information, the release curve may deviate from the ideal shape. For example, if the operator makes smaller releases in the early part of the hydrograph, it is likely that the reservoir will completely fill before point D. In that eventuality, the operator will be forced to make releases in excess of the safe carrying capacity of the downstream channel thereby causing flood damages. Conversely, if at the beginning of a flood event, the operator starts making higher releases in the expectation of a major flood and such a flood does not occur, the reservoir may not fill to the desired level by the end of the filling season.

The efficiency function of a flood control reservoir can be defined by Nagy et al. (2002):

$$K = \mathrm{f}(Q_d{}^{\max}) \tag{11.12}$$

where K represents the storage capacity to ensure that the maximum outflow $Q_d^{\max}$ from the reservoir will be lower than the maximum upstream flood. At the starting point of the efficiency function curve is a reservoir with zero capacity which will not modify the inflows. At the other end is a reservoir whose capacity is equal to the volume of the entire flood flow (V):

$$K = V, \text{ and } Q_d{}^{\max} = 0 \tag{11.13}$$

Nagy et al. (2002) have described a method to determine the required capacity of a storage reservoir using the efficiency function and a given reliability. Before discussing the approaches for flood control operation of a reservoir, a technique that is used to allocate storage space for flood control in a multipurpose reservoir is discussed.

11.6.1 Flood Control Reservation Diagram

To allocate storage space in a multipurpose reservoir, the flood control diagram is especially useful. The curves of this diagram define the amount of vacant storage space to be kept available to control the floods of known or expected magnitude over a specified time interval. In calculating the amount of space required, it is assumed that releases in excess of channel capacity will not be made.

The vacant space required for flood control in a reservoir on any date is the difference in the volume of inflow between the date in question and the date of maximum storage (end of filling season) and the volume of water released from the reservoir. Usually the time step size of the analysis is a week or a fortnight. To derive curves for the first day of the various time steps during the filling season, the historical flow record is routed through the reservoir. The amount of storage space required to control runoff after the date in question is plotted for each year. The positive values indicate the vacant space that must be kept available in addition to the minimum reservation for flood control. The years having negative values of required space indicate that the flood could have been controlled with less than the minimum reservation for flood control. A line is then fitted through the points for the date in question.

The flood control space requirement can be linearly related with the remaining season runoff. The generalized equation is:

$$Y = mX + C_t \tag{11.14}$$

where Y is the required space in volume units, m is the tangent of the straight line, X is the remaining season runoff in volume unit, and C_t is the ordinate intercept (volume units) which is a function of the time (day of the filling season). If the flood control space given by Eq. (11.14) is not available in the reservoir on the given date, one must make sufficient releases so that enough vacant space is created. Depending on the inflows, this release may or may not exceed the normal outflow required to meet the conservation demands.

It is important to note that the flood control reservation chart does not indicate the size of the release to be made. The decision concerning the rate of release will depend on the actual inflows. If the flood control space required is violated marginally and the inflows are not likely to increase rapidly for some time, the release may be made to gradually vacate the encroached space. This will save water, particularly if subsequent runoff is small. If, however, the inflows are likely to be large, the reservoir should be drawn down rapidly.

As the filling season progresses, the runoff likely to be generated between the current time and the end of filling season decreases and thereby, the space required for flood control also decreases. It implies that the reservoir should be gradually filled so that the objective of having the reservoir full at the end of the flood season can be met. In other words, the amount of joint storage space committed to flood control decreases while that for conservation increases as the season progresses. At the end of the flood season, the entire joint-use storage space is committed to conservation.

11.6.2 Strategies to Reservoir Operation During Floods

Two strategies are common in controlling flood peaks by reservoirs. The first consists of operating the reservoir to reduce every flood peak by the maximum possible amount. In the second approach, attention is focused on moderation of larger floods; the smaller flood peaks are not given much attention. In any case, it would be ideal that the reservoir is operated such that the release is always less than the safe carrying capacity of the downstream channel. Since substantial water may be temporarily stored in the reservoir, the water level rises above the full reservoir level (FRL). After the flood has peaked, the reservoir is gradually bought back to FRL.

There can be several possible approaches for regulating an incoming flood. This classification is based on the peak of the incoming flood.

Regulation Based on the Maximum Use of Available Space in Each Flood Event

This type of regulation aims at reducing damaging stages at locations sought to be protected as much as possible during each flood with the judicious use of the available space. The possibility of having an appreciable portion of the flood control storage capacity already filled up before the occurrence of a large subsequent flood is disregarded.

The regulation to obtain the maximum benefits during ordinary floods can be successful depending on the ability to forecast flow conditions at the reservoir and flood-prone areas below. This requires adequate network of hydrologic stations and capability to properly evaluate weather forecasts. This method is helpful in cases where the available flood control storage is insufficient to control larger floods which occur less frequently. The disadvantage is that the available protection is limited if a dangerous flood occurs after most of the available capacity has been utilized to regulate lesser floods.

Regulation Based on the Control of Project Design Flood

In a project which has a flood regulation capacity based on the control of the project design flood, the regulation may be based on the assumption that each incoming flood might develop as design flood. The release rates are so established that all the flood control storage capacity is utilized if the current flood turns out to be the same as the project design flood.

As the project design flood occurrence is an unusual event, a schedule for its regulation will normally afford a satisfactory moderation of most floods. However, with this strategy, a less satisfactory regulation of lesser floods (which occur more frequently) can result at times.

Regulation with Combination Method

The best overall plan of operation often results by adopting a combination of the above two methods. For instance, to protect an agricultural area, a regulation plan for the maximum damage reduction during the main farming season may be desired, but it may not be the most advantageous when the fields are empty. As the flood waters carry nutrients, there is a view that controlled flooding of agricultural areas may not be all that bad. Nevertheless, it is desirable to reserve the reservoir storage to provide assured flood protection to an important town or areas (e.g., an industrial estate). In such cases, a schedule of releases to assure greater control of major floods at the expense of less regulation of moderate floods may be more desirable.

Flood Control Rule Curves

For the multipurpose reservoirs, wherein flood control is the main purpose, some storage space is allocated for flood control and is specified by rule curves (Loucks and van Beek, 2017). This space is typically kept vacant to moderate the incoming floods. The size of this space is based on the volume of design flood, the spillway release capacity, and the safe carrying capacity at the downstream damage center. The incoming floods are moderated by utilizing this space; it is emptied after a flood event to be ready for a subsequent flood. This practice is based on the premise that damage-causing releases should be avoided till the available storage is utilized.

A limitation of this approach is its rigidity during real-time operation especially with environmental constraints. Notwithstanding the limitations, rule curves offer a backup even when advanced decision-making tools are present. Improvements in the way rule curves are developed and implemented would be helpful in flood control regulation.

11.6.3 Predepletion of Reservoirs

In some situations, it is not desirable to allow the reservoir level to rise above FRL for some reasons, for example, the land near the periphery could not be acquired. In that situation, the outlet capacity should be large enough to control the rise of the reservoir water level. Alternatively, the reservoir level is lowered before the arrival of the flood by making anticipatory prereleases and the level is brought back to FRL after the flood peak has passed (see Fig. 11.13). This strategy involves the use of inflow forecasts and the confidence of the operator in making prereleases, and therefore, depends on the reliability and timely availability of forecasts. This aspect is further discussed in a later section.

In some reservoirs, the space available for flood control is small compared to the volume of the design flood. In such situations also, predepletion proves to be useful for flood moderation. Consider that the reservoir is at or below FRL at the beginning of the flood control operation. The operator can lower the reservoir level by prereleases to create the storage space for flood moderation. During the passage of the flood (see Fig. 11.14), the reservoir is regulated such that the maximum water level (MWL) is below MWL. Thereafter, releases are made such that the reservoir level comes back to the

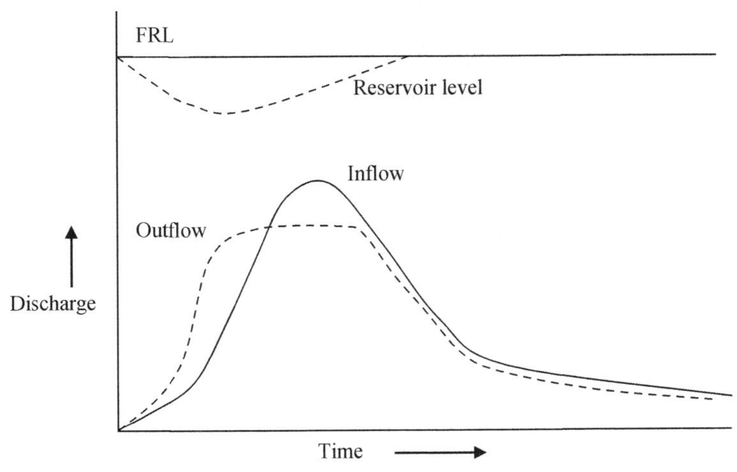

FIGURE 11.13

Flood moderation through reservoir predepletion.

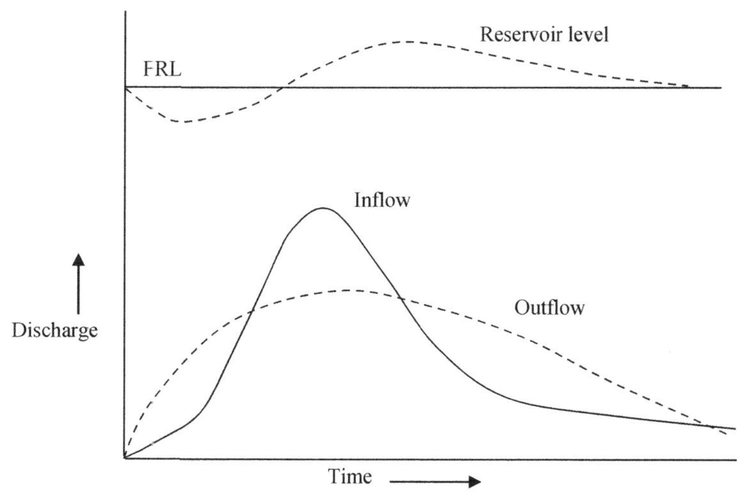

FIGURE 11.14

Flood moderation through predepletion and use of flood control space.

desired elevation. Thus, the flood is moderated using the space created by prereleases and the space between FRL and MWL. If not constrained by other factors, this approach perhaps involves making the most efficient use of the reservoir storage space.

The important considerations in predepletion are discussed below.

When the Predepletion Should Commence?

The predepletion should begin as soon as the (expected) inflows exceed a specified value. This value depends on the reservoir and catchment characteristics. A typical value could be about 10% of the minimum of the outlet capacity at FRL and the safe carrying capacity of the downstream channel.

How Much the Reservoir Level Should Be Depleted?

The extent of predepletion will depend on the current reservoir level, the shape of elevation-capacity curve, the inflow and their rate of increase, and the rainfall in the catchment. The net rainfall in the catchment provides an estimate of the volume of inflows expected in the reservoir. If the reservoir is predepleted by this much volume, one can be reasonably sure to refill the reservoir back to the original level.

At what Stage After Passage of Flood, Should the Reservoir Be Filled Back?

It would be necessary to have an estimate of the net rainfall in the catchment at each time step (usually 1–3 h). This would give the volume of expected inflows. After the passage of peak, the reservoir refilling should commence whenever the expected inflows are a little more than the volume required to refill the reservoir up to the desired level.

11.6.4 Normal and Emergency Operation

In many cases, the current hydrologic situation in terms of observed values of relevant hydrometeorological variables are not available at the reservoir control centers. In such situations, the decision-making process should be based on indicators which the operator at the dam site can easily observe. These are: (a) the current reservoir level and its rate of rise, and (b) the inflows and their trend (whether these are increasing or decreasing).

The operation of a reservoir for flood control usually begins as soon as the water level in the reservoir exceeds some specified level, such as FRL, or the inflow rate is greater than a threshold. The operation policy for flood control regulation could be based on the concept of a cut-off reservoir-level, cut-off inflow magnitude, and the nature of inflow (i.e., rising or falling). While developing the regulation policy, it is assumed that the operator has the following objectives:

(a) to attain FRL at the earliest and to maintain it;
(b) to control the rise of reservoir level above FRL to the extent possible;
(c) to utilize the reservoir storage capacity fully before releasing water at rates exceeding the safe channel capacity; and
(d) to release water at rates above the nondamaging discharge in the downstream reaches for the least possible time.

A procedure for operation of flood control reservoirs that is based on two variables has been found to be satisfactory in most cases. The first variable, the *reservoir critical level (RCL)* is a level between FRL and MWL. If, during the passage of a flood wave, the reservoir level rises above RCL, it indicates that the normal operation procedure has not been able to contain floods within the desired range. This could be due to limitations of storage capacity or outlet release capacity or because the magnitude of the flood is higher than anticipated. As a very high water level in the reservoir can endanger the safety of the dam, it is necessary that the reservoir level be brought down below RCL at the earliest. To do that, the operator may be forced to make releases which may cause damages in the downstream reaches. The *critical flood inflow (CFI)* may also be defined in a similar way and indicates abnormally high inflows to the reservoir. If the actual inflows are greater than CFI at any stage, and the reservoir level is above RCL, the emergency operation schedule should be followed without delay.

The RCL and CFI depend on the reservoir capacity, particularly between FRL and MWL, the capacity of spillway and other outlet works, safe carrying capacity of downstream channel, the volume under the design flood hydrograph, and the peak of design flood. No simple analytical technique is available to determine RCL and CFI for a reservoir. These can be determined by simulation through a trial-and-error approach. As a rough guide, the initial value of RCL may be taken as half way between FRL and MWL and CFI equal to the safe carrying capacity of the downstream channel. Simulation of the operation of the reservoir is then carried out with either the design flood hydrograph or the hydrograph of a major flood that has occurred in the past. The detailed working table should be examined to flag the periods during which the reservoir level was undesirably high and the releases were more than the safe carrying capacity of the downstream channel. The parameters RCL and CFI are now tuned to best attain the stated objectives. Normally, convergence is rapidly achieved. In case, a system of reservoirs is analyzed, this procedure should be carried out first for individual reservoirs. After these parameters have been obtained for each reservoir, the entire system is studied. At this stage, the parameters for individual reservoirs may have to be changed a little bit so that the overall performance of the system is the best.

Now, the reservoir operation scenario can be classified into two categories: *normal operation* and *emergency operation*.

Normal Operation

The normal operation of the reservoir for flood control begins as soon as the first signs of an impending flood event are noticed. Specifically, the normal operation policy is initiated when:

(a) the reservoir level is between FRL and RCL and the inflow rate is less than CFI; or
(b) the reservoir level is between FRL and RCL, the inflow rate is greater than CFI, and it is decreasing; or
(c) the reservoir level is greater than RCL, the inflow rate is less than CFI, and it is decreasing.

Under normal operation, water is released at a rate which is less than or equal to the safe carrying capacity of the downstream channel. The aim is to bring the reservoir back to FRL at the earliest so that the next flood, if any, may be moderated.

Emergency Operation

This mode of operation is invoked when the flood build-up is bigger than anticipated and the normal operation has failed to control it. Either the reservoir level is already high or the inflows are big, indicating the likelihood of an extreme flood. Under the emergency operation, the safety of a dam becomes important. The emergency operation policy is followed when:

(a) the reservoir level is greater than RCL and the inflow rate is less than CFI but it is increasing; or
(b) the reservoir level is greater than RCL, inflow rate is greater than CFI, and it is decreasing; or
(c) at any reservoir level, inflow rate is more than CFI and it is increasing.

If the release at the rate of safe downstream channel capacity is likely to cause overtopping, the release is made equal to the outlet capacity at the current elevation. The objective is to bring the reservoir down to safer level at the earliest and ensure safety of the dam. The minimum rate at which water should be released from the reservoir under emergency conditions is the lesser of the inflow rate and the outlet capacity. A software on the above lines was reported by Jain and Goel (1999).

11.6.5 Flood Control Operation of a Multireservoir System

A reservoir system may have single-purpose reservoirs in series, in parallel, or a mixed configuration of multipurpose reservoirs.

Flood Control Reservoirs in Series

Since the objective of flood control is to transmit the water from the system at the earliest while causing the least damage, the operation policy is opposite to the conservation operation policy. Therefore, in a series of reservoirs serving solely for flood control, any flow which does not cause flood damage should be allowed to pass at the earliest. If the excess water is to be stored, the headwater reservoirs should be filled first and the downstream reservoirs are to be emptied first. Since the closest upstream reservoir to a location to be protected has the greatest flood control capability and most of the area to be protected lies in the lower reaches of a basin, this approach appears to be the best strategy for flood management.

Flood Control Reservoirs in Parallel

The usual approach for flood control operation of parallel reservoirs is to maintain the reservoirs in balance in terms of occupied capacities and flood flow from respective catchment areas. These reservoirs are operated for flood control such that the combined release plus the flow from the intermediate catchment is below the safe carrying capacity of the channel at the damage center and if this is not possible, the flow in excess of the safe channel capacity is exceeded by the smallest amount and for the least possible time. If the releases are to be increased, the increment is higher for the reservoir whose flood control storage occupation is the most, or which is likely to receive higher inflow. The situation is reverse when the releases are to be curtailed.

The objective behind balancing of parallel reservoirs is to logically and properly use the volume of flood control storage available, while maximizing nondamaging releases from the system. The principle of balancing the flood control storage on parallel reservoirs could be developed on the same lines as the space rule.

The USACE procedure to allocate flood control space between two parallel reservoirs (see Fig. 11.8) with a common downstream flood damage center is as follows:

1. Route the reservoir design flood or other observed major floods for reservoir 1 while making releases at maximum nondamaging rate. Allow reservoir 2 to make the remaining releases up to the maximum nondamaging level. Plot the space required in reservoir 1 versus the total space required.
2. Perform the same analysis in reservoir 2. Plot the space required at reservoir 2 versus the total space required.
3. The ratio for balancing flood storage between the two reservoirs should lie between these two curves.

11.6.6 Flood Control Operation of a Multipurpose Reservoir

Permanent allocation of space exclusively for flood control at the top of the conservation pool becomes necessary in those multipurpose reservoirs for which flood control is one of the main purposes and floods can be experienced at any time of the year. The size of flood reservation may vary according to the magnitude of expected floods. The flood storage space allocation at different times of the year is so determined that incoming floods would be absorbed or mitigated to the maximum degree. The floods should be so regulated that the downstream damage does not exceed permissible limits even if a big flood comes. If the reservoir is in a region where floods are experienced only in a particular season, the allocation of space for flood management varies during different parts of the flood season, depending on the magnitude of likely floods. After the flood season is over, this space may be used to store water for other uses.

A multipurpose reservoir whose primary objective is flood protection should be operated to provide maximum moderation to potentially dangerous floods. Water can be stored in the joint use space of a reservoir for other purposes as long as the flood control operation is not hindered. Before the onset of a flood, the main decision demanding the operator's attention is the amount of flood control space that ought to be available at any time and the size of the releases required to create additional space, if necessary.

The reservoir operation should also include the periodic assessment of future incoming volumes based on rainfall and other data gathered from the rain gauge stations in the catchment. The frequency of such a review depends on the catchment size and storm characteristics. The release decisions are also influenced by intuition, experience, and judgment of the operator.

The broad reservoir operation approaches/strategies adopted for flood mitigation in a multipurpose scenario can be classified into three categories, viz.: the flood control rule curves (discussed earlier), the concept of flood limited water level (FLWL), and systems analysis techniques (Jain et al., 2023).

Flood Limited Water Level

Reservoir operation during the flood season should aim to moderate floods while conserving water to meet dry season demands. The maximum allowable water level in the reservoir to ensure adequate vacant space for flood control during the high flow season is termed the FLWL (Li et al., 2010). The FLWL concept was developed in China where a distinct flood season is experienced. However, it can be applied to any region which has a distinct high flow season and the reservoirs follow a drawdown-refill cycle. Somewhat similar is the concept of Wurbs and Cabezas (1987), redistributing reservoir storage capacity in Brazos Basin, USA, based on the seasonal variation of floods properties.

The storage space between FLWL and the MWL or FRL is used to store flood water (Fig. 11.15). FLWL concept governs trade-off between flood control and water conservation (Yun and Singh, 2008). FLWL, computed using the annual design storm, does not vary within the flood season.

If the reservoir water level at the beginning of flood season is above FLWL, it is lowered to FLWL, and near the end of the flood season, the reservoir is filled up to FRL. The use of single-valued FLWL, also termed the annual FLWL, is suboptimal in some cases. If the flood season begins late and the reservoir level is low, demands may be met partially till inflows increase sufficiently. If flood season

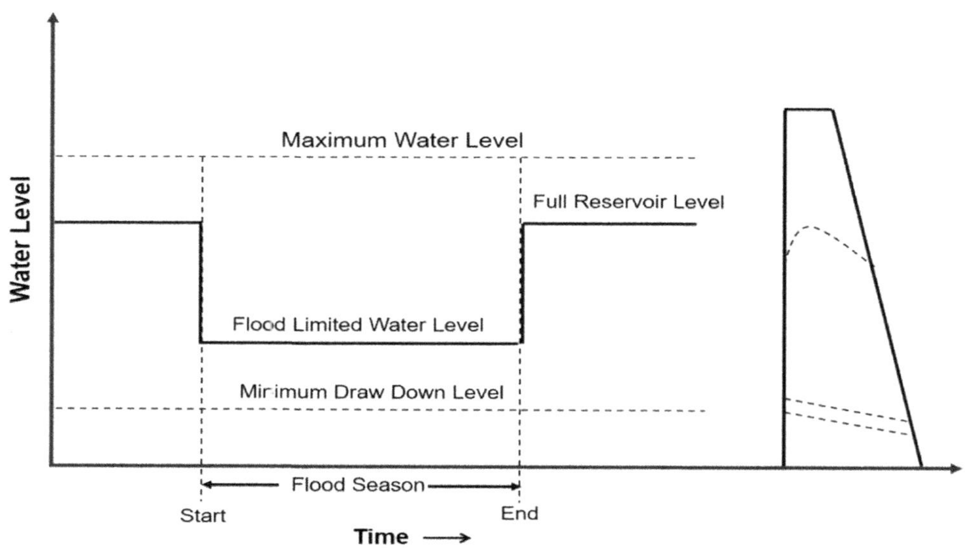

FIGURE 11.15

Depiction of flood limited water level.

ends before the usual dates, reservoir may not be filled to the desired level and shortages may be faced. Hence, flood mitigation policies have to be devised considering variability in timing, magnitude, and frequency of flood flows.

Two pioneering improvements in FLWL approach are the multiple duration FLWL (MDFLWL, also termed as seasonal FLWL), and the dynamic FLWL (DFLWL) (Yun and Singh, 2008). In case of MDFLWL, instead of deriving one design storm for the whole flood season, design storms for smaller durations, such as a month, are derived and used to determine monthly limited water level.

Fig. 11.16 shows the monthly and annual duration limited water levels. MDFLWL is commonly higher than the annual FLWL implying that more water is conserved and the demands are served better. Since occurrence of a large flood event early in the flood season reduces the occurrence probability of another large event in later parts of the flood season and vice versa, the limited water levels in the subsequent periods can be raised for that year by using conditional probabilities of occurrence of large storms in the reservoir's catchment. The limiting water levels in the later segments do not require adjustment if no large floods occur in the initial segment.

DFLWL method considers revising the limited water levels dynamically in light of new data. DFLWL conserves more water in the reservoir without harming its effectiveness in flood control and water shortages due to delays in flood season or short flood season can also be addressed. Some studies have recommended the use of monthly limited water levels in reservoir operation in place of annual FLWL.

Reservoir operation by using the FLWL gives high priority to low probability flood events and neglects the inflow uncertainty. Researchers have successfully integrated the forecast information in the reservoir operation within the flood season, leading to DFLWL during real-time operation.

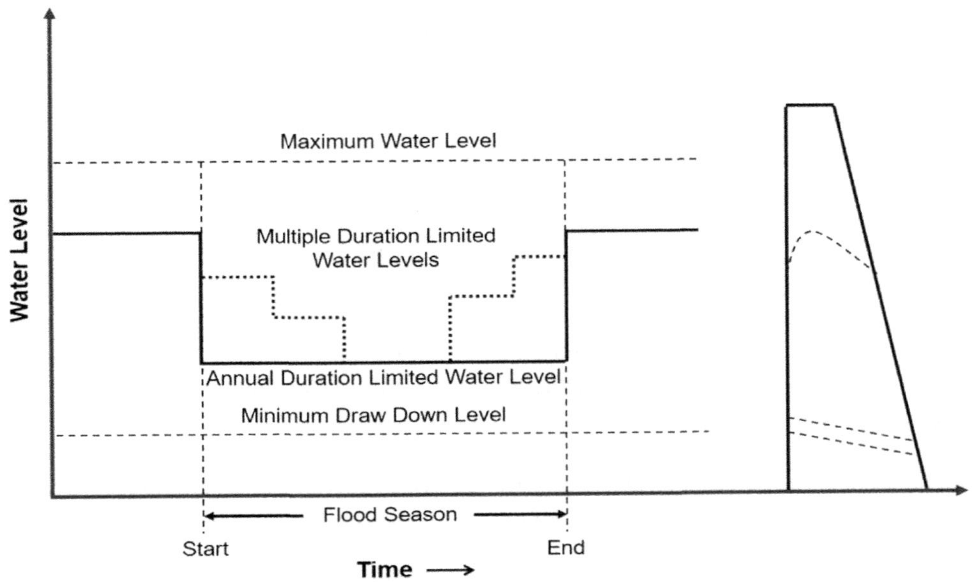

FIGURE 11.16

Depiction of monthly and annual duration limited water levels.

Dynamic updating of FLWL in the flood season provides better water utilization. A direct improvement to the MDFLWL/seasonal FLWL is to optimize the seasonal water levels. Jiang et al. (2015) identified the seasonal FLWL while considering forecast error as a fuzzy variable and using a credibility-based chance constrained simulation. Incorporation of seasonal flood risk analysis and analysis of hydrologic, hydraulic, and reservoir conditions along with forecast information allows higher water conservation levels in the subseasons.

The characteristics of 2011 flood in the Chao Phraya river basin, Thailand (Komori et al., 2012), contradicts the supposition of conditional probability of occurrence of successive large events in the multiple flood subseasons. This is because the high rainfalls in the season occurred due to multiple typhoons initially followed by an above normal monsoon rainfall in the year. When the downstream flooding in the basin could be attributed to low terrain slope, poor inflow forecast led to bad flood mitigation at the reservoirs, leading to levee breaks. Another notable aspect is the lack of flood operation strategies for an early start and late ending of flood season. Hence, it is better to devise flood operation policies considering within year and across year variability of floods. Such an adaptive reservoir operation strategy along with reliable forecast of precipitation and evaporation helps in real-time operation of reservoirs for flood mitigation.

The DFLWL approach can also be implemented to derive joint operation schemes for cascade reservoirs. Large size problems can be tackled by using aggregation-decomposition (AD) technique and dynamic programming (DP). Floodplains of some rivers can temporarily hold substantial quantity of water, and hence joint operation of reservoirs and floodplains can be highly beneficial without additional flood risks.

Hedging in Reservoir Operation for Flood Control

Hedging is rationing of water supplied for various demands. Its rationality is that the expected marginal benefit from storage must be the same as the marginal benefit from release. When applying the hedging concept for flood control, storage space to moderate floods is considered a scare resource which should be judiciously allocated between the current and future time periods.

Hedging involves accepting small current losses so as to avoid larger future losses. Fig. 11.17 depicts an operating policy based on the hedging concept in the backdrop of the SLOP. The reservoir releases (R) depend upon the expected water availability during the decision period. From time "D" onwards on x-axis, the operator may keep reservoir sufficiently vacant by progressively increasing releases, thereby minimizing flood risk. The downstream flood risk is avoided by making releases below safe river capacity (horizontal line at R). The parameters in a two-point linear/nonlinear hedging rule are the starting and ending water availability (SWA and EWA), respectively. Predepletion of storage before a flood increases vacant storage and increases the ability to regulate large future floods. This is akin to hedging. Reliable forecasts help in successful implementation of hedging.

Hedging is difficult in large flood events owing to big flood volume and is of little help in small floods. The optimal hedging releases are useful for medium floods to minimize the overall risk.

Although the downstream flood protection is enhanced by prereleases from the reservoir, excessive releases increase downstream flood risk and may also increase water shortage if postflood inflows are insufficient. With perfect forecast of incoming storm and known damage function, optimal prereleases could be made. Hui et al. (2016) attempted to minimize the overall losses that arise due to prereleases under various cases of storm forecasts and damage functions. Hua et al. (2020) extended the prerelease concept considering the dry days between flood occurrences. Depending upon the forecast lead time

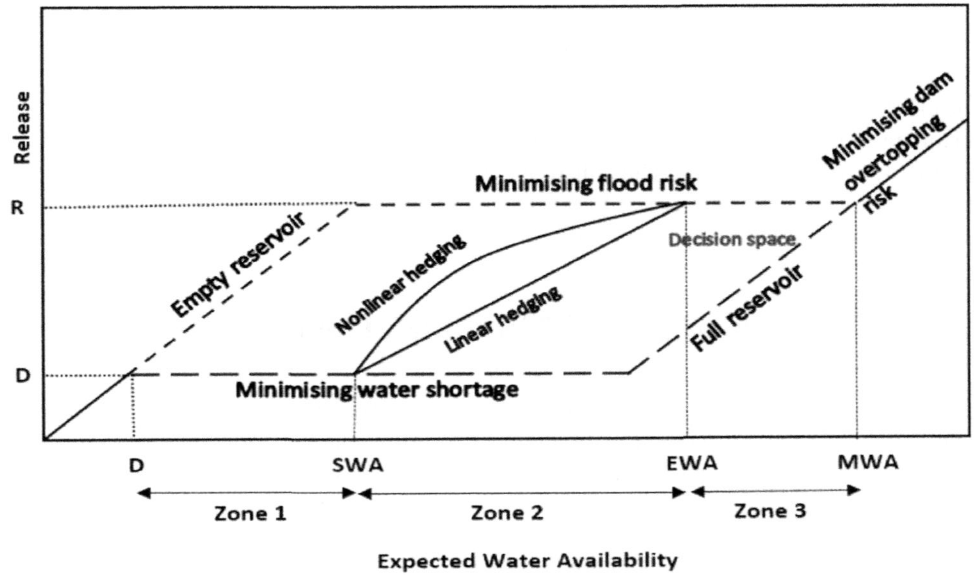

FIGURE 11.17

Depiction of hedging rules for reservoir operation for flood control.

and the expected length of dry days, releases were made either to vacate the space above the upper bound of dynamic control of FLWL (DC-FLWL) or to generate hydropower.

The mathematical underpinning for a two-stage hedging operation, to seek trade-off between water conservation in stage 1 (see Eq. 11.15) and flood risk in stage 2 (see Eq. 11.16), is presented here. They considered flood risk associated with inflow uncertainty to be zero in stage 1 as the forecast error in the near future is less, and the water conservation loss in stage 2 is taken as 1 as the target level at the end of flood season is set to the lower FLWL. The objective functions (OFs) for stage 1 and 2 are given as:

$$g_{w,1}(W_1) = \left(\frac{S_r - S_1}{S_r - S_{\min}}\right)^m = \left(\frac{W_r - W_1}{W_r}\right)^m \tag{11.15}$$

$$g_{f,2}(\delta, \sigma) = \int_{\delta}^{+\infty} h(\varepsilon)\mathrm{d}\varepsilon \tag{11.16}$$

In Eq. (11.15), S_r denotes reservoir storage capacity for water supply and W_r is corresponding water availability, $S_{\min}$ is the storage corresponding to lower bound of FLWL, and W_1 is the water availability in Stage 1. Further, W_1 is the difference between S_1 and $S_{\min}$ and is denoted as "excess storage" carried over to Stage 2 as water conservation. The shape of the water conservation OF is guided by factor m; $m > 1$ is the condition for a convex $g_{w,1}$. In Eq. (11.16), $h(\varepsilon)$ is the probability density function of forecast error ε with forecast uncertainty σ. The downstream flood risk is brought into the OF $g_{f,2}$ through δ.

Wan et al. (2016) derived hedging refill strategies by using KKT conditions to solve a two-stage optimization problem and the result showed that the refill operation is optimal when the marginal loss of conservation in the current period equals the marginal flood damage in the future period. Implementation of the methodology highlighted its practical use as it invoked gradual refilling of reservoir using forecasts.

The efficacy of flood control operation largely depends on the accuracy of inflow forecast, available flood storage space in the reservoir and downstream channel, acceptable risk, and decision maker's preferences for other water uses. Thus, there is need for a comprehensive, dynamic hedging strategy that helps managers attain the largest benefits, without compromising on dam safety. Currently, potential of hedging is underutilized and more research on this topic is needed.

11.7 SYSTEMS ENGINEERING TECHNIQUES FOR RESERVOIR MANAGEMENT

Determination of the reservoir operation policy to efficiently manage available water is a complex problem because it involves random hydrologic events. Many attempts have been made to solve this problem using optimization and simulation models. Reservoir optimization models allow the user to generate operating decisions that are optimal in some "measurable" sense. By making a number of runs of a simulation model with alternative decision policies, a (near) optimal solution can be reached. Maass et al. (1962), Loucks et al. (1981), Wurbs (1996), Loucks and van Beek (2017) and many others have discussed optimization and simulation models and their underlying differences at length. Wurbs (1996) has presented an annotated bibliography of optimization and simulation models.

11.7.1 Optimization

The linear programming (LP) and DP optimization techniques have been extensively used in water resources. Loucks et al. (1981) have illustrated applications of LP, nonlinear programming (NLP), and DP to water resources. Many reviews of applications of systems techniques for water resources problems have been published from time to time, such as those by Yakowitz (1982), Yeh (1985), Simonovic (1992), and Wurbs (1993).

After little modifications, the model for design of a reservoir that was presented in Section 10.10.4 can be used to determine an operation policy. The modified formulation is:

$$\text{Max } D \tag{11.17}$$

$$S_t + I_t - R_t = S_{t+1} \quad t = 1, 2, \ldots n \tag{11.18}$$

$$R_t \geq \alpha_i D, \quad t = 1, 2, \ldots n \tag{11.19}$$

$$S_t \leq C, \quad t = 1, 2, \ldots n \tag{11.20}$$

$$S_t, C, D, R_t \geq 0, \quad t = 1, 2, \ldots n \tag{11.21}$$

This formulation does not consider evaporation loss from the reservoir which could be significant in arid climates. An approximate method to include evaporation losses involves establishing a linear relation between the average reservoir storage for the period and surface area A_t:

$$A_t = a + b(S_t + S_{t-1})/2 \tag{11.22}$$

The volume of water lost during the period (m^3) will be the product of area and the depth of evaporation (d_t). After incorporating evaporation loss, the constraint given by Eq. (11.18) becomes

$$S_t + I_t - R_t - a \times d_t - b \times d_t \times S_t/2 - b \times d_t \times S_{t-1}/2 = S_{t+1}, t = 1, 2, \ldots n \tag{11.23}$$

Note that in Eq. (11.23), R_t includes release as well as spill. The generation of hydropower is also considered in a similar manner. The average head (h_t) in period t is expressed as a function of the average storage

$$h_t = c + d \times (S_t + S_{t-1})/2 \tag{11.24}$$

The power generated during the period is

$$P_t = 9.817\ T\eta\ R_t\ [c + d(S_t + S_{t-1})/2] \tag{11.25}$$

$$\text{or } P_t = \alpha\ R_t\ c + \alpha\ R_t\ d\ S_t/2 + \alpha\ R_t\ d\ S_{t-1}/2 \tag{11.26}$$

where $\alpha = 9.817\ T\eta$. This equation can be used with the formulation given by Eqs. (11.17) to (11.21). It is well known that LP cannot be directly used to determine the maximum firm power. The model has to be repeatedly solved due to the nonlinear nature of Eq. (11.25). The basic model described here can be extended to a system of reservoirs. ReVelle (1999) has described many such formulations for a variety of purposes.

Young (1967) was the first to propose the use of a linear regression procedure to derive general operating rules from deterministic optimization. The approach that he had used is called "Monte Carlo DP." Basically his method is to generate, for the river in question, a number of series of synthetic annual streamflow sequences using the Monte Carlo technique. For each of these, he used a DP formulation with a forward computation procedure. The optimum policies obtained for each of the synthetic streamflow sequences were then used in regression analysis in an attempt to determine the causal factor influencing the optimal policy. This method appears to be sound and is computationally efficient. The results are a good approximation of the true optimal policy.

A programming model for the design of multireservoir flood control system was developed by Windsor (1975). Karamouz and Houck (1987) derived general operating rules using deterministic DP and regression (DPR). The DPR model incorporates a multiple linear regression procedure suggested by Bhaskar and Whitlach (1980). Rules for the operation of multireservoir systems can also be developed by SDP which requires an explicit characterization of streamflow probabilities and a loss function. This approach has been used by Butcher (1971), Loucks et al. (1981), and numerous others. SDP approaches to deriving operating rules suffer from large computational needs for large problems. Attempts to reduce computational requirements by increasing the coarseness of storage and flow discretizations lead to approximate results. A comparison of SDP with other operating rules by Karamouz and Houck (1987) and Johnson et al. (1991) shows that well-designed conventional operating rules may perform equally well or sometimes better than SDP.

Optimization models are frequently used in reservoir operation studies employing flow forecast as input. Datta and Burges (1984) derived a short-term operation policy for multipurpose reservoirs from an optimization model with the objective of minimizing short-term losses. They examined the sensitivity of various performance criteria for the operation of a single reservoir to the accuracy of forecast streamflow volumes. The study revealed that when there is a trade-off incurring one unit of storage deviation and one unit of release deviation from respective target values, the optimized solution depends on uncertain future streamflow as well as the shape of the loss function.

The application of optimization models for multiple-reservoir operation is bit cumbersome. Difficulties in application include model development, trained manpower, cost of solution, an adequate inclusion of uncertain future hydrologic conditions, inability to identify and quantify all relevant objectives, and the need for better interaction with the user. However, this should not be interpreted as a judgment against optimization techniques which are very useful tools for water resources analyst. Another approach which is being used these days to account for stochasticity of inflows is the fuzzy logic programming. The fuzzy set theory was introduced by Zadeh (1965). In a fuzzy set of objects, there is no sharp boundary between those elements that belong to the class and those that do not. The membership function assigns a grade of membership, varying from zero to one to each object. Jairaj and Vedula (2000) applied this approach to multireservoir optimization.

11.7.2 Simulation

Since it is not possible to do experiments with a real reservoir, mathematical simulation models are developed and used in studies. Experiments can be conducted using these models to provide insights into the problem. For reservoir operation, the model studies bring into focus certain aspects of operation which serve to improve the manager's ability to control the system wisely. The simulation models associated with reservoir operation include the mass-balance computation of reservoir inflows, outflows, and changes in storage. They may also provide an economic evaluation of flood damages, hydropower benefits, irrigation benefits, and other similar characteristics. The simulation technique has provided a bridge from early analytical tools for analyses of reservoir systems to complex general-purpose packages. According to Simonovic (1992), the concepts inherent in simulation are easier to understand and communicate than other modeling concepts.

Simulation models can provide more realistic and detailed representation of reservoir systems and their operations (such as detailed responses of individual reservoirs and channels or the effects of certain time-varying phenomena). They also allow added flexibility in deriving responses which cannot always be readily defined in economic terms (recreational benefits, preservation of fish and wildlife, etc.). The time required to prepare inputs, run models, and other computational demands of simulation is much lower than those of optimization models. The simulation results will readily bring out the trade-offs in case of multiple objectives. The ability is helpful in arriving at the best policy of system operation. Practical real-time operation requires the specification of reservoir operating rules which is much simpler in simulation.

A number of general-purpose computer software are available which can be used to do analysis related to planning, design, and operation of reservoirs. Most of the software can be run on microcomputers which are widely available these days. Moreover, once the data required for particular software have been prepared for a problem, it is easy to modify the same and hence the consequences of various alternative design/operation decisions can be quickly evaluated.

Examples of simulation date back to the early 1950s. Perhaps, the first contributions were made by the Harvard Water Program (Maass et al., 1962). Probably one of the most popular and widely used generalized reservoir system simulation models is the HEC-5 model developed by the Hydrologic Engineering Center (Feldman, 1981; Wurbs, 1996). Some other well-known simulation models are: the Acres model (Sigvaldason, 1976); the Streamflow Synthesis and Reservoir Regulation (SSARR) model (USACE, 1987), the Interactive River System Simulation (IRIS) model (Loucks et al., 1989), and the Water Rights Analysis Package (WRAP) (Wurbs et al., 1993). Lund and Ferriera (1996) studied the Missouri River reservoir system and found simulation models to be superior to classical regression techniques for inferring and refining operating rules derived from deterministic DP. Jain and Goel (1997) have presented a generalized simulation model for conservation operation of a reservoir system based on rule curves. Despite the availability of several generalized models, the need to develop simulation models for specific reservoir system is usually present as each reservoir system has some unique features.

Multireservoir simulation models used to assess the impact of various operating policies are useful only if the voluminous output from all the various runs can be compared and evaluated. The analyst computes the means and variances and the time distribution of reservoir performance indicators, such as reservoir storage volumes, releases, associated benefits or losses, and these can be used for policy evaluation and comparison. The evaluation can also make use of concepts, such as system reliability, resilience, and vulnerability [see Chapter 8]. Simulation models for reservoir operation are of value for aiding in assessing possible impacts of alternative operating policies and for forecasting the future state of the system, given a specific operating policy and predicted hydrologic scenarios.

Steps for Application of a Simulation Model

The steps to perform a simulation study of a reservoir system are as follows:

(a) Prepare the diagram of the system showing names of reservoirs and diversion weirs/barrages, their location, and the length and direction of rivers and tributaries.
(b) Collect general details about the operation like the number of control locations in the system, initial month, day and hour, and the total number of periods of operation.
(c) Assign numbers to all the control points (storage reservoir, diversion weir, barrage, etc.) starting from the upstream node. Some models may require special numbering system.
(d) Collect general details about each location which include maximum capacity up to the FRL, initial storage, elevation-area-capacity table, demands, minimum release to be made in the downstream channel, and the evaporation depths for all the months of the year.
(e) For each structure, calculate the local flow coming from the free catchment area at that structure for all the periods of operation. If routing is to be performed, the values of the required parameters are to be provided.
(f) Simulate the operation of the system. An examination of the working tables of the various reservoirs, performance statistics, and graphs of relevant variables, such as release, reservoir storage, and demand, will show the improvement that can be made in the operation policy.
(g) Modify the input operation policy and run the model again. Repeat this procedure till the desired results are obtained.

The performance statistics that are typically used in step (f) include time and volume reliabilities, the largest shortage and spill, the frequency of spill, and the maximum number of periods of consecutive shortages.

Although simulation and optimization are different approaches, many mathematical models employ both. While converging to the optimal solution, optimization models reproduce behavior of the system by a simulation model. It is often impossible to consider all the complexities of a system in an optimization model, but a simulation model can handle most such complexities. At times, it is best to use optimization and simulation together to harness the strengths of both techniques. An optimization model may be used for preliminary screening of alternatives, and the short-listed alternatives may be investigated in detail by a simulation model (Wurbs, 2005).

Labadie (2004) reviewed the applications of optimization methods to reservoir systems. He also provided an overview of multiobjective and heuristic optimization procedures, besides the application of machine learning (ML) algorithms (neural network and fuzzy theory) in reservoir operation. Generalized formulations of optimization models for flood control problems are described in Appendix A.

Using systems approach, a number of generalized mathematical models that analyze the river-reservoir system include the tools developed by The Hydrologic Engineering Center (HEC), U.S. Army Corps of Engineers, such as the Hydrologic Modeling System (HEC-HMS), River Analysis System (HEC-RAS), Reservoir System Simulation (HEC-ResSim), and Flood Damage Reduction Analysis (HEC-FDA) (https://www.hec.usace.army.mil/). The Corps Water Management System links HEC-HMS, HEC-RAS, and HEC-ResSim (Sensoy et al., 2018). Other models include MIKE 11 (https://www.mikepoweredbydhi.com), NIH_ReSyP (NIH_*Re*servoir *Sy*stems *P*ackage) developed by the National Institute of Hydrology of India (Goel et al., 2020), RiverWare developed at University of Colorado (http://cadswes.colorado.edu/) (Zagona et al., 2001), and MODSIM (Labadie, 2006).

11.7.3 Applications of Systems Analysis Techniques for Flood Management

The decision-making concerning reservoir flood management involves risk. Jain et al. (1992) suggested a methodology with explicit risk consideration to develop flood season operation policy for a reservoir by maximizing storage at the end of the flood season, ensuring that the risk of an overflow is kept below the acceptable limit. Jain et al. (1998) divided the reservoir flood control space in two parts: lower and upper. When reservoir level is in lower part, incoming flood is regulated by following the normal flood regulation policy. If the reservoir level rises to the upper part and safety of the dam is threatened, attempts should be made to lower the level at the earliest. The current reservoir level and nature of inflow (increasing or decreasing) determines which operation mode to follow. LeNgo et al. (2007) derived optimal flood rule curves considering the downstream nondamaging discharge conditions. The suitability of the operating policies can be evaluated based on indices related to the flood mitigation efficiency, for example, peak reservoir release, MWL, mean number of gate maneuvers, and power generation (Bianucci et al., 2015).

System dynamics approach to integrated flood management has been applied considering structural and nonstructural flood mitigation measures. The structural measures are the dams and watershed management, and the nonstructural measures included determination of the reservoir level at the start and safe reservoir release at the start and end of the flood. Zhao et al. (2022) proposed and employed a novel artificial neural network (ANN)-based framework to develop intensity duration frequency (IDF) curves and flood risk change maps in the changing climate and socioeconomic scenarios for Phnom Penh, Cambodia. They emphasized on the joint implementation of mitigation (structural measure) and adaptation strategies (nonstructural measures) to attain best and sustainable results. Their adaptation strategies are flexible and can be extended to other geographic locations and also put forward the expertise of the policymaker for efficient and effective countermeasures.

Due to climate and land use-land cover (LULC) changes, reservoir inflow and demand properties may change and previously developed rule levels may not remain optimal. Hence, "dynamic rule curves" may be adopted wherein the system under study is reoptimized for every operational period by using ensemble streamflow and forecasts of water demand to produce ensemble of optimized reservoir storages. The optimized storage values are used to produce rule levels for each operational period as a function of system state and streamflow forecast. An advantage of this method is that systems self-adjust as the climate evolves.

To harness the advantages of both simulation and optimization, many researches have used S–O approaches for the design of flood control reservoirs or to develop operation policies. The decision variables in the optimization formulation are either the reservoir releases or the parameters of the operation policy which are obtained by optimizing a single objective (flood mitigation) or multiple objectives (flood mitigation and other purposes). Finding efficient operating policies for a complex multireservoir system with simulation alone is challenging. Combined use of simulation and optimization gives a framework to efficiently complete this task (Rani and Moreira, 2010). The basic idea behind coupled S–O is: an "outer loop" finds optimum values for the decision variables by using an approximate system configuration and an "inner loop" simulates operation by using more realistic configuration, determines the values for the uncertain variables, and completes statistical evaluation, and returns values that the "outer loop" search process can use. The process is repeated iteratively till the best possible outcome is obtained.

S–O approach has been used to modify the design rule curves by considering the altered reservoir inflow conditions due to climate change or the ecological flow requirements. Further, direct interpolation of monthly flood rule curves to daily scale was found to increase the flood risk. Hence, daily rule curves should be developed independently.

The objectives of reservoir flood control operation, viz., minimizing the maximum reservoir water level as well as downstream water levels can be efficiently achieved through the use of advanced optimization algorithms such as multiobjective evolutionary algorithms (MOEAs). The advantage of MOEAs over traditional multicriteria methods is that these could produce a set of nondominated solutions in a single run. Combination of conventional decomposition techniques with MOEAs (MOEA/D) has received wide application. MOEA/D decomposes the target multiobjective optimization problem into multiple subproblems by using aggregation approaches and then optimizes them simultaneously by using an evolutionary algorithm. Such models were found to be suitable to incorporate preference information (weights) (MOEA/D-PWA) which could refine the solution search.

Dynamic changes occurring in the physical river-reservoir system, such as the hydrometeorological conditions in the catchment, the vacant storage for flood mitigation, the river stage at the damage center, and the available river capacities below the bank-full stage, influence the reservoir operation decisions. An adaptive S–O framework, in which the flood hydrograph was divided into three phases according to their slope in the rising and recession limbs, has been used in a multireservoir system to consider dynamic changes.

11.7.4 Operation of a Network of Multipurpose Reservoirs

The development of operation policy for multipurpose multireservoir systems is very challenging because of nonlinearity, high dimensionality, correlated decision variables, and conflicting objectives. Different reservoirs receive different regulated and unregulated inflows, and have different storage capacities and different vacant storages. However, by skillful regulation, these factors can be turned to advantage and a high-peak hydrograph can be turned into an outflow with smaller peaks. Lund and Guzman (1999) provided an overview of operating policies for network of multipurpose reservoirs and also highlighted the opportunities for developing rules for water allocation and storage for various system configurations and uses.

A widely studied strategy for flood control operation, the DC-FLWL has been extended to multireservoir systems. In network application, DC-FLWL is simultaneously derived for all the reservoirs in the system. The optimization algorithms to arrive at the optimal DC-FLWL for each reservoir in a system include progressive optimality algorithm (POA), NGSA II, and stochastic programming. Studies have confirmed that jointly developed operation rules for multireservoir systems provide better flood protection than the individual flood operation rules, and water conservation in joint operation is higher. A concern with S—O framework is that the dimensionality problem increases with the number of reservoirs.

"Curse of dimensionality" is a common problem in studies dealing with multireservoir systems. In the AD approach, reservoirs are aggregated into a single virtual reservoir whose operation is optimized and the decisions are distributed among the individual reservoirs. The AD approach was used to derive optimal joint flood control rule curves for a parallel multireservoir system in the Lijiang river basin by Pan et al. (2020) who used the nested annealing accuracy penalty function—adaptive multivariant particle swarm optimization (AAPF-AMPSO) algorithm to maximize the peak reduction. Decomposition of the aggregated results was based on the ratio of the flood storage capacity in individual reservoirs to the aggregated single reservoir. The application of this technique resulted in higher reduction of flood peaks. One way to reduce the number of reservoirs to handle is by identifying a set of "effective" reservoirs that are jointly operated; "non-effective" reservoirs were handled separately.

DP has been widely employed for optimizing reservoir operation rules as it can efficiently handle discrete variables and nonconvex, noncontinuous, and nondifferentiable OFs. However, the curse of dimensionality limits its use in optimization of multireservoir systems. To overcome this curse, modified DP algorithms like the discrete differential DP (DDDP) and the POA have been used. Zhou et al. (2018) used DP-POA in a two-stage model to derive optimal rules for a cascade of reservoirs with multiple control points by sequentially progressing from upstream to downstream.

Hui and Lund (2015) investigated the impact of hydrograph shapes in arriving at optimal operation of parallel reservoirs. Four basic hydrograph shapes were considered and two useful concepts that relate peak flow reduction and additional storage requirements were defined: flood storage efficiency (FSE) (the peak reduction per unit storage) and marginal flood storage efficiency (MFSE, first derivative of FSE). Optimal flood storage allocation was derived for various combinations of storms. A conclusion was that flood storage be allotted such that the MFSE is equally distributed among the reservoirs and the series of storms.

Zhang et al. (2016) explicitly considered the movement of flood waves in the downstream river channel. A hybrid algorithm, termed as the POA and successive approximation (POA-SA) was used to solve the optimization problem.

Postevent analysis of Aug 2018 Kerala (India) floods that affected large population and damaged infrastructure and environment (Sudheer et al., 2019) is interesting from systems analysis perspective. One of the worst affected basins, the Periyar river basin (PRB), has 17 interconnected dams/barrages. Authors analyzed reservoir operation to examine the possibility of improvements and to find whether predepletion of PRB reservoirs would have helped in higher flood attenuation. They concluded that releases from major reservoirs had a minor role in flooding in the PRB. Most of the flow that caused damages came from the areas which did not have any reservoir. Thus, better reservoir regulation would have helped in limited reduction of flood peaks; improved decisions were possible if reliable rainfall forecasts at higher lead time were available.

Although many reservoir operation studies have been conducted employing systems analysis tools, the largest size of systems considered so far is about 14 reservoirs. More research is needed to develop methodologies for the regulation of larger reservoir systems. To manage large water resources systems, the selected model should be able to efficiently handle complex systems. Further, the success of flood control operation by reservoirs critically depends on accurate rainfall prediction at higher lead times.

11.7.5 Network Flow Models

The network flow models have been discussed in Chapter 5. Most models developed for water resources systems use the "out-of-kilter" algorithm (OKA) to find the optimum flow in the network. This algorithm solves a special class of LP problems, each of which can be represented as a "capacitated network," that is, as a series of nodes and interconnecting arcs. The objective is expressed as the minimum collective cost of flows through all arcs, subject to two types of constraints. The first type is simply the mass balance equation at each node. The second type of constraints state that every arc flow must be within some specified lower and upper limits. Fortunately, many water resource problems can be transposed into an equivalent network representation. Storage changes in reservoirs during individual time periods and changes in system operation through a sequence of time periods can also be represented effectively. It is also easy to assign priorities to various uses of water. An attractive feature of the OKA is its computational efficiency. The *surface water allocation model AL-V* developed by Martin (1982) uses OKA for optimal water allocation.

11.7.6 Linear Decision Rule

A linear decision rule (LDR) relates releases from a reservoir to storage content and decision parameters. The concept of a decision rule in reservoir planning and operation was proposed by Young (1967) who found that LDRs provided as good or better fit to stochastic input data as more complicated rules when quadratic loss functions are associated with release deficits. Extensive work on LDR in water resource management was published in the 1970s in a series of papers beginning with ReVelle et al. (1969). This rule is very useful in formulations that involve chance-constraints. In the simplest form for reservoir management, LDR assumes that the release is a linear function of reservoir storage:

$$R_t = S_t - b_t \tag{11.27}$$

in which R_t denotes the release during time period t, S_t denotes the storage at the beginning of time period t, and b_t is the decision parameter that is determined by optimizing a criterion function. Note that reservoir losses are not explicitly considered herein. Numerous investigators have modified, extended, or applied this rule to a variety of reservoir optimization problems. Loucks and Dorfman (1975) suggested the following LDR for the initial reservoir storage in period $t+1$:

$$S_{t+1} = \lambda_t I_t + b_t \tag{11.28}$$

in which I_t is the inflow during period t, and λ_t is a known coefficient ($0 \le \lambda_t \le 1$). When this decision rule is substituted in the storage continuity equation, the LDR for reservoir releases in each within-year period t is obtained:

$$R_t = S_t + (1 - \lambda_t)I_t - b_t \tag{11.29}$$

In this rule, coefficient λ_t can be interpreted as a weight given to inflow in the current period while determining the release. As λ_t is increased from 0 to 1, the rule becomes more conservative and more reservoir capacity is required to meet commitments. The LDR by ReVelle et al. (1969) is a special case of Eq. (11.29) when λ_t equals 1:

$$R_t = S_t - b_t = I_{t-1} + b_{t-1} - b_t \tag{11.30}$$

If a reservoir is operated by following this decision rule, the releases at the beginning of a time period can only be determined if $0 \le b_t \le S_t$ since the initial storage S_t will be known at the beginning of period t. If $b_t < 0$ or if $b_t > S_t$, then the release will be determined in part by the current inflow I_t, which will not be known at the beginning of period t (Loucks and Dorfman, 1975).

A less conservative decision rule is obtained when λ_t is equal to 0 in Eq. (11.29):

$$R_t = S_t + I_t - b_t = I_t + b_{t-1} - b_t \tag{11.31}$$

and

$$S_{t+1} = b_t$$

This rule equates b_t with the random end of the period storage volume and thus does not allow a release commitment at the beginning of the period. In Eq. (11.27) (obtained by setting λ_t equal to 1), release is a function of the past inflow I_{t-1}, while in Eq. (11.31) where $\lambda_t = 0$, release depends on the current inflow I_t. ReVelle and Gundelach (1975) proposed an LDR that is a function of current storage and past inflows:

$$R_t = S_{t-1} + \beta_t I_t - \beta_{t-1} I_{t-1} + \ldots + \beta_{t-k} I_{t-k} + b_t \tag{11.32}$$

in which constants β_t, β_{t-1}, β_{t-k} are to be determined. This extension considers incorporation of the stochastic nature of the streamflow process into an LDR formulation. This rule results in smaller reservoirs than the original LDR with an objective of minimizing the reservoir storage capacity subject to several performance criteria (Gundelach and ReVelle, 1975).

LDR has been a subject of considerable controversy and is said to produce conservative results. The conservative nature of LDR originates from (Joeres et al., 1981): (1) inappropriate comparisons of the multiobjective LDR model with a single-purpose "yield maximizing design methods," (2) inappropriate comparisons of noncomparable operating modes, and (3) the assumption in LDR that streamflows in successive periods are independent of each other. Loucks and Dorfman (1975) mentioned that the conservative nature of LDR arises because the rule itself is an additional operating constraint in the system. They suggest that while this technique may be suitable for screening studies, it is not satisfactory to derive optimal operating policies for reservoir(s). ReVelle (1999) has provided many extensions of the basic LP model given in Section 11.7.1 for the allocation of reservoir services among water supply, flood control, and hydropower using chance-constrained programming (discussed in Section 4.6.1). For example, the chance-constrained model to determine the maximum firm release R_F from a reservoir will be:

$$\text{Max } R_F \tag{11.33}$$

$$R_F + S_{t+1} - S_t + L_t = I_t \quad t = 1, 2, \ldots n \tag{11.34}$$

$$R_F - S_t \le I_t^{\alpha} \quad t = 1, 2, \ldots n \tag{11.35}$$

$$S_n \geq S_0 \quad t = 1, 2, \ldots n \tag{11.36}$$

$$S_t \leq C, \quad t = 1, 2, \ldots n \tag{11.37}$$

$$S_t, C, R_F, L_t \geq 0, \quad t = 1, 2, \ldots n \tag{11.38}$$

where L_t is spill from the reservoir during period t. The constraint of Eq. (11.35) ensures that storage at the end of period t is greater than or equal to zero with probability α. $I^{\alpha}{}_t$ is the inflow for the month t that is exceeded with a probability α, also known as the (1-α)-percentile flow for month t (see Fig. 11.18). The constraint of Eq. (11.36) ensures that the storage at the end of the computation horizon is not greater than the beginning storage or the total volume of release should not exceed the total inflows. Note that the stochasticity of inflows can be taken into account either through the chance-constrained models or through synthetic inflow sequences. For correct answers, it is necessary that the synthetic inflows properly reproduce the critical flows of the record or the persistence of low flows is maintained. In the chance-constrained programming, these low flows are accounted for through $I^{\alpha}{}_t$.

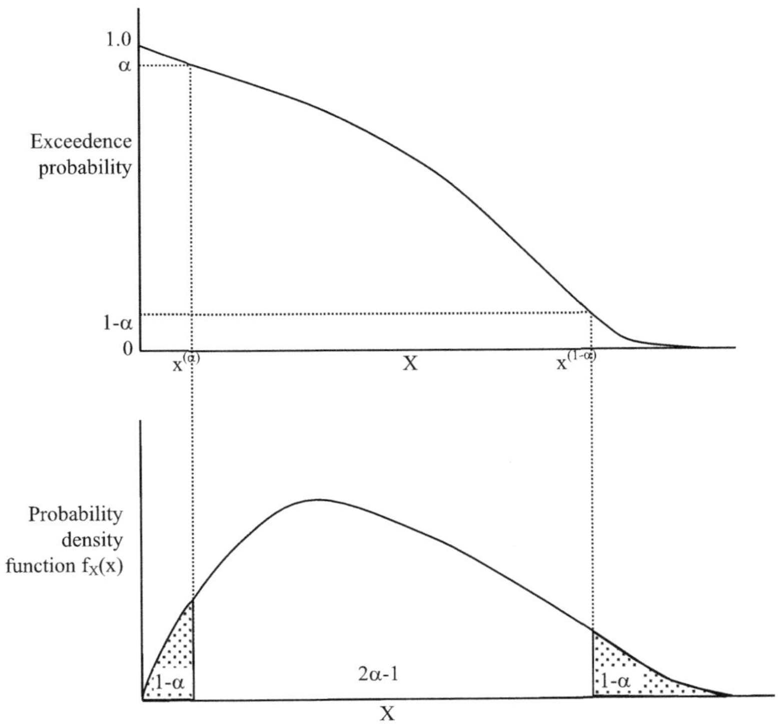

FIGURE 11.18

Probability distributions of random variable *X*. The probability of exceedance of $x^{(\alpha)}$ is α.

11.7.7 Recomposition-Decomposition Approach

In large river basins where many developmental activities may be going on, it may be cumbersome to carry out the systems study in one mathematical operation. A system has various components and each of them has a different function to perform. Therefore, it can be expected that the models for the various components could be different. In other words, there is no single universal model governing all components and functions.

A meaningful study of such a system can be carried out by decomposing it into smaller subsystems, each of them being within manageable limits for systems analysis. Each subsystem can be optimized from a functional point of view. Toward the end, these subsystems can be recomposed into the whole system. This approach has many advantages.

Hall and Dracup (1970) proposed that a water resources system could be decomposed into 5 subsystem on the basis of purposes, viz., watershed subsystem (WSSS), river regulation subsystem (RRSS), water distribution subsystem (WDSS), water use subsystem (WUSS), and water waste regulation subsystem (WWRSS). The WSSS is normally managed for purposes other than water resources. This subsystem usually affects the flow of the water in rivers and sediment into reservoirs. The RRSS consists of reservoir storages (both surface reservoir and ground water reservoirs) for the purpose of converting the stochastic flow into assured and regulated flow. WDSS consists of river reaches and canal systems. This starts from a terminus of RRSS to a place of the use of water. WUSS comprises the fields in the command receiving water from the distribution system. WWRSS consists of all collection and treatment works involved in removing waste or unused water from WUSS and for treating it to the extent required for ultimate disposal or reuse.

A somewhat similar strategy that has been proposed in the literature is referred to as "linear dynamic decomposition programming." In this approach, DP is used for optimizing individual reservoirs and LP for combining the reservoirs collectively into an integrated optimization model. The approach uses dual variables from the LP solution to constrain the DP solution. In turn, the DP solution provides column vectors for the LP matrix. An optimal solution is obtained after a series of iterations back and forth between the LP and DP portions of the model.

A multireservoir system may have many small reservoirs whose ability to influence the system operation and the impact on the overall performance may be marginal. Many times, the desired data about these small reservoirs are not available. In studies dealing with such systems, it is common to aggregate some of the reservoirs to reduce computational or data demands of mathematical models. Usually, reservoirs in series are easier to aggregate and disaggregate than reservoirs in parallel.

Example 11.2. This example illustrates the rule curves for Panchet dam, India. Several reservoirs were constructed in the Damodar valley (India) in the 1950s for the purpose of flood control, irrigation, power generation, and municipal and industrial water supply. The major dams are Tilaiya, Konark, Maithon, and Panchet. Consider, for illustration, the Panchet dam whose drainage area is 10961 sq. km and the average annual runoff is 4541 Mm^3. The Panchet dam has a gross storage capacity of 1476 Mm^3 at an elevation 135.67 m (445 ft). The top of the conservation pool is at an elevation 125 m (410 ft) at which the storage capacity is 392 Mm^3. Thus, the storage space of 1084 Mm^3 is available for flood control. The maximum observed flood is 8558 cumec, while the spillway design flood is 17853 cumec. The installed capacity of power plant is 40 MW. The rule curves for Panchet multipurpose reservoir are shown in Fig. 11.19.

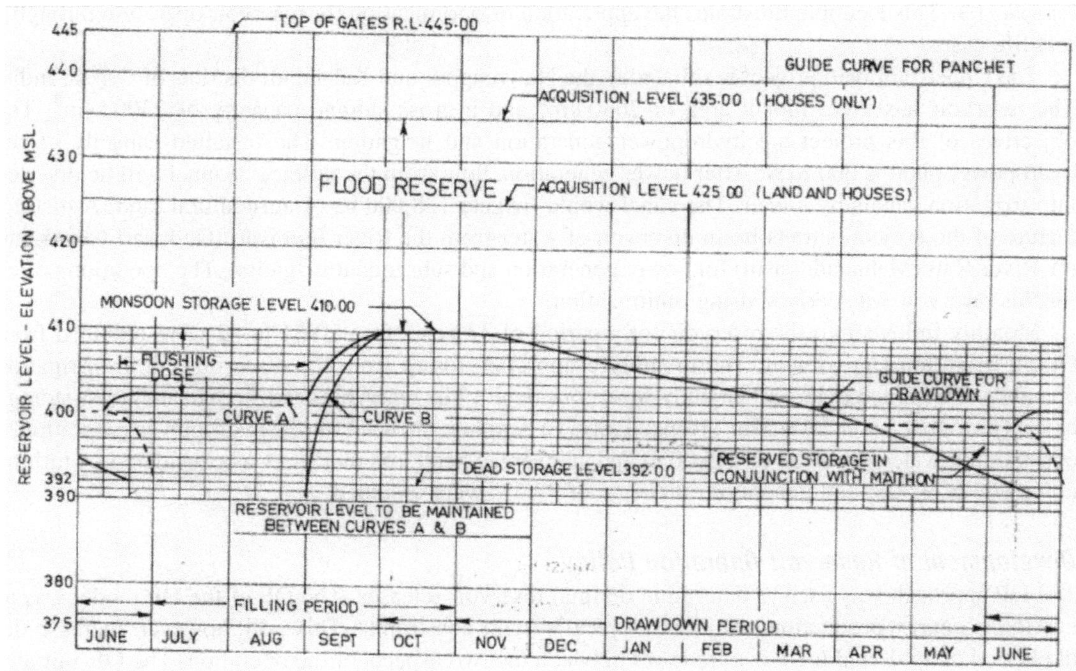

FIGURE 11.19

Rule curve for Panchet reservoir, Damodar valley (India).

According to the guidelines, the reservoir level may be drawn down to 119.5 m to meet the power requirements in June. On the first day of July, the Panchet reservoir should be kept as near an elevation 121.95 m (400 ft) as possible to ensure generation of hydropower. The reservoir level should be kept between curves A and B (Fig. 11.19) during July, Aug, and closer to curve A during Sep. The rule curves are to be followed such that the reservoir attains a level of 125 m on 1st Oct.

Whenever the water level rises above 125 m, flood control operation commences and this operation ceases as soon as the level comes down to 125 m. This reservoir along with another reservoir, namely Maithon, is operated to control floods at a downstream point by trying to keep the maximum flow of less than 7084 cumec at that point. The emergency operation for flood control is triggered when half of the flood control space in the reservoir is occupied. The outflow from the dam is made equal to inflow when 100% flood control space is occupied.

The water available from monsoon storage and dry season flows are utilized to meet the committed requirements of irrigation, industrial and municipal water supply, and hydropower needs during the drawdown season. The reservoir operation guidelines prepared by the Damodar Valley Corporation also recommend that before the commencement of monsoon, inspection and operation of gates should be made for smooth operation during monsoon. The communication system should also be checked for trouble-free service.

Example 11.3. This example illustrates the application of optimization for reservoir operation through a real-life case.

The Upper Indravati project is situated in the Nowrangpur and Kalahandi districts of Orissa, India. The reservoir has a catchment area of 2630 km^2 and a gross storage capacity of 2300 Mm^3. The objectives of this project are hydropower generation and irrigation. The installed capacity of the hydropower plant is 600 MW. After power generation, flow from the tailrace channel will be diverted into irrigation canals by a weir. The canal would irrigate 128,000 ha of agricultural land. A distinct feature of the project is trans-basin diversion of water from the River Indravati (Godavari basin) into the River Hati (Mahanadi basin) for power generation and subsequent irrigation. The operation policy for this reservoir was derived using optimization.

Monthly inflows into the reservoir for a period of 32 years from 1951 to 82 were obtained from Orissa Irrigation Department. The monthly evaporation losses from the reservoir and the irrigation demands determined in an earlier study were considered. The reservoir dead storage and gross storage capacity are 814.5 and 2300 Mm^3, respectively. To compute the head for the hydropower generation, a constant tail water elevation of 265.00 m and an average head loss of 12 m was considered. A turbine efficiency of 92.0% and generator efficiency of 97.5% were assumed.

Development of Reservoir Operation Policy

The DP approach was used to determine optimal reservoir releases. The OF of the DP model was to maximize energy production, subject to typical system constraints. This will, however, increase the likelihood of spill, and there is a trade-off between the two aspects of the operation. The OF, initially chosen, was a weighted sum of hydropower generation and saving water in storage:

$$\text{Max}\sum_{t=1}^{N}\left(\frac{H_t * \text{rel}_t}{10^6}+\frac{\text{Sav}_t}{10^4}\right) \tag{11.39}$$

where H_t represents the average effective head (m) during time period t (average pool elevation-tail water level), rel_t is the release (Mm^3) during period t, N is the number of time periods within the optimization horizon, and Sav_t is the average storage for the period t (Mm^3).

Constraints

Two types of constraints were considered: those which represent the inherent system characteristics and will not change during optimization and others which are loss or penalty functions. These constraints are discussed below.

i. Mass-balance or continuity equation

$$S_{t+1} = S_t + \text{inf}_t - \text{elos}_t - \text{rel}_t - \text{spil}_t \quad t = 1, 2, \ldots, N \tag{11.40}$$

where S_t is the initial reservoir storage, inf_t is the reservoir inflow, elos_t is the evaporation loss, and spil_t is the spill from the reservoir, all for time period t.

ii. Storage S_t can vary only between the maximum and minimum storage bounds:

$$S_{\min} \le S_t \le S_{\max} \quad t = 1, 2, \ldots, N \tag{11.41}$$

iii. Hydropower generation is proportional to the release and the operating head and is given by

$$\text{gen}_t = c \times 9.8 \times \text{rel}_t \times H_t \times \eta \quad t = 1, 2, \ldots, N \tag{11.42}$$

where gen_t is the energy generated in period t, c is a constant for converting release in MCM to cumec, and η is the efficiency of power plant.

iv. The hydropower generation is limited by the installed capacity:

$$\text{gen}_t \leq \text{Installed capacity}, t = 1, 2, \ldots, N \tag{11.43}$$

v. The penalty for not meeting irrigation demands:

$$\text{ben}_t = -\{\text{rel}_t - \text{dem}_t\}^2 / 10^8 \quad t = 1, 2, \ldots, N \tag{11.44}$$

where ben_t and dem_t are the benefit and the demand during time period t, respectively.

The problem is one of adopting the appropriate OF and identifying a suitable loss function which can accommodate both the objectives, viz., maximization of power and minimization of irrigation deficit. This was finalized after a trial-and-error procedure by first choosing a set of functions, then running the DP model, and finally evaluating the model performance as per certain criteria and by repeating the entire procedure after altering the functions. The DDDP approach was used to determine the optimal releases.

Sensitivity Analysis of the Objective Function

As the hydropower generation is a function of head as well as discharge, a term for higher storage (i.e., higher pool elevation) was added in some runs. A penalty was also imposed for deficits. Several combinations of OF and penalty were examined. The criteria adopted for the performance appraisal of various DP models were: total generation during the optimization period, irrigation deficit, and spill. A summary of the results obtained from some competing model runs is given in Table 11.4. From this table, the following inferences can be drawn about the Indravati system:

Table 11.4 DP performance appraisal for alternative objectives and penalty functions.

SN	Objective function	Penalty function	Spill Mm^3	Generation (MW)	Deficit (Mm^3)	Mean square deficit (Mm^3)
1	$(S_{av} + H_{av} \times \text{rel})/10^6$	$-(\text{rel - dem})^2/10^{11}$	3524	100,376	2590	4737
2	$(S_{av} + H_{av} \times \text{rel})/10^6$	If water is available & rel < dem: -ben- $[H_{av} \times (\text{rel-dem})^2]/10^6$ else $-(\text{rel-dem})^2/10^{11}$	2512	100,969	5253	9081
3	$(S_{av} + H_{av} \times \text{rel})/10^6$	- ben $-H_{av} \times (\text{rel-dem})^2/10^6$	2265	101,217	2309	2484
4	$S_{av}/10^4 + (H_{av} \times \text{rel})/10^6$	$-(\text{rel - dem})^2/10^8$	3795	99,641	2326	2363
5	$(H_{av} \times \text{rel})/10^6$	$-(\text{rel - dem})^2/10^8$	3607	99,914	2326	2362
6	$(H_{av} \times \text{rel})/10^6$	- ben $-H_{av} \times (\text{rel-dem})^2/10^6$	2242	101,248	2309	2484

In runs 1 and 2, only penalty function is different. For case 2, spill is less, power generation is about the same but deficit is very large. The penalty function forced releases in the current period if water was available and this caused large deficits in latter periods.

Comparison of runs 1, 2, and 3 shows that the penalty function of run 3 is more appropriate as the spill is least, generation is maximum, and the deficit is least, although OF is the same. The term for storage in the OF of run 4 has more weight than has run 5. This results in higher amounts of spill and reduction in power generation, whereas the deficit remains unaffected. The OF for runs 5 and 6 is the same, but the penalty function for run 6 is dependent on head also. In run 6, the spill and deficit are least and power generation is the maximum.

Overall, the set-up of run 6 appears to be the best. Regarding DDDP, it was observed that by increasing the number of iterations and successively reducing the corridor width, the total deficit is distributed among a larger number of time periods with smaller deficits. Further details of this study are available in Jain et al. (1999).

11.8 REAL-TIME RESERVOIR OPERATION

Generally, the reservoir operation policy is developed taking into account the demands of the past and using data from historical or synthetic time series of hydrological variables. But the probability that an actual event will occur in the same way as prior events of the same type is small. In a flood event, reservoir inflows and flows in downstream river can rapidly change and the operators need to take, implement, and update decisions accounting for changed situation. A reservoir system can be efficiently operated if the time interval between the occurrence of an event and the execution of the control adapted for that event is short. In real-time operation, the release decisions are based on short-term information. The definition of short-term varies in accordance with the purpose. If the reservoir is operated for flood control, the short term may refer to (multi) hourly operation and if it is serving conservation purposes, the short term may be a day or longer.

The term *real-time control* denotes the execution of a decision process concurrently with a physical system such that the results of the analysis based on on-line data are available in time to usefully control the physical system. Here on-line implies that the data about the system are received without any delay as the events take place and are immediately used. In real-time reservoir operation, the release decisions for a finite future time horizon are taken, based on the condition of the reservoir at that instant when these decisions are to be taken and the forecast about the likely inflows/demands over this time horizon, if available. After a certain time, interval, new information about the reservoir state becomes available, the forecasts are updated, and the decisions are modified in light of these.

Real-time operation is especially suitable during floods where the catchment response changes rapidly, decisions have to be taken quickly and adapted frequently. A model of the system is developed in which release is a decision variable. A forecasting algorithm is used to provide inflow forecast for a finite number of future time periods based on the present state of the system as well as its past behavior. A mathematical model uses forecasts to determine the optimum amount of water to be released from the reservoir. Although the optimum releases are determined for a finite number of future time periods, they are implemented only for the immediately next time period. After this period, the next set of observations becomes available and the entire process is repeated. The control process can also use the

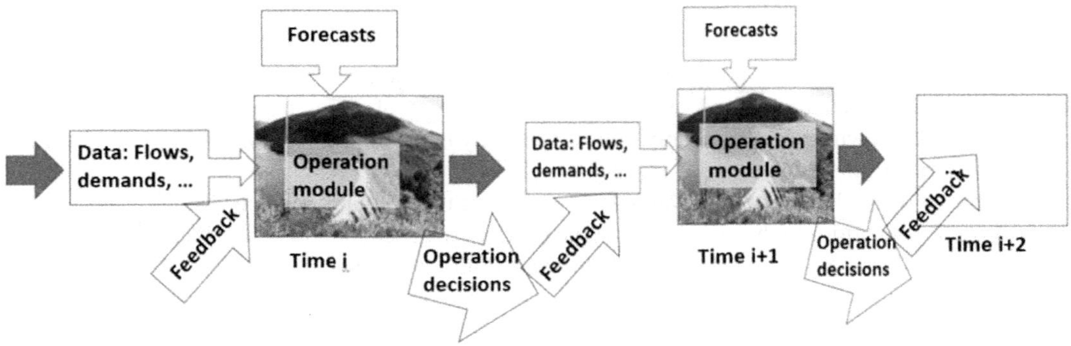

FIGURE 11.20

Adaptive feedback control of a reservoir using inflow forecasts.

information about river flows at critical locations while taking a decision. This process of control of a system is known as adaptive control—the decision is adapted based on the feedback received from the system (Fig. 11.20).

11.8.1 Real-Time Flow Forecasting

Flow forecasting (FF) is an important application of hydrology. The benefits from real-time operation of a reservoir can be substantially increased if good forecasts are available. The main components involved in the inflow forecasting are weather forecasting, rainfall-runoff modeling, and channel routing. For a forecast in proper time, the transmission of information and processing of data has to be done rapidly by employing computers.

FF is the estimation of future water levels or flows at a single or multiple sites of a river system for different lead times. It is one the most challenging and difficult problems in hydrology. When reliable forecasts of incoming floods are available sufficient in advance, projects can be operated efficiently. Forecasts of floods are commonly prepared by using a hydrologic model. Depending upon the catchment features, size, climate, etc., the hydrologic model may be a flood routing model, a rainfall-runoff model, a statistical model, etc. The typical inputs to a flow forecast model are observed precipitation and/or flows. If precipitation forecasts are available in real-time, better flow forecasts with higher lead times can be prepared. Point measurements of precipitation are commonly used to compute spatial average values. These days, such measurements may be supplemented by data obtained from weather radars which produce spatial estimates of precipitation and also describe storm movement. Forecast-based operation of reservoirs provides better flood mitigation, reduces water deficit, and meets ecological demands. A key requirement in real-time reservoir operation is to update release decisions based on the current and forecasted inputs to the reservoir. Fig. 11.20 shows a schematic of real-time reservoir operation by using inflow forecasts.

Real-time FF involves the estimation of discharges in a river some period prior to its occurrence. The forecast lead-time proves useful in mitigating some of the adverse effects of flooding. The forecast lead-time depends on catchment characteristics which affect the time taken by the catchment to transform rainfall into discharge at a point. The utility of a forecast depends on the accuracy and timeliness. Clearly, underprediction of a flood event can lead to a dangerous situation and may result in loss of life and damage to property in the affected areas. On the other hand, overprediction of a flood event will result in avoidable and unnecessary high releases of water from the dam, evacuation of people from the downstream area likely to be affected by floods, unnecessary flood fighting measures, and panic among the people.

A forecast would be valuable if available much before the event takes place. The forecast lead-time is the advance time prior to the occurrence of the event over which the forecast is issued. In general, the longer the forecast lead-time, the greater will be its utility. However, the errors in forecasts increase progressively with increase in the lead-time. Thus, there is a need for proper balance between the accuracy and the lead-time, although a certain minimum lead-time (depending on the local conditions) is necessary to organize flood fighting measures. Normally, a lead-time of at least 10–12 h is necessary to organize a meaningful flood fighting measure.

The field of FF is vast and highly developed. A detailed discussion is beyond the scope of this book, and a large volume of reference material is available, e.g., Jain et al. (2018).

11.8.2 Special Considerations in Real-Time Operation

The necessity for real-time operation arises from the fact that the inflows to reservoirs are random in nature and uncertain. To know the actual inflows to the reservoir and to forecast future inflows, adequate data collection and transmission network is required. During periods of high flows, operation of gates becomes an important aspect and needs to be given due consideration. In case, where it may not be possible to fully absorb the flood in the reservoir, efficient information dissemination system is required to warn the people downstream well in advance.

Data-acquisition system: A sophisticated hydrologic model and a fast computer system serve their useful purpose only if the data acquisition system is reliable and fast. The exchange of data between widely separated data observing equipment and forecasting center is called telemetry. In this system, usually no human element is involved and this completely eliminates the human error and reduces the time of observation and transmission of data. Chapter 2 provides a detailed discussion on data acquisition and processing. A successful application of the real-time operation requires a good telemetry system.

Downstream conditions: A reservoir may have other reservoirs downstream of it, an area to be protected immediately downstream of it, or an area to be protected far away from it. In the latter case, there will be a large uncontrolled catchment between the reservoir and the damage centers. While making release from the reservoir, it will be important to ensure that the peaks of the flow from the uncontrolled area and the reservoir release do not occur at the same time at the damage center.

The nondamaging channel capacity at major damage centers downstream of the dam is a primary consideration in sizing the flood control storage and regulation. Usually, the nondamaging discharge tends to reduce with time. Before the commencement of flood season, a survey of the downstream area must be made to ascertain any changes in the channel carrying capacity. If there are significant changes in this capacity, these will have to be incorporated in the operation procedure.

Regulation of outlet structures: The outlet capacity is an important factor in flood control operation of a reservoir. This capacity is crucial in cases where inflows expected could be more than the outlet capacity because in such cases the safety of the dam itself may be under threat. It is not only the total outlet capacity but also the capacity at various levels of a reservoir that influences the operation decisions. There may be large capacity sluices and crest spillway with or without gates. The sluices which can operate at lower levels of reservoir are effective in operation where prereleases from the dam are required to be made. However, the discharging capacity of sluices varies with the square root of head and they are not very effective at higher heads. Spillways are more effective in disposing flood waters because the discharge varies with head raised to the power of 1.5.

The spillway gates can be operated manually or mechanically. In the manual operation, the gate movement is slow. The gate operation in mechanical system is rapid and a uniform rate of opening can be maintained. If there are rapid variations of inflow or if some emergency situation arises, gates may have to be operated frequently and the time taken for opening/closing of gates becomes important. In case of mechanical operation, stand-by arrangements must be made for the supply of electric power. When operating outlets, the rate of change in release and the reservoir level should be within permissible limits.

Upstream conditions: Backwater surface profiles must be studied to determine acquisition and easement requirements upstream of the reservoir. Sedimentation deposition will adversely affect these areas, and this must be anticipated and evaluated. Upstream conditions determine the levels up to which a reservoir can be filled during floods.

Jain et al. (2018) reviewed techniques and applications of flood forecasting and highlighted the need for computationally efficient models that can adequately consider catchment properties and their variabilities, can efficiently update forecasts, can compute uncertainties in forecasts, and incorporate them in decisions and warnings. Ensemble forecasts that rely on a set of likely forecasts, help produce better forecasts, and provide estimates of uncertainty in outcomes. A hydrologic model forced by an ensemble of precipitation forecast provides a band of reservoir inflows and estimates of forecast uncertainty. Wu et al. (2020) reviewed the current status and future opportunities in ensemble flood forecasting. Many of the works reviewed by them have focused on research questions related to ensemble modeling techniques or their use to simulate past flood events. Greater application of ensemble forecasting helps in improved flood control by reservoirs.

Connaughton et al. (2014) explored the role of inflow forecasting by comparing the efficacy of three flood operating rules in controlling the peak of outflow and provided insight into the role of inflow forecasting and damage functions in reservoir operation. They used four hydrograph shapes (triangular, flood pulse, abrupt wave, and broad peak) and three operation rules: (i) the "Minimize Flood Peak" rule (uses vacant reservoir storage to control flood), (ii) the "Minimize Flooding Frequency" rule (reservoir release is kept below the safe carrying capacity of the downstream river to the extent possible), and (iii) the "Short Forecast Peak Minimization" rule (use existing flood storage capacity to minimize peak over a short future horizon). In this rule, release R_t is calculated as

$$R_t = \left(\int_t^{t+f} I_t(\tau)\mathrm{d}\tau - S_t \right) / f \tag{11.45}$$

where S_t is unfilled storage capacity at time t, f is the forecast period, and I_t are the forecasted flows. Eq. (11.45) is basically the Pack Rule of reservoir operation. Without using inflow forecast, the flood storage capacity may fill up before the arrival of flood peak and moderation would be less. However, flow forecast errors and uncertainty pose a challenge in the reservoir operation. If the forecast errors are heteroscedastic (variance increases as inflows increase), forecast uncertainty would be large when the inflow is large. Hence, a larger part of storage would have to be reserved to control floods. But, when forecasted inflow is comparatively small, due to heteroscedastic forecast errors, forecast uncertainty will be smaller. Thus, reliable forecast horizons (FHs) should be identified for an efficient forecast-based flood control operation.

Traditionally, lumped catchment models are used to prepare reservoir inflow forecasts, but at times, these models are unable to provide the desired accuracy. This has promoted the use of physically based distributed hydrological model (PBDHM) to forecast inflows. However, research is needed to overcome issues concerning calibration and forecast updating while using PBDHM. In future, PBDHM will likely be increasingly used in reservoir inflow forecasting. The performance of hydrological models for forecasting could also be improved through data assimilation techniques, leading to their wider use in real-time operations.

In the past few decades, ML methods have been developed to simulate the complex processes of flood formation. Popular ML techniques include ANNs, adaptive neuro-fuzzy inference systems (ANFISs), multilayer perceptron (MLP), long short-term memory (LSTM), wavelet neural networks, and support vector machines (SVMs). ML models can learn the nonlinear rainfall-runoff process based on historical data and do not require detailed knowledge of the underlying processes. Mosavi et al. (2018) reviewed the state-of-the-art of ML models in flood forecasting. ANNs and deep networks are the most popular ML tools in water resources because of their accuracy, fault tolerance capacity, and parallel processing abilities. The use of ML methods for hydrological modeling and forecasting is expected to grow with time.

Despite numerous benefits by the use of forecasts for reservoir operation and flood management, reliable short-term forecasts are not widely available and used in reservoir regulation, particularly in developing countries. Key reasons behind this are lack of skills and expertise, lack of data, and lack of infrastructure/logistics. It would be hugely rewarding to overcome these limitations. Pappenberger et al. (2015) estimated potential monetary benefits of early flood warnings in Europe by using flood damage cost data and found that the benefits were to the tune of 400 Euro for every 1 Euro invested.

11.8.3 Real-Time Flood Control Operation of Reservoirs

Real-time flood control operation is accomplished by integrating: (i) data collection, validation and processing, (ii) FF, (iii) flood reservoir operation, and (iv) information query. Researchers have attempted real-time reservoir operation for flood mitigation by integrating a flood forecasting and an optimization model. Some of the optimization models that have been used to make reservoir release decision during the forecast period are mixed integer linear programming (MILP), genetic algorithms, preemptive goal programming, and genetic programming (GP). Yang et al. (2019) used the recurrent neural networks (RNNs) to simulate reservoir operations. Researchers have also integrated various system simulation models with optimization models for real-time reservoir operation. Evolutionary algorithms like GA have also found wide application in real-time reservoir operation for flood mitigation due to their suitability in coupling with hydrologic modeling software. Che and Mays (2015, 2017) developed and illustrated the applicability of an elaborate real-time optimal operation model for a multireservoir system that consists of (i) a short-term rainfall forecasting module, (ii) HEC-HMS for rainfall-runoff modeling, (iii) HEC-RAS for unsteady river flow routing, and (iv) GA-based optimization module to decide on reservoir releases. The optimization algorithm was developed by using MATLAB, which also served to link the hydrologic models. Applications of the model to a hypothetical system showed their utility and suitability, but problem may arise in practical applications when a key module fails to give realistic output.

Since a number of factors and data are involved in real-time operation, a decision support system (DSS) that logically organizes data and models can be of great value. Dam managers are motivated to use DSS by their successful application in many engineering/management tasks and wider availability of computers with user-friendly graphic interfaces. A spatial DSS (SDSS) for real-time flood control in a multipurpose, multireservoir system integrates several subsystems: real-time data monitoring, database management, geographic information, and S—O. An ANN was employed to develop spatially distributed flood forecasts in real-time (which are regularly updated). Many studies describe the development and application of DSS to real-life problems with the aid of remote sensing, GIS, graphical tools integrated to multiple criteria decision-making. Sensoy et al. (2018) combined Supervisory Council and Data Acquisition (SCADA) system with Numerical Weather Prediction (NWP) data, HEC-HMS, HEC-ResSim, and HEC-RAS models to create a DSS for real-time reservoir operation and reported improved system performance by use of this DSS.

Although computer models are of immense value in real-time operation, seasoned operators cross-check the results with their experience and in-situ data (say, verify the inflow rate with rate of change of reservoir water level). Undoubtedly, decisions are to be taken quickly, but risk of wrong decisions taken in haste is to be avoided. Spillway gate operation is a critical component of real-time operation. It is a specialized task and the sequence and rate of opening the gates is project-specific.

11.8.4 Information Dissemination

Whenever high outflow from the dam is likely, it is the duty of the in-charge of reservoir operation to alert the concerned authorities. The information must also be communicated to civil authorities, revenue authorities, police, and general public for taking precautionary measures in respect of alerting and evacuating the people in the area likely to be affected.

To take full advantage of real-time operation, a good information dissemination system is necessary. In real-time operation, advance information can be given about the likely outflow and if the public is informed in time, the flood damage can be significantly reduced. The organizations responsible for issuing flood warning and flood fighting should be informed about the likely high releases from the dam as early as possible so that the required action is planned and activities set into operation with least possible delay. They should also be kept informed of the development of the high releases and any change in the present as well as anticipated future situation. The information is provided by the flood forecasting authorities in the form of *bulletins*. These bulletins must be very clear and should include necessary details so that a realistic picture of the incoming danger is depicted. There should be arrangement to double-check the information supplied with clear-cut responsibilities in every office authorized to issue such bulletins. It is necessary to avoid dissemination of wrong information including even the inadvertent mistakes because hardly any time is left for review between dissemination of information and triggering a chain of follow-up activities.

11.8.5 Advantages of Real-Time Operation

The advantages of real-time operation over the conventional methods are as follows:

- It is highly flexible compared to the conventional methods because the current state of the system and forecast of future inflows are into account while taking a decision.
- Real-time operation is the most realistic operation for reservoirs as operation decisions are frequently updated with the availability of new information.
- In emergencies, such as floods, it gives high lead-time to the authorities to take precautionary measures in respect of alerting and evacuating the area likely to be affected.

11.9 UNCERTAINTIES AND RISKS IN FLOOD CONTROL OPERATION OF RESERVOIRS

In reservoir operation for flood mitigation, risks arise due to the uncertainties in current and predicted inflows, remaining flood cushion in the reservoir, or the downstream channel characteristics. Reservoir regulation also involves a small risk of dam overtopping and break. These uncertainties put the operators in a risky situation where flood mitigation or conservation targets or the safety of the dam may be compromised. The flood mitigation decisions depend on the flood damage risk estimated as the probability of reservoir releases exceeding the conveyance capacity of river at the control point and the dam overtopping risk which is the safety margin relating dam crest height to the/likely reservoir water level while routing the floods. The impact of streamflow forecasting on flood control operations depends on factors such as reservoir size, the lead-time over which reliable forecasts can be made, the ability of the decision-making process to utilize the forecasts, the reservoir capacity-inflow ratio, the spillway release capacity, and safe downstream river carrying capacity.

The uncertainty analysis methods include the first-order second-moment method (FOSM), the point estimation methods, the Latin hypercube sampling (LHS), and the Monte-Carlo simulation (MCS). The analytical methods are simple but may get cumbersome with increasing number of random variables. Lack of reliable data and round-off errors affect the risk estimates which are mostly based on empirical estimates of the moments of random variables. The LHS and MCS methods overcome such shortcomings and also yield output distribution. The mean, the variance, and the distribution of the output obtained by using the LHS is nearly unbiased and requires much less effort for the same level of precision, compared to the conventional MCS methods. Impact of various uncertainty factors on reservoir operation can also be quantified by using the Shannon entropy concept.

Errors in inflow forecast are a major source of uncertainty in reservoir operation, and their effect on flood operation is crucial. Stochastic simulation of inflows or ensemble forecasts are adopted to describe uncertainty in reservoir inflow and local flows at the downstream control point. To reduce the risk due to inflow forecast errors, a reliable FH based on various levels of streamflow forecast uncertainties for different lead times can be defined. The other kinds of uncertainties that invoke flood risks are associated with the water level-storage curve, flood hydrograph shape, reservoir operation parameters, wind wave effect, and the climate induced uncertainties. The various uncertainties can be jointly considered for evaluating the upstream and downstream flood risks.

Studies have computed the risk in reservoir operation for flood mitigation in a multireservoir system also. The FHs of each reservoir in a multireservoir system could be different, whereas their operation horizons (OHs) are the same. Gong et al. (2020) developed a two-stage model in which a multivariate autoregressive model was used to derive multinormal distribution of multireservoir storage during the first stage of FH. For the time between FH and OH, the flood risk was computed by routing the design floods through the reservoirs. Total risk during the OH was the combined risk at all the stages, and the operation corresponding to minimum total flood risk was optimal operation for the entire system.

Wang et al. (2021) examined the uncertainties and evaluated the associated risk in a multi-reservoir system. They performed a detailed evaluation of the risk that arises due to the uncertainty in streamflow series, flood forecast error, water-level storage curve error, and reservoir discharge capacity error independently. The probability of failure is the number of times the loading (maximum flood storage capacity, the largest peak discharge at the downstream control point, and the flood diversion volumes in the flood detention basins) exceeds the resistance (safe storage capacity, safe water release, and safe flood diversion). Consequences of failure are quantified using the conditional value-at risk index. Their study is a comprehensive evaluation of uncertainties and associated risk in a multireservoir flood control system located in the upper Yangtze river basin in China.

To summarize, uncertainties arise in flood control operation of reservoirs due to imperfect knowledge about the system, inputs (including forecasts), and downstream conditions. Uncertainty analysis should be a necessary component of real-time reservoir operation. Every uncertainty analysis method has its own hypothesis, advantages, and limitations. There is big scope for research concerning uncertainty analysis in a multireservoir system. The temporal and spatial joint modeling of multiple uncertain factors followed by risk evaluation of the reservoir flood mitigation system could ensure better performance. The uncertainty arising due the variable length of the forecast period in different reservoirs in a system has seldom been considered. Structural measures for flood disaster risk reduction such as dams and embankments provide a feel of safety to the people residing in their vicinity. Since the structures may fail, this sense of security is perceived to be "false" and frequently the "residual risk" is underestimated. Thus, there is a need of increasing flood risk awareness and accounting for "residual flood risk" related with the structural flood control measures.

11.10 DEVELOPMENT OF OPERATING RULES FOR SABARMATI SYSTEM

For judicious regulation of water resources in the Sabarmati basin, operation policies are required for conservation and flood control purposes for the various hydraulic structures. A detailed description of the Sabarmati basin was given in Chapter 1. A line diagram of this system is given in Fig. 11.21. The following discussion illustrates the application of systems analysis techniques to develop operation procedures for the reservoirs in this system. It also brings out the difficulties that are faced while applying the techniques to a system where the data availability situation is somewhat unsatisfactory.

The aim of the conservation regulation of this system is to meet the demands in the best manner and avoid severe scarcity of water in the basin. In the flood season, the Dharoi reservoir is to be operated such that the total flow at Ahmedabad city does not exceed the safe carrying capacity of the river (14,160 cumec). However, the safety of the dam is to be given the top priority. The maximum discharge capacity of spillways at the Dharoi dam is 16,992 cumec.

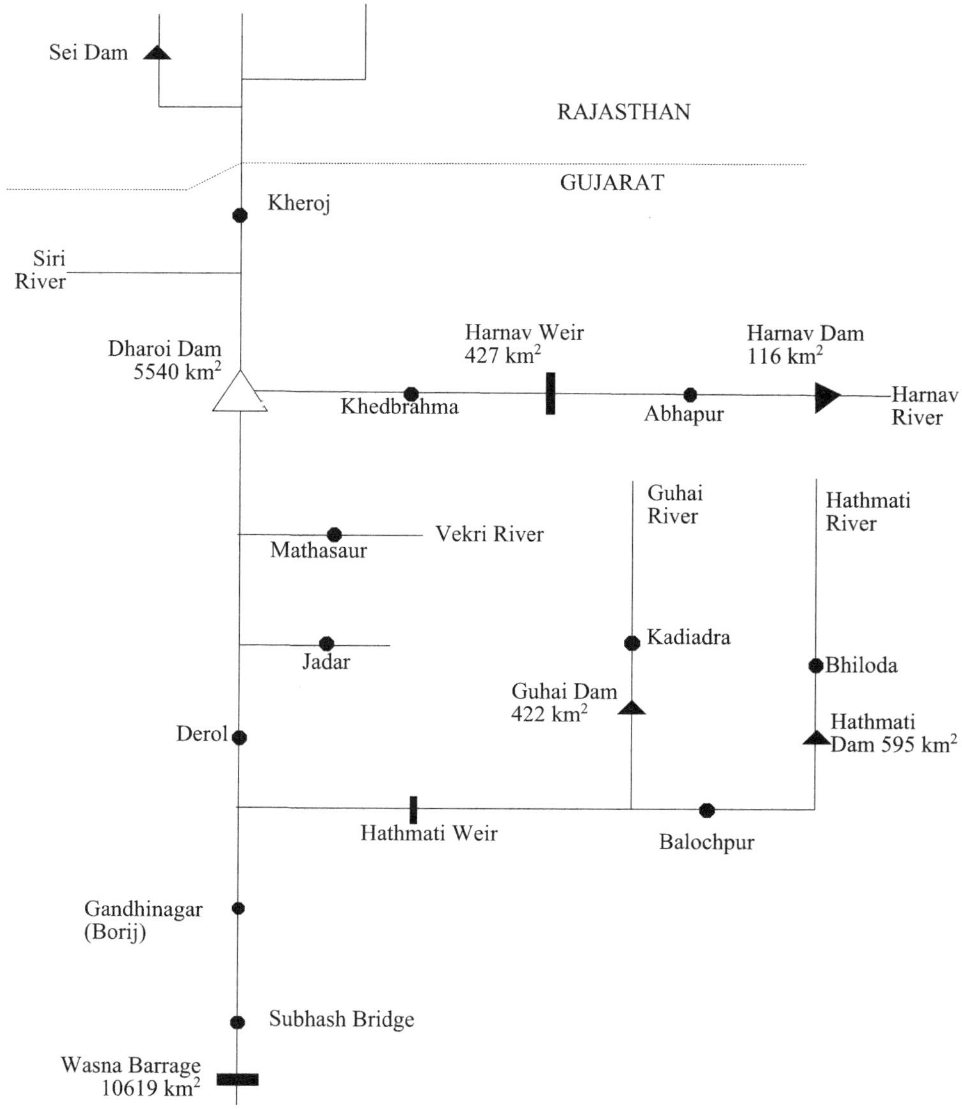

FIGURE 11.21

Line diagram of the Sabarmati basin up to Wasna Barage (Ahmedabad).

The simulation approach using historic observed flows was adopted to derive the regulation policy for various impoundments in the Sabarmati basin. Rule curves were derived for all the reservoirs serving various purposes. For flood regulation, the modeling of flow in the Sabarmati River at Ahmedabad was carried out and the release policy for the Dharoi dam was developed using the design flood hydrograph.

11.10.1 Solution Approach

A rule curve-based operation procedure was adopted for the conservation regulation of this system. Initial rule curves were derived for different reservoirs and these were refined using simulation.

A monthly time step was chosen to simulate the operation of the system. The operation of the system was simulated in accordance with the trial policy and calculates the release for different purposes, spill (if any), and the evaporation losses for each reservoir. During simulation, it takes into account the flow coming from upstream structures, the diversion of flow, and the release for meeting downstream demands. The output of the program includes a detailed working table and the time and volumetric reliability of each structure for the trial policy. It also calculates the number of critical failure months (release less than 75% of the target demand).

Since the flood control capacity of the structures in the Hathmati system was limited, the entire intermediate catchment (between the Dharoi dam and the Ahmedabad city) was assumed to be uncontrolled. For developing flood regulation procedure for the Sabarmati system, the modeling of flow at Ahmedabad was carried out. The flow at Ahmedabad was split into two components: the routed release from the Dharoi reservoir (which can be controlled) and the flow from the intermediate catchment (which cannot be controlled). Using the information about the releases from the Dharoi reservoir for the past few flood events, flow routing was carried out using the Muskingum–Cunge routing procedure (Ponce, 1989), and the various parameters of the method were calibrated. The flow from the intermediate catchment was modeled using the unit hydrograph approach.

This reservoir moderates the inflow in the space between the FRL at 189.59 m and the MWL at 193.60 m. The maximum observed flood at the Dharoi dam site was estimated to be 14,158 cumec. In India, the design flood for major dams, such as Dharoi, is developed using the probable maximum precipitation (PMP) approach. The design flood hydrograph for the Dharoi reservoir was used for developing the flood regulation procedure. The design storm is convoluted with the unit hydrograph for the catchment, and the base flow is added to obtain the design flood hydrograph. The peak of the design flood hydrograph is 27,180 cumec and its volume is 3095 Mm^3. It was felt that the flood regulation policy developed using the design flood hydrograph would be quite appropriate. A flood regulation simulation analysis was carried out for the Dharoi reservoir using different normal and emergency conditions under various scenarios of the downstream safe channel capacity.

11.10.2 Data Availability and Processing

Discharge is gauged in the basin at eight locations; the frequency of observation being hourly (during flood season) or once/twice/four-times a day. A sufficiently long inflow series at a structure was necessary for the simulation study. Usually, data for the period 1967–1993 were available. However, the virgin flows at some nodes were not available and the flows were estimated by adjusting the flow data at the nearest location in proportion to the catchment area, assuming that the catchment yield per unit area is the same. The target monthly demands, normal monthly evaporation depths, and storage details for all the structures were collected from the concerned operating authorities.

A preliminary analysis of the Hathmati subbasin flow data revealed that on the average, the annual flows in this basin were less than the annual target demands. Because of the deficit of water availability, any structure in this subsystem (Hathmati dam, Guhai dam, and Hathmati weir) could not be used to supply water at the Ahmedabad/Wasna barrage. It was also not possible to divert water from any other structure to this subbasin. Hence, the Hathmati subsystem was studied as an independent unit. The concurrent local flow data at the Guhai dam, the Hathmati dam, and the Hathmati weir were processed for the period 1964 to 1993. For the Dharoi subsystem, consisting of the Harnav dam, the Dharoi dam, and the Harnav weir, the average annual flows were more than the annual target demands. Hence, it was decided to meet some requirements of the Wasna barrage from the Dharoi reservoir. The common period of flow observation for the Dharoi subbasin was from 1967 to 1993. Since for integrated operation, a common period of observation is necessary, the simulation analysis for the whole system was carried out for the period 1967 to 1993.

For a satisfactory flood regulation study, the modeling of flow in the Sabarmati river basin using short interval data (say, hourly) was necessary. Such data at various locations were available concurrently only after the year 1982. Those events were considered for which the hourly discharge data at various gauging stations were available concurrently with the hourly rainfall data at various self-recording rain gauge stations. The hourly discharge data of Derol, Gandhinagar, and Ahmedabad and the releases from the Dharoi reservoir were utilized for routing studies. Rainfall data of 11 rain gauge stations were used to obtain the average hourly rainfall in the intermediate basin. Seven flood events, having consistent and concurrent short interval rainfall and discharge data, were used. These events were: July 23–26, 1982; Aug 11–20, 1983; Aug 4–7, 1984; Aug 11–15, 1984; July 19–22, 1988; Aug 4–6, 1988; and Aug 24–26, 1990.

11.10.3 Integrated Conservation Operation of the System

In the ranking of conservation demands, the domestic and industrial water supply demand is given the top priority, while the irrigation demand is given the next priority. For conservation purposes, the system was to be operated such that: (a) the water supply demand could be met to the extent possible; and (b) the available irrigation water must be managed such that severe water shortages and subsequent crop failures can be minimized.

Derivation of Initial Rule Curves

More than 80% of the annual rainfall in India occurs in the four monsoon months from June to Sep. The reservoirs are filled during this period, and the stored water is used throughout the water year. Using the power transformation approach, monthly inflows for 50%, 75%, and 90% dependability were estimated for each reservoir. Depending on the number, nature, and priority of conservation demands, different rule curves were derived for each reservoir. For a reservoir serving for irrigation only, two rule curves, namely an upper rule curve and a lower rule curve, were derived. For a reservoir serving for water supply demand also, an additional rule curve for water supply was developed. The upper rule curve levels specify the maximum storage by month to which a reservoir should be filled if there is sufficient inflow. Lowering the upper rule curve below FRL provides storage for flood retention during the monsoon season, although it may affect the performance of the reservoir for conservation demands. Initially, the upper rule curve was set at FRL from June to Sep to ensure reservoir filling. In the months after the monsoon season, upper rule levels were calculated using 50% probable inflow, full target demands, evaporation losses, and uncontrolled spills from the reservoir using the forward computation:

$$\text{Storage}_{\text{end}} = \text{Storage}_{\text{begining}} + \text{Inflow} - \text{Demand} - \text{Evaporation} - \text{Spills} \tag{11.46}$$

The evaporation losses were considered at normal monthly rate over the surface area of the reservoir corresponding to a particular elevation.

The middle/lower (whichever is applicable) rule curve is critical for irrigation demands and was derived for the situation when water is scarce and not all demands can be met throughout the water year. These were derived by assuming that the reservoir level will reach the dead storage level by the end of May. To compute them, 75% dependable inflow, full target demands, evaporation losses, and uncontrolled spills from the reservoir were considered and backward calculations were carried out starting from the end of May as:

$$\text{Storage}_{\text{begining}} = \text{Storage}_{\text{end}} - \text{Inflow} + \text{Demand} + \text{Evaporation} + \text{Spills} \tag{11.47}$$

If the reservoir falls below this level in a particular month, the supply for irrigation must be suitably curtailed so that the reduced release can be made for a longer duration, thus avoiding severe crop failures. It is proposed that below this level, releases will be made to satisfy at the most 75% of the irrigation demands and full water supply demands (if any).

The lower rule curve levels (in reservoirs with water supply demand for domestic and industrial purposes) are critical for higher priority water supply demand. When the reservoir level falls below the lower rule level, the supply for irrigation must be completely curtailed and the release made only for the essential demands. This level was calculated using the approach mentioned for the middle rule level except that the 90% dependable inflow, only water supply demands, evaporation losses, and uncontrolled spills were considered. The trial rule curves were derived for all the reservoirs in a similar way.

Since the Dharoi reservoir serves irrigation and water supply demands, three rule curves were derived: upper rule curve, middle rule curve for full target demands, and lower rule curve for water supply demands only. The Hathmati reservoir serves irrigation demands of its own command area and the higher priority irrigation demands of the command area of the Hathmati weir. Hence, three rule curves were derived to serve two demands of different priority. Since Hathmati dam is ungated and the reservoir is filled to FRL before any spill, there was no need to derive the upper rule curve and FRL at this reservoir corresponds to the upper rule level in all months. The middle rule curve was derived for the total irrigation demand, while the lower rule curve was derived for the irrigation demands of the Hathmati weir. The other two reservoirs (Harnav and Guhai) have only irrigation demands in their own command area. So only two rule curves (upper and lower) were sufficient for these reservoirs.

Simulation of the System Operation

To derive the optimum rule curve reservoir levels (those giving the highest monthly time reliability with the least number of critical failure months), a detailed operation table was utilized. This table included the starting storage, virgin inflow, inflow from any upstream structure, demands from the structure, actual evaporation losses based on initial and final water spread areas, release for different demands, spills, and final elevation at each of the reservoirs for each month. For diversion structures, the simulation table included initial storage (in case of associated small storage structures), virgin inflow, inflow from upstream structures, demands at the structure, actual releases made, spills, and the final storage. Initially, the upper rule curve levels were kept unchanged. Based on the observations from the simulation operation table, the initial lower rule curve levels were modified as long as the number of failure months (release <100% of the target demand) could be reduced without increasing the number of critical failure months (release <75% of the target demand). A number of simulation runs were made, and the lower and the middle rule curve levels were tuned. Next, the trial upper rule curve levels were lowered in the four-monsoon months till there was no increase in the number of failure months.

Integrated operation of the Sabarmati system was simulated for the period 1967 to 1993. Based on the results from a number of trial runs, it was found that in the Hathmati subbasin, the Hathmati dam could be operated to serve for a maximum of 70% of the target demand of the Hathmati weir. At the Hathmati weir, there exist two small storage structures, Limla dam and Karol dam with capacities of 10.28 and 7.5 M Cum, respectively, to store excess water. This water can serve the demands in the command area of the Hathmati weir. Before releasing the excess water in the downstream main Hathmati River, these dams are filled to their capacities. The operating authorities at the Dharoi reservoir retain storage reserve equivalent to one-year water supply demand to cater for the eventuality of a monsoon failure in the following year. Thus, water is not released for irrigation from the Dharoi reservoir below the reservoir level 180.69 m. Trial simulation also demonstrated that only 50% of the Wasna barrage demands could be met from the Dharoi reservoir.

In the Dharoi subsystem, the monthly reliabilities of the Harnav and the Dharoi reservoirs for full assured supply were estimated to be 81.2% and 86.1%, respectively. For 75% assured irrigation supply, the monthly reliabilities for these structures were 93.8% and 90.1%, respectively. The monthly reliability of the Dharoi reservoir for water supply demand was 96.9%. The capacity of the Harnav reservoir (19.97 M Cum) is not adequate to support even the water supply demands at Dharoi. Furthermore, in periods of water scarcity at the Dharoi reservoir, water was also scarce in the Harnav dam. Hence, the Harnav dam may be operated to meet only its own demands. For the Harnav reservoir, in the span of 27 years, there were only 20 months when the release was less than 75% of the target demand. Most of these months occurred during the three acute drought years (1969, 1986, and 1987). Similarly, for the Dharoi reservoir, five years (1969, 1972, 1974, 1986, and 1987) were acute drought years in which 29 of the 32 critical failures occurred. Simulation showed that the present policy of the Dharoi operators of not supplying water for irrigation below 180.69 m (Dead Storage Level 175.87 m) could be relaxed in June and July as this improves the reliability of the reservoir with no increase in the number of months of water supply failure.

The simulation of the Hathmati subbasin revealed that it is better to operate the Hathmati dam to meet 70% of the target Hathmati weir demand. The monthly reliability of the Guhai dam, the Hathmati dam, and the Hathmati weir for meeting full target demand came out to be 61.4%, 54.6%, and 59.6%, respectively. The years 1969, 1974, 1979, 1980, 1985 to 1990 and 1992 were severe drought years for this subbasin and most of the failures occurred in these years. It was found that there is scarcity of water in this region and either target demands must be reduced and/or water conservation practices must be adopted.

As discussed earlier, the reservoir system operating policy meets the demands for as long a duration as possible. However, as soon as a shortage of water is anticipated at any time, the supply is curtailed so that the reduced supply could be maintained throughout the crop period. Simulation showed that 75% of the target demands could be met most of the time. The final rule curves for the Dharoi reservoir are shown in Fig. 11.22.

The upper rule curve implies that excess water above this level should be spilled from the reservoir to make some room for flood attenuation. This aspect of the system operation will be

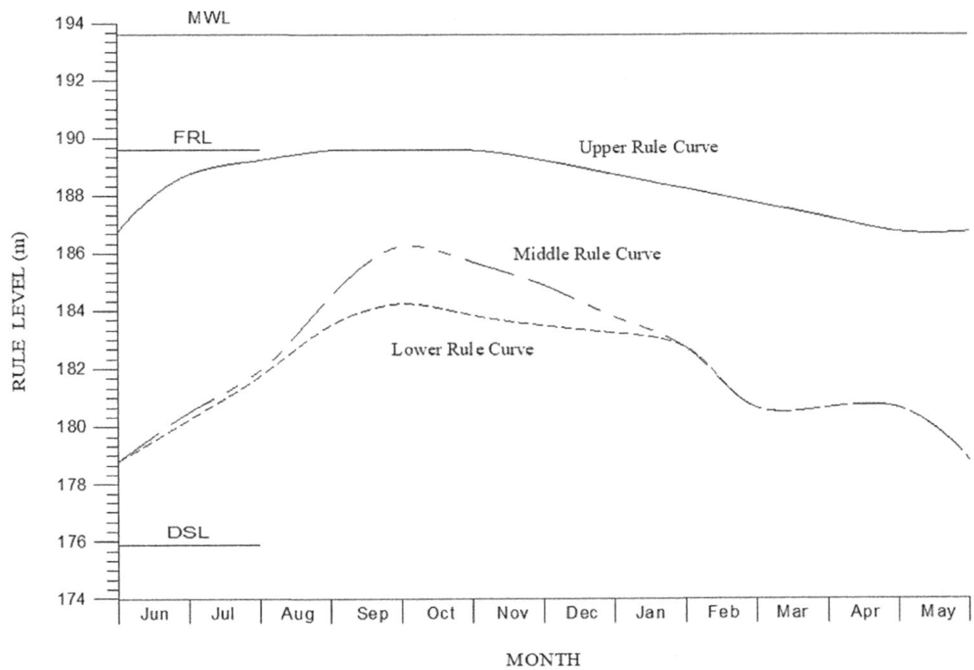

FIGURE 11.22

Developed rule curves for the Dharoi reservoir.

addressed in the next section. The middle rule curve implies a critical situation in the Dharoi reservoir for meeting full target demands in the remaining part of the water year. If the reservoir level drops below the middle rule curve, the authorities should initiate measures to avoid severe or sudden crop failure. The lower rule curve signifies the start of a critical water supply demand event (the highest priority demand). In this situation, releases for all other purposes should be curtailed to ensure that this highest priority demand is met to the fullest possible extent. When operating a reservoir using rule curves, it is advisable to periodically review and update the previous decision for the remaining duration of that month in light of new information, such as increased inflows, reduced demands, and so forth.

11.10.4 Flood Control Operation of the System

Reservoirs are frequently operated to control flooding at locations downstream of the dam. If the potential damage center is located far away downstream of the dam, a significant portion of the river flow at that location may be due to the contribution of the intermediate catchment area. In such cases, it is necessary to determine the likely flow at the damage center due to the rainfall occurring in the intermediate catchment and the release from the upstream reservoir. Thus, modeling the flow at the damage center involves two steps: first, channel routing to analyze the effect of the release from the dam site; and second, the rainfall-runoff modeling of the intermediate catchment.

The Dharoi reservoir is the only major flood control storage in the Sabarmati basin. Since the flood retention capacity of the hydraulic structure in the Hathmati subbasin was limited, the entire intermediate catchment (between the Dharoi dam and the Ahmedabad city) was assumed to be uncontrolled. In the Dharoi reservoir, space above FRL (189.59 m) is exclusively for flood control. Therefore, it is proposed that this reservoir be operated for flood control only at elevations above 189.59 m. While deciding the release from the Dharoi reservoir, the operator must know the maximum safe release that can be made from the reservoir such that the flow at the Ahmedabad city does not exceed the desired limits.

Development of Flood Regulation Policy for Dharoi Reservoir

Simulation was adopted to derive the flood regulation policy. Various policies of the flood regulation were tried using different scenarios of the downstream safe channel capacity and different normal/emergency conditions in the reservoir. An exhaustive simulation analysis for the reservoir was carried out using the design flood hydrograph.

Through simulation, it was found that by keeping the emergency level at 191.00 m, the safe channel capacity at 11,500 cumec, and the spillway release capacity at 90% of the maximum (16,990 cumec), it was possible to restrict the reservoir level below MWL (193.60 m). The analysis also suggests that as soon as a flood is anticipated, releases at a rate equal to the minimum of the safe channel capacity and the spillway release capacity at the current reservoir level, but more than the inflow rate must be made. This will create an extra empty space in the reservoir for flood attenuation.

The operation scenario of the Dharoi reservoir for flood regulation was classified into two categories: normal operation and emergency operation. Normal operation policy is applicable in any of the following conditions: (i) reservoir level is below 191.00 m and the inflow rate less than 16,992 cumec, (ii) reservoir level is above 191.00 m and the inflow rate is less than 16,992 cumec and it is decreasing. Under normal conditions, the maximum release from the reservoir would be equal to 50%–80% of the safe carrying capacity of the downstream channel. This way, some additional empty space can be created in the reservoir for flood absorption. However, if the inflow starts decreasing, the release should be curtailed till the reservoir is filled to FRL. In case of (ii), the release should be such that the water level falls below the emergency level at the earliest.

The emergency operation is activated in any of the following conditions: (i) reservoir level is above 191.00 m and either the inflow rate <16,992 cumec but increasing, or the inflow rate >16,992 cumec although decreasing; and (ii) at any reservoir level, the inflow rate is more than 16,992 cumec and it is increasing. Under emergency conditions, the release from the reservoir (may be more than the safe channel capacity) should be such that the reservoir level can be brought down to a safer level at the earliest, thereby avoiding overtopping of MWL. The minimum release rate under emergency conditions is the smaller of the safe channel capacity and the spillway release capacity.

With the developed flood regulation policy, it was found from the simulation analysis that by using the design flood hydrograph as inflow, the safe channel capacity of 9000 cumec and the maximum emergency release as 13,900 cumec, the reservoir level could be maintained below MWL. Thus, as soon as a flood is anticipated in the reservoir, release at rates greater than the inflow rate but less than the safe channel capacity should be started. If the conditions in the reservoir are in the normal range,

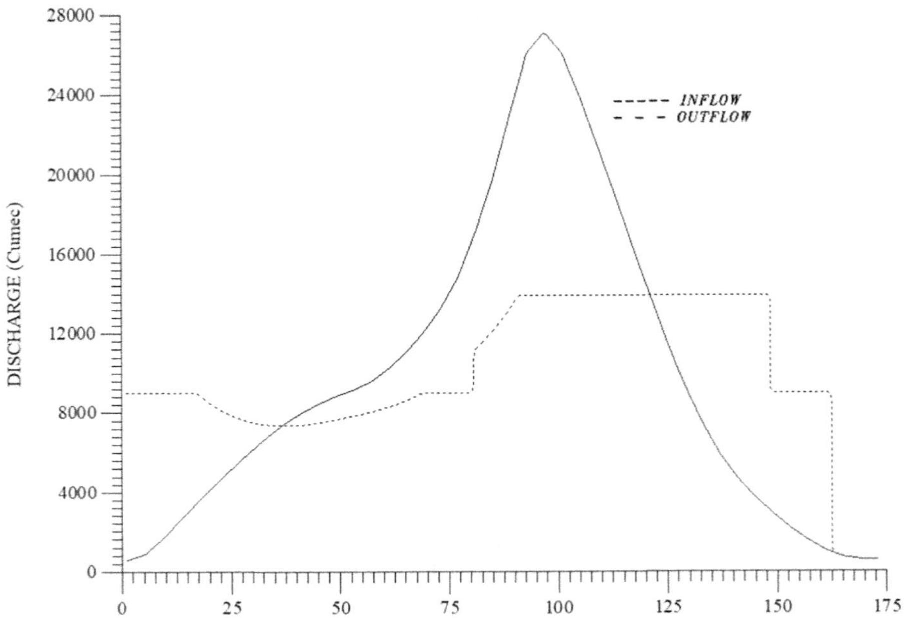

FIGURE 11.23

Regulation of design flood hydrograph using the developed policy.

normal operation policy must be adopted. However, if the conditions in the reservoir reach the emergency state, emergency operation rules should be followed so that the safety of the structure can be ensured even if that entails exceeding channel capacity. The regulation of the design flood hydrograph through the Dharoi reservoir with the developed policy is presented in Fig. 11.23.

The likely hydrograph at Ahmedabad can be estimated by routing the Dharoi releases down the river up to Ahmedabad and adding the contribution of the intermediate catchment to it.

Routing of Reservoir Releases

Using the data about the releases from the Dharoi reservoir for the past flood events, flow routing in the Sabarmati River was carried out from Dharoi/Derol up to Ahmedabad. The Muskingum–Cunge method (Ponce, 1989) of flood routing was used in this study. Two main reasons for adopting this technique were: (a) the simplicity of the method, and (b) this method makes use of topographic and cross-section information which was available for the present study.

Let the time be discretized in steps of Δt duration and the distance along the river in steps of Δx length. Following this method, if the inflow to the river reach at time $n\Delta t$ is denoted by I_n, the outflow Q_{n+1} at time $(n+1)\Delta t$ is given by

$$Q_{n+1} = C_o I_{n+1} + C_1 I_n + C_2 Q_n \tag{11.48}$$

Define the cell Reynolds number $D = q_o/(S_o\, c\, \Delta x)$, and the Courant number $C = c\, \Delta t/\Delta x$, where S_o is the channel bottom slope, q_o is the reference flow per unit channel width, and c is the wave celerity

($c = \beta V$, β is the rating exponent and V is the average velocity corresponding to peak flow). The coefficients in Eq. (11.48) are:

$$C_o = \frac{-1 + C + D}{1 + C + D}$$
$$C_1 = \frac{1 + C - D}{1 + C + D} \quad (11.49)$$
$$C_2 = \frac{1 - C + D}{1 + C + D}$$

The methodology would be applicable, subject to the satisfaction of the criteria: $T_r S_o V_o / D_o \geq 85$, where T_r is the time to peak of the inflow hydrograph, V_o is the average flow velocity, and D_o is the average flow depth. The value of Δt is chosen such that $\Delta t \leq \Delta x / c$ and $\Delta t \leq T_r/5$. The value of Δx should be chosen such that $q_o \Delta t / (0.25\ S_o) \geq \Delta x \geq q_o / (S_o c)$. To ensure numerical stability, the values of Δx and Δt should satisfy condition $C + D \geq 1$.

During a flood, the time available to the dam operator for deciding various parameters of the Muskingum–Cunge routing procedure is short. For this reason, estimates of different parameters corresponding to various peak discharges were developed. The number of reaches to be considered for routing was found to vary inversely with the discharge in the river. From Dharoi to Ahmedabad, the number of required reaches was found to be in the range from 13 to 35. The top width of the river, the flow area, and the rating exponent were also calibrated for different peak discharge values. The time step (Δt) equal to half-hour was found suitable for satisfying various conditions as mentioned above. Using a half-hourly time step, space step (Δx) values are calculated. Using these values, one can estimate the routed flow at Ahmedabad corresponding to any trial release from the Dharoi reservoir.

Flood routing was carried out for those events for which the concurrent flow data were available at various gauging stations. For routing between Derol and Ahmedabad, the observed hourly discharge data at Derol, Gandhinagar, and Ahmedabad were used. The Hathmati River is the only major tributary that joins the Sabarmati River downstream of, but very near to, the Derol gauging site. Thus, a large volume of lateral inflow enters the river Sabarmati between Derol and Ahmedabad. There is no gauging site on the Hathmati River near its confluence with Sabarmati. The lateral inflow hydrograph from the Hathmati River was estimated by routing the flow at Derol up to Gandhinagar and then subtracting the routed flow from the observed flow at Gandhinagar.

For the flood events for which the release data from the Dharoi reservoir were available, the flow from the intermediate catchment was calculated by routing the release from Dharoi dam up to Ahmedabad and then subtracting the routed flow from the observed flow at Ahmedabad. For the events when the reservoir release was nil, the entire flow at Ahmedabad was assumed to be the contribution from the intermediate catchment. These computed flows were used for rainfall-runoff modeling of the intermediate catchment.

Modeling of the Intermediate Catchment

The catchment area between the Dharoi dam and Ahmedabad city, including the Hathmati subbasin, is 5079 sq. km. To model the contribution from a catchment of this size, it would be appropriate to carry a detailed modeling by subdividing it into subcatchments, but the required short-term rainfall and discharge data of sufficient duration for each subcatchment were not available. In view of limited data, it was decided to develop a unit hydrograph for the whole intermediate catchment. Release data from the Dharoi reservoir were available for seven events, and these were used for modeling the intermediate catchment. The average hourly rainfall for the intermediate catchment was estimated using the Thiessen polygon method. Since the entire intermediate catchment was modeled as a single unit, the total lateral flow hydrograph at Ahmedabad, computed during the routing analysis, was used as the observed flow.

The unit hydrograph was developed by using Snyder's synthetic unit hydrograph method. The average hourly rainfall over the intermediate catchment was estimated as discussed above. The volumes of the observed lateral flow and the observed rainfall were computed and the runoff coefficient corresponding to different rainfall depths was calibrated. The runoff coefficient for the intermediate catchment was found to be in the range from 0.06 to 0.15. A synthetic unit hydrograph was developed for the intermediate basin and convoluted with the effective hourly rainfall. The observed and computed hydrographs were compared, and the unit hydrograph was modified until a satisfactory response was achieved.

Software to Calculate Safe Release

A computer program was developed to calculate the flow hydrograph at Ahmedabad city resulting from the release from the dam and the rainfall in the intermediate catchment. The program uses the Muskingum—Cunge routing procedure and the unit hydrograph for this purpose. This program can assist the operator at the dam site in deciding the safe release from the reservoir at any time. The input to the program includes the past half-hourly reservoir releases, present trial release, and the average rainfall in the intermediate catchment during all the past hours of the current flood event. The past reservoir releases and past average rainfall depths are stored in relevant files. The various calibrated parameters (space step, number of reaches, top channel width, flow area, wave celerity, and runoff coefficient) have been specified in the program. The output from the program is the release from the current time period.

The data availability was the main constraint in the application of improved techniques to this real-life problem. In addition, the accuracy and authenticity of the data collection system created additional problems in successfully modeling the Sabarmati river basin system.

11.11 CLOSURE

Due to the increase in population and urbanization, the conservation demands as well as damage potential of floods are increasing day-by-day, but it is not easy to create new reservoirs because of social and environmental ramifications. Therefore, it is essential to operate the existing reservoirs as efficiently as possible. Real-time operation is an efficient way of operating a reservoir system in which the control decisions are made on the basis of prevailing conditions of the system and the forecast about the likely inflow in the reservoir. For real-time operation of a reservoir, automatic telemetry system is essential for direct transmission of data at a regular interval to the forecasting station from where forecasts are issued. This hydrological forecasting is used as input in the operation model to find the optimized value of the release from the reservoir.

Many smaller reservoirs are operated using simple "common-sense" rules such as keeping reservoirs full for conservation purposes or empty to control floods. Rules for major projects are derived using systems analysis techniques. However, in general, simulation appears to be the technique that is most common in deriving procedures that are used in practical situations. Despite significant developments in the reservoir operation area, many reservoirs are still operated using procedures that have not been developed "scientifically."

A lot of work on real-time operation is going on and efforts are on to develop models which can capture the details of the system in as much detail as possible. Computation time is an important aspect in real-time operation as the 'lead time'" of a forecast is affected by it. Due to the increasing availability of high capability and low-cost computers with telemetry networks and remote control features, the hardware required for this purpose is easily available. A number of models using different optimization techniques, such as LP, DP, and goal programming have been developed and applied to various real-life problems in the form of case studies. A considerable amount of software is also available. A number of DSSs have also been developed which can assist the operator in making better decisions.

APPENDIX A

Generalized Formulations of Optimization Models

Reservoir operation during flood season is best carried out by accounting for dynamic changes that occur in the physical system as the flood progresses in time. The changes include the hydrometeorological conditions in the upstream watershed, the space available in the reservoir for flood control, the flow conditions, and residual safe river capacities available at the damage centers (Prakash et al., 2015). Reservoir operation for flood control can be modeled through a predefined rule curve or within a real-time framework.

Real-time reservoir operation model may have a variety of OFs and constraints to represent the physical system in the formulation of operational framework. Every period of operation is characterized by a number of constraints. The total number of constraints may run into a few thousands when the larger period of operation is considered. The formulation becomes complex as the storage-elevation relationship and outlet discharge function are nonlinear. In real-time flood control operation, the optimizer has to access the reservoir routing and channel routing modules at every time step of the operation. Further, conditional decisions on release from the reservoir are to be included in the framework. Therefore, the operational framework is a highly constrained nonlinear optimization problem and requires careful solution strategy.

A key objective during flood control operation is minimizing total flood damages which is usually defined as a function of water surface elevations in flood-prone areas or reservoir itself. Here, we have categorized generic formulation of OF in flood control operation by considering five OFs. The first four OFs deal with reductions in total excess volume and flood peak mitigation at the reservoir or the control point or both. The fifth OF deals with minimizing the absolute deviation of the actual reservoir water elevation at the end of the flood horizon from the prefixed target water surface elevation. Typical OFs and constraints considered by researchers are as follows (Prakash et al., 2015):

Objective Functions

Objective 1: Maximize the cumulative volume of water released from the reservoir during the flood OH. This OF aims to make the reservoir as empty as possible thereby ensuring safety of the dam against a succeeding flood.

$$\text{Maximize} \sum_{t=1}^{T_R} O_t - \sum_{t=1}^{T_R} I_t \tag{A.1}$$

where I_t is the reservoir inflow; O_t is the outflow from the reservoir; T_R is the flood OH at the reservoir; and t is the time index varying from 1 to T_R.

Objective 2: Minimize the cumulative volume in excess of the specified channel capacity at the downstream control point during the flood OH. This objective target to minimize the flooding at the control point by storing water in the reservoir.

$$\text{Minimize} \sum_{t=1}^{T_C} [\max(0, Q_{Ct} - CC)] \tag{A.2}$$

where Q_{Ct} is the flow at the control point at time t; CC is the specified channel capacity; T_C is the flood OH at the control point; $T_C \geq T_R$

Objective 3: Maximize the flood peak attenuation at the reservoir. This objective tries to achieve maximum flood mitigation by reducing the peak flood discharge from the reservoir.

$$\text{Maximize } (I_p - O_p) \tag{A.3}$$

where I_p is the peak inflow to the reservoir, and O_p is the peak reservoir outflow.

Objective 4: Maximize flood peak attenuation at the control point. This objective target to reduce the flood peak discharge at the control point thereby minimizing flood damage in the vicinity of the control point.

$$\text{Maximize } (I_p - Q_{Cp}) \tag{A.4}$$

where Q_{Cp} is the peak flow at the control point.

Objective 5: Minimize the deviation of the reservoir water surface elevation at the end of flood OH from the desired target water surface elevation at the reservoir. This manager may set low target reservoir water level at the end of operation to accommodate any succeeding flood event or high to conserve water for future use.

$$\text{Minimize } (|E_{\text{targ}} - E_{TR}|) \tag{A.5}$$

where E_{targ} is the target reservoir water level; E_{TR} is the actual water level at the end of the OH (T_R).

Constraints

The constraints arising in the above formulation are storage continuity for the reservoir, upper limit for release from outlets, upper limit on rate of change of releases through spillway and bottom outlet, upper and lower limits of the reservoir storage at specified time, and the channel routing equation.

$$S_t = S_{t-1} + \left(\frac{I_{t-1} + I_t}{2}\Delta t\right) - \left(\frac{O_{t-1} + O_t}{2}\Delta t\right) \tag{A.6}$$

$O_t = f_1(H_t)$, where $H_t = f_2(S_t)$

$$R_t^b \leq R_{t\,\max}^b \tag{A.7}$$

$$O_t = R_t^b + W_t \tag{A.8}$$

If $R_t^b \geq R_{t-1}^b$ then,

$$R_t^b - R_{t-1}^b \leq RI \tag{A.9}$$

If $R_t^b \leq R_{t-1}^b$ then,

$$R_{t-1}^b - R_t^b \leq RD \tag{A.10}$$

$$S_{\min} \leq S_t \leq S_{\max} \tag{A.11}$$

$$Q_{Ct} = C_0 O_t + C_1 O_{t-1} + C_2 O_{t-1} + LF_t \tag{A.12}$$

where S_t is the reservoir storage; Δt is the routing time interval; W_t is the uncontrolled spillway discharge from the reservoir; R_t^b is the release through the bottom sluice; $R_{t\,\max}^b$ is the maximum release possible through the bottom sluice; RI is the upper limit of hourly rate of increase in the release from the bottom outlet; RD is the upper limit of the hourly rate of decrease in the release from the bottom outlet; $S_{\min}$ is the inactive storage; $S_{\max}$ maximum reservoir storage; Q_{Ct} is the control point flow; LF_t is the local flow contribution at the control point at time t; C_0, C_1, C_2 are the routing coefficients; and Q_t, Q_{t-1}, Q_{t-1} are the reservoir outflows at time $t, t-1$ and $t-2$, respectively.

Not all OFs would be activated simultaneously during optimization; a combination of one or more of these OFs can be used depending on the physical state of the river–reservoir system, as the flood passes (Chou and Wu, 2015). Out of the five OFs defined above, the variants of Eqs. (A.1) and (A.2), which avoid dam overtopping without exceeding the safe carrying capacity of the channel at the downstream control point(s) are found to be employed widely and are defined as in Eqs. (A.13) and (A.14).

$$\text{Min}\, f_1(Q) = \min\{\max(Z_t)\}; \quad t = 1, 2, \ldots, T_R \tag{A.13}$$

where Z_t is the water level in the reservoir for tth time period.

$$\text{Min}\, f_2(Q) = \min\{\max(Q_t)\} \tag{A.14}$$

where Q_t is the reservoir flood release for tth time period.

It is also to be noted here that the above OFs (Eqs. A.1–A.5) are developed with an assumption that there is prior knowledge of the entire flood hydrograph. The equations, hence, would require suitable modification to ensure adaptation in a real-time scenario. In the real-time operation of a river-reservoir system, where the main objective is to control the flood flows and flood elevation at various locations, Eqs. (A.2) and (A.4) could be modified as (Che and Mays, 2017) in Eq. (A.15).

$$\text{Minimize}\left(\sum_{i=1}^{N}\sum_{t=1}^{T_C}[C_i Q_{i,t}]\right) \tag{A.15}$$

where $Q_{i,t}$ is the time series of flow rate vectors of the control points, $i, (1 \leq i \leq N)$, in the river-reservoir system, and C_i is the penalty coefficient at the control points.

In addition to the operational constraints defined in Eqs. (A.6)–(A.11), in a real-time simulation optimization framework, the hydrologic constraints are also to be defined which depend on the rainfall-runoff relationship of the watershed systems. The hydrologic constraint defined in Che and Mays (2017) is as given in Eq. (A.16).

$$h(P_{i,t}, H_{i,t}, Q_{i,t}) = 0 \quad \text{(A.16)}$$

where $P_{i,t}$ is the matrix of precipitation data of the watershed, $H_{i,t}$ denotes the rainfall losses in the system, and $Q_{i,t}$ is the discharge matrix of the system. While using a systems analysis tool like the HEC-HMS for the rainfall-runoff transformations, the constraint is solved implicitly within the model for each time step t.

Real-time flood operation differs from normal operation of reservoirs in the sense that for flood operation, hydraulic constraints are also used in addition to operation constraints. The hydraulic constraints are the equality constraints consisting of the equations that describe the flow in the system. Hydraulic constraints are defined by the Saint–Venant equations for one-dimensional gradually varied unsteady flow and other relationships such as upstream, downstream, and internal boundary conditions and initial conditions that describe the flow in the different components of a river reservoir system.

In case of one-dimensional gradually varying unsteady flow simulation of the downstream channel, the hydrologic routing constraint in Eq. (A.12) has to be replaced by hydraulic constraints defined by the *Saint–Venant* equations (Che and Mays, 2017) which could be generally represented as Eq. (A.17).

$$g(h_{i,t}, Q_{i,t}) = 0 \quad \text{(A.17)}$$

where $h_{i,t}$ is the water elevation in the system. The constraint could be implicitly solved within a river analysis tool, for example, the HEC-RAS.

APPENDIX B

Monte Carlo Simulation for Reservoir Operation

Reliability concepts are commonly applied to the analysis and design of water resources systems. Three statistical indices, *reliability, resilience*, and *vulnerability,* discussed in Chapter 8 have found numerous applications, for example, Vogel et al. (1999) and Jain and Bhunya (2008). Use of long data series to compute these indices is recommended. If the length of hydrologic data series is inadequate, Monte Carlo Simulations (MCS) (see Chapter 5 and Singh et al., 2007), can be used to compute performance indices.

In MCS, random values of each stochastic variable are generated according to their respective probability distributions and the operation of the system is simulated. By repeating this process a large number of times, the statistical properties and the probability distribution of the model output can be determined. Techniques to generate random variables that follow a variety of probability distributions are described by Singh et al. (2007). The accuracy of MCS results depends on the number of simulations performed. For models with a large number of parameters and for which events with low probabilities (<0.1) of occurrence are of interest, thousands of simulations are required. MCS is a flexible and robust method for a variety of problems, particularly when computing power is not a limitation.

Here, we describe the use of the MCS in a real-life problem using the data of Dharoi reservoir, India. Statistical parameters of the observed inflows at Dharoi were used to generate synthetic sequences of monthly flows by using short-memory (SM) and long-memory (LM) models. SM implies that the impact of a particular value of the time series on the future values becomes negligible after a short time. LM implies that such influence on future values is felt for a long time. In this study, the monthly Thomas–Fiering model (Bras and Rodriguez-Iturbe, 1985) was employed to generate flows with short memory.

LM or persistence in hydrological time series is manifested by the tendency of wet years (when flows are larger than the mean) or dry years (when flows are smaller than the mean) to cluster together. Named after the engineer who discovered this phenomenon, the *Hurst phenomenon* is displayed by the data of a wide variety of natural processes such as river flows, precipitation, and width of tree rings. For a standard normal independent series, H is about 0.5 but Hurst found that the average value of H for time series of geophysical processes was 0.73. The value of the Hurst exponent over 0.5 is a measure of memory in a time series.

To generate a flow series by following the LM concept, the value of Hurst exponent H of the observed series is computed. For the inflows to the Dharoi reservoir, the Hurst exponent H turned out to be 0.7445. The fractional Gaussian noise (fGn) model was used to reproduce the Hurst phenomenon. The correlation function, ρ for the fGn processes is expressed as

$$\rho(j,H) = 0.5\left[|j+1|^{2H} - 2|j|^{2H} + |j-1|^{2H}\right] \cong H(2H-1)j \tag{B.1}$$

where j is the lag and H is the Hurst exponent.

Annual flows generated by a symmetric moving average approach (Koutsoyiannis, 2002) were disaggregated into monthly flows by following a linear disaggregation model. Five hundred sequences, each of 200-year length representing inflows to the Dharoi reservoir, were generated, and reservoir operation was simulated using three rule curves. For each sequence, several performance indices were computed, viz., time and volume reliability for municipal and industrial (M&I) water supply and time and volume reliability for irrigation demands.

Results of Monte Carlo Simulations

Table 2 summarizes the results of the various runs using the data generated by LM and SM models. Fig. 11.2 shows the time and volume reliabilities for M&I and irrigation demands for LM and SM models. In the case of the LM runs, the time and volume reliabilities for M & I demands are very high, the mean being 0.904 and 0.914 and the maximum is very close to unity in both cases. There are large fluctuations in these reliabilities across the various MCS runs as evidenced by a large coefficient of variation (CV) of 0.09 and 0.08 for time and volume reliabilities, respectively.

In the case of SM runs, the average time and volume reliabilities are 0.987 and 0.991, respectively, with corresponding CV equal to 0.006 and 0.005. On comparing the results of LM and SM, it can be observed that the variation in reliabilities in the case of the LM model is higher. Further, LM gives a mean resilience of 0.23 while the same for SM is higher at 0.43. The average vulnerability in LM is 1.99, which is a little higher than 1.63 for SM.

Table B.1 Summary of the results of runs using data generated by LM and SM models.

Inflow generation process	Statistics	M&I demands			
		Time Rel.	Vol. Rel.	Resilience	Vulnerability
Long-memory model (LM)	Max	0.9990	0.9990	0.6000	2.4280
	Min	0.4060	0.4400	0.1360	0.8570
	Mean	0.9042	0.9144	0.2279	1.9941
	SD	0.0831	0.0773	0.0589	0.1456
	CV	0.0919	0.0846	0.2585	0.0730
Short-memory model (SM)	Max	0.9990	1.0000	1.0010	2.2630
	Min	0.9690	0.9740	0.2220	0.5750
	Mean	0.9866	0.9909	0.4326	1.6284
	SD	0.0057	0.0048	0.1083	0.2948
	CV	0.0058	0.0048	0.2504	0.1810

Similar analysis was completed for irrigation demands, and resilience and vulnerability were computed. One of the conclusions is that the variations of the indices using LM runs were higher than the SM runs. This was observed for M&I water supply and irrigation demands. More details of the study can be seen in Jain and Bhunya (2008).

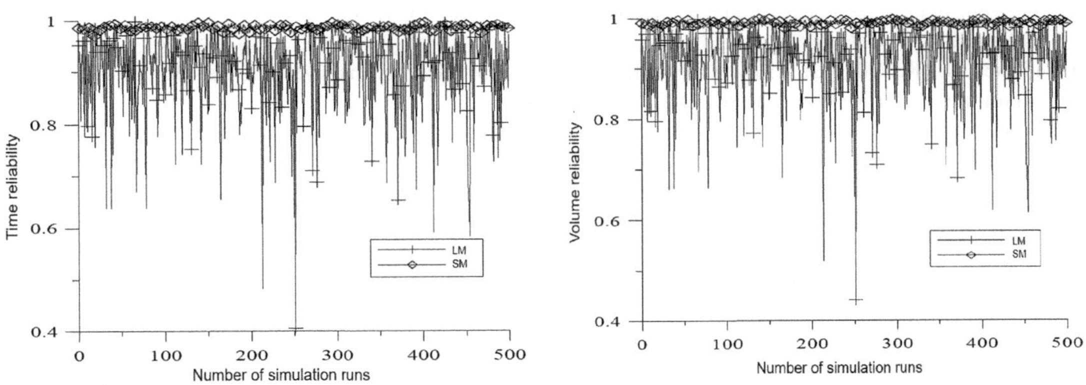

FIGURE B.1

Time and volume reliabilities in the case of LM and SM using MCS.

References

Bhaskar, N.R., Whitlach, E.E., 1980. Derivation of monthly reservoir release policies. Water Resources Research 16 (6), 987–993. https://doi.org/10.1029/WR016i006p00987.

Bianucci, P., Sordo-Ward, Á., Moralo, J., Garrote, L., 2015. Probabilistic-multiobjective comparison of user-defined operating rules. Case study: hydropower dam in Spain. Water 7 (3), 956–974.

Biswas, A.K., 1991. Water for sustainable development in the 21st century: a global perspective. Water International 6 (4), 219–224.

Butcher, W., 1971. Stochastic dynamic programming for optimum reservoir operation. Water Resources Bulletin 7 (1), 115–123.

Bras, R.L., Rodriguez-Iturbe, I., 1985. Random Functions in Hydrology. Dover Publications, New York, USA.

Che, D., Mays, L.W., 2015. Development of an optimization/simulation model for real-time flood-control operation of river-reservoirs systems. Water Resources Management 29 (11), 3987–4005.

Che, D., Mays, L., 2017. Application of an optimization/simulation model for real-time flood-control operation of river-reservoirs systems. Water Resources Management 31 (7).

Chou, F.N.F., Wu, C.W., 2015. Stage-wise optimizing operating rules for flood control in a multi-purpose reservoir. Journal of Hydrology 521, 245–260.

Connaughton, J., King, N., Dong, L., Ji, P., Lund, J., 2014. Comparing simple flood reservoir operation rules. Water 6 (9), 2717–2731.

Datta, B., Burges, S.J., 1984. Short-term single, multiple purpose reservoir operation. Water Resources Research 20 (9), 1167–1176.

Feldman, A.D., 1981. HEC models for water resources system simulation: theory and experience. In: Chow, V.T. (Ed.), Advances in Hydroscience, vol. 12. Academic Press, New York.

Fontane, D., Labadie, J.W., Loftis, B., 1981. Optimal control of reservoir discharge quality through selective withdrawal. Water Resources Research 17 (6), 1594–1604.

Goel, M.K., Jain, S.K., Rani, D., Subrahmanyam, G.V., Visweswararao, M., 2020. Developing operation procedures for individual reservoirs in a large multistate River Basin in context of tribunal awards. Journal of Water Resources Planning and Management 146 (8), 05020013.

Gong, Y., Liu, P., Zhang, J., Liu, D., Zhang, X., Zhang, X., 2020. Considering different streamflow forecast horizons in the quantitative flood risk analysis for a multi-reservoir system. Reliability Engineering & System Safety 204, 107128.

Gundelach, J., ReVelle, C., 1975. Linear decision rules in reservoir management and design 5:A general algorithm. Water Resources Research 11 (2), 204–207.

Hall, W.A., Dracup, J.A., 1970. Water Resources System Engineering. McGraw-Hill Book Company, New York.

Hua, L., Wan, X., Wang, X., Zhao, F., Zhong, P.A., Liu, M., Yang, Q., 2020. Floodwater utilization based on reservoir pre-release strategy considering the worst-case scenario. Water 12 (3), 892.

Hui, R., Lund, J.R., 2015. Optimal flood pre-releases—flood hedging for a single reservoir. In: World Environmental and Water Resources Congress 2015, pp. 2192–2201.

Hui, R., Lund, J., Zhao, J., Zhao, T., 2016. Optimal pre-storm flood hedging releases for a single reservoir. Water Resources Management 30 (14), 5113–5129.

Jain, S.K., Goel, M.K., 1997. Software for Reservoir Analysis, Report No. UM-1/96-97. National Institute of Hydrology, Roorkee.

Jain, S.K., Goel, M.K., 1999. Flood Control Regulation of a Multi-Reservoir System. National Institute of Hydrology, Roorkee.

Jain, S.K., Yoganarasimhan, G.N., Seth, S.M., 1992. A risk-based approach for flood control operation of a multipurpose reservoir. JAWRA: Journal of the American Water Resources Association 28 (6), 1037–1043.

Jain, S.K., Goel, M.K., Agarwal, P.K., 1998. Reservoir operation studies of Sabarmati system, India. Journal of Water Resources Planning and Management 124 (1), 31–37.

Jain, S.K., Goel, M.K., Agarwal, P.K., 1998. Reservoir operation studies for Sabarmati system, India. Journal of Water Resources Planning and Management, ASCE 124 (1), 31–38.

Jain, S.K., Das, A., Srivastava, D.K., 1999. Application of ANN for reservoir inflow prediction and operation. Journal of Water Resources Planning and Management, ASCE 125 (5), 263–271.

Jain, S.K., Bhunya, P.K., 2008. Reliability, resilience and vulnerability of a multipurpose storage reservoir. Hydrological Sciences Journal 53 (2), 434–447.

Jain, S.K., Mani, P., Jain, S.K., Prakash, P., Singh, V.P., Tullos, D., Dimri, A.P., 2018. A brief review of flood forecasting techniques and their applications. International Journal of River Basin Management 16 (3), 329–344.

Jain, S.K., Shilpa, L.S., Natyan, D., Sudheer, K.P., 2023. State-of-the-art review: operation of multi-purpose reservoirs during flood season. Journal of Hydrology 618 (2023), 129165.

Jairaj, P.G., Vedula, S., 2000. Multireservoir system optimization using fuzzy mathematical programming. Water Resources Management 14, 457–472.

James, L.D., Lee, R.R., 1971. Economics of Water Resources Planning. McGraw-Hill Book Co., New York.

Jiang, Z., Sun, P., Ji, C., Zhou, J., 2015. Credibility theory based dynamic control bound optimization for reservoir flood limited water level. Journal of Hydrology 529, 928–939.

Joeres, E.F., Seus, G.J., Engelmann, H.M., 1981. The linear decision rules (LDR) reservoir problem with correlated inflows 1: model development. Water Resources Research 17 (1), 18–24.

Johnson, S.A., Stedinger, J.R., Staschus, K., 1991. Heuristic operating policies for reservoir system simulation. Water Resources Research 27 (6), 673–685.

Karamouz, M., Houck, M.H., 1987. Comparison of stochastic and deterministic dynamic programming for reservoir operating rule generation. Water Resources Bulletin 23 (1), 1–9.

Komori, D., Nakamura, S., Kiguchi, M., Nishijima, A., Yamazaki, D., Suzuki, S., Kawasaki, A., Oki, K., Oki, T., 2012. Characteristics of the 2011 Chao Phraya river flood in Central Thailand. Hydrological Research Letters 6 (2012). https://doi.org/10.3178/hrl.6.41.

Kuiper, E., 1965. Water Resources Development. Butterworths and Company Ltd., London.

Koutsoyiannis, D., 2002. The Hurst phenomenon and fractional Gaussian noise made easy. Hydrological Sciences Journal 47 (4), 573–596.

Labadie, J.W., 2004. Optimal operation of multireservoir systems: state-of-the-art review. Journal of Water Resources Planning and Management 130 (2), 93–111.

Labadie, J.W., 2006. MODSIM: Decision Support System for Integrated River Basin Management. Diss. International Environmental Modeling and Software Society.

LeNgo, L., Madsen, H., Rosbjerg, D., 2007. Simulation and optimisation modelling approach for operation of the Hoa Binh reservoir, Vietnam. Journal of Hydrology 336 (3–4), 269–281.

Li, X., Guo, S., Liu, P., Chen, G., 2010. Dynamic control of flood limited water level for reservoir operation by considering inflow uncertainty. Journal of Hydrology 391 (1–2), 124–132.

Loucks, D.P., Dorfman, P., 1975. An evaluation of some linear decision rules in chance constrained models for reservoir planning and operation. Water Resources Research 11 (6), 777–782.

Loucks, D.P., Eelco, van B., 2017. Water Resource Systems Planning and Management: An Introduction to Methods, Models, and Applications. Jointly Published by Deltares. UNESCO-IHE, and Springer Nature, Switzerland.

Loucks, D.P., Sigvaldason, O.T., 1980. Operations research in multiple reservoir operation. In: Yaron, D., Tapiero, C. (Eds.), Operations Research in Agriculture and Water Resources. North Holland Publishing Company.

Loucks, D.P., Stedinger, J.R., Haith, D.A., 1981. Water Resources Systems Planning and Analysis. Prentice Hall Inc., New Jersey.

Loucks, D.P., Salewicz, K.A., Taylor, M.R., 1989. IRIS: An Interactive River System Simulation Model, General Introduction and Description. International Institute for Applied Systems Analysis, Laxenburg, Austria.

Lund, J.R., Ferreira, I., 1996. Operating rule optimization for the Missouri River reservoir system. Journal of Water Resources Planning and Management, ASCE 122 (4), 287–295.

Lund, J.R., Guzman, J., 1999. Derived operating rules for reservoirs in series or in parallel. Journal of Water Resources Planning and Management, ASCE 125 (3), 143–153.

Maass, A., Hufschmidt, M.M., Dorfman, R., Thomas, H.A., Marglin, S.A., Fair, G.M., 1962. Design of Water Resource Systems. Harvard University Press, USA.

Martin, Q.W., 1982. Multivariate Simulation and Optimization Model (SIM-V), Program Documentation and User's Manual. UM-38. Texas Department of Water Resources, Austin, USA.

Mosavi, A., Ozturk, P., Chau, K.-wing, 2018. Flood prediction using machine learning models: literature review. Water 10, 1536. https://doi.org/10.3390/w10111536.

Nagy, I.V., Asante-Duah, K., Zsuffa, I., 2002. Hydrological Dimensioning and Operation of Reservoirs. Water Science and Technology Library, Kluwer Academic Publishers, Dordrecht.

Pan, Z., Chen, L., Teng, X., 2020. Research on joint flood control operation rule of parallel reservoir group based on aggregation–decomposition method. Journal of Hydrology 590, 125479.

Pappenberger, F., Cloke, H.L., Parker, D.J., Wetterhall, F., Richardson, D.S., Thielen, J., 2015. The monetary benefit of early flood warnings in Europe. Environmental Science & Policy 51, 278–291.

Ponce, V.M., 1989. Engineering Hydrology: Problems and Practices. Prentice-Hall, Inc., Englewood Cliffs, NJ.

Prakash, O., Srinivasan, K., Sudheer, K.P., 2015. Adaptive multi-objective simulation–optimization framework for dynamic flood control operation in a river–reservoir system. Hydrology Research 46 (6), 893–911.

Rani, D., Moreira, M.M., 2010. Simulation–optimization modeling: a survey and potential application in reservoir systems operation. Water Resources Management 24 (6), 1107–1138.

ReVelle, C., 1999. Optimizing Reservoir Resources, Including a New Model for Reservoir Reliability. John Wiley and Sons, New York.

ReVelle, C., Gundelach, J., 1975. Linear decision rule in reservoir management and design, 4. A rule that minimizes output variance. Water Resources Research 11 (2), 197–203.

ReVelle, C., Joers, E., Kirby, W., 1969. The linear decision rule in reservoir management and design, 1. Development of the stochastic model. Water Resources Research 5 (4), 767–777.

Şensoy, A., Uysal, G., Şorman, A.A., 2018. Developing a decision support framework for real-time flood management using integrated models. Journal of Flood Risk Management 11, S866–S883.

Sheer, D., 1986. Dan Sheer's Reservoir Operating Guidelines. School of Civil and Environmental Engineering, Cornell University, Ithaca, N.J.

Sigvaldason, O.T., 1976. A simulation model for operating a multi-purpose multi-reservoir system. Water Resources Research 12 (2), 263–278.

Simonovic, S.P., 1992. Reservoir system analysis: closing gap between theory and practice. Journal of Water Resources Planning and Management, ASCE 118 (3), 262–280.

Sudheer, K.P., Bhallamudi, S.M., Narasimhan, B., Thomas, J., Bindhu, V.M., Vema, V., Kurian, C., 2019. Role of dams on the floods of Aug 2018 in Periyar River Basin, Kerala. Current Science 116 (5), 00113891.

Singh, V.P., Jain, S.K., Tyagi, A., 2007. Risk and Reliability Analysis: A Handbook for Civil and Environmental Engineers. American Soc. of Civil Engineers, USA.

USACE, 1985. Engineering Manual: Engineering and Design, Hydropower. (1985). EM 1110-2-1701. U.S. Army Corps of Engineers, Washington, D.C.

USACE, 1987. SSARR Users Manual: Streamflow Synthesis and Reservoir Regulation. US Army Corps of Engineers, Portland, Oregon, USA.

Vogel, R.M., Lane, M., Ravindiran, R.S., Kirshen, P., 1999. Storage reservoir behavior in the United States. J. Water Resour. Plan. Manage. ASCE 125 (5), 245–254.

Wan, W., Zhao, J., Lund, J.R., Zhao, T., Lei, X., Wang, H., 2016. Optimal hedging rule for reservoir refill. Journal of Water Resources Planning and Management 142 (11), 04016051.

Wang, Q., Zhou, J., Dai, L., Huang, K., Zha, G., 2021. Risk assessment of multireservoir joint flood control system under multiple uncertainties. Journal of Flood Risk Management 14 (4), e12740.

Windsor, J.S., 1975. A programming model for the design of multireservoir flood control systems. Water Resources Research 10 (1), 30–35.
Wu, W., Emerton, R., Duan, Q., Wood, A.W., Wetterhall, F., Robertson, D.E., 2020. Ensemble flood forecasting: current status and future opportunities. WIREs Water 2020 (7), e1432.
Wurbs, R.A., 1993. Reservoir system simulation and optimization models. Journal of Water Resources Planning and Management, ASCE 119 (4), 455–472.
Wurbs, R.A., 1996. Modeling and Analysis of Reservoir System Operations. Prentice Hall, New York.
Wurbs, R.A., 2005. Comparative evaluation of generalized river/reservoir system models. Technical Report No. 282. Texas Water Resources Institute, The Texas A& M University System, College Station, TX.
Wurbs, R.A., Cabezas, L.M., 1987. Analysis of reservoir storage reallocations. Journal of Hydrology 92, 77–95.
Wurbs, R.A., Dunn, D.D., Walls, W.B., 1993. Water Rights Analysis Program (TAMU WRAP), Model Description and Users Manual, Tech. Rep. 146. Texas Water Resources Institute, College Station, Texas, USA.
Yakowitz, S., 1982. Dynamic programming applications in water resources. Water Resources Research 18 (4), 673–696.
Yang, S., Yang, D., Chen, J., Zhao, B., 2019. Real-time reservoir operation using recurrent neural networks and inflow forecast from a distributed hydrological model. Journal of Hydrology 579, 124229.
Yeh, W.W.-G., 1985. Reservoir management and optimization models: a state-of-the-art review. Water Resources Research 21 (12), 1797–1818.
Young, G.K., 1967. Finding reservoir operating rules. Journal of the Hydraulics Division, ASCE 93 (HY6), 297–321.
Yun, R., Singh, V.P., 2008. Multiple duration limited water level and dynamic limited water level for flood control, with implications on water supply. Journal of Hydrology 354 (1–4), 160–170.
Zadeh, L.A., 1965. Fuzzy sets. Information and Control 8, 338–353.
Zagona, E.A., Fulp, T.J., Shane, R., Magee, T., Goranflo, H.M., 2001. Riverware: a generalized tool for complex reservoir system modeling 1. JAWRA Journal of the American Water Resources Association 37 (4), 913–929.
Zhang, W., Liu, P., Chen, X., Wang, L., Ai, X., Feng, M., Liu, Y., 2016. Optimal operation of multi-reservoir systems considering time-lags of flood routing. Water Resources Management 30 (2), 523–540.
Zhao, W., Kinouchi, T., Ang, R., Zhuang, Q., 2022. A framework for quantifying climate-informed heavy rainfall change: implications for adaptation strategies. Science of the Total Environment 835, 155553.
Zhou, C., Sun, N., Chen, L., Ding, Y., Zhou, J., Zha, G., Yang, X., 2018. Optimal operation of cascade reservoirs for flood control of multiple areas downstream: a case study in the upper Yangtze river basin. Water 10 (9), 1250.

Further Reading

Bhasker, N.R., Whitlatch, E.E., 1987. Comparision of reservoir linear operation rules using linear and dynamic programming. Water Resources Bulletin 23 (6), 1027–1036.
Biswas, A.K. (Ed.), 1976. Systems Approach to Water Management. McGraw Hill Book Company, New York.
Eastman, J., ReVelle, C., 1973. Linear decision rule in reservoir management and design, 3. Direct capacity determination and intraseasonal constraints. Water Resources Research 9 (1), 29–43.
Goulter, I.C., Tai, F.K., 1985. Practical implications in the use of stochastic dynamic programming for reservoir operation. Water Resources Bulletin 21 (1), 65–74.
HEC, 1976. Flood Control by Reservoirs. Hydrologic Engineering Methods for Water Resources Development, vol. 7. Hydrologic Engineering Center, Davis, USA.
HEC, 1982. HEC - 5 Simulation of Flood Control and Conservation System. Program Users Manual, Hydrologic Engineering Center, Davis, USA.
Kite, G.W., 1977. Frequency and Risk Analysis in Hydrology. Water Resources Publications, USA.

Lettenmaier, D.P., Wood, E.F., 1993. Hydrological forecasting. In: Maidment, D.R. (Ed.), Handbook of Hydrology. McGraw-Hill Book Company, New York.

Loucks, D.P., Sigvaldason, O.T., 1982. Multiple-reservoir operation in North America. In: Kaczmarek, Z., Kindler, J. (Eds.), The Operation of Multiple Reservoir Systems. International Institute for Applied Systems Analysis, Laxenburg, Austria.

Marien, J.L., Damazio, J.M., Costa, F.S., 1994. Building flood control rule curves for multipurpose multireservoir systems using controllability conditions. Water Resources Research 30 (4), 1135–1144.

McMahon, T.A., Mein, R.G., 1986. River and Reservoir Yield. Water Resources Publications, Colorado, USA.

Oliveira, R., Loucks, D.P., 1997. Operating rules for multireservoir systems. Water Resources Research 33 (4), 839–852.

ReVelle, C., Kirby, W., 1970. Linear decision rule in reservoir management and design, 2. Performance optimization. Water Resources Research 6 (4), 1033–1044.

Stedinger, J.R., Sule, B.F., Pei, D., 1983. Multiple reservoir system screening models. Water Resources Research 19 (6), 1383–1393.

Unny, T.E., McBean, E.A. (Eds.), 1982. Decision Making for Hydrosystems: Forecasting and Operation. Water Resources Publications, Littletown, USA.

Yeh, W.W.-G., Becker, L., 1982. Multiobjective analysis of multireservoir operation. Water Resources Research 18 (5), 1326–1336.

CHAPTER

Reservoir Sedimentation

12

Chapter outline

Water Resources Systems Planning and Management. https://doi.org/10.1016/B978-0-12-821349-0.00016-2

The objectives of this chapter are:

- to explain the problem of reservoir sedimentation and its consequences,
- to explain methods and models of estimation of sediment yield of watersheds,
- to explain the conventional and advanced techniques of reservoir surveys, and
- to explain methods of managing reservoir sedimentation.

Soil erosion is the detachment and transportation of the soil. It is a universal and natural phenomenon. The most important agents that cause soil erosion are precipitation, runoff, wind, and activities of living beings. The eroded soil is carried into water courses by flood and storm waters resulting in tremendous sediment movement. Every stream carries some sediments in suspension and moves larger particles as bed load down to reservoirs, lakes, estuaries, bays, and oceans. The impact of sediment erosion, transport, and deposition is widespread.

Uncontrolled deforestation, forest fires, grazing, improper method of tillage, and unwise agricultural and land use practices accelerate soil erosion resulting in a large increase of sediment inflow into streams. The deposition of sediment in channels or reservoirs creates a variety of problems, such as raising of stream beds, increasing flood heights, choking of navigation channels, and, of course, depletion of capacity in storage reservoirs.

The sediment content in rivers varies from month to month. While it is negligible in winter and summer months, it attains the maximum in the flood season. Some rivers indeed carry very heavy sediment load. According to Alam (2001), the total quantity of sediment transported annually to the sea by rivers of the world is about 2×10^{10} tons or about 13.5 km^3 in terms of volume. Assuming that all this sediment enters into the reservoirs of the world, it would take about 481 years to fill up the estimated 6500 km^3 of the storage volume available. However, the sediment source and reservoir locations are not uniformly distributed. Experts fear that the loss could be even higher and faster if the forecasts on climate change prove to be true and the rates of deforestation in many parts of the world are not controlled. The severity of storms and rains is likely to increase as a result of global warming and this may accelerate the natural erosion rates in catchments. Global warming is also likely to exaggerate the extremes in precipitation patterns, thereby making it even more vital that the available storage capacity of reservoirs is maintained. The rates of erosion from hillsides, planted with crops, are about 100 times higher than from the same land covered with trees.

Suspended sediment load of selected rivers of the world is given in Table 12.1.

The sedimentation problem is quite severe in some countries. China has more than 80,000 reservoirs that annually lose about 2.3% of storage capacity due to sedimentation. The sediment transport features for some important rivers are given in Table 12.2.

Soil erosion is considerably high in arid climates. In India, about 5333 million tonnes (16.35 t/ha) of soil is detached annually due to agriculture and associated activities, and about 29% of this is carried away by rivers into the sea. Nearly 10% of it is deposited in reservoirs resulting in loss of 1%–2% of the storage capacity (Dhruva Narayana, 1995). While the rivers in the Indian peninsula, such as the Krishna and the Godavari, carry about 100 ppm (parts per million), the silt carried by the Ganga often exceeds 2000 ppm. In another North Indian river, Kosi, the silt content is much larger, being 3310 ppm. Garde and Kothyari (1986) have studied sediment yield estimation and have presented the average erosion rates for large river catchments in India (see Table 12.3).

Each sediment particle being transported by flow is affected by two dynamic forces: a horizontal component acting in the direction of flow and a vertical component due to gravity; there is also a force of water turbulence. Since the specific gravity of soil materials is about 2.65, the particles of suspended

Table 12.1 Sediment load of selected rivers.

S. no	Rivers and country	Average annual suspended sediment load	
		10^6 tons	Tons/sq. km
1.	Yellow, China	2080	2910
2.	Ganges, India	1600	620
3.	Brahmaputra, Bangladesh	800	309
4.	Yangtze, China	550	212
5.	Indus, Pakistan	480	185
6.	Amazon, Brazil	400	154
7.	Mississippi, United States	344	132
8.	Irrawaddy, Burma	330	127
9.	Missouri, United States	240	92
10.	Kosi, India	190	73

Adapted from Holeman (1968).

Table 12.2 Sediment transport features of selected rivers.

River	Drainage area (km^2)	Average sediment concentration (kg/m^3)	Erosion modulus ($t/km^2/yr$)
World rivers			
Nile	2,978,000	1.25	37
Missouri	1,370,000	3.54	159
Colorado	637,000	27.5	212
Indus	969,000	2.49	449
Irrawaddy	430,000	0.70	695
Brahmaputra	666,000	1.89	1090
Red	119,000	1.06	1092
Ganges	955,000	3.92	1519
Rivers in China			
Huaihe	261,500	0.46	153
Liaohe	166,300	6.86	240
Pearl	355,000	0.35	260
Yangtze	1,807,200	0.54	280
Haihe	50,800	60.8	1944
Yellow	752,400	37.6	2480

Adapted from Qiang and Dai (1980).

Table 12.3 Average erosion rates for large river catchments in India.

River	No. of points at which erosion rates were considered	Catchment area in 10^4 km^2	Erosion rate ton/km^2-yr estimated by Garde and Kothyari (1986)
Ganga	23	86.15	1969.0
Indus	4	32.13	1942.5
Brahmaputra	5	18.71	1891.0
Mahanadi	11	14.16	1287.0
Sabarmati	7	2.17	1277.0
Cauvery	5	8.79	1214.0
Krishna	7	25.0	1191.0
Godavari	10	31.28	954.0
Tapi	10	6.69	935.0
Narmada	10	9.88	906.0
Mahi	3	3.76	820.0
Luni	1	0.001	250.0

sediment tend to settle at the channel bottom, but upward currents in the turbulent flow counteract the gravitational settling. The sediment inflow and outflow in the natural river reaches is mostly in balance.

A dam on a stream channel changes the hydraulic characteristics of flow and its sediment transport capacity. As the reservoir width is much bigger than the river channel width, the velocity of flow entering into it decreases tremendously. At the same time, there is a dampening of water turbulence. All these factors contribute to make the flow unable to transport all the sediment particles and the particles begin to deposit. First, the larger suspended particles and most of the bed load is deposited at the mouth of the reservoir. The smaller particles remain in suspension for a long time and some may pass the dam with water discharged through sluices, turbines, or the spillways. The deposition of coarse sediments reduces the reservoir storage and channel conveyance for water supply, irrigation, and navigation and causes extensive disturbance to streams. Suspended sediments reduce the water clarity and sunlight penetration, thereby affecting the biotic life. The settlement of sediments at the bottom of water bodies buries and kills the vegetation and changes the ecosystem.

12.1 RESERVOIR SEDIMENTATION

The accumulation of sediments is one of the principal factors that threaten the longevity of river valley projects. In fact, sometimes a project is not constructed just because the silting rate is so high that the reservoir will fill up before the investment is fully recovered. The problems of concern for planners are the rapidity of reservoir sediment deposition and the time that will elapse before the use of the reservoir storage capacity is seriously impaired. In general, a dam designer needs to determine:

(a) the volume of sediments that will accumulate in the reservoir each year,
(b) the distribution of sediments in the reservoir,
(c) the aggradation above a reservoir, and
(d) the reservoir trap efficiency.

The ultimate destiny of all reservoirs is to be filled with sediments. Reservoir planning must include consideration of the probable rate of sedimentation to determine whether the useful life of the proposed reservoir will be sufficient to warrant its construction. If the sediment inflow is large compared to the reservoir capacity, special care is needed in design and operation, otherwise the useful life of the reservoir may be short. There are instances of reservoirs being filled-up within a few years of their operation. A small water-supply reservoir in the United States of America was filled with sediment during the first year after its completion. Morris and Fan (1998) quote many interesting examples. The Sanmenxia dam, constructed during 1957–60, was the first major dam on the middle reaches of the Yellow River. In the first 18 months after the dam closure, 1.8 billion metric tons of sediment accumulated in the reservoir, representing a trap efficiency of 93%. The sediment deposits were also found to raise the bed elevations and flood levels in the Yellow River as far as 260 km upstream of the dam. The Xinghe reservoir in the Shaanxi province took 2 years to construct but only 1 year to fill with sediment. The 21.8 Mm^3 Laoying reservoir in Shaanxi province silted up even before the irrigation canal was completed. The 76 m high Warsak dam on the Kabul River in Pakistan lost 18% of its storage volume in the very first year of operation (Nagy et al., 2002).

The right approach to solve reservoir sedimentation problem has the following three components: (a) collection and analysis of field data, (b) setting up of appropriate models, and (c) development of an operational policy for the reservoir. When different operation modes are adopted for a reservoir, deposition and scour may differ considerably. To foresee what changes are likely to occur so that remedial measures could be taken as early as possible, a reliable prediction is needed before the decision is made.

The sediment deposits in a reservoir can be divided into three groups: topset beds, foreset beds, and bottomset beds as shown in Fig. 12.1. The topset beds are composed of large size sediment deposits, but one may also find fine particles. These extend up to the point where the backwater curve ends. The downstream limit of the topset bed corresponds to the downstream limit of the bed material transport in the reservoir. These deposits cause a minor reduction in the reservoir storage capacity. Foreset deposits represent the face of the delta deposit advancing in the reservoir toward the dam. It is a transition zone having steeper slopes and decreasing grain size. The bottomset beds consist of fine sediments which are deposited beyond the delta by turbidity currents or nonstratified flow. Note that this particle distribution may change due to reservoir drawdown, slope failures, and extreme floods. In a reservoir with significant water level fluctuations, the nature of deposition will depend on these fluctuations because these can move the topset and foreset beds further downstream fairly quickly. During a major earthquake, sediment deposits may be subjected to liquefaction and may move abruptly and destroy water intake towers and other structures in the reservoir (Alam, 2001).

For proper allocation of the storage space and management of a reservoir, knowledge about the sediment deposition pattern in various zones is essential. It is essential to periodically conduct surveys and assess the sedimentation rate in a reservoir. With the correct knowledge of the sedimentation processes taking place in a reservoir, remedial measures can be undertaken well in advance.

12.1.1 Problems Due to Reservoir Sedimentation

The sediments deposited in the reservoir are an indication of the sediment yield of the entire catchment. The problems due to sedimentation can take place upstream, downstream, and in the reservoir.

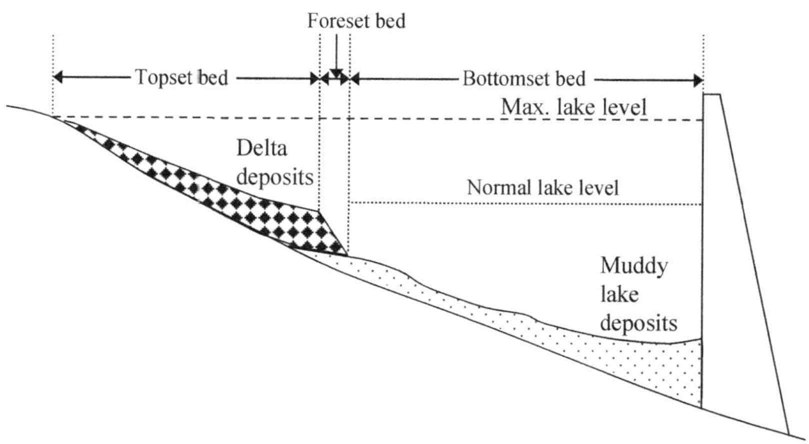

FIGURE 12.1

Sediment deposit zones in a reservoir.

The pattern of deposition depends on several factors, such as size and texture of sediment particles, characteristics of reservoir outlets, size and shape of the reservoir, and its operation schedule. Generally, coarser sediments deposit first in the reservoir headwaters and finer particles are transported by density currents down to the dam.

Sedimentation reduces the storage capacity of reservoirs and thereby, their ability to conserve water for various intended purposes. Consequently, the frequency and magnitude of failures increases. Sediment deposition may also hamper the operation of outlet structures. Due to sediments brought by a flood, the outlet structures of the Guayabal irrigation dam in Puerto Rico remained buried for several days and it was not possible to deliver water for irrigation. Sedimentation also reduces the survival of aquatic species. It results in increased evaporation due to the higher exposure area of the water. In one instance, it was found that the additional loss through evapotranspiration due to increased vegetation was about 10% of the annual supply.

12.1.2 Factors Influencing Reservoir Sedimentation

Sediment transport by rivers varies from near zero during dry weather to extremely large quantities during major floods. Hence, it is difficult to predict the sediment accumulation during a short period of time. The main source of knowledge of the reservoir sedimentation rates are surveys of sediment accumulation in reservoirs that have been studied for many years. These surveys indicate the specific weight of the settled sediments and the percentage of entering sediment that is deposited in the reservoir. Further, the sediment accumulation during a period of a few years may not indicate the long-term sedimentation rate.

The two dominant factors that influence the rate of silting in any storage reservoir are: (a) capacity to inflow (CI) ratio and (b) sediment content in the water flowing in. The other factors that affect the long-term loss of storage capacities are the texture and size of the sediment, trap efficiency, size, shape, and length of the reservoir, and the method of reservoir operation. The CI ratio is the ratio of reservoir storage capacity to mean annual inflow. A reservoir having this ratio more than 50% is considered hydrologically large and may have significant carry-over. If the CI ratio is large, the trap efficiency will also be large. Note that the sediment inflow depends on the catchment area too. All other things remaining the same, a dam of the same capacity in the upper catchment will have a higher rate of silting compared to a dam lower down the valley.

The two principal factors mentioned above have a complete range of interplay. A reservoir having a small CI ratio and small sediment inflow and the other having a large CI ratio and large sediment inflow may have more or less the same average annual percentage loss of capacity. With a high CI ratio and high sediment content in inflow, a high rate of silting can be expected. On the other hand, a high CI ratio and low sediment content in inflows will result in a small rate of silting.

The detention period denotes the time required to replenish water in the reservoir. It is the ratio of reservoir storage capacity and the inflow rate over a specified duration. For a reservoir, this ratio varies with season. In a hydrologically small reservoir, the detention period during a big flood may be of the order of a few hours, while during a dry season, it can be up to several months. A hydrologically small reservoir will have a short detention period, and the flood water will not stay in the reservoir for a long time.

To evaluate the effect of each of these factors and to serve as a guide for future planning, systematic capacity surveys of reservoirs should be undertaken at regular intervals. It further helps in planning corrective measures by way of the catchment area treatment, if surveys reveal abnormal deviations. The characteristics of sediments, particularly the particle size distribution, help determine their unit weight and location of deposition within the reservoir.

12.1.3 Trap Efficiency

The trap efficiency of a reservoir is the ratio of sediment retained in the reservoir to the sediment brought into it. Thus, it is the percentage of the total incoming sediment retained in the reservoir. The trap efficiency primarily depends on the sediment characteristics (particle size distribution and the behavior of the finer fractions under varying concentration, temperature, etc.), the detention time of inflow, method of operation, and age of reservoir. The detention time depends on: (a) the CI ratio, (b) the shape of the reservoir basin, and (c) the type of outlets and operation schedule. Clearly, greater the period of retention (similar to detention period above) in a given pool, lower the transit velocity and turbulence, the higher will be the percentage of deposition of incoming sediment. The *sedimentation index* is the ratio of the period of retention to the mean water velocity through the reservoir. The *period of retention* is equal to the reservoir capacity (m^3) divided by the average daily inflow to the reservoir (in m^3/s). A small reservoir on a large stream passes most of its inflow so quickly that finer sediments do not settle but are discharged downstream. A larger reservoir, on the other hand, may retain water for several years and the outflow from it may be completely devoid of suspended sediment. The trap efficiency of a reservoir decreases with age as the reservoir capacity is reduced by the sediment accumulation and complete filling may require a very long time.

The CI ratio is an indicator of the period of retention. The reservoirs with CI ratios less than unity are called seasonal storage while those with CI ratio over unity are known as carry over reservoirs. When CI ratio >1, water is rarely spilled from the reservoir. Taking into consideration the effect of seepage and evaporation losses, the trap efficiency of such reservoirs will obviously be close to 100%.

When the estimated sediment accumulation is a substantial percentage of the reservoir capacity, it may be necessary to analyze the trap efficiency for some incremental periods of the reservoir life. Theoretically, the reservoir trap efficiency will progressively decrease, once storage has begun. However, it is generally not practical to analyze the trap efficiency by increments of less than five years. While allocating space for dead storage during the planning stages, the trap efficiency is usually considered about 90%.

The trap efficiency can be computed from the inflow and outflow data of sediments. The outflow of sediments can be assessed from the observations downstream to the dam immediately after the outlets. Empirical relations have also been derived based on the trap efficiency actually observed.

Brune (1953) analyzed data from 44 reservoirs in the United States of America, 40 being normal ponded reservoirs with catchment areas varying from 0.098 to 478,000 km^2 and the CI ratio ranging from 0.0016 to 2.05. Besides normal ponded reservoirs, the analysis included two desilting basins and two semidry reservoirs. Desilting basins are shallow reservoirs normally constructed at a point where the stream gradient suddenly decreases. The average effective depth of such a reservoir is generally less than 1.2 m. The semidry reservoirs are those which are not allowed to fill the available storage capacity due to the nonacquisition of the full land and thereby restricting the operation of the dam. Besides the trap efficiency, data were also collected on the capacity, annual inflow, shape of the reservoir basin, method of operation, location and characteristics of outlets, and density-current, if any.

This analysis brought out that the laws of sediment deposition are the same for all types of reservoirs, and the factors influencing the trap efficiency are the same irrespective of the size of the reservoir. The observations showed that it is probable that reservoirs having a very low CI ratio may alternately fill and scour, depending on the stream flow conditions, and may have a trap efficiency of zero or less during periods of scour. In many dry and semidry reservoirs, the available capacity is not filled but is operated with a small storage so that most of the sediment flows out of the dam without being trapped. Thus, although the CI ratio may be high, the trap efficiency can be low. Even in carry-over reservoirs, if the operation is adjusted so as to allow a major portion of the inflow through the outlets, the trap efficiency may be brought to nearly 60% with a CI ratio of 1.7. Under normal operating conditions, the trap efficiency would be of the order of 90%.

Brune (1953) presented a set of envelope curves between CI ratio and trap efficiency, shown in Fig. 12.2. This figure shows a semilogarithmic curvilinear relation between trap efficiency and the CI ratio. Note that the data from reservoirs in the United States were used to develop these graphs. Murthy (1977) summarized the important conclusions on the trap efficiency:

1. The CI ratio shows a good correlation with the reservoir trap efficiency.
2. Although reservoirs are unlikely to have a trap efficiency of zero or 100%, under actual field conditions, trap efficiencies of zero or 100% are sometimes found.
3. Efforts at sluicing or venting sediment from reservoirs vary in effectiveness. Proper planning and correct timings of venting operations to intercept gravity underflows can treble or quadruple the amount of sediment cleared from a reservoir.
4. Desilting basins, largely because of their shape, have much higher trap efficiencies than do normal ponded reservoirs. For desilting basins not equipped with mechanical removal of sediment, trap efficiencies above 90% appear to prevail with a CI ratio as low as 0.02. With mechanical removal of sediment, such trap efficiencies may be found in even lower CI ratio ranges of as low as 0.001.

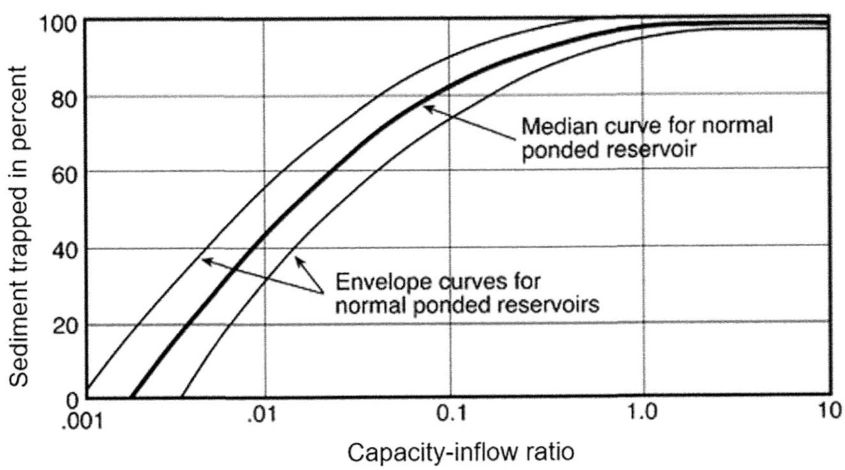

FIGURE 12.2

Reservoir trap efficiency curves by Brune (1953).

5. Semidry reservoirs may be expected to have much lower trap efficiencies than do normal ponded reservoirs. Even carry-over storage reservoirs, if operated so as to allow for large flows of water to pass unrestricted through the dam, may have trap efficiencies in the range of 60% rather than above 90%, as would be expected with normal operation.

Table 12.4 summarizes the trap efficiency and other relevant data for some Indian reservoirs as reported by Murthy (1977).

A relationship between sediment release efficiency and sedimentation index was developed by Churchill (1948). The Churchill's curve is shown in Fig. 12.3.

Table 12.4 Capacity, CI ratio, and trap efficiency of a few Indian reservoirs.

Name	Capacity (10^6 m^3)	CI ratio	Trap efficiency
Matatila	1132.7	0.187	67–90
Hirakud	8100	0.2	65–90
Gandhi Sagar	4700	0.66	100
Bhakra	9800	0.66	99

Courtesy Murthy (1977).

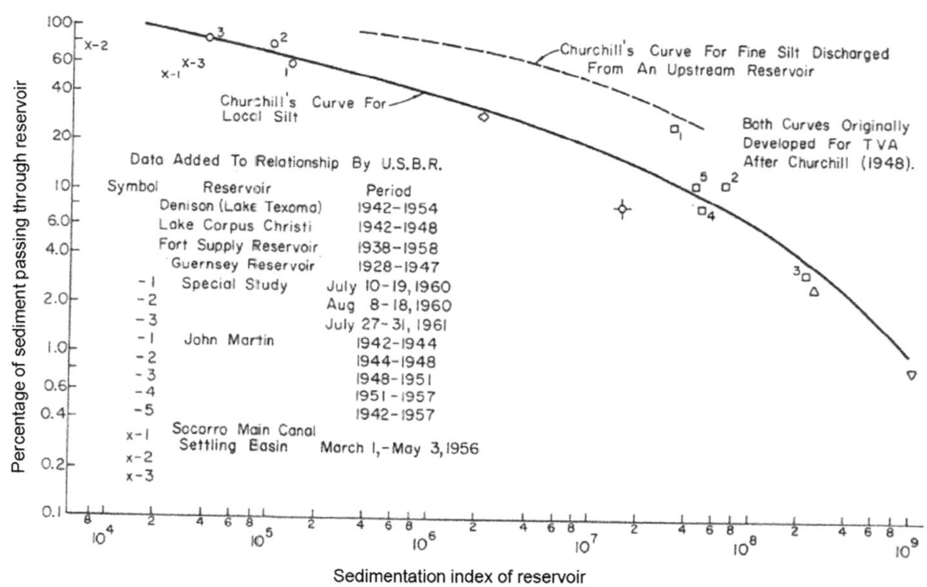

FIGURE 12.3

Reservoir trap efficiency curve by Churchill.

Courtesy Borland (1971).

12.1.4 Sedimentation and Life of a Reservoir

The term *life of a reservoir* appears to be a misnomer, since the reservoirs do not have a single well-defined life which denotes two functional states: *ON* and *OFF*. Rather they show a gradual degradation of performance. Sedimentation and the consequent reduction of capacity is a gradual process, which can be classified in various phases. From the point of view of operation, it is important to find when the sedimentation reaches the extent that the satisfaction of the purpose of the reservoir begins to suffer. At this stage, may be lesser command area can be serviced or lesser power can be generated. The flood control pool is at the top of the conservation pool, and this purpose begins to suffer after the sediments largely occupy the conservation pool. Thus, the "life" of a reservoir for different purposes will be affected at different times. Murthy (1977) defined the following terms connected to the life of a reservoir.

Useful life: It is the period during which the capacity occupied by sediments does not prevent the reservoir from serving its intended primary purpose. It is said to be over when the reservoir cannot meet the intended commitments.

Economic life: This is determined by the point of time after which the effect of various factors, such as physical deterioration by sedimentation, changing requirements for project services, etc., cause the operating costs of the reservoir to exceed the additional benefits from its continuation. In other words, this is the period during which it can be operated with economic efficiency.

Usable life: This is the period of time over which the reservoir can continue to serve some of the purposes, although to a limited extent, even after the expiration of its economic life, singly or in conjunction with additional facilities created for the purpose.

Design life: This is the period that is adopted for economic analysis. It is either the useful life or shorter of the expected economic life or fixed span of life 50/100 years (according to the practice of the agency owning the project) keeping various criteria in view.

Full life: It is the number of years required for the reservoir capacity to be fully depleted by sedimentation.

Half-life: This is the time over which half of the reservoir storage capacity is expected to be occupied by sediments. Note that the trap efficiency decreases with storage capacity and hence the entire storage will not be filled in time which is twice the half life. Usually, a half-life is the time after which the adverse effects of sedimentation begin to affect the operation of a conservation reservoir.

According to the approach incorporated in the Indian Standard IS: 12182 (1987), the life of a reservoir has five phases. The end of Phase-I is said to occur at the end of the period in which the reservoir is capable of yielding the full planned benefits. The Phase-II would depict a period when the operation of the reservoir is also trouble free, in regard to sedimentation, although the efficiency of the reservoir gradually reduces, and management measures to adjust to the reduction are required. The Phase-III would be a period of troubled operation, and unless some new engineering solutions are implemented, the project may have to be given up in phase-IV or phase-V.

12.1.5 Allocation of Space for Sediments

The most common procedure to deal with the sediment problem is to designate a portion of the reservoir capacity as sediment storage. This, however, is a negative approach that in no way reduces the sediment accumulation but merely postpones the date when it becomes serious. Since sediment is deposited all through the reservoir, the allocation for sediment storage cannot be exclusively the dead storage but must also include some useful storage. Actually, the reservoir sedimentation cannot be totally prevented, but it can be retarded. One way of doing this is to select a site where the sediment inflow is naturally low. Some basins are more prolific sources of sediment than others because of soil type, land slopes, vegetal cover, and rainfall characteristics. If an alternative site exists, prolific sediment sources should be avoided.

After a site has been selected, the reservoir capacity should be made large enough so that its useful life is sufficient to warrant the project. Although the trap efficiency of a large reservoir is high, it does not increase linearly, and the useful life of a large reservoir is longer than that of a small reservoir, if all other factors remain constant.

12.2 LOSS OF STORAGE CAPACITY

After completion of a dam, a backwater region is formed. Sediment coming from upstream starts to deposit and the storage capacity of the reservoir decreases as time passes. According to Mahmood (1987), world wide, reservoirs are losing storage at an annual rate equivalent to 1% of the storage capacity which amounts to about 65 km^3 per year. The replacement cost of this storage is indeed very high. According to estimates by Crowder (1987), the rate of loss of the reservoir storage in the United States is about 0.22% per year which is equivalent to 2020 million m^3 per year. Based on the weighted average data of 144 reservoirs in India, the annual loss of the gross storage is about 0.44%.

12.2.1 Rate of Loss of Storage Capacity

Approximately 8060 Mm^3 of sediments were accumulated in the reservoirs, and 19.2% of the design capacity was lost, even though most of the reservoirs had been in operation for less than 20 years. Considering the fact that the amount of material washed out from a basin is a function of the erosion rate and drainage area of the basin, the following empirical formula was derived (Deyi and Fan, 1991):

$$R_s = 0.0002G^{0.95}(F/S)^{0.8} \tag{12.1}$$

in which R_s is the % average loss rate of reservoir capacity per year; G is the average soil erosion from the basin in $t/km^2/year$; F is the drainage area in m^2; and S is the reservoir capacity in m^3.

Observations show that in the reservoirs which have a small sluicing capacity with respect to normal floods and which have no reservoirs above them, the siltation rate is comparatively high in the first 15–20 years and thereafter it falls off and may ultimately become negligible. From the data of reservoir capacity surveys, Shangle (1991) found that the sedimentation rates in major reservoirs (storage >100 million m^3) in India that have completed more than 50 years of their useful life varied from 0.30 to 4.89 Ha-m/100 sq. km/year. The rate for those major reservoirs that have completed less than 50 years of their useful life was found to vary from 0.34 to 27.85 Ha-m/100 sq. km/year. The data given in Table 12.5 also show that the rate of siltation of a reservoir falls with time.

A possible explanation is that the obstruction by the dam causes the dips and flanks of the storage basin to fill up with silt in early years. A stage reaches when the river section adjusts itself to carry the normal discharge and the disposal of suspended load in the area of the reservoir is harmonized

Table 12.5 Rate of siltation of some Indian reservoirs.

S.no.	Name of reservoir (State)	First period of 10 years (A)	Rate of siltation	Last period of 10 years (B)	Rate of siltation	Total period between midpoints of A and B	% decrease in rate of siltation in the total period	Percentage decrease in rate siltation per year
1.	Panchet Hill (Bihar)	1956–66	0.973	1986–96	0.313	30	67.83	2.261
2.	Maithon (Bihar)	1955–65	1.170	1984–94	1.132	29	03.24	0.117
3.	Pong (HP)	1974–84	2.558	1988–98	1.350	14	47.22	3.383
4.	Tungabhadra (Karnataka)	1953–63	0.602	1983–93	0.226	30	62.46	2.082
5.	Hirakud (Orissa)	1967–77	0.657	1984–94	0.562	17	14.46	0.851
6.	Bhakra (Punjab)	1958–68	0.633	1988–98	0.663	30	–	–
7.	Lower Bhavani (TN)	1953–63	0.306	1973–83	0.246	20	19.60	0.980
8.	Vaigai (TN)	1958–68	0.409	1973–83	0.380	15	07.10	0.473
9.	Matatila (UP)	1956–66	0.849	1984–94	0.340	28	59.95	2.141
10.	Dhukwan (UP)	1907–17	0.042	1970–80	0.012	63	71.43	1.138

Courtesy Central Water Commission, New Delhi, India.

with the condition of the flow. Besides, the progressive development of deltas above the reservoir helps in trapping of some of the silt load. Shrinkage and settlement of deposited silt also takes place with time due to superimposed loads of additional silt load. This results in reduction in silt volume thereby reducing the sedimentation rate. However, a complete explanation of this behavior is not available.

The prediction of sediments which are likely to be deposited over a time horizon is necessary for many purposes. There are many empirical relations in the literature to predict the number of years in which the reservoir will fill completely. This number will depend on the reservoir capacity and variables on which the sediment yield depends. For example, Garde (1995) provides the following equation to estimate the number of years (T_0) in which the reservoir will completely fill if all the sediment entering into the reservoir stays there:

$$T_0 = 3.789 \times 10^{-7} A^{0.886}\, p^{2.869} \,/\, \left(C^{1.771}\, {D_d}^{1.819}\, {F_c}^{8.678}\right) \tag{12.2}$$

where A is the catchment area in km^2, p is annual precipitation in cm, C is the reservoir capacity in Mm^3, D_d is the drainage density in km^{-1}, and F_c is the erosion factor given as

$$F_c = (0.2A_1 + 0.4A_2 + 0.6A_3 + 0.8A_4 + A_5) \,/\, (A_1 + A_2 + A_3 + A_4 + A_5) \tag{12.3}$$

where $A_1 \ldots A_5$ are the areas (in km^2) of closed and dense forest, unclassed forest, arable area, scrub and grass area and waste area, respectively. Murthy (1977) proposed the following equation to relate the sediment load in dead storage as a percentage of the total load, S, versus the dead storage as a percentage of the total capacity C_1 for four types of reservoirs:

$$S = K{C_1}^N \tag{12.4}$$

The values of K and N for the four types of reservoirs are:

Type	Description	K	N
I	Lake	3.39	0.78
II	Flood plain in foot hill	9.33	0.56
III	Hill	25.12	0.35
IV	Gorge	32.36	0.30

It must be emphasized that such methods do not consider all the variables on which the deposition pattern depends and hence, give only approximate results. These relations for regions other than those for which data were used in developing them should be applied with caution.

12.2.2 Unit-Weight of Deposited Sediments

The sediment load obtained by the measurements of suspended silt from the streams is usually expressed in weight and to convert it to space occupied, the weight-volume relationship has to be established. The unit-weight of sediments is the dry weight (kg) per unit volume (cubic meter) of the material. The unit-weight of a reservoir deposits varies widely, and the density of the deposited sediment has been observed to vary from 500 to 2000 kg/m^3. If unit-weight estimates are wrong, the silt load will not be correctly estimated. Therefore, the correct assessment of the density of deposited sediment is necessary. The following classification of sediment according to size is normally used.

Sediment type	Size range (mm)
Clay	Less than 0.004
Silt	0.004–0.0625
Sand	0.0625–2.0

Important factors that influence the unit weight of the deposited sediment are the manner in which the reservoir is operated, the texture and size of sediment particles, and the compaction or consolidation rate. Other factors, such as density currents, thalweg slope, and the effect of vegetation in head reaches of the reservoir, are less influential. The reservoir operation is the most influential of these factors. If a reservoir is operated, lowering its levels from time to time, the deposited sediment gets exposed to sun and air and gets dense. In case of detention basins where flood is temporarily held and evacuated as early as possible, there will be considerable time gap between floods so that the sediment deposited by previous floods gets dried up and is consolidated before the next flood. The degree of consolidation depends on the weight of the overlying material, its exposure, sediment size, and time. The reservoir which is always filled up has a low density of deposit. Power and irrigation reservoirs belong to the intermediate class. In multipurpose reservoirs which are operated depending on the various requirements, the determination of sediment density becomes complicated. Generally, lower densities are observed in the vicinity of the dam under submerged conditions, while higher densities are noticed in the upstream portions of reservoirs.

The size of the incoming sediment particles has a significant effect on the unit-weight. Sediment deposits composed of silt and sand will have a higher unit-weight than those in which clay predominates. Based on the results of the unit-weight and size distribution analysis of 1300 samples, Lara and Pemberton of U.S.B.R. developed a method to estimate the initial unit-weight of sediment deposits when the particle size of the incoming sediments and the proposed reservoir operation schemes are known. Reservoir operations were classified according to different types as follows:

Type	Reservoir operation
I	Sediment always submerged or nearly submerged,
II	Normally moderate to considerable reservoir drawdown,
III	Reservoir normally empty,
IV	River bed sediments.

After the reservoir type has been selected, the unit-weight of the sediment deposits can be estimated using the following equation:

$$\gamma = 16.05\ (W_c P_c + W_m P_m + W_s P_s) \tag{12.5}$$

where γ is unit-weight in kg/m^3; P_c, P_m, P_s are percentages of clay, silt, and sand of the incoming sediment respectively; and W_c, W_m, W_s are the coefficients of clay, silt, and sand, respectively, which may be obtained from the following table (Murthy, 1977):

Reservoir type	W_c	W_m	W_s
I	26	70	97
II	35	71	97
III	40	73	97
IV	60	73	97

Example 12.1. The particle size analysis for a type I reservoir shows 23% clay, 40% silt, and 37% sand. Estimate the unit weight of sediments.

Solution. The unit weight is computed as

$$\gamma = 16.05[26(0.23) + 70(0.40) + 97(0.37)] = 16.05(5.98 + 28.00 + 35.89)$$
$$= 1121.4 \text{ kg/m}^3$$

12.2.3 Aggradation and Degradation

A river reach is in equilibrium when the sediment load entering into it is equal to that going out. If, due to some reason, the sediment entering is more than sediment leaving, the balance of sediment is deposited in the reach. Aggradation refers to the up-rising of the riverbed to a new elevation and profile due to sediment deposition. The major consequences of reservoir aggradation are:

- delta deposits leading to reduced channel capacities;
- the rise of the backwater profile of the channel upstream from the reservoir, thus creating problems for riparian villages besides being eyesores;
- adverse environmental effects, such as formation of stagnant pools in adjacent lands; and
- deterioration of water channels due to larger sediment concentration, and infestations of phreatophytes, such as salt cedars.

It is necessary to determine the elevation up to which sediments will accumulate so as to fix the location of undersluices and other outlet works. If the reservoir is to be used for recreation, then the location of these facilities should be decided keeping in view the sediment accumulation.

The water released from a dam is relatively free of sediments and has the capacity to erode and transport sediments. If the downstream channel consists of loose material, the same is eroded by this water resulting in lowering of the bed level. This is known as degradation and in some cases, this may extend for hundreds of kilometers downstream. This may disturb the river environment and may cause other problems in the downstream areas. Garde and Rangaraju (1977) cite many examples of such problems. The Islam barrage on River Sutlej failed due to degradation. A subsidiary weir had to be constructed downstream of the Naga Hemadi barrage on the Nile River to control degradation. Due to degradation below the Hoover dam on the Colorado River, 150 million cubic yards of material had been removed from the channel over a distance of 150 km during 1935–51. Degradation may also endanger the foundations of bridges.

12.2.4 Distribution of Sediments in Reservoirs

An understanding of the pattern or profile of sedimentation in the reservoir helps in predicting the extent to which services will be affected at various times and the remedial actions to be taken. In planning and design stages, the designer is interested to know up to what height the sediment will accumulate in a given period to fix up the sill elevation of the outlets and the penstocks gate elevation, etc. The pattern is also needed to mark the region where delta would be formed, and backwater levels will increase. The backwater levels are important particularly if the reach happens to be in a developing area. Finally, the pattern is necessary to locate spots for recreational facilities, such as swimming and boating.

A commonly used empirical method to estimate the new reservoir profile is discussed next.

12.2.5 Empirical Area Reduction Method

Based on the elevation-capacity characteristics, reservoirs are classified into four types, namely, (a) gorge, (b) hill, (c) flood plain-foot hill, and (d) lake. The empirically derived sediment distribution curves are used to distribute the sediment throughout the reservoir section. The *empirical area reduction* method, as revised by Lara (1962), is the most popularly used method to predict the new reservoir bed profile. It is necessary to first estimate the amount of sediment deposited in the reservoir. This method is illustrated using an example drawn from Murthy (1977). The example data in Table 12.6 pertain to a reservoir whose original capacity was 8253.3 ha-m. The top of the conservation zone was 15.43 m above the original bed near the dam. A sediment survey showed that the sediment deposition was 1143.9 ha-m.

The steps of computations are as follows.

1. Plot the depth versus the original reservoir capacity on a double log paper with capacity on the x-axis and depth on the y-axis. Fit a straight line to the data and compute its slope m. If a single line does not represent the behavior, take the slope of the portion where the water level lies most of the time. On the basis of m, the reservoir is classified in the following classes:

m	Reservoir shape	Type
3.5–4.5	Lake	I
2.5–3.5	Floodplain–foothill	II
1.5–2.5	Hill and gorge	III
1.0–1.5	Gorge	IV

The type curves developed by USBR are shown in Fig. 12.4. From the curve of Type I reservoir, it is clear that about 50% sediments accumulate in 30% (70%–100%) of the shallow depth zone; thus in the lake–type reservoir, more sediments are deposited in the shallow water region. In contrast, in the gorge type reservoir, about 50% sediments accumulate in 20% (0%–20%) depth and such reservoirs have a propensity to accumulate sediments in deeper zones. The sediment distribution depends on the operation of the reservoir, and from this consideration, Morris and Fan (1998) classify the reservoir as stable pool, moderate drawdown, considerable drawdown, or normally empty. They have provided the following table to determine the weighted reservoir type by giving equal weight to the shape and operation:

Table 12.6 Illustrative example of empirical area reduction method.

Original survey data				Relative		Computed sediment distribution			Revised	
Elevation h (m)	Area A_h (10^4 m^2)	Capacity V_h (10^6 m^3)	F value	Depth P (%)	Area a	Area (10^4 m^2)	Volume increment (10^6 m^3)	Cumulative volume (10^6 m^3)	Area (10^4 m^2)	Capacity (10^6 m^3)
1	2	3	4	5	6	7	8	9	10	11
575.95	1394.14	8253.30		1.000	0.000	0.000	0.00	1143.90	1394.14	7109.40
574.55	1197.87	6442.74		0.909	0.875	72.44	54.12	1089.78	1125.43	5352.96
573.02	1006.45	4771.17		0.180	1.113	91.86	124.84	964.93	914.59	3806.23
571.50	833.65	3372.66		0.711	1.230	101.57	146.98	817.95	732.07	2554.71
569.97	679.06	2226.30		0.613	1.277	105.62	157.44	660.51	573.44	1565.79
568.45	513.95	1316.10		0.514	1.267	104.41	159.90	500.61	409.54	815.49
566.93	366.64	651.90	0.087	0.415	1.210	99.95	155.59	345.01	266.69	306.88
565.55	191.41	268.63	0.296	0.326	1.117	92.27	131.61	213.40	99.15	55.22
564.18	79.72	95.81	0.852	0.237	0.975	79.72	117.59	95.81	0	0
562.35	19.42	10.0	3.785	0.119	–	19.42	85.97	9.84	0	0
560.52	0	0	0	0	–	0	9.84	0	0	0

Adapted from Murthy (1977).

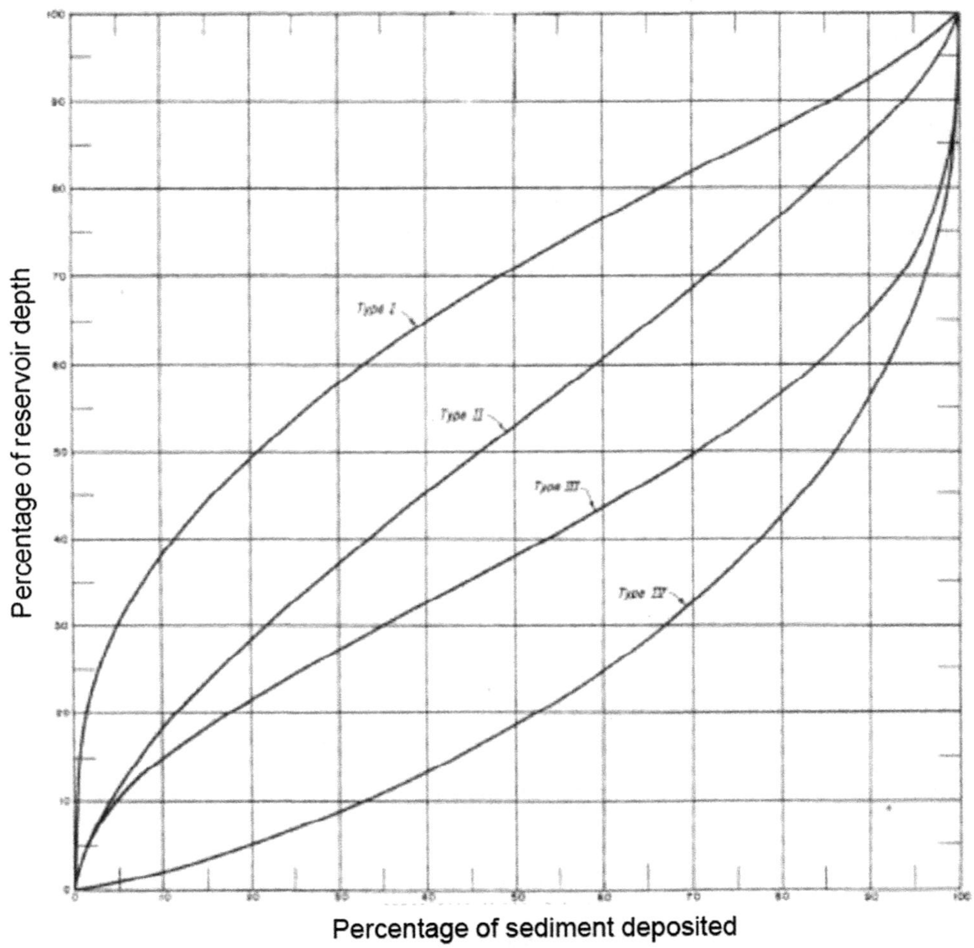

FIGURE 12.4

Reservoir storage design curves.

Courtesy Strand (1977).

Reservoir operation	Operational class	Shape class	Weighted class
Sediment submerged (continuous high pool level)	I	I	I
		II	I or II
		III	II
Moderate drawdown	II	I	I or II
		II	II
		III	II or III

Continued

Considerable drawdown	III	I	II
		II	II or III
		III	III
Normally empty	IV	All	IV

The type should be selected by giving due importance to that aspect of the reservoir, that is, shape or operation, whichever has more influence on sedimentation. In most river basins, the grain size distribution is not an important factor in influencing sediment distribution. Only in those cases where there is a choice between two type classes, the following table can be used to finalize the type:

Predominant grain size	Type
Sand or coarser	I
Silt	II
Clay	III

The reservoir under consideration is a Type II reservoir.

2. The original elevation-capacity data are used to compute the values of F (column 4) at different reservoir elevations in the deeper part (where elevation is lesser):

$$F = (S - V_h) / HA_h \quad (12.6)$$

where S is the total sediment deposition (1143.9 ha-m), H is the original depth of reservoir below the conservation pool (15.43 m), and V_h and A_h are capacity and area at elevation h.

3. The decimal values of relative depth are calculated using:

$$p = (h - h_{min}) / H \quad (12.7)$$

where h_{min} is the original bottom elevation (560.52 m) of the dam.

4. Plot the F and p values on the type curve graph. The intersection of this curve with the curve representing the reservoir type defines the point known as the new *zero elevation* (NZE). In this case, the intersection point p_0 is 0.237 and the NZE.
   $h_0 = p_oH + h_{min} = 0.237 * 15.43 + 560.52 = 564.177$ m. From the original curve, the corresponding area A_0 is 79.72 ha.
5. To estimate the sediment distribution within various zones, the value of the relative sediment area a at each depth p (column 6) is computed as:

$$a = bp^c (1 - p)^d \quad (12.8)$$

The values of b, c, and d depend on the type of the reservoir as given below:

Type	b	c	d
I	5.047	1.85	0.36
II	2.487	0.57	0.41

III	16.967	1.15	2.32
IV	1.486	−0.25	1.34

For this dam (Type II) and p = 0.237, the value of a (at p_0) is 0.979.

6. The area correction factor is $A_0/a_{p0} = 79.72 * 10^4/0.979 = 81.43 * 10^4$ m^2.
7. The area at each reservoir elevation (column 7) occupied by the sediment is obtained by multiplying the relative sediment area (column 6) by the area correction factor. Note that below NZE, the areas in columns 7 and 2 are equal.
8. For each elevation above NZE, the sediment volume (column 8) is computed by the end area method. Below the NZE, the sediment takes the entire space and hence the sediment volume equals the reservoir capacity.
9. The cumulative volume of the sediment deposited (column 9) is obtained by summing the values in column 8. The total volume of sediment should be approximately equal to the value used in the beginning of computations.
10. The revised area and volumes (columns 9 and 10) are found by subtracting column 7 from the original area (column 2) and column 9 from the original volume (column 3).

These computations can be easily carried out using an electronic spreadsheet.

12.2.6 Economics of Reservoir Sedimentation

The consequences of reservoir sedimentation ultimately lead to a gradual reduction in benefits from the reservoir. The extent of loss depends on the type and nature of purposes being served and the rate of loss of the storage capacity. Gunatilake and Gopalakrishnan (1999) worked out the cost of reservoir sedimentation in Mahaweli reservoirs in Sri Lanka. The total cost was composed of the loss of irrigable area, hydropower production loss, cost of water purification, and fisheries yield loss. The total cost of sedimentation was estimated as \$838040 as of 1993 (see Table 12.7) and it is likely to increase to

Table 12.7 Cost of sedimentation in Mahaweli reservoirs.

Year	Irrigable area loss (\$000s)	Hydropower production loss (\$000s)	Cost of water purification (\$000s)	Fisheries yield loss (\$000s)	Total cost of sedimentation (\$000s)
1993	62.14	453.87	198.17	119.90	838.04
2002	133.09	885.91	236.83	119.90	1380.47
2012	211.92	1365.95	288.69	119.90	1992.25
2022	290.76	1846.00	351.92	119.90	2615.61
2032	369.59	2326.05	428.98	119.90	3253.10
2042	448.42	2806.09	522.93	119.90	7604.71
Present value	2693.73 (10.2%)	17592.75 (66.6%)	4230.29 (116.0%)	1189.85 (7.2%)	26406.62 (100%)

Gunatilake and Gopala-krishnan (1999). © Taylor & Francis Ltd. (www.tandf.co.uk/journals). Used by permission.

\$7604710 within a 50-year period. The expenditure on this project is nearly \$1 billion. The present values were calculated using a 6% discount rate. Clearly, the loss of hydropower generation capacity is the most significant category of the cost of sedimentation in the Mahaweli reservoirs.

12.3 SEDIMENT YIELD OF WATERSHEDS

The major causes of erosion of the sediment that enters the reservoirs are rainfall, discharge, or natural geological reasons. The erosion rate is expressed in terms of the mass of soil removed per unit area per unit time (ton/km^2/year). The soil erosion, transport, and deposition processes are important for design and management of WRD projects as well as for structures which interact with water, for example, bridges. The rocks may change in character, break, decay, and turn into soil by chemical and mechanical actions known as *weathering*. The process of loosening and removal of soil and rock from any part of the earth surface is known as *erosion*. The human activities influence hydrological processes in a watershed and can accelerate or decelerate erosion.

Soil erosion caused by water can be divided into the following categories: sheet erosion (due to forces of rain drop impact and surface runoff), gully erosion (by small channels of about 15–20 cm deep), channel erosion (of banks and beds by perennial or intermittent streams), flood erosion (due to the flow of flood water on plains), and mass movement (landslides, slope failures, avalanches, etc.).

The delivery of eroded material from the place of origin to any downstream point is a complex process conditioned by variations in watersheds characteristics, such as size, topography, slope, land cover, degree of channelization, hydrologic and hydraulic factors, etc. The sediment content of the flow is a result of many interrelated factors of which the following are the most significant:

(a) the source and character of runoff,
(b) the aerial extent and density of vegetative cover on the watershed,
(c) susceptibility of soils and valley alluvium to erosion,
(d) the hydraulic efficiency of the drainage system, and
(e) the aerial extent and density of vegetative cover on the watershed.

All eroded material does not enter a stream system. Some detached particles travel for a short distance and get deposited for want of sufficient overland or channel flow. Some may travel downstream and get lodged in the vegetation on the banks. Some may be carried downstream only to be deposited in the plains. Thus, all the sediment produced by a watershed may not be delivered at a downstream point. Measurements show that as little as 5% and as much as 100% of the materials eroded in some watersheds may be delivered to a downstream point. The total amount of eroded material that passes through a section, such as a gauging site, is the *sediment yield* at that point. The quantity of sediment delivered to a reservoir depends on the rate of gross erosion in the watershed and the ability of the stream system to transport eroded material to the reservoir. The yield per unit area is termed as *specific sediment yield*. The sediment that enters into a reservoir is the sediment yield of the catchment and not the total eroded matter. The sediment production rate is worked out by dividing the annual production by the watershed area and is normally expressed as tonnes per unit drainage area per year.

As a general rule, the average rate of sediment production decreases as the size of the drainage area increases, just as runoff per unit area decreases with the increase of drainage area. The main cause is that larger the watershed, the greater the opportunity for deposition between the points of origin and the reservoir. Besides, the larger the watershed, the lesser is the variation between the rates. Generally, very high and very low production of sediments per unit area are found in small watersheds. This is because a very small watershed may be entirely forested as well as entirely cultivated. The watershed with forests may produce small amounts of sediment, while the cultivated watersheds produce very high amounts. In larger watersheds, land use tends toward greater uniformity with less variation between rates of sediment production.

The fraction of eroded sediment that is delivered at a point is known as the sediment-delivery ratio. Higher sediment-delivery ratio is associated with smaller catchments. As one moves upstream, the basin area decreases and the topographic factors that promote sediment delivery become more intensified resulting in a higher sediment delivery ratio. The sediment delivery ratio (D_R) is mathematically expressed as

$$D_R = S_y/E_g \tag{12.9}$$

where S_y is the sediment yield and E_g is the gross erosion. In general, the sediment discharged to large rivers is usually less than one-fourth of that eroded. ASCE (1975) have tabulated sediment delivery ratio and drainage area which show a near straight line relation when plotted on a semi-log graph (Fig. 12.5). The equation of the best-fit line is:

$$D^R = 5.07 \ln A + 36.86 \tag{12.10}$$

where A is catchment area (km^2).

Besides area, catchment physiography, transport system, and texture of eroded material influence this ratio. The sediment delivery ratio is small for big catchments with mild slopes and is large for small catchments with steep slopes. The effect of slope can be expressed through *relief length ratio* (R) as:

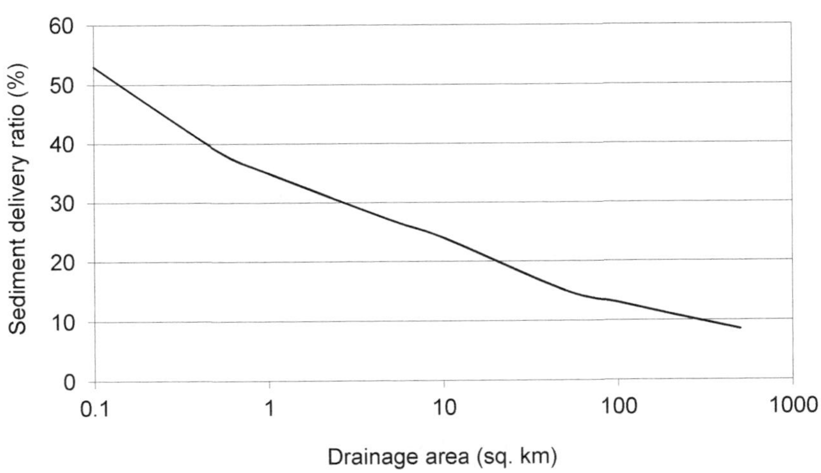

FIGURE 12.5

Variation of sediment delivery ratio with drainage area.

$$R = h/L \tag{12.11}$$

where h is the relief of watershed between the minimum and the maximum elevation (m), and L is the maximum length of watershed (m). The relief length ratio can be determined from topographic maps. Many workers have developed regression equations relating sediment yield with factors to account for climate and vegetative growth, topography, and soil properties (see Singh, 1992).

12.3.1 Methods of Sediment Yield Determination

A huge amount of research effort has been spent on understanding and modeling the soil erosion process. Many scientific studies have been conducted on experimental watersheds to assess the effect of land use on soil loss. The land uses which have been studied include forests, grasslands, agricultural lands, fallow lands, ravine lands, bare lands, and horticultural lands.

Approaches ranging from empirical methods to detailed physical model have been used to predict reservoir sedimentation. The choice of a prediction method largely depends on the objectives of the study; it may vary in different stages (planning, feasibility study, design, and operation). As such, there is no best method that can be used for all the river basins. Nowadays, mathematical modeling has become a widely used tool. These methods to determine sediment yield are briefly summarized in what follows.

12.3.2 Comparison with Nearby Watersheds and Reservoirs

If sufficient data about the watershed in question are not available, the annual sediment yield rate (per unit of drainage area) of another watershed of similar characteristics (physiography, climatology) can form an initial estimate for the project watershed. Field inspections of the watershed will disclose the main sources of sediment, such as sheet erosion, gullying, flood erosion due to deforestation, and stream channel erosion.

If there is a reservoir in a nearby watershed and its sediment-deposition data are available for many years, the sediment yield rate of that basin can be estimated by considering the trap efficiency of the reservoir. However, the results of nearby basins are comparable only if the catchment properties, size, and the hydrometeorological characteristics are identical. It may be necessary to adjust the yield rate to account for variations in the drainage characteristics.

The data obtained from surveys of the reservoirs in the region can be plotted against catchment area and a regression relation can be developed. The data can also be used to prepare sediment yield maps. However, such maps should be used with caution because while preparing such maps, a large area with wide variations in factors that affect soil erosion is divided into a smaller number of supposedly homogenous areas. Nonetheless, these maps are useful when there is very little or no data about the basin in question.

12.3.3 Stream Gaging

This method involves setting up a stream gaging site and carrying out sediment measurement along with gauge and discharge observations. This aspect is discussed in Chapter 2. It is important to have long-term sediment data because sediment load varies widely from season to season.

The sediment inflow into the reservoir including the bed load, and the outflow from it need to be measured at all significant points of entry and exit. The difference gives the quantity deposited during the period of analysis. The points of inflow measurement should be sufficiently close to the reservoir periphery. Care must be taken while outflow sampling because it should be done before the flow meets an erodable channel downstream.

12.3.4 Mathematical Modeling of Reservoir Sedimentation

The equilibrium sediment transport model is adopted in most mathematical models. In the equilibrium sediment transport model, the difference between the instantaneous sediment concentration and sediment carrying capacity is neglected. If this difference is taken into account, the approach belongs to the nonequilibrium sediment transport model. For coarse sediment particles, equilibrium sediment transport model can be used, but for very fine sediment particles, nonequilibrium model better reflects the reality. Unit sediment graphs, which follow the same concept as the unit hydrograph, are explained by Singh (1992).

Mathematical models of sediment transport and deposition solve a system of governing equations. These are: the equation of continuity for water, momentum equation, equation of continuity for sediment, sediment transport law (e.g., a power law), and resistance law (Manning's or Chezy's law). The various mathematical models basically differ in the use of sediment transport and resistance law, the nature and type of terms that are included in momentum equation, and methods of solution of the equations. Before discussing a few models, the universal soil loss equation and its variants are discussed. This is perhaps the most widely used method of estimating soil erosion.

12.3.5 Universal Soil Loss Equation

This is an empirical equation which was developed in the 1960s by Wischmeier and Smith (1965) to predict long-term interrill and rill erosion rates based on analysis of data of a large number of experimental plots in the United States. This equation has, however, been used world wide in varying climatic, geologic, and landuse scenarios. Interrill erosion is a process of soil detachment by the impact of raindrops, transport by shallow sheet flow, and delivery to rill channels. Rill erosion is the erosion of sediment by concentrated flow. Rills carry flow from interrill areas as well as the rain that directly falls on them. The USLE is written as:

$$A = R \times K \times L \times S \times C \times P \quad (12.12)$$

where A is the soil loss per unit area, expressed in the units of K and period selected for R; R is the rainfall and runoff erosivity factor; K is the soil erodibility factor; L is the slope length factor; S is the slope steepness factor; C is the crop management factor; and P is the support practice factor. The word "universal" is used probably because the equation considers the five principle factors which influence soil loss: K, R, LS, C, and P.

In the 1980s, the USLE was revised to incorporate additional research and technology developed. This resulted in a new equation called the revised USLE or RUSLE (Renard et al., 1994). RUSLE maintains the basic structure of USLE, but the algorithms used to calculate the individual factors have been changed significantly. The estimation of the factors has been computerized to assist their determination.

A brief description of the various factors of Eq. (12.12) follows.

R-Factor: The *rainfall-runoff erosivity factor* in RUSLE is calculated as the product of storm kinetic energy times the maximum 30-min storm depth and summed for all storms in a year. The R-factor represents the input that drives the sheet and rill erosion processes. The differences in R values represent differences in erosivity of the climate. Part of the R-factor calculation involves a seasonal distribution to permit weighting of the soil erodibility value, K, and the cover-management factor, C. To facilitate these calculations, climate data files have been developed (called a city code) for climatically homogeneous areas.

K-Factor: This soil erodibility factor is a measure of the inherent erodibility of a given soil under the standard condition of the unit USLE plot maintained in continuous fallow. Soils with high sand and clay contents have lower values and soils with high silt contents have higher values. In RUSLE, K also varies seasonally which is a major change over the USLE procedure. Experimental data show that K is not constant but varies with season, being highest in early spring and lowest in mid-fall or when the soil is frozen.

L- and S-Factor: The estimation of the length-slope factor is somewhat subjective, because the choice of a slope length involves judgment; different users choose different slope lengths for similar situations. RUSLE includes improved guides for choosing the slope length values to give greater consistency among users. Regarding the L-factor, the soil loss is less sensitive to the slope length than to any other USLE factor. For typical slope conditions, a 10% error in the slope length measurement results in a 5% error in the computed soil loss. RUSLE uses four separate slope length relationships. These are functions of slope steepness as in USLE, and of the susceptibility of the soil to rill erosion relative to interrill erosion. For a given slope and its length, the LS factor can be computed as:

$$LS = (\lambda/72.6)^m \left(65.41 \sin^2\theta + 4.56 \sin\theta + 0.065\right) \tag{12.13}$$

where λ is the slope length in feet; θ is the angle of slope; and m = 0.5 if the percent slope is 5 or more, = 0.4 on slopes of 3.5%–4.5%, = 0.3 on slopes of 1%–3%, and 0.2 on uniform gradients of less than 1%. Soil loss is much more sensitive to changes in slope steepness than to changes in slope length. Thus, special attention should be given to obtaining good estimates of slope steepness. RUSLE has a more nearly linear slope steepness relationship and also provides a slope steepness relationship for short slopes subject primarily to interrill erosion.

C-Factor: The vegetative cover factor is perhaps the most important RUSLE factor because it represents conditions that can most easily be managed to reduce erosion. The values of C can vary from near zero for a very well protected soil to 1.5 for a finely tilled, ridged surface that produces much runoff and leaves the soil highly susceptible to rill erosion.

Values of C are a weighted average of the soil loss ratios (SLRs) that represent the soil loss for a given condition at a given time, to that of the unit plot (a unit plot is one maintained in clean-tilled fallow). Thus, SLRs vary during the year as soil and cover conditions change. To compute C, SLRs are weighted according to the distribution of erosivity during a year. In RUSLE, a subfactor method is used to compute SLRs as a function of four subfactors given as

$$C = PLU \,.\, CC \,.\, SC \,.\, SR \tag{12.14}$$

where PLU is the prior land use, CC is the crop canopy, SC is the surface or ground cover (including erosion pavement), and SR is the surface roughness.

P-factor: The erosion control practice factor mainly represents how surface conditions affect flow paths and flow hydraulics. For example, with contouring, runoff flows around the slope in channels formed by tillage. The grade and flow velocities could be much lower than in up-and-down hill flow paths. Of all the factors, the values for the P-factor are the least reliable. There are many interacting variables that determine the effect of contouring. The size of storm, antecedent soil water, and tillage are some of these variables that interact in such a way that a contouring factor may vary widely from storm to storm and field to field; these interactions have made it difficult to document in the limited number of field studies dealing with contouring. Likewise, identifying these subtle characteristics in the field is difficult when applying RUSLE. Thus, the P-factor values represent broad, general effects of such practices as contouring.

The RUSLE P-factors are treated as the product of subfactors computed based on practices applied to the landscape. In RUSLE, extensive data (both field and model) have been analyzed to reevaluate the effect of contouring. The results have been interpreted to give factor values for contouring as a function of ridge height, furrow grade, and climatic erosivity. New P-factor values for the effect of terracing account for grade along the terrace, while a broader array of strip cropping conditions are considered in RUSLE. Finally, P-factors in RUSLE have been developed to reflect conservation practices on range lands. The practices require estimates of surface roughness and runoff reduction.

Although the universal soil loss equation has been widely applied all over the world and for catchments of widely varying sizes, the results may be erroneous unless the model is applied with care and the parameters are adapted to local conditions. Note that no physical processes are simulated in this model and the antecedent conditions are not considered. Therefore, the model is not well suited to predict soil loss from individual events. Further, this model was developed using the data of soils on mild slopes and hence its application to soils on steep slopes should be with caution.

Example 12.2. A catchment is located in foothills of the Himalayas and has clayey soil with $K = 0.33$. The average slope is 2.41% and the slope length is 393 ft (120 m). It has subhumid temperate climate with annual rainfall of 1705 mm, rainfall-runoff factor about 350, crop-management factor $C = 0.3$, and $P = 0.2$. Calculate the average annual soil erosion.

Solution. For a slope length of 393 ft, and $\theta = 2.17°$, $m = 0.3$. Using Eq. (12.12), $LS = 0.35$. Hence, the average annual soil loss per unit catchment area is:

$$A = 345 \times 0.33 \times 0.35 \times 0.30 \times 0.20 = 2.39\ \text{t/ha/year}.$$

12.3.6 HEC-6 Model

The HEC-6 model was developed at the Hydrologic Engineering Centre of the U.S. Army Corps of Engineers, and this description is mainly based on the users' manual for the model software (HEC, 1995). HEC-6 is a one-dimensional steady flow model designed to analyze scour and deposition by modeling the interaction between the water-sediment mixture, sediment material forming the stream's boundary, and the hydraulics of flow. It simulates the ability of the stream to transport sediment and considers the full range of conditions embodied in Einstein's bed load function plus silt and clay transport and deposition, armoring and the destruction of the armor layer. The model subdivides channel cross-section into two parts—a part which has a movable bed, and that which does not; the boundary between these parts remains fixed. The entire movable bed part of the cross section is moved vertically up and down. The model does not account for density currents and secondary currents.

The reservoir deposition can be analyzed to determine both the volume and location. The degradation of the streambed downstream from a dam can also be determined. Long-term trends of scour or deposition in a stream channel, for instance, those that would result from modifying frequency and duration of the water discharge or stage or from encroaching of flood plains, can be simulated. The HEC-6 program can be used to assess the influence that dredging has on the rate of deposition, scour during floods, the impact of a reservoir, etc., on the water surface profile and the water depth.

The basis for simulating the movable bed is the solution of the continuity equation for sediment material (the Exner equation):

$$\frac{\partial G}{\partial x} + B_0 \frac{\partial Y_S}{\partial(DD)} = 0 \tag{12.15}$$

where G is the sediment load in m^3/day, DD is the duration of time step, Y_S is the depth of sediment deposit in the control volume, x is the distance along the channel, and B_0 is the width of deposit (movable bed). This equation is expressed in finite difference form for point *P* using the notations shown in Fig. 12.6.

$$\frac{G_u - G_d}{0.5(L_d + L_u)} + \frac{B_{sp}\left(Y'_{sp} - Y_{sp}\right)}{DD} = 0 \tag{12.16}$$

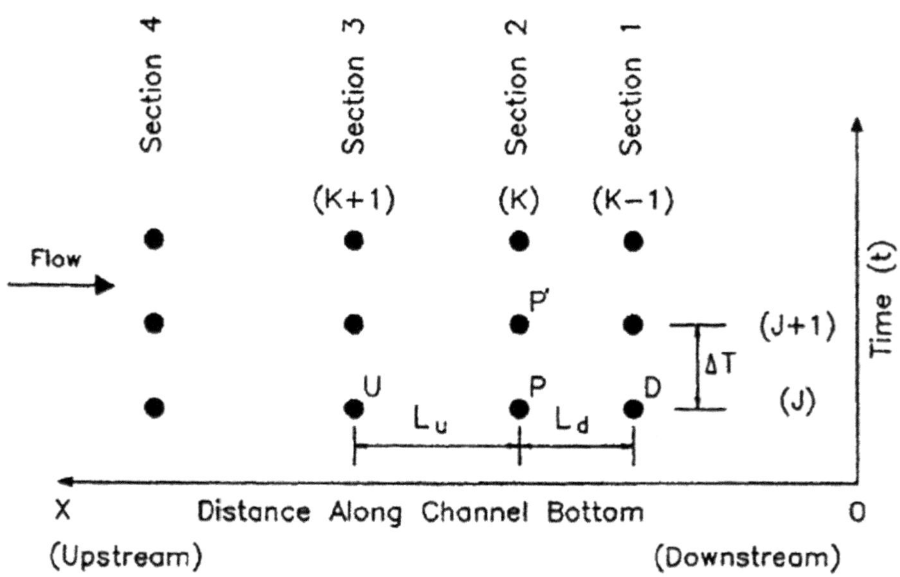

FIGURE 12.6

Computation grid of HEC-6.

Courtesy HEC (1991).

$$Y'_{sp} = Y_{sp} - \frac{DD}{0.5B_{sp}} \cdot \frac{G_u - G_d}{L_d + L_u} = 0 \tag{12.17}$$

where B_{sp} is the width of the movable bed at point P, G_u and G_d are sediment loads at the upstream and downstream cross sections, respectively, L_u and L_d are the upstream and downstream reach lengths, respectively, Y_{sp} and Y'_{sp} are the depths of sediment before and after time step, respectively. The initial depth of bed material at point P defines the initial values of Y_{sp}. The sediment load G_u is the amount of sediment, by grain size, entering the control volume from the upstream control volume. For the upstream-most reach, this is the inflowing boundary condition provided by the user. The sediment leaving the control volume, G_d, becomes the G_u for the next downstream control volume.

The sediment load, G_d, is calculated from the transport capacity at point P, the sediment inflow, and availability of material in the bed and armoring. The difference between G_d and G_u is the amount of material deposited or scoured in the reach between points D and U during the time step and is converted to a change the in bed elevation using Eq. (12.17).

The time step of fraction of a day is typical for large water discharges; it may be appropriate to use several days or even months for low flows. It is important that each time interval be short enough so that changes in bed elevation due to scour or deposition during that time interval do not significantly influence the transport capacity by the end of the time interval. Regarding the amount of change in bed elevation that can be tolerated in one time step, a value equal to 0.3 m or 10% of the water depths whichever is less, gives good results. The gradation of the bed material is recalculated during the time interval because the amount of material transported is very sensitive to the gradation of the bed material.

The basic hydraulic parameters needed to calculate sediment transport capacity are velocity, depth, width, and slope—all of which come from water surface profile calculations. The one-dimensional energy equation, shown below, is solved using the standard step method, and the above hydraulic parameters are calculated at each cross section (see Fig. 12.7):

$$WS_2 + \frac{\alpha_2 V_2^2}{2g} = WS_1 + \frac{\alpha_1 V_1^2}{2g} + h_e \tag{12.18}$$

$$h_e = h_f + h_0 \tag{12.19}$$

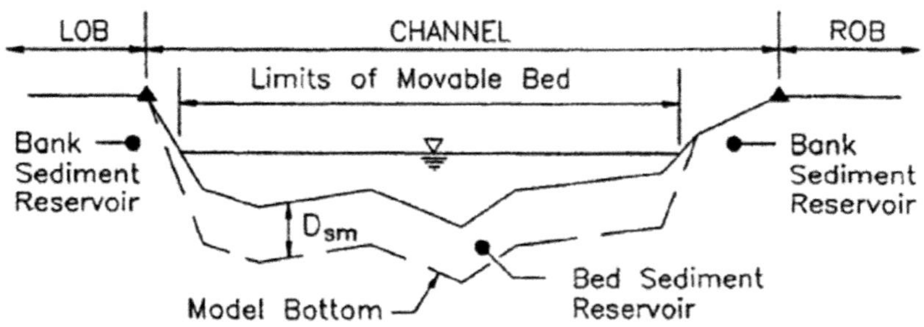

FIGURE 12.7

Cross-section representation in HEC-6.

Courtesy HEC (1991).

The energy loss term, h_e, in Eq. (12.18) is composed of friction loss, h_f, and form losses, h_0. Only the contraction and expansion losses are considered in the form loss term.

Further details of the model are available in Thomas (1979) and HEC (1991).

12.3.7 The WEPP Model

The Water Erosion and Prediction Project (WEPP) model was developed as a cooperative effort of four organizations with the leading role for the US Department of Agriculture. The main aim was to employ the current knowledge to develop a model as an alternative to USLE. The model carries out a simulation of the physical processes that cover erosion. There are three versions of the model: mountain slope, watershed, and grid. The concepts of stochastic weather generation, infiltration theory, hydrology, hydraulics, soil physics, plant science, and erosion mechanism have been used in this model. The mountain slope model simulates the erosion process from different types of land uses and treatment. The watershed version is a catchment model which includes the slope model and estimates the sediment delivery to channels. The sediment that is loaded is routed to the catchment outlet by simulating the process of erosion and transportation. The grid version is used for large areas which need not be within one watershed. The model uses a daily time step and simulates the process of plant growth, soil properties, and hydrologic processes. A conceptual representation of a hillslope and a watershed for application of WEPP are shown in Figs. 12.8 and 12.9.

The WEPP watershed model is made up of four major components: hillslope, channel, impoundment, and irrigation. The hillslope component is the WEPP hillslope model which calculates erosion and deposition on rill and interrill flow areas. It can consider processes, such as infiltration, sediment transport and deposition, evaporation, transpiration, snow melt, residue and canopy effects on soil detachment, and contour effects. It is able to account for spatial and temporal variations in topography, surface roughness, soil properties, crops, and landuse on hillslopes. The channel component calculates erosion and deposition within concentrated flow areas which can be represented

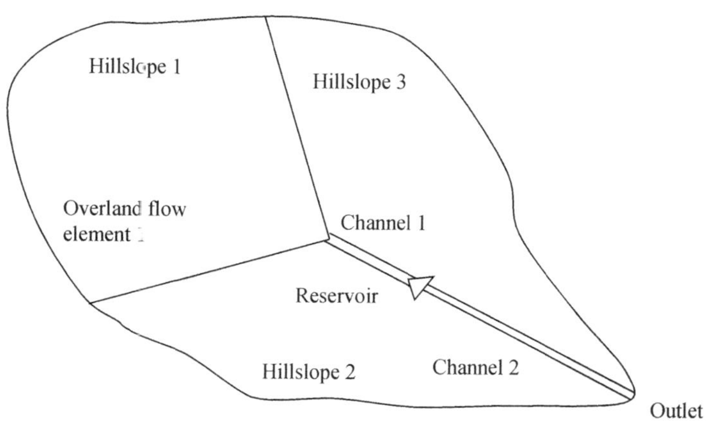

FIGURE 12.8

Representation of watershed in WEPP erosion model.

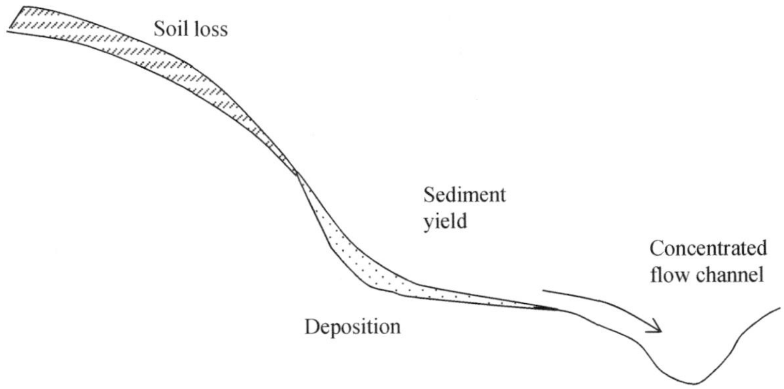

FIGURE 12.9

Representation of hillslope in WEPP erosion model.

as permanent channels or ephemeral gullies. The impoundment component calculates the deposition of sediment within terrace impoundments and stock tanks. The irrigation component calculates erosion and deposition on border irrigation areas.

A watershed must be represented by at least one hillslope element. In the WEPP watershed model, hillslope elements can contain up to 10 overland flow subelements which may represent changes in cropping patterns (strip cropping), soil variation in the downslope direction, different land use patterns, or changes in grazing intensities. A hillslope element can drain into a channel either at the headwaters or laterally or into an impoundment. A channel element can receive water and sediment input from hillslope elements, upstream channel elements (up to three channel elements), or an impoundment. An impoundment element can receive input from hillslope or channel elements.

Specifically, the model considers three erosion processes: detachment, transportation, and deposition. The channel erosion and grid erosion is modeled using hydraulic concepts. In case there is a reservoir within the catchment, the same can also be considered. A brief description of the model is given here. For details, reference may be made to Lane and Nearing (1989). The WEPP erosion model uses a steady-state sediment continuity equation to describe the downslope movement of suspended sediment in a rill:

$$dG / dx = D_f + D_i \tag{12.20}$$

where x (m) represents the distance downslope, G (kg/s/m) is the sediment load, D_i (kg/s/m^2) is the lateral sediment flow from the interrill areas, and D_f (kg/s/m^2) is the rill erosion or deposition rate. Interrill sediment delivery, and D_i, is considered to be independent of x. Rill erosion, D_f, is positive for detachment and negative for deposition.

Interrill erosion is conceptualized as a process of sediment delivery to rills, whereby the interrill sediment is either carried off the hillslope by the flow in the rill or deposited in the rill. Sediment delivery from the interrill areas is considered to be proportional to the square of rainfall intensity, with the constant of proportionality being the interrill erodibility parameter. The function for interrill sediment delivery also includes terms to account for ground and canopy cover effects, which are discussed below.

The net soil detachment in rills is calculated for the case when hydraulic shear stress exceeds the critical shear stress of the soil and when sediment load is less than the sediment transport capacity. For the case of rill detachment,

$$D_f = D_c(1 - G/T_c) \tag{12.21}$$

where D_c is the detachment capacity by flow and T_c (kg/s/m) is the sediment transport capacity in the rill. Rill detachment is considered to be zero when the hydraulic shear stress is less than critical shear strength of the soil. The net deposition is computed when the sediment load, G, is greater than the sediment transport capacity, T_c. For the case of deposition

$$D_f = (V_f/q)(T_c - G) \tag{12.22}$$

where V_f (m/s) is the effective fall velocity for the sediment, and q (m^2/s) is the discharge per unit width. The overland flow hydrograph is developed by assuming broad, uniform sheet flow. Once the unsteady flow calculations are made to get the runoff peak rate and duration, quasi–steady-state flow is assumed at the peak rate and is partitioned into broad sheet flow for interrill erosion and concentrated flow for rill erosion. The kinematic wave equations for one-dimensional overland flow are:

$$\partial h/\partial t + \partial q/\partial x = r - f = v \tag{12.23}$$

$$q = \alpha h^{3/2} \tag{12.24}$$

where h is the local flow depth (m); t is the time (s); q is the discharge per unit width (m^2/s); x is the distance down the plane (m); r is the rainfall intensity (m/s); f is the infiltration rate (m/s); v is excess rainfall rate (m/s); and α is the depth–discharge coefficient ($m^{0.5}$/s).

Prediction of the effects of land use and management practices on erosion control are perhaps the most important part of an erosion prediction tool if the purpose is to plan land and farm management systems to control erosion. Farmers can control soil loss through residue management and tillage practices. In the WEPP erosion model, the interrill sediment delivery is adjusted to account for the effects of ground cover, dead roots, live roots, and canopy cover.

The parameters used by the hydrology and erosion components of the model that must be input by the user include soil conditions for the day of the rainfall event, crop canopy, surface residue, days since last disturbance, surface random roughness, oriented roughness, etc. WEPP model may also be executed in a single-storm mode, although it is very effective when used as a continuous simulation model. Surface residue, for example, plays an important role in terms of predicting the amount of soil lost during a given rainfall event. An erosion model may use a plant growth and residue decay model to estimate the amount of crop residue present on the soil surface for each day through the year.

The output of the continuous simulation model represents the time-integrated estimate of erosion. In nature as well as in the model predictions, a large percentage of erosion occurs due to a small percentage of rainfall events. The model simulates erosion for some number of years and sums the total soil loss over those years for each point on the hillslope to obtain the average annual values of erosion along the hillslope. The model calculates both detachment and deposition. It predicts where deposition begins and/or ends on a hillslope, which may vary from storm to storm. Certain points on the hillslope may experience detachment during some rainfall events and deposition during other events. In this case, the output represents an average of the erosion events.

The WEPP landscape profile version model requires four input data files: a soil file, a slope profile file, a crop management file, and a climate file. The model requires huge data on various variables for a good implementation; the required climatic data include rainfall, its peak and duration, temperature data, wind velocity and direction, solar radiation, snowfall, and snowmelt. The user must have file building tools and access to appropriate soil, tillage implement, plant, and climate databases in order to build the four data files. Since this is a tedious task, expert systems have been developed for the assistance of a user. The theory of the model has been described in detail by Finkner et al. (1989) and USDA (1989).

12.4 RESERVOIR SURVEYS

Sediments accumulated in an existing reservoir can be determined by periodically running a sediment survey of the reservoir. It is a direct measurement procedure to assess the volume of deposit along with its pattern in the reservoirs. Other valuable information gathered in these surveys includes how the sediment deposits are distributed in the reservoir. Sediment data collected during the surveys are analyzed to determine the specific weights of the deposits, their grain size distribution, sediment accumulation rates, and trap efficiency. The recent advances in technology have considerably reduced the efforts in reservoir surveys and analysis of data. Note that the maximum information about the reservoir bed profile will be obtained when the reservoir water level is high.

The frequency of surveying the reservoirs depends on the sediment accumulation rate. Reservoirs that have high accumulation rates are surveyed more often than those with lower rates. The cost of running a survey also plays a critical part in deciding their frequency. Generally, the reservoirs are surveyed every 3–10 years. Special circumstances may necessitate a change in the established schedule. For example, a reservoir might be surveyed after a major flood that has carried heavy sediment load in the reservoir. A survey may also be run following the closure of a major dam upstream in the same catchment since the reduction in the free drainage area leads to a reduction in the sediment accumulation rate of the downstream reservoir. The volume of the sediment that has accumulated in a reservoir is computed by subtracting the revised capacity from the original capacity at a reference reservoir elevation (usually the FRL). Since this is the difference of two large numbers, an error, even by a few percentages in either of the two numbers will significantly influence the results.

The advantages of reservoir surveys are:

1. The reservoir survey can be less costly than continuous sediment measurement at several locations in the catchment.
2. The accuracy of these surveys is usually very high, particularly if advanced equipment are used.
3. It is possible to estimate the total sediment load (bed and suspended load) being carried by the river.
4. The survey can be carried out at any convenient time to get the total sedimentation after the last survey.
5. The time required for a survey can be considerably shortened with the use of advanced equipment.

There are some limitations of the reservoir sedimentation survey.

1. The unit weight of sediment is required to estimate sediment yield. This weight is estimated using samples from selected locations within the reservoir. Usually only limited samples are taken and thus spatial variation may not be properly estimated. Furthermore, due to compaction, the weight changes with time and this may introduce errors in the results.
2. Such surveys do not provide any information about the variation of sediment yield with time and give only the total sediments accumulated since the last survey. This information can only be obtained by gauging.
3. This method does not provide subcatchment wise sediment yield which can be obtained by sediment sampling of different streams.
4. This approach is not very effective where sedimentation is small, as the errors of measurement may mask the true sedimentation rates.
5. To find the total sediment inflow, sediment outflow data is also needed.

It is essential to have accurate map of the reservoir at an appropriate scale, for example, 1:10,000 scale prior to commencement of the hydrographic survey. The important reservoir features, such as the FRL along the periphery, position of dam, outlets, location of inflowing streams, etc., should be precisely marked on the map. Other topographic details, such as position of islands, permanent structures, bridges, roads, villages, etc., are also recorded. It is also necessary to mark control points in the study reach prior to commencement of the survey. Horizontal and vertical control points are fixed at a suitable interval (say 5 km in horizontal and a few meters in vertical) on the circumference of the reservoir. After fixing the control points in the outer boundary of the study reach, x-sections are planned at a suitable interval depending on the reservoir size.

The contour and range methods are two basic techniques of the reservoir survey. In some situations, a combination of both is used. The selection of a method depends on the quantity and distribution of sediment indicated by field inspections, shape of the reservoir, purpose of the survey, and desired accuracy.

12.4.1 Contour Survey Method

This is a very accurate approach to obtain the complete profile of the reservoir bed. The contour method of survey is generally helpful for all types of reservoir shapes. The capacity of the reservoir at the time of survey is computed, based on depth measurements over the reservoir bed. The general methods of contour survey are grid contouring, radial contouring, and circular contouring.

Using the level data gathered during the survey, a contour map of the reservoir is prepared. By planimetering the successive areas enclosed by the contours starting from the lowest contour, a table of elevation and submergence area is prepared. The storage capacity between successive elevations (ΔV) can be worked out by the Prismoidal formula:

$$\Delta V = \Delta H(A_1 + A_2 + \sqrt{A_1 * A_2}) / 3 \tag{12.25}$$

where ΔH is the elevation difference between adjacent contours, and A_1 and A_2 are the areas enclosed by the contours. The cumulative value of the capacity is obtained by starting from the lowest elevation and adding the successive values and the elevation-capacity table is obtained. The difference in the capacity between two surveys indicates the loss of capacity due to sediment deposition during the intervening period.

12.4.2 Range Survey Method

This method is widely used and enables estimation of sediment accumulation with minimum of field data. The range method is more suitable for reservoirs with relatively straight reaches. This method of capacity survey consists of leveling or sounding along a fixed set of ranges or cross-section lines across the reservoir which may be resurveyed at a predetermined interval later on. The survey data forms the basis for computation of volume contents of the reservoirs. A suitable combination of range and contour survey may also be justified in some cases.

The object is to compute the end areas at different cross sections and carry out volumetric computation on that basis. Nowadays, many advanced tools are available to help in the surveys. The position of the surveys can be accurately determined through the Global Positioning System (GPS) described below. The range layout can be easily planned using software packages; many such commercial packages are available.

A typical pattern of range locations is shown in Fig. 12.10. The spacing between range lines need not be constant; it should be closer in the area of special interest or the zone where higher sedimentation is expected. The range lines should be perpendicular to the reservoir axis. The ranges should be set up across the mouth of each major arm of the reservoir as well as each major inflowing channel. The extreme end where sediment deposition begins should also be covered. It has been recommended that the minimum of three ranges should be marked. Morris and Fan (1998) have suggested the following formula as a rough guide to decide the number of range lines:

$$\text{Number of range lines} = 2.942\ A^{0.3652} \tag{12.26}$$

where A is the reservoir surface area in ha. A number of methods are available to process the data from range surveys. The methods used in contour surveys can also be applied.

12.4.3 Instruments for Reservoir Survey

Hydrographic surveys of reservoirs can be conducted either by a conventional method with plane table, sextant, range finders, and sounding pole or by using automated computerized technology comprising positioning system, depth measuring system, and data acquisition and analysis system. The equipment required to run a survey consist of land surveying instruments, sonic sounders, boats, and a variety of auxiliary equipment. Two-way radios are needed when it is necessary to maintain communication between shore and boat parties. Currently, portable survey equipment and measurement devices are available which can measure and store data about the x, y, and z coordinates of the reservoir bottom in real-time on a computer media for subsequent processing. Software to generate contour maps or range profiles including 3-D maps from these data are widely available.

Global Positioning System

The United States Department of Defence conceived and installed a radionavigation "ranging" or "distance measurement" system using a constellation of satellites. This system was named as Global Positioning System (GPS). It is a satellite surveying system to provide, using the known positions of satellites in space, precise location of unknown positions on land, sea, and in the air. GPS reached full operational capability in July 1995.

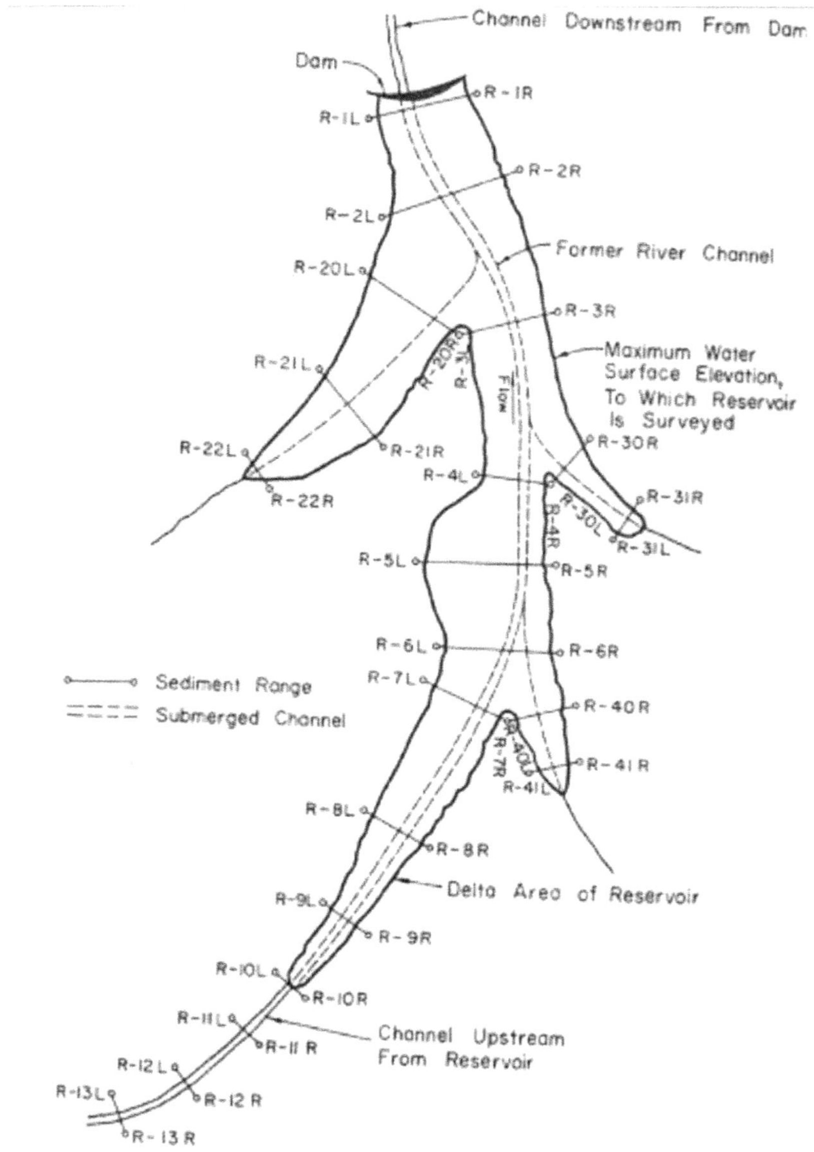

FIGURE 12.10

A layout of range lines for reservoir surveys.

Borland (1971).

GPS consists of three segments: space, control, and user.

The space segment consists of 24 operational satellites in six circular orbits 20,200 km (10,900 NM) above the earth at an inclination angle of 55° with a 12-h period. The satellites are spaced in orbit so that at any time a minimum of six satellites will be in view to users anywhere in the world. The satellites continuously broadcast position and time data to users throughout the world.

The control segment consists of a master control station in Colorado Springs, with five monitor stations and three ground antennas located throughout the world. The monitor stations track all GPS satellites in view and collect ranging information from the satellite broadcasts. The information collected from each of the satellites is sent back to the master control station, which computes extremely precise satellite orbits. The information is then formatted into updated navigation messages for each satellite.

The user segment consists of the receivers, processors, and antennas that allow land, sea, or airborne operators to receive the GPS broadcasts. The user's receiver measures the time delay for the signal to reach the receiver, which is the direct measure of the apparent range to the satellite and is used to compute their precise position, velocity, and time.

GPS works on the principle of resection, that is, locating the position of the surveyor with reference to known positions in space. The known positions refer to those of 24 satellites that continuously transmit microwave radio signals on two frequencies. GPS signals contain three key information, viz., satellite identification, precise time of the atomic clocks on the satellite, and orbital location of the satellite. To fix the position of the GPS receiver in space, the values of three unknowns, that is, x, y, and z are to be determined in an earth-centered cartesian coordinate system. Additionally, the fourth unknown parameter is the time offset between precise atomic clocks on-board satellites and the clock on the GPS receiver. To solve for these four unknowns, ranging to at least four satellites is required.

GPS provides two levels of service—a Standard Positioning Service (SPS) for general public use and an encoded Precise Positioning Service (PPS) primarily intended for use by the Department of Defense. SPS coverage is continuous and worldwide, and the signal accuracy is intentionally degraded to protect the US national security interests. This process, called selective availability (SA), controls the availability of the system's full capabilities. The SPS available to civilian users normally gives a 100-m horizontal accuracy (95% of the time after SA). The vertical accuracy is about 1.5 times worse than horizontal, due to the satellite geometry.

Differential GPS (DGPS) is a means of correcting for some system errors by using the errors observed at a known location to correct the readings of a roving receiver. The basic concept is that the reference station "knows" its position and determines the difference between that known position and the position as determined by a GPS receiver. This error measurement is then passed to the roving receiver which can adjust its indicated position to compensate.

The differential reference station computes the errors in the pseudorange measurements for each satellite in view separately and broadcasts the error and other system status information. A differential beacon receiver receives and decodes this information and sends it to the "differential ready" GPS receiver. The GPS receiver combines this information with the individual pseudorange measurements it makes, before calculating the position. DGPS eliminates the error introduced by SA and errors caused by variations in the ionosphere. DGPS systems can give coordinates within 3 m, or so.

The use of GPS has significantly cut down the time to complete a survey. It eliminates the need for intervisibility between receivers. There is no need for extra stations or line-of-sight visibility. Thus, control can be established in a short period of time even over a large area. Fog and rain do not affect data transmission and so little time is lost during inclement weather conditions. Work can be carried out at night also when atmospheric conditions are most favourable for GPS observations. While the application of GPS opens a whole new range of possibilities in survey planning and cost reduction, there are a few problems too. Signals from satellites are influenced by foliage and the system does not work well in the forested area. In urban areas with high-rise buildings, the reflected signals lead to multipath problems. It may also be noted that high solar radiation can cause disturbance to radio signals from satellites. This topic has been covered in an earlier chapter.

Depth Measuring Equipment

The main component of depth-measuring unit is an echo sounder. The echo sounder transmits the sound pulses downward into the water by a transducer. The echo reflected from the bed is also received by the echo sounder. The time interval between the emission of the sound pulse and its return as an echo is used to estimate the depth of the water. The echo sounder is capable of recording a continuous profile of the reservoir bed. Typically, the accuracy of the order of several cm can be obtained. Dual frequency echo sounder provides a higher accuracy, particularly in the cases where the reservoir bed is soft.

Sediment Samplers

Samples of the reservoir deposits are taken during surveys to determine specific weights and grain size distribution. A variety of samplers are used. Physical samples taken with core-type samplers are analyzed in the laboratory for gradation and specific unit weight. Radioactive probes can measure only in-situ wet bulk densities. The specific weight data are used to compute the mean specific weight of the sediment. The size distribution of the sample material is obtained from mechanical analyses. The size distribution data can be used to compute specific weights of sediment deposits. Generally, at least one sample is collected at each range. The total number of samples depends on the reservoir size, the type and texture of the inflowing sediments, and the location and number of inflowing rivers.

Software

Many computer software packages are available to process hydrographic survey data. A versatile software can interface a series of echo sounders and position-fixing systems. The depth recorded by the echo sounder and its position determined by the GPS can be logged simultaneously by the software and the logged data are stored in the computer for postprocessing.

12.5 ASSESSMENT OF RESERVOIR SEDIMENTATION USING REMOTE SENSING

Remote sensing techniques, enabling acquisition and analysis of synoptic data over a broad spectral range, are an alternative to conventional methods of data acquisition and processing. The advantage of satellite data over conventional sampling procedures includes repetitive coverage of a given area every few weeks, the availability of a synoptic view which is unobtainable by conventional methods, and almost instantaneous spatial data over the areas of interest. The remote sensing analysis is highly cost effective and requires lesser time as compared to conventional methods. Spatial, spectral, and temporal attributes of satellite data provide invaluable synoptic and timely information regarding the water spread area.

During the planning and design phases of a dam, contour maps of the reservoir area are carefully prepared. Due to deposition of sediments, the reservoir water spread area at various elevations goes on decreasing. A greater deposition of sediments at an elevation causes a greater decrease in this area. In the remote sensing approach, a series of imageries covering the range of reservoir water levels are obtained. These imageries are analyzed to compute the number of contiguous reservoir water pixels in each. Multiplying the number of water pixels with the area of a pixel gives the water spread area of the reservoir at the time of satellite overpass.

Most reservoirs have annual drawdown and refill cycles. The actual water surface elevation in the reservoir at the time of satellite pass can be obtained from the dam authorities. An analysis of a series of imageries will give water spread of the reservoir at various elevations over the operation range. The reservoir capacity between two levels can be computed by the trapezoidal or prismoidal formula and the elevation-capacity table can be prepared. A comparison of this table with a previous table yields the capacity lost during the intervening period. The reservoir water spread area is determined by analyzing the satellite imagery. The reduction in the water spread area with time helps in determining the sediment distribution and deposition pattern in a reservoir. This information can be used to quantify the rate of reservoir sedimentation.

It is important to note that the amount of sediments deposited below the lowest observed water level cannot be determined through remote sensing. Thus, it is not possible to estimate the actual sedimentation rate in the whole reservoir. It is only possible to calculate the sedimentation rate within the operational zone of the reservoir. However, to operate the reservoir, the capacity of the live storage zone and the pattern of sediment deposition within this zone is important.

12.5.1 Identification of Water Spread Area

The reflectance characteristics for vegetation, soil, and water were presented in Fig. 3.5. In the visible region of the spectrum (0.4–0.7 μm), the transmittance of water is significant and the absorptance and reflectance are low; the reflectance scarcely rises above 5%. The absorptance of water rises rapidly in the near-infrared region (NIR) (0.77–0.86 μm) where both the reflectance and transmittance are low. The transmittance of visible radiation through water implies that if depth is shallow, the radiation is reflected by the bottom of the water body, transmitted back through the water, and detected by the sensor. In such cases, it may not be clear from the visible bands whether there is a thin water layer above the water surface. To resolve this, the image in the NIR band must be inspected as a submerged surface will not be detected in this portion because of the lack of transmittance. At NIR wavelengths, water apparently acts as a black body absorber and the boundary between the water and other surface features is quite prominent.

The reflectance from the wetland along the reservoir periphery may be quite similar to the reflectance from the adjacent shallow water. The reservoir water may be heavily laden with silt. It is also possible that a pixel at the soil–water interface may represent mixed conditions (some part water and other part soil). To differentiate water pixels from the adjacent wetland pixels, comparative analysis of the digital numbers in different bands is carried out. The behavior of the reflectance curves of water and soil is different from the blue band (0.53–0.59 μm) onwards. Beyond the blue band, with increase in wavelength, water reflectance curves show downward trend while soil curves show an upward trend. This characteristic can be used to differentiate the water pixels from the peripheral wetland pixels. The variation of soil reflectance with moisture content and the reflectance of water in different conditions is demonstrated in Figs. 12.11 and 12.12, respectively.

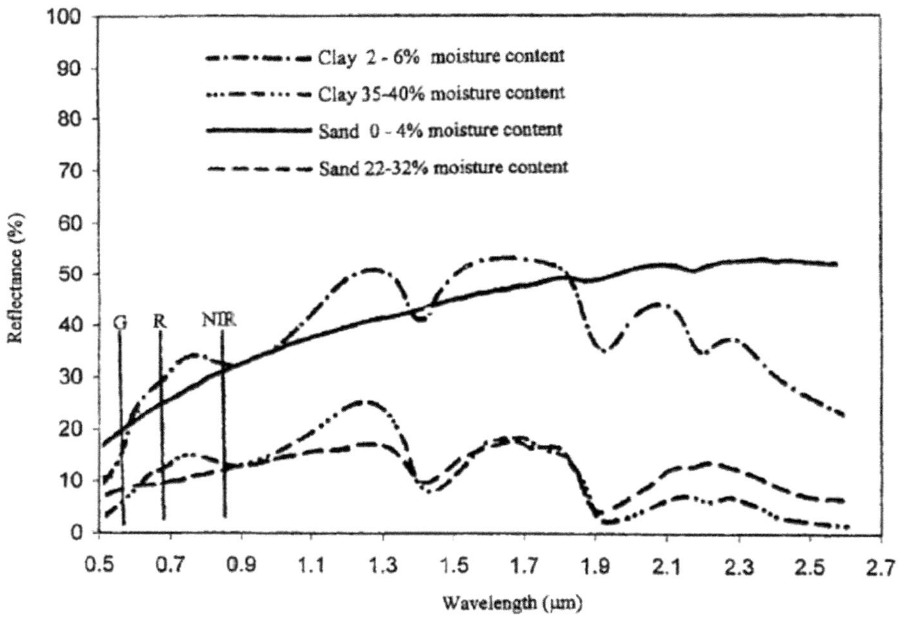

FIGURE 12.11

Variation of soil reflectance with moisture content.

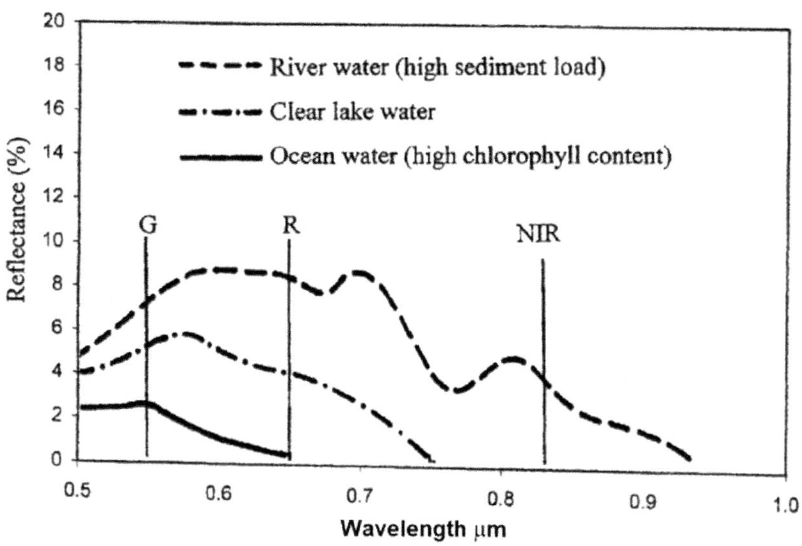

FIGURE 12.12

Reflectance of water in different conditions.

12.5.2 Analysis of Imageries

The first step of analysis is to select the period whose data is to be used. It is a good idea to choose the year corresponding to the maximum variation in the elevation of the reservoir water level and consequently, the water spread area. Large water level variations will be noticed in a wet year followed by a dry year. Multispectral data are required for the identification of water pixels and to differentiate the water pixels from the peripheral wetland pixels. It is necessary to ascertain that good-quality cloud-free satellite data are available. Besides, there might be some other reasons to select the period of analysis. It is also desirable to use high-resolution data for better results. The data of a number of satellites are available these days, and a choice is usually made based on the frequency of satellite pass, spatial resolution, and cost considerations. These days, satellite data are mostly supplied on CD-ROM and Internet is being increasingly used for this purpose.

In digital image processing, the information of different spectral bands can be utilized. The information on the pixels covered by clouds can be extracted indirectly and noise in the data can be removed. A number of commercial software are available for digital image processing. The imagery needs to be imported in the software system before analysis can commence. While using the temporal satellite data of the same area, it is necessary to geo-reference the imageries acquired at different times. The geo-referenced imageries can be overlaid and changes in the water spread area can be detected. Geo-referencing also helps to manipulate the information below the clouds and under the noise pixels. One of the imageries which is sharp, clear, and cloud- and noise-free is chosen as the base (master). The imageries of other dates are considered slaves and geo-referenced with the master. Although the reservoir area may be covered in a small part of the scene, the full scene should be utilized for geo-referencing to improve accuracy. Clearly identifiable features, such as crossing of rivers, roads or lineaments, sharp turns in the rivers, bridges, the rock outcrops, are selected as control points. At least 10 control points should be selected. The geo-referencing statistics is examined and the points which generated large errors are edited/deleted or replaced by other points so as to obtain satisfactory results. Typically, the final error should be less than the size of a pixel.

Depending on the areal extent and spatial resolution, the file size of each scene can be very large. Since the area of interest is only the reservoir area, the reservoir water spread area and its surrounding can be extracted from the full scene before proceeding with analysis. This will result in less consumption of disk space, easy handling of files, and reduction in the analysis time. This will also reduce the efforts for display and editing files. The RS packages contain utilities for this purpose. For example, ERDAS/IMAGINE has a utility named *area of interest* (AOI). A polygon covering the reservoir spread area and some area adjacent to it is constructed. The data corresponding to this AOI polygon are saved in a new file.

Identification of Water Pixels

Many techniques are available to demarcate water pixels. Density slicing of the NIR band is one such method. Although most of the water pixels can be separated out by density slicing, it may fail under certain conditions. The sliced pixels may include some saturated soil pixels also since the reflectance value of the saturated soil is very low in the NIR band. Supervised classification is another approach. Although clearly distinguishable water pixels could be easily separated out by this technique, sometimes it is difficult to provide accurate training sets for peripheral pixels. Another approach is to apply a model that uses multispectral data and tests multiple conditions to ascertain whether a pixel represents water or not. Most modern packages have a provision to write algorithms to differentiate water pixels by processing the data of multiple bands.

After the water spread area is separated out, the resulting imagery can be compared with the NIR imagery and the standard FCC. There is a possibility of interpretation error because of the presence of mixed pixels along the periphery of the spread area. However, depending on the area covered by the water or soil in a mixed pixel, classification of some pixels as pure water and some as pure soil can mutually counterbalance the effect of misclassification to some extent. Note that the estimation of sedimentation by remote sensing is highly sensitive to determination of the water spread area. The data of high-resolution sensors help to reduce the error in remote sensing analysis.

Accounting for Cloud Effect, Noise, and Tails

If the imagery has clouds, their shadows might fall over the reservoir area and its periphery. It is necessary to determine whether the pixels occupied by clouds and shadows correspond to water or not. If clouds and shadows are present over the reservoir area or around the periphery in an imagery taken during the draw-down cycle, the imagery for the next cloud-free date is examined. If the area covered by the cloud in a particular imagery has water at the same location on the next date's imagery, the pixels below the cloud are classified as water pixels. The reason is that the reservoir water surface area decreases with time during the draw-down cycle and so the pixels having water on a given date will also have water on the previous date.

Some pixels may be affected by noise in the data and are edited in a similar way using the imagery of previous or subsequent days. Due to the presence of local depressions and islands around the reservoir periphery, a few water pixels might be present near the reservoir area. Such pixels that do not form part of the continuous water spread should be removed. Many streams join the reservoir from different directions around the periphery. Beyond a certain point, these do not form a part of the reservoir. The imagery is edited to suitably remove such tails.

The number of water pixels in an imagery can be obtained from the image histogram. The water spread area is calculated by multiplying the number of pixels by the area of one pixel. The reservoir capacity between two consecutive contours (ΔV) can be computed using the trapezoidal formula (Eq. 12.25). The contours can also be used to prepare the DEM of the area. The DEM of two different dates can be compared to determine the depth of sediment deposition at various points.

12.5.3 Case Study—Assessment of Sedimentation in Dharoi Reservoir

Multitemporal data acquired by the LISS-II sensor (having a resolution of 36.25 m) of IRS-1A satellite were used by Goel and Jain (1996) to assess sedimentation in Dharoi reservoir. In India, more than 80% of the annual rainfall is received during the four monsoon months of June to September. Therefore, the water level in a reservoir can be expected to be highest after the monsoon season after which it gradually depletes till onset of the next monsoon season. After carefully examining the availability of good quality data, the data of eight dates were chosen to capture the maximum variation in the water level: May 02, 1988; October 03, 1988; October 12, 1989; November 25, 1989; January 30, 1990; February 21, 1990; March 15, 1990; and April 28, 1990.

The presence of sediments in water changes the backscattering characteristics of water. Suspended particles tend to increase the total scatter and backscatter, reduce the average path length, and consequently change the spectral distribution of light. Thus, turbid water is more reflective than clear water. The measured signal at any wavelength depends on particle size and concentration. Band 1 (0.45–0.52 μm) and band 2 (0.53–0.59 μm) of the LISS-II sensor provide information about the suspended sediments in water. Band 3 (0.62–0.68 μm) and 4 (0.77–0.86 μm) do not show a clear pattern of suspended sediments due to limited penetration power.

At greater wavelengths, water apparently behaves as a black body absorber and the reflected solar energy is very low. In band 4 (0.77–0.86 μm) of IRS-1A, sunlight is almost completely absorbed by water and the boundary between water and other surface features is clear. Hence, band 4 was used to calculate the water surface area for satellite data of different dates. To differentiate water pixels from the adjacent wetland pixels, a comparative analysis of the digital numbers in different bands was carried out. The behavior of the reflectance curves of water and soil/vegetation is different from band 2 onwards. Beyond band 2, as wavelength increases, water reflectance curves show a downward trend while soil/vegetation curves show an upward trend. This characteristic was used to differentiate water pixels in band 4 and density slicing was employed to obtain the water surface areas for the eight different dates. Only continuous extents of water pixels were considered to calculate the area. Water surface areas were obtained by multiplying the number of water pixels by the area of individual pixels.

The volume of the reservoir between two consecutive levels was calculated using the prismoidal formula (Eq. 12.25). The volume of water below the lowest observed level (172.90 m) was assumed to be the same as at the time of construction of dam. By successive addition, the volume up to the highest observed level (187.98 m) was determined. Original and revised elevation-capacity curves for the Dharoi reservoir are given in Fig. 12.13. It was found that about 57.27 million m^3 of sediments were deposited in the reservoir over a span of 14 years. Assuming a uniform rate of sedimentation, the average sedimentation rate works out to 4.091 million m^3/year.

Recently, better algorithms than density slicing that use data of different spectral bands have been suggested. A *normalized difference water index* (NDWI) was proposed by Goel et al. (2002):

$$\mathrm{NDWI} = (\mathrm{B2} - \mathrm{B3})/(\mathrm{B2} + \mathrm{B3}) \tag{12.27}$$

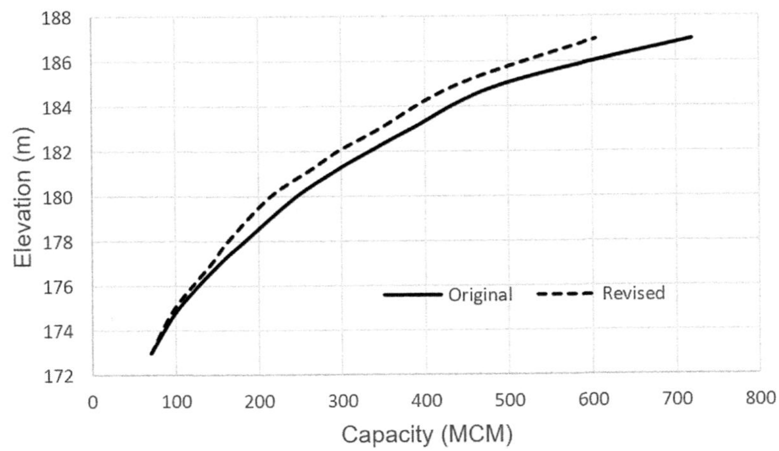

FIGURE 12.13

Original and revised elevation-capacity curves for the Dharoi reservoir.

where B2 and B3 represent reflectivity in bands 2 and 3, respectively. Normally, NDWI for water pixels is either equal to or greater than a threshold value. In addition, a model can be applied to check the DN values of different bands to ascertain whether a pixel represents water or not. The biggest advantage of using a model is that it avoids the necessity of subjectively selecting different limits in different images as required in density slicing. The results of assessment of reservoir sedimentation using satellite data have also been given by Goel and Jain (1998) and Gupta (1999).

12.6 METHODS TO CONTROL SEDIMENT INFLOW INTO A RESERVOIR

The problem of reservoir sedimentation is complex but manageable. The problem can be controlled to a large extent by judicious design, construction, and management of reservoirs. The Yellow River in China is notorious for very high volumes of sediments in its water. The Sanmenxia dam was the first dam built in the middle reach of this river. Immediately after the impounding commenced, about 1.8 billion metric tonnes of sediments accumulated during the first 18 months. This represented a trap efficiency of 93%. To achieve the balance between sediment inflow and outflow, the dam was extensively reconstructed by providing high-capacity bottom outlets and reservoir operation was substantially changed. As a result, sediment balance was achieved in 1970 and Sanmenxia was the first major reservoir in the world where this was accomplished.

There are various ways to manage the sedimentation problem, and the effectiveness of an approach depends on the site conditions. Thus, it is not possible to identify a technique which will work everywhere. Broadly, the methods to control the reservoir sedimentation are as follows:

- Control the sediment inflow into the reservoir,
- Do not allow the entering sediment to settle in the reservoir, and
- Remove the settled sediment from the reservoir.

Regarding the first approach, there are three ways to prevent the sediment from entering the reservoir: (i) construct reservoirs away from streams, (ii) place physical barriers in the way of sediment movement, and (iii) manage watersheds to check soil erosion. A discussion on these follows.

12.6.1 Off-stream Reservoirs

If the site conditions permit, a reservoir can be constructed away from the main stream. Water of low sediment concentration is diverted from the main river to the reservoir as shown in Fig. 12.14. The flows containing high sediment load are excluded from diversion. There are many methods of sediment exclusion from the diverted water. These basically work on the premise that the distribution of sediment is not uniform across the cross-section in a channel and water is, therefore, withdrawn from the zones that contain a lower sediment concentration. An additional advantage of an off-stream reservoir is that only a desired amount of flow is diverted to it and, therefore, its capacity can be limited. Furthermore, this reservoir does not require a large spillway and hence the construction cost can be considerably small.

12.6.2 Check Dams

A check dam is a small dam of a few meters height which is constructed across a stream in the headwater region to reduce flow velocity and control channel erosion. Due to fall in velocity, most of

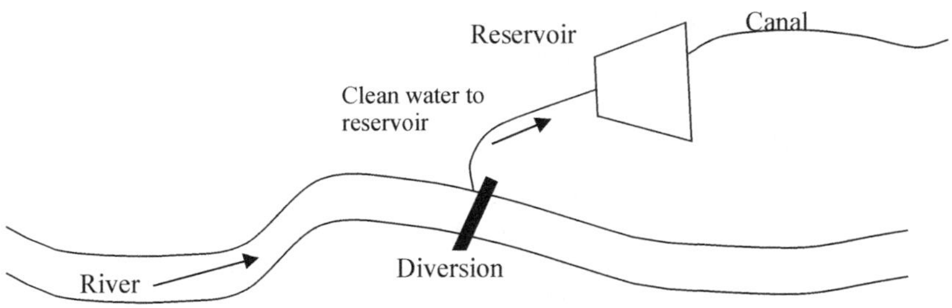

FIGURE 12.14

An off-stream reservoir.

the sediment is deposited behind the check dam and relatively clear water comes out downstream. These dams can be highly effective in controlling reservoir sedimentation. The cost of a small check dam is not very high. Some saving is possible by using locally available construction material. Therefore, before constructing a series of such dams, the watershed should be thoroughly investigated to identify the most suitable location for such structures. These dams are more efficient for sediment trapping if they are spaced farther apart. These are not a long-term solution to the problem because they do not control the sediment erosion and once the reservoir behind the check dam is filled with sediments, it no longer serves the purpose.

12.6.3 Watershed Management to Reduce Soil Erosion

This is one of the widely followed methods and is the only effective alternative in many instances. The term watershed management has wide implications and its objectives may be to reduce soil erosion, increase infiltration, and improve water yield and quality. The programs may include point treatment as well as treatment of nonpoint sources. The watershed management programs covering large areas are expensive and the benefits are available only after some time lag. These programs, besides being beneficial for reservoir sedimentation, yield many other advantages. The techniques, such as contour bunding and afforestation, were successfully applied in the catchment of Tungabhadra dam in India and siltation rate was reduced to 1/3 of the original value. Large-scale soil conservation measures in the middle reaches of the Yellow River in China reduced the sediment loads of tributaries, such as Wuding, Fenhe, Wingshui, to 50% of their original value.

Soil conservation is an effective method to prevent movement of soil particles and transport of sediment to the reservoir. Before any major soil conservation program is launched, the underlying physical process should be understood. A good approach to reduce soil erosion is to plan erosion control techniques specifically designed for the soil, climate, and topography of the basin. Agricultural activities should be planned by providing suitable cover to erodible soils. The protection of land through forest and agriculture is a cheap alternative in the long term. A good strategy is to increase the vegetation cover to the extent possible, increase infiltration, and reduce the highly erosive concentrated flows. Terracing, limiting slope length, and steepness can control concentrated flows.

Unless a watershed management program is properly planned and executed, the benefits may be small despite large investments. Some areas, for example, the Himalayas, are highly erosive due to geological and climate factors and it may, therefore, not be always possible to appreciably reduce soil erosion. In any case, one may not expect complete stoppage of soil erosion in any region.

12.6.4 Vegetative Measures

These techniques basically make use of vegetation or mulch to protect soil by diminishing the erosive forces. Extensive and sustained efforts over a long period of time are required to control soil loss if sizeable land area is degraded. Vegetation is more effective in controlling distributed flows; this measure in isolation may not be able to have any worthwhile control on erosion due to concentrated flow. For best results, the choice of vegetation should be a function of climate, soil, and topography.

Forests for erosion control: The erosion of soil in a forested area is less as the tree canopy intercepts rainfall and the drops fall down slowly. To achieve a significant reduction in erosion, the forests planted for soil and water conservation should cover large area, have multiple levels, a dense canopy, and a high density. The trees selected should be able to grow under prevailing climatic conditions, thin top soil, strong winds, and should be fast growing. Their root system should be highly developed to consolidate the soil. Their fallen leaves can be used as forage, fertilizer, and fuel. The trees which give other useful products, such as oil, are preferred in most cases.

Highly resistant shrubs are usually planted in arid areas, while arbors are common in humid areas. Depending on climate, a mixture of arbors, shrubs, coniferous trees, and broadleaf trees may be planted to increase the ground coverage. The management of forests and croplands requires significant efforts in the beginning but is cheap in the long-term besides additional benefits. These measures improve the environment quality, availability of food and fodder as well as increase in the farmers' income. Most forests are self-renewing and require little maintenance after initial years.

Farming practices: Land is cultivated without due attention to soil and water conservation in many hilly catchments. Due to the high population density in some regions, even those lands are being cultivated which are not suitable for agriculture. The practice of agriculture along the contours is known as contouring. This is an old concept. The design of terraces depends on climate, soil, crops, and farming system. It is very effective in reducing surface runoff and formation of rills and gullies. Strip cropping involves planting alternate strips of different crops which may follow the contour pattern. In this arrangement, some part of the field is always covered with a crop.

The common methods for land preparation are level terraces, level ditches, and fish-scale pits. Level terraces are built along contours by balanced cutting and filling. The exterior edge of the terrace should be higher than the interior edge by 9–18 cm to conserve moisture. Usually, the width of terrace is 0.6–1.2 m and the height 1.2–1.8 m.

Highly resistant grasses: Grasses are an important component in vegetative measures for soil and water conservation due to fast growth. These are also planted for the supply of forage and fuel. Grass seeds should be highly resistant to adverse natural conditions. Land preparation is necessary before seeding of grass, and some manure and fertilizer are applied. The required thickness of covering to suit local conditions should be ascertained.

Sediment trapping by vegetative screens: Vegetative screens considerably reduce the flow velocity and cause the sediment to deposit behind them. If such screens, a few km wide are put before the reservoir periphery, they prevent the sediment from entering the reservoir. The effectiveness of the screen depends upon its spatial extent.

12.6.5 Engineering Measures

These include slope surface engineering works, gully engineering works, and works for controlling slope disintegration.

Slope surface engineering works: Terracing is an important measure to raise grain yield, to prevent soil erosion, to preserve soil fertility, and to maintain long-term stable production. There are three kinds of terraces: (a) bench terraced farmlands; (b) sloping terraced farmland; (c) combination level terraced farmlands and natural slope land.

Bench terrace is the basic type of farmland in mountains. A bench terrace with its level platform and projected or ridged rim may hold rainwater for irrigation. Proper drainage should be provided in rainy areas. Under sloping or retention terraced farmlands, only riser dikes are built and no land leveling is made. The land surface will be flattened gradually by deep ploughing over the years. The spacing between riser dikes varies with the natural slope. In parallel with the siltation of sloping terraces, the riser dikes are made higher. The sloping terrace is less effective than the bench terrace in both soil and water conservation. It is mostly adopted in regions where the per capita land availability is high.

Gully engineering works: The check dams are the most effective gully engineering works. Their purpose is to retain silt, to fix gully bed, to raise erosion datum, to stop downward cutting of gully bed, or to prevent slope disintegration. In addition, storage of excess runoff in suitably located farm ponds also helps in soil conservation.

12.6.6 Watershed Prioritization

Since watershed treatment requires huge resources, it is desirable to target these programs in the areas that produce large quantities of sediments. Alternately, the treatment programs can be taken up in phases; zones that produce more sediments are attended first and so on. An assessment of soil loss from subwatersheds is made and a priority ranking is prepared.

In view of extensive data requirements of sediment yield models, an alternate is to use topographical/morphometric, soil, and vegetation data for qualitative estimation of erosion. Manavalan et al. (1993) presented an empirical model named *WA*tershed *R*esponse *A*nalysis (WARA) model (see Fig. 12.15) that uses data of slope angle, drainage density, soil, and vegetation to study erosion susceptibility and surface runoff potential of a watershed. This model lacks a rigorous theoretical basis. A key feature of this study was that satellite imageries were used to prepare thematic input maps. Thematic maps were analyzed to assign weights to subcatchments and these formed the basis for prioritization. Typical input theme layers are: watershed boundary, drainage, slope, vegetation, and soil texture. The preparation of thematic layers is discussed below. The number of categories into which the layers are divided depends on the variation of the relevant property.

Vegetation map: The presence of vegetation reduces the sediment detachment and transportation. In a multispectral satellite image, a normalized difference vegetation index (NDVI) is an indicator of vegetation condition. The NDVI is computed as:

$$\mathrm{NDVI} = (\mathrm{NIR} - \mathrm{RED})/(\mathrm{NIR} + \mathrm{RED}) \tag{12.28}$$

where NIR and RED are reflectances in near infrared (say 0.77–0.86 μm range) and red (0.62–0.68 μm) bands in an imagery. The normalization minimizes the effect of the illumination

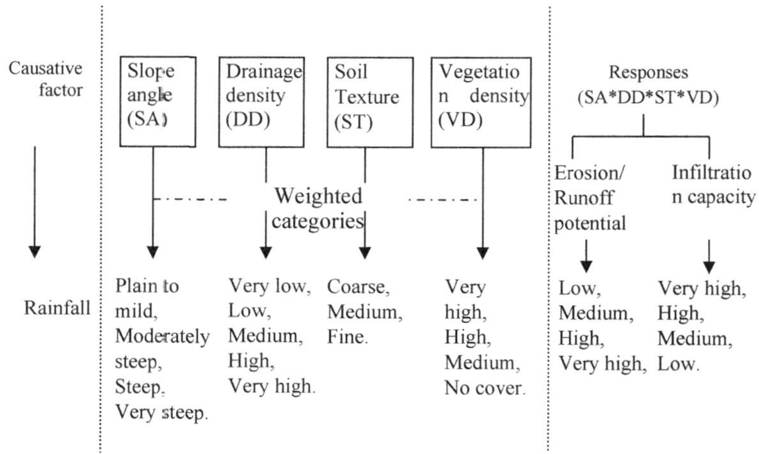

FIGURE 12.15

Watershed response analysis model.

Adapted from Manavalan et al. (1993).

geometry as well as surface topography although it does not eliminate the additive effects due to atmospheric attenuation. The NDVI image can be density sliced in vegetation classes and relative weights are given to each vegetation category to calculate the area-weighted value of the vegetation. The *area weighted vegetation* (AWV) for a watershed having five classes is calculated as:

$$AWV = (A_1 * V_1 + \ldots + A_5 * V_5) / (A_1 + \ldots + A_5) \tag{12.29}$$

where $A_1, \ldots, A_5$ are the areas under each vegetation category, and $V_1, \ldots, V_5$ are the weights for each vegetation category. An example of relative weights is given in Table 12.8 which is based on the reasoning that the watershed with a denser vegetation will have lesser erosion and must be given low priority in soil conservation measures and vice versa.

Soil brightness index maps: Physical properties of the soil affect the extent to which it can be detached, dispersed, and transported. The soil brightness index (SBI) indicates the susceptibility of soil to erosion. Tone and texture are the major soil properties which affect its brightness. To compute SBI, it is necessary to transform the data into different axes. For Landsat MSS data, *Tasseled Cap*

Table 12.8 Weights for vegetation categories.

Vegetation density	Weight
Very high	5
High	4
Medium	3
Small	2
Very less or no vegetation	1

transformation is widely used because it captures the greatest amount of data variability in fewest features and enhances data interpretability. The maximum variation can be captured in brightness and greenness, the two major axes of the transformed data set. The transformation used by Sharma et al. (1990) to calculate SBI using Indian Remote Sensing Satellite (IRS)-1B, LISS-II data is:

$$SBI = 0.2623 * B1 + 0.6432 * B2 + 0.6302 * B3 + 0.3471 * B4 \quad (12.30)$$

where B1, B2, B3, and B4 are bands 1, 2, 3, and 4 of the sensor, respectively. Similar to NDVI, the SBI image for a watershed is classified into different classes and relative weights are given to each SBI category. Similar to vegetation, the area-weighted soil (AWS_o) values are calculated for all watersheds. High AWS_o is given higher weight and vice versa.

Slope map: Slope is an important factor governing soil erosion in a watershed. Steeper the slope, more will be the erosion. The slope map of a watershed can be prepared from a contour map. Since a watershed contains many slope categories, the area-weighted slope (AWS) can be calculated using the following equation for a 5-category case:

$$AWS = (A_1 * wS_1 + \ldots + A_5 * wS_5) / (A_1 + \ldots + A_5) \quad (12.31)$$

where AWS is the area-weighted slope, A_1, …, A_5 are the areas under slope categories, wS_1, …, wS_5 are the weights for slope categories. The slope image is classified into different categories and weights are assigned.

Morphological/topographical parameters: Morphological and topographical parameters can be obtained from topographic maps at an appropriate scale. For stream ordering, the Strahler system, which is a modified form of Horton's method, can be used. Using the GIS database, morphological characteristics of watersheds like the drainage density, form factor, circulatory ratio, and elongation ratio can be estimated and a DEM can be generated.

The drainage pattern is an indicator of flow characteristics. More the overland flow, more will be the silt load traveling to the channel. Drainage density (D_d) is the ratio of total length of streams to the total drainage area and is expressed in length per unit area:

$$D_d = \text{Total stream length of all order} - \text{streams}/\text{Watershed area} \quad (12.32)$$

A higher drainage density implies a higher number of streams per unit area and thus a rapid storm response which is conducive for higher erosion. The drainage map can be prepared from topographic maps. The shape of a watershed also governs the movement of silt—more elongated a watershed is, less is the possibility of silt reaching the outlet. A circular watershed has less time of concentration and more possibility of silt load reaching the outlet point. The form factor describes the shape of a watershed and indicates its erosion potential; higher form factor induces sediment delivery of lesser erosion and vice versa. The form factor is calculated as:

$$\text{Form Factor} = \text{Area}/L^2 \quad (12.33)$$

where L is the length of basin. The form factor values can also be classified in five classes and weights assigned for different ranges of form factors.

Prioritization: The process of prioritization is integration of all the themes and weights. Based on the evaluation of different parameters as described above, priority values can be assigned to the subwatersheds. Higher the sum weight, more prone is the subwatershed to soil erosion and higher should be the priority for its treatment.

12.7 SEDIMENT ROUTING

Sediment routing is an effective method to control reservoir sedimentation. It includes the methods to manage the hydraulic behavior of the reservoir and its geometry so as to allow the maximum sediments to pass through the storage. The sediment load transported by stream varies with time; it is the highest during the period of intense precipitation and is very small during low flows. Therefore, to ensure that the least amount of sediment gets deposited in the reservoir, it is important to manage sediment-laden flow differently from the flow containing small amounts of sediments. The concepts of sediment routing were developed in China and the guiding principle is: *discharge the muddy water, impound the clear water.* An advantage of sediment routing is that the characteristics of the sediment being transported are not significantly changed. In this sense, it is the most environmental friendly approach.

The sediment routing techniques can be classified into two groups: sediment pass-through and sediment by-pass. Sediment is passed through the reservoir by either drawing down the reservoir water level or venting the turbid density currents.

12.7.1 Reservoir Drawdown

In the reservoir drawdown approach, the water level is bought down so as to pass the turbid flow through the reservoir without deposition. Typically, the reservoir level is brought down in the beginning of a flood event because the rising limb of the hydrograph carries larger amounts of sediments than does the falling limb. Since a significant amount of water may be lost during drawdown, it is important that the reservoir is filled up in the falling stage of the hydrograph and the purposes that it is required to serve do not suffer. In some cases, reservoir drawdown for sediment management is incorporated in the rule curve and it becomes an integral part of reservoir operation.

12.7.2 Density Currents

A density current is defined as the gravity flow of fluid under, over, or through another fluid of approximately equal density or the density of which differs by a small amount from that of the other fluid. Further, it is essential that two fluids are miscible and that the density difference be due to differences in temperature, and/or sediment concentration of the two fluids, but independent of pressure, density, and elastic properties of the fluid. Density currents are generated when sediment-laden water enters a relatively still water body. A sediment-laden inflow with density greater than the reservoir water descends to the lower water layers and moves toward the dam. The density current venting approach (Fig. 12.16) provides a clear unhindered path to sediment-laden flows which are released through the low-level outlets of the dam giving least opportunity to the current to dump sediments.

Density currents that transport sediments from upstream parts of the reservoir to downstream parts closer to the dam are important for sedimentation management. Upon entering the reservoir, the inflowing muddy water may plunge into the clear still water of the lake and travel below the surface due to higher density. Partial mixing may occur due to turbulence. As this current loses velocity, the ability to carry sediments reduces and coarser particles are deposited. Depending on the residual sediment concentration, the current may dissipate or continue to move forward. After reaching the dam, the current (including the sediments) can be let out through low-level outlets. The outlets should be located at the right elevation and operated to provide an unhindered path to the density current

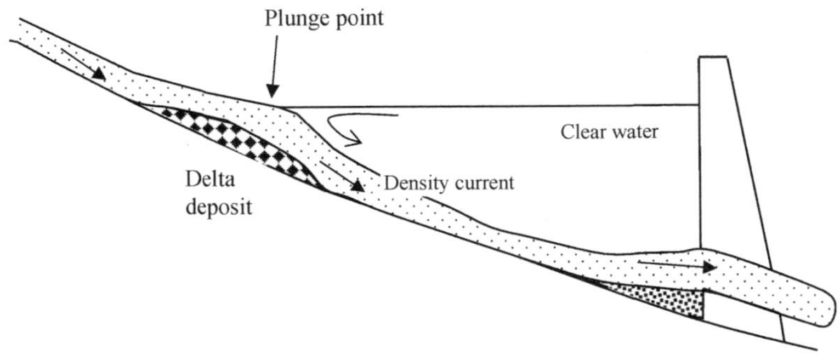

FIGURE 12.16

Venting of density current through bottom outlet in a reservoir.

before it has an opportunity to settle down in the reservoir. This approach has been followed in many reservoirs, for example, Elephant Butte and Lake Mead reservoirs in the United States, and Guanting, Yongding, and Sanmenxia reservoirs in China.

The plunge point or the place where sediment laden flow plunges underneath the clear water reservoir is at a section where (Garde, 1995):

$$U_o / [(\Delta\gamma_o / \gamma)gh_o] = 0.60 \tag{12.34}$$

where U_o and h_o are the velocity and depth of flow at the plunge section; γ is the unit weight of sediment laden flow; and $\Delta\gamma_0$ is the difference in unit weights of sediment laden flow and water in reservoir. The velocity of density current (U') is given by

$$U' = [(8\Delta\gamma/f\gamma) * gqS_0]^{1/3} \tag{12.35}$$

where f is Darcy–Weisbach resistance coefficient including the resistance at the bottom and the interface; q is the discharge of density current per unit width; and S_o is the stream slope.

The venting efficiency is the ratio of inflowing and outflowing turbidity currents. It depends on the length of the reservoir: in reservoirs of up to about 1 km length, close to 100% efficiency has been obtained. The efficiency decreases substantially beyond the reservoir length of 10 km. This efficiency also decreases as the average outflow to inflow ratio decreases. Morris and Fan (1998) cite the cases where it has been possible to release more than half of the total sediment load in an individual turbid current by proper operation. This approach is considered to be environment-friendly, since the sediment-related properties of outflows are close to those of inflows.

12.7.3 Sediment By-pass

This is an intuitively appealing arrangement which basically prohibits sediment laden flow from entering the reservoir. It consists of a diversion structure upstream of the dam (Fig. 12.17) by means of which the flows with heavy sediment concentration are diverted to a by-pass channel or conduit which joins the main river downstream of the dam. As a result, the flows entering the reservoir are relatively clean and naturally, the sedimentation problem will not be serious. This scheme has been used for the Tezden reservoir in the (former) USSR and the Amsteg reservoir in Switzerland.

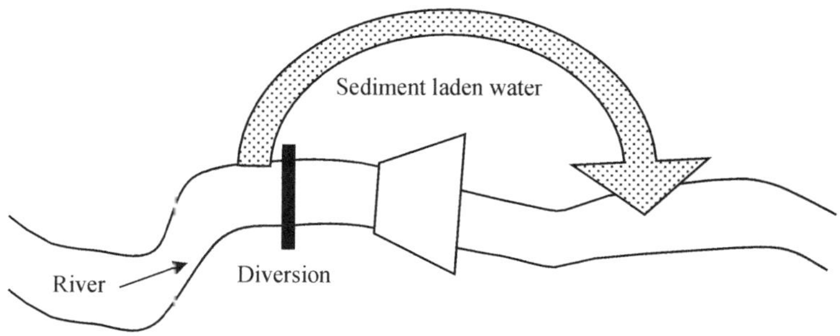

FIGURE 12.17

Sediment by-pass.

12.8 RECOVERY OF STORAGE CAPACITY

Two methods of recovering the storage capacity of a reservoir are briefly described in what follows.

12.8.1 Flushing

The aim of flushing is to establish for a short time the same flow conditions in the reservoir area (at least the channel portion) as had existed before the impoundment began. Emptying and flushing operations may be used in reservoirs where a balance between deposition and erosion cannot be obtained. This enables the river current to erode some of the deposited sediment and flush it out of the reservoir through low-level outlets. The channel area thus scoured is the regain of the storage capacity. The efficiency of sediment flushing depends on the topography of the reservoir, the capacity and elevation of outlets, and the characteristics of the inflowing sediments. For the best results, the flushing sluices should be located as deep in the dam as possible and should be sufficiently wide so that the minimum backwater is produced. Furthermore, two sluices side-by-side are preferred to sluices at different levels. The deposition on the flood plains is not affected by the flushing operation. The flushing operation lasts from a few hours to several days. Note that the lowering of the reservoir level adversely affects conservation purposes such as power generation and irrigation. Therefore, this approach is particularly suitable for reservoirs whose CI ratio is small (usually less than 0.3) since these are quickly refilled after flushing is over. According to Garde (1995), not more than 10%−15% of sediment can be removed in this manner under the most favorable conditions.

Normally the flushing is carried out once every year. It has three stages. During the draw-down stage, the reservoir level is lowered by releasing water through the under sluices or hydropower plants for some time. In the final drawdown phase, the water level is rapidly lowered using the bottom outlets and it lasts over a few hours. In the erosion stage, the riverine flow is established along the length of the reservoir and the flow erodes the sediment in the channel cross-section due to high velocity. The width of erosion depends upon magnitude of flows. If the flows are large, a wider cross-section of the channel is eroded, otherwise the erosion may be confined to a narrow region. This stage may last over several days. In the third and final stage when the storage recovery is satisfactory, the bottom outlets are closed to refill the reservoir. Flushing is most effective when the low level outlets are placed near the original bed of the river and reservoir is completely emptied. In many reservoirs, flushing is the only viable means of controlling sedimentation. For instance, a number of sediment management strategies including watershed management, construction of check dams, and dredging were considered for the Tarbela Dam in Pakistan. However, none of these was found to be effective because the dam is located on the Indus River in a tectonically active region of the Himalayas. Due to very large volume of sediment being transported, dredging was an expensive proposition.

Flushing efficiency (F_e) is defined as the ratio of deposited volume that is eroded to the water volume used during flushing over the specified time intervals. Let inflow and outflow water volumes be V_i and V_O (m^3), inflow and outflow total sediment concentration be C_i and C_O (kg/m^3), and the bulk density of the deposited material ρ (kg/m^3). Then, the flushing efficiency is:

$$F_e = (V_o C_o - V_i C_i) / (V_o \rho) \tag{12.36}$$

Typical values of flushing efficiency vary in the range 0.006–0.12. Morris and Fan (1998) describe the details of flushing operation at the Cachi, Gebidam, and Sanmenxia reservoirs. As a result of flushing, the trap efficiency of Cachi hydropower reservoir constructed on Reventzon River in Costa Rica was reduced from 82% to 27%. This approach is also in use in the reservoirs of Damodar valley system in India.

Flushing is not considered environmentally friendly because a large amount of accumulated sediment is released during a short period when the natural flow may be smaller and this may cause problems in the downstream areas. Further, the characteristics of the sediment released from the reservoir are significantly different from those of the incoming flows. The release of large amounts of sediments in a shorter time can choke irrigation canals and is harmful to the biological life in the downstream channel.

12.8.2 Dredging

Dredging is the process of removing the sediment from a water body (reservoir or channel), transporting, and depositing it at another location far away. Generally, dredging is an expensive means of recovering the storage capacity unless the deposits removed can be used for beneficial purposes. However, with the increase in water demand and reduction in construction of new projects, it is expected that the number of incidents where dredging is taken up will increase. Dredging may be resorted to when other methods of sediment management are not viable or successful.

Dredging is usually focused on small areas in the reservoir, for example, the intake structure, the regions that are being used for recreational purposes, due to cost considerations. Of course, there are problems in locating suitable sites to place the excavated material. Ideally such sites should be near the reservoir to reduce the cost of transportation. The sediment that is removed may be put in the river downstream of the dam or may be dumped in a depression away from the river. If the material is dumped in the downstream river channel, it is carried away by the river flow. Some coarse sediments may be used as construction material.

While planning a dredging program, the location and depth from which the sediment is to be removed, the volume to be cleared, and grain size distribution of the sediment should be ascertained by sampling. The equipment to be used, rating of pumps, and other considerations critically depend on it. The dredging equipment to be employed depends on the depth of excavation, the distance of dumpsite, and the elevation difference between the excavation area and dump site. The means of transporting the sediment depends on the dredging technique. Trucks may be used in dry excavation, while pipelines are employed to move slurry in wet dredging. The dredging operations should be timed when the fluctuations in the reservoir water level are small and gradual.

Dredging can be carried out without affecting the normal operation of the reservoir. If the area to be cleared is normally dry, the earth moving equipment can be deployed. Sometimes, the reservoir is specifically emptied for this purpose. However, unless the area is totally dry for some time, such excavation is problematic, because the heavy earth moving vehicles may get bogged down in swamps or may skid on clayey soils. The cost of dry excavation depends on the volume and type of material to be handled, the distance to the dump site, and the elevation differences between the excavation and the dump sites.

Hydraulic and mechanical dredges are employed for wet dredging. In hydraulic dredging, the sediment is excavated, mixed with water to form slurry which is transported to the dump site. Mechanical dredges use small buckets mounted on a chain to dig, lift, and transport the excavated material.

From the point of view of convenience, the discharge of dredged material downstream to the dam is the easiest option. However, this option just transfers the problems of sediments from one place to another. The sediment that is moved by the river may again cause problem in the downstream areas, may be deposited in canals or damage water supply networks. Therefore, before depositing the sediment downstream to the dam, its consequences should be properly assessed. Sometimes the dredged sediment is dumped in a dry valley. As far as possible, such sediment should not be deposited in wetlands and should be ensured that the future use of the dumped area does not get hampered. As the availability of dump sites is limited in most cases, dredging is not a preferred or permanent solution to the problem.

12.9 CASE STUDY: SEDIMENTATION MANAGEMENT IN SANMENXIA RESERVOIR, CHINA

The following description is based on Wang et al. (2005) and other sources.

Yellow River Basin is the second largest river basin in China, covering a drainage area of 752,000 km^2 and the length of the river is 5464 km. Yellow river flows through mostly arid and semiarid area. This river is known for high soil erosion. The annual sediment load at Sanmenxia station is 1.6×10^9 tons which makes it first in terms of sediment yield. Most of the sediment load consists of silt with a median diameter of about 0.03 mm.

In the past, the Yellow River has caused the highest flood disasters in China than any other river. Therefore, control of floods in this basin by means of a dam has been a dream of Chinese people. Based on a report outlining comprehensive river basin plan for the Yellow river, construction of multipurpose Sanmenxia Dam, serving for flood control, hydropower, irrigation, navigation, and ice jam control, was taken up in 1957. The dam was completed in 1960. It was the first dam ever built on the Yellow River. The project consists of the Sanmenxia dam, outlets, and a power plant with an installed capacity of 116 MW. The maximum reservoir level had been set at elevation of 335 m, proving for a total storage capacity of 9.75×10^9 m^3. The drainage area at the dam is 688,400 km^2, which is 92% of the total catchment area of the Yellow River. Importantly, 89% of the runoff and 98% of the sediment generated in the basin enters into the reservoir. Fig. 12.18 shows the map of Yellow River basin; the location of Sanmenxia reservoir is marked.

FIGURE 12.18

Map of Yellow river basin, showing the location of Sanmenxia reservoir.

Courtesy Wang et al. (2005).

Sanmenxia reservoir faced serious sedimentation problems soon after the impoundment of water. After completion of the dam, the reservoir was operated as a storage basin, but during the first 18 months of operation, 93% of the incoming sediment was trapped in the reservoir, leading to loss of storage capacity, backwater deposition, and increased flood risk. To address the problems, the operation procedure was changed to store only floodwaters in flood season. But the capacity of outlets to release water was limited, it did not help in reducing sedimentation. By October 1964, the accumulated sediment deposition below the elevation of 330 m reached about 3.75×10^9 m^3, which was a 62.9% loss of the corresponding storage capacity.

Toward the end of 1964, a decision was taken to reconstruct outlet structures to increase their discharge capacity. This was attained in two stages. After the completion of the two stages of remodeling, the releasing capacity of all the outlets increased to 9319 m^3/s at an elevation of 315 m. After remodeling, the dam could discharge sediment-laden water without detaining significant amounts of flood water. In addition, the operation of the dam was also modified to achieve sediment balance (Fig. 12.19). Three modes of operation were developed:

(a) Storage: This mode was employed during September 1960 to March 1962, when the reservoir was operated at high storage level throughout the year, according to the original project design.

(b) The second mode is flood detention which was applied from March 1962 to October 1973, during which the reservoir was used for flood detention and sediment sluicing, water being

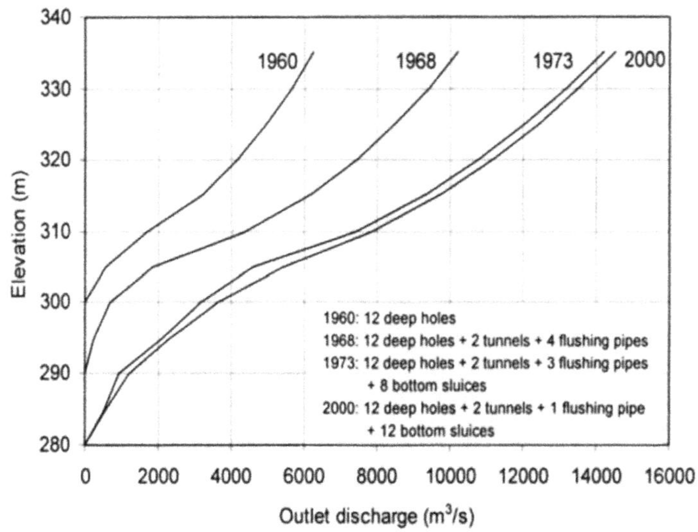

FIGURE 12.19

Progressive increase in the discharge capacity of the outlets (excluding the penstocks) of Sanmenxia Dam.

Courtesy Wang et al. (2005).

released without restrictions. The reservoir was operated at a low storage level throughout the year, detaining floods only during flood seasons and sluicing sediment with the largest possible discharges.

(c) The third mode is controlled release. This mode has been used after November 1973, to store relatively clear water in nonflood seasons (November–June) and discharge turbid water in flood seasons (July–October). In this period, the reservoir has been operated at a high operational level, with the maximum level not exceeding a value between 322 and 326 m, to store water for irrigation and hydropower generation in nonflood seasons; and at a low storage level of 305 m during flood seasons, allowing low-head hydropower generation. In times of flood discharge greater than 2500 m^3/s, all the outlets were to be opened to sluice as much sediment as possible.

Dam operator realized that sedimentation in the reservoir was different, corresponding to different operational modes and the outlet discharge capacities. Because the reservoir was operated at a low storage level and the discharge capacity was increased in the period of flood detention, the reservoir behavior changed from deposition to scouring. In the period of controlled release, deposition occurred during nonflood seasons while scouring occurred in flood seasons, maintaining a sediment balance in the reservoir within a water year or in several years. Within the range of operation levels below 323 m, the capacity of 1.0×10^9 m^3 is available for controlling medium floods in the flood season whenever necessary.

The Three Gorges Project, built on the Yangtze River, was also designed based on the experience gained in Sanmenxia reservoir.

12.9.1 Lessons Learned

The Yellow River is deficient in water and abundant in sediment. Many other rivers display similar behavior to different extents. Because of the regulation of water in large reservoirs and the utilization of water resources for irrigation in recent years, the annual runoff and its temporal distribution within a year have been greatly modified.

To manage sedimentation in Sanmenxia reservoir, outlet structures were rebuilt and their discharge capacity was significantly increased so that the sediment-laden flood flows can rapidly pass through the reservoir without allowing the sediment particles to get deposited. Dam operation was also changed and the sediment transport capacity was utilized to flush and move sediments out of the reservoir. Dam operators followed the strategy of storing clear water in nonflood seasons and discharging turbid water in flood seasons; this has proved to be very effective in managing water and sediment in the reservoir. Despite the unexpected sedimentation problems, the dam is still providing the basin with flood control, irrigation, and hydropower generation, even though some benefits are lower than the original design. Further adjustments of the operating rules will continue to accommodate the changing inflow conditions into the reservoir that result from basin development.

Importantly, the experiments at Sanmenxia Dam served as a real-life field scale study of water and sediment management in silt-laden rivers. Experience gained from the reconstruction and operation of Sanmenxia Dam is valuable to achieve sustainable use of dams by solving sediment problems of large-sized reservoirs built on sediment-laden rivers. To that end, it is essential to correctly estimate the long-term trend of sediment inflow to the reservoir. Enough outlet capacity at various levels, especially at low levels, in a dam is a fundamental requirement to facilitate the release of flood flows and sediment removal for long-term reservoir sustainability. It is necessary to regulate both water and sediment by using the concept of controlled release for dam operation to maintain a balance between sediment inflow and outflow and to conserve the useful storage capacity of the reservoir.

Sediment particles entering the Sanmenxia Reservoir are mainly the size of silt. Hence the experience gained at Sanmenxia is typically applicable to large reservoirs in the middle and lower reaches of large rivers. Upstream reservoirs have a more aggravated sedimentation problem due to the significant quantities of coarse material which tends to deposit at the head of the reservoir and the solutions that worked at Sanmenxia may not work in those cases.

12.10 CLOSURE

Sedimentation of reservoirs is a cause of concern in many regions of the world. In these areas, natural processes are predominantly adverse. However, dams are required to be constructed to meet various demands as well as control flooding. It is necessary that remedial design and management measures are taken so that the life of artificial reservoirs is prolonged and they do not silt up rapidly.

References

Alam, S., 2001. A critical evaluation of sedimentation management design practice. International Journal on Hydropower and Dams 8 (1), 54–59.

ASCE, 1975. Sedimentation Engineering Manual. American Society of Civil Engineers, New York.

Borland, W.M., 1971. Reservoir sedimentation. In: Shen, H.W. (Ed.), River Mechanics, vol II. H. W. Shen, Colorado, USA.

Brune, G.M., 1953. Trap efficiency of reservoirs. Transaction American Geophysical Union 34 (3), 407–418.

Churchill, M.A., 1948. Discussion of "Analysis and use of reservoir sedimentation data. In: Gottschalk, L.C. (Ed.), Federal Inter-agency Sedimentation Conference, Colorado, 1947, pp. 139–140.

Crowder, B.M., 1987. Economic cost of reservoir sedimentation: a regional approach to estimating cropland erosion damages. Journal of Soil and Water Conservation 42 (3), 194–197.

Deyi, W., Fan, J., 1991. Method of preserving reservoir capacity. In: Lecture Notes of Regional Training Course on Reservoir Sedimentation and Control. Central Water Commission, New Delhi.

Dhruva Narayana, V.V., 1995. Research in Soil and Water Conservation in India with Special Emphasis on Watershed Management. Scientific Contribution No. INCOH/SAR-5/95. National Institute of Hydrology, Roorkee.

Finkner, S.C., Nearing, G.R., Foster, G.R., Gilley, J.E., 1989. A simplified equation for modeling sediment transport capacity. Transactions ASAE 32 (5), 355–359.

Garde, R.J., 1995. Reservoir Sedimentation. State of Art Report No. INCOH/SAR-6/95. National Institute of Hydrology, Roorkee.

Garde, R.J., Kothyari, U.C., 1986. Erosion in Indian catchments. In: Proceedings of 3rd International Symposium on River Sedimentation, Mississippi, USA.

Garde, R.J., Rangaraju, K.G., 1977. Mechanics of Sediment Transportation and Alluvial Stream Problems. Wiley Eastern Ltd., New Delhi.

Goel, M.K., Jain, S.K., 1996. Evaluation of reservoir sedimentation using multi-temporal IRA-1A LISS-II data. Asian-Pacific Remote Sensing and GIS Journal 8 (2), 39–43.

Goel, M.K., Jain, S.K., 1998. Reservoir Sedimentation Study for Ukai Dam Using Satellite Data. Report No. UM-1/97-98. NIH, Roorkee.

Goel, M.K., Jain, S.K., Agrawal, P.K., 2002. Assessment of sediment deposition rate in Bargi reservoir using digital image processing. Hydrological Science Journal 47, Special Issue, S81–S92.

Gupta, S.C., 1999. Status paper on reservoir sedimentation assessment using remote sensing techniques. In: Proceedings of the National Workshop on Reservoir Sedimentation Assessment Using Remote Sensing Data, May 7–8. NIH, Roorkee.

Gunatilake, H.M., Gopalakrishnan, C., 1999. The economics of reservoir sedimentation: a case study of Mahaweli reservoirs in Sri Lanka. Water Resources Development 15 (4), 511–526.

HEC, 1991. HEC – 6: Scour and Deposition in Rivers and Reservoirs. User's Manual. Hydrologic Engineering Center, Davis.

HEC, 1995. HEC-6 Reservoir Sediment Control Applications. Hydrologic Engineering Center, Davis, California, United States.

Holeman, J.N., 1968. The sediment yield of major rivers of the world. Water Resources Research 4 (4), 737–747.

Lane, L.J., Nearing, M.A. (Eds.), 1989. Water Erosion and Prediction Project Landscape Profile Model Documentation, Report No. 2. National Soil Erosion Research Laboratory, Purdue University, USA.

Lara, J.M., 1962. Revision of Procedures to Compute Sediment Distribution in Large Reservoirs. US Bureau of Reclamation, Colorado, USA.

Mahmood, K., 1987. Reservoir Sedimentation: Impact, Extent, Mitigation. Technical Report No. 71. The World Bank, Washington, USA.

Manavalan, P., Krishnamurthy, J., Manikiam, B., Adiga, S., Radhakrishnan, K., Chadrasekhar, M.G., 1993. Watershed response analysis using digital data integration technique. Advances in Space Research 13 (5), 177–180.

Morris, G.L., Fan, J., 1998. Reservoir Sedimentation Handbook. McGraw Hill Book Company, New York.

Murthy, B.N., 1977. Life of Reservoirs. Technical Report No. 19. Central Board of Irrigation and Power, New Delhi.

Nagy, I.V., Asante-Duah, K., Zsuffa, I., 2002. Hydrological Dimensioning and Operation of Reservoirs. Water Science and Technology Library, Kluwer Academic Publishers, Dordrecht.

Qiang, N., Dai, D., 1980. The problems of river sedimentation and the present status of its research in China. In: Proceedings of the International Symposium on River Sedimentation, Beijing.

Renard, K.G., Laften, J.M., Foser, G.R., McCool, D.K., 1994. The revised universal soil loss equation. In: Lal, R. (Ed.), Soil Erosion Research Methods. Soil and Water Conservation Society, Ankeny, IA.

Shangle, A.K., 1991. Reservoir sedimentation – status in India. In: Jalvigyan Sameeksha, vol V. INCOH, National Institute of Hydrology, Roorkee, pp. 63–70.

Sharma, S.A., Bhatt, H.P., Ajai, 1990. Generation of brightness and greenness transformations for IRS LISS II data. Photonirvachak 18 (3), 25–31.

Singh, V.P., 1992. Elementary Hydrology. Prentice Hall, New Jersey.

Strand, R.I., 1977. Sedimentation. In: Design of Small Dams. Bureau of Reclamation, U.S. Department of the Interior, Washington.

Thomas, W.A., 1979. Computer modeling of rivers: HEC 1-6. In: Shen, H.W. (Ed.), Modeling of Rivers. Wiley-Interscience, New York.

USDA, 1989. USDA-Water Erosion Prediction Project. NSERL Report No. 2. USDA Agricultural Research Service, Indiana.

Wang, G., Wu, B., Wang, Z.-Y., 2005. Sedimentation problems and management strategies of Sanmenxia Reservoir, Yellow River, China. Water Resources Research 41, W09417. https://doi.org/10.1029/2004WR003919.

Wischmeier, W.H., Smith, D.D., 1965. Predicting Rainfall-Erosion Losses from Cropland East of the Rocky Mountains. Agricultural Handbook No. 282. US Department of Agriculture, Washington.

Further Reading

Harmancioglu, N.B., Singh, V.P., Alpaslan, M.N., 1998. Environmental Data Management. Kluwer Academic Publishers, Dordrecht.

Lara, J.M., Pemberton, E.L., 1963. Initial unit weight of deposited sediments. In: Proceedings of Federal Inter-agency Sedimentation Conference. USDA-ARS Miscellaneous Publication 970, USA.

Milliman, J.D., Meade, R.H.(, 1983. World –wide delivery of sediments to the oceans. The Journal of Geology 91 (1), 1–21.

Tejwani, K.G., 1984. Reservoir sedimentation in India – its causes, control and future course of action. Water International 9 (4), 150–154.

Tejwani, K.G., 1987. Sedimentation of reservoirs in the Himalayan region. Mountain Research and Development 7 (3), 323–327.

CHAPTER

Water Quality Modeling

13

Chapter outline

Water Resources Systems Planning and Management. https://doi.org/10.1016/B978-0-12-821349-0.00017-4

The objectives of this chapter are:

- to introduce water quality concepts, monitoring, and standards,
- to provide a detailed discussion on river water quality modeling,
- to explain basics of lake water quality modeling, and
- to cover catchment and ground water quality models.

The ancient river valley civilizations that prospered in India, Egypt, and Mesopotamia were aware of the importance of the quality of water, although at that time water quality problems were not of serious nature. In recent years, with the growth of population and industrial activity, use of chemical fertilizers in agriculture, and use of chemicals in our daily lives, the quality of water has deteriorated at many places and has become an important issue all over the world. Clearly, water quality is closely linked to water use and to the state of economic development of the society. Although various countries have developed water quality standards for different purposes which are being enforced, it has not yet been possible to provide water of desired quality to all people. The quality of water is deteriorating sharply in many regions of the world and there are cases of poisoning of water by toxic chemicals, such as arsenic. According to the World Health Organization (WHO) estimates, more than one billion people drink unsafe water and 2.4 billion (40% of the human race) are without adequate sanitation. Every year, nearly 3.4 million people, mostly children, unnecessarily die due to water-related diseases (more than one million from malaria alone).

In view of the central role of supply of clean water in social health, organizations such as WHO strongly urge several basic measures, including purifying water (say, chlorination), and improving hygiene, as immediate means of improving social well-being in developing countries. Chlorination, for example, is a proven means of ridding piped water of disease-causing microorganisms. A good example of successful chlorination is to be found in the Maldives where a national control program used it in wells and in oral rehydration salts against diarrhea. According to WHO, 20 years after the program started, all of the Maldives islands have their own community rainwater collection tanks, and deaths from diarrhea are virtually unknown. Elsewhere, good water management has almost eradicated guinea worm, a disfiguring, disabling disease which afflicted 50 million people in Africa and Asia in the mid-1900s. By 1999 that number had fallen to below 100,000. But poor irrigation water management has, in sharp contrast, led to a huge spread of schistosomiasis (snail fever) in areas of the world where it never existed before. According to the WHO, an estimated 200 million people are infected today with schistosomiasis.

During the past 50 years, there has been a strong emphasis on medical interventions, including, for example, drug use and this has tended to reduce the attention and priority given to safe water supply and adequate sanitation. With the growing resistance to antibiotics, insecticides, and standard drugs, health authorities now understand the limitations of a purely medical approach. That makes safe water and sanitation more important than ever. Maintaining the quality of waters is an important aspect of sustainable development. Before discussing water quality, it is helpful to recall the physical properties of water.

13.1 RELEVANT PROPERTIES OF WATER

Two atoms of hydrogen and one atom of oxygen join together to form one molecule of water (H_2O). A noteworthy property of water is that it is denser than ice at its melting point. It is because of this property that ice floats on the surface of lakes and reservoirs in winter and the deeper layers of water are in liquid form and the aquatic life survives in these layers. Another noteworthy feature of water is its thermal properties. It has the highest specific heat (4.18 J/g/deg) of any known substance which means that the temperature of a large water body changes slowly. A very large amount of heat is needed for evaporation of water because its latent heat of vaporization is about 600 times larger than its heat capacity (the heat required to raise water temperature by 1°C). This, according to Maidment (1993), makes evaporation the dominant component of the energy balance associated with water cycle and about 23% of the solar radiation reaching the earth surface is absorbed by evaporating water. Vaporization of water also partly helps maintain the temperature of human body which contains about 70% water by weight.

The main cause of quality-related problems is the molecular structure of water that makes water a nearly universal solvent. Due to this structure, it dissolves the largest number of substances, many of which are harmful leading to water quality problems. Water has a high surface tension which makes raindrops acquire the shape of spherical drops and there is capillary rise in soil zone.

The temperature of water is an important physical variable that affects its density, viscosity, surface tension, vapor pressure, solubility, etc. The temperature of a water body depends on the location, season, source, and depth of sampling point. Normally, the temperature of water in surface bodies varies from 0 to 35°C. Selected physical, chemical, and biological characteristics of surface and ground waters are given in Table 13.1.

The range of a few key water quality parameters and their typical values are given in Table 13.2.

Table 13.1 Physical, chemical, and biological characteristics of various water sources.

	Water source	
Characteristics	**Typical surface water**	**Typical groundwater**
Suspended solids, mg/L	>50	–
Dissolved solids, mg/L	<100	>100
Temperature, °C	0.5–30	2.7–25
Turbidity (JTU)	10–50	–
Chemicals: inorganic matter		
Alkalinity, mg/L as $CaCO_3$	<100	>100
Hardness, mg/L as $CaCO_3$	<100	>100
Chlorides, mg/L	50	200
Heavy metals, mg/L	–	0.5
Nitrogen, mg/L	<10	<10

Continued

Table 13.1 Physical, chemical, and biological characteristics of various water sources.—cont'd

	Water source	
Characteristics	**Typical surface water**	**Typical groundwater**
Organic, mg/L	5	–
Nitrate, mg/L	<5	5
pH, (units)	6–9	6.5–8
Chemicals: organic matter		
Total organic carbon (TOC), mg/L	<5	–
Pesticides, mg/L	<0.1	–
Phenols, mg/L	<0.001	–
Surfactants, mg/L	<0.5	<0.5
Oxygen, mg/L	7.5	7.5
Bacteria, MPN/100 mL	<2000	<100
Viruses, plaque-forming units (pfu)	<10	<1

Adapted from Malina (1996).

Table 13.2 Typical value and range of parameters of river water quality.

Water quality parameter	Typical value	Observed range
Inorganic carbon (mg/L)	50	5–250
Total organic carbon (mg/L)	1–10	0.01–40
Dissolved organic carbon (mg/L)	1–6	0.3–32
Total organic matter (mg/L)	2–20	0.02–80
Biochemical oxygen demand (5 day) (mg/L)	3–15	1–100
Temperature	10–20	0–30
Dissolved oxygen	4–9	0–20

Compiled from McCutcheon et al. (1993), Biswas (1981), and others.

13.2 WATER QUALITY MONITORING

Fundamental to water resources, environmental, and ecosystem management is the assessment of the degree of water pollution which is done using modeling techniques. However, modeling requires continual monitoring of both water quantity and quality. The term monitoring indicates long-term, standardized measurement and observation of the environment in order to define the status and trends in the uses of resources. In survey, a finite duration (usually short) intensive program is launched to measure and observe the quality of aquatic environment for a specific purpose. The term surveillance is used for a continuous specific measurement and observation for the purpose of water quality measurement and operational activities. Note that all the three activities require collection of data and therefore sometimes the term monitoring is used to denote all three. Usually the value of a water quality variable is a function of space and time. Therefore, one can write

$$c = f\,(x, y, z, t) \tag{13.1}$$

The quality of a water body may show rapid variations. In general, the variations are more in river, somewhat less in lakes, and much less in aquifers. Fig. 13.1 shows a schematic of facets of water quality monitoring. Monitoring and modeling of water quality variations require, inter alia, a detailed knowledge of hydraulics and hydrology of the water body. In case a program is launched from an operational point of view, the periods of worse conditions, such as summers when the flows are small, and the periods when the concentration of pollutants is likely to be highest, should be given more attention.

Guidelines regarding frequency of observation for trend monitoring were suggested by Chapman (1992). The characteristics of spatial and temporal variations in water quality, following Chapman (1992), are given in Table 13.3. The design of any water quality monitoring program should be periodically examined. The review should take into account new approaches, instruments, techniques, etc. The observers who are deployed for water quality samples and analysis must have the requisite background and training, since these are specialized activities.

The Global Water Quality database and information system GEMStat is hosted, operated, and maintained by the International Centre for Water Resources and Global Change (ICWRGC) in Koblenz, Germany (https://gemstat.org). It operates within the framework of the GEMS/Water Programme of the United Nations Environment Programme (UNEP), and in cooperation with the Federal Institute of Hydrology. GEMStat hosts water quality data of ground and surface waters providing a global overview of the condition of water bodies and the trends at global, regional and local levels. The database is growing and currently, it contains > 18 million entries for rivers, lakes, reservoirs, wetlands

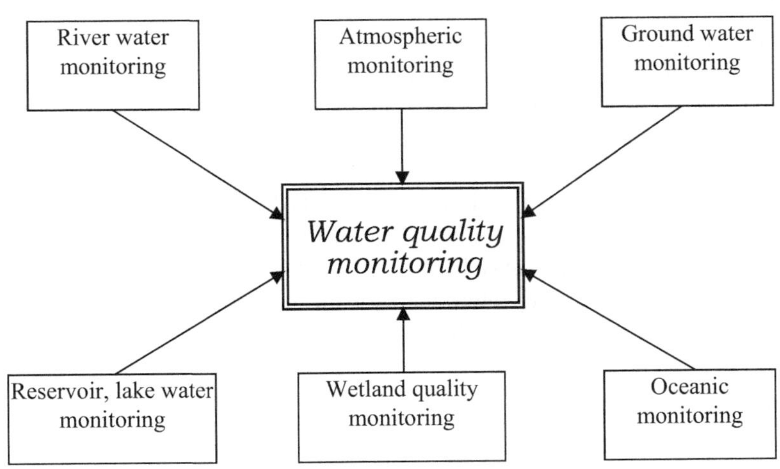

FIGURE 13.1

Facets of water quality monitoring.

Table 13.3 Spatial and temporal variations in water quality variables.

Rivers	Lakes and reservoirs	Ground water
Spatial variations (x—longitudinal direction; y—transverse direction; z—vertical direction)		
In fully mixed narrow rivers—variability only in x. Wide rivers—variability in x and y, particularly downstream of sewage outfall or junction with a tributary.	High variability in z for most lakes. High variability in x, y, and z in irregularly shaped lakes. In reservoirs, variation also depends on purposes and operation.	Commonly, high variability in x and y. In multilayer aquifers and in the unsaturated zone, high variability in x, y, and z may be present.
Temporal variations		
Variation is closely related to discharge, but a unique relation may not exist. High sediment load during floods. High BOD, low DO, higher concentration of pollutants during low flows.	Thermal stratification influenced by weather changes. Variations depend upon residence time.	Low temporal variation. Definite trend can be seen if the aquifer is hydraulically connected to a river or lake that receives pollutants.

and groundwater systems from 94 countries and more than 15,000 stations. Overall, data is available for the time period from 1965 to 2022 and more than 200 parameters (see data section for more information).

13.2.1 Monitoring Network

The location of sampling stations largely depends on the purpose of study. In general, sampling stations may be divided into two categories: i) basic stations and ii) auxiliary stations. Basic stations are usually located at the mouth of main streams and major tributaries, downstream of river development projects, at or near hydrometric stations, gauge discharge sites, at state boundaries, upstream and downstream of waste outfalls, industrial and urban centers, and at points of water use. Auxiliary stations are established to investigate the effect of pollutants discharged into a stream, determination of assimilation capacity of stream, and similar studies with a common objective. These stations are purposely related to each other and may be moved to another place or operated temporarily. Data collected at these stations are used to interpolate stream quality at other points or predict the water quality described under a set of given conditions.

13.2.2 Sampling Program

Water quality samples should be taken at places where the composition is expected to be homogeneous over the cross-section. Usually there will be no difficulty in locating such points in effluents discharged from pipes or small open channels. Sampling points in rivers should be well away from any possible disturbing influence, such as pools, stagnant zones, heavy growth of weed or sewage fungus, or points where ground water enters, unless it is desired to specifically study their effects on water quality.

Sampling frequency largely depends on the purpose of network, relative importance of the station, variability of the data, and accessibility of the station. The workload involved and the financial commitment for a decided frequency of sampling must match the available resources. At newly started stations, samples may be collected at a higher frequency so that within 2–3 years, a sufficient number of observations are available for statistical analysis to determine mean, variations, cycles, and trends in the relevant parameters used in monitoring water quality. The frequency of sampling may be changed after such an evaluation. For basic station networks operated to collect baseline data, the sampling frequency may be from 3 to 4 times per year to monthly. In any event, at least one sample should be taken in each season. When stations are operated in relation to specific use, the frequency of sampling will be governed by that use. The sampling of raw water for water works may be carried out daily. Weekly or bimonthly samples may be collected at a sampling station which is maintained downstream of a waste outfall to monitor the effectiveness of the wastewater management program and its effect on the stream. Where the stream water is used for swimming and other recreational purposes, sampling may be confined only to the season of use.

The details of water quality monitoring critically depend on the level of deterioration of the water body. Clearly, greater the deterioration of the water body is, more intense—in terms of the number of parameters and frequency of observation—the monitoring is. The aim of such monitoring may be to identify the sources and causes of pollution, to decide a program of treatment, or to assess the effectiveness of remedial action.

Critical parameters: The choice of water quality characteristics that are to be measured must necessarily be determined by local circumstances. In general, it will be necessary to compromise between the number of parameters measured and the marginal difference that information on each parameter makes to subsequent management decisions. The number of organic constituents present in water could be very large, but it would be impractical to measure them all on a routine basis. Instead, it will be necessary to rely on nonspecific tests, such as chemical oxygen demand (COD) or total organic carbon (TOC), to indicate the general level of organic pollution, together with a few specific tests for selected pollutants with particularly harmful properties.

As a general idea, the parameters to be measured may include the following: TOC, biochemical oxygen demand (BOD), cyanide, pesticides, suspended solids, nitrogen, fluoride, cadmium, chromium, copper, lead, nickel, zinc, mercury, boron, dissolved oxygen (DO), pH value, and coliform bacteria. If the objective is to develop a model which would permit the evaluation of the influence of pollution control measures on the cost of treatment of public water supply, additional information might be needed on total hardness, alkalinity, calcium hardness, sulfate, phosphate, sodium, potassium, etc.

Microplastics: As the use of plastic materials is increasing, there is rising prevalence of plastic material in water. In particular, microplastics are small plastic pieces which are less than five millimeters long. There are a variety of sources of microplastics including breaking of larger plastic pieces, resin pellets used to manufacture plastic, the microbeads (small plastic beads used in health and beauty products). Microplastics can pose many health hazards to humans. Microplastics can be absorbed, ingested, or inhaled into the human body through the skin, gastrointestinal system, or lungs. Inside the body, they can physically block the digestive system, stimulate the mucous membrane, and may as well damage it. When the size of microplastics becomes smaller than 1 micrometer, they form nanoplastics, which are ultrafine plastics. Such particles can pass through the primary tissue barrier in the body and penetrate the capillary blood vessel through the blood stream. Over a period of time, they can cause much harm.

13.2.3 Water Quality Standards

Whether water from a given source can be used for a specific purpose critically depends on its quality. Even though adequate water may be available from a source, if the quality is not up to the mark, either it cannot be used or it has to be treated before use. Water quality standards are legally enforceable criteria that specify parameters, such as temperature, concentration of various pollutants, etc., whose adherence makes water suitable for a given purpose. Clearly, the permissible limits depend on the intended use of a source, for example, water from a particular source may be good for irrigation, but it may not be suitable for drinking.

The continued advances in research on water quality and the growing concern with risks to health have caused various governments and international bodies to draw standards of water quality for various purposes. The WHO had issued water quality guidelines 1971. These were later revised by following a philosophy that emphasizes a risk-benefit approach in the formulation and enforcement of national standards. This new approach was contained in the publication *Guidelines for drinking-water quality*, issued by World Health Organization (1996). Research findings and field experience continue to expand the knowledge base on water and health. The latest WHO guidelines and standards are available on its website at (https://www.who.int/teams/environment-climate-change-and-health/water-sanitation-and-health/water-safety-and-quality/drinking-water-quality-guidelines) and in (WHO, 2022). Table 13.4 gives the bacteriological quality of drinking water, and Table 13.5 lists substances and parameters in drinking water that may give rise to complaints from consumers. The various national governments have drawn standards for use within the territory of their jurisdiction.

Table 13.4 Bacteriological quality of drinking water.[a]

Organisms	Guideline value
All water intended for drinking	
Escherichia coli or thermotolerant coliform bacteria[b,c]	Must not be detectable in any 100-mL sample
Treated water entering the distribution system	
E. coli or thermotolerant coliform bacteria[b]	Must not be detectable in any 100-mL sample
Total coliform bacteria	Must not be detectable in any 100-mL sample
Treated water in the distribution system	
E. coli or thermotolerant coliform bacteria[b]	Must not be detectable in any 100-mL sample
Total coliform bacteria	Must not be detectable in any 100-mL sample. In the case of large supplies, where sufficient samples are examined, must not be present in 95% of samples taken throughout any 12-month period

[a]*Immediate investigative action must be taken if either* E. coli *or total coliform bacteria are detected. The minimal action in the case of total coliform bacteria is repeat sampling; if these bacteria are detected in the repeat sample, the cause must be determined by immediate further investigation.*

[b]*Although* E. coli *is the more precise indicator of fecal pollution, the count of thermotolerant coliform bacteria is an acceptable alternative. If necessary, proper confirmatory tests must be carried out. Total coliform bacteria are not acceptable indicators of the sanitary quality of rural water supplies, particularly in tropical areas where many bacteria of no sanitary significance occur in almost all untreated supplies.*

[c]*It is recognized that, in the great majority of rural water supplies in developing countries, fecal contamination is widespread. Under these conditions, the national surveillance agency should set medium-term targets for the progressive improvement of water supplies, as recommended in Volume 3 of* Guidelines for drinking-water quality.

Source: WHO website www.who.int.

Table 13.5 Substances and parameters in drinking water that may give rise to complaints from consumers.

	Levels likely to give rise to consumer complaints[a]	Reasons for consumer complaints
Physical parameters		
Colour	15 TCU[b]	Appearance
Taste and odor	–	Should be acceptable
Temperature	–	Should be acceptable
Turbidity	5 NTU[c]	Appearance; for effective terminal disinfection, median turbidity = 1 NTU, single sample = 5 NTU
Inorganic constituents		
Aluminum	0.2 mg/L	Depositions, discoloration
Ammonia	1.5 mg/L	Odor and taste
Chloride	250 mg/L	Taste, corrosion
Copper	1 mg/L	Staining of laundry and sanitary ware (health-based provisional guideline value 2 mg/L)
Hardness	–	High hardness: scale deposition, scum formation low hardness: possible corrosion
Hydrogen sulfide	0.05 mg/L	Odor and taste
Iron	0.3 mg/L	Staining of laundry and sanitary ware
Manganese	0.1 mg/L	Staining of laundry and sanitary ware (health-based guideline value 0.5 mg/L)
Dissolved oxygen	–	Indirect effects
Ph	–	Low pH: corrosion High pH: taste, soapy feel Preferably <8.0 for effective disinfection with chlorine
Sodium	200 mg/L	Taste
Sulfate	250 mg/L	Taste, corrosion
Total dissolved solids	1000 mg/L	Taste
Zinc	3 mg/L	Appearance, taste
Organic constituents		
Toluene	24–170 μg/L	Odor, taste (health-based guideline value 700 μg/L)
Xylene	20–1800 μg/L	Odor, taste (health-based guideline value 500 μg/L)
Ethylbenzene	2–200 μg/L	Odor, taste (health-based guideline value 300 μg/L)
Styrene	4–2600 μg/L	Odor, taste (health-based guideline value 20 μg/L)

Continued

Table 13.5 Substances and parameters in drinking water that may give rise to complaints from consumers.—cont'd

	Levels likely to give rise to consumer complaints[a]	Reasons for consumer complaints
Monochlorobenzene	10–120 μg/L	Odor, taste (health-based guideline value 300 μg/L)
1,2-Dichlorobenzene	1–10 μg/L	Odor, taste (health-based guideline value 1000 μg/L)
1,4-Dichlorobenzene	0.3–30 μg/L	Odor, taste (health-based guideline value 300 μg/L)
Trichlorobenzenes (total)	5–50 μg/L	Odor, taste (health-based guideline value 20 μg/L)
Synthetic detergents	–	Foaming, taste, odor
Disinfectants and disinfectant by-products		
Chlorine	500–1000 μg/L	Taste and odor (health-based guideline value 5 μg/L)
Chlorophenols		
2-Chlorophenol	0.1–10 μg/L	Taste, odor
2,4-Dichlorophenol	0.3–40 μg/L	Taste, odor
2,4,6-Trichlorophenol	2–300 μg/L	Taste, odor (health-based guideline value 200 μg/L)

[a]*The levels indicated are not precise numbers. Problems may occur at lower or higher values according to local circumstances. A range of taste and odor threshold concentrations is given for organic constituents.*
[b]*TCU, true color unit.*
[c]*NTU, nephelometric turbidity unit.*
Source: WHO Website www.who.int.

13.2.4 Water Quality-Based Classification of River

Rivers are generally classified on the basis of annual average discharge in various categories: large rivers >1000 m^3/s, rivers 150–1000 m^3/s, streams 5–150 m^3/s, and small stream <5 m^3/s. Note that these distinctions do not consider annual variations of river flow. In subtropical and arid zones, rivers often range from zero discharge during the dry season to large discharge during wet periods.

Knowledge of the natural water quality of a stream, its self-purification capacity, and the effect of various wastes on its ecosystem is necessary for a planned development of its uses. While the overlying strata protect the ground water, a stream is highly vulnerable to anthropogenic activities in its basin. Also, they are one of the most extensively exploited water resources. Although water does not exist in streams in its pure chemical form, it may be classified as unpolluted in view of its beneficial use. A polluted stream tends to regain its natural quality with time as the water flows in it. Various physical, chemical, and biological phenomena are responsible for this self-purification. Some conservative pollutants may persist for a long time before they are diluted to an insignificant level or are removed from the liquid phase due to physicochemical reactions of precipitation and adsorption.

One of the most efficient and effective ways to present numerical results is through the use of graphs and maps. The visualization of findings in the form of maps provides an easy and rapid comprehension of the situation. Additionally, such maps may show the location of human settlements, industrial plants, major water outfalls and intakes, water purification plants, power plants, etc. A band, the width of which indicates the mean discharge, represents the river course. Colors indicate the quality of the water. Usually the following color code is used:

Blue:	Category I, no pollution to slight pollution
Green:	Category II, moderate pollution
Yellow:	Category III, heavy pollution
Red:	Category IV, excessive pollution
Black:	Category V, zone of devastation.

13.3 RIVER WATER QUALITY MODELING

Water quality modeling has witnessed a rapid development in the past 50 years and as a result a multitude of mathematical models have been developed (Stolarska and Skrzypski, 2012). These models have been applied to determine changes in water quality in rivers, canals, lakes, reservoirs, wetlands, and bays. The changes in water quality are caused by wastewater discharge, dumping of industrial waste, chemical pollution from agricultural lands, acid rain, urban waste heaped as landfill, and so on. The models are vital for integrated water resources management, for they aid in analyzing different pollution scenarios, assessing potential impacts of human activities, determining the fate of pollutants in water and assessing their related hazards, computing the effect of changes in the aquatic environment, setting limits for pollutant discharge, identifying the location of sources of pollution, deciphering the causes of water quality deterioration in streams or other water bodies, evaluating environmental impact, and selecting an optimal approach for sustainable development. Thus, water quality models are essentially decision support tools.

Water quality models can be broadly classified into laboratory and mathematical models (UNESCO, 2005). Our focus in this chapter is on mathematical models which are computer simulation models. These models can be classified, depending on the complexity based on dimensionality, theoretical basis, type of pollutants, area of application, nature, spatial detail, linearity, time variability or state of flow, and data requirement. Models can be 1D, 2D, or 3D, depending on dimensionality; physically based, conceptual, or empirical, depending on theory; conservative or nonconservative, physical, chemical, or biological, depending on the pollutant type; watershed, groundwater, river, canal, lake, reservoir, wetland, or integrated, depending on the application; deterministic, stochastic, or mixed, depending on the nature; lumped or distributed, depending on spatial detail; linear, quasilinear, or nonlinear, depending on linearity; steady state or unsteady state, depending on time variability; and extensive, moderate, or low, depending on data requirements (Tsakiris and Alexakis, 2012).

Mathematical river water quality models consist of a set of governing dynamic equations that describe hydrologic, thermal, and biochemical processes that take place in the riverine ecosystem. These are the equations of conservation of mass, momentum, and energy. However, because of incomplete understanding of some of the processes as well as different levels of detail, a whole lot of different models are available. Mostly one-dimensional models are used in river studies; higher dimensional models have been mainly employed to study lakes and estuaries. The earliest attempt to model water quality in rivers was made by Streeter and Phelps in 1925. Since that time, many additions have been made to improve the Streeter–Phelps model.

River water quality modeling has two functions: (1) forecasting the developments in a basin and consequent effects on water quality, and (2) forecasting of changes in the concentration of pollutants within the stream channel. The second component has two problems. The first is the prediction problem, that is, the solute concentration at a specific point in time and space are known and the solute concentrations at various times at various locations downstream are to be determined. The second objective involves formulating across the problem in a descriptive manner, that is, a time series of solute concentrations along the stream is given and it is required to find what the dominant processes are that control the solute concentration.

Forecasting the concentration of contaminants which are stable in nature is simple, since their concentration changes only due to dilution or evaporation. Concentrations of many biologically stable organic and inorganic substance change due to precipitation, sedimentation, adsorption and chemical reactions with other substances. These reactions are influenced by several factors, such as pH, temperature, bed characteristics, etc., and they are to be considered separately for respective mixtures of substances and local conditions.

Similar to hydrographs, the plot of concentration of pollutants with respect to time is known as a pollutograph. The graph of pollutant load (concentration) multiplied by flow rate/time is known as loadograph. Pollutants may enter into a water body from a point source or a nonpoint source. As the name suggests, a point source indicates that the pollutants enter the water body through a well-defined outlet or locations as happens with municipal or industrial waste. Other examples of point sources include discharges and spills that occur due to accidents, for example, when transporters of hazardous materials release contaminants into a water body. The pollutants from a nonpoint source come from flow distributed over large land surfaces; a typical example is runoff from agricultural areas. Nonpoint sources of solutes are distributed along the watercourse. Sediment and associated pollutants also originate from nonpoint sources and result from the interplay of a large number of factors, such as climate, precipitation (amount, intensity, and distribution), geology, soil properties, land use, and streamflow characteristics. Note that this differentiation between point and nonpoint is a bit confusing because if the waste from a large city is collected to a sewer system and discharged into a river at a well-defined outlet, it will be termed as point source pollution, although the catchment producing waste is spread over a large area.

Pollution sources may also be roughly classified according to the time-pattern of outflow. Some sources continuously discharge waste into a water body, while others may do it instantaneously or intermittently. Continuous sources dump pollutants into the water body over a long period of time. A common example of a continuous source is a municipal wastewater treatment plant. Although mass-loading rates may vary in time, most of these treatment plants continuously discharge effluents into the water body. It needs to be highlighted that the treatment of concentrated wastes is easier than diluted wastes.

The waste from an instantaneous source enters the water body over a very short period of time. Although truly instantaneous sources are hard to find, situations do arise in which solutes are added to the river over time intervals that are indeed very small relative to the time span of interest. The addition of pollutant to a water body in one gulp is one example of an instantaneous source. Another example is an accidental spill from a tanker where contaminants enter a river or lake in just a few minutes.

The riverine ecosystem can be schematically represented as a food web (Fig. 13.2). Organic debris of natural origin (leaves, humus, etc.) constitute nutrients for bacteria. The bacterial population degrades organic substances into inorganic compounds, mainly nitrates, phosphates, and carbon dioxide (self-purification process) in aerobic conditions. These inorganic compounds are the principal nutrients for phytoplankton, which consumes them to make biomass through the process of photosynthesis in the presence of sun light. A useful by-product of photosynthesis is oxygen gas. Another source of oxygen in water is diffusion at the free surface, known as superficial reaeration. Oxygen that is made available through these two processes counterbalances the oxygen deficit due to bacterial activity.

The biochemical and chemical processes of a river are significantly influenced by hydraulic and thermal conditions. The main hydraulic variables that influence water quality are flow velocity, depth, and discharge. Obviously, an increase of discharge increases the dilution rate and decreases pollutant concentration. An increase of flow velocity reduces biochemical self-purification, since the travel time between two river sections reduces. The amount of sediment load being transported by a river depends on flow velocity. Another important consequence of velocity variations is turbulence in flow. An increase in the turbulence definitely increases reaeration, because there is better intake and mixing of oxygen at choppy layer at the surface. Due to this reason, considerably higher aeration takes place in shallow, fast flowing streams. In many water-treatment processes, artificial aeration is an important activity. Besides aeration, turbulence also influences the ecological interactions within the food web;

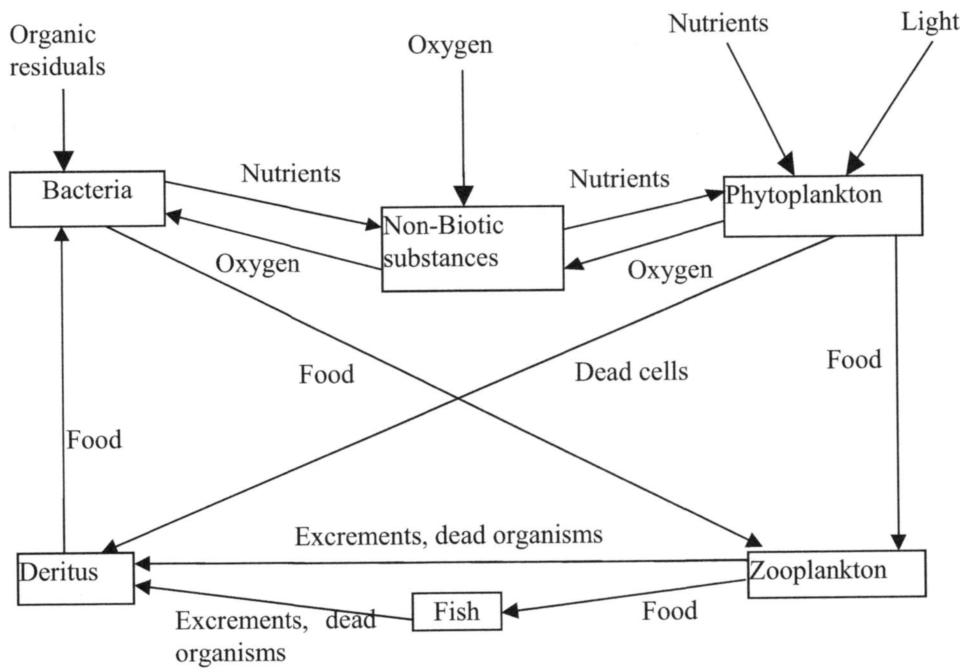

FIGURE 13.2

Simplified representation of aquatic ecosystem as a food web.

After Gandolfi et al. (1996).

the efficiency of bacterial degradation processes which involve exoenzyme drops if turbulence increases. As far as the flow depth is concerned, the penetration of sunlight is small at greater depths and this causes a slowdown in photosynthetic processes.

Clearly, high organic loads cause a strong depletion of the DO in water. Sometimes this causes severe damages to higher-order species and produces considerable amounts of mineral nutrients. Mineral nutrients may also directly enter into a river by various sources, such as treatment plant outflows, scour and leaching from agricultural lands, or some industrial processes. In rivers with low flowing velocities, the time of residence is quite large and in such cases, high contents of mineral nutrients can be assimilated by algae, giving rise to intensive algal blooms. As a consequence, the mass of phytoplankton may block the penetration of sunlight in deeper layers. This may cause massive death of organisms, which in turn causes oxygen consumption and a further increase of mineral nutrients. In this way, a degenerative phenomenon, known as *eutrophic cycle*, may be established in the water body.

Among nondegradable materials, heavy metals, even at relatively small concentrations, may seriously influence the bacterial metabolism. At times, this may lead to the death of bacteria. Oily substances form a thin film on the water surface, thus reducing reaeration. Penetration of sunlight at deeper layers can be reduced due to high concentration of suspended solids.

Regarding thermal conditions, unless the temperature is very high, biochemical processes accelerate with increase in temperature. The solubility of gases in water decreases with the rise in temperature. A direct fall out is the slower intake of oxygen through superficial reaeration and higher BOD.

13.3.1 Components of a River Water Quality Model

In a detailed mathematical model of river water quality, all the biochemical, hydraulic, and thermal phenomena must be described by means of differential equations. Gandolfi et al. (1996) suggested that the water quality model must be composed by three submodels: biochemical, hydraulic, and thermal submodels, and these could be coupled together as shown in Fig. 13.3a. Evidently, there will be interactions among the state variables of these submodels, although some of the interactions will be

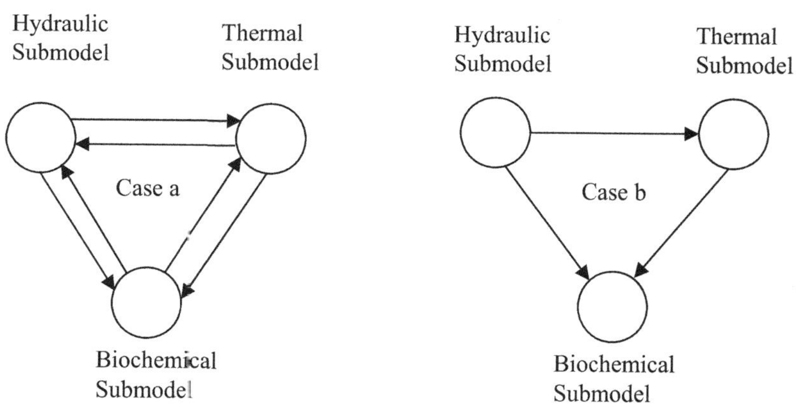

FIGURE 13.3

Water quality model structure: (a) the general case; (b) when minor interactions are ignored.

Source: Gandolfi et al. (1996).

stronger and some weaker. In particular, the influence of temperature variations (in the range of values normally encountered in most natural rivers) on the hydraulic conditions is negligible. Based on this reasoning, the structure of the water quality model can be simplified as shown in Fig. 13.3b. These three submodels can be sequentially solved, beginning with the hydraulic submodel and ending with the biochemical one.

13.3.2 Hydraulic and Thermal Models

In one-dimensional unsteady flow process, two important variables are the flow depth h (or the area of flow cross-section A which is a function of h) and the mean velocity v (or discharge). The two pertinent equations are the mass balance and momentum equations. The St. Venant equations are the general form of these equations.

In a thermal model, one state variable, that is, temperature, is adequate to provide the requisite information. The temperature is an input to the biochemical model and in view of this and observations about temperature above, one may skip the thermal model.

13.3.3 Biochemical Model

A large number of variables are involved in the biochemical model. The number of agents that are involved in the self-purification processes is so huge that it would be impractical, because of logistics and computational problems, to introduce a state variable corresponding to each of them. If one were to introduce all the interactions of a real-life case, the resulting model would be unmanageable. This calls for a need to reduce the number of variables either by substitution or by grouping similar variables and then letting one variable represent each group.

A pertinent variable that is a representative of the water quality status or the self-purification process in a river is the DO concentration. A detailed discussion on DO can be found in Section 13.4. BOD is the measure of the amount of oxygen needed for complete biochemical oxidation of all the matter present in a unit volume of water. Therefore, BOD is the aggregated variable representative of all the biochemically oxidizable compounds in that matter. Not all water quality problems require highly sophisticated models. In fact, many water quality decisions can be taken, based on representative indicators, such as BOD.

It needs to be emphasized here that any biochemical model is a highly simplified representation of the processes that actually take place in the water body. Most parameters of the equations cannot be measured in the field and, therefore, these models require careful calibration. Of course, one needs to have sufficient field data to set-up the model.

13.3.4 Geochemical Processes

The discussion so far focused on physical processes that aid and affect solute transport. Although these processes play a large role in determining the fate of solutes, chemical and biological processes may have equally important impact on solute transport. Some of the basic geochemical reactions affecting solutes in natural waters are explained in what follows.

A dissolved chemical species may take part in two types of chemical reactions: homogeneous and heterogeneous. The reactions in which dissolved species interact with species of the same phase are known as homogeneous reactions. These reactions are termed as homogeneous, because the species involved are in the same (dissolved) form. Some examples of homogeneous reactions are: acid-base, complexation, and hydrolysis reactions.

The second type of reactions in which a dissolved species may take part are heterogeneous reactions. These reactions involve species from more than one phase. Thus, a heterogeneous reaction may involve the interaction of a dissolved phase species with a species in gas or solid phase. For example, the DO concentrations in water may be dependent on the oxygen in the atmosphere over the surface water body. If the water body is undersaturated in terms of DO concentrations, oxygen will enter the water body through the air/water interface and the concentrations of DO will increase. Conversely, if water becomes oversaturated with oxygen due to some reason, oxygen will escape to the atmosphere through the process of degassing and the concentrations of DO will decrease. Other examples of heterogeneous reactions include solid-phase reactions, such as precipitation/dissolution and sorption.

There are two fundamental approaches to study geochemical reactions in natural waters. The appropriate approach for a given problem depends on the rates at which relevant reactions occur, relative to the time scale of interest. If reaction rates are "slow" relative to the study time scale, a kinetic approach is employed, wherein the dynamics of the system are studied. If reaction rates are relatively "fast," an equilibrium approach is implemented to determine the steady-state composition of waters. Other factors, such as data availability, also play a role in determining the appropriate approach.

All geochemical reactions require some finite amount of time, τ_c, for completion. The required amount of time is dependent on the rate of reaction, r, such that τ_c and r are inversely related (Runkel and Bencala, 1995). A kinetic approach is often implemented when τ_c exceeds the time period of observation, τ_0. For example, τ_c for many biologically mediated oxidation/reduction reactions is on the order of days to years. If the interest is in the state of the system over the course of a day, a kinetic approach is needed to describe how the concentrations of the chemical species change over time. In this case, differential equations may be written to describe the temporal variation in species concentrations and kinetic rate constants (r) may be empirically determined. The equations may be solved to yield the species concentrations over time.

13.3.5 Sorption/Desorption

Sorption is the process in which a dissolved species becomes associated with a solid surface. This is an important process that controls the concentrations of solutes in natural waters. The dissolved species that takes part in a sorption reaction is known as a *sorbate*, and the solid species with which it gets associated with is known as the *sorbent*. In case the dissolved species penetrates the sorbent, the process is known as *absorption*. The term *adsorption* is used to describe the interaction of the dissolved species with the surface or interface of the sorbent (Weber et al., 1991).

Sorption may have a significant effect on dissolved inorganic species in water. Morel and Hering (1993) noted that trace elements, such as zinc and copper, are known to sorb to hydrous ferric oxides. Kuwabara et al. (1984) found copper to sorb to the bed of a small stream.

There is close coupling between physical and geochemical processes in a river. The distribution of a solute mass among three phases, viz., dissolved, solid, and gas, is determined by geochemical reactions and in this way, they affect the amount of mass that is available for physical transport. The physical transport mechanisms affect the location of solute mass. This is subjected to various geochemical actions as it moves in water.

The coupling between transport and chemistry may be illustrated by examining a conceptual classification of geochemistry based on homogenous and heterogeneous reactions. This distinction between reaction types becomes important when considered within the context of physical transport. To begin with, heterogeneous reactions are divided into three subclasses. In heterogeneous I reactions, a dissolved species interacts with a mobile solid species. Examples of heterogeneous I reactions include the formation of a precipitate in the water column due to oversaturated conditions, and the sorption of a dissolved species onto a particulate solid within the water column. Heterogeneous II reactions are those in which a dissolved species interacts with a stationary solid species. An example of a heterogeneous II reaction is the sorption of a dissolved species onto a stationary solid such as particles coating the streambed or debris in the channel. Finally, in heterogeneous III reactions, a dissolved species interacts with a gas phase, such as the degassing of DO to the atmosphere.

13.3.6 Pollutant Concentration and Load

Concentration is the primary measure of the quantity of a constituent in a fluid environment and is defined as

$$C = M/V \tag{13.2}$$

where C is the concentration (mg/L), m is the mass of constituent (mg), and V is the volume of fluid (L). Note that the density of water is nearly 1.0 g/cm^3 and hence the units of mg/L and g/m^3 are numerically equivalent to *parts per million* (ppm) by mass in water. For variables that are not measured in mass units, concentration may be defined in terms of numbers. For example, bacteria are often measured as *most probable number* (MPN) per unit volume.

The impact of constituents on a water body is influenced by both concentration and load. The term *load* may mean either the total mass M in a volume V of water (see Eq. 13.2)

$$M = CV \tag{13.3}$$

or the mass flow rate L (mass/time) in water flowing with a discharge Q (m^3/s):

$$L = CQ \tag{13.4}$$

13.3.7 Transport of Solutes in Rivers

A *solute* can be defined as any substance or entity that is transported downstream by the flowing waters. Thus, solutes may be pollutants, such as pesticides and hydrocarbons, or naturally occurring substances, such as dissolved gases, nutrients, and trace elements. Study of the processes affecting solutes is important, because pollutants may pose a threat to public health when the water of the affected body is used for drinking or recreational purposes. The impact of pollutants on the aquatic organisms that live in the stream ecosystem is also an important issue. The solutes that occur naturally are also of interest, because they often interact with pollutants and change their toxicity. The concentrations of naturally occurring solutes are also affected by human activities and in many cases this significantly degrades the quality of water body. Nutrients also play a role in eutrophication—the process in which abnormally high nutrient concentrations cause excessive plant growth.

The fate and transport of solutes in rivers depends on a number of processes. The most obvious and perceptible are the physical processes which cause solutes to move downstream and mix with other constituents in water. The residence time of a given solute in the system under study is often determined by physical transport characteristics. The additional processes that influence solute concentrations are the chemical, biological, and geochemical reactions that take place in the water body. The residence time effectively determines the time scales over which chemical and biological processes have an effect. All these factors make a close and complex coupling among the various processes involved in solute transport.

Consider that a dose of pollutant is released at the center of a free-flowing stream. A few important terms and processes related to its transport are defined below.

Flux: Flux is the rate of flow of mass or energy per unit area normal to the direction of flow. The transport of a constituent is measured by the flux; the common units are quantity per unit area per unit time. If the constituent is measured in units of mass, a typical unit of flux is g/m^2s. The movement and concentration of a pollutant is affected by two important transport mechanisms: advection and dispersion.

Advection: Advection, also known as convection, is a bodily transport of the constituent due to the motion of the fluid. The constituent moves with the fluid velocity without change in concentration. The flux F due to advection is the product of velocity and concentration:

$$F = UC \tag{13.5}$$

where U is the velocity (m/s) and C is the concentration in terms of quantity per unit volume (gm/m^3).

Advection does not alter the shape (distribution of concentration vs. distance) of the constituent distribution as long as the velocity distribution remains uniform. Fig. 13.4a shows a case of pure advection. A dose of a dye is released into water at time t_0. At time greater than t_0, the location of dye changes as it moves in the downstream direction with water, but its shape and concentration remain unaltered.

Dispersion: As the constituent moves in the downstream direction, small-scale mixing causes it to spread out, thereby increasing the volume and decreasing the concentration of the constituent-containing water. This process is called dispersion. Sometimes the terms dispersion and diffusion are used interchangeably. See Holley (1996) for a detailed treatment of the subject. Dispersion is attributed to both molecular diffusion and velocity variations caused by shear stress. Molecular diffusion takes place due to random motion of particles. This is a significant mixing process in still water or laminar flow, such as a calm lake or ground water. In natural streams, the role of molecular diffusion is less and the effect of velocity variation becomes the dominant mixing mechanism. It is well known that in a river, the flow velocity increases with distance from the stream bed and is maximum at some point close to the water surface. In the lateral direction, the velocity is the maximum at the center and reduces as the bank is approached. These velocity variations cause considerable mixing as different particles move with different velocities.

The flux due to diffusion is given by Fick's law of diffusion. According to this law, the mass flux due to molecular diffusion is proportional to the concentration gradient. Thus, flux in the x direction would be:

$$F_x = -E_x \frac{\partial C}{\partial x} \tag{13.6}$$

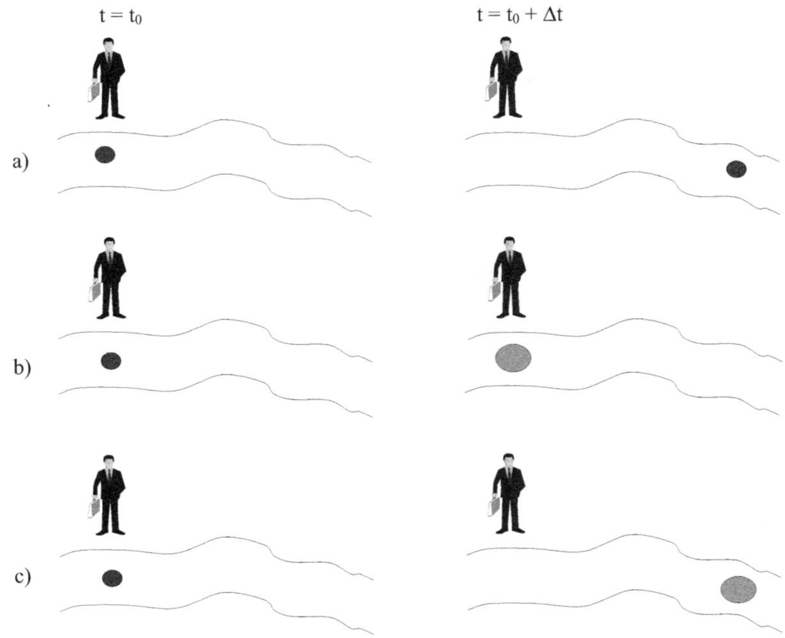

FIGURE 13.4

Effects of advection and dispersion on a constituent. (a) The case of advection only; the location of the constituent changes with time, but its volume is constant; (b) the case of dispersion only; the center of the constituent is stationary, but the volume increases; (c) combined effect of advection and dispersion; the location and volume of the constituent changes with time.

After Runkel and Bencala (1995).

where concentration C is expressed in units of quantity per unit volume and E_x is the longitudinal dispersion coefficient or diffusivity, with units of L^2/T. The negative sign indicates a positive flux in the direction of negative gradient (i.e., in the direction of decreasing concentration). When dispersion is the only transport process, the behavior of a constituent is shown in Fig. 13.4b. Here, the center of the constituent mass remains at the same place, but the volume of constituent-containing water tends to increase with time.

The behavior of the constituent mass due to the combined effect of advection and dispersion is shown in Fig. 13.4c. Here, the location of the center moves downstream with time and its concentration goes on reducing with the passage of time.

In surface waters, most diffusion is by the process of turbulent diffusion. Therefore, in practice, diffusion often is used to account for all unknown factors in a problem, including undefined velocity fields, trapping of the constituents along boundaries, secondary currents, density effects, etc. If velocity fluctuations due to turbulence could be described exactly as a function of time and space, then turbulent transport could be analyzed as an advective process. Unfortunately, turbulent fluctuations ordinarily can only be described statistically. Although Eq. (13.6) is convenient and widely used to describe turbulent transport, the diffusivities are usually unknown and must be either estimated by tracer method or empirical data. Otherwise, it becomes a calibration parameter for the site.

Dispersion is one of many processes which can reduce the concentration of contaminants transported by groundwater. It is a physical phenomenon of major importance which affects contaminant concentration, as these materials travel in groundwater systems. This process will not only tend to mix contaminated flows with uncontaminated groundwater leading to reductions in concentration by dilution but will also result in contaminants spreading longitudinally and transversely forming a typical plume. This process also results in the contaminants arriving at a distant location earlier than predicted by flow models which do not account for dispersion. It is important to note that the concentration of the contaminant that "arrives early" will be less than the concentration reported by the flow models.

The general term dispersion refers to both the process of mechanical mixing during fluid advection and molecular diffusion due to the thermal-kinetic energy of the contaminant material. Diffusion, which is driven by concentration differences, is a dispersion process of importance only at low velocities. Dispersion due to mechanical mixing during fluid advection is referred to as hydraulic dispersion.

Hydraulic dispersion is generally separated into microscopic and macroscopic levels, as shown in Fig. 13.5. Hydraulic dispersion is the spreading of distribution of contaminant material in groundwater systems and results from inherent heterogeneity of soil matrix geometry in all natural soil systems. Microscopic dispersion results from the numbers of pore pathways available to a slug of groundwater as well as the hydraulics of flow around individual soil particles. Macroscopic dispersion applies to the

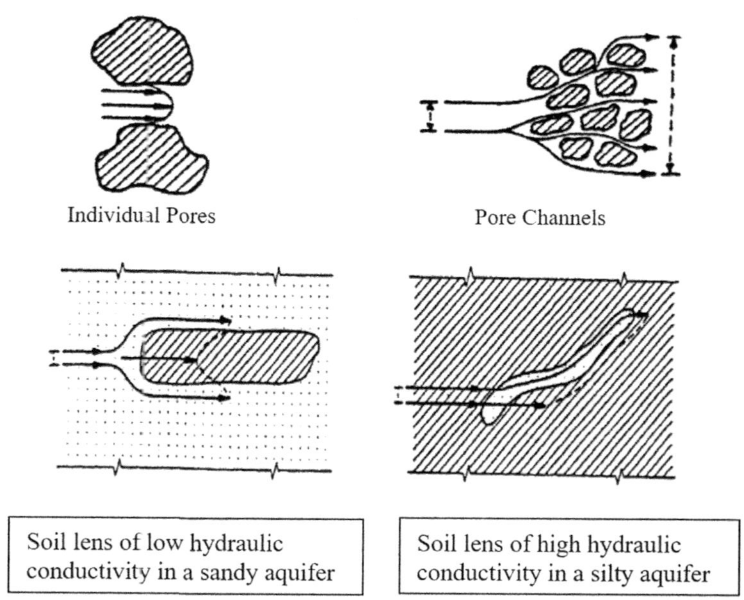

FIGURE 13.5

Types of dispersion in ground water flow: (a) microscopic dispersion and (b) macroscopic dispersion.

impact of small soil bodies or lenses present in larger soil bodies of significantly different hydraulic conductivity. This condition can lead to wide ranges in time of travel for contaminants depending on what route the ground water carrier takes.

Hydrodynamic dispersion is the macroscopic outcome of the actual movements of individual tracer particles through the pores and includes two processes. One mechanism is mechanical dispersion which depends on both the flow of fluid and the characteristics of porous medium through which the flow takes place. The process of water movement through saturated porous media involves both transport and adsorption of fluid. Advective and dispersive transports are the mass movement mechanisms associated with hydrodynamic dispersion. It is generally assumed that the amount of material transferred parallel to any given direction is the sum of the advective and dispersive mass transport components. Advective or mechanical mass transport is attributed to the variation of local microscopic velocity in the porous medium matrix. The dispersive transport phenomenon, or the so-called physicochemical dispersion or molecular diffusion, is caused by the existence of concentration.

13.3.8 Governing Advective-Diffusion Equation

From a spatial perspective, solutes enter surface waters through point and nonpoint sources. From the standpoint of environmental monitoring, point sources are relatively easy to quantify. Mass loading from these sources may be estimated by measuring the flow and solute concentration associated with a plant's effluent:

$$W = Q_e C_e \tag{13.7}$$

where W is the mass loading rate [M/T], Q_e is the volumetric flow rate for the point source, and C_e is the solute concentration in the effluent. In practice, calculations of mass loading rates are often considerably more complex than a straightforward application of Eq. (13.7) because Q_e and C_e vary with time.

Since nonpoint sources of solutes are distributed along the watercourse, these are often diffuse, in that mass entering at any one point in space is relatively small, yet the aggregate mass loading rate is significant. An example of a nonpoint source is agricultural runoff that enters a stream as overland flow. During precipitation, runoff from cultivated fields may contain pesticide residues and fertilizer as well as suspended sediments that sorb contaminants. Loading due to this type of nonpoint source is represented by a lateral inflow term in the general transport equation. Another example of a nonpoint source is that of acid rain or acid deposition that arises due to combustion of fossil fuels. This loading may impact a watershed, leading to the acidification of surface waters.

The mass loading rates for nonpoint sources are often difficult to measure because the loads are not associated with a specific point in space and there might be large spatial variations. The pollutant load that may enter as lateral flow in the river may have considerable variation in the longitudinal (along the flow) direction. In the case of acid deposition, an estimate of mass loading rates can be made by analyzing the rain samples collected at gauges that are installed at different locations in the study area. Here also, considerable variation in loading rates through the focus area can be expected. Solute sources and loading rates also vary with respect to time. These factors necessitate a carefully designed sampling program to monitor nonpoint sources of pollution.

An equation describing the spatial and temporal effects of advection and dispersion on solute concentration is developed using the law of conservation of mass. This law ensures that the mass is neither created nor destroyed and that the change of mass in a unit control volume of water is equal to the difference between the mass flowing in and leaving the control volume:

$$\text{Time Rate of Change of Mass} = \text{Mass Inflow} - \text{Mass Outflow} \tag{13.8}$$

where each term in the equation is expressed in terms of mass per unit time. The governing equation for conservation of mass of a constituent can be derived by equating the change of mass in a control volume to the sum of the net (advective plus diffusive) flux through the control volume plus sources and sinks. The general three-dimensional form of the law in cartesian coordinates is

$$\frac{\partial C}{\partial t} + u\frac{\partial C}{\partial x} + v\frac{\partial C}{\partial y} + w\frac{\partial C}{\partial z} = \frac{\partial}{\partial x}\left(E_x\frac{\partial C}{\partial x}\right) + \frac{\partial}{\partial y}\left(E_y\frac{\partial C}{\partial y}\right) + \frac{\partial}{\partial z}\left(E_Z\frac{\partial C}{\partial z}\right) - KC \tag{13.9}$$

in which u, v, and w are the velocity components in the x, y, z directions, respectively; C is the concentration in turbulent flow; E_x, E_y, E_z are anisotropic (a function of direction), nonhomogeneous (a function of location) turbulent diffusivities in the x, y, z directions; and first-order decay (with coefficient K) is assumed. Eq. (13.9) can be applied to two-dimensional problems by averaging in one coordinate direction. For example, if averaging is in the z direction, the two-dimensional, anisotropic, nonhomogeneous form in cartesian coordinates is

$$\frac{\partial(hC)}{\partial t} + \frac{\partial(uhC)}{\partial x} + \frac{\partial(vhC)}{\partial y} = \frac{\partial}{\partial x}\left(hE'_x\frac{\partial C}{\partial x}\right) + \frac{\partial}{\partial y}\left(hE'_y\frac{\partial C}{\partial y}\right) - hKC \tag{13.10}$$

For this equation, the concentration and u and v velocities (in the x and y directions) are vertically averaged over the variable depth h(x,y). Thus, E'_x and E'_y include a diffusive mixing component due to shear-flow dispersion. Eq. (13.10) can be simplified if the depth is constant in space and time, since h can then be eliminated from each term.

The one-dimensional form of the advective-dispersion equation commonly applied in streams is

$$\frac{\partial C}{\partial t} + \frac{\partial(uC)}{\partial x} = \frac{1}{A}\frac{\partial}{\partial x}\left(AE_L\frac{\partial C}{\partial x}\right) - KC \tag{13.11}$$

In Eq. (13.11), $U = Q/A$ is the average longitudinal velocity, A(x) is the cross-sectional area at any location, and $E_L(x)$ is the longitudinal dispersion coefficient. The advection-dispersion equation describes the spatial and temporal variation in solute concentration C:

$$\frac{\partial C}{\partial t} = -U\frac{\partial C}{\partial x} + D_L\frac{\partial^2 C}{\partial x^2} \tag{13.12}$$

where U is the advective velocity [L/T], and D_L is the longitudinal dispersion coefficient [L^2/T]. Coefficients U and D_L are assumed to be constant in space and time.

Although the variables of river water quality vary in three spatial dimensions, for practical purposes, a one-dimensional description is generally adequate. It is assumed that the variables show appreciable variation only along the length of flow and that the variation in respect of depth and width of the cross-section is negligible. The requirements of data and other resources of higher dimensional models are large in comparison with one-dimensional models.

Gandolfi et al. (1996) described a general form of the one-dimensional balance equation, valid for the generic component P which can be any water quality constituent or the biomass of a population whose concentration is p. Let t be the time, l the spatial coordinate, v the mean velocity of fluid in the cross section of area A, and D the longitudinal dispersion coefficient. Then, the equation becomes

$$\frac{\partial(Ap)}{\partial t}+\frac{\partial(Avp)}{\partial l}-\frac{\partial}{\partial l}\left[AD\frac{\partial p}{\partial l}\right]=E_p+AI_p \tag{13.13}$$

where E_p represents the external sources and sinks of the constituent p, and I_p represents all the phenomena that take place inside the river and influence the concentration p (e.g., decay, sedimentation). The first term on the left hand side represents the variation of p with respect to time as seen by a stationary observer at the river bank, while the second and third terms represent, respectively, the convective and diffusive components of the transport process. All the variables that appear in Eq. (13.13) are functions of space and time. The equations of the three submodels can be obtained by substituting p with the appropriate variable and defining the source and sink terms.

A multitude of analytical solutions is available for Eqs. (13.11) to (13.13). One taxonomic breakdown considering the similarities and structure is to consider instantaneous and continuous sources (in time), and point (three-dimensional), line (two-dimensional), and plane (one-dimensional) injections in space. For homogeneous, anisotropic turbulence, the solutions may be reduced to the Gaussian form.

For solutions, a point source means that the contaminant release of mass M is concentrated at a single point and can diffuse in all three coordinate directions. A line source means that the contaminant release is spread uniformly over the length h of a line and diffuses only in the x-y plane perpendicular to the line. A plane source means that the contaminant release is spread uniformly over a plane surface of area A and diffuses only in the x direction perpendicular to the plane.

13.4 MODELING OF OXYGEN IN RIVERS

Since oxygen is crucial for all biological life and is an indicator of water quality, its modeling is of central importance in river water quality studies. A detailed treatment of the topic is in order.

13.4.1 Dissolved Oxygen

The biological life that is present in water, including the organisms that are responsible for the self-purification processes, depends on the DO for survival. Oxygen is not much soluble in water, and the DO content of natural waters varies with temperature, atmospheric pressure, dissolved solids, turbulence, the photosynthetic activity of algae and plants, and atmospheric pressure. The solubility of oxygen decreases as temperature and salinity increase. In fresh waters, DO at 1 atmospheric pressure ranges from 15 mg/L at 0°C to 6 mg/L at 40°C. Huber (1993) has provided tables of DO saturation concentration at sea level at various temperatures.

Variations in DO occur seasonally, or even during a day as a function of temperature and biological activity. Biological respiration, including that related to decomposition processes, reduces DO concentrations. Waste discharges high in organic matter and nutrients can lead to decreases in DO concentrations as a result of the increased microbial activity during the degradation of the organic matter. As the oxygen solubility is less when the water temperature is high, the DO in water is deficient during summer season when temperatures are high and flow in the river is small.

The waste material present in water is degraded by two processes: aerobic and anaerobic. In aerobic processes, oxygen is used for the oxidation of organic matter and the end products are relatively harmless. In anaerobic processes, slow degradation without oxygen takes place and the end products are unwanted or even harmful. Thus, it is important that wastes are degraded through aerobic route. Since the solubility of oxygen in water is low, it is desirable that wastes are treated before they are dumped in a water body so that the quality of water does not deteriorate below a certain minimum level at any place.

The level of DO indicates the degree of pollution by organic matter, the destruction of organic substances, and the level of self-purification of the water. Concentrations below 5 mg/L may adversely affect the functioning and survival of biological communities and may lead to the death of most fish below 2 mg/L.

Determination of the DO concentration is important in water quality management, since oxygen is either involved in or influences nearly all chemical and biological processes in water bodies. There are two main methods to determine DO. The titration method involves the chemical fixation of oxygen in a water sample collected in an air-tight bottle. Fixation is carried out in the field and titration in the laboratory. The method is time-consuming but can give a high degree of accuracy. It is suitable for most kinds of sources of water. The alternative oxygen probe method is quick and can be used in-situ or for continuous monitoring, although its accuracy is somewhat less.

DO is of limited use as an indicator of pollution in groundwater and is not useful to evaluate the use of groundwater for normal purposes. In addition, the determination of DO in groundwater requires special equipment and is, therefore, not widely carried out.

13.4.2 Biochemical Oxygen Demand

The amount of oxygen consumed by indigenous microorganisms in water, while degrading under aerobic conditions is known as BOD. It is defined as the amount of oxygen required for the aerobic microorganisms present in the sample to oxidize the organic matter to a stable inorganic form. BOD is an important indicator of the status of quality of water. BOD tests are routinely carried out to determine the quality of water body. Sometimes, the COD is determined which is an indirect, although somewhat inaccurate, indicator of BOD. The BOD tests take several days, while COD tests can be performed in a few hours.

BOD is an approximate measure of the amount of biochemically degradable organic matter present in a water sample. The method is subject to various complicating factors, such as the oxygen demand resulting from the respiration of algae in a sample and the possibility of oxidation of ammonia. The presence of toxic substances in a sample may affect microbial activity leading to a reduction in the measured BOD.

Unpolluted waters typically have BOD values of 2 mg/L or less. Typical BOD concentrations for rivers are 2–15 mg/L; values up to 65 mg/L have been observed in practice. Those rivers that receive wastewaters may have considerably higher BOD particularly near the point of wastewater inflow. BOD of the raw sewage may be about 600 mg/L, whereas it should come down to 20–100 mg/L after treatment, depending on the treatment process followed.

Standard laboratory tests are used to determine BOD by measuring the amount of oxygen consumed after incubating the sample in the dark at 20°C temperature for a specific period of time. Usually this period is five days, and this is denoted by BOD_5. The oxygen consumption is determined from the difference between the DO concentrations in the sample before and after the incubation period.

BOD_5 can be calculated as $BOD_5 = D_o - D_1$ in which D_o and D_1 are the DO concentrations (mg/L) at time 0 and 5 days, respectively. BOD_5 test can also be used to characterize municipal and industrial wastewaters. However, in these tests, in addition to dilution of water, acclimated seed organisms, nutrients, and the presence or absence of toxic substances must be considered. As mentioned above, the samples are incubated at 20°C in standard BOD_5 analyses. However, if the samples are incubated at some other temperature, the rate coefficient can be adjusted using the relationship:

$$K_5 = K_{20}\theta^{(r-20)} \tag{13.14}$$

where $\theta = 1.024$ for pure water. The BOD remaining at time t is

$$L_t = L^* 10^{-kt} \tag{13.15}$$

The amount of BOD that has been exerted (satisfied) at time t, y, is

$$y = L - L_t \tag{13.16}$$

where L is the ultimate BOD (mg/L). Therefore,

$$y = L\left(1 - 10^{-kt}\right) \tag{13.17}$$

Recently, online, continuous BOD meters for wastewater plants have been introduced. Some of these systems have good reproducibility and correlation with standard five-day BOD tests. Standard features include automatic line washing, sensor autocalibration, and automatic control of sample volume for three months of maintenance-free operation. Since the correlated BOD of a sample can be measured in as little as 20 min, the instrument enables users to monitor BOD trends before a wastewater plant problem gets out of hand, evaluate the toxicity of influents and their effect on wastewater treatment plant loading, and monitor effluent quality. BOD_5 levels from 0.5 to 10,000 mg/L can be measured.

The instrument used is basically an online bioreactor in which a suspension of microorganisms (biomass) is aerated until it reaches the endogenous respiration stage. When a wastewater sample is added, the microorganisms begin to degrade it rapidly, causing an increase in the oxygen uptake rate (OUR) and a decrease in DO compared to the levels during endogenous respiration. When the organic matter in the sample is consumed, the microorganisms return to the endogenous stage.

13.4.3 Chemical Oxygen Demand

COD is a measure of oxygen equivalent of organic matter in a water sample that is susceptible to oxidation by a strong chemical oxidant. COD is widely used as a measure of the susceptibility to oxidation of the organic and inorganic materials present in water bodies and in the municipal and industrial wastes. The COD test of natural water yields the total quantity of oxygen that is required for oxidation of a waste to carbon dioxide and water (McCutcheon et al., 1993). In a BOD test, only biologically reactive carbon is oxidized, while in a COD test, all organic matter is converted to carbon dioxide. The test for COD does not identify the oxidizable material or differentiate between the organic material and inorganic material present. Similarly, it does not indicate the TOC present. Consequently, the COD values are higher than BOD. Nevertheless, COD is a useful variable that can be rapidly measured; the COD test can be performed in 3 h against five days required for a BOD_5 test.

The COD concentrations observed in surface water resources typically range from 20 mg/L or less in unpolluted waters to greater than 200 mg/L in waters receiving effluents. Industrial wastewaters may have COD ranging from 100 mg/L to 60,000 mg/L (Chapman, 1992).

13.4.4 Reaeration

Reaeration is the physical absorption of oxygen from the atmosphere by water. It occurs at the air–water interface if a nonequilibrium condition between air phase and the water phase exists for oxygen. Reaeration is the dominant natural means by which a water body may recover DO concentrations. From the point of view of water quality management, it is important that physical reaeration process taking place in a channel is clearly understood and the reaeration amount is correctly estimated. The theoretical background related to reaeration coefficient that controls this process is available in many studies including Rao (1999). As per two-film theory, mass transfer occurs through the gas and liquid interfaces until a dynamic equilibrium is established. The rate of transfer of oxygen from the atmosphere to the body of the liquid is generally proportional to the difference between the existing concentration C and the equilibrium or saturation concentration C* of oxygen in the liquid (Rao, 1999). Mathematically, it can be written as

$$\frac{dm}{dt} = K_L A(C^* - C) \tag{13.18}$$

where m is the mass of oxygen, K_L is the coefficient of diffusion of oxygen in the liquid, and A is the area through which oxygen is diffused. Also,

$$\frac{dm}{dt} = V_1 \frac{dC}{dt} \tag{13.19}$$

where V_1 is the volume of liquid. Thus, using Eqs. (13.18) and (13.19), one obtains

$$\frac{dC}{dt} = K_L \left(\frac{A}{V_1}\right)(C^* - C) \tag{13.20}$$

Parameter $K_L(A/V_1)$ is generally denoted by $(K_2)_T$ and is widely termed as the oxygen transfer or reaeration coefficient at temperature T (°C). The value of $(K_2)_T$ is related to the value $(K_2)_{20}$ as follows:

$$(K_2)_{20} = (K_2)_T / \theta^{(T-20)} \tag{13.21}$$

The measurement of the reaeration coefficient requires considerable effort, both in the field and in laboratory. The required time and funds may not be available in many cases and therefore, many efforts have been made to relate the reaeration coefficient with channel characteristics. Besides the theoretical approaches to define the reaeration coefficient, a large number of predictive equations have been developed to relate the reaeration coefficient K_2 with the mean flow velocity (V), shear stress velocity (V^*), depth of flow (H), Froude number (F_r), and the channel bed slope (S). Moog and Jirka (1998) concluded that slope is an essential component of reaeration equations and proposed two predictive equations, based on slope criteria. These are:

$$K_2 = 1740 V^{0.46} S^{0.79} H^{0.74} \quad \text{for } S > 0.00 \tag{13.22}$$

$$K_2 = 5.59 S^{0.16} H^{0.73} \quad \text{for } S < 0.00 \tag{13.23}$$

where V is the velocity of stream water (m/s), H is the flow depth (m), S is the slope, and F_r is the Froude number. Thackston and Dawson (2001), while recalibrating reaeration equations, observed that reaeration equations should not be applied for very high Froude numbers, where the influence of turbulent water is prevalent. Maier and Dandy (1996) used ANNs to predict water quality parameters.

13.4.5 Modeling of Dissolved Oxygen

The pair of differential equations that provide mass balances for BOD and oxygen deficit in a stream segment with constant flow and geometry is (Chapra, 1996):

$$U\frac{dL}{dx} = -k_r L \tag{13.24}$$

$$U\frac{dD}{dx} = -k_a D + k_d L \tag{13.25}$$

where U is the average stream velocity (m/d), L is the carbonaceous BOD concentration (mg/L), x is the distance downstream from the discharge point (m), k_r is the total removal rate of BOD by decomposition and settling (1/d), D is the DO deficit (mg/L), k_a is the aeration rate (1/d), and k_d is the removal rate of BOD by decomposition (1/d). The deficit is related to the oxygen concentration c by

$$c = c_s - D \tag{13.26}$$

where c_s is the saturation concentration for DO (mg/L).

Eqs. (13.24) and (13.25) are written for a steady-state condition and simulate the spatial distribution of oxygen below a treatment plant. Variations in time are not considered in this simple representation. Further, these ignore longitudinal dispersion and sources and sinks of oxygen, such as photosynthesis, respiration, and sediment oxygen demand. These equations can be solved for the BOD and DO concentrations below a single point source of sewage as

$$L = L_0 \exp[-(k_r/U)x] \tag{13.27}$$

and

$$c = c_s - D_0 \exp[-(k_a/U)x] - [\exp\{-(k_r/U)x\} - \exp\{-(k_a/U)x\}] * k_d L_0/(k_a - k_r) \tag{13.28}$$

where L_o and D_o are, respectively, the concentrations of BOD and oxygen deficit at the mixing point where the waste is discharged.

Assume that the flow in a river is Q_u and BOD L_u when a sewer with flow Q_s and BOD L_s meets it (see Fig. 13.6). After complete mixing, the resulting BOD L_0 can be obtained by a simple mass balance:

$$L_0 = (Q_u L_u + Q_s L_s)/(Q_u + Q_s) \tag{13.29}$$

As the flow moves downstream in the river, BOD is subject to advection and diffusion/dispersion. A sink term ($-K_1L$) is included in the advective-diffusion Eq. (13.15) with C replaced by L, and other source/sink terms can be added optionally. For the steady-state condition, $\partial L/\partial t = 0$ in Eq. (13.15), and the ordinary differential equation is readily solved to give

$$L(x) = L_0 e^{mx} \tag{13.30}$$

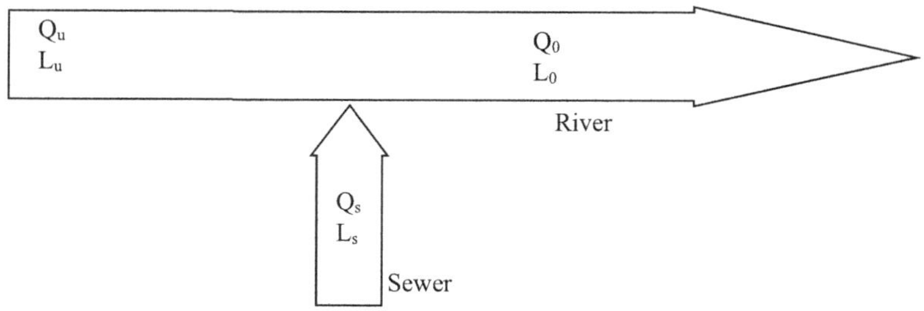

FIGURE 13.6

Schematic of a point source discharging in a river.

where L(x) is the BOD distribution downstream of the initial condition L, and the exponent m is

$$m = -K_1 / U \tag{13.31}$$

Since travel time t = x/U, Eq. (13.24) can be simplified as

$$L(x) = L_0 \exp(-K_1 t) \tag{13.32}$$

in which L is a simple exponential decay as a function of t.

In the DO balance, the term $-K_1L(x)$ is a sink and reaeration at air—water interface is a source. Terms can be added to represent other sources and sinks due to photosynthesis, sediment oxygen demand, etc. Considering only the BOD sink term and the reaeration source term in one-dimensional advective-diffusive equation (neglecting dispersion), the steady-state DO deficit D(x) (Medina et al., 1981) is

$$D\backslash(x) = [K_1L_0/(K_2 - K_1)]^*[e^{mx} - e^{rx}] + D_0\, e^{rx} \tag{13.33}$$

where $D(x) = [C_s - C(x)]$, C_s is the saturation DO concentration, D_o is the initial, well-mixed deficit at $x = 0$, $r = K_2/U$, and K_2 is the reaeration coefficient. Note that the effect of dispersion is important for estuaries where the flow velocity is low but dispersion can be neglected for rivers.

Eq. (13.33) describes the classical DO *sag curve*, due to the typical shape of the curve shown in Fig. 13.7. As the pollutants enter the river, oxygen is consumed in degradation and therefore the deficit

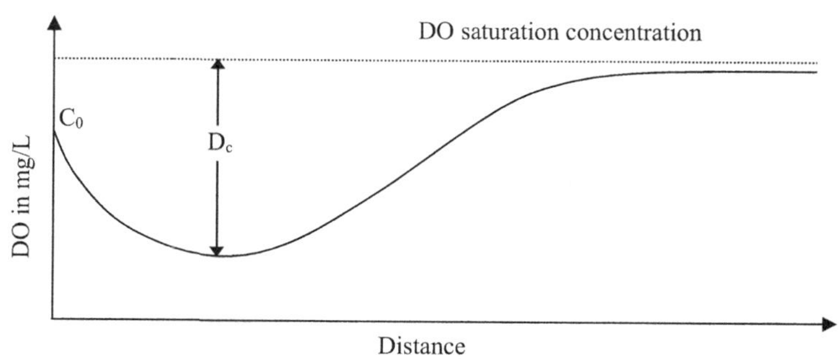

FIGURE 13.7

DO sag curve.

increases. The oxygen from atmosphere enters the river through reaeration at the water—air interface. DO goes on decreasing as long as the BOD oxidation exceeds reaeration till a minimum concentration or maximum deficit D_C is reached. From this point onwards, the reaeration process dominates and DO increases until saturation is reached. In this way, a "sag" is produced in the DO curve. Note that this description is a highly simple case and there may be additional BOD sources or DO sinks in a real-life problem.

13.5 CATCHMENT-SCALE WATER QUALITY MODELS

Sediment and sediment-associated pollutants that reach a water body originate largely from nonpoint sources in the catchment and result from a complex interplay of numerous forces. Of special importance are (1) climate, in general, and the nature, amount, and intensity of precipitation, in particular; (2) orientation, degree, and length of slopes; (3) geology and soil types; (4) land use; (5) condition and density of the channel system; (6) particle settling velocity; and (7) streamflow regime. The uncertainty in estimation of pollutants largely depends on the variability of these factors.

The term *loading function* describes calculation procedures to estimate the average inflow of nonpoint pollutants or chemical loads over a time period or a storm event from an individual land use category. A number of different loading functions have been developed and are in use. These functions are used only for preliminary or rough calculations. For detailed and improved estimates, simulation models are used which perform hydrologic, soil erosion and transportation, and chemical/biological pollutant computations over certain time interval. The resulting values for each variable of interest, such as runoff, sediment, and pollutant load, can be analyzed to get the desired information.

The development and application of catchment-scale water quality models have been largely driven by the need for tools that can be used to evaluate two issues. These are nonpoint source pollution effects (i.e., exposure and/or loadings assessment), and chemical management plans, such as agricultural best management plans for fertilizers and pesticides or, more recently, the total maximum daily loads (TMDLs), including both point and nonpoint contaminant sources. Catchment scale water quality models must represent dominant processes that determine the interaction between the following state variables: water, temperature, sediment, DO, nutrients (nitrogen and phosphorus species) and bacteria (coliforms), pesticides and other toxic organics, metals, and selected biological variables (phytoplankton, benthic algae, zooplankton, herbivorous fish, predatory fish).

With the exception of biological variables, all the other above-listed variables can be critical to modeling water quality in the presence of nonpoint sources, both urban and nonurban. Temperature is important in modeling all hydrodynamic, water quality, and soil processes. Estimation and prediction of sediment transport and deposition is useful for many applications related to design and management. DO is a useful indicator of water quality and its modeling is needed to support analyses of nutrients and biological parameters. The important nitrogen species are ammonium, ammonia, nitrite, nitrate, and organic nitrogen. The phosphorous species that must be considered are orthophosphorus and organic phosphorus. Additional chemicals that should be modeled by a comprehensive water quality model include pesticides and other toxic organics, and metals. Note that these two categories include a large number of abiotic variables and hence a generalized approach is necessary. The general physical and chemical processes important for fate and transport are adsorption/desorption, diffusion, atmospheric deposition, volatilization, and chemical reactions/degradation. The processes that need to be considered for biological variables depend on the trophic level.

A large number of computer models are available for modeling of water quality of a catchment. Tetra Tech, Inc. (2018) evaluated 18 water quality models, based on their maintenance and support status. These models are currently used to support the development of TMDL plans, assess permit conditions for national pollutant discharge elimination system facilities, evaluate policy and water quality effects of potential actions, and perform other water quality analyses. These models include AGNPS, AQUATOX, BASINS, BATHUB, CE-QUAL-ICM, CE-QUAL-W2, CORMIX, EFDC, HAQWS, HSPF, LSPC, QUAL2KW, SPPAROW, SWAT, SWMM, SWTOOLBOX, VISUAL PLUMES, and WASP.

AGNPS (Agricultural Non-Point Source Pollution Model) evaluates the effect of predominantly agricultural management decisions that impact water, sediment, and chemical loadings within a watershed system. AQUATOX—an ecological model—predicts the impacts of various environmental pollutants on aquatic ecosystems, aquatic organisms, including fish, invertebrates, and aquatic plants. BASINS (Better Assessment Science Integration Point and Nonpoint Source) interfaces with HSPF, SWAT, and AQUATOX and is designed for watershed and water quality studies by local, state, and regional entities. BATHTUB is designed for the simulation of morphologically complex lakes and reservoirs. CE-QUAL-ICM is a three-dimensional eutrophication model and simulates biochemical cycles. CE-QUAL-W2 is a two-dimensional (vertical and longitudinal) hydrodynamic and water quality model. CORMIX is a mixing zone model and decision support system for environmental impact assessment. EFDC (Environmental Fluid Dynamics Code) is a linked hydrodynamic and water quality modeling system, including modules for sediment transport and fate and transport of toxic contaminants. HAWQS (Hydrologic and Water Quality System) uses the Soil and Water Assessment Tool (SWAT) as a core model and is an interactive water quantity and quality model.

HSPF (Hydrologic Simulation Program Fortran) is a model for the simulation of watershed hydrology and water quality for both conventional and toxic chemicals. LSPC (Loading Simulation Program C++) is also a model for the simulation of watershed hydrology and water quality for both conventional and toxic chemicals. QUAL2KW is a one-dimensional river and water quality model and simulates temperature, DO, nutrients, pH, periphyton, macrophytes, phytoplankton, and sediment diagenesis. SPARROW (Spatially Referenced Regressions on Watershed Attributes) model relates water quality with local landscape characteristics. SWAT (Soil and Water Assessment Tool) predicts the effect of management decisions on water, sediment, nutrient, and pesticide yields. SWMM (Storm Water Management Model) simulates runoff quantity and quality from urban and semiurban watershed. VISUAL PLUMES is a computer model that estimates the dilution of outfall discharges. WASP (Water Quality Analysis Simulation Program) models aquatic systems, including both the water column and the underlying benthos.

Donigian et al. (1995) carried out a comprehensive study of a number of models. Among the models for full-scale simulation for urban areas, four models, namely, DR3M-QUAL, HSPF, STORM, and SWMM, were found to be the best. DR3M-QUAL is a version of the USGS Distributed Routing Rainfall Runoff Model that includes quality simulation (Alley and Smith, 1982). Runoff generation and subsequent routing use the kinematic wave method, and parameter estimation assistance is included in the model. The Hydrological Simulation Program—Fortran (HSPF) is the culmination of hydrologic routines that originated with the Stanford Watershed Model in 1966 and eventually incorporated many nonpoint source modeling efforts (Bicknell et al., 1993). This model has been widely used for nonurban nonpoint source modeling. It incorporates field scale models for nonpoint loadings into a catchment and basin-scale analysis framework that includes fate and transport in one-dimensional stream channels. Several flow routing and water quality options are available in the model. The Storage, Treatment, Overflow, Runoff Model (STORM) developed by the Hydrologic Engineering Center (HEC, 1977) was among the first continuous simulation models in urban hydrology.

The Storm Water Management Model (SWMM) was originally developed as a single-event model specifically for the analysis of combined sewer outflow. Its capabilities were later enhanced, and the Version 4 of the model (Huber and Dickinson, 1988) performs both continuous and single-event simulation. It can simulate backwater, surcharging, pressure flow, and looped connections (by solving the complete dynamic wave equations) and has a variety of options for quality simulation, including traditional buildup and washoff formulations as well as rating curves and regression techniques. *Buildup* is a term that represents all of the complex spectrum of dry-weather processes that occur in an urban area between storms, including deposition, wind erosion, street cleaning, etc. All such processes lead to an accumulation of solids and other pollutants that are then *washed off* during storm events.

The MOUSE (Modeling of Urban Sewers) model was developed by the Danish Hydraulic Institute, Denmark. Included in the package are modules for the generation of runoff from rainfall, sewer routing (the S11S model), and a simple quality routine that uses the constant concentration approach (Jacobsen et al., 1984).

Chemicals, Runoff, and Erosion from Agricultural Management Systems (CREAMS) was developed by the U.S. Department of Agriculture—Agricultural Research Service (Knisel, 1980) for the analysis of agricultural best management practices for pollution control. CREAMS is a field scale model that uses separate hydrology, erosion, and chemistry submodels connected together by pass files. Runoff volume, peak flow, infiltration, evapotranspiration, soil water content, and percolation are computed on a daily basis. Daily erosion and sediment yield, including particle size distribution, are estimated at the edge of the field. Plant nutrients and pesticides are simulated and storm load and average concentrations of sediment-associated and dissolved chemicals are determined in runoff, sediment, and percolation through the root zone. CREAMS can also simulate user-defined management activities, such as aerial spraying or soil incorporation of pesticides, animal waste management, and agricultural best management practices.

The Areal Non-point Source Watershed Environment Response Simulation (ANSWERS) is an event based, distributed parameter model capable of predicting the hydrologic and erosion response of agricultural watersheds (Beasley and Huggins, 1981). The application of ANSWERS requires that the catchment be subdivided into a grid of square elements, each of which should be small enough so that all important parameter values within its boundaries are uniform. Within each element, the model simulates the processes of interception, infiltration, surface storage, surface flow, subsurface drainage, sediment drainage, and sediment detachment, transport, and deposition. The output of one element then becomes an input source to an adjacent element.

The surface water quality models can be classified in two categories: *far-field* models which simulate quality in the entire sections of a water body and mixing zone models which simulate *near-field* dilution processes. The far-field models can be subdivided on the basis of water body type, that is, rivers, lakes and reservoirs, and estuaries. The models to simulate water quality defer from one another mainly in terms of their representation of four key attributes: dimensionality, time, hydrodynamics and pollutant loadings.

Dimensionality determines how a model represents spatial features of the water body. Most river and stream models are one-dimensional and can be applied to branching systems. Two-dimensional models can cover either longitudinal and lateral (X/Y) or longitudinal and vertical (X/Z) dimensions. "Box" models can be applied in 1, 2, or 3 dimensions.

Time determines how a model can represent the water body dynamics. Steady-state models predict concentrations that do not vary in time. These are useful primarily for rivers under low-flow design conditions. Both dynamic and quasidynamic solutions predict concentrations that vary with time. Quasidynamic solutions allow some major forcing functions, such as flow, loading, or solar effects on photosynthesis, to vary with time.

All receiving water quality models require information on the movement of water. Some provide this information in the input data set. These models are steady or quasidynamic in nature. Others require input information on flow or velocity; still others can either accept input flows or be linked to simulated hydrodynamics. Some water quality models provide for the hydrodynamic calculations internally. Most, however, require linkage to an external hydrodynamics file.

Almost all models simulate advection and dispersion. Most 1-D riverine models do not require dispersion because most rivers are not highly dispersive and the model network and solution techniques introduce some degree of numerical dispersion. All models allow the user to input steady pollutant loads. Some allow the specification of variable loads from the input data set. A few models provide internal or external linkage to nonpoint source loading simulations.

One of the most widely used models is QUAL2E (Brown and Barnwell, 1987). The enhanced model is capable of simulating several water quality constituents. A finite difference scheme is used to solve the one-dimensional advection-dispersion mass transport and reaction equation. The model can predict DO, CBOD, temperature, and phytoplankton dynamics as affected by organic material and nutrients. It has nine state variables: total algal biomass, BOD, DO, ammonia, nitrites, nitrates, organic nitrogen, organic phosphorus, and orthophosphates.

13.5.1 Impacts of Climate Change on Water Quality

It is now widely accepted fact that climate change is unavoidable. Climate change is likely to impact not only temperature, precipitation and meteorological variables but also the quality of water. Rising water temperatures will impact dissolved gases in water and the chemical reactions in aquatic ecosystems. Changes in precipitation intensity and patterns would also alter soil erosion, sediment transport, morphology of rivers, and habitats of aquatic life. Increased instances of (flash) flooding would lead to higher discharges entering in the rivers and lakes from urban areas. Invasion by alien species is likely to increase and the migration of species is also going to change due to changes in flow patterns and velocities, as well as construction of dams and barrages. As the use of fertilizers and pesticides increases, flow of nutrients to water bodies will rise, increasing the potential for toxic algal blooms. Thus, in future, water managers will have to tackle a range of new problems, arising due to water contamination caused by the existing and emerging pollutants and climate change. Sixth Assessment Report of IPCC (Caretta et al., 2022) has covered this topic in greater details and has noted that evidence of global-scale changes in water quality due to climate change is limited as of now because many studies are isolated and have limited regional coverage.

13.6 WATER QUALITY IN LAKES AND RESERVOIRS

A lake is a partially enclosed inland body of fresh water surrounded by land. Lakes vary in size from many thousands of square kilometres in area and many meters in depth to only a few square kilometres and depths of less than 10 m. Lakes are used for many purposes including municipal and industrial water supply, recreation, navigation, fishing, power generation, etc. During warm weather, lakes are heated by the incident solar energy. During cold season, they loose heat to the atmosphere and thereby cool down. It is well known that warm water is lighter and therefore, it floats on colder denser water. The maximum density of water occurs at 4°C. The result is thermal stratification of lakes.

On the basis of thermal characteristics, lakes are usually divided into three zones: epilimnion, hypolimnion, and metalimnion. The *epilimnion* zone is the upper layer of the lake consisting of the warmer, lighter water. The downward movement of this water requires displacement of the denser and colder waters in the lower region of the lake and there is, therefore, a thermal resistance to mixing. Warm waters in the epilimnion zone are circulated by winds and do not go far below the surface but move along the top of the cold water zone and then returns to the surface. The *hypolimnion* zone is composed of the noncirculating cold mass of water in the bottom layers of the lake during the stagnant period. The hypolimnion zone has little or no opportunity to gain heat from the sun or oxygen from the atmosphere during the warm weather seasons. The *metalimnion* zone, known as *thermocline*, is the transition zone between the epilimnion and hypolimnion and is the area in which steep temperature gradients are found. Because of the difference in densities of epilimnion and hypolimnion waters, the thermocline acts as a barrier to the downward movement of the lighter waters. The various chemical and biological processes that take place in a lake are significantly influenced by this thermal stratification.

13.6.1 Differences Between Lakes and Reservoirs

Although the lakes and reservoirs appear to be similar on a casual look, there are many fundamental differences between them and these also influence the chemical and biological processes that take place. The differences in capacity-inflow ratio and water level fluctuations in lakes and reservoirs are largely responsible for this difference in behavior. A small residence time means that many species do not have an opportunity to reproduce in most reservoirs. The important differences between natural lakes and man-made reservoirs are summarized in Table 13.6.

Age wise, natural lakes are usually very old; some of them date back to thousands of years, while most large reservoirs were made during the past century. Both natural lakes and artificial reservoirs receive a wide variety of sediments and nutrients as inputs which get accumulated and deposited. There are large variations in deposition rates, depending on the catchment properties as well as the properties of water body (including regulation in the case of reservoirs). In general, the rate of sedimentation in man-made reservoirs is much higher than in natural lakes. The process of sedimentation leads to a gradual reduction in the storage capacity (see Chapter 12) and over time, the lake or reservoir may ultimately reduce to a marshy land. Many programs have been launched world over to recover the storage capacity by the removal of sediments and biomass through mechanical means.

Geometrically, lakes and reservoir are quite different and the reason lies in their origin. Natural lakes are formed at local depressions and are typically circular or oval shaped. The central portion of a

Table 13.6 Important differences between natural lakes and man-made reservoirs.

Feature	Natural lake	Man-made reservoir
Age	1000s of years	100s of years
Capacity-inflow ratio	Large	Wide range
Water level fluctuations	Small	Usually large and seasonal
Location of maximum depth	Commonly near the center	Close to the dam
Source of inflows	Surface and subsurface	Predominantly surface
Outlet of water	Surface and subsurface	Almost totally surface
Catchment: water surface area	Small	Large
Shape	Mostly oval or circular	Usually linear or dendritic
Sediment and nutrient loading	Low	High
Biodiversity	Higher	Lower
Primary productivity	Lower	Higher
Water quality gradients	Concentric	Longitudinal

lake is the deepest. In contrast, the site for an artificial reservoir is decided based on many factors; a deeper reservoir will have less surface area and less loss of water due to evaporation. Such a reservoir is created where river slopes are high. The upstream end of a reservoir has shallow depth and the maximum depth is near the dam. Since most of the flow enters at the farthest upstream end and leaves at the downstream end, there is a strong longitudinal flow and density currents may also be present.

Some of the products of biological processes settle and combine with the inorganic components of sediments present at the lake bottom. A variety of pollutants, such as pesticides, trace elements, metals, nonbiodegradable substances, etc., also enter the lake and are responsible for many problems.

13.6.2 Chemical Considerations

The most serious challenges to the use of lakes from a water quality angle have been *eutrophication* and inflow of toxic chemicals. Naturally, eutrophication or "lake enrichment" has been one of the most intensely studied subjects in water quality modeling. A water-quality analysis of lakes commonly involves questions of deleterious effects due to increased nutrient (nitrogen and phosphorous) supply. The deleterious effects are increased life and growth of a lake's biota, especially algae and macrophytes (large aquatic plants) with a consequent increase in turbidity and color, possible reduction in DO, and change in nature of the fish population. The degree of eutrophication is largely dependent on nutrient concentration in the lake waters. As summarized in Table 13.7, in-lake nitrogen and phosphorus concentrations are a reflection of the trophic status of a lake.

Some authors have defined the trophic status of lakes on the basis of phosphorus concentrations alone. Alternately, indices have also been suggested to denote the trophic state of lakes. Reckhow and

Table 13.7 Trophic status of lakes.

Water-quality variable	Oligotrophic	Mesotrophic	Eutrophic
Total phosphorus, g/L	<10	10–20	>20
Chlorophyll a, g/L	<4	4–10	>10
Secchi depth, m	>4	2–4	<2
Hypolimnetic oxygen, % saturation	>80	10–80	<10

Sources: NAS/NAE (1972), USEPA (1974).

Chapra (1983) described the use of a regression equation that relates the values of Secchi disk readings and concentrations of phosphorus, nitrogen, and chlorophyll a. Considerable efforts are made by lake and reservoir managers to control nutrient levels to reduce vegetational growth and thereby enhance the use of lakes. Nutrients in this context are usually considered to include phosphorous, nitrogen, carbon, and silica in various chemical forms.

Nitrogen and phosphorus are required for growth of algae. On the basis of their relative presence, the trophic status of a lake is determined. Phosphorus is of paramount importance in lake eutrophication. According to Thomann and Mueller (1987), when the ratio of the total nitrogen (N) to the total phosphorus (P) is less than 13, lakes are nitrogen-limited. When this N/P ratio is in the range 13–21, the lakes are known as nutrient-balanced and if N/P > 21, then it is phosphorus-limited. The term "phosphorus-limited" implies that additional phosphorus is required to produce additional algal growth (Huber 1993). Most lakes are phosphorus-limited, especially those for which nonpoint-source runoff is the dominant source of phosphorus. Note that these limits (13 and 21) are approximate only and depend on algal species. The clear lakes with low biological productivity and poorer in nutrient materials and organic life are termed as *oligotrophic*. The lakes with high biological productivity are referred to as *eutrophic*. Such lakes are rich in nutritive materials and contain an abundance of plankton organisms, shore vegetation, and animal life. In between fall *mesotrophic* lakes which have an intermediate level of biological productivity.

Lakes contain a wide variety of physical and chemical substances. These range from floating debris to suspended materials, dissolved inorganic matter, nutrients, metals, organics, and DO. Of these, dissolved matter, suspended solids, nutrients, and DO are among the most important constituents in assessing the productivity of lakes. Dissolved matter or total dissolved solids is a measure of the total inorganic substances dissolved in water. These substances include major chemical ions, such as calcium, magnesium, sodium, potassium, carbonates, sulfates, and chlorides as well as dissolved metals. Suspended solids in the lake water are, for the most part, finely suspended particles of insoluble material, including sand, silt, clay, debris from vegetative growth, algae, chlorophyll, and other substances. These materials originate from shore erosion through wind and wave action, turbidity inputs, biological activities and from pollution sources. Nutrients play a very significant role in the life of a lake.

DO is an important factor in the health of lakes. It is essential to the production and support of biological life in lake waters and necessary to the decomposition and decay of organic wastes and deceased organisms. Oxygen is consumed by the respiration of plants and animals, by bacterial decomposition of organic matter, and by the chemical oxidation of waste substances. It plays an important role in the organic cycle of lakes.

13.6.3 Biological Considerations

Biological substances in lakes consist of different life groups of organisms, including bacteria, fungi, phytoplankton, zooplankton, benthic fauna, aquatic plant life, and fish. Each of these organisms is an essential part of the biological community in lakes and each is critical to the overall balance and stability of lake environment. Out of these, the phytoplankton and zooplankton, benthic fauna, and fish have been most commonly studied.

Lake stratification is driven by seasonally variable forces: solar energy, inflowing water temperature, and variable suspended solid concentrations. If temperature profiles are measured year round, the temperate zone in lakes and reservoirs will demonstrate a strong summer stratification due to solar heating of surface waters, but isothermal conditions occur during winter, culminating in density gradients and allowing the entire water column to mix vertically, a process called turnover. During turnover, nutrients which have accumulated in the deeper water are returned to the surface where they can promote algal growth in the summer as more solar radiation is received, daylight hours increase, temperatures rise, and ice covers melt away.

Continuous cycling of carbon between inorganic and organic forms is attained by the use of solar energy in photosynthesis. The process converts the solar energy into chemical energy. The process of photosynthesis can be expressed by the following chemical equation:

$$6CO_2 + 6H_2O \underset{\text{Respiration}}{\overset{\text{Photosynthesis}}{\Leftrightarrow}} C_6H_{12}O_6 + 6O_2 \tag{13.34}$$

13.6.4 Lake Mass Balance

A simple mass balance equation for a lake can be written as (Young and Dolan, 1995):

$$V\frac{dC}{dt} = \sum_i W_{TRIB,i} + \sum_j W_{PTS,j} + \sum_k W_{NPS,k} + W_{ATM} - J_{SED}A_s - J_{EVAP}A - QC - KCV \tag{13.35}$$

The lake has volume, V, an outflow rate of Q, a contaminant of concern with water column concentration, C. It receives loads of the contaminant from tributaries $\Sigma W_{TRIB,i}$, direct point sources $\Sigma W_{PTS,j}$, external direct nonpoint sources $\Sigma W_{NPS,k}$, and atmospheric sources ΣW_{ATM}. The lake loses mass due to the net balance of sedimentation, resuspension, diffusion, and bioturbation over the sediment with surface area A_s, (J_{SED} A_S), by evaporation/volatilization over the water surface with area A, ($J_{SED}A$), which react within the water column by a first-order process with rate law $-kC$.

When contaminant transport occurs by pure advective processes, the average load, L [MT^{-1}], delivered to a receiving water during some specific time interval, t [T], through an interface of area, A [L^2], for a contaminant that is carried by water flowing at an instantaneous velocity, v [LT^{-1}], and concentration, C [ML^{-3}], may be described by the equation:

$$S = \frac{1}{\Delta t}\int_{\Delta t}\int_{A} \nu\{A(\tau),\tau\}C\{A(\tau),\tau\}dAd\tau \tag{13.36}$$

If contaminant transport occurs by purely diffusive or dispersive processes along the instantaneous concentration gradient normal to the interface, dC/dz [ML^{-4}], with a transport coefficient, D_z [L^2T^{-1}], then Eq. (13.37) may be used to represent the average load to the receiving water:

$$S = -\frac{1}{\Delta t}\int_{\Delta t}\int_{A} D_z\{A(\tau),\tau\}\frac{dC\{A(\tau),\tau\}}{dz}dAd\tau \tag{13.37}$$

Eqs. (13.30) and (13.31) are one-dimensional, idealized, deterministic simplifications of the complex, stochastic situation that exists and determines loads during contaminant transport. If it were possible to monitor continuously in time and space and with absolute accuracy all boundaries of a receiving water body, then one could compute loads without uncertainty, at least in theory. Even if monitoring with that intensity were technologically feasible, however, it would be impractical. Nevertheless, these equations can serve to illustrate the major sources of uncertainty that affect estimates of chemical loading and other forcing functions of interest for aquatic system modeling. In particular, any condition or change in conditions that alters any of the variables in Eqs. (13.36) and (13.37) will affect, in turn, the resulting estimate of load. By corollary, uncertainty in the variables of Eqs. (13.36) and (13.37) will propagate through the relationship to affect the load estimate. Any quantity, therefore, that produces uncertainty in flow velocities, concentrations, dispersion or diffusion coefficients, or concentration gradients can lead to uncertainty in contaminant loading estimates.

The dominant source of input to water and chemicals to most lakes is surface inflows and outflows. Depending on the factors such as the behavior of inflowing streams and their number, accuracy required, and funds available, a suitable monitoring program is devised. The commonly used methods for estimating the inflow of chemical and sediments include measurements of their concentrations at discrete times and estimating the total loading by multiplying the concentration with discharge. Although this method is most popular, it may give erroneous results, particularly if large amounts of chemicals enter water bodies during a short period of high flows. An improved approach is to develop a relationship between flow and pollutant concentration and then estimate loadings using these relations.

An important way of exchange of water and chemicals from a lake is through atmosphere. Rain and precipitation that directly fall on the lake surface constitutes input, while evaporation is the process through which the outflow of water from a lake takes place. Chemicals enter the lake through deposition of particles and absorption of gases to the water surface. Depending on the circumstances and location, atmospheric loading can be an important source of chemicals, particularly for the lakes that have large surface areas. Sometimes, chemical composition of precipitation is determined to estimate chemical loading. However, this method may not give reliable results for which on-site monitoring using collectors is necessary. These collectors collect precipitation and other deposition to the lake. Data of pan evaporimeters are most commonly used to estimate evaporation.

Seepage through subsurface processes is another source of exchange for lakes. If groundwater contains large concentrations of constituent matters, seepage inflows can be an important source of chemical inputs. Seepage is commonly estimated as the residual term in the lake water balance equation. However, all the errors of estimation in individual components of water balance will be aggregated in seepage terms, making it error prone and will not give correct results when there is inflow as well as outflow of water and chemicals through seepage. In some basins, due to geologic formations, there might be inflow due to seepage at one end and outflow at the other end. The water balance approach will clearly give wrong results in such cases. In the flownet approach, a network of piezometers is set up in the catchment of the lake to determine the water table gradients. The inflows and outflows can be estimated using governing groundwater equations.

13.7 GROUNDWATER QUALITY

Groundwater has long been regarded as the best resource of water for any type of use. It is used for different purposes, the major ones being community water supply, agriculture, and industrial processes. Each type of use requires certain water quality criteria which determine whether groundwater in question is suitable for the purpose. Although it is generally well protected from contaminating influences, the very uses for which it is deployed are causing its degradation. In some cases, excessive abstraction of groundwater has caused a gradual degeneration and a number of serious pollution problems in large groundwater bodies. The need to conserve vitally important aquifers as raw water sources calls for careful management of groundwaters with respect to their quantity and quality. In general, groundwater contamination is irreversible, that is, once it is polluted, it is difficult to restore the quality over a short span of time.

Rocks and soils are the two main components which influence the groundwater quality. The interaction of rocks and water is a complex process due to a great variety of rocks and environmental conditions. Soil and rock basically can be distinguished from one another in that soil contains organic and inorganic constituents. Inorganic constituents in soil range from 90% to 95%, while rock is almost completely inorganic. Soil is the result of the total interaction of water, air, climate, plant, and animal organisms with rocks.

Groundwater, although protected by the soil cover, is subject to quality changes as a result of activities of man on the overlying cover. The most important pollution sources include:

- domestic waste water infiltrated into the aquifer through cesspools or septic tanks;
- municipal sewage (due to leaks in sewerage system) or percolation from waste ponds, etc.;
- leachates from garbage dumps and sanitary landfills;
- industrial wastes from mining, refineries and oil industries, metal processing, and other industries;
- cooling water infiltration through cooling water recharge wells;
- accidental discharges through petroleum products;
- irrigation-return flows; and
- artificial recharge with treated sewage.

The self-purification of groundwater occurs due to a variety of physical, chemical, and biological processes. Physical processes include dispersion and filtration. Dispersion causes dilution of wastes and filtration favors reduction in the amounts of substances associated with colloidal or larger-sized particles. The geochemical processes include complexation, acid-base titrations, oxidation-reduction, precipitation-solution, and adsorption-desorption. Biological processes include decay and respiration, cell synthesis, etc.

Unlike surface water pollution, the damage to groundwater is not readily "visible" and may not be detected until it is seriously polluted. Analysis of any groundwater problem requires sound knowledge of the geological and hydrological conditions of the aquifer that is subject to pollution. The basic network of groundwater observation points consists usually of production wells and piezometers. In most cases, however, there might be a need for additional monitoring points for groundwater levels and selected quality parameters.

Sampling and analysis of groundwater can be executed in two ways: (1) direct measurement of groundwater quality in test hole. For this purpose, a probe is lowered or permanently installed in the observation well. Pollution indicators, such as electrical conductivity and temperature, can be easily measured this way. (2) Water samples are pumped from the observation well. If accurate measurements of pH, redox potential, and electrical conductivity are desired, it will be necessary to analyze these parameters at the site. For other parameters, samples can be preserved by adding an appropriate preservative and analysis can be carried out in the laboratory. The selection of analytical parameters has to be made according to the objectives of the monitoring program.

13.7.1 Models for Groundwater Quality

Groundwater modeling is concerned with simulating the hydrologic behavior of subsurface systems. These days, it is the preferred method to understand the movement of water and the pollutants and for regulatory purposes. Contaminant transport models are a step beyond flow models. These models include all the considerations incorporated in flow models plus relationships which are designed to track the contaminants of interest and determine the change in their concentration with time. The continuity equations for contaminant transport simulations include not only terms for dispersion and flow but also other processes, such as chemical and biological reactions which quantify the expected changes in contaminant concentrations with time, as this material travels through the soil system of interest. The controlling concept is that the total mass is always accounted for. Models that are used to predict the groundwater contaminant transport can be classified into three categories:

(a) *Advection models*: These models define the movement of contaminants as a result of groundwater flow only. A slug of water carrying contaminants moves through the soil system along with groundwater flow. Contaminants are transported with no change in concentration with distance.

(b) *Advection-dispersion models*: When the concept of dispersion is introduced into the model, a term is included which provides for dispersion related mixing and spreading and leads to time-related changes in contaminant concentration. The dispersion term takes into consideration molecular diffusion, microscopic dispersion, and macroscopic dispersion. Generally, because of the scale of applications in terms of land area involved and relatively high flow velocities, molecular diffusion is of small consequence compared to micro- and macrodispersion.

(c) *Advection-dispersion-chemical/biological reaction models*: Another step in model sophistication is the inclusion of the effects of reactions which change the concentration of transported contaminants. The reactions may be chemical or biological and can be incorporated into advection-only models or advection-dispersion models. Because of the current lack of knowledge regarding subsurface reaction kinetics, only chemical processes, such as ion exchange and adsorption, have been considered in most applications.

An additional class involves coupling of geochemical models with groundwater flow models. Such models are complex and were developed to study the chemistry of natural waters. These are not designed for application to contaminant transport problems. Their applications have been limited to simulate the evolution of groundwater quality along regional groundwater flow paths in systems dominated by calcium-magnesium-sulfate reactions. The application of this modeling approach to meet industry needs appears to be of limited value.

Consider the transport of a contaminant in a saturated flow through a porous medium. The term contaminant refers to any species of interest in a solution. Symbol C denotes the concentration of a contaminant, that is, mass of contaminant per unit volume of solution. It is assumed that the porous medium is homogeneous and isotropic with respect to dispersivity, the flow regime is laminar, and in general, variations in contaminant concentration cause changes in the density and viscosity of the liquid. These, in turn, affect the flow regime. The equation describing the mass transport and dispersion of dissolved chemical constituents in a saturated porous medium may be written as

$$\left[\frac{\partial C}{\partial t}+\frac{\partial}{\partial x_i}(V_i C)\right]-\left[\frac{\partial}{\partial x_i}\left(D'_{ij}\frac{\partial C}{\partial x_j}\right)+q'_c\right]=0 \quad i,j=1,2,3\ldots \tag{13.38}$$

where D'_{ij} is the coefficient of hydrodynamic dispersion, V_i is the component of seepage velocity, q'_c is the mass flux of source or sink, and x_i are the Cartesian coordinates. The theoretical basis and the derivation of the diffusion-convection equation are discussed in detail by Bear (1979). In Eq. (13.38), the first term represents the time rate of change of contaminant concentration. The second term describes the advective transport of C in the x_i-direction, which is proportional to seepage velocity. The third term is the transport (redistribution) of C due to dispersion and molecular diffusion. Finally, the last term represents the time rate of production or decay of C.

The advective dispersion equation is a nonlinear partial differential equation of parabolic type. The relation is nonlinear because of the advective term, and because of the transport coefficient which is a function of the dependent variable V. The advective term is nonsymmetric and has been a principal source of difficulty in the numerical solution of the advective dispersion equation. Bobba and Singh (1995) have described the initial and boundary conditions for this equation and have discussed the many finite difference and finite element schemes that are used in contaminant transport models.

Among the models for groundwater quality, perhaps the most popular is the MODLFOW model. MODFLOW is a MODular 3-dimensional finite difference groundwater FLOW model developed by McDonald and Harbough (1988) of the USGS, USA. Langevin, et al. (2022) provide thė documentation for MODFLOW 6, released in 2022. It simulates steady and nonsteady flow in three dimensions for an irregularly shaped flow system in which the aquifer layer can be confined, unconfined, or a combination of confined and unconfined. Flow from external sources, such as flow to wells, areal recharge, evapotranspiration, flow to drains, and flow through river, can be simulated.

MODFLOW uses a modular structure wherein similar program functions are grouped together. The modular structure consists of a main program and a large number of independent subroutines called "modules." The modules, in turn, have been grouped into "packages." Each package deals with a specific aspect of the hydrological system to be simulated. For example, the option "well package" simulates the effect of wells, the "river package" simulates the effect of river, etc.

The three-dimensional unsteady movement of groundwater of constant density through porous earth material in a heterogeneous anisotropic medium can be described by the following partial differential equation:

$$\frac{\partial}{\partial x}\left(K_{xx}\frac{\partial h}{\partial x}\right)+\frac{\partial}{\partial y}\left(K_{yy}\frac{\partial h}{\partial y}\right)+\frac{\partial}{\partial z}\left(K_{zz}\frac{\partial h}{\partial z}\right)-W=S_s\frac{\partial h}{\partial t} \tag{13.39}$$

where K_{xx}, K_{yy}, K_{zz} denote the hydraulic conductivities along major axes [LT^{-1}], h is the potentiometric head [L], W is the volumetric flux per unit volume and represents sources and/or sinks of water [T_{-1}], S_s is the specific storage of the porous material [L^{-1}], and t is time [T]. Possible inflow/outflow terms (W) are recharge from rainfall, artificial recharge through wells, pumping through wells, evapotranspiration loss, recharge from river/canal cells, outflow into a river/canal cell, inflow/outflow across a boundary cell, outflow through drains, spring flow, etc.

MT3D is a Modular 3-Dimensional solute Transport model for simulating changes in concentration of *single species* miscible contaminants in groundwater considering advection, dispersion, and some *simple chemical reactions* with various types of boundary conditions and external sources or sinks in groundwater systems. The model was developed by Zheng (1992).

MT3D is based on a modular structure to permit simulation of transport components independently or jointly. It interfaces directly with the USGS finite-difference groundwater flow model MODFLOW for the head solution and supports all the hydrologic and discretization features of MODFLOW. It has been accepted by practitioners and researchers alike and applied in numerous field-scale studies throughout the world. The modular structure of the MT3D transport model makes it possible to simulate advection, dispersion, source/sink mixing, and chemical reactions independently.

13.7.2 Model Selection

There can be different criteria for selecting water quality models, such as purpose, availability of quality data, type of waterbody, model complexity, model capability, model applicability, cost, familiarity, easy access to program source code, support for help, and popularity of the model (Kayode and Muthukrishna, 2018). Once a model is selected, it has to be calibrated and validated. The U.S. Environmental Protection Agency (Grimsrud et al., 1976) outlined the process of model selection and criteria with four levels of selection phases. The phases of model selection include Phase I (model applicability test), Phase II (cost constraint test), Phase III (performance index rating; simplified), and Phase IV (performance index rating; advanced). The first two phases help eliminate unsuitable models and the last two phases aid the ranking of the remaining models.

The elimination of models depends on the suitability of model for the objective of application qualified by the type of waterbody, input data, time variability, space variability, discretization, special features, pollutants, driving factors, and boundary conditions. If all the models are excluded in the first phase, then the user has to search either for a new model, reexamine the applicability constraints, or relax some of the constraints. Before water quality simulation, suitable models should be selected by inspecting the type of pollutant problem affecting the water system, determining the cause of water pollution, identifying the best management practice solutions, modeling objectives, and available resources.

13.8 CLOSURE

Water quantity and quality are an integral part of the natural hydrologic environment. These two processes are in continuous dynamic interactions. Heavy dependence of the modern society on chemical products is fraught with environmental degradation, health hazard, and pollution of precious natural resources. Therefore, proper assessment, development, and management of water resources require full understanding of the environmental processes and their interactions. Historically, water quantity has been the governing factor in determining water use. However, in a world that is increasingly aware and conscious of development that does not damage the environment, the modeling of water quality has occupied the central stage in water resources utilization. A large number of water quality models are available these days and provide critical inputs in decision-making.

After obtaining good understanding of the causes behind water quality degradation and the likely future trends, the next step would be to create infrastructure to treat waste water before it is disposed in the receiving water bodies. It would be good to consider a range of options to treat water – nature-based solutions to treatment plants.

References

Alley, W.M., Smith, P.E., 1982. Distributed Routing Rainfall-Runoff Model Version II. USGS Open File Report 82-344. Gulf Coast Hydroscience Center, NSTL Station, MS.

Bear, J., 1979. Hydraulics of Ground Water. McGraw-Hill, New York.

Beasley, D.B., Huggins, L.F., 1981. ANSWERS Users Manual. U.S.EPA, Region V, Chicago, IL. EPA-905/9-82-001.

Bicknell, B.R., Imhoff, J.C., Kittle, J.L., Donigian Jr., A.S., Johanson, R.C., 1993. Hydrological Simulation Program FORTRAN (HSPF): User's Manual for Release 10. U.S. Environmental Protection Agency, Athens, GA. EPA-600/R-93/174.

Biswas, A.K. (Ed.), 1981. Models for Water Quality Management. McGraw-Hill International Book Company, New York.

Bobba, A.G., Singh, V.P., 1995. Groundwater contamination modeling. In: Singh, V.P. (Ed.), Environmental Hydrology. Kluwer Academic Publishers, Dordrecht.

Brown, L.C., Barnwell, T.O., 1987. The Enhanced Stream Water Quality Model QUAL2E and QUAL2E-UNCAS: Documentation and User Manual. U.S. EPA, Athens, GA. EPA-600/3-87/007.

Caretta, A.M.M.A., Arfanuzzaman, R.B.M., Morgan, S.M.R., Kumar, M., 2022. In: Pörtner, H.-O., Roberts, D.C., Tignor, M., Poloczanska, E.S., Mintenbeck, K., Alegría, A., Craig, M., Langsdorf, S., Löschke, S., Möller, V., Okem, A., Rama, B. (Eds.), Climate Change 2022: Impacts, Adaptation, and Vulnerability. Contribution of Working Group II to the Sixth Assessment Report of the Intergovernmental Panel on Climate Change. Cambridge University Press, Cambridge, UK and New York, NY, USA, pp. 551–712. https://doi.org/10.1017/9781009325844.006.

Chapman, D. (Ed.), 1992. Water Quality Assessments. Published E and FN Spon on behalf of UNESCO, WHO and UNEP, London.

Chapra, S.C., 1996. Rivers and streams. In: Mays, L.W. (Ed.), Water Resources Handbook. McGraw-Hill, New York.

Donigian Jr., A.A., Imhoff, J.C., Ambrose, R.B., 1995. Modeling watershed water quality. In: Singh, V.P. (Ed.), Chapter 11 in Environmental Hydrology. Kluwer Academic Publishers, Amsterdam.

Gandolfi, C., Kraszewski, A., Sessa, R.S., 1996. River water quality modeling. In: Singh, V.P. (Ed.), Environmental Hydraulics. Kluwer Academic Publishers, Dordrecht.

Grimsrud, G.P., Finnemore, E.J., Owen, H.J., 1976. Evaluation of Water Quality Models: A Management Guide for Planners; Office of Air, Land and Water Use. Office of Research and Development, US Environmental Protection Agency, Washington, DC, USA.

HEC, 1977. Storage, Treatment, Overflow, Runoff Model, STORM, User's Manual. Hydrologic Engineering Center, Davis, CA.

Holley, E.R., 1996. Diffusion and dispersion. In: Singh, V.P., Hager, W.H. (Eds.), Environmental Hydraulics. Kluwer Academic Publishers, Dordrecht.

Huber, W.C., 1993. Contaminant transport in surface water. In: Maidment, D.R. (Ed.), Handbook of Hydrology. McGraw-Hill Inc., New York.

Huber, W.C., Dickinson, R.E., 1988. Storm Water Management Model User's Manual. U.S. Environmental Protection Agency, Athens, GA. EPA/600/3-88/001a (NTIS PB88-236641/AS), Version 4.

Jacobsen, P., Harremoes, P., Jakobsen, C., 1984. In: Balmer, P., Malmqvist, P.-A., Sjoberg, A. (Eds.), The Danish Stormwater Modelling Package: The SVK-System. Proc. Third Int. Conference on Urban Storm Drainage, vol. 2. Chalmers University, Goteborg, Sweden, pp. 453–562.

Kayode, O.O., Muthukrishna, V.K., 2018. Assessment of some existing water quality models. Nature Environment and Pollution Technology 17 (3), 939–948.

Knisel, W. (Ed.), 1980. CREAMS: A Field-Scale Model for Chemicals, Runoff, and Erosion-From Agricultural Management Systems. U.S. Department of Agriculture. Conservation Research Report No. 26, Washington, DC.

Kuwabara, J.S., Leland, H.V., Bencaa, K.E., 1984. Copper transport along a Sierra Nevada stream. Journal of Environmental Engineering, ASCE 110 (3), 646–655.

Langevin, C.D., Provost, A.M., Panday, S., Hughes, J.D., et al., 2022. Documentation for the MODFLOW 6 Groundwater Transport Model. U.S. Geological Survey Techniques and Methods, A61, 6. U.S. Geological Survey, USA, p. 56. https://doi.org/10.3133/tm6A61.

Maidment, D.R., 1993. Handbook of Hydrology. McGraw-Hill Book Company, New York.

Maier, H.R., Dandy, G.C., 1996. The use of artificial neural networks for the prediction of water quality parameters. Water Resources Research 32 (4), 1013–1022.

Malina, J.F., 1996. Water quality. In: Mays, L.W. (Ed.), Water Resources Handbook. McGraw-Hill Inc., New York.

McCutcheon, S.C., Martin, J.L., Barnwell, T.O., 1993. Water quality. In: Maidment, D.R. (Ed.), Handbook of Hydrology. McGraw-Hill Inc., New York.

McDonald, M.D., Harbaugh, A.W., 1988. A Modular Three-Dimensional Finite Difference Groundwater Flow Model. U.S. Geological Survey, Open-File Report, pp. 83–875.

Medina, M., Huber, W.C., Hinney, J.P., 1981. Modeling stormwater storage/treatment transients: Applications. Journal of the Environmental Engineering Division, ASCE 107 (4), 799–816. https://doi.org/10.1061/JEEGAV.0001199.

Moog, D.B., Jirka, G.H., 1998. Analysis of reaeration equations using mean multiplicative error. Journal of Environmental Engineering, ASCE 112 (2), 104–110.

Morel, F.M.M., Hering, J.G., 1993. Principles and Applications of Aquatic Chemistry. Wiley Interscience, New York.

Nas/Nae, 1972. Water Quality Criteria. National Academy of Science/National Academy of Engineering, Washington, D.C.

Rao, A.R., 1999. Prediction of reaeration rates in square, stirred tanks. Journal of Environmental Engineering, ASCE 125 (3), 215–223.

Reckhow, K.H., Chapra, S.C., 1983. Engineering Approaches for Lake Management, Volume 1: Data Analysis and Empirical Modeling. Butterworth, Mass.

Runkel, R.L., Bencala, K.E., 1995. Transport of reacting solutes in rivers and streams. In: Singh, V.P. (Ed.), Environmental Hydrology. Kluwer Academic Publishers, Dordrecht.

Stolarska, A.Z., Skrzypski, J., 2012. Review of mathematical models of water quality. Ecological Chemical Engineering S 19 (2). https://doi.org/10.2478/v10216-011-0015-x.

Tetra Tech Inc, 2018. Assessment of Surface Water Model Maintenance and Support Status. Environmental Protection Agency, Washington, D.C. EPA/600/R-18/220.

Thackston, E.L., Dawson, J.W., 2001. Recalibration of a reaeration equation. Journal of Environmental Engineering, ASCE 127 (4), 317–321.

Thomann, R.V., Mueller, J.A., 1987. Principles of Surface Water Quality Modeling and Control. Harper and Row Publishers, New York.

Tsakiris, G., Alexakis, D., 2012. Water quality models: an overview. European Water 37, 33–46.

UNESCO, 2005. Water Resources Systems Planning and Management. Retrieved August 5, 2019, From. https://ecommons.cornell.edu/bitstream/handle/1813/2804/12_chapter12.pdf?sequence=9&isAllowed=y.

USEPA, 1974. The Relationships of Phosphorus and Nitrogen to the Trophic State of Northeast and North-central Lakes and Reservoirs. National Eutrophication Survey Working Paper No. 23. U.S. Environmental Protection Agency, Washington D.C.

Weber, W.J., McGinley, P.M., Katz, L.E., 1991. Sorption Phenomena in Subsurface Systems: Concepts, Models and Effects on Contaminant Fate and Transport, Water Research, pp. 499–528, 1991.

WHO, 2022. Guidelines for Drinking-Water Quality: Fourth Edition Incorporating the First and Second Addenda. World Health Organization, Geneva.

World Health Organization, 1996. Guidelines for Drinking-Water Quality, vol. 2. Health criteria and other supporting information, World Health Organization, Geneva.

Young, T.C., Dolan, D.M., 1995. Assessment and control of loading uncertainty for managing eutrophication and toxic chemical fate in lakes. In: Singh, V.P. (Ed.), Chapter 10 in Environmental Hydrology. Kluwer Academic Publishers, Amsterdam.

Zheng, C., 1992. MT3D: A Modular Three-Dimensional Transport Model, Version 1.5, Documentations and Users' Guide. S.S. Papadopulos and Associates, Inc.

Further Reading

Bartram, J., Balance, R. (Eds.), 1996. Water Quality Monitoring: A Practical Guide to the Design and Implementation of Freshwater Quality Studies and Monitoring Programmes. Published by E and FN Spon on behalf of UNEP and WHO, London.

Helmer, R., Hespanhol, I. (Eds.), 1997. Water Pollution Control: A Guide to the Use of Water Quality Management Principles. Published by E and FN Spon on behalf of UNEP, Water Supply and Sanitation Collaborative Council and WHO, London.

CHAPTER

Climate Change and Its Impacts on Water Resources

14

Chapter outline

Water Resources Systems Planning and Management. https://doi.org/10.1016/B978-0-12-821349-0.00021-6

The objectives of this chapter are to:

- Explain the causes behind climate change;
- Explain the methodology to study the likely impacts of climate change on water resources;
- Explain the techniques to downscale climate data;
- Explain the likely impacts of climate change on various components of the water cycle.

14.1 INTRODUCTION

Climate of a region/place represents the long-term (typically more than 30 years) average of weather. Climate is the outcome of a complex system where a range of meteorological variables vary with time. Thus, climate can be said to be "average weather" or the long-term regional or global average of temperature, humidity and rainfall patterns over seasons, years or decades (https://climateknowledgeportal.worldbank.org/overview). These meteorological variables are typically temperature, precipitation, and wind. The period over which the average is taken is 30 years, as suggested by the World Meteorological Organization. Fig. 14.1 shows a pictorial view of the global climate system.

Climate change (CC) is a very common word in the present-day world. It refers to a change in either the average state or in other statistics (such as standard deviation, the occurrence of extremes, etc.) of climate. The change should persist for a long period, normally decades or longer. Further, the change should be statistically significant. Climate change in Intergovernmental Panel on Climate Change (IPCC) usage refers to any change in climate over time, whether due to natural variability or as a result of human activity. This definition is different than the United Nations Framework Convention on Climate Change (UNFCCC) definition, which defines climate change as "a change of climate which is attributed directly or indirectly to human activity that alters the composition of the global atmosphere and which is in addition to natural climate variability observed over comparable time periods." Climate change is now regarded as a major global environmental problem.

The IPCC is a United Nations body, created in 1988 by the WMO and the United Nations Environment Programme (UNEP). The objective of the IPCC is to provide governments with scientific information that can be used by them to develop climate policies. The IPCC has constituted three Working Groups that focus on different aspects of climate change: Working Group I—Physical Science Basis of Climate Change, Working Group II—Climate Change Impacts, Adaptation and Vulnerability, and Working Group III—Mitigation of Climate Change. After every 5–6 years, IPCC brings out comprehensive assessment reports (ARs), which provide a review and the latest information about the past, current, and future climate based on a review of published material. An elaborate procedure is followed to distill the information. Recently, in 2021–22, IPCC released the sixth assessment report (AR6). The IPCC does not conduct its own research. More information about IPCC can be found at https://www.ipcc.ch/.

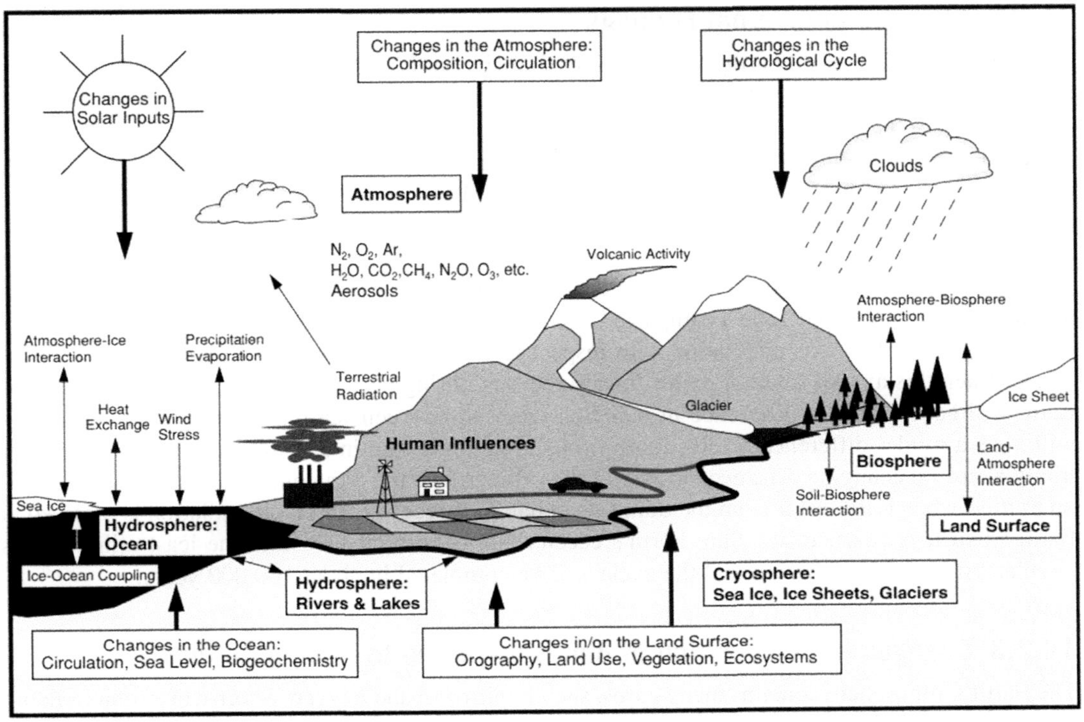

FIGURE 14.1

Schematic view of the components of the global climate system (*bold*), their processes and interactions (*thin arrows*), and some aspects that may change (*bold arrows*).

Reproduced from Baede et al. (2001). Copyright © IPCC. Used with permission.

14.2 CAUSES AND EVIDENCE OF CLIMATE CHANGE

Climate of Earth is changing due to: (1) natural internal processes, (2) external (to the Earth) forcings, (3) changes in the contents of the atmosphere due to human activities, or (4) changes in land use and land cover. Until recent times, climate forcings, such as changes in solar radiation reaching the Earth and emissions of volcanic dust, were entirely natural. However, in recent times, human activities have created additional forcings in the climate system, such as increases in greenhouse gases in the atmosphere that are causing warming.

14.2.1 Natural Internal Processes

Internal forcing mechanisms of CC, as the name suggests, operate from inside the climate system. For example, some volcanic eruptions release large volumes of ash, aerosols, and gases that intercept solar radiation reaching the Earth. Global climate responds to such time-varying forcings and undergoes changes. However, this type of change is different from internal climate variability. The El Niño–Southern Oscillation (ENSO) phenomenon is an example of internal variability of climate.

14.2.2 Changes in External Forcings

Solar radiation received on Earth is changing very slowly. Earth's orbit around the Sun and the tilt of its rotational axis are changing very slowly due to gravitational forces exerted by the movement of other planets of the solar system itself in the Galaxy. "Milankovitch cycles" is the collective name for natural cycles in Earth's movement; the name is used to honor the Serbian scientist Milutin Milankovitch who discovered these cycles operating in nature. Changes in Milankovitch cycles cause extremely slow and long-term climate change. Here, three orbital variations are important. First, the tilt of Earth's axis with respect to the plane of orbit or obliquity varies between 22.1 and 24.5 degrees over a period of about 41,000 years. This change in tilt does not change the total solar radiation received on Earth but its distribution in space and time is impacted. Second, the axial precession is the slow shift in the orientation of Earth's axis of rotation with respect to stars. This cycle is completed in about 26,000 years. When this axis of rotation points toward the Sun during perihelion, one of the hemispheres of Earth will have a greater difference in radiation received between seasons but the other hemisphere will have a milder difference. Third, Earth orbits around the Sun in an elliptical path whose eccentricity keeps on changing. The eccentricity of the orbit around the Sun controls the radiation received on Earth—when Earth is far from the Sun, less radiation is received and more radiation is received on Earth when it is closer to the Sun. Earth's eccentricity is currently close to the least elliptic (most circular); it is slowly reducing and the cycle will be completed in about 100,000 years.

14.2.3 Changes in Composition of Atmosphere due to Human Actions

The Earth's atmosphere contains many gases, such as nitrogen and oxygen. Short-wave solar radiation passes through the atmosphere and reaches the land surface. Passing through the atmosphere, a part of this radiation is reflected back into space, some part is absorbed by the atmosphere, and the remaining radiation reaches the land surface, which either reflects it or absorbs it. The land surface emits radiation toward space as long-wave and part of it is absorbed by some gases in the atmosphere. Some atmospheric gases radiate back part of the long-wave radiation to the Earth and this radiation warms the lower atmosphere. In this way, greenhouse gases modify the heat balance of the Earth by keeping long-wave radiation that would otherwise have returned to space (Fig. 14.2). This effect is called the greenhouse effect. The gases whose presence causes this are known as greenhouse gases (GHGs). The main greenhouse gases in the atmosphere are carbon dioxide (CO_2), nitrous oxide (N_2O), methane (CH_4), water vapor, and ozone (O_3).

In earlier times, the concentration of GHGs in the atmosphere was low but after the industrial revolution (it began about 150 years ago), human actions have released large quantities of GHGs into the atmosphere. Recently released Assessment Report#6 by IPCC (Arias et al., 2021) states that "it is unequivocal that the increase of CO_2, methane (CH_4), and nitrous oxide (N_2O) in the atmosphere over the industrial era is the result of human activities and that human influence is the main driver of many changes observed across the atmosphere, ocean, cryosphere and biosphere." The report goes on to say that "for the decade 2011−20, the increase in global surface temperature since 1850−1900 is assessed to be 1.09 [0.95 to 1.20]°C." The historical time series data of temperature and CO_2 levels were compiled, and based on the reconstructions of past climate series, IPCC showed that the concentrations of CO_2 in the atmosphere and global surface temperature are strongly coupled (see Fig. 14.3).

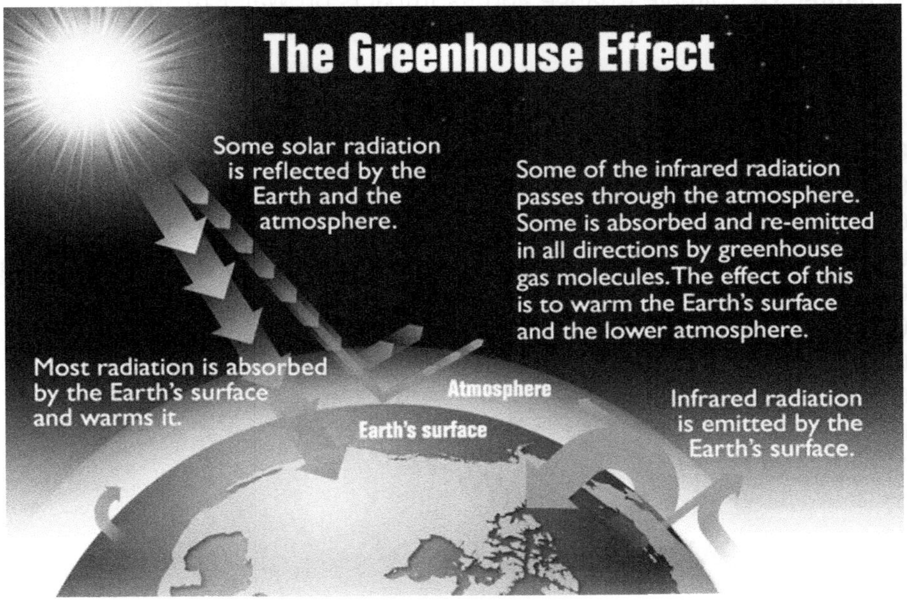

FIGURE 14.2

Illustration of the greenhouse effect.

Reproduced from https://www.epa.gov/climatechange-science/basics-climate-change.

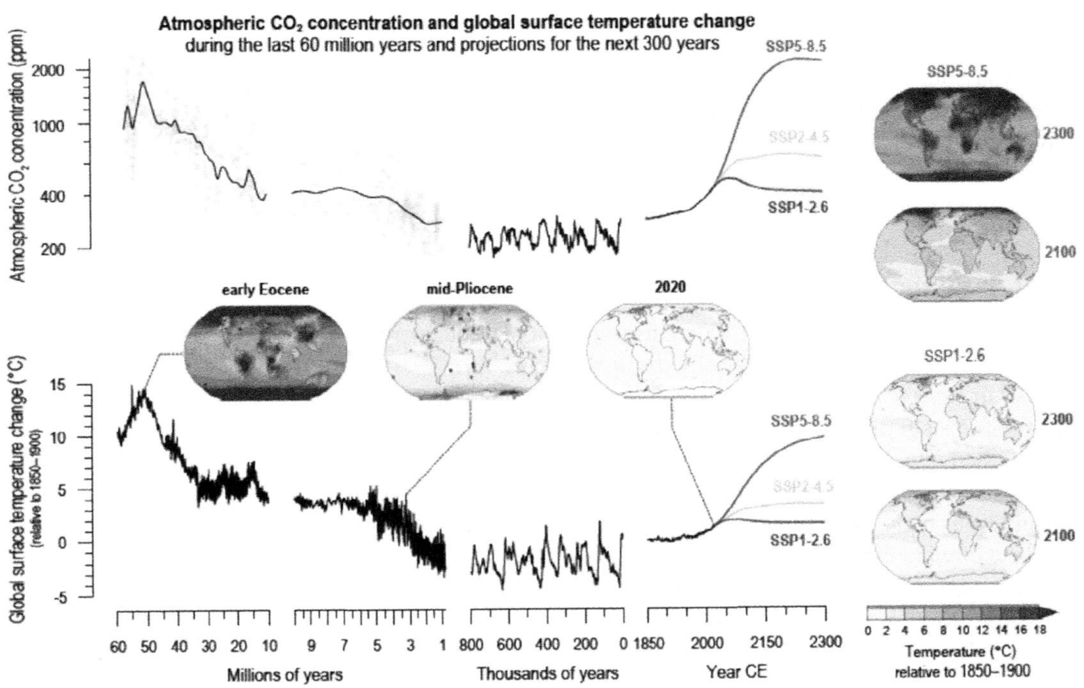

FIGURE 14.3

Changes in the atmospheric CO_2 and global surface temperature (relative to 1850–1900) from the deep past to the next 300 years.

Courtesy: Arias et al. (2021). Copyright © IPCC. Used with permission.

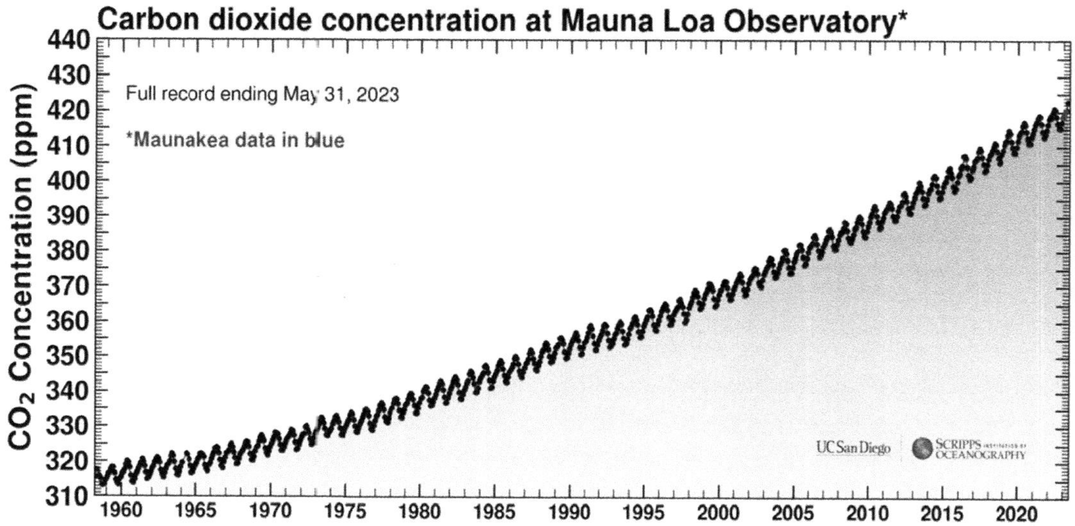

FIGURE 14.4

Concentration of CO_2 at the Mauna Loa observatory, 1960–2023.

Courtesy: https://keelingcurve.ucsd.edu/.

The concentrations of various greenhouse gases in the atmosphere of Earth are continuously rising, mainly arising from the combustion of fossil fuels and land-use changes. Fig. 14.4 shows the so-called Keeling curve, showing the concentration of CO_2 at the famous Mauna Loa observatory from 1960 onwards. This data can be considered to be representative of the world. It can be seen that the concentration of CO_2 is monotonically increasing. The cited website also contains CO_2 concentration curves from the past 800 thousand years till now.

14.2.4 Changes in Land Use—Land Cover

Land use and land cover (LULC) changes play a significant role in climate change. Major LULC changes and their impacts are briefly discussed here.

Deforestation: Trees absorb CO_2 to produce carbohydrates by photosynthesis. However, men have cleared forests for agriculture or urbanization and harvested timber. As a result, large amounts of CO_2 are released into the atmosphere and the ability to absorb CO_2 from the atmosphere is also reduced. So, deforestation leads to increased warming besides damage to ecosystems, reduction in biodiversity, less groundwater recharge, and higher soil erosion. Planting trees in deforested or degraded areas can help mitigate climate change.

Agriculture: Expansion of agricultural areas often comes with deforestation and increased GHG emissions. Moreover, farmers use fertilizers in agriculture that releases nitrous oxide (N2O), a GHG. Sustainable agricultural practices and more efficient land use can help mitigate these impacts.

Urbanization: Conversion of natural landscapes into built-up areas reduces infiltration and ET. Concrete and asphalt surfaces absorb and retain heat, resulting in higher temperatures compared to adjacent rural (green) areas and lead to the "urban heat island (UHI)" effect. UHI may also impact the local weather patterns. Buildings also impact wind patterns. Conversion of land for nonagricultural uses can cause soil degradation and loss of fertile land. To address the LULC impacts on climate change effectively, it is important to consider LULC changes under broader strategies for sustainable development.

14.3 APPROACH TO STUDY CLIMATE CHANGE IMPACTS ON WATER RESOURCES

An essential climate variable (ECV) is a physical, chemical or biological variable or a group of linked variables that critically contributes to the characterization of Earth' s climate. Data pertaining to ECVs provide the empirical evidence required to understand and predict how the climate evolves. These data also guide mitigation and adaptation measures and help assess risks and attribution of climate events. ECVs have been categorized in many groups, e.g., atmosphere, land, and ocean. Some data relevant for water sector are: precipitation, pressure (surface), surface radiation budget, surface wind speed and direction, temperature (near surface), water vapor, river discharge, groundwater volume change, levels, recharge; water quality, lake water level, extent, water temperature; soil moisture, area covered by snow, snow depth, snow water equivalent, glacier area, mass balance; albedo, and land cover.

We now discuss future climate scenarios.

14.3.1 Future Climate Scenarios and Their Purpose

Future GHG emissions will depend on the evolution of the socioeconomic system, which is very complex and dynamic. Its evolution is influenced by driving forces, such as industrial, demographic, transportation, and socioeconomic developments, and the adoption of new technologies. The future evolution of climate, of course, is highly uncertain. Scenarios are alternative paths describing the way the future might develop, depending upon the interaction of global, regional, environmental, and economic factors. Scenarios assist in climate modeling and in the assessment of the impacts of climate change, adaptation, and mitigation. Since the future is highly uncertain, it is difficult to say which emission path the climate will follow at various times in the future.

Based on a detailed appraisal of driving forces and emissions, alternative modeling approaches, and after wide consultations and feedback, IPCC (2000) released the "Special Report on Emissions Scenarios (SRES)," describing four narrative storylines. These are depicted and labeled as A1, A2, B1, and B2 in Fig. 14.5. Each of the four storylines represents different combinations of demographic, technological, social, economic, and environmental developments.

As shown in Fig. 14.6, a total of 40 SRES scenarios were developed by six modeling teams. All the scenarios are equally likely, and no probability of occurrence has been assigned to anyone. These 40 scenarios have been subdivided into six scenario groups: three groups from the A1 storyline and one group each from storylines A2, B1, and B2. Each of these four storylines represents different paths and a wide range of developments in demographic, social, economic, technological, and environmental conditions. All the scenarios emerging from the same storyline constitute a scenario "family." Within each family and group of scenarios, there is a harmonized scenario, marked as "HS." Further, "OS" scenarios denote uncertainties in driving forces beyond those in the harmonized scenarios. Scenarios exclude "surprise" or "disaster" situations.

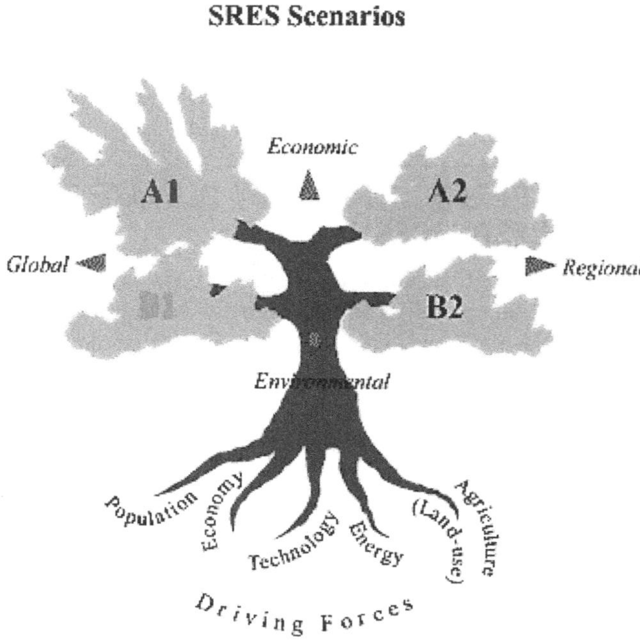

FIGURE 14.5

Schematic illustration of SRES scenarios. The four scenario "families" are illustrated, very simplistically, as branches of a two-dimensional tree.

Reproduced from Nakicenovic et al. (2000). Copyright © IPCC. Used with permission.

According to the A1 family, the future world would see very rapid economic growth, where new and/or more efficient technologies would be introduced at a fast pace, but population growth would be low. The A1 scenario family consists of four groups that follow different directions of energy generation technology. There are subsets to the A1 family: A1FI—emphasizes the use of fossil fuels, A1B—places balanced emphasis on the use of all energy sources, and in A1T, energy sources are mainly nonfossil. The world of the A2 scenario family would be very heterogeneous; the focus would be on self-reliance and maintaining regional identity. Regional fertility trends converge very slowly, which will result in population growth at a high rate. The world will see fragmented growth; economic development will depend on regions. In comparison to other storylines, per capita economic development and technical advancement will be fragmented and slower.

The B1 family assumes that the world will be convergent and the growth of population will be low but economic structures will change rapidly. There will be reductions in material intensity; resource-efficient technologies will be introduced, and shift toward a service and information economy is envisaged. According to the B2 family, the world will have an emphasis on indigenous solutions to achieve sustainable economic, social, and environmental development. In comparison to the B1 and A1 storylines, the world of the B2 family is likely to see a moderate population increase, midway levels of economic development, and less fast but varied technology advancement. The B1 scenario concentrates on local and regional levels but is also concerned with social equality and conservation of the environment. Fig. 14.6 provides details about the scenarios and their classification.

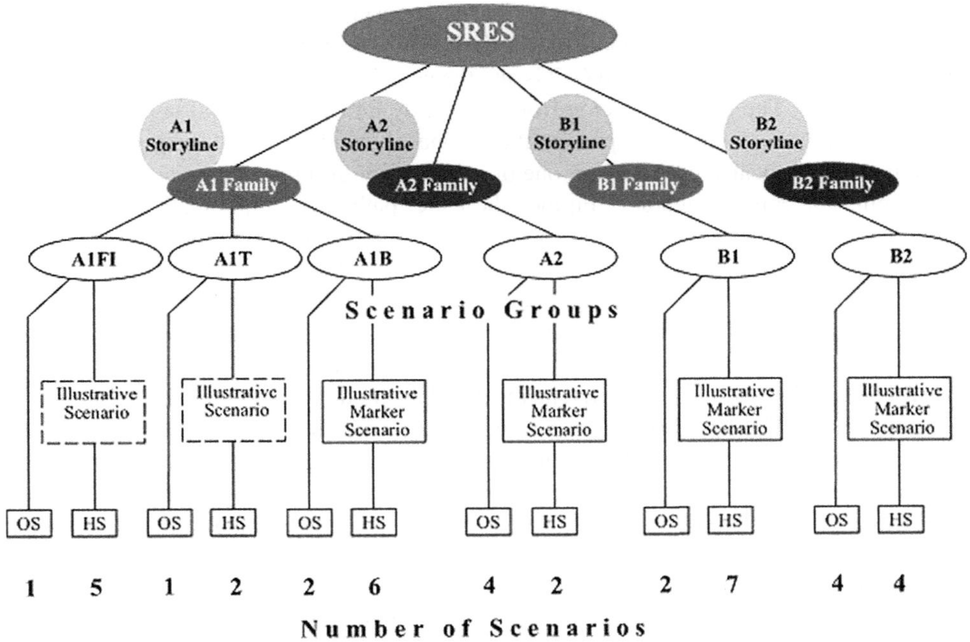

FIGURE 14.6

Schematic illustration of SRES scenarios. Four qualitative storylines yield four sets of scenarios called "families": A1, A2, B1, and B2. Altogether 40 SRES scenarios have been developed by six modeling teams. All are equally valid with no assigned probabilities of occurrence. The set of scenarios consists of six scenario groups drawn from the four families: one group each in A2, B1, B2, and three groups within the A1 family, characterizing alternative developments of energy technologies: A1FI (fossil fuel intensive), A1B (balanced), and A1T (predominantly nonfossil fuel). Within each family and group of scenarios, some share "harmonized" assumptions on global population, gross world product, and final energy. These are marked as "HS" for harmonized scenarios. "OS" denotes scenarios that explore uncertainties in driving forces beyond those of the harmonized scenarios. The number of scenarios developed within each category is shown. For each of the six scenario groups, an illustrative scenario (which is always harmonized) is provided. Four illustrative marker scenarios, one for each scenario family, were used in draft form in the 1998 SRES open process and are included in revised form in this report. Two additional illustrative scenarios for the groups.

Courtesy: Nakicenovic et al. (2000). Copyright © IPCC. Used with permission.

14.3.2 Representative Concentration Pathways

Representative concentration pathways (RCPs) are a group of scenarios that are input to climate models (van Vuuren et al., 2011). RCPs follow the concept of radiative forcing (Watts/sq. m), which is additional energy taken up by the Earth's atmosphere due to the increased GHG effect. It is the difference in the amount of energy that moves in the atmosphere and that goes back to outer space, compared to the situation before the industrial period. Global temperature increases as radiative forces increase. Both the positive forcing from GHGs and the negative forcing from aerosols contribute to the net radiative forcing. The positive forcing from CO_2 is by far the most important element.

The name RCP was adopted to emphasize the rationale behind the use of these scenarios. RCPs are representative, which means that these are one of a number of different scenarios that have analogous radiative forcing and emission features. Further, these are "pathways" implying that their main purpose is to provide projections of concentrations of atmospheric GHG as a function of time. Pathway emphasizes that a particular long-term concentration or radiative forcing outcome is not of interest, but the trajectory is also important. The magnitude of changes (GHG concentrations, LULC, and other changes) depends on population and economic growth, technology change, innovation, urbanization, etc.

Each RCP was developed by a different independent modeling group, and these RCPs are not directly comparable. The key differences between RCPs and SRES scenarios are that no fixed assumptions are associated with any RCP concerning population growth, economic development, or technology. The same level of radiative forcing is possible in the event of many different socioeconomic future conditions. This allows researchers to test a range of permutations of climate policies and social, technological, and economic policies. For instance, adverse climatic impacts of increased energy consumption could be overcome by higher use of energy generated by renewable sources. Four RCPs have been developed and are characterized by approximately total radiative forcing in the year 2100 relative to 1750 as shown in Table 14.1 which has been compiled from various reports.

Each RCP contains spatially resolved data sets of emissions of air pollutants from different sectors, land use changes, and specifies annual GHG concentrations and emissions up to 2100. Approximately, CO_2 concentrations by the year 2100 for RCPs are: RCP2.6—reaching 421 ppm, RCP4.5—reaching 538 ppm, RCP6.0—reaching 670 ppm, and RCP 8.5—reaching 936 ppm. Fig. 14.7 shows evolution of radiate forcings till 2100 for various RCPs.

14.3.3 Shared Socioeconomic Pathways

Along with RCPs, scientists were also working to find how socioeconomic factors might change in the future, depending upon population growth, economic growth, education, urbanization, and the technology development. This led to the creation of shared socioeconomic pathways (SSPs) that look at five different ways in which the world might develop in the future. Thus, SSPs are scenarios of

Table 14.1 Four RCPs and corresponding total radiative forcing.

Scenario	Description
RCP2.6	A mitigation scenario of very low radiative forcing of 2.6 W/m^2 by the year 2100
RCP4.5	A stabilization scenario, radiative forcing of 4.5 W/m^2 by the year 2100
RCP6.0	A stabilization scenario, radiative forcing of 6.0 W/m^2 by the year 2100
RCP8.5	A scenario of very high GHG emissions, radiative forcing of 8.5 W/m^2 by the year 2100

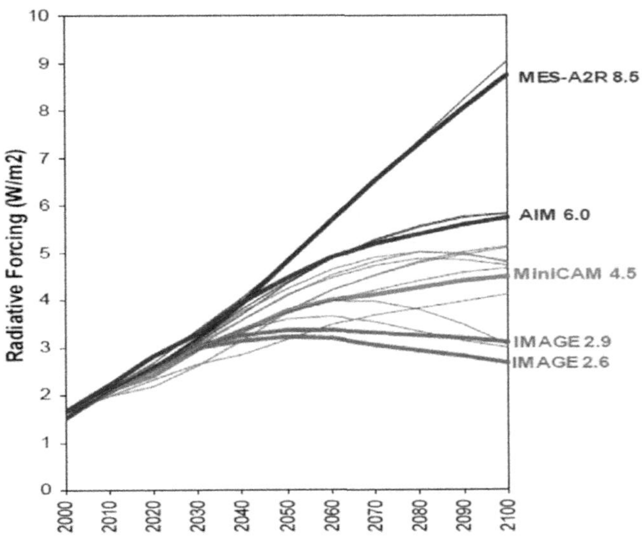

FIGURE 14.7

Radiative forcing compared to preindustrial (left panel) and energy and industry CO_2 emissions for the RCP candidates (colored lines).

Reproduced from Moss et al. (2008). Copyright © IPCC. Used with permission.

projected socioeconomic changes in the world up to the year 2100; these are used to determine emissions of GHG scenarios under different climate policies. The five SSP scenarios are:

- SSP1: sustainability (taking the green road, low mitigation and adaptation challenges),
- SSP2: middle of the road (medium challenges to mitigation and adaptation are faced),
- SSP3: regional rivalry (a rocky road, high challenges to mitigation and adaptation),
- SSP4: inequality (a road divided, low challenges to mitigation, high challenges to adaptation), and
- SSP5: fossil-fueled development (taking the highway, high challenges to mitigation, low challenges to adaptation).

The SSPs describe alternative socio-economic development narratives. These are a qualitative portrayal of the narratives; in quantitative terms, they provide data on populations, urbanization, and per capita GDP. Details of the five SSPs compiled from IPCC reports and other sources follow.

SSP1: Sustainability

The world shifts gradually, but widely, toward a more sustainable path; development is more comprehensive, respecting safe environmental limits.

SSP2: Middle of the Road

The world follows a trajectory in which social, economic, and technological trends do not shift significantly from historical trends. Development and income growth is uneven; some nations make good progress; some others achieve low growth.

SSP3: Regional Rivalry

The world will see growing nationalism, increased competition, and regional conflicts. Countries focus more on national and regional security. Nations will try to achieve food and energy security within their territory. Investments in education, capacity building, and technology would decline. Economic development would be slow, people are more materialistic, and inequalities may increase with time.

SSP4: Inequality

This scenario will see rising inequalities. Investments in human resources, environment protection, and conservation would be highly uneven. Gap between developed and developing nations will widen. Social cohesion would be less and conflict and discontent become increasingly common. The sectors using high-technology see rapid advancements. Environmental regulations and policies are concerned with regional problems in middle- and high-income communities.

SSP5: Fossil-Fueled Development

The SSP5 baseline scenarios point to very high amounts of use of fossil fuels, up to a doubling of global food demand, and up to a tripling of energy demand and GHG emissions by the year 2100. Currently, SSP5 are the only scenarios that would result in a high radiative forcing pathway.

SSP Temperature Projections from the IPCC Sixth Assessment Report

The IPCC AR6 assessed the projected warming levels for five scenarios based on the SSP framework. The scenarios have been named corresponding to the SSP on which they are based (SSP1−SSP5) and are combined with the likely amount of radiative forcing (1.9−8.5 W/m^2) for the year 2100. Table 14.2 shows likely warming for near-, mid-, and long-term for various SSPs.

AR6 of IPCC did not estimate the likelihood of various scenarios. However, a study has suggested that SSP5-8.5 is highly unlikely, SSP3-7.0 is unlikely, and SSP2-4.5 is the likely scenario (https://en.wikipedia.org/wiki/Shared_Socioeconomic_Pathways).

14.3.4 Global Climate Models

The planetary scale movements or "circulation" of air and water in the climate system is governed by radiative transfer, the Earth's rotation, gravity, surface friction, thermodynamics, etc. As controlled experiments involving the climate system are impossible, mathematical models of the climate have been developed to simulate the complex climatic processes. The development of climate models has a long history, beginning with very simple models. The land surface, the carbon cycle, forests, and many other components of the climate system were all included in climate models by the 2010s. Climate simulations offer a tool to investigate the various relationships and consequences of factors including solar irradiance, aerosols, and global climate models (GHGs).

The best tools now available to simulate the global climate and make future projections are the general circulation models, also known as global climate models (GCMs). Physical processes occurring in the atmosphere, ocean, and cryosphere and on land surface are represented in GCMs in detail. GCMs are the most sophisticated tools for simulating how the global climate system will react to rising GHG concentrations. Fig. 14.8 gives a generalized diagram of a GCM. Information produced by GCM is used in impact analysis, mitigation, and adaptation.

As seen in the figure, GCMs represent the climate system by using a three-dimensional grid over the Earth. Typical horizontal grid side ranges between 200 and 600 km. There may be 10−20 vertical layers in the atmosphere. The ocean part may have up to 30 vertical layers. The grid size of GCMs depends on a number of factors that include the solution algorithm, computational resources, time, and finances available.

Table 14.2 Changes in global surface temperature, which are assessed based on multiple lines of evidence, for selected 20-year time periods and the five illustrative emissions scenarios considered. Temperature differences relative to the average global surface temperature of the period 1850–1900 are reported in °C. This includes the revised assessment of observed historical warming for the AR5 reference period 1986–2005, which in AR6 is higher by 0.08 [−0.01 to +0.12] °C than in AR5. Changes relative to the recent reference period 1995–2014 may be calculated approximately by subtracting 0.85°C, the best estimate of the observed warming from 1850–1900 to 1995–2014.

	Near term, 2021–40		Mid-term, 2041–60		Long term, 2081–2100	
Scenario	Best estimate (°C)	Very likely range (°C)	Best estimate (°C)	Very likely range (°C)	Best estimate (°C)	Very likely range (°C)
SSP1-1.9	1.5	1.2–1.7	1.6	1.2–2.0	1.4	1.0–1.8
SSP1-2.6	1.5	1.2–1.8	1.7	1.3–2.2	1.8	1.3–2.4
SSP2-4.5	1.5	1.2–1.8	2.0	1.6–2.5	2.7	2.1–3.5
SSP3-7.0	1.5	1.2–1.8	2.1	1.7–2.6	3.6	2.8–4.6
SSP5-8.5	1.6	1.3–1.9	2.4	1.9–3.0	4.4	3.3–5.7

IPCC (2021). Copyright © IPCC. Used with permission.

Note that the GCM grid size mentioned may not be good enough for some reasons. GCMs with relatively coarse resolution cannot properly represent the finer topographic and other details and the climate, particularly in places where orography is complex, land surface cover is heterogeneous, small islands, mountains, or coastlines. Many physical processes, for example, those associated with clouds, take place at small spatial and temporal scales and cannot be properly modeled by coarse-scale GCMs. Further, at coarse discretization, extreme events, such as thunderstorms, cyclones, or intense rainfalls/cloud bursts, are not correctly captured. One way to overcome this limitation is to average the relevant properties over a larger scale in a technique known as parameterization. However, this introduces a source of uncertainty in GCM-based simulation of likely future climate. Many other sources of uncertainty also exist. For these reasons, the responses to the same forcings GCMs may be simulated differently by different GCMs. The response depends on the way the processes and responses are modeled by a particular GCM.

Climate models developed at different places are different in many ways, such as grid type, resolution, and parameterizations. To address the problem of comparing their results, scientists have developed model intercomparison projects (MIPs) prescribing standard experiment designs, time periods, output variables, etc. These experiments are named in their serial improving stages, for example, coupled model intercomparison project 5 (CMIP5). The latest project in the series is CMIP6.

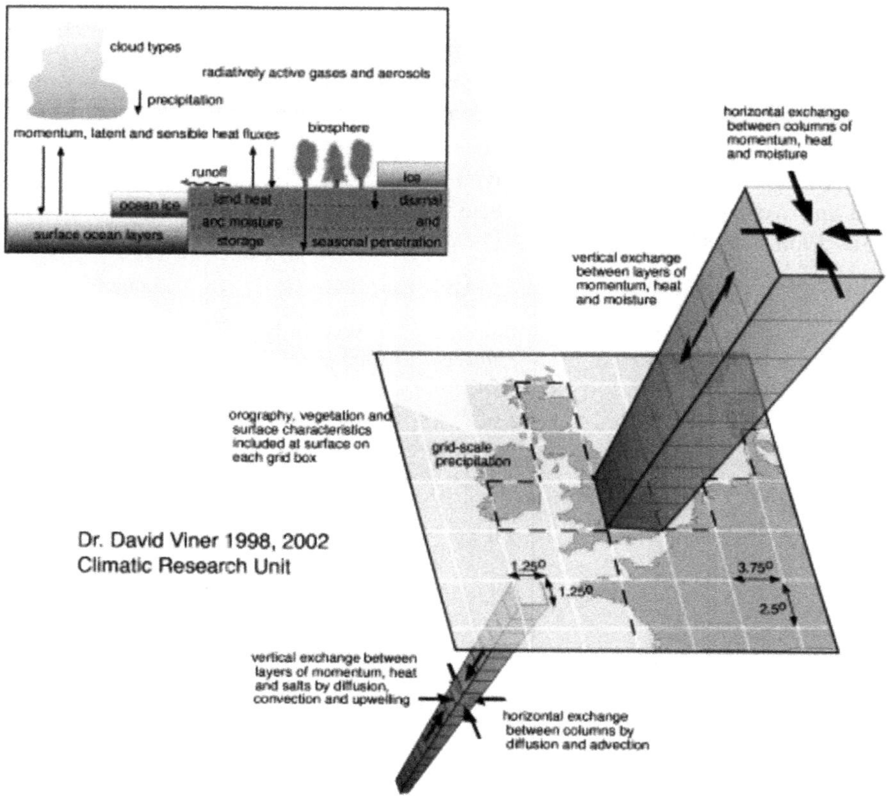

FIGURE 14.8

A conceptual diagram of a GCM. Courtesy

https://www.ipcc-data.org/guidelines/pages/gcm_guide.html. Copyright © IPCC. Used with permission.

Historically, GCMs have been the main source to provide data to construct climate scenarios and continue to deliver data for assessments of climate change at different scales from local to global. One way to use the GCM outputs is by employing a technique known as downscaling.

14.4 DOWNSCALING CLIMATE DATA

In climate change studies, the temporal scales range from a short time interval, say 10 min (flood studies) to a year or more. Similarly, spatial dimension may span from a few hectares (urban and rural watersheds) to thousands of square km (in the case of large river basins). GCMs are not able to simulate well the behavior of regional/local climate at time and space scales that are required for hydrological studies. In simple words, GCM outputs are commonly at resolutions that are too big for typical studies to determine climate change impacts. Thus, there is a need to scale down GCM outputs to regional/local/station scales.

Most impact assessment models work with data at a spatial resolution of 50 km or less. Clearly, some way is desirable to get the data at that spatial resolution. Downscaling is carried out to derive local and/or regional scale data from larger-scale data. In downscaling climate projections, the coarse-resolution outputs of GCMs are used to derive finer spatial and temporal scale data that is required to assess impacts at local and regional scales. Downscaling attempts to reduce the gap between the data that the climate models can provide and that needed by the impact assessment community (Fig. 14.9).

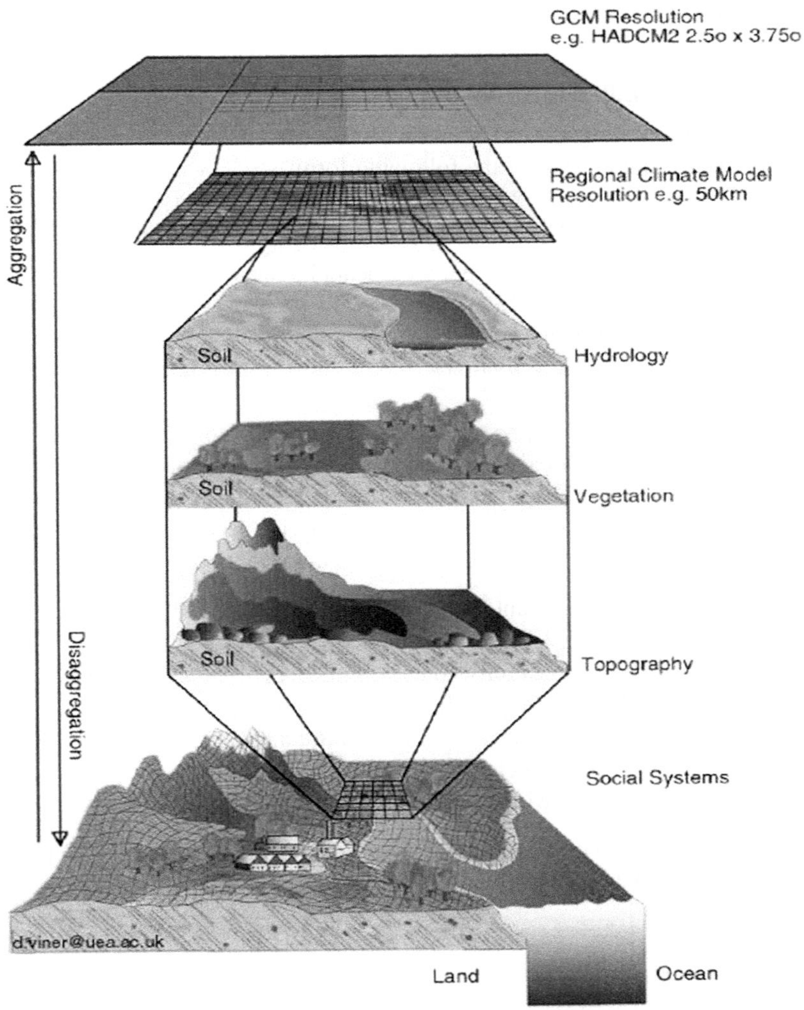

FIGURE 14.9

A schematic illustrating the general approach to downscaling.

Courtesy: USAID (2014).

Downscaling assumes that regional climate is directed by the climate on larger scales, say regional or even global. The regional climate results from the interaction of overall atmospheric and oceanic circulation and region-specific factors, such as topography, LULC, land–ocean interaction, and geographic location. In downscaling, data at a large scale are brought "down" to smaller scales.

Downscaling procedures can be classified into two broad categories: (1) statistical (or empirical) downscaling (SD) that utilizes relationships between observed large-scale climatological variables and observed local-scale variables, derived from the two data sets, and (2) dynamical downscaling (DD) that involves the use of high-resolution regional modeling to determine the impacts of large-scale climate processes on the climate at regional or local scales.

14.4.1 Statistical (or Empirical) Downscaling

Statistical/empirical downscaling tries to determine the local-scale data from the large-scale data by establishing a functional relationship of the following form:

$$\text{Regional/local climate response } R = f\,(\text{large-scale external forcing } E) \tag{14.1}$$

Here, R are the predictands (the set of regional/local climate variables that are to be predicted), E are the predictors (the set of large-scale external variables), and f a stochastic and/or deterministic function which maps E to R. In general, function f is unknown and is determined empirically from observed and/or modeled data sets. Sometimes, sets R and E may contain the same variables but on different spatial or temporal scales.

The concept of "downscaling" expressed by Eq. (14.1) does not mean that the regional climate variables could be uniquely computed by the large-scale state. For the given large-scale states, the corresponding regional states may be quite different. In other words, the regional climate is viewed as a process whose outcome is largely conditioned upon the driving large-scale climate.

While using SD downscaling, three assumptions are made:

1. The predictor variables are relevant for the task and are realistically simulated by the GCM(s) at the scale under consideration.
2. The identified predictor variables fully contain signatures of climate change for the study area.
3. The local-scale surface variables are a function of synoptic forcing.
4. The transfer function f will remain valid under future altered climatic conditions. This assumption, cannot be proved beforehand.
5. The observed data contains a wide range of past climatic variations.

Statistical (or empirical) downscaling (SD) methodologies themselves can be classified into three groups based on the computational procedures used: weather typing, stochastic weather generators, and regression.

Weather Typing

Here, the atmospheric circulation pattern is divided into limited classes by using these approaches. Next, weather types are simulated by using stochastic models; rainfall or other hydrometeorological processes are simulated by using weather types, and the relationship between the likelihood of rainfall and weather type is established by using conditional probabilities. Weather typing techniques have the capacity to produce lengthy data sequences at a location based on limited previous data and take into account links between large-scale climate and local weather patterns. The possibility that precipitation variations brought on by weather pattern changes may not match those generated by the host GCM is a drawback of such techniques.

Weather Generators

These are statistical models of observed sequences of weather variables. In a way, weather generators are random number generators whose outputs look like daily weather data at a given place. Depending upon the approach adopted to model daily precipitation, two basic types of daily weather generators are used: (1) the Markov chain approach, and (2) the spell-length approach. The Markov chain approach employs a random process to determine, conditional upon the state of the preceding day, if a day at a station can be considered as rainy or dry. If the day turns out to be rainy, the amount of rainfall is determined from a probability distribution.

A stochastic weather generator (WG) generates a time series of weather data for a place using the statistical features of observed weather at that place. WG models commonly follow two steps. In the first step, daily precipitation is modeled, and in the second step, the other variables of interest (e.g., daily maximum/minimum temperature, solar radiation, humidity, wind speed, etc.) are generated, conditional on the occurrence of precipitation.

Regression-Based Downscaling Methods

The downscaling techniques based on regression mainly depend on the causal connections between predictors and predictands. The key benefit of these techniques is that they are less computationally complex and simpler than previous downscaling strategies. In addition, these techniques are common among researchers. Of course, these techniques should be used only when it is possible to find trustworthy predictor-predictand correlations.

Building multiple regression models connecting atmospheric grid point values to surface variables is the most straightforward strategy here. Piecewise linear or nonlinear interpolation can be used as an alternative to linear regression. Artificial intelligence techniques, which have proven to be effective tools for nonlinear mapping, are becoming popular for this task.

Validation

Validation aims to ascertain that the model set up for downscaling and the chosen predictors are able to correctly represent the behavior of variables under investigation. The conventional validation approach is to calibrate/train a downscaling model by using a subset of available observed data and then assess the performance of the developed model by comparing its predictions with independent observed values. Statistical indices and graphical plots are used to evaluate the model performance.

Scientists have used a range of predictands, for example, temperature, precipitation, snowpack, and monthly or annual climatic statistics. While choosing predictors in SD, one needs to consider those variables also that will impact the predictand in the future. For example, if it is found that in the future, precipitation may depend on humidity also, it would be advisable to include humidity in the set of predictors. Note that a predictor may not be significant when developing the downscaling function in the present climates, but it might be important in the changed conditions. Among the SD techniques that have been used to establish relations between local weather and large-scale GCMs produced variables, the Statistical Downscaling Model (SDSM) and the stochastic weather generator LARS-WG are popular.

The use of downscaled data from GCM simulations for impact assessment by hydrological models has been identified as a source of large uncertainties. A way out is to apply ensembles of GCMs and downscaling methods to reduce uncertainties.

14.4.2 Dynamical Downscaling

In dynamical downscaling, outputs of a GCM are used as initial and boundary conditions and inputs for more detailed modeling of a limited geographic area with higher resolution in space and time. Physical processes in the limited area are simulated by using regional climate models (RCMs) such that there is consistency between the simulations of GCM and RCM.

An RCM covers a small portion of the entire world with a high spatial resolution (usually less than 50 km). RCMs are also capable of resolving small surface characteristics. An RCM and a GCM are relatively similar; both often include representations of the atmosphere and land surface, as well as the key climate system processes (such as clouds, radiation, rainfall, and soil hydrology). RCMs must be driven at their limits by large-scale fields (such as wind fields, temperature, water vapor, and atmospheric pressure), which vary over time. This is because RCMs are limited area models. This information originates from either GCM or observational analysis. The idea of a nested regional climate model is shown in Fig. 14.10.

Numerous surface features, such as complicated orographies, coasts, tiny islands, and peninsulas, can be better represented in an RCM. The majority, if not all, of the interactions, feedbacks, and processes between the many elements of the climate system that are represented by GCMs are represented by RCMs.

These days, a wide variety of RCMs developed at various modeling facilities worldwide are available. Due to variations in model formulation and small-scale internal variability produced by the RCM, various RCMs produce distinct scenarios of local weather for a given boundary forcing. Through the intercomparison of several RCMs, there has recently been a significant effort to quantify uncertainty in regional climate change. RCM simulations with grid sizes of 5 km or smaller are being developed at some modeling centers; recently, RCM simulations with grid size of less than 20 km have been performed for specific regions.

Developing, setting up, and using an RCM over a specific region requires considerable effort from a team of experienced climate modelers. Like GCMs, RCMs require large computing facilities. Due to these factors, many developing countries are not able to produce climate predictions and scenarios tailored to their needs. To promote this topic in a coordinated manner, an initiative known as CORDEX (Coordinated Regional Downscaling Experiment) is running. CORDEX is sponsored by the World Climate Research Program (WCRP) with the aim to develop an improved framework for generating regional-scale climate projections for impact assessment and adaptation studies. A key topic in the RCM research is to develop convection-permitting models that are useable for resolutions of a few kilometers. Giorgi (2019) has provided a comprehensive review and status of RCMs.

14.4.3 Advantages and Limitations of RCMs

RCMs simulate local/regional climate more realistically: Where terrain is flat for extensive areas which are away from coast, the coarse resolution GCM may work well. But when the region has

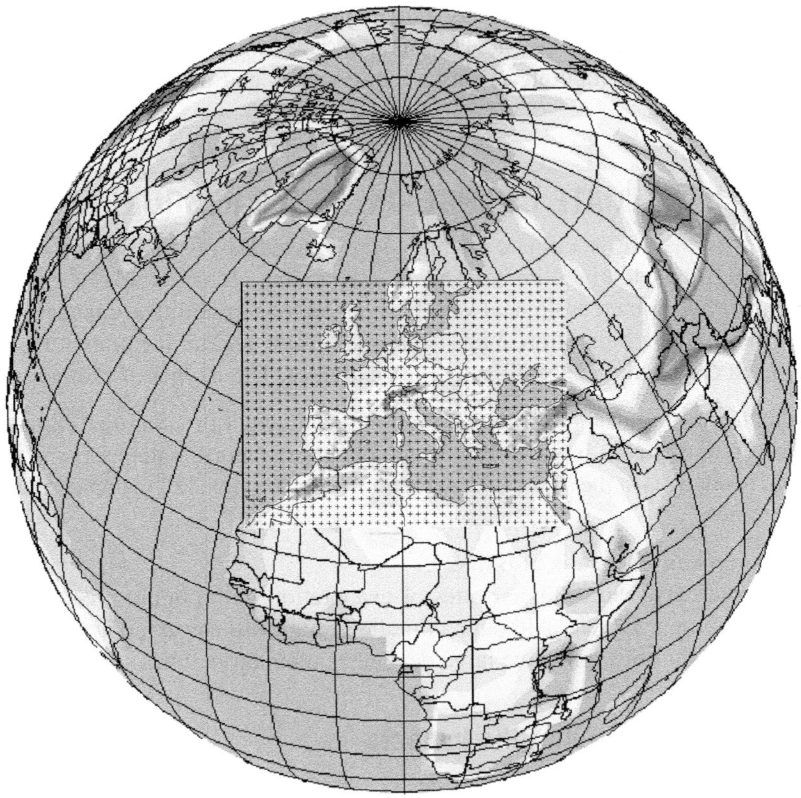

FIGURE 14.10

Nested regional climate model.

Reproduced from Giorgi (2019).

undulating topography on small scales that are represented by fine-resolution data, RCM results may be much better than GCM results.

RCMs represent smaller islands: With coarse resolution data, small islands cannot be represented and hence their climate is not correctly simulated. This drawback can be overcome in an RCM.

RCMs predict climate change with more detail: The finer spatial scale, where mountains are correctly represented, helps in the correct modeling of wind flows and other associated variables.

RCMs are much better at simulating extreme weather: Extremes of weather, for example, heavy rainfall events, cyclones, thunderstorms, are likely to be modeled much better by RCMs than by GCMs.

Limitations: RCMs, of course, cannot provide a solution to all the problems concerning downscaling. Predictions from an RCM are dependent on how correctly the GCM has simulated a broader climate. Resources and expertise needed to develop and run an RCM are much less than that for a GCM but these are not available at many places.

Comparisons of DD and SD techniques for climate impact indicate that neither of these techniques is consistently better. However, SD methods have some advantages over DD methods. SD methods have relatively inexpensive computing resource requirements and these can be tailored to meet the needs of the study at hand.

14.5 IMPACT OF CLIMATE CHANGE ON WATER RESOURCES

Water resources are sensitive to climate change. Many elements of the hydrological cycle will be directly or indirectly impacted by climate change. A changing climate is likely to cause changes in shifts in the mean of hydro-meteorological variables, increased variabilities, changes in the symmetry of distribution; and frequency, duration, intensity, spatial extent, and timing of extreme weather events (see Fig. 14.11) and can result in exceptionally high extremes. Three types of changes are shown here:

1. effects due to shifting of the whole distribution toward right or a warmer climate;
2. effects due to an increase in variability of temperature; average value remains the same;
3. effects due to changes in the shape of distribution; here, the right tail becomes heavier, resulting in hotter extremes.

More than one type of change is also likely.

A warmer climate may lead to higher rates of evaporation, more liquid precipitation, and intensification of the hydrological cycle. Although climatic change will cause impacts globally, it is likely to severely affect the countries that have large dependence on agriculture.

14.5.1 Methodology to Predict the Impacts

To identify the initiatives that the water sector needs to undertake to overcome the adverse impacts of climate change on water resources, it is important to determine how climate change will impact various hydrological variables and processes. To that end, the following methodology can be followed:

- Use GCMs to simulate future climate under the chosen climate by choosing the appropriate GHG emission scenario or SSP.
- Downscale GCM outputs to the scales at data will be used by the hydrological model.
- Employ hydrologic model(s) to determine how climate change will cause changes in the hydrological variables at the regional and/or local scale.
- Use the outputs from the hydrologic models as inputs to water management models that inform how the performance of hydro-infrastructure such as a hydro-power plant will change in the changed scenario and if required, suggest changes in management.

The changes to the hydrological cycle due to climate change will impact the availability of water for society in terms of quantity, quality, and accessibility. We now discuss the likely impacts of climate changes.

14.5.2 Impacts on Quantity of Water

Many countries of the world are already facing water stress due to rising irrigation and energy demands, increasing industrialization, industrial pollution, sewage, growing population, etc. The stress is

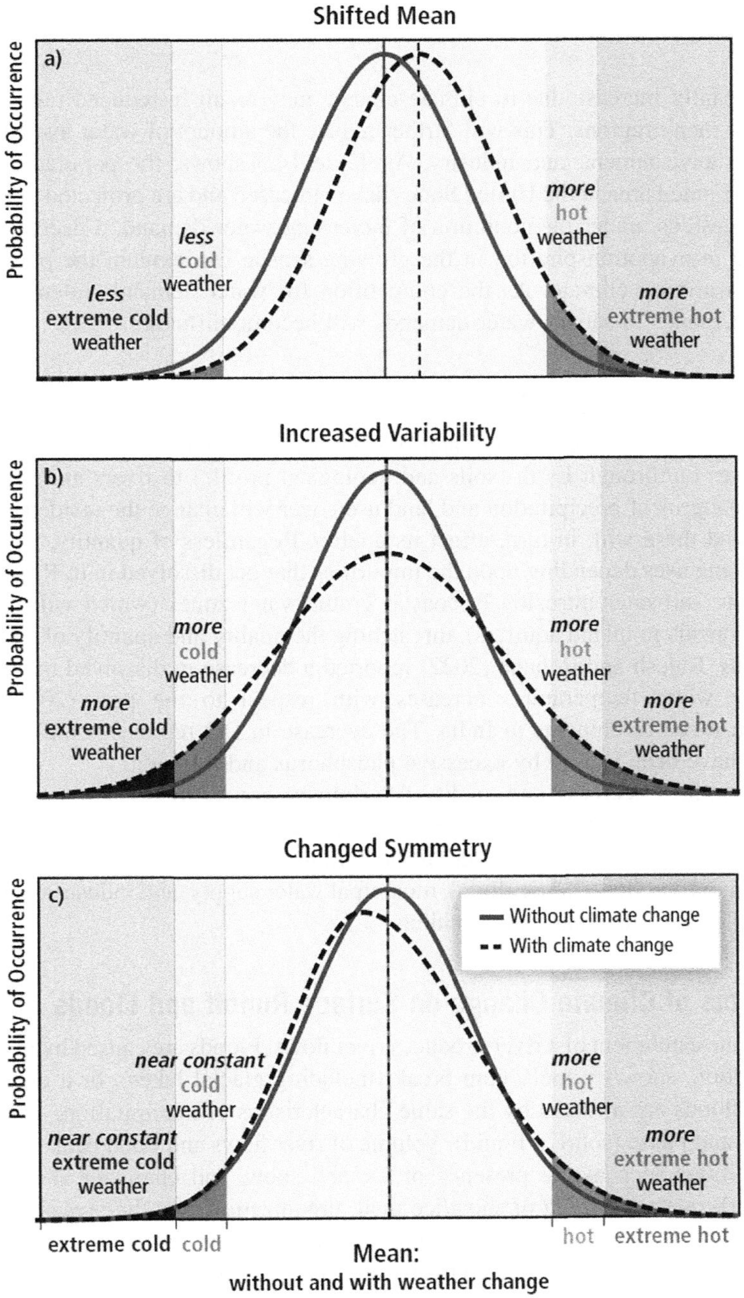

FIGURE 14.11

The effect of changes in temperature distribution on extremes. Different changes in temperature distributions between present and future climate and their effects on extreme values of the distributions: (a) effects of a simple shift of the entire distribution toward a warmer climate; (b) effects of an increased temperature variability with no shift of the mean; and (c) effects of an altered shape of the distribution, in this example an increased asymmetry toward the hotter part of the distribution.

IPCC (2012). Copyright © IPCC. Used with permission.

likely to significantly increase due to climate change may result in reduced rainfall and increasing temperatures for many regions. This will further reduce the amount of water available for household use, agriculture, environment, and industry. As Table 14.3 shows, the population, electricity consumption, and irrigated area in the United States have increased and are projected to increase under the A1B scenario. Besides, under the condition of increasing water demand, a decrease in precipitation and an increase in evapotranspiration in the growing season can explain the potential water stress impacted by a warming climate. As the competition for water demands intensifies under climate change, the governance to satisfy water demands will become difficult.

14.5.3 Impacts on Quality of Water

Water captures most of its geochemical and biochemical substance after it falls on the Earth from clouds and moves on/through Earth (soils and geological profile) to rivers and sea. Changes in the amount and/or patterns of precipitation and land use/cover will change the residence time of water in the catchment and these will, in turn, affect its quality. Regardless of quantity, water could become unsuitable for some uses depending upon the impurities that get dissolved in it. Rising seawater levels will lead to more saltwater intrusion in coastal groundwater zones (which will be accentuated by increased withdrawals in inland aquifers), threatening the quality and quantity of freshwater access to large populations. Rajesh and Rehana (2022) reported a decrease in dissolved oxygen (DO) concentration as river water temperature increases with respect to the past (2001–15) and future (2071–2100) in seven catchments in India. The decrease in DO represents the degraded quality of water that may have been caused by excessive phosphorus and nitrogen.

As water quantities decrease and quality deteriorates, competition for remaining resources will become more intense. Agricultural and domestic water demands are likely to rise during hooter and drier periods. In many countries, agriculture is the dominant user of water. In the event of falling water tables and reducing river flows, agricultural, municipal water supply, and industrial needs will compete with each other, raising the chances of conflicts.

14.5.4 Impacts of Climate Change on Surface Runoff and Floods

Precipitation in the catchment of a river produces river flows. Floods are caused by intense and/or long-lasting precipitation, snow/ice melt, dam break (including glacial lakes), or a combination thereof. River flow and floods are affected by the same characteristics of precipitation, such as duration, intensity, amount, and phase (solid or liquid). Volume of river flows and flood peaks are also affected by catchment conditions such as the presence of ice and snow, soil character and status (antecedent moisture content), rate and timing of snow/ice melt, urbanization, and the presence of reservoirs and diversions (Bates et al., 2008). In this section, the focus is on spatial, temporal, and seasonal changes in flows and peak discharges in rivers related to climate change. Climate change may change many of the factors affecting floods (e.g., precipitation, snow/ice cover, soil moisture distribution, sea level, glacial lakes, vegetation) and thus may consequently change the characteristics of floods.

Observed records of floods at river gauge stations are limited worldwide in spatial and temporal coverage. Very few river gauging stations have data series whose length is more than 50 years; even fewer have data series of length more than 75 years. Among these data, when one tries to locate the stations where flows have not been impacted by upstream withdrawals/regulation and land use

Table 14.3 Past and projected drivers of water use and meteorologic conditions in the United States.

Year	Population (millions)	Electricity consumption (kWh/p day)	Irrigated area in east (ha/10^6)	Irrigated area in west (ha/10^6)	Mean annual temperature (°C)	Growing season precipitation (cm)	Growing season mean ETp (mm/day)
2005	294	32.3	6.1	18.5	11.8	44.4	5.4
2010	307	32.6	6.5	18.2	12.0	45.4	5.5
2020	334	33.0	7.2	17.9	12.5	44.3	5.5
2030	361	33.2	7.6	17.7	12.8	45.3	5.6
2040	389	33.3	7.8	17.5	13.4	42.4	5.9
2050	417	33.4	8.0	17.3	13.7	44.1	6.2
2060	444	33.5	8.1	17.1	13.6	45.2	5.9
2070	461	33.5	8.1	17.0	14.4	42.8	6.3
2080	480	33.5	8.2	16.9	14.3	44.8	6.2
2090	499	33.5	8.2	16.8	15.0	43.6	6.4

Courtesy: Brown et al. (2013).

changes, one is left with data of only limited stations. Further, restrictions in data access and data sharing protocols prevent systematic and comprehensive assessments of trends in streamflow in many regions of the world. Literature on the impact of climate change on riverine floods is also scarce.

Many river systems are no longer in their natural state. Hydro-infrastructure such as reservoirs and diversions regulate river flows/floods and may also affect flows in downstream rivers. Many large dams operated to meet irrigation demands may have resulted in significant land use change, which in turn may affect runoff generation in a catchment. Thus the river flow and floods behavior may alter due to a number of reasons and it is difficult to separate changes in the river flows that are caused by climate change from those caused by other factors. Seneviratne et al. (2021) also note that one cannot always find a one-to-one correspondence between an extreme precipitation event and a flood event, or between changes in extreme precipitation and changes in floods. The reason is that floods are affected by multiple factors, besides heavy precipitation. Here, the focus is on the changes that can be attributed to changes in climate.

Many published studies that have tried to estimate trends in streamflow have used stream gauging records from stations thought to be minimally affected by human interventions such as upstream diversions/regulation, groundwater withdrawal, and land use change. Hence, the trends determined in these studies can be attributed to variability induced by changes in climate.

Research does not show clear and widespread evidence of climate-driven observed changes in the magnitude or frequency of floods at the global level based on instrumental records, and there is thus low confidence regarding the magnitude and frequency and even the sign of these changes.

In places where seasonal snow storage and melting play a significant role in the annual runoff, the hydrologic regime is affected by changes in temperature. In a warmer world, a smaller portion of precipitation falls as snow and the winter snow melts earlier in spring, resulting in a shift in peak river runoff to winter and early spring. There is evidence for an earlier occurrence of spring peak flows in snowmelt- and glacier-fed rivers. The observed trends toward earlier timing of snowmelt-driven flows in the western United States since 1950 are detectably different from natural variability (Barnett et al., 2008; Hidalgo et al., 2009). Thus, observed warming over several decades, attributable to anthropogenic forcing, has been linked to earlier spring peak flows in snowmelt- and glacier-fed rivers. It is, however, unclear if atmospheric warming has affected the magnitude of snowmelt peak flows, but warming may result either in an increase in spring peak flows where winter snow depth increases (Meehl et al., 2007) or a decrease in spring peak flows because of decreased snow cover and amount.

In cold regions where snowmelt dominates the flow regime responding to warming, the seasonality of floods has changed. There is low confidence about peak flow trends over past decades on the global scale, but some regions have experienced increases in peak flows.

Recent studies have detected the influence of anthropogenically induced climate change in variables that affect floods, such as mean precipitation, heavy precipitation, and snowpack, though a direct statistical link between anthropogenic climate change and trends in the magnitude and frequency of floods is still not established. There is still a lack of studies identifying the influence of anthropogenic climate change on rain-generated peak streamflow trends because of limited observed streamflow data, uncertainty in the data, and weak signals in the data. Sharma et al. (2018) noted that scientists have evidence of increasing precipitation extremes. However, the flood data do not show corresponding increases in flooding. Possible reasons could be a reduction in antecedent soil moisture, decreasing storm extent, and a decrease in snowmelt. Thus, understanding the link between changes in precipitation and changes in flooding is a challenge for the hydrologic community.

Table 14.4 Trend test analysis for the frequency of flood events in the global and each latitudinal sectors. Significance level is 0.05. √ and × represent significant trend and insignificant trend, respectively.

	Frequency of flood events (1985–2015)				
Spatial scale	**Total flood events**	**Kendalls's τ**	**Sen's slope**	***P*-value**	**Trend**
Global	4311	0.26	2.12	0.0429	√
Mid-latitudes (north)	1077	0.22	0.5	0.086	×
Subtropics (north)	856	0.032	0.048	0.8115	×
Tropics	2020	0.4	1.74	0.0016	√
Subtropics (south)	210	0.366	0.22	0.0038	√
Mid-latitudes (south)	59	0.327	0.083	0.0077	√

Najibi and Devineni (2018).

It is likely that the frequency of floods and droughts will increase during the 21 century due to projected climate change, which would enhance the severity of water extreme events. Changes in the amount of rainfall, patterns, and intensity would affect streamflow. Najibi and Devineni (2018) analyzed the frequency and duration of floods by the use of global flood database of the Dartmouth Flood Observatory (DFO) to examine the existence of trends during 1985–2015 at global and latitudinal scales, as shown in Tables 14.4 and 14.5. Except for the middle latitudes and subtropics in the northern hemisphere, the frequency and duration of floods showed statistically significant increasing trends. Flood and drought management schemes have to be planned keeping in view the increase in severity of floods and droughts. It would be prudent to incorporate the possible effects of climate change in the design and management of water resources projects.

Table 14.5 Trend test analysis for the median of flood durations in the global and each latitudinal sectors. Significance level is 0.05. √ and × represent significant trend and insignificant trend, respectively.

	Median of flood durations (1985–2015)				
Spatial scale	**Maximum flood durations in the period (days)**	**Kendalls's τ**	**Sen's slope**	***P*-value**	**Trend**
Global	168	0.484	0.125	~0.000	√
Mid-latitudes (north)	131	0.2667	0.0909	0.0346	√
Subtropics (north)	122	0.3097	0.125	0.0141	√
Tropics	168	0.4473	0.15	0.0004	√
Subtropics (south)	93	0.3312	0.1667	0.0088	√
Mid-latitudes (south)	21	0.3613	0.2105	0.0034	√

Reproduced from Najibi and Devineni (2018).

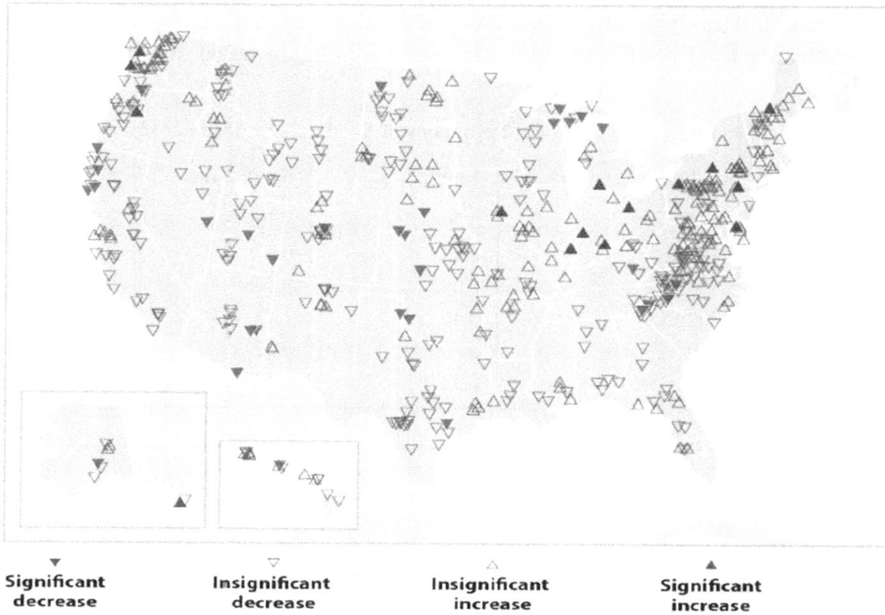

FIGURE 14.12

Changes in the magnitude of flooding events in rivers in the United States between 1965 and 2015.

Fig. 14.12, credited to US EPA (https://www.epa.gov/climate-indicators/climate-change-indicators-river-flooding), shows changes in the magnitude of flooding events in rivers in the United States during 1965–2015. Blue upward-pointing triangles show stations where floods have become bigger; brown downward-pointing triangles show stations where floods have become lesser. The color-filled triangles represent locations where the change in floods was found to be statistically significant. The first impression from this figure is that the river flows and floods will rise at some places and fall at some other places. In the central and western parts of the United States, there are more instances of floods becoming smaller.

Summarizing the findings for IPCC AR6, Seneviratne et al. (2021) conclude that there is low confidence in concluding that humans have caused changes in high river flows on a global scale. This low confidence arises because the number of studies is limited, the results of the studies are different, and modeling has large uncertainties.

14.5.5 Impacts of Climate Change on Droughts

In addition to floods, climate change will also have significant impacts on droughts. The Fifth Assessment of the Intergovernmental Panel on Climate Change (IPCC AR5) reports that the global mean temperature has been increasing since the late 1900s (Randall et al., 2013). The increase in global mean temperature increases potential evapotranspiration as well as high-intensity precipitation with greater intermittence. Further, global arid regions are expected to increase ~10% by 2100. In the latter half of the 21st century, more severe and more widespread droughts will occur in most of the

Americas, Australia, and Africa, south of Europe, and Southeast Asia (Dai, 2011). Kogan et al. (2020) calculated the global drought area and intensity in three categories (severe-to-exceptional, extreme-to-exceptional, and exceptional) during 1981–2018 using the mean weekly vegetation health index (VHI). They showed the rate of trend change of global drought area based on intensity ranges for the four different periods (1981–2018, 1998–2018, 1998–2014, and 1981–97). The percentage change was calculated as the relative difference between the beginning and end of the period divided by the beginning. The period 1998–2018, which included an abrupt temperature rise, showed a larger range of trend change rate than the period 1998–2014. It would be interesting to study how global warming affects drought.

Global warming impacts the ocean–atmosphere–land interaction and internal atmospheric variability. The increase in global mean temperature increases atmospheric water demand (AWD) and intensifies land–atmosphere feedback. With the increase in AWD and the intermittent period of precipitation, the dry period gets longer and drought becomes more likely. On the other hand, increased global mean temperature results in increased melting of the Himalayan snowpack more and reduced albedo during spring and summer, thus amplifying soil moisture and streamflow and reducing sensible heat. Therefore, the low thermal land–sea contrast makes the monsoon system weak, reducing precipitation under global warming. Under global warming, drought causative mechanisms are more apt to frequent and severe drought occurrences.

14.5.6 Impacts of Climate Change on Groundwater

Groundwater is the principal source of drinking water in rural areas in many countries. Groundwater regime is also expected to be severely impacted by climate change. As rainfall intensities increase, water will stay on the catchment surface for shorter periods and the opportunities for recharge will be fewer. As urbanization increases, more ground surfaces will become impervious, leading to reduced infiltration of water. Climate change is also expected to impact groundwater regime due to changes in precipitation patterns and higher evapotranspiration.

The projected global warming resulting in accelerated melting/shrinking of glaciers and sea level rise will adversely impact water balance in various catchments. Increased intrusion of saltwater in coastal aquifers is going to increase salinity in ground waters along coastal plains. Increased frequency and severity of floods may adversely affect groundwater quality in inland alluvial aquifers since floods frequently carry polluted water.

Fig. 14.13 shows the decreasing trend of yearly cumulative groundwater loss in the Central Valley in California, the United States. Groundwater losses were calculated by combining the USGS's Central Valley Hydrologic Model (CVHM) and analysis of data collected by GRACE/FO satellites. The black line represents the overall groundwater depletion estimated by combining the CVHM and GRACE from 1962 to 2021 (Liu et al., 2022). Similar groundwater depletion can be seen in many other parts of the world. Drop in groundwater levels causes depleted baseflows in the rivers and associated harms.

14.5.7 Impacts of Climate Change on Hydropower Generation

The basic elements of potential water power are river or stream flow and available head or fall through which stream flow may be utilized in the development of power (power plant). Hydropower projects may be classified into two major types: storage-based projects and run-of-river (ROR) projects.

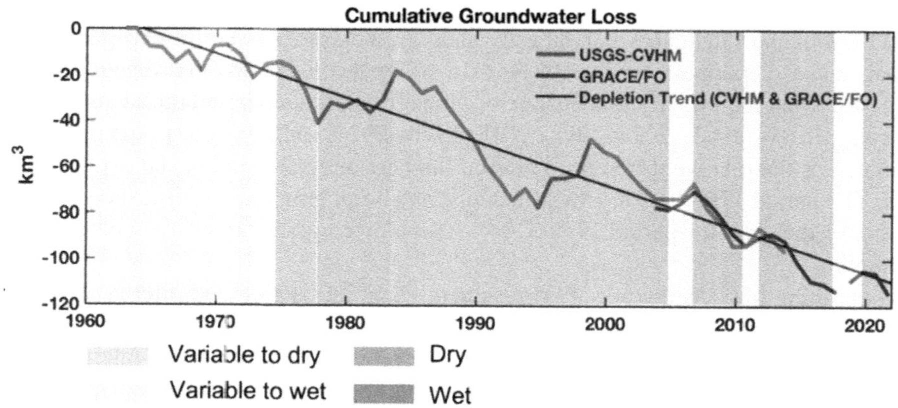

FIGURE 14.13

Annually cumulative groundwater losses in the Central Valley.

Source: Liu et al. (2022).

Storage-based hydropower projects can better address increased seasonality in inflow series and will be less vulnerable to climate change.

The following framework can be used to study the impact of CC on hydropower generation:

1. Collect hydrological and meteorological data for the study area along with features of the reservoir and hydropower plant.
2. Carry out hydrological modeling of the catchment area of the hydropower plant to simulate inflows to the project. The model should be able to realistically account for the changes in the meteorological and other inputs to the catchment.
3. Construct the likely future climate scenarios in terms of precipitation, temperature, and other meteorological variables.
4. Using the model calibrated for the catchment, determine the inflows to the reservoir under different climatic scenarios.
5. Simulate the operation of the hydropower plant to determine firm energy, firm power, or hydroelectric energy generated by the project under different scenarios along with associated reliabilities.
6. Intercomparison of the results of different scenarios will yield the impact of CC change on hydropower generation by a project.

Climate change may impact hydropower generation through changed inflow volumes and their distribution. If the volume of inflows is reduced, power generation will be less and vice versa. Further, if the bulk of the inflows is received in fewer months, the chances of spill will be more, and less energy will be generated. Fig. 14.14 shows the monthly hydroelectricity generation in the Pacific Northwest and California. Compared to 10-year average monthly values (2011–20), Pacific Northwest (California) generated hydroelectricity below the 10-year average overall during 2019–21 (2020–21) due to drought. The historic drought in 2021 led to 14% (48%) below normal for Pacific Northwest (California). Climate change could also induce a mismatch in timing between energy generation and

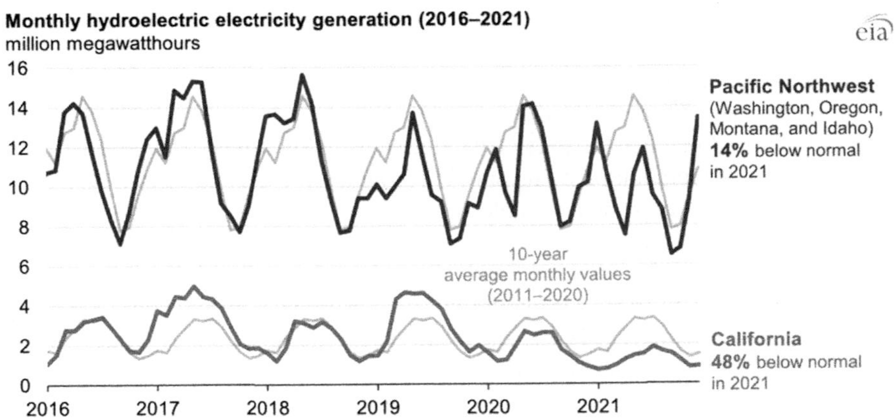

FIGURE 14.14

Monthly hydroelectric generation in Pacific Northwest and California.

Courtesy: U.S. Energy Information Administration, https://www.eia.gov/.

demand. Additionally, higher inflows in the reservoirs during the wet season could lead to greater spillage and less overall energy generation. These and the other potential effects can be studied using an optimization/simulation model of the catchment and reservoir system.

14.6 CLIMATE CHANGE: ADAPTATION AND MITIGATION

IPCC defines adaptation as the "Adjustment in natural or human systems in response to actual or expected climatic stimuli or their effects, which moderates harm or exploits beneficial opportunities." (https://archive.ipcc.ch/publications_and_data/ar4/wg2/en/annexessglossary). Climate adaptation is the ability of a system to acclimate to climate change so the possible harms are lessened, opportunities are seized, and the harmful effects are addressed. Adaptation comes in a variety of forms. A proactive or anticipatory adaptation is one that occurs before the effects of climate change are actually felt. Autonomous or spontaneous adaptation is a nonconscious response to climatic stimuli. Additionally, an adaptation that occurs due to a conscious policy choice, based on the knowledge that the circumstances have changed or will soon change and a decision needs to be made to return to a desired condition is known as planned adaptation. Adaptive capacity is the ability of a system to adapt to CC to mitigate potential harm, grab opportunities, or deal with the consequences. A well-planned adaptation program can lessen the vulnerability by way of strengthening the existing strategies to cope with harmful inputs.

According to IPCC (Klein et al., 2007), mitigation is: "an anthropogenic intervention to reduce the sources or enhance the sinks of greenhouse gases (GHG)." Thus, climate mitigation is any action taken to eliminate or lessen the risk to human life and properties due to CC. While mitigation addresses the causes of climate change, adaptation addresses the effects. The water sector has a somewhat limited role in mitigation.

In general, with more mitigation, the impacts to which we will have to adjust, and the risks for which we will have to prepare, will be less. A greater climatic change results from less mitigation, necessitating more adaptation. There is, therefore, an urgency surrounding GHG emission reductions for which a coordinated set of measures in a larger strategy is required. Climate mitigation and adaptation should not be considered as alternatives to one another and should be considered in tandem. Mitigation strategies can impact the management of water resources and, at the same time, water management policies and practices impact GHG emissions, while mitigation efforts of other sectors impact water supplies. Adaptation planning should recognize the uncertainties in CC scenarios, and adaptation plans themselves must be adaptable.

Many harmful and widespread impacts may arise due to climate change including damages to crops, falling energy generation, increased flood damages, and deteriorating water quality and health of ecosystems. The outcome of various studies points toward urgency in taking up adaptation measures since any delay will result in increased costs and make people more vulnerable. Hydro-infrastructure is crucial for not only achieving water, food, and environmental security but also for minimizing damages to life and property during disasters. It is, therefore, appropriate that we urgently review water resources development (WRD) pathways and policies and update these in view of new and emerging challenges. The development should have an appropriate mix of gray, blue, and green infrastructure.

With the increase in warming, risks to ecosystems, society, and infrastructure will increase everywhere. In the presence of other global trends such as population growth, rapid urbanization, land degradation, overuse of resources, biodiversity loss, poverty, and inequity, reducing the climate-change-induced risk is more complicated. We require unified solutions that integrate the strategies to reduce climate risks (adaptation) with actions to reduce GHG emissions (mitigation).

14.6.1 Climate Resilient Development

To comprehensively and sustainably address the adverse impacts of climate change, development actions should be climate resilient. Climate Resilient Development (CRD) combines strategies to adapt to climate change with actions to reduce greenhouse gas emissions to support sustainable development for everyone (Schipper et al., 2022). IPCC suggests that action to implement the CRD concept should start now because decision-making is already challenging at current global warming levels. If temperatures exceed 2°C of warming, CRD will become impossible in some regions of the world.

CRD actions should be scalable and well aligned with adequate investment. Well-planned CRD pathways will surely reduce exposure and vulnerability to climate hazards, will promote biodiversity, reduce GHG emissions, increase water, food, and energy security, and will lead to sustainable urban and rural development. As each country has different capacities, resources, and opportunities for CRD, the best CRD path will also be different.

CRD is the result of the choices and paths that the society adopts for climate risk reduction, emissions reduction, and sustainable development. To address the risks and impacts of climate change successfully, commitment from a diverse group of stakeholders is needed. Leadership is required from not only policy-makers but all concerned in the government, private sector, and civil society. As an example, consider that agriculture practices and crop grown are to be changed. This will require efforts from concerned government officers, regulators, farmers, input suppliers, and buyers who purchase, process, transport, and distribute the produce. If all these actors are associated since the inception stage, the changes would not meet much resistance and the solutions would lead to the desired reduction in consumption of water and fertilizer, reduced emissions of GHGs, and enhanced farmer's income.

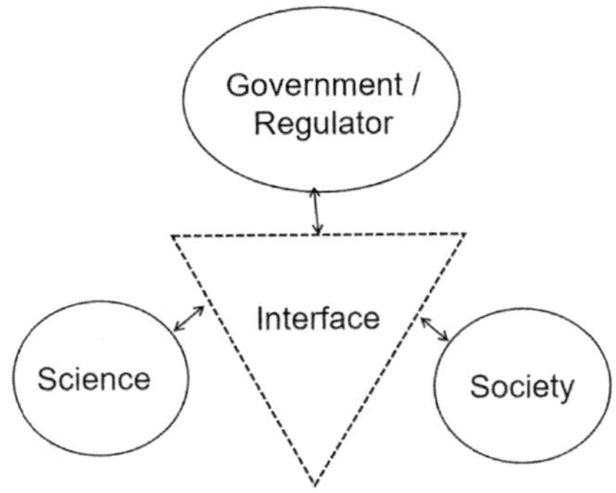

FIGURE 14.15

The trialogue model of interaction.

CRD is effective when governments at all levels work with society and the actions are driven by science. Fig. 14.15 shows the Trialogue Model of interaction that can be effectively adopted for CRD.

To create a climate-resilient, sustainable world, fundamental changes are required in the way society functions. As the world is changing and will continue to change, CRD offers ways to usher the desired changes by reducing climate risk and inequities. Development that is climate resilient is already challenging when the warming has not yet crossed the threshold of 1.5°C; it will become progressively more difficult if and when warming crosses 2°C. In some regions, for example, low-lying coastal cities, small islands, and some mountain areas, it will not be possible to live if warming exceeds 2°C. This is the reason why urgent actions are required to keep warming below 2°C. Table 14.6 lists the required adaptation actions for selected sectors and likely changes in climatic features.

It can be seen that many adaptation solutions are no-regret actions. Further, for the best results, water conservation actions should be supplemented with actions to reduce water usage. Adaptation to CC is indispensable and urgent since the climate is already changing in many respects and the impacts of mitigation actions will take a long time before the outcomes are seen. In addition, even if the mitigation efforts are implemented widely and persistently, climate will continue to change throughout this century and beyond. CC impacts on WR should be considered together with existing stresses and adaptation should be coordinated with other water management measures and be integrated into an overall strategy to adapt to global changes.

14.7 UNCERTAINTY IN IMPACT ASSESSMENTS

Uncertainty in projected changes in the climatic variables is large. As noted earlier, GCM projections are used to determine the likely future climate and the choice of GCMs is the largest source of uncertainties in hydrological projections at the catchment/river-basin scale. Uncertainties from emission scenarios and downscaling methods are also relevant but are of less importance (Prudhomme and Davies, 2009). After downscaling, hydrological models are used to determine the variables of concern and this step also introduces some uncertainties. The relative weightage of GCM selection,

Table 14.6 Required adaptation actions for selected sectors and likely changes in climatic features.

Feature/Sector	Likely change	Adaptation actions
Meteorology and climate	Rising temperatures	Enhance water conservation—catch the rain at the right place!
	Higher variability in precipitation, both spatial and temporal	Conserve water to maximize the amount of utilizable water.
	Changes in other meteorological variables	Employ the right mix of solutions to conserve water.
	Less solid precipitation in areas receiving snowfall	
Disasters	More extreme precipitation, cloudbursts, leading to human casualties, damages to crops, threats to infrastructure, bridges, roads,	Weather and flow forecasting
	more frequent and severe floods and droughts	Demarcate flood plains and implement zoning regulations
	Higher risks of compound events	Strengthen measurement networks
		Review and revise hydrological design standards including those for stormwater drainage
		Low impact development
		Better maintenance of urban/rural drainages
		Research for improved understanding of hydrological response and forecasting
Agriculture	Rising temperature	Providing irrigation cover to more rainfed agriculture areas
	Changing precipitation patterns and variability	Demand management by scientific cropping and efficient crop water management
	Damages to crops due to more intense rains	Precise application of fertilizers and pesticides based on soil and crop health
	Damages to crops due to severe and extended droughts	Better drainage in agricultural fields
Cryosphere	Less snowfall	Strengthen monitoring, prediction, and forecasting of snowfall
	More intense snow storms	Conserve the precipitation/water that glaciers would have held
	Reduction in snow/glacier cover	Monitor vulnerable glacier lakes and issue GLOF warnings
	Increased instances of glacier lake outburst floods (GLOFs)	Infrastructure to absorb/channelize incoming floods to protect vulnerable areas
		Flood plain zoning and regulation

Table 14.6 Required adaptation actions for selected sectors and likely changes in climatic features.—cont'd

Feature/Sector	Likely change	Adaptation actions
Groundwater	Reduced groundwater recharge due to more short-duration and intense rainfall events Degrading groundwater quality due to influx of polluted water Sea water ingress in coastal aquifers	Control groundwater withdrawal, more so in critical and overexploited places Control dumping of untreated waste in ponds, lakes, rivers, and aquifers Managed aquifer recharge (MAR) Restore ponds and wetlands and store water in them for recharge and for environmental improvements Control groundwater withdrawals near coasts
Water quality	Deteriorating water quality due to rising temperatures and dumping untreated or partially treated waste in rivers Improper solid waste management Threats to water supply systems due to poor quality water	Avoid dumping untreated or partially treated waste in ponds, lakes, rivers, and aquifers More STPs to treat waste water, dual pipelines to supply gray waters for specific purposes Compliance of water and sanitation rules by industries Check on the usage of fertilizers and pesticides in agriculture
Ecosystems	Ecosystem degradation due to changing climate Changes in the composition of aquatic ecosystems Opportunistic invaders in lakes and reservoirs	Soil and water conservation Afforestation Slope stabilization Nexus-based planning considering hydrology, ecology, geology Development subject to carrying capacity of a region Take decisions based on cumulative impact assessment
Health	Malnutrition owing to water and food shortages Nonavailability of safe drinking water during disasters Resurgence of water-related vectorborne diseases New breeding places for mosquitoes and other disease-transmitting insects	Take steps to provide water-food-energy security to the people and protect livelihoods Improved crop water management Follow the scientific crop calendar Improved sanitation in public places and households

downscaling, bias correction, and the hydrological models employed will depend, in addition to the factors mentioned, on the selected region/catchment.

Downscaling and bias correction are also a source of uncertainty in rain-dominated catchments (van Pelt et al., 2009). There is considerable uncertainty in the projections of changes in flood, particularly about magnitude and frequency. Due to this, the confidence in the future changes in the magnitude and frequency of floods derived from modeling and simulations is low. However, from the physical reasons, if rainfall increases in a catchment, it is likely to increase local floods unless there is a major change in other causative factors.

References

Arias, P.A., Bellouin, N., Coppola, E., Jones, R.G., Krinner, G., Marotzke, J., Naik, V., Palmer, M.D., Plattner, G.-K., Rogelj, J., Rojas, M., Sillmann, J., Storelvmo, T., Thorne, P.W., Trewin, B., Achuta Rao, K., Adhikary, B., Allan, R.P., Armour, K., Bala, G., Barimalala, R., Berger, S., Canadell, J.G., Cassou, C., Cherchi, A., Collins, W., Collins, W.D., Connors, S.L., Corti, S., Cruz, F., Dentener, F.J., Dereczynski, C., Di Luca, A., Diongue Niang, A., Doblas-Reyes, F.J., Dosio, A., Douville, H., Engelbrecht, F., Eyring, V., Fischer, E., Forster, P., Fox-Kemper, B., Fuglestvedt, J.S., Fyfe, J.C., Gillett, N.P., Goldfarb, L., Gorodetskaya, I., Gutierrez, J.M., Hamdi, R., Hawkins, E., Hewitt, H.T., Hope, P., Islam, A.S., Jones, C., Kaufman, D.S., Kopp, R.E., Kosaka, Y., Kossin, J., Krakovska, S., Lee, J.-Y., Li, J., Mauritsen, T., Maycock, T.K., Meinshausen, M., Min, S.-K., Monteiro, P.M.S., Ngo-Duc, T., Otto, F., Pinto, I., Pirani, A., Raghavan, K., Ranasinghe, R., Ruane, A.C., Ruiz, L., Sallée, J.-B., Samset, B.H., Sathyendranath, S., Seneviratne, S.I., Sörensson, A.A., Szopa, S., Takayabu, I., Tréguier, A.-M., van den Hurk, B., Vautard, R., von Schuckmann, K., Zaehle, S., Zhang, X., Zickfeld, K., 2021. Technical summary. In: Masson-Delmotte, V., Zhai, P., Pirani, A., Connors, S.L., Péan, C., Berger, S., Caud, N., Chen, Y., Goldfarb, L., Gomis, M.I., Huang, M., Leitzell, K., Lonnoy, E., Matthews, J.B.R., Maycock, T.K., Waterfield, T., Yelekçi, O., Yu, R., Zhou, B. (Eds.), Climate Change 2021: The Physical Science Basis. Contribution of Working Group I to the Sixth Assessment Report of the Intergovernmental Panel on Climate Change. Cambridge University Press, Cambridge, United Kingdom and New York, NY, USA, pp. 33–144. https://doi.org/10.1017/9781009157896.002.

Baede, A.P.M., Ahlonsou, E., Ding, Y., Schimel, D., 2001. The climate system: an overview. In: Houghton, J.T., Ding, Y., Griggs, D.J., Noguer, M., van der Linden, P.J., Dai, X., Maskell, K., Johnson, C.A. (Eds.), Climate Change 2001: The Scientific Basis. Contribution of Working Group I to the Third Assessment Report of the Intergovernmental Panel on Climate Change. Cambridge University Press, Cambridge, United Kingdom and New York, NY, USA, p. 881.

Barnett, T.P., Pierce, D.W., Hidalgo, H.G., Bonfils, C., Santer, B.D., Das, T., Bala, G., Wood, A.W., Nozawa, T., Mirin, A.A., Cayan, D.R., Dettinger, M.D., 2008. Human induced changes in the hydrology of the western United States. Science 319 (5866), 1080–1083.

Bates, B.C., Kundzewicz, Z.W., Wu, S., Palutikof, J.P., et al., 2008. Climate change and water. Technical Paper of the Intergovernmental Panel on Climate Change. IPCC Secretariat, Intergovernmental Panel on Climate Change, Geneva.

Brown, T.C., Foti, R., Ramirez, J.A., 2013. Projected freshwater withdrawals in the United States under a changing climate. Water Resources Research 49 (3), 1259–1276. https://doi.org/10.1002/wrcr.20076.

Dai, A., 2011. Drought under global warming: a review. Wiley Interdisciplinary Reviews: Climate Change 2 (1), 45–65.

Giorgi, F., 2019. Thirty years of regional climate modeling: where are we and where are we going next? Journal of Geophysical Research: Atmospheres 124, 5696–5723. https://doi.org/10.1029/2018JD030094.

Hidalgo, H.G., Das, T., Dettinger, M.D., Cayan, D.R., Pierce, D.W., Barnett, T.P., Bala, G., Mirin, A., Wood, A.W., Bonfils, C., Santer, B.D., Nozawa, T., 2009. Detection and attribution of streamflow timing changes to climate change in the western United States. Journal of Climate 22 (13), 3838–3855.

IPCC, 2000. Special Report Emissions Scenarios. Intergovernmental Panel on Climate Change. Cambridge University Press, Cambridge, United Kingdom and New York, NY, USA.

IPCC, 2012. Summary for policymakers. In: Field, C.B., Barros, V., Stocker, T.F., Qin, D., Dokken, D.J., Ebi, K.L., Mastrandrea, M.D., Mach, K.J., Plattner, G.-K., Allen, S.K., Tignor, M., Midgley, P.M. (Eds.), Managing the Risks of Extreme Events and Disasters to Advance Climate Change Adaptation, A Special Report of Working Groups I and II of the Intergovernmental Panel on Climate Change. Cambridge University Press, Cambridge, UK, and New York, NY, USA, pp. 1–19.

IPCC, 2021. Summary for policymakers. In: Masson-Delmotte, V., Zhai, P., Pirani, A., Connors, S.L., Péan, C., Berger, S., Caud, N., Chen, Y., Goldfarb, L., Gomis, M.I., Huang, M., Leitzell, K., Lonnoy, E., Matthews, J.B.R., Maycock, T.K., Waterfield, T., Yelekçi, O., Yu, R., Zhou, B. (Eds.), Climate Change 2021: The Physical Science Basis. Contribution of Working Group I to the Sixth Assessment Report of the Intergovernmental Panel on Climate Change. Cambridge University Press, Cambridge, United Kingdom and New York, NY, USA, pp. 3–32. https://doi.org/10.1017/9781009157896.001.

Klein, R.J.T., Huq, S., Denton, F., Downing, T.E., Richels, R.G., Robinson, J.B., Toth, F.L., 2007. Interrelationships between adaptation and mitigation. In: Parry, M.L., Canziani, O.F., Palutikof, J.P., van der Linden, P.J., Hanson, C.E. (Eds.), Climate Change 2007: Impacts, Adaptation and Vulnerability. Contribution of Working Group II to the Fourth Assessment Report of the Intergovernmental Panel on Climate Change. Cambridge University Press, Cambridge, UK, pp. 745–777.

Kogan, F., Guo, W., Yang, W., 2020. Near 40-year drought trend during 1981-2019 earth warming and food security. Geomatics, Natural Hazards and Risk 11 (1), 469–490. https://doi.org/10.1080/19475705.2020.1730452.

Liu, P.W., Famiglietti, J.S., Purdy, A.J., Adams, K.H., McEvoy, A.L., Reager, J.T., et al., 2022. Groundwater depletion in California's Central Valley accelerates during mega drought. Nature Communications 13 (1), 7825. https://doi.org/10.1038/s41467-022-35582-x.

Meehl, G.A., Stocker, T.F., Collins, W.D., Friedlingstein, P., Gaye, A.T., Gregory, J.M., Kitoh, A., Knutti, R., Murphy, J.M., Noda, A., Raper, S.C.B., Watterson, I.G., Weaver, A.J., Zhao, Z.C., 2007. Global climate projections. In: Solomon, S., Qin, D., Manning, M., Chen, Z., Marquis, M., Averyt, K.B., Tignor, M., Miller, H.L. (Eds.), Climate Change 2007: The Physical Science Basis. Contribution of Working Group I to the Fourth Assessment Report of the Intergovernmental Panel on Climate Change. Cambridge University Press, Cambridge, UK, and New York, NY, pp. 747–845.

Moss, R., Babiker, M., Sander, B., Calvo, E., Carter, T., Edmonds, J., Elgizouli, I., Emori, S., Lin, E., Hibbard, K., Jones, R., Kainuma, M., Kelleher, J., Lamarque, J.F., Manning, M., Matthews, B., Meehl, J., Meyer, L., Mitchell, J., Nakicenovic, N., O'Neill, B., Pichs, R., Riahi, K., Rose, S., Paul, R., Stouffer, R., van Vuuren, D., Weyant, J., Wilbanks, T., Pascal van Ypersele, J., Zurek, M., 2008. Towards New Scenarios for Analysis of Emissions, Climate Change, Impacts, and Response Strategies. Intergovernmental Panel on Climate Change, Geneva, p. 132.

Najibi, N., Devineni, N., 2018. Recent trends in the frequency and duration of global floods. Earth System Dynamics 9 (2), 757–783. https://doi.org/10.5194/esd-9-757-2018.

Nakicenovic, N., Joseph, A., Davis, G., de Vries, B., Fenhann, J., Gaffin, S., Gregory, K., Griibler, A., Jung, T.Y., Kram, T., La Rovere, E.L., Michaelis, L., Mori, S., Morita, T., Pepper, W., Pitcher, H., Price, L., Riahi, K., Alexander, R., Rogner, H.-H., Sankovski, A., Schlesinger, M., Shukla, P., Smith, S., Swart, R., van Rooijen, S., Victor, N., Zhou, D., 2000. Special Report on Emissions Scenarios. The Press Syndicate of the University of Cambridge, UK.

Prudhomme, C., Davies, H., 2009. Assessing uncertainties in climate change impact analyses on the river flow regimes in the UK. Part 2: future climate. Climatic Change 93, 197–222. https://doi.org/10.1007/s10584-008-9461-6.

Rajesh, M., Rehana, S., 2022. Impact of climate change on river water temperature and dissolved oxygen: Indian riverine thermal regimes. Scientific Reports 12 (1), 9222. https://doi.org/10.1038/s41598-022-12996-7.

Randall, M.R., Rojas, M., Sabine, C., Shindell, D., Talley, L.D., Vaughan, D.G., Xie, S.-P., 2013. Technical summary. In: Stocker, T.F., Qin, D., Plattner, G.-K., Tignor, M., Allen, S.K., Boschung, J., Nauels, A., Xia, Y., Bex, V., Midgley, P.M. (Eds.), Climate Change 2013: The Physical Science Basis. Contribution of Working Group I to the Fifth Assessment Report of the Intergovernmental Panel on Climate Change. Cambridge University Press, Cambridge, United Kingdom and New York, NY, USA.

Schipper, E.L.F., Revi, A., Preston, B.L., Carr, E.R., Eriksen, S.H., Fernandez-Carril, L.R., Glavovic, B.C., Hilmi, N.J.M., Ley, D., Mukerji, R., Muylaert de Araujo, M.S., Perez, R., Rose, S.K., Singh, P.K., 2022. Climate resilient development pathways. In: Pörtner, H.-O., Roberts, D.C., Tignor, M., Poloczanska, E.S., Mintenbeck, K., Alegría, A., Craig, M., Langsdorf, S., Löschke, S., Möller, V., Okem, A., Rama, B. (Eds.), Climate Change 2022: Impacts, Adaptation and Vulnerability. Contribution of Working Group II to the Sixth Assessment Report of the Intergovernmental Panel on Climate Change. Cambridge University Press, Cambridge, UK and New York, NY, USA, pp. 2655–2807. https://doi.org/10.1017/9781009325844.027.

Seneviratne, S.I., Zhang, X., Adnan, M., Badi, W., Dereczynski, C., Di Luca, A., Ghosh, S., Iskandar, I., Kossin, J., Lewis, S., Otto, F., Pinto, I., Satoh, M., Vicente-Serrano, S.M., Wehner, M., Zhou, B., 2021. Weather and climate extreme events in a changing climate. In: Masson-Delmotte, V., Zhai, P., Pirani, A., Connors, S.L., Péan, C., Berger, S., Caud, N., Chen, Y., Goldfarb, L., Gomis, M.I., Huang, M., Leitzell, K., Lonnoy, E., Matthews, J.B.R., Maycock, T.K., Waterfield, T., Yelekçi, O., Yu, R., Zhou, B. (Eds.), Climate Change 2021: The Physical Science Basis. Contribution of Working Group I to the Sixth Assessment Report of the Intergovernmental Panel on Climate Change. Cambridge University Press, Cambridge, United Kingdom and New York, NY, USA, pp. 1513–1766. https://doi.org/10.1017/9781009157896.013.

Sharma, A., Wasko, C., Lettenmaier, D.P., 2018. If precipitation extremes are increasing, why aren't floods? Water Resources Research 54, 8545–8551. https://doi.org/10.1029/2018WR023749.

U.S. Energy Information Administration. https://www.eia.gov/todayinenergy/detail.php?id=51839.

USAID, 2014. A Review of Downscaling Methods for Climate Change Projections. United States Agency for International Development, Burlington, Vermont.

van Pelt, S.C., Kabat, P., ter Maat, H.W., van den Hurk, B.J.J.M., Weerts, A.H., 2009. Discharge simulations performed with a hydrological model using bias corrected regional climate model input. Hydrology and Earth System Sciences 13, 2387–2397. http://www.hydrol-earth-syst-sci.net/13/2387/2009/.246.

van Vuuren, D.P., Edmonds, J., Kainuma, M., Riahi, K., Thomson, A., Hibbard, K., Hurtt, G.C., Kram, T., Krey, V., Lamarque, J.-F., Masui, T., Meinshausen, M., Nakicenovic, N., Smith, S.J., Rose, S.K., 2011. The representative concentration pathways: an overview. Climatic Change 109. https://doi.org/10.1007/s10584-011-0148-z.

Further Reading

Hannaford, J., Marsh, T.J., 2008. High-flow and flood trends in a network of undisturbed catchments in the UK. International Journal of Climatology 28 (10), 1325–1338.

IPCC, 2001. In: Houghton, J.T., Ding, Y., Griggs, D.J., Noguer, M., van der Linden, P.J., Dai, X., Maskell, K., Johnson, C.A. (Eds.), Climate Change 2001: The Scientific Basis. Contribution of Working Group I to the Third Assessment Report of the Intergovernmental Panel on Climate Change. Cambridge University Press, Cambridge, United Kingdom and New York, NY, USA, p. 881.

IPCC, 2014. In: Core Writing Team, Pachauri, R.K., Meyer, L.A. (Eds.), Climate Change 2014: Synthesis Report. Contribution of Working Groups I, II and III to the Fifth Assessment Report of the Intergovernmental Panel on Climate Change. IPCC, Geneva, Switzerland, p. 151.

Milly, P.C., Dunne, K.A., Vecchia, A.V., 2005. Global pattern of trends in streamflow and water availability in a changing climate. Nature 438 (7066), 347–350. https://doi.org/10.1038/nature04312.

Zhao, C., Liu, B., Piao, S., Wang, X., Lobell, D.B., Huang, Y., Asseng, S., 2017. Temperature increase reduces global yields of major crops in four independent estimates. Proceedings of the National Academy of Sciences 114 (35), 9326–9331. https://doi.org/10.1073/pnas.1701762114.

CHAPTER

River Basin Planning and Management

15

Chapter outline

Water Resources Systems Planning and Management. https://doi.org/10.1016/B978-0-12-821349-0.00002-2

Water is the driver of nature.

Leonardo da Vinci.

Water is pivotal to our environment, and it influences and shapes the landscape. The sustenance of life and economic and social development are not possible without sufficient water of the right quality. River basins possess freshwater and are most important from the point of view of water resources development and management. The availability of water at a given place depends on the properties of its upstream basin. Most of the consumptive water use, such as irrigation, and nonconsumptive uses, such as hydropower production, recreation, and navigation, take place in the basin itself. A basin also receives most of the return flow from irrigation and wastewater. Due to these reasons, river basins are the most important management units. The ecosystem approach to water management is a somewhat related concept that aims to integrate social, economic, and environmental interests within the wider framework of the basin. The basic motivation behind this is that people depend on their ecosystem and, therefore, the capacity of ecosystem to deliver goods and services should be maintained over a longer timeframe.

Undoubtedly, river basins are among the earth's most productive ecosystems. Freshwater ecosystems provide a number of goods and services to global, regional, national, and local economies. Besides water, goods and benefits derived from these systems include fish, timber, fuel, biodiversity, wildlife, and fertile lands. A diverse range of industries, such as agriculture, tourism, fisheries, forestry, and construction, benefit both directly and indirectly from freshwater ecosystems. There is also an intrinsic social value through their links to aesthetic, cultural, and heritage aspects. Clearly, freshwater ecosystems generate multiple and wide-ranging economic benefits.

Rivers provide a range of services, including water supply; waste assimilation; fisheries; energy production; flood attenuation; spiritual, cultural, and social benefits; and habitat that supports a multitude of ecosystems. In general, demands on rivers exceed their natural capabilities because of overexploitation, pollution, alien infestation, floodplain changes, and habitat destruction as a result of improper decision-making and inadequate management, poor planning. To protect and restore rivers, proper planning becomes a must and constitutes the foundation of river management.

The array of goods and services detailed above can be classified into two groups: public and private. Public goods and services are those whose provision is "nonexclusive" and "nondivisible." This means that once they are provided, anyone can benefit from their provision without diminution of their availability to others. Many of the quality-related facets of goods and services derived from a freshwater ecosystem may be construed as public goods. These include water quality, storage and purification, groundwater recharge, flood control, storm protection, nutrient retention, and microclimate changes. If public goods are managed properly for one user, the entire society enjoys the benefits. Conversely, when these are poorly managed, the whole society suffers due to mismanagement.

In contrast to public goods, the provision of private goods and services is exclusive, that is, once these are provided to one person, that individual holds them exclusively and the quantity available to others gets reduced by an equal amount. Many of the quantity-related facets of goods and services from freshwater ecosystems are of this nature. For example, once a particular quantity of water is given to a user or group for consumption, it becomes his exclusive property and nobody else can claim a right on it. The public or private nature of a freshwater ecosystem's goods and services determines how it can be managed in the most efficient way. Private goods are best managed by free market-based mechanisms because such markets provide many incentives for efficient management.

The main cause of damage to freshwater systems, as for any other system, is their overexploitation. It may be through activities, such as excessive water extraction or excessive dumping of waste in them. The resulting damages to these systems are termed as *externalities* and are not included in the traditional benefit-cost analysis. A sustainable use of freshwater systems requires that all users must be made responsible for bearing their share of the cost of maintaining freshwater ecosystems. This philosophy is commonly known as the *polluter pays* principle.

All of these considerations indicate that decentralization of the administration of freshwater systems is necessary for efficiently managing them. However, one comes across many instances of centralized control and decision-making, which introduces more complexities by virtue of the range of problems with which a centralized management is concerned.

The demand and use of water have rapidly increased after the middle of the 20th century. In many basins, water resources are overexploited, but the demands are still increasing. The consequent reduction in the capacity to meet different demands has resulted in conflicts between different water uses and between upstream and downstream uses. An integrated approach, covering all waters of the basin (surface and subsurface) is necessary to solve these conflicts. It should consider temporal and spatial distribution of the water quantity and quality; and the interaction of water with land, vegetation, and other resources. Ideally, it should also ensure the integration of social, economic, legal, political, and administrative issues.

Sustainable river basin management (RBM), which is the basic objective, requires a sound understanding of water resources systems and their internal relations (groundwater, surface water; quantity and quality; biotic components; upstream and downstream interactions). The water systems should be studied and managed as part of the broader environment and in relation to socioeconomic demands and potentials, with the due influence of the political and cultural settings. The water itself should be seen as a social, environmental, and economic resource, and each of these three aspects must be recognized in the decision-making process. The management of water resources cannot be addressed in isolation; it is necessary to consider the functioning of ecosystems simultaneously at different hierarchical levels, in both space and time. This involves planning and management interventions at local levels (e.g., field, farm, and village) as well as at regional levels (e.g., catchment and river basins).

Sustainable RBM also requires cooperation and political commitment within a country and among the basin nations in the case of international basins. In many river basins, pressures on the environment have reached or surpassed the levels that may be sustainable. Consequently, vulnerability from extreme natural events has increased, and conflicts between different water uses and between upstream and downstream uses are increasing. The capacity of many basins to meet growing social demands, including basic human needs, such as drinking water, is decreasing rapidly. But the basic water needs of people and ecosystems have to be fulfilled, first. This requires that the essential ecological and physical processes should be protected and the harmful effects of overexploitation of the sources of water should be minimized.

15.1 DEFINITION AND SCOPE OF RIVER BASIN MANAGEMENT

A river basin can be defined as *the geographical area demarcated by the topographic limits of the system of waters, including surface and subsurface water, flowing into a common point* (the groundwater may, however, not exactly follow this rule). The boundaries of most river basins are clearly defined by topography, although there may be instances with ill-defined boundaries. Some rivers have a shared delta and the basin areas in flat topography may not be distinct. In a basin, there are strong interactions between land and water resources, between groundwater and surface water, and between water quantity and quality. In a nutshell, a river basin is a coherent system of interacting and interdependent elements.

River basin management (RBM) is defined as *the management of water resources of a basin as part of the natural ecosystem and in relation to their socioeconomic setting*. The term integrated water resources management (IWRM) is also frequently used to convey similar concepts. The term "integrated" was used by Downs et al. (1991) for an intermediate stage in which more than one sectoral interest is linked at both operational and strategic levels. The schemes that approach the basin as an energy or ecosystem are termed "holistic." However, there is no unanimity in the use of these terms. Sometimes holistic is used for the actions at the strategic level while integrated is for those at the operational level. Note, however, that the term RBM does not imply that river basins are closed systems or the only relevant geographical areas. Other units, such as administrative areas, are also frequently used as management units. Nevertheless, river basins are the most logical and important units that should be carefully managed for the benefit of all concerned.

Another perspective of integration is planning and management of both surface water and groundwater considering both water quality and quantity as well as the impacts of land use and vegetation management on natural resources and people in a participative way, and the adoption of new resource use policy can be referred to as IWRM. Integration extends to economic, social, and environmental considerations, accounting for not only financial and economic costs and benefits of natural resources management decisions, but also social and environmental costs and benefits to attain an acceptable balance among all of them.

The unified management of a river basin by a single authority looks intuitively appealing too. Such a body can best prepare comprehensive basin-wide development and management plans, can realistically look at the "big-picture," implement the best available technology, and deploy appropriate models and decision support systems. However, due to a wide range of associated tasks, many river basin organizations become unwieldy and have their offices at many locations. Care is necessary to ensure that additional complexities and communication gaps are not introduced just because of their size and the range of issues that they have to handle. If there is no proper coordination among the various wings in a large basin organization, the basic aim of unified decision-making, nonredundancy, and transparency gets defeated.

Broadly, the aim of RBM is to ensure the use of water and allied resources of a river basin in a sustainable manner. For example, the stated aim of the Mekong Commission (www.mrcmekong.org) is *to promote and coordinate sustainable management and development of water and related resources for the countries' mutual benefit and the people's well-being by implementing strategic programs and activities and providing scientific information and policy advice*. As the capacity of a river basin to serve various uses is limited, priorities have to be assigned to different uses. Considerations of a hierarchy of needs suggest that the basic human requirements—water supply for daily necessities, including basic hygiene—have to be, first, taken care of. The next level of priorities depends on the natural, social, and economic conditions in the pertinent basin and the values and occupations of its population.

The need for RBM arises because the noncoordinated use of resources is inefficient as well as damaging. A key characteristic of RBM is satisfactory conflict resolution. Despite the intrinsic appeal and obvious benefits, RBM is not practiced in a proper way in many countries. The main reasons are that there is no river basin organization, the basins fall under incompatible administrative units, or there is not enough will for RBM.

15.1.1 Scope of RBM

To understand the intricacies of RBM, it is useful to distinguish six different activities: planning, construction, operation, monitoring, analysis, and decision-making. The river basins and their users are directly affected by operation and management actions (see Fig. 15.1). Of course, at each stage, the basin managers receive feedback from the basin and its users. Planning and construction are the primary means to install facilities for operation and management. As shown in the figure, monitoring and analysis provide inputs for planning, construction, and operation.

RBM affects river basins in many ways. It may alter the natural physical processes in a river basin by constructing structures to store and carry water; regulate the use of water with the help of allocation

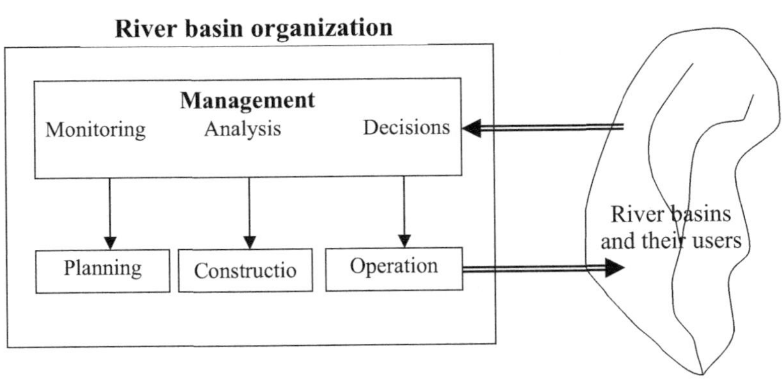

FIGURE 15.1

Activities in river basin management.

rules, prices, water rights, and permits; and apply economic instruments, such as taxes and subsidies to control the water usage. Different inputs are necessary to apply these instruments, such as money, personnel, legal, and appropriate policy directives. RBM may also induce change in the behavior of users by penalizing or allowing/encouraging certain activities. An effective RBM will require a mix of instruments depending on circumstances. For long-term sustenance, it is necessary to build in-house capacity by training staff and keep the general public well informed. Table 15.1 gives an overview of different types of operational RBM instruments.

The scope of RBM also contains a number of related activities, viz., public participation, professional cooperation between related organizations, and international cooperation. Because of this wide view, RBM has a broader scope than does traditional water management. In fact, it covers all human activities that use or affect freshwater systems; many of these are beyond the scope of this book. The planning process was the theme of Chapter 9 and analytical tools have been discussed in Chapters 3–8. Decision support systems will be covered in Section 15.4. Three important topics, viz., operation, water rights, and water charges, will be covered in what follows.

Table 15.1 Activities and instruments of operational RBM.

Activities	Characteristics	Instruments
Structural	Direct interference and control of water flow	Dams for water regulation Weirs for river flow diversion Embankments to confine streamflows River training works Canals, tunnels, and pipes for water conveyance Pumping from aquifers, artificial recharge Catchment treatment Interbasin transfer
Regulation	Influences the users by encouraging or penalizing some activities	Rules and regulations Water rights and permits Monitoring and enforcement
Economic instruments	Influence water use by means of financial (dis)incentives	Charges (taxes, levies, etc.) Subsidies Tradable water use and pollution rights
Communication and awareness raising	Influences users by providing information	Public involvement Seminars, lectures, open house Exhibitions
Financing	Supports the various activities by providing finances	Funding desirable activities
Institutional support	Supports the previous instruments by providing other necessary resources (personnel, legal competencies, policy directives)	Capacity building Legislation Extension services

Adapted from Mostert et al. (1999).

15.1.2 Operations

As the infrastructure starts coming up, it is to be put to designated beneficial uses. The term operations (some authors use operational management) implies regulation of facilities; and application of economic, legal, and policy instruments. As shown in Fig. 15.1, operations have a direct influence on river basins and their users. Operations are carried out using the analysis and decision support systems (DSS) that are set up specifically for this purpose. These tools and techniques have been discussed in detail in the preceding chapters. As far as regulation of structures is concerned, the most important component is reservoir operation. An extensive amount of work has been done on this as explained in Chapter 11.

15.1.3 Water Charges

An important and very sensitive operational issue in RBM relates to water charges. Charges are an effective and efficient means to finance resource development activities, minimize wastage, and control pollution. If water charges are based on the full cost of providing services, it is easier to attract and involve the private sector. However, caution is to be exercised here, as water is also a social good and, therefore, it should not be so expensive that the poor are not able to afford it. From a management point of view, very high charges are difficult to enforce and there is a resistance to pay. Moreover, high water charges can substantially reduce margins on agricultural products and can have a cascading effect on other commodities. Inputs to agriculture are subsidized to varying degrees in most countries and there is always political opposition and resistance to high water charges. Nevertheless, water plays a key role in many economic developments, and by virtue of that, it has a high value, although it is politically difficult to reflect this value in water transactions.

Water can be charged in two ways: based on the actual usage or on a lump-sum basis. The charges that are based on actual water use certainly help reduce water use and wastage to a certain extent. The extent of the impact of the price depends on the price elasticity or the sensitivity of water use to the costs of use. It is generally low in the case of drinking water use and high for irrigation water. When charges are on a lump-sum basis, wastage is high and recovery of full cost of providing services is difficult.

Ideally, water rates should be based on the opportunity cost of the water use (the value of the next best alternative use). If this is not feasible, charges should be fixed to at least recover the cost of providing related services, such as domestic water supply, and waste-water treatment. This will ensure that enough funds are available for operation and maintenance. Sometimes the aim could also be to raise money for expansion of services and reduce wastage. By equating prices with their marginal cost, the full benefit of those goods to society will be reflected in equilibrium. Users facing such prices will base consumption decisions on the real economic cost of providing the goods and so should reach the socially optimal level of consumption (Dinar et al., 1997). The fixation of water charges is somewhat easier in the case of water supply and irrigation, while the evaluation of the benefits of drainage and flood control is more complicated. Note that irrigation services are heavily subsidized in many countries. Of late, many countries have privatized water services or are in the process of doing so with a view to improve efficiency.

From practical considerations, marginal cost pricing is hard to implement, as it is usually difficult to derive the marginal cost curves. While trying to incorporate the scarcity value of water, it is unlikely that authorities will be able to devote the time or money necessary to find the true marginal cost of provision (Dinar et al., 1997). Additionally, the marginal cost varies over time; this means that prices would constantly need readjusting to take into account the seasonal availability cycle, and the level of demand. The potential magnitude of this difference is illustrated in Zimbabwe, where tariffs to farmers in the various provinces have varied from Z$9 per cubic meter to Z$278 (Winpenny, 1994). Recall that equity issues are one of the primary concerns of the government of Zimbabwe. It is possible that in a dry spell the marginal cost of provision would rise so much that it would become too expensive for lower-income users to afford. In fact, charges are more often based on the average prices combined with the subsidies to promote equity issues. In many countries, consumers pay much lower prices than necessary to achieve the economic efficiency of allocation.

The scarcity value of water in arid climates can be appreciated by noting that the currency of Botswana, a semiarid country in Africa (mean annual rainfall in the range of 650–250 mm) is Pula. In the local Setswana language, the word Pula means rain!

Maintenance of the infrastructure usually is the first casualty in the case of a shortage of funds. Due to the nonrecovery of operational expenses, the quality of services at many places has degraded and these have not expanded to the desired extent. Poor users could be levied lower charges, but the economic viability of providing services should not be undermined. Ironically, if the services are not up to the desired level, it is the poor people who are the first and the worst sufferers. A possible solution is to link the rates with the ability to pay. Evidently, fixing water prices is not a technical or economic issue; it is more a socio-political issue.

15.1.4 Water Rights

There are two types of water rights: ownership rights and the right to use water. Water bodies can be government-owned, privately owned, or the property of a whole community. The ownership status depends on the national legal system and the type of water body (navigable or nonnavigable river, ponds, and groundwater). The practice of riparian rights is followed in many countries and under this, landowners along natural water bodies have certain minimum rights to water use. The mechanism that is commonly followed in developing countries is through legislative authorization of major water-related bodies. The nongovernmental organizations have a major role in advocating and providing services in water resources. Regarding the groundwater rights, mostly the land owner the has right to pump water beneath it.

According to the riparian doctrine, water rights are a component of the property interest that arises from the ownership of the land bordering a natural water course and include the right to make reasonable use of water on riparian lands. As "reasonable" is a relative term, the riparian right is commonly not fixed in magnitude and can vary with time. However, it must be compatible with other uses relying on the same source of water. Commonly, the maximum extent of the riparian land is the boundary of a stream catchment.

Water rights should be flexible and responsive to changing circumstances at both national and international levels. If the government is the owner of water, the use rights are generally granted by a designated agency by means of permits, concessions, etc. Private ownership, however, does not mean that the owner can use the water as he pleases; he is not supposed to encroach upon the rights of others, his ownership right might be limited by law and permits may be needed for specific uses. If the water is seen as the property of a whole community or as incapable of being owned by anyone, the water use is often regulated by the government. However, in many places, local communities of users have their own bodies to manage water use.

When individuals own a confirmed right to a proportion of the resource, they have an economic incentive to exploit that resource efficiently. If a user has the assurance that he can continue to use the resource, he will invest in maintaining and running the system. The major issues concerning the rights of privately owned waters relate to their flexibility and the extent to which the ownership right can be limited. In the case of water use rights, the major issues are whether these rights are granted for a specific period or in perpetuity and under which conditions they can be revised. A relatively high degree of flexibility—combined with existing rights and if necessary, compensation—seems essential for effective RBM in a changing world.

If the rights are appropriately determined and allocated, the aggregate level of consumption will gradually reach an optimal level. Thus, first, the socially optimal level of resource use must be determined and then divided into appropriate rights to ensure that the total use is within limits. For the rights to be acceptable, it is essential that the historical pattern of appropriation is taken into account and is not drastically disturbed and no one should be made worse off. If the allocation of rights leads to a situation in which some existing users are unable to use their share, the equity criterion has not been met. Many customary rights dictate the kind of equipment that can be used rather than the volume of water that must be extracted. The right to pollute can also be issued to ensure that the concentration of waste in the water body does not exceed the optimal level. As the type of pollutant, location of the deposition, and the kind of use all affect the health of the ecosystem, this heterogeneity must be taken into account.

A careful identification of the exact entitlement and responsibilities associated with the rights and long-term repercussions is essential. For instance, a requirement that the failure "to use a water right for a set number of years can lead to the forfeiture or abandonment" of rights would at first sight seem to prevent players from claiming rights speculatively. In reality, the limit can reduce the time horizon of the holder and encourage him to just use water to ensure his entitlement (Anderson and Hill, 1997).

The discussion so far should not give an impression that rights alone will ensure an efficient use of water resources. Water and land resources in a basin are owned by various categories of people and institutions. Water resources ownership is sometimes linked to land ownership. In any case, a clear definition of ownership is essential for sustainable management.

To achieve an economically efficient outcome for society, it is essential that the rights to use the freshwater ecosystem's goods belong to those activities that have the highest marginal social benefit. In a majority of cases, the numbers of users, the heterogeneity of their interests, the asymmetry of information available to the agency determining the rights, and the need to ensure an equitable allocation results in an initially suboptimal allocation of rights. A free market for rights allows low-value users to exchange permits with high-value users, get the market price for the goods transferred and be fully compensated for the loss of the right. For a market in permits to allocate rights effectively, the prior agreement of rights and their clear definition are essential for any trade. Second, rights must be valid for a period long enough so that owners can assign a value to them. Information about the exact nature of rights and the rules of trading must be clear so that both buyers and sellers understand the full implications of the situation (Dinar et al., 1997).

The concept of *water banks* has been used in the United States as a method to match sellers with excess rights to buyers who can exchange the rights for a limited period of time. In this concept, water is bought from farmers and then resold "to those who have the most critical needs" (Le-Moigne, 1994). The state of California has used water banks extensively in years of drought to reallocate rights to scarce water resources. The efficiency of the system has improved over time using the experience gained (Dinar et al., 1997).

Wurbs (2008) has developed the Texas A&M University Water Rights Analysis Package (TAMUWRAP) to handle the prior appropriation ("first in time, first in right") system of legal water rights prevalent in the western United States. This model is capable of simulating river flows, multireservoir operations, and diversions to multiple uses with a set of priority allocations, each defined by annual volume, priority number, type of use, and optional water rights group. TAMUWRAP has been applied to the Brazos River system of 12 reservoirs in Texas, where over 1300 permit holders are allowed to use water.

15.2 PLANNING AND RIVER BASIN MANAGEMENT

Plans and policies have an important role to support RBM. Planning helps assess the present situation in the basin, the situation desired, the gap between the two, and how to close the gap. It helps orient operational management and set priorities. Second, it is often not possible or effective to do policy analysis and organize public participation for each individual operational decision. In these cases, planning may offer a framework and focus. Third, open and participatory planning processes will result in more public support or acceptance of the resulting plan/policy. Planning processes can bring different river basin managers into discussion with each other and the resulting plans and policies can act as common focal points.

Planning must consider tradeoffs among economic, social, and environmental objectives as well as between current and future demands; environmental water needs and ecosystem functioning; interactions among hydrological, ecological, economical, and social systems; uncertainty in future development and climate considering alternative hydro-economic scenarios; prioritization of demands considering economic development, social justice, and environmental protection (Pegram et al., 2013).

Planning is an important and often indispensable means to utilize water resources and in the operation of the projects. Planning has four related functions:

(a) To assess the current situation (including the identification of conflicts and priorities), formulate visions, set goals and targets, and thus orient operation and management,
(b) To provide a framework for public participation and feedback,
(c) To increase the legitimacy and mobilize public acceptance, and
(d) To facilitate the interaction among concerned organizations and stakeholders.

The topic of water resources planning has been dealt with in detail in Chapter 9. This section focuses on some aspects of planning that are relevant to RBM.

15.2.1 Planning Process

Although planning requires extensive technical and scientific information, it is not a purely technical or scientific exercise. It must have a human touch, as it ultimately affects humans. The following are the important activities of planning for RBM:

1. Identification of the need, scope, and geographical coverage of the area;
2. Analysis of the institutional framework for RBM, identification of decisions that are to be taken, and the bodies who have to take these decisions;

3. Identification of the main stakeholders, their likings, and expectations;
4. Preparation of a blueprint, describing the scope of planning; identify different phases and groups to be involved in each phase; and prepare a flowchart of activities;
5. Formulation of a plan and its approval; and
6. Implementation of the plan.

Note that planning is an iterative process—the first plans are not always the best. But after a plan is ready, it can be widely circulated and suggestions for improvements can be invited. Most people are able to react in clear terms when a concrete proposal is presented to them. A plan may also encourage planners to look for ways to overcome too restrictive constraints that might be limiting the overall development.

As an aid in river basin planning and management, electronic spreadsheet-based simulation models have been developed recently. In these models, the formulas governing the working of the system are entered into the cells of the spreadsheet. The basic data of the system (such as configuration, inflows) might be fixed and the user can vary pertinent parameters, such as the amount of water withdrawn for municipal uses, power generation, the price of water, and the quality of returned water. After defining a new scenario, the operation of the system can be simulated in a few seconds and the revised tables and graphs are readily available. Such models are of significant value in getting a "feel" for the system and the interaction of its various components. These are also helpful as a teaching aid in the classroom through role-playing. Roles can be assigned to different students—someone can assume the role of water supply utility, another of wastewater treatment utility, the third of RM, and so on. Operation policies and charges for services can be fixed through negotiations and the sensitivity can be examined.

15.2.2 River Basin Planning Systems

Plans and policies relevant to RBM can differ on many dimensions: policy sectors, geographical scope, etc. What types of plans are needed in a specific situation depends on a number of factors, such as the most important policy issues; whether river basins are located in one, two, or more jurisdictions; the funds that can reasonably be spent on planning, etc. These factors differ from country to country and from basin to basin.

RBM is the management of water systems as part of the broader natural environment and in relation to their socioeconomic environment. Consequently, river basin planning should ensure consideration of interrelations within water systems (surface and groundwater, quantity and quality), the interrelations between climate, land, and water; and interrelations between complete river basins and their socioeconomic environment.

The types of plans needed depend on the need for different functions that plans can perform. For instance, if in a basin there is an urgent need of providing drinking water to a rapidly growing city, there may be no need for integrated strategic planning that requires an overall description of the basin and sets long-term goals. Generally, the number of plans should be small, especially in countries and basins with a scarcity of competent technical personnel. If too much planning is going on at the same time, too few resources may be available for each planning exercise and coordination between plans can become a problem.

15.2.3 Implementation of Integrated Approach

Experiences of different countries on the best practice of IBRM point to five common attributes (World bank, 2006):

1. Establishment of a basin-wide strong, clear, and flexible institutional framework.
2. Allowing for an equal partnership among all member administrative units supported by legislation, regulation, or decree. The administrative units operate within the basin and develop an integrated natural resource policy agenda and clear financing and budgeting systems for the range of basin-wide activities.
3. Strategic assessment of water and related resources, including all types of basin data and their attributes.
4. Development of policies, strategies, decisions, and projects in an integrated manner, recognizing the holistic and interactive way of natural resources.
5. Incorporation of community and stakeholder participation and their views into the planning and management processes.
6. Establishment of a system for sustainable river basin management, including ongoing monitoring and auditing, and basin-wide institutional arrangements achieving the goals and objectives set by governments.

15.2.4 River Basin Organization

A river basin organization must be flexible, evolving with time and changing conditions. It must consider political realities and key drivers promoting change, such as natural disasters, conflict in water sharing, severely poor regulation and management, or severe resource degradation. These drivers provide opportunities for the promotion of river basin management. The organization should undertake five categories of functions (World Bank, 2006):

1. Collection and processing of water and natural resources data, systems modeling, and water and natural resources planning.
2. Project feasibility, design, implementation, operation and maintenance, and fundraising.
3. Allocation and monitoring of water and natural resources (both quality and quantity) shares, and cost sharing.
4. Development of policy and strategy for social, economic, and environmental considerations, and community awareness and participation.
5. Monitoring water use and shares, monitoring of pollution and environmental conditions, river basin organization partners providing overview and review, monitoring and assessing the health of natural resources, and monitoring the sustainability of natural resources management.

15.3 INTEGRATED WATER RESOURCES MANAGEMENT

A river basin system can be classified in three components (McKinney et al., 1999): (1) source components, such as rivers, canals, reservoirs, and aquifers; (2) demand components which could be off-stream (irrigation fields, industrial plants, and cities) and in-stream (hydropower, recreation, environment); and (3) intermediate components, such as treatment plants and water reuse and recycling facilities. A schematic diagram of the components of a river basin system is given in Fig. 15.2. This figure shows the water supply system (groundwater and surface water), the delivery system (canal network), the system of water users (agricultural, municipal, and industrial), and the drainage collection system (surface and subsurface). The upper bound of a river basin is the atmosphere through which mass (e.g., precipitation) and energy (e.g., radiation) exchange takes place. These exchanges have important influences on the hydrologic processes that take place in a basin. Besides natural influences, human activities have significant bearing on the state of resources in a basin. This influence is exerted through artificial interventions (impoundments, diversions, deforestation, etc.), and the use of water for consumptive uses, such as municipal and agricultural, affecting its availability; and application of chemicals and fertilizers thereby affecting its quality. An integrated basin management model must take into account both the quality and quantity aspects.

Integrated water resources management (IWRM) has been defined in many ways. According to The Technical Advisory Committee of the Global Water Partnership (TAC, 2000) *IWRM is a process that promotes the coordinated development and management of water, land, and related resources, to maximize the resultant economic and social welfare in an equitable manner without compromising the sustainability of vital ecosystems*. This definition emphasizes that IWRM is not a goal in itself but it

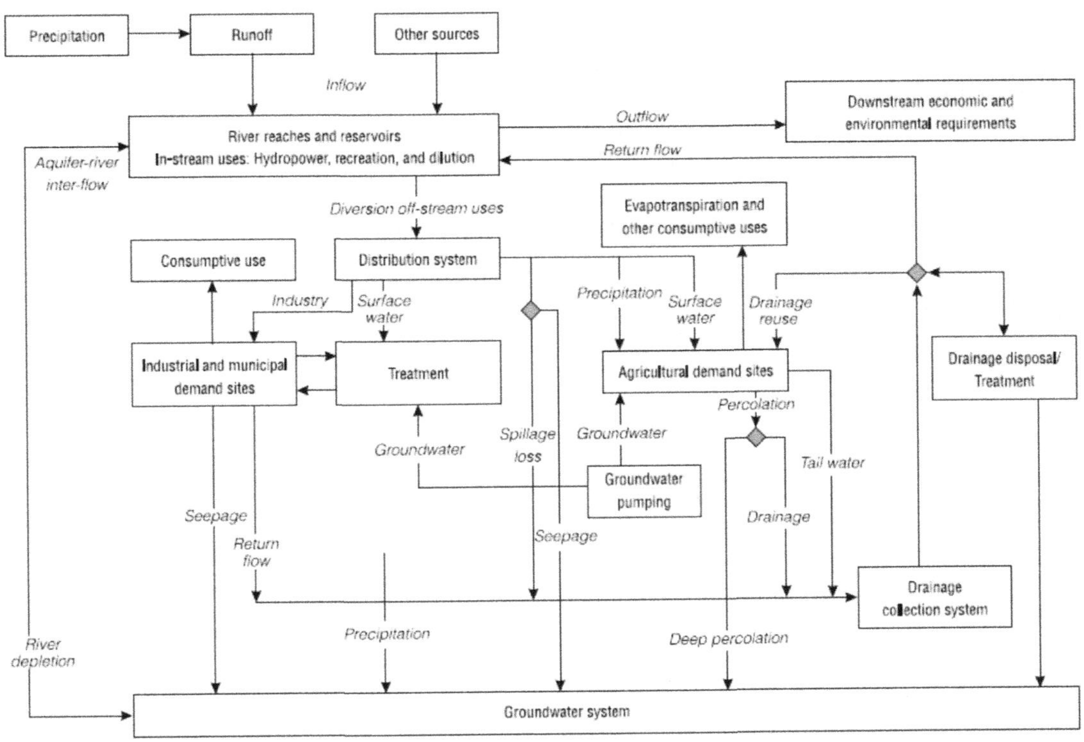

FIGURE 15.2

Schematic representation of river basin processes.

Reproduced from McKinney et al. (1999). Copyright International Water Management Institute. Used by permission.

should be viewed as a process of balancing and making trade-offs between different goals in an informed way. In this process, there are two fundamental categories of integration, the natural system and the human system. The water managers face a variety of challenges, circumstances differ greatly among and within countries, and policies and practices that are acceptable in one place may not be appropriate for another. Therefore, IWRM is not a blueprint, nor does it come with an instruction manual valid for all eventualities. IWRM stresses the interrelationship of actions of different types, working at different levels of influence. Moreover, water cannot be taken in isolation and water policies must also take account of other sectoral policies, in particular land use.

A conceptual view of IWRM has been presented by van Beek (2002) who termed it as a "structured process of policy analysis." This view, shown in Fig. 15.3, emphasizes the integrating character of IWRM among three systems: natural, socioeconomic, and institutional. Integration in the natural system in itself has many components, the first being the integration of land and water. The storage and consumption of water depend on the properties of land and vegetation cover. Traditionally, water management tends to pay more attention to *blue water*, water that can be extracted from surface bodies and aquifers. However, equally important is the management of *green water*, water stored in plants. An integrated management of green and blue water can result in substantial savings of water, higher use efficiency, and higher crop yields. The other important components requiring integration are water quality and quantity, upstream and downstream interests, and fresh and saline water management. Faced with such a range of issues, IWRM was viewed as an integrating handle by TAC (2000), as shown in Fig. 15.4. In a river basin, water storage/utilization in upstream areas changes the patterns of floods and recharge in the downstream reaches and water use upstream influences the quality and quantity of water available to downstream stakeholders. All this is likely to give birth to conflicts. Sometimes, these conflicts become quite bitter, if the upstream and downstream areas fall under different political entities (recall the example of the Middle East region in Chapter 1). An integrated management recognizes these physical and social linkages and attempts to manage them properly.

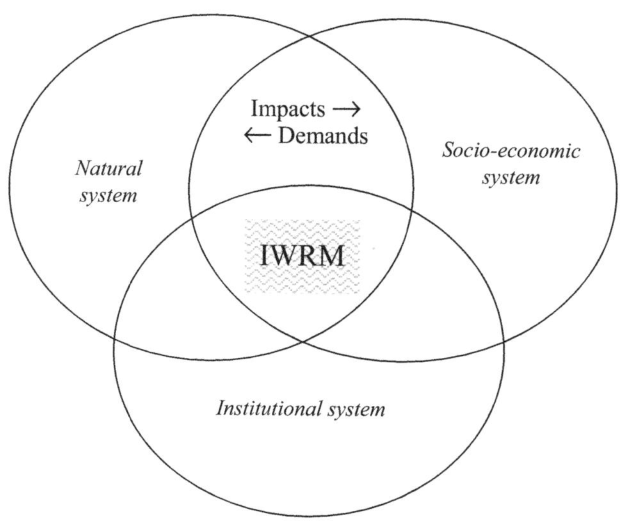

FIGURE 15.3

A conceptual view of IWRM, showing interactions among related systems.

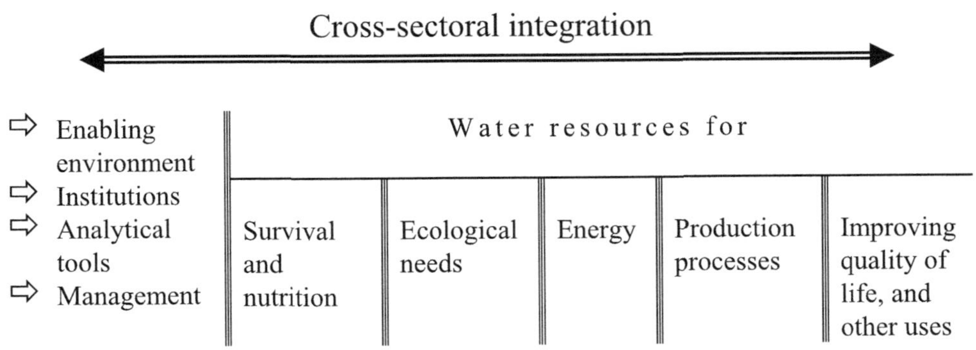

FIGURE 15.4

The concept of IWRM as an integrating handle, growing from subsectoral to cross-sectoral management.

Adapted from TAC (2000).

Water resources management modeling of a river basin system should include not only natural and physical processes, but artificial "hardware" (physical infrastructure projects) and "software" (management policies) as well. This classification of management components in two parts was discussed in Chapter 1. An ideal management model also needs some submodel of human behavior in response to policy initiatives. In a simple way, this can be introduced through the price elasticity of demand coefficients.

Alleviation of poverty is the key issue in most developing countries and it is closely linked to the provision of good quality water and sanitation facilities to the poor. These services should be priced so that the charges do not put an unreasonable burden on them but at the same time, the wastages are minimum.

In a basin, water is used for in-stream purposes, including hydropower generation, recreation, and waste dilution, as well as off-stream purposes that are differentiated into agricultural water uses and municipal and industrial (M&I) water uses. The steps for the management of water in a basin are depicted in Fig. 15.5 on the same lines as given in Section 9.4. IWRM is concerned not only with the management of various projects in a basin but also with the utilization of water for consumptive uses, nonconsumptive uses, and in-stream uses. The quality of water that is returned to the system after various uses should conform to the standards laid down.

Following van Beek (2002), an IWRM study can be divided into three phases: inception, development, and selection. During inception or initial analysis, the objectives of the study are defined; and the parameters, such as the base year, the time horizon, the growth rate, the discount rate, the boundaries of analysis, and its details are fixed. The commonly studied scenarios include business as usual (BAU), the possibility of large-scale use of new technology, or any other setting that may be appropriate in the prevalent socio-political environment. The development phase consists of extensive data collection, model set-up, and bottleneck analysis. About 10–15 strategies may be examined in the final section phase which includes impact assessment, scenario analysis, and sensitivity analysis. The indicator 6.5.1 of SDG 6 tracks the degree of IWRM implementation by assessing the four key dimensions of IWRM: (a) enabling environment, (b) institutions and participation, (c) management instruments, and (d) financing. Sustainable IWRM is vital for long-term social, economic and

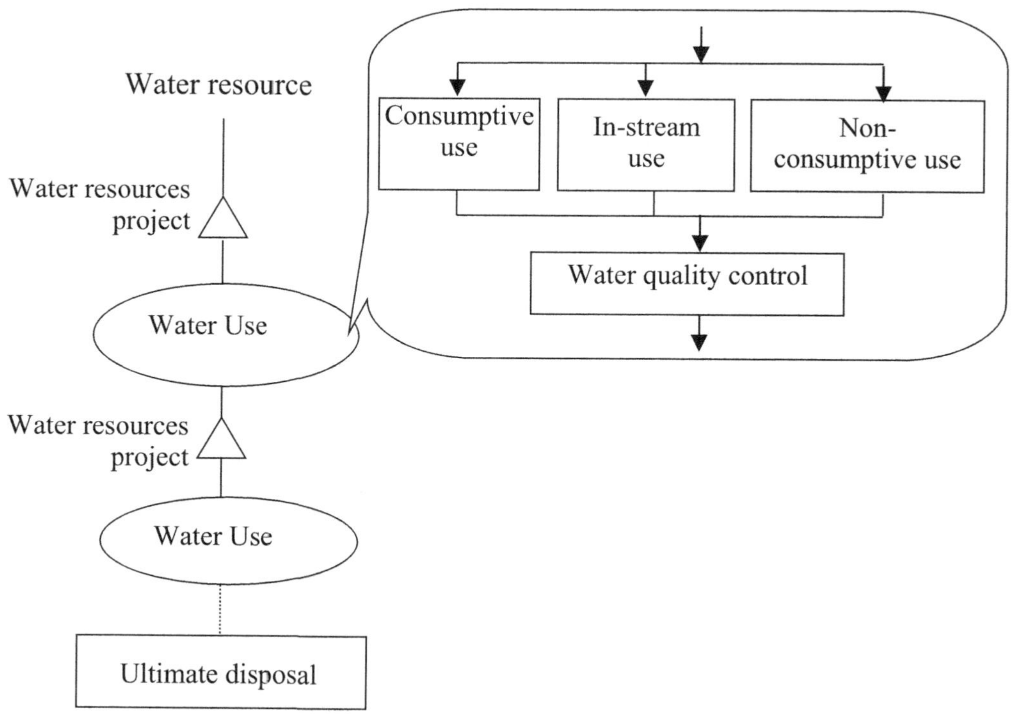

FIGURE 15.5

Steps in water resources management.

environmental well-being (UNEP, 2021). UNEP (2021) presents the global status on IWRM and acceleration needs to achieve target 6.5 by 2030, based on the latest data on indicator 6.5.1.

15.3.1 Conjunctive Use of Surface and Ground Water

As rivers and aquifers are two interrelated sources of water in an area, it is rational to manage them jointly. A joint use of surface and subsurface water is necessary for cost-effective environmentally friendly water management in a stream-aquifer system. The term *conjunctive use* of water denotes coordinated use of surface and groundwater in space and time. Thus, when surface water is scarce, groundwater is utilized (subject to availability) to meet the demands and when surface water is in excess, groundwater is recharged. The main advantage of conjunctive use is an overall increase in benefits and reduction in adverse affects due to nonsustainable use of either of the resources. Such management is also necessary to control water logging and soil salinity and increase irrigation efficiency.

Traditionally, surface water is considered as a source of energy, while groundwater requires energy to pump it out. Given this, a coordinated use of the two resources is an attractive proposition. Another major advantage of conjunctive use is the higher reliability of meeting demands and the reduction in the required storage capacity of surface reservoirs. This can be possible because the underground aquifers are also used as storage. An added advantage of aquifer storage is that there is no loss of water due to evaporation and the quality of water is also better than surface storage. Despite many obvious advantages of conjunctive use, such practices are not very widely followed. In many places in the world, wells are drilled by farmers to supplement irrigation without any centralized plan. Although farmers may be using both sources, such a utilization is not truly a conjunctive use system because there is no monitoring on withdrawal and no planned scheme for recharge.

The conjunctive use also yields greater flexibility in the choice of source to meet demands. If water from a source is not suitable for some demands, the alternate source can be tapped. The drainage problems are in conjunctive use systems, because groundwater is pumped in the locations where the water table is at a shallow depth. Note that if aquifers have consolidated due to severe overdraft of groundwater, it may not be possible to restore the status by artificial recharge.

There are three possible strategies in a conjunctive use system: (a) conjunctive use in time, (b) conjunctive use in space, and (c) a combination of (a) and (b). In the first case, surface water is used when it is available in sufficient quantities and groundwater is used during lean flow season. In case of (b), parcels of lands are assigned to surface and groundwater resources, depending upon topography, hydrogeology, vulnerability to water logging, etc. However, in this system, enforcement may be difficult particularly if the prices of surface and subsurface water have large differences. The third strategy is undoubtedly the best, as it exploits the advantages of the first and second strategy.

The water resources literature contains numerous models and studies on the conjunctive use of surface and groundwater resources. Gorelick (1983), Willis and Yeh (1987), and Coe (1990), among others, provide excellent reviews of integrated water quantity and quality management modeling in aquifer and stream-aquifer systems. Young and Bredehoeft (1972) used a detailed hydrologic simulation model in conjunction with a net benefit optimization model to address a conjunctive use problem in Colorado, USA. They found that centrally controlled groundwater development would probably lead to greater net benefits than would unregulated development. However, usually, groundwater is pumped both by government-owned and privately owned wells and there is simply no regulation on its use.

Conjunctive use is an area in which LP and DP techniques have been applied extensively. Billib et al. (1995) developed a multistep modeling approach for a multiobjective decision problem of conjunctive use. The system that they studied had a surface water reservoir with a hydropower plant, an aquifer, an artificial recharge area, pumping fields, and a distribution system for five irrigation areas. Their formulation also considered irrigation, hydropower production, water supply, as well as water quality maintenance. The authors applied a three-step procedure to combine the short-term (hydrologic year) decision with the multiyear analysis.

Wong et al. (1997) presented a methodology to determine the multiperiod optimal conjunctive use of surface water and groundwater with water quality constraints. The methodology included three models: a two-dimensional groundwater flow model, a two-dimensional contaminant transport model, and a nonlinear optimization model. The flow and contaminant transport models were solved separately. Based on the results, a drawdown limit and a concentration limit were established in the optimization model to determine, for each time period, the water supply from the surface water source, the groundwater source, and an imported source. Conjunctive use modeling was an important component of the Ganga-Brahmaputra study that was described in Section 9.12.

15.3.2 Concept of Water-Energy-Food-Ecosystem Nexus

Dwindling quantities of available water and degrading quality of raw water sources are adversely impacting sustainable access to water for a large section of society. Water security is the capacity of a population to safeguard sustainable access to adequate quantities of water of acceptable quality for sustaining livelihoods, human well-being, and socioeconomic development (UN Water, 2013). Water security also requires that protection be provided against water-borne pollution and disasters related to water. Related to water security is the concept of food security. According to the declaration of the World Summit on Food Security (2009), food security arises when all people, at all times, have physical, social, and economic access to sufficient, safe, and nutritious food. The food should be according to the preferences and needs of the people. Food security has four pillars: food availability, physical and economic access to food, stability of supply and access, and food utilization.

Another related concept is energy security which is the uninterrupted availability of energy sources to a country or a region at an affordable price (IEA, 2022). Energy security is important at long and short-term horizons. Long-term energy security is necessary for the supply of energy for economic development and the ecosystem. Short-term energy security requires that the energy production systems can supply energy as per the short-term fluctuations in demand. Finally, ecosystem security is a subcategory of the broader concept of human security. The Human Development Report 1994 by the United Nations Development Program (UNDP, 1994) identified environmental security together with food, health, personal, community, economic, and political security as the key components of human security. Ecosystem security comprises compliance, conservation, pollution prevention, restoration, and application of environmental security technology. It denotes the state of protection of the natural environment of a region from the threats arising from human and natural impacts on the environment (Zurlini and Muller, 2008).

Current approaches for planning and management in water-energy-food-ecosystem (WEFE) sectors appear to be unsustainable and suboptimal because of silo-based approaches that ignore interconnectedness across these interdependent sectors. A nexus approach that considers the interactions and interdependence among the sectors helps overcome weaknesses of silo-based approaches to better address synergies and trade-offs.

The WEFE nexus concept refers to the interconnectedness of these four elements and how they affect each other (UNECE, 2022). Water, food, and energy are essential for human survival and economic development, and the health of our ecosystems plays a critical role in maintaining the availability and sustainability of these resources. The WEFE nexus emphasizes the importance of considering the interdependence of these elements and the need for integrated, holistic approaches to managing them to optimize equitable economic and social welfare and environmental sustainability.

Fig. 15.6 presents a conceptual view of the WEFE Nexus. It shows that there are interconnections and interdependencies among these sectors. Therefore, the actions taken for utilization or management in one of these sectors are likely to impact the other sectors at various spatial and temporal scales.

The availability of water, food, energy, and environmental resources varies considerably across regions. Given that WEFE sectors have many interconnections and interdependencies, decisions taken in one of these sectors have impacts on the other sectors. For example, water availability and uses impact the options for energy generation, food production, and the state of the ecosystem. In many

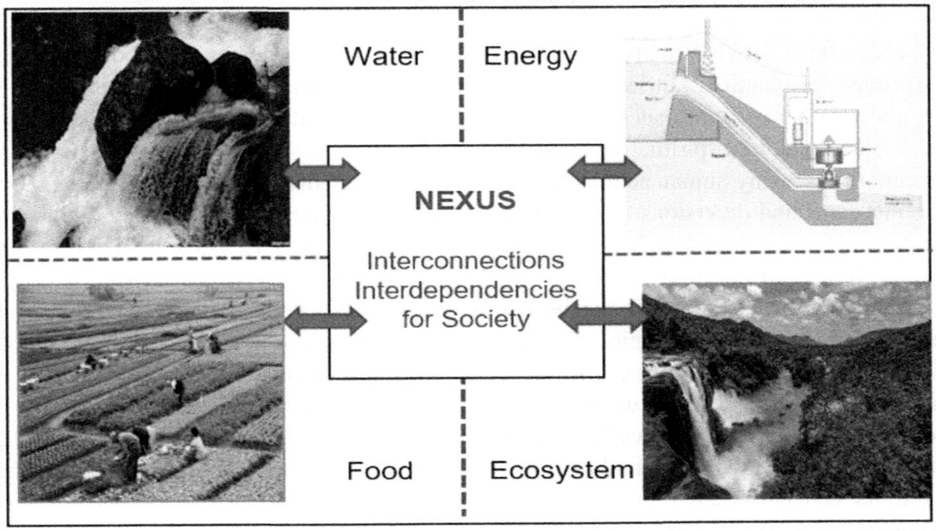

FIGURE 15.6

A conceptual view of the WEFE Nexus.

cases, almost the entire river flow in the lean season is diverted to irrigate farms, leaving little flow in the river, and consequently, the river ecosystem begins to deteriorate. Similarly, while the use of groundwater for irrigation has helped self-sufficiency in the food sector in many countries, this has placed a high burden on energy utilities as electricity is provided at subsidized rates. In turn, this is leading to overexploitation of groundwater which is detrimental to water, energy, and ecosystem sectors. Declining water tables are also resulting in contamination of groundwater and degradation of ecosystems.

With increasingly connected WEFE systems and an increasing number of policy/programs, largely working in silos, there is a need to identify, assess, prioritize, and scale the most suitable WEFE nexus management interventions. The nexus approach allows accounting for these synergies and tradeoffs to support more sustainable, equitable, and economically optimal outcomes. Hoff (2011) identified some key benefits of adopting a nexus approach: increased resources productivity, use of waste as a resource in multiuse systems, stimulating development through economic incentives, governance, institutions, and policy coherence, benefiting from productive ecosystems, integrated poverty alleviation and green growth, and capacity building and awareness raising.

Many ills in water management can be substantially overcome by practicing the nexus framework in the design/implementation of policies and management of projects. Jain et al. (2023) reviewed key Indian government policies and programmes in the WEFE sector to assess the synergies and trade-offs among them and found that the policies of different sectors give inadequate consideration to the impacts of decisions in a sector on the other related sectors. They have also discussed the key challenges and barriers in implementing the nexus concept and how to overcome them.

15.3.3 Environmental Flows and Their Assessment

Healthy ecosystems provide many services to society, such as water, wood, food, and fuel. They support primary production, nutrient cycling, and soil formations. Healthy ecosystems also regulate climate, and weather extremes such as droughts and floods. Healthy ecosystems also provide social, cultural, recreational, and spiritual benefits.

In recent times, many human actions have damaged ecosystems. River flows are being modified by constructing dams and diversions to provide water for agriculture, city water supply, industries, and navigation, and for flood control. In addition, land use land cover (LULC) changes have also impacted the amount, time, and duration of river flows. Increasing withdrawals are resulting in significantly reduced river flows. Many people living close to the river directly depend on rivers for their livelihood. In many countries, numerous religious and cultural ceremonies are held near river. Nature has been providing ecosystems services or benefits for "free" to all since time immemorial. But, when the river flows are drastically reduced and water is polluted, most of these benefits vanish.

The concept of minimum flows in the rivers evolved in the 1970s when people realized that to sustain human well-being, health and integrity of the entire ecosystem including rivers is paramount. This led to the concept of environmental flows (EFs). EFs are defined as: "environmental flows (EFs) are the quantity, timing, duration, frequency and quality of flows required to sustain freshwater, estuarine and near-shore ecosystems and the human livelihoods and well-being that depend on them" (Arthington, 2012). This definition of EFs is contained in the Brisbane Declaration (2007). EFs are critical in maintaining river health, economic development, and poverty alleviation. These are a way to balance the use of river water to meet the needs of society and the needs of ecosystems.

For sustainable management and keeping the rivers healthy, it is necessary that EF requirements are met. Following this mechanism, flows of specified magnitudes, at specified timings and durations are present in the river. EFs help sustain the aquatic habitats and ecosystem processes in a river. A related term "ecological flows" is interpreted as the flows that are required to sustain the flora and fauna present in the river and their ecological functions. Some scientists use the term "flows" instead of "flow" because the term "flow" refers to a single quantity whereas "flows" refers to the whole flow regime which changes with time.

Importance of Environmental Flows

An ecosystem consists of biotic and abiotic parts. Biotic parts include plants, animals, and microorganisms living side by side in the nonliving abiotic (physical) parts. A river reach is an aquatic (water-based) ecosystem but its catchment is a terrestrial (earth-based) ecosystem. Although ecosystems provide a wide range of valuable services to society, they are frequently neglected while allocating water. In some cases, almost all water is used by society, leaving nothing for the ecosystem itself. This leads to severe ecosystem degradation. These days, we have a growing need to safeguard ecosystems while utilizing water for societal needs.

The livelihood of a large section of society, especially in villages, depends directly on the ecosystem services that rivers provide. A river transports water, energy, nutrients, sediments, and aquatic biota, mostly from upstream to downstream; at times fishes move in the reverse direction also. Thus, a river provides longitudinal connectivity in the catchment. It also provides lateral connectivity between the main channel and flood plains/wetlands. Lateral and longitudinal connectivity helps the movement of water, food, nutrients, and carbon across the landscape. Jain and Acreman (2017) have covered the topic of environmental flows in detail.

Methods for Environmental Flows Assessment

Over the past 40 years, many methods have been developed to estimate EF for a given river stretch. These methods range from quick and simple approaches to resource-intensive, complex approaches. In earlier times, the main objective of environmental flow assessment (EFA) was to determine the minimum flow required in a river to ensure the survival of one or a few fish species. They respond to changes that impact their habitats and show a consistent ecological response. Therefore, the composition and status of the fish community are commonly used to determine and validate EF regimes. Of course, while computing EFs, a better approach is to take care of the entire ecosystem.

In response to growing requirements, data availability, expertise, time, and other resources, several methodologies have been developed to compute EFs. The methodologies range from simple, desktop approaches to complex approaches that require intensive resources. The methodologies can be divided into four major groups: (1) hydrological methods, (2) hydraulic rating methods, (3) habitat simulation methods, and (4) holistic methods. A brief description of the methods follows.

Hydrological Methods

Hydrology data are the basic data for water resources management. The river flow regime is the key driving force that influences the ecology of rivers. Indices based on streamflow data are useful indicators of the river ecosystem. Hence, many initial estimates of EFs utilized hydrological indices.

Tennant (1976) analyzed data from a number of river gauging sites in the mid-western United States and divided the entire range of mean annual flow (MAF) at a site into several ecologically relevant ranges. These were used to specify EFs:

	Recommended flow regime (%)	
Flow category	**October to March**	**April to September**
Flushing/maximum	200	200
Optimum	60–100	60–100
Outstanding	40	60
Excellent	30	50
Good	20	40
Fair/degrading	10	30
Poor/minimum	10	10
Severe degradation	<10	<10

Although Tennant (1976) developed indices for certain climatic regions in North America, these have been used in other countries. Modifications to Tennant's method have been suggested (Tennant, 1976). Tessmann (1980) considered natural variations in flow on a monthly basis to determine the flow thresholds and the resulting approach is called the Modified Tennant Method or Tessman Method.

In some countries, minimums for rivers have been specified. As per the freshwater fishing law in France, the residual flows in the bypassed section of a river must be at least 1/40 of the average flow for existing schemes and 1/10 for new schemes (Souchon and Keith, 2001). In the United Kingdom, an index based on flow, Q_{95} (which is the flow that is equaled or exceeds 95% of the time) often forms the basis to define EFs. In Greece, the stipulation is that at least 1/3 of the average summer flow of the river should be reserved for the environment. In South Africa, a hydrological index, CV/BFI is used to determine EFs. Here CV is the coefficient of variation of river flows (CV) and BFI (base flow index) is the proportion of total flow that is base flow.

Range of Variability Approach

Richter et al. (1997) developed the range of variability approach (RVA) in which various aspects of the variability of flow are indicated by computing 32 indices. These indices are calculated from natural daily flows on an annual basis. They reflect different aspects of flow variability. Some scientists note that RVA uses too many indices; many indices are likely to be correlated or their values may not differ much. Note that indices are computed solely from hydrological data and do not consider ecology data.

Hydrology and Hydraulics-Based Methods

Hydrologic- and hydraulic (HH)-based methods are a refinement of hydrologic methods to estimate EFs. The HH methods find and use the relations between river flow and channel hydraulic characteristics, for example, flow depth, wetted perimeter, and flow velocity to compute EF. A relationship between hydraulic habitat and discharge is used to derive the EF. An adequate number of channel cross-sections should be considered to ensure that the longitudinal hydraulic variability of stream reaches is accounted. A breakpoint on the Q-habitat response curve is considered as a threshold below. If discharge falls below the breakpoint, the habitat quality is likely to be significantly degraded and thus, the discharge at the breakpoint is the required EF.

Hydraulic Rating Methods

Hydraulic rating methods (HRMs) use hydrologic, hydraulic, and ecological data and the opinions of experts to arrive at EFs. HRMs use indices of habitat conditions (morphological and ecological data) in the river to compute EF. Easily measurable hydraulic characteristics of a river are water depth, velocity, wetted perimeter, etc. Wetted perimeter of the river's cross-section is a useful surrogate for habitat quality for the target biota. Methods based on wetted perimeter use the relationship among discharge, cross-section properties, habitat quality, and wetted perimeter to arrive at the value of EF.

Hydro-Biology Methods

Hydro-biology methods use hydrological, hydraulic, and biological data and habitat simulation models. These methods are also called holistic methods because they consider almost all the factors that govern EFs. Ecosystem constituents that are commonly used in such approaches include channel morphology, hydraulic habitat, geomorphology, water quality, vegetation, fish, and other biological life. In fact, these methods are frameworks that combine hydrological, hydraulic, and habitat simulation models and explicitly adopt a comprehensive ecosystem-based approach for EF determination.

Building Block Method

The building block method (BBM) is based on the hypothesis that the survival and growth of riverine communities depend on different elements (high, medium low) of flows which can be termed as building blocks of the flow regime (King and Louw, 1998). Hence, it is important to preserve the key features of the natural flow regime, including variation to keep the river ecosystem in healthy condition. When combined together, these building blocks create a flow regime that enables the river to reach a prespecified condition.

The flow blocks comprise low flows as well as higher flows required for channel maintenance, breeding stimuli, fish movement, and for connectivity with floodplains. Each flow component may vary with season, and the flow regime may be defined for "normal years" and "dry years." Table 15.2 lists typical building blocks (see Acreman et al., 2009).

Habitat Simulation Methods

The next stage of improvement in EF estimation is to analyze the suitability of the habitat of indicator species as a function of river flow properties. The resulting methods that are based on simulating the response of species have been called habitat simulation methods (HSM). To apply HSMs, detailed hydrological, hydraulic, ecological, and biological data are collected. According to Tharme (2003), these techniques attempt to assess EFRs by analyzing the quantity and suitability of instream habitats, that are available to target species in different flow regimes. Fish is the most common indicator species for river habitat.

The Physical Habitat Simulation Model (PHABSIM) was developed by Bovee (1986). The PHABSIM model has been designed to predict the microhabitat (depth, velocities, channel indices) conditions in rivers as a function of streamflow, and the relative suitability of those conditions to aquatic life (www.fort.usgs.gov/products/software/phabsim/). PHABSIM deals with the physical habitat results of various flow regimes.

Case Study

This section presents a case study of EF assessment for a stretch of the Yamuna River, India. The material presented in this section has been largely drawn from Jain et al. (2022). This river originates from the Yamunotri Glacier in the lower Himalayas in the Uttarakhand state, India at an altitude of 6390 m. In the upper stretch, the Yamuna River flows through Uttarakhand, Himachal Pradesh, and Haryana states. Fig. 15.7 shows the location of the Yamuna River basin. At the Hathnikund barrage (Yamuna Nagar district, Haryana state), river water is diverted into two canals for irrigation: the Western Yamuna Canal (WYC) and the Eastern Yamuna Canal (EYC).

Downstream of Hathnikund barrage, about 10 cumecs of water is released in the Yamuna river to meet the demand of the areas adjacent to the river in Haryana and Uttar Pradesh states. However, this amount is inadequate as evidenced by the poor state of the ecosystem in this reach. In this case study, EF was estimated for the river reach between Hathnikund and Okhla barrage (Fig. 15.7).

Yamuna supports the livelihood of a large number of people living in its basin states. To fulfill the growing water needs, increasingly larger volumes of water are being withdrawn from the river. During

Table 15.2 List of building blocks making up an environmental flow regime.

Building block	Purpose
Low flows	Habitat for juveniles and prevention of invasive species
Maintenance flows	Stimulate species migration, spawning, and dispersal
Freshets	Stimulate species migration, spawning, and dispersal
Small floods	Sort river sediments, connect river and floodplain habitats
Large floods	Remove undesired species, maintain channel structure and evolution

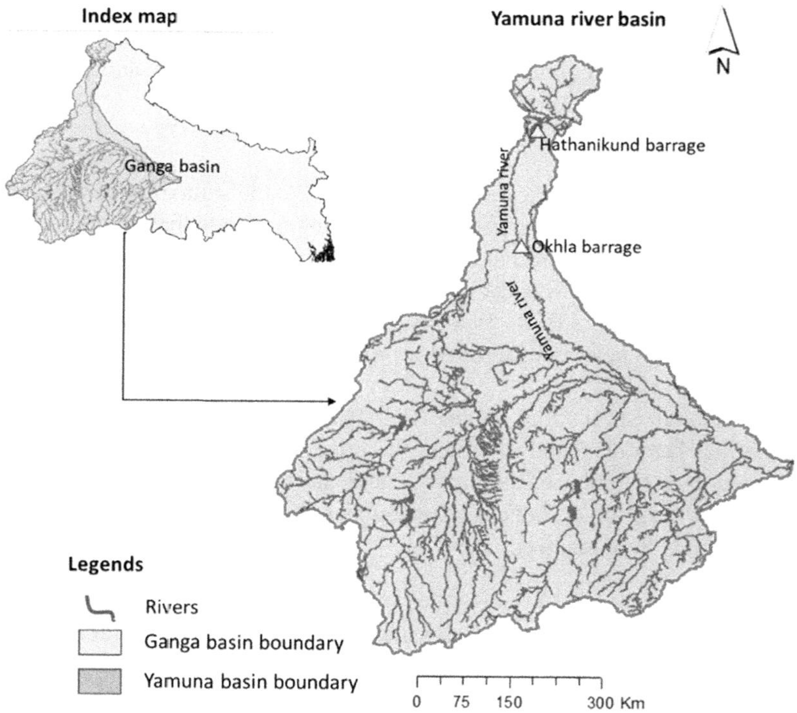

FIGURE 15.7

Map showing the upper Yamuna basin and the river stretch between Hathnikund and Okhla barrage.

the low flow season, a very small quantity of water flows in the river reaches between Hathnikund to Okhla barrage. Further, large quantities of treated/untreated sewage and industrial effluent are released into the river from the towns and industries located on its banks. This has adversely affected the health of the river as well as of the people living in the region. EFs flows are necessary for Yamuna to regain its health and remain a healthy living river.

Integrated hydrologic and hydrodynamic modeling was carried out to compute EFs in the study reach and to determine the releases required from the Hathnikund barrage in the river to maintain requisite EFs. As limited information about fish species in the study reach was present, field surveys were conducted to identify fish species. Surveys revealed that 40 species of fishes were present in the river stretch upstream of Delhi. The identified indicator species, Bangana dero and Raiamas bola, were found to be thriving well in the river.

The values of depth suitable for indicator aquatic species were converted into flow values for the entire hydrologic regime by using the HEC-RAS 1D model. In addition, the required releases from Hathnikund barrage to maintain the minimum required flows at various sites were estimated by using the results of the calibrated SWAT model. The computed releases were modified to mimic the natural variability of natural flows and are shown in Fig. 15.8. The recommended releases are about 30% of the inflows at Hathnikund.

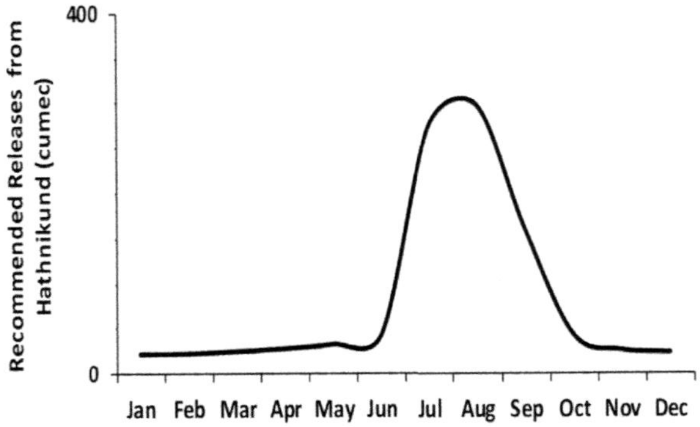

FIGURE 15.8

Recommended releases from Hathnikund barrage or environmental flow for the study reach.

Implementation of Environmental Flows

Implementation of EFs is challenging due to many reasons. Different regulators look after different facets of the environment, for example, rivers, forests, wetlands, fishery, and wildlife. The importance of these is different in different segments of the river. The waters of many rivers are overallocated, and it is a challenge to find water for EFs in an overallocated river because the allocation of some current uses will have to be curtailed. This causes economic loss and opposition. In some cases, water for EFs was found by purchasing water rights, modifying dam operations, and increasing water use efficiency in agriculture.

The final decision on EFs is a societal choice supported by scientific studies. Many countries have limited experience in implementing EFs and therefore, adaptive management may be followed. Here, EFs are assessed and implemented, outcomes are monitored, and feedback is used to revise assessments. When the experience is limited, a good strategy is to adopt a phased approach. Many programs have begun with simple approaches and evolved as more data and experience became available. In recent times, allocation of water for the environment has been mainstreamed in many countries across the world and the topic of environmental flows is a part of river basin modeling.

15.3.4 Models for Integrated Water Resources Management

An early program wherein the concept of river basin modeling was introduced was the Harvard Water Program (Maass et al., 1962). The development and application of mathematical models saw rapid growth in the 1970s and the 1980s with the advent of computers. There is a long list of models that address a wide range of water resources problems—many of these are problem-specific and many are general purpose. The first generation of river basin models mainly focused on hydraulic and hydrologic aspects, such as flood routing and reservoir regulation. Side by side, models for water quality simulation and sediment transport were also developed. The Streamflow Synthesis and Reservoir Regulation (SSARR) model of USACE (1987) was a widely used model in the 1980s. The SIMYLD-II model of Texas Water Development Board (1972), USA, was based on network flow programming techniques to simulate a river-reservoir-diversion system. The HEC-5 model of the Hydrologic Engineering Center is widely used to simulate the operation of a system of reservoirs.

The models that are able to consider both hydrologic and water quality aspects could be labeled as second-generation models. With the wider availability of personal computers and the use of graphical user interfaces, the models began to employ interactive analysis and graphical display of results. The Interactive River-Aquifer Simulation (IRAS) model by Loucks et al. (1995) simulates flows, storage, water quality, hydropower, and energy for pumping in an interdependent surface water-groundwater system. It made extensive use of graphics capabilities in system simulation. The later improvements in the model included sediment transport modules, interfacing with a watershed runoff component, or other user-defined modules. This model has been used in many countries.

Basin-scale models that require hydrologic, crop, and economic input data and simulate the behavior of various hydrologic, water quality, economic, or other variables under a fixed set of water allocation and infrastructure management policies are being increasingly used to assess the performance of water resources systems. A useful outcome of the simulation of the operation of a water resources system under a range of conditions is the identification of system components that are likely to fail. The models that use detailed hydrometeorological input data can also be employed to assess the system performance under various scenarios of climate change, and changing demands, such as those due to population growth, change in command areas, and cropping patterns.

The European Hydrological System (SHE) model has been developed as a joint effort by the Institute of Hydrology in Great Britain, SOGREAH (France), and the Danish Hydraulic Institute (DHI) (Abbott and Refsgaard, 1996). SHE is a distributed and physically based modeling system for describing the major flow processes of the entire land phase of the hydrologic cycle. Note that this is not a system management model but results from such models can be useful in RBM. MIKE SHE is a version of this model that is supported by DHI (1995) and it has a number of add-on modules for specific problems, like water quality, soil erosion, or irrigation. MIKE SHE is being used by several academic and research organizations and consulting companies. However, large requirements of input data, computer resources, and trained staff mean that such models are mostly applied by major government and academic/research institutions or consulting firms. The philosophy and role of distributed hydrological models in water resources management have been described by Abbott and Refsgaard (1996).

Increasingly, the water quality simulation capability is a standard feature of river basin models. Early water quality models were one-dimensional and could compute only temporal variation of relatively few water quality variables, such as temperature, dissolved oxygen (DO), and biochemical oxygen demand (BOD). Subsequent models that accounted for spatial variability were one-dimensional and allowed for the simulation of more complex variables subject to adsorption or decay processes, such as nutrients and coliforms. Recently, three-dimensional, time-dependent models incorporating more realistic descriptions of processes affecting water quality have appeared. A widely applied water quality model is the Enhanced Stream Water Quality Model (QUAL2E) of the United States Environmental Protection Agency (EPA, 1998). QUAL2E simulates temperature, DO, BOD, chlorophyll A, nitrogen (organic, ammonia, and nitrate), phosphorus (organic and inorganic), and coliforms in addition to constituents with user-defined decay properties. It is a widely used tool as far as water quality modeling is concerned. The Water Quality for River Reservoir Systems (WQRRS) developed by the Hydrologic Engineering Center (USACE,1998) simulates DO, total dissolved solids, P, ammonia, nitrate, alkalinity, total carbon, organic constituents, and a range of aquatic biota.

The third generation of models refers to interactive models that are supported by graphical user interfaces, GIS for input and analysis of spatial data, and screen display of results. These models are gradually becoming common in river basin simulation. The WaterWare model was developed by a consortium of European Union-sponsored research institutes under a collaborative research program Eureka EU 487 (Jamieson and Fedra, 1996a). WaterWare has a GIS component and modules for expert systems a two-dimensional, finite-difference groundwater model. It has modular architecture and components for demand forecasting, water resources planning, and ground and surface water pollution. WaterWare has been applied to the Thames River basin in the United Kingdom, and the Rio Lerma basin in Mexico (Jamieson and Fedra, 1996b).

The Tennessee Valley Authority (TVA) Environment and River Resource Aid (TERRA) model is a reservoir and power generation operations management tool linked to a local area network for the real-time functioning of the complex TVA system (Reitsma et al., 1994). A unique feature of TERRA is that it manages hydro-meteorological input and processed output data for a range of users with different levels of security access. The TERRA model has been developed specifically for the TVA system and, therefore, cannot be applied to other river basins without modifications.

RIBASIM (River Basin Simulation Model) developed by Delft Hydraulics (2002), the Netherlands, is a software for river basin simulation and modeling. This software is capable of simulating the behavior of river basins under various hydrological conditions. The model is a comprehensive and flexible tool that links hydrological inputs at various locations with specific water uses in the basin. It is also capable of evaluating a number of alternatives related to infrastructure, operational, and demand management through an advanced DSS. RIBASIM has facilities to link to a GIS. The important modules of RIBASIM are:

WADIS is a generic district water balance model. It calculates the water demand or water surplus of districts (hydrological units, i.e., watersheds or parts thereof). It makes use of other models, such as AGWAT, to calculate the water demand of certain use categories.

AGWAT is a generic agricultural water demand and impacts model that can be linked to RIBASIM/WADIS as a subroutine.

HYMOS is a hydrological data processing and analysis system.

The MIKE BASIN is another water resources management tool developed by DHI. It is structured as a network model in which rivers and their main tributaries are represented by a network consisting of branches and nodes. MIKE BASIN uses a graphical user interface with a linkage to ArcView GIS. The model output includes information on the performance of each individual reservoir and irrigation scheme within the simulation period, illustrating the frequency and magnitude of water shortages. The combined effect of selected schemes on river flows can also be handled through simulation of the time series of the river flow at all nodes (DHI, 1997, 1998).

McKinney et al. (1999) have grouped the basin scale models into two types: models that simulate water resources behavior in accordance with a defined set of rules governing water allocations and infrastructure operations and models that optimize and select allocations and infrastructure based on an objective function. In their opinion, the assessment of system performance can be best addressed with simulation models while the optimization models are more useful if the main goal is the improvement of system performance.

15.3.5 Sustainable Development Goals

Sustainable development goals (SDGs) were adopted by all UN Members in 2015 to end poverty, protect the planet, and ensure that all people enjoy peace and prosperity by 2030. There are 17 SDGs:

1. No poverty
2. Zero hunger
3. Good health and well-being
4. Quality education
5. Gender equality
6. Clean water and sanitation
7. Affordable and clean energy
8. Decent work and economic growth
9. Industry, innovation, and infrastructure
10. Reduced inequality
11. Sustainable cities and communities
12. Responsible consumption and production
13. Climate action
14. Life below water
15. Life on land
16. Peace and justice strong institutions
17. Partnerships to achieve the goal

The SDGs are pictorially shown in Fig. 15.9. The 17 SDGs to transform the world are integrated. Action in one area will affect outcomes in others.

SDG 6 is focused on water; it aims to "ensure availability and sustainable management of water and sanitation for all." Its subgoals are concerned with different facets of water management. SDG 6 has six outcome-oriented targets, which are to be achieved by 2030. These include: (1) safe and affordable drinking water; (2) end open defecation and provide access to sanitation, and hygiene, (3) improve water quality, wastewater treatment, and safe reuse, (4) increase water-use efficiency and ensure freshwater supplies, (5) implement IWRM, protect and (6) restore water-related ecosystems. Progress toward the targets will be measured by using 11 indicators.

Subgoals of SDG 6 are shown in Fig. 15.10. SDG 11.5, water-related disasters, is also closely associated with developments in the water sector.

It is seen that water forms an important factor in many other SDGs, notably SDG-2, 7, 13, 3, 7, 11, and 14. A concerted action to meet the SDGs will lead to sustainable development and avoid unintended harm to other sectors while managing our water resources.

Seventeen Sustainable Development Goals (SDGs) were adopted by all UN Members in 2015 to end poverty, protect the planet, and ensure that all people enjoy peace and prosperity by 2030.

15.4 DECISION SUPPORT SYSTEMS

The management of natural resources requires an integration of large volumes of disparate information from diverse sources. A framework is required to couple this information with efficient tools for the assessment and evaluation that allow for broad, interactive participation in planning and decision-

FIGURE 15.9

The 17 sustainable development goals.

Source: https//www.un.org/sustainabledevelopment/ and "The content of this publication has not been approved by the United Nations and does not reflect the views of the United Nations or its officials or Member States."

making processes and effective methods of communicating results to a broader audience. Better and useful information needs to be made available to a larger number of participants in more open and participatory decision-making if information is to be effectively integrated into decision-making processes. It is a challenge to integrate new information technologies with traditional methods of analysis and to put these tools to work in practice. A decision support system (DSS) helps in attaining this objective. The integration of techniques, such as database management, GIS, simulation and optimization models, interactive, graphical user interfaces, animated graphics, hypertext, and multimedia systems, has the necessary power and flexibility to support environmental planning and management (Fedra, 1994).

Advances in information technology have made it possible to easily access large volumes of information and databases. As the people involved in decision-making may include analysts, technical managers, policymakers, as well as the affected public, a new paradigm of man-machine systems is needed to handle the various phases of the problem definition and solving. The information being provided should be adequate and understandable to all those involved in the decision-making process. An information system that can cater to all these needs must be well conceived with due attention to psychological, cognitive, and institutional aspects. Mallach (1994) has provided a discussion on the human decision-making process.

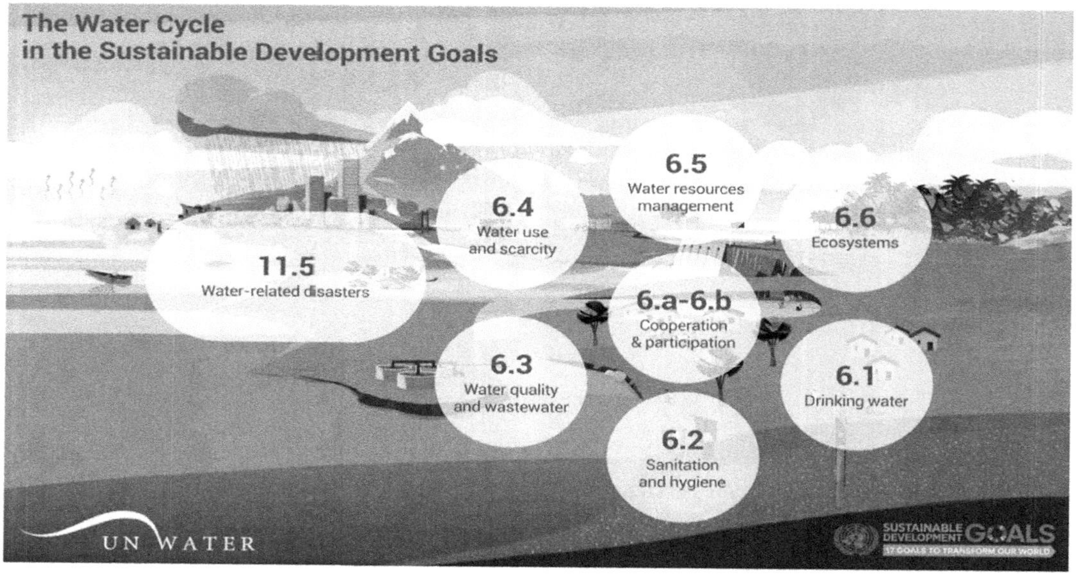

FIGURE 15.10

Subgoals of SDG 6.

Source: https://www.un.org/sustainabledevelopment/ and "The content of this publication has not been approved by the United Nations and does not reflect the views of the United Nations or its officials or Member States."

Nowadays, almost everyone who is involved in water resources planning and management uses mathematical models, typically to estimate the inputs to the system, to understand the system in a better way, or to examine the consequences of a decision or policy. However, many decision-makers are not able to effectively use the models either because the inputs are not readily available, or cannot be put in the desired format, they are not in a position to interpret the output, etc. An effective and widespread application of models requires the development of user-friendly interfaces so that a bigger group of users utilize them to obtain the desired information.

15.4.1 Definition and Objectives

"Computer-based models together with their interactive interfaces are typically called decision support systems (DSSs)" (Loucks, 1995). Morton (1971) viewed a DSS as "an aid for those management problems that are large, unstructured, … and that involve management judgment." The typical user of a DSS might be a decision-maker who may want to view a problem from various perspectives and solve it rapidly or those who require results to make an informed decision. The common objective of all DSSs is to provide timely information that supports decision-making. Note that time is critical here. Decision-makers need information when the opportunity to use that information exists, for any information provided thereafter is of not much use. This need is the key consideration that motivated the development of DSSs.

The key to useful computer-based decision support is integration. It implies that in a real-world application, several sources of information or databases, more than one problem representation or model, and a multifaceted and problem-oriented user interface, ought to be combined in a common framework to provide realistic, timely, and useful information. At the level of data and background information, numerous and often incompatible, noncommensurate data from disparate sources have to be compiled together. The user should be provided with processed data of controlled quality. The objective of a computer-based DSS for water resources management is to improve planning and decision-making processes by providing useful and scientifically sound information to the user.

15.4.2 Need and Types of DSS

The need for DSSs probably came from two main reasons. First, there is the necessity for a wider practical application of systems analysis models. An important reason behind the scanty application of these models is that real-world problems are often big in size, complex, and less structured. An application of models may require properly organizing a lot of input data. Often those responsible for handling real-world problems are not skilled in using computers and models. If these modeling tools are to be used effectively on a broad scale, the field personnel need support, help, and a framework in which these can be easily applied. DSSs are supposed to help in this.

The second reason is the need for timely information to arrive at decisions. In fact, the information needs in decision-making process are the key motivators for the development of DSSs. The decision-making can be improved by a judicious pooling of humans and machines. Loucks (1995) presented an illustration (see Fig. 15.11) of objectives and information characteristics associated with various levels of decision-making.

Depending upon the purpose that they serve, DSSs can be divided into seven types (Mallach, 1994). The *file drawer systems* retrieve data from a database. The *data analysis systems* additionally perform some analysis of the data. An *analysis information system* can combine information from several files. In *accounting models*, the calculations are performed using the data from that time period. A computerized spreadsheet is an example of accounting models. *Representational models* forecast the future effects of a decision. An *optimization system* selects the best among several alternatives. *Suggestion systems* can provide an optimal solution when decisions are highly structured. In some ways, these systems are close to expert systems discussed in Chapter 3.

15.4.3 Components of a DSS

The design of a DSS largely depends on the purpose(s) that it is to serve. The information needs of a decision-maker depend on the issues being addressed and the level of decision-making required. Therefore, the DSS developers need to know what information is used in decision-making and the appropriate format. The analysis tools in a DSS include models of environmental, economic, and social processes. Typical components of a DSS are as follows:

(a) Tools to help in system design and determine operation policies;
(b) Optimization and simulation models to determine values of decision variables or system performance indicators, given inputs and constraints;
(c) Algorithms to calibrate environmental models;
(d) Empirical models that can be used for quick calculations with limited data;
(e) Geographic information systems (see Section 3.2), for analyses and display of spatial data;

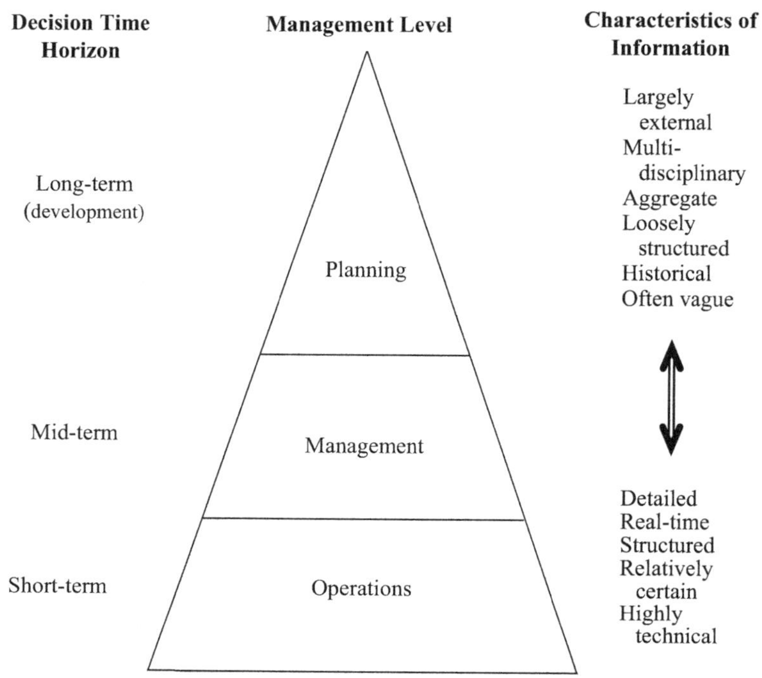

FIGURE 15.11

Decision pyramid and information characteristics associated with various types of decisions.

Adapted from Loucks (1995).

(e) Knowledge-based expert systems (see Section 3.4) that can process rules and data to draw conclusions and can provide an explanation of how those conclusions were reached;
(f) Management information systems (databases and analysis tools); and
(g) Statistical, graphical, and spreadsheet software for data analyses and display.

The input and output data of a DSS can be in a wide variety of forms. Most commonly, the interactive input is in terms of numbers that might be obtained through a series of questions. In a GUI, the user may click one of several buttons to indicate his choice. The output can be tables, text, time- or space-series graphs, maps, and video animation.

The input data to a model may be: control parameters, such as time step size; global parameters, such as the density of water; dynamic data, such as the time series of river stages; and spatially distributed data, such as the land use map. These data will come from numerous sources, in different formats, and with different quality. Their integration into one unifying information system requires a number of tools to process the original data and store it in a common format. A database management system, discussed in Chapter 2, is necessary for this purpose.

Input data preparation is often the main effort while applying a model. Hence, the integration of databases and models, which allows users to automatically retrieve and load input data for modeling, is a natural step. Computerized databases of water resources, and social and economic variables are now increasingly available in digital form and many of these can be accessed through the Internet. Some organizations maintain databases of time series and the user has to only specify which series is needed and its duration. A user can obtain data from a computer database, irrespective of his or her database's location, can analyze the data, and share the results in the text as well as graph forms with other users. These capabilities clearly provide tremendous options for information processing and communication.

Mathematical models are an important component of a DSS. The commonly used mathematical models include optimization, simulation, statistical models, decision analyses, genetic algorithms, and neural networks. The choice of a particular model depends on a number of factors, such as the type of problem, data available, personal preference, and the result that the decision-maker is looking for. Often, an integrated use of different methods, for example, simulation and optimization, is also made. The simulation and optimization models have been discussed in detail in Chapter 5.

15.4.4 Designing a DSS

It is a good idea to spend some time to get an understanding of the main objectives of a DSS, the issues to be addressed, the information needs, and preferences of the potential users. Involving one or two key people from the user organization also helps in development efforts. Many useful tips on the requirements of users are difficult to specify beforehand and get clarified during product development. Without close association of the users in various stages of the DSS development, the chances of implementation and real use of the system are considerably diminished. Though the association of the users is easier said than done, the benefits are worth the effort. This interaction also makes the users acquainted with the system and increases their confidence in the results.

Before commencing the development of a DSS, extensive discussions with the users of the concerned organization help understand their mental model and expectations from the DSS. Potential users can be approached to define the various tasks that they perform, the kind of help they expect from the DSS, the type of information needed in their analysis and how this is gathered, etc. A review and evaluation of existing DSSs should be taken up to learn their capabilities, including their model types and programming features, data input and display capabilities, documentation, model calibration and verification, interfaces, and reasons behind their success or failure.

The next stage is to translate these ideas into a framework or system architecture. Only after this, the development of various modules may be taken up. This approach will lead to a flexible and open architecture, which will be easy to expand as the need arises. A flexible DSS architecture developed by Jamieson and Fedra (1996a) is shown in Fig. 15.12. The main advantage here is that models of various complexities can be developed and incorporated into the system.

Evidently, many persons will be involved in the programming of such a large system which requires different kinds of expertise. The team to work on visual and user interfaces will be different from those working on mathematical models; these groups can work in parallel and the development work can simultaneously proceed on many fronts. A good practice is to clearly define the variables of each module and insert sufficient comments in the code. This is helpful when a new person has to work with a program written by someone else. After each module is developed, it is first independently tested. Next, it is integrated into the system and then tested again. A good strategy is to add the components one by one and test the whole system after a new component is added. Special care is needed for data flow among the various modules. Keeping a log of the changes that have been made and preparing detailed notes for programmers obviates many future headaches. A detailed manual

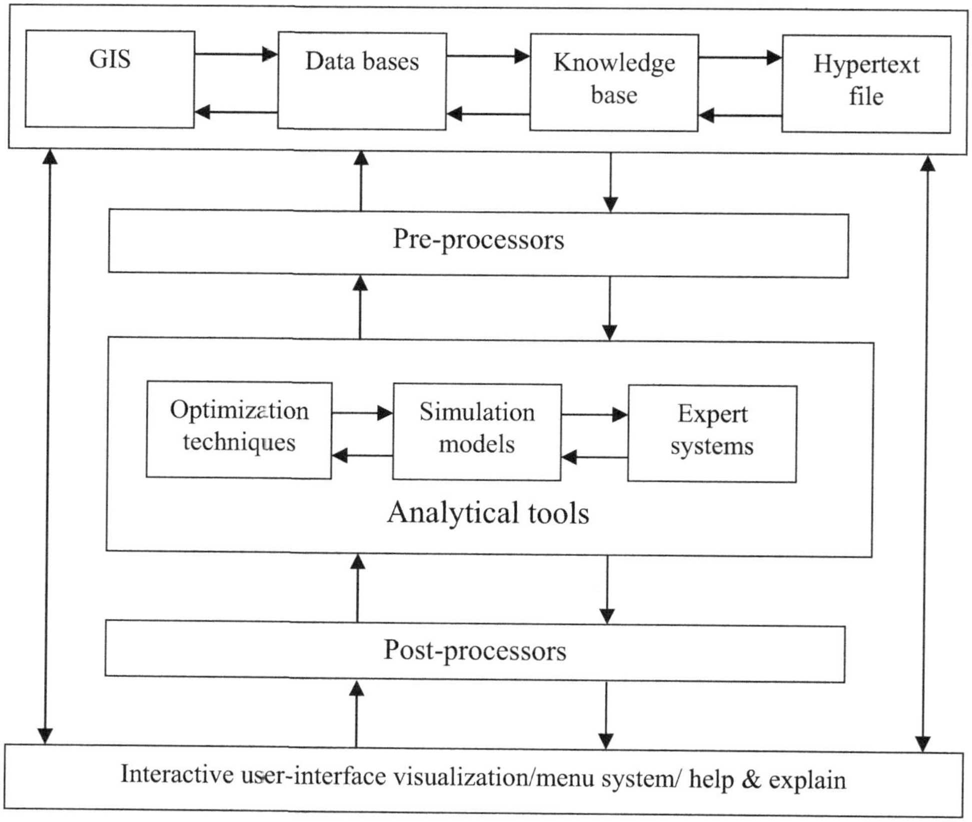

FIGURE 15.12

The DSS system architecture of Jamieson and Fedra (1996a).

should be prepared for general users with illustrative examples. Besides the hard copy manuals, online help should also be provided. As a DSS may have to be implemented on different hardware, it would be better if programming is hardware independent. Many current programming languages, such as Java, are highly portable.

When a workable prototype of the DSS is ready, it should be given to the actual users to carry out real work and their reactions should be observed and followed up. The feedback of these users should result in modifications and improvements. The development is an iterative process—designing, coding, testing, feedback, and improving. The process may never end as long as a particular DSS is in use.

Even with all user-friendly features, a certain amount of training for users is necessary. Training is an essential component of any large software package, and it should be structured in such a way that a new user is brought to a reasonable level within a period of 1 week or so. The DSS documentation and training should cater to the requirements of the end users and for those who may be responsible for programming modifications and extension. The users' guide can also be online, that is, readable on-screen when the system is being used and should have a search facility using key words. One may never lose sight of the fact that the user is the best judge of suitability or otherwise of various modeling assumptions, interface design, and implementation. Therefore, aspects, such as user training, data entry, maintenance of the systems, and its adaptations and updates, should be given due attention at the design stage itself.

While the earlier DSSs used to run on a standalone system, these days client/server computing is in vogue, because it offers many advantages. This configuration is also advantageous for DSSs that help a group of people to make decisions jointly or group DSSs. Finally, like any other product, the measure of the success of a DSS is in its usage and the additional benefits that arise from its implementation.

User Interface

Interaction is a central feature of any man-machine system and poorly designed interfaces are responsible for the rejection of many versatile and powerful packages. A good interface allows the user to define and explore a problem incrementally in response to questions from the system user. Graphical displays are preferred media to communicate complex information. A good interface communicates the results to the users in a form that they can easily understand and facilitates interactions between the user and the computer. The visual interfaces also make it simpler to interpret the model output.

Many real-world planning or management concepts, such as risk or reliability, are rather abstract. Such concepts are better understood through graphical representation. Visualization is very effective to communicate and understand large amounts of highly structured information and helps in the intuitive understanding of processes, interdependencies, and spatial and temporal patterns.

Extensive guidelines for the development of DSS user interfaces are available (Mallach, 1994). It is a good practice to ensure consistency of data entry, display, control actions, and operational procedures. Repeated entries of long data should always be avoided. An interesting, informative, and enjoyable to use DSS will surely have a greater value to the user. A good interface saves the user's time, is versatile, and easy to use, and the user is able to recall how to use the system after he has not used it for sometime, provides on-line help, is adaptable to the users' needs, and should be interesting to use. Colors can give an attractive view to an interface.

The central idea to develop and implement a DSS is to support the users in the management and decision-making process. Many of the potential users of DSSs would be experts or people with many years of practical experience, and it would be advantageous to include their expertise and experience in the system. Another advantage of involving the users in design and development is that they begin to consider themselves to be part of the system and the problems of user acceptance are automatically solved. The users can also provide valuable suggestions in the design of interfaces, and the input screens can be the same as the hardcopy forms that are used in their offices. For example, in a case in India, the acceptability of a data entry system was found to increase manifold when the screen layouts were changed to resemble the hardcopy forms that the users were familiar with.

The development of a user interface is a simpler task these days due to numerous interactive graphics-based toolkits that eliminate the need to write modules for the graphical user interface (GUI). For ease of updating and portability, the interface should be an independent module of DSS. This allows easy customization to meet the needs of different users who may be using different languages, and having different skills and preferences. In fact, many popular operating systems, such as Windows, and software packages, such as word processors now have options to customize the pull-down menus. Software tool kits to assist in developing object-oriented programs and visual interactive interfaces are widely available these days.

The development of the user interface is a process of incremental improvements. Typically, a simple working skeleton is developed and the views of users on it are obtained (Loucks, 1995). The design is then modified, based on their inputs, and the process is repeated. One of the ways to get the reaction of users is the thinking-aloud method. While using the DSS, the user is asked to speak out his thoughts, what he is trying to do, what questions or confusion arise, what he expects will happen next, etc. The observations of the user and his reactions are noted by an observer. Such studies help pinpoint problems in users' understanding of DSSs, to identify problems, detect any annoying or confusing stage, and learn the users' feeling about the DSS.

15.4.5 Applications

Many DSSs are being applied to a wide variety of problems. The growth in the DSS development and use has been substantial in the water resources fields (e.g., Labadie, 1989; Loucks and da Costa, 1991; Santos, 1991). Some versatile tools have been described by Fedra (1992) and Diersch and Grundler (1993). Other studies describing the implementation of DSSs include a program for integrating hydraulic models with a supervisory control and data acquisition (SCADA) simulator in Illinois (Schulte and Malm, 1993). DSSs for simulating river system design and operation include those developed by Andreu et al. (1996) and Basson et al. (1994), the Center for Advanced Decision Support for Water and Environmental Systems (1992), Ford (1990), HEC (1993), Kuczera (1990, 1993), Loucks et al. (1995), and Randall et al. (1995). Wardropper and Brookfield (2022) have provided a comprehensive discussion on water decision support systems.

The WaterWare DSS was developed by Jamieson and Fedra (1996a) as a tool for river basin planning. It can address issues, such as what, where, and when new resources should be developed; formulating strategies for river and groundwater pollution control programmes; assessing the environmental impact of water-related developments; determining the limits of sustainable development; and evaluating the impact of new environmental legislation. This software integrates the capabilities of GISs, database management systems (DBMS), modeling techniques, optimization procedures, and expert systems. The system architecture shown in Fig. 15.8 depicts an open, modular system with different degrees and mechanics of coupling at various levels of integration. The basic architecture comprises: (1) a main program to coordinate individual tasks and provide access through a menu of options; (2) a GIS to store, display, and analyze geographical data (this could be a commercial product or user-developed model); (3) a DBMS to provide access to nonspatial data (this also could be a commercial or a custom-designed dedicated system); (4) simulation, optimization, expert system models which can access data from both the GIS and DBMS; (5) a set of pre- and postprocessors, which support mainly editing of input data and visualization of model output; (6) a user interface with access to different functional components of the system and help files; and (7) a set of utility functions which assist in data preparation and management tasks. A river-network editor can be used to configure the physical system.

The WaterWare system has a spectrum of analytical models of different sophistication giving the user a choice depending on his requirements. This package has been successfully applied to several systems. Jamieson and Fedra (1996b) have described studies for the Thames River in England and the Rio Lerma basin in Mexico. These two basins had different types of problems; the Larma basin application was concerned with the planning of basin-wide infrastructure, while the Thames was an operation study. Such application results have clearly brought out the possible range of problems of water resources to which DSS can be applied. A special issue of the Journal of Hydrology published in 2022 on "Decision-support systems for water management" contains many articles on this topic.

The current technology allows integrating multimedia into a DSS, which improves the presentation of information, online help, and interactive tutorials. In a multimedia environment, the user can also see the consequences of various decisions through video animation. Now that DSSs and the requisite computer hardware are widely available, the attention is to use the power of these tools to achieve higher levels of system performance.

15.5 INSTITUTIONAL ASPECTS OF BASIN MANAGEMENT

A vast and well-organized institutional structure is necessary to carry out the various tasks of RBM. Institutional structures consist of formal and informal working rules. The *operational rules* provide a framework for operations (sometimes also known as operational management), for example, policy directives regarding the storage and withdrawal of water from reservoirs and aquifers. The institutional structure for RBM should facilitate the necessary coordination within the water management sector and between the water management sector and other sectors, such as land use and environment to achieve sustainable water use and maintain the balance of the river system. The institutional structure should also be a means of empowerment. All stakeholders, including economic interest groups, local communities, environmental NGOs, and women should be encouraged to play an active role in RBM. Policy formulation, mediatory, regulatory, and other management tasks should be well defined, clearly allocated, and made transparent. As RBM is often characterized by parochial interests and intractable problems, leadership and political commitment are essential to achieve progress.

The infrastructure requirements depend on the tasks to be performed in RBM. Three major tasks of an RBM organization are planning, construction, and operation. The infrastructure and expertise needed for these are completely different. In light of this, normally the organizations have separate wings headed by an experienced officer to look after each of these. Although planning receives much attention in the initial years, it is in fact a continuous activity. If the requisite expertise is not available within the organization, it is necessary to involve experts from other institutes, sometimes through commissioned studies. The construction activities are the most important usually during the initial years when infrastructure is being developed and a large share of funds as well as manpower is allocated for the same. This, however, diminishes appreciably after the infrastructure is in place and this wing may be later responsible for routine maintenance jobs. In addition, the RBM organization may also be responsible for the enforcement and implementation of decisions including legal issues.

15.5.1 Models of RBM

Normally, catchment boundaries do not coincide with political and social boundaries. Many human boundaries exist within and across a catchment, such as individual farms, villages, ethnic groups, and provincial boundaries. This "mismatch" between a basin perspective and a sociopolitical perspective has an important place in RBM. Mostert (1998a) has discussed three different models for RBM. The first is the *hydrological model* in which the organizational structure for water management is based on hydrological boundaries, that is, subbasins and so on. In the hydrological model, administrative boundaries coincide with hydrological boundaries and there is the least chance of upstream-downstream conflicts. At the topmost level, the entity that is responsible for overall management is the "river basin authority." Although this model is highly suitable for water management, it may isolate water management from other relevant policy sectors, and intersectoral coordination may become a problem. Moreover, this model implies centralization of water management activities, and it will not be preferred in countries with decentralized water management. This model is also not likely to suit international river basins. Due to these reasons, the hydrological model of RBM is a good option for smaller national basins only.

In the *administrative model,* the zones of control follow the political/administrative boundaries. Various bodies, viz., provincial, municipal, etc., may be assigned responsibilities for water management in the area of their jurisdiction. Clearly, in this model, the water management responsibility is not based on hydrological boundaries. An advantage of this model is that relevant policy sectors, such as land-use planning, can be kept together and thus there can be better intersectoral coordination. However, there is a risk of upstream-downstream conflicts and the lack of a platform or mechanism to discuss these problems. The *coordinated model* is a mixture of hydrological model and administrative model. In this setup, water management is performed following the administrative model but there are river basin commissions with a coordinating task. These commissions also provide a platform to discuss and resolve conflicts. They may consist of representatives of different bodies and ensure coordination in water planning and management.

15.5.2 Decentralization and Privatization

Some tasks related to RBM, such as irrigation water management and water quality control, require at-site monitoring and decentralized management. There are several advantages to decentralization, the main advantage being that it brings the government as close as possible to individual citizens. Second, decentralized administration is better informed about local circumstances. It also allows for local participation in the decisions that can be tailored to suit local circumstances and preferences.

Decentralization is also possible for tasks with a supralocal character, provided the decentralized governments cooperate with each other or are supervised by a higher-level government. Supervision also improves the enforcement of regulations in case the decentralized governments have too close relations with the persons and organizations they have to regulate. Decentralization is not possible for tasks, such as establishing the institutional structure and formulating policies that apply to a large region. Decentralization may also not be possible if the decentralized governments lack the necessary management capacity. This could be overcome by local capacity building and advisory services by specialized central governments. However, decentralized governments have superior information on local conditions because of their (usually) closer contact with the population.

Another form of decentralized management involves the participation of users in the supply of freshwater ecosystem goods at the local level. *User associations* are becoming popular in India and other parts of Asia and these help to ensure efficient allocation of water resources at the local level. Communication between farmers and governmental officials greatly increases after the formation of user associations. In addition, cooperation among users themselves reduced conflicts over water use. User associations are particularly effective where an efficient allocation of freshwater ecosystem goods requires intra- rather than intersectoral reallocation (Dinar et al., 1997).

Privatization is frequently advanced as a solution to overcome bureaucracy, and inefficiency; the basic premise being that it leads to efficient services. Private companies have to reduce costs and make profits to survive in business. However, there may not be many competitors because of the involvement of higher costs and expertise. The basic purpose of privatization may be defeated in the absence of competition. To overcome this, it would be desirable to have a strong regulator who can control prices and quality of services. The regulator may, additionally, monitor the environmental parameters because many private agencies are not keen to improve or maintain the quality of the environment.

Privatization is possible for specific services only; most commonly privatized services include the construction and operation of dams, powerplants, water supply and wastewater treatment infrastructure, supplying materials and equipment, and maintenance works. Of late, projects are being handed over to private companies on the basis of build–operate–transfer (BOT) of facilities. Some functions, such as regulatory and policy-making, have to necessarily remain with the government. In fact, private parties with departmental supervision frequently do construction. In many countries, projects like hydropower generation are being offered on a build–own–operate–transfer basis. Often, the infrastructure remains in the hands of the government but the maintenance is on a contract basis. Finally, private firms may even own, operate, and maintain the infrastructure.

Another management option is to assign the responsibilities of water services through public limited companies. This option is midway between the above two options. These companies can have flexibility in decision-making and raising money through various financial instruments. Their efficiency can be as good as a private firm and they can pay adequate attention to environmental aspects too. However, care is to be taken to ensure that they do not fall under the clutches of bureaucracy and become another organ of it.

Local institutions and NGOs also play an important role in managing river basins. Their involvement may prevent misuse and wastage. There are also instances of beneficiaries organizing themselves to develop or manage a particular resource and prevent overexploitation. Based on their intimate knowledge of the resource, they may devise rules to limit use and may set up self-financing systems, self-monitoring systems, and conflict-resolution procedures.

Ten regional water authorities were created in England and Wales in 1974, dealing with both water utility and river basin functions (Merrett, 1997). In 1989, the utility functions of the water providers were transferred to independently regulated private agencies, while the environmental functions, for example, flood defense and pollution control, remained in the control of a public agency. The UK experience shows that the benefits of privatizing water utilities were limited. First, the main operational efficiency gains within regional water authorities came before they were sold, to encourage the initial privatization (Kinnersley, 1993). Second, the backlog of capital spending and the EU's environmental legislation necessitated a large capital-spending program, driving up overhead costs and pushing the rate of increase in charges well above the inflation rate (Merrett, 1997). In addition, the rationalization of the utilities involved large-scale job losses. Shareholders seemed to gain to a large extent and the proportionate increase in salaries to directors was much higher than to other workers. As a result, the problems that privatization had aimed to solve were replaced by another set of problems (Kinnersley, 1993).

15.5.3 Monitoring and Analysis

Monitoring of critical indicators of basin state is an important task of RBM. In many river basins, routine hydrometeorological observations are made by automatic equipment and these are communicated to a control station in real time. In some countries, the real-time data of a few variables, such as river stage and water temperature, are available on the Internet in real time.

With the advancements in information technology, it is possible to combine online measurements with computer models to formulate forecasts. Flood early warning systems have been installed in many important basins in the world. Another possibility is to issue early warnings for water quality indices and alert the public in events, such as accidental spills. Yet another step forward is the automatic operation of controls, such as canal gates, pumps, and sluices. A preprogrammed computer that receives input data as measured by sensors at regular intervals can control the operation of these devices. As a result, human operators are relieved of some of their routine work, and the risk of human errors is eliminated.

Most of the efforts related to the development of solutions to RBM problems have been directed to specific cases. Usually, softwares are developed in response to specific requirements. The efforts and challenges in developing general purpose and comprehensive RBM tools are considerably more. It is rather difficult to find a funding agency that may be willing to commit funds for an effort that is not immediately rewarding and well focused. Further, the diversity of the RBM problems can be very large, and extensive data of several basins with divergent settings will be required to test such software. It is unlikely that such data will be available in a single organization.

Generally, there is a lack of data to fully describe and understand the complex processes that take place in a river basin and their interaction with the socioeconomic system. In the absence of these, it is an enormous task to incorporate all considerations in a comprehensive tool at the river basin level. An important aspect is that many basin processes operate at various levels and scales. This calls for the development of tools for different geographical scales and levels. Many such tools have been described in earlier chapters of this book. RBM also has an element of negotiation and the decisions taken may be dependent on the negotiation skills and strength of concerned parties. Due to these reasons, the decision taken may not be consistent over a large basin and may differ from one decision-maker to another.

15.5.4 Practical Aspects of RBM

It is one thing to know how river basins should be managed, it is another thing to actually manage them. As RBM involves conflicting interests, naturally, there is always a trade-off. The important question is how to achieve the best trade-off. Given the complexities and diversities of problems, at best some guidelines can be issued. But any guideline will be implemented only if it is "realistic." The guidelines should reflect the differing hydrological, socioeconomic, and cultural contexts and should be technically/scientifically sound.

Without a basic national legislative framework, RBM is not likely to succeed. For example, it should be clear who has jurisdiction over different waters. The concerned organization should also have adequate authority, support, and power to enforce rules. Plans lose their worth unless these are implemented in a timely fashion. The purpose of plans is to orient operational RBM and improve its effectiveness. Implementation is, therefore, a pivotal step in any RBM strategy.

Any strategy can benefit greatly from a feedback and periodic evaluation. Were the targets, and goals attained? Were the planned measures all implemented, and if not, why not? Were the measures effective? Could the planning process have been better, and were there any crucial bottlenecks in the legislative and organizational framework? Evaluations, such as these, can provide valuable input for a new round and contribute to a continuing improvement of RBM.

15.5.5 Role of Financiers

Many water resources projects require large financial outlays and it may be necessary to seek loans or funding from international financiers. As the funds available with these financiers are limited and many of them are under obligation to advance the policies of their promoters, the proposed projects are scrutinized before funds are allocated. If the project fails to meet its norms, the funds may be denied till the design or operation is suitably modified. The adverse impacts of a project on the environment are viewed very seriously these days and it is difficult to find a reputed funding organization willing to support a project which does not pass strict environmental tests. International funding agencies are sometimes able to significantly influence management policies by providing funds to the activities that are designed and operated by following relevant international treaties and agreements. Of late, most international agencies treat water as an economic commodity and emphasize on active participation of stakeholders. For example, according to the new policy of the World Bank, water is treated as an economic good. The World Bank (1993) also emphasizes decentralized management and active participation by stakeholders.

15.5.6 Cooperation Among Basin Management Organizations

New knowledge and technology are necessary to improve any set-up including RBM. The issues involved in technology and knowledge transfer and research cooperation are as diverse as the types of technologies and knowledge. Transfer of knowledge and technology are best attained in response to the needs in the receiving organization and should match its capacity to absorb. In practice, the technology transfer might also be motivated by the interests of the providers of a specific technology rather than by the needs of the recipient. One possibility to avoid this trap is through cooperation among the organizations responsible for basin management.

The cooperative activities include joint site visits, discussions and presentations, and exchange of information and staff. Usually, the aim is mutual learning with respect to the operational, policy, and institutional aspects of RBM. To this end, short seminars are also organized where people from various organizations including academic institutes can interact with each other. Such cooperation always leads to the development of improved expertise in the participating organizations. However, the ultimate outcome depends on the quality, interests, and motivation of the personnel involved.

15.5.7 Some Important River Basins Organizations

A large number of river basin organizations are functional all over the world. The Tennessee Valley Authority (TVA) in the United States (www.tva.gov) is one of the earliest river basin authorities. In its early years, TVA initiated a wide range of regional planning and development activities, including afforestation, extension programs for improved soil and land use management, and community development.

In the United States, the Mississippi River Commission (MRC) was created by the Act of Congress on June 28, 1879. The general duties of the MRC include the recommendations of policy and work programs, the study of and reporting upon the necessity for modifications or additions to the flood control and navigation project, recommendation upon any matters authorized by law, inspection trips, and holding public hearings. The work of the MRC is directed by its president and carried out by the six Army Engineer Districts in St. Paul, Rock Island, St. Louis, Memphis, Vicksburg, and New Orleans. Current activities of the MRC are in three broad categories: (1) general investigations to determine needed improvements, (2) construction of new facilities, and (3) maintenance and operation of existing systems. Included in its responsibilities are the main river from Cairo, Ill., to Head of

Passes, and the basins of St. Francis, Tensas, Yazoo, Atchafalaya, Lower Red, Lower Arkansas, Lower White, and West Tennessee rivers. The address of the website of the MRC is http://www.mvd.usace.army.mil/MRC.

The Thames Water Authority in the United Kingdom (www.thameswater.co.uk) is a classic example of integrated river basin management. The authority supports about 3500 abstractions—1200 for agriculture, 500 for domestic water supplies, and 1800 for industrial and other uses. The river receives industrial effluents at 6500 locations and effluents from sewage treatment works at 450 locations. Besides, the river is used for fishing and boating. The river flows are regulated and managed to ensure that discharges do not pollute water supplies and abstractions do not affect the level of the river to the extent that it puts at risk natural life or the enjoyment of those who use the river for recreation.

The Murray-Darling Basin Commission of Australia (www.mdbc.gov.au) has been managing the waters of one of the very dry basins in the world successfully and has contributed considerably to the economic development of that continent. In France, river basin management is primarily based on the water laws of 1964 and 1992. The law of 1964 divided France into six river basins, created River Basin Committees and Water Agencies, and a system of financial management. The 1992 law recognizes water as a single unitary resource irrespective of its physical and geographical distinctions. The management of water is done in the framework of a river basin. The French system is basically founded on the following features.

(i) The water user pays for the water he is using and the polluter pays for the water quality deterioration he is causing. The system takes into account the capacity to pay for each category of users (domestic, industrial, and farmers for irrigation).
(ii) The water resources development and management are financially self-sustaining. The water charges are proposed by the Agency's Board of Directors and later on agreed upon by the River Basin Committee. As elected representatives are also members of the River Basin Committee, there is general consensus on the water charges and recovery of water charges is satisfactory.
(iii) More than 90% of the money collected is afterward redistributed under the form of financial assistance (loans and grants) either for pollution control action or for the development of water resources and their sustainability.

In Asia, the Mekong River basin covering the whole of the Lao PDR and Cambodia, one-third of Thailand and two-fifths of Vietnam, is under a Mekong Committee. This was established by the ESCAFE now ESCAP (Economic and Social Commission for Asia and Pacific) in 1957 with its secretariat office in Bangkok and it coordinates the work of collection of basic data and river basin planning. The Mekong River Commission was created by four basin countries, namely, Cambodia, Laos, Thailand, and Vietnam, under an agreement on the cooperation for the sustainable development of the Mekong basin. In a study on the Mekong basin, Chenoweth et al. (2001) found that the Mekong River Commission lacked the institutional capability to achieve integrated management of water resources of the basin. Good organizational capacity and a sufficiently strong environment are prerequisites for IRBM.

In China, the 1988 water law requires that basin plans should serve as the basis for water development, utilization, and prevention of damage. There are seven commissions covering the six major river basins and one lake basin. These are central agencies having planning and regulatory functions under the Ministry of Water. The Yellow River Conservancy Commission has the additional responsibility of flood management in the lower Yellow River basin and operation of all the reservoirs. It has considerable financial strength and autonomy as a consequence of water and power receipts from the operation of projects. However, major basins are not managed by a single agency.

The International Commission for the Protection of the Danube River (ICPDR) is a transnational body, which has been established to implement the Danube River Protection Convention (https://www.icpdr.org/main/icpdr/about-us). ICPDR works to ensure the sustainable and equitable use of waters in the Danube River Basin. The work of the ICPDR is based on the Danube River Protection Convention (DRPC), the major legal instrument for cooperation and transboundary water management in the Danube River Basin.

The Congo basin covers an area of about 3.8 million km^2. Congo River has an average annual discharge of 41,000 m^3/s which is second in the world after the Amazon River. Congo basin straddles 10 countries and has more than 25,000 km of navigable waterways. CICOS (Commission Internationale du Bassin Congo-Oubangui-Sangha) is an intergovernmental organization charged with the promotion of inland waterway navigation and integrated management of water resources in the Congo basin (www.cicos.int). Six countries are members of CICOS: Angola, Cameroon, Central African Republic, Congo, Democratic Republic of Congo, and Gabon.

The Mahaweli Authority in Sri Lanka is a body for both the development and management of major storage and irrigation projects. It is also provided with a secretariat for planning and other water management activities.

India is a union of states and most of India's river basins are interstate in nature. India's constitution provides power to the states to develop water resources within their boundaries, subject to the parliament empowering union government to regulate and develop interstate rivers to the extent to which such regulation and development are declared by the parliament by law to be expedient in the public interest. The National Water Policy adopted in 2002, among other things, recommends:

> *Water resources development and management will have to be planned for a hydrological unit such as drainage basin as a whole or for a sub-basin, multisectorally, taking into account surface and ground water for sustainable use incorporating quantity and quality aspects as well as environmental considerations. All individual developmental projects and proposals should be formulated and considered within the framework of such an overall plan keeping in view the existing agreements/awards for a basin or a subbasin so that the best possible combination of options can be selected and sustained.*

Among the major Indian river basin organizations, the Bhakra Beas Management Board (bhakra.nic.in) manages waters in the Sutlej basin in the Himalayas. The Damodar Valley Corporation (https://www.dvc.gov.in/) manages most of the projects in the Damodar basin and the Brahmaputra Board has been set up for the systematic exploitation of water resources of this basin. The Narmada Control Authority (NCA) (http://nca.gov.in) has been set up under the final orders and decision of the Narmada Water Disputes Tribunal as a machinery for implementation of its directions regarding utilization of the water resources of the Narmada River.

15.6 PUBLIC INVOLVEMENT

Public involvement (PI) plays an important role in planning and policy-making. This topic has also been discussed in Section 9.7, so this discussion will be limited to the aspects that are relevant to RBM. Increasingly, PI is being viewed as a means of improving the quality, effectiveness, and acceptability of the decision-making process by involving the stakeholders in decision-making. Another view could be that PI is a legal right or a means to empower individuals and social groups. Three components that are considered pillars of PI are: access to information, involvement in the decision-making process, and access to justice.

Governments of many countries increasingly endorse the view that the public is to be involved in the environmental decision-making process. This idea is expressed in international policy documents, such as the Rio Declaration and Agenda 21. The Dublin Statement on Water and Sustainable Development specifically echoes this concept for freshwater resources management.

The Rio Principle 10 says: "Environmental issues are best handled with the participation of all concerned citizens, at the relevant level ...." Note that the Principle does not guide as to how one should establish the "relevant level." This level may be the local, the river basin, the national, or the international level. For international rivers, in addition to PI at the intranational level, the need for PI at the international level may also be felt. Many times, there are conflicting water use interests in different basin countries, for example, an industrial plant in the upstream country may be discharging wastewater into the river and the water from this river might be used for municipal needs in a downstream country. Before attempting PI at this level, it will be useful to attempt PI at the national level. There might be large differences in social, cultural, languages, etc. among nations and hence in the perceptions of the people. Therefore, the involvement of the public at the international level is still a more complex activity. For many international basins, a river basin commission has been established where important decisions are taken.

Concerning the first pillar of PI—access to information—the information that can be disseminated is often categorized into different groups. Possible categories are: (1) official (approved) information, including international agreements, action plans, and programs; (2) working documents and drafts; (3) hydrological and water demand/use data, maps; and (4) financial and personnel information. Of course, some of these data may be classified and not available to everyone. It should be ensured that all nonclassified information is accessible on a nondiscriminatory basis and the user should be charged minimal or only reproduction expenses.

PI, as a means of community development, is closely related to decentralization. The participation of the public to improve the quality and efficiency of decision-making is the most relevant reason behind PI. The public can come up with information that would otherwise not be available and with innovative solutions. PI in the decision-making process enhances the legitimacy of the process and the acceptance by the public of the resulting decisions and often costly and time-consuming litigation can be prevented. Despite all the benefits, PI is easier said than done. Recall that it involves a lot of interaction with the public at large at various levels and the entire process will have to be carefully handled. If PI is to realize its potential, a number of issues need to be addressed. The important ones are discussed below (see also Section 9.7).

15.6.1 Approaches for PI

Different approaches are appropriate for different target groups. The type of information to be provided and the method of delivery for NGOs which may have some professionals also would be quite different than for local communities. The approach will also be guided by the type of issues at stake—whether any controversial issue is involved or not and how much emotionally surcharged the environment is. The social and cultural background of the participants is also very important. For instance, in a culture where consulting the public is seen as a sign of weakness on the side of the leaders, the usual "Western" methods of PI could be a kind of political suicide and other methods have to be devised (Mostert et al., 1999). Whenever the specific methods chosen allow for large numbers of participants, the public should be able to select itself for involvement—after it gets sufficient information in an appropriate form.

PI should be organized early enough so that the feedback could be used to improve decisions, but not too early because different plans and ideas should be specific enough to interest the public. A possibility is to organize PI in different phases and target different segments of the public: in early phases (semi-) professional NGOs, and in a later phase the local population and individuals. In any case, it should never appear that the decisions have already been made and the public has been called just to communicate the same. The river basin authorities may make the meetings with the public a regular feature. They may also earmark a day as an "open-house" day and organize activities, such as seminars, and question-answer sessions, by involving the public.

15.6.2 Information Dissemination and Follow-Up

If the decision-makers do not follow up on the outcomes of PI sincerely, the net result could be that it will lose the interest of people, legitimacy, and acceptance. Also, once people become disenchanted, it will be very difficult to involve them again. In this respect, a legal-administrative approach to PI could be useful. This entails a legal requirement to publish and react to the comments received (an action-taken report) in combination with access to justice.

The use of the Internet has proved to be a valuable tool for making information widely accessible. Many river commissions already have a homepage where the approved information can be viewed. For example, the Mekong River is an international river in Asia and the homepage of Mekong Commission (www.mrcmekong.org) provides a lot of information about this basin. The website has a separate section "information resources" with subsections on publications, data, and maps. Likewise, the homepage of the Murray Darling Basin Commission (www.mdbc.gov.au) contains links to information about the basin, natural resources management, action room, newsroom, and a tour of the basin. The website of the Bhakra Beas Management Board (www.bhakra.nic.in) provides a historical account of the projects in the basin, the power generation and reservoir operation data, technical papers, and tender announcements. It is helpful to see the "what's new" section on websites to know the latest interesting developments.

The meetings of the plenary body and/or subsidiary bodies of some river basin commissions are open to all. However, in most cases, the general public cannot participate in the meetings, but sometimes NGOs can get observer status. Their admittance is usually made dependent on some criteria of recognition. Pragmatic modalities are found in those situations where public involvement is not formally or insufficiently provided for. Such modalities include (1) representatives of NGOs as members or experts in the national delegation; (2) involvement of NGOs in national preparatory meetings for the plenary meetings of the joint body and/or its subsidiary bodies; and (3) special consultative meetings with NGOs organized by the river commission. Moreover, several river basin commissions invite NGO members to their meetings as experts. Involvement as an observer or expert in meetings of the commission automatically involves access to information that otherwise might not have been disclosed.

15.7 INTERBASIN WATER TRANSFER

When there is a shortage of water in an area with respect to demands, two options are available: supply management and demand management. In supply management, steps are taken to increase the availability of water and the means that can be adopted include interbasin transfer of water, artificial recharge, and desalinization of water.

The term *water transfer* refers to the transport of water through engineering structures, usually across river basins for some beneficial purpose. Interbasin Water Transfer (IBWT) is one of the possible solutions of water deficiency and is somewhat similar to other alternatives, such as dams, desalination, and groundwater extraction. The interbasin water transfer involves the transportation of surplus water from a basin to another basin that is deficient in water. The starting point of interbasin water transfer is an unsustainable situation in the receiving basin in the sense that it suffers from recurrent water shortages. If the surplus and deficient basins are not near each other, which is quite often the case, this will involve the transfer of water over large distances, sometimes of the order of thousands of kilometers. IBWT is an ancient approach and under certain conditions, it is a rational and often indispensable measure. In fact, most water resources development projects involve some kind of water transfer, though usually over short distances. Diversion of water by IBWT increases the resilience of the water system and decreases the risk of shortages. The most common purpose of IBWT projects is the water demand of agricultural areas or megacities. Typically, large-distance water transfer is carried out to improve national/regional economy, self-sufficiency in national/regional agricultural outputs and to remove regional disparities in development activities. As the human settlements are not always near the places of water occurrence or the available water may be inadequate to meet all demands at a place, waters have been transferred from one basin to another since time immemorial.

The special attributes of long-distance water transfer are:

1. Large amounts of water involved in transfer, often exceeding one billion cubic meters per year.
2. Large distances of water transfer, often exceeding 500 km.
3. Large costs of infrastructure and the possibility of extensive and irreversible environmental consequences.
4. Significant influence on the economy of the receiving area.

Most commonly, canals are used to carry water from one basin to another; tunnels and pipelines are used to negotiate ridges. However, the final selection depends on the topography of the area, climate, soil properties en route, and the volume of water involved. Many interbasin transfer projects involve the pumping of water in some stretches when a mountain is to be crossed and the construction of a tunnel is not feasible. To the extent possible, a gravity flow system is preferred over a system that involves pumping, even if a slightly longer route is to be taken. The running cost of a system with pumping is significantly higher and this places an additional burden on the infrastructure, particularly in those countries which are deficient in electric energy. Additional maintenance problems arise if the water that is to be pumped contains sediments because these cause much wear and tear. GISs are being increasingly used these days to finalize the route of the transfer link.

15.7.1 Planning for IBWT Projects

Large-scale interbasin water transfer has been an essential component of water management for a long time in many parts of the world and will remain so in the future. Generally, the water being exported from a basin must be surplus after meeting all the needs of the basin in the foreseeable future. The requirement of the water importing basin should be minimized by water-saving measures, but without impairing the efficiency of water uses.

Before any large-scale water transfer project is taken up, it would be helpful to mull over the following questions:

- Is water transfer the only option to overcome the present and likely problems?
- Is water transfer the most efficient alternative?
- What are the tradeoffs involved in the water transfer?
- Is the requisite institutional and infrastructural support available?

As per the practice being followed in India, if the water available in a basin is more than the demands that are likely to arise in the foreseeable future (time span of 25 years or so), then this basin is considered as a water surplus basin. The volume of water over and above the projected demands is labeled as surplus for that basin and this can be made available for the transfer to other deficient basins.

An important issue in IBWT is the sharing of water resources between the donor and receiving basin. Sustainable development in both basins should be practiced through shared economic and social benefits from the project. In the donor basin, water transfer must not have negative impacts on the sustainability of water use. Water transfer agreements should take care of monitoring and periodical assessment, with the possibility of adjustments of mutual obligations. If water transfer involves two countries, it may hopefully contribute to wider political cooperation between the countries.

15.7.2 Evaluation of IBWT Projects

IBWT projects are generally cost-effective solutions and technical problems are seldom the limiting factor except for those projects that involve long distances of transfer. The water transfer projects should always be compared with other water management instruments. As these projects may also involve a large-scale population displacement, adequate compensation to the project-affected persons must be an essential component of the project proposal. The necessity or otherwise of an IBWT project can be evaluated by the following criteria:

1. The recipient basin must have a substantial deficit in meeting the present or projected future water demands after considering alternative water supply sources and all reasonable measures for reducing the water demand.
2. The future development of the donor basin must not be substantially constrained by water scarcity.
3. An IBWT project should be taken up after a comprehensive environmental impact assessment (EIA) indicates that it will not substantially degrade environmental quality within the area of origin or area of delivery.
4. A comprehensive assessment of socio-cultural impacts must indicate a reasonable degree of certainty that the project will not cause sociocultural problems in the donor or recipient basins.
5. The net benefits from the transfer must be shared equitably between the donor and recipient basins.

EIA is a necessary step in the evaluation of any major IBWT project. However, due to serious and long-term environmental implications, such considerations should be an important part, and not just an addendum or formality for project clearance. Importantly, EIA should not be viewed as an obstacle to the project—the environmental assessment may direct to initiate IBWT. Due to water transfer, the ecological balance in the recipient basin may improve, and the transfer of water can help sustain cultural and emotional values that are associated with a water body. Environmental norms widely differ between countries—stringent norms are followed in developed countries, while developing countries give preference to economic and social progress. The environmental aspects of water resources projects are discussed in Chapter 7.

While trying to identify the environmental impacts, a distinction has to be made between impacts in the exporting area, the water importing area, and the transfer path. It is necessary to consider impacts on water quantity, water quality, micro and macroclimate changes, impacts on soil erosion, sedimentation, etc. These projects are also likely to have a significant influence on the regional economy, agricultural production, energy availability, and employment as well as the impact on aquatic life. As the techniques and methodologies for the evaluation of the impact of large-scale water transfer are not fully developed, it is essential to carry out monitoring on a continuous basis.

Sometimes EIA is limited to only water quality studies, but the concept includes the protection of biodiversity, the possible transfer of species between the two basins, and the impact of the new water on existing ecosystems in the receiving basin. While there are no generally accepted quantitative indicators of the required in-stream water quality and flow regimes from an ecological perspective, the attempt is to maintain or recreate the natural flow regime to the extent possible. Environmental, aesthetic, and human interests in the affected regions must be respected. In certain cases, large-scale project initiatives had to be abandoned mainly because of environmental implications, and also owing to political and economic developments in the region.

The institutional support includes relevant laws, policies, and administrative setup. An analysis of past projects shows that these could be implemented without much fuss and opposition because these were small, not technically complex, and the finances involved were not large. Many of these projects were confined to a single administrative unit and the environmental issues were either not raised at that time or the awareness was limited. However, current projects are facing considerable opposition because the technical problems are complex, the size of the project is large, and the environmental and social issues have gained significant importance. Many such projects involve more than one state of a country and some of them are international in nature, such as projects in North America.

In many instances, IBWT projects are the cheapest and most effective solution to overcome the water deficit in an area. The necessity of these schemes may become even more crucial if the crisis due to the hindrance of economic growth on account of water deficiency becomes serious. Some aspects that may further enhance the attraction of these projects in the future include technological advances that reduce costs of water transport, lessening of adverse socioenvironmental consequences, and changes in climate that make more water available in donor regions.

15.7.3 Examples of IBWT Projects

Long-distance interbasin transfer of water is a concept that has been in practice for a long time. In arid and semiarid regions, IBWT is sometimes crucial to alleviate acute water shortages and to strengthen the resilience of water systems in case of droughts. The Western Yamuna Canal and the Agra Canal in India were built about five centuries ago and carried water from the Himalayas to the distant parts of Punjab, U.P., and Rajasthan. The Kurnool Cuddappah Canal (constructed in 1860–70) and Periyar Vaigai (constructed in 1896) are other examples of interbasin water transfers executed in India in the

19th century. The Indira Gandhi Canal Project diverts water from the Himalayas to the deserts of Rajasthan. This scheme comprises a large multipurpose project constructed across the Beas River at Pong, a barrage at Harike, and a grand canal system. The Sardar Sarovar project (discussed in Chapter 7) involves the transfer of water from the Narmada River in central India to Saurashtra and Kutch in western regions.

In many countries, interbasin water transfer projects are among the most controversial issues of water resources utilization. Nevertheless, these projects are proposed and taken up because they often offer the most attractive solution to a given water problem. These projects are often necessary because, in many instances, the regions where water resources are abundant are not necessarily the regions where most of the population resides or where the industrial or agricultural activities are concentrated. For example, about 60% of water in Canada flows toward the north while 90% of the population and majority of industries are concentrated within 300 km of its border with the United States (Sewell, 1985). Some existing and under-construction IBWT projects in Canada include Kemano, Churchill Diversion, Welland Canal, James Bay, and Churchill Falls.

The Lingua Canal (completed in 214 BCE) and the Grand Canal (completed in 605 AD) are two examples of IBWT from ancient China. Recently completed projects in China include the Biliuha-Dalian interbasin water supply system and the transbasin transfer of water of the Luhana River to Tiajian and Tengshan. The southern part of China is abundant in water resources, whereas the northern part is water deficient. The basin of Huang He, Huai He, and Hai He rivers suffer from water deficiency. The Chang Jiang (Yangtze) River basin is abundant in water resources and this will be the main basin to supply water to North China. The South–North Water Transfer Project is a mega project in China to transfer about 44.8 billion cubic meters of water annually from the Chang Jiang River in southern China to the arid and industrialized north through three routes: (1) The Eastern Route through the course of the Grand Canal; (2) The Central Route from the upper reaches of Han River (a tributary of Yangtze River) via the Grand Aqueduct to Beijing and Tianjin; and (3) The Western Route which goes from three tributaries of Yangtze River. The eastern route component was to be taken up first in view of inadequate water requirements. In the middle route component, a large amount of water from Han Jiang and Chang Jiang (middle reaches) rivers will be transferred. Diversion of Quiantang River water and diversion of Yellow River surpluses are other ambitious proposed projects.

The major water transfer projects that have been proposed for the United States include the North American Water and Power Alliance, the Texas Water Plan, and the California State Water Project. The Texas Water Plan envisages the redistribution of water in Texas and New Mexico to meet the needs of the year 2020. The California State Water Project, the first phase of which was completed in 1973, provides for the diversion of 4 cubic km of flow from better-watered northern California to the drier central and southern parts of the state. The conveyance system comprises a 715 km California Aqueduct, a complex system of lined and unlined canals, pumping stations, siphons, and tunnels. The lift involved is nearly 1000 m (IWRS, 1996). Similarly, the waters of the Colorado River (an international river between the United States and Mexico) are being diverted outside the basin to the Imperial Valley in California. In Mexico, the project for the water supply of Mexico City through the transfer of groundwater from the Lerma basin was completed in 1958. The Mahaveli-Ganga project of Sri Lanka includes several interbasin transfer links.

The Great Man-Made River (GMMR) is a network of pipes to supply fresh water pumped from the Nubian Sandstone Aquifer System in Libya. It is the world's largest water supply project. GMMR utilizes a pipeline system that pumps water from the aquifers in south Libya to cities in north Libya including Tripoli and Benghazi. GMMR provides 70% of all freshwater used in Libya. GMMR is the largest (2820 km) underground network of pipes in the world that supplies 6.5 million m^3 of fresh water every day. The Libyan president Gaddafi termed it as the "Eighth Wonder of the World." GMMR was conceived in the late 1960s, approval to fund it was granted by the Libyan government in Oct. 1983, and work on the project began in 1984. The total cost of the project was estimated at about US$25 billion. GMMR extracts fossil groundwater from deep aquifers and has been criticized for environmental considerations.

A notable interbasin transfer scheme executed in the former USSR is the Irtysh Karganda scheme in central Kazakhistan. The link canal is about 450 km long with a maximum capacity of 75 m^3/s and the lift involved is 14–22 m. There is another plan to transfer 90,000 million m^3 from the north-flowing river to the area in the south. Other proposals include partial redistribution of water resources of northern rivers and lakes of the European part to the Caspian Sea basin involving two million hectare-m of water. More examples include IBWT for urban drinking water supply in Spain, France, and Germany; for irrigation, drinking water and hydropower in India, navigation and environmental aspects in Europe, and for environmental improvement in the United States and Australia. Glubev and Biswas (1985) and IWRA (1986) contain many case studies of IBWT projects.

15.8 MANAGEMENT OF INTERNATIONAL RIVER BASINS

About half of the land area of the earth is occupied by international river basins. The management of international river basins poses unique problems. The international basins are usually less homogenous and there are wide variations in socioeconomic, cultural, language, and administrative matters. The consequences of an action in the headwater regions in a country may significantly influence the conditions in the downstream areas which may be in a different country. Given this, it is necessary to have a high level of cooperation and coordination in international basins to prevent or solve upstream–downstream conflicts. In cases where a river basin falls under the jurisdiction of three or even more states, the need for multilateral cooperation is even greater. Mostert et al. (1999) have suggested the approach of "lowest common denominator" in the management of international basins. They explain that few obligations can be imposed on countries without their own consent and, therefore, many international agreements simply reflect the commonalities in the national policies of the concerned states. They have also presented nine mechanisms to overcome the lowest common denominator problem. These are given in Table 15.3.

River basin treaties and other forms of international cooperation should reflect the relevant principles of international law, primarily the principles of equitable and reasonable use, the obligation not to cause significant harm, and the duty to notify and exchange information. International river basin authorities with decision-making and enforcement powers may be a good option for specific operational tasks, such as the restoration of water quality, shipping, and the joint operation and management of infrastructure.

15.8.1 International River Basin Organizations

The basic premise behind the management of international river basins is that the interest of the international community is supreme, superseding the interest of individual countries. There are a number

Table 15.3 Nine mechanisms for reaching international agreements that go beyond the lowest common denominator.

Mechanism	Explanation
1. Issue linkage	It implies that a contentious issue on which national interests conflict (e.g., upstream–downstream conflict) is linked to another issue where the distribution of (perceived) costs and benefits is the reverse. Solving such issues simultaneously can result in a net gain for all parties involved, thus overcoming the conflict of interests. The second issue might be either an RBM issue or a totally different issue, but the former is usually more effective because on both issues the same parties are involved and costs and benefits fall on the same groups.
2. Diffuse reciprocity	A situation in which countries accept less favorable agreements to keep good relations and create a "reservoir of goodwill" from which they can draw in the future.
3. Side payments	These are payments—directly or through increased subsidies or reduced contributions—in return for a concession. Side payments will be most effective in cases of agreements affecting the economy or the finances of countries. They will be less effective when deeply held values or basic human needs are involved and can be experienced as bribery. Moreover, side payments for pollution reduction conflicts with the polluter pays principle.
4. Large geographical scope	Strict national environmental standards may limit the competitiveness of industry in a basin, but the effects are much smaller if several countries adopt similar standards for their whole territory.
5. Appealing goals/ mobilizing vision	Ambitious agreements can also be reached if they contain goals or a vision of the future that is attractive for large sections of society in the countries concerned. Such goals and such a vision can act as a form of awareness raising. Moreover, they can implicitly incorporate forms of issue linkage and diffuse reciprocity.
6. Slack cutting	It occurs when national government bodies use international agreements for introducing a more ambitious policy domestically or for promoting the enforcement of existing.
7. Intended noncompliance	It refers to the fact that countries may be willing to accept ambitious international agreements if they expect that the agreements will not be enforced. Obviously, agreements reached in this way are usually not implemented.
8. Unforeseen consequences	Ambitious agreements can also be reached if their consequences are not foreseen. Negotiators might be too confident about their national situation and assume too easily that no adaptation will be necessary. Furthermore, international courts may give unexpectedly strict interpretations to agreements. Finally, the negotiators may be inexperienced or the time to study proposals may simply be lacking, especially in case of last-minute changes.
9. Majority voting	In some rare cases, international agreements are the result of majority voting. In these cases, the more conservative countries can be overruled, at least in theory. However, the more conservative countries can link the issue to another issue where their cooperation is needed, either because unanimity is required for that issue or to obtain a majority.

Adapted from Mostert et al. (1999).

of bi- and multilateral agreements among the basin countries for the management of international river basins. The cooperation in the management of international basins depends on the mutual relations of the countries involved. If the relations are good, for example, the Danube River in Europe whose basin is shared by a large number of countries, problems can be easily resolved through across-the-table discussions and meetings. However, in the case the basin states do not have a good understanding, cooperation in the management of basins is the first casualty. There are a number of bi- and multilateral agreements among the basin countries for the management of international river basins.

A major aspect of many river basin treaties is the establishment of a river basin organization. There are two types of national river basin organizations: river basin commissions with a primarily coordinating task and river basin authorities with decision-making and policing powers. The same type of organizations can be found in international basins.

The importance of international river basin commissions has been widely recognized. River basin commissions may coordinate monitoring and research efforts, add legitimacy to the monitoring and research results, and in this way provide a common, generally agreed upon factual basis for management. Furthermore, they offer the basin states a platform for coordinating their policy and management. River basin commissions can also prepare RBM plans and programs, but these have to be adopted by the basin countries. River basin commissions may also oversee the implementation of plans and programs if the basin countries agree to these. Finally, river basin commissions can play a significant role in resolving river-related international conflicts. They constitute a relatively informal forum for discussion, may help in selecting fact-finders and arbitrators, or may even do fact-finding or act as arbitrators themselves.

Mutual understanding, trust, and information sharing are the basis for international cooperation. Technical cooperation involving the collection and dissemination of information promotes the acceptance of this information by all basin states and stimulates mutual understanding and trust. In times of international conflicts, at least technical cooperation should be maintained. Several mechanisms could be used to overcome conflicting (upstream–downstream) interests. Contentious international issues could be linked with other issues. Moreover, countries may accept less favorable agreements in the expectation that other countries will do the same in the future ("diffuse reciprocity"). In some cases, financial compensation from the benefiting country to the country having to incur costs could be justified, provided the polluter-pays principle is respected.

International river basin commissions can perform many useful functions in the management of international basins, such as coordination of research and monitoring, coordination of river basin management between participating basin states, planning, compliance monitoring, and conflict resolution. International river basin commissions are almost indispensable for international basins. States sharing several international waters may also establish joint water commissions.

For river basins falling within one jurisdiction, intersectoral planning offers good opportunities for intersectoral coordination. The unrestricted exchange of data and knowledge is a prerequisite to efficient management and cooperation in both national and international river basins. Monitoring data collected with public funds should be publicly available and easily accessible, nationally and internationally.

The importance of mechanisms for promoting river basin cooperation is becoming more widely recognized and implemented and is reflected in support for the International Network of River Basin Organizations (INBO). About 102 organizations in 42 countries are members of this network. At the

International Conference on Water and Sustainable Development, held in Paris in 1998, INBO recommended the establishment of a legal framework that takes into account five objectives regarding river basin management that should be organized:

1. On the relevant scale of large river basins and aquifers;
2. With the participation in decision-making of the local authorities concerned, different categories of users and associations for environmental protection besides the appropriate governmental administrations;
3. Based on master plans that define the long-term objectives to be achieved as regards water resources management in times of scarcity or flooding and that enable the management of users integrated into land use planning and preserve the quality of ecosystems;
4. Within priority investment programs that result from these master plans;
5. With the mobilization of appropriate financial resources, based on the polluter-pays principle and user-pays systems.

15.8.2 International Initiatives for Freshwater Management

The first step in cooperation among riparian countries is technical cooperation, which may begin with the sharing of information and joint monitoring of developmental activities. This can be followed by cooperation on more substantive issues and the exchange of ideas at a higher level. This matter has also been the theme of many international conferences. The first global conference which paid specific attention to fresh water issues was held in Mar del Plata in 1977. It was recognized and emphasized here that people have a right to water for their basic needs.

Among the United Nations agencies, UNESCO has a long-term program in the water sector in the form of the *International Hydrological Program* (IHP). IHP is a vehicle through which countries can upgrade their knowledge of the water cycle and thereby increase their capacity to better manage and develop their water resources. It aims at the improvement of the scientific and technological basis for the development of methods for the rational management of water resources. The Program started as the International Hydrological Decade (IHD, 1965–74) and was followed as a long-term program executed in phases of a 6-year duration. IHP-I lasted from 1975 to 1980. IHP-II, on the other hand, was of a shorter duration (1981–83). The IHD was mainly research-oriented. IHP-I, which followed on from the IHD, maintained much of the same research orientation. However, the next phases were oriented to include practical aspects of hydrology and water resources. Hence, IHP-II (1981–83) and IHP-III (1984–89) were planned under the theme "Hydrology and the scientific bases for rational water resources management." The theme for IHP-IV (1980–95) was "Hydrology and water resources: sustainable development in a changing environment," while the theme of the fifth phase (1996–2001) was "Hydrology and water resources development in a vulnerable environment." In recognition of the shift in thinking about water from fragmented compartments of scientific inquiry to a more holistic integrated approach, the general theme for IHP-VI (2002–07) has been defined as "*Water interactions: systems at risk and social challenges*."

The relevant IHP-VI topics on hydrologic research, water resources management, and education are framed under five themes, with the transition and interaction from the global scale to the watershed scale being the overall driving force for consideration of the complex relationships between water and society and the overall need for knowledge, information and technology transfer: (1) global changes and water resources; (2) integrated watershed and aquifer dynamics; (3) land habitat hydrology; (4) water and society; (5) water education and training. Two crosscutting program components: FRIEND (Flow Regimes for International Experimental and Network Data) and HELP (Hydrology for Environment, Life and Policy) have been identified that, through their operational concept, interact with all themes.

IHP constitutes a framework for applied research and education in the field of hydrology and water management. It is a dynamic concept whose aim is to improve links between research, application, and education and to promote scientific and educational activities. Further information about IHP is available at the IHP website: www.unesco.org/water/ihp.

The World Meteorological Organization (WMO) has many programs related to water resources and has brought out a large number of publications that are extensively referred to by water resources professionals. Many other agencies of the United Nations, notably, the Food and Agriculture Organization (FAO), and the United Nations Environment Program (UNEP), are actively involved in the water sector. The websites of these agencies (see Appendix B) contain a large number of useful publications and data that are of immense use to practitioners.

15.9 CLOSURE

The United Nations observed the decade 1981–90 as the *International Decade on Water Supply and Sanitation*. The aim was to provide access to clean drinking water and sanitation for all people by the year 1990, a target that could be achieved only to a limited extent. Two important conferences were held in 1992: International Conference on Water and the Environment at Dublin and the UN Conference on Environment and Development at Rio de Janeiro. A noteworthy outcome of the Rio Conference was the Agenda 21 which has provided a program of action for attaining sustainable development. Chapter 18 of Agenda 21, entitled "Protection of the Quality and Supply of Freshwater Resources: Application of Integrated Approaches to the Development, Management and use of Water Resources" is relevant to the management of freshwater resources. The main objectives enunciated in it are: (1) access should be ensured for all people to safe and sufficient water supplies, or at least water supplies to meet the basic drinking and food-growing requirements; (2) public participation and management at the lowest appropriate level should be enhanced; and (3) integrated development and management of water resources should be attained. The year 2002 was declared as *the International Year of Mountains* and the year 2003 is declared as *the International Year of Freshwater.*

Decision-making for hydrosystems is gradually becoming more complex. The last few decades have witnessed that the role of social and political factors is becoming equally important as the technical inputs. Seven strategic priorities for water and energy resources development have been advanced by Asmal (2002). The charges for various uses should be fixed in a manner that leads to savings and the polluter should pay for the damage caused by him. It will be necessary that the development alternatives are arrived at after a comprehensive examination of all options, these should have the acceptance of the public, and the benefits be shared fairly. The touchstone for future development strategies will essentially be sustainability, poverty alleviation, participatory decision-making, and conflict avoidance. The water resources professionals will have to provide solutions that are scientifically sound and socially acceptable.

References

Abbott, M.B., Refsgaard, J.C., 1996. Distributed Hydrological Modelling. Water Science and Technology Library # 22. Kluwer Academic Publishers, Dordrecht.

Acreman, M., Aldrick, J., Binnie, C., Black, A., Cowx, I., Dawson, H., Dunbar, M., Extence, C., Hannaford, J., Harby, A., Holmes, N., Jarritt, N., Old, G., Peirson, G., Webb, J., Wood, P., 2009. Environmental flows from dams: the water framework directive. Engineering Sustainability 162 (1), 13–22. https://doi.org/10.1680/ensu.2009.162.1.13. March 2009.

Anderson, T., Hill, P. (Eds.), 1997. Water-Marketing: The Next Generation. Rowman and Littlefield, London.

Andreu, J., Capilla, J., Sanchis, E., 1996. AQUATOOL, a generalized decision-support system for water-resources planning and operational management. Journal of Hydrology 177 (3–4), 269–291.

Arthington, A.H., 2012. Environmental Flows: Saving Rivers in the Third Millennium. University of California Press, Berkeley, p. 424.

Asmal, K., 2002. Parting the waters. Journal of Water Resources Planning and Management, ASCE 128 (2), 87–90.

Basson, M.S., Allen, R.B., Pegram, G.G.S., van Rooyen, J.A., 1994. Probabilistic Management of Water Resources and Hydropower Systems. Water Resources Publications, Colorado, USA.

Billib, M.H.A., Boochs, P.W., Matheja, A., Rusteberg, B., 1995. Interactive management of a conjunctive use system considering quality aspects. In: Proceedings of an International Symposium Held at Boulder, Colorado, July 1–14, IAHS Publication No. 231, pp. 273–281.

Bovee, K.D., 1986. Development and evaluation of habitat suitability criteria for use in instream flow incremental methodology. U.S. Fish and Wildlife Service Biological Report 86 (7) (U.S. Fish and Wildlife Service).

Center for Advanced Decision Support for Water and Environmental Systems, 1992. River Simulation System (RSS) User's Manual, Programming Guide, and Tutorial. University of Colorado, Boulder, Colorado.

Chenoweth, J.L., Malano, H., Bind, J.F., 2001. Integrated river basin management in multi-jurisdictional river basin: the case study of Mekong river basin. Water Resources Development 17 (3), 365–377.

Coe, J.J., 1990. Conjunctive use – advantages, constraints, and examples. Journal of Irrigation and Drainage Engineering, ASCE 116 (3), 427–443.

DHI, 1995. MIKE SHE. Short Description. Danish Hydraulic Institute, Horsholm, Denmark.

DHI, 1997. MIKE BASIN – A Tool for River Planning and Management. Danish Hydraulic Institute, Horsholm, Denmark.

DHI, 1998. http://www.dhi.dk/mikebasn/index.htm.

Diersch, H.J.G., Grundler, R., 1993. GIS-based ground water flow and transport modeling – the simulation system FEFLOW. In: International Conference HYDROGIS-93, Vienna, Austria.

Dinar, A., Rosegrant, M., Meinzen-Dick, R., 1997. Water Allocation Mechanisms: Principles and Examples. The World Bank, Washington, DC.

Downs, P.W., Gregory, K.J., Brookes, A., 1991. How integrated is river basin management? Environmental Management 15 (3), 299–309.

EPA, 1998. United States Environmental Protection Agency. http://www.epa.gov/docs/QUAL2E_WINDOWS/metadata.txt.html.

Fedra, K., 1992. Advanced Computer Applications, Options. International Institute for Applied Systems Analysis, Laxenburg, Austria, pp. 4–14.

Fedra, K., Weigkricht, E., Winkelbauer, L., 1992. Chapter 17: Decision support and information systems for regional development planning. In: Vasko, T. (Ed.), Problems of Economic Transition: Regional Development in Central and Eastern Europe. Avebury, Aldershot, United Kingdom, pp. 216–242.

Fedra, K., 1994. GIS and Environmental Modeling. RR-94-2. International Institute for Applied Systems Analysis, Laxenburg, Austria. Reprinted from. In: Goodchild, M.F., Parks, B.O., Steyaert, L.T. (Eds.), Environmental Modeling with GIS. Oxford University Press, pp. 35–50.

Ford, D.T., 1990. Reservoir storage reallocation analysis with PC. Journal of Water Resources Planning and Management, ASCE 116 (3), 402–416.

Golubev, G.N., Biswas, A.K. (Eds.), 1985. Large-scale Water Transfers: Emerging Environmental and Social Issues, UNEP Water Resources Series, vol. 7. Tycooly Publishing Ltd., Oxford, UK.

Gorelick, S.M., 1983. A review of distributed parameter groundwater management modeling methods. Water Resources Research 19 (2), 305–319.

HEC, 1993. HEC-PRN Simulation Model, User's Manual. Hydrologic Engineering Center U.S. Army Corps of Engineers, Davis, California.

Hoff, H., 2011. Understanding the Nexus. In: Background Paper for the Bonn 2011 Conference: The Water, Energy and Food Security Nexus. Stockholm Environment Institute, Stockholm.

IEA, 2022. Energy Security. Available from: https://www.iea.org/topics/energy-security.

IWRA, 1986. Proceedings of the IWRA Seminar on Inter-basin Water Transfer, Held at Beijing, China. Hosted by Chinese Hydraulic Engineering Society.

IWRS, 1996. Interbasin transfers of water for national development – problems and prospects. In: Theme Paper Prepared for Water Resources Day 1996. Indian Water Resources Society, Roorkee.

Jain, S.K., Acreman, M.C., 2017. Chapter 134: Environmental flows. In: Singh, V.P. (Ed.), Handbook of Applied Hydrology. McGraw-Hill, New York.

Jain, S.K., Mishra, P.K., Sharma, A., 2022. Chapter 1: Environmental flows and their assessment. In: Mukherjee, A. (Ed.), Riverine Systems – Understanding the Hydrological, Hydrosocial and Hydro-Heritage Dynamics. Capital Publishing Company, New Delhi, India.

Jain, S.K., Sikka, A.K., Alam, M.F., 2023. Water-energy-food-ecosystem nexus in India—A review of relevant studies, policies, and programmes. Frontiers in Water 5, 1128198. https://doi.org/10.3389/frwa.2023.1128198.

Jamieson, D.G., Fedra, K., 1996a. The 'Water Ware' decision-support system for river-basin planning 1. Conceptual design. Journal of Hydrology 177 (3–4), 163–175.

Jamieson, D.G., Fedra, K., 1996b. The 'Water Ware' decision-support system for river-basin planning 3. Example applications. Journal of Hydrology 177 (3–4), 199–211.

King, J.M., Louw, D., 1998. Instream flow assessments for regulated rivers in South Africa using the Building Block Methodology. Aquatic Ecosystems Health and Restoration 1, 109–124.

Kinnersley, D., 1993. Privatization and the water environment in England. In: Le-Moigne, G. (Ed.), Country Experiences with Water Resources Management: Economic, Institutional, Technological and Environmental Issues. The World Bank, Washington, DC.

Kuczera, G., 1990. WATN-ET, Generalized Water Supply Simulation Using Network Linear Programming. Department of Civil Engineering and Surveying, University of Newcastle, NSW, Australia.

Kuczera, G., 1993. Network linear programming codes in water supply headworks modeling. Journal of Water Resources Planning and Management, ASCE 119 (3), 412–417.

Labadie, J.W., Brazil, L.E., Corbu, I., Johnson, L.E. (Eds.), 1989. Computerized Decision Support Systems for Water Managers. American Society of Civil Engineers, New York.

Le-Moigne, G., Subramanian, A., Mei, X., Giltner, A. (Eds.), 1994. A Guide for the Formulation of Water Resources Strategy. World Bank Technical Paper No. 263. The World Bank, Washington, DC.

Loucks, D.P., da Costa, J.R. (Eds.), 1991. Decision Support Systems. Water Resources Planning, NATO Series G, vol. 26. Springer-Verlag, Berlin.

Loucks, D.P., 1995. Developing and implementing decision support systems: a critique and a challenge. Water Resources Bulletin 31 (4), 571–582.

Loucks, D.P., French, R.N., Taylor, M.R., 1995. Interactive River-Aquifer Simulation, Program Description and Operation. School of Civil and Environmental Engineering, Cornell University, Ithaca, New York.

Maass, A., 1962. Design of Water Resource Systems. New Techniques for Relating Economic Objectives, Engineering Analysis, and Governmental Planning. Harvard University Press, Cambridge, Massachusetts.

Mallach, E.G., 1994. Understanding Decision Support Systems and Expert Systems. Irwin, Burr Ridge, Illinios.

McKinney, D.C., Cai, X., Rosegrant, M.W., Ringler, C., Scott, C.A., 1999. Modeling Water Resources Management at the Basin Level: Review and Future Directions. SWIM Paper 6. International Water Management Institute, Colombo, Sri Lanka.

Merrett, S., 1997. Introduction to the Economics of Water Resources. UCL Press, London.

Morton, M.S., 1971. Management Support Systems. Computer-Based Support for Decision Making. Division of Research, Harvard University, Cambridge, Massachusetts.

Mostert, E., 1998. River basin management in the European Union; how it is done and how it should be done. European Water Management 1 (No. 3), 26–35.

Mostert, E., van Beek, E., Bouman, N.W.M., Hey, E., Savenije, H.H.G., Thissen, W.A.H., 1999. River basin management and planning. In: Keynote Paper for International Workshop on River Basin Management. The Hague.

Pegram, G., Li, Y., Quesne, T.L., Speed, R., Li, J., Shen, F., 2013. River Basin Planning: Principles, Procedures and Approaches for Strategic River Basin Planning. UNESCO, Paris.

Randall, D., Link, G.W., Cleland, L., Kuehne, C.S., Sheer, D.P., 1995. A Water Supply Planning Simulation Model Using Mixed Integer Linear Programming Engine. Mirneo, Water Resources Management, Inc., Columbia, Maryland.

Reitsma, R., Ostrowski, P., Wehrend, S., 1994. Geographically distributed decision support: the TVA TERRA System. In: Fontaine, D.G., Tuvel, H.N. (Eds.), Water Policy and Management. Solving the Problems. American Society of Civil Engineers, New York.

Richter, B.D., Baumgartner, J.V., Wigington, R., Braun, D.P., 1997. How much water does a river need? Freshwater Biology 37 (1), 231–249.

Santos, M.A., 1991. Decision Support Systems in Water Resources. Laboratoiio Nacional de Engenharia Civil, Lisboa, Portugal.

Schulte, A.M., Malm, A.P., 1993. Integrating hydraulic modeling and SCADA systems for system planning and control. Journal of the American Water Works Association 85 (7), 62–66.

Sewell, W.R.D., 1985. Inter-basin water diversions: Canadian experiences and perspectives. In: Golubev, G.N., Biswas, A.K. (Eds.), Large-scale Water Transfers: Emerging Environmental and Social Issues, UNEP Water Resources Series, vol. 7. Tycooly Publishing Ltd., Oxford, UK.

Souchon, Y., Keith, P., 2001. Freshwater fish habitat: science, management and conservation in France. Aquatic Ecosystem Health & Management 4, 401–412.

TAC, 2000. Integrated Water Resources Management. Background Paper No. 4. Technical Advisory Committee of Global Water Partnership, Stockholm, Sweden.

Tennant, D.L., 1976. Instream flow regimes for fish. Wildlife, Recreation and Related Environment Resources, Fisheries 1 (4), 6–10.

Tessmann, S., 1980. Environmental assessment, technical appendix e in environmental use sector reconnaissance elements of the westerndakotas region of south dakota study. South Dakota State University, Water Resources Institute, Brookings, South Dakota, USA.

Texas Water Development Board, 1972. Economic Optimization and Simulation Techniques for Management of Regional Water Resource Systems: River Basin Simulation Model SIMYLD-II. Program Description. Texas Water Development Board, Austin, Texas.

Tharme, R.E., 2003. A global perspective on environmental flow assessment: emerging trends in the development and application of environmental flow methodologies for rivers. River Research and Applications 19, 397–442.

UNDP (United Nations Development Programme), 1994. Human Development Report 1994: New Dimensions of Human Security. New York.

UNECE, 2022. Water-Food-Energy-Ecosystem Nexus. United Nations Economic Commission for Europe (UNECE). Available from: https://unece.org/environment-policy/water/areas-work-convention/water-food-energy-ecosystem-nexus#:~:text=The%20%E2%80%9Cnexus%E2%80%9D%20term%20in%20the,human%20activities%20ultimately%20depend%20upon.

UN Water, 2013. What Is Water Security? Available from. https://www.unwater.org/publications/what-water-security-infographic.

UNEP, 2021. Progress on Integrated Water Resources Management. Tracking SDG 6 series: global indicator 6.5.1 updates and acceleration needs. United Nations Environment Programme, Kenya.

USACE, 1987. SSARR Users Manual: Streamflow Synthesis and Reservoir Regulation. US Army Corps of Engineers, Portland, Oregan, USA.

USACE, 1998. http://www.wrc-hec.usace.army.mil/software/software.html.

van Beek, E., 2002. Lecture Notes of IWRM Course at the National Water Academy, Pune, India.

Wardropper, C., Brookfield, A., 2022. Decision-support systems for water management. Editorial. Journal of Hydrology 610, 127928.

Willis, R.L., Yeh, W.W.-G., 1987. Groundwater Systems Planning and Management. Prentice-Hall, Englewood Cliffs, New Jersey.

Winpenny, J., 1994. Managing Water as an Economic Resource. ODI. Routledge, London.

Wong, H.S., Sun, N., Yeh, W.W.-G., 1997. Optimization of conjunctive use of surface water and groundwater with water quality constraints. In: Water Resources Planning and Management Division, ASCE. Proceedings of Annual Conference, Houston.

World Bank, 1993. Water Resources Management; A World Bank Policy Paper. The World Bank, Washington, DC.

World Bank, 2006. Integrated River Basin Management: From Concept to Good Practice. Brief Note 1, Washington, DC.

World Summit on Food Security, 2009. Draft Declaration of the World Summit on Food Security. Food and Agriculture Organization of United Nations, Rome. Available from: http://www.fao.org/fileadmin/templates/wsfs/Summit/Docs/Declaration/WSFS09_Draft_Declaration.pdf.

Wurbs, R.A., 2008. Water rights analysis package (WRAP) modeling system reference manual, Technical Report No. TR-255. Texas Water Resources Institute, College Station, Texas, USA.

Young, R.A., Bredehoeft, J.D., 1972. Digital computer simulation for solving management problems of conjunctive ground-water and surface water systems. Water Resources Research 8, 533–556.

Zurlini, G., Muller, F., 2008. Environmental security. In: Jørgensen, S.E., Fath, B.D. (Eds.), Systems Ecology. Vol. [2] of Encyclopedia of Ecology, 5 vols. Elsevier, Oxford, pp. 1350–1356.

Further Reading

Fedra, K., Loucks, D.P., 1985. Interactive computer technology for planning and policy modeling. Water Resources Research 21 (2), 114–122.

Goodchild, M.F., Parks, B.O., Steyaert, L.T. (Eds.), 1994. Environmental Modeling with GIS. Oxford University Press.

Lattenmaier, D.P., McCabe, G., Stakhiv, E.Z., 1996. Chapter 29: Global climate change: effect on hydrologic cycle. In: Mays, L.W. (Ed.), Water Resources Handbook. McGraw-Hill Inc., New York.

Loucks, D.P., Kindler, J., Fedra, K., 1985. Interactive water resources modeling and model use: an overview. Water Resources Research 21 (2), 95–102.

Mimikou, M.A., 1995. Climatic change, in environmental hydrology. In: Singh, V.P. (Ed.), Water Science and Technology Library # 15. Kluwer Academic Publishers, Dordrecht.

Mostert, E., 1998b. A Framework for Conflict Resolution. Water International, December.

Mostert, E. (Ed.), 1999. River Basin Management and Planning; Institutional Structures, Approaches and Results in Five European Countries and Six International Basins. RBA Series on River Basin Administration. Research Report No. 10. RBA Centre, Delft, The Netherlands.

Schmid, C.F., 1983. Statistical Graphics: Design Principles and Practices. John Wiley and Sons, New York.

Sheer, D.P., Randall, D., 1989. Methods for evaluating the potential impacts of global climate change: case studies of the State of California and Atlanta, Georgia. In: Smith, J.B., Tirpak, D.A. (Eds.), The Potential Effects of Global Climate Change on the United States, Appendix A: Water Resources. Report EPA-230-05-89-051. U.S. Environmental Protection Agency, 2-1:2-28.

UNCED, 1992. The Rio Declaration on Environment and Development, United Nations Conference on Environment and Development, Rio de Janeiro, 3–14 June.

UNCED (United Nations Conference on Environment and Development), 1998. Agenda 21, Chapter 18. Final, Advanced Version. United Nations Environment Programme.

United Nations, 1992. The Dublin Statement and Report of the Conference. International Conference on Water and the Environment: Development Issues for the 21st Century, 26–31 January, Dublin, Ireland. United Nations Administrative Committee on Co-ordination Inter-Secretariat Group for Water Resources, Geneva.

Appendix A: Conversion Factors

Conversion factors—length

	cm	m	km	inch	foot	mile
cm	1	0.01	1×10^{-5}	0.3937	0.032808	6.2137×10^{-6}
m	100	1	0.001	39.37	3.2808	6.2137×10^{-4}
km	10^5	1000	1	39,370	3280.8	0.62137
inch	2.54	0.0254	2.54×10^{-5}	1	0.083333	1.5783×10^{-5}
foot	30.48	0.3048	3.048×10^{-4}	12	1	1.8939×10^{-4}
mile	1.6093×10^5	1609.3	1.6093	63,360	5280	1
Example to use this table: 1 mile = 1.6093 km.						

Conversion factors—area

	square meter	hectare	square km	square foot	acre	square mile
square meter	1	10^{-4}	10^{-6}	10.764	2.4711×10^{-4}	3.861×10^{-7}
hectare	10^4	1	0.01	107,639	2.4711	0.003861
square kilometer	10^6	100	1.0764×10^7	247.11	247.11	0.38610
square foot	0.092903	9.2903×10^{-6}	9.2903×10^{-4}	1	2.2957×10^{-5}	3.5870×10^{-8}
acre	4046.9	0.40469	0.0040469	43.560	1	0.0015625
square mile	2.590×10^6	259.0	2.59	2.7878×10^7	640	1

Conversion factors—volume

	Liter	m^3	ha-m	ft^3	acre-ft
liter	1	0.001	10^{-7}	0.035315	8.1071×10^{-7}
m^3	1000	1	10^{-4}	35.315	8.11071×10^{-4}
hectare meter	10^7	10^4	1	353,147	8.1071
ft^3	28.317	0.028317	2.8317×10^{-6}	1	2.2957×10^{-5}
acre-foot	1233.5×10^3	1233.5	0.12335	43,560	1

Multiplying factors and their values

Name of multiplying factor	Numerical value
peta	10^{15}
tera	10^{12}
giga	10^{9}
mega	10^{6}
kilo	10^{3}
hecto	10^{2}
deca	10^{1}
deci	10^{-1}
centi	10^{-2}
milli	10^{-3}
micro	10^{-6}
nano	10^{-9}
pico	10^{-12}
femto	10^{-15}

OTHER UNITS

1 pound (lb) = 453.6 gm
1 ton (tonne) = 1000 kg
1 horsepower = 745.7 watt
1 nautical mile = 1.151 miles = 1.852 km
1 US gallon = 3.7854 liter
$e = 2.71828$
$\log10\ e = 0.43429$
$\log e\ 10 = \ln 10 = 2.30259$

Index

Note: 'Page numbers followed by "*f*" indicate figures and "*t*" indicate tables.'

D

E

F

G

H

I

J

K

M

N

O

Q

R

S

T

U

Y

Z

9780128213490